To the mathematics instructors of Edgewood High School, the University of Redlands, and California State University at Los Angeles, whose love of their discipline and dedication to their students continue to be an inspiration to me

To Karl Jacobs, a fine mentor and college president

AST

RDG

Books in the Tussy and Gustafson Series

In paperback:

Basic Mathematics for College Students Third Edition
Student edition: ISBN 0-534-42223-3
Instructor's edition: ISBN 0-534-42224-1

Prealgebra Third Edition
Student edition: ISBN 0-534-40280-1
Instructor's edition: ISBN 0-534-40281-X

Developmental Mathematics
Student edition: ISBN 0-534-99776-7
Instructor's edition: ISBN 0-534-99775-5

Introductory Algebra Third Edition
Student edition: ISBN 0-534-40735-8
Instructor's edition: ISBN 0-534-40736-6

Intermediate Algebra Third Edition
Student edition: ISBN 0-534-49394-7
Instructor's edition: ISBN 0-534-49395-5

In hardcover:

Elementary Algebra Third Edition
Student edition: ISBN 0-534-41914-3
Instructor's edition: ISBN 0-534-41915-1

Intermediate Algebra Third Edition
Student edition: ISBN 0-534-41923-2
Instructor's edition: ISBN 0-534-41924-0

Elementary and Intermediate Algebra Third Edition
Student edition: ISBN 0-534-41932-1
Instructor's edition: ISBN 0-534-41933-X

THIRD EDITION

Intermediate Algebra

Alan S. Tussy

Citrus College

R. David Gustafson

Rock Valley College

THOMSON

™

BROOKS/COLE

Australia • Canada • Mexico • Singapore • Spain • United Kingdom • United States

THOMSON
™
BROOKS/COLE

Intermediate Algebra, Third Edition
Alan S. Tussy and R. David Gustafson

Executive Editor: Jennifer Laugier

Development Editor: Kirsten Markson

Assistant Editor: Rebecca Subity

Editorial Assistant: Sarah Woicicki

Technology Project Manager: Earl Perry

Marketing Manager: Greta Kleinert

Marketing Assistant: Jessica Bothwell

Advertising Project Manager: Bryan Vann

Project Manager, Editorial Production: Hal Humphrey

Art Director: Vernon Boes

Print/Media Buyer: Barbara Britton

Permissions Editor: Kiely Sisk

Production Service: Helen Walden

Text Designer: Diane Beasley

Art Editor: Helen Walden

Photo Researcher: Helen Walden

Copy Editor: Helen Walden

Illustrator: Lori Heckelman

Cover Designer: Cheryl Carrington

Cover Image: © CORBIS

Cover Printer: Quebecor World/Dubuque

Compositor: G & S Typesetters, Inc.

Printer: Courier/Kendallville

Printed in the United States of America
4 5 6 7 08 07

For more information about our products, contact us at:
Thomson Learning Academic Resource Center
1-800-423-0563

For permission to use material from this text or product, submit a request online at **http://www.thomsonrights.com**.

Any additional questions about permissions can be submitted by email to **thomsonrights@thomson.com**.

Library of Congress Control Number: 2004118197
Student Edition: ISBN-13: 978-0-495-18889-6
ISBN-10: 0-495-18889-1
Annotated Instructor's Edition: ISBN 0-534-49395-5

Thomson Higher Education
10 Davis Drive
Belmont, CA 94002-3098
USA

Asia (including India)
Thomson Learning
5 Shenton Way #01-01
UIC Building
Singapore 068808

Australia/New Zealand
Thomson Learning Australia
102 Dodds Street
Southbank, Victoria 3006
Australia

Canada
Thomson Nelson
1120 Birchmount Road
Toronto, Ontario M1K 5G4
Canada

UK/Europe/Middle East/Africa
Thomson Learning
High Holborn House
50/51 Bedford Row
London WC1R 4LR
United Kingdom

Latin America
Thomson Learning
Seneca, 53
Colonia Polanco
11560 Mexico D.F.
Mexico

Spain (including Portugal)
Thomson Paraninfo
Calle Magallanes, 25
28015 Madrid, Spain

CONTENTS

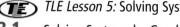

8 Quadratic Equations, Functions, and Inequalities 569

9 Exponential and Logarithmic Functions 644

10 Conic Sections; More Graphing   733

For the Instructor

Algebra is a language in its own right. The purpose of this textbook is to teach students how to read, write, and think mathematically using the language of algebra. It presents all of the topics associated with a second course in algebra. *Intermediate Algebra,* Third Edition, employs a variety of instructional methods that reflect the recommendations of NCTM and AMATYC. You will find extensive opportunities for skill practice and the well-defined pedagogy that are hallmarks of a traditional approach to teaching mathematics. You will also find that we emphasize the reasoning, modeling, and communication skills that are a part of today's mathematics reform movement.

The third edition retains the basic philosophy of the second edition. However, we have made several improvements as a direct result of the comments and suggestions we received from instructors and students. Our goal has been to make the book more enjoyable to read, easier to understand, and more relevant.

New to This Media Edition

New features make this the most student relevant and engaging edition of this popular text.

Integration with Intermediate AlgebraNow™: This new Media Edition features access to Intermediate AlgebraNow, a powerful online learning companion. Icons throughout the text indicate opportunities to reinforce concepts with online tutorials.

New Chapter Openers with TLE Labs: TLE (The Learning Equation) is interactive courseware that uses a guided inquiry approach to teaching developmental math concepts. Each chapter opener integrates a TLE Lab with a real-world application, which offers an opportunity to seamlessly integrate TLE lessons into your course. For more information on TLE, instructors can see the Preview at the beginning of their book.

Check Your Knowledge: New pretests at the beginning of each chapter are helpful in gauging a student's knowledge base for the upcoming chapter. Instructors can assign the pretest to see how well students are prepared for the material in the chapter, and customize subsequent lessons based on the needs of their students. Students can also take the pretest themselves as a warm-up for the chapter and to structure review. Answers to the pretests appear at the back of the book.

Study Skills Workshop: This complete mini-course in mathematics study skills helps students and instructors tackle the problem of lack of preparedness and inadequate study habits in an organized series of lessons. Each chapter opens with a one-page Study Skills Workshop sequenced to address relevant study skills issues as the students move through the course. For example, students learn how to use a calendar to schedule study times in the first lesson, best practices for study groups around the time of a midterm exam, and how to effectively study for a final in one of the last lessons. This helpful reference can be used in the classroom or assigned as homework.

Think It Through: Each chapter contains one or two of these features, which make the connection between mathematics and student life. These problems are student-relevant and require mathematics skills from the chapter to be applied to a real-life situation. Topics include tuition costs, statistics about college life, job opportunities, and many more topics directly connected to the student experience.

A colorful new design visually organizes information on the page.

▊ Proven Features of *Intermediate Algebra*

- The authors' proven five-step problem-solving strategy teaches students to analyze the problem, form an equation, solve the equation, state the conclusion, and check the result. In a step-by-step manner, this approach clarifies the thought process and mathematical skills necessary to solve a wide variety of problems. As a result, students' confidence is increased and their problem-solving abilities are strengthened.

- STUDY SETS are found at the end of every section and feature a unique organization, tailored to improve students' ability to read, write, and communicate mathematical ideas, thereby approaching topics from a variety of perspectives. Each comprehensive STUDY SET is divided into seven parts: VOCABULARY, CONCEPTS, NOTATION, PRACTICE, APPLICATIONS, WRITING, and REVIEW.

  - VOCABULARY, NOTATION, and WRITING problems help students improve their ability to read, write, and communicate mathematical ideas.

  - The CONCEPT problems section in the STUDY SETS reinforces major ideas through exploration and fosters independent thinking and the ability to interpret graphs and data.

  - PRACTICE problems in the STUDY SETS provide the necessary drill for mastery while the APPLICATIONS provide opportunities for students to deal with real-life situations. Each STUDY SET concludes with a REVIEW section that consists of problems randomly selected from previous sections.

- SELF CHECK problems, adjacent to most worked examples, reinforce concepts and build confidence. The answer to each Self Check follows the problem to give students instant feedback.

- The KEY CONCEPT section is a one-page review, found at the end of each chapter, which revisits the importance of the role the concept plays in the overall picture.

- REAL-LIFE APPLICATIONS are presented from a number of disciplines, including science, business, economics, manufacturing, entertainment, history, art, music, and mathematics.

- CALCULATOR SNAPSHOT sections introduce keystrokes and show how scientific calculators can be used to solve application problems, for instructors who wish to integrate calculators into their course.

- CUMULATIVE REVIEW EXERCISES at the end of every chapter except Chapter 1 help students retain what they have learned in prior chapters.

For detailed information on the ancillary resources available for this text, instructors can see the Preview section at the beginning of their book.

▊ Acknowledgments

We are grateful to the following people who reviewed this manuscript and the other manuscripts in the paperback series at various stages of development. They all had valuable suggestions that have been incorporated into the text.

The following people reviewed the first and second editions:

Linda Beattie
Western New Mexico University

Julia Brown
Atlantic Community College

Linda Clay
Albuquerque TVI

John Coburn
Saint Louis Community College–
Florissant Valley

Sally Copeland
Johnson County Community College

Ben Cornelius
Oregon Institute of Technology

James Edmondson
Santa Barbara Community College

David L. Fama
Germanna Community College

Barbara Gentry
Parkland College

Laurie Hoecherl
Kishwaukee College

Judith Jones
Valencia Community College

Therese Jones
Amarillo College

Joanne Juedes
University of Wisconsin–Marathon
County

Dennis Kimzey
Rogue Community College

Sally Lesik
Holyoke Community College

Elizabeth Morrison
Valencia Community College

Jan Alicia Nettler
Holyoke Community College

Scott Perkins
Lake–Sumter Community College

Angela Peterson
Portland Community College

J. Doug Richey
Northeast Texas Community College

Angelo Segalla
Orange Coast College

June Strohm
Pennsylvania State Community College–
DuBois

Rita Sturgeon
San Bernardino Valley College

Jo Anne Temple
Texas Technical University

Sharon Testone
Onondaga Community College

Marilyn Treder
Rochester Community College

Thomas Vanden Eynden
Thomas More College

The following people reviewed the series in preparation for the third edition:

Cedric E. Atkins
Mott Community College

William D. Barcus
SUNY, Stony Brook

Kathy Bernunzio
Portland Community College

Girish Budhwar
United Tribes Technical College

Sharon Camner
Pierce College–Fort Steilacoom

Robin Carter
Citrus College

Ann Corbeil
Massasoit Community College

Carolyn Detmer
Seminole Community College

Maggie Flint
Northeast State Technical Community
College

Charles Ford
Shasta College

Michael Heeren
Hamilton College

Monica C. Kurth
Scott Community College

Sandra Lofstock
St. Petersberg College–Tarpon Springs
Center

Marge Palaniuk
United Tribes Technical College

Jane Pinnow
University of Wisconsin–Parkside

Eric Sims
Art Institute of Dallas

Annette Squires
Palomar College

Lee Ann Spahr
Durham Technical Community College

John Strasser
Scottsdale Community College

Stuart Swain
University of Maine at Machias

Celeste M. Teluk
D'Youville College

Sven Trenholm
Herkeimer County Community College

Stephen Whittle
Augusta State University

Mary Lou Wogan
Klamath Community College

Without the talents and dedication of the editorial, marketing, and production staff of Brooks/Cole, this revision of *Intermediate Algebra* could not have been so well accomplished. We express our sincere appreciation for the hard work of Bob Pirtle, Jennifer Laugier, Helen Walden, Lori Heckleman, Vernon Boes, Diane Beasley, Sarah Woicicki, Greta Kleinert, Jessica Bothwell, Bryan Vann, Kirsten Markson, Rebecca Subity, Hal Humphrey, Ana deArmas, Christine Davis, Diane Koenig, Earl Perry, and G&S Typesetters for their help in creating the book. Special thanks to David Casey of Citrus College for his hard work on the pretests and to Sheila Pisa for writing the excellent Study Skills Workshops.

Alan S. Tussy
R. David Gustafson

For the Student

Success in Algebra

To be successful in mathematics, you need to know how to study it. The following checklist will help you develop your own personal strategy to study and learn the material. The suggestions below require some time and self-discipline on your part, but it will be worth the effort. This will help you get the most out of the course.

As you read each of the following statements, place a check mark in the box if you can truthfully answer Yes. If you can't answer Yes, think of what you might do to make the suggestion part of your personal study plan. You should go over this checklist several times during the semester to be sure you are following it.

Preparing for the Class

❑ I have made a commitment to myself to give this course my best effort.
❑ I have the proper materials: a pencil with an eraser, paper, a notebook, a ruler, a calculator, and a calendar or day planner.
❑ I am willing to spend a minimum of two hours doing homework for every hour of class.
❑ I will try to work on this subject every day.
❑ I have a copy of the class syllabus. I understand the requirements of the course and how I will be graded.
❑ I have scheduled a free hour after the class to give me time to review my notes and begin the homework assignment.

Class Participation

❑ I know my instructor's name.
❑ I will regularly attend the class sessions and be on time.
❑ When I am absent, I will find out what the class studied, get a copy of any notes or handouts, and make up the work that was assigned when I was gone.

❏ I will sit where I can hear the instructor and see the board.

❏ I will pay attention in class and take careful notes.

❏ I will ask the instructor questions when I don't understand the material.

❏ When tests, quizzes, or homework papers are passed back and discussed in class, I will write down the correct solutions for the problems I missed so that I can learn from my mistakes.

Study Sessions

❏ I will find a comfortable and quiet place to study.

❏ I realize that reading a math book is different from reading a newspaper or a novel. Quite often, it will take more than one reading to understand the material.

❏ After studying an example in the textbook, I will work the accompanying Self Check.

❏ I will begin the homework assignment only after reading the assigned section.

❏ I will try to use the mathematical vocabulary mentioned in the book and used by my instructor when I am writing or talking about the topics studied in this course.

❏ I will look for opportunities to explain the material to others.

❏ I will check all my answers to the problems with those provided in the back of the book (or with the *Student Solutions Manual*) and resolve any differences.

❏ My homework will be organized and neat. My solutions will show all the necessary steps.

❏ I will work some review problems every day.

❏ After completing the homework assignment, I will read the next section to prepare for the coming class session.

❏ I will keep a notebook containing my class notes, homework papers, quizzes, tests, and any handouts — all in order by date.

Special Help

❏ I know my instructor's office hours and am willing to go in to ask for help.

❏ I have formed a study group with classmates that meets regularly to discuss the material and work on problems.

❏ When I need additional explanation of a topic, I use the tutorial videos and the interactive CD, as well as the Web site.

❏ I make use of extra tutorial assistance that my school offers for mathematics courses.

❏ I have purchased the *Student Solutions Manual* that accompanies this text, and I use it.

To follow each of these suggestions will take time. It takes a lot of practice to learn mathematics, just as with any other skill.

No doubt, you will sometimes become frustrated along the way. This is natural. When it occurs, take a break and come back to the material after you have had time to clear your thoughts. Keep in mind that the skills and discipline you learn in this course will help make for a brighter future. Good luck!

Intermediate AlgebraNow Can Help You Succeed in Math

Intermediate AlgebraNow is a powerful online learning companion that helps you gauge your unique study needs and provides you with a personalized learning plan. The resources in Intermediate AlgebraNow give you all the learning tools you need to master core concepts. Intermediate AlgebraNow and this new edition of Tussy and

Gustafson's *Intermediate Algebra* enhance each other, providing you with a seamless learning system. Completely integrated with the textbook, icons in the text direct you to the online tutorials and TLE Labs found in Intermediate AlgebraNow.

The Intermediate AlgebraNow system consists of three powerful, easy-to-use assessment components:

1. **What Do I Know?**
   Take a diagnostic pre-test to find out.

2. **What Do I Need to Learn?**
   Your personalized learning plan helps you focus on the areas you need to study.

3. **What Have I Learned?**
   Take a post-test to see how your understanding of key concepts has improved; results can be e-mailed to your instructor.

■ What You'll Find in Your Personalized Learning Plan

Text-specific tutorials, live online tutoring, and TLE online labs provide you with multiple tools to help you get a better grade.

Text-Specific Tutorials

The view of a tutorial looks like this. To navigate between chapters and sections, use the drop-down menu below the top navigation bar. This will give you access to the study activities available for each section.

Math Toolbar

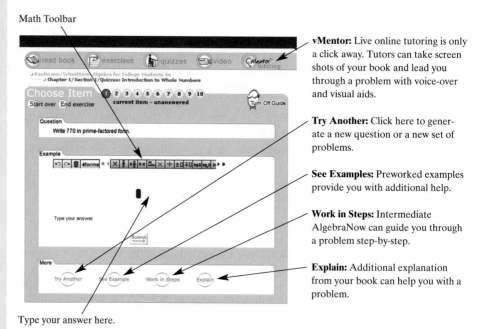

vMentor: Live online tutoring is only a click away. Tutors can take screen shots of your book and lead you through a problem with voice-over and visual aids.

Try Another: Click here to generate a new question or a new set of problems.

See Examples: Preworked examples provide you with additional help.

Work in Steps: Intermediate AlgebraNow can guide you through a problem step-by-step.

Explain: Additional explanation from your book can help you with a problem.

Type your answer here.

Online Tutoring with vMentor

Access to live online tutors and support through vMentor™. To access vMentor while you are working in the Exercises or Tutorial areas, click on the **vMentor Tutoring** button at the top right of the navigation bar above the problem or exercise.

Next, click on the **vMentor** button; you will be taken to a Web page that lists the steps for entering a vMentor classroom. If you are a first-time user of vMentor, you might need to download Java software before entering the class for the first class. You can either take an Orientation Session or log in to a vClass from the links at the bottom of the opening screen.

All vMentor Tutoring is done through a vClass, an Internet-based virtual classroom that features two-way audio, a shared whiteboard, chat, messaging, and experienced tutors.

You can access vMentor during the following times:

Monday through Thursday:

5 p.m. to 12 a.m. Pacific Time

6 p.m. to 1 a.m. Mountain Time

7 p.m. to 2 a.m. Central Time

8 p.m. to 3 a.m. Eastern Time

Sunday:

12 p.m. to 12 a.m. Pacific Time

1 p.m. to 1 a.m. Mountain Time

2 p.m. to 2 a.m. Central Time

3 p.m. to 3 a.m. Eastern Time

If you need additional help using vMentor, you can access the Participant Quick Reference Guide at this Web site: **http://www.elluminate.com/support/docs/Elive_Participant_Quick_Reference_Guide_6.0.pdf.**

TLE Online Labs

Use TLE Online Labs to explore and reinforce key concepts introduced in this text. These electronic labs give you access to additional instruction and practice problems, so you can explore each concept interactively, at your own pace. Not only will you be better prepared, but you will perform better in the class overall.

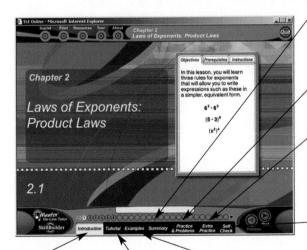

Summary: The Summary revisits the problem presented in the Introduction and encourages you to apply the mathematics you learned in the Tutorial and Examples.

Practice & Problems: Practice & Problems presents up to 25 questions organized in four or five categories.

Extra Practice: Extra practice presents questions like those in the Examples. After each question you have the option to try again, see the answer, see a sample solution, or try another question of the same type.

Self-Check: Self-Check presents up to 10 dynamically generated questions. To complete a lesson you must obtain the minimum standard (about 70%).

Introduction: Each lesson opens with objectives and prerequisites and provides brief instructions on using TLE.

Tutorial: The Tutorial provides the main instruction for the lesson. Hint and Success Tips teach strategies that can be used to solve the problem.

Examples: The examples expand on what you learned in the Tutorial. A hidden picture is progressively revealed as you complete each example.

■ Accessing Intermediate AlgebraNow through 1pass

Registering with the PIN Code on the 1pass Card *Situation:* Your instructor has not given you a PIN code for an online course, but you have a textbook with a 1pass PIN code. With 1pass, you have one simple PIN code access to all media resources associated with your textbook. Please refer to your 1pass card for a complete list of those resources.

Initial Log-in

To access your web gateway through 1pass:

1. Check the outside of your textbook to see if there is an additional 1pass card.
2. Take this card (and the additional 1pass card if appropriate) and go to http://1pass.thomson.com.
3. Type in your 1pass access code (or codes).
4. Follow the directions on the screen to set up your personal username and password.
5. Click through to launch your personal portal.
6. Access the media resources associated with your text . . . all the resources are just one click away.
7. Record your username and password for future visits and be sure to use the same username for all Thomson Learning resources.

For tech support, contact us at 1 (800) 423-0563.

> You will be asked to enter a valid e-mail address and password. Save your password in a safe place. You will need them to log in the next time you use 1pass. Only your e-mail address and password will allow you to reenter 1pass.

Subsequent Log-in

1. Go to **http://1pass.thomson.com.**
2. Type your e-mail address and password (see boxed information above) in the "Existing Users" box; then click on **Login.**

APPLICATIONS INDEX

Examples that are applications are shown with boldface page numbers.

Exercises that are applications are shown with lightface page numbers.

A Review of Basic Algebra

INTERMEDIATE
Algebra $f(x)$ **Now** ™

Throughout the chapter, this icon introduces resources on the Intermediate AlgebraNow Web site, accessed through **http://1pass .thomson.com**, that will

- Help you test your knowledge of the material with a pre-test and a post-test

- Provide a personalized learning plan targeting areas you should study

© Getty/Brand X Pictures

Today, banks and lending institutions offer a variety of financial services. To manage our money wisely, we need to be familiar with the terms and conditions of our checking accounts, credit cards, and loans. Financial transactions are described using whole numbers, fractions, and decimals. These numbers belong to a set that we call the *real numbers*.

To learn more about real numbers and how they are used in the financial world, visit *The Learning Equation* on the Internet at http://tle.brookscole.com. (The log-in instructions are in the Preface.) For Chapter 1, the online lesson is:

- *TLE* Lesson 1: The Real Numbers

Check Your Knowledge

1. A(n) _____ is a mathematical sentence that contains an = symbol.

2. The _____ of a geometric figure is the distance around it. The _____ of a figure is the amount of surface it encloses. The _____ of a solid figure is its capacity.

3. All points on the number line represent the set of _____ numbers.

4. Terms with exactly the same variables raised to exactly the same powers are called _____ terms.

5. The sum of two numbers is 12. If one number is x, write an expression that represents the other number.

6. Consider the set $\left\{ -\pi, -\dfrac{2}{3}, 0.125, -2, 0.\overline{9}, \dfrac{24}{5}, \sqrt{3} \right\}$.

   a. Graph the set on the number line.

   b. Which of the numbers are integers?

   c. Which of the numbers are rational?

   d. Which of the numbers are irrational?

Evaluate:

7. $-|3 - 5|$

8. $3 - 2^2$

9. $(-5)^3$

10. $\dfrac{1}{2} \div \left(-\dfrac{3}{2} \right)$

11. $\dfrac{3}{5} - \dfrac{7}{8}$

12. $27 - 5[3 - 2(7 - \sqrt{4})]$

13. $\sqrt{\dfrac{64}{49}}$

14. $\dfrac{-b - \sqrt{b^2 - 4ac}}{2a}$, for $a = 1$, $b = 2$, and $c = -3$

15. a. Solve $F = \dfrac{9}{5}C + 32$ for C.   b. Find C if $F = -40$.

16. What property of the real numbers justifies each statement?
    a. $-2(x - y) = -2x + 2y$
    b. $2x + (3x + 4) = (2x + 3x) + 4$

17. Simplify: $-6\left(7 + \dfrac{y}{36} - \dfrac{x}{2} \right)$.

18. Solve: $3x - 7 = 2(x - 2)$.   19. Solve: $\dfrac{y + 2}{2} - \dfrac{y - 3}{3} = \dfrac{5y + 4}{6}$.

20. Consider the set $\{4, 7, 12, 8, 12, 19, 15\}$.
    a. Find the mean.   b. Find the median.   c. Find the mode.

21. George needs 36 pounds of a coffee bean mixture worth $8.00 per pound. How many pounds each of $12.00 per pound coffee beans and $6.00 per pound coffee beans should he mix together?

22. Two trains are 340 miles apart, and one is traveling 15 miles per hour faster than the other. They travel toward each other and meet in 4 hours. Find the speed of each train.

Study Skills Workshop

PREPARING FOR SUCCESS

When starting any class, especially a math class, it's important to allot enough time to successfully complete your work. Being prepared can make you more confident and relaxed so you can concentrate on learning new material and recollecting what you have learned in past courses. Early in the course, make a study calendar, collect class materials, familiarize yourself with your textbook, and locate areas for quiet study.

Attend Class. Attending class every meeting is one of the most important things that you can do to succeed. Your instructor explains material and also may give extra information about topics that are not in your book, or may make changes in homework assignments or test dates. Think of your class as a big support group — it provides a structure for your studies and also helps to keep you accountable for your learning.

Time Management. Set aside enough time for homework and study. Two hours outside of class for every hour inside is a general rule of thumb for how much time you should devote to study. If your class meets for 5 hours per week, expect to spend 10 hours per week on homework. This is only a general rule — you might need to spend more time depending on the length of time since your last math class and the grade that you received, and the other classes you are taking. The best way to ensure that you have allotted enough time is to make up a calendar that shows all of your time commitments. Write in class times and study times for all of your classes, as well as other regularly scheduled events (work, church activities, etc.) Be sure to schedule some time to get rest and have fun, too. You can record this information using a regular calendar, a day planner, a PDA (personal digital assistant), or your computer.

Materials. By your second class meeting you should have all of the supplies that your instructor requires: textbook, notebook, pencils, and calculator are the usual materials, but check your syllabus if in doubt. Divide your notebook into three sections: one for class notes, one for returned homework assignments, and one for returned tests. This will allow you to keep track of graded assignments for future study or in case you have a discrepancy in grades. Bring all of your supplies to each class meeting. Take notes at every class meeting, and use them when going over homework and when studying for tests.

A Good Study Environment. Find a place where you can work efficiently, spread out a little, and be relatively free from interruptions. Some people prefer total silence in their study environment, while others like to have some background noise. Your study place may be in your home or in another location, like the school library. Certain places may work well at some times but not others.

▇ ASSIGNMENT

1. Download a calendar online at series.brookscole.com/tussypaperback, or make your own calendar with class times and study times for each course as well as times for work and other essential activities (e.g., church activities, social obligations, time with children, etc.). You may also want to schedule additional time to study a week before a test. Also include time for physical exercise and rest — this is important to reduce the effects of stress that school brings.
2. Download and print out the Course Information Sheet online at series.brookscole. com/tussypaperback, or make a list of the following items:
   - Instructor's name, office location and hours, phone number, email address
   - Test dates, if scheduled
   - The work that determines your course grade (homework, tests, class participation, etc.)
3. Make or find area(s) for study that are accessible during your study times.

Algebra is a mathematical language that can be used to solve many types of problems.

1.1   The Language of Algebra

- Variables, algebraic expressions, and equations • Verbal models • Constructing tables
- Graphical models • Formulas

Algebra is the result of contributions from many cultures over thousands of years. The word *algebra* comes from the title of the book *Ihm Al-jabr wa'l muqābalah*, written by the Arabian mathematician al-Khwarizmi around A.D. 800. Using the vocabulary, symbols, and notation of algebra, we can mathematically **model** the real world. In this section, we review some of the basic components of the language of algebra.

■ Variables, algebraic expressions, and equations

A rental agreement for a banquet hall is shown in Figure 1-1.

Rental Agreement
ROYAL VISTA BANQUET HALL
Wedding Receptions•Dances•Reunions•Fashion Shows

Rented To_____Date_____
Lessee's Address_____

Rental Charges
- $100 per hour
- Nonrefundable $200 cleanup fee

Terms and conditions
Lessor leases the undersigned lessee the above described property upon the terms and conditions set forth on this page and on the back of this page. Lessee promises to pay rental cost stated herein.

FIGURE 1-1

In the agreement, we see that two operations must be performed to calculate the cost of renting the hall.

- First, we must *multiply* the hourly rental cost of $100 by the number of hours the hall is to be rented.
- To that result, we must then *add* the cleanup fee of $200.

In words, we can describe the process as follows.

| The cost of renting the hall | is | 100 | times | the number of hours it is rented | plus | 200. |

We can describe this procedure more concisely by using *variables* and mathematical symbols. A **variable** is a letter that is used to stand for a number. If we let C stand for the cost of renting the hall and h stand for the number of hours it is rented, the words can be translated to form a *mathematical model*.

The cost of renting the hall	is	100	times	the number of hours it is rented	plus	200.
C	=	100	·	h	+	200

The statement $C = 100 \cdot h + 200$ is called an *equation.* The = symbol indicates that two quantities, C and $100 \cdot h + 200$, are equal.

> ### Equations
> An **equation** is a mathematical sentence that contains an = symbol.

Here are some examples of equations.

$$2 + 3 = 5 \qquad 3x - 2 = 4x + 10 \qquad P = a + b + c$$

In the equation $C = 100 \cdot h + 200$, the variable h is multiplied by 100. When we multiply a variable by a number, we can omit the symbol for multiplication. Thus, the equation $C = 100 \cdot h + 200$ can be written as $C = 100h + 200$.

On the right-hand side of $C = 100h + 200$, the notation $100h + 200$ is called an **algebraic expression,** or more simply, an **expression.**

> ### Algebraic expressions
> Variables and/or numbers can be combined with the operations of addition, subtraction, multiplication, division, raising to a power, and finding a root to create **algebraic expressions.**

Here are some examples of algebraic expressions.

$5a - 12$	This expression involves the operations of multiplication and subtraction.
$\dfrac{50 - y}{3y^3}$	This expression involves the operations of subtraction, multiplication, division, and raising to a power.
$\sqrt{a^2 + b^2}$	This expression involves the operations of addition, raising to a power, and finding a root.

◼ Verbal models

Table 1-1 lists some words and phrases that are often used in mathematics to denote the operations of addition, subtraction, multiplication, and division.

Addition +	Subtraction −	Multiplication ·	Division ÷
added to	subtracted from	multiplied by	divided by
plus	difference	product	quotient
the sum of	less than	times	ratio
more than	decreased by	percent (or fraction) of	half
increased by	reduced by	twice	into
greater than	minus	triple	per

TABLE 1-1

In the banquet hall example, the equation $C = 100h + 200$ was used to describe a procedure to calculate the cost of renting the hall. Using vocabulary from Table 1-1, we can write a **verbal model** that also describes this procedure. One such model is:

The cost (in dollars) of renting the hall is the *product* of 100 and the number of hours it is rented, *increased by* 200.

Self Check 1
After winning a lottery, three friends split the prize equally. Each person had to pay $2,000 in taxes on his or her share. Write a mathematical and a verbal model that relate the amount of each person's share, after taxes, to the amount of the lottery prize.

Answers $S = \dfrac{p}{3} - 2,000$; each person's share (in dollars), after taxes, is the quotient of the lottery prize and 3, decreased by 2,000

EXAMPLE 1   The cost to have a dinner catered is $6 per person. For groups of more than 200, a $100 discount is given. Write a mathematical and a verbal model that describe the relationship between the catering cost and the number of people being served for groups larger than 200.

Solution   To find the catering cost C (in dollars) for groups larger than 200, we need to *multiply* the number n of people served by $6 and then *subtract* the $100 discount.

The catering cost	is	6	times	the number of people served	minus	100.

Using variables, the mathematical model is:

$C = 6n - 100$

A verbal model is:

The catering cost (in dollars) is the *product* of 6 and the number of people served, *decreased by* 100.

Constructing tables

In the banquet hall example, the equation $C = 100h + 200$ can be used to determine the cost of renting the banquet hall for any number of hours.

Self Check 2
Find the cost of renting the hall for 6 hours and for 7 hours. Write the results in a table.

Answer

h	C
6	800
7	900

EXAMPLE 2   Find the cost of renting the hall for 3 hours and for 4 hours.

Solution   To write the results in a table, we choose appropriate column headings: h for the number of hours the hall is rented and C for the cost (in dollars) to rent the hall. Then we enter the number of hours of each rental time in the left column.

h	C
3	500
4	600

Next, we use the equation $C = 100h + 200$ to find the total rental cost for 3 hours and for 4 hours.

$C = 100h + 200$ $C = 100h + 200$
$C = 100(3) + 200$ Substitute 3 for h. $C = 100(4) + 200$ Substitute 4 for h.
$\quad = 300 + 200$ Multiply. $\quad = 400 + 200$ Multiply.
$\quad = 500$ $\quad = 600$

Finally, we enter these results in the right-hand column of the table: $500 for a 3-hour rental and $600 for a 4-hour rental.

Graphical models

The cost of renting the banquet hall for various lengths of time can also be presented graphically. The following **bar graph** in Figure 1-2(a) has a **horizontal axis** labeled "Number of hours the hall is rented." The **vertical axis** of the graph, labeled "Cost to rent the hall ($)," is scaled in units of 50 dollars. The bars directly over each of the times (1, 2, 3, 4, 5, 6, and 7 hours) extend to a height that gives the corresponding cost to rent the hall. For example, if the hall is rented for 5 hours, the height of the bar indicates that the cost is $700.

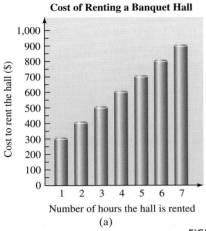
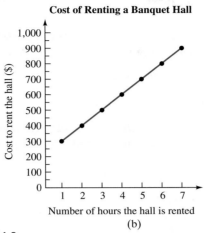

FIGURE 1-2

A **line graph** can also be used to show the rental costs. This type of graph consists of a series of dots drawn at the correct height, connected with line segments. (See Figure 1-2(b).) Not only does the line graph present all the information contained in the bar graph, but it also provides additional information. We can use the line graph to find the cost of renting the banquet hall for a length of time not shown in the bar graph.

EXAMPLE 3 Use the line graph in Figure 1-2(b) to determine the cost of renting the hall for $4\frac{1}{2}$ hours.

Solution In Figure 1-3, we locate $4\frac{1}{2}$ on the horizontal axis of the line graph and draw a line straight up to intersect the graph. From the point of intersection with the graph, we draw a horizontal line to the left that intersects the vertical axis. On the vertical axis, we can read that the rental cost is $650 for $4\frac{1}{2}$ hours.

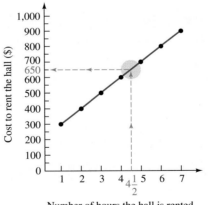

FIGURE 1-3

The video icons (see below) show which examples are taught on tutorial video tapes or disks.

INTERMEDIATE
Algebra *f(x)* Now™

Self Check 3
Use Figure 1-3 to find the cost of renting the hall for $6\frac{1}{2}$ hours.

Answer $850

■ Formulas

Equations that express a relationship between two or more quantities, represented by variables, are called **formulas.** Formulas are used in many fields, such as automotive technology, economics, medicine, retail sales, and banking.

EXAMPLE 4 Use variables to express each relationship as a formula.

a. The distance in miles traveled by a vehicle is the product of its average rate of speed in mph and the time in hours it travels at that rate.

b. The sale price of an item is the difference between the regular price and the discount.

INTERMEDIATE
Algebra *f(x)* Now™

Self Check 4
Express each relationship as a formula:
a. The retail price of an item is the sum of its wholesale cost and the markup.

b. The simple interest earned by a deposit is the product of the principal, the annual rate of interest, and the time (in years).

Answers a. $r = c + m$,
b. $I = Prt$

Solution

a. The word *product* indicates multiplication. If we let d stand for the distance traveled in miles, r for the vehicle's average rate of speed in mph, and t for the length of time traveled in hours, we can write the formula as

$$d = rt$$

b. The word *difference* indicates subtraction. If we let s stand for the sale price of the item, p for the regular price, and d for the discount, we have

$$s = p - d$$

The **perimeter** of a geometric figure is a measure of the distance around it. Table 1-2 shows the formulas for the perimeter P of several geometric figures.

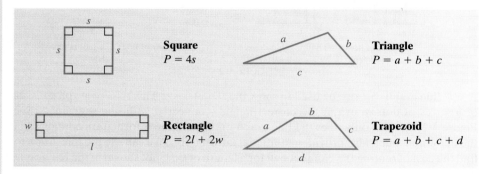

Square
$P = 4s$

Triangle
$P = a + b + c$

Rectangle
$P = 2l + 2w$

Trapezoid
$P = a + b + c + d$

TABLE 1-2

Self Check 5
Find the amount of fencing needed to enclose a triangular lot with sides that are 150 ft, 205.5 ft, and 165 ft long.

Answer 520.5 ft

EXAMPLE 5   Find the number of feet of redwood edging needed to outline a square flower bed having sides that are 6.5 feet long.

Solution   To find the amount of redwood edging needed, we need to find the perimeter of the square flower bed.

$P = 4s$ This is the formula for the perimeter of a square.

$P = 4(\mathbf{6.5})$ Substitute 6.5 for s, the length of one side of the square.

$ = 26$ Perform the multiplication.

26 feet of redwood edging is needed to outline the flower bed.

Section 1.1 STUDY SET

■ **VOCABULARY**   *Fill in the blanks.*

1. An _____ is a mathematical sentence that contains an = symbol.

2. A _____ is a letter that is used to stand for a number.

3. Variables and/or numbers can be combined with mathematical operations to create _____.

4. Phrases such as *the sum, increased by,* and *more than* are used to indicate the operation of _____.

5. A _____ is an equation that expresses a relationship between two or more quantities represented by variables.

6. Phrases such as *the difference, decreased by,* and *less than* are used to indicate the operation of _____.

7. Words such as *of, product,* and *twice* indicate the operation of _____.

8. The distance around a geometric figure is its _____.

CONCEPTS *Classify each of the following as an expression or an equation.*

9. $x - 5$

10. $x - 5 = 5$

11. $P = a + b + c$

12. $\dfrac{x + y}{8}$

13. $d = rt$

14. Prt

15. $2x^2$

16. $T = 6n - 400$

17. a. What type of graph is shown below?

    b. What units are used to scale the horizontal axis? The vertical axis?

    c. Estimate the height of the candle after it has burned for $3\frac{1}{2}$ hours. For 8 hours.

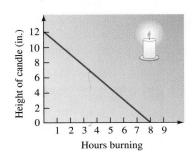

Hours burning

18. a. What type of graph is shown below?

    b. What units are used to scale the horizontal axis? The vertical axis?

    c. In what year was the average expenditure on auto insurance the least? Estimate the amount. In what year was it the greatest? Estimate the amount.

U.S. Average Consumer Expenditures on Auto Insurance

Source: Insurance Information Institute

Translate each verbal model to a mathematical model. (Answers may vary depending on the variables chosen.)

19. The cost each semester is $13 times the number of units taken plus a student services fee of $24.

20. The yearly salary is $25,000 plus $75 times the number of years of experience.

21. The quotient of the number of clients and 75 gives the number of social workers needed.

22. The difference between 500 and the number of people in a theater gives the number of unsold tickets.

23. Each test score was increased by 15 points to give a new adjusted test score.

24. The weight of a super-size order of French fries is twice that of a regular-size order.

25. The product of the number of boxes of crayons in a case and 12 gives the number of crayons in a case.

26. The perimeter of an equilateral triangle can be found by tripling the length of one of its sides.

Use the data in each table to find a formula that mathematically describes the relationship between the two quantities.

27.

Tower height t (ft)	Height of base h (ft)
15.5	5.5
22	12
25.25	15.25
45.125	35.125

28.

Seasonal employees s	Employees e
25	75
50	100
60	110
80	130

NOTATION

29. Algebraic expressions are combinations of variables, numbers, and the operations of mathematics. For the expression $2D + 4$,

    a. What variable is used?

    b. What numbers are used?

    c. What operations are used?

30. Consider the equation $T = 3r - 15$.

    a. How many variables does the equation contain?

    b. What expression forms the right-hand side of the equation?

    c. What operations are indicated on the right-hand side of the equation?

PRACTICE   *Use the given equation (formula) to complete each table.*

31. $c = \dfrac{p}{12}$

Number of packages p	Cartons c
24	
72	
180	

32. $y = 100c$

Number of centuries c	Years y
1	
6	
21	

33. $n = 22.44 - K$

K	n
0	
1.01	
22.44	

34. $y = x + 15$

x	y
0	
15	
30	

35. The lengths of the two parallel sides of a trapezoid are 10 inches and 15 inches. The other two sides are each 6 inches long. Find the perimeter of the trapezoid.

36. Find the perimeter of a rectangular-shaped playground that is 200 feet long and 75 feet wide.

37. Find the perimeter of a square quilt that has sides 2 yards long.

38. Find the perimeter of a triangular postage stamp with sides 1.8, 1.8, and 1.5 centimeters long.

APPLICATIONS

39. CARPET CLEANING See the following ad.

Rent the in-home
Carpet Cleaning System
Do it yourself and save!
Safe, effective
Costs only $10 an hour
plus $20 for supplies

  a. Write a formula that states the relationship between the cost C of renting the carpet-cleaning system and the number of hours h it is rented.

b. Use your result from part a to complete the table below, and then draw a line graph.

h	C
1	
2	
3	
4	
8	

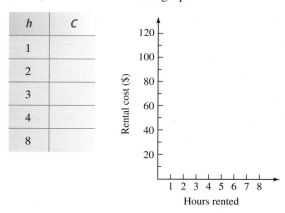

40. UTILITY VEHICLES What geometric concept applies when finding the length of the plastic trim around the cargo area floor mat shown in the illustration? Estimate the amount of trim used.

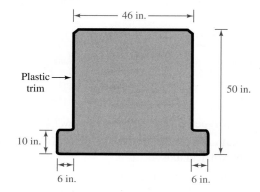

41. CARPENTRY A miter saw can pivot 180° to make angled cuts on molding. (See the illustration on the next page.) The formula that relates the angle measure s on the scrap piece of molding and the angle measure f on the finish piece of molding is $s = 180 - f$. Complete the following table and then draw a line graph.

f	s
30	
45	
90	
135	
150	

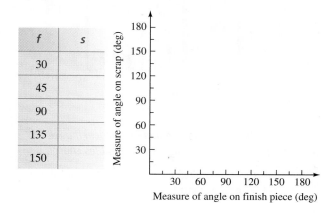

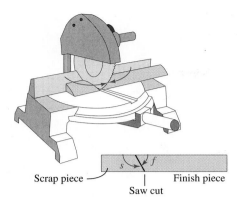

Scrap piece — | — Finish piece
Saw cut

s f

42. PRODUCTION PLANNING Use the following diagram to complete four formulas involving r that could be used by planners to order the necessary number of oak mounting plates p, bar holders b, chrome bars c, and mounting screws s for a production run that will produce r towel racks.

$$p = \quad \quad b = \quad r \quad \quad c = \quad \quad s = \quad r$$

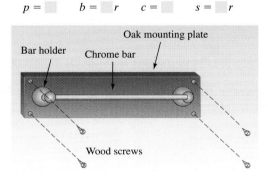

Oak mounting plate

Bar holder   Chrome bar

Wood screws

■ **WRITING**

43. Explain the difference between an expression and an equation. Give examples.

44. Name four ways in which numerical relationships were described in this section. Give examples.

45. Use each word below in a sentence so that a mathematical operation is indicated. If you are unsure of the meaning of a word, look it up in a dictionary.

quadrupled	*deleted*	*bisected*
confiscated	*annexed*	*docked*

46. Consider the formula $h = 7d$, where d is the age of a dog and h is the equivalent human age of the dog. Explain what information is given by the formula.

1.2 The Real Number System

- Natural numbers, whole numbers, and integers • Rational numbers • Irrational numbers
- Real numbers • The real number line • Inequality symbols • Absolute value

In this section, we will define each type of number used in this course. Then we will show that together, they form a collection of numbers called the *real numbers*.

■ Natural numbers, whole numbers, and integers

The graph in Figure 1-4 shows the daily low temperatures for Anchorage, Alaska, for the first seven days of January 1998. On the horizontal axis, the numbers 1, 2, 3, 4, 5, 6, and 7 have been used to denote the calendar days of the month. This collection of numbers is called a **set,** and the members (or **elements**) of the set can be listed within **braces** { }.

{1, 2, 3, 4, 5, 6, 7}

This is read as "the set containing the elements 1, 2, 3, 4, 5, 6, and 7."

Each of these numbers also belongs to a more extensive set of numbers that we use to count with, called the *natural numbers*.

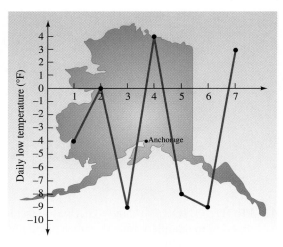

FIGURE 1-4

Natural numbers

The set of **natural numbers** is $\{1, 2, 3, 4, 5, 6, 7, 8, 9, 10, \ldots\}$.

The three dots used in the definition above indicate that the list of natural numbers continues on forever.

Two important sets of numbers included within the natural numbers are the prime numbers and the composite numbers.

Prime numbers

The **prime numbers** are the natural numbers greater than 1 that are divisible only by themselves and 1.

Composite numbers

The **composite numbers** are the natural numbers greater than 1 that are not prime numbers.

The first ten prime numbers are 2, 3, 5, 7, 11, 13, 17, 19, 23, and 29. The first ten composite numbers are 4, 6, 8, 9, 10, 12, 14, 15, 16, and 18. Since every prime number is a natural number, we say that the set of prime numbers is a **subset** of the set of natural numbers. Similarly, the set of composite numbers is a subset of the natural numbers.

! COMMENT Every composite number can be written as the product of prime numbers. For example,

$$6 = 3 \cdot 2, \qquad 25 = 5 \cdot 5, \qquad \text{and} \qquad 168 = 7 \cdot 3 \cdot 2 \cdot 2 \cdot 2$$

The natural numbers together with 0 make up the set of **whole numbers.**

Whole numbers

The set of **whole numbers** is $\{0, 1, 2, 3, 4, 5, 6, 7, 8, 9, 10, \ldots\}$.

Since every natural number is also a whole number, the set of natural numbers is a subset of the set of whole numbers.

The graph in Figure 1-4 contains both **positive numbers,** numbers greater than 0, and **negative numbers,** numbers less than 0. For example, on January 7, the low temperature in Anchorage was 3° F (3 degrees above zero). On January 3, the low temperature was −9° F (9 degrees below zero). On January 2, the low temperature was 0° F. Zero is neither positive nor negative. The numbers used to represent the temperatures in the graph are members of a set of numbers called the *integers*.

Integers

The set of **integers** is $\{\ldots, -4, -3, -2, -1, 0, 1, 2, 3, 4, \ldots\}$.

Integers that are divisible by 2 are called **even integers,** and integers that are not divisible by 2 are called **odd integers.**

Even integers: . . . , −6, −4, −2, 0, 2, 4, 6, . . .

Odd integers: . . . , −5, −3, −1, 1, 3, 5, . . .

Since every whole number is also an integer, the set of whole numbers is a subset of the set of integers.

Rational numbers

In this course, we will work with positive and negative fractions. For example, if money is deposited in a savings account for 7 months, we say it is deposited for $\frac{7}{12}$ of a year. If a tank loses 40 gallons of water over a 3-minute period, we can describe the change per minute in the number of gallons of water in the tank using the fraction $-\frac{40}{3}$.

We will also work with mixed numbers. For instance, we might speak of $5\frac{7}{8}$ cups of flour or of a river that is $3\frac{1}{2}$ feet below flood stage ($-3\frac{1}{2}$ ft). Fractions and mixed numbers belong to the set of numbers called the **rational numbers.** A rational number is any number that can be written as a fraction with an integer numerator and a nonzero integer denominator.

Rational numbers

A **rational number** is any number that can be written as $\frac{a}{b}$, where a and b represent integers and $b \neq 0$.

Some examples of rational numbers are $\frac{7}{2}$, $\frac{-40}{3}$ (which can be written $-\frac{40}{3}$), and $\frac{25}{99}$.

The mixed numbers $5\frac{7}{8}$ and $-3\frac{1}{2}$ are also rational numbers, because they can be written as fractions with integer numerators and nonzero integer denominators:

$$5\frac{7}{8} = \frac{47}{8} \qquad \text{and} \qquad -3\frac{1}{2} = -\frac{7}{2} = \frac{-7}{2}$$

All integers are rational numbers, because they can be written as fractions with a denominator of 1. For example, $-4 = \frac{-4}{1}$ and $0 = \frac{0}{1}$. Therefore, the set of integers is a subset of the rational numbers.

In this course, we will also work with decimals. For example,

- The interest rate of a loan was 11% or 0.11.
- In baseball, the distance from home plate to second base is 127.279 feet.
- The third-quarter loss for a business was −2.7 million dollars.

Terminating decimals such as 0.11, 127.279, and -2.7 are rational numbers, since they can be written as fractions with integer numerators and nonzero integer denominators.

$$0.11 = \frac{11}{100} \qquad 127.279 = 127\frac{279}{1,000} = \frac{127,279}{1,000} \qquad -2.7 = -2\frac{7}{10} = \frac{-27}{10}$$

Examples of **repeating decimals** are 0.333. . . and 4.252525. . . . The three dots indicate that the digits continue in the pattern shown. Any repeating decimal can be expressed as a fraction with an integer numerator and a nonzero integer denominator. For example, 0.333. . . $= \frac{1}{3}$ and 4.252525. . . $= 4\frac{25}{99} = \frac{421}{99}$. Since every repeating decimal can be written as a fraction, repeating decimals are also rational numbers.

Rational numbers

The set of **rational numbers** is the set of all terminating and all repeating decimals.

INTERMEDIATE
Algebra $f(x)$ **Now**™

Self Check 1

Change each fraction to a decimal and determine whether it terminates or repeats:

a. $\dfrac{25}{990}$

b. $\dfrac{47}{50}$

Answers a. $\frac{25}{990} = 0.02\overline{5}$, repeating decimal, **b.** $\frac{47}{50} = 0.94$, terminating decimal

EXAMPLE 1   Change each fraction to a decimal and determine whether the decimal terminates or repeats:

a. $\dfrac{4}{5}$ and **b.** $\dfrac{17}{6}$.

Solution

a. To change $\frac{4}{5}$ to a decimal, we divide the numerator by the denominator.

$$\begin{array}{r} .8 \\ 5\overline{)4.0} \\ \underline{4\,0} \\ 0 \end{array}$$ Write a decimal point and a 0 to the right of 4.

In decimal form, $\frac{4}{5}$ is 0.8. This is a terminating decimal.

b. To change $\frac{17}{6}$ to a decimal, we use a calculator to divide 17 by 6 and obtain 2.8333. . . . This repeating decimal can be written as $2.8\overline{3}$, where the **overbar** indicates that the 3 repeats.

▮ Irrational numbers

Some decimals neither terminate nor repeat. One example is the decimal that is represented by the Greek letter π (read as "pi").

$$\pi = 3.141592654. . .$$

This nonterminating, nonrepeating decimal cannot be written as a fraction with an integer numerator and denominator. Therefore, π is not a rational number; it is an *irrational number*.

Irrational numbers

Irrational numbers are nonterminating, nonrepeating decimals.

Many geometric formulas involve π. For example, to find the distance around a circle, called its **circumference,** we multiply the diameter of the circle by π. See Figure 1-5(a).

Another irrational number is $\sqrt{2}$, which is the length of a diagonal of a square that has sides of length 1. See Figure 1-5(b).

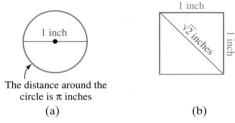

The distance around the
circle is π inches

(a)   (b)

FIGURE 1-5

We can use a calculator to approximate irrational numbers such as π and $\sqrt{2}$.

Approximating irrational numbers   CALCULATOR SNAPSHOT

To find an approximate value of π with a scientific calculator, we press the $\boxed{\pi}$ key.

$\boxed{\pi}$ (You may have to use a $\boxed{\text{2nd}}$ or $\boxed{\text{Shift}}$ key first.)

$$\boxed{3.141592654}$$

We see that $\pi \approx 3.141592654$. (Read $\approx$ as "is approximately equal to.") Rounded to the nearest thousandth, $\pi = 3.142$.

To approximate $\sqrt{2}$, we enter 2 and press the $\boxed{\sqrt{}}$ key (the square root key).

$2\boxed{\sqrt{}}$ $\boxed{1.414213562}$

We see that $\sqrt{2} \approx 1.414213562$. To the nearest hundredth, $\sqrt{2} = 1.41$.

To approximate π and $\sqrt{2}$ with a graphing calculator, we enter these numbers and press these keys.

$\boxed{\text{2nd}} \; \pi \boxed{\text{ENTER}}$

$\boxed{\text{2nd}} \boxed{\sqrt{}} 2 \boxed{)} \boxed{\text{ENTER}}$

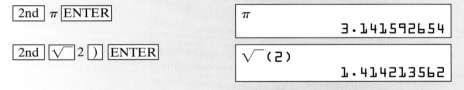

Other examples of irrational numbers are

$\sqrt{97} = 9.848857802...$

$-\sqrt{7} = -2.64575131...$ This is a negative irrational number.

$2\pi = 6.283185307...$ 2π means $2 \cdot \pi$.

Not all square roots are irrational numbers. When we simplify square roots such as $\sqrt{9}$, $\sqrt{36}$, and $\sqrt{400}$, it is apparent that they are rational numbers: $\sqrt{9} = 3$, $\sqrt{36} = 6$, and $\sqrt{400} = 20$.

! COMMENT Don't make the mistake of classifying a number such as 4.12122122212222... as a repeating decimal. Although this number exhibits a pattern, no single block of digits repeats forever. Therefore, it is a nonterminating, nonrepeating decimal, which is an irrational number.

▮ Real numbers

The set of rational numbers together with the set of irrational numbers form the set of **real numbers.** This means that every real number can be written as either a terminating, a repeating, or a nonterminating, nonrepeating decimal. Thus, the set of real numbers is the set of all decimals.

> **The real numbers**
>
> A **real number** is any number that is either a rational or an irrational number.
> All the points on a number line represent the set of real numbers.

Figure 1-6 shows how the sets of numbers introduced in this section are related; it also gives some specific examples of each type of number. Note that a number can belong to more than one set. For example, −6 is an integer, a rational number, and a real number.

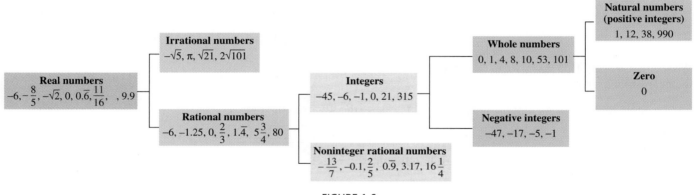

FIGURE 1-6

 INTERMEDIATE
Algebra *f(x)* Now™

Self Check 2
Use the instructions for Example 2 with the following set:
$$\left\{-\pi, -5, 3.4, \sqrt{19}, 1, \frac{16}{5}, 9.\overline{7}\right\}$$

Answers natural numbers: 1;
whole numbers: 1; integers: −5,
1; rational numbers: −5, 3.4, 1, $\frac{16}{5}$,
$9.\overline{7}$; irrational numbers: $-\pi$, $\sqrt{19}$;
real numbers: all

EXAMPLE 2   Which numbers in the following set are natural numbers, whole numbers, integers, rational numbers, irrational numbers, and real numbers?

$$\left\{\frac{5}{8}, -0.03, 45, -9, \sqrt{7}, 5\frac{2}{3}, 0, -1.727227222.\;.\;.\,, 0.\overline{25}\right\}$$

Solution

Natural numbers: 45

Whole numbers: 45, 0

Integers: 45, −9, 0

Rational numbers: $\frac{5}{8}$, 45, −9, $5\frac{2}{3}$, and 0 are rational numbers because each of them can be expressed as a fraction: $45 = \frac{45}{1}$, $-9 = \frac{-9}{1}$, $5\frac{2}{3} = \frac{17}{3}$, and $0 = \frac{0}{1}$. The terminating decimal −0.03 and the repeating decimal $0.\overline{25}$ are also rational numbers.

Irrational numbers: The nonterminating, nonrepeating decimals $\sqrt{7} = 2.645751311\ldots$ and $-1.727227222\ldots$ are irrational numbers.

Real numbers: $\frac{5}{8}$, −0.03, 45, −9, $\sqrt{7}$, $5\frac{2}{3}$, 0, −1.727227222. . . , $0.\overline{25}$

The real number line

We can illustrate real numbers using the **number line** shown in Figure 1-7. To each real number, there corresponds a point on the line and to each point on the line, there corresponds a number, called its **coordinate.** The point labeled 0 is called the **origin.**

The figure shows the **graph** of −3 and 4. As we move from left to right on the number line, the coordinates of the points get larger. Since the graph of 4 is to the right of the graph of −3, we know that 4 is greater than −3 and that −3 is less than 4.

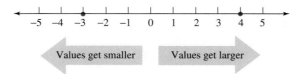

FIGURE 1-7

EXAMPLE 3 Graph the set on a number line.

$$\left\{-\frac{8}{3},\ -1.1,\ 0.\overline{56},\ \frac{\pi}{2},\ -\sqrt{15},\ 2\sqrt{2}\right\}$$

Solution
To help locate the graph of each number, we make some observations.

- Expressed as a mixed number, $-\frac{8}{3} = -2\frac{2}{3}$.
- Since −1.1 is less than −1, its graph is to the left of −1.
- $0.\overline{56} = 0.565656\ldots \approx 0.6$ Read ≈ as "is approximately equal to."
- From a calculator, $\frac{\pi}{2} \approx 1.6$.
- From a calculator, $-\sqrt{15} \approx -3.9$.
- $2\sqrt{2}$ means $2 \cdot \sqrt{2}$. From a calculator, $2\sqrt{2} \approx 2.8$.

Self Check 3
Graph the set on a number line.
$$\left\{\pi,\ -2.\overline{1},\ \sqrt{3},\ \frac{11}{4},\ -0.9\right\}$$

Answer

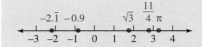

Inequality symbols

The **inequality symbols** shown in Table 1-3 are used to compare real numbers.

Symbol	Read as	Examples
≠	"is not equal to"	$6 \neq 9$ and $0.33 \neq \frac{3}{5}$
<	"is less than"	$\frac{22}{3} < \frac{23}{3}$ and $-7 < -6$
>	"is greater than"	$19 > 5$ and $\frac{1}{2} > 0.3$
≤	"is less than or equal to"	$3.5 \leq 3.\overline{5}$ and $1\frac{4}{5} \leq 1.8$
≥	"is greater than or equal to"	$29 \geq 29$ and $-15.2 \geq -16.7$

TABLE 1-3

It is always possible to write an inequality with the inequality symbol pointing in the opposite direction. For example,

$$-3 < 4 \qquad \text{is equivalent to} \qquad 4 > -3$$
$$5.3 \geq 2.9 \qquad \text{is equivalent to} \qquad 2.9 \leq 5.3$$

INTERMEDIATE
Algebra *(f(x))* **Now**™

Self Check 4
Use one of the symbols $\geq$ or $\leq$ to make each statement true:

a. $\dfrac{2}{3}$ ☐ $\dfrac{4}{3}$

b. $8\dfrac{1}{2}$ ☐ 8.4

Answers **a.** $\leq$, **b.** $\geq$

EXAMPLE 4   Use one of the symbols $>$ or $<$ to make each statement true:

a. -24 ☐ -25 and **b.** $\dfrac{3}{4}$ ☐ 0.76.

Solution
a. Since -24 is to the right of -25 on the number line, $-24 > -25$.

b. If we express the fraction $\frac{3}{4}$ as a decimal, we can easily compare it to 0.76. Since $\frac{3}{4} = 0.75$, we know that $\frac{3}{4} < 0.76$.

▨ Absolute value

In Figure 1-8, we can see that -3 and 3 are both a distance of 3 units away from zero on the number line. Because of this, we say that -3 and 3 are **opposites** or **additive inverses**.

Parentheses are used to express the opposite of a negative number. For example, the opposite of -3 is written as $-(-3)$. Since -3 and 3 are the same distance from zero, the opposite of -3 is 3. Symbolically, this can be written $-(-3) = 3$. This leads us to the following conclusion.

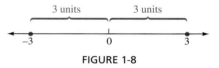

FIGURE 1-8

The double negative rule

If a is any real number, then $-(-a) = a$.

The **absolute value** of any real number is the distance between the number and zero on a number line. To indicate absolute value, the number is inserted between two vertical bars. For example, the points shown in Figure 1-8 with coordinates of 3 and -3 both lie 3 units from zero. Thus, $|3| = |-3| = 3$.

The absolute value of a number is defined more formally as follows.

Absolute value

For any real number a, $\begin{cases} \text{If } a \geq 0, \text{ then } |a| = a. \\ \text{If } a < 0, \text{ then } |a| = -a. \end{cases}$

❗ COMMENT Since absolute value expresses distance, the absolute value of a number is always positive or zero, but never negative.

INTERMEDIATE
Algebra *f(x)* **Now**™

EXAMPLE 5   Find the value of each expression: **a.** $|34|$,   **b.** $\left|-\dfrac{4}{5}\right|$,   **c.** $|0|$, and   **d.** $-|-1.8|$.

Solution

a. Since 34 is a distance of 34 from 0 on a number line, $|34| = 34$.

b. $-\dfrac{4}{5}$ is a distance of $\dfrac{4}{5}$ from 0 on a number line. Therefore, $\left|-\dfrac{4}{5}\right| = \dfrac{4}{5}$.

c. $|0| = 0$.

d. $-|-1.8| = -(1.8)$ Find $|-1.8|$ first: $|-1.8| = 1.8$.

          $= -1.8$ Rewrite without parentheses.

Self Check 5
Find the value of each expression:
a. $|-9.6|$
b. $-|-12|$
c. $\left|\dfrac{3}{2}\right|$

Answers **a.** 9.6,   **b.** -12,   **c.** $\dfrac{3}{2}$

Section 1.2   STUDY SET

INTERMEDIATE
Algebra *f(x)* **Now**™

VOCABULARY   *Fill in the blanks.*

1. A _____ number is any number that can be written as a fraction with an integer numerator and a nonzero integer denominator.

2. The _____ numbers are the natural numbers greater than 1 that are divisible only by themselves and 1.

3. The _____ _____ of any real number is the distance between the number and zero on a number line.

4. Together, the set of rational numbers and the set of irrational numbers form the set of _____ numbers.

5. _____ numbers are nonterminating, nonrepeating decimals.

6. _____ numbers are greater than 0, and _____ numbers are less than 0.

7. A _____ number is a natural number greater than 1 that is not prime.

8. The distance around a circle is called its _____.

CONCEPTS   *Which elements of the following set belong to each category?*

$$\left\{-3, -\tfrac{8}{5}, 0, \tfrac{2}{3}, 1, 2, \sqrt{3}, \pi, 4.75, 9, 16.\overline{6}\right\}$$

9. Natural numbers

10. Whole numbers

11. Integers

12. Rational numbers

13. Irrational numbers

14. Real numbers

15. Even natural numbers

16. Odd integers

17. Prime numbers

18. Composite numbers

19. Odd composite numbers

20. Odd prime numbers

Determine whether each number is a repeating or a nonrepeating decimal, and whether it is a rational or an irrational number.

21. $0.090090009\ldots$

22. $0.\overline{09}$

23. $5.41414141\ldots$

24. $1.414213562\ldots$

25. Show that each of the following numbers is a rational number by expressing it as a fraction with an integer numerator and a nonzero integer denominator.

$$7, \ -7\tfrac{3}{5}, \ 0.007, \ 700.1$$

26. Determine whether each statement is true or false.

   a. All prime numbers are odd numbers.

   b. $6 \geq 6$

   c. 0 is neither even nor odd.

   d. If x is negative, $|x| = -x$.

27. Find x if $|x| = 3.5$.

28. Name two numbers that are 6 units away from -2 on the number line.

29. The diagram below can be used to show how the natural numbers, whole numbers, integers, rational numbers, and irrational numbers make up the set of real numbers. If the natural numbers can be represented as shown, label each of the other sets.

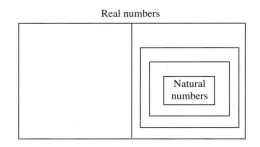

30. The formula $C = \pi D$ gives the circumference C of a circle, where D is the length of its diameter. Find the circumference of the wedding ring. Give an *exact* answer and then an *approximate* answer, rounded to the nearest hundredth of an inch.

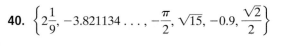

NOTATION   *Fill in the blanks.*

31. The symbol $<$ means "_____."

32. $|-2|$ is read as "the _____ value ____ -2."

33. The symbols $\{\ \}$ are called _____.

34. The symbol $\geq$ means "_____."

PRACTICE   *Change each fraction into a decimal and classify it as a terminating or a repeating decimal.*

35. $\dfrac{7}{8}$

36. $\dfrac{8}{3}$

37. $-\dfrac{11}{15}$

38. $-\dfrac{19}{16}$

Graph each set on the number line.

39. $\left\{ -\dfrac{5}{2}, -0.1, 2.142765\ldots, \dfrac{\pi}{3}, -\sqrt{11}, 2\sqrt{3} \right\}$

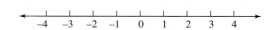

40. $\left\{ 2\dfrac{1}{9}, -3.821134\ldots, -\dfrac{\pi}{2}, \sqrt{15}, -0.9, \dfrac{\sqrt{2}}{2} \right\}$

41. $\{ 3.\overline{15}, \dfrac{22}{7}, 3\dfrac{1}{8}, \pi, \sqrt{10}, 3.1 \}$

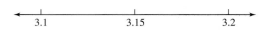

42. $\{ -0.\overline{331}, -0.331, -\dfrac{1}{3}, -\sqrt{0.11} \}$

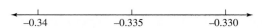

43. The set of prime numbers less than 8

44. The set of integers between -7 and 0

45. The set of odd integers between 10 and 18

46. The set of composite numbers less than 10

47. The set of positive odd integers less than 12

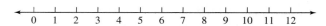

48. The set of negative even integers greater than -7

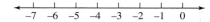

Insert either a $<$ or a $>$ symbol to make a true statement.

49. $8 \quad 9$

50. $9 \quad 0$

51. $-(-5) \quad -10$

52. $|-3| \quad -(-6)$

53. $-7.999 \quad -7.1$

54. $4\dfrac{1}{2} \quad \dfrac{7}{2}$

55. $6.\overline{1} \quad 6$

56. $-6.07 \quad -\dfrac{17}{6}$

Write each statement with the inequality symbol pointing in the opposite direction.

57. $19 > 12$

58. $-3 \geq -5$

59. $-6 \leq -5$

60. $-10 < 0$

Find the value of each expression.

61. $|20|$

62. $|-20|$

63. $-|-6|$

64. $-|-8|$

65. $|-5.9|$

66. $-|1.\overline{27}|$

67. $\left|\dfrac{5}{4}\right|$

68. $\left|-\dfrac{5}{16}\right|$

APPLICATIONS

69. DRAFTING Express each dimension in the following drawing of a bracket as a four-place decimal.

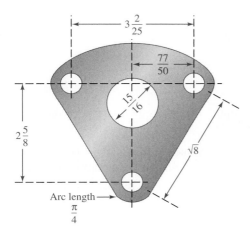

70. pH SCALE The pH scale, shown below is used to measure the strength of acids and bases (alkalines) in chemistry. It can be thought of as a number line. On the scale, graph and label each pH measurement given in the table.

Solution	pH
Seawater	8.5
Cola	2.9
Battery acid	1.0
Milk	6.6
Blood	7.4
Ammonia	11.9
Saliva	6.1
Oven cleaner	13.2
Black coffee	5.0
Toothpaste	9.9
Tomato juice	4.1

Strong acid

Increasing acidity

Neutral

Increasing alkalinity

Strong base

0 1 2 3 4 5 6 7 8 9 10 11 12 13 14

71. HELICOPTERS Refer to the illustration. How far does a point on the tip of a rotor blade travel when it makes one complete revolution? Round to the nearest hundredth of a foot. (*Hint:* Use the formula $C = \pi D$.)

72. TIRES The distance a tire rolls in one revolution can be found by computing the circumference of the circular tire. How far will the tire shown below roll in one revolution? Round to the nearest tenth of an inch. (*Hint:* Use the formula $C = \pi D$.)

One revolution

WRITING

73. Explain why the whole numbers are a subset of the integers.

74. What is a real number? Give examples.

75. Explain why there are no even prime numbers greater than 2.

76. Explain why every integer is a rational number, but not every rational number is an integer.

REVIEW

77. Is $\dfrac{3x - 4}{2}$ an equation or an expression?

78. Translate into mathematical symbols: The weight of an object in ounces is 16 times its weight in pounds.

Complete each table.

79. $T = x - 1.5$

x	T
3.7	
10	
30.6	

80. $j = 3m$

m	j
0	
15	
300	

1.3 Operations with Real Numbers

- Rules for adding real numbers • A rule for subtracting real numbers
- Rules for multiplying real numbers • Rules for dividing real numbers
- Raising a real number to a power • Finding a square root of a real number
- Order of operations • Evaluating algebraic expressions • Area and volume

Six operations can be performed with real numbers: addition, subtraction, multiplication, division, raising to a power, and finding a root. In this section, we will review the rules for performing each of these operations. We will also discuss how to evaluate numerical expressions involving several operations.

Rules for adding real numbers

When two numbers are added, we call the result their **sum.** The rules for adding two real numbers are as follows.

> **Adding two real numbers**
>
> *With like signs:* Add the absolute values of the numbers and keep the common sign.
>
> *With unlike signs:* Subtract the absolute values of the numbers (the smaller from the larger) and keep the sign of the number with the larger absolute value.

Self Check 1
Add:
a. $-34 + 25$
b. $-70.4 + (-21.2)$
c. $\dfrac{7}{4} + \left(-\dfrac{3}{2}\right)$

Answers **a.** -9, **b.** -91.6,
c. $\dfrac{1}{4}$

EXAMPLE 1   Find each sum: **a.** $-5 + (-3)$,   **b.** $8.9 + (-5.1)$,   and
c. $-\dfrac{13}{15} + \dfrac{3}{5}$.

Solution
a. $-5 + (-3) = -8$ Because the numbers have like signs, add their absolute values, 5 and 3, to get 8. Keep the common negative sign.

b. $8.9 + (-5.1) = 3.8$ Since the numbers have unlike signs, subtract their absolute values, 5.1 from 8.9, to get 3.8. Because 8.9 has the larger absolute value, the answer is positive.

c. $-\dfrac{13}{15} + \dfrac{3}{5} = -\dfrac{13}{15} + \dfrac{9}{15}$ Express $\dfrac{3}{5}$ in terms of the lowest common denominator, 15: $\dfrac{3}{5} = \dfrac{3 \cdot 3}{5 \cdot 3} = \dfrac{9}{15}$.

$= -\dfrac{4}{15}$ Subtract the absolute values. $\dfrac{9}{15}$ from $\dfrac{13}{15}$, to get $\dfrac{4}{15}$. Keep the $-$ sign, because $-\dfrac{13}{15}$ has the larger absolute value.

A rule for subtracting real numbers

When two numbers are subtracted, we call the result their **difference.** To find a difference, we can change the subtraction into an equivalent addition. For example, the subtraction $7 - 4$ is equivalent to the addition $7 + (-4)$, because they have the same answer:

$$7 - 4 = 3 \quad \text{and} \quad 7 + (-4) = 3$$

This suggests that to subtract two numbers, we can change the sign of the number being subtracted and add.

Subtracting two real numbers

If a and b represent real numbers, then $a - b = a + (-b)$.

This rule indicates that *subtraction is the same as adding the opposite of the number to be subtracted.*

EXAMPLE 2   Find each difference: **a.** $12 - 4$,   **b.** $-1.3 - 5.5$,   and
c. $-\frac{14}{3} - \left(-\frac{7}{3}\right)$.

Solution   In each case, we change the sign of the number that is being subtracted and add.

a. $12 - 4 = 12 + (-4)$

  Here, 4 is being subtracted, so we change the sign of 4 and add.

  $= 8$

  Use the rule for adding two numbers with unlike signs.

b. $-1.3 - 5.5 = -1.3 + (-5.5)$ Change the sign of 5.5 and add.

  $= -6.8$ Use the rule for adding two numbers with like signs.

c. $-\dfrac{14}{3} - \left(-\dfrac{7}{3}\right) = -\dfrac{14}{3} + \dfrac{7}{3}$ Change the sign of $-\frac{7}{3}$ and add.

  $= -\dfrac{7}{3}$ Use the rule for adding two numbers with unlike signs.

INTERMEDIATE
Algebra *f(x)* **Now**™

Self Check 2

Subtract:

a. $-15 - 4$

b. $-12.1 - (-7.6)$

c. $\dfrac{5}{9} - \dfrac{7}{9}$

Answers **a.** -19, **b.** -4.5,

c. $-\dfrac{2}{9}$

■ Rules for multiplying real numbers

When two numbers are multiplied, we call the numbers **factors** and the result their **product.** The rules for multiplying two real numbers are as follows.

Multiplying two real numbers

With like signs: Multiply their absolute values. The answer is positive.

With unlike signs: Multiply their absolute values. Make the answer negative.

EXAMPLE 3   Find each product: **a.** $4(-7)$,   **b.** $-5.2(-3)$,   and
c. $-\dfrac{7}{9}\left(\dfrac{3}{16}\right)$.

Solution

a. $4(-7) = -28$

  Multiply the absolute values of 4 and -7: $4 \cdot 7 = 28$. Since the signs of 4 and -7 are unlike, make the answer negative.

b. $-5.2(-3) = 15.6$

  Multiply the absolute values: $5.2 \cdot 3 = 15.6$. Since the signs of -5.2 and -3 are like, the answer is positive.

c. $-\dfrac{7}{9}\left(\dfrac{3}{16}\right) = -\dfrac{7 \cdot 3}{9 \cdot 16}$

  Multiply the numerators and multiply the denominators. Since the signs of the factors are unlike, the answer is negative.

  $= -\dfrac{7 \cdot \overset{1}{\cancel{3}}}{\underset{1}{\cancel{3}} \cdot 3 \cdot 16}$

  Simplify the fraction: Factor 9 as $3 \cdot 3$ and remove the common factor of 3 in the numerator and denominator.

  $= -\dfrac{7}{48}$

  Multiply in the numerator and denominator.

INTERMEDIATE
Algebra *f(x)* **Now**™

Self Check 3

Multiply:

a. $(-6)(5)$

b. $(-4.1)(-8)$

c. $\left(\dfrac{4}{3}\right)\left(-\dfrac{1}{8}\right)$

Answers **a.** -30, **b.** 32.8,

c. $-\dfrac{1}{6}$

Self Check 4
Find each product:
a. $-2(-30)(-5)$
b. $-0.4(2)(100)(-2)$

EXAMPLE 4   Find each product: **a.** $-5(12)(-1)$ and
b. $-1.2(-10)(4)(-3)$.

Solution   In each case, we perform the multiplications working from left to right.

a. $-5(12)(-1) = -60(-1)$ Multiply -5 and 12. The product of two numbers with unlike signs in negative.

$= 60$ The product of two numbers with like signs is positive.

b. $-1.2(-10)(4)(-3) = 12(4)(-3)$ Multiply -1.2 and -10. The product of two numbers with like signs is positive.

$= 48(-3)$ Multiply 12 and 4. The product is positive.

$= -144$ The product of two numbers with unlike signs is negative.

Answers **a.** -300,   **b.** 160

■ Rules for dividing real numbers

When two numbers are divided, we call the result their **quotient.** The rules for dividing two real numbers are as follows.

> **Dividing two real numbers**
> *With like signs:* Divide their absolute values. The answer is positive.
>
> *With unlike signs:* Divide their absolute values. Make the answer negative.

INTERMEDIATE
Algebra *f(x)* **Now**™

Self Check 5
Divide:
a. $\dfrac{55}{-5}$
b. $\dfrac{-7.2}{-6}$

EXAMPLE 5   Find each quotient: **a.** $\dfrac{-44}{11}$ and   **b.** $\dfrac{-2.7}{-9}$.

Solution

a. $\dfrac{-44}{11} = -4$ Divide the absolute values of -44 and 11: $\frac{44}{11} = 4$. Since the signs of -44 and 11 are unlike, make the answer negative.

b. $\dfrac{-2.7}{-9} = 0.3$ Divide the absolute values: $\frac{2.7}{9} = 0.3$. Since the signs of -2.7 and -9 are like, the answer is positive.

Answers **a.** -11,   **b.** 1.2

To divide two fractions, we multiply the first fraction by the **reciprocal** of the second fraction. In symbols, if a, b, c, and d represent real numbers and b, c, and d are not 0, then

$$\frac{a}{b} \div \frac{c}{d} = \frac{a}{b} \cdot \frac{d}{c} \qquad \frac{d}{c} \text{ is the reciprocal of } \frac{c}{d}.$$

INTERMEDIATE
Algebra *f(x)* **Now**™

Self Check 6
Divide:
a. $-\dfrac{7}{8} \div \dfrac{2}{3}$
b. $-\dfrac{1}{10} \div (-5)$

EXAMPLE 6   Find each quotient: **a.** $\dfrac{2}{3} \div \left(-\dfrac{3}{5}\right)$ and   **b.** $-\dfrac{1}{2} \div (-6)$.

Solution

a. $\dfrac{2}{3} \div \left(-\dfrac{3}{5}\right) = \dfrac{2}{3} \cdot \left(-\dfrac{5}{3}\right)$ Multiply by the reciprocal of $-\dfrac{3}{5}$, which is $-\dfrac{5}{3}$.

$= -\dfrac{10}{9}$ The factors have unlike signs, so the answer is negative.

b. $-\dfrac{1}{2} \div (-6) = -\dfrac{1}{2} \cdot \left(-\dfrac{1}{6}\right)$ Multiply by the reciprocal of -6, which is $-\dfrac{1}{6}$.

$\qquad\qquad\quad = \dfrac{1}{12}$ The factors have like signs, so the answer is positive.

Answers **a.** $-\dfrac{21}{16}$, **b.** $\dfrac{1}{50}$

■ Raising a real number to a power

Exponents indicate repeated multiplication. For example,

$3^2 = 3 \cdot 3$ Read 3^2 as "3 to the second power" or "3 squared."

$(-9.1)^3 = (-9.1)(-9.1)(-9.1)$ Read $(-9.1)^3$ as "-9.1 to the third power" or "-9.1 cubed".

$\left(\dfrac{2}{3}\right)^4 = \left(\dfrac{2}{3}\right)\left(\dfrac{2}{3}\right)\left(\dfrac{2}{3}\right)\left(\dfrac{2}{3}\right)$ Read $\left(\dfrac{2}{3}\right)^4$ as "$\dfrac{2}{3}$ to the fourth power."

These examples suggest the following definition.

Natural-number exponents

If n represents a natural number, then

$$\overbrace{x^n = x \cdot x \cdot x \cdot \,\cdots\, \cdot x}^{n \text{ factors of } x}$$

The **exponential expression x^n** is called a **power of x,** and we read it as "x to the nth power." In this expression, x is called the **base,** and n is called the **exponent.** A natural-number exponent tells how many times the base of an exponential expression is to be used as a factor in a product.

Base $\rightarrow x^n \leftarrow$ Exponent

EXAMPLE 7   Find each power: **a.** $(-2)^5$,   **b.** $\left(\dfrac{3}{4}\right)^2$,   and   **c.** 0.1 cubed.

Solution   In each case, we use the fact that an exponent tells how many times the base is to be used as a factor in a product.

a. $(-2)^5 = (-2)(-2)(-2)(-2)(-2) = -32$ The base is -2. The exponent is 5.

b. $\left(\dfrac{3}{4}\right)^2 = \dfrac{3}{4}\left(\dfrac{3}{4}\right) = \dfrac{9}{16}$ The base is $\dfrac{3}{4}$. The exponent is 2.

c. 0.1 cubed means $(0.1)^3$.   The base is 0.1. The exponent is 3.

$(0.1)^3 = (0.1)(0.1)(0.1) = 0.001$

INTERMEDIATE
Algebra $f(x)$ **Now**™
Self Check 7
Find each power:
a. $(-3)^3$
b. $(0.8)^2$
c. 2^4
d. $\dfrac{7}{5}$ squared

Answers **a.** -27, **b.** 0.64, **c.** 16, **d.** $\dfrac{49}{25}$

❗ COMMENT   Although the expressions $(-3)^2$ and -3^2 look alike, they are not. In $(-3)^2$, the base is -3. In -3^2, the base is 3. The $-$ sign in front of 3^2 means the negative of 3^2. When we evaluate them, we see that the results are different:

$(-3)^2 = (-3)(-3)$ $-3^2 = -(3 \cdot 3)$
$\qquad\;\; = 9$ $\qquad\; = -9$

■ Finding a square root of a real number

Since the product $3 \cdot 3$ can be denoted by the exponential expression 3^2, we say that 3 is squared. The opposite of squaring a number is called finding its **square root.**

All positive numbers have two square roots: one positive and one negative. For example, the two square roots of 9 are 3 and -3. The number 3 is a square root of 9, because $3^2 = 9$, and -3 is a square root of 9, because $(-3)^2 = 9$.

The symbol $\sqrt{}$, called a **radical symbol,** is used to represent the positive (or *principal*) square root of a number.

> **Principal square root**
>
> A number b is a square root of a if $b^2 = a$.
>
> If $a > 0$, the expression $\sqrt{a}$ represents the **principal** (or positive) **square root** of a. The principal square root of 0 is 0: $\sqrt{0} = 0$.

The principal square root of a positive number is always positive. Although 3 and -3 are both square roots of 9, only 3 is the principal square root. The symbol $\sqrt{9}$ represents 3. To represent -3, we place a $-$ sign in front of the radical symbol:

$$\sqrt{9} = 3 \qquad \text{and} \qquad -\sqrt{9} = -3$$

INTERMEDIATE
Algebra *f(x)* **Now**™

Self Check 8

Find each square root:
a. $\sqrt{64}$, **b.** $-\sqrt{100}$, **c.** $\sqrt{\dfrac{4}{25}}$,
d. $\sqrt{1}$, **e.** $\sqrt{0.81}$, and
f. $-\sqrt{400}$.

Answers **a.** 8, **b.** -10, **c.** $\frac{2}{5}$,
d. 1, **e.** 0.9, **f.** -20

EXAMPLE 8 Find **a.** $\sqrt{121}$, **b.** $-\sqrt{49}$, **c.** $\sqrt{\dfrac{1}{4}}$, and **d.** $\sqrt{0.09}$.

Solution

a. $\sqrt{121} = 11$, because $11^2 = 121$.

b. Since $\sqrt{49} = 7$, we have $-\sqrt{49} = -7$.

c. $\sqrt{\dfrac{1}{4}} = \dfrac{1}{2}$, because $\left(\dfrac{1}{2}\right)^2 = \dfrac{1}{4}$.

d. $\sqrt{0.09} = 0.3$, because $(0.3)^2 = 0.09$.

■ Order of operations

We will often have to **evaluate** expressions involving more than one operation. For example, consider $3 + 2 \cdot 5$. We can evaluate it in two different ways. We can perform the addition first and then the multiplication. Or we can perform the multiplication first and then the addition. However, we get different results.

Method 1: **Perform the addition first**	**Method 2:** **Perform the multiplication first**
$3 + 2 \cdot 5 = 5 \cdot 5$ Add 3 and 2 first.	$3 + 2 \cdot 5 = 3 + 10$ Multiply 2 and 5 first.
$= 25$ Perform the multiplication.	$= 13$ Perform the addition.

$$\longleftarrow \text{Different results} \longrightarrow$$

This example shows the need to establish an order of operations. To avoid the possibility of obtaining two answers, we will use the following set of priority rules.

Rules for the order of operations

1. Perform all calculations within parentheses and other grouping symbols, following the order listed in steps 2–4 and working from the innermost pair to the outermost pair.

2. Evaluate all exponential expressions (powers) and roots.

3. Perform all multiplications and divisions as they occur from left to right.

4. Perform all additions and subtractions as they occur from left to right.

When all grouping symbols have been removed, repeat steps 2–4 to complete the calculation.

If a fraction bar is present, evaluate the expression above the bar (the *numerator*) and the expression below the bar (the *denominator*) separately. Then perform the division indicated by the fraction bar, if possible.

Grade Point Average (GPA)

THINK IT THROUGH

"In considering all of the factors that are important to employers as they recruit students in colleges and universities nationwide, college major, grade point average, and work related experience usually rise to the top of the list."
Mary D. Feduccia, Ph.D., Career Services Director, Louisiana State University

A grade point average (GPA) is a weighted average based on the grades received and the number of units (credit hours) taken. A GPA for one semester (or term) is defined as:

> *the quotient of the sum of the grade points earned for each class and the sum of the number of units taken. The number of grade points earned for a class is the product of the number of units assigned to the class and the value of the grade received in the class.*

1. Use the table of grade values to compute the GPA for the student whose semester grade report is shown. Round to the nearest hundredth.

Grade	Value
A	4
B	3
C	2
D	1
F	0

Class	Units	Grade
Geology	4	C
Algebra	5	A
Psychology	3	C
Spanish	2	B

2. If you were enrolled in school last semester (or term), list the classes taken, units assigned, and grades received in a grade report table like the one shown in Exercise 1. Then calculate your GPA.

To evaluate $3 + 2 \cdot 5$ correctly, we must apply the rules for the order of operations. Since the expression does not contain grouping symbols or exponential expressions, we follow steps 3 and 4 of the previous list.

$$3 + \mathbf{2 \cdot 5} = 3 + \mathbf{10} \qquad \text{Ignore the addition for now and multiply 2 and 5.}$$
$$= 13 \qquad \text{Perform the addition.}$$

Using the rules for the order of operations, we see that the correct answer is 13.

INTERMEDIATE
Algebra *f(x)* **Now™**

Self Check 9
Evaluate each expression:
a. $-9 + 2(-4)^2$ and
b. $20 \div (-5) - (-6)(-5) + (-12)$

EXAMPLE 9   Evaluate each expression: **a.** $-5 + 4(-3)^2$ and
b. $-10 \div 5 - 5(3) + 6$.

Solution
a. Although the expression contains parentheses, there are no operations to perform within the parentheses. So we proceed with steps 2, 3, and 4 of the rules for the order of operations.

$$-5 + 4(-\mathbf{3})^2 = -5 + 4(\mathbf{9}) \qquad \text{First, evaluate the power: } (-3)^2 = 9.$$
$$= -5 + 36 \qquad \text{Perform the multiplication.}$$
$$= 31 \qquad \text{Perform the addition.}$$

b. Since the expression does not contain any powers, we perform the multiplications and divisions, working from left to right.

$$-\mathbf{10 \div 5} - 5(3) + 6 = -\mathbf{2} - 5(3) + 6 \qquad \text{Perform the division: } -10 \div 5 = -2.$$
$$= -2 - 15 + 6 \qquad \text{Perform the multiplication.}$$
$$= -17 + 6 \qquad \text{Working from left to right, perform the subtraction: } -2 - 15 = -17.$$
$$= -11 \qquad \text{Perform the addition.}$$

Answers **a.** 23,   **b.** -46

Grouping symbols serve as mathematical punctuation marks. They help determine the order in which an expression is evaluated. Examples of grouping symbols are parentheses ( ), brackets [ ], and the fraction bar ———.

Self Check 10
Evaluate each expression:
a. $(5 - 3)^3 - 40$
b. $-3[-2(5^3 - 3) + 4] - 1$

EXAMPLE 10   Evaluate each expression: **a.** $3 - (4 - 8)^2$ and
b. $2 + 3[-2 - 8(4 - 3^2)]$.

Solution
a. We begin by performing the operation within the parentheses.

$$3 - (\mathbf{4 - 8})^2 = 3 - (-\mathbf{4})^2 \qquad \text{Perform the subtraction: } 4 - 8 = -4.$$
$$= 3 - 16 \qquad \text{Evaluate the power: } (-4)^2 = 16.$$
$$= -13 \qquad \text{Perform the subtraction.}$$

b. First, we perform the work within the innermost grouping symbols (the parentheses), using the rules for the order of operations.

$$2 + 3[-2 - 8(4 - \mathbf{3^2})] = 2 + 3[-2 - 8(4 - \mathbf{9})] \qquad \text{Find the power: } 3^2 = 9.$$
$$= 2 + 3[-2 - 8(-5)] \qquad \text{Perform the subtraction: } 4 - 9 = -5.$$

Next, we work within the brackets.

$$= 2 + 3[-2 - (-40)] \qquad \text{Perform the multiplication: } 8(-5) = -40.$$
$$= 2 + 3(-2 + 40)$$

Since only one set of grouping symbols was needed, we wrote $-2 + 40$ within parentheses.

$$= 2 + 3(38) \quad \text{Add: } -2 + 40 = 38.$$
$$= 2 + 114 \quad \text{Perform the multiplication.}$$
$$= 116 \quad \text{Perform the addition.}$$

Answers **a.** -32, **b.** 719

INTERMEDIATE
Algebra *f(x)* **Now**™

EXAMPLE 11   Evaluate: $-5|-45 + 30|$.

Self Check 11
Evaluate: $2|-25 - (-6)(3)|$.

Solution $-5|-45 + 30|$ means $-5 \cdot |-45 + 30|$. Since absolute value symbols are grouping symbols, we perform the operation within them first.

$$-5|{-45} + 30| = -5|{-15}| \quad \text{Perform the addition within the absolute value symbol.}$$
$$= -5 \cdot 15 \quad \text{Find the absolute value: } |-15| = 15.$$
$$= -75 \quad \text{Perform the multiplication.}$$

Answer   14

Order of operations

CALCULATOR SNAPSHOT

Calculators are programmed to follow the rules for the order of operations. For example, when computing $3 + 2 \cdot 5$, both scientific and graphing calculators give the correct answer, 13.

3 [+] 2 [×] 5 [=]

| 13 |

3 [+] 2 [×] 5 [ENTER]

```
3+2*5
        13
```

Both types of calculators use left and right parentheses keys [(] [)] when grouping symbols are needed. To evaluate $3 - (4 - 8)^2$, we enter these numbers and press these keys.

3 [−] [(] 4 [−] 8 [)] [x^2] [=]

| −13 |

3 [−] [(] 4 [−] 8 [)] [x^2] [ENTER]

```
3-(4-8)²
        -13
```

Both types of calculators require that we group the terms in the numerator together and the terms in the denominator together when calculating the value of an expression such as $\frac{200 + 120}{20 - 16}$.

[(] 200 [+] 120 [)] [÷] [(] 20 [−] 16 [)] [=]

| 80 |

[(] 200 [+] 120 [)] [÷] [(] 20 [−] 16 [)] [ENTER]

```
(200+120)/(20-16
)
        80
```

If parentheses aren't used when finding $\frac{200 + 120}{20 - 16}$, you will obtain an incorrect result of 190. That is because the calculator will interpret the entry as $200 + \frac{120}{20} - 16$.

Evaluating algebraic expressions

Recall that an algebraic expression is a combination of variables and numbers with the operations of arithmetic. To **evaluate an algebraic expression,** we substitute specific numbers for the variables and simplify.

INTERMEDIATE
Algebra *f(x)* Now™

Self Check 12
If $r = 2$, $s = -5$, and $t = 3$, evaluate:

a. $-\dfrac{1}{3}s^3t$

b. $\dfrac{\sqrt{-5s}}{(s+t)r^2}$

EXAMPLE 12   If $a = -2$, $b = 9$, and $c = -1$, evaluate

a. $-\dfrac{1}{2}a^2$ and   **b.** $\dfrac{-a\sqrt{b} + 3c^3}{c(c-b)}$.

Solution
a. We substitute -2 for a and simplify.

$$-\frac{1}{2}a^2 = -\frac{1}{2}(-2)^2 \qquad \text{Substitute } -2 \text{ for } a. \text{ Write parentheses around } -2 \text{ so that it is squared.}$$

$$= -\frac{1}{2}(4) \qquad \text{Evaluate the power: } (-2)^2 = 4.$$

$$= -2 \qquad \text{Perform the multiplication.}$$

b. $\dfrac{-a\sqrt{b} + 3c^3}{c(c-b)} = \dfrac{-(-2)\sqrt{9} + 3(-1)^3}{-1(-1-9)}$ Substitute -2 for a, 9 for b, and -1 for c.

$$= \frac{-(-2)(3) + 3(-1)}{-1(-10)} \qquad \begin{array}{l}\text{In the numerator, evaluate the square root}\\\text{and the power: } \sqrt{9} = 3 \text{ and } (-1)^3 = -1.\\\text{In the denominator, perform the subtraction.}\end{array}$$

$$= \frac{2(3) + 3(-1)}{-1(-10)} \qquad \text{In the numerator, simplify: } -(-2) = 2.$$

$$= \frac{6 + (-3)}{10} \qquad \begin{array}{l}\text{In the numerator, perform the}\\\text{multiplications.}\\\text{In the denominator, perform the}\\\text{multiplication.}\end{array}$$

$$= \frac{3}{10} \qquad \text{In the numerator, perform the addition.}$$

Answers **a.** 125,   **b.** $-\dfrac{5}{8}$

Area and volume

The **area** of a two-dimensional geometric figure is a measure of the surface it encloses. Table 1-4 shows the formulas for the area A of several geometric figures.

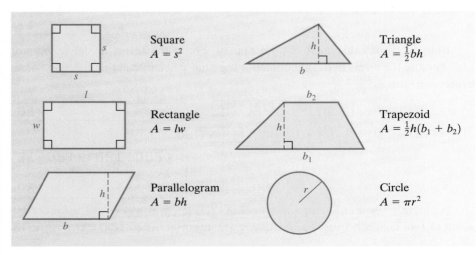

Square
$A = s^2$

Triangle
$A = \frac{1}{2}bh$

Rectangle
$A = lw$

Trapezoid
$A = \frac{1}{2}h(b_1 + b_2)$

Parallelogram
$A = bh$

Circle
$A = \pi r^2$

TABLE 1-4

EXAMPLE 13 Find the amount of skin covered by a rectangular bandage $\frac{5}{8}$ inches wide and $3\frac{1}{2}$ inches long.

Solution To find the amount of skin covered by the bandage, we need to find its area.

$$A = lw \qquad \text{The formula for the area of a rectangle.}$$

$$A = 3\frac{1}{2}\left(\frac{5}{8}\right) \qquad \text{Substitute } 3\frac{1}{2} \text{ for } l \text{ and } \frac{5}{8} \text{ for } w.$$

$$A = \frac{7}{2}\left(\frac{5}{8}\right) \qquad \text{Write } 3\frac{1}{2} \text{ as an improper fraction.}$$

$$A = \frac{35}{16}$$

The bandage covers $\frac{35}{16}$ or $2\frac{3}{16}$ in.^{2} (square inches) of skin.

Self Check 13
A solar panel is in the shape of a trapezoid. Its upper and lower bases measure $53\frac{1}{2}$ centimeters and $79\frac{1}{2}$ centimeters, respectively, and its height is 47 centimeters. In square centimeters, how large a surface do the sun's rays strike?

Answer $3,125\frac{1}{2}$ cm^2

The **volume** of a three-dimensional geometric figure is a measure of its capacity. Table 1-5 shows the formulas for the volume V of several geometric figures.

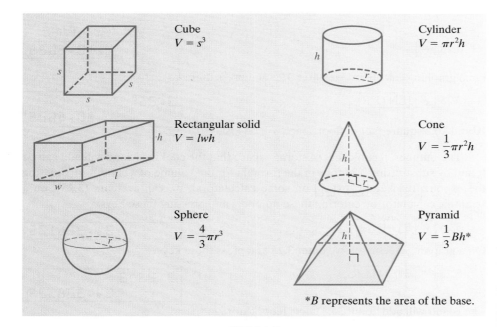

Cube
$V = s^3$

Cylinder
$V = \pi r^2 h$

Rectangular solid
$V = lwh$

Cone
$V = \frac{1}{3}\pi r^2 h$

Sphere
$V = \frac{4}{3}\pi r^3$

Pyramid
$V = \frac{1}{3}Bh*$

*B represents the area of the base.

TABLE 1-5

EXAMPLE 14 Find the amount of sand in the hourglass shown in Figure 1-9 on the next page.

Solution The sand is in the shape of a cone. The radius of the cone is one-half the diameter of the base of the hourglass, and the height of the cone is one-half the height of the hourglass. To find the amount of sand, we substitute 1 for r and 2.5 for h in the formula for the volume of a cone.

Self Check 14
Find the volume of a drinking straw that is 250 millimeters long with an inside diameter of 6 millimeters.

$$V = \frac{1}{3}\pi r^2 h$$

$$V = \frac{1}{3}\pi (1)^2 (2.5)$$

$$V = \frac{2.5\pi}{3} \qquad \text{Evaluate the power: } (1)^2 = 1. \text{ Then multiply.}$$

$$V \approx 2.617993878 \qquad \text{Use a calculator.}$$

There are about 2.6 in.^{3} (cubic inches) of sand in the hourglass.

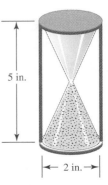

5 in.

2 in.

Answer about 7,069 mm^3

FIGURE 1-9

CALCULATOR SNAPSHOT   **The squaring and exponential keys**

A homeowner plans to install the cooking island shown in Figure 1-10. To find the number of square feet of kitchen floor space that will be lost, we substitute 3.25 for s in the formula for the area of a square, $A = s^2$. Using the squaring key $\boxed{x^2}$ on a scientific calculator, we can evaluate $(3.25)^2$ by entering these numbers and pressing these keys.

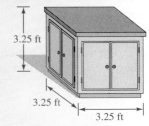

3.25 ft

3.25 ft

3.25 ft

FIGURE 1-10

$$3.25 \; \boxed{x^2} \qquad\qquad \boxed{10.5625}$$

On a graphing calculator, we enter 3.25 and press these keys.

$$3.25 \; \boxed{x^2} \; \boxed{\text{ENTER}} \qquad\qquad \boxed{\begin{array}{l} 3.25^2 \\ \quad 10.5625 \end{array}}$$

About 10.6 square feet of floor space will be lost.

The number of cubic feet of storage space that the cooking island will add can be found by substituting 3.25 for s in the formula for the volume of a cube, $V = s^3$. Using the exponential key $\boxed{y^x}$ ($\boxed{x^y}$ on some calculators), we can evaluate $(3.25)^3$ on a scientific calculator by entering these numbers and pressing these keys.

$$3.25 \; \boxed{y^x} \; 3 \; \boxed{=} \qquad\qquad \boxed{34.328125}$$

On a graphing calculator, we enter 3.25 and press these keys.

$$3.25 \; \boxed{\wedge} \; 3 \; \boxed{\text{ENTER}} \qquad\qquad \boxed{\begin{array}{l} 3.25^\wedge 3 \\ \quad 34.328125 \end{array}}$$

The island will add about 34.3 cubic feet of storage space.

Section 1.3 STUDY SET

INTERMEDIATE
Algebra $f(x)$ Now™

VOCABULARY *Fill in the blanks.*

1. When we add two numbers, the result is called the _____. When we subtract two numbers, the result is called the _____.

2. When we multiply two numbers, the result is called the _____. When we divide two numbers, the result is called the _____.

3. To _____ an algebraic expression, we substitute values for the variables and then apply the rules for the order of operations.

4. In the expression $9 + 6[22 - (6 - 1)]$, the _____ are the innermost grouping symbols, and the brackets are the _____ grouping symbols.

5. 6^2 can be read as "six _____," and 6^3 can be read as "six _____."

6. 4^5 is the fifth _____ of four.

7. In the exponential expression x^2, x is the _____, and 2 is the _____.

8. An _____ is used to represent repeated multiplication.

9. Subtraction is the same as adding the _____ of the number being subtracted.

10. The principal square _____ of 16 is 4.

CONCEPTS

11. Consider the expression $6 + 3 \cdot 2$.

   a. In what two different ways might we evaluate the given expression?

   b. Which result from part a is correct and why?

12. a. What operations does the expression $60 - (-9)^2 + 5(-1)$ contain?

   b. In what order should they be performed?

13. What are we finding when we calculate

   a. the amount of surface a circle encloses?

   b. the capacity of a cylinder?

   c. the distance around a rectangle

14. Rewrite each subtraction as addition of the opposite. Then evaluate the expression.

   a. $-16 - 1$

   b. $-16 - (-1)$

   c. $-16 - (-4) - 2$

NOTATION   *Complete each solution.*

15. $60 - 20 \cdot 2^3 = 60 - 20()$

$ = 60 - $

$ = -100$

16. $9 + 8[(1 + 4) \cdot 5] = 9 + 8[() \cdot 5]$

$ = 9 + 8()$

$ = 9 + $

$ = 209$

17. What is the name of the symbol $\sqrt{}$?

18. What is the one number that a fraction cannot have as its denominator?

19. a. In the expression $(-6)^2$, what is the base?

   b. In the expression -6^2, what is the base?

20. Translate each expression into symbols, and then evaluate on it.

   a. Negative four squared

   b. The opposite of four squared

PRACTICE   *Perform the operations.*

21. $-3 + (-5)$

22. $-2 + (-8)$

23. $-7.1 + 2.8$

24. $3.1 + (-5.2)$

25. $-3 - 4$

26. $-11 - (-17)$

27. $-3.3 - (-3.3)$

28. $0.14 - (-0.13)$

29. $-2(6)$

30. $-3(-7)$

31. $-0.3(5)$

32. $-0.4(-0.6)$

33. $-5(6)(-2)$

34. $-9(-1)(-3)$

35. $\dfrac{-8}{4}$

36. $\dfrac{-16}{-4}$

37. $\dfrac{1}{2} + \left(-\dfrac{1}{3}\right)$

38. $-\dfrac{3}{4} + \left(-\dfrac{1}{5}\right)$

39. Subtract $-\dfrac{3}{5}$ from $\dfrac{1}{2}$.

40. Subtract $\dfrac{11}{13}$ from $\dfrac{1}{26}$.

41. $\left(-\dfrac{3}{5}\right)\left(\dfrac{10}{7}\right)$

42. $\left(-\dfrac{6}{7}\right)\left(-\dfrac{5}{12}\right)$

43. $-\dfrac{16}{5} \div \left(-\dfrac{10}{3}\right)$

44. $-\dfrac{5}{24} \div \dfrac{10}{3}$

Evaluate each expression.

45. 12^2

46. 9^2

47. -5^2

48. $(-5)^2$

49. $4 \cdot 2^3$

50. $(4 \cdot 2)^3$

51. $(1.3)^2$

52. $\left(\dfrac{3}{5}\right)^2$

53. $\sqrt{64}$

54. $\sqrt{121}$

55. $-\sqrt{\dfrac{9}{16}}$

56. $-\sqrt{0.16}$

57. $3 - 5 \cdot 4$

58. $12 - 2 \cdot 3$

59. $(-3 - \sqrt{25})^2$

60. $4^2 - (-2)^2$

61. $2 + 3\left(\dfrac{25}{5}\right) + (-4)$

62. $(-2)^3\left(\dfrac{-6}{-2}\right)(-1)$

63. $\dfrac{-\sqrt{49} - 3^2}{2 \cdot 4}$

64. $\dfrac{1}{2}\left(\dfrac{1}{8}\right) + \left(-\dfrac{1}{4}\right)^2$

65. $-2|4 - 8|$

66. $|\sqrt{49} - 8(4 - 7)|$

67. $(4 + 2 \cdot 3)^4$

68. $|9 - 5(1 - 8)|$

69. $3 + 2[-1 - 4(5)]$

70. $-3[5^2 - (7 - 3)^2]$

71. $3 - [3^3 + (3 - 1)^3]$

72. $3|-(3 \cdot 5 - 2 \cdot 6)^2|$

73. $\dfrac{|-25| - 2(-5)}{2^4 - 9}$

74. $\dfrac{2[-4 - 2(3 - 1)]}{3(3)(2)}$

75. $\dfrac{3[-9 + 2(7 - 3)]}{(8 - 5)(9 - 7)}$

76. $\dfrac{5 \cdot 4 \cdot 3 \cdot 2 \cdot 1}{1 \cdot 2 \cdot 3 \cdot 4}$

77. $\dfrac{(6 - 5)^4 + 21}{|(\sqrt{16})^2 - 27|}$

78. $\dfrac{3(3,246 - 1,111)}{561 - 546}$

79. ▦ $54^3 - 16^4 + 19(3)$

80. ▦ $\dfrac{36^2 - 2(48)}{(25)^2 - \sqrt{105,625}}$

Evaluate each expression for the given values.

81. $-\dfrac{2}{3}a^2$ for $a = -6$

82. $\left(-\dfrac{2}{3}a\right)^2$ for $a = -6$

83. $\dfrac{y_2 - y_1}{x_2 - x_1}$ for $x_1 = -3, x_2 = 5, y_1 = 12, y_2 = -4$

84. $P_0\left(1 + \dfrac{r}{k}\right)^{kt}$ for $P_0 = 500, r = 4, k = 2, t = 3$

85. $(x + y)(x^2 - xy + y^2)$ for $x = -4, y = 5$

86. $\dfrac{-b + \sqrt{b^2 - 4ac}}{2a}$ for $a = 1, b = 2, c = -3$

87. $\dfrac{x^2}{a^2} + \dfrac{y^2}{b^2}$ for $x = -3, y = -4, a = 5, b = -5$

88. $\dfrac{n}{2}[2a_1 + (n - 1)d]$ for $n = 50, a_1 = -4, d = 5$

89. $\sqrt{(x_2 - x_1)^2 + (y_2 - y_1)^2}$ for $x_1 = -2, x_2 = 4$, $y_1 = 4, y_2 = -4$

90. $\dfrac{|Ax_0 + By_0 + C|}{\sqrt{A^2 + B^2}}$ for $A = 3, B = 4, C = -5$, $x_0 = 2, y_0 = -1$

▦ *Find each area to the nearest tenth.*

91. The area of a triangle with a base of 2.75 centimeters (cm) and a height of 8.25 cm

92. The area of a circle with a radius of 5.7 meters

▦ *Find each volume to the nearest hundredth.*

93. The volume of a rectangular solid with dimensions of 2.5 cm, 3.7 cm, and 10.2 cm

94. The volume of a pyramid whose base is a square with each side measuring 2.57 cm and with a height of 12.32 cm

95. The volume of a sphere with a radius of 5.7 meters

96. The volume of a cone whose base has a radius of 5.5 in. and whose height is 8.52 in.

▦ **APPLICATIONS**   *For some exercises, a calculator will be helpful.*

97. ALUMINUM FOIL Find the number of *square feet* of aluminum foil on a roll if the dimensions printed on the box are $8\frac{1}{3}$ yards $\times$ 12 inches.

98. HOCKEY A goal is scored in hockey when the puck, a vulcanized rubber disk 2.5 cm (1 in.) thick and 7.6 cm (3 in.) in diameter, is driven into the opponent's goal. Find the volume of a puck in cubic centimeters and cubic inches. Round to the nearest tenth.

99. PAPER PRODUCTS When folded, the paper sheet shown on the next page forms a rectangular-shaped envelope. The formula

$$A = \dfrac{1}{2}h_1(b_1 + b_2) + b_3h_3 + \dfrac{1}{2}b_1h_2 + b_1b_3$$

gives the amount of paper (in square units) used in the design. Explain what each of the four terms in the formula finds. Then evaluate the formula for $b_1 = 6$, $b_2 = 2$, $b_3 = 3$, $h_1 = 2$, $h_2 = 2.5$, and $h_3 = 3$. All dimensions are in inches.

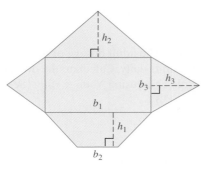

100. BONDS The illustration shows the annual net cash flow in the U.S. bond market for 1993–2002. A positive number indicates increased investment in bonds as compared to the previous year; a negative number indicates a decrease. What was the net cash flow into the bond market for the decade?

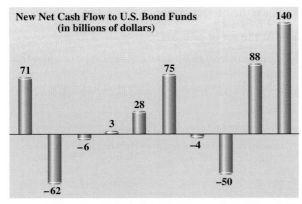

Source: Investment Company Institute

101. ACCOUNTING On a financial balance sheet, debts (negative numbers) are denoted within parentheses. Assets (positive numbers) are written without parentheses. What is the 2004 fund balance for the preschool whose financial records are shown in the next column?

Community Care Preschool Balance Sheet, June, 2004	
Fund balances	
Classroom supplies	$ 5,889
Emergency needs	927
Holiday program	(2,928)
Insurance	1,645
Janitorial	(894)
Licensing	715
Maintenance	(6,321)
BALANCE	?

102. TEMPERATURE EXTREMES The highest and lowest temperatures ever recorded in several cities are shown below. List the cities in order, from the smallest to the largest range in temperature extremes.

City	Extreme temperatures	
	Highest	Lowest
Atlanta, Georgia	105	−8
Boise, Idaho	111	−25
Helena, Montana	105	−42
New York, New York	107	−3
Omaha, Nebraska	114	−23

103. ICE CREAM If the two equal-sized scoops of ice cream shown in the illustration melt completely into the cone, will they overflow the cone?

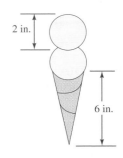

104. TENNIS In men's tennis, blazing ace serves often make the volleys between the players too brief. One suggested remedy is to use a bigger ball. The illustration shows the standard size tennis ball and a proposed larger ball. By how much do their volumes differ?

Standard size

Oversized ball

105. PODIUMS The illustration shows the dimensions of the trapezoidal-shaped top of a podium. Find its area in square inches.

1 ft 2 in.

11 in.

1 ft 8 in.

106. PHYSICS Waves are motions that carry energy from one place to another. The illustration shows an example of a wave called a *standing wave*. What is the difference in the height of the crest of the wave and the depth of the trough of the wave?

Crest

0.8 meter

0.4

0.0

Trough

107. PEDIATRICS Young's rule is used by some doctors to calculate dosage for infants and children:

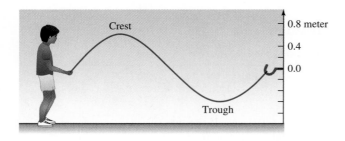

$$\frac{\text{Age of child}}{\text{Age of child} + 12} \left(\begin{array}{c} \text{average} \\ \text{adult dose} \end{array} \right) = \text{child's dose}$$

The following syringe shows the adult dose of a certain medication. Use Young's rule to determine the dosage for a 6-year-old child. Then use an arrow to locate the dosage on the calibration.

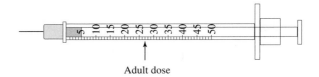

5   10 15 20 25 30 35 40 45 50

Adult dose

108. DOSAGES The adult dosage of procaine penicillin is 300,000 units daily. Calculate the dosage for a 12-year-old child using Young's rule. (See Exercise 103.)

WRITING

109. Explain what the statement $x - y = x + (-y)$ means.

110. Explain why rules for the order of operations are necessary.

111. See the illustration. It would be incorrect to say that the area of the square is 25^2 ft. Explain why.

5 ft

5 ft

112. When evaluating expressions, should multiplications be performed before divisions? Explain your answer.

REVIEW

113. What two numbers are a distance of 5 away from -2 on the number line?

114. Place the proper symbol ($>$ or $<$) in the blank: $-4.6 \ \rule{1em}{0.1em} \ -4.5$.

115. List the set of integers.

116. Translate into mathematical symbols: ten less than twice x.

117. True or false: The real numbers is the set of all decimals.

118. True or false: Irrational numbers are nonterminating, nonrepeating decimals.

1.4 Simplifying Algebraic Expressions

- Properties of real numbers   • Properties of 0 and 1   • More properties of real numbers
- Simplifying algebraic expressions   • The distributive property   • Combining like terms

Suppose we are asked to find the total amount of the checks that have been recorded in the check register shown in Figure 1-11 on the next page. As the notes in the figure point out, we can simplify the computation by adding the numbers in a different order from the way they are entered. The *commutative* and *associative* properties of addition guarantee that we will obtain the same result whether we add them in their original order or in this more convenient way. In this section, we will introduce some important properties of real numbers and see how they can be used to simplify expressions.

Number	Date	Description of Transaction	Payment/Debit	
101	3/6	DR. OKAMOTO, DDS	$64	00
102	3/6	UNION OIL CO.	$25	00
103	3/8	STATER BROS.	$16	00
104	3/9	LITTLE LEAGUE	$75	00

Add $64.00 and $16.00 to get $80.00.

Add $25.00 and $75.00 to get $100.00.

Now add the two subtotals to get the total dollar amount of the checks: $80.00 + $100.00 = $180.00.

FIGURE 1-11

Properties of real numbers

When working with real numbers, we will use the following properties.

Properties of real numbers

If a, b, and c represent real numbers, then we have:

The associative properties of addition and multiplication

$$(a + b) + c = a + (b + c) \qquad (ab)c = a(bc)$$

The commutative properties of addition and multiplication

$$a + b = b + a \qquad ab = ba$$

The *associative properties* enable us to group the numbers in a sum or a product any way that we wish and get the same result.

EXAMPLE 1 Evaluate $(14 + 94) + 6$ in two ways.

Solution

$$(\mathbf{14 + 94}) + 6 = \mathbf{108} + 6 \qquad \text{Work within the parentheses first.}$$
$$= 114$$

To evaluate the expression in a second way, we apply the associative property of addition.

$$(14 + 94) + 6 = 14 + (\mathbf{94 + 6}) \qquad \text{Use parentheses to group 94 with 6.}$$
$$= 14 + \mathbf{100} \qquad \text{Perform the addition within the parentheses.}$$
$$= 114$$

Notice that the results are the same.

COMMENT Subtraction and division are not associative, because different groupings give different results. For example,

$$(\mathbf{8 - 4}) - 2 = 4 - 2 = 2 \qquad \text{but} \qquad 8 - (\mathbf{4 - 2}) = 8 - 2 = 6$$
$$(\mathbf{8 \div 4}) \div 2 = 2 \div 2 = 1 \qquad \text{but} \qquad 8 \div (\mathbf{4 \div 2}) = 8 \div 2 = 4$$

INTERMEDIATE
Algebra *f(x)* **Now**™

Self Check 1
Evalute $2 \cdot (50 \cdot 37)$ in two ways.

Answer 3,700

The *commutative properties* enable us to add or multiply two numbers in either order and obtain the same result. Here are two examples.

$$3 + (-5) = -2 \quad \text{and} \quad -5 + 3 = -2$$
$$-2.6(-8) = 20.8 \quad \text{and} \quad -8(-2.6) = 20.8$$

! COMMENT Subtraction and division are not commutative, because performing these operations in different orders will give different results. For example,

$$8 - 4 = 4 \quad \text{but} \quad 4 - 8 = -4$$
$$8 \div 4 = 2 \quad \text{but} \quad 4 \div 8 = \frac{1}{2}$$

Properties of 0 and 1

The real numbers 0 and 1 have important special properties.

Properties of 0 and 1

Additive identity: The sum of 0 and any number is the number itself.

$$0 + a = a + 0 = a$$

Multiplicative identity: The product of 1 and any number is the number itself.

$$1 \cdot a = a \cdot 1 = a$$

Multiplication property of 0: The product of any number and 0 is 0.

$$a \cdot 0 = 0 \cdot a = 0$$

For example,

$$7 + 0 = 7, \quad 1(5.4) = 5.4, \quad \left(-\frac{7}{3}\right)1 = -\frac{7}{3}, \quad \text{and} \quad -19(0) = 0$$

! COMMENT An error is often made when subtracting a number from 0. In the problem $0 - 8$, the answer is not 8. By writing the subtraction as addition of the opposite, we see that the result is -8.

$$0 - 8 = 0 + (-8) = -8$$

More properties of real numbers

If the sum of two numbers is 0, they are called **additive inverses, negatives,** or **opposites** of each other. For example, 6 and -6 are additive inverses, because $6 + (-6) = 0$.

The additive inverse property

For every real number a, there is a real number $-a$ such that

$$a + (-a) = -a + a = 0$$

If the product of two numbers is 1, the numbers are called **multiplicative inverses** or **reciprocals** of each other.

The multiplicative inverse property

For every nonzero real number a, there exists a real number $\dfrac{1}{a}$ such that

$$a \cdot \frac{1}{a} = \frac{1}{a} \cdot a = 1$$

Some example of reciprocals (multiplicative inverses) are

- 5 and $\dfrac{1}{5}$ are reciprocals, because $5\left(\dfrac{1}{5}\right) = 1.$

- $\dfrac{3}{2}$ and $\dfrac{2}{3}$ are reciprocals, because $\dfrac{3}{2}\left(\dfrac{2}{3}\right) = 1.$

- -0.25 and -4 are reciprocals, because $-0.25(-4) = 1.$

The reciprocal of 0 does not exist, because $\frac{1}{0}$ is undefined.

Recall that when a number is divided by 1, the result is the number itself, and when a number is divided by itself, the result is 1.

Division properties

Division by 1: For any real number a, $\dfrac{a}{1} = a.$

Division of a number by itself: For any nonzero real number a, $\dfrac{a}{a} = 1.$

There are three types of division involving zero that we will encounter in mathematics. To help explain the concept of division of zero, we consider the division $\frac{0}{5} = ?$ and its equivalent multiplication fact $5(?) = 0.$

Multiplication fact	**Division fact**
$5(?) = 0$	$\dfrac{0}{5} = \mathbf{0}$
↑	↑
This must be 0 if the product is to be 0.	So the quotient is 0.

This example suggests that the quotient of zero and any nonzero number is zero.

To illustrate that division by zero is not possible, we consider the division $\frac{5}{0} = ?$ and its equivalent multiplication fact $0(?) = 5.$

Multiplication fact	**Division fact**
$0(?) = 5$	$\dfrac{5}{0} = $ undefined
↑	↑
There is no number that gives 5 when multiplied by 0.	There is no quotient.

This example illustrates that the quotient of any number divided by zero is *undefined.*

To show that division of zero by zero doesn't have a precisely determined result, we consider $\frac{0}{0} = ?$ and its equivalent multiplication fact $0(?) = 0.$

Multiplication fact	**Division fact**
$0(?) = 0$	$\dfrac{0}{0} = \text{indeterminate}$
↑	↑
Any number multiplied by 0 gives 0.	We cannot determine this — it could be any number.

This example illustrates that zero divided by zero is *indeterminate*.

Division with 0

Division of 0: For any nonzero real number a, $\dfrac{0}{a} = 0$.

Division by 0: For any nonzero real number a, $\dfrac{a}{0}$ is undefined.

Division of 0 by 0: $\dfrac{0}{0}$ is indeterminate.

▮ Simplifying algebraic expressions

To **simplify algebraic expressions,** we use properties of the real numbers to write the expressions in a less complicated form. As an example, let's consider the expression $6(5x)$ and simplify it.

$$\begin{aligned}
6(5x) &= 6 \cdot (5 \cdot x) &\quad 6(5x) = 6 \cdot 5x \text{ and } 5x = 5 \cdot x. \\
&= (6 \cdot 5) \cdot x &\quad \text{Use the associative property of multiplication to group 5 with 6.} \\
&= 30x &\quad \text{Perform the multiplication within the parentheses.}
\end{aligned}$$

Since $6(5x) = 30x$, we say that $6(5x)$ simplifies to $30x$.

INTERMEDIATE
Algebra *f(x)* **Now**™

Self Check 2
Simplify:
a. $14 \cdot 3s$

b. $-1.6b(3t)$

c. $-\dfrac{2}{3}x(-9)$

EXAMPLE 2 Simplify: **a.** $9(10t)$, **b.** $-5.3r(-2s)$, and **c.** $-\dfrac{21}{2}a\left(\dfrac{1}{3}\right)$.

Solution

a. $9(10t) = (9 \cdot 10)t$ — Use the associative property of multiplication to regroup the factors.

$= 90t$ — Perform the multiplication within the parentheses.

b. $-5.3r(-2s) = [-5.3(-2)](r \cdot s)$ — Use the commutative and associative properties to group the numbers and group the variables.

$= 10.6rs$ — Perform the multiplications.

c. $-\dfrac{21}{2}a\left(\dfrac{1}{3}\right) = -\dfrac{21}{2}\left(\dfrac{1}{3}\right)a$ — Use the commutative property of multiplication to change the order of the factors a and $\frac{1}{3}$.

$= -\dfrac{7}{2}a$ — Multiply: $-\dfrac{21}{2} \cdot \dfrac{1}{3} = -\dfrac{21 \cdot 1}{2 \cdot 3} = -\dfrac{7 \cdot \overset{1}{\cancel{3}} \cdot 1}{2 \cdot \underset{1}{\cancel{3}}} = -\dfrac{7}{2}$.

Answers a. $42s$, **b.** $-4.8bt$,
c. $6x$

▮ The distributive property

The **distributive property** enables us to evaluate many expressions involving a multiplication and an addition. For example, let's consider $4(5 + 3)$, which can be evaluated in two ways.

Method 1: Rules for the Order of Operations

We compute the sum within the parentheses first.

$$4(5 + 3) = 4(8) \qquad \text{Perform the addition within the parentheses first.}$$
$$= 32 \qquad \text{Perform the multiplication.}$$

Method 2: The Distributive Property

We distribute the multiplication by 4 across the 5 and the 3, find each product separately, and add the results.

$$4(5 + 3) = 4 \cdot 5 \quad + \quad 4 \cdot 3 \qquad \begin{array}{l}\text{To apply the distributive property,} \\ \text{we multiply each number within the} \\ \text{parentheses by the factor outside the} \\ \text{parentheses.}\end{array}$$

First product   Second product

$$= 20 \quad + \quad 12 \qquad \text{Perform the multiplications first.}$$
$$= 32 \qquad \text{Perform the addition.}$$

Notice that each method gives a result of 32. We now state the distributive property in symbols.

> **The distributive property of multiplication over addition**
>
> If a, b, and c represent real numbers,
>
> $$a(b + c) = ab + ac$$

We can use the distributive property to *remove the parentheses* when an algebraic expression is multiplied by a quantity. For example, to remove the parentheses in the expression $5(x + 2)$, we proceed as follows:

$$5(x + 2) = 5 \cdot x + 5 \cdot 2 \qquad \text{Distribute the multiplication by 5.}$$
$$= 5x + 10 \qquad \text{Perform the multiplications.}$$

Since subtraction is the same as adding the opposite, the distributive property also holds for subtraction.

$$5(x - 2) = 5 \cdot x - 5 \cdot 2 \qquad \text{Distribute the multiplication by 5.}$$
$$= 5x - 10 \qquad \text{Perform the multiplications.}$$

INTERMEDIATE
Algebra $f(x)$ Now™

EXAMPLE 3   Use the distributive property to remove parentheses: **a.** $6(a + 9)$ and   **b.** $-15(4b - 1)$.

Solution

a. $6(a + 9) = 6 \cdot a + 6 \cdot 9 \qquad \text{Distribute the multiplication by 6.}$
$\qquad\qquad = 6a + 54 \qquad \text{Perform the multiplications.}$

b. $-15(4b - 1) = -15(4b) - (-15)(1) \qquad \text{Distribute the multiplication by } -15.$
$\qquad\qquad\qquad = -60b - (-15) \qquad \text{Perform the multiplications.}$
$\qquad\qquad\qquad = -60b + 15 \qquad \text{Add the opposite of } -15, \text{ which is 15.}$

Self Check 3
Use the distributive property to remove parentheses:
a. $9(r + 4)$
b. $-11(-3x - 5)$

Answers **a.** $9r + 36$,
b. $33x + 55$

! COMMENT The distributive property cannot be applied to all expressions that contain parentheses. For example, the distributive property does not apply to

$$5(12x) \qquad \text{or} \qquad 5(-4 \cdot y) \qquad \text{Here, a product is multiplied by 5.}$$

However, the distributive property does apply to

$$5(12 + x) \qquad \text{and} \qquad 5(-4 - y) \qquad \text{Here, a sum and a difference are multiplied by 5.}$$

A more general form of the distributive property is the **extended distributive property.**

$$a(b + c + d + e + \cdots) = ab + ac + ad + ae + \cdots$$

Self Check 4
Remove parentheses:
$\dfrac{1}{3}(-6t + 3s - 9)$

EXAMPLE 4   Remove parentheses: $-0.5(7 - 5y + 6z)$.

Solution

$$-0.5(7 - 5y + 6z)$$
$$= -0.5(7) - (-0.5)(5y) + (-0.5)(6z) \qquad \text{Distribute the multiplication by } -0.5.$$
$$= -3.5 - (-2.5y) + (-3z) \qquad \text{Perform the multiplications.}$$
$$= -3.5 + 2.5y - 3z \qquad \text{Subtracting } -2.5y \text{ is the same as adding } 2.5y. \text{ Adding } -3z \text{ is the same as subtracting } 3z.$$

Answer $-2t + s - 3$

Since multiplication is commutative, we can write the distributive property in the following forms.

$$(b + c)a = ba + ca, \qquad (b - c)a = ba - ca, \qquad (b + c + d)a = ba + ca + da$$

INTERMEDIATE
Algebra $f(x)$ **Now**™

Self Check 5
Remove parentheses:
$(-5s + 4t)(-10)$

EXAMPLE 5   Remove parentheses: $(-8 - 3y)(-30)$.

Solution

$$(-8 - 3y)(-30) = -8(-30) - 3y(-30) \qquad \text{Distribute the multiplication by } -30.$$
$$= 240 - (-90y) \qquad \text{Perform the multiplications.}$$
$$= 240 + 90y \qquad \text{Add the opposite of } -90y.$$

Answer $50s - 40t$

To use the distributive property to simplify $-(x + 3)$, we note that the negative sign in front of the parentheses represents the number -1.

The $-$ sign represents -1.
$$\downarrow \qquad\qquad \downarrow$$
$$-(x + 3) = -1(x + 3)$$
$$= -1(x) + (-1)(3) \qquad \text{Distribute the multiplication by } -1.$$
$$= -x + (-3) \qquad \text{Perform the multiplications.}$$
$$= -x - 3$$

INTERMEDIATE
Algebra $f(x)$ **Now**™

Self Check 6
Simplify: $-(-27k + 15)$.

EXAMPLE 6   Simplify: $-(-21 - 20m)$.

Solution

$$-(-21 - 20m) = -1(-21 - 20m) \qquad \text{Write the } - \text{ sign in front of the parentheses as } -1.$$
$$= -1(-21) - (-1)(20m) \qquad \text{Distribute the multiplication by } -1.$$
$$= 21 - (-20m) \qquad \text{Perform the multiplications.}$$
$$= 21 + 20m$$

Answer $27k - 15$

Combining like terms

Addition signs separate algebraic expressions into parts called **terms.** For example, the expression $3x^2 + 2x + 4$ has three terms: $3x^2$, $2x$, and 4. A term may be

- a number (called a **constant**); examples are -6, 45.7, 35, and $\frac{2}{3}$.

- a variable or a product of variables (which may be raised to powers); examples are x, bh, s^2, Prt, and a^3bc^4.

- a product of a number and one or more variables (which may be raised to powers); examples are $3x$, $-7y$, $2.5y^2$, and $\pi r^2 h$.

Since subtraction can be written as addition of the opposite, the expression $6a - 5b$ can be written in the equivalent form $6a + (-5b)$. We can then see that the algebraic expression $6a - 5b$ contains two terms, $6a$ and $-5b$.

The **numerical coefficients** (or simply the **coefficients**) of the terms of the expression $x^3 - 5x^2 - x + 28$ are 1, -5, -1 and 28, respectively.

Terms with exactly the same variables raised to exactly the same powers are called **like terms** or **similar terms.** Any constants in an expression are considered to be like terms.

$5x$ and $6x$ are like terms.	$27x^2y^3$ and $-326x^2y^3$ are like terms.
$4x$ and $-17y$ are unlike terms, because they have different variables.	$15x^2y$ and $6xy^2$ are unlike terms, because the variables have different exponents.

If we are to add (or subtract) quantities, they must have the same units. For example, we can add dollars to dollars and inches to inches, but we cannot add dollars to inches. The same is true when working with terms of an algebraic expression. They can be added or subtracted only when they are like terms.

This expression can be simplified, because it contains like terms.	This expression cannot be simplified, because its terms are not like terms.
$5x + 6x$	$5x + 6y$
↑ ↑	↑ ↑
Like terms	Unlike terms
The variable parts are the same.	The variable parts are not the same.

Simplifying the sum or difference of like terms is called **combining like terms.** To simplify expressions with like terms, we use the distributive property. For example,

$$5x + 6x = (5 + 6)x \qquad \text{and} \qquad 32y - 16y = (32 - 16)y$$
$$= 11x \qquad\qquad\qquad\qquad\qquad = 16y$$

These examples suggest the following rule.

Combining like terms

To add or subtract like terms, combine their coefficients and keep the same variables with the same exponents.

EXAMPLE 7   Simplify each expression: **a.** $-8f + (-12f)$,   **b.** $0.56s^3 - 0.2s^3$, and   **c.** $-\frac{1}{2}ab + \frac{1}{3}ab$.

Solution

a. $-8f + (-12f) = -20f$ Add the coefficients of the like terms: $-8 + (-12) = -20$. Keep the variable f.

b. $0.56s^3 - 0.2s^3 = 0.36s^3$ Subtract: $0.56 - 0.2 = 0.36$. Keep s^3.

INTERMEDIATE
Algebra $f(x)$ **Now**™

Self Check 7
Simplify by combining like terms:
a. $5k + 8k$

b. $-600a^2 - (-800a^2)$

c. $\dfrac{2}{3}xy - \dfrac{3}{4}xy$

Answers **a.** $13k$, **b.** $200a^2$,

c. $-\dfrac{1}{12}xy$

c. $-\dfrac{1}{2}ab + \dfrac{1}{3}ab = -\dfrac{1 \cdot 3}{2 \cdot 3}ab + \dfrac{1 \cdot 2}{3 \cdot 2}ab$ Express each fraction in terms of the LCD, 6.

$= -\dfrac{3}{6}ab + \dfrac{2}{6}ab$ Perform the multiplications.

$= -\dfrac{1}{6}ab$ Add the coefficients: $-\dfrac{3}{6} + \dfrac{2}{6} = -\dfrac{1}{6}$. Keep ab.

INTERMEDIATE
Algebra $f(x)$ **Now**™

Self Check 8
Simplify: $8R + 7r - 14R - 21r$

Answer $-6R - 14r$

EXAMPLE 8   Simplify: $9b - B - 14b + 34B$.

Solution   This expression has four terms. The uppercase B and the lowercase b are different variables. So the first and third terms are like terms, and the second and fourth terms are like terms.

$9b - B - 14b + 34B = -5b + 33B$ Combine like terms: $9b - 14b = -5b$ and $-B + 34B = 33B$.

INTERMEDIATE
Algebra $f(x)$ **Now**™

Self Check 9
Simplify: $-5(y - 4) + 2(4y + 8)$

Answer $3y + 36$

EXAMPLE 9   Simplify: $9(x + 1) - 3(7x - 1)$.

Solution   We apply the distributive property and then combine like terms.

$9(x + 1) - 3(7x - 1) = 9x + 9 - 21x + 3$ Use the distributive property twice.

$= -12x + 12$ Combine like terms: $9x - 21x = -12x$ and $9 + 3 = 12$.

Section 1.4 STUDY SET

INTERMEDIATE
Algebra $f(x)$ **Now**™

VOCABULARY   *Fill in the blanks.*

1. _____ terms are terms with exactly the same variables raised to exactly the same powers.

2. To add or subtract like terms, combine their _____ and keep the same variables and exponents.

3. A number or the product of numbers and variables is called a _____.

4. $\dfrac{1}{3}$ and 3 are _____, because $\dfrac{1}{3} \cdot 3 = 1$.

5. To _____ algebraic expressions, we use properties of real numbers to write the expressions in a less complicated form.

6. The _____ property tells us how to multiply 5 and $(x + 7)$: $5(x + 7) = 5x + 35$.

7. Division of a nonzero number by 0 is _____.

8. The _____ of the term $-8c$ is -8.

CONCEPTS

9. **a.** Using the variables x, y, and z, write the associative property of addition.

   b. Using the variables x and y, write the commutative property of multiplication.

   c. Using the variables r, s, and t, write the distributive property.

10. **a.** What is the additive identity?

    b. What is the multiplicative identity?

    c. Simplify: $-(-10)$.

11. What number should be

    a. subtracted from 5 to obtain 0?

    b. added to 5 to obtain 0?

12. By what number should

    a. 5 be divided to obtain 1?

    b. 5 be multiplied to obtain 1?

13. Give the reciprocal.

    a. $\dfrac{15}{16}$ **b.** -20

    c. 0.5 **d.** x

14. Evaluate each expression.

    a. $0 + 15$ **b.** $-3 + 0$

    c. $0 - 4$ **d.** $0 - (-50)$

15. Simplify each expression.

    a. $0 + 2x$ **b.** $-3a + 0$

    c. $0 - 9t$ **d.** $0 - (-22x)$

16. Does the distributive property apply?

    a. $2(3)(5)$ **b.** $2(3 \cdot 5)$

    c. $2(3x)$ **d.** $2(x - 3)$

17. Consider the expression $2x^2 - x + 6$.

    a. What are the terms of the expression?

    b. Give the coefficient of each term.

18. Which properties of real numbers involve changing *order* and which involve changing *grouping*?

Determine whether the terms are like terms. If they are, combine them.

19. $2x, 6x$ **20.** $-3x, 5y$

21. $-5xy, -7yz$ **22.** $-3t^2, 12t^2$

23. $3x^2, -5x^2$ **24.** $5y^2, 7xy$

25. $xy, 3xt$ **26.** $-4x, -5x$

▌ NOTATION

27. In $-(x - 7)$, what does the negative sign in front of the parentheses represent?

28. Perform each division, if possible.

    a. $\dfrac{0}{8}$ **b.** $\dfrac{8}{0}$

▌ PRACTICE *Fill in the blanks by using the given property of the real numbers.*

29. $3 + 7 =$ _____
Commutative property of addition

30. $2(5 \cdot 97) =$ _____
Associative property of multiplication

31. $3(2 + d) =$ _____
Distributive property

32. $1 \cdot y =$ _____
Commutative property of multiplication

33. $c + 0 =$ _____
Additive identity property

34. $-4(x - 2) =$ _____
Distributive property and simplifying

35. $25 \cdot \dfrac{1}{25} =$ _____
Multiplicative inverse property

36. $z + (9 - 27) =$ _____
Commutative property of addition

37. $8 + (7 + a) =$ _____
Associative property of addition

38. _____ $\cdot 3 = 3$
Multiplicative identity property

39. $(x + y)2 =$ _____
Commutative property of multiplication

40. $h + (-h) =$ _____
Additive inverse property

Evaluate each side of the equation to show that the same result is obtained. Identify the property of real numbers that is being illustrated.

41. $(37.9 + 25.2) + 14.3 = 37.9 + (25.2 + 14.3)$

42. $7.1(3.9 + 8.8) = 7.1 \cdot 3.9 + 7.1 \cdot 8.8$

43. $2.73(4.534 + 57.12) = 2.73 \cdot 4.534 + 2.73 \cdot 57.12$

44. $(6.789 + 345.1) + 27.347 = (345.1 + 6.789) + 27.347$

Use the distributive property to remove the parentheses.

45. $-4(t - 3)$ **46.** $-4(-t + 3)$

47. $-(t - 3)$ **48.** $-(-t + 3)$

49. $\dfrac{2}{3}(3s - 9)$ **50.** $\dfrac{1}{5}(5s - 15)$

51. $0.7(s + 2)$ **52.** $2.5(6s - 8)$

53. $3\left(\dfrac{4}{3}x - \dfrac{5}{3}y + \dfrac{1}{3}\right)$

54. $6\left(-\dfrac{4}{3} + \dfrac{7}{6}s + \dfrac{16}{3}t\right)$

55. $(9m + n)8$ **56.** $(-6s + t)5$

57. $(-4r - 3s)(-6)$ **58.** $(-9x - y)(-8)$

Simplify each expression.

59. $9(8m)$

60. $12n(4)$

61. $5(-9q)$

62. $-7(2t)$

63. $(-5p)(-6b)$

64. $(-7d)(-7e)$

65. $-5(8r)(-2y)$

66. $-7s(-4t)(-1)$

67. $3x + 15x$

68. $12y - 17y$

69. $18x^2 - 5x^2$

70. $37x^2 + 3x^2$

71. $-9x + 9x$

72. $-26y + 26y$

73. $-b^2 + b^2$

74. $-3c^3 + 3c^3$

75. $8x + 5x - 7x$

76. $-y + 3y + 6y$

77. $3x^2 + 2x^2 - 5x^2$

78. $8x^3 - x^3 + 2x^3$

79. $3.8h - 0.7h$

80. $-5.7m + 5.3m$

81. $\frac{2}{5}ab - \left(-\frac{1}{2}ab\right)$

82. $-\frac{3}{4}st - \frac{1}{3}st$

83. $\frac{3}{5}t + \frac{1}{3}t$

84. $\frac{3}{16}x - \frac{5}{4}x$

85. $4(y + 9) - 8y$

86. $-(4 + z) + 2z$

87. $2z + 5(z - 4)$

88. $12(2m + 11) - 11$

89. $8(2c + 7) - 2(c - 3)$

90. $9(z + 3) - 5(3 - z)$

91. $2x^2 + 4(3x - x^2) + 3x$

92. $3p^2 - 6(5p^2 + p) + p^2$

93. $-(a + 2) - (a - b)$

94. $3z - 2(y - z) + y$

95. $-3(p - 2) + 2(p + 3) - 5(p - 1)$

96. $5(q + 7) - 3(q - 1) - (q + 2)$

▮ APPLICATIONS

97. PARKING AREAS A restaurant parking lot is shown in the next column.

 a. Express the area of the entire parking lot as the product of its length and width.

 b. Express the area of the entire lot as the sum of the areas of the self-parking space and the valet parking space.

 c. Write an equation that shows that your answers to parts a and b are equal. What property of real numbers is illustrated by this example?

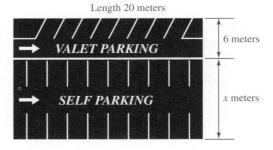

Length 20 meters

VALET PARKING   6 meters

SELF PARKING   x meters

98. CROSS SECTIONS When the steel casting shown below is cut down the middle, we see that it has a uniform cross section consisting of two identical trapezoids. What is the area of the cross section? (The measurements are in inches.)

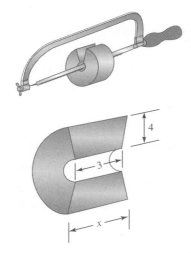

4

3

x

99. GIFT WRAPPING How much ribbon is needed to wrap the package below if x inches are used to make the bow?

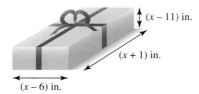

$(x - 11)$ in.

$(x + 1)$ in.

$(x - 6)$ in.

100. AMERICAN RED CROSS In 1891, Clara Barton founded the Red Cross. Its symbol is a white flag with a red cross. In the illustration, each side of the cross has length x centimeters. Write an algebraic expression for

 a. the perimeter of the cross.

 b. the area of the cross.

x

■ WRITING

101. Explain why the distributive property does not apply to $6(2 \cdot x)$.

102. In each case, explain what you can conclude about one or both of the numbers described below.

    a. When the two numbers are added, the result is 0.

    b. When the two numbers are subtracted, the result is 0.

    c. When the two numbers are multiplied, the result is 0.

    d. When the two numbers are divided, the result is 0.

103. What are like terms?

104. Use each of the words *commute, associate,* and *distribute* in a sentence in which the context is nonmathematical.

■ REVIEW   *Evaluate each expression.*

105. $-5.6 - (-5.6)$

106. $-(-4 \cdot 2)^3$

107. $\left(-\dfrac{3}{2}\right)\left(\dfrac{7}{12}\right)$

108. $-\sqrt{\dfrac{9}{16}}$

109. $(4 + 2 \cdot 3)^3$

110. $-3|4 - 8|$

111. $\dfrac{-\sqrt{64} - 5^2}{2 \cdot 4 + 3}$

112. $\dfrac{1}{2} - \left(-\dfrac{4}{5}\right)$

1.5 Solving Linear Equations and Formulas

- Solutions of equations • Linear equations • Properties of equality
- Solving linear equations • Simplifying expressions to solve equations
- Identities and contradictions • Solving formulas

In one method to solve problems, we begin by letting a variable stand for an unknown quantity. We then write an equation involving the variable to describe the situation. Finally, we perform a series of steps to find the value represented by the variable. The process of determining the value (or values) represented by a variable is called *solving the equation.* In this section, we will discuss strategy to solve one basic type of equation called a *linear equation.*

■ Solutions of equations

An **equation** is a statement indicating that two quantities are equal. The equation $2 + 4 = 6$ is true, and the equation $2 + 5 = 6$ is false. If an equation contains a variable (say, x), it can be either true or false, depending on the value of x. For example, if x is 1, the equation $7x - 3 = 4$ is true.

$$7x - 3 = 4$$

$7(\mathbf{1}) - 3 \stackrel{?}{=} 4$ Substitute 1 for x. At this stage, we don't know whether the left- and right-hand sides of the equation are equal, so we use an "is possibly equal to" symbol $\stackrel{?}{=}$.

$7 - 3 \stackrel{?}{=} 4$ Perform the multiplication.

$4 = 4$ We obtain a true statement.

Since 1 makes the equation true, we say that 1 *satisfies* the equation. However, the equation is false for all other values of x.

The set of numbers that satisfy an equation is called its **solution set.** The elements of the solution set are called **solutions** or **roots** of the equation. Finding the solution set of an equation is called **solving the equation.**

EXAMPLE 1   Determine whether 2 is a solution of $3x + 2 = 2x + 5$.

Solution   We substitute 2 for x wherever it appears in the equation and see whether it satisfies the equation.

INTERMEDIATE
Algebra $f(x)$ **Now**™

Self Check 1
Is -5 a solution of $2x - 5 = 3x$?

$$3x + 2 = 2x + 5 \qquad \text{This is the original equation.}$$
$$3(\mathbf{2}) + 2 \overset{?}{=} 2(\mathbf{2}) + 5 \qquad \text{Substitute 2 for } x.$$
$$6 + 2 \overset{?}{=} 4 + 5 \qquad \text{Perform the multiplication.}$$
$$8 = 9 \qquad \text{Perform the addition.}$$

Since $8 = 9$ is a false statement, the number 2 does not satisfy the equation. It is not a solution.

Answer yes

Linear equations

We are not usually told the solutions of an equation — we need to find them ourselves. In this text, you will learn how to solve many different types of equations. The easiest to solve are **linear equations.**

> **Linear equations**
>
> A **linear equation in one variable** (say, x) is any equation that can be written in the form
>
> $$ax + b = c \qquad (a, b, \text{ and } c \text{ represent real numbers and } a \neq 0)$$

Some examples of linear equations in one variable are

$$-2x - 8 = 0 \qquad 3(a + 2) = 20 \qquad 4b - 7 + 2b = 1 + 2b + 8$$

Linear equations are also called **first-degree equations,** because the highest power on the variable is 1.

Properties of equality

When solving linear equations, the objective is to *isolate* the variable on one side of the equation. We achieve this by undoing the operations performed on the variable. As we undo the operations, we produce a series of simpler equations, all having the same solutions. Such equations are called **equivalent equations.**

> **Equivalent equations**
>
> Equations with the same solutions are called **equivalent equations.**

The solution of the equation $x = 2$ is obviously 2, because replacing x with 2 yields a true statement, $2 = 2$. The equation $x + 4 = 6$ also has a solution of 2. Since $x = 2$ and $x + 4 = 6$ have the same solution, they are equivalent equations.

The following properties are used to isolate a variable on one side of an equation.

> **Properties of equality**
>
> If a, b, and c represent real numbers and $a = b$, then
>
> $$a + c = b + c \qquad \textbf{(Addition property of equality)}$$
> $$a - c = b - c \qquad \textbf{(Subtraction property of equality)}$$
>
> If $c \neq 0$, then
>
> $$ca = cb \qquad \textbf{(Multiplication property of equality)}$$
> $$\frac{a}{c} = \frac{b}{c} \qquad \textbf{(Division property of equality)}$$

> **Properties of equality**
>
> In words,
>
> *If any quantity is added to (or subtracted from) both sides of an equation, a new equation is formed that is equivalent to the original equation.*
>
> *If both sides of an equation are multiplied (or divided) by the same nonzero quantity, a new equation is formed that is equivalent to the original equation.*

Solving linear equations

EXAMPLE 2 Solve: $-2x - 8 = 0$.

Solution We note that x is multiplied by -2 and then 8 is subtracted from that product. To isolate x on the left-hand side of the equation, we proceed as follows.

- To undo the subtraction of 8, we add 8 to both sides.
- To undo the multiplication by -2, we divide both sides by -2.

$$-2x - 8 \mathbf{+ 8} = 0 \mathbf{+ 8} \qquad \text{Add 8 to both sides.}$$

$$-2x = 8 \qquad \text{Simplify both sides of the equation.}$$

$$\frac{-2x}{\mathbf{-2}} = \frac{8}{\mathbf{-2}} \qquad \text{Divide both sides by } -2.$$

$$x = -4 \qquad \text{Simplify both sides of the equation. } x \text{ is isolated.}$$

Check: We substitute -4 for x to verify that it satisfies the original equation.

$$-2x - 8 = 0$$

$$-2(\mathbf{-4}) - 8 \stackrel{?}{=} 0 \qquad \text{Substitute } -4 \text{ for } x.$$

$$8 - 8 \stackrel{?}{=} 0 \qquad \text{Perform the multiplication.}$$

$$0 = 0 \qquad \text{Perform the subtraction.}$$

Since we obtain a true statement, -4 is the solution of $-2x - 8 = 0$, and the solution set is $\{-4\}$.

INTERMEDIATE
Algebra *f(x)* **Now**™
Self Check 2
Solve: $-3a + 15 = 0$.

Answer 5

INTERMEDIATE
Algebra *f(x)* **Now**™
Self Check 3
Solve: $\frac{2}{3}b - 3 = -15$.

EXAMPLE 3 Solve: $\frac{3}{4}y = -7$.

Solution On the left-hand side, y is multiplied by $\frac{3}{4}$. We can undo the multiplication by dividing both sides by $\frac{3}{4}$. Since division by $\frac{3}{4}$ is equivalent to multiplication by its reciprocal, we can isolate y by multiplying both sides by $\frac{4}{3}$.

$$\frac{3}{4}y = -7$$

$$\frac{4}{3}\left(\frac{3}{4}y\right) = \frac{4}{3}(-7) \qquad \text{Use the multiplication property of equality: Multiply both sides by the reciprocal of } \frac{3}{4}, \text{ which is } \frac{4}{3}.$$

$$\left(\frac{4}{3} \cdot \frac{3}{4}\right)y = \frac{4}{3}(-7) \qquad \text{Use the associative property of multiplication to regroup.}$$

$$1y = \frac{4}{3}(-7) \qquad \text{The product of a number and its reciprocal is 1: } \frac{4}{3} \cdot \frac{3}{4} = 1.$$

$$y = -\frac{28}{3} \qquad \text{On the right-hand side, perform the multiplication.}$$

Check: $\quad \dfrac{3}{4}y = -7 \qquad$ This is the original equation.

$$\dfrac{3}{4}\left(-\dfrac{28}{3}\right) \stackrel{?}{=} -7 \qquad \text{Substitute } -\dfrac{28}{3} \text{ for } y.$$

$$-\dfrac{\overset{1}{3}\cdot\overset{1}{4}\cdot 7}{\underset{1}{4}\cdot\underset{1}{3}} \stackrel{?}{=} -7 \qquad \begin{array}{l}\text{Multiply the numerators and the denominators. Then simplify}\\ \text{the fraction: Factor 28 and remove the common factors in the}\\ \text{numerator and denominator.}\end{array}$$

$$-7 = -7 \qquad \text{Simplify the left-hand side.}$$

The solution is $-\dfrac{28}{3}$, and the solution set is $\left\{-\dfrac{28}{3}\right\}$.

Answer -18

The equation in Example 3 can be solved using an alternate two-step approach.

$$\dfrac{3}{4}y = -7$$

$$4\left(\dfrac{3}{4}y\right) = 4(-7) \qquad \text{Multiply both sides by 4 to undo the division by 4.}$$

$$3y = -28 \qquad \text{Simplify: } 4\left(\dfrac{3}{4}y\right) = \dfrac{4}{1}\left(\dfrac{3}{4}y\right) = \dfrac{\overset{1}{4}\cdot 3}{1\cdot\underset{1}{4}}y = 3y.$$

$$\dfrac{3y}{3} = \dfrac{-28}{3} \qquad \text{Divide both sides by 3 to undo the multiplication by 3.}$$

$$y = -\dfrac{28}{3}$$

◼ Simplifying expressions to solve equations

To solve more complicated equations, we often need to apply the distributive property and combine like terms.

INTERMEDIATE
Algebra *f(x)* Now™

Self Check 4
Solve: $-2(x + 3) = 18$.

EXAMPLE 4 Solve: $3(a + 2) = 20$.

Solution

$$3(a + 2) = 20$$

$$3a + 6 = 20 \qquad \text{Distribute the multiplication by 3.}$$

$$3a + 6 - 6 = 20 - 6 \qquad \text{To undo the addition of 6, subtract 6 from both sides.}$$

$$3a = 14 \qquad \text{Simplify each side of the equation.}$$

$$\dfrac{3a}{3} = \dfrac{14}{3} \qquad \text{To undo the multiplication by 3, divide both sides by 3.}$$

$$a = \dfrac{14}{3} \qquad \text{Simplify the left-hand side.}$$

Check: $3(a + 2) = 20$ This is the original equation.

$$3\left(\frac{14}{3} + 2\right) \stackrel{?}{=} 20 \quad \text{Substitute } \frac{14}{3} \text{ for } a.$$

$$3\left(\frac{14}{3} + \frac{6}{3}\right) \stackrel{?}{=} 20 \quad \text{Get a common denominator: } 2 = \frac{6}{3}.$$

$$3\left(\frac{20}{3}\right) \stackrel{?}{=} 20 \quad \text{Add the fractions: } \frac{14}{3} + \frac{6}{3} = \frac{20}{3}.$$

$$20 = 20 \quad \text{Simplify the left-hand side.}$$

The solution is $\frac{14}{3}$, and the solution set is $\left\{\frac{14}{3}\right\}$.

Answer -12

INTERMEDIATE
Algebra *f(x)* **Now**™

EXAMPLE 5   Solve: $4b - 7 + 2b = 1 + 2b + 8$.

Self Check 5
Solve:
$-6t - 12 - 6t = 1 + 2t - 5.$

Solution   First, we combine like terms on each side of the equation.

$$4b - 7 + 2b = 1 + 2b + 8$$
$$6b - 7 = 2b + 9 \quad \text{Combine like terms: } 4b + 2b = 6b \text{ and } 1 + 8 = 9.$$

We note that terms involving b appear on both sides of the equation. To isolate b on the left-hand side, we need to eliminate $2b$ on the right-hand side.

$$6b - 7 = 2b + 9$$
$$6b - 7 - 2b = 2b + 9 - 2b \quad \text{Subtract } 2b \text{ from both sides.}$$
$$4b - 7 = 9 \quad \text{Combine like terms: } 6b - 2b = 4b \text{ and } 2b - 2b = 0.$$
$$4b - 7 + 7 = 9 + 7 \quad \text{To undo the subtraction of 7, add 7 to both sides.}$$
$$4b = 16 \quad \text{Simplify each side of the equation.}$$
$$b = 4 \quad \text{Divide both sides by 4.}$$

Check: $4b - 7 + 2b = 1 + 2b + 8$ This is the original equation.
$$4(4) - 7 + 2(4) \stackrel{?}{=} 1 + 2(4) + 8 \quad \text{Substitute 4 for } b.$$
$$16 - 7 + 8 \stackrel{?}{=} 1 + 8 + 8 \quad \text{Perform each multiplication.}$$
$$17 = 17 \quad \text{Simplify each side.}$$

The solution is 4, and the solution set is {4}.

Answer $-\dfrac{4}{7}$

In general, to solve linear equations in one variable, we follow these steps.

Solving linear equations in one variable

1. If the equation contains fractions, multiply both sides of the equation by a number that will eliminate the denominators.

2. Use the distributive property to remove all sets of parentheses and then combine like terms.

3. Use the addition and subtraction properties of equality to get all variable terms on one side of the equation and all constant terms on the other side. Combine like terms, if necessary.

4. Use the multiplication and division properties of equality to make the coefficient of the variable equal to 1.

5. Check the result by replacing the variable with the proposed solution and verifying that the number satisfies the equation.

EXAMPLE 6   Solve: $\dfrac{1}{3}(6x - 15) = \dfrac{3}{2}(x - 2) + 2$.

Solution

Step 1: One way to clear the equation of fractions is to multiply both sides by the least common denominator (LCD) of $\frac{1}{3}$ and $\frac{3}{2}$. The LCD of these fractions is the smallest number that can be divided by both 2 and 3 exactly. That number is 6.

$$\frac{1}{3}(6x - 15) = \frac{3}{2}(x - 2) + 2$$

$$6\left[\frac{1}{3}(6x - 15)\right] = 6\left[\frac{3}{2}(x - 2) + 2\right]$$

To eliminate the fractions, multiply both sides by the LCD, 6.

$$2(6x - 15) = 6 \cdot \frac{3}{2}(x - 2) + 6 \cdot 2$$

On the left-hand side, perform the multiplication: $6 \cdot \frac{1}{3} = 2$. On the right-hand side, distribute the multiplication by 6.

$$2(6x - 15) = 9(x - 2) + 12$$

Perform the multiplications on the right-hand side.

Step 2: We use the distributive property to remove parentheses and then combine like terms.

$$12x - 30 = 9x - 18 + 12$$

On the left-hand side, distribute the multiplication by 2. On the right-hand side, distribute the multiplication by 9.

$$12x - 30 = 9x - 6$$

Combine like terms: $-18 + 12 = -6$.

Step 3: We use the addition and subtraction properties of equality by subtracting $9x$ from both sides and by adding 30 to both sides.

$$12x - 30 - 9x + 30 = 9x - 6 - 9x + 30$$

$$3x = 24$$

On each side, combine like terms.

Step 4: The coefficient of the variable x is 3. To undo the multiplication by 3, we divide both sides by 3.

$$\frac{3x}{3} = \frac{24}{3}$$ Divide both sides by 3.

$$x = 8$$ Perform the divisions.

Step 5: We check by substituting 8 for x in the original equation and simplifying:

$$\frac{1}{3}(6x - 15) = \frac{3}{2}(x - 2) + 2$$

$$\frac{1}{3}[6(\mathbf{8}) - 15] \stackrel{?}{=} \frac{3}{2}(\mathbf{8} - 2) + 2$$

$$\frac{1}{3}(48 - 15) \stackrel{?}{=} \frac{3}{2}(6) + 2$$

$$\frac{1}{3}(33) \stackrel{?}{=} 9 + 2$$

$$11 = 11$$

The solution is 8, and the solution set is {8}.

EXAMPLE 7 Solve: $\dfrac{x+2}{5} - 4x = \dfrac{8}{5} - \dfrac{x+9}{2}$.

Solution

$$\dfrac{x+2}{5} - 4x = \dfrac{8}{5} - \dfrac{x+9}{2}$$

$$10\left(\dfrac{x+2}{5} - 4x\right) = 10\left(\dfrac{8}{5} - \dfrac{x+9}{2}\right)$$ To eliminate the fractions, multiply both sides by the LCD, 10.

$$10\cdot\dfrac{x+2}{5} - 10\cdot 4x = 10\cdot\dfrac{8}{5} - 10\cdot\dfrac{x+9}{2}$$ On each side, distribute the multiplication by 10.

$$2(x+2) - 40x = 2(8) - 5(x+9)$$ Perform each multiplication by 10.

$$2x + 4 - 40x = 16 - 5x - 45$$ On each side, distribute.

$$-38x + 4 = -5x - 29$$ On each side, combine like terms.

$$-33x = -33$$ Add $5x$ to both sides. Subtract 4 from both sides. These steps can be done in your head — we don't show them.

$$x = 1$$ Divide both sides by -33. This step can also be done in your head.

Check: $\dfrac{x+2}{5} - 4x = \dfrac{8}{5} - \dfrac{x+9}{2}$

$$\dfrac{1+2}{5} - 4(1) \overset{?}{=} \dfrac{8}{5} - \dfrac{1+9}{2}$$ Substitute 1 for x.

$$\dfrac{3}{5} - 4 \overset{?}{=} \dfrac{8}{5} - 5$$

$$\dfrac{3}{5} - \dfrac{20}{5} \overset{?}{=} \dfrac{8}{5} - \dfrac{25}{5}$$ Write 4 as $\dfrac{20}{5}$ and 5 as $\dfrac{25}{5}$.

$$-\dfrac{17}{5} = -\dfrac{17}{5}$$ Perform each subtraction.

Since 1 satisfies the equation, it is the solution.

Self Check 7
Solve:
$\dfrac{a+3}{2} + 2a = \dfrac{3}{2} - \dfrac{a+27}{5}$.

Answer -2

EXAMPLE 8 Solve: $-35.6 = 77.89 - x$.

Solution

$$-35.6 = 77.89 - x$$

$$-35.6 - \mathbf{77.89} = 77.89 - x - \mathbf{77.89}$$ Subtract 77.89 from both sides.

$$-113.49 = -x$$ Simplify each side of the equation.

$$-113.49 = -1x$$ $-x = -1x$.

$$\dfrac{-113.49}{-1} = \dfrac{-1x}{-1}$$ To isolate x, divide both sides by -1.

$$113.49 = x$$ Simplify each side of the equation.

$$x = 113.49$$

Check to see that 113.49 satisfies the original equation.

Self Check 8
Solve: $-1.3 = -2.6 - x$.

Answer -1.3

For more complicated equations involving decimals, we can multiply both sides of the equation by a power of 10 to clear the equation of decimals.

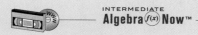

EXAMPLE 9   Solve: $0.04(12) + 0.01x = 0.02(12 + x)$.

Solution   The equation contains the decimals 0.04, 0.01, and 0.02. Multiplying both sides by 100 changes the decimals in the equation to integers, which are easier to work with.

$$0.04(12) + 0.01x = 0.02(12 + x)$$

$$\mathbf{100}[0.04(12) + 0.01x] = \mathbf{100}[0.02(12 + x)]$$
To make 0.04, 0.01, and 0.02 integers, multiply both sides by 100.

$$\mathbf{100} \cdot 0.04(12) + \mathbf{100} \cdot 0.01x = \mathbf{100} \cdot 0.02(12 + x)$$
On the left-hand side, distribute the multiplication by 100.

$$4(12) + 1x = 2(12 + x)$$
Perform the three multiplications by 100.

$$48 + x = 24 + 2x$$
Distribute.

$$48 + x - 24 - x = 24 + 2x - \mathbf{24} - \mathbf{x}$$
Subtract 24 and x from both sides.

$$24 = x$$
Simplify each side.

$$x = 24$$

Check by substituting 24 for x in the original equation.

▌Identities and contradictions

The equations discussed so far are called **conditional equations.** For these equations, some numbers satisfy the equation and others do not. An **identity** is an equation that is satisfied by every number for which both sides of the equation are defined.

Self Check 10
Solve: $3(a + 4) + 5 = 2(a - 1) + a + 19$ and give the solution set.

EXAMPLE 10   Solve: $-2(x - 1) - 4 = -4(1 + x) + 2x + 2$.

Solution

$$-2(x - 1) - 4 = -4(1 + x) + 2x + 2$$

$$-2x + 2 - 4 = -4 - 4x + 2x + 2$$
Use the distributive property.

$$-2x - 2 = -2x - 2$$
On each side, combine like terms.

$$-2 = -2$$
Add $2x$ to both sides.

The terms involving x drop out. The resulting true statement indicates that the original equation is true for every value of x. The solution set is the set of real numbers denoted $\mathbb{R}$. The equation is an identity.

Answer all real numbers

A **contradiction** is an equation that is never true.

EXAMPLE 11   Solve: $-6.2(-x - 1) - 4 = 4.2x - (-2x)$.

Solution

$$-6.2(-x - 1) - 4 = 4.2x - (-2x)$$

$6.2x + 6.2 - 4 = 4.2x + 2x$	On the left-hand side, distribute -6.2. On the right-hand side, write the subtraction as addition of the opposite.
$6.2x + 2.2 = 6.2x$	On each side, combine like terms.
$6.2x + 2.2 - \mathbf{6.2x} = 6.2x - \mathbf{6.2x}$	Subtract $6.2x$ from both sides.
$2.2 \neq 0$	Simplify each side.

The terms involving x drop out. The resulting false statement indicates that no value for x makes the original equation true. The solution set contains no elements and can be denoted as the **empty set {   }** or the **null set $\varnothing$.** The equation is a contradiction.

Self Check 11
Solve: $3(a + 4) + 2 = 2(a - 1) + a + 19$ and give the solution set.

Answer no solution

Solving formulas

To solve a formula for a variable means to isolate that variable on one side of the equation and have all other quantities on the other side. We can use the skills discussed in this section to solve many types of formulas for a specified variable.

EXAMPLE 12   Solve $A = \frac{1}{2}bh$ for h.

Solution

$$A = \frac{1}{2}bh$$

$2A = bh$	To eliminate the fraction, multiply both sides by 2.
$\dfrac{2A}{b} = h$	To isolate h, divide both sides by b.
$h = \dfrac{2A}{b}$	Write the equation with h on the left-hand side.

Self Check 12
Solve $A = \frac{1}{2}bh$ for b.

Answer $b = \dfrac{2A}{h}$

INTERMEDIATE
Algebra $f(x)$ Now™

EXAMPLE 13   For simple interest, the formula $A = P + Prt$ gives the amount of money in an account at the end of a specific time. A represents the amount, P the principal, r the rate of interest, and t the time. We can solve the formula for t as follows:

Solution

$$A = P + Prt$$

$A - P = Prt$	To isolate the term involving t, subtract P from both sides.
$\dfrac{A - P}{Pr} = t$	To isolate t, divide both sides by Pr.
$t = \dfrac{A - P}{Pr}$	Write the equation with t on the left-hand side.

Self Check 13
Solve $A = P + Prt$ for r.

Answer $r = \dfrac{A - P}{Pt}$

Self Check 14

Solve $S = \dfrac{180(t - 2)}{7}$ for t.

EXAMPLE 14   The formula $F = \frac{9}{5}C + 32$ converts degrees Celsius to degrees Fahrenheit. Solve the formula for C.

Solution

$$F = \frac{9}{5}C + 32$$

$$F - 32 = \frac{9}{5}C \qquad \text{To isolate the term involving } C, \text{ subtract 32 from both sides.}$$

$$\frac{5}{9}\left(F - 32\right) = \frac{5}{9}\left(\frac{9}{5}C\right) \qquad \text{To isolate } C, \text{ multiply both sides by } \frac{5}{9}.$$

$$\frac{5}{9}(F - 32) = C \qquad \text{On the right-hand side: } \frac{5}{9} \cdot \frac{9}{5} = 1 \text{ and } 1C = C.$$

$$C = \frac{5}{9}(F - 32)$$

Answer $t = \dfrac{7S + 360}{180}$ or

$t = \dfrac{7S}{180} + 2$

To convert degrees Fahrenheit to degrees Celsius, we can use the formula $C = \frac{5}{9}(F - 32)$.

Section 1.5   STUDY SET

VOCABULARY   *Fill in the blanks.*

1. An _____ is a statement that two quantities are equal.

2. If two equations have the same solution set, they are called _____ equations.

3. If a number is substituted for a variable in an equation and the equation is true, we say that the number _____ the equation.

4. An _____ is an equation that is true for all values of its variable.

5. An equation that is not true for any values of its variable is called a _____.

6. $3x + 1 = 10$ is a linear _____ in one variable.

CONCEPTS   *Fill in the blanks.*

7. If a, b, and c are real numbers, and $a = b$, then $a + c = b + \boxed{}$ and $a - c = b - \boxed{}$.

8. If a, b, and c are real numbers, and $a = b$, then $c \cdot a = \boxed{} \cdot b$ and $\dfrac{a}{c} = \boxed{}$ $(c \neq 0)$

9. **a.** Simplify: $5y + 2 - 3y$.

   b. Solve: $5y + 2 - 3y = 8$.

   c. Evaluate $5y + 2 - 3y$ for $y = 8$.

10. **a.** In $-5b + 3 = -18$, what can we do to both sides to undo the addition of 3?

    b. In $-5b = -18$, what can we do to both sides to undo the multiplication by -5?

11. Suppose you solve a linear equation in one variable, the variable drops out, and you obtain $8 = 8$. What is the solution set?

12. Suppose you solve a linear equation in one variable, the variable drops out, and you obtain $8 = 7$. What is the solution set?

13. When solving $\dfrac{x + 1}{3} - \dfrac{2}{15} = \dfrac{x - 1}{5}$, why would we multiply both sides by 15?

14. When solving $1.45x - 0.5(1 - x) = 0.7x$, why would we multiply both sides by 100?

NOTATION   *Complete each solution.*

15.
$$-2(x + 7) = 20$$
$$\boxed{} - 14 = 20$$
$$-2x - 14 + \boxed{} = 20 + \boxed{}$$
$$-2x = \boxed{}$$
$$\dfrac{-2x}{\boxed{}} = \dfrac{34}{\boxed{}}$$
$$x = -17$$

16. $\dfrac{d}{3} + 4 = 1$

$\dfrac{d}{3} + 4 - \blacksquare = 1 - \blacksquare$

$\dfrac{d}{3} = \blacksquare$

$\blacksquare\left(\dfrac{d}{3}\right) = \blacksquare(-3)$

$d = -9$

17. Fill in the blanks.

a. $-x = \blacksquare\, x$

b. $\dfrac{2t}{3} = \blacksquare\, t$

18. When checking a solution of an equation, the symbol $\overset{?}{=}$ is used. What does it mean?

PRACTICE   *Determine whether 5 is a solution of each equation.*

19. $3x + 2 = 17$

20. $7x - 2 = 53 - 5x$

21. $3(2m - 3) = 15$

22. $\dfrac{3}{5}p - 5 = -2$

Solve each equation.

23. $\dfrac{x}{4} = 7$

24. $-\dfrac{x}{6} = 8$

25. $-\dfrac{4}{5}s = 16$

26. $-3 = -\dfrac{9}{8}s$

27. $2x - 1 = 0$

28. $3x - 4 = 0$

29. $5y + 6 = 0$

30. $7y + 3 = 0$

31. $8x = x$

32. $-z = 5z$

33. $0 = -x$

34. $-\dfrac{2}{3}x = 0$

35. $2x + 2(1) = 6$

36. $3x - 4(1) = 8$

37. $4(3) + 2y = -6$

38. $5(2) + 10y = -10$

39. $200 = 34 - t$

40. $8 - x = -12$

41. $1.6a = 4.032$

42. $0.52 = 0.05y$

43. $3x + 1 = 3$

44. $8k - 2 = 13$

45. $3(k - 4) = -36$

46. $4(x + 6) = 84$

47. $4j + 12.54 = 18.12$

48. $9.8 - 15r = -15.7$

49. $4a - 22 - a = -2a - 7$

50. $a + 18 = 5a - 3 + a$

51. $2(2x + 1) = x + 15 + 2x$

52. $-2(x + 5) = x + 30 - 2x$

53. $2(a - 5) - (3a + 1) = 0$

54. $8(3a - 5) - 4(2a + 3) = 12$

55. $9(x + 2) = -6(4 - x) + 18$

56. $3(x + 2) - 2 = -(5 + x) + x$

57. $\dfrac{1}{2}x - 4 = -1 + 2x$

58. $2x + 3 = \dfrac{2}{3}x - 1$

59. $\dfrac{b}{2} - \dfrac{b}{3} = 4$

60. $\dfrac{w}{2} + \dfrac{w}{3} = 10$

61. $\dfrac{a + 1}{3} + \dfrac{a - 1}{5} = \dfrac{2}{15}$

62. $\dfrac{2z + 3}{3} + \dfrac{3z - 4}{6} = \dfrac{z - 2}{2}$

63. $\dfrac{5a}{2} - 12 = \dfrac{a}{3} + 1$

64. $5 - \dfrac{x + 2}{3} = 7 - x$

65. $\dfrac{3 + p}{3} - 4p = 1 - \dfrac{p + 7}{2}$

66. $\dfrac{4 - t}{2} - \dfrac{3t}{5} = 2 + \dfrac{t + 1}{3}$

67. $0.45 = 16.95 - 0.25(75 - 3x)$

68. $0.02x + 0.0175(15{,}000 - x) = 277.5$

69. $0.04(12) + 0.01t - 0.02(12 + t) = 0$

70. $0.25(t + 32) = 3.2 + t$

Solve each equation. If the equation is an identity or a contradiction, so indicate.

71. $4(2 - 3t) + 6t = -6t + 8$

72. $2x - 6 = -2x + 4(x - 2)$

73. $3(x - 4) + 6 = -2(x + 4) + 5x$

74. $2(x - 3) = \dfrac{3}{2}(x - 4) + \dfrac{x}{2}$

75. $2y + 1 = 5(0.2y + 1) - (4 - y)$

76. $-3x = -2x + 1 - (5 + x)$

Solve each formula for the indicated variable.

77. $V = \dfrac{1}{3}Bh$ for B

78. $A = \dfrac{1}{2}bh$ for b

79. $I = Prt$ for t

80. $E = mc^2$ for m

81. $P = 2l + 2w$ for w

82. $T - W = ma$ for W

83. $A = \dfrac{1}{2}h(B + b)$ for B

84. $l = a + (n - 1)d$ for n

85. $y = mx + b$ for x

86. $\lambda = Ax + AB$ for B

87. $\bar{v} = \dfrac{1}{2}(v + v_0)$ for v_0

88. $l = a + (n - 1)d$ for d

89. $S = \dfrac{a - \ell r}{1 - r}$ for ℓ

90. $s = \dfrac{1}{2}gt^2 + vt$ for g

91. $S = \dfrac{n(a + l)}{2}$ for l

92. $K = \dfrac{Mv_o^2}{2} + \dfrac{Iw^2}{2}$ for I

APPLICATIONS

93. CONVERTING TEMPERATURES In preparing an American almanac for release in Europe, editors need to convert temperature ranges for the planets from degrees Fahrenheit to degrees Celsius. Solve the formula $F = \frac{9}{5}C + 32$ for C. Then use your result to make the conversions for the data shown in the Table. Round to the nearest degree.

Planet	High °F	Low °F	High °C	Low °C
Mercury	810	−290		
Earth	136	−129		
Mars	63	−87		

94. THERMODYNAMICS In thermodynamics, the Gibbs free-energy function is given by the formula $G = U - TS + pV$. Solve for S.

95. WIPER DESIGN The following illustration shows the area cleaned by a windshield wiper assembly. The area is given by the formula

$$A = \dfrac{d\pi(r_1^2 - r_2^2)}{360}$$

Engineers have determined the amount of windshield area that needs to be cleaned by the wiper for two different vehicles. Solve the equation for d and use your result to find the number of degrees d the wiper arm must swing in each case. Round to the nearest degree.

Vehicle	Area cleaned	d (deg)
Luxury car	513 in.²	
Sport utility vehicle	586 in.²	

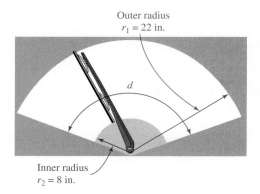

Outer radius $r_1 = 22$ in.

d

Inner radius $r_2 = 8$ in.

96. ELECTRONICS The figure below is a schematic diagram of a resistor connected to a voltage source of 60 volts. As a result, the resistor dissipates power in the form of heat. The power P lost when a voltage E is placed across a resistance R (in ohms) is given by the formula

$$P = \dfrac{E^2}{R}$$

Solve for R. If P is 4.8 watts and E is 60 volts, find R.

Battery　　$E = 60$ v　　Resistor

97. CHEMISTRY LABS In chemistry, the ideal gas law equation is $PV = nR(T + 273)$, where P is the pressure, V the volume, T the temperature, and n the number of moles of a gas. R is a constant, 0.082. Solve the equation for n. Then use your result and the data from the student lab notebook in the illustration on the next page to find the value of n to the nearest thousandth for trial 1 and trial 2.

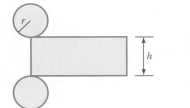

Ideal gas law Lab #1		Betsy Kinsell Chem 1 Section A	
Data:	Pressure (Atmosph.)	Volume (Liters)	Temp (°C)
Trial 1	0.900	0.250	90
Trial 2	1.250	1.560	−10
R = 0.082 (Constant)			

98. INVESTMENTS An amount P, invested at a simple interest rate r, will grow to an amount A in t years according to the formula $A = P(1 + rt)$. Solve for P. Suppose a man invested some money at 5.5%. If he had \$6,693.75 on deposit after 5 years, what amount did he originally invest?

99. COST OF ELECTRICITY The cost of electricity in a city is given by the formula $C = 0.07n + 6.50$, where C is the cost and n is the number of kilowatt hours used. Solve for n. Then see the illustration and find the number of kilowatt hours used each month by the homeowner whose checks to pay the monthly electric bills are shown.

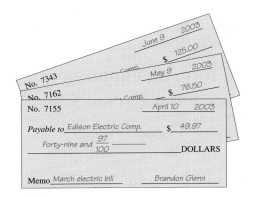

100. COST OF WATER A monthly water bill in a certain city is calculated by using the formula $n = \frac{5{,}000C - 17{,}500}{6}$, where n is the number of gallons used and C is the monthly cost. Solve for C and compute the bill for quantities of 500, 1,200, and 2,500 gallons.

101. SURFACE AREA To find the amount of tin needed to make the coffee can shown in the next column, we use the formula for the surface area of a right circular cylinder,

$$A = 2\pi r^2 + 2\pi rh$$

Solve the formula for h.

102. CARPENTRY A regular polygon has n equal sides and n equal angles. The measure a of an interior angle in degrees is given by $a = 180\left(1 - \frac{2}{n}\right)$. Solve for n. How many sides does the outdoor bandstand design shown below have if the performance platform is a regular polygon with interior angles measuring 135°?

WRITING

103. What does it mean to *solve an equation?*

104. Why doesn't the equation $x = x + 1$ have a real-number solution?

105. What is an identity? Give an example.

106. When solving a linear equation in one variable, the objective is to isolate the variable on one side of the equation. What does that mean?

REVIEW *Simplify each expression.*

107. $-(4 + t) + 2t$

108. $12(2r + 11) - 11 - 3$

109. $4(b + 8) - 8b$

110. $-2(m - 3) + 8(2m + 7)$

111. $3.8b - 0.9b$ **112.** $0 - 0.6p$

113. $\frac{3}{5}t + \frac{2}{5}t$ **114.** $-\frac{3}{16}x - \frac{5}{16}x$

1.6   Using Equations to Solve Problems

- A problem-solving strategy   • Translating words to form an equation
- Number–value problems   • Drawing diagrams   • Geometry problems
- Using formulas to solve problems

One objective of this course is to improve your problem-solving skills. In the next two sections, you will have the opportunity to do that as we discuss how to use equations to solve many types of problems.

■ A problem-solving strategy

The key to problem solving is understanding the problem and devising a plan for solving it. The following list provides a strategy for solving problems.

> **Problem solving**
>
> 1. ***Analyze the problem*** by reading it carefully to understand the given facts. What information is given? What vocabulary is used? What are you asked to find? Often a diagram or table will help you visualize the facts of the problem.
> 2. ***Form an equation*** by picking a variable to represent the numerical value to be found. Then express all other unknown quantities as expressions involving that variable. Key words or phrases can be helpful. Finally, translate the words of the problem into an equation.
> 3. ***Solve the equation.***
> 4. ***State the conclusion.***
> 5. ***Check the result*** in the words of the problem.

■ Translating words to form an equation

In order to solve problems, which are almost always given in words, we must translate those words into mathematical symbols. In the next example, we use translation to write an equation that mathematically models the situation.

INTERMEDIATE
Algebra $f(x)$ **Now**™

EXAMPLE 1   Leading U.S. employers.   In 2003, McDonald's and Wal-Mart were the nation's top two employers. Their combined work forces totaled 2,900,000 people. If Wal-Mart employed 100,000 fewer people than McDonald's, how many employees did each company have?

Analyze the problem

- The phrase *combined work forces totaled 2,900,000* suggests that if we add the number of employees of each company, the result will be 2,900,000.
- The phrase *Wal-Mart employed 100,000 fewer people than McDonald's* suggests that the number of employees of Wal-Mart can be found by subtracting 100,000 from the number of employees of McDonald's.
- We are to find the number of employees of each company.

Form an equation

If we let x = the number of employees of McDonald's, then $x - 100,000$ = the number of employees of Wal-Mart. We can now translate the words of the problem into an equation.

The number of employees of McDonald's	plus	the number of employees of Wal-Mart	is	2,900,000
x	$+$	$x - 100{,}000$	$=$	$2{,}900{,}000$

Solve the equation

$$x + x - 100{,}000 = 2{,}900{,}000$$
$$2x - 100{,}000 = 2{,}900{,}000 \qquad \text{Combine like terms.}$$
$$2x = 3{,}000{,}000 \qquad \text{Add 100,000 to both sides.}$$
$$x = 1{,}500{,}000 \qquad \text{Divide both sides by 2.}$$

Recall that x = the number of employees of McDonald's. To find the number of employees of Wal-Mart, we evaluate $x - 100{,}000$ for $x = 1{,}500{,}000$.

$$x - 100{,}000 = \mathbf{1{,}500{,}000} - 100{,}000$$
$$= 1{,}400{,}000$$

State the conclusion

In 2003, McDonald's had 1,500,000 employees and Wal-Mart had 1,400,000 employees.

Check the result

Since $1{,}500{,}000 + 1{,}400{,}000 = 2{,}900{,}000$, and since 1,400,000 is 100,000 less than 1,500,000, the answers check.

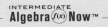

INTERMEDIATE
Algebra $f(x)$ Now™

EXAMPLE 2 **Travel promotions.** The price of a 7-day Alaskan cruise, normally $2,752 per person, is reduced by $1.75 per person for large groups traveling together. How large a group is needed for the price to be $2,500 per person?

Analyze the problem

For each member of the group, the cost is reduced by $1.75. For a group of 20 people, the $2,752 price would be reduced by 20($1.75) = $35.

The per-person price of the cruise = $2{,}752 - 20(\$1.75)$

For a group of 30 people, the $2,752 cost would be reduced by 30($1.75) = $52.50.

The per-person price of the cruise = $2{,}752 - 30(\$1.75)$

Form an equation

We can describe this situation in words. If we let x = the group size necessary for the price of the cruise to be $2,500 per person, we can translate the words into an equation.

The price of the cruise	is	$2,752	minus	the number of people in the group	times	$1.75.
$2{,}500$	$=$	$2{,}752$	$-$	x	$\cdot$	1.75

Solve the equation

$$2{,}500 = 2{,}752 - 1.75x$$

$$2{,}500 - \mathbf{2{,}752} = 2{,}752 - 1.75x - \mathbf{2{,}752} \qquad \text{Subtract 2,752 from both sides.}$$

$$-252 = -1.75x \qquad\qquad\qquad \text{Simplify each side.}$$

$$144 = x \qquad\qquad\qquad\qquad \text{Divide both sides by } -1.75.$$

State the conclusion

If 144 people travel together, the price will be $2,500 per person.

Check the result

For 144 people, the cruise cost of $2,752 will be reduced by 144($1.75) = $252. If we subtract, $2,752 − $252 = $2,500. The answer checks.

■ Number–value problems

If a problem deals with quantities that have a value, we must distinguish between the *number of* and the *value of* the unknown quantity. For problems such as these, we will use the relationship

Number · value = total value

EXAMPLE 3   Portfolio analysis.   A college foundation owns stock in Coca-Cola (selling at $58 per share), Boeing (selling at $65 per share), and CIGNA (selling at $125 per share). The foundation owns an equal number of shares of Coca-Cola and Boeing stock, but five times as many shares of CIGNA stock. If this portfolio is worth $523,600, how many shares of each stock does the foundation own?

Analyze the problem

The value of the Coca-Cola stock plus the value of the Boeing stock plus the value of the CIGNA stock must equal $523,600. We need to find the number of shares of each of these stocks that the foundation has in its portfolio.

Form an equation

If we let x = the number of shares of Coca-Cola stock, then x also represents the number of shares of Boeing stock. If the foundation owns five times as many shares of CIGNA stock as of Coca-Cola or Boeing stock, $5x$ = the number of shares of CIGNA. The value of the shares of each company's stock would be the *product* of the number of shares of that stock and its per-share value. See Figure 1-12.

Stock	Number of shares ·	Value per share =	Total value of the stock
Coca-Cola	x	58	$58x$
Boeing	x	65	$65x$
CIGNA	$5x$	125	$125(5x)$

FIGURE 1-12

We can now form the equation.

The value of Coca-Cola stock	plus	the value of Boeing stock	plus	the value of CIGNA stock	is	the total value of all of the stock.
58x	+	65x	+	125(5x)	=	523,600

Solve the equation

$$58x + 65x + 125(5x) = 523,600$$

$$58x + 65x + 625x = 523,600 \quad \text{125(5x) = 625x.}$$

$$748x = 523,600 \quad \text{Combine like terms on the left-hand side.}$$

$$x = 700 \quad \text{Divide both sides by 748.}$$

State the conclusion

The foundation owns 700 shares of Coca-Cola, 700 shares of Boeing, and 5(700) = 3,500 shares of CIGNA.

Check the result

The value of 700 shares of Coca-Cola stock is 700($58) = $40,600. The value of 700 shares of Boeing stock is 700($65) = $45,500. The value of 3,500 shares of CIGNA is 3,500($125) = $437,500. The sum is $40,600 + $45,500 + $437,500 = $523,600. The answers check.

▌Drawing diagrams

When solving problems, diagrams are often helpful, because they allow us to visualize the facts of the problem.

INTERMEDIATE
Algebra $f^{(x)}$ Now™

EXAMPLE 4   Triathlons.   A triathlon includes swimming, long-distance running, and cycling. The long-distance run is 11 times longer than the distance the competitors swim. The distance they cycle is 85.8 miles longer than the distance they run. Overall, the competition covers 140.6 miles. Find the length of each part of the triathlon and round each length to the nearest tenth of a mile.

Analyze the problem

The entire course covers a distance of 140.6 miles. We note that the distance the competitors run is related to the distance they swim, and the distance they cycle is related to the distance they run.

Form an equation

If x = the distance the competitors swim, then $11x$ = the length of the long-distance run, and $11x + 85.8$ = the distance they cycle. From the diagram in Figure 1-13 on the next page, we see that the sum of the individual parts of the triathlon must equal the total distance covered.

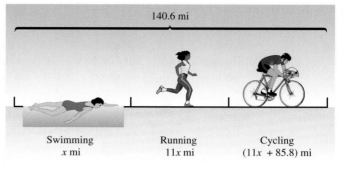

FIGURE 1-13

The distance they swim	plus	the distance they run	plus	the distance they cycle	is	the total length of the course.
x	$+$	$11x$	$+$	$11x + 85.8$	$=$	140.6

Solve the equation

$$x + 11x + 11x + 85.8 = 140.6$$

$$23x + 85.8 = 140.6 \qquad \text{Combine like terms.}$$

$$23x = 54.8 \qquad \text{Subtract 85.8 from both sides.}$$

$$x \approx 2.382608696 \qquad \text{Divide both sides by 23.}$$

State the conclusion

To the nearest tenth, the distance the competitors swim is 2.4 miles. The distance they run is $11x$, or approximately $11(2.382608696) = 26.20869565$ miles. To the nearest tenth, that is 26.2 miles. The distance they cycle is $11x + 85.8$, or approximately $26.20869565 + 85.8 = 112.0086957$ miles. To the nearest tenth, that is 112.0 miles.

Check the result

If we add the lengths of the three parts of the triathlon and round to the nearest tenth, we get 140.6 miles. The answers check.

▮ Geometry problems

Sometimes a geometric fact is helpful in solving a problem. Figure 1-14 shows several geometric figures. A **right angle** is an angle whose measure is 90°. A **straight angle** is an angle whose measure is 180°. An **acute angle** is an angle whose measure is greater than 0° and less than 90°. An angle whose measure is greater than 90° and less than 180° is called an **obtuse angle.**

If the sum of two angles equals 90°, the angles are called **complementary,** and each angle is called the **complement** of the other. If the sum of two angles equals 180°, the angles are called **supplementary,** and each angle is the **supplement** of the other.

A **right triangle** is a triangle with one right angle. An **isosceles triangle** is a triangle with two sides of equal measure that meet to form the **vertex angle.** The angles opposite the equal sides, called the **base angles,** are also equal. An **equilateral triangle** is a triangle with three equal sides and three equal angles.

For any triangle, the sum of the measures of its angles is 180°.

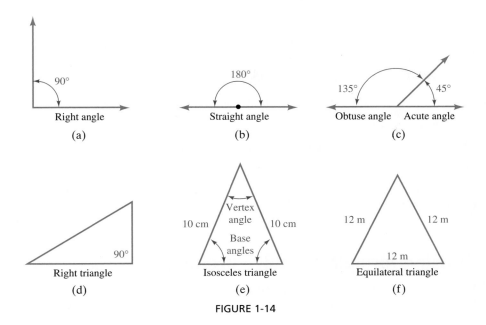

FIGURE 1-14

EXAMPLE 5   Flag designs.

The flag of Guyana, a republic on the northern coast of South America, is one isosceles triangle superimposed over another isosceles triangle on a field of green, as shown in Figure 1-15. The measure of a base angle of the larger triangle is 14° more than the measure of a base angle of the smaller triangle. The measure of the vertex angle of the larger triangle is 34°. Find the measure of each base angle of the smaller triangle.

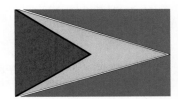

FIGURE 1-15

Analyze the problem

We are working with isosceles triangles. Therefore, the base angles of the smaller triangle have the same measure, and the base angles of the larger triangle have the same measure.

Form an equation

If we let x = the measure in degrees of one base angle of the smaller isosceles triangle, then the measure of its other base angle is also x. (See Figure 1-16.) The measure of a base angle of the larger isosceles triangle is $x + 14°$, since its measure is 14° more than the measure of a base angle of the smaller triangle. We are given that the vertex angle of the larger triangle measures 34°.

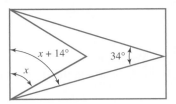

FIGURE 1-16

The sum of the measures of the angles of any triangle (in this case, the larger triangle) is 180°.

The measure of one base angle	plus	the measure of the other base angle	plus	the measure of the vertex angle	is	180°.
$x + 14$	$+$	$x + 14$	$+$	34	$=$	180

Solve the equation

$$x + 14 + x + 14 + 34 = 180$$
$$2x + 62 = 180 \quad \text{Combine like terms.}$$
$$2x = 118 \quad \text{Subtract 62 from both sides.}$$
$$x = 59 \quad \text{Divide both sides by 2.}$$

State the conclusion

The measure of each base angle of the smaller triangle is 59°.

Check the result

If x is 59, then $x + 14 = 73$. The sum of the measures of each base angle and the vertex angle of the larger triangle is $73° + 73° + 34° = 180°$. The answer checks.

▍Using formulas to solve problems

When preparing to write an equation to solve a problem, the given facts of the problem often suggest a formula that can be used to model the situation mathematically.

INTERMEDIATE
Algebra $f(x)$ Now™

EXAMPLE 6   **Designing a kennel.**   A man has a 50-foot roll of fencing to make a rectangular-shaped kennel. If he wants the kennel to be 6 feet longer than it is wide, find its dimensions.

Analyze the problem

The perimeter P of the rectangular kennel is 50 feet. Recall that the formula for the perimeter of a rectangle is $P = 2l + 2w$. We need to find its length and width.

Form an equation

We let $w = $ the width of the kennel. Then the length, which is 6 feet more than the width, is represented by the expression $w + 6$. See Figure 1-17.

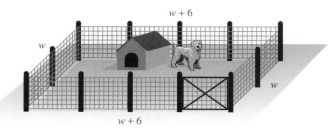

FIGURE 1-17

We can now form the equation by substituting 50 for P, $w + 6$ for the length, and w for the width in the formula for the perimeter of a rectangle.

$$P = 2l + 2w$$
$$50 = 2(w + 6) + 2w$$

Solve the equation

$$50 = 2(w + 6) + 2w$$
$$50 = 2w + 12 + 2w \quad \text{Use the distributive property.}$$
$$50 = 4w + 12 \quad \text{Combine like terms.}$$
$$38 = 4w \quad \text{Subtract 12 from both sides.}$$
$$9.5 = w \quad \text{Divide both sides by 4.}$$

State the conclusion

The width of the kennel is 9.5 feet. The length would be 6 feet more than this, or 15.5 feet.

Check the result

If a rectangle has a width of 9.5 feet and a length of 15.5 feet, its length is 6 feet more than its width, and the perimeter is 2(9.5) feet + 2(15.5) feet = 50 feet.

Section 1.6 STUDY SET

■ **VOCABULARY** *Fill in the blanks.*

1. An _____ angle has a measure of more than 0° and less than 90°.
2. A _____ angle is an angle whose measure is 90°.
3. If the sum of the measures of two angles equals 90°, the angles are called _____ angles.
4. If the sum of the measures of two angles equals 180°, the angles are called _____ angles.
5. If a triangle has a right angle, it is called a _____ triangle.
6. If a triangle has two sides of equal length, it is called an _____ triangle.
7. The sum of the measures of the _____ of a triangle is 180°.
8. An _____ triangle has three sides of equal length and three angles of equal measure.

■ **CONCEPTS**

9. The unit used to measure the intensity of sound is called the *decibel*. Translate the comments in the right-hand column of the following table into mathematical symbols to complete the decibel column.

	Decibels	Compared to conversation
Conversation	d	—
Vacuum cleaner		15 decibels more
Circular saw		10 decibels less than twice
Jet takeoff		20 decibels more than twice
Whispering		10 decibels less than half
Rock band		Twice the decibel level

10. The following table shows the four types of problems an instructor put on a history test.
    a. Complete the table.
    b. Which type of question appears the most on the test?
    c. Write an algebraic expression that represents the total number of points on the test.

Type of question	Number	Value	Total value
Multiple choice	x	5	
True/false	$3x$	2	
Essay	$x - 2$	10	
Fill-in	x	5	

11. The illustration below shows the three pieces that fit together to make a flute.
    a. What is the length of the shortest piece?
    b. What is the length of the longest piece?
    c. Write an algebraic expression that represents the length of the flute.

12. For each illustration below, what geometric concept is illustrated?

Radiation warning

13. Write an algebraic expression that represents the length in inches of the head of the tennis racquet shown below.

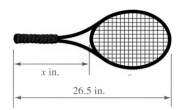

x in.

26.5 in.

14. The surface area of Earth is 510,066,000 square kilometers (km^2). If x represents the number of km^2 covered by water, what would $510,066,000 - x$ represent?

APPLICATIONS

15. CEREAL SALES In 2001, the two top-selling cereals were General Mills' Cheerios and Kellogg's Frosted Flakes, with combined sales of $1,003 million. Frosted Flakes sales were $324 million less than sales of Cheerios. What were the 2001 sales for each brand?

16. FILMS As of March 2004, Denzel Washington's three top-grossing films, *Remember the Titans, The Pelican Brief,* and *Crimson Tide,* had earned $307.8 million. If *Remember the Titans* earned $14.8 million more than *The Pelican Brief,* and if *The Pelican Brief* earned $9.4 million more than *Crimson Tide,* how much did each film earn as of that date?

17. SPRING TOURS A group of junior high students will be touring Washington, DC. Their chaperons will have the $1,810 cost of the tour reduced by $15.50 for each student they personally supervise. How many students will a chaperon have to supervise so that his or her cost to take the tour will be $1,500?

18. MACHINING Each pass through a lumber plane shaves off 0.015 inch of thickness from a board. How many times must a board, originally 0.875 inch thick, be run through the planer if a board of thickness 0.74 inch is desired?

19. MOVING EXPENSES To help move his furniture, a man rents a truck for $29.25 per day plus 15¢ per mile. If he has budgeted $75 for transportation expenses, how many miles will he be able to drive the truck if the move takes 1 day?

20. COMPUTING SALARIES A student working for a delivery company earns $47.50 per day plus $1.75 for each package she delivers. How many deliveries must she make each day to earn $100 a day?

21. VALUE OF AN IRA In an Individual Retirement Account (IRA) valued at $53,900, a couple has 500 shares of stock, some in Big Bank Corporation and some in Safe Savings and Loan. If Big Bank sells for $115 per share and Safe Savings sells for $97 per share, how many shares of each does the couple own?

22. ASSETS OF A PENSION FUND A pension fund owns 2,000 fewer shares in mutual stock funds than mutual bond funds. Currently, the stock funds sell for $12 per share, and the bond funds sell for $15 per share. How many shares of each does the pension fund own if the value of the securities is $165,000?

23. SELLING CALCULATORS Last month, a bookstore ran the ad shown below. Sales of $4,980 were generated, with 15 more graphing calculators sold than scientific calculators. How many of each type of calculator did the bookstore sell?

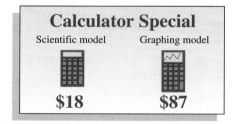

Calculator Special

Scientific model   Graphing model

$18 $87

24. SELLING GRASS SEED A seed company sells two grades of grass seed. A 100-pound bag of a mixture of rye and Kentucky bluegrass sells for $245, and a 100-pound bag of bluegrass sells for $347. How many bags of each are sold in a week when the receipts for 19 bags are $5,369?

25. WOODWORKING The carpenter saws a board that is 22 feet long into two pieces. He wants one piece to be 1 foot longer than twice the length of the shorter piece. Find the length of each piece.

22 ft

26. STATUE OF LIBERTY From the foundation of the large pedestal on which it sits to the top of the torch, the Statue of Liberty National Monument measures 305 feet. The pedestal is 3 feet taller than the statue. Find the height of the pedestal and the height of the statue.

27. NURSING The illustration below shows the angle a needle should make with the skin when administering a certain type of intradermal injection. Find the measure of both angles labeled in the illustration.

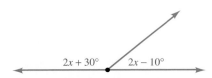

28. SUPPLEMENTARY ANGLES Refer to the illustration below and find *x*.

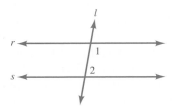

29. ARCHITECTURE Because of soft soil and a shallow foundation, the Leaning Tower of Pisa in Italy is not vertical, as shown below. How many degrees from vertical is the tower?

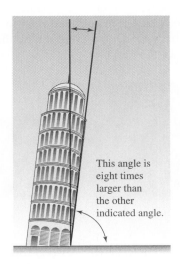

This angle is eight times larger than the other indicated angle.

30. STEPSTOOLS The sum of the measures of the three angles of any triangle is 180°. In the illustration in the next column, the measure of ∠2 (angle 2) is 10° larger than the measure of ∠1. The measure of ∠3 is 10° larger than the measure of ∠2. Find the measure of each angle.

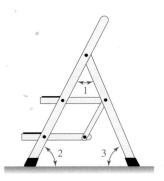

31. SUPPLEMENTARY ANGLES AND PARALLEL LINES In the illustration below, lines *r* and *s* are cut by a third line *l* to form ∠1 (angle 1) and ∠2. When lines *r* and *s* are parallel, ∠1 and ∠2 are supplementary. If ∠1 = *x* + 50°, ∠2 = 2*x* − 20°, and lines *r* and *s* are parallel, find *x*.

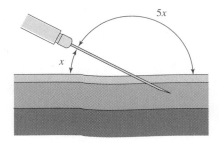

32. ANGLES AND PARALLEL LINES In the illustration below, *r* ∥ *s* (read as "line *r* is parallel to line *s*"), and *a* = 103. Find *b*, *c*, and *d*. (*Hint:* See Exercise 31.)

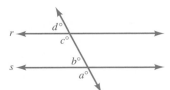

33. VERTICAL ANGLES When two lines intersect, as in the illustration, four angles are formed. Angles that are side-by-side, such as ∠1 (angle 1) and ∠2, are called **adjacent angles.** Angles that are nonadjacent, such as ∠1 and ∠3 or ∠2 and ∠4, are called **vertical angles.** From geometry, we know that if two lines intersect, vertical angles have the same measure. If ∠1 = 3*x* + 10° and ∠3 = 5*x* − 10°, find *x*.

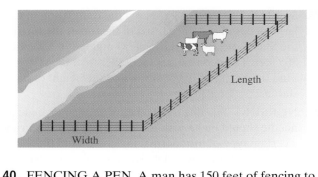

34. ANGLES AND PARALLEL LINES In the illustration, $r \parallel s$ (read as "line r is parallel to line s"), and $b = 137$. Find a and c.

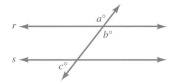

35. QUADRILATERALS The sum of the angles of any four-sided figure (called a *quadrilateral*) is 360°. The quadrilateral shown below has two equal base angles. Find their measures.

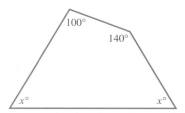

36. HEIGHT OF A TRIANGLE If the height of a triangle with a base of 8 inches is tripled, its area is increased by 96 square inches. Find the height of the triangle.

37. GOLDEN RECTANGLES Throughout history, most artists and designers have felt that rectangles having a length 1.618 times as long as their width have the most visually attractive shape. Such rectangles are known as *golden rectangles*. Measure the length and width of the following rectangles. Which of the rectangles is closest to being a golden rectangle?

i.
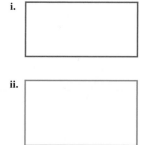

ii.

38. QUILTING A woman is planning to make a quilt in the shape of a golden rectangle. (See Exercise 37.) She has exactly 22 feet of a special lace that she plans to sew around the edge of the quilt. What should the length and width of the quilt be? Round both answers up to the nearest hundredth.

39. FENCING A PASTURE A farmer has 624 feet of fencing to enclose the pasture shown in the next column. Because a river runs along one side, fencing will be needed on only three sides. Find the dimensions of the pasture if its length is double its width.

40. FENCING A PEN A man has 150 feet of fencing to build the pen shown below. If one end is a square, find the outside dimensions.

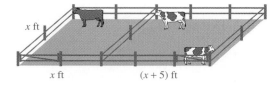

41. ENCLOSING A SWIMMING POOL A woman wants to enclose the pool shown below and have a walkway of uniform width all the way around. How wide will the walkway be if the woman uses 180 feet of fencing?

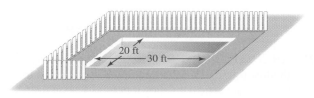

42. SOLAR HEATING One solar panel in the illustration below is to be 3 feet wider than the other. To be equally efficient, they must have the same area. Find the width of each.

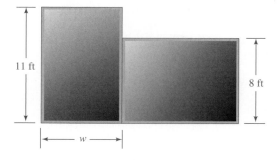

43. MAKING FURNITURE A woodworker wants to put two partitions crosswise in a drawer that is 28 inches deep, as shown on the next page. He wants to place the partitions so that the spaces created increase by 3 inches from front to back. If the thickness of each partition is $\frac{1}{2}$ inch, how far from the front end should he place the first partition?

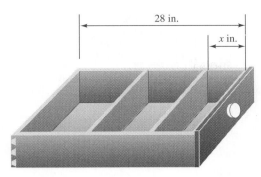

28 in.

x in.

44. BUILDING SHELVES A carpenter wants to put four shelves on an 8-foot wall so that the five spaces created decrease by 6 inches as we move up the wall. If the thickness of each shelf is $\frac{3}{4}$ inch, how far will the bottom shelf be from the floor?

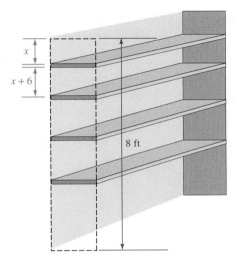

x

x + 6

8 ft

WRITING

45. Briefly explain what should be accomplished in each of the steps (*analyze, form, solve, state,* and *check*) of the problem-solving strategy used in this section.

46. Write a problem that can be represented by the following verbal model and equation.

Measure of 1st angle	plus	measure of 2nd angle	plus	measure of 3rd angle	is 180°.
x	+	$2x$	+	$x + 10$	= 180

REVIEW

47. When expressed as a decimal, is $\frac{7}{9}$ a terminating or a repeating decimal?

48. Solve: $x + 20 = 4x - 1 + 2x$.

49. List the integers.

50. Solve: $2x + 2 = \frac{2}{3}x - 2$.

51. Evaluate $2x^2 + 5x - 3$ for $x = -3$.

52. Solve $T - R = ma$ for R.

1.7 More Problem Solving

- Percent problems • Statistics problems • Investment problems
- Uniform motion problems • Mixture problems

In this section, we will again use equations as we solve a variety of problems.

Percent problems

Numeric information is often expressed using percents. Percent means per one hundred. We can use the **percent formula** (Amount = percent · base) to solve many types of percent problems.

EXAMPLE 1 **Gold mining.** Use the data in Figure 1-18 (on the next page) about South Africa, the world's largest producer of gold, to determine the total world gold production in 2002.

Analyze the problem

In the **circle graph,** we see that the 388.5 metric tons of gold produced by South Africa was 15% of the world's production in 2002.

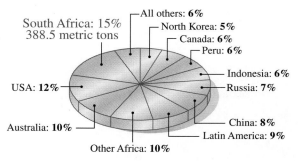

South Africa: 15%
388.5 metric tons

All others: **6%**
North Korea: **5%**
Canada: **6%**
Peru: **6%**

Indonesia: **6%**
Russia: **7%**

USA: **12%**

China: **8%**
Latin America: **9%**

Australia: **10%**

Other Africa: **10%**

Source: World Gold Council

FIGURE 1-18

Form an equation

Let x = world gold production (in metric tons) for 2002. First, we write a percent sentence using the given data. Then we translate to form an equation.

| 388.5 | is | 15% | of | what? |

| 388.5 | = | 15% | · | x |

The amount is 388.5, the percent is 15%, and the base is x.

Solve the equation

$388.5 = 15\% \cdot x$

$388.5 = 0.15x$ Write 15% as a decimal: 15% = 0.15.

$\dfrac{388.5}{0.15} = \dfrac{0.15\,x}{0.15}$ Divide both sides by 0.15.

$2{,}590 = x$ Divide.

State the conclusion

The world produced 2,590 metric tons of gold in 2002.

Check the result

If 2,590 metric tons of gold were produced, then the 388.5 metric tons produced by South Africa were $\frac{388.5}{2,590} = 0.15 = 15\%$ of the world's production. The answer checks.

When the regular price of merchandise is reduced, the amount of reduction is called **markdown** (or discount).

| Sale price | = | regular price | − | markdown |

Usually, the markdown is expressed as a percent of the regular price.

| Markdown | = | percent of markdown | · | regular price |

EXAMPLE 2   Wedding gowns.   At a bridal shop, a wedding gown that normally sells for $397.98 is on sale for $265.32. Find the percent of markdown.

Analyze the problem

In this case, $265.32 is the sale price, $397.98 is the regular price, and the markdown is the *product* of $397.98 and the percent of markdown.

Form an equation

We let $r =$ the percent of markdown, expressed as a decimal. We then substitute $265.32 for the sale price and $397.98 for the regular price in the formula.

Sale price	is	regular price	minus	markdown.
265.32	=	397.98	−	$r \cdot 397.98$

Solve the equation

$$265.32 = 397.98 - r \cdot 397.98$$

$$265.32 = 397.98 - 397.98r \qquad \text{Rewrite } r \cdot 397.98 \text{ as } 397.98r.$$

$$-132.66 = -397.98r \qquad \text{Subtract 397.98 from both sides.}$$

$$\frac{-132.66}{-397.98} = \frac{-397.98r}{-397.98} \qquad \text{Divide both sides by } -397.98.$$

$$0.333333\ldots = r \qquad \text{Perform the divisions.}$$

$$33.3333\ldots\% = r \qquad \text{To write the decimal as a percent, multiply } 0.333333\ldots \text{ by 100 and insert a \% sign.}$$

State the conclusion

The percent of markdown on the wedding gown is $33.3333\ldots\%$, or $33\frac{1}{3}\%$.

Check the result

The markdown is $33\frac{1}{3}\%$ of $397.98, or $132.66. The sale price is $397.98 − $132.66, or $265.32. The answer checks.

Percents are often used to describe how a quantity has changed. To describe such changes, we use **percent increase** or **percent decrease.**

EXAMPLE 3   Movie theaters.   See Figure 1-19. Use the data to determine the percent increase in the number of movie theater screens in the United States from 1990 to 2002.

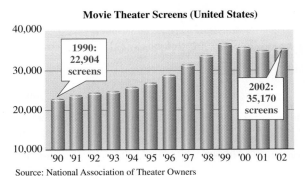

Movie Theater Screens (United States)

Source: National Association of Theater Owners

FIGURE 1-19

Analyze the problem

To find the percent increase, we first find the *amount of increase* by subtracting the number of screens in 1990 from the number of screens in 2002.

$$35{,}170 - 22{,}904 = 12{,}266$$

Form an equation

Next, we find what percent of the original 22,904 screens the 12,266 increase represents. We let x = the unknown percent and translate the words into an equation.

12,266	is	what percent	of	22,904?

$$12{,}266 \quad = \quad x \quad \cdot \quad 22{,}904$$

The amount is 12,266, the percent is x, and the base is 22,904.

Solve the equation

$$12{,}266 = x \cdot 22{,}904$$
$$12{,}266 = 22{,}904x$$
$$\frac{12{,}266}{22{,}904} = \frac{22{,}904x}{22{,}904}$$ Divide both sides by 22,904.

$$0.535539644\ldots = x$$ Perform the divisions.
$$53.5539644\ldots\% \approx x$$ Write the decimal as a percent.
$$54\% \approx x$$ Round to the nearest one percent.

State the conclusion

There was a 54% increase in the number of movie screens in the United States from 1990 to 2002.

Check the result

A 50% increase from 22,904 screens would be approximately 11,000 additional screens. It seems reasonable that 12,266 more screens would be a 54% increase.

THINK IT THROUGH

Fastest Growing Occupations

"With the Baby Boomer generation aging, more medical professionals will be required to manage the health needs of the elderly."
Jonathan Stanewick, Compensation Analyst, Salary.com, 2004

The table below shows predictions by the U.S. Department of Labor, Bureau of Labor Statistics, of the fastest growing occupations for the years 2002–2012. Which occupation has the greatest percent increase?

Occupation	Number of jobs in 2002	Predicted number of jobs in 2012	Education or training required
Home health aide	580,000	859,000	On-the-job training/AA degree
Medical assistant	365,000	579,000	On-the-job training/AA degree
Physician assistant	63,000	94,000	Bachelor's degree/ Masters degree
Social service assistant	305,000	454,000	On-the-job training/AA degree
Systems, data analyst	186,000	292,000	Bachelor's degree

Statistics problems

Statistics is a branch of mathematics that deals with the analysis of numerical data. Three types of averages are commonly used in statistics as measures of central tendency of a distribution (collection) of numbers: the **mean,** the **median,** and the **mode.**

Mean, median, and mode

The **mean** $\bar{x}$ of a collection of values is the sum S of those values divided by the number of values n.

$$\bar{x} = \frac{S}{n} \quad \text{Read } \bar{x} \text{ as "x bar."}$$

The **median** of a collection of values is the middle value. To find the median,

1. Arrange the values in increasing order.

2. If there are an odd number of values, choose the middle value.

3. If there are an even number of values, add the middle two values and divide by 2.

The **mode** of a collection of values is the value that occurs most often.

EXAMPLE 4 Physiology experiments.

As part of a project for a physiology class, a student measured 10 people's reaction times to a visual stimulus. Their reaction times, in seconds, are listed below. Find **a.** the mean, **b.** the median, and **c.** the mode of the distribution.

0.29, 0.22, 0.19, 0.36, 0.28, 0.23, 0.16, 0.28, 0.33, 0.26

Solution

a. To find the mean, we add the values and divide by the number of values, which is 10.

$$\bar{x} = \frac{0.29 + 0.22 + 0.19 + 0.36 + 0.28 + 0.23 + 0.16 + 0.28 + 0.33 + 0.26}{10}$$

$$= 0.26 \text{ second}$$

b. To find the median, we first arrange the values in increasing order:

0.16, 0.19, 0.22, 0.23, 0.26, 0.28 , 0.28, 0.29, 0.33, 0.36

Because there are an even number of measurements, the median will be the sum of the middle two values, 0.26 and 0.28, divided by 2. Thus, the median is

$$\text{Median} = \frac{0.26 + 0.28}{2} = 0.27 \text{ second}$$

c. Since the time 0.28 second occurs most often, it is the mode.

EXAMPLE 5 Bank charges.

When the average (mean) daily balance of a customer's checking account falls below $500 in any week, the bank assesses a $15 service charge. What minimum balance will the account shown in Figure 1-20 on the next page need to have on Friday to avoid the service charge?

Security Savings

☐ Weekly Statement ☐

Acct: 201-234-002   Type: checking

Day	Date	Daily balance	Comments
Mon	3/11	$730.70	
Tue	3/12	$350.19	
Wed	3/13	–$50.19	overdrawn
Thu	3/14	$275.55	
Fri	3/15		

FIGURE 1-20

Analyze the problem

We can find the average (mean) daily balance for the week by adding the daily balances and dividing by 5. We want the mean to be $500 so that there is no service charge.

Form an equation

We will let x = the minimum balance needed on Friday. Then we translate the words into mathematical symbols.

The sum of the five daily balances	divided by	5	is	$500.

$$\frac{730.70 + 350.19 + (-50.19) + 275.55 + x}{5} = 500$$

Solve the equation

$$\frac{730.70 + 350.19 + (-50.19) + 275.55 + x}{5} = 500$$

$$\frac{1{,}306.25 + x}{5} = 500 \qquad \text{Combine like terms in the numerator.}$$

$$5\left(\frac{1{,}306.25 + x}{5}\right) = 5(500) \qquad \text{Multiply both sides by 5 to clear the equation of the fraction.}$$

$$1{,}306.25 + x = 2{,}500$$

$$x = 1{,}193.75 \qquad \text{Subtract 1,306.25 from both sides.}$$

State the conclusion

On Friday, the account balance needs to be $1,193.75 to avoid a service charge.

Check the result

Check the result by adding the five daily balances and dividing by 5.

▌ Investment problems

The money an investment earns is called *interest*. **Simple interest** is computed by the formula $I = Prt$, where I is the interest earned, P is the principal (amount invested), r is the annual interest rate, and t is the length of time the principal is invested.

EXAMPLE 6 **Interest income.** To protect against a major loss, a financial analyst suggests a diversified plan for a client who has $50,000 to invest for 1 year.

1. Alco Development, Inc. Builds mini-malls. High yield: 12% per year. Risky!

2. Certificate of deposit (CD). Insured, safe. Low yield: 4.5% annual interest.

If the client puts some money in each investment and wants to earn $3,600 in interest, how much should be invested at each rate?

Analyze the problem

In this case, we are working with two investments made at two different rates for 1 year. If we add the interest from the two investments, the sum should equal $3,600.

Form an equation

If we let x = the number of dollars invested at 12%, the interest earned is $I = Prt = \$x(12\%)(1) = \$0.12x$. If x is invested at 12%, there is $\$(50,000 - x)$ to invest at 4.5%, which will earn $\$0.045(50,000 - x)$ in interest. These facts are listed in Figure 1-21.

	P ·	r ·	t =	I
Alco Development, Inc.	x	0.12	1	$0.12x$
Certificate of deposit	$50,000 - x$	0.045	1	$0.045(50,000 - x)$

FIGURE 1-21

The sum of the two amounts of interest should equal $3,600. We now translate the words of the problem into an equation.

The interest earned at 12%	plus	the interest earned at 4.5%	is	the total interest earned.
$0.12x$	+	$0.045(50,000 - x)$	=	3,600

Solve the equation

$$0.12x + 0.045(50,000 - x) = 3,600$$

$\mathbf{1,000}[0.12x + 0.045(50,000 - x)] = \mathbf{1,000}(3,600)$ To eliminate the decimals, multiply both sides by 1,000.

$120x + 45(50,000 - x) = 3,600,000$ Distribute the 1,000 and simplify both sides.

$120x + 2,250,000 - 45x = 3,600,000$ Distribute 45.

$75x + 2,250,000 = 3,600,000$ Combine like terms.

$75x = 1,350,000$ Subtract 2,250,000 from both sides.

$x = 18,000$ Divide both sides by 75.

State the conclusion

$18,000 should be invested at 12% and $(50,000 - 18,000) = \$32,000$ should be invested at 4.5%.

Check the result

The annual interest on $18,000 is $0.12(\$18,000) = \$2,160$. The interest earned on $32,000 is $0.045(\$32,000) = \$1,440$. The total interest is $\$2,160 + \$1,440 = \$3,600$. The answers check.

Uniform motion problems

Problems that involve an object traveling at a constant rate for a specified period of time over a certain distance are called **uniform motion** problems. To solve these problems, we use the formula $d = rt$, where d is distance, r is rate, and t is time.

EXAMPLE 7 Travel times.

After visiting her grandparents' farm, a girl returns home, 385 miles away. To split up the drive, the parents and grandparents start at the same time and drive toward each other, planning to meet somewhere along the way. If the parents travel at an average rate of 60 mph and the grandparents at 50 mph, how long will it take them to reach each other?

Analyze the problem

The vehicles are traveling toward each other, as shown in Figure 1-22(a). We know the rates the cars are traveling (60 mph and 50 mph). We also know that they will travel for the same amount of time.

Form an equation

We can let t = the time that each vehicle travels. Then the distance traveled by the parents is $60t$ miles, and the distance traveled by the grandparents is $50t$ miles. This information is organized in the table in Figure 1-22(b). The sum of the distances traveled by the parents and grandparents is 385 miles.

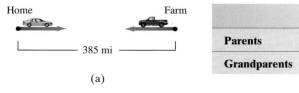

	r	·	t	=	d
Parents	60		t		60t
Grandparents	50		t		50t

(a)   (b)

FIGURE 1-22

The distance the parents travel	plus	the distance the grandparents travel	is	the distance between the child's home and the grandparents' farm.
60t	+	50t	=	385

Solve the equation

$$60t + 50t = 385$$

$110t = 385$ Combine like terms.

$t = 3.5$ Divide both sides by 110.

State the conclusion

The parents and grandparents will meet in $3\frac{1}{2}$ hours.

Check the result

The parents travel $3.5(60) = 210$ miles. The grandparents travel $3.5(50) = 175$ miles. The total distance traveled is $210 + 175 = 385$ miles. The answer checks.

! **COMMENT** When using the formula $d = rt$, make sure that the units match. For example, if the rate is given in miles per *hour*, the time must be expressed in *hours* and the distance in miles.

Mixture problems

We now discuss two types of mixture problems. In the first example, a *dry mixture* of a specified value is created from two differently priced components.

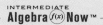

EXAMPLE 8   Mixing nuts.   The owner of a candy store notices that 20 pounds of gourmet cashews are getting stale. They did not sell because of their high price of $12 per pound. The owner decides to mix peanuts with the cashews to lower the price per pound. If peanuts sell for $3 per pound, how many pounds of peanuts must be mixed with the cashews to make a mixture that could be sold for $6 per pound?

Analyze the problem

To solve this problem, we will use the formula $v = np$, where v represents value, n represents the number of pounds, and p represents the price per pound.

Form an equation

We can let x = the number of pounds of peanuts to be used. Then $20 + x$ = the number of pounds in the mixture. We enter the known information in the table in Figure 1-23. The value of the cashews plus the value of the peanuts will be equal to the value of the mixture.

	n	$\cdot$ p	$=$ v
Cashews	20	12	240
Peanuts	x	3	$3x$
Mixture	$20 + x$	6	$6(20 + x)$

FIGURE 1-23

Next we translate the verbal model into mathematical symbols.

The value of the cashews	plus	the value of the peanuts	is	the value of the mixture.
240	$+$	$3x$	$=$	$6(20 + x)$

Solve the equation

$$240 + 3x = 6(20 + x)$$
$$240 + 3x = 120 + 6x \qquad \text{Distribute the multiplication by 6.}$$
$$120 = 3x \qquad \text{Subtract } 3x \text{ and 120 from both sides.}$$
$$40 = x \qquad \text{Divide both sides by 3.}$$

State the conclusion

The owner should mix 40 pounds of peanuts with the 20 pounds of cashews.

Check the result

The cashews are valued at $12(20) = $240, and the peanuts are valued at $3(40) = $120. The mixture is valued at $6(60) = $360. Since the value of the cashews plus the value of the peanuts equals the value of the mixture, the answer checks.

In the next example, a *liquid mixture* of a desired strength is to be made from two solutions with different concentrations.

EXAMPLE 9 Milk production.

Owners of a dairy find that milk with a 2% butterfat content is their best seller. Suppose the dairy has large quantities of whole milk having a 4% butterfat content and milk having a 1% butterfat content. How much of each type of milk should be mixed to obtain 120 gallons of milk that is 2% butterfat?

Analyze the problem

We are to find the amount of 4% milk to mix with 1% milk to get 120 gallons of a milk that has a 2% butterfat content. In Figure 1-24(a), if we let g = the number of gallons of the 4% milk used in the mixture, then $120 - g$ = the number of gallons of the 1% milk needed to obtain the desired concentration.

Form an equation

The amount of butterfat in a tank is the *product* of the percent butterfat and the number of gallons of milk in the tank. In the first tank shown in Figure 1-24(a), 4% of the g gallons, or $0.04g$ gallons, is butterfat. In the second tank, 1% of the $(120 - g)$ gallons, or $0.01(120 - g)$ gallons, is butterfat. Upon mixing, the third tank will have $0.02(120)$ gallons of butterfat in it. These results are recorded in the last column of the table in Figure 1-24(b).

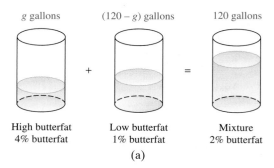

g gallons	(120 − g) gallons	120 gallons
High butterfat 4% butterfat	Low butterfat 1% butterfat	Mixture 2% butterfat

(a)

	Gallons of milk	. Percent butterfat =	Gallons of butterfat
High butterfat	g	0.04	$0.04g$
Low butterfat	$120 - g$	0.01	$0.01(120 - g)$
Mixture	120	0.02	$0.02(120)$

(b)

FIGURE 1-24

The sum of the amounts of butterfat in the first two tanks should equal the amount of butterfat in the third tank.

The amount of butterfat in g gallons of 4% milk	plus	the amount of butterfat in $(120 - g)$ gallons of 1% milk	is	the amount of butterfat in 120 gallons of the mixture.
$0.04g$	$+$	$0.01(120 - g)$	$=$	$0.02(120)$

Solve the equation

$$0.04g + 0.01(120 - g) = 0.02(120)$$

$$4g + 1(120 - g) = 2(120) \qquad \text{Multiply both sides by 100 to clear the equation of decimals.}$$

$$4g + 120 - g = 240$$

$$3g + 120 = 240 \qquad \text{Combine like terms.}$$

$$3g = 120 \qquad \text{Subtract 120 from both sides.}$$

$$g = 40 \qquad \text{Divide both sides by 3.}$$

State the conclusion

40 gallons of 4% milk and $120 - 40 = 80$ gallons of 1% milk should be mixed to get 120 gallons of milk with a 2% butterfat content.

Check the result

The 40 gallons of 4% milk contains $0.04(40) = 1.6$ gallons of butterfat, and 80 gallons of 1% milk contains $0.01(80) = 0.8$ gallons of butterfat—a total of $1.6 + 0.8 = 2.4$ gallons of butterfat. The 120 gallons of the 2% mixture contains 2.4 gallons of butterfat. The answers check.

Section 1.7   STUDY SET

VOCABULARY   *Fill in the blanks.*

1. When an investment is made, the amount of money invested is called the _____.

2. The value that occurs the most in a distribution of numbers is called the _____.

3. The middle value of a distribution of numbers is called the _____.

4. The _____ of several values is the sum of those values divided by the number of values.

5. In the statement "10 is 20% of 50," 10 is the _____, and 50 is the _____.

6. When the regular price of an item is reduced, the amount of reduction is called the _____.

CONCEPTS

7.

> Total Paid Circulation
> *Seventeen* Magazine
> **1995:** 2,172,923   **2002:** 2,445,539

   a. Find the *amount* of increase in circulation of *Seventeen* magazine.

   b. Fill in the percent sentence that can be used to find the percent of increase in circulation.

   _____ is _____ % of _____ ?

8. Complete the following table for each 1-year investment.

Account	Principal	Rate	Time	Interest earned
CD	$1,500	6%	1	
Bonds	$x	5.65%	1	
Stocks	$(850 − x)	7%	1	

9. a. If $5,000 of $30,000 is invested at 5%, how much is left to be invested at another rate?

   b. If $x of $30,000 is invested at 5%, how much is left to be invested at another rate?

10. Complete the following table, given the speed of light and sound through air and water.

	Rate	Time	Distance
Light (air)	186,224 mi/sec	60 sec	
Light (water)	140,060 mi/sec	60 sec	
Sound (air)	1,088 ft/sec	x sec	
Sound (water)	4,870 ft/sec	$(x - 3)$ sec	

11. Complete the following table.

	Pounds	Price	Value
M & M plain	30	2.49	
M & M peanut	p	2.79	
Mixture	$p + 30$	2.59	

12. Complete the following table that could be used to solve this problem: How many pints of punch from the orange cooler must be mixed with the entire contents of the blue cooler to get a 12% punch mixture?

	Amount	·	Strength	=	Pure concentrate
Too strong					
Too weak					
Mixture					

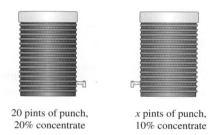

20 pints of punch, x pints of punch,
20% concentrate   10% concentrate

NOTATION   *Translate each statement into mathematical symbols.* **Do not solve the resulting equation.**

13. What number is 5% of 10.56?

14. 16 is what percent of 55?

15. 32.5 is 74% of what number?

16. What is 83.5% of 245?

Complete each solution.

17. Solve: $0.09x + 0.08(2{,}000 - x) = 400$.

$$\boxed{}\,[0.09x + 0.08(2{,}000 - x)] = \boxed{}\,(400)$$
$$\boxed{} + 8(2{,}000 - x) = \boxed{}$$
$$9x + \boxed{} - 8x = 40{,}000$$
$$x = 24{,}000$$

18. Solve: $0.2(5) + 0.6l = 0.4(5 + l)$.

$$\boxed{}\,[0.2(5) + 0.6l] = \boxed{}\,[0.4(5 + l)]$$
$$\boxed{}\,(5) + 6l = \boxed{}\,(5 + l)$$
$$10 + 6l = 20 + \boxed{}$$
$$\boxed{} = 10$$
$$l = 5$$

19. What formula is used to solve simple interest problems?

20. What formula is used to solve uniform motion problems?

21. What formula is used to solve dry mixture problems?

22. Give the percent formula.

APPLICATIONS

23. ENERGY In 2002, the United States alone accounted for 23.5% of the world's total energy consumption, using 97.6 quadrillion British thermal units (Btu). What was the world's energy consumption in 2002?

24. SPACE FLIGHTS Yuri Gagarin of the Soviet Union was the first person in space (April 12, 1961). Alan Shepard was the first American in space (May 5, 1961). Shepard's 15-minute flight was only 13.9% as long as Gagarin's. How long did Gagarin's space flight last? Round to the nearest minute.

25. BUYING A WASHER AND A DRYER Find the percent of markdown of the sale shown below.

One-Day Sale!

Regularly $726   Now only $580.80

Washer/ Dryer

26. BUYING FURNITURE A bedroom set regularly sells for $983. If it is on sale for $737.25, what is the percent of markdown?

27. FLEA MARKETS A vendor sells tool chests at a flea market for $65. If she makes a profit of 30% on each unit sold, what does she pay the manufacturer for each tool chest? (*Hint:* The retail price = the wholesale price + the markup.)

28. MANAGING BOOKSTORES A bookstore sells a textbook for $39.20. If the bookstore makes a profit of 40% on each sale, what does the bookstore pay the publisher for each book? (*Hint:* The retail price = the wholesale price + the markup.)

29. IMPROVING PERFORMANCE The following graph shows how the installation of a special computer chip increases the horsepower of a truck. What is the percent increase in horsepower for the engine running at 4,000 revolutions per minute (rpm)? Round to the nearest percent.

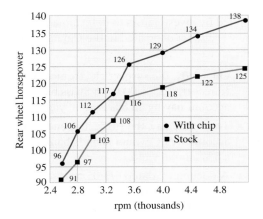

30. GREENHOUSE GASES The total U.S. greenhouse gas emissions in 2000 were 1,906 million metric tons carbon equivalent. In 2001, they were 1,883 million metric tons carbon equivalent. What percent of decrease is this? Round to the nearest tenth of one percent.

31. BROADWAY SHOWS Complete the following table to find the percent increase in attendance at Broadway shows for each season compared to the previous season. Round to the nearest tenth of one percent.

Season	Broadway attendance	% increase
2000–01	11.89 million	—
2001–02	10.95 million	
2002–03	11.42 million	

Source: LiveBroadway.com

32. DRIVE-INS The number of drive-in movie theaters in the United States (see the illustration in the next column) peaked in 1958. Since then, the numbers have steadily declined. Determine the percent of decrease in the number of drive-ins. Round to the nearest one percent.

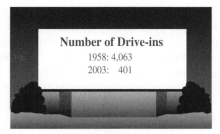

Source: Drive-in Theater Owners Association

33. FUEL EFFICIENCY The table below lists the 10 most fuel-efficient gasoline-powered cars produced in 2001, based on manufacturer's estimated city and highway average miles per gallon (mpg). Find the mean, median, and mode of both sets of data.

Model	mpg city	mph hwy
Honda Insight	61	68
Toyota Prius	52	45
Suzuki Swift	36	42
Toyota Echo	34	41
Chevrolet Prizm	32	41
Toyota Corolla	32	41
Honda Civic	32	39
Mitsubishi Mirage	32	39
Suzuki Esteem	30	37
Saturn SL	29	40

Based on data from the U.S. Department of Energy

34. SPORT FISHING The following report lists the fishing conditions at Pyramid Lake for a Saturday in January. Find the median and the mode of the weights of the striped bass caught at the lake.

Pyramid Lake—Some striped bass are biting but are on the small side. Striking jigs and plastic worms. Water is cold: 38°. Weights of fish caught (lb): 6, 9, 4, 7, 4, 3, 3, 5, 6, 9, 4, 5, 8, 13, 4, 5, 4, 6, 9

35. JOB TESTING To be accepted into a police training program, a recruit must have an average (mean) score of 85 on a battery of four tests. If a candidate scored 78 on the oral test, 91 on the physical fitness test, and 87 on the psychological test, what is the lowest score she can obtain on the written test and still be accepted into the program?

36. GAS MILEAGES The city mileage estimates of Ford models selling for $25,000–$30,000 are listed in the following table. Suppose management wants to add a new model in this price range. What mileage should the new model get if the city mileage average (mean) for this group must then be 18.5 mpg?

Model	City mileage (mpg)
Explorer	14.5
F-Series (P/U)	15.0
Mustang	20.5
Taurus	19.0

37. INVESTING Based on the information in the following table, a woman invested $12,000, some in a money market account and the rest in a 5-year CD. How much was invested in each account if the income from both investments is $1,060 per year?

First Republic Savings and Loan	
Account	Rate
NOW	5.5%
Savings	7.5%
Money Market	8.0%
Checking	4.0%
5-year CD	9.0%

38. ENTREPRENEURS Last year, a women's professional organization made two small-business loans totaling $28,000 to young women beginning their own businesses. The money was lent at 7% and 10% simple interest rates. If the annual income the organization received from these loans was $2,560, what was each loan amount?

39. INHERITANCES Paula split an inheritance between two investments, one a certificate of deposit paying 7% annual interest, and the other a promising biotech company offering an annual return of 10%. She invested twice as much in the 10% investment as she did in the 7% investment. If her combined annual income from the two investments was $4,050, how much did she inherit?

40. INCOME TAX RETURNS On the following federal income tax form, a taxpayer forgot to write in the amount of interest income he earned for the year. From what is written on the form, determine the

amount of interest earned from each investment and the amount he invested in stocks.

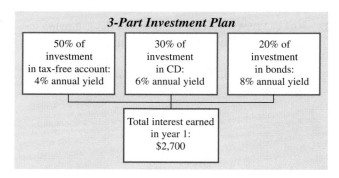

41. MONEY-LAUNDERING Use the evidence compiled by investigators to determine how much money a suspect deposited in the Cayman Islands bank.

- On 6/1/04, the suspect electronically transferred $300,000 to a Swiss bank account paying an 8% annual yield.
- That same day, the suspect opened another account in a Cayman Islands bank that offered a 5% annual yield.
- A document, dated 6/3/05, was seized during a raid of the suspect's home. It stated, "The total interest earned in one year from the two overseas accounts was 7.25% of the total amount deposited."

42. FINANCIAL PRESENTATIONS A financial planner showed her client the investment plan below. Find the total amount the client will have to invest to earn $2,700 in interest.

3-Part Investment Plan

50% of investment in tax-free account: 4% annual yield	30% of investment in CD: 6% annual yield	20% of investment in bonds: 8% annual yield

Total interest earned in year 1: $2,700

43. TRAVEL TIMES A man called his wife from his office to tell her that they needed to switch vehicles so that he could use the family van to pick up some building materials after work. The wife then left their home, traveling toward his office in their van at 35 mph. At the same time, the husband left his office in his car, traveling toward their home at 45 mph. If his office is 20 miles from their home, how long will it take them to meet so that they can switch vehicles?

44. **AIR TRAFFIC CONTROL** An airplane leaves Los Angeles bound for Caracas, Venezuela, flying at an average rate of 500 mph. At the same time, another airplane leaves Caracas bound for Los Angeles, averaging 550 mph. If the airports are 3,675 miles apart, when will the air traffic controllers have to make the pilots aware that the planes are passing each other?

45. **CYCLING** A cyclist leaves his training base for a morning workout, riding at the rate of 18 mph. One hour later, her support staff leaves the base in a car going 45 mph in the same direction. How long will it take the support staff to catch up with the cyclist?

46. **RUNNING MARATHONS** Two marathon runners leave the starting gate, one running 12 mph and the other 10 mph. If they maintain the pace, how long will it take for them to be one-quarter of a mile apart?

47. **RADIO COMMUNICATIONS** At 2 P.M., two military convoys leave Eagle River, Wisconsin, one headed north and one headed south. The convoy headed north averages 50 mph, and the convoy headed south averages 40 mph. They will lose radio contact when the distance between them is more than 135 miles. When will this occur?

48. **SEARCH AND RESCUE** Two search-and-rescue teams leave base camp at the same time, looking for a lost child. The first team, on horseback, heads north at 3 mph, and the other team, on foot, heads south at 1.5 mph. How long will it take them to search a distance of 18 miles between them?

49. **JET SKIING** A jet ski can go 12 mph in still water. If a rider goes upstream for 3 hours against a current of 4 mph, how long will it take the rider to return? (*Hint:* Upstream speed is (12 − 4) mph; how far can the rider go in 3 hours?)

50. **PHYSICAL FITNESS** For her workout, Sarah walks north at the rate of 3 mph and returns at the rate of 4 mph. How many miles does she walk if the round trip takes 3.5 hours?

51. **SWEET AND SOUR** The owner of a candy store wants to make a 30-pound mixture of two candies to sell for $2 per pound. If red licorice bits sell for $1.90 per pound and lemon gumdrops sell for $2.20 per pound, how many pounds of each should be used?

52. **HEALTH FOODS** A pound of dried pineapple bits sells for $6.19, a pound of dried banana chips sells for $4.19, and a pound of raisins sells for $2.39 a pound. Two pounds of raisins are to be mixed with equal amounts of pineapple and banana to create a trail mix that will sell for $4.19 a pound. How many pounds of pineapple and banana chips should be used?

53. **GARDENING** A wholesaler of premium organic planting mix notices that the retail garden centers are not buying her product because of its high price of $1.57 per cubic foot. She decides to mix sawdust with the planting mix to lower the price per cubic foot. If the wholesaler can buy the sawdust for $0.10 per cubic foot, how many cubic feet of each must be mixed to have 6,000 cubic feet of planting mix that could be sold to retailers for $1.08 per cubic foot?

54. **METALLURGY** A one-ounce commemorative coin is to be made of a combination of pure gold, costing $380 dollars an ounce, and a gold alloy that costs $140 dollars an ounce. If the cost of the coin is to be $200, and 500 are to be minted, how many ounces of gold and gold alloy are needed to make the coins?

55. **DILUTING SOLUTIONS** In the illustration below, how much water should be added to 20 ounces of a 15% solution of alcohol to dilute it to a 10% solution?

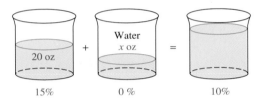

56. **INCREASING CONCENTRATION** The beaker shown below contains a 2% saltwater solution.
    a. How much water must be boiled away to increase the concentration of the salt solution from 2% to 3%?
    b. Where on the beaker will the new water level be?

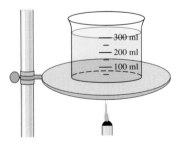

57. **DAIRY FOODS** Cream is approximately 22% butterfat. How many gallons of cream must be mixed with milk testing at 2% butterfat to get 20 gallons of milk containing 4% butterfat?

58. **LOWERING FAT** How many pounds of extra-lean hamburger that is 7% fat must be mixed with 30 pounds of lean hamburger that is 15% fat to obtain a mixture that is 10% fat?

WRITING

59. If a car travels at 60 mph for 30 minutes, explain why the distance traveled is *not* $60 \cdot 30 = 1{,}800$ miles.

60. If a mixture is to be made from solutions with concentrations of 12% and 30%, can the mixture have a concentration less than 12% or greater than 30%? Explain.

61. Write a mixture problem that can be represented by the following verbal model and equation.

The value of the regular coffee	plus	the value of the gourmet coffee	equals	the value of the blend.
$4x$	+	$7(40 - x)$	=	$5(40)$

62. Write a uniform motion problem that can be represented by the following verbal model and equation.

The distance traveled by the 1st train	plus	the distance traveled by the 2nd train	equals	330 miles.
$45t$	+	$55t$	=	330

REVIEW   *Solve each equation.*

63. $9x = 6x$

64. $7a + 2 = 12 - 4(a - 3)$

65. $\dfrac{8(y - 5)}{3} = 2(y - 4)$

66. $\dfrac{t - 1}{3} = \dfrac{t + 2}{6} + 2$

KEY CONCEPT

Let $x =$

In Chapter 1, we have discussed one of the most important problem-solving techniques used in algebra. In this method, the first step is to let a variable represent the unknown quantity. Then we use the variable in writing an equation that mathematically describes the situation in question. Finally, we solve the equation to find the value represented by the variable and state the solution called for in the problem. Let's review some of the key parts of this problem-solving method.

Analyzing the Problem

Always begin problem solving by reading the problem carefully. What facts are given? What are we asked to find? Can we infer anything? In Exercises 1 and 2, what do we know and what are we asked to find?

1. **GEOGRAPHY** Of the 48 contiguous states, 4 more lie east of the Mississippi River than lie west of the Mississippi. How many states are west of the Mississippi River?

2. **GEOMETRY** In a right triangle, the measure of one acute angle is 5° more than twice that of the other angle. Find the measure of the smallest angle.

Letting a Variable Represent an Unknown Quantity

We usually let the variable represent what we are asked to find. In Exercises 3 and 4, what quantity should the variable represent? State your response in the form "Let $x = \ldots$."

3. **CAMPING** To make anchor lines for a tent, a 60-foot rope is cut into four pieces, each successive piece twice as long as the previous one. Find the length of each anchor line.

4. **INSURANCE COVERAGE** While waiting for his van to be repaired, a man rents a car for $12 per day and 10 cents per mile. His insurance company will pay up to $100 of the rental fee. If he needs the car for 2 days, how many miles of driving will his policy cover?

Forming an Equation

Several methods can be used to help form an equation.

5. **TRANSLATIONS** For each phrase, what operation is indicated?
   a. less than
   b. of
   c. increased by
   d. ratio

6. **FORMULAS** What formula is suggested by each type of problem?
   a. Uniform motion
   b. Simple interest
   c. Dry mixture
   d. Perimeter of a rectangle

7. **TABLES** Complete the table below. What equation is suggested?

	Amount	% alcohol	Amount alcohol
Spray 1	15	0.15	
Spray 2	x	0.50	
Mixture	$15 + x$	0.40	

8. **DIAGRAMS** What equation is suggested by the diagram below?

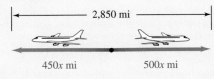

$2{,}850$ mi

$450x$ mi $500x$ mi

ACCENT ON TEAMWORK

SECTION 1.1

PATTERNS   Examine the series of figures below.

a. Draw the next three figures.
b. Complete the table.

Blocks in the bottom row	1	2	3	4	5	6
Total number of blocks						

Examine the series of figures below.

a. Draw the next three figures.
b. Complete the table.

Triangles in the bottom row					
Total number of triangles					

SECTION 1.2

REPEATING DECIMALS   When we change $\frac{1}{3}$ to a decimal, the result is the repeating decimal 0.333 . . . = 0.3. Find a fraction whose decimal equivalent has
a. a two-digit block that repeats.
b. a three-digit block that repeats.
c. a four-digit block that repeats.
d. a five-digit block that repeats.

SECTION 1.3

ORDER OF OPERATIONS   Insert one pair of parentheses in the expression

$$3 - 2 \cdot 3^3 - 3 + 1$$

so that when it is evaluated, the result will be
a. −55   **b.** −49
c. −47   **d.** −44

SECTION 1.4

PROPERTIES OF REAL NUMBERS   Use the numbers 100 and 10 to show that there is no commutative property of subtraction or of division. Use 100, 10, and 2 to show that there is no associative property of subtraction or of division.

SECTION 1.5

SUBTRACTION PROPERTY OF EQUALITY   Borrow a scale and some weights from your school's science department and use them as part of a class presentation to model how the subtraction property of equality is used to solve the equation $x + 2 = 5$.

SECTION 1.6

GOLDEN RECTANGLE   See Exercise 37 in Study Set 1.6. Cut out a set of six rectangles, each with the same width but with different lengths. One of the rectangles should have a length 1.618 times its width; this is a golden rectangle. Paste the rectangles on a piece of poster board randomly. Survey 50 people. Ask them to pick the rectangle that is "the most visually appealing." See whether the golden rectangle is picked most often. Make a table showing the results.

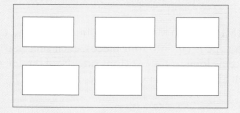

SECTION 1.7

MIXTURE PROBLEM   Make a 10% solution of lemonade by mixing 1 paper cup of lemonade concentrate with 9 paper cups of water in a pitcher. In a similar manner, using a 4-to-6 ratio, make a 40% solution in another pitcher. Then use algebra to determine how many ounces of the 40% lemonade must be mixed with 8 ounces of the 10% lemonade to get a 20% solution. Measure out the appropriate amounts and make the 20% mixture. Taste each of the three solutions. Can you tell the difference in the concentrations?

CHAPTER REVIEW

| SECTION 1.1 | *The Language of Algebra* |

CONCEPTS

A *variable* is a letter that stands for a number.

An *equation* is a mathematical sentence that contains an = symbol.

Formulas are equations that express a known relationship between two or more quantities.

REVIEW EXERCISES

Translate each verbal model into a mathematical model.

1. The cost C (in dollars) to rent t tables is $15 more than the product of $2 and t.

2. A rectangle has an area of 25 in.² The length of the rectangle is the quotient of its area and its width.

3. The waiting period for a business license is now 3 weeks less than it used to be.

4. To determine the proper cooking time for prime rib, a cookbook suggests using the formula $T = 30p$, where T is the cooking time in minutes and p is the weight of the prime rib in pounds. Use this formula to complete the table.

p	T
6.0	
6.5	
7.0	
7.5	
8.0	

Bar graphs and *line graphs* are used to describe numerical relationships.

5. *Use the data from the table in Exercise 4 to draw each type of graph.*

   a. Bar graph

   b. Line graph

The *perimeter* of a geometric figure is the distance around it.

6. GRAND OPENINGS The owner of a new business wants to frame the first dollar bill her business ever received. How long a piece of molding will she need if a dollar is 2.625 inches wide and 6.125 inches long?

| SECTION 1.2 | *The Real Number System* |

Natural numbers:
  {1, 2, 3, . . .}

Whole numbers:
  {0, 1, 2, 3, . . .}

Integers:
  {. . . , −3, −2, −1, 0, 1, 2, 3, . . .}

List the numbers in $\{-5, 0, -\sqrt{3}, 2.4, 7, -\frac{2}{3}, -3.\overline{6}, \pi, \frac{15}{4}, 0.13242368. . .\}$ *that belong to the following sets.*

7. Natural numbers

8. Whole numbers

9. Integers

10. Rational numbers

11. Irrational numbers

12. Real numbers

13. Negative numbers

14. Positive numbers

15. Prime numbers

16. Composite numbers

17. Even integers

18. Odd integers

Prime numbers:
$\{2, 3, 5, 7, 11, 13, \ldots\}$

Composite numbers:
$\{4, 6, 8, 9, 10, 12, \ldots\}$

Integers divisible by 2 are *even integers.* Integers not divisible by 2 are *odd integers.*

Rational numbers are numbers that can be written as $\frac{a}{b}$, where a and b are integers and $b \neq 0$. They form the set of all terminating and all repeating decimals.

Irrational numbers are numbers that can be written as nonterminating, nonrepeating decimals.

A *real number* is any number that is either a rational or an irrational number. All points on the number line represent the set of real numbers.

For any real number x:
$$\begin{cases} \text{If } x \geq 0, \text{ then } |x| = x. \\ \text{If } x < 0, \text{ then } |x| = -x. \end{cases}$$

Use one of the symbols $>$ or $<$ to make each statement true.

19. $-16 \quad\quad -17$

20. $-(-1.8) \quad\quad 2\frac{1}{2}$

Determine whether each statement is true or false.

21. $23.000001 \geq 23.1$

22. $-11 \leq -11$

23. Graph the prime numbers between 20 and 30 on the number line.

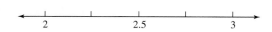

24. Graph the set $\{2.75, 2.\overline{3}, \sqrt{7}, \frac{8}{3}, \frac{3\pi}{4}\}$ on the number line.

Find the value of each expression.

25. $|-18|$

26. $-|6.26|$

SECTION 1.3   *Operations with Real Numbers*

Adding real numbers: With like signs, add the absolute values and keep the common sign. With unlike signs, subtract the absolute values and keep the sign of the number that has the greatest absolute value.

Subtracting real numbers:
$$x - y = x + (-y)$$

Multiplying and dividing real numbers: With like signs, multiply (or divide) their absolute values. The answer is positive. With unlike signs, multiply (or divide) their absolute values. The answer is negative.

Perform the operations.

27. $-3 + (-4)$

28. $-70.5 + 80.6$

29. $-\dfrac{1}{2} - \dfrac{1}{4}$

30. $-6 - (-8)$

31. $(-4.2)(-3.0)$

32. $-\dfrac{1}{10} \cdot \dfrac{5}{16}$

33. $\dfrac{-2.2}{-11}$

34. $-\dfrac{9}{8} \div 21$

35. $15 - 25 - 23$

36. $-3.5 + (-7.1) + 4.9$

37. $-3(-5)(-8)$

38. $-1(-1)(-1)(-1)$

x^n is a *power of x*. x is the *base*, and n is the *exponent*.

$$\overbrace{x^n = x \cdot x \cdot x \cdot \cdots \cdot x}^{n \text{ factors of } x}$$

A number b is a *square root* of a if $b^2 = a$. $\sqrt{a}$ represents the *principal* (positive) *square root* of a.

Order of operations:

1. Work from the innermost pair to the outermost pair of grouping symbols in the following order.

2. Evaluate all powers and roots.

3. Perform all multiplications and divisions, working from left to right.

4. Perform all additions and subtractions, working from left to right.

When the grouping symbols have been removed, repeat the steps above to finish the calculation. In a fraction, simplify the numerator and denominator separately, and then simplify the fraction.

To *evaluate an algebraic expression,* substitute values for the variables and then apply the rules for the order of operations.

The *area* of a figure is the amount of surface it encloses. The *volume* of solid is its capacity.

Evaluate each expression.

39. $(-3)^5$

40. $\left(-\dfrac{2}{3}\right)^2$

41. 0.3 cubed

42. -5^2

Evaluate each expression.

43. $\sqrt{4}$

44. $-\sqrt{100}$

45. $\sqrt{\dfrac{9}{25}}$

46. $\sqrt{0.64}$

Evaluate each expression.

47. $-6 + 2(-5)^2$

48. $\dfrac{-20}{4} - (-3)(-2)(-\sqrt{1})$

49. $4 - (5 - 9)^2$

50. $4 + 6[-1 - 5(25 - 3^3)]$

51. $2|-1.3 + (-2.7)|$

52. $\dfrac{(7 - 6)^4 + 32}{36 - (\sqrt{16} + 1)^2}$

53. $(-10)^3\left(\dfrac{-6}{-2}\right)(-1)$

54. $-(-2 \cdot 4)^2$

Evaluate the algebraic expression for the given values of the variables.

55. $(x + y)(x^2 - xy + y^2)$ for $x = -2$ and $y = 4$

56. $\dfrac{-b - \sqrt{b^2 - 4ac}}{2a}$ for $a = 2, b = -3,$ and $c = -2$

57. SAFETY CONES Find the area covered by the square rubber base if its sides are 10 inches long.

58. SAFETY CONES Find the volume of the cone that is centered atop the base. Round to the nearest tenth.

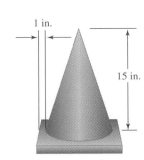

Simplifying Algebraic Expressions

Properties of real numbers:

1. *Associative properties:*
   $(a + b) + c = a + (b + c)$
   $(ab)c = a(bc)$

2. *Commutative properties:*
   $a + b = b + a$
   $ab = ba$

3. *Distributive property:*
   $a(b + c) = ab + ac$

4. 0 is the *additive identity:*
   $a + 0 = 0 + a = a$

5. 1 is the *multiplicative identity:*
   $1 \cdot a = a \cdot 1 = a$

6. *Multiplication property of* 0:
   $a \cdot 0 = 0 \cdot a = 0$

7. $-a$ is the *negative* (or *additive inverse*) of a:
   $a + (-a) = 0$

8. If $a \neq 0$, then $\frac{1}{a}$ is the *reciprocal* (or *multiplicative inverse*) of a:
   $a \cdot \dfrac{1}{a} = \dfrac{1}{a} \cdot a = 1$

Fill in the blanks by applying the indicated property of the real numbers.

59. $3(x + 7) =$ _____
Distributive property
(and simplify)

60. $t \cdot 5 =$ ____
Commutative property
of multiplication

61. $-x + x =$ ____
Additive inverse property

62. $(27 + 1) + 99 =$ _____
Associative property of addition

63. $\frac{1}{8} \cdot 8 =$ ____
Multiplicative inverse property

64. $0 + m =$ ____
Additive identity property

65. ____ $\cdot 9.87 = 9.87$
Multiplicative identity property

66. $5(-9)(0)(2{,}345) =$ ____
Multiplication property of 0

67. $(-3 \cdot 5)2 =$ _____
Associative property
of multiplication

68. $(t + z) \cdot t =$ _____
Commutative property
of addition

Perform each division, if possible.

69. $\dfrac{102}{102}$

70. $\dfrac{0}{6}$

71. $\dfrac{-25}{1}$

72. $\dfrac{5.88}{0}$

To *simplify algebraic expressions*, we use properties of real numbers to write them in a less complicated form.

Terms with exactly the same variables raised to exactly the same powers are called *like* (or *similar*) *terms*.

To add or subtract *like terms*, combine their coefficients and keep the same variables with the same exponents.

Use the distributive property to remove the parentheses.

73. $8(x + 6)$

74. $-6(x - 2)$

75. $-(-4 + 3y)$

76. $(3x - 2y)1.2$

77. $10(2c^2 - c + 1)$

78. $\dfrac{2}{3}(3t + 9)$

Simplify each expression.

79. $8(6k)$

80. $(-7x)(-10y)$

81. $-9(-3p)(-7)$

82. $15a + 30a + 7$

83. $3g^2 - 3g^2$

84. $-m + 4(m - 12)$

85. $\dfrac{7}{8}x + \dfrac{1}{8}x$

86. $21.45l - 45.99l$

87. $5t^3 + 6t^2 - 2t^2 + t^3$

88. $8(2h + 9) - 5(h - 1)$

Solving Linear Equations and Formulas

The set of numbers that satisfy an equation is called its *solution set*.

Determine whether -6 *is a solution of each equation.*

89. $6 - x = 2x + 24$

90. $\dfrac{5}{3}(x - 3) = -12$

Properties of equality:
If $a = b$, then

$$a + c = b + c$$

$$a - c = b - c$$

$$ca = cb \ (c \neq 0)$$

$$\frac{a}{c} = \frac{b}{c} \ (c \neq 0)$$

To solve a linear equation in one variable:

1. Clear the equation of any fractions.

2. Remove all parentheses and combine like terms.

3. Get all variables on one side and all numbers on the other. Combine like terms.

4. Make the coefficient of the variable 1.

5. Check the result.

An *identity* is an equation that is satisfied by every number for which both sides are defined. A *contradiction* is an equation that is never true.

To *solve a formula* for a variable, isolate that variable on one side and isolate all other quantities on the other side.

Solve each equation and give the solution set.

91. $\dfrac{x}{5} = -45$

92. $t - 3.67 = 4.23$

93. $0.0035 = 0.25g$

94. $0 = x + 4$

Solve each equation.

95. $5x + 12 = 0$

96. $-3x - 7 + x = 6x + 20 - 5x$

97. $4(y - 1) = 28$

98. $2 - 13(x - 1) = 4 - 6x$

99. $\dfrac{8(x - 5)}{3} = 2(x - 4)$

100. $\dfrac{3y}{4} - 14 = -\dfrac{y}{3} - 1$

101. $-k = -0.06$

102. $\dfrac{5}{4}p = -10$

103. $\dfrac{4t + 1}{3} - \dfrac{t + 5}{6} = \dfrac{t - 3}{6}$

104. $33.9 - 0.5(75 - 3x) = 0.9$

Solve each equation. If the equation is an identity or a contradiction, so state.

105. $2(x - 6) = 10 + 2x$

106. $-5x + 2x - 1 = -(3x + 1)$

Solve each formula for the indicated variable.

107. $V = \pi r^2 h$ for h

108. $Y + 2g = m$ for g

109. $\dfrac{T}{6} = \dfrac{1}{6}ab(x + y)$ for x

110. $V = \dfrac{4}{3}\pi r^3$ for r^3

| **SECTION 1.6** | *Using Equations to Solve Problems* |

Problem-solving strategy:

1. Analyze the problem.

2. Form an equation.

3. Solve the equation.

4. State the conclusion.

5. Check the result.

Number · value = total value

111. AIRPORTS The world's two busiest airports are Hartsfield Atlanta International and Chicago O'Hare International. Together they served 143 million passengers in 2002, with Atlanta handling 10 million more than O'Hare. How many passengers did each airport serve?

112. TUITION A private school reduces the monthly tuition cost of $245 by $5 per child if a family has more than one child attending the school. Write an algebraic expression that gives the monthly tuition cost per child for a family having c children.

113. TREASURY BILLS What is the face value of five $1,000 T-bills? What is the face value of x $1,000 T-bills?

114. CABLE TV A 186-foot television cable is to be cut into four pieces. Find the length of each piece if each successive piece is 3 feet longer than the previous one.

115. TOOLING The illustration shows the angle at which a drill is to be held when drilling a hole into a piece of aluminum. Find the measures of both angles labeled in the illustration.

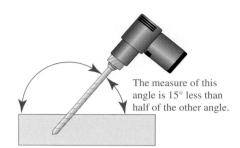

The measure of this angle is 15° less than half of the other angle.

More Problem Solving

Interest = Principal · rate · time

Percent means per one hundred.

Amount = percent · base

116. INVESTMENTS Sally has $25,000 to invest. She invests some money at 10% interest and the rest at 9%. If her total annual income from these two investments is $2,430, how much does she invest at each rate?

117. a. Determine the percent of increase in the number of Toyota Camrys sold in 2002 compared to 2001. Round to the nearest tenth of one percent.

 b. Determine the percent of decrease in the number of Honda Accords sold in 2002 compared to 2001. Round to the nearest tenth of one percent.

The Two Top-Selling Passenger Cars in the U.S.

	2002	
1. Toyota Camry		434,145
2. Honda Accord		398,980
	2001	
1. Honda Accord		414,718
2. Toyota Camry		390,449

Source: *The World Almanac 2004*

$Mean = \dfrac{\text{sum of the values}}{\text{number of values}}$.

The *mode* is the value that occurs most often.

118. SCHOLASTIC APTITUDE TEST The mean SAT math test scores of college-bound seniors for the years 1993–2003 are listed below. Find the mean, median, and mode. Round to the nearest one point.

1993	1994	1995	1996	1997	1998	1999	2000	2001	2002	2003
500	499	504	505	505	505	505	505	506	504	507

The *median* is the middle value after the data have been ordered.

Distance = rate · time

119. PAPARAZZI A celebrity leaves a nightclub in his car and travels at 1 mile per minute (60 mph) trying to avoid a tabloid photographer. One minute later, the photographer leaves the nightclub on his motorcycle, traveling at 1.5 miles per minute (90 mph) in pursuit of the celebrity. How long will it take the photographer to catch up with the celebrity?

120. PEST CONTROL How much water must be added to 20 gallons of a 12% pesticide/water solution to dilute it to an 8% solution?

$v = np$, where v is the value, n is the number of pounds, and p is the price per pound.

121. CANDY SALES Write an algebraic expression that gives the value of a mixture of 3 pounds of Tootsie Rolls with x pounds of Bit O'Honey, if the mix is to sell for $3.95 per pound.

CHAPTER 1

Translate each verbal model into a mathematical model.

1. Each test score T was increased by 10 points to give a new adjusted test score s.

2. The area A of a triangle is the product of one-half the length of the base b and the height h.

Consider the set $\{-2, \pi, 0, -3\frac{3}{4}, 9.2, \frac{14}{5}, 5, -\sqrt{7}\}$.

3. Which numbers are integers?

4. Which numbers are rational numbers?

5. Which numbers are irrational numbers?

6. Which numbers are real numbers?

Graph each set on the number line.

7. $\left\{\dfrac{7}{6}, \dfrac{\pi}{2}, 1.8234503\ldots, \sqrt{3}, 1.\overline{91}\right\}$

8. The prime numbers less than 12

Evaluate each expression.

9. $-|8|$

10. $|-5.5|$

11. $7 - (-5.3)$

12. $-\dfrac{5}{3}\left(-\dfrac{4}{25}\right)$

13. $\dfrac{1}{2} - \left(-\dfrac{3}{5}\right)$

14. $(-4)^3$

15. $\dfrac{2[-4 - 2(3 - 1)]}{3(\sqrt{9})(2)}$

16. $7 + 2[-1 - 4(5)]$

17. Evaluate the expression for $a = 2$, $b = -3$, and $c = 4$.

$$\dfrac{-3b + a}{ac - b}$$

18. PEDIATRICS Some doctors use Young's rule in calculating dosage for infants and children.

$$\dfrac{\text{Age of child}}{\text{Age of child} + 12}\left(\dfrac{\text{average}}{\text{adult dose}}\right) = \text{child's dose}$$

The adult dose of Achromycin is 250 mg. What is the dose for an 8-year-old child?

Determine which property of real numbers justifies each statement.

19. $3 + 5 = 5 + 3$

20. $x(yz) = (xy)z$

21. $a(b + c) = ab + ac$

22. $d \cdot \dfrac{1}{d} = 1$

Simplify each expression.

23. $-y + 3y + 9y$

24. $5(-9q)$

25. $-(4 + t) + t$

26. $12\left(-\dfrac{4}{3} + \dfrac{1}{6}a + \dfrac{3}{2}b\right)$

Solve each equation.

27. $9(x + 4) + 4 = 4(x - 5)$

28. $\dfrac{y - 1}{5} + 2 = \dfrac{2y - 3}{3}$

29. $6 - (x - 3) - 5x = 3[1 - 2(x + 2)]$

Refer to the data in the following table.

U.S. Unemployment Rate (1990–2002)												
'90	'91	'92	'93	'94	'95	'96	'97	'98	'99	'00	'01	'02
5.6	6.8	7.5	6.9	6.1	5.6	5.4	4.9	4.5	4.2	4.0	4.7	5.8

Source: U.S. Department of Labor

30. Find the mean. Round to the nearest tenth.

31. Find the median.

32. Find the mode.

33. Solve $P = L + \dfrac{s}{f}i$ for i.

34. CALCULATORS The viewing window of a calculator has a perimeter of 26 centimeters and is 5 centimeters longer than it is wide. Find the dimensions of the window.

35. WOMEN'S TENNIS Determine the percent of increase in prize money earned by Serena Williams in 2002 compared to 2001. Round to the nearest one percent.

> **Serena Williams — WTA Tennis Tour**
> Prize Money 2002: $3,935,668
> Prize Money 2001: $2,136,263

Source: AcemanTennis.com

36. INVESTING An investment club invested part of $10,000 at 9% annual interest and the rest at 8%. If the annual income from these investments was $860, how much was invested at 8%?

37. GOLDSMITHING How many ounces of a 40% gold alloy must be mixed with 10 ounces of a 10% gold alloy to obtain an alloy that is 25% gold?

38. What does it mean when we say that 3 is a solution of the equation $2x - 3 = x$?

Graphs, Equations of Lines, and Functions

© Getty Images / Thinkstock

INTERMEDIATE
Algebra $f(x)$ **Now™**

Throughout the chapter, this icon introduces resources on the Intermediate AlgebraNow Web site, accessed through **http://1pass .thomson.com**, that will

- Help you test your knowledge of the material with a pre-test and a post-test

- Provide a personalized learning plan targeting areas you should study

TLE An increasing number of people today are choosing to lease rather than buy a car. When deciding whether to lease or buy, one needs to weigh the advantages and disadvantages of each option. Graphs can be helpful in making these comparisons. In this chapter, we will draw graphs that display paired data using a rectangular coordinate system. This type of graph can be used to show the annual costs of leasing verses buying a vehicle over an extended period of time.

To learn more about the rectangular coordinate system, visit *The Learning Equation* on the Internet at http://tle.brookscole.com. (The log-in instructions are in the Preface.) For Chapter 2, the online lessons are:

- *TLE* Lesson 2: The Rectangular Coordinate System
- *TLE* Lesson 3: Rate of Change and the Slope of a Line
- *TLE* Lesson 4: Function Notation

Check Your Knowledge

1. In the ordered pair $(-5, 13)$, the number -5 is called the _____ coordinate and 13 is called the _____ coordinate.

2. The graph of an equation is the graph of all points (x, y) on the rectangular coordinate system whose coordinates _____ the equation.

3. Any equation whose graph is a straight line is called a _____ equation.

4. Two lines are _____ when their slopes are negative reciprocals. Two lines are _____ when they have the same slope.

5. The set of all output values of a function is called the _____ and the set of all inputs is called the _____ of the function.

6. Use the graph in the illustration to answer the following questions.

   a. What are the values of x when $y = 0$?

   b. What is the y-intercept of the graph?

   c. What is the minimum y value?

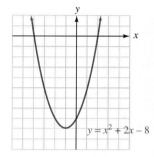

$y = x^2 + 2x - 8$

PROBLEM 6

7. Find the midpoint of the line segment joining $(2, -7)$ and $(-5, 2)$.

8. Find the x- and y-intercepts of the graph of $2x + 3y = 6$. Then graph the equation.

Fnd the slope of each line, if possible.

9. The graph of $x = 3$

10. The line through $(2, -3)$ and $(-3, 2)$

11. The graph of $4x - 2y = 7$

12. Find the slope of the line shown in the illustration.

13. Write an equation of the line with slope -2 that passes through $(2, 7)$. Give the answer in slope–intercept form.

14. Write an equation of the line that passes through the points $(2, 1)$ and $(-1, 4)$. Give the answer in general form.

15. Write an equation of the line that passes through the origin and is perpendicular to the graph of $y = x + 2$. Give the answer in slope–intercept form.

16. Does $y^2 = x$ define y to be a function of x?

17. Find the domain and range of the function $f(x) = -\dfrac{1}{x}$.

18. Consider $f(x) = 2x^2 + 3x + 2$.

   a. Find $f(-1)$.   b. Find $f(0)$.

19. Which of the following are graphs of functions?

   a.   b.   c.

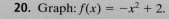

20. Graph: $f(x) = -x^2 + 2$.

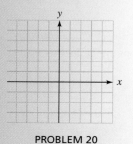

PROBLEM 8

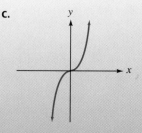

PROBLEM 12

PROBLEM 20

Study Skills Workshop

GETTING THE MOST FROM YOUR TEXTBOOK AND RELATED TECHNOLOGY

Your textbook and its accompanying software are more than just a source of homework problems. Proper use of these materials both before and after concepts are presented in class will help you in completing homework assignments and in understanding your lectures better. Because there are many concepts from Introductory Algebra that are learned in greater depth in this course, there is a great deal of built-in review in your Intermediate Algebra text. Making use of the textbook that you used in your previous math course can also help you in preparing for this class. If you did not keep your previous textbook or cannot remember which book you used, obtain a copy of an Introductory Algebra textbook from the college library or bookstore.

Before Lecture. Look in your textbook for the concepts that you will be covering in your next class meeting. Are there topics that match these concepts in your Introductory Algebra text, or are familiar because you learned them before? If so, look at the homework assignments at the end of the sections in both books. Are there some problems that look the same? If so, do you remember how to do these problems? If you can't remember, look for worked examples in your old text that might jog your memory. Are there some problems that are different? How are they different? By pinpointing the differences, you can make yourself alert for topics that your instructor will be presenting in class.

After Lecture. Read through the text, paying special attention to definitions and worked examples. For each example, try to fill in the missing steps (if there are any) and see whether you can supply a reason for each step in the problem. After you have looked at all of the examples, start your homework assignment. If you have trouble with any of these problems, see whether you can find an example in your text that resembles your problem. In doing this, you may find a clue to help where you became stuck.

Technology. Some people prefer to use the computer, rather than the textbook, for reviewing problems. If you purchased a new text, a CD-ROM was shrink-wrapped with the book that allows you to practice problems like those in your homework sections. Material of this type can also be found at the iLrn Web site, www.iLrn.com. A passcode that allows you to access iLrn was also included in the purchase of your new textbook. Log on to the program and click around to familiarize yourself with the tutorial section of the program.

ASSIGNMENT

1. Find a chapter and section in an Introductory Algebra book that looks similar to Chapter 2 in this text. Compare homework problems in your Chapter 2 assignment to homework problems in the introductory text. Make a list of the following things:
   a. Name and author of the introductory text,
   b. Five homework problems in the introductory text that look similar to five problems in your current homework assignment,
   c. Problems in your current assignment that are different from those in your introductory book. How are they different?
2. Log on to the iLrn Web site and locate both tutorial and quiz problems that relate to your current Chapter 2 assignment. What are the advantages to working problems online? What disadvantages did you find?

Many relationships between two quantities can be described by using a table, a graph, or an equation.

2.1 The Rectangular Coordinate System

- The rectangular coordinate system   • Graphing mathematical relationships
- Reading graphs • Step graphs • The midpoint formula

It is often said that a picture is worth a thousand words. In this chapter, we will show how numerical relationships can be described by mathematical pictures called *graphs*.

The rectangular coordinate system

Many cities are laid out on a rectangular grid. For example, on the east side of Rockford, Illinois, all streets run north and south, and all avenues run east and west. (See Figure 2-1.) If we agree to list the street numbers first, every address can be identified by using an ordered pair of numbers. If Jose Quevedo lives on the corner of Third Street and Sixth Avenue, his address is given by the ordered pair (3, 6).

This is the street. ⟶ ↓ ↓ ⟵ This is the avenue.
(3, 6)

If Lisa has an address of (6, 3), we know that she lives on the corner of Sixth Street and Third Avenue. From the figure, we can see that

- Bob Anderson's address is (4, 1).
- Rosa Vang's address is (7, 5).
- The address of the store is (8, 2).

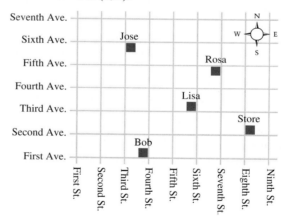

FIGURE 2-1

The idea of associating an ordered pair of numbers with points on a grid is attributed to the 17th-century French mathematician René Descartes. The grid is often called a **rectangular coordinate system,** or **Cartesian coordinate system** after its inventor.

In general, a rectangular coordinate system is formed by two intersecting perpendicular number lines. (See Figure 2-2 on the next page.)

- The horizontal number line is usually called the **x-axis.**
- The vertical number line is usually called the **y-axis.**

The positive direction on the *x*-axis is to the right, and the positive direction on the *y*-axis is upward.

The point where the axes cross is called the **origin.** This is the 0 point on each axis. The two axes form a **coordinate plane** and divide it into four regions called **quadrants,** which are numbered using Roman numerals, as shown in Figure 2-2.

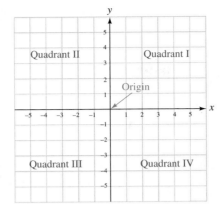

Every point on a coordinate plane can be identified by a pair of real numbers x and y, written as (x, y). The first number in the pair is the **x-coordinate,** and the second number is the **y-coordinate.** The numbers are called the **coordinates** of the point. Some examples of such pairs are $(-4, 6)$, $(2, 3)$, and $(6, -4)$.

FIGURE 2-2

$$(6, -4)$$

The x-coordinate is listed first. ——↑ ↑—— The y-coordinate is listed second.

The process of locating a point in the coordinate plane is called **graphing** or **plotting** the point. In Figure 2-3(a), we use red arrows to graph the point with coordinates $(6, -4)$. Since the x-coordinate, 6, is positive, we start at the origin and move 6 units to the right along the x-axis. Since the y-coordinate, -4, is negative, we then move *down* 4 units, and draw a dot. This locates the point $(6, -4)$, which lies in quadrant IV.

In the figure, blue arrows are used to show how to plot $(-4, 6)$. We start at the origin, move 4 units to the *left* along the x-axis, and then 6 units *up* and draw a dot. This locates the point $(-4, 6)$, which lies in quadrant II.

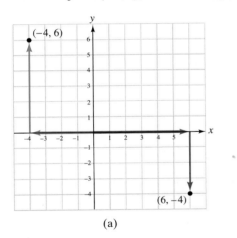

(a)

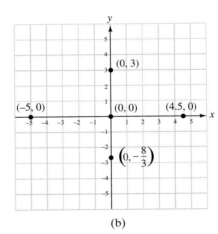

(b)

FIGURE 2-3

! **COMMENT** Note that the point $(-4, 6)$ is not the same as the point $(6, -4)$. This illustrates that the order of the coordinates of a point is important. This is why we call the pairs **ordered pairs.**

In Figure 2-3(b), we see that the points $(-5, 0)$, $(0, 0)$, and $(4.5, 0)$ lie on the x-axis. In fact, every point with a y-coordinate of 0 lies on the x-axis. Note that the coordinates of the origin are $(0, 0)$.

In Figure 2-3(b), we also see that the points $\left(0, -\frac{8}{3}\right)$, $(0, 0)$, and $(0, 3)$ lie on the y-axis. In fact, every point with an x-coordinate of 0 lies on the y-axis.

A point may lie in one of the four quadrants or it may lie on one of the axes, in which case the point is not considered to be in any quadrant. For points in quadrant I, the *x*- and *y*-coordinates are positive. Points in quadrant II have a negative *x*-coordinate and a positive *y*-coordinate. In quadrant III, both coordinates are negative. In quadrant IV, the *x*-coordinate is positive and the *y*-coordinate is negative.

! COMMENT If no scale is indicated on the axes, we assume that they are scaled in units of 1.

EXAMPLE 1 **Halley's comet.** Halley's comet passes Earth every 76 years as it travels in an elliptical orbit about the sun. Figure 2-4 shows its approximate path. Use the graph to determine the comet's position for the years 1912, 1930, 1948, 1966, 1978, and the most recent time it passed by the Earth, 1986.

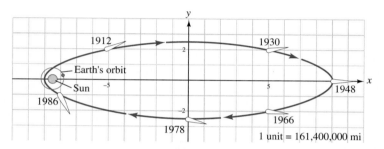

FIGURE 2-4

Solution To find the coordinates of each position, we start at the origin and move left or right along the *x*-axis to find the *x*-coordinate and then up or down to find the *y*-coordinate.

Year	Position of comet on graph	Coordinates
1912	5 units to the *left*, then 2 units *up*	$(-5, 2)$
1930	5 units to the *right*, then 2 units *up*	$(5, 2)$
1948	9 units to the *right*, no units *up* or *down*	$(9, 0)$
1966	5 units to the *right*, then 2 units *down*	$(5, -2)$
1978	No units *left* or *right*, then 2.5 units *down*	$(0, -2.5)$
1986	8 units to the *left*, then 1 unit *down*	$(-8, -1)$

Graphing mathematical relationships

Every day, we deal with quantities that are related.

- The distance we travel depends on how fast we are going.
- Your test score depends on the amount of time you study.
- The height of a toy rocket depends on the time since it was shot into the air.

We can use graphs to visualize relationships between two quantities. For example, suppose that we know the height of a toy rocket at 1-second intervals from 0 second to 6 seconds. We can list this information in the table shown in Figure 2-5. Each entry in the table represents an ordered pair, where the *x*-coordinate is the time (in seconds) since the rocket was shot into the air, and the *y*-coordinate is the height of the rocket (in feet).

Time (in seconds)	Height of rocket (in feet)		
0	0	→	(0, 0)
1	80	→	(1, 80)
2	128	→	(2, 128)
3	144	→	(3, 144)
4	128	→	(4, 128)
5	80	→	(5, 80)
6	0	→	(6, 0)
↑	↑		↑
x-coordinate	*y*-coordinate		The data in the table can be expressed as ordered pairs.

FIGURE 2-5

The information in the table can be used to construct a graph that shows the relationship between the height of the rocket and the time since it was shot into the air. To get the graph in Figure 2-6, we plot the seven ordered pairs shown in the table and draw a smooth curve through the data points.

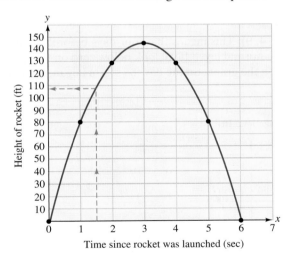

This graph shows the height of the rocket in relation to the time since it was shot into the air. It does not show the path of the rocket.

FIGURE 2-6

From the graph, we can see that the height of the rocket increases as the time increases from 0 second to 3 seconds. Then the height decreases until the rocket hits the ground in 6 seconds. We can also use the graph to make observations about the height of the rocket at other times. For example, the dashed blue lines on the graph show that in 1.5 seconds, the height of the rocket will be approximately 108 feet.

▮ Reading graphs

As we have seen, graphs can be used to describe how quantities change with time. From such a graph, we can determine when a quantity is increasing and when it is decreasing.

Self Check 2

Refer to the graph in Figure 2-7 to answer the following questions.

a. When was the water at the normal level?

b. By how many feet did the water level rise during the storm?

c. After the storm ended and the water level began to fall, how long did it take for the water level to return to normal?

EXAMPLE 2   Water management.   The graph in Figure 2-7 shows the water level of a reservoir before, during, and after a storm. On the *x*-axis, zero represents the day the storm began. On the *y*-axis, zero represents the normal water level that operators try to maintain.

a. In anticipation of the storm, operators released water to lower the level of the reservoir. By how many feet was the water lowered prior to the storm?

b. After the storm ended, on what day did the water level begin to fall?

c. When was the water level 2 feet above normal?

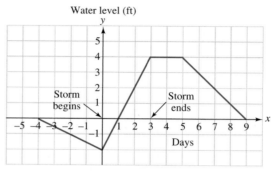

FIGURE 2-7

Solution

a. The graph starts at the point $(-4, 0)$. This means that four days before the storm began, the water level was at the normal level. If we look below zero on the *y*-axis, we see that the point $(0, -2)$ is on the graph. So the day the storm began, the water level had been lowered 2 feet.

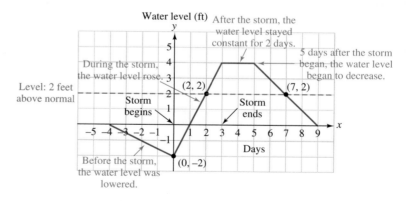

FIGURE 2-8

b. If we look at the *x*-axis, we see that the storm lasted 3 days. From the third to the fifth day, the water level remained constant, 4 feet above normal. The graph does not begin to decrease until day 5.

Answers a. 4 days before the storm began and 1 day and 9 days after the storm began, **b.** 6 ft,   **c.** 4 days

c. We can draw a horizontal line passing through 2 on the *y*-axis. This line intersects the graph in two places—at the points $(2, 2)$ and $(7, 2)$. This means that 2 days and 7 days after the storm began, the water level was 2 feet above normal.

EXAMPLE 3   Sunburn.   The ultraviolet index shown in Figure 2-9 is printed in the weather section of a newspaper each day. For a given UV level and skin sensitivity, the graph gives the number of minutes it takes for a person exposed to the sun to get sunburned. How long will it take a sunbather with "most sensitive" skin to get sunburned on a Los Angeles beach?

Self Check 3
See Figure 2-9. In Las Vegas, how long will it take a person with "least sensitive" skin to get sunburned?

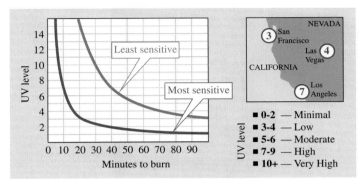

FIGURE 2-9

Solution   The UV level for Los Angeles is 7. We locate 7 on the vertical axis of the graph and extend a horizontal line that intersects the graph for the "most sensitive" skin type. The horizontal line intersects the graph at the point (10, 7). This means that the sunbather will burn in 10 minutes of exposure to the sun unless he or she applies sunscreen.

Answer 70 min.

Step graphs

The graph in Figure 2-10 shows the cost of renting a rototiller for different periods of time. The horizontal axis is labeled with the variable d. This reinforces the fact that it is associated with the number of *days* the rototiller is rented. The vertical axis is labeled with the variable c, for *cost*. In this case, ordered pairs on the graph will be of the form (d, c). For example, the point $(3, 50)$ on the graph tells us that the cost of renting a rototiller for 3 days is $50. The cost of renting the rototiller for more than 4 days up to 5 days is $70. Since the jumps in cost form steps in the graph, we call the graph a **step graph.**

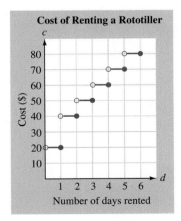

FIGURE 2-10

The solid dot at the end of each step indicates the rental cost for 1, 2, 3, 4, and 5 days. Each open circle indicates that that point is not on the graph.

Self Check 4
Referring to Figure 2-11, find the cost of renting the rototiller for $2\frac{1}{2}$ days.

EXAMPLE 4   Rental costs.   Use the graph to answer the following questions.
a. Find the cost of renting the rototiller for 2 days.   **b.** Find the cost of renting the rototiller for $5\frac{1}{2}$ days.   **c.** How long can you rent the rototiller if you have budgeted $60 for the rental?   **d.** Is the cost of renting the rototiller the same each day?

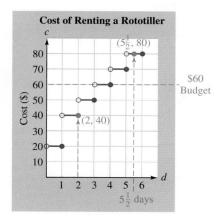

FIGURE 2-11

Solution
a. We locate 2 on the *d*-axis and move up to locate the point on the graph directly above the 2. Since that point has coordinates (2, 40), a two-day rental costs $40.

b. We locate $5\frac{1}{2}$ on the *d*-axis and move straight up to locate the point on the graph with coordinates $(5\frac{1}{2}, 80)$, which indicates that a $5\frac{1}{2}$-day rental would cost $80.

c. We draw a horizontal line through the point labeled 60 on the *c*-axis. Since this line intersects one of the steps of the graph, we can look down to the *d*-axis to find the *d*-values that correspond to a *c*-value of 60. we see that the rototiller can be rented for more than 3 and up to 4 days for $60.

d. The cost each day is not the same. If we look at how the *c*-coordinates change, we see that the first-day rental fee is $20. The second day, the cost jumps another $20. The third day, and all subsequent days, the cost jumps $10.

Answer $50

We have seen that points in the coordinate plane can be identified using ordered pairs of real numbers. There is a formula we can use to find the coordinates of the midpoint of a line segment joining two points in the plane.

■ The midpoint formula

If point *M* in Figure 2-12(a) on the next page lies midway between point *P* with coordinates $(-2, 5)$ and point *Q* with coordinates $(4, -2)$, it is called the **midpoint** of line segment *PQ*. To find the coordinates of point *M*, we find the mean (average) of the *x*-coordinates and the mean of the *y*-coordinates of points *P* and *Q*.

$$x = \frac{-2 + 4}{2} \qquad \text{and} \qquad y = \frac{5 + (-2)}{2}$$

$$= \frac{2}{2} \qquad\qquad\qquad = \frac{3}{2}$$

$$= 1$$

Thus, point *M* has coordinates $\left(1, \frac{3}{2}\right)$.

To distinguish between the coordinates of two general points on a line segment, we often use **subscript notation.** In Figure 2-12(b), point $P(x_1, y_1)$ is read as "point *P* with coordinates of x sub 1 and y sub 1," and point $Q(x_2, y_2)$ is read as "point Q with coordinates of x sub 2 and y sub 2." Using this notation, we can write the midpoint formula in the following way.

The midpoint formula

The **midpoint** of a line segment with endpoints at (x_1, y_1) and (x_2, y_2) has coordinates

$$\left(\frac{x_1 + x_2}{2}, \frac{y_1 + y_2}{2} \right)$$

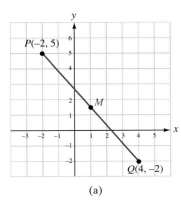
(a)

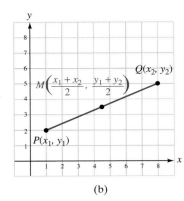
(b)

FIGURE 2-12

! COMMENT Note the difference between x^2 and x_2. In the notation x^2, the exponent 2 is a *superscript,* used to represent repeated multiplication: $x^2 = x \cdot x$. In the notation x_2, the 2 is a *subscript.* In this section, it is used to represent the x-coordinate of a point.

EXAMPLE 5 The midpoint of the line segment joining $P(-5, -3)$ and $Q(x_2, y_2)$ is the point $M(-1, 2)$. Find the coordinates of point Q.

Solution We can let $P(x_1, y_1) = P(-5, -3)$ and $M(x_M, y_M) = M(-1, 2)$, where x_M represents the x-coordinate of point M, and y_M represents the y-coordinate of point M. We can then find the coordinates of point Q using the midpoint formula.

$$x_M = \frac{x_1 + x_2}{2} \quad \text{and} \quad y_M = \frac{y_1 + y_2}{2}$$

$$-1 = \frac{-5 + x_2}{2} \qquad\qquad 2 = \frac{-3 + y_2}{2}$$

$$-2 = -5 + x_2 \qquad\qquad 4 = -3 + y_2 \quad \text{Multiply both sides by 2.}$$

$$3 = x_2 \qquad\qquad\qquad 7 = y_2$$

Since $x_2 = 3$ and $y_2 = 7$, the coordinates of point Q are $(3, 7)$.

Self Check 5
If the midpoint of a segment PQ is $M(-2, 5)$ and one endpoint is $Q(6, -2)$, find the coordinates of point P.

Answer $(-10, 12)$

Section 2.1 STUDY SET

INTERMEDIATE
Algebra $f(x)$ Now™

▮ VOCABULARY *Fill in the blanks.*

1. The pair of numbers $(6, -2)$ is called an _____ pair.

2. In the ordered pair $(-2, -9)$, the number -9 is called the ____ coordinate.

3. The point with coordinates $(0, 0)$ is the _____.

4. The x- and y-axes divide the coordinate plane into four regions called _____.

5. Ordered pairs of numbers can be graphed on a _____ coordinate system.

6. The process of locating a point on a coordinate plane is called _____ the point.

7. If a point is midway between two points P and Q, it is called the _____ of segment PQ.

8. If a line segment joins points P and Q, points P and Q are called _____ of the segment.

▮ CONCEPTS *Fill in the blanks.*

9. To plot the point $(6, -3.5)$, we start at the _____ and move 6 units to the _____ and then 3.5 units _____.

10. To plot the point $\left(-6, \frac{3}{2}\right)$, we start at the _____ and move 6 units to the _____ and then $\frac{3}{2}$ units _____.

11. In which quadrant do points with a negative x-coordinate and a positive y-coordinate lie?

12. In which quadrant do points with a positive x-coordinate and a negative y-coordinate lie?

13. Use the graph to complete the table.

x	y
0	
1	
2	
	3

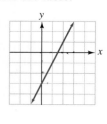

14. Use the graph to complete the table.

x	y
-2	
-1	
	-2
1	
2	

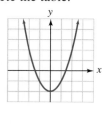

15. How many midpoints does a segment have?

16. The x-coordinate of the midpoint of the line segment joining (x_1, y_1) and (x_2, y_2) is _____, and the y-coordinate is _____.

▮ NOTATION

17. For the ordered pair (t, d), which variable is associated with the horizontal axis?

18. Do these ordered pairs name the same point?
$$\left(5.25, -\tfrac{3}{2}\right), \left(5\tfrac{1}{4}, -1.5\right), \left(\tfrac{21}{4}, -1\tfrac{1}{2}\right)$$

19. How do you read the expression "x_1"?

20. Explain the difference between x^2 and x_2.

▮ PRACTICE *Plot each point on the rectangular coordinate system.*

21. $(4, 3)$

22. $(-2, 1)$

23. $(3.5, -2)$

24. $(-2.5, -3)$

25. $(5, 0)$

26. $(-4, 0)$

27. $\left(\frac{8}{3}, 0\right)$

28. $\left(0, \frac{10}{3}\right)$

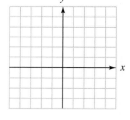

Give the coordinates of each point shown.

29. A

30. B

31. C

32. D

33. E

34. F

35. G

36. H

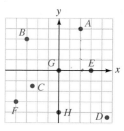

37. SUBMARINES The graph on the next page shows the depths of a submarine at certain times.

a. Where is the sub when $t = 2$?

b. What is the sub doing as t increases from $t = 2$ to $t = 3$?

c. How deep is the sub when $t = 4$?

d. How large an ascent does the sub begin to make when $t = 6$?

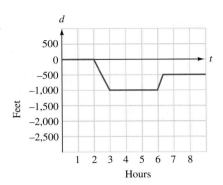

38. AIRPLANES The following graph shows the altitudes of a plane at certain times.

a. Where is the plane when $t = 0$?

b. What is the plane doing as t increases from $t = 1$ to $t = 2$?

c. What is the altitude of the plane when $t = 2$?

d. How much of a descent does the plane begin to make when $t = 4$?

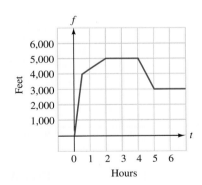

39. PETROLEUM Refer to the following graph.

a. When did imports first surpass production?

b. Estimate the difference in U.S. petroleum imports and production for 2002.

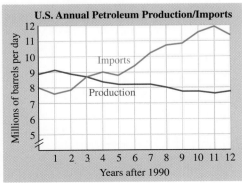

U.S. Annual Petroleum Production/Imports

Source: United States Department of Energy

40. TRACK AND FIELD Refer to the following graph.

a. Which runner ran faster at the start of the race?

b. Which runner stopped to rest first?

c. Which runner dropped the baton and had to go back and get it?

d. At what times was runner 1 stopped and runner 2 running?

e. Describe what was happening at time D.

f. Which runner won the race?

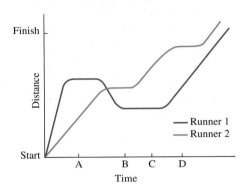

Find the midpoint of line segment PQ.

41. $P(0, 0), Q(6, 8)$

42. $P(10, 12), Q(0, 0)$

43. $P(6, 8), Q(12, 16)$

44. $P(10, 4), Q(2, -2)$

45. $P(2, 4), Q(5, 8)$

46. $P(5, 9), Q(8, 13)$

47. $P(-2, -8), Q(3, 4)$

48. $P(-5, -2), Q(7, 3)$

49. $Q(-3, 5), P(-5, -5)$

50. $Q(2, -3), P(4, -8)$

51. If $M(-2, 3)$ is the midpoint of segment PQ and the coordinates of P are $(-8, 5)$, find the coordinates of Q.

52. If $M(6, -5)$ is the midpoint of segment PQ and the coordinates of Q are $(-5, -8)$, find the coordinates of P.

53. If $M(-7, -3)$ is the midpoint of segment PQ and the coordinates of Q are $(6, -3)$, find the coordinates of P.

54. If $M\left(\frac{1}{2}, -2\right)$ is the midpoint of segment PQ and the coordinates of P are $\left(-\frac{5}{2}, 5\right)$, find the coordinates of Q.

APPLICATIONS

55. ROAD MAPS Road maps have a built-in coordinate system to help locate cities. Use the map below to find the coordinates of these cities in South Carolina: Jonesville, Easley, Hodges, and Union. Express each answer in the form (number, letter).

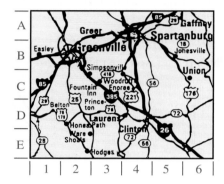

56. HURRICANES A coordinate system that designates the location of places on the surface of the Earth uses a series of latitude and longitude lines, as shown below.

  a. If we agree to list longitude first, what are the coordinates of New Orleans, expressed as an ordered pair?

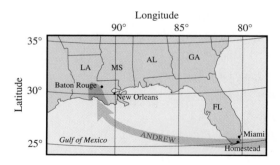

  b. In August 1992, Hurricane Andrew destroyed Homestead, Florida. Estimate the coordinates of Homestead.

  c. Estimate the coordinates of where the hurricane hit Louisiana.

57. EARTHQUAKES The map in the next column shows the area where damage was caused by an earthquake.

  a. Find the coordinates of the epicenter (the source of the quake).

  b. Was damage done at the point (4, 5)?

  c. Was damage done at the point $(-1, -4)$?

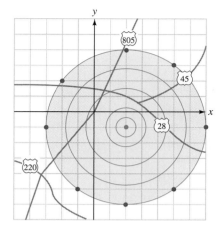

58. GEOGRAPHY The illustration shows a cross-sectional profile of the Sierra Nevada mountain range in California.

  a. Estimate the coordinates of blue oak, sagebrush scrub, and tundra using an ordered pair of the form (distance, elevation).

  b. The *treeline* is the highest elevation at which trees grow. Estimate the treeline for this mountain range.

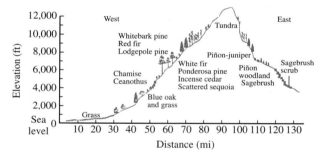

59. GOLF Refer to the graph below. Tiger Woods rallied in the final round of the 2000 AT&T Pebble Beach National Pro-Am golf tournament to overtake the leader, Matt Gogel. (In golf, the player with the score that is the farthest *under* par is the winner.)

  a. At the beginning of the final round, by how many strokes did Gogel lead Woods?

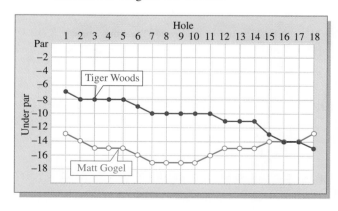

b. What was the largest lead that Gogel had over Woods in the final round?

c. On what hole did Woods tie up the match?

d. On what hole did Woods take the lead?

60. HURRICANES The following illustration shows the scale used to rate the intensity of hurricanes. Use the graph to categorize the hurricanes listed in the table below.

Name/Year	Location	Wind Speed	Category
Luis/1995	Leeward Isl.	138 mph	
Fran/1996	North Carolina	115 mph	
Danny/1997	Alabama	80 mph	
Mitch/1998	Caribbean	178 mph	
Bonnie/1998	North Carolina	110 mph	

Saffir-Simpson Hurricane Scale

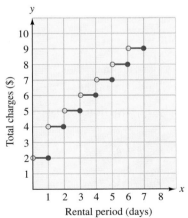

61. VIDEO RENTALS The charges for renting a video are shown in the graph.

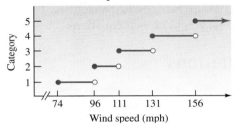

a. Find the charge for a 1-day rental.

b. Find the charge for a 2-day rental.

c. Find the charge if the video is kept for 5 days.

d. Find the charge if the video is kept for a week.

62. POSTAGE RATES The graph shown in the next column gives the first-class postage rates for mailing parcels weighing up to 5 ounces.

a. Find the cost to mail a 3-oz letter.

b. Find the difference in cost for a 2.75-oz letter and a 3.75-oz letter.

c. What is the heaviest letter that can be mailed first class for 97¢?

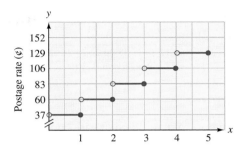

63. BOEING C-17 Engineers use a coordinate system with three axes when designing airplanes. As shown, in the illustration, the *x*-axis is used to describe left/right on the airplane, the *y*-axis forward/backward, and the *z*-axis up/down. Any point on the airplane can be described by an *ordered triple* of the form (x, y, z). The coordinates of three points on the plane are $(0, 181, 56)$, $(-46, 48, 19)$, and $(84, 94, 24)$. Which highlighted part of the plane corresponds with which ordered triple?

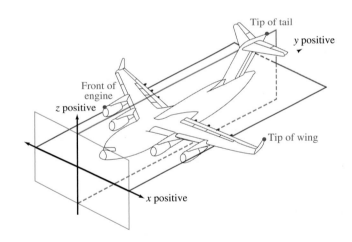

64. ROAST TURKEY The thawing guidelines that appear on the label of a frozen turkey are listed in the following table. In the illustration on the next page, draw a step graph that represents these instructions.

Size	Refrigerator thawing
10 lb to just under 18 lb	3 days
18 lb to just under 22 lb	4 days
22 lb to just under 24 lb	5 days
24 lb to just under 30 lb	6 days

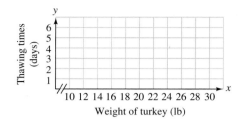

WRITING

65. Explain how to plot the point with coordinates of $(-2, 5)$.

66. Explain why the coordinates of the origin are $(0, 0)$.

REVIEW   *Evaluate each expression.*

67. $-5 - 5(-5)$

68. $(-5)^2 + (-5)$

69. $\dfrac{-3 + 5(2)}{9 + 5}$

70. $|-1 - 9|$

71. Solve: $-4x + 0.7 = -2.1$.

72. Solve $P = 2l + 2w$ for w.

2.2 Graphing Linear Equations

- Graphing linear equations   •   The intercept method   •   Horizontal and vertical lines
- Linear models

In this section, we will discuss equations that contain two variables. Such equations are used to describe algebraic relationships between two quantities. To see a picture of these relationships, we can construct graphs of their equations.

Graphing linear equations

The equation $y = -\frac{1}{2}x + 4$ contains the variables x and y. The solutions of this equation can be written as ordered pairs of real numbers. For example, the ordered pair $(-4, 6)$ is a solution, because the equation is satisfied when $x = -4$ and $y = 6$.

$$y = -\frac{1}{2}x + 4$$

$$6 \overset{?}{=} -\frac{1}{2}(-4) + 4 \quad \text{Substitute } -4 \text{ for } x \text{ and } 6 \text{ for } y.$$

$$6 \overset{?}{=} 2 + 4 \qquad\qquad \text{Perform the multiplication: } -\frac{1}{2}(-4) = 2.$$

$$6 = 6 \qquad\qquad\quad \text{This is a true statement.}$$

This pair and others that satisfy the equation are listed in the **table of solutions** shown in Figure 2-13.

$y = -\dfrac{1}{2}x + 4$		
x	**y**	**(x, y)**
−4	6	(−4, 6)
−2	5	(−2, 5)
0	4	(0, 4)
2	3	(2, 3)
4	2	(4, 2)

Note that we choose x-values that are multiples of the denominator, 2. This makes the computations easier when multiplying the x value by $-\frac{1}{2}$ to find the corresponding y-value.

Choose values for x.   Compute each y-value.   Write each solution as an ordered pair.

FIGURE 2-13

The **graph of the equation** $y = -\frac{1}{2}x + 4$ is the graph of all points (x, y) on the rectangular coordinate system whose coordinates satisfy the equation.

EXAMPLE 1   Graph: $y = -\frac{1}{2}x + 4$.

Solution   To graph the equation, we plot the five ordered pairs listed in the table in Figure 2-13. These points lie on the straight line shown in Figure 2-14. In fact, if we were to plot more pairs that satisfied the equation, it would become obvious that the resulting points will all lie on the line.

When we say that the graph of an equation is a line, we imply two things:

1. Every point with coordinates that satisfy the equation will lie on the line.

2. Any point on the line will have coordinates that satisfy the equation.

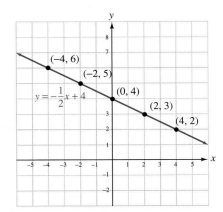

FIGURE 2-14

When the graph of an equation is a straight line, we call the equation a **linear equation.** Linear equations can be written in the form $Ax + By = C$, called **general form,** where A, B, and C represent numbers (called **constants**) and x and y are variables.

EXAMPLE 2   Graph: $3x + 2y = 12$.

Solution   We can pick values for either x or y, substitute them into the equation, and solve for the other variable. For example, if $x = 2$,

$$3x + 2y = 12$$
$$3(2) + 2y = 12 \qquad \text{Substitute 2 for } x.$$
$$6 + 2y = 12 \qquad \text{Perform the multiplication.}$$
$$2y = 6 \qquad \text{Subtract 6 from both sides.}$$
$$y = 3 \qquad \text{Divide both sides by 2.}$$

The ordered pair $(2, 3)$ satisfies the equation. If $y = 6$,

$$3x + 2y = 12$$
$$3x + 2(6) = 12 \qquad \text{Substitute 6 for } y.$$
$$3x + 12 = 12 \qquad \text{Perform the multiplication.}$$
$$3x = 0 \qquad \text{Subtract 12 from both sides.}$$
$$x = 0 \qquad \text{Divide both sides by 3.}$$

INTERMEDIATE
Algebra $f(x)$ **Now**™

Self Check 1

Complete the table of solutions for $y = 2x - 3$ and graph the equation.

x	y
-1	
0	
1	
2	
3	

Answer

x	y
-1	-5
0	-3
1	-1
2	1
3	3

$y = 2x - 3$

INTERMEDIATE
Algebra $f(x)$ **Now**™

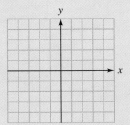

Self Check 2

Graph: $2x - 3y = -6$.

A second ordered pair that satisfies the equation is $(0, 6)$.

These pairs and others that satisfy the equation are shown in the table in Figure 2-15. After we plot each pair, we see that they all lie on a line. The graph of the equation is the line shown in the figure.

Answer

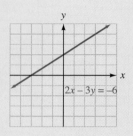

$3x + 2y = 12$		
x	y	(x, y)
-2	9	$(-2, 9)$
0	6	$(0, 6)$
2	3	$(2, 3)$
4	0	$(4, 0)$
6	-3	$(6, -3)$

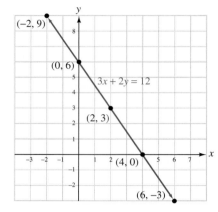

FIGURE 2-15

Generating tables of solutions

If an equation in x and y is solved for y, we can use a graphing calculator to generate tables of solutions like the one shown in Figure 2-15. Several brands of graphing calculators are available, and each one has its own sequence of keystrokes to make tables. The instructions in this discussion are for a TI-83 Plus graphing calculator. For specific details about other brands, please consult the owner's manual.

To construct a table of solutions for $3x + 2y = 12$, we first solve for y.

$$3x + 2y = 12$$

$$2y = -3x + 12 \qquad \text{Subtract } 3x \text{ from both sides.}$$

$$y = -\frac{3}{2}x + 6 \qquad \text{Divide both sides by 2 and simplify.}$$

To enter $y = -\frac{3}{2}x + 6$, we press $\boxed{Y =}$ and enter $-(3/2)x + 6$, as shown in Figure 2-16(a). (Ignore the subscript 1 on y; it is not relevant at this time.)

To enter the x-values that are to appear in the table, we press $\boxed{\text{2nd}}$ $\boxed{\text{TBLSET}}$ and enter the first value for x on the line labeled TblStart =. In Figure 2-16(b), -2 has been entered on this line. Other values for x that are to appear in the table are determined by setting an **increment value** on the line labeled $\triangle$Tbl =. Figure 2-16(b) shows that an increment of 2 was entered. This means that each x-value in the table will be 2 units larger than the previous x-value.

The final step is to press the keys $\boxed{\text{2nd}}$ $\boxed{\text{TABLE}}$. This will display a table of solutions, as shown in Figure 2-16(c). The table contains all of the entries that we obtained in Figure 2-15, as well as two additional solutions: $(8, -6)$ and $(10, -9)$.

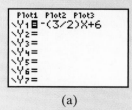

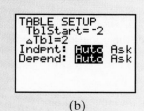

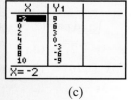

(a)   (b)   (c)

FIGURE 2-16

◼ The intercept method

In Example 2, the graph intersected the *y*-axis at the point with coordinates $(0, 6)$ (called the **y-intercept**) and intersected the *x*-axis at the point with coordinates $(4, 0)$ (called the **x-intercept**). In general, we have the following definitions.

> **Intercepts of a line**
>
> The **y-intercept** of a line is the point $(0, b)$, where the line intersects the *y*-axis. To find *b*, substitute 0 for *x* in the equation of the line and solve for *y*.
>
> The **x-intercept** of a line is the point $(a, 0)$, where the line intersects the *x*-axis. To find *a*, substitute 0 for *y* in the equation of the line and solve for *x*.

EXAMPLE 3   Use the *x*-and *y*-intercepts to graph $2x - 5y = 10$.

Solution   To find the *y*-intercept, we substitute 0 for *x* and solve for *y*:

$$2x - 5y = 10$$
$$2(\mathbf{0}) - 5y = 10 \qquad \text{Substitute 0 for } x.$$
$$-5y = 10 \qquad \text{Perform the multiplication: } 2(0) = 0.$$
$$y = -2 \qquad \text{Divide both sides by } -5.$$

The *y*-intercept is the point $(0, -2)$. To find the *x*-intercept, we substitute 0 for *y* and solve for *x*:

$$2x - 5y = 10$$
$$2x - 5(\mathbf{0}) = 10 \qquad \text{Substitute 0 for } y.$$
$$2x = 10 \qquad \text{Perform the multiplication: } 5(0) = 0.$$
$$x = 5 \qquad \text{Divide both sides by 2.}$$

The *x*-intercept is the point $(5, 0)$.

Although two points are enough to draw the line, it is a good idea to find and plot a third point as a check. To find the coordinates of a third point, we can substitute any convenient number (such as -5) for *x* and solve for *y*:

$$2x - 5y = 10$$
$$2(\mathbf{-5}) - 5y = 10 \qquad \text{Substitute } -5 \text{ for } x.$$
$$-10 - 5y = 10 \qquad \text{Perform the multiplication.}$$
$$-5y = 20 \qquad \text{Add 10 to both sides.}$$
$$y = -4 \qquad \text{Divide both sides by } -5.$$

The line will also pass through the point $(-5, -4)$.

A table of solutions and the graph of $2x - 5y = 10$ are shown in Figure 2-17.

$2x - 5y = 10$		
x	*y*	*(x, y)*
0	−2	$(0, -2)$
5	0	$(5, 0)$
−5	−4	$(-5, -4)$

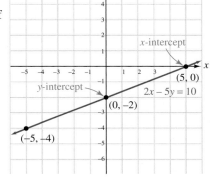

FIGURE 2-17

INTERMEDIATE
Algebra *f(x)* **Now**™

Self Check 3
Find the *x*- and *y*-intercepts and graph $5x + 15y = -15$.

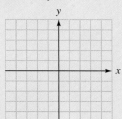

Answer

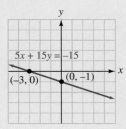

■ Horizontal and vertical lines

Equations such as $y = 3$ and $x = -2$ are linear equations, because they can be written in the form $Ax + By = C$.

$y = 3$ is equivalent to $0x + 1y = 3$

$x = -2$ is equivalent to $1x + 0y = -2$

Algebra $f(x)$ **Now™**

Self Check 4

Graph $x = 4$ and $y = -3$ on one set of coordinate axes.

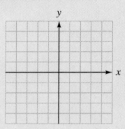

EXAMPLE 4   Graph: **a.** $y = 3$ and   **b.** $x = -2$.

Solution

a. Since the equation $y = 3$ does not contain x, the numbers chosen for x have no effect on y. The value of y is always 3. (See the table below on the left.)

   After plotting the ordered pairs shown in the table on the left in Figure 2-18, we see that the graph is a horizontal line, parallel to the x-axis, with a y-intercept of $(0, 3)$. The line has no x-intercept.

b. Since the equation $x = -2$ does not contain y, the value of y can be any number. After plotting the ordered pairs shown in the table on the right below, we see that the graph is a vertical line, parallel to the y-axis, with an x-intercept of $(-2, 0)$. The line has no y-intercept.

$y = 3$		
x	y	(x, y)
-3	3	$(-3, 3)$
0	3	$(0, 3)$
2	3	$(2, 3)$
4	3	$(4, 3)$

$x = -2$		
x	y	(x, y)
-2	-2	$(-2, -2)$
-2	0	$(-2, 0)$
-2	2	$(-2, 2)$
-2	6	$(-2, 6)$

The value of x →
can be any number.

← The value of y can be
any number.

Answer

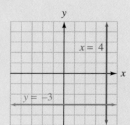

FIGURE 2-18

The results of Example 4 suggest the following facts.

> **Horizontal and vertical lines**
>
> If a and b represent real numbers, then
>
> The graph of the equation $x = a$ is a vertical line with x-intercept at $(a, 0)$.
>
> The graph of the equation $y = b$ is a horizontal line with y-intercept at $(0, b)$.

! COMMENT As shown in Figure 2-19, the linear equation that describes the *x*-axis is $y = 0$, and the linear equation that describes the *y*-axis is $x = 0$.

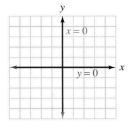

FIGURE 2-19

Linear models

In the next examples, we will see how linear equations can model real-life situations. In each case, the equations describe a *linear relationship* between two quantities; when they are graphed, the result is a line. We can make observations about what has happened in the past and what might take place in the future by carefully inspecting the graph.

EXAMPLE 5   U.S. labor statistics.   The linear equation $p = 0.6t + 38$ models the percent of women 16 years or older who were part of the civilian labor force for each of the years 1960–2000. In the equation, *t* represents the number of years after 1960, and *p* represents the percent. Graph this equation.

Solution   The variables *t* and *p* are used in the equation. We will associate *t* with the horizontal axis and *p* with the vertical axis. Ordered pairs will be of the form (t, p).

To graph the equation, we pick three values for *t*, substitute them into the equation, and find each corresponding value of *p*. The results are listed in the following table.

For $t = 0$	**For $t = 10$**	**For $t = 20$**
(The year 1960)	(The year 1970)	(The year 1980)
$p = 0.6t + 38$	$p = 0.6t + 38$	$p = 0.6t + 38$
$p = 0.6(0) + 38$	$p = 0.6(10) + 38$	$p = 0.6(20) + 38$
$p = 38$	$p = 6 + 38$	$p = 12 + 38$
	$p = 44$	$p = 50$

The pairs (0, 38), (10, 44), and (20, 50) satisfy the equation. Next, we plot these points and draw a line through them. From the graph, we see that there has been a steady increase in the percent of the female population 16 years or older that is part of the labor force.

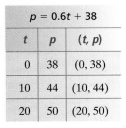

$p = 0.6t + 38$		
t	*p*	(t, p)
0	38	(0, 38)
10	44	(10, 44)
20	50	(20, 50)

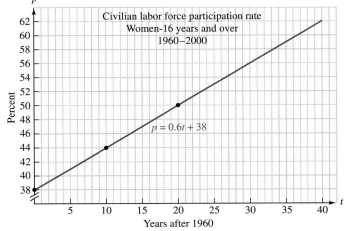

Source: Bureau of Labor Statistics

FIGURE 2-20

INTERMEDIATE
Algebra *f(x)* **Now**™

Self Check 5

a. Use the equation $p = 0.6t + 38$ to determine the percent of women 16 years or older who were part of the labor force in 1975.

b. Use the graph to determine the percent of women who were part of the labor force in 2000.

Answers **a.** 47%, **b.** 62%

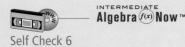

Self Check 6

a. Referring to Example 6, use the equation $y = -950x + 6,750$ to determine when the copier will be worth $3,900.

b. Use the graph in Figure 2-21 to determine when the copier will be worth $2,000.

EXAMPLE 6 Linear depreciation. A copy machine that has been purchased for $6,750 is expected to depreciate (lose value) according to the formula $y = -950x + 6,750$, where y is the value of the copier after x years. When will the copier have no value?

Solution   The copier will have no value when y is 0. To find x when $y = 0$, we substitute 0 for y and solve for x.

$$y = -950x + 6,750$$
$$0 = -950x + 6,750$$
$$-6,750 = -950x \qquad \text{Subtract 6,750 from both sides.}$$
$$7.105263158 = x \qquad \text{Divide both sides by } -950.$$

The copier will have no value in about 7.1 years.

The equation $y = -950x + 6,750$ is graphed in Figure 2-21. Important information can be obtained from the intercepts of the graph.

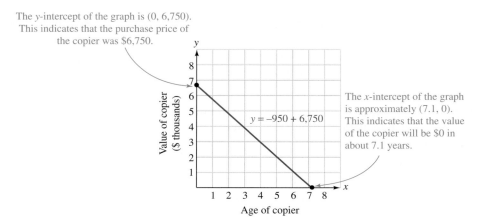

The y-intercept of the graph is $(0, 6,750)$. This indicates that the purchase price of the copier was $6,750.

The x-intercept of the graph is approximately $(7.1, 0)$. This indicates that the value of the copier will be $0 in about 7.1 years.

Answers a. 3 years, **b.** 5 years

FIGURE 2-21

CALCULATOR SNAPSHOT   **Graphing lines**

We have graphed linear equations by finding solutions, plotting points, and drawing lines through those points. Graphing is much easier if we use a graphing calculator.

Window settings

Graphing calculators have a window to display graphs. To see the proper picture of a graph, we must decide on the minimum and maximum values for the x- and y-coordinates. A window with standard settings of

$$\text{Xmin} = -10 \qquad \text{Xmax} = 10 \qquad \text{Ymin} = -10 \qquad \text{Ymax} = 10$$

will produce a graph where the values of x and the values of y are between -10 and 10, inclusive. We use the notation $[-10, 10]$ to describe each of these intervals.

Graphing lines

To graph $3x + 2y = 12$, we must first solve the equation for y.

$$3x + 2y = 12$$

$2y = -3x + 12$ Subtract $3x$ from both sides.

$y = -\dfrac{3}{2}x + 6$ Divide both sides by 2 and simplify.

Next, we press $\boxed{Y=}$ and enter the right-hand side of the equation after the symbol $Y_1 =$. (See Figure 2-22(a).) We then press the $\boxed{\text{GRAPH}}$ key to get the graph shown in Figure 2-22(b). To show more detail, we can draw the graph in a different window. A window with settings of $[-1, 5]$ for x and $[-2, 7]$ for y will give the graph shown in Figure 2-22(c).

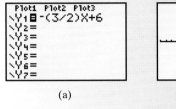
     (a)   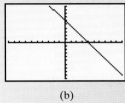 (b)   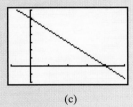 (c)

FIGURE 2-22

Finding the coordinates of a point on the graph

If we reenter the standard window settings of $[-10, 10]$ for x and $[-10, 10]$ for y, press $\boxed{\text{GRAPH}}$, and then press the $\boxed{\text{TRACE}}$ key, we get the display shown in Figure 2-23(a). The y-intercept of the graph is highlighted by the flashing cursor, and the x- and y-coordinates of that point are given at the bottom of the screen. We can use the $\boxed{\blacktriangleright}$ and $\boxed{\blacktriangleleft}$ keys to move the cursor along the line to find the coordinates of any point on the line. For example, after pressing the $\boxed{\blacktriangleright}$ key 12 times, we will get the display in Figure 2-23(b).

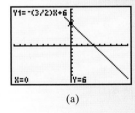 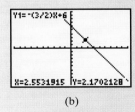 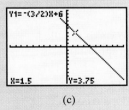
    (a)   (b)   (c)

FIGURE 2-23

To find the y-coordinate of any point on the line, given its x-coordinate, we press $\boxed{\text{2nd}}$ $\boxed{\text{CALC}}$ and select the *value* option. We enter the x-coordinate of the point and press $\boxed{\text{ENTER}}$. The y-coordinate is then displayed. In Figure 2-23(c), 1.5 was entered for the x-coordinate, and its corresponding y-coordinate, 3.75, was found.

The *table* feature, discussed on page 114, gives us a third way of finding the coordinates of a point on the line.

Determining the *x*-intercepts of a graph

To determine the x-intercept of the graph of $y = -\frac{3}{2}x + 6$, we can use the *zero* option, found under the CALC menu. (Be sure to reenter the standard window settings for x and y before using CALC.) After we guess left and right bounds, as shown in Figure 2-24(a), the cursor automatically moves to the x-intercept of the graph when we press $\boxed{\text{ENTER}}$. Figure 2-24(b) shows how the coordinates of the x-intercept are then displayed at the bottom of the screen.

We can also use the *trace* and *zoom* features to determine the *x*-intercept of the graph of $y = -\frac{3}{2}x + 6$. After graphing the equation using the standard window settings, we press TRACE. Then we move the cursor along the line toward the *x*-intercept until we arrive at a point with the coordinates shown in Figure 2-24(c). To get better results, we press ZOOM, select the *zoom* in option, and press ENTER to get a magnified picture. We press TRACE again and move the cursor to the point with coordinates shown in Figure 2-24(d). Since the *y*-coordinate is nearly 0, this point is nearly the *x*-intercept. We can achieve better results with more zooms and traces.

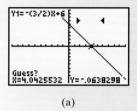
(a)

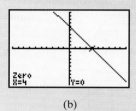
(b)

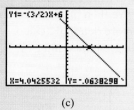
(c)

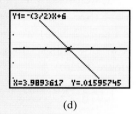
(d)

FIGURE 2-24

Section 2.2 STUDY SET

VOCABULARY   *Fill in the blanks.*

1. The graph of an equation is the graph of all points (x, y) on the rectangular coordinate system whose coordinates _____ the equation.

2. Any equation whose graph is a straight line is called a _____ equation.

3. The point where a graph intersects the *x*-axis is called the _____. The point where a graph intersects the *y*-axis is called the _____.

4. A solution of an equation in two variables is an _____ _____ of numbers that make a true statement when substituted into the equation.

5. The graph of any equation of the form $x = a$ is a _____ line.

6. The graph of any equation of the form $y = b$ is a _____ line.

CONCEPTS

7. Determine whether the given ordered pair is a solution of $y = -5x - 2$.

   a. $(-1, 3)$ **b.** $(3, -13)$

8. Consider the linear equation $6x - 4y = -12$.

   a. Find the *x*-intercept of its graph.

   b. Find the *y*-intercept of its graph.

   c. Does its graph pass through $(2, 6)$?

9. A table of solutions for a linear equation is given below. From the table, determine the *x*-intercept and the *y*-intercept of the graph of the equation.

x	y	(x, y)
−6	0	(−6, 0)
−4	1	(−4, 1)
−2	2	(−2, 2)
0	3	(0, 3)
2	4	(2, 4)

10. Assume that $a \neq 0$ and $b \neq 0$. On which axis does each point lie?

    a. $(0, b)$ **b.** $(a, 0)$

11. See the illustration.

    a. What is the *x*-intercept and what is the *y*-intercept of the line?

    b. If the coordinates of point *M* are substituted into the equation of the line that is graphed here, will a true or a false statement result?

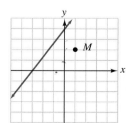

12. Use the graph shown on the right to determine three solutions of $2x + 3y = 9$.

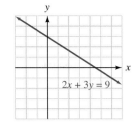

$2x + 3y = 9$

13. 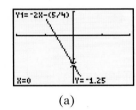 A graphing calculator display is shown here. It is a table of solutions for which one of the following linear equations?

$$y = -2x - 1$$
$$y = -3x - 1$$
$$y = -4x - 1$$

14. The graphing calculator displays below show the graph of $y = -2x - \frac{5}{4}$.

a. In illustration (a), what important feature of the line is highlighted by the cursor?

b. In illustration (b), what important feature of the line is highlighted by the cursor?

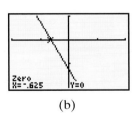   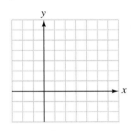

(a)   (b)

NOTATION *Complete each solution.*

15. Verify that $(-3, -1)$ is a solution of $2x + 2y = -8$.

$$2x + 2y = -8$$
$$2(\quad) + 2(-1) \overset{?}{=} -8$$
$$-6 + (\quad) \overset{?}{=} -8$$
$$-8 = -8$$

16. To find the coordinates of a point on the graph of $5x + 2y = 10$, choose $x = 1$ and find y.

$$5x + 2y = 10$$
$$5(\ ) + 2y = 10$$
$$5 + \quad = 10$$
$$2y = \quad$$
$$y = \frac{5}{2}$$

The point $\quad$ is on the graph of $5x + 2y = 10$.

17. The graph of the equation $x = 0$ is which axis?

18. The graph of the equation $y = 0$ is which axis?

PRACTICE *Complete each table.*

19. $y = -x + 4$

x	y
−1	
0	
2	

20. $y = x - 2$

x	y
−2	
0	
4	

21. $y = -\frac{1}{3}x - 1$

x	y
−3	
0	
3	

22. $y = -\frac{1}{2}x + \frac{5}{2}$

x	y
−1	
3	
5	

Use the results from Problems 19–22 to graph each equation.

23. $y = -x + 4$

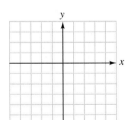

24. $y = x - 2$

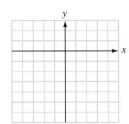

25. $y = -\frac{1}{3}x - 1$

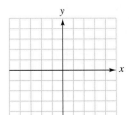

26. $y = -\frac{1}{2}x + \frac{5}{2}$

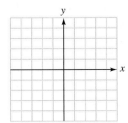

Construct a table of solutions and graph the equation.

27. $y = x$

28. $y = -2x$

29. $y = -3x + 2$

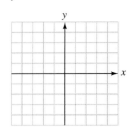

30. $y = 2x - 3$

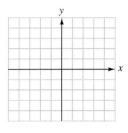

39. $-3y + 2 = 5$

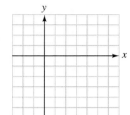

40. $-2x + 3 = 11$

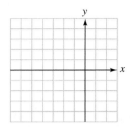

31. $y = 3 - x$

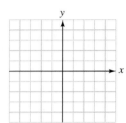

32. $y = 5 - x$

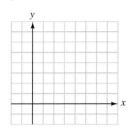

Graph each equation using the intercept method. Label the intercepts on each graph.

41. $3x + 4y = 12$

42. $4x - 3y = 12$

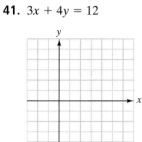

33. $y = \dfrac{x}{4} - 1$

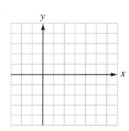

34. $y = -\dfrac{x}{4} + 2$

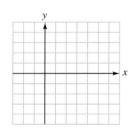

43. $3y = 6x - 9$

44. $2x = 4y - 10$

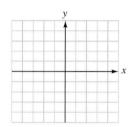
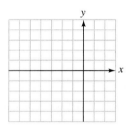

35. $x = 3$

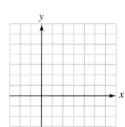

36. $y = -4$

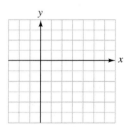

45. $2y + x = -2$

46. $4y + 2x = -8$

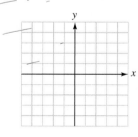
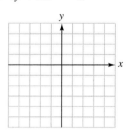

Write each equation in $y = b$ or $x = a$ form and graph it.

37. $y - 2 = 0$

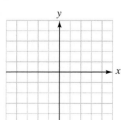

38. $x + 1 = 0$

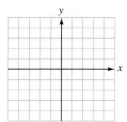

47. $3x + 4y - 8 = 0$

48. $-2y - 3x + 9 = 0$

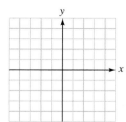
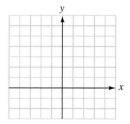

49. $3x = 4y - 11$ **50.** $-5x + 3y = 11$

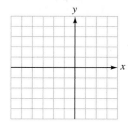

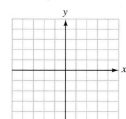

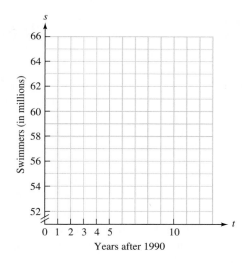

Years after 1990

Use a graphing calculator to graph each equation, and find the *x*-coordinate of the *x*-intercept to the nearest hundredth.

51. $y = 3.7x - 4.5$ **52.** $y = \dfrac{3}{5}x + \dfrac{5}{4}$

53. $1.5x - 3y = 7$ **54.** $0.3x + y = 7.5$

APPLICATIONS

55. HOURLY WAGES The following table gives the amount *y* (in dollars) that a student can earn for working *x* hours. Plot the ordered pairs and estimate how much the student will earn for working 8 hours.

x	2	4	6
y	15	30	45

56. VALUE OF A CAR The following table shows the value *y* (in dollars) of a car that is *x* years old. Plot the ordered pairs and estimate the value of the car when it is 4 years old.

x	0	1	3
y	15,000	12,000	6,000

57. SWIMMING Surveys done by the National Sporting Goods Association have found that interest in swimming has been declining. The equation $s = -0.9t + 65.5$ is a linear model that gives the approximate number of people who went swimming during a given year. *s* is the annual number of swimmers (in millions), and *t* is the number of years since 1990. Graph the equation in the illustration.

a. What information about the number of swimmers can be obtained from the *s*-intercept of the graph?

b. From the graph, estimate the number of swimmers in 1998.

58. LIVING LONGER According to the National Center for Health Statistics, life expectancy in the United States is increasing. The equation $y = 0.13t + 74$ is a linear model that approximates life expectancy; *y* is the number of years of life expected for a child born *t* years after 1980. Graph the equation in the illustration. From the graph, estimate the life expectancy of someone born in 1998.

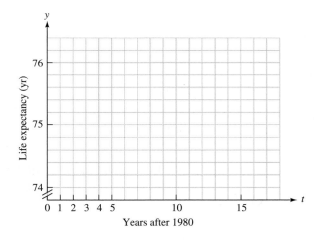

Years after 1980

59. HOUSE APPRECIATION A house purchased for $125,000 is expected to appreciate (gain value) according to the formula $y = 7,500x + 125,000$, where *y* is the value of the house after *x* years. Find the value of the house 5 years later.

60. DEMAND EQUATIONS The number of television sets that consumers buy depends on price. The higher the price, the fewer TVs people will buy. The equation that relates price to the number of TVs sold at that price is called a **demand equation.** If the demand equation for a 13-inch TV is $p = -\dfrac{1}{10}q + 170$, where *p* is the price and *q* is the number of TVs sold at that price, how many TVs will be sold at a price of $150?

61. CAR DEPRECIATION A car purchased for $17,000 is expected to depreciate (lose value) according to the formula $y = -1,360x + 17,000$. When will the car have no value?

62. SUPPLY EQUATIONS The number of television sets that manufacturers produce depends on price. The higher the price, the more TVs manufacturers will produce. The equation that relates price to the number of TVs produced at that price is called a **supply equation.** If the supply equation for a 13-inch TV is $p = \frac{1}{10}q + 130$, where p is the price and q is the number of TVs produced for sale at that price, how many TVs will be produced if the price is $150?

63. BUYING TICKETS Tickets to a circus cost $10 each from Ticketron plus a $2 service fee for each block of tickets.

    a. Write a linear equation that gives the cost c when t tickets are purchased.

    b. Complete the table and graph the equation.

    c. Use the graph to estimate the cost of buying 6 tickets.

t	c
1	
2	
3	
4	

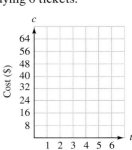

Cost ($)

64 56 48 40 32 24 16 8

1 2 3 4 5 6 t

Number of tickets

64. TELEPHONE COSTS In a community, the monthly cost of local telephone service is $5 per month, plus 25¢ per call.

    a. Write a linear equation that gives the cost c for a person making n calls.

    b. Complete the table and graph the equation.

    c. Use the graph to estimate the cost of service in a month when 20 calls were made.

n	c
4	
8	
12	
16	

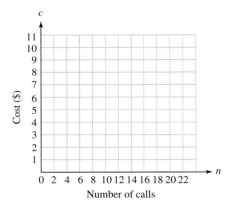

Cost ($)

11 10 9 8 7 6 5 4 3 2 1

0 2 4 6 8 10 12 14 16 18 20 22 n

Number of calls

WRITING

65. Explain how to graph a line using the intercept method.

66. When graphing a line by plotting points, why is it a good practice to find three solutions instead of two?

REVIEW

67. List the prime numbers between 10 and 30.

68. Write the first ten composite numbers.

69. In what quadrant does the point $(-2, -3)$ lie?

70. What is the formula that gives the area of a circle?

71. Simplify: $-4(-20s)$.

72. Approximate π to the nearest thousandth.

73. Simplify: $-(-3x - 8)$.

74. Simplify: $\frac{1}{3}b + \frac{1}{3}b + \frac{1}{3}b$.

2.3 Rate of Change and the Slope of a Line

- Average rate of change
- Slope of a line
- Interpretation of slope
- Horizontal and vertical lines
- Slopes of parallel lines
- Slopes of perpendicular lines

Our world is one of constant change. In this section, we will show how to describe the amount of change in one quantity in relation to the amount of change in another by finding an *average rate of change*.

▋ Average rate of change

The line graphs in Figure 2-25 are models that approximate the number of daily morning newspapers and the number of evening newspapers published in the United States for the years 1990–1999. From the graph, we can see that the number of morning newspapers increased and the number of evening newspapers decreased from 1990 to 1999.

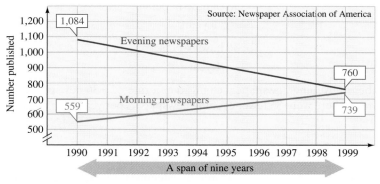

Based on data from *Editor & Publisher*

FIGURE 2-25

If we want to know the rate at which the number of morning newspapers increased or the rate at which the number of evening newspapers decreased over this period, we can do so by finding an **average rate of change.** To find an average rate of change, we compare the change in the number of newspapers published to the length of time in which that change took place, using a **ratio.**

Ratios and rates

A **ratio** is the quotient of two numbers or two quantities with the same units. In symbols, if a and b are two numbers, the ratio of a to b is $\frac{a}{b}$. Ratios that are used to compare quantities with different units are called **rates.**

In Figure 2-25, we see that in 1990, the number of morning newspapers published was 559. In 1999, the number grew to 739. This is a change of $739 - 559$ or 180 over a 9-year time span. So we have

Average rate of change $= \dfrac{\text{change in number of morning newspapers}}{\text{change in time}}$ The rate of change is a ratio that involves units.

$$= \frac{180 \text{ newspapers}}{9 \text{ years}}$$

$$= \frac{\overset{1}{\cancel{9}} \cdot 20 \text{ newspapers}}{\underset{1}{\cancel{9}} \text{ years}} \qquad \text{Factor 180 as } 9 \cdot 20 \text{ and simplify: } \tfrac{9}{9} = 1.$$

$$= \frac{20 \text{ newspapers}}{1 \text{ year}}$$

The number of morning newspapers published in the United States increased, on average, at a rate of 20 newspapers per year (written as 20 newspapers/year) from 1990 through 1999.

In Figure 2-25 shown above, we see that in 1990 the number of evening newspapers published was 1,084. In 1999, the number fell to 760. To find the change, we subtract: $760 - 1{,}084 = -324$. The negative result indicates a decline in the number of evening newspapers over the 9-year time span. So we have

$$\text{Average rate of change} = \frac{\text{change in number of evening newspapers}}{\text{change in time}}$$

$$= \frac{-324 \text{ newspapers}}{9 \text{ years}}$$

$$= \frac{-36 \cdot \overset{1}{\cancel{9}} \text{ newspapers}}{\underset{1}{\cancel{9}} \text{ years}} \qquad \text{Factor } -324 \text{ as } -36 \cdot 9 \text{ and simplify: } \tfrac{9}{9} = 1.$$

$$= \frac{-36 \text{ newspapers}}{1 \text{ year}}$$

The number of evening newspapers changed at a rate of -36 newspapers/year. That is, on average, there were 36 fewer evening newspapers per year, every year, from 1990 through 1999.

■ Slope of a line

In the newspaper example, we measured the steepness of each line in Figure 2-25 to determine the average rates of change. In doing so, we found the **slope** of each line. The slope of a nonvertical line is a number that measures the line's steepness.

To calculate the slope of a line (usually denoted by the letter m), we must first pick two points on the line. To distinguish between the coordinates of two points, we use **subscript notation.** The first point can be denoted as (x_1, y_1), and the second point as (x_2, y_2). After picking two points on the line, we write the ratio of the vertical change to the corresponding horizontal change as we move from one point to the other. The following formula shows how we compute the slope of a line.

Slope of a line

The **slope** of a line passing through points (x_1, y_1) and (x_2, y_2) is

$$m = \frac{\text{change in } y}{\text{change in } x} = \frac{y_2 - y_1}{x_2 - x_1} \quad \text{where } x_2 \neq x_1$$

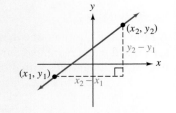

Self Check 1

Find the slope of the line passing through the points $(-3, 6)$ and $(4, -8)$.

EXAMPLE 1   Find the slope of the line shown in Figure 2-26 on the next page, passing through $(-2, 4)$ and $(3, -4)$.

Solution   We can let $(x_1, y_1) = (-2, 4)$ and $(x_2, y_2) = (3, -4)$. Then

$$x_1 = -2 \qquad y_1 = 4 \qquad \text{and} \qquad x_2 = 3 \qquad y_2 = -4$$

$$m = \frac{\text{change in } y}{\text{change in } x}$$

$$m = \frac{y_2 - y_1}{x_2 - x_1} \qquad \text{This is the slope formula.}$$

$$= \frac{-4 - 4}{3 - (-2)} \qquad \text{Substitute } -4 \text{ for } y_2, 4 \text{ for } y_1, 3 \text{ for } x_2, \text{ and } -2 \text{ for } x_1.$$

$$= \frac{-8}{5}$$

$$= -\frac{8}{5}$$

The slope of the line is $-\frac{8}{5}$.

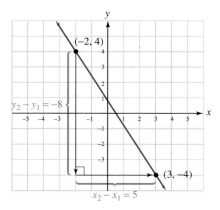

FIGURE 2-26

When calculating slope, it does not matter which point is (x_1, y_1) and which point is (x_2, y_2). We would have obtained the same result in Example 1 if we had let $(x_1, y_1) = (3, -4)$ and $(x_2, y_2) = (-2, 4)$.

$$m = \frac{y_2 - y_1}{x_2 - x_1} = \frac{4 - (-4)}{-2 - 3} = \frac{8}{-5} = -\frac{8}{5}$$

❗ COMMENT When using the slope formula, be careful to subtract the y-coordinates and the x-coordinates in the same order. For example, in Example 1 with $(x_1, y_1) = (-2, 4)$ and $(x_2, y_2) = (3, -4)$, it would be incorrect to write either of the following.

This is $y_2 - y_1$.
↓

$$m = \frac{-4 - 4}{-2 - 3}$$
↑
This is $x_1 - x_2$. The subtraction is not in the same order.

This is $y_1 - y_2$. The subtraction is not in the same order.
↓

$$m = \frac{4 - (-4)}{3 - (-2)}$$
↑
This is $x_2 - x_1$.

The change in y (often denoted as Δy and read as "delta y") is the **rise** of the line between two points on the line. The change in x (often denoted as Δx and read as "delta x") is the **run.** Using this terminology, we can define slope as the ratio of the rise to the run:

$$m = \frac{\Delta y}{\Delta x} = \frac{\text{rise}}{\text{run}} \quad \text{where} \ \Delta x \neq 0$$

EXAMPLE 2   Find the slope of the line determined by $3x - 4y = 12$.

Solution   We first find the coordinates of two points on the line.

- If we choose $x = 0$, then $y = -3$. The point $(0, -3)$ is on the line.
- If we choose $y = 0$, then $x = 4$. The point $(4, 0)$ is on the line.

We then refer to Figure 2-27 on the next page and find the slope of the line passing through $(0, -3)$ and $(4, 0)$ by substituting 0 for y_2, -3 for y_1, 4 for x_2, and 0 for x_1 in the formula for slope.

Self Check 2
Find the coordinates of two other points on the line $3x - 4y = 12$. Then calculate the slope of the line.

$$m = \frac{\text{rise}}{\text{run}}$$

$$= \frac{y_2 - y_1}{x_2 - x_1}$$

$$= \frac{0 - (-3)}{4 - (0)}$$

$$= \frac{3}{4}$$

The slope of the line is $\frac{3}{4}$.

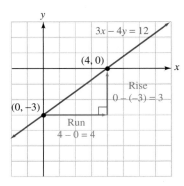

FIGURE 2-27

Answer $\dfrac{3}{4}$

! COMMENT The answers to Example 2 and the Self Check illustrate an important fact. When finding the slope of a line, any pair of points on the line can be used and the result will be the same.

Interpretation of slope

For applied problems, slope can be thought of as the average rate of change in one quantity per unit change in another quantity.

Self Check 3

What information is given by the c-intercept of the graph in Figure 2-28?

EXAMPLE 3   Carpeting cost.   A store sells a carpet for \$25 per square yard, plus a \$20 delivery charge. The total cost c of n square yards is given by the following formula.

Total cost	=	cost per square yard	·	the number of square yards	+	the delivery charge.
c	=	25	·	n	+	20

Graph this equation and interpret the slope of the line.

Solution   The total cost c depends on the number n of square yards of carpet purchased. This relationship is described by $c = 25n + 20$. We can graph the equation on a coordinate system with a horizontal n-axis and a vertical c-axis. Figure 2-28 shows a table of solutions and the graph.

$c = 25n + 20$		
n	c	(n, c)
10	270	$(10, 270)$
30	770	$(30, 770)$
40	1,020	$(40, 1,020)$

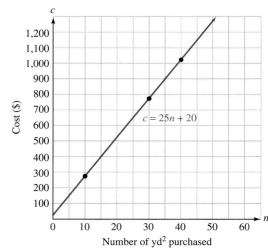

FIGURE 2-28

If we pick the points (30, 770) and (40, 1,020) to find the slope, we have

$$m = \frac{\Delta c}{\Delta n}$$ Read as "m equals delta c over delta n."

$$= \frac{c_2 - c_1}{n_2 - n_1}$$ This is the form the slope formula takes when working with ordered pairs of the form (n, c).

$$= \frac{1,020 - 770}{40 - 30}$$ Substitute 1,020 for c_2, 770 for c_1, 40 for n_2, and 30 for n_1.

$$= \frac{250}{10}$$

$$= 25$$

The slope of 25 is the price (in dollars) per square yard of the carpet.

Answer The c-coordinate of the c-intercept is the delivery charge, $20.

INTERMEDIATE
Algebra *f(x)* **Now**™

EXAMPLE 4   Building stairs.

The slope of a staircase is defined to be the ratio of the total rise to the total run, as shown in the illustration. What is the slope of the staircase?

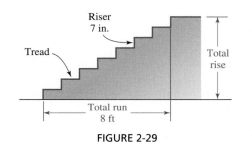

Riser 7 in.

Tread

Total rise

Total run 8 ft

FIGURE 2-29

Self Check 4
Find the slope of the staircase if the riser height is changed to 6.5 inches.

Solution   Since the design has eight 7-inch risers, the total rise is $8 \cdot 7 = 56$ inches. The total run is 8 feet, or 96 inches. With these quantities expressed in the same units, we can now form their ratio.

$$m = \frac{\text{total rise}}{\text{total run}}$$

$$= \frac{56}{96}$$

$$= \frac{7}{12}$$ Simplify the fraction: $\dfrac{56}{96} = \dfrac{\overset{1}{\cancel{8}} \cdot 7}{\underset{1}{\cancel{8}} \cdot 12} = \dfrac{7}{12}$.

The slope of the staircase is $\dfrac{7}{12}$.

Answer $\dfrac{13}{24}$

▮ Horizontal and vertical lines

If (x_1, y_1) and (x_2, y_2) are points on a horizontal line as shown in Figure 2-30(a) on the next page, then $y_1 = y_2$, and the numerator of the fraction

$$\frac{y_2 - y_1}{x_2 - x_1}$$ On a horizontal line, $x_2 \neq x_1$.

is 0. Thus, the value of the fraction is 0, and the slope of the horizontal line is 0.

If (x_1, y_1) and (x_2, y_2) are two points on a vertical line as shown in Figure 2-30(b), then $x_1 = x_2$, and the denominator of the fraction

$$\frac{y_2 - y_1}{x_2 - x_1} \qquad \text{On a vertical line, } y_2 \neq y_1.$$

is 0. Since the denominator of a fraction cannot be 0, a vertical line has no defined slope.

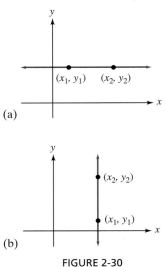

FIGURE 2-30

Slopes of horizontal and vertical lines

Horizontal lines (lines with equations of the form $y = b$) have a slope of 0.

Vertical lines (lines with equations of the form $x = a$) have no defined slope.

THINK IT THROUGH   Community Colleges

The community college has maintained a unique role as a vital component of postsecondary education in America. And now that role is on the ascent.
Arthur M. Cohen, Director Educational Resources Information Center, UCLA, 2002

The graph below shows that the number of community colleges in the United States has steadily increased since 1900. In which decade was the rate of increase in community colleges the greatest? Find that rate of increase.

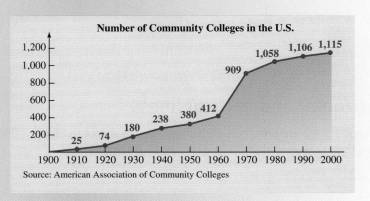

If a line rises as we follow it from left to right, as in Figure 2-31(a), its slope is positive. If a line drops as we follow it from left to right, as in Figure 2-31(b), its slope is negative. If a line is horizontal, as in Figure 2-31(c), its slope is 0. If a line is vertical, as in Figure 2-31(d), it has no defined slope.

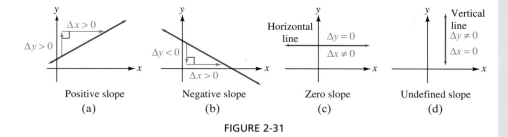

Positive slope (a)   Negative slope (b)   Zero slope (c)   Undefined slope (d)

FIGURE 2-31

Slopes of parallel lines

To see a relationship between parallel lines and their slopes, we refer to the parallel lines l_1 and l_2 shown in Figure 2-32, with slopes of m_1 and m_2, respectively. Because right triangles ABC and DEF are similar, it follows that

$$m_1 = \frac{\Delta y \text{ of } l_1}{\Delta x \text{ of } l_1}$$

$$= \frac{\Delta y \text{ of } l_2}{\Delta x \text{ of } l_2} \qquad \text{Since the triangles are similar, corresponding sides of } \Delta ABC \text{ and } \Delta DEF \text{ are proportional: } \frac{CB}{BA} = \frac{FE}{ED}.$$

$$= m_2$$

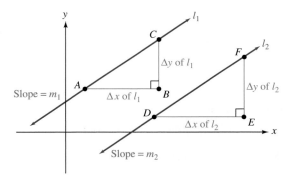

FIGURE 2-32

Thus, if two nonvertical lines are parallel, they have the same slope. It is also true that when two lines have the same slope, they are parallel.

Slopes of parallel lines

Nonvertical parallel lines have the same slope, and different lines having the same slope are parallel.

Since vertical lines are parallel, lines with no defined slope are parallel.

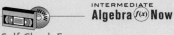

Self Check 5

Determine whether the line that passes through $(4, -8)$ and $(1, -2)$ is parallel to a line with a slope of 2.

EXAMPLE 5   Determine whether the line that passes through $(-6, 2)$ and $(3, -1)$ is parallel to a line with a slope of $-\frac{1}{3}$.

Solution   We can use the slope formula to find the slope of the line that passes through $(-6, 2)$ and $(3, -1)$.

$$m = \frac{y_2 - y_1}{x_2 - x_1}$$

$$m = \frac{-1 - 2}{3 - (-6)} \quad \text{Substitute } -1 \text{ for } y_2, 2 \text{ for } y_1, 3 \text{ for } x_2, \text{ and } -6 \text{ for } x_1.$$

$$= \frac{-3}{9}$$

$$= -\frac{1}{3}$$

Answer They are not parallel.

Both lines have a slope of $-\frac{1}{3}$, and therefore they are parallel.

▧ Slopes of perpendicular lines

The two lines shown in Figure 2-33 meet at right angles and are called **perpendicular lines.** In the figure, the symbol ⌐ is used to denote a right angle. Each of the four angles that are formed has a measure of 90°.

The product of the slopes of two (nonvertical) perpendicular lines is -1. For example, the perpendicular lines shown in Figure 2-33 have slopes of $\frac{3}{2}$ and $-\frac{2}{3}$. If we find the product of their slopes, we have

$$\frac{3}{2}\left(-\frac{2}{3}\right) = -\frac{6}{6} = -1$$

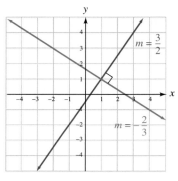

FIGURE 2-33

Two numbers whose product is -1, such as $\frac{3}{2}$ and $-\frac{2}{3}$, are called **negative reciprocals.**

> **Slopes of perpendicular lines**
>
> If two nonvertical lines are perpendicular, their slopes are negative reciprocals.
>
> If the slopes of two lines are negative reciprocals, the lines are perpendicular.

We can also state the fact given above symbolically: If the slopes of two nonvertical lines are m_1 and m_2, then the lines are perpendicular if

$$m_1 \cdot m_2 = -1 \quad \text{or} \quad m_2 = -\frac{1}{m_1}$$

Because a horizontal line is perpendicular to a vertical line, a line with a slope of 0 is perpendicular to a line with no defined slope.

Self Check 6

In Figure 2-34, is the line l_1 perpendicular to the line l_3?

EXAMPLE 6   Are the lines l_1 and l_2 shown in Figure 2-34 on the next page perpendicular?

Solution   We find the slopes of the lines and see whether they are negative reciprocals.

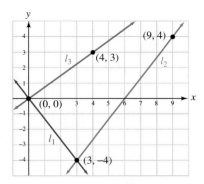

FIGURE 2-34

$$\text{Slope of line } l_1 = \frac{y_2 - y_1}{x_2 - x_1} \qquad \text{and} \qquad \text{Slope of line } l_2 = \frac{y_2 - y_1}{x_2 - x_1}$$

$$= \frac{-4 - 0}{3 - 0} \qquad\qquad\qquad = \frac{4 - (-4)}{9 - 3}$$

$$= -\frac{4}{3} \qquad\qquad\qquad\quad = \frac{8}{6}$$

$$\qquad\qquad\qquad\qquad\qquad\quad = \frac{4}{3}$$

Since their slopes are not negative reciprocals $\left(-\frac{4}{3} \cdot \frac{4}{3} \neq -1 \right)$, the lines are not perpendicular.

Answer yes

Section 2.3 **STUDY SET**

VOCABULARY *Fill in the blanks.*

1. _____ is defined as the change in y divided by the change in x.

2. A slope is an average _____ of change.

3. The _____ in x (denoted as Δx) is the horizontal run of the line between two points on the line.

4. The change in y (denoted as Δy) is the vertical _____ of the line between two points on the line.

5. $\frac{7}{8}$ and $-\frac{8}{7}$ are negative _____.

6. _____ lines have the same slope. The slopes of _____ lines are negative reciprocals.

CONCEPTS

7. See the illustration in the next column.

  a. Which line is horizontal? What is its slope?

  b. Which line is vertical? What is its slope?

c. Which line has a positive slope? What is it?

d. Which line has a negative slope? What is it?

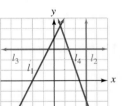

8. See the illustration.

  a. Find the slopes of lines l_1 and l_2. Are they parallel?

  b. Find the slopes of lines l_2 and l_3. Are they perpendicular?

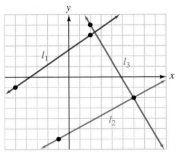

9. THE RECORDING INDUSTRY The following graphs are models that approximate the number of CDs and cassettes that were shipped for sale from 1990 to 1999.

a. What was the rate of increase in the number of CDs shipped?

b. What was the rate of decrease in the number of cassettes shipped?

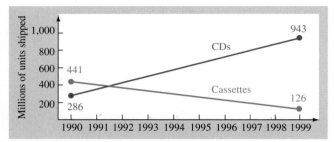

Source: *Statistical Abstract of the United States* (2003)

10. HALLOWEEN A couple kept records of the number of trick-or-treaters who came to their door on Halloween night. Find the slope of the line. What information does the slope give?

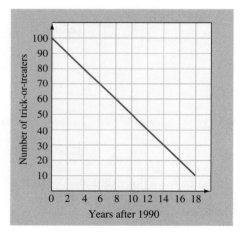

11. a. Determine the slope of the line shown below.

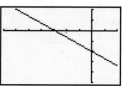

b. Determine the slope of the line shown.

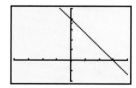

12. A table of solutions for a linear equation is shown. Find the slope of the graph of the equation.

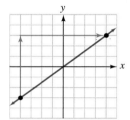

NOTATION

13. What formula is used to find the slope of a line?

14. Explain the difference between x^2 and x_2.

15. See the illustration.

a. What is Δy?

b. What is Δx?

c. What is $\dfrac{\Delta y}{\Delta x}$?

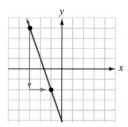

16. See the illustration.

a. What is Δy?

b. What is Δx?

c. What is $\dfrac{\Delta y}{\Delta x}$?

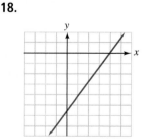

PRACTICE   *Find the slope of each line.*

17.

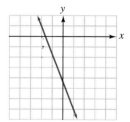

18.

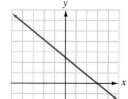

19.

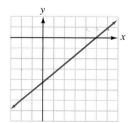

20.

Find the slope of the line that passes through the given points, if possible.

21. $(0, 0), (3, 9)$

22. $(9, 6), (0, 0)$

23. $(-1, 8), (6, 1)$

24. $(-5, -8), (3, 8)$

25. $(3, -1), (-6, 2)$

26. $(0, -8), (-5, 0)$

27. $(7, 5), (-9, 5)$

28. $(2, -8), (3, -8)$

29. $(-7, -5), (-7, -2)$

30. $(3, -5), (3, 14)$

31. $(a, b), (b, a)$

32. $(a, b), (-b, -a)$

Find the slope of the line determined by each equation.

33. $3x + 2y = 12$

34. $2x - y = 6$

35. $3x = 4y - 2$

36. $x = y$

37. $y = \dfrac{x - 4}{2}$

38. $x = \dfrac{3 - y}{4}$

39. $4y = 3(y + 2)$

40. $x + y = \dfrac{2 - 3y}{3}$

Determine whether the lines with the given slopes are parallel, perpendicular, or neither.

41. $m_1 = 3, m_2 = -\dfrac{1}{3}$

42. $m_1 = \dfrac{1}{4}, m_2 = 4$

43. $m_1 = 4, m_2 = 0.25$

44. $m_1 = -5, m_2 = -\dfrac{1}{0.2}$

45. $m_1 = \dfrac{1}{a}, m_2 = a$

46. $m_1 = a, m_2 = -\dfrac{1}{a}$

Determine whether the line that passes through the two given points is parallel or perpendicular (or neither) to a line with a slope of -2.

47. $(3, 4), (4, 2)$

48. $(6, 4), (8, 5)$

49. $(-2, 1), (6, 5)$

50. $(3, 4), (-3, -5)$

51. $(5, 4), (6, 6)$

52. $(-2, 3), (4, -9)$

APPLICATIONS

53. LANDING PLANES A jet descends in a stairstep pattern, as shown below. The required elevations of the plane's path are given. Find the slope of the descent in each of the three parts of its landing that are labeled. Which part is the steepest?

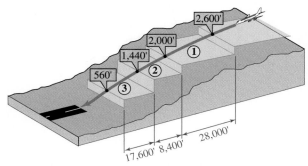

Based on data from *Los Angeles Times* (August 7, 1997), p. A8

54. DROP IN PRICES The price of computers has been dropping for the past ten years. If a desktop PC cost $5,700 ten years ago, and the same computing power cost $400 two years ago, find the rate of decrease per year. (Assume a straight-line model.)

55. MAPS Topographic maps have contour lines that connect points of equal elevation on a mountain. The vertical distance between contour lines in the illustration is 50 feet. Find the slope of the west face and the east face of the mountain peak.

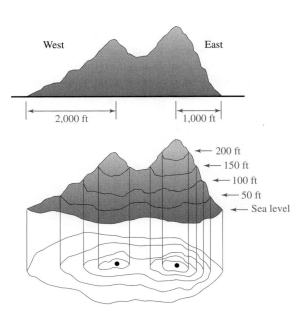

56. SKIING The men's giant slalom course shown below is longer than the women's course. Does this mean that the men's course is steeper? Use the concept of the slope of a line to explain.

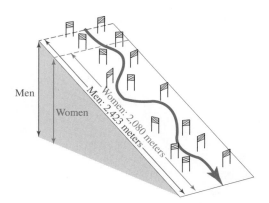

57. ROAD SIGNS Find the slope of the road shown below. Use this information to complete the road warning sign for truckers by expressing the slope as a percent. (*Hint:* 1 mi = 5,280 ft.)

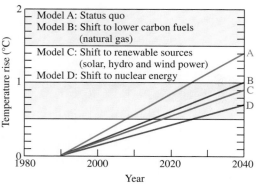

58. GREENHOUSE EFFECT The following graphs are estimates of future average global temperature rise due to the greenhouse effect. Assume that the models are straight lines. Estimate the average rate of change of each model. Express your answers as fractions.

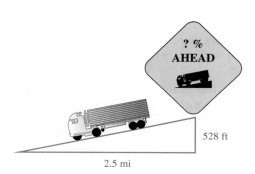

Based on data from *The Blue Planet* (Wiley, 1995)

59. DECK DESIGN See the illustration below. Find the slopes of the cross-brace and the supports. Is the cross-brace perpendicular to either support?

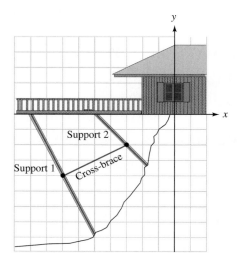

60. AIR PRESSURE Air pressure, measured in units called **Pascals** (Pa), decreases with altitude. Find the rate of change in Pascals for the fastest and the slowest decreasing steps of the graph in the illustration.

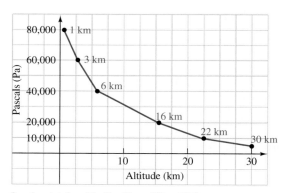

Based on data from *The Blue Planet* (Wiley, 1995)

WRITING

61. POLITICS The illustration on the next page shows how federal Medicare spending would have continued if the Republican-sponsored Balanced Budget Act hadn't become law in 1997. Write a brief statement explaining why Democrats could argue that the budget act "cut spending." Then write a brief statement explaining why Republicans could respond by saying, "There was no cut in spending — only a reduction in the rate of growth of spending."

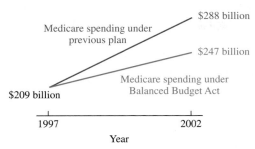

$288 billion

Medicare spending under previous plan

$247 billion

$209 billion

Medicare spending under Balanced Budget Act

1997   2002

Year

Based on information supplied by Congressman David Drier's office

62. NUCLEAR ENERGY Since 1998, the number of nuclear reactors licensed for operation in the United States has remained about the same. Knowing this, what can be said about the rate of change in the number of reactors since 1998? Explain your answer.

63. Explain why a vertical line has no defined slope.

64. Explain how to determine from their slopes whether two lines are parallel, perpendicular, or neither.

REVIEW

65. HALLOWEEN CANDY A candy maker wants to make a 60-pound mixture of two candies to sell for $2 per pound. If black licorice bits sell for $1.90 per pound and orange gumdrops sell for $2.20 per pound, how many pounds of each should be used?

66. MEDICATIONS A doctor prescribes an ointment that is 2% hydrocortisone. A pharmacist has 1% and 5% concentrations in stock. How many ounces of each should the pharmacist use to make a 1-ounce tube?

67. READING GRAPHS In the illustration below, d is the distance a person has walked after t hours. How many hours did it take for the person to walk 10 miles?

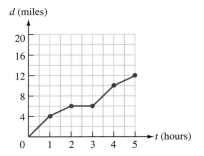

d (miles)

t (hours)

68. CIRCLE GRAPHS In the illustration, each part of the circle represents the amount of money spent in each of five categories. Approximately what percent was spent on rent?

Monthly Expenses of Joe Sigueri

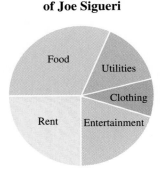

Food

Utilities

Clothing

Rent

Entertainment

2.4 Writing Equations of Lines

- Point–slope form of the equation of a line
- Slope–intercept form of the equation of a line
- Using slope as an aid in graphing
- Parallel and perpendicular lines
- Straight-line depreciation
- Curve fitting

We have seen that linear relationships are often presented in graphs. In this section, we begin a discussion of how to write an equation to model a linear relationship.

Point–slope form of the equation of a line

Suppose that line l in Figure 2-35 on the next page has a slope of m and passes through (x_1, y_1). If (x, y) is a second point on line l, we have

$$m = \frac{y - y_1}{x - x_1}$$

or if we multiply both sides by $x - x_1$, we have

(1) $y - y_1 = m(x - x_1)$

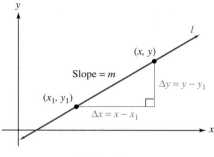

FIGURE 2-35

Because Equation 1 displays the coordinates of the point (x_1, y_1) on the line and the slope m of the line, it is called the **point–slope form** of the equation of a line.

> **Point–slope form**
>
> The equation of the line that passes through (x_1, y_1) and has slope m is
>
> $$y - y_1 = m(x - x_1)$$

INTERMEDIATE
Algebra *f(x)* **Now™** ———

Self Check 1
Write an equation of the line that has a slope of $\frac{5}{4}$ and passes through $(4, 10)$.

EXAMPLE 1 Write an equation of the line that has a slope of $-\frac{2}{3}$ and passes through $(-4, 5)$.

Solution We substitute $-\frac{2}{3}$ for m, -4 for x_1, and 5 for y_1 into the point–slope form and simplify.

$$y - y_1 = m(x - x_1) \qquad \text{This is the point–slope form.}$$

$$y - 5 = -\frac{2}{3}[x - (-4)] \quad \text{Substitute } -\frac{2}{3} \text{ for } m, -4 \text{ for } x_1, \text{ and 5 for } y_1.$$

$$y - 5 = -\frac{2}{3}(x + 4) \qquad \text{Simplify the expression within the brackets.}$$

$$y - 5 = -\frac{2}{3}x - \frac{8}{3} \qquad \text{Distribute the multiplication by } -\frac{2}{3}.$$

$$y = -\frac{2}{3}x + \frac{7}{3} \qquad \text{To solve for } y, \text{ add 5 in the form of } \frac{15}{3} \text{ to both sides and simplify.}$$

Answer $y = \frac{5}{4}x + 5$

The equation of the line is $y = -\frac{2}{3}x + \frac{7}{3}$.

INTERMEDIATE
Algebra *f(x)* **Now™**

Self Check 2
Write an equation of the line passing through $(-2, 5)$ and $(4, -3)$.

EXAMPLE 2 Write an equation of the line passing through $(-5, 4)$ and $(8, -6)$.

Solution First we find the slope of the line.

$$m = \frac{y_2 - y_1}{x_2 - x_1} \qquad \text{This is the slope formula.}$$

$$= \frac{-6 - 4}{8 - (-5)} \quad \text{Substitute } -6 \text{ for } y_2, 4 \text{ for } y_1, 8 \text{ for } x_2, \text{ and } -5 \text{ for } x_1.$$

$$= -\frac{10}{13}$$

Since the line passes through $(-5, 4)$ and $(8, -6)$, we can choose either point and substitute its coordinates into the point–slope form. If we choose $(-5, 4)$, we substitute -5 for x_1, 4 for y_1, and $-\frac{10}{13}$ for m and proceed as follows.

$$y - y_1 = m(x - x_1) \qquad \text{This is the point–slope form.}$$

$$y - 4 = -\frac{10}{13}[x - (-5)] \qquad \text{Substitute } -\frac{10}{13} \text{ for } m, -5 \text{ for } x_1, \text{ and 4 for } y_1.$$

$$y - 4 = -\frac{10}{13}(x + 5) \qquad \text{Simplify the expression within the brackets.}$$

$$y - 4 = -\frac{10}{13}x - \frac{50}{13} \qquad \text{Distribute the multiplication by } -\frac{10}{13}.$$

$$y = -\frac{10}{13}x + \frac{2}{13} \qquad \text{To solve for } y, \text{ add 4 in the form of } \frac{52}{13} \text{ to both sides and simplify.}$$

The equation of the line is $y = -\frac{10}{13}x + \frac{2}{13}$.

Answer $y = -\frac{4}{3}x + \frac{7}{3}$

■ Slope–intercept form of the equation of a line

Since the y-intercept of the line l shown in Figure 2-36 is the point $(0, b)$, we can write its equation by substituting 0 for x_1 and b for y_1 in the point–slope form and simplifying.

$$y - y_1 = m(x - x_1)$$

$$y - b = m(x - 0)$$

$$y - b = mx$$

$$\textbf{(2)} \qquad y = mx + b \qquad \text{To solve for } y, \text{ add } b \text{ to both sides.}$$

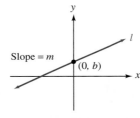

Slope $= m$ $(0, b)$

FIGURE 2-36

Because Equation 2 displays the slope m and the y-coordinate b of the y-intercept, it is called the **slope–intercept form** of the equation of a line.

> **Slope–intercept form**
>
> The equation of the line having slope m and y-intercept $(0, b)$ is
>
> $$y = mx + b$$

INTERMEDIATE
Algebra $f(x)$ Now™

EXAMPLE 3 Write an equation of the line that has a slope of 4 and passes through $(5, 9)$.

Solution Since we are given that $m = 4$ and that $(5, 9)$ satisfies the equation, we can substitute 5 for x, 9 for y, and 4 for m in the equation $y = mx + b$ and solve for b.

$$y = mx + b \qquad \text{This is the slope–intercept form.}$$

$$9 = 4(5) + b \qquad \text{Substitute 9 for } y, 4 \text{ for } m, \text{ and 5 for } x.$$

$$9 = 20 + b \qquad \text{Perform the multiplication.}$$

$$-11 = b \qquad \text{To solve for } b, \text{ subtract 20 from both sides.}$$

Because $m = 4$ and $b = -11$, the equation is $y = 4x - 11$.

Self Check 3
Write an equation of the line that has a slope of -2 and passes through $(-2, 8)$.

Answer $y = -2x + 4$

When an equation describing a linear relationship between two quantities is written in slope–intercept form, two pieces of information about the relationship are easily seen. As an example, let's consider the equation $L = -0.05t + 7.25$. If we begin with a pencil 7.25 inches long, this linear model gives the new length L in inches of the pencil after it has been inserted into a sharpener and the handle turned t times. (See Figure 2-37.)

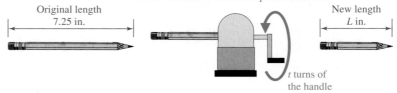

FIGURE 2-37

The value of m (in this case, -0.05) gives the change in the length of the pencil for one turn of the handle. Because the slope is negative, we know that the length of the pencil *decreases* by 0.05 inch for each turn of the handle. The value of b (in this case, 7.25) tells us that before any turns were made (when $t = 0$), the length of the pencil was 7.25 inches.

$$L = -0.05t + 7.25$$

The slope is the rate of change of the length of the pencil.

The intercept is the original length of the pencil.

Using slope as an aid in graphing

It is easy to graph a linear equation when it is written in slope–intercept form. For example, to graph $y = \frac{4}{3}x - 2$, we note that $b = -2$ and that the y-intercept is $(0, b) = (0, -2)$. (See Figure 2-38.)

Because the slope of the line is $\frac{\Delta y}{\Delta x} = \frac{4}{3}$, we can locate another point Q on the line by starting at point $P(0, -2)$ and counting 3 units to the right (run) and 4 units up (rise). The change in x from point P to point Q is $\Delta x = 3$, and the corresponding change in y is $\Delta y = 4$. The line through points P and Q is the graph of the equation.

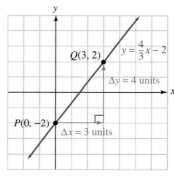

FIGURE 2-38

Self Check 4
Find the slope and the y-intercept of the line with the equation $3x - 2y = -4$. Then graph the line.

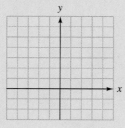

EXAMPLE 4   Find the slope and the y-intercept of the line with the equation $2x + 3y = -9$. Then graph the line.

Solution   We write an equation in the form $y = mx + b$ to find the slope m and the y-intercept $(0, b)$.

$2x + 3y = -9$ This is the given equation in general form.

$3y = -2x - 9$ Subtract $2x$ from both sides.

$\dfrac{3y}{3} = \dfrac{-2x}{3} - \dfrac{9}{3}$ To solve for y, divide both sides by 3.

$y = -\dfrac{2}{3}x - 3$ Simplify both sides. We see that $m = -\dfrac{2}{3}$ and $b = -3$.

The slope of the line is $-\frac{2}{3}$, which can be expressed as $\frac{-2}{3}$. After plotting the y-intercept, $(0, -3)$, we move 2 units downward (rise) and then 3 units to the right (run). This locates a second point on the line, $(3, -5)$. From this point, we move another 2 units downward and 3 units to the right to locate a *third point* on the line, $(6, -7)$. Then we draw a line through the points to obtain the graph shown in the figure.

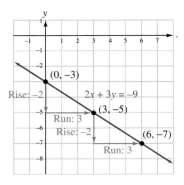

FIGURE 2-39

Answer $m = \frac{3}{2}, (0, 2)$

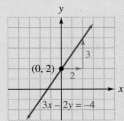

■ Parallel and perpendicular lines

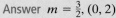
EXAMPLE 5   Show that the lines represented by $4x + 8y = 10$ and $2x = 12 - 4y$ are parallel.

Solution   We solve each equation for y to see that the lines are distinct and that their slopes are equal.

$$4x + 8y = 10 \qquad\qquad 2x = 12 - 4y$$
$$8y = -4x + 10 \qquad\qquad 4y = -2x + 12$$
$$y = -\frac{1}{2}x + \frac{5}{4} \qquad\qquad y = -\frac{1}{2}x + 3$$

Since the values of b in these equations are different $\left(\frac{5}{4} \text{ and } 3\right)$, the lines are distinct. Since the slope of each line is $-\frac{1}{2}$, they are parallel.

Self Check 5
Are the lines represented by $3x - 2y = 4$ and $2x = 5(y + 1)$ parallel?

Answer no

EXAMPLE 6   Show that the lines represented by $4x + 8y = 10$ and $4x - 2y = 21$ are perpendicular.

Solution   We solve each equation for y to see that the slopes of their straight-line graphs are negative reciprocals.

$$4x + 8y = 10 \qquad\qquad 4x - 2y = 21$$
$$8y = -4x + 10 \qquad\qquad -2y = -4x + 21$$
$$y = -\frac{1}{2}x + \frac{5}{4} \qquad\qquad y = 2x - \frac{21}{2}$$

Since the slopes are negative reciprocals $\left(-\frac{1}{2} \text{ and } 2\right)$, the lines are perpendicular.

Self Check 6
Are the lines represented by $3x + 2y = 6$ and $2x - 3y = 6$ perpendicular?

Answer yes

Self Check 7
Write an equation of the line that is parallel to the line $y = 8x - 3$ and passes through the origin.

EXAMPLE 7   Write an equation of the line that passes through $(-2, 5)$ and is parallel to the line $y = 8x - 3$.

Solution   Since the slope of the line given by $y = 8x - 3$ is the coefficient of x, the slope is 8. The desired equation is to have a graph that is parallel to the graph of $y = 8x - 3$. Its slope must also be 8.

We substitute -2 for x_1, 5 for y_1, and 8 for m in the point–slope form and simplify.

$$y - y_1 = m(x - x_1)$$
$$y - 5 = 8[x - (-2)] \quad \text{Substitute 5 for } y_1, 8 \text{ for } m, \text{ and } -2 \text{ for } x_1.$$
$$y - 5 = 8(x + 2) \quad \text{Simplify the expression within the brackets.}$$
$$y - 5 = 8x + 16 \quad \text{Distribute the multiplication by 8 and simplify.}$$
$$y = 8x + 21 \quad \text{Add 5 to both sides.}$$

Answer $y = 8x$

The equation is $y = 8x + 21$.

When asked to *write the equation of a line,* determine what you know about the graph of the line; its slope, its y-intercept, points it passes through, and so on. Then substitute the appropriate numbers into one of the following forms of a linear equation.

Forms for the equation of a line	
General form of a linear equation	$Ax + By = C$ A and B cannot both be 0.
Slope–intercept form of a linear equation	$y = mx + b$ The slope is m, and the y-intercept is $(0, b)$.
Point–slope form of a linear equation	$y - y_1 = m(x - x_1)$ The slope is m, and the line passes through (x_1, y_1).
A horizontal line	$y = b$ The slope is 0, and the y-intercept is $(0, b)$.
A vertical line	$x = a$ There is no defined slope, and the x-intercept is $(a, 0)$.

Straight-line depreciation

For tax purposes, many businesses use *straight-line depreciation* to find the declining value of aging equipment.

EXAMPLE 8   Accounting.   After purchasing a new drill press, a machine shop owner had his accountant prepare a depreciation worksheet for tax purposes. See the illustration.

Depreciation Worksheet

Drill press . . . . $1,970 (new)

Salvage value . . . . . . $270 (in 10 years)

a. Assuming straight-line depreciation, write an equation that gives the value v of the drill press after x years of use.

b. Find the value of the drill press after $2\frac{1}{2}$ years of use.

c. What is the economic meaning of the v-intercept of the line?

d. What is the economic meaning of the slope of the line?

Solution

a. The facts presented in the worksheet can be expressed as ordered pairs of the form

$$(x, v)$$

number of years of use ⟵⎤ ⎡⟶ value of the drill press

- When purchased, the new $1,970 drill press had been used 0 years: (0, 1,970).
- After 10 years of use, the value of the drill press will be $270: (10, 270).

A simple sketch showing these ordered pairs and the line of depreciation is helpful in visualizing the situation.

Since we know two points that lie on the line, we can write its equation using the point-slope form. First, we find the slope of the line.

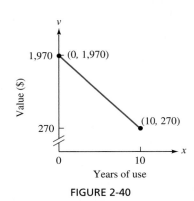

FIGURE 2-40

$$m = \frac{v_2 - v_1}{x_2 - x_1} \qquad \text{This is the slope formula written in terms of } x \text{ and } v.$$

$$= \frac{270 - 1{,}970}{10 - 0} \qquad (x_1, v_1) = (0, 1{,}970) \text{ and } (x_2, v_2) = (10, 270).$$

$$= \frac{-1{,}700}{10}$$

$$= -170$$

To find the equation of the line, we substitute -170 for m, 0 for x_1, and $1{,}970$ for v_1 in the point–slope form and simplify.

$$v - v_1 = m(x - x_1) \qquad \text{This is the point–slope form written in terms } x \text{ and } v.$$

$$v - 1{,}970 = -170(x - 0)$$

$$v = -170x + 1{,}970 \qquad \text{This is the straight-line depreciation equation.}$$

The value v of the drill press after x years of use is given by the linear model $v = -170x + 1{,}970$.

b. To find the value of the drill press after $2\frac{1}{2}$ years of use, we substitute 2.5 for x in the depreciation equation and find v.

$$v = -170x + 1{,}970$$

$$= -170(2.5) + 1{,}970$$

$$= -425 + 1{,}970$$

$$= 1{,}545$$

In $2\frac{1}{2}$ years, the drill press will be worth $1,545.

c. From the sketch, we see that the v-intercept of the graph of the depreciation line is (0, 1,970). This gives the original cost of the drill press, $1,970.

d. Each year, the value of the drill press decreases by $170, because the slope of the line is -170. The slope of the line is the *annual depreciation rate*.

Curve fitting

In statistics, the process of using one variable to predict another is called **regression.** For example, if we know a man's height, we can usually make a good prediction about his weight, because taller men tend to weigh more than shorter men.

The table in Figure 2-41(a) shows the results of sampling twelve men at random and recording the height h and weight w of each. In Figure 2-41(b), the ordered pairs (h, w) from the table are plotted to form a **scatter diagram.** Notice that the data points fall more or less along an imaginary straight line, indicating a linear relationship between h and w.

Height h in.	66	67	68	68	70	70	71	72	73	74	75	75
Weight w lb	145	150	150	165	180	165	175	200	190	190	205	215

(a)

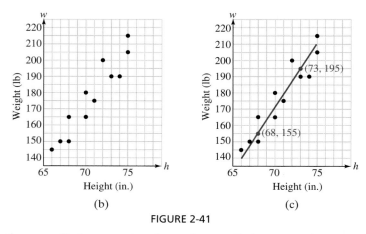

(b)   (c)

FIGURE 2-41

To write a *prediction equation* (sometimes called a *regression equation*) that relates height and weight, we must find the equation of the line that comes closer to all of the data points in the scatter diagram than any other possible line. In statistics, there are exact methods to find this equation; however, they are beyond the scope of this book. In this course, we will draw "by eye" a line that we feel best fits the data points.

In Figure 2-41(c), a straight edge was placed on the scatter diagram and a line was drawn that seemed to best fit all of the data points. Note that it passes through (68, 155) and (73, 195). To write the equation of that line, we first need to find its slope.

$$m = \frac{w_2 - w_1}{h_2 - h_1}$$ This is the slope formula written in terms of h and w.

$$= \frac{195 - 155}{73 - 68}$$ Choose $(h_1, w_1) = (68, 155)$ and $(h_2, w_2) = (73, 195)$.

$$= \frac{40}{5}$$

$$= 8$$

We then use the point–slope form to find the equation of the line. Since the line passes through (68, 155) and (73, 195), we can use either one to write its equation.

$w - w_1 = m(h - h_1)$ This is the point–slope form written in terms of h and w.

$w - 155 = 8(h - 68)$ Choose (68, 155) for (h_1, w_1).

$w - 155 = 8h - 544$ Distribute the multiplication by 8.

$w = 8h - 389$ To solve for w, add 155 to both sides.

The equation of the line that was drawn through the data points in the scatter diagram is $w = 8h - 389$. We can use this equation to predict the weight of a man who is 72 inches tall.

$w = 8h - 389$

$w = 8(72) - 389$ Substitute 72 for h.

$w = 576 - 389$

$w = 187$

We predict that a 72-inch-tall man chosen at random will weigh about 187 pounds.

Section 2.4 STUDY SET

■ VOCABULARY *Fill in the blanks.*

1. The point–slope form of the equation of a line is _____ .

2. The _____ form of the equation of a line is $y = mx + b$.

3. Two lines are _____ when their slopes are negative reciprocals.

4. Two lines are _____ when they have the same slope.

■ CONCEPTS

5. If you know the slope of a line, is that enough information about the line to write its equation?

6. If you know a point that a line passes through, is that enough information about the line to write its equation?

7. The line graphed on the right passes through the point $(-2, -3)$. Find its slope. Then write its equation. Express your answer in point–slope form.

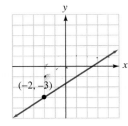

8. For the line graphed on the right, find the slope and the y-intercept. Then write the equation of the line. Express your answer in slope–intercept form.

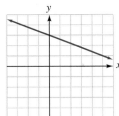

9. When the graph of the line $y = -\frac{2}{3}x + 1$ is drawn, what slope and y-intercept will the line have?

10. When the graph of the line $y - 3 = -\frac{2}{3}(x + 1)$ is drawn, what slope will it have? What point does the equation indicate it will pass through?

11. Do the equations $y - 2 = 3(x - 2)$, $y = 3x - 4$, and $3x - y = 4$ all describe the same line.

12. See the linear model below.

    a. What information does the y-intercept give?

    b. What information does the slope give?

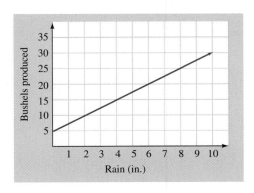

13. When each equation is graphed, what will the y-intercept be?

    a. $y = 2x$ b. $x = -3$

14. When each equation is graphed, what will be the slope of the line?

    a. $y = -x$ b. $x = -3$

15. 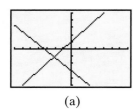 The two lines graphed in illustration (a) below appear to be perpendicular. Their equations are displayed in illustration (b). Are they perpendicular? Explain.

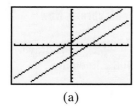

(a)   (b)

16. 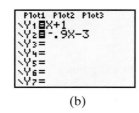 The two lines graphed in illustration (a) below appear to be parallel. Their equations are displayed in illustration (b). Are they parallel? Explain.

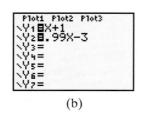

(a)   (b)

NOTATION *Complete each solution.*

17. Write $y + 2 = \frac{1}{3}(x + 3)$ in slope–intercept form.

$$y + 2 = \frac{1}{3}(x + 3)$$

$$y + 2 = \boxed{} + 1$$

$$y + 2 - \boxed{} = \frac{1}{3}x + 1 - \boxed{}$$

$$y = \frac{1}{3}x - \boxed{}$$

$$m = \boxed{} , b = \boxed{}$$

18. Write an equation of the line that has slope -2 and passes through the point $(3, 1)$.

$$y - y_1 = m(x - x_1)$$

$$y - \boxed{} = -2(x - \boxed{})$$

$$y - 1 = \boxed{} + 6$$

$$y = -2x + 7$$

PRACTICE *Use point–slope form to write an equation of the line with the given properties. Then write each equation in slope–intercept form.*

19. $m = 5$, passing through $(0, 7)$

20. $m = -8$, passing through $(0, -2)$

21. $m = -3$, passing through $(2, 0)$

22. $m = 4$, passing through $(-5, 0)$

Use point–slope form to write an equation of the line passing through the two given points. Then write each equation in slope–intercept form.

23. $(0, 0), (4, 4)$

24. $(-5, 5), (0, 0)$

25.

x	y
3	4
0	-3

26.

x	y
4	0
6	-8

27.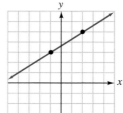

28.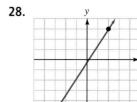

Use the slope–intercept form to write an equation of the line with the given properties.

29. $m = 3, b = 17$

30. $m = -2, b = 11$

31. $m = -7$, passing through $(7, 5)$

32. $m = 3$, passing through $(-2, -5)$

33. $m = 0$, passing through $(2, -4)$

34. $m = -7$, passing through the origin

35. passing through $(6, 8)$ and $(2, 10)$

36. passing through $(-4, 5)$ and $(2, -6)$

Write each equation in slope–intercept form. Then find the slope and the y-intercept of the line determined by the equation.

37. $3x - 2y = 8$

38. $-2x + 4y = 12$

39. $-2(x + 3y) = 5$

40. $5(2x - 3y) = 4$

Find the slope and y-intercept and use them to graph the line.

41. $y = x - 1$

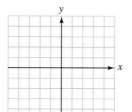

42. $y = -x + 2$

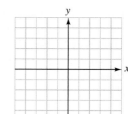

43. $y = \dfrac{2}{3}x + 2$

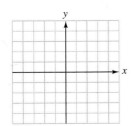

44. $y = -\dfrac{5}{4}x + \dfrac{5}{2}$

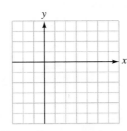

45. $4y - 3 = -3x - 11$

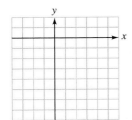

46. $-2x + 4y = 12$

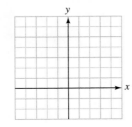

Determine whether the graphs of each pair of equations are parallel, perpendicular, or neither.

47. $y = 3x + 4, y = 3x - 7$

48. $y = 4x - 13, y = \dfrac{1}{4}x + 13$

49. $x + y = 2, y = x + 5$

50. $x = y + 2, y = x + 3$

51. $3x + 6y = 1, y = \dfrac{1}{2}x$

52. $2x + 3y = 9, 3x - 2y = 5$

53. $y = 3, x = 4$

54. $y = -3, y = -7$

Write an equation of the line that passes through the given point and is parallel to the given line. Write the answer in slope–intercept form.

55. $(0, 0), y = 4x - 7$

56. $(0, 0), x = -3y - 12$

57. $(2, 5), 4x - y = 7$

58. $(-6, 3), y + 3x = -12$

59. $(4, -2), x = \dfrac{5}{4}y - 2$

60. $(1, -5), x = -\dfrac{3}{4}y + 5$

Write an equation of the line that passes through the given point and is perpendicular to the given line. Write the answer in slope–intercept form.

61. $(0, 0), y = 4x - 7$

62. $(0, 0), x = -3y - 12$

63. $(2, 5), 4x - y = 7$

64. $(-6, 3), y + 3x = -12$

65. $(4, -2), x = \dfrac{5}{4}y - 2$

66. $(1, -5), x = -\dfrac{3}{4}y + 5$

APPLICATIONS *In Exercises 67–70, assume straight-line depreciation or straight-line appreciation.*

67. BIG-SCREEN TVs Find a linear depreciation equation for the TV in the want ad shown.

For Sale: 3-year-old 45-inch TV, with matrix surround sound & picture within picture, remote. $1,750 new. Asking $800. Call 875-5555. Ask for Mike.

68. SALVAGE VALUE A truck was purchased for $19,984. Its salvage value at the end of 8 years is expected to be $1,600. Find the depreciation equation.

69. ART In 1987, the painting *Rising Sunflowers* by Vincent van Gogh sold for $36,225,000. Suppose that an art appraiser expected the painting to double in value in 20 years. Let x represent the time in years after 1987. Find an appreciation equation.

70. REAL ESTATE LISTINGS Use the information given in the description of the property below to write an appreciation equation for the house.

Vacation Home
$122,000
Only 2 years old

• Great investment property!
• Expected to appreciate $4,000/yr

Sq ft: 1,635	Fam rm: yes	Den: no
Bdrm: 3	Ba: 1.5	Gar: enclosed
A/C: yes	Firepl: yes	Kit: built-ins

71. CRIMINOLOGY City growth and the number of burglaries are related by a linear equation. Records show that 575 burglaries were reported in a year when the local population was 77,000 and that the rate of increase in the number of burglaries was 1 for every 100 new residents.

a. Using the variables p for population and B for burglaries, write an equation (in slope–intercept form) that police can use to predict future burglary statistics.

b. How many burglaries can be expected when the population reaches 110,000?

72. CABLE TV Since 1990, when the average monthly basic cable TV rate in the United States was $15.81, the cost has risen by about $1.52 a year.

a. Write an equation in slope–intercept form to predict cable TV costs in the future. Use t to represent time in years after 1990 and C to represent the average basic monthly cost.

b. If the equation in part a were graphed, what would be the meaning of the C-intercept and the slope of the line?

73. PSYCHOLOGY EXPERIMENTS The scattergram below shows the performance of a rat in a maze.

a. Draw a line through (1, 10) and (19, 1). Write its equation using the variables t and E. In psychology, this equation is called the *learning curve* for the rat.

b. What does the slope of the line tell us?

c. What information does the x-intercept of the graph give?

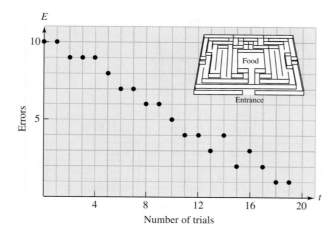

Number of trials

74. UNDERSEA DIVING The illustration below shows that the pressure p that divers experience is related to the depth d of the dive. A linear model can be used to describe this relationship.

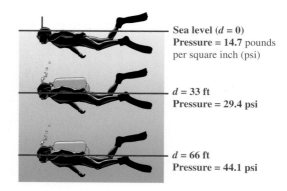

a. Write a linear model in slope–intercept form.

b. Pearl and sponge divers often reach depths of 100 feet. What pressure do they experience? Round to the nearest tenth.

c. Scuba divers can safely dive to depths of 250 feet. What pressure do they experience? Round to the nearest tenth.

75. WIND-CHILL A combination of cold and wind makes a person feel colder than the actual temperature. The table shows what temperatures of 35°F and 15°F feel like when a 15-mph wind is blowing. The relationship between the actual temperature and the wind-chill temperature can be modeled with a linear equation.

a. Write an equation that models this relationship. Answer in slope–intercept form.

b. What information is given by the y-intercept of the graph of the equation found in part a?

Actual temperature	Wind-chill temperature
35°F	16°F
15°F	−11°F

76. **DRAFTING** The illustration on the next page shows a computer-generated drawing of an airplane part. When the designer clicks the mouse on a line on the drawing, the computer finds the equation of the line. Use a calculator to determine whether the angle where the weld is to be made is a right angle.

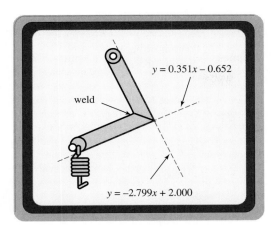

$$y = 0.351x - 0.652$$

weld

$$y = -2.799x + 2.000$$

WRITING

77. Explain how to find the equation of a line passing through two given points.

78. Explain what m, x_1, and y_1 represent in the point–slope form of the equation of a line.

79. Explain what m and b represent in the slope–intercept form of the equation of a line.

80. Linear relationships between two quantities can be described by an equation or a graph. Which do you think is the more informative? Why?

REVIEW

81. INVESTMENTS Equal amounts are invested at 6%, 7%, and 8% annual interest. The three investments yield a total of $2,037 annual interest. Find the total amount of money invested.

82. MIXING COFFEES To make a mixture of 80 pounds of coffee worth $272, a grocer mixes coffee worth $3.25 a pound with coffee worth $3.85 a pound. How many pounds of cheaper coffee should the grocer use?

83. Solve: $3x = 2x$.

84. Find the area of a rectangle that has a perimeter of 32 centimeters and a length of 12 centimeters.

2.5 Introduction to Functions

- Representing functions • Function notation • The graph of a function
- Finding the domain and range of a function • The vertical line test • Linear functions

In Chapter 1, we examined the rental hall agreement shown below. We determined that the cost C of renting the hall depended on the number of hours h the hall was to be rented. We described this dependence in four ways: using words, using an equation, using a table, and using a graph. See the next page.

In this example, a specific rule assigns to each number of hours h the hall is rented a unique number C, the cost to rent the hall. Correspondences of this type are called **functions.** In this section, we introduce the concept of function and discuss a special notation that is used when working with functions.

Rental Agreement
ROYAL VISTA BANQUET HALL
Wedding Receptions•Dances•Reunions•Fashion Shows

Rented To_____Date_____
Lessee's Address_____

Rental Charges
• $100 per hour
• Nonrefundable $200 cleanup fee

Terms and conditions
Lessor leases the undersigned lessee the above described property upon the terms and conditions set forth on this page and on the back of this page. Lessee promises to pay rental cost stated herein.

Words

| The cost to rent the hall | is | 100 | times | the number of hours it is rented | plus | 200. |

Equation

$C = 100h + 200$

Table

h	C
1	300
2	400
3	500
4	600
5	700
6	800
7	900

Graph

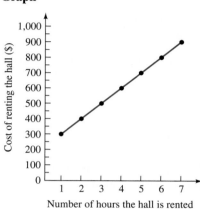

FIGURE 2-42

■ Representing functions

If x and y are real numbers, an equation in x and y determines a relationship between the values of x and y. To see how, we consider the equation $y = \frac{1}{2}x + 3$. To find the value of y (called an **output value**) that corresponds to $x = 4$ (called an **input value**), we multiply 4 by $\frac{1}{2}$ and then add 3.

$$y = \frac{1}{2}x + 3$$

$$y = \frac{1}{2}(4) + 3 \qquad \text{Substitute the input value of 4 for } x.$$

$$= 2 + 3$$

$$= 5 \qquad\qquad \text{The output value is 5.}$$

The ordered pair $(4, 5)$ satisfies the equation and shows that a y-value of 5 corresponds to an x-value of 4. This ordered pair and others that satisfy the equation appear in the table shown in Figure 2-43. The graph of the equation also appears in the figure.

$y = \frac{1}{2}x + 3$

x	y	
−2	2	2 corresponds to an x-value of −2.
0	3	3 corresponds to an x-value of 0.
2	4	4 corresponds to an x-value of 2.
4	5	5 corresponds to an x-value of 4.
6	6	6 corresponds to an x-value of 6.

We choose the inputs. ↑   We compute each output. ↑

FIGURE 2-43

To see how the table determines the relationship, we simply find an input in the x column and then read across to find the corresponding output in the y column. For example, if we select 2 as an input value, we get 4 as an output value. Thus, a y-value of 4 corresponds to an x-value of 2.

To see how the graph determines the relationship, we draw a vertical and a horizontal line through any point (say, point P) on the graph, as shown in Figure 2-43. Because these lines intersect the x-axis at 4 and the y-axis at 5, the point $P(4, 5)$ associates 5 on the y-axis with 4 on the x-axis. This shows that a y value of 5 is assigned to an x-value of 4.

When a relationship is described by an equation, by a table, using words, or with a graph, and only one y-value corresponds to each x-value, we call the relationship (or correspondence) a **function.** The set of input values x is called the **domain** of the function, and the set of output values y is called the **range.** Since the value of y usually depends on the number x, we call y the **dependent variable** and x the **independent variable.**

Functions

A **function** is a correspondence between a set of input values x (called the **domain**) and a set of output values y (called the **range**), where exactly one y-value in the range is assigned to each number x in the domain.

EXAMPLE 1   Does $y = 2x - 3$ define y to be a function of x? If so, illustrate the function with a table and a graph.

Solution   For $y = 2x - 3$ to define a function, every input number x must determine one output value of y. To find y in the equation $y = 2x - 3$, we multiply x by 2 and then subtract 3. Since this arithmetic gives one result, each choice of x determines one value of y. Thus, the equation defines y to be a function of x.

A table of values and the graph appear in Figure 2-44.

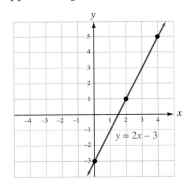

$y = 2x - 3$	
x	y
-4	-11
-2	-7
0	-3
2	1
4	5
6	9

FIGURE 2-44

Self Check 1
Does $y = -2x + 3$ define y to be a function of x?

Answer yes

As you will see in the next example, not every equation in two variables defines a function.

EXAMPLE 2   Does $y^2 = x$ define y to be a function of x?

Solution   For a function to exist, each input x must determine one output y. If we let $x = 16$, for example, y could be either 4 or -4, because $4^2 = 16$ and $(-4)^2 = 16$. Since more than one value of y is determined when $x = 16$, the equation does not represent a function.

Self Check 2
Does $|y| = x$ define y to be a function of x? Explain your answer.
Answer No; if $x = 2$, y could be either 2 or -2.

Function notation

There is a special notation that we will use to denote functions.

> **Function notation**
>
> The notation $y = f(x)$ denotes that the variable y is a function of x.

The notation $y = f(x)$ is read as "y equals f of x." Note that y and $f(x)$ are two notations for the same quantity. Thus, the equations $y = 4x + 3$ and $f(x) = 4x + 3$ are equivalent. We read $f(x) = 4x + 3$ as "f of x is equal to $4x$ plus 3."

This is the variable used to represent input values.
$$f(x) = 4x + 3$$
This is the name of the function.

This expression shows how to obtain an output value from a given input value.

! COMMENT The notation $f(x)$ does not mean "f times x."

The notation $y = f(x)$ provides a way of denoting the value of y (the dependent variable) that corresponds to some number x (the independent variable). For example, if $y = f(x)$, the value of y that is determined by $x = 3$ is denoted by $f(3)$.

INTERMEDIATE
Algebra ⓕ⁽ˣ⁾ Now™

Self Check 3
If $f(x) = -2x - 1$, find:
a. $f(2)$
b. $f(-3)$
c. $f(-t)$

EXAMPLE 3 Let $f(x) = 4x + 3$. Find **a.** $f(3)$, **b.** $f(-1)$, **c.** $f(0)$, and **d.** $f(r + 1)$.

Solution In each case, we will substitute the value for x within the parentheses of the function notation and in the algebraic expression $4x + 3$. Then we evaluate the expression.

a. To find $f(3)$, we replace x with 3:

$$f(x) = 4x + 3$$
$$f(3) = 4(3) + 3$$
$$= 12 + 3$$
$$= 15$$

b. To find $f(-1)$, we replace x with -1:

$$f(x) = 4x + 3$$
$$f(-1) = 4(-1) + 3$$
$$= -4 + 3$$
$$= -1$$

c. To find $f(0)$, we replace x with 0:

$$f(x) = 4x + 3$$
$$f(0) = 4(0) + 3$$
$$= 3$$

d. To find $f(r + 1)$, we replace x with $r + 1$:

$$f(x) = 4x + 3$$
$$f(r + 1) = 4(r + 1) + 3$$
$$= 4r + 4 + 3$$
$$= 4r + 7$$

Answers a. -5, **b.** 5,
c. $2t - 1$

To see why function notation is helpful, we consider the following equivalent sentences:

1. For the equation $y = 4x + 3$, find the value of y when x is 3.
2. For the function $f(x) = 4x + 3$, find $f(3)$.

Statement 2, which uses $f(x)$ notation, is much more concise.

EXAMPLE 8 **Manicurists.** A recent graduate of a cosmetology school rents a station from the owner of a beauty salon for $18 a day. She expects to make $12 profit from each customer she serves. Write a linear function describing her daily income if she serves *c* customers per day. Then graph the function.

Solution The manicurist makes a profit of $12 per customer, so if she serves *c* customers a day, she will make $12*c*. To find her income, we must *subtract* the $18 rental fee she pays from the profit. Therefore, the income function is $I(c) = 12c - 18$.

The graph of this linear function, shown in Figure 2-51, is a line with slope 12 and intercept $(0, -18)$. Since the manicurist cannot have a negative number of customers, we do not extend the line into quadrant III.

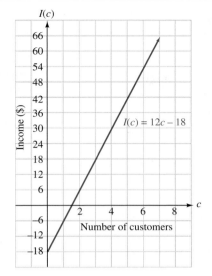

FIGURE 2-51

Evaluating functions

CALCULATOR SNAPSHOT

We can use a graphing calculator to find the value of a function for different input values. For example, to find the income earned by the manicurist in Example 8 for different numbers of customers, we first graph the income function $I(c) = 12c - 18$ as $y = 12x - 18$, using window settings of $[0, 10]$ for x and $[0, 100]$ for y. To find her income when she serves seven customers, we trace and move the cursor until the x-coordinate on the screen is nearly 7, as in Figure 2-52(a). From the screen, we can read that her income is about $66.25.

We can also use the *table* feature to evaluate a function. For example, after entering $Y_1 = 12x - 18$ to represent $I(c) = 12c - 18$, and after adjusting the *table set* (TBLSET), we press $\boxed{2\text{nd}}$ $\boxed{\text{TABLE}}$ to get the display shown in Figure 2-52(b). The next-to-last line of the table indicates that $I(6) = 54$. That is, if she serves 6 customers, her income will be $54.

With some graphing calculator models, we can evaluate a function by entering function notation. For example, to find the income earned by the manicurist if she serves 15 customers, we can use the following steps on a TI-83 Plus graphing calculator.

With $I(c) = 12c - 18$ entered as $Y_1 = 12x - 18$, we call up the home screen by pressing $\boxed{2\text{nd}}$ $\boxed{\text{QUIT}}$. Then we enter $\boxed{\text{VARS}}$ $\boxed{\blacktriangleright}$ $\boxed{1}$ $\boxed{\text{ENTER}}$. The symbolism Y_1 will be displayed. Next, we enter the input value 15 within parentheses and press $\boxed{\text{ENTER}}$. In Figure 2-52(c), we see that $Y_1(15) = 162$. That is, $I(15) = 162$. The manicurist will earn $162 if she serves 15 customers in one day.

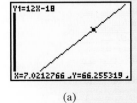
(a)

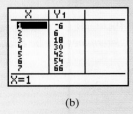
(b)

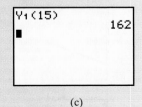
(c)

FIGURE 2-52

Section 2.5 STUDY SET

VOCABULARY   *Fill in the blanks.*

1. A _____ is a correspondence between a set of input values and a set of output values, where each input value determines _____ output value.

2. When graphing a function, input values are associated with the _____ axis and output values with the _____ axis.

3. For a function, the set of all output values is called the _____ of the function.

4. For a function, the set of all inputs is called the _____ of the function.

5. For $f(x) = 2x - 9$, the independent variable is ▢.

6. The _____ line test can be used to tell whether the graph of an equation determines a function.

7. For $y = -x + 3$, the dependent variable is ▢.

8. Any substitution for x in $f(x) = 5x - 4$ is called an _____ value.

CONCEPTS

9. Fill in the blank so that the following statements ask for the same thing.

  1. For the equation $y = -5x + 1$, find the value of y when $x = -1$.

  2. For the function $f(x) = -5x + 1$, find ▢.

10. For the function $f(x) = \dfrac{1}{x + 4}$, why isn't -4 in the domain of f?

11. Consider the graph of the function on the right.

  a. Label each arrow in the illustration with the appropriate term: *domain* or *range*.

  b. Give the domain and range.

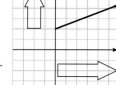

12. Use the graph of function f on the right to find each of the following.

  a. $f(-2)$

  b. $f(0)$

  c. $f(1)$

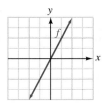

NOTATION   *Complete each solution.*

13. If $f(x) = x^2 - 3x$, find $f(-5)$.

$$f(x) = x^2 - 3x$$
$$f(-5) = (-5)^2 - 3()$$
$$= + 15$$
$$= 40$$

14. If $g(x) = \dfrac{2 - x}{6}$, find $g(8)$.

$$g(x) = \dfrac{2 - x}{6}$$
$$g(8) = \dfrac{2 - }{6}$$
$$= \dfrac{}{6}$$
$$= -1$$

15. Complete the sentence: $f(5) = 6$ is read "f _____ 5 is 6."

16. 📟 The illustration below shows a table of values generated by a graphing calculator for a function f. Find the following.

  a. $f(-1)$ **b.** $f(3)$

X	Y1
-2	10
-1	1
0	-4
1	-5
2	-2
3	5
4	16

X= -2

PRACTICE   *Determine whether the equation determines y to be a function of x.*

17. $y = 2x + 3$ **18.** $y = 4x - 1$

19. $y = 2x^2$ **20.** $y^2 = x + 1$

21. $y = 3 + 7x^2$ **22.** $y^2 = 3 - 2x$

23. $x = |y|$ **24.** $y = |x|$

Find $f(3)$ and $f(-1)$.

25. $f(x) = 3x$

26. $f(x) = -4x$

27. $f(x) = 2x - 3$

28. $f(x) = 3x - 5$

29. $f(x) = 7 + 5x$

30. $f(x) = 3 + 3x$

31. $f(x) = 9 - 2x$

32. $f(x) = 12 + 3x$

Find $g(2)$ and $g(3)$.

33. $g(x) = x^2$

34. $g(x) = x^2 - 2$

35. $g(x) = x^3 - 1$

36. $g(x) = x^3$

37. $g(x) = (x + 1)^2$

38. $g(x) = (x - 3)^2$

39. $g(x) = 2x^2 - x$

40. $g(x) = 5x^2 + 2x$

Find $h(2)$ and $h(-2)$.

41. $h(x) = |x| + 2$

42. $h(x) = |x| - 5$

43. $h(x) = x^2 - 2$

44. $h(x) = x^2 + 3$

45. $h(x) = \dfrac{1}{x + 3}$

46. $h(x) = \dfrac{3}{x - 4}$

47. $h(x) = \dfrac{x}{x - 3}$

48. $h(x) = \dfrac{x}{x^2 + 2}$

Complete each table.

49. $f(t) = |t - 2|$

t	f(t)
−1.7	
0.9	
5.4	

50. $f(r) = -2r^2 + 1$

Input	Output
−1.7	
0.9	
5.4	

51. $g(x) = x^3$

Input	Output
$-\frac{3}{4}$	
$\frac{1}{6}$	
$\frac{5}{2}$	

52. $g(x) = 2\left(-x - \dfrac{1}{4}\right)$

x	g(x)
$-\frac{3}{4}$	
$\frac{1}{8}$	
$\frac{5}{2}$	

Find $g(w)$ and $g(w + 1)$.

53. $g(x) = 2x$

54. $g(x) = -3x$

55. $g(x) = 3x - 5$

56. $g(x) = 2x - 7$

Find the domain of each function.

57. $g(x) = \dfrac{x}{6 - x}$

58. $H(x) = \dfrac{4}{3 - 2x}$

59. $s(x) = |x - 7|$

60. $t(x) = \left|\dfrac{2x}{3} + 1\right|$

61. $f(x) = x^2$

62. $g(x) = x^3$

63. $s(x) = 3x + 6$

64. $h(x) = \dfrac{4}{5}x - 8$

Find the domain and range of each function.

65. $\{(-2, 3), (4, 5), (6, 7)\}$

66. $\{(0, 2), (1, 2), (3, 4)\}$

67. $f(x) = \dfrac{1}{x - 4}$

68. $f(x) = \dfrac{5}{x + 1}$

Use the vertical line test to determine whether the given graph represents a function.

69.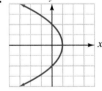

70.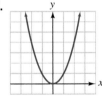

71.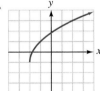

72.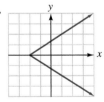

73.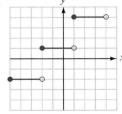

74.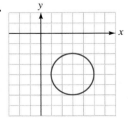

75. 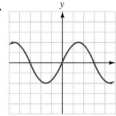   **76.**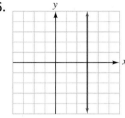

Graph each function and find the domain and range.

77. $f(x) = 2x - 1$ **78.** $f(x) = -x + 2$

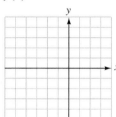   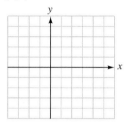

79. $f(x) = \dfrac{2}{3}x - 2$ **80.** $f(x) = -\dfrac{3}{2}x - 3$

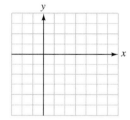   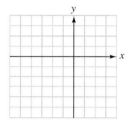

Determine whether each equation defines a linear function.

81. $y = 3x^2 + 2$ **82.** $y = \dfrac{x - 3}{2}$

83. $y = x$ **84.** $y = 3x^3 - 4$

APPLICATIONS

85. DECONGESTANTS The temperature in degrees Celsius that is equivalent to a temperature in degrees Fahrenheit is given by the linear function $C(F) = \frac{5}{9}(F - 32)$. Use this function to find the temperature range, in degrees Celsius, at which a bottle of Dimetapp should be stored. The label directions are shown below.

> **DIRECTIONS:** Adults and children 12 years of age and over: Two teaspoons every 4 hours. DO NOT EXCEED 6 DOSES IN A 24-HOUR PERIOD. Store at a controlled room temperature between 68°F and 77°F.

86. BODY TEMPERATURES The temperature in degrees Fahrenheit that is equivalent to a temperature in degrees Celsius is given by the linear function $F(C) = \frac{9}{5}C + 32$. Convert each of the temperatures in the following excerpt from *The Good Housekeeping Family Health and Medical Guide* to degrees Fahrenheit. (Round to the nearest degree.)

> In disease, the temperature of the human body may vary from about 32.2°C to 43.3°C for a time, but there is grave danger to life should it drop and remain below 35°C or rise and remain at or above 41°C.

87. CONCESSIONAIRES A baseball club pays a peanut vendor $50 per game for selling bags of peanuts for $1.75 each.

  a. Write a linear function that describes the income the vendor makes for the baseball club during a game if she sells b bags of peanuts.

  b. Find the income the baseball club will make if the vendor sells 110 bags of peanuts during a game.

88. NEW HOME CONSTRUCTION In a proposal to some prospective clients, a housing contractor listed the following costs.

  - Fees, permits, miscellaneous $12,000
  - Construction, per square foot $75

  a. Write a linear function that the clients could use to determine the cost of building a home having f square feet.

  b. Find the cost to build a home having 1,950 square feet.

89. EARTH'S ATMOSPHERE The illustration shows a graph of the temperatures of the atmosphere at various altitudes above Earth's surface. The temperature is expressed using the Kelvin scale, which is used in scientific work.

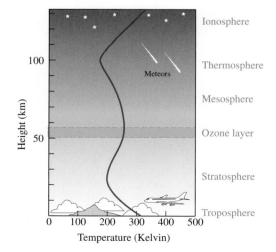

a. Estimate the coordinates of three points on the graph that have an *x*-coordinate of 200.

b. Explain why this is not the graph of a function.

90. CHEMICAL REACTIONS When students in a chemistry laboratory mixed solutions of acetone and chloroform, they found that heat was immediately generated. As time went by, the mixture cooled down. The illustration below shows a graph of data points of the form (time, temperature) taken by the students during the experiment.

a. The linear function $T(t) = -\frac{t}{240} + 30$ models the relationship between the elapsed time *t* since the solutions were combined and the temperature $T(t)$ of the mixture. Graph the function in the illustration.

b. Predict the temperature of the mixture immediately after the two solutions are combined.

c. Is $T(180)$ more or less than the temperature recorded by the students for $t = 300$ on the graph?

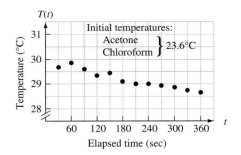

91. INCOME TAXES The function

$$T(a) = 700 + 0.15(a - 7,000)$$

(where *a* is adjusted gross income) is a model of the instructions given on the first line of the tax rate Schedule X shown.

a. Find $T(25,000)$ and interpret the result.

Schedule X–Use if your filing status is **Single**			2003
If your adjusted gross income is: *Over —*	*But not over —*	Your tax is	*of the amount over —*
$ 7,000	$28,400	$ 700 + 15%	$ 7,000
$28,400	$68,800	$3,910 + 25%	$28,400

b. Write a function that models the second line on Schedule X.

92. COST FUNCTIONS An electronics firm manufactures tape recorders, receiving $120 for each recorder it makes. If *x* represents the number of recorders produced, the income received is determined by the *revenue function* $R(x) = 120x$. The manufacturer has fixed costs of $12,000 per month and variable costs of $57.50 for each recorder manufactured. Thus, the *cost function* is $C(x) = 57.50x + 12,000$. How many recorders must the company sell for revenue to equal cost? (*Hint:* Set $R(x) = C(x)$.)

93. BALLISTICS The height of a toy rocket shot from the ground straight upward is given by the function $f(t) = -16t^2 + 256t$.

a. Find the height of the rocket 3 seconds after it is shot.

b. Find $f(16)$. Interpret the result.

94. PLATFORM DIVING
The number of feet a diver is above the surface of the water is given by the function $h(t) = -16t^2 + 16t + 32$, where *t* is the elapsed time in seconds after the diver jumped. Find the height of the diver for the times shown in the table.

t	$h(t)$
0	
0.5	
1.5	
2.0	

WRITING

95. Give four ways in which a function can be described. Which do you think is the most informative? Why?

96. Explain why we can think of a function as a machine.

REVIEW *Show that each number is a rational number by expressing it as a ratio of two integers.*

97. $-3\frac{3}{4}$

98. 4.7

99. 0.333. . .

100. $-0.\overline{6}$

2.6 Graphs of Functions

- Graphs of nonlinear functions   •   Translations of graphs
- Reflections of graphs   •   Solving equations graphically

Many real-world situations can be modeled by linear functions. When these functions are graphed, we can learn information by examining the resulting line. In this section, we will discuss three more types of functions. Like linear functions, they can be used as mathematical models. But unlike linear functions, their graphs are not straight lines. Therefore, they are called **nonlinear functions.**

Graphs of nonlinear functions

The first nonlinear function we will discuss is $f(x) = x^2$, called the **squaring function.**

INTERMEDIATE
Algebra $f(x)$ **Now™**

Self Check 1
Graph $g(x) = x^2 - 2$. Find the domain and the range. Compare the graph to the graph of $f(x) = x^2$.

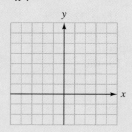

Answers D: the set of real numbers, R: the set of all real numbers greater than or equal to -2; the graph has the same shape but is 2 units lower

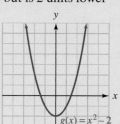

EXAMPLE 1   **The squaring function.**   Graph the function $f(x) = x^2$ and find the domain and range.

Solution   We substitute values for x in the equation and compute the corresponding values of $f(x)$. For example, if $x = -3$, we have

$$f(x) = x^2$$
$$f(-3) = (-3)^2 \quad \text{Substitute } -3 \text{ for } x.$$
$$= 9$$

Since $f(-3) = 9$, the ordered pair $(-3, 9)$ lies on the graph of f. In a similar manner, we find the corresponding values of $f(x)$ for other x-values and list the ordered pairs in the table of values. Then we plot the points and draw a smooth curve through them to get the graph, called a **parabola.**

$f(x) = x^2$		
x	**f(x)**	**(x, f(x))**
-3	9	$(-3, 9)$
-2	4	$(-2, 4)$
-1	1	$(-1, 1)$
0	0	$(0, 0)$
1	1	$(1, 1)$
2	4	$(2, 4)$
3	9	$(3, 9)$

Choose values for x. ⬉   ↑ Compute each $f(x)$.

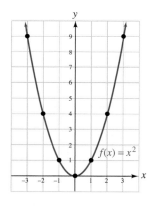

FIGURE 2-53

From the graph, we can see that x can be any real number. This means that the domain of the squaring function is the set of real numbers. We can also see that y is always positive or zero. This means that the range is the set of nonnegative real numbers.

EXAMPLE 2   The cubing function.   Graph the function $f(x) = x^3$ and find the domain and range.

Solution   We substitute values for x in the equation and compute the corresponding values of $f(x)$. For example, if $x = -2$, we have

$$f(x) = x^3$$
$$f(-2) = (-2)^3 \quad \text{Substitute } -2 \text{ for } x.$$
$$= -8$$

Since $f(-2) = -8$, the ordered pair $(-2, -8)$ lies on the graph of f. In a similar manner, we find the corresponding values of $f(x)$ for other x-values and list the ordered pairs in the table. Then we plot the points and draw a smooth curve through them to get the graph.

Self Check 2
Graph $g(x) = x^3 + 1$. Find the domain and the range. Compare the graph to the graph of $f(x) = x^3$.

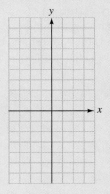

$f(x) = x^3$		
x	$f(x)$	$(x, f(x))$
-2	-8	$(-2, -8)$
-1	-1	$(-1, -1)$
0	0	$(0, 0)$
1	1	$(1, 1)$
2	8	$(2, 8)$

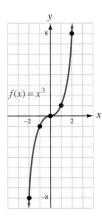

FIGURE 2-54

Answers D: the set of real numbers, R: the set of real numbers; the graph has the same shape but is 1 unit higher

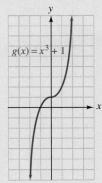

From the graph, we can see that x can be any real number. This means that the domain of the cubing function is the set of real numbers. We can also see that y can be any real number. This means that the range is the set of real numbers.

A third nonlinear function is $f(x) = |x|$, called the **absolute value function.**

EXAMPLE 3   The absolute value function.   Graph the function $f(x) = |x|$ and find the domain and range.

Solution   We substitute values for x in the equation and compute the corresponding values of $f(x)$. For example, if $x = -3$, we have

$$f(x) = |x|$$
$$f(-3) = |-3| \quad \text{Substitute } -3 \text{ for } x.$$
$$= 3$$

Since $f(-3) = 3$, the ordered pair $(-3, 3)$ lies on the graph of f. In a similar manner, we find the corresponding values of $f(x)$ for other x-values and list the ordered pairs in the table on the next page. Then we plot the points and connect them to get the graph.

Self Check 3
Graph $g(x) = |x - 2|$. Find the domain and the range. Compare the graph to the graph of $f(x) = |x|$.

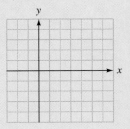

Answers D: the set of real numbers, R: the set of nonnegative real numbers; the graph has the same shape but is 2 units to the right

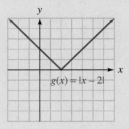

$f(x) = \lvert x \rvert$		
x	$f(x)$	$(x, f(x))$
-3	3	$(-3, 3)$
-2	2	$(-2, 2)$
-1	1	$(-1, 1)$
0	0	$(0, 0)$
1	1	$(1, 1)$
2	2	$(2, 2)$
3	3	$(3, 3)$

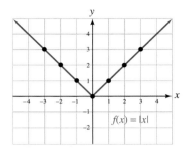

FIGURE 2-55

From the graph, we can see that x can be any real number. So the domain of the absolute value function is the set of real numbers. We can also see that y is always positive or zero. This means that the range is the set of nonnegative real numbers.

CALCULATOR SNAPSHOT   Graphing functions

We can graph nonlinear functions with a graphing calculator. For example, to graph $f(x) = x^2$ in a standard window of $[-10, 10]$ for x and $[-10, 10]$ for y, we press $\boxed{Y=}$, and then enter the function by typing x and then $\boxed{x^2}$. Finally, we press the $\boxed{\text{GRAPH}}$ key to obtain the graph shown in Figure 2-56(a).

To graph $f(x) = x^3$, we enter the function by typing x^3 and then press the $\boxed{\text{GRAPH}}$ key to obtain the graph shown in Figure 2-56(b). To graph $f(x) = \lvert x \rvert$, we enter the function by selecting abs from the NUM option within the MATH menu, typing x, and pressing the $\boxed{\text{GRAPH}}$ key to obtain the graph shown in Figure 2-56(c).

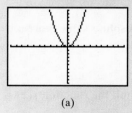   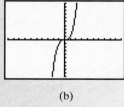   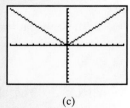

(a)   (b)   (c)

FIGURE 2-56

When using a graphing calculator, we must be sure that the viewing window does not show a misleading graph. For example, if we graph $f(x) = \lvert x \rvert$ in the window $[0, 10]$ for x and $[0, 10]$ for y, we will obtain a misleading graph that looks like a line. See Figure 2-57. This is not correct. The proper graph is the V-shaped graph shown in Figure 2-56(c). One of the challenges of using graphing calculators is finding an appropriate viewing window.

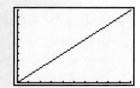

FIGURE 2-57

■ Translations of graphs

Examples 1, 2, and 3 and their Self Checks suggest that the graphs of different functions may be identical except for their positions in the *xy*-plane. For example, Figure 2-58 shows the graph of $f(x) = x^2 + k$ for three different values of *k*. If $k = 0$, we get the graph of $f(x) = x^2$. If $k = 3$, we get the graph of $f(x) = x^2 + 3$, which is identical to the graph of $f(x) = x^2$ except that it is shifted 3 units upward. If $k = -4$, we get the graph of $f(x) = x^2 - 4$, which is identical to the graph of $f(x) = x^2$ except that it is shifted 4 units downward. These shifts are called **vertical translations.**

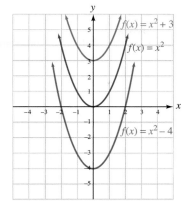

FIGURE 2-58

In general, we can make these observations.

Vertical translations

If *f* is a function and *k* represents a positive number, then

- The graph of $y = f(x) + k$ is identical to the graph of $y = f(x)$ except that it is translated *k* units upward.
- The graph of $y = f(x) - k$ is identical to the graph of $y = f(x)$ except that it is translated *k* units downward.

INTERMEDIATE
Algebra *(f(x))* **Now**™

EXAMPLE 4   Graph: $g(x) = |x| + 2$.

Self Check 4
Graph: $g(x) = |x| - 3$.

Solution   The graph of $g(x) = |x| + 2$ will be the same V-shaped graph as $f(x) = |x|$, except that it is shifted 2 units up. The graph appears in Figure 2-59.

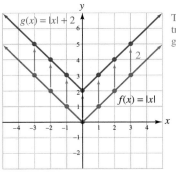

To graph $g(x) = |x| + 2$, translate each point on the graph of $f(x) = |x|$ up 2 units.

FIGURE 2-59

Answer

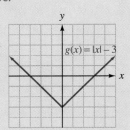

Figure 2-60 shows the graph of $f(x) = (x + h)^2$ for three different values of h. If $h = 0$, we get the graph of $f(x) = x^2$. The graph of $f(x) = (x - 3)^2$ is identical to the graph of $f(x) = x^2$ except that it is shifted 3 units to the right. The graph of $f(x) = (x + 2)^2$ is identical to the graph of $f(x) = x^2$ except that it is shifted 2 units to the left. These shifts are called **horizontal translations.**

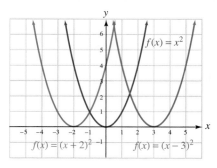

FIGURE 2-60

In general, we can make these observations.

Horizontal translations

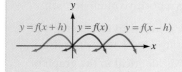

If f is a function and h is a positive number, then

- The graph of $y = f(x - h)$ is identical to the graph of $y = f(x)$ except that it is translated h units to the right.

- The graph of $y = f(x + h)$ is identical to the graph of $y = f(x)$ except that it is translated h units to the left.

Self Check 5
Graph: $g(x) = (x - 2)^2$.

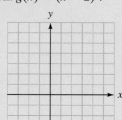

Answer

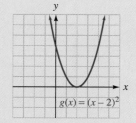

EXAMPLE 5   Graph: $g(x) = (x + 3)^3$.

Solution   The graph of $g(x) = (x + 3)^3$ will be the same shape as the graph of $f(x) = x^3$ except that it is shifted 3 units to the left. The graph appears in Figure 2-61.

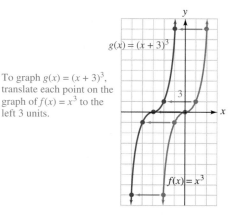

To graph $g(x) = (x + 3)^3$, translate each point on the graph of $f(x) = x^3$ to the left 3 units.

FIGURE 2-61

EXAMPLE 6 Graph: $g(x) = (x - 3)^2 + 2$.

Solution   In this example, two translations are made to a basic graph. We can graph this equation by translating the graph of $f(x) = x^2$ to the right 3 units and then 2 units up, as shown in Figure 2-62.

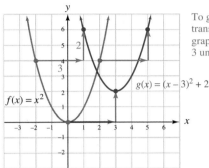

To graph $g(x) = (x - 3)^2 + 2$, translate each point on the graph of $f(x) = x^2$ to the right 3 units and then 2 units up

FIGURE 2-62

Self Check 6
Graph: $f(x) = |x + 2| - 3$.

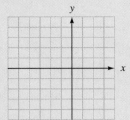

Answer

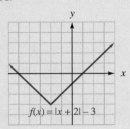

$f(x) = |x + 2| - 3$

Reflections of graphs

Figure 2-63 shows a table of values for $f(x) = x^2$ and for $g(x) = -x^2$. We note that for a given value of x, the corresponding y-values in the tables are opposites. When graphed, we see that the $-$ in $g(x) = -x^2$ has the effect of flipping the graph of $f(x) = x^2$ over the x-axis so that the parabola opens downward. We say that the graph of $g(x) = -x^2$ is a **reflection** of the graph of $f(x) = x^2$ about the x-axis.

	$f(x) = x^2$			$g(x) = -x^2$	
x	$f(x)$	$(x, f(x))$	x	$g(x)$	$(x, g(x))$
-2	4	$(-2, 4)$	-2	-4	$(-2, -4)$
-1	1	$(-1, 1)$	-1	-1	$(-1, -1)$
0	0	$(0, 0)$	0	0	$(0, 0)$
1	1	$(1, 1)$	1	-1	$(1, -1)$
2	4	$(2, 4)$	2	-4	$(2, -4)$

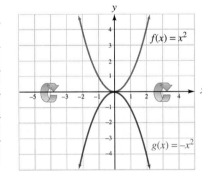

FIGURE 2-63

EXAMPLE 7 Graph: $g(x) = -x^3$.

Solution   To graph $g(x) = -x^3$, we use the graph of $f(x) = x^3$ from Example 2. First, we reflect the portion of the graph of $f(x) = x^3$ in quadrant I to quadrant IV, as shown in Figure 2-64 on the next page. Then we reflect the portion of the graph of $f(x) = x^3$ in quadrant III to quadrant II.

Self Check 7
Graph: $f(x) = -|x|$.

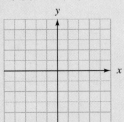

Answer

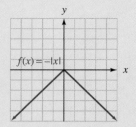

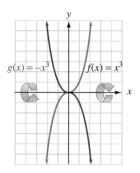

FIGURE 2-64

Reflection of a graph

The graph of $y = -f(x)$ is the graph of $y = f(x)$ reflected about the x-axis.

Solving equations graphically

Some of the graphing concepts discussed in this chapter can be used to solve equations. For example, the solution of $-2x - 4 = 0$ is the number x that will make y equal to 0 in the equation $y = -2x - 4$. To find this number, we inspect the graph of $y = -2x - 4$ and locate the point on the graph that has a y-coordinate of 0. In Figure 2-65, we see that the point is $(-2, 0)$, which is the x-intercept of the graph. We can conclude that the x-coordinate of the x-intercept, $x = -2$, is the solution of $-2x - 4 = 0$.

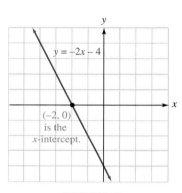

FIGURE 2-65

Check:
$$-2x - 4 = 0$$
$$-2(-2) - 4 \stackrel{?}{=} 0$$
$$4 - 4 \stackrel{?}{=} 0$$
$$0 = 0$$

Another approach that can be used to solve equations involves a graphical determination of the value of the variable that will make the sides of the equation equal.

Self Check 8
Solve $2x + 4 = 2$ graphically.

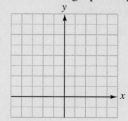

Answer −1

EXAMPLE 8   Solve $2x + 4 = -2$ graphically.

Solution   The graphs of $y = 2x + 4$ and $y = -2$ are shown in Figure 2-66. To solve the equation $2x + 4 = -2$, we need to find the value of x that makes $2x + 4$ equal -2. The point of intersection of the graphs is $(-3, -2)$. This tells us that if $x = -3$, the expression $2x + 4$ equals -2. So the solution of $2x + 4 = -2$ is $x = -3$.

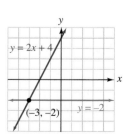

FIGURE 2-66

To solve more complex equations by graphing, we can use a graphing calculator.

Solving equations graphically

To solve $2(x - 3) - 6 = -6x$ with a graphing calculator, we add $6x$ to both sides so that the right-hand side of the equation is 0.

$$2(x - 3) - 6 = -6x$$
$$2(x - 3) - 6 + 6x = 0 \qquad \text{Add } 6x \text{ to both sides.}$$

Then we enter the left-hand side of the equation into a graphing calculator in the form $y = 2(x - 3) - 6 + 6x$, as shown in Figure 2-67(a). Note that it is not necessary to simplify the expression $2(x - 3) - 6 + 6x$.

Next, we enter window settings of $[-5, 5]$ for x and $[-5, 5]$ for y and select the ZERO feature found under the CALC menu. After we guess left and right bounds and press $\boxed{\text{ENTER}}$, the cursor automatically locates the x-intercept of the graph and displays its coordinates. (See Figure 2-67(b).) The x-coordinate of the x-intercept of the graph is called a **zero** of $y = 2(x - 3) - 6 + 6x$. The zero (in this case, 1.5) is the solution of $2(x - 3) - 6 = -6x$.

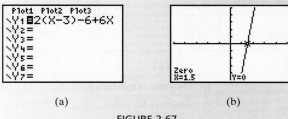

(a)   (b)

FIGURE 2-67

To solve the equation $2(x - 3) + 3 = 7$ by the method of Example 8 using a graphing calculator, we graph the left-hand side and the right-hand side of the equation in the same window by entering

$$Y_1 = 2(x - 3) + 3$$
$$Y_2 = 7$$

Figure 2-68(a) shows the graphs, generated using settings of $[-10, 10]$ for x and $[-10, 10]$ for y.

The coordinates of the point of intersection of the graphs can be determined using the INTERSECT feature found on most graphing calculators. With this feature, the cursor automatically highlights the intersection point, and the x- and y-coordinates are displayed. Consult your owner's manual for specific instructions concerning this feature.

In Figure 2-68(b), we see that the point of intersection is $(5,7)$, which indicates that 5 is a solution of $2(x - 3) + 3 = 7$.

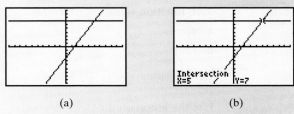

(a)   (b)

FIGURE 2-68

Section 2.6 STUDY SET

INTERMEDIATE
Algebra *f(x)* Now™

VOCABULARY *Fill in the blanks.*

1. The function $f(x) = x^2$ is called the _____ function.

2. The function $f(x) = x^3$ is called the _____ function.

3. The function $f(x) = |x|$ is called the _____ _____ function.

4. Functions whose graphs are not straight lines are called _____ functions.

5. Shifting the graph of an equation up or down is called a _____ translation.

6. Shifting the graph of an equation to the left or to the right is called a horizontal _____.

7. The graph of $f(x) = -x^2$ is a _____ of the graph of $f(x) = x^2$ about the x-axis.

8. The set of _____ real numbers is the set of real numbers greater than or equal to 0.

CONCEPTS *Fill in the blanks.*

9. The graph of $f(x) = (x + 4)^3$ is the same as the graph of $f(x) = x^3$ except that it is shifted ▨ units to the _____.

10. The graph of $f(x) = x^3 - 2$ is the same as the graph of $f(x) = x^3$ except that it is shifted ▨ units _____.

11. The graph of $f(x) = x^2 + 5$ is the same as the graph of $f(x) = x^2$ except that it is shifted ▨ units _____.

12. The graph of $f(x) = |x - 5|$ is the same as the graph of $f(x) = |x|$ except that it is shifted ▨ units to the _____.

13. Use the graphs on the right to solve each equation.

  a. $3x - 2 = 4$

  b. $3x - 2 = -2$

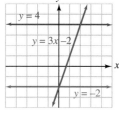

14. Use the graphs on the right to solve each equation.

  a. $-3x + 2 = x - 2$

  b. $-x - 4 = x - 2$

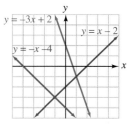

15. 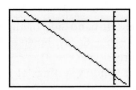 Use the graph of $y = 2(x - 6) - 4x$, shown below, to solve $2(x - 6) - 4x = 0$.

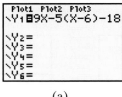

16. **a.** Use the information in the given calculator displays to solve $9x - 5(x - 6) - 18 = 0$.

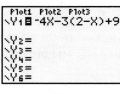   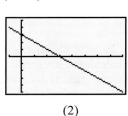

    (a)   (b)

  b. Use the information in the given calculator displays to solve $-4x - 3(2 - x) = -9$.

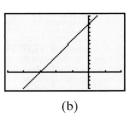   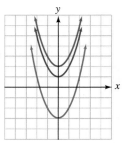

    (1)   (2)

17. The illustration on the right shows the graph of $f(x) = x^2 + k$, for three values of k. What are the three values?

18. Use the graph on the right to find each function value.

  a. $f(-2)$

  b. $f(0)$

  c. $f(1)$

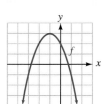

19. Use the graph on the right to find each function value.

  a. $g(-2)$

  b. $g(1)$

  c. $g(2.5)$

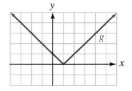

20. 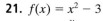 Use a graphing calculator to sketch the reflection of the graph shown below.

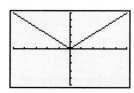

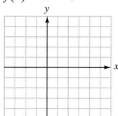 Graph each function using window settings of [−4, 4] for x and [−4, 4] for y. The graph is not what it appears to be. Pick a better viewing window and find a better representation of the true graph.

29. $f(x) = x^2 + 8$ **30.** $f(x) = x^3 - 8$

PRACTICE *Graph each function by plotting points. Give the domain and range.* **Check your work with a graphing calculator.**

31. $f(x) = |x + 5|$ **32.** $f(x) = |x - 5|$

21. $f(x) = x^2 - 3$ **22.** $f(x) = x^2 + 2$

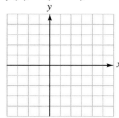   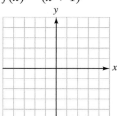

33. $f(x) = (x - 6)^2$ **34.** $f(x) = (x + 9)^2$

35. $f(x) = x^3 + 8$ **36.** $f(x) = x^3 - 12$

23. $f(x) = (x - 1)^3$ **24.** $f(x) = (x + 1)^3$

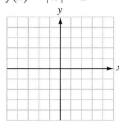   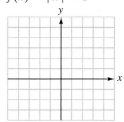

For each function, first sketch the graph of its associated function, $f(x) = x^2$, $f(x) = x^3$, or $f(x) = |x|$. Then draw each graph using a translation or a reflection.

37. $f(x) = x^2 - 5$ **38.** $f(x) = x^3 + 4$

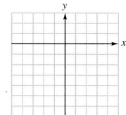   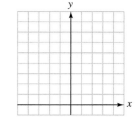

25. $f(x) = |x| - 2$ **26.** $f(x) = |x| + 1$

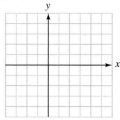   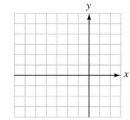

39. $f(x) = (x - 1)^3$ **40.** $f(x) = (x + 4)^2$

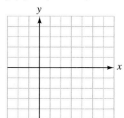   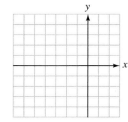

27. $f(x) = |x - 1|$ **28.** $f(x) = |x + 2|$

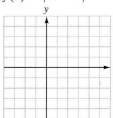   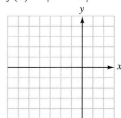

41. $f(x) = |x - 2| - 1$ **42.** $f(x) = (x + 2)^2 - 1$

43. $f(x) = (x + 1)^3 - 2$

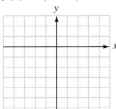

44. $f(x) = |x + 4| + 3$

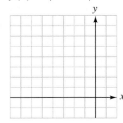

45. $f(x) = -x^3$

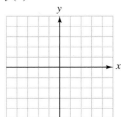

46. $f(x) = -|x|$

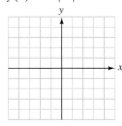

47. $f(x) = -x^2$

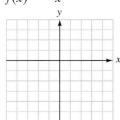

48. $f(x) = -(x + 1)^2$

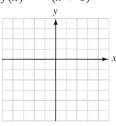

Use a graphing calculator to solve each equation.

49. $4(x - 1) = 3x$

50. $4(x - 3) - x = x - 6$

51. $11x + 6(3 - x) = 3$

52. $2(x + 2) = 2(1 - x) + 10$

APPLICATIONS

53. OPTICS The law of reflection states that the angle of reflection is equal to the angle of incidence. What function studied in this section mathematically models the path of the reflected light beam with an angle of incidence measuring 45°?

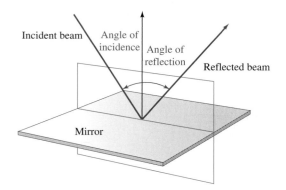

54. BILLIARDS In the illustration below, a rectangular coordinate system has been superimposed over a billiard table. Write a function that mathematically models the path of the ball that is shown banking off of the far cushion.

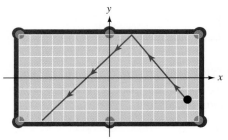

55. CENTER OF GRAVITY As a diver performs a $1\frac{1}{2}$-somersault in the tuck position, her center of gravity follows a path that can be described by a graph shape studied in this section. What graph shape is that?

56. FLASHLIGHTS Light beams coming from a bulb are reflected outward by a parabolic mirror as parallel rays.

   a. The cross-sectional view of a parabolic mirror is given by the function $f(x) = x^2$ for the following values of x: $-0.7, -0.6, -0.5, -0.4, -0.3, -0.2,$ $-0.1, 0, 0.1, 0.2, 0.3, 0.4, 0.5, 0.6, 0.7$. Sketch the parabolic mirror using the graph below.

   b. From the lightbulb filament at $(0, 0.25)$, draw a line segment representing a beam of light that strikes the mirror at $(-0.4, 0.16)$ and then reflects outward, parallel to the y-axis.

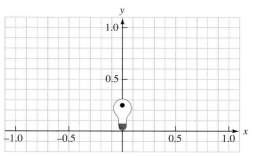

WRITING

57. Explain how to graph a function by plotting points.

58. Explain why the correct choice of window settings is important when using a graphing calculator.

59. Explain how to solve the equation $2x + 6 = 0$ graphically.

60. What does it mean when we say that the domain of a function is the set of all real numbers?

REVIEW *Solve each formula for the indicated variable.*

61. $T - W = ma$ for W

62. $a + (n - 1)d = \ell$ for n

63. $s = \dfrac{1}{2}gt^2 + vt$ for g

64. $e = mc^2$ for m

65. BUDGETING Last year, Rock Valley College had an operating budget of \$4.5 million. Due to salary increases and a new robotics program, the budget was increased by 20%. Find the operating budget for this year.

66. In the illustration, the line passing through points R, C, and S is parallel to line segment AB. Find the measure of $\angle ACB$. (Read $\angle ACB$ as "angle ACB." *Hint:* Recall from geometry that alternate interior angles have the same measure.)

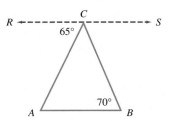

KEY CONCEPT

Functions

In Chapter 2, we introduced one of the most important concepts in mathematics, that of a **function.** Functions are used to describe relationships where one quantity depends on another.

1. Fill in the blanks to make the statements true.

   a. A function is a _____ between a set of _____ values x (called the domain) and a set of output values y (called the _____), where exactly _____ y-value in the range is assigned to each number x in the _____.

   b. Since the value of y usually depends on the number x, we call y the _____ variable and x the _____ variable.

2. We can think of a function as a machine. Using the words *input, output, domain,* and *range,* explain how the function machine shown below works.

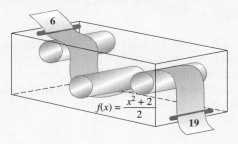

$$f(x) = \frac{x^2 + 2}{2}$$

Four Ways to Represent a Function

Functions can be described in words, with an equation, with a table, or with a graph.

3. x and y are related by the following correspondence: To find the value of y, double the value of x and then add 3. Find the value of y that corresponds to $x = -10$.

4. The area of a circle is the product of π and the radius squared. Use the variables A and r to describe this relationship with an equation.

5. Use the table below to determine the height of a projectile 1.5 seconds after it was shot vertically into the air.

Time t (seconds)	Height h (feet)
0	0
0.5	28
1.0	48
1.5	60
2.0	64

6. Use the graph below to determine what output value the function f assigns to an input of 1.

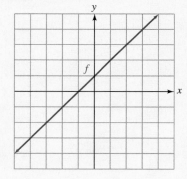

Function Notation

There is a special notation used with functions. The notation $y = f(x)$ denotes that the variable y is a function of x.

7. Write the equation in Problem 3 using function notation. Find $f(0)$.

8. Use function notation to represent the relationship described in Problem 4.

9. If the function $h(t) = -16t^2 + 64t$ gives the height of the projectile described in Problem 5, find $h(4)$ and interpret the result.

10. Use the graph in Problem 6 to determine $f(-3)$ and $f(3)$.

ACCENT ON TEAMWORK

SECTION 2.1

CITY MAP Get a map of the city in which you live and use a black marker to draw a rectangular coordinate system on the map. The size of the grid you use will depend on the size of the map. The origin of the coordinate system should be your place of residence. Determine the coordinates of ten important locations in your city, such as the library, fire stations, parks, schools, and city hall.

SECTION 2.2

GRAPHING LINES Draw rectangular coordinate systems with 1-inch grids on two pieces of white poster board. On the first poster board, graph $y = mx + 2$ for $m = 1, 2, 3, 4$, and 5. Label each line with its respective equation. At the bottom of the poster board, write your observations about how the graph changes for different values of m. On the second poster board, graph $y = 2x + b$ for $b = 1, 2, 3, 4$, and 5. At the bottom of the poster board, write your observations about how the graph changes for different values of b.

SECTION 2.3

MEASURING SLOPE Use a ruler and a level to find the slopes of five ramps or inclines by measuring $\frac{\text{rise}}{\text{run}}$, as shown. Record your results in a chart of the form shown below. List the examples in order, from that with the smallest slope to that with the largest slope.

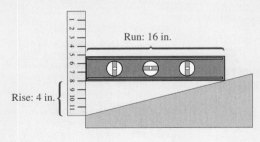

Object/location	Slope
Ramp outside cafeteria	$\dfrac{\text{Rise}}{\text{Run}} = \dfrac{4 \text{ in.}}{16 \text{ in.}} = \dfrac{1}{4}$

SECTION 2.4

A LINEAR MODEL For this project, you will need a 5 gallon pail, a yardstick, a watch that shows seconds, and access to a garden hose.

Tape the yardstick to the pail as shown below. Turn on the water, leaving it running at a constant rate, and begin to fill the pail. Keep track of the time in seconds it takes for the water in the pail to reach heights of 2, 4, 6, 8, 10, and 12 inches.

Draw a coordinate system, with the horizontal axis representing time in seconds and the vertical axis the height of the water in inches. Then plot your data as ordered pairs of the form (time, height). Draw the line that best fits the data and write the equation of the line. Use the equation to predict the height the column of water would be if it were to run for 24 hours, which is 86,400 seconds.

SECTION 2.5

LINEAR FUNCTIONS Think of a real-life situation like that given in Exercise 87 of Study Set 2.5 and write a linear function that describes it. Graph the function and explain the importance of the slope of the line and its y-intercept.

SECTION 2.6

GRAPHING FUNCTIONS Draw a rectangular coordinate system using a 1-inch grid on a piece of white poster board. Graph $y = k|x|$ for $k = \frac{1}{2}, 1, 2$, and 3. Label each graph with its respective equation. At the bottom of the poster board, write your observations about how the graph changes for different values of k.

PARABOLAS Borrow some calculus books from your mathematics instructor, or find them in your school library. Look up the word *parabola* in the index and find some real-life applications of parabolas mentioned in these books. Write a brief description of each use, and include a sketch of the application.

CHAPTER REVIEW

SECTION 2.1	*The Rectangular Coordinate System*

CONCEPTS

A *rectangular coordinate system* is formed by two intersecting perpendicular number lines called the *x-axis* and the *y-axis,* which divide the plane into four *quadrants.*

The process of locating a point in the coordinate plane is called *plotting* or *graphing* that point.

Data in a two-column table can be expressed as ordered pairs.

REVIEW EXERCISES

Plot each point on the rectangular coordinate system shown on the right.

1. $(0, 3)$

2. $(-2, -4)$

3. $\left(\frac{5}{2}, -1.75\right)$

4. the origin

5. $(2.5, 0)$

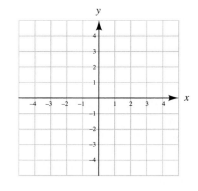

6. Use the graph on the right to complete the table.

x	y	(x, y)
−3		(−3,)
−2		(−2,)
	3	(, 3)
0		(0,)
1		(1,)
	0	(, 0)
	−5	(, −5)

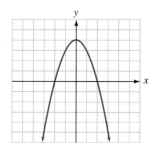

Graphs can be used to visualize relationships between two quantities.

The graph on the right shows how the height of the water in a flood control channel changed over a 7-day period.

7. Describe the height of the water at the beginning of day 2.

8. By how much did the water level increase or decrease from day 4 to day 5?

9. During what time period did the water level stay the same?

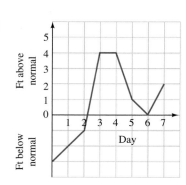

AUCTIONS The dollar increments used by an auctioneer during the bidding process depend on what initial price the auctioneer began with for the item. See the step graph.

10. What increments are used by the auctioneer if the bidding on an item began at $150?

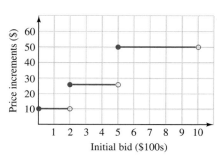

11. If the first bid on an item being auctioned is \$750, what will be the next price asked for by the auctioneer?

Midpoint formula: The midpoint of a line segment with endpoints at (x_1, y_1) and (x_2, y_2) is the point M with coordinates

$$\left(\frac{x_1 + x_2}{2}, \frac{y_1 + y_2}{2} \right)$$

12. Find the midpoint of the line segment joining $(8, -2)$ and $(-4, 6)$.

13. The midpoint of a line segment has coordinates of $\left(-\frac{1}{2}, \frac{1}{2}\right)$. If one endpoint has coordinates of $(-3, 7)$, what are the coordinates of the other endpoint?

| SECTION 2.2 | *Graphing Linear Equations* |

The *graph of an equation* is the graph of all points on the rectangular coordinate system whose coordinates satisfy the equation.

Graph each equation.

14. $y = 3x + 4$

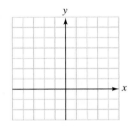

15. $y = -\frac{1}{3}x - 1$

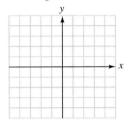

To find the *y-intercept* of a line, substitute 0 for x in the equation and solve for y. To find the *x-intercept* of a line, substitute 0 for y in the equation and solve for x.

Graph each equation using the intercept method.

16. $2x + y = 4$

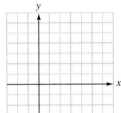

17. $3x - 4y - 8 = 0$

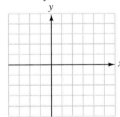

The graph of the equation $x = a$ is a *vertical line* with x-intercept at $(a, 0)$.

The graph of the equation $y = b$ is a *horizontal line* with y-intercept at $(0, b)$.

Graph each equation.

18. $y = 4$

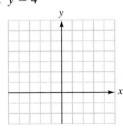

19. $x = -2$

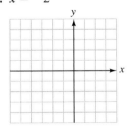

Complete the table of solutions for each equation.

20. $y = -3x$

x	y
−3	
0	
3	

21. $y = \frac{1}{2}x - \frac{5}{2}$

x	y
−3	
0	
3	

The *slope* of a nonvertical line is defined to be

$$m = \frac{\text{rise}}{\text{run}} = \frac{\Delta y}{\Delta x}$$

The *slope formula:* If $(x_2 \neq x_1)$,

$$m = \frac{y_2 - y_1}{x_2 - x_1}$$

The slope of a line gives the *average rate of change.*

22. Find the slope of lines l_1 and l_2 in the illustration.

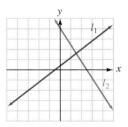

23. U.S. VEHICLE SALES On the graph in the illustration, draw a line through the points $(0, 21.2)$ and $(20, 48.6)$. Use this linear model to estimate the rate of increase in the market share of minivans, sport utility vehicles, and light trucks over the years 1980–2000.

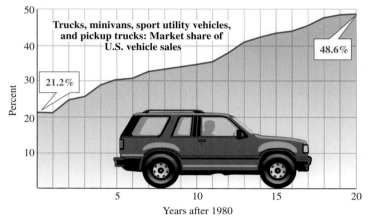

Source: American Automotive Association and U.S. Bureau of Economic Analysis

Horizontal lines have a slope of 0. Vertical lines have no defined slope.

Find the slope of the line passing through the given points.

24. $(2, 5)$ and $(5, 8)$

25. $(3, -2)$ and $(-6, 12)$

26. $(-2, 4)$ and $(8, 4)$

27. $(-5, -4)$ and $(-5, 8)$

Find the slope of the graph of each equation, if one exists.

28. $y = \frac{2}{3}x + 18$

29. $4x + 2y = 8$

30. $x = 10$

31. $y = 7$

Parallel lines have the same slope. The slopes of two nonvertical *perpendicular lines* are negative reciprocals.

Determine whether the lines with the given slopes are parallel, perpendicular, or neither.

32. $m_1 = 4, m_2 = -\frac{1}{4}$

33. $m_1 = 0.5, m_2 = \frac{1}{2}$

Equations of a line:
Point–slope form:

$$y - y_1 = m(x - x_1)$$

Write an equation of the line with the given properties. Express the result in slope–intercept form.

34. Slope of 3; passing through $(-8, 5)$

35. Passing through $(-2, 4)$ and $(6, -9)$

Slope–intercept form:

$$y = mx + b$$

General form:

$$Ax + By = C$$

Write an equation of the line with the given properties. Write the answer in general form.

36. Passing through $(-3, -5)$; parallel to the graph of $3x - 2y = 7$

37. Passing through $(-3, -5)$; perpendicular to the graph of $3x - 2y = 7$

38. Write $3x + 4y = -12$ in slope–intercept form. Give the slope and y-intercept of the graph of the equation. Then use this information to graph the line.

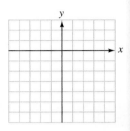

Many real-life situations can be modeled by linear equations.

39. DEPRECIATION A business purchased a copy machine for \$8,700 and will depreciate it on a straight-line basis over the next 5 years. At the end of its useful life, it will be sold as scrap for \$100. Find its depreciation equation.

SECTION 2.5   *Introduction to Functions*

A *function* is a correspondence between a set of input values x and a set of output values y, where exactly one value of y in the *range* is assigned to each number x in the *domain*.

Determine whether each equation determines y to be a function of x.

40. $y = 6x - 4$

41. $y = 4 - x^2$

42. $y^2 = x$

43. $|y| = x$

The notation $y = f(x)$ denotes that the variable y (the *dependent* variable) is a function of x (the *independent* variable).

Assume that $f(x) = 3x + 2$ *and* $g(x) = \dfrac{x^2 - 4x + 4}{2}$ *and find each value.*

44. $f(-3)$

45. $g(8)$

46. $g(-2)$

47. $f(t)$

The *domain* of a function is the set of input values. The *range* is the set of output values.

Find the domain and range of each function.

48. $f(x) = 4x - 1$

49. $f(x) = x^2 + 1$

50. $f(x) = \dfrac{4}{2 - x}$

51. $y = -|4x|$

The *vertical line test* can be used to determine whether a graph represents a function.

Use the vertical line test to determine whether each graph represents a function.

52.

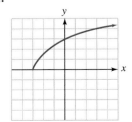

53.

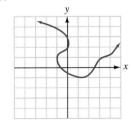

VIDEOS A video production company charges a $175 setup fee and $105 per hour to tape a wedding and reception.

54. Write a function that gives the cost to tape a wedding and reception that lasts x hours.

55. Use your answer to Exercise 54 a to find the cost to tape a wedding and reception lasting 6 hours.

Determine which are linear functions.

56. $f(x) = 3x + 2$

57. $f(x) = x^2 - 25$

Graphs of Functions

Graphs of *nonlinear functions* are not straight lines.

The *squaring function:*

$f(x) = x^2$

The *cubing function:*

$f(x) = x^3$

The *absolute value function:*

$f(x) = |x|$

A *horizontal translation* shifts a graph left or right. A *vertical translation* shifts a graph upward or downward. A *reflection* "flips" a graph about the x-axis.

Graph each function.

58. $f(x) = x^2 - 3$

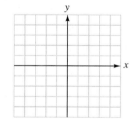

59. $f(x) = (x - 2)^3 + 1$

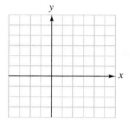

60. Graph $f(x) = |x + 2|$, $g(x) = |x - 2|$, and $h(x) = -|x|$ on the same coordinate system.

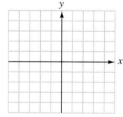

61. Use the graph to find each function value.

a. $f(-2)$ **b.** $f(3)$

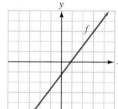

Use the graphs below to solve each equation.

62. $2(2 - x) + x = 0$

63. $2(2 - x) + x = 5$

64. $2(2 - x) + x = x$

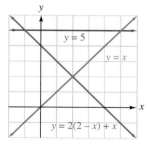

Refer to the graph below, which shows the height of an object at different times after it was shot straight up into the air.

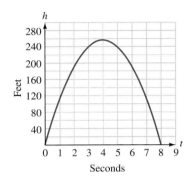

1. How high was the object 3 seconds into the flight?

2. At what times was the object about 110 feet above the ground?

3. What was the maximum height reached by the object?

4. How long did the flight take?

5. Find the coordinates of the midpoint of the line segment joining $(-2, -5)$ and $(7, 8)$.

6. Graph: $y = -\dfrac{2}{3}x - 2$.

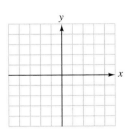

7. Find the x- and y-intercepts of the graph of $2x - 5y = 10$. Then graph the equation.

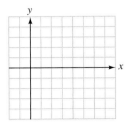

8. Graph: $y = -2$.

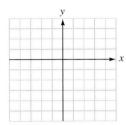

9. Find the slope of the line shown below.

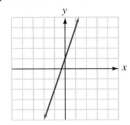

10. Find the rate of change of the temperature for the period of time shown in the graph below.

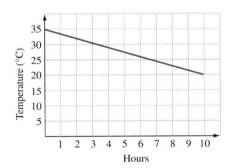

Find the slope of each line, if possible.

11. The line that passes through $(-2, 4)$ and $(6, 8)$

12. The graph of $2x - 3y = 8$

13. The graph of $x = 12$

14. The graph of $y = 12$

15. Write an equation of the line that has slope of $\dfrac{2}{3}$ and passes through $(4, -5)$. Give the answer in slope–intercept form.

16. Write an equation of the line that passes through $(-2, 6)$ and $(-4, -10)$. Give the answer in general form.

17. Find the slope and the y-intercept of the graph of $-2x + 6 = 6y + 15$.

18. Determine whether the graphs of $4x - y = 12$ and $y = \frac{1}{4}x + 3$ are parallel, perpendicular, or neither.

19. Write an equation of the line that passes through the origin and is parallel to the graph of $y = \frac{3}{2}x - 7$.

20. ACCOUNTING   After purchasing a new color copier, a business owner had his accountant prepare a depreciation worksheet for tax purposes. (See the illustration.)

   a. Assuming straight-line depreciation, write an equation that gives the value v of the copier after x years of use.

   b. If the depreciation equation is graphed, explain the significance of its v-intercept.

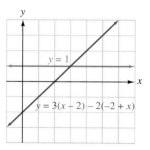

Depreciation Worksheet

Color copier $4,000
(new)

Salvage value $400
(in 6 years)

21. Does $|y| = x$ define y to be a function of x?

22. Does the table define y as a function of x?

x	y
−3	4
4	−3
1	4
2	5

23. Find the domain and range of the function $f(x) = |x|$.

24. Find the domain and range of the function $f(x) = x^3$.

Let $f(x) = 3x + 1$ and $g(x) = x^2 - 2x - 1$. Find each value.

25. $f(3)$

26. $g(0)$

27. $f\left(\dfrac{2}{3}\right)$

28. $g(r)$

Determine whether each graph represents a function.

29.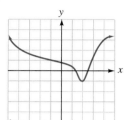

30.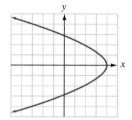

31. Graph: $f(x) = x^2 + 3$.   32. Graph: $g(x) = -|x + 2|$.

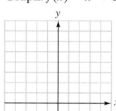   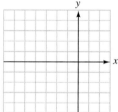

33. See the illustration below. Explain how to solve $3(x - 2) - 2(-2 + x) = 0$ graphically. What is the solution?

34. See the illustration below. Explain how to solve $3(x - 2) - 2(-2 + x) = 1$ graphically. What is the solution?

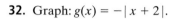

$y = 1$

$y = 3(x - 2) - 2(-2 + x)$

35. Describe four different ways to represent a function.

36. Explain why the graph of a circle does not represent a function.

List the elements of $\left\{-2, 0, 1, 2, \frac{13}{12}, 6, 7, \sqrt{5}, \pi\right\}$ that belong to the following set.

1. Natural numbers

2. Whole numbers

3. Rational numbers

4. Irrational numbers

5. Negative numbers

6. Real numbers

7. Prime numbers

8. Composite numbers

9. Even numbers

10. Odd numbers

Evaluate each expression.

11. $-|5| + |-3|$

12. $\dfrac{|-5| + |-3|}{-|4|}$

13. $2 + 4 \cdot 5$

14. $\dfrac{8 - 4}{2 - 4}$

15. $-\dfrac{16}{5} \div \left(-\dfrac{10}{3}\right)$

16. $\dfrac{(9 - 8)^4 + 21}{3^3 - (\sqrt{16})^2}$

Evaluate each expression for $x = 2$ and $y = -3$.

17. $-x - 2y$

18. $\dfrac{x^2 - y^2}{2x + y}$

Determine which property of real numbers justifies each statement.

19. $(a + b) + c = a + (b + c)$

20. $3(x + y) = 3x + 3y$

21. $(a + b) + c = c + (a + b)$

22. $(ab)c = a(bc)$

Simplify each expression.

23. $12y - 17y$

24. $-7s(-4t)(-1)$

25. $3x^2 + 2x^2 - 5x^2$

26. $-(4 + z) + 2z$

Solve each equation.

27. $2x - 5 = 11$

28. $\dfrac{2x - 6}{3} = x + 7$

29. $4(y - 3) + 4 = -3(y + 5)$

30. $2x - \dfrac{3(x - 2)}{2} = 7 - \dfrac{x - 3}{3}$

31. $-3 = -\dfrac{9}{8}s$

32. $0.04(24) + 0.02x = 0.04(12 + x)$

Solve each formula.

33. $S = \dfrac{n(a + l)}{2}$ for a

34. $A = \dfrac{1}{2}h(b_1 + b_2)$ for h

35. **INVESTMENTS** A woman invested part of $20,000 at 6% and the rest at 7%. If her annual interest is $1,260, how much did she invest at 6%?

36. **DRIVING RATES** John drove to a distant city in 5 hours. When he returned, there was less traffic, and the trip took only 3 hours. If he drove 26 mph faster on the return trip, how fast did he drive each way?

37. Graph: $2x - 3y = 6$.

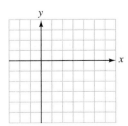

38. Find the slope of the line shown below.

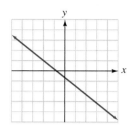

39. Write an equation of the line passing through $(-2, 5)$ and $(8, -9)$. Give the answer in slope–intercept form.

40. Write an equation of the line passing through $(-2, 3)$ and parallel to the graph of $3x + y = 8$. Give the answer in slope–intercept form.

Let $f(x) = 3x^2 + 2$ and $g(x) = -2x - 1$. Find each value.

41. $f(-1)$ **42.** $g(0)$

43. $g(-2)$ **44.** $f(-r)$

Graph each equation and determine whether it is a function. If it is a function, give the domain and range.

45. $y = -x^2 + 1$

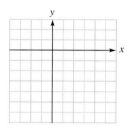

46. $y = |x - 3|$

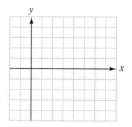

47. See the illustration below. Explain why there is not a linear relationship between the height of the antenna and its maximum range of reception.

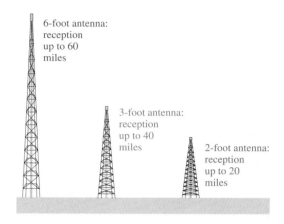

6-foot antenna: reception up to 60 miles

3-foot antenna: reception up to 40 miles

2-foot antenna: reception up to 20 miles

48. Explain how to use the following illustration to solve each equation. Give each solution.

  a. $-7 - 5(x - 1) + 4x = 0$

  b. $-7 - 5(x - 1) + 4x = -4$

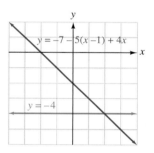

$y = -7 - 5(x - 1) + 4x$

$y = -4$

Systems of Equations

INTERMEDIATE
Algebra ⨍(x) Now™

Throughout the chapter, this icon introduces resources on the Intermediate AlgebraNow Web site, accessed through http://1pass .thomson.com, that will

- Help you test your knowledge of the material with a pre-test and a post-test

- Provide a personalized learning plan targeting areas you should study

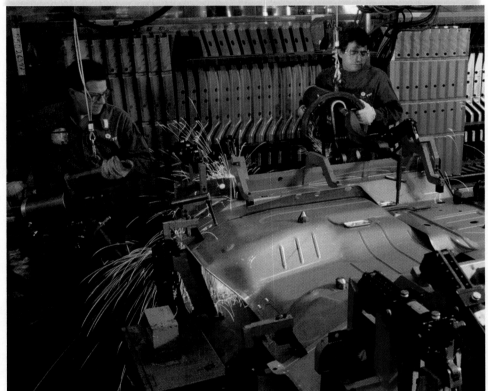

© Chuck Savage/Getty Images

Manufacturing companies regularly consider replacing old machinery with newer, more efficient models to increase productivity and expand profits. Financial decisions such as this can be made by writing and then graphing a system of equations to compare the costs associated with the purchase of new equipment to the costs of continuing with the equipment currently in use.

To learn more about solving systems of linear equations graphically, visit *The Learning Equation* on the Internet at http://tle.brookscole.com. (The log-in instructions are in the Preface.) For Chapter 3, the online lesson is:

- *TLE* Lesson 5: Solving Systems of Equations by Graphing

Check Your Knowledge

1. If a system of equations has one or more solutions, it is called a _____ system. A system of equations that has no solutions is called an _____ system.

2. If two equations have different graphs, they are called _____ equations. Two equations with the same graph are called _____ equations.

3. When solving a system of two linear equations by the graphing method, we look for the point of _____ of the two lines.

4. A _____ is a rectangular array of numbers.

5. _____ rule uses determinants to solve systems of linear equations.

6. Is $(5, 1)$ a solution of $\begin{cases} x + 2y = 7 \\ y = 2x - 4 \end{cases}$?

7. Solve $\begin{cases} x - y = 3 \\ 3x + y = 1 \end{cases}$ by graphing.

8. Use substitution to solve $\begin{cases} x + y = 5 \\ x - 2y = 2 \end{cases}$.

9. Use addition to solve $\begin{cases} 2x + 3y = 7 \\ 3x + 2y = 3 \end{cases}$.

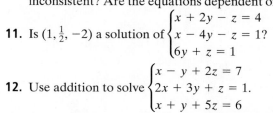

PROBLEM 7

10. Is the system of equations $\begin{cases} 4x - 2y = 6 \\ y = 2x - 3 \end{cases}$ consistent or inconsistent? Are the equations dependent or independent?

11. Is $(1, \frac{1}{2}, -2)$ a solution of $\begin{cases} x + 2y - z = 4 \\ x - 4y - z = 1 \\ 6y + z = 1 \end{cases}$?

12. Use addition to solve $\begin{cases} x - y + 2z = 7 \\ 2x + 3y + z = 1 \\ x + y + 5z = 6 \end{cases}$.

Write a system of equations to solve each problem.

13. Sandra invested part of $15,000 at 4% interest and the rest at 6%. If the amount of income after 1 year from these investments was $820, how much did Sandra invest at each rate?

14. Harry has available a 5% sugar solution and a 10% solution. How much of each should he combine to get 20 gallons of a 7% sugar solution?

Use matrices to solve each system of equations.

15. $\begin{cases} x - y = 5 \\ 2x + y = 1 \end{cases}$

16. $\begin{cases} x + y - z = 2 \\ 2x + y + z = -2 \\ x + 2y + 3z = 1 \end{cases}$

Evaluate each determinant.

17. $\begin{vmatrix} 1 & -1 \\ 2 & 5 \end{vmatrix}$

18. $\begin{vmatrix} 1 & 1 & 0 \\ 2 & 0 & -2 \\ 1 & 2 & 1 \end{vmatrix}$

19. Use Cramer's rule to solve the system. $\begin{cases} -x + y = 3 \\ 2x + y = 0 \end{cases}$.

20. Solve the following system for y only, using Cramer's rule:
$$\begin{cases} x + y + z = -2 \\ 2x + 3y - z = 1 \\ y - z = 1 \end{cases}$$

Study Skills Workshop

LEARNING STYLES AND PERSONAL STRENGTHS

The process of getting an education is more than a chance to learn new subject matter. It is also an opportunity to learn new things about yourself: how well you react to new situations, adapt to change, and accept responsibility. Some personal traits that will affect your learning are learning style, anxiety about math, attitude, and willingness to take responsibility for your learning. Identifying your unique learning style and personal strengths and weaknesses can help you adapt your study methods to give you the best chance of success in your courses.

Learning Style. Knowing the type of learner you are can help you determine the best study techniques for you. Visual-verbal learners learn best by reading and writing; a good strategy for them in studying is to rewrite notes and examples. Auditory learners learn best by listening, so making audiotapes of important concepts may be their best study strategy. Kinesthetic learners like to move and do things with their hands, such as using models or 3-D applications to represent math concepts, so incorporating these types of activities in studying may be helpful.

Attitude. Attitudes about your chance for success, learning in general, and math in particular will all affect how well you complete your course. Keeping a positive attitude, seeking help when it is needed, and believing in your own ability to learn new material are crucial to your success in your math course.

Math and Test Anxiety. Sometimes students feel a great deal of anxiety when faced with taking tests or feel stressed when asked to do any kind of work related to numbers or math problems. These counterproductive feelings can be overcome with extra preparation, support services, relaxation techniques, and even hypnotherapy.

Taking Responsibility for Your Learning. Often students complain that "bad teaching" has lead to failure in math classes. However, according to Dr. Benjamin Bloom, quality of instruction is only responsible for 25% of your overall achievement in a class (B. Bloom, 1976, *Human Characteristics and School Learning,* McGraw-Hill). Your personal characteristics, including study habits, are at least as important as the instruction you receive. This means that you are really in control of your outcome for this class! You can enhance your knowledge of math by reviewing problems that you learned in introductory courses, by adhering to your study schedule, and by seeking help in the Tutorial Center, math lab, study group, or from your instructor when you need it.

ASSIGNMENT

1. Take a learning skills inventory test such as that found at http://www. metamath. com/multiple/multiple_choice_questions.cgi to determine what type of learner you are. How will you use this information to help you succeed in class?
2. Have your past experiences in math courses been good or bad? Why? Do you feel that you suffer from math anxiety or test anxiety? If so, schedule an appointment at your school's counseling office to find ways of dealing with this problem.
3. Find out whether your school offers tutorial services and/or a math lab or learning center and its hours of operation. How do you go about using these services?
4. Ask your instructor whether you can pass around (before or after class) a sign-up sheet for students interested in participating in a study group. Have people write their names, phone number or email address, and times that they are available for study. Distribute copies to interested classmates at your next class meeting.
5. Repeat Exercise 1 from the previous study skills lesson, but relate it to material found in Chapter 3 from this text.

To solve many problems, we must use two and sometimes three variables. This requires that we solve a system of equations.

3.1 Solving Systems by Graphing

- The graphing method   • Consistent systems   • Inconsistent systems
- Dependent equations

The red line in Figure 3-1 shows the cost for a company to produce a given number of skateboards. The blue line shows the revenue the company will earn for selling a given number of those skateboards. The graph offers the company important financial information.

- The production costs exceed the revenue earned if fewer than 400 skateboards are sold. In this case, the company loses money.

- The revenue earned exceeds the production costs if more than 400 skateboards are sold. In this case, the company makes a profit.

- Production costs equal revenue earned if exactly 400 skateboards are sold. This fact is indicated by the point of intersection of the two lines, (400, 20,000), which is called the **break-even point.**

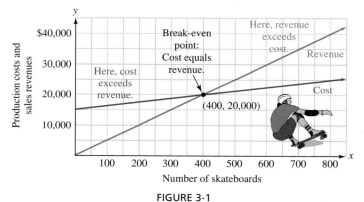

FIGURE 3-1

This example shows that information can be learned by finding the point of intersection of two lines. In this section, we will discuss how to use the *graphing method* to solve systems of two equations, each with two variables.

The graphing method

We have discussed linear equations in two variables, x and y. We found that such equations had infinitely many solutions (x, y), and that we could graph them on a rectangular coordinate system. In this chapter, we will discuss **systems of linear equations** involving two or three equations. A left brace { is often used when writing a system of equations.

In the pair of equations

$$\begin{cases} x + 2y = 4 \\ 2x - y = 3 \end{cases}$$ (called a system of two linear equations)

there are infinitely many pairs (x, y) that satisfy the first equation and infinitely many pairs (x, y) that satisfy the second equation. However, there is only one pair (x, y) that satisfies both equations. The process of finding this pair is called *solving the system.*

To solve a system of two equations in two variables by graphing, we use the following steps.

The graphing method

1. On a single set of coordinate axes, graph each equation.

2. Find the coordinates of the point (or points) where the graphs intersect. These coordinates give the solution of the system.

3. If the graphs have no point in common, the system has no solution.

4. Check the proposed solution in both of the original equations.

▌ Consistent systems

When a system of equations (as in Example 1) has a solution, the system is called a **consistent system.**

INTERMEDIATE
Algebra $f(x)$ **Now**™

EXAMPLE 1 Solve the system: $\begin{cases} x + 2y = 4 \\ 2x - y = 3 \end{cases}$.

Self Check 1

Solve: $\begin{cases} x - 3y = -5 \\ 2x + y = 4 \end{cases}$.

Solution We graph both equations on one set of coordinate axes, as shown in Figure 3-2.

Although infinitely many ordered pairs (x, y) satisfy $x + 2y = 4$, and infinitely many ordered pairs (x, y) satisfy $2x - y = 3$, only the coordinates of the point where the graphs intersect satisfy both equations. From the graph, it appears that the intersection point has coordinates $(2, 1)$. To verify that it is the solution, we substitute 2 for x and 1 for y in both equations and verify that $(2, 1)$ satisfies each one, as shown on the next page.

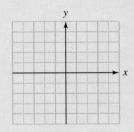

$x + 2y = 4$		
x	y	(x, y)
4	0	$(4, 0)$
0	2	$(0, 2)$
-2	3	$(-2, 3)$

$2x - y = 3$		
x	y	(x, y)
$\frac{3}{2}$	0	$(\frac{3}{2}, 0)$
0	-3	$(0, -3)$
-1	-5	$(-1, -5)$

Use the intercept method to graph each line.

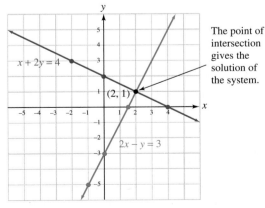

The point of intersection gives the solution of the system.

FIGURE 3-2

Answer (1, 2)

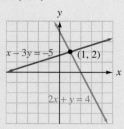

Check:

The first equation	The second equation
$x + 2y = 4$	$2x - y = 3$
$2 + 2(1) \stackrel{?}{=} 4$	$2(2) - 1 \stackrel{?}{=} 3$
$2 + 2 \stackrel{?}{=} 4$	$4 - 1 \stackrel{?}{=} 3$
$4 = 4$	$3 = 3$

Since (2, 1) makes both equations true, it is the solution of the system.

Inconsistent systems

When a system has no solution (as in Example 2), it is called an **inconsistent system.**

INTERMEDIATE
Algebra *f(x)* **Now**™

Self Check 2

Solve: $\begin{cases} 3y - 2x = 6 \\ 2x - 3y = 6 \end{cases}$.

EXAMPLE 2   Solve $\begin{cases} 2x + 3y = 6 \\ 4x + 6y = 24 \end{cases}$ if possible.

Solution   Using the intercept method, we graph both equations on one set of coordinate axes, as shown in Figure 3-3.

$2x + 3y = 6$		
x	**y**	**(x, y)**
3	0	(3, 0)
0	2	(0, 2)
−3	4	(−3, 4)

$4x + 6y = 24$		
x	**y**	**(x, y)**
6	0	(6, 0)
0	4	(0, 4)
−3	6	(−3, 6)

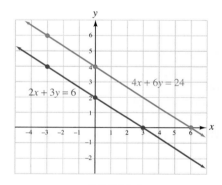

FIGURE 3-3

In this example, the graphs are parallel, because the slopes of the two lines are equal and they have different y-intercepts. We can see that the slope of each line is $-\frac{2}{3}$ by writing each equation in slope–intercept form.

$$2x + 3y = 6 \qquad\qquad 4x + 6y = 24$$
$$3y = -2x + 6 \qquad\qquad 6y = -4x + 24$$
$$y = -\frac{2}{3}x + 2 \qquad\qquad y = -\frac{2}{3}x + 4$$

Since the graphs are parallel lines, the lines do not intersect, and the system does not have a solution. It is an inconsistent system.

Answer no solutions

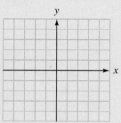

Dependent equations

When the equations of a system of two equations in two variables have different graphs (as in Examples 1 and 2), the equations are called **independent equations.** Two equations with the same graph are called **dependent equations.**

Bachelor's Degrees

"Women now earn more associate's, bachelor's and master's degrees than their male counterparts. Women also earned nearly half of the Ph.D.s as well as professional degrees, which include medical, law and dental degrees."
CNN/Money, April 27,2004

Examine the graph below. Determine the point of intersection of the graph and explain its importance. (*Note:* The *y*-axis does not need to be scaled for you to answer this question.)

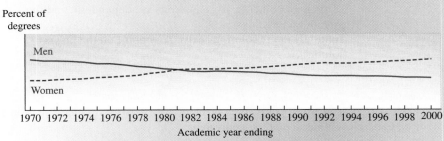

Source: U.S. Department of Education, National Center for Education Statistics

---

EXAMPLE 3   Solve the system: $\begin{cases} y = \frac{1}{2}x + 2 \\ 2x + 8 = 4y \end{cases}$.

Solution   We graph each equation on one set of coordinate axes, as shown in Figure 3-4. Since the graphs coincide (are the same), the system has infinitely many solutions. Any ordered pair (x, y) that satisfies one equation also satisfies the other. From the graph shown in Figure 3-4, we see that $(-4, 0)$, $(0, 2)$, and $(2, 3)$ are three of the infinitely many solutions. Because the two equations have the same graph, they are dependent equations.

Graph by using the slope and *y*-intercept.

$$y = \frac{1}{2}x + 2$$
$$m = \frac{1}{2} \qquad b = 2$$
Slope $= \frac{1}{2}$ *y*-intercept: $(0, 2)$

Graph by using the intercept method.

2x + 8 = 4y		
x	**y**	**(x, y)**
−4	0	(−4, 0)
0	2	(0, 2)
2	3	(2, 3)

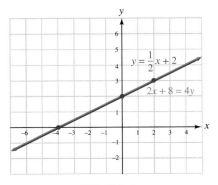

FIGURE 3-4

INTERMEDIATE
Algebra $f(x)$ **Now**™

Self Check 3

Solve: $\begin{cases} 2x - y = 4 \\ y = 2x - 4 \end{cases}$.

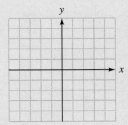

Answer There are infinitely many solutions; three of them are $(0, -4)$, $(2, 0)$, and $(4, 4)$.

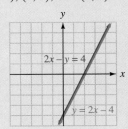

We now summarize the possibilities that can occur when two linear equations, each in two variables, are graphed.

Solving a system of equations by the graphing method

If the lines are different and intersect, the equations are independent, and the system is consistent.
One solution exists. It is the point of intersection.

If the lines are different and parallel, the equations are independent, and the system is inconsistent.
No solution exists.

If the lines coincide, the equations are dependent, and the system is consistent.
Infinitely many solutions exist. Any point on the line is a solution.

If each equation in one system is equivalent to a corresponding equation in another system, the systems are called **equivalent.**

INTERMEDIATE
Algebra $f(x)$ **Now™**

Self Check 4

Solve: $\begin{cases} \dfrac{5}{2}x - y = 2 \\ x + \dfrac{1}{3}y = 3 \end{cases}$

by the graphing method.

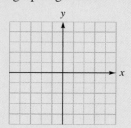

Answer $(2, 3)$

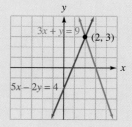

EXAMPLE 4 Solve the system: $\begin{cases} \dfrac{3}{2}x - y = \dfrac{5}{2} \\ x + \dfrac{1}{2}y = 4 \end{cases}$.

Solution We multiply both sides of $\frac{3}{2}x - y = \frac{5}{2}$ by 2 to eliminate the fractions and obtain the equation $3x - 2y = 5$. We multiply both sides of $x + \frac{1}{2}y = 4$ by 2 to eliminate the fractions and obtain the equation $2x + y = 8$.

The new system

$$\begin{cases} 3x - 2y = 5 \\ 2x + y = 8 \end{cases}$$

is equivalent to the original system and is easier to solve, since it has no fractions. If we graph each equation in the new system, as in Figure 3-5, it appears that the lines intersect at $(3, 2)$. Verify that $(3, 2)$ is the solution by substituting 3 for x and 2 for y in each equation of the original system.

3x − 2y = 5		
x	y	(x, y)
$\frac{5}{3}$	0	$(\frac{5}{3}, 0)$
0	$-\frac{5}{2}$	$(0, -\frac{5}{2})$
1	−1	$(1, -1)$

2x + y = 8		
x	y	(x, y)
4	0	$(4, 0)$
0	8	$(0, 8)$
1	6	$(1, 6)$

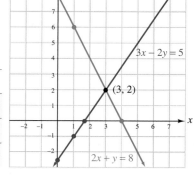

FIGURE 3-5

! COMMENT When checking a solution of a system of equations, always substitute the values of the variables into the original equations.

Solving systems by graphing

The graphing method has limitations. First, the method is limited to equations with two variables. Systems with three or more variables cannot be solved graphically. Second, it is often difficult to find exact solutions graphically. However, graphing calculators enable us to get very good approximations of such solutions.

For example, to solve the system

$$\begin{cases} 3x + 2y = 12 \\ 2x - 3y = 12 \end{cases}$$

with a graphing calculator, we first solve each equation for y so that we can enter the equations into the calculator. After solving for y, we obtain the equivalent system:

$$\begin{cases} y = -\frac{3}{2}x + 6 \\ y = \frac{2}{3}x - 4 \end{cases}$$

If we use window settings of $[-10, 10]$ for x and $[-10, 10]$ for y, the graphs of the equations will look like those in Figure 3-6(a).

We can find the intersection of the two lines by using the INTERSECT feature found under the CALC menu. After graphing the lines and using INTERSECT, we obtain the display shown in Figure 3-6(b), which shows the approximate coordinates of the point of intersection to be $(4.62, -0.92)$.

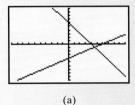   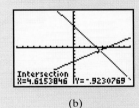

(a)   (b)

FIGURE 3-6

Section 3.1 STUDY SET

VOCABULARY   *Fill in the blanks.*

1. $\begin{cases} x + 2y = 4 \\ 2x - y = 3 \end{cases}$ is called a _____ of linear equations.

2. When a system of equations has one or more solutions, it is called a _____ system.

3. If a system has no solutions, it is called an _____ system.

4. If two equations have different graphs, they are called _____ equations.

5. Two equations with the same graph are called _____ equations.

6. When solving a system of two linear equations by the graphing method, we look for the point of _____ of the two lines.

CONCEPTS

7. Refer to the illustration. Determine whether a true or a false statement would be obtained when the coordinates of

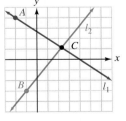

a. point A are substituted into the equation for line l_1.

b. point B are substituted into the equation for line l_1.

c. point C are substituted into the equation for line l_1.

d. point C are substituted into the equation for line l_2.

8. Refer to the illustration.

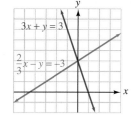

a. How many ordered pairs satisfy the equation $3x + y = 3$? Name three.

b. How many ordered pairs satisfy the equation $\frac{2}{3}x - y = -3$? Name three.

c. How many ordered pairs satisfy both equations? Name it or them.

9. a. The intercept method can be used to graph the equation $2x - 4y = -8$. Complete the following table.

x	y	(x, y)
	0	
0		
2		

b. What is the x-intercept of the graph of $2x - 4y = -8$? What is the y-intercept?

10. a. To graph $y = 3x + 1$, we can pick three numbers for x and find the corresponding values of y. Complete the following table.

x	y	(x, y)
−1		
0		
2		

b. We can also graph $y = 3x + 1$ if we know the slope and the y-intercept of the line. What are they?

11. Write a system of two linear equations that has

a. only one solution, (2, 3).

b. an infinite number of solutions.

c. no solution.

12. 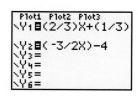 Estimate the solution of the system of linear equations shown in the display below. Then check your answer.

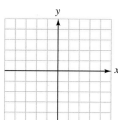

NOTATION *Fill in the blanks.*

13. A left _____ { is often used when writing a system of equations.

14. The solution of a system of two linear equations in two variables is written as an ordered _____.

PRACTICE *Determine whether the ordered pair is a solution of the system of equations.*

15. $(1, 2)$; $\begin{cases} 2x - y = 0 \\ y = \frac{1}{2}x + \frac{3}{2} \end{cases}$

16. $(-1, 2)$; $\begin{cases} y = 3x + 5 \\ y = x + 4 \end{cases}$

17. $(2, -3)$; $\begin{cases} y + 2 = \frac{1}{2}x \\ 3x + 2y = 0 \end{cases}$

18. $(-4, 3)$; $\begin{cases} 4x - y = -19 \\ 3x + 2y = -6 \end{cases}$

Solve each system by graphing, if possible. If a system is inconsistent or if the equations are dependent, so indicate.

19. $\begin{cases} x + y = 6 \\ x - y = 2 \end{cases}$
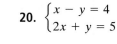

20. $\begin{cases} x - y = 4 \\ 2x + y = 5 \end{cases}$
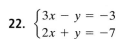

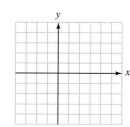
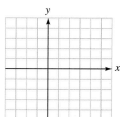

21. $\begin{cases} 2x + y = 1 \\ x - 2y = -7 \end{cases}$

22. $\begin{cases} 3x - y = -3 \\ 2x + y = -7 \end{cases}$

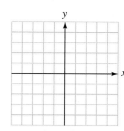
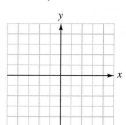

23. $\begin{cases} 3x - 2y = 0 \\ 2x + 3y = 0 \end{cases}$
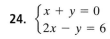

24. $\begin{cases} x + y = 0 \\ 2x - y = 6 \end{cases}$

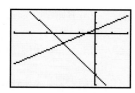
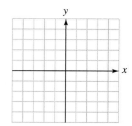

25. $\begin{cases} 4x - 3y = 5 \\ 2x + y = 0 \end{cases}$

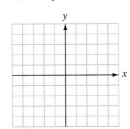

26. $\begin{cases} 4x + y = 0 \\ x - 2y = 0 \end{cases}$

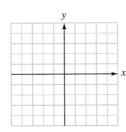

35. $\begin{cases} y = 3 \\ x = 2 \end{cases}$

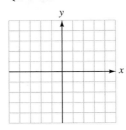

36. $\begin{cases} 2x + 3y = -15 \\ 2x + y = -9 \end{cases}$

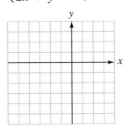

27. $\begin{cases} 3x + y = 3 \\ 3x + 2y = 0 \end{cases}$

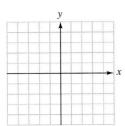

28. $\begin{cases} 2x + 2y = -1 \\ 3x + 4y = 0 \end{cases}$

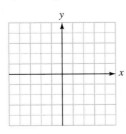

37. $\begin{cases} x = \dfrac{11 - 2y}{3} \\ y = \dfrac{11 - 6x}{4} \end{cases}$

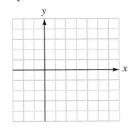

38. $\begin{cases} x = \dfrac{1 - 3y}{4} \\ y = \dfrac{12 + 3x}{2} \end{cases}$

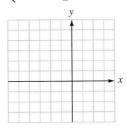

29. $\begin{cases} x = 13 - 4y \\ 3x = 4 + 2y \end{cases}$

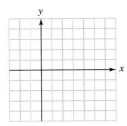

30. $\begin{cases} 3x = 7 - 2y \\ 2x = 2 + 4y \end{cases}$

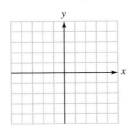

39. $\begin{cases} y = -\frac{5}{2}x + \frac{1}{2} \\ 2x - \frac{3}{2}y = 5 \end{cases}$

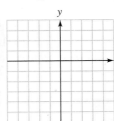

40. $\begin{cases} \frac{5}{2}x + 3y = 6 \\ y = -\frac{5}{6}x + 2 \end{cases}$

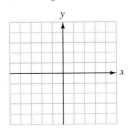

31. $\begin{cases} x = 3 - 2y \\ 2x + 4y = 6 \end{cases}$

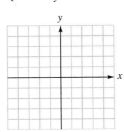

32. $\begin{cases} 3x = 5 - 2y \\ 3x + 2y = 7 \end{cases}$

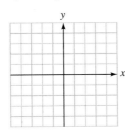

41. $\begin{cases} x = \dfrac{5y - 4}{2} \\ x - \dfrac{5}{3}y + \dfrac{1}{3} = 0 \end{cases}$

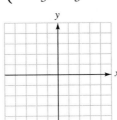

42. $\begin{cases} 2x = 5y - 11 \\ 3x = 2y \end{cases}$

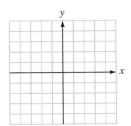

33. $\begin{cases} x = 2 \\ y = -\frac{1}{2}x + 2 \end{cases}$

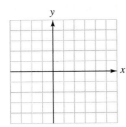

34. $\begin{cases} y = -2 \\ y = \frac{2}{3}x - \frac{4}{3} \end{cases}$

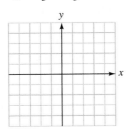

43. $\begin{cases} x = -\frac{3}{2}y \\ x = \frac{3}{2}y - 2 \end{cases}$

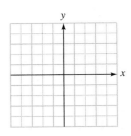

44. $\begin{cases} 4x = 3y - 1 \\ 3y = 4 - 8x \end{cases}$

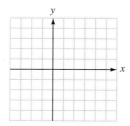

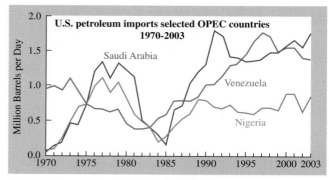

U.S. petroleum imports selected OPEC countries 1970-2003

Based on data from the Energy Information Administration

Use a graphing calculator to solve each system. Give all answers to the nearest hundredth.

45. $\begin{cases} y = 3.2x - 1.5 \\ y = -2.7x - 3.7 \end{cases}$

46. $\begin{cases} y = -0.45x + 5 \\ y = 5.55x - 13.7 \end{cases}$

47. $\begin{cases} 1.7x + 2.3y = 3.2 \\ y = 0.25x + 8.95 \end{cases}$

48. $\begin{cases} 2.75x = 12.9y - 3.79 \\ 7.1x - y = 35.76 \end{cases}$

APPLICATIONS

49. MAPS See the following illustration. Name the cities that lie along Interstate 40. Name the cities that lie along Interstate 25. What city lies on both interstate highways?

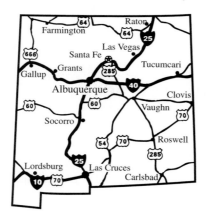

50. PETROLEUM IMPORTS See the graph in the next column.

   a. In which years did the United States import approximately the same amount of petroleum from each of the countries?

   b. For each answer to part a, estimate the amount of petroleum that was imported from each country (in barrels per day).

51. HEARING TESTS See the illustration below. At what frequency and decibel level were the hearing test results the same for the left and right ear? Write your answer as an ordered pair.

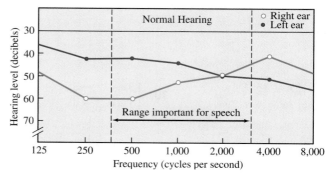

52. BUSINESS Estimate the break-even point (where cost = revenue) on the graph below. Then explain why is it called the break-even point.

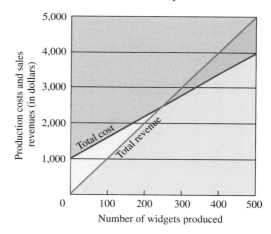

53. SUPPLY AND DEMAND The demand function, graphed on the next page, describes the relationship between the price x of a certain camera and the demand for the camera.

   a. The supply function, $S(x) = \frac{25}{4}x - 525$, describes the relationship between the price x of the camera and the number of cameras the manufacturer is willing to supply. Graph this function.

b. For what price will the supply of cameras equal the demand?

c. As the price of the camera is increased, what happens to supply and what happens to demand?

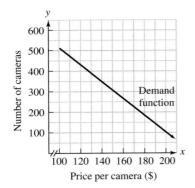

54. COST AND REVENUE The function $C(x) = 200x + 400$ gives the cost for a college to offer x sections of an introductory class in CPR (cardiopulmonary resuscitation). The function $R(x) = 280x$ gives the amount of revenue the college brings in when offering x sections of CPR.

a. Find the break-even point (where cost = revenue) by graphing each function on the same coordinate system.

b. How many sections does the college need to offer to make a profit on the CPR training course?

55. NAVIGATION The paths of two ships are tracked on the same coordinate system. One ship is following a path described by the equation $2x + 3y = 6$, and the other is following a path described by the equation $y = \frac{2}{3}x - 3$.

a. Is there a possibility of a collision?

b. What are the coordinates of the danger point?

c. Is a collision a certainty?

56. AIR TRAFFIC CONTROL Two airplanes flying at the same altitude are tracked using the same coordinate system on a radar screen. One plane is following a path described by the equation $y = \frac{2}{5}x - 2$, and the other is following a path described by the equation $2x = 5y + 7$. Is there a possibility of a collision?

WRITING

57. Suppose the solution of a system of two linear equations is $\left(\frac{14}{5}, -\frac{8}{3}\right)$. Knowing this, explain any drawbacks with solving the system by the graphing method.

58. Can a system of two linear equations have exactly two solutions? Why or why not?

REVIEW Let $f(x) = -x^3 + 2x - 2$ and $g(x) = \frac{2 - x}{9 + x}$. Find each value.

59. $f(-1)$ **60.** $f(10)$

61. $g(2)$ **62.** $g(-20)$

63. Determine the domain and range of $f(x) = x^2 - 2$.

64. Find the slope of the line passing through the points $(-4, 8)$ and $(3, 8)$.

65. The area of the square on the right is 81 square centimeters. Find the area of the shaded triangle.

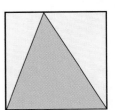

66. If the area of the circle on the right is 49π square centimeters, find the area of the square.

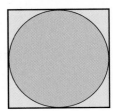

3.2 Solving Systems Algebraically

- The substitution method
- The addition method
- An inconsistent system
- A system with infinitely many solutions
- Problem solving

The graphing method provides a way to visualize the process of solving systems of equations. However, it can sometimes be difficult to determine the exact coordinates of the point of intersection. In this section, we will discuss two other methods, called the *substitution* and the *addition* methods, that can be used to find the exact solutions of systems of equations.

▮ The substitution method

To solve a system of two equations in two variables by the **substitution method,** we follow these steps.

> **The substitution method**
>
> 1. If necessary, solve one equation for a variable — preferably one with a coefficient of 1 or −1. We call the equation found in step 1 the **substitution equation.**
> 2. Substitute the resulting expression for that variable into the other equation and solve it.
> 3. Find the value of the other variable by substituting the value of the variable found in step 2 into the substitution equation.
> 4. State the solution.
> 5. Check the proposed solution in both of the original equations. Write the solution as an ordered pair.

INTERMEDIATE
Algebra $f(x)$ Now™

Self Check 1

Solve: $\begin{cases} x + 3y = 9 \\ 2x - y = -10 \end{cases}$.

EXAMPLE 1   Solve the system: $\begin{cases} 4x + y = 13 \\ -2x + 3y = -17 \end{cases}$.

Solution

Step 1: We solve the first equation for y, because y has a coefficient of 1.

$$4x + y = 13$$
$$y = -4x + 13 \quad \text{To isolate } y, \text{ subtract } 4x \text{ from both sides.}$$
This is the substitution equation.

Step 2: We then substitute $-4x + 13$ for y in the second equation of the system. This step will eliminate the variable y from that equation. The result will be an equation containing only one variable, x.

$$-2x + 3y = -17 \quad \text{This is the second equation of the system.}$$
$$-2x + 3(\mathbf{-4x + 13}) = -17 \quad \text{Substitute } -4x + 13 \text{ for } y. \text{ The variable } y \text{ is eliminated from the equation.}$$
$$-2x - 12x + 39 = -17 \quad \text{Distribute the multiplication by 3.}$$
$$-14x = -56 \quad \text{To solve for } x, \text{ first combine like terms and then subtract 39 from both sides.}$$
$$x = 4 \quad \text{Divide both sides by } -14.$$

Step 3: To find y, we substitute 4 for x in the substitution equation and simplify:

$$y = -4x + 13$$
$$= -4(\mathbf{4}) + 13 \quad \text{Substitute 4 for } x.$$
$$= -3$$

Step 4: The solution is $(4, -3)$. The graphs of the equations of the given system would intersect at the point $(4, -3)$.

Step 5: To verify that this result satisfies both equations, we substitute 4 for x and −3 for y into the original equations of the system and simplify.

Check: **The first equation**   **The second equation**

$$4x + y = 13$$

$$4(4) + (-3) \stackrel{?}{=} 13$$

$$16 - 3 \stackrel{?}{=} 13$$

$$13 = 13$$

$$-2x + 3y = -17$$

$$-2(4) + 3(-3) \stackrel{?}{=} -17$$

$$-8 - 9 \stackrel{?}{=} -17$$

$$-17 = -17$$

Since $(4, -3)$ satisfies both equations of the system, it checks.

Answer $(-3, 4)$

INTERMEDIATE
Algebra *f(x)* **Now**™

EXAMPLE 2   Solve the system: $\begin{cases} \dfrac{2}{9}x - \dfrac{2}{9}y = \dfrac{2}{3} \\ 0.1x = 0.2 - 0.1y \end{cases}$.

Self Check 2

Solve: $\begin{cases} \dfrac{x}{8} + \dfrac{y}{4} = \dfrac{1}{2} \\ 0.01y = -0.02x + 0.04 \end{cases}$.

Solution   First we find an equivalent system without fractions or decimals. To do this, we multiply both sides of the first equation by 9, which is the lowest common denominator of the fractions in the equation. Then we multiply both sides of the second equation by 10.

(1)
(2) $\begin{cases} 2x - 2y = 6 \\ x = 2 - y \end{cases}$ This is the substitution equation.

Since the variable x is isolated in Equation 2, we will substitute $2 - y$ for x in Equation 1. This step will eliminate x from Equation 1, leaving an equation containing only one variable, y. We then solve for y.

$$2x - 2y = 6 \qquad \text{This is Equation 1.}$$

$$2(2 - y) - 2y = 6 \qquad \text{Substitute } 2 - y \text{ for } x.$$

$$4 - 2y - 2y = 6 \qquad \text{Distribute the multiplication by 2.}$$

$$-4y = 2 \qquad \begin{array}{l}\text{To solve for } y, \text{ combine like terms and then subtract 4 from} \\ \text{both sides.}\end{array}$$

$$y = -\frac{1}{2} \qquad \text{Divide both sides by } -4 \text{ and then simplify the fraction.}$$

We can find x by substituting $-\frac{1}{2}$ for y in the substitution equation and simplifying:

$$x = 2 - y$$

$$x = 2 - \left(-\frac{1}{2}\right) \qquad \text{Substitute } -\frac{1}{2} \text{ for } y.$$

$$= 2 + \frac{1}{2}$$

$$= \frac{5}{2} \qquad\qquad 2 + \frac{1}{2} = \frac{4}{2} + \frac{1}{2} = \frac{5}{2}.$$

The solution is $\left(\frac{5}{2}, -\frac{1}{2}\right)$. Verify that this solution satisfies both equations in the original system.

Answer $\left(\dfrac{4}{3}, \dfrac{4}{3}\right)$

■ The addition method

Another method for solving a system of linear equations is the **addition method.** In this method, we combine the equations in a way that will eliminate the terms involving one of the variables.

> **The addition method**
> 1. Write both equations of the system in general form: $Ax + By = C$.
> 2. Multiply the terms of one or both of the equations by constants chosen to make the coefficients of x (or y) differ only in sign.
> 3. Add the equations and solve the resulting equation, if possible.
> 4. Substitute the value obtained in step 3 into either of the original equations and solve for the remaining variable.
> 5. State the solution obtained in steps 3 and 4.
> 6. Check the proposed solution in both of the original equations. Write the solution as an ordered pair.

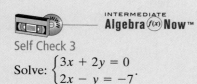

INTERMEDIATE
Algebra $f(x)$ Now™

Self Check 3

Solve: $\begin{cases} 3x + 2y = 0 \\ 2x - y = -7 \end{cases}$.

EXAMPLE 3   Solve the system: $\begin{cases} 4x + y = 13 \\ -2x + 3y = -17 \end{cases}$.

Solution

Step 1: This is the system discussed in Example 1. In this example, we will solve it by the addition method. Since both equations are already written in general form, step 1 is unnecessary.

Step 2: We note that the coefficient of x in the first equation is 4. If we multiply both sides of the second equation by 2, the coefficient of x in that equation will be -4. Then the coefficients of x will differ only in sign.

$$\begin{cases} 4x + y = 13 \\ -4x + 6y = -34 \end{cases}$$

Step 3: When these equations are added, the terms involving x drop out (or are eliminated), and we get an equation that contains only the variable y. We then proceed by solving for y.

$$\begin{aligned} 4x + y &= 13 \\ + \underline{-4x + 6y} &= \underline{-34} \\ 7y &= -21 \\ y &= -3 \end{aligned}$$

Add the like terms, column by column: $4x + (-4x) = 0$, $y + 6y = 7y$, and $13 + (-34) = -21$.

To solve for y, divide both sides by 7.

Step 4: To find x, we substitute -3 for y in either of the original equations and solve for x. If we use the first equation, we have

$$\begin{aligned} 4x + y &= 13 \\ 4x + (\mathbf{-3}) &= 13 \qquad \text{Substitute } -3 \text{ for } y. \\ 4x &= 16 \qquad \text{To solve for } x, \text{ add 3 to both sides.} \\ x &= 4 \qquad \text{Divide both sides by 4.} \end{aligned}$$

Step 5: The solution is $(4, -3)$.

Answer $(-2, 3)$

Step 6: The check was completed in Example 1.

INTERMEDIATE
Algebra $f(x)$ Now™

Self Check 4

Solve: $\begin{cases} 4(2x - y) = 18 \\ 3(x - 3) = 2y - 1 \end{cases}$.

EXAMPLE 4   Solve the system: $\begin{cases} 4x = 3(2 + y) \\ 3(x - 10) = -2y \end{cases}$.

Solution   To use the addition method, we must write each equation in general form. In each case, the first step is to use the distributive property to remove the parentheses.

The first equation

$$4x = 3(2 + y)$$
$$4x = 6 + 3y$$
$$4x - 3y = 6$$

The second equation

$$3(x - 10) = -2y$$
$$3x - 30 = -2y$$
$$3x + 2y = 30$$

We now solve the equivalent system

(1) $\begin{cases} 4x - 3y = 6 \\ 3x + 2y = 30 \end{cases}$
(2)

Since the coefficients of y already have opposite signs, we choose to eliminate y. To make the y-terms drop out when we add the equations, we multiply both sides of Equation 1 by 2 and both sides of Equation 2 by 3 to get

$$\begin{cases} 8x - 6y = 12 \\ 9x + 6y = 90 \end{cases}$$

When these equations are added, the y-terms drop out, and we get

$17x = 102$ $8x + 9x = 17x, -6y + 6y = 0,$ and $12 + 90 = 102.$

$x = 6$ To solve for x, divide both sides by 17.

To find y, we can substitute 6 for x in any equation that contains both variables. It appears the computations will be simplest if we use Equation 2.

$$3x + 2y = 30$$
$3(6) + 2y = 30$ Substitute 6 for x.
$18 + 2y = 30$ Perform the multiplication.
$2y = 12$ Subtract 18 from both sides.
$y = 6$ Divide both sides by 2.

The solution is $(6, 6)$. Check this result.

Answer $\left(1, -\dfrac{5}{2}\right)$

■ An inconsistent system

EXAMPLE 5   Solve the system: $\begin{cases} y = 2x + 4 \\ 8x - 4y = 7 \end{cases}$ if possible.

Solution   Because the first equation is already solved for y, we use the substitution method.

$8x - 4y = 7$ This is the second equation.
$8x - 4(2x + 4) = 7$ Substitute $2x + 4$ for y.

We then solve this equation for x:

$8x - 8x - 16 = 7$ Use the distributive property to remove parentheses.
$-16 \neq 7$ Combine like terms.

The x-terms drop out. The false result, $-16 = 7$, indicates that the equations in the system are independent and that the system is inconsistent. Since the system has no solution, the graphs of the equations in the system will be parallel.

INTERMEDIATE
Algebra *(f(x))* Now™
Self Check 5

Solve: $\begin{cases} x = -2.5y + 8 \\ y = -0.4x + 2 \end{cases}$.

Answer no solution

A system with infinitely many solutions

Self Check 6

Solve: $\begin{cases} x - \frac{5}{2}y = \frac{19}{2} \\ -\frac{2}{5}x + y = -\frac{19}{5} \end{cases}$.

EXAMPLE 6   Solve the system: $\begin{cases} 4x + 6y = 12 \\ -2x - 3y = -6 \end{cases}$.

Solution   Since the equations are written in general form, we use the addition method. We copy the first equation and multiply both sides of the second equation by 2 to get

$$\begin{array}{r} 4x + 6y = 12 \\ -4x - 6y = -12 \\ \hline \end{array}$$

After adding the left-hand sides and the right-hand sides, we get

$$0x + 0y = 0$$
$$0 = 0$$

Here, both the x- and y-terms drop out. The true statement $0 = 0$ indicates that the equations are dependent and that the system is consistent with infinitely many solutions.

Note that the equations of the system are equivalent, because when the second equation is multiplied by -2, it becomes the first equation. The graphs of these equations would coincide. Any ordered pair that satisfies one of the equations also satisfies the other. To find some solutions, we can substitute 0, 3, and -3 for x in either original equation to obtain $(0, 2)$, $(3, 0)$, and $(-3, 4)$.

Answer There are infinitely many solutions; three of them are $(2, -3)$, $(12, 1)$, and $\left(\frac{19}{2}, 0\right)$.

Problem solving

To solve problems using two variables, we follow the same problem-solving strategy discussed previously, except that we form two equations instead of one.

EXAMPLE 7   **Wedding pictures.**   In Figure 3-7, a professional photographer offers two different packages for wedding pictures. Use the information in the figure to determine the cost of an 8 × 10-in. and a 5 × 7-in. photograph.

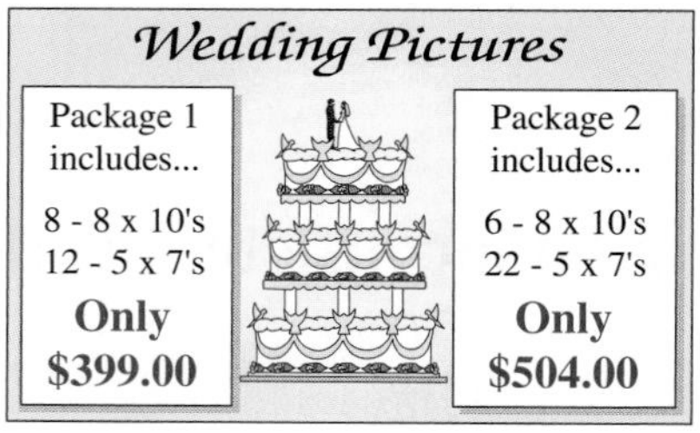

Wedding Pictures

Package 1 includes...

8 - 8 x 10's
12 - 5 x 7's
Only
$399.00

Package 2 includes...

6 - 8 x 10's
22 - 5 x 7's
Only
$504.00

FIGURE 3-7

Analyze the problem

From the figure, we see that eight 8 × 10 and twelve 5 × 7 pictures cost $399, and six 8 × 10 and twenty-two 5 × 7 pictures cost $504. We need to find the cost of an 8 × 10 and a 5 × 7 photograph.

Form two equations

We can let x = the cost of an 8 × 10 photograph and let y = the cost of a 5 × 7 photograph. For the first package, the cost of eight 8 × 10 pictures is $8 \cdot \$x = \$8x$, and the cost of twelve 5 × 7 pictures is $12 \cdot \$y = \$12y$. For the second package, the cost of six 8 × 10 pictures is $\$6x$, and the cost of twenty-two 5 × 7 pictures is $\$22y$. To find x and y, we must write and solve two equations.

The cost of eight 8 × 10 photographs	plus	the cost of twelve 5 × 7 photographs	is	the value of the first package.
$8x$	$+$	$12y$	$=$	399

The cost of six 8 × 10 photographs	plus	the cost of twenty-two 5 × 7 photographs	is	the value of the second package.
$6x$	$+$	$22y$	$=$	504

Solve the system

To find the cost of the 8 × 10 and the 5 × 7 photographs, we can solve the following system:

$$\begin{cases} 8x + 12y = 399 \\ 6x + 22y = 504 \end{cases}$$

We will use the addition method to solve this system. To make the x-terms drop out, we multiply both sides of the first equation by 3. Then we multiply both sides of the second equation by -4. We then add the resulting equations and solve for y:

$$\begin{array}{rl} 24x + 36y = & 1,197 \\ -24x - 88y = & -2,016 \\ \hline -52y = & -819 \\ y = & 15.75 \end{array}$$

Add like terms, column by column. The x-terms drop out.

Divide both sides by -52.

To find x, we substitute 15.75 for y in the first equation of the original system and solve for x:

$$\begin{aligned} 8x + 12y &= 399 \\ 8x + 12(\mathbf{15.75}) &= 399 \qquad \text{Substitute 15.75 for } y. \\ 8x + 189 &= 399 \qquad \text{Perform the multiplication.} \\ 8x &= 210 \qquad \text{Subtract 189 from both sides.} \\ x &= 26.25 \qquad \text{Divide both sides by 8.} \end{aligned}$$

State the conclusion

The cost of an 8 × 10 photo is $26.25, and the cost of a 5 × 7 photo is $15.75.

Check the result

If the first package contains eight 8 × 10 and twelve 5 × 7 photographs, the value of the package is 8($26.25) + 12($15.75) = $210 + $189 = $399. If the second package contains six 8 × 10 and twenty-two 5 × 7 photographs, the value of the package is 6($26.25) + 22($15.75) = $157.50 + $346.50 = $504. The answers check.

INTERMEDIATE
Algebra $f(x)$ Now™

EXAMPLE 8 Water treatment.

A technician determines that 50 fluid ounces of a 15% muriatic acid solution needs to be added to the water in a swimming pool to kill a growth of algae. If the technician has 5% and 20% muriatic solutions on hand, how many ounces of each must be combined to create the 15% solution?

Analyze the problem

We need to find the number of ounces of a 5% solution and the number of ounces of a 20% solution that must be combined to obtain 50 ounces of a 15% solution.

Form two equations

We can let x = the number of ounces of the 5% solution and let y = the number of ounces of the 20% solution that are to be mixed. (See Figure 3-8(a).) Then the amount of muriatic acid in the 5% solution is $0.05x$ ounces, and the amount of muriatic acid in the 20% solution is 0.20y ounces. The sum of these amounts is also the amount of muriatic acid in the final mixture, which is 15% of 50 ounces. This information is shown in the table in Figure 3-8(b).

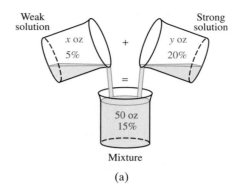

Mixture

(a)

Solution	Ounces	· % acid	= Amount of acid
Weak	x	0.05	0.05x
Strong	y	0.20	0.20y
Mixture	50	0.15	0.15(50)

↑   ↑

One equation comes from the information in this column.   Another equation comes from the information in this column.

(b)

FIGURE 3-8

The facts of the problem provide the following two equations:

The number of ounces of 5% solution	plus	the number of ounces of 20% solution	is	the total number of ounces in the 15% mixture.
x	+	y	=	50

The acid in the 5% solution	plus	the acid in the 20% solution	is	the acid in the 15% mixture.
0.05x	+	0.20y	=	0.15(50)

Solve the system

To find out how many ounces of each are needed, we solve the following system:

$$\begin{cases} x + y = 50 \\ 0.05x + 0.20y = 7.5 \end{cases} \quad \text{Simplify the right-hand side: } 0.15(50) = 7.5.$$

To solve this system by substitution, we can solve the first equation for y:

$$x + y = 50$$

$$y = 50 - x \quad \text{Subtract } x \text{ from both sides. This is the substitution equation.}$$

Then we substitute $50 - x$ for y in the second equation of the system and solve for x.

$$0.05x + 0.20y = 7.5$$

$$0.05x + 0.20(\mathbf{50 - x}) = 7.5 \qquad \text{Substitute } 50 - x \text{ for } y.$$

$$5x + 20(50 - x) = 750 \qquad \text{Multiply both sides by 100.}$$

$$5x + 1{,}000 - 20x = 750 \qquad \text{Use the distributive property to remove parentheses.}$$

$$-15x = -250 \qquad \text{Combine like terms and subtract 1,000 from both sides.}$$

$$x = \frac{-250}{-15} \qquad \text{Divide both sides by } -15.$$

$$x = \frac{50}{3} \qquad \text{Simplify: } \frac{250}{15} = \frac{\overset{1}{\cancel{5}} \cdot 50}{\underset{1}{\cancel{5}} \cdot 3}.$$

To find y, we can substitute $\frac{50}{3}$ for x in the substitution equation.

$$y = 50 - x$$

$$= 50 - \frac{\mathbf{50}}{\mathbf{3}} \qquad \text{Substitute } \frac{50}{3} \text{ for } x.$$

$$= \frac{100}{3} \qquad \text{Express 50 as } \frac{150}{3} \text{ and then subtract.}$$

State the conclusion

To obtain 50 ounces of a 15% solution, the technician must mix $\frac{50}{3}$ or $16\frac{2}{3}$ ounces of the 5% solution with $\frac{100}{3}$ or $33\frac{1}{3}$ ounces of the 20% solution.

Check the result

We note that $16\frac{2}{3}$ ounces of solution plus $33\frac{1}{3}$ ounces of solution equals the required 50 ounces of solution. We also note that 5% of $16\frac{2}{3} \approx 0.83$ and 20% of $33\frac{1}{3} \approx 6.67$, giving a total of 7.5, which is 15% of 50. The answers check.

INTERMEDIATE
Algebra $f(x)$ Now™

EXAMPLE 9   Parallelograms.

Refer to the parallelogram shown in Figure 3-9 and find the values of x and y.

Solution   To solve this problem, we will use two important facts about parallelograms.

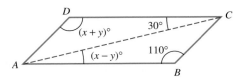

FIGURE 3-9

- When a diagonal intersects two parallel sides of a parallelogram, pairs of *alternate interior angles* have the same measure. In Figure 3-9, $\angle BAC$ and $\angle DCA$ are alternate interior angles and therefore have the same measure. Thus, $(x - y)° = 30°$.
- *Opposite angles* of a parallelogram have the same measure. Since $\angle B$ and $\angle D$ in Figure 3-9 are opposite angles of the parallelogram, $(x + y)° = 110°$.

We can form the following system of equations and solve it by addition.

$$
\begin{aligned}
x - y &= 30 \\
\underline{x + y} &= \underline{110} \\
2x &= 140 \qquad \text{Add the equations. The } y\text{-terms drop out.} \\
x &= 70 \qquad \text{Divide both sides by 2.}
\end{aligned}
$$

We can substitute 70 for x in the second equation of the system and solve for y.

$$x + y = 110$$
$$\mathbf{70} + y = 110 \quad \text{Substitute 70 for } x.$$
$$y = 40 \quad \text{Subtract 70 from both sides.}$$

Thus, x is 70 and y is 40.

Running a machine involves both *setup costs* and *unit costs*. Setup costs include the cost of preparing a machine to do a certain job. The costs to make one item are unit costs. They depend on the number of items to be manufactured, including costs of raw materials and labor.

EXAMPLE 10 Break point.

The setup cost of a machine that makes wooden coathangers is $400. After setup, it costs $1.50 to make each hanger (the unit cost). Management is considering the purchase of a new machine that can manufacture the same type of coathanger at a cost of $1.25 per hanger. If the setup cost of the new machine is $500, find the number of coathangers that the company would need to manufacture to make the cost the same using either machine. This is called the **break point.**

Analyze the problem

We are to find the number of coathangers that will cost equal amounts to produce on either machine. The machines have different setup costs and different unit costs.

Form two equations

The cost C_1 of manufacturing x coathangers on the machine currently in use is $1.50x + $400 (the number of coathangers manufactured times $1.50, plus the setup cost of $400). The cost C_2 of manufacturing the same number of coathangers on the new machine is $1.25x + $500. The break point occurs when the costs to make the same number of hangers using either machine are equal ($C_1 = C_2$).

If $x =$ the number of coathangers to be manufactured, the cost C_1 using the machine currently in use is

The cost of using the current machine	is	the cost of manufacturing x coathangers	plus	the setup cost.
C_1	$=$	$1.5x$	$+$	400

The cost C_2 using the new machine to make x coathangers is

The cost of using the new machine	is	the cost of manufacturing x coathangers	plus	the setup cost.
C_2	$=$	$1.25x$	$+$	500

Solve the system

To find the break point, we must solve the system $\begin{cases} C_1 = 1.5x + 400 \\ C_2 = 1.25x + 500 \end{cases}$.

Since the break point occurs when $C_1 = C_2$, we can substitute $1.5x + 400$ for C_2 in the second equation to get

$$1.5x + 400 = 1.25x + 500$$
$$1.5x = 1.25x + 100 \qquad \text{Subtract 400 from both sides.}$$
$$0.25x = 100 \qquad \text{Subtract } 1.25x \text{ from both sides.}$$
$$x = 400 \qquad \text{Divide both sides by 0.25.}$$

State the conclusion

The break point is 400 coathangers.

Check the result

To make 400 coathangers, the cost on the current machine would be $\$400 + \$1.50(400) = \$400 + \$600 = \$1,000$. The cost using the new machine would be $\$500 + \$1.25(400) = \$500 + \$500 = \$1,000$. Since the costs are equal, the break point is 400. The answer checks.

Section 3.2 STUDY SET

VOCABULARY   *Fill in the blanks.*

1. $Ax + By = C$ is the _____ form of a linear equation.

2. In the equation $x + 3y = -1$, the x-term has an understood coefficient of ▢.

3. When we add the two equations of the system $\begin{cases} x + y = 5 \\ x - y = -3 \end{cases}$, the y-terms are _____

4. To solve $\begin{cases} y = 3x \\ x + y = 4 \end{cases}$, we can _____ $3x$ for y in the second equation.

CONCEPTS

5. If the system $\begin{cases} 4x - 3y = 7 \\ 3x - 2y = 6 \end{cases}$ is to be solved using the addition method, by what constants should each equation be multiplied if

   a. the x-terms are to drop out?

   b. the y-terms are to drop out?

6. If the system $\begin{cases} 4x - 3y = 7 \\ 3x + y = 6 \end{cases}$ is to be solved using the substitution method, what variable in what equation would it be easier to solve for?

7. Can the system $\begin{cases} 2x + 5y = 7 \\ 4x - 3y = 16 \end{cases}$ be solved more easily by the substitution or the addition method?

8. Given the equation $3x + y = -4$.

   a. Solve for x.

   b. Solve for y.

   c. Which variable was easier to solve for? Explain why.

9. The substitution method was used to solve three systems of linear equations. The results after y was eliminated and the remaining equation was solved for x are listed below. Match each result with a possible graph of the system from the illustration.

   a. $-2 = 3$ b. $x = 3$ c. $3 = 3$

   Possible graphs

   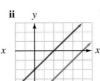

   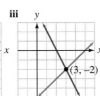

10. Consider the system $\begin{cases} \frac{2}{3}x - \frac{y}{6} = \frac{16}{9} \\ 0.03x + 0.02y = 0.03 \end{cases}$.

    a. What algebraic step should be performed to clear the first equation of fractions?

    b. What algebraic step should be performed to clear the second equation of decimals?

PRACTICE   *Solve each system by substitution, if possible. If a system is inconsistent or if the equations are dependent, so indicate.*

11. $\begin{cases} y = x \\ x + y = 4 \end{cases}$

12. $\begin{cases} y = x + 2 \\ x + 2y = 16 \end{cases}$

13. $\begin{cases} x = 2 + y \\ 2x + y = 13 \end{cases}$

14. $\begin{cases} x = -4 + y \\ 3x - 2y = -5 \end{cases}$

15. $\begin{cases} x + 2y = 6 \\ 3x - y = -10 \end{cases}$

16. $\begin{cases} 2x - y = -21 \\ 4x + 5y = 7 \end{cases}$

17. $\begin{cases} \frac{3}{2}x + 2 = y \\ 0.6x - 0.4y = -0.4 \end{cases}$

18. $\begin{cases} 2x - \frac{5}{2} = y \\ 0.04x - 0.02y = 0.05 \end{cases}$

Solve each system by addition, if possible. If a system is inconsistent or if the equations are dependent, so indicate.

19. $\begin{cases} x - y = 3 \\ x + y = 7 \end{cases}$

20. $\begin{cases} x + y = 1 \\ x - y = 7 \end{cases}$

21. $\begin{cases} 2x + y = -10 \\ 2x - y = -6 \end{cases}$

22. $\begin{cases} x + 2y = -9 \\ x - 2y = -1 \end{cases}$

23. $\begin{cases} 2x + 3y = 8 \\ 3x - 2y = -1 \end{cases}$

24. $\begin{cases} 5x - 2y = 19 \\ 3x + 4y = 1 \end{cases}$

25. $\begin{cases} 4(x - 2) = -9y \\ 2(x - 3y) = -3 \end{cases}$

26. $\begin{cases} 2(2x + 3y) = 5 \\ 8x = 3(1 + 3y) \end{cases}$

Solve each system by any method, if possible. If a system is inconsistent or if the equations are dependent, so indicate.

27. $\begin{cases} 3x - 4y = 9 \\ x + 2y = 8 \end{cases}$

28. $\begin{cases} 3x - 2y = -10 \\ 6x + 5y = 25 \end{cases}$

29. $\begin{cases} 2(x + y) + 1 = 0 \\ 3x + 4y = 0 \end{cases}$

30. $\begin{cases} 5x + 3y = -7 \\ 3(x - y) - 7 = 0 \end{cases}$

31. $\begin{cases} 0.16x - 0.08y = 0.32 \\ 2x - 4 = y \end{cases}$

32. $\begin{cases} 0.6y - 0.9x = -3.9 \\ 3x - 17 = 4y \end{cases}$

33. $\begin{cases} x = \frac{3}{2}y + 5 \\ 2x - 3y = 8 \end{cases}$

34. $\begin{cases} x = \frac{2}{3}y \\ y = 4x + 5 \end{cases}$

35. $\begin{cases} 0.5x + 0.5y = 6 \\ \frac{x}{2} - \frac{y}{2} = -2 \end{cases}$

36. $\begin{cases} \frac{x}{2} - \frac{y}{3} = -4 \\ \frac{x}{2} + \frac{y}{9} = 0 \end{cases}$

37. $\begin{cases} \frac{3}{4}x + \frac{2}{3}y = 7 \\ \frac{3}{5}x - \frac{1}{2}y = 18 \end{cases}$

38. $\begin{cases} \frac{2}{3}x - \frac{1}{4}y = -8 \\ 0.5x - 0.375y = -9 \end{cases}$

39. $\begin{cases} \frac{3x}{2} - \frac{2y}{3} = 0 \\ \frac{3x}{4} + \frac{4y}{3} = \frac{5}{2} \end{cases}$

40. $\begin{cases} \frac{3x}{5} + \frac{5y}{3} = 2 \\ \frac{6x}{5} - \frac{5y}{3} = 1 \end{cases}$

41. $\begin{cases} 12x - 5y - 21 = 0 \\ \frac{3}{4}x - \frac{2}{3}y = \frac{19}{8} \end{cases}$

42. $\begin{cases} 4y + 5x - 7 = 0 \\ \frac{10}{7}x - \frac{4}{9}y = \frac{17}{21} \end{cases}$

The following systems involve variables other than x and y. When writing the solution as an ordered pair, write the values for the variables in alphabetical order.

43. $\begin{cases} \frac{3}{2}p + \frac{1}{3}q = 2 \\ \frac{2}{3}p + \frac{1}{9}q = 1 \end{cases}$

44. $\begin{cases} a + \frac{b}{3} = \frac{5}{3} \\ \frac{a + b}{3} = 3 - a \end{cases}$

45. $\begin{cases} \frac{m - n}{5} + \frac{m + n}{2} = 6 \\ \frac{m - n}{2} - \frac{m + n}{4} = 3 \end{cases}$

46. $\begin{cases} \frac{r - 2}{5} + \frac{s + 3}{2} = 5 \\ \frac{r + 3}{2} + \frac{s - 2}{3} = 6 \end{cases}$

•

Solve each system. To do this, substitute a for $\frac{1}{x}$ and b for $\frac{1}{y}$ and solve for a and b. Then find x and y using the fact that $a = \frac{1}{x}$ and $b = \frac{1}{y}$.

47. $\begin{cases} \frac{1}{x} + \frac{1}{y} = \frac{5}{6} \\ \frac{1}{x} - \frac{1}{y} = \frac{1}{6} \end{cases}$

48. $\begin{cases} \frac{1}{x} + \frac{1}{y} = \frac{9}{20} \\ \frac{1}{x} - \frac{1}{y} = \frac{1}{20} \end{cases}$

49. $\begin{cases} \frac{1}{x} + \frac{2}{y} = -1 \\ \frac{2}{x} - \frac{1}{y} = -7 \end{cases}$

50. $\begin{cases} \frac{3}{x} - \frac{2}{y} = -30 \\ \frac{2}{x} - \frac{3}{y} = -30 \end{cases}$

APPLICATIONS   *Use two variables to solve each problem.*

51. ADVERTISING Use the information in the fee schedule shown to find the cost of a 15-second and a 30-second radio commercial on radio station KLIZ.

| ADVERTISE YOUR COMPANY ON THE RADIO KLIZ 1250 AM | Plan 1: Four 30-second spots, six 15-second spots Cost: $6,050 Plan 2: Three 30-second spots, five 15-second spots Cost: $4,775 |

52. TEMPORARY HELP A law firm had to hire several workers to help finish a large project. From the billing records shown on the next page, determine the daily fee charged by the employment agency for a clerk-typist and for a computer programmer.

TEMPORARY EMPLOYMENT, INC.
We meet your employment needs!

Billed to: _Archer Law Offices_ Attn: _B. Kinsell_

Day	Position/Employee Name	Total cost
Mon. 3/22	*Clerk-typists:* K. Amad, B. Tran, S. Smith *Programmers:* T. Lee, C. Knox	$685
Tues. 3/23	*Clerk-typists:* K. Amad, B. Tran, S. Smith, W. Morada *Programmers:* T. Lee, C. Knox, B. Morales	$975

53. PETS According to the Pet Food Institute, in 2003 there were an estimated 135 million dogs and cats in the United States. If there were 15 million more cats than dogs, how many of each type of pet were there in 2003?

54. ELECTRONICS In the illustration, two resistors in the voltage divider circuit have a total resistance of 1,375 ohms. To provide the required voltage, R_1 must be 125 ohms greater than R_2. Find both resistances.

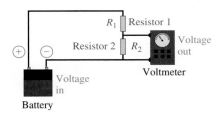

55. FENCING A FIELD The rectangular field shown below is surrounded by 72 meters of fencing. If the field is partitioned as shown, a total of 88 meters of fencing is required. Find the dimensions of the field.

56. GEOMETRY In a right triangle, one acute angle is 15° greater than two times the other acute angle. Find the difference between the measures of the angles.

57. BRACING The bracing of a basketball backboard shown in the next column forms a parallelogram. Find the values of *x* and *y*.

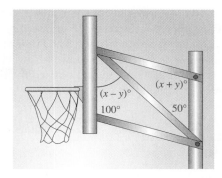

58. TRAFFIC SIGNALS In the illustration below, brace A and brace B are perpendicular. Find the values of *x* and *y*.

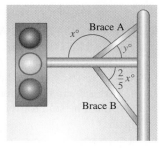

59. INVESTMENT CLUBS Part of $8,000 was invested by an investment club at 10% interest and the rest at 12%. If the annual income from these investments is $900, how much was invested at each rate?

60. RETIREMENT INCOME A retired couple invested part of $12,000 at 6% interest and the rest at 7.5%. If their annual income from these investments is $810, how much was invested at each rate?

61. TV NEWS A news van and a helicopter left a TV station parking lot at the same time, headed in opposite directions to cover breaking news stories that were 145 miles apart. If the helicopter had to travel 55 miles farther than the van, how far did the van have to travel to reach the location of the news story?

62. DELIVERY SERVICES A delivery truck travels 50 miles in the same time that a cargo plane travels 180 miles. The speed of the plane is 143 mph faster than the speed of the truck. Find the speed of the delivery truck.

63. PRODUCTION PLANNING A bicycle manufacturer builds racing bikes and mountain bikes, with the per unit manufacturing costs shown below. The company has budgeted $31,800 for labor and $26,150 for materials. How many bicycles of each type can be built?

Model	Cost of materials	Cost of labor
Racing	$110	$120
Mountain	$140	$180

64. FARMING A farmer keeps some animals on a strict diet. Each animal is to receive 15 grams of protein and 7.5 grams of carbohydrates. The farmer uses two food mixes, with nutrients as shown below. How many grams of each mix should be used to provide the correct nutrients for each animal?

Mix	Protein	Carbohydrates
Mix A	12%	9%
Mix B	15%	5%

65. RECORDING COMPANIES Three people invest a total of $105,000 to start a recording company that will produce reissues of classic jazz. Each release will be a set of 3 CDs that will retail for $45 per set. If each set can be produced for $18.95, how many sets must be sold for the investors to make a profit?

66. MACHINE SHOPS Two machines can mill a brass plate. One machine has a setup cost of $300 and a cost per plate of $2. The other machine has a setup cost of $500 and a cost per plate of $1. Find the break point.

67. PUBLISHING A printer has two presses. One has a setup cost of $210 and can print the pages of a certain book for $5.98. The other press has a setup cost of $350 and can print the pages of the same book for $5.95. Find the break point.

68. MIXING CANDY How many pounds of each candy shown below must be mixed to obtain 60 pounds of candy that would be worth $4.50 per pound?

69. COSMETOLOGY A beauty shop specializing in permanents has fixed costs of $2,101.20 per month. The owner estimates that the cost for each permanent is $23.60, which covers labor, chemicals, and electricity. If her shop can give as many permanents as she wants at a price of $44 each, how many must be given each month to break even?

70. PRODUCTION PLANNING A paint manufacturer can choose between two processes for manufacturing house paint, with monthly costs as shown below. Assume that the paint sells for $18 per gallon.

Process	Fixed costs	Unit cost (per gallon)
A	$32,500	$13
B	$80,600	$5

a. Find the break even point where production costs equal revenue earned for process A.

b. Find the break even point for process B.

c. If expected sales are 7,000 gallons per month, which process should the company use?

71. DERMATOLOGY Tests of an antibacterial face-wash cream showed that a mixture containing 0.3% Triclosan (active ingredient) gave the best results. How many grams of cream from each tube shown below should be used to make an equal-size tube of the 0.3% cream?

72. MIXING SOLUTIONS How many ounces of the two alcohol solutions shown below must be mixed to obtain 100 ounces of a 12.2% solution?

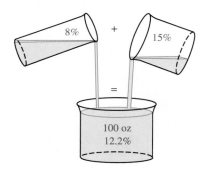

73. RETOOLING A manufacturer of automobile water pumps is considering retooling for one of two manufacturing processes, with monthly fixed costs and unit costs as indicated in the table. Each water pump can be sold for $50.

Process	Fixed costs	Unit cost
A	$12,390	$29
B	$20,460	$17

a. Find the break even point where production costs equal revenue earned for process A.

b. Find the break even point for process B.

c. If expected sales are 550 per month, which process should be used?

74. SALARY OPTIONS A sales clerk can choose from two salary plans: a straight 7% commision, or $150 + 2% commission. How much would the clerk have to sell for each plan to produce the same monthly paycheck?

WRITING

75. Which method would you use to solve the system $\begin{cases} 4x + 6y = 5 \\ 8x - 3y = 3 \end{cases}$? Explain why.

76. Which method would you use to solve the system $\begin{cases} x - 2y = 2 \\ 2x + 3y = 11 \end{cases}$? Explain why.

77. When solving a problem using two variables, why must we write two equations?

78. Write a problem that can be solved by solving the system $\begin{cases} x + y = 36 \\ \$1.29x + \$2.29y = \$72.44 \end{cases}$.

79. Write a problem to fit the information given in the table.

Solution	Ounces	% insecticide	Amount of insecticide
Weak	x	0.02	$0.02x$
Strong	y	0.10	$0.10y$
Mixture	80	0.07	0.07(80)

80. Make a table like the one below, listing one advantage and one disadvantage for each of the methods that can be used to solve a system of two linear equations.

Method	Advantage	Disadvantage
Graphing		
Substitution		
Addition		

REVIEW *Find the slope of each line.*

81.

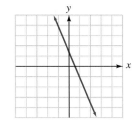

82.

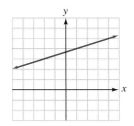

83. The line passing through $(0, -8)$ and $(-5, 0)$

84. The line with equation $y = -3x + 4$

85. The line with equation $4x - 3y = -3$

86. The line with equation $y = 3$

3.3 Systems of Three Equations

- Solving systems of three equations in three variables • Consistent systems
- An inconsistent system • Systems with dependent equations
- Problem solving • Curve fitting

In the preceding sections, we solved systems of two linear equations in two variables. In this section, we will solve systems of three linear equations in three variables by using a combination of the addition method and the substitution method. We will then use that procedure to solve problems involving three variables.

Solving systems of three equations in three variables

We now extend the definition of a linear equation to include equations of the form $Ax + By + Cz = D$, where A, B, C, and D represent real numbers. The solution of a system of three linear equations in three variables is an **ordered triple** of numbers. For example, the solution of the system

$$\begin{cases} 2x + 3y + 4z = 20 \\ 3x + 4y + 2z = 17 \\ 3x + 2y + 3z = 16 \end{cases}$$

is the triple $(1, 2, 3)$, since each equation is satisfied if $x = 1$, $y = 2$, and $z = 3$.

$2x + 3y + 4z = 20$	$3x + 4y + 2z = 17$	$3x + 2y + 3z = 16$
$2(1) + 3(2) + 4(3) = 20$	$3(1) + 4(2) + 2(3) = 17$	$3(1) + 2(2) + 3(3) = 16$
$2 + 6 + 12 = 20$	$3 + 8 + 6 = 17$	$3 + 4 + 9 = 16$
$20 = 20$	$17 = 17$	$16 = 16$

The graph of an equation of the form $Ax + By + Cz = D$ is a flat surface called a **plane.** A system of three linear equations with three variables is consistent or inconsistent, depending on how the three planes corresponding to the three equations intersect. Figure 3-10 illustrates some of the possibilities.

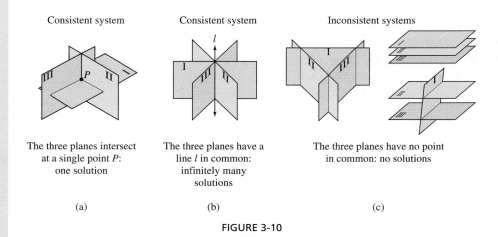

Consistent system	Consistent system	Inconsistent systems
The three planes intersect at a single point P: one solution	The three planes have a line l in common: infinitely many solutions	The three planes have no point in common: no solutions
(a)	(b)	(c)

FIGURE 3-10

To solve a system of three linear equations in three variables, we follow these steps.

Solving three equations in three variables

1. Write each equation of the system in **general form:** $Ax + By + Cz = D$.
2. Pick any two equations and eliminate a variable.
3. Pick a different pair of equations and eliminate the same variable as in step 2.
4. Solve the resulting pair of two equations in two variables.
5. To find the value of the third variable, substitute the values of the two variables found in step 4 into any equation containing all three variables and solve the equation.
6. Check the proposed solution in all three of the original equations. Write the solution as an ordered triple.

■ Consistent systems

Recall that when a system has a solution, it is called a **consistent system.**

EXAMPLE 1 Solve the system: $\begin{cases} 2x + y + 4z = 12 \\ x + 2y + 2z = 9 \\ 3x - 3y - 2z = 1 \end{cases}$.

Self Check 1

Solve: $\begin{cases} 2x + y + 4z = 16 \\ x + 2y + 2z = 11 \\ 3x - 3y - 2z = -9 \end{cases}$.

Solution

Step 1: Since the equations are written in general form, we proceed to step 2.

(1) $\begin{cases} 2x + y + 4z = 12 \\ x + 2y + 2z = 9 \\ 3x - 3y - 2z = 1 \end{cases}$ To clarify the solution process, we number each equation.
(2)
(3)

Step 2: If we pick Equations 2 and 3 and add them, the variable z is eliminated.

(2) $\quad x + 2y + 2z = \ 9$
(3) $\quad \underline{3x - 3y - 2z = \ 1}$
(4) $\quad 4x - \ y \quad\quad = 10$ This equation does not contain z.

Step 3: We now pick a different pair of equations (Equations 1 and 3) and eliminate z again. If each side of Equation 3 is multiplied by 2 and the resulting equation is added to Equation 1, z is eliminated.

(1) $\quad 2x + \ y + 4z = 12$
$\quad\quad \underline{6x - 6y - 4z = \ 2}$ Multiply both sides of Equation 3 by 2.
(5) $\quad 8x - 5y \quad\quad = 14$ This equation does not contain z.

Step 4: Equations 4 and 5 form a system of two equations in two variables, x and y.

(4) $\begin{cases} 4x - y = 10 \\ 8x - 5y = 14 \end{cases}$
(5)

To solve this system, we multiply Equation 4 by -5 and add the resulting equation to Equation 5 to eliminate y:

$\quad -20x + 5y = -50$ Multiply both sides of Equation 4 by -5.
(5) $\quad \underline{\ \ 8x - 5y = \ \ 14}$
$\quad -12x \quad\quad = -36$

$\quad\quad x = 3$ To find x, divide both sides by -12.

To find y, we substitute 3 for x in any equation containing x and y only (such as Equation 5) and solve for y:

(5) $\quad 8x - 5y = 14$
$\quad 8(3) - 5y = 14$ Substitute 3 for x.
$\quad 24 - 5y = 14$ Simplify.
$\quad -5y = -10$ Subtract 24 from both sides.
$\quad y = 2$ Divide both sides by -5.

Step 5: To find z, we substitute 3 for x and 2 for y in any equation containing x, y, and z (such as Equation 1) and solve for z:

(1) $\quad 2x + y + 4z = 12$
$\quad 2(3) + 2 + 4z = 12$ Substitute 3 for x and 2 for y.
$\quad 8 + 4z = 12$ Simplify.
$\quad 4z = 4$ Subtract 8 from both sides.
$\quad z = 1$ Divide both sides by 4.

$\left. 4a + 2b - 6c = -5 \right.$

Solution We can multiply the first equation of the system by 2 and add the resulting equation to the second equation to eliminate b:

$\quad 4a + 2b - 6c = -6$ Multiply both sides of the first equation by 2.
$\quad \underline{3a - 2b + 4c = \ \ 2}$
(1) $\quad 7a \quad\quad - 2c = -4$

Now add the second and third equations of the system to eliminate b again:

$\quad 3a - 2b + 4c = \ \ 2$
$\quad \underline{4a + 2b - 6c = -7}$
(2) $\quad 7a \quad\quad - 2c = -5$

The solution of the system is $(x, y, z) = (3, 2, 1)$. Because this system has a solution, it is a consistent system.

Answer $(1, 2, 3)$

Step 6: Verify that these values satisfy each equation in the original system.

When one or more of the equations of a system is missing a term, the elimination of a variable that is normally performed in Step 2 of the solution process can be skipped.

INTERMEDIATE
Algebra *f(x)* **Now**™

Equations 1 and 2 form the system

(1) $\begin{cases} 7a - 2c = -4 \\ 7a - 2c = -5 \end{cases}$
(2)

Since $7a - 2c$ cannot equal both -4 and -5, the system is inconsistent and has no solution.

Answer no solution

Systems with dependent equations

When the equations in a system of two equations in two variables are dependent, the system has infinitely many solutions. This is not always true for systems of three equations in three variables. In fact, a system can have dependent equations and still be inconsistent. Figure 3-11 illustrates the different possibilities.

Consistent system

Inconsistent system

When three planes coincide, the equations are dependent, and there are infinitely many solutions.

When three planes intersect in a common line, the equations are dependent, and there are infinitely many solutions.

When two planes coincide and are parallel to a third plane, the system is inconsistent, and there are no solutions.

(a)

(b)

(c)

FIGURE 3-11

INTERMEDIATE
Algebra *f(x)* **Now**™

Self Check 4

Solve: $\begin{cases} 3x + 2y + z = -1 \\ 2x - y - z = 5 \\ 5x + y = 4 \end{cases}$.

EXAMPLE 4   Solve the system: $\begin{cases} 3x - 2y + z = -1 \\ 2x + y - z = 5 \\ 5x - y = 4 \end{cases}$.

Solution   We can add the first two equations to get

$$\begin{array}{rcl} 3x - 2y + z &=& -1 \\ 2x + y - z &=& 5 \\ \hline \textbf{(1)} \quad 5x - y &=& 4 \end{array}$$

Since Equation 1 is the same as the third equation of the system, the equations of the system are dependent, and there will be infinitely many solutions. From a graphical perspective, the equations represent three planes that intersect in a common line, as shown in Figure 3-11(b).

To write the general solution of this system, we can solve Equation 1 for y to get

$$5x - y = 4$$
$$-y = -5x + 4 \qquad \text{Subtract } 5x \text{ from both sides.}$$
$$y = 5x - 4 \qquad \text{Multiply both sides by } -1.$$

We can then substitute $5x - 4$ for y in the first equation of the system and solve for z to get

$$\begin{array}{ll} 3x - 2y + z = -1 & \\ 3x - 2(5x - 4) + z = -1 & \text{Substitute } 5x - 4 \text{ for } y. \\ 3x - 10x + 8 + z = -1 & \text{Use the distributive property to remove parentheses.} \\ -7x + 8 + z = -1 & \text{Combine like terms.} \\ z = 7x - 9 & \text{Add } 7x \text{ and } -8 \text{ to both sides.} \end{array}$$

Since we have found the values of y and z in terms of x, every solution of the system has the form $(x, 5x - 4, 7x - 9)$, where x can be any real number. For example,

If $x = 1$, a solution is $(1, 1, -2)$. $5(1) - 4 = 1$, and $7(1) - 9 = -2$.

If $x = 2$, a solution is $(2, 6, 5)$. $5(2) - 4 = 6$, and $7(2) - 9 = 5$.

If $x = 3$, a solution is $(3, 11, 12)$. $5(3) - 4 = 11$, and $7(3) - 9 = 12$.

Answer There are infinitely many solutions. A general solution is $(x, 4 - 5x, -9 + 7x)$. Three solutions are $(1, -1, -2)$, $(2, -6, 5)$, and $(3, -11, 12)$.

Problem solving

INTERMEDIATE
Algebra $f(x)$ **Now** ™

EXAMPLE 5 Manufacturing. A company makes three types of hammers, which are marketed as "good," "better," and "best." The cost of manufacturing each type of hammer is $4, $6, and $7, respectively, and the hammers sell for $6, $9, and $12. Each day, the cost of manufacturing 100 hammers is $520, and the daily revenue from their sale is $810. How many hammers of each type are manufactured?

Analyze the problem

We need to find how many of each type of hammer are manufactured daily. We must write three equations to find three unknowns.

Form three equations

If we let x = the number of good hammers, y = the number of better hammers, and z = the number of best hammers, we know that

The total number of hammers is $x + y + z$.

The cost of manufacturing the good hammers is $4x$ ($4 times x hammers).

The cost of manufacturing the better hammers is $6y$ ($6 times y hammers).

The cost of manufacturing the best hammers is $7z$ ($7 times z hammers).

The revenue received by selling the good hammers is $6x$ ($6 times x hammers).

The revenue received by selling the better hammers is $9y$ ($9 times y hammers).

The revenue received by selling the best hammers is $12z$ ($12 times z hammers).

We can assemble the facts of the problem to write three equations.

The number of good hammers	plus	the number of better hammers	plus	the number of best hammers	is	the total number of hammers.
x	$+$	y	$+$	z	$=$	100

The cost of good hammers	plus	the cost of better hammers	plus	the cost of best hammers	is	the total cost.
$4x$	$+$	$6y$	$+$	$7z$	$=$	520

The revenue from good hammers	plus	the revenue from better hammers	plus	the revenue from best hammers	is	the total revenue.
$6x$	$+$	$9y$	$+$	$12z$	$=$	810

Solve the system

We must now solve the system

(1) $\begin{cases} x + y + z = 100 \\ 4x + 6y + 7z = 520 \\ 6x + 9y + 12z = 810 \end{cases}$
(2)
(3)

If we multiply Equation 1 by -7 and add the result to Equation 2, we get

$$-7x - 7y - 7z = -700 \quad \text{Multiply both sides of Equation 1 by } -7.$$

(2) $\underline{\quad 4x + 6y + 7z = \quad 520}$
(4) $\quad -3x - \quad y \qquad\quad = -180$

If we multiply Equation 1 by -12 and add the result to Equation 3, we get

$$-12x - 12y - 12z = -1{,}200 \quad \text{Multiply both sides of Equation 1 by } -12.$$

(3) $\underline{\quad 6x + \quad 9y + 12z = \qquad 810}$
(5) $\quad -6x - \quad 3y \qquad\quad = \quad -390$

If we multiply Equation 4 by -3 and add it to Equation 5, we get

$$9x + 3y = \quad 540 \quad \text{Multiply both sides of Equation 4 by } -3.$$

(5) $\underline{-6x - 3y = -390}$
$\quad 3x \qquad\quad = \quad 150$
$\qquad\quad x = 50 \qquad \text{To find } x, \text{ divide both sides by } 3.$

To find y, we substitute 50 for x in Equation 4:

$$\begin{aligned} -3x - y &= -180 \\ -3(\mathbf{50}) - y &= -180 \qquad \text{Substitute 50 for } x. \\ -150 - y &= -180 \qquad -3(50) = -150. \\ -y &= -30 \qquad \text{Add 150 to both sides.} \\ y &= 30 \qquad \text{Divide both sides by } -1. \end{aligned}$$

To find z, we substitute 50 for x and 30 for y in Equation 1:

$$\begin{aligned} x + y + z &= 100 \\ \mathbf{50} + \mathbf{30} + z &= 100 \\ z &= 20 \qquad \text{Subtract 80 from both sides.} \end{aligned}$$

State the conclusion

The company manufactures 50 good hammers, 30 better hammers, and 20 best hammers each day.

Check the result

Check the proposed solution in each equation in the original system.

INTERMEDIATE
Algebra $f(x)$ Now™

■ Curve fitting

EXAMPLE 6 Finding the equation of a parabola. The equation of a parabola opening upward or downward is of the form $y = ax^2 + bx + c$. Find the equation of the parabola shown in Figure 3-12 on the next page by determining the values of a, b, and c.

Solution   Since the parabola passes through the points $(-1, 5)$, $(1, 1)$, and $(2, 2)$, each pair of coordinates must satisfy the equation $y = ax^2 + bx + c$. If we substitute the x- and y-coordinates of each point into the equation and simplify, we obtain the following system of three equations in three variables.

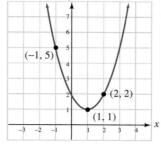

(1) $\begin{cases} a - b + c = 5 \quad & \text{Substitute } (-1, 5) \text{ into} \\ & y = ax^2 + bx + c \text{ and simplify.} \\ \\ a + b + c = 1 \quad & \text{Substitute } (1, 1) \text{ into} \\ & y = ax^2 + bx + c \text{ and simplify.} \\ \\ 4a + 2b + c = 2 \quad & \text{Substitute } (2, 2) \text{ into} \\ & y = ax^2 + bx + c \text{ and simplify.} \end{cases}$

(2)

(3)

FIGURE 3-12

If we add Equations 1 and 2, we obtain

$$\begin{array}{r} a - b + c = 5 \\ a + b + c = 1 \\ \hline 2a + 2c = 6 \end{array}$$

(4)

If we multiply Equation 1 by 2 and add the result to Equation 3, we get

$$\begin{array}{r} 2a - 2b + 2c = 10 \\ 4a + 2b + c = 2 \\ \hline 6a + 3c = 12 \end{array}$$

(5)

We can then divide both sides of Equation 4 by 2 to get Equation 6 and divide both sides of Equation 5 by 3 to get Equation 7. We now have the system

(6) $\begin{cases} a + c = 3 \\ 2a + c = 4 \end{cases}$

(7)

To eliminate c, we multiply Equation 6 by -1 and add the result to Equation 7. We get

$$\begin{array}{r} -a - c = -3 \\ 2a + c = 4 \\ \hline a = 1 \end{array}$$

To find c, we can substitute 1 for a in Equation 6 and find that $c = 2$. To find b, we can substitute 1 for a and 2 for c in Equation 2 and find that $b = -2$.

After we substitute these values of a, b, and c into the equation $y = ax^2 + bx + c$, we have the equation of the parabola.

$$y = ax^2 + bx + c$$
$$y = 1x^2 - 2x + 2$$
$$y = x^2 - 2x + 2$$

Section 3.3   STUDY SET

VOCABULARY   *Fill in the blanks.*

1. $\begin{cases} 2x + y - 3z = 0 \\ 3x - y + 4z = 5 \\ 4x + 2y - 6z = 0 \end{cases}$ is called a _____ of three linear equations.

2. If the first two equations of the system in Exercise 1 are added, the variable y is _____.

3. The equation $2x + 3y + 4z = 5$ is a linear equation in _____ variables.

4. The graph of the equation $2x + 3y + 4z = 5$ is a flat surface called a _____.

5. When three planes coincide, the equations of the system are _____ , and there are infinitely many solutions.

6. When three planes intersect in a line, the system will have _____ many solutions.

CONCEPTS

7. For each graph of a system of three equations, tell whether the solution set contains one solution, infinitely many solutions, or no solution.

a. 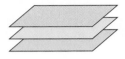   b.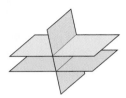

8. Consider the system: $\begin{cases} -2x + y + 4z = 3 & \textbf{(1)} \\ x - y + 2z = 1 & \textbf{(2)} \\ x + y - 3z = 2 & \textbf{(3)} \end{cases}$

  a. What is the result if Equation 1 and Equation 2 are added?

  b. What is the result if Equation 2 and Equation 3 are added?

  c. What variable was eliminated in the steps performed in parts a and b?

NOTATION

9. Write the equation $3z - 2y = x + 6$ in $Ax + By + Cz = D$ form.

10. Fill in the blank to make a true statement: Solutions of a system of three equations in three variables, x, y, and z, are written in the form (x, y, z) and are called ordered _____ .

PRACTICE *Determine whether the given ordered triple is a solution of given system.*

11. $(2, 1, 1)$, $\begin{cases} x - y + z = 2 \\ 2x + y - z = 4 \\ 2x - 3y + z = 2 \end{cases}$

12. $(-3, 2, -1)$, $\begin{cases} 2x + 2y + 3z = -1 \\ 3x + y - z = -6 \\ x + y + 2z = 1 \end{cases}$

Solve each system. If a system is inconsistent or if the equations are dependent, so indicate.

13. $\begin{cases} x + y + z = 4 \\ 2x + y - z = 1 \\ 2x - 3y + z = 1 \end{cases}$ **14.** $\begin{cases} x + y + z = 4 \\ x - y + z = 2 \\ x - y - z = 0 \end{cases}$

15. $\begin{cases} 2x + 2y + 3z = 10 \\ 3x + y - z = 0 \\ x + y + 2z = 6 \end{cases}$ **16.** $\begin{cases} x - y + z = 4 \\ x + 2y - z = -1 \\ x + y - 3z = -2 \end{cases}$

17. $\begin{cases} b + 2c = 7 - a \\ a + c = 8 - 2b \\ 2a + b + c = 9 \end{cases}$ **18.** $\begin{cases} 2a = 2 - 3b - c \\ 4a + 6b + 2c - 5 = 0 \\ a + c = 3 + 2b \end{cases}$

19. $\begin{cases} 2x + y - z = 1 \\ x + 2y + 2z = 2 \\ 4x + 5y + 3z = 3 \end{cases}$ **20.** $\begin{cases} 4x + 3z = 4 \\ 2y - 6z = -1 \\ 8x + 4y + 3z = 9 \end{cases}$

21. $\begin{cases} a + b + c = 180 \\ \frac{a}{4} + \frac{b}{2} + \frac{c}{3} = 60 \\ 2b + 3c - 330 = 0 \end{cases}$ **22.** $\begin{cases} 2a + 3b - 2c = 18 \\ 5a - 6b + c = 21 \\ 4b - 2c - 6 = 0 \end{cases}$

23. $\begin{cases} 0.5a + 0.3b = 2.2 \\ 1.2c - 8.5b = -24.4 \\ 3.3c + 1.3a = 29 \end{cases}$ **24.** $\begin{cases} 4a - 3b = 1 \\ 6a - 8c = 1 \\ 2b - 4c = 0 \end{cases}$

25. $\begin{cases} 2x + 3y + 4z = 6 \\ 2x - 3y - 4z = -4 \\ 4x + 6y + 8z = 12 \end{cases}$ **26.** $\begin{cases} x - 3y + 4z = 2 \\ 2x + y + 2z = 3 \\ 4x - 5y + 10z = 7 \end{cases}$

27. $\begin{cases} x + \frac{1}{3}y + z = 13 \\ \frac{1}{2}x - y + \frac{1}{3}z = -2 \\ x + \frac{1}{2}y - \frac{1}{3}z = 2 \end{cases}$ **28.** $\begin{cases} x - \frac{1}{5}y - z = 9 \\ \frac{1}{4}x + \frac{1}{5}y - \frac{1}{2}z = 5 \\ 2x + y + \frac{1}{6}z = 12 \end{cases}$

APPLICATIONS

29. MAKING STATUES An artist makes three types of ceramic statues at a monthly cost of $650 for 180 statues. The manufacturing costs for the three types are $5, $4, and $3. If the statues sell for $20, $12, and $9, respectively, how many of each type should be made to produce $2,100 in monthly revenue?

30. POTPOURRI The owner of a home decorating shop wants to mix dried rose petals selling for $6 per pound, dried lavender selling for $5 per pound, and buckwheat hulls selling for $4 per pound to get 10 pounds of a mixture that would sell for $5.50 per pound. She wants to use twice as many pounds of rose petals as lavender. How many pounds of each should she use?

31. NUTRITION A hospital dietitian is to design a meal that will provide a patient with exactly 14 grams (g) of fat, 9 g of carbohydrates, and 9 g of protein. She is to use a combination of the three foods listed in the table below. If one ounce (oz) of each of the foods has the nutrient content shown, how many ounces of each should be used?

Food	Fat	Carbohydrates	Protein
A	2 g/oz	1 g/oz	2 g/oz
B	3 g/oz	2 g/oz	1 g/oz
C	1 g/oz	1 g/oz	2 g/oz

32. NUTRITIONAL PLANNING One serving of each of three foods has the vitamin and mineral content shown below. How many servings of each must be used to provide exactly 22 milligrams (mg) of niacin, 12 mg of zinc, and 20 mg of vitamin C?

Vitamin/Mineral Content of 1 Serving			
Food	Niacin	Zinc	Vitamin C
A	1 mg	1 mg	2 mg
B	2 mg	1 mg	1 mg
C	2 mg	1 mg	2 mg

33. CHAINSAW SCULPTING A north woods sculptor carves three types of statues with a chainsaw. The number of hours required for carving, sanding, and painting a totem pole, a bear, and a deer are shown in the table. How many of each should be produced to use all available labor hours?

	Totem pole	Bear	Deer	Time available
Carving	2 hr	2 hr	1 hr	14 hr
Sanding	1 hr	2 hr	2 hr	15 hr
Painting	3 hr	2 hr	2 hr	21 hr

34. MAKING CLOTHES A clothing manufacturer makes coats, shirts, and slacks. The time required for cutting, sewing, and packaging each item is shown in the table. How many of each should be made to use all available labor hours?

	Coats	Shirts	Slacks	Time available
Cutting	20 min	15 min	10 min	115 hr
Sewing	60 min	30 min	24 min	280 hr
Packaging	5 min	12 min	6 min	65 hr

35. EARTH'S ATMOSPHERE Use the information in the circle graph to determine what percent of Earth's atmosphere is nitrogen, is oxygen, and is other gases.

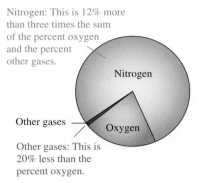

Nitrogen: This is 12% more than three times the sum of the percent oxygen and the percent other gases.

Other gases: This is 20% less than the percent oxygen.

36. NFL RECORDS Jerry Rice, who played with the San Francisco 49ers and the Oakland Raiders, holds the all-time record for touchdown passes caught. Here are some interesting facts about this feat.

- He caught 30 more TD passes from Steve Young than he did from Joe Montana.
- He caught 39 more TD passes from Joe Montana than he did from Rich Gannon.
- He caught a total of 156 TD passes from Young, Montana, and Gannon.

Determine the number of touchdown passes Rice has caught from Young, from Montana, and from Gannon as of 2003.

37. GRAPHS OF SYSTEMS Explain how each illustration could be thought of as an example of the graph of a system of three linear equations. Then describe the solution, if any.

a.

b.

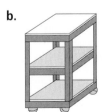

c.

d.

38. ZOOLOGY An X ray of a laboratory mouse revealed a cancerous tumor located at the intersection of the coronal, sagittal, and transverse planes, as shown below. From this description, would you expect the tumor to be at the base of the tail, on the back, in the stomach, on the tip of the right ear, or in the mouth of the mouse?

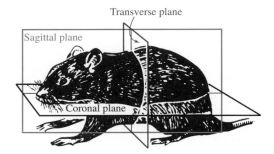

39. ASTRONOMY Comets have elliptical orbits, but the orbits of some comets are so vast that they are indistinguishable from parabolas. Find the equation of the parabola that closely describes the orbit of the comet shown below.

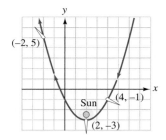

40. CURVE FITTING Find the equation of the parabola.

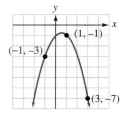

41. WALKWAYS A circular sidewalk is to be constructed in a city park. The walk is to pass by three particular areas of the park, as shown in the graph in the next column. If an equation of a circle is of the form $x^2 + y^2 + Cx + Dy + E = 0$, find the equation that describes the path of the sidewalk by determining C, D, and E.

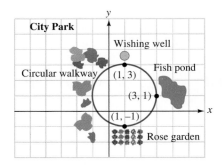

42. CURVE FITTING The equation of a circle is of the form $x^2 + y^2 + Cx + Dy + E = 0$. Find the equation of the circle shown by determining C, D, and E.

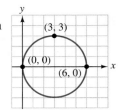

43. TRIANGLES The sum of the angles in any triangle is 180°. In triangle ABC, $\angle A$ is 100° less than the sum of $\angle B$ and $\angle C$, and $\angle C$ is 40° less than twice $\angle B$. Find the measure of each angle.

44. QUADRILATERALS The sum of the angles of any four-sided figure is 360°. In the quadrilateral shown, the measures of $\angle A$ and $\angle B$ are the same, $\angle C$ is 20° greater than $\angle A$, and $\angle D$ measures 40°. Find the measure of each angle.

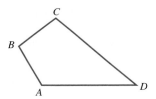

45. INTEGER PROBLEMS The sum of three integers is 48. If the first integer is doubled, the sum is 60. If the second integer is doubled, the sum is 63. Find the integers.

46. INTEGER PROBLEMS The sum of three integers is 18. The third integer is four times the second, and the second integer is 6 more than the first. Find the integers.

WRITING

47. Explain how a system of three equations in three variables can be reduced to a system of two equations in two variables.

48. What makes a system of three equations in three variables inconsistent?

REVIEW *Graph each function.*

49. $f(x) = |x|$

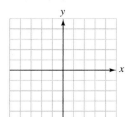

50. $g(x) = x^2$

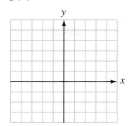

51. $h(x) = x^3$

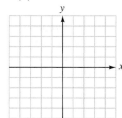

52. $S(x) = x$

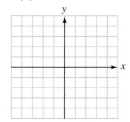

3.4 Solving Systems Using Matrices

- Matrices • Augmented matrices • Gaussian elimination
- Solving a system of three equations • Inconsistent systems and dependent equations

In this section, we will discuss another method for solving systems of linear equations. This technique uses a mathematical tool called a *matrix* in a series of steps that are based on the addition method.

Matrices

Another method of solving systems of equations involves rectangular arrays of numbers called **matrices** (plural for **matrix**).

> **Matrices**
>
> A **matrix** is any rectangular array of numbers arranged in rows and columns, written within brackets.

Some examples of matrices are

$$A = \begin{bmatrix} 1 & -3 & 8 \\ 2 & 5 & -1 \end{bmatrix} \begin{matrix} \leftarrow \text{Row 1} \\ \leftarrow \text{Row 2} \end{matrix}$$

$$\begin{matrix} \uparrow & \uparrow & \uparrow \\ \text{Column} & \text{Column} & \text{Column} \\ 1 & 2 & 3 \end{matrix}$$

$$B = \begin{bmatrix} 1 & 4 & -2 & -4 \\ 6 & -2 & 6 & 1 \\ 3 & 8 & -3 & 12 \end{bmatrix} \begin{matrix} \leftarrow \text{Row 1} \\ \leftarrow \text{Row 2} \\ \leftarrow \text{Row 3} \end{matrix}$$

$$\begin{matrix} \uparrow & \uparrow & \uparrow & \uparrow \\ \text{Column} & \text{Column} & \text{Column} & \text{Column} \\ 1 & 2 & 3 & 4 \end{matrix}$$

The numbers in each matrix are called **elements.** Because matrix A has two rows and three columns, it is called a 2×3 matrix (read "2 by 3" matrix). Matrix B is a 3×4 matrix (three rows and four columns).

Augmented matrices

To show how to use matrices to solve systems of linear equations, we consider the system

$$\begin{cases} x - y = 4 \\ 2x + y = 5 \end{cases}$$

which can be represented by the following matrix, called an **augmented matrix:**

$$\begin{bmatrix} 1 & -1 & \vdots & 4 \\ 2 & 1 & \vdots & 5 \end{bmatrix}$$

Each row of the augmented matrix represents one equation of the system. The first two columns of the augmented matrix are determined by the coefficients of x and y in the equations of the system. The last column is determined by the constants in the equations.

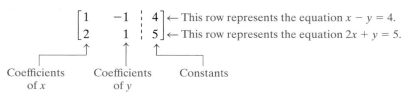

$\begin{bmatrix} 1 & -1 & \vdots & 4 \\ 2 & 1 & \vdots & 5 \end{bmatrix}$ ← This row represents the equation $x - y = 4$.
← This row represents the equation $2x + y = 5$.

Coefficients   Coefficients   Constants
of x of y

Self Check 1

Represent each system using an augmented matrix:

a. $\begin{cases} 2x - 4y = 9 \\ 5x - y = -2 \end{cases}$

b. $\begin{cases} a + b - c = -4 \\ -2b + 7c = 0 \\ 10a + 8b - 4c = 5 \end{cases}$

Answers

a. $\begin{bmatrix} 2 & -4 & \vdots & 9 \\ 5 & -1 & \vdots & -2 \end{bmatrix}$,

b. $\begin{bmatrix} 1 & 1 & -1 & \vdots & -4 \\ 0 & -2 & 7 & \vdots & 0 \\ 10 & 8 & -4 & \vdots & 5 \end{bmatrix}$

EXAMPLE 1   Represent each system of equations using an augmented matrix:

a. $\begin{cases} 3x + y = 11 \\ x - 8y = 0 \end{cases}$ and **b.** $\begin{cases} 2a + b - 3c = -3 \\ 9a + 4c = 2 \\ a - b - 6c = -7 \end{cases}$.

Solution

a. $\begin{cases} 3x + y = 11 & \leftrightarrow \\ x - 8y = 0 & \leftrightarrow \end{cases}$ $\begin{bmatrix} 3 & 1 & \vdots & 11 \\ 1 & -8 & \vdots & 0 \end{bmatrix}$

b. $\begin{cases} 2a + b - 3c = -3 & \leftrightarrow \\ 9a + 4c = 2 & \leftrightarrow \\ a - b - 6c = -7 & \leftrightarrow \end{cases}$ $\begin{bmatrix} 2 & 1 & -3 & \vdots & -3 \\ 9 & 0 & 4 & \vdots & 2 \\ 1 & -1 & -6 & \vdots & -7 \end{bmatrix}$

Gaussian elimination

To solve a 2×2 system of equations by **Gaussian elimination,** we transform the augmented matrix into the following matrix, which has 1's down its main diagonal and a 0 below the 1 in the first column.

$$\begin{bmatrix} 1 & a & \vdots & b \\ 0 & 1 & \vdots & c \end{bmatrix} \quad a, b, \text{ and } c \text{ represent real numbers}$$

Main diagonal

To write the augmented matrix in this form, we use three operations called **elementary row operations.**

Elementary row operations

Type 1: Any two rows of a matrix can be interchanged.
Type 2: Any row of a matrix can be multiplied by a nonzero constant.
Type 3: Any row of a matrix can be changed by adding a nonzero constant multiple of another row to it.

- A type 1 row operation corresponds to interchanging two equations of the system.
- A type 2 row operation corresponds to multiplying both sides of an equation by a nonzero constant.
- A type 3 row operation corresponds to adding a nonzero multiple of one equation to another.

None of these row operations will change the solution of the given system of equations.

EXAMPLE 2   Consider the matrices:

$$A = \begin{bmatrix} 2 & 4 & \vdots & -3 \\ 1 & -8 & \vdots & 0 \end{bmatrix}, \quad B = \begin{bmatrix} 1 & -1 & \vdots & 2 \\ 4 & -8 & \vdots & 0 \end{bmatrix}, \quad C = \begin{bmatrix} 2 & 1 & -8 & \vdots & 4 \\ 0 & 1 & 4 & \vdots & -2 \\ 0 & 0 & -6 & \vdots & 24 \end{bmatrix}.$$

a. Interchange rows 1 and 2 of matrix A.

b. Multiply row 3 of matrix C by $-\frac{1}{6}$.

c. To the numbers in row 2 of matrix B, add the results of multiplying each number in row 1 by -4.

Solution

a. Interchanging the rows of matrix A, we obtain $\begin{bmatrix} 1 & -8 & \vdots & 0 \\ 2 & 4 & \vdots & -3 \end{bmatrix}$.

b. We multiply each number in row 3 of matrix C by $-\frac{1}{6}$. Rows 1 and 2 remain unchanged.

$$\begin{bmatrix} 2 & 1 & -8 & \vdots & 4 \\ 0 & 1 & 4 & \vdots & -2 \\ 0 & 0 & 1 & \vdots & -4 \end{bmatrix}$$ We can represent the instruction to multiply the third row by $-\frac{1}{6}$ with the symbolism $-\frac{1}{6}R_3$.

c. If we multiply each number in row 1 of matrix B by -4, we get

$$-4 \quad 4 \quad -8$$

We then add these numbers to row 2. (Note that row 1 remains unchanged.)

$$\begin{bmatrix} 1 & -1 & \vdots & 2 \\ 4 + (-4) & -8 + 4 & \vdots & 0 + (-8) \end{bmatrix}$$ We can abbreviate this procedure using the notation $-4R_1 + R_2$, which means "Multiply row 1 by -4 and add the result to row 2."

After simplifying, we have the matrix

$$\begin{bmatrix} 1 & -1 & \vdots & 2 \\ 0 & -4 & \vdots & -8 \end{bmatrix}$$

Self Check 2

Refer to Example 2.

a. Interchange the rows of matrix B.

b. To the numbers in row 1 of matrix A, add the results of multiplying each number in row 2 by -2.

c. Interchange rows 2 and 3 of matrix C.

Answers **a.** $\begin{bmatrix} 4 & -8 & \vdots & 0 \\ 1 & -1 & \vdots & 2 \end{bmatrix}$,

b. $\begin{bmatrix} 0 & 20 & \vdots & -3 \\ 1 & -8 & \vdots & 0 \end{bmatrix}$,

c. $\begin{bmatrix} 2 & 1 & -8 & \vdots & 4 \\ 0 & 0 & -6 & \vdots & 24 \\ 0 & 1 & 4 & \vdots & -2 \end{bmatrix}$

We now solve a system of two linear equations using the **Gaussian elimination** process, which involves a series of elementary row operations.

EXAMPLE 3   Solve: the system $\begin{cases} 2x + y = 5 \\ x - y = 4 \end{cases}$.

Solution   We can represent the system with the following augmented matrix:

$$\begin{bmatrix} 2 & 1 & \vdots & 5 \\ 1 & -1 & \vdots & 4 \end{bmatrix}$$

Self Check 3

Solve: $\begin{cases} 3x - 2y = -5 \\ x - y = -4 \end{cases}$.

First, we want to get a 1 in the top row of the first column where the red 2 is. This can be achieved by applying a type 1 row operation: Interchange rows 1 and 2.

$$\begin{bmatrix} 1 & -1 & \vdots & 4 \\ 2 & 1 & \vdots & 5 \end{bmatrix}$$ Interchanging row 1 and row 2 can be abbreviated as $R_1 \leftrightarrow R_2$.

To get a 0 under the 1 in the first column, we use a type 3 row operation. To row 2, we add the results of multiplying each number in row 1 by -2.

$$\begin{bmatrix} 1 & -1 & \vdots & 4 \\ 0 & 3 & \vdots & -3 \end{bmatrix}$$ $-2R_1 + R_2.$

To get a 1 in the bottom row of the second column, we use a type 2 row operation: Multiply row 2 by $\frac{1}{3}$.

$$\begin{bmatrix} 1 & -1 & \vdots & 4 \\ 0 & 1 & \vdots & -1 \end{bmatrix}$$ $\frac{1}{3}R_2.$

This augmented matrix represents the equations

$$1x - 1y = 4$$
$$0x + 1y = -1$$

Writing the equations without the coefficients of 1 and -1, we have

(1) $x - y = 4$

(2) $y = -1$

From Equation 2, we see that $y = -1$. We can back substitute -1 for y in Equation 1 to find x.

$$x - y = 4$$
$$x - (-1) = 4 \quad \text{Substitute } -1 \text{ for } y.$$
$$x + 1 = 4 \quad -(-1) = 1.$$
$$x = 3 \quad \text{Subtract 1 from both sides.}$$

The solution of the system is $(3, -1)$. Verify that this ordered pair satisfies the original system.

Answer $(3, 7)$

In general, if a system of linear equations has a single solution, we can use the following steps to solve the system using matrices.

Solving systems of linear equations using matrices

1. Write an augmented matrix for the system.

2. Use elementary row operations to transform the augmented matrix into a matrix with 1's down its main diagonal and 0's under the 1's.

3. When step 2 is complete, write the resulting system. Then use back substitution to find the solution.

4. Check the proposed solution in each equation of the original system.

■ Solving a system of three equations

To show how to use matrices to solve systems of three linear equations containing three variables, we consider the system

$$\begin{cases} x - 2y - z = 6 \\ 2x + 2y - z = 1 \\ -x - y + 2z = 1 \end{cases}$$

which can be represented by the augmented matrix

$$\begin{bmatrix} 1 & -2 & -1 & \vdots & 6 \\ 2 & 2 & -1 & \vdots & 1 \\ -1 & -1 & 2 & \vdots & 1 \end{bmatrix}$$

To solve a 3×3 system by Gaussian elimination, we transform the augmented matrix into a matrix with 1's down its main diagonal and 0's below its main diagonal.

$$\begin{bmatrix} 1 & a & b & \vdots & c \\ 0 & 1 & d & \vdots & e \\ 0 & 0 & 1 & \vdots & f \end{bmatrix} \quad a, b, c, \dots, f \text{ represent real numbers}$$

Main diagonal

INTERMEDIATE
Algebra $f(x)$ Now™

EXAMPLE 4 Solve the system: $\begin{cases} 3x + y + 5z = 8 \\ 2x + 3y - z = 6 \\ x + 2y + 2z = 10 \end{cases}$.

Solution This system can be represented by the augmented matrix shown on the left below. To get a 1 in the first column in place of the red 3, we perform a type 1 row operation and interchange rows 1 and 3, as shown on the right.

Self Check 4

Solve: $\begin{cases} 2x - y + z = 5 \\ x + y - z = -2 \\ -x + 2y + 2z = 1 \end{cases}$.

$$\begin{bmatrix} 3 & 1 & 5 & \vdots & 8 \\ 2 & 3 & -1 & \vdots & 6 \\ 1 & 2 & 2 & \vdots & 10 \end{bmatrix} \qquad \begin{bmatrix} 1 & 2 & 2 & \vdots & 10 \\ 2 & 3 & -1 & \vdots & 6 \\ 3 & 1 & 5 & \vdots & 8 \end{bmatrix} \quad R_1 \leftrightarrow R_2$$

To get a 0 under the 1 in the first column in place of the red 2, we perform a type 3 row operation: Multiply row 1 by -2 and add the results to row 2. Row 1 remains the same. (See below left.)

To get a 0 under the 0 in the first column in place of the red 3, we perform another type 3 row operation: Multiply row 1 by -3 and add the results to row 3. Row 1 remains the same. (See below right.)

$$\begin{bmatrix} 1 & 2 & 2 & \vdots & 10 \\ 0 & -1 & -5 & \vdots & -14 \\ 3 & 1 & 5 & \vdots & 8 \end{bmatrix} \quad -2R_1 + R_2 \qquad \begin{bmatrix} 1 & 2 & 2 & \vdots & 10 \\ 0 & -1 & -5 & \vdots & -14 \\ 0 & -5 & -1 & \vdots & -22 \end{bmatrix} \quad -3R_1 + R_3$$

To get a 1 under the 2 in the second column in place of the red -1, we perform a type 2 row operation: Multiply row 2 by -1. (See below left.)

To get a 0 under the 1 in the second column in place of the red -5, we perform a type 3 row operation: Multiply row 2 by 5 and add the results to row 3. Row 2 remains the same. (See below right.)

$$\begin{bmatrix} 1 & 2 & 2 & \vdots & 10 \\ 0 & 1 & 5 & \vdots & 14 \\ 0 & -5 & -1 & \vdots & -22 \end{bmatrix} \quad -1R_2. \qquad \begin{bmatrix} 1 & 2 & 2 & \vdots & 10 \\ 0 & 1 & 5 & \vdots & 14 \\ 0 & 0 & 24 & \vdots & 48 \end{bmatrix} \quad 5R_2 + R_3$$

To get a 1 under the 5 in the third column in place of the red 24, we perform a type 2 row operation: Multiply row 3 by $\frac{1}{24}$.

$$\begin{bmatrix} 1 & 2 & 2 & \vdots & 10 \\ 0 & 1 & 5 & \vdots & 14 \\ 0 & 0 & 1 & \vdots & 2 \end{bmatrix} \quad \frac{1}{24}R_3$$

The final matrix represents the system

$$\begin{cases} 1x + 2y + 2z = 10 \\ 0x + 1y + 5z = 14 \\ 0x + 0y + 1z = 2 \end{cases}$$ which can be written without the coefficients of 0 and 1 as $\begin{cases} x + 2y + 2z = 10 \quad \textbf{(1)} \\ y + 5z = 14 \quad \textbf{(2)} \\ z = 2 \quad \textbf{(3)} \end{cases}$

From Equation 3, we can read that z is 2. To find y, we back substitute 2 for z in Equation 2 and solve for y:

$$y + 5z = 14 \qquad \text{This is Equation 2.}$$
$$y + 5(2) = 14 \qquad \text{Substitute 2 for } z.$$
$$y + 10 = 14$$
$$y = 4 \qquad \text{Subtract 10 from both sides.}$$

Thus, y is 4. To find x, we back substitute 2 for z and 4 for y in Equation 1 and solve for x:

$$x + 2y + 2z = 10 \qquad \text{This is Equation 1.}$$
$$x + 2(4) + 2(2) = 10 \qquad \text{Substitute 2 for } z \text{ and 4 for } y.$$
$$x + 8 + 4 = 10$$
$$x + 12 = 10$$
$$x = -2 \qquad \text{Subtract 12 from both sides.}$$

Thus, x is -2. The solution of the given system is $(-2, 4, 2)$. Verify that this ordered triple satisfies each equation of the original system.

Answer $(1, -1, 2)$

Inconsistent systems and dependent equations

In the next example, we consider a system with no solution.

INTERMEDIATE
Algebra *f(x)* **Now**™

Self Check 5

Solve: $\begin{cases} 4x - 8y = 9 \\ x - 2y = -5 \end{cases}$.

EXAMPLE 5 Using matrices, solve the system: $\begin{cases} x + y = -1 \\ -3x - 3y = -5 \end{cases}$.

Solution This system can be represented by the augmented matrix

$$\begin{bmatrix} 1 & 1 & \vdots & -1 \\ -3 & -3 & \vdots & -5 \end{bmatrix}$$

Since the matrix has a 1 in the top row of the first column, we proceed to get a 0 under it by multiplying row 1 by 3 and adding the results to row 2.

$$\begin{bmatrix} 1 & 1 & \vdots & -1 \\ 0 & 0 & \vdots & -8 \end{bmatrix} \qquad 3R_1 + R_2.$$

This matrix represents the system

$$\begin{cases} x + y = -1 \\ 0 + 0 = -8 \end{cases}$$

This system has no solution, because the second equation is never true. Therefore, the system is inconsistent. It has no solutions.

Answer no solution

In the next example, we consider a system with infinitely many solutions.

INTERMEDIATE
Algebra *f(x)* **Now**™

Self Check 6

Solve: $\begin{cases} 5x - 10y + 15z = 35 \\ -3x + 6y - 9z = -21 \\ 2x - 4y + 6z = 14 \end{cases}$.

EXAMPLE 6 Using matrices, solve the system: $\begin{cases} 2x + 3y - 4z = 6 \\ 4x + 6y - 8z = 12 \\ -6x - 9y + 12z = -18 \end{cases}$.

Solution This system can be represented by the augmented matrix

$$\begin{bmatrix} 2 & 3 & -4 & \vdots & 6 \\ 4 & 6 & -8 & \vdots & 12 \\ -6 & -9 & 12 & \vdots & -18 \end{bmatrix}$$

To get a 1 in the top row of the first column, we multiply row 1 by $\frac{1}{2}$.

$$\begin{bmatrix} 1 & \frac{3}{2} & -2 & \vdots & 3 \\ 4 & 6 & -8 & \vdots & 12 \\ -6 & -9 & 12 & \vdots & -18 \end{bmatrix} \quad \frac{1}{2}R_1.$$

Next, we want to get 0's under the 1 in the first column. This can be achieved by multiplying row 1 by -4 and adding the results to row 2, and multiplying row 1 by 6 and adding the results to row 3.

$$\begin{bmatrix} 1 & \frac{3}{2} & -2 & \vdots & 3 \\ 0 & 0 & 0 & \vdots & 0 \\ 0 & 0 & 0 & \vdots & 0 \end{bmatrix} \quad \begin{matrix} -4R_1 + R_2. \\ 6R_1 + R_3. \end{matrix}$$

The last matrix represents the system

$$\begin{cases} x + \frac{3}{2}y - 2z = 3 \\ 0x + 0y + 0z = 0 \\ 0x + 0y + 0z = 0 \end{cases}$$

If we clear the first equation of fractions, we have the system

$$\begin{cases} 2x + 3y - 4z = 6 \\ 0 = 0 \\ 0 = 0 \end{cases}$$

This system has dependent equations and infinitely many solutions. Solutions of this system would be any triple (x, y, z) that satisfies the equation $2x + 3y - 4z = 6$. Two such solutions would be $(0, 2, 0)$ and $(1, 0, -1)$.

Answer There are infinitely many solutions — any triple satisfying the equation $x - 2y + 3z = 7$.

Section 3.4 STUDY SET

INTERMEDIATE
Algebra *f(x)* Now™

VOCABULARY   *Fill in the blanks.*

1. A _____ is a rectangular array of numbers.

2. The numbers in a matrix are called its _____.

3. A 3×4 matrix has 3 _____ and 4 _____.

4. Elementary _____ operations are used to produce new matrices that lead to the solution of a system.

5. A matrix that represents the equations of a system is called an _____ matrix.

6. The augmented matrix $\begin{bmatrix} 1 & 3 & \vdots & -2 \\ 0 & 1 & \vdots & 4 \end{bmatrix}$ has 1's down its main _____.

CONCEPTS

7. For each matrix, tell the number of rows and the number of columns.

a. $\begin{bmatrix} 4 & 6 & \vdots & -1 \\ \frac{1}{2} & 9 & \vdots & -3 \end{bmatrix}$ **b.** $\begin{bmatrix} 1 & -2 & 3 & \vdots & 1 \\ 0 & 1 & 6 & \vdots & 4 \\ 0 & 0 & 1 & \vdots & \frac{1}{3} \end{bmatrix}$

8. For each augmented matrix, give the system of equations it represents.

a. $\begin{bmatrix} 1 & 6 & \vdots & 7 \\ 0 & 1 & \vdots & 4 \end{bmatrix}$

b. $\begin{bmatrix} 2 & -2 & 9 & \vdots & 1 \\ 3 & 1 & 1 & \vdots & 0 \\ 2 & -6 & 8 & \vdots & -7 \end{bmatrix}$

9. Write the system of equations represented by the augmented matrix. Then use back substitution to find the solution.

$$\begin{bmatrix} 1 & -1 & \vdots & -10 \\ 0 & 1 & \vdots & 6 \end{bmatrix}$$

10. Write the system of equations represented by the augmented matrix. Then use back substitution to find the solution.

$$\begin{bmatrix} 1 & -2 & 1 & \vdots & -16 \\ 0 & 1 & 2 & \vdots & 8 \\ 0 & 0 & 1 & \vdots & 4 \end{bmatrix}$$

11. Matrices were used to solve a system of two linear equations. The final matrix is shown here. Explain what the result tells about the system.

$$\begin{bmatrix} 1 & 2 & \vdots & -4 \\ 0 & 0 & \vdots & 2 \end{bmatrix}$$

12. Matrices were used to solve a system of two linear equations. The final matrix is shown here. What does the result tell about the equations?

$$\begin{bmatrix} 1 & 2 & \vdots & -4 \\ 0 & 0 & \vdots & 0 \end{bmatrix}$$

▌ NOTATION

13. Consider the matrix $A = \begin{bmatrix} 3 & 6 & -9 & \vdots & 0 \\ 1 & 5 & -2 & \vdots & 1 \\ -2 & 2 & -2 & \vdots & 5 \end{bmatrix}$.

   a. Explain what is meant by $\frac{1}{3}R_1$. Then perform the operation on matrix A.

   b. Explain what is meant by $-R_1 + R_2$. Then perform the operation on the answer to part a.

14. Consider the matrix $B = \begin{bmatrix} -3 & 1 & \vdots & -6 \\ 1 & -4 & \vdots & 4 \end{bmatrix}$.

   a. Explain what is meant by $R_1 \leftrightarrow R_2$. Then perform the operation on matrix B.

   b. Explain what is meant by $3R_1 + R_2$. Then perform the operation on the answer to part a.

Complete each solution to solve the system.

15. Solve: $\begin{cases} 4x - y = 14 \\ x + y = 6 \end{cases}$.

$$\begin{bmatrix} 4 & & \vdots & 14 \\ 1 & 1 & \vdots & 6 \end{bmatrix}$$

$$\begin{bmatrix} & 1 & \vdots & 6 \\ 4 & -1 & \vdots & 14 \end{bmatrix} \quad R_1 \leftrightarrow R_2.$$

$$\begin{bmatrix} 1 & 1 & \vdots & 6 \\ 0 & & \vdots & -10 \end{bmatrix} \quad -4R_1 + R_2.$$

$$\begin{bmatrix} 1 & 1 & \vdots & 6 \\ 0 & 1 & \vdots & \end{bmatrix} \quad -\frac{1}{5}R_2.$$

This matrix represents the system
$$\begin{cases} x + y = 6 \\ = 2 \end{cases}$$
The solution is (, 2).

16. Solve: $\begin{cases} 2x + 2y = 18 \\ x - y = 5 \end{cases}$.

$$\begin{bmatrix} 2 & 2 & \vdots & 18 \\ & -1 & \vdots & 5 \end{bmatrix}$$

$$\begin{bmatrix} 1 & 1 & \vdots & 9 \\ & -1 & \vdots & 5 \end{bmatrix} \quad \frac{1}{2}R_1.$$

$$\begin{bmatrix} 1 & 1 & \vdots & 9 \\ 0 & -2 & \vdots & -4 \end{bmatrix} \quad -R_1 + R_2.$$

$$\begin{bmatrix} 1 & 1 & \vdots & 9 \\ 0 & 1 & \vdots & \end{bmatrix} \quad -\frac{1}{2}R_2.$$

This matrix represents the system
$$\begin{cases} x + y = \\ y = 2 \end{cases}$$
The solution is (, 2).

▌ PRACTICE   *Use matrices to solve each system. If the equations of a system are dependent or if a system is inconsistent, so indicate.*

17. $\begin{cases} x + y = 2 \\ x - y = 0 \end{cases}$ **18.** $\begin{cases} x + y = 3 \\ x - y = -1 \end{cases}$

19. $\begin{cases} 2x + y = 1 \\ x + 2y = -4 \end{cases}$ **20.** $\begin{cases} 5x - 4y = 10 \\ x - 7y = 2 \end{cases}$

21. $\begin{cases} 2x - y = -1 \\ x - 2y = 1 \end{cases}$ **22.** $\begin{cases} 2x - y = 0 \\ x + y = 3 \end{cases}$

23. $\begin{cases} 3x + 4y = -12 \\ 9x - 2y = 6 \end{cases}$ **24.** $\begin{cases} 2x - 3y = 16 \\ -4x + y = -22 \end{cases}$

25. $\begin{cases} x + y + z = 6 \\ x + 2y + z = 8 \\ x + y + 2z = 9 \end{cases}$ **26.** $\begin{cases} x - y + z = 2 \\ x + 2y - z = 6 \\ 2x - y - z = 3 \end{cases}$

27. $\begin{cases} 3x + y - 3z = 5 \\ x - 2y + 4z = 10 \\ x + y + z = 13 \end{cases}$ **28.** $\begin{cases} 2x + y - 3z = -1 \\ 3x - 2y - z = -5 \\ x - 3y - 2z = -12 \end{cases}$

29. $\begin{cases} 3x - 2y + 4z = 4 \\ x + y + z = 3 \\ 6x - 2y - 3z = 10 \end{cases}$ **30.** $\begin{cases} 2x + 3y - z = -8 \\ x - y - z = -2 \\ -4x + 3y + z = 6 \end{cases}$

31. $\begin{cases} 2a + b + 3c = 3 \\ -2a - b + c = 5 \\ 4a - 2b + 2c = 2 \end{cases}$ **32.** $\begin{cases} 3a + 2b + c = 8 \\ 6a - b + 2c = 16 \\ -9a + b - c = -20 \end{cases}$

33. $\begin{cases} x - 3y = 9 \\ -2x + 6y = 18 \end{cases}$ **34.** $\begin{cases} -6x + 12y = 10 \\ 2x - 4y = 8 \end{cases}$

35. $\begin{cases} 4x + 4y = 12 \\ -x - y = -3 \end{cases}$ **36.** $\begin{cases} 5x - 15y = 10 \\ 2x - 6y = 4 \end{cases}$

37. $\begin{cases} 6x + y - z = -2 \\ x + 2y + z = 5 \\ 5y - z = 2 \end{cases}$ **38.** $\begin{cases} 2x + 3y - 2z = 18 \\ 5x - 6y + z = 21 \\ 4y - 2z = 6 \end{cases}$

39. $\begin{cases} 2x + y - z = 1 \\ x + 2y + 2z = 2 \\ 4x + 5y + 3z = 3 \end{cases}$ **40.** $\begin{cases} x - 3y + 4z = 2 \\ 2x + y + 2z = 3 \\ 4x - 5y + 10z = 7 \end{cases}$

41. $\begin{cases} 5x + 3y = 4 \\ 3y - 4z = 4 \\ x + z = 1 \end{cases}$ **42.** $\begin{cases} y + 2z = -2 \\ x + y = 1 \\ 2x - z = 0 \end{cases}$

43. $\begin{cases} x - y = 1 \\ 2x - z = 0 \\ 2y - z = -2 \end{cases}$ **44.** $\begin{cases} x + y - 3z = 4 \\ 2x + 2y - 6z = 5 \\ -3x + y - z = 2 \end{cases}$

<hr>

APPLICATIONS   *Remember these facts from geometry: The sum of the measures of complementary angles is 90°, the sum of the measures of supplementary angles is 180°, and the sum of the measures of the interior angles of a triangle is 180°.*

45. One angle measures 46° more than the measure of its complement. Find the measure of each angle.

46. One angle measures 14° more than the measure of its supplement. Find the measure of each angle.

47. In the illustration, $\angle B$ measures 25° more than the measure of $\angle A$, and the measure of $\angle C$ is 5° less than twice the measure of $\angle A$. Find the measure of each angle of the triangle.

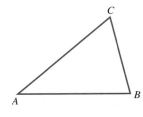

48. In the illustration, $\angle A$ measures 10° less than the measure of $\angle B$, and the measure of $\angle B$ is 10° less than the measure of $\angle C$. Find the measure of each angle of the triangle.

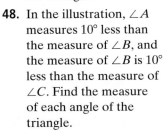

Write a system of equations to solve each problem. Then use matrices to solve it.

49. PHYSICAL THERAPY   After an elbow injury, a volleyball player has restricted movement of her arm. Her range of motion (angle 1) is 28° less than angle 2. Find the measure of each angle.

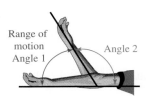

50. PIGGY BANKS   When a child breaks open her piggy bank, she finds a total of 64 coins, consisting of nickels, dimes, and quarters. The total value of the coins is $6. If the nickels were dimes, and the dimes were nickels, the value of the coins would be $5. How many nickels, dimes, and quarters were in the piggy bank?

51. THEATER SEATING   The illustration shows the cash receipts from two sold-out performances of a play and the ticket prices. Find the number of seats in each of the three sections of the 800-seat theater.

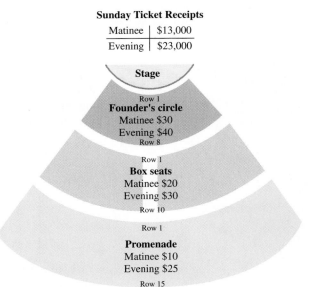

52. ICE SKATING Three circles were traced out by a figure skater during her performance. If the centers of the circles are the given distances apart, find the radius of each circle.

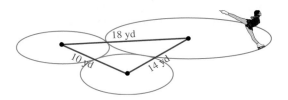

WRITING

53. Explain what is meant by the phrase *back substitution*.

54. Explain how a type 3 row operation is similar to the addition method of solving a system of equations.

REVIEW

55. What is the formula used to find the slope of a line, given two points on the line?

56. What is the form of the equation of a horizontal line? Of a vertical line?

57. What is the point–slope form of the equation of a line?

58. What is the slope–intercept form of the equation of a line?

3.5 Solving Systems Using Determinants

- Determinants • Evaluating a determinant
- Using Cramer's rule to solve a system of two equations
- Using Cramer's rule to solve a system of three equations

In this section, we will discuss another method for solving systems of linear equations. With this method, called *Cramer's rule,* we work with combinations of the coefficients and the constants of the equations written as *determinants.*

Determinants

An idea closely related to the concept of matrix is the **determinant.** A determinant is a number that is associated with a **square matrix,** a matrix that has the same number of rows and columns. For any square matrix A, the symbol $|A|$ represents the determinant of A. To write a determinant, we put the elements of a square matrix between two vertical lines.

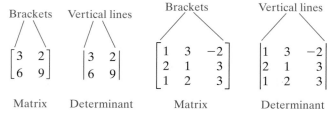

Like matrices, determinants are classified according to the number of rows and columns they contain. The determinant on the left is a 2×2 determinant. The other is a 3×3 determinant.

■ Evaluating a determinant

The determinant of a 2 × 2 matrix is the number that is equal to the product of the numbers on the main diagonal minus the product of the numbers on the other diagonal.

$$\begin{vmatrix} a & b \\ c & d \end{vmatrix}$$
Main diagonal

$$\begin{vmatrix} a & b \\ c & d \end{vmatrix}$$
Other diagonal

Value of a 2 × 2 determinant

If a, b, c, and d represent numbers, the **determinant** of the matrix $\begin{bmatrix} a & b \\ c & d \end{bmatrix}$ is

$$\begin{vmatrix} a & b \\ c & d \end{vmatrix} = ad - bc$$

INTERMEDIATE
Algebra $f(x)$ Now™

EXAMPLE 1   Find each value: **a.** $\begin{vmatrix} 3 & 2 \\ 6 & 9 \end{vmatrix}$ and   **b.** $\begin{vmatrix} -5 & \frac{1}{2} \\ -1 & 0 \end{vmatrix}$.

Self Check 1

Evaluate: $\begin{vmatrix} 4 & -3 \\ 2 & 1 \end{vmatrix}$.

Solution   From the product of the numbers along the main diagonal, we subtract the product of the numbers along the other diagonal.

a. $\begin{vmatrix} 3 & 2 \\ 6 & 9 \end{vmatrix} = 3(9) - 2(6)$

$= 27 - 12$

$= 15$

b. $\begin{vmatrix} -5 & \frac{1}{2} \\ -1 & 0 \end{vmatrix} = -5(0) - \frac{1}{2}(-1)$

$= 0 + \frac{1}{2}$

$= \frac{1}{2}$

Answer 10

A 3 × 3 determinant is evaluated by **expanding by minors.**

Value of a 3 × 3 determinant

$$\begin{vmatrix} a_1 & b_1 & c_1 \\ a_2 & b_2 & c_2 \\ a_3 & b_3 & c_3 \end{vmatrix} = a_1 \overset{\text{Minor of } a_1}{\begin{vmatrix} b_2 & c_2 \\ b_3 & c_3 \end{vmatrix}} - b_1 \overset{\text{Minor of } b_1}{\begin{vmatrix} a_2 & c_2 \\ a_3 & c_3 \end{vmatrix}} + c_1 \overset{\text{Minor of } c_1}{\begin{vmatrix} a_2 & b_2 \\ a_3 & b_3 \end{vmatrix}}$$

To find the minor of a_1, we cross out the elements of the determinant that are in the same row and column as a_1:

$$\begin{vmatrix} a_1 & b_1 & c_1 \\ a_2 & b_2 & c_2 \\ a_3 & b_3 & c_3 \end{vmatrix} \qquad \text{The minor of } a_1 \text{ is } \begin{vmatrix} b_2 & c_2 \\ b_3 & c_3 \end{vmatrix}.$$

To find the minor of b_1, we cross out the elements of the determinant that are in the same row and column as b_1:

$$\begin{vmatrix} a_1 & b_1 & c_1 \\ a_2 & b_2 & c_2 \\ a_3 & b_3 & c_3 \end{vmatrix} \qquad \text{The minor of } b_1 \text{ is } \begin{vmatrix} a_2 & c_2 \\ a_3 & c_3 \end{vmatrix}.$$

To find the minor of c_1, we cross out the elements of the determinant that are in the same row and column as c_1:

$$\begin{vmatrix} a_1 & b_1 & c_1 \\ a_2 & b_2 & c_2 \\ a_3 & b_3 & c_3 \end{vmatrix} \qquad \text{The minor of } c_1 \text{ is } \begin{vmatrix} a_2 & b_2 \\ a_3 & b_3 \end{vmatrix}.$$

INTERMEDIATE
Algebra $f(x)$ Now™

Self Check 2

Evaluate: $\begin{vmatrix} 2 & -1 & 3 \\ 1 & 2 & -2 \\ 3 & 1 & 1 \end{vmatrix}$.

EXAMPLE 2 Find the value of $\begin{vmatrix} 1 & 3 & -2 \\ 2 & 1 & 3 \\ 1 & 2 & 3 \end{vmatrix}$.

Solution We evaluate this determinant by expanding by minors along the first row of the determinant.

$$\begin{array}{ccc} & \text{Minor} & \text{Minor} & \text{Minor} \\ & \text{of 1} & \text{of 3} & \text{of } -2 \\ & \downarrow & \downarrow & \downarrow \end{array}$$

$$\begin{vmatrix} 1 & 3 & -2 \\ 2 & 1 & 3 \\ 1 & 2 & 3 \end{vmatrix} = 1\begin{vmatrix} 1 & 3 \\ 2 & 3 \end{vmatrix} - 3\begin{vmatrix} 2 & 3 \\ 1 & 3 \end{vmatrix} + (-2)\begin{vmatrix} 2 & 1 \\ 1 & 2 \end{vmatrix}$$

$$= 1(3 - 6) - 3(6 - 3) - 2(4 - 1) \qquad \text{Evaluate each } 2 \times 2 \text{ determinant.}$$

$$= 1(-3) - 3(3) - 2(3)$$

$$= -3 - 9 - 6$$

$$= -18$$

Answer 0

We can evaluate a 3×3 determinant by expanding it along any row or column. To determine the signs between the terms of the expansion of a 3×3 determinant, we use the following array of signs.

> **Array of signs for a 3 × 3 determinant**
>
> $$\begin{array}{ccc} + & - & + \\ - & + & - \\ + & - & + \end{array}$$
>
> This array of signs is commonly referred to as the **checkerboard pattern.**

INTERMEDIATE
Algebra $f(x)$ Now™

Self Check 3

Evaluate $\begin{vmatrix} 1 & 3 & -2 \\ 2 & 1 & 3 \\ 1 & 2 & 3 \end{vmatrix}$ by expanding along the last column.

EXAMPLE 3 Evaluate the determinant $\begin{vmatrix} 1 & 3 & -2 \\ 2 & 1 & 3 \\ 1 & 2 & 3 \end{vmatrix}$ by expanding on the middle column.

Solution This is the determinant of Example 2. To expand it along the middle column, we use the signs of the middle column of the array of signs:

$$\begin{array}{ccc} \text{Minor} & \text{Minor} & \text{Minor} \\ \text{of 3} & \text{of 1} & \text{of 2} \\ \downarrow & \downarrow & \downarrow \end{array}$$

$$\begin{vmatrix} 1 & 3 & -2 \\ 2 & 1 & 3 \\ 1 & 2 & 3 \end{vmatrix} = -3\begin{vmatrix} 2 & 3 \\ 1 & 3 \end{vmatrix} + 1\begin{vmatrix} 1 & -2 \\ 1 & 3 \end{vmatrix} - 2\begin{vmatrix} 1 & -2 \\ 2 & 3 \end{vmatrix}$$

Use the middle column of the checkerboard pattern:
$$\begin{array}{ccc} + & - & + \\ - & + & - \\ + & - & + \end{array}$$

$$= -3(6 - 3) + 1[3 - (-2)] - 2[3 - (-4)] \qquad \text{Evaluate each } 2 \times 2 \text{ determinant.}$$

$$= -3(3) + 1(5) - 2(7)$$

$$= -9 + 5 - 14$$

$$= -18$$

Answer −18

As expected, we get the same value as in Example 2.

Evaluating determinants

It is possible to use a graphing calculator to evaluate determinants. For example, to evaluate the determinant in Example 3, we first enter the matrix by pressing the $\boxed{\text{MATRIX}}$ key, selecting EDIT, and pressing the $\boxed{\text{ENTER}}$ key. Next, we enter the dimensions and the elements of the matrix to get Figure 3-13(a). We then press $\boxed{\text{2nd}}$ $\boxed{\text{QUIT}}$ to clear the screen, press $\boxed{\text{MATRIX}}$, select MATH, and press 1 to get Figure 3-13(b). We then press $\boxed{\text{MATRIX}}$, select NAMES, press 1, and press $\boxed{)}$ and $\boxed{\text{ENTER}}$ to get the value of the determinant. Figure 3-13(c) shows that the value of the determinant is -18.

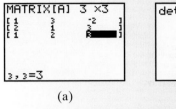
(a)

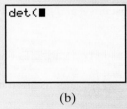
(b)

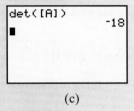
(c)

FIGURE 3-13

Using Cramer's rule to solve a system of two equations

The method of using determinants to solve systems of linear equations is called **Cramer's rule,** named after the 18th-century mathematician Gabriel Cramer. To develop Cramer's rule, we consider the system

$$\begin{cases} ax + by = e \\ cx + dy = f \end{cases}$$

where x and y are variables and a, b, c, d, e, and f are constants.

If we multiply both sides of the first equation by d and multiply both sides of the second equation by $-b$, we can add the equations and eliminate y:

$$\begin{aligned} adx + bdy &= ed \\ -bcx - bdy &= -bf \\ \hline adx - bcx &= ed - bf \end{aligned}$$

To solve for x, we use the distributive property to write $adx - bcx$ as $(ad - bc)x$ on the left-hand side and divide each side by $ad - bc$:

$$(ad - bc)x = ed - bf$$

$$x = \frac{ed - bf}{ad - bc} \qquad \text{where } ad - bc \neq 0$$

We can find y in a similar manner. After eliminating the variable x, we get

$$y = \frac{af - ec}{ad - bc} \qquad \text{where } ad - bc \neq 0$$

Determinants provide an easy way of remembering these formulas. Note that the denominator for both x and y is

$$\begin{vmatrix} a & b \\ c & d \end{vmatrix} = ad - bc$$

The numerators can be expressed as determinants also:

$$x = \frac{ed - bf}{ad - bc} = \frac{\begin{vmatrix} e & b \\ f & d \end{vmatrix}}{\begin{vmatrix} a & b \\ c & d \end{vmatrix}} \quad \text{and} \quad y = \frac{af - ec}{ad - bc} = \frac{\begin{vmatrix} a & e \\ c & f \end{vmatrix}}{\begin{vmatrix} a & b \\ c & d \end{vmatrix}}$$

If we compare these formulas with the original system

$$\begin{cases} ax + by = e \\ cx + dy = f \end{cases}$$

we note that in the expressions for x and y on the preceding page, the denominator determinant is formed by using the coefficients a, b, c, and d of the variables in the equations. The numerator determinants are the same as the denominator determinant, except that the column of coefficients of the variable for which we are solving is replaced with the column of constants e and f.

Cramer's rule for two equations in two variables

The solution of the system $\begin{cases} ax + by = e \\ cx + dy = f \end{cases}$ is given by

$$x = \frac{D_x}{D} = \frac{\begin{vmatrix} e & b \\ f & d \end{vmatrix}}{\begin{vmatrix} a & b \\ c & d \end{vmatrix}} \quad \text{and} \quad y = \frac{D_y}{D} = \frac{\begin{vmatrix} a & e \\ c & f \end{vmatrix}}{\begin{vmatrix} a & b \\ c & d \end{vmatrix}}$$

If every determinant is 0, the system is consistent, but the equations are dependent. The system has infinitely many solutions.

If $D = 0$ and D_x or D_y is nonzero, the system is inconsistent and does not have a solution.

Self Check 4

Solve $\begin{cases} 2x - 3y = -16 \\ 3x + 5y = 14 \end{cases}$
using Cramer's rule.

EXAMPLE 4 Use Cramer's rule to solve $\begin{cases} 4x - 3y = 6 \\ -2x + 5y = 4 \end{cases}$.

Solution The value of x is the quotient of two determinants, D and D_x. The denominator determinant D is made up of the coefficients of x and y:

$$D = \begin{vmatrix} 4 & -3 \\ -2 & 5 \end{vmatrix}$$

To solve for x, we form the numerator determinant D_x from D by replacing its first column (the coefficients of x) with the column of constants (6 and 4).

To solve for y, we form the numerator determinant D_y from D by replacing its second column (the coefficients of y) with the column of constants (6 and 4).

To find the values of x and y, we evaluate each determinant:

$$x = \frac{D_x}{D} = \frac{\begin{vmatrix} 6 & -3 \\ 4 & 5 \end{vmatrix}}{\begin{vmatrix} 4 & -3 \\ -2 & 5 \end{vmatrix}} = \frac{6(5) - (-3)(4)}{4(5) - (-3)(-2)} = \frac{30 + 12}{20 - 6} = \frac{42}{14} = 3$$

$$y = \frac{D_y}{D} = \frac{\begin{vmatrix} 4 & 6 \\ -2 & 4 \end{vmatrix}}{\begin{vmatrix} 4 & -3 \\ -2 & 5 \end{vmatrix}} = \frac{4(4) - 6(-2)}{14} = \frac{16 + 12}{14} = \frac{28}{14} = 2$$

Answer $(-2, 4)$

The solution of this system is $(3, 2)$. Verify that $x = 3$ and $y = 2$ satisfy both equations.

EXAMPLE 5   Use Cramer's rule to solve $\begin{cases} 7x = 8 - 4y \\ 2y = 3 - \frac{7}{2}x \end{cases}$.

Solution   We multiply both sides of the second equation by 2 to eliminate the fraction and write the system in the form

$$\begin{cases} 7x + 4y = 8 \\ 7x + 4y = 6 \end{cases}$$

When we attempt to use Cramer's rule to solve this system for x, we obtain

$$x = \frac{D_x}{D} = \frac{\begin{vmatrix} 8 & 4 \\ 6 & 4 \end{vmatrix}}{\begin{vmatrix} 7 & 4 \\ 7 & 4 \end{vmatrix}} = \frac{8}{0} \quad \text{which is undefined}$$

Since the denominator determinant D is 0 and the numerator determinant D_x is not 0, the system is inconsistent. It has no solutions.

We can see directly from the system that it is inconsistent. For any values of x and y, it is impossible that 7 times x plus 4 times y could be both 8 and 6.

▮ Using Cramer's rule to solve a system of three equations

Cramer's rule can be extended to solve systems of three linear equations in three variables.

Cramer's rule for three equations in three variables

The solution of the system $\begin{cases} ax + by + cz = j \\ dx + ey + fz = k \\ gx + hy + iz = l \end{cases}$ is given by

$$x = \frac{D_x}{D}, \qquad y = \frac{D_y}{D}, \qquad \text{and} \qquad z = \frac{D_z}{D}$$

where

$$D = \begin{vmatrix} a & b & c \\ d & e & f \\ g & h & i \end{vmatrix} \qquad D_x = \begin{vmatrix} j & b & c \\ k & e & f \\ l & h & i \end{vmatrix}$$

$$D_y = \begin{vmatrix} a & j & c \\ d & k & f \\ g & l & i \end{vmatrix} \qquad D_z = \begin{vmatrix} a & b & j \\ d & e & k \\ g & h & l \end{vmatrix}$$

If every determinant is 0, the system is consistent, but the equations are dependent. The system has infinitely many solutions.

If $D = 0$ and D_x or D_y or D_z is nonzero, the system is inconsistent and does not have a solution.

EXAMPLE 6   Use Cramer's rule to solve $\begin{cases} 2x + y + 4z = 12 \\ x + 2y + 2z = 9 \\ 3x - 3y - 2z = 1 \end{cases}$.

Solution The denominator determinant D is the determinant formed by the coefficients of the variables. The numerator determinants, D_x, D_y, and D_z, are formed by replacing the coefficients of the variable being solved for by the column of constants. We form the quotients for x, y, and z and evaluate each determinant by expanding by minors about the first row:

$$x = \frac{D_x}{D} = \frac{\begin{vmatrix} 12 & 1 & 4 \\ 9 & 2 & 2 \\ 1 & -3 & -2 \end{vmatrix}}{\begin{vmatrix} 2 & 1 & 4 \\ 1 & 2 & 2 \\ 3 & -3 & -2 \end{vmatrix}}$$

$$= \frac{12\begin{vmatrix} 2 & 2 \\ -3 & -2 \end{vmatrix} - 1\begin{vmatrix} 9 & 2 \\ 1 & -2 \end{vmatrix} + 4\begin{vmatrix} 9 & 2 \\ 1 & -3 \end{vmatrix}}{2\begin{vmatrix} 2 & 2 \\ -3 & -2 \end{vmatrix} - 1\begin{vmatrix} 1 & 2 \\ 3 & -2 \end{vmatrix} + 4\begin{vmatrix} 1 & 2 \\ 3 & -3 \end{vmatrix}}$$

$$= \frac{12(2) - 1(-20) + 4(-29)}{2(2) - 1(-8) + 4(-9)}$$

$$= \frac{-72}{-24}$$

$$= 3$$

$$y = \frac{D_y}{D} = \frac{\begin{vmatrix} 2 & 12 & 4 \\ 1 & 9 & 2 \\ 3 & 1 & -2 \end{vmatrix}}{\begin{vmatrix} 2 & 1 & 4 \\ 1 & 2 & 2 \\ 3 & -3 & -2 \end{vmatrix}}$$

$$= \frac{2\begin{vmatrix} 9 & 2 \\ 1 & -2 \end{vmatrix} - 12\begin{vmatrix} 1 & 2 \\ 3 & -2 \end{vmatrix} + 4\begin{vmatrix} 1 & 9 \\ 3 & 1 \end{vmatrix}}{-24}$$

$$= \frac{2(-20) - 12(-8) + 4(-26)}{-24}$$

$$= \frac{-48}{-24}$$

$$= 2$$

$$z = \frac{D_z}{D} = \frac{\begin{vmatrix} 2 & 1 & 12 \\ 1 & 2 & 9 \\ 3 & -3 & 1 \end{vmatrix}}{\begin{vmatrix} 2 & 1 & 4 \\ 1 & 2 & 2 \\ 3 & -3 & -2 \end{vmatrix}}$$

$$= \frac{2\begin{vmatrix} 2 & 9 \\ -3 & 1 \end{vmatrix} - 1\begin{vmatrix} 1 & 9 \\ 3 & 1 \end{vmatrix} + 12\begin{vmatrix} 1 & 2 \\ 3 & -3 \end{vmatrix}}{-24}$$

$$= \frac{2(29) - 1(-26) + 12(-9)}{-24}$$

$$= \frac{-24}{-24}$$

$$= 1$$

Answer $(2, -2, 3)$

The solution of this system is $(3, 2, 1)$.

Section 3.5   STUDY SET

INTERMEDIATE
Algebra *f(x)* Now™

VOCABULARY   *Fill in the blanks.*

1. $\begin{vmatrix} 2 & 1 \\ -6 & 1 \end{vmatrix}$ is a 2×2 _____.

2. A _____ matrix has the same number of rows and columns.

3. The _____ of b_1 in $\begin{vmatrix} a_1 & b_1 & c_1 \\ a_2 & b_2 & c_2 \\ a_3 & b_3 & c_3 \end{vmatrix}$ is $\begin{vmatrix} a_2 & c_2 \\ a_3 & c_3 \end{vmatrix}$.

4. In $\begin{vmatrix} 7 & -3 \\ 1 & 2 \end{vmatrix}$, 7 and 2 lie along the main _____.

5. A 3×3 determinant has 3 _____ and 3 _____.

6. _____ rule uses determinants to solve systems of linear equations.

CONCEPTS   *Fill in the blanks.*

7. If the denominator determinant D for a system of equations is zero, the equations of the system are _____ or the system is _____.

8. To find the minor of 5, we _____ _____ the elements of the determinant that are in the same row and column as 5.

$$\begin{vmatrix} 3 & 5 & 1 \\ 6 & -2 & 2 \\ 8 & -1 & 4 \end{vmatrix}$$

9. What is the value of $\begin{vmatrix} a & b \\ c & d \end{vmatrix}$?

10. $\begin{vmatrix} 5 & 1 & -1 \\ 8 & 7 & 4 \\ 9 & 7 & 6 \end{vmatrix} = -1\begin{vmatrix} 8 & 7 \\ 9 & 7 \end{vmatrix} - 4\begin{vmatrix} 5 & 1 \\ 9 & 7 \end{vmatrix}$
$\qquad\qquad + 6\begin{vmatrix} 5 & 1 \\ 8 & 7 \end{vmatrix}$

In evaluating this determinant, about what row or column was it expanded?

11. What is the denominator determinant D for the system $\begin{cases} 3x + 4y = 7 \\ 2x - 3y = 5 \end{cases}$?

12. What is the denominator determinant D for the system $\begin{cases} x + 2y = -8 \\ 3x + y - z = -2 \\ 8x + 4y - z = 6 \end{cases}$?

13. For the system $\begin{cases} 3x + 2y = 1 \\ 4x - y = 3 \end{cases}$, $D_x = -7$, $D_y = 5$ and $D = -11$. What is the solution of the system?

14. For the system $\begin{cases} 2x + 3y - z = -8 \\ x - y - z = -2 \\ -4x + 3y + z = 6 \end{cases}$
$D_x = -28$, $D_y = -14$, $D_z = 14$, and $D = 14$. What is the solution?

NOTATION   *Complete the evaluation of each determinant.*

15. $\begin{vmatrix} 5 & -2 \\ -2 & 6 \end{vmatrix} = 5(\ ) - (-2)(-2)$
$\qquad\qquad = \ \boxed{} - 4$
$\qquad\qquad = 26$

16. $\begin{vmatrix} 2 & 1 & 3 \\ 3 & 4 & 2 \\ 1 & 5 & 3 \end{vmatrix}$

$= 2\begin{vmatrix} 4 & \boxed{\ } \\ 5 & 3 \end{vmatrix} - \boxed{\ }\begin{vmatrix} 3 & 2 \\ \boxed{\ } & 3 \end{vmatrix} + 3\begin{vmatrix} 3 & 4 \\ 1 & \boxed{\ } \end{vmatrix}$
$= 2(\boxed{\ } - 10) - 1(9 - \boxed{\ }) + 3(15 - \boxed{\ })$
$= 2(2) - 1(\boxed{\ }) + \boxed{\ }(11)$
$= 4 - 7 + \boxed{\ }$
$= 30$

PRACTICE   *Evaluate each determinant.*

17. $\begin{vmatrix} 2 & 3 \\ -2 & 1 \end{vmatrix}$

18. $\begin{vmatrix} 3 & -2 \\ -2 & 4 \end{vmatrix}$

19. $\begin{vmatrix} -1 & 2 \\ 3 & -4 \end{vmatrix}$

20. $\begin{vmatrix} -1 & -2 \\ -3 & -4 \end{vmatrix}$

21. $\begin{vmatrix} 10 & 0 \\ 1 & 20 \end{vmatrix}$

22. $\begin{vmatrix} 1 & 15 \\ 15 & 0 \end{vmatrix}$

23. $\begin{vmatrix} -6 & -2 \\ 15 & 4 \end{vmatrix}$

24. $\begin{vmatrix} 3 & -2 \\ 12 & -8 \end{vmatrix}$

25. $\begin{vmatrix} 1 & 2 & 0 \\ 0 & 1 & 2 \\ 0 & 0 & 1 \end{vmatrix}$

26. $\begin{vmatrix} -1 & 2 & 1 \\ 2 & 1 & -3 \\ 1 & 1 & 1 \end{vmatrix}$

27. $\begin{vmatrix} 1 & -2 & 3 \\ -2 & 1 & 1 \\ -3 & -2 & 1 \end{vmatrix}$

28. $\begin{vmatrix} 1 & 1 & 2 \\ 2 & 1 & -2 \\ 3 & 1 & 3 \end{vmatrix}$

29. $\begin{vmatrix} 1 & 0 & 1 \\ 0 & 1 & 0 \\ 1 & 1 & 1 \end{vmatrix}$

30. $\begin{vmatrix} 3 & 5 & 1 \\ 6 & -2 & 2 \\ 8 & -1 & 4 \end{vmatrix}$

31. $\begin{vmatrix} 1 & 2 & 1 \\ -3 & 7 & 3 \\ -4 & 3 & -5 \end{vmatrix}$

32. $\begin{vmatrix} 1 & 4 & 7 \\ 2 & 5 & 8 \\ 3 & 6 & 9 \end{vmatrix}$

Use Cramer's rule to solve each system of equations, if possible. If a system is inconsistent or if the equations are dependent, so indicate.

33. $\begin{cases} x + y = 6 \\ x - y = 2 \end{cases}$ **34.** $\begin{cases} x - y = 4 \\ 2x + y = 5 \end{cases}$

35. $\begin{cases} 2x + 3y = 0 \\ 4x - 6y = -4 \end{cases}$ **36.** $\begin{cases} 4x - 3y = -1 \\ 8x + 3y = 4 \end{cases}$

37. $\begin{cases} 3x + 2y = 11 \\ 6x + 4y = 11 \end{cases}$ **38.** $\begin{cases} 5x + 6y = 12 \\ 10x + 12y = 24 \end{cases}$

39. $\begin{cases} y = \frac{-2x + 1}{3} \\ 3x - 2y = 8 \end{cases}$ **40.** $\begin{cases} 2x + 3y = -1 \\ x = \frac{y - 9}{4} \end{cases}$

41. $\begin{cases} x + y + z = 4 \\ x + y - z = 0 \\ x - y + z = 2 \end{cases}$ **42.** $\begin{cases} x + y + z = 4 \\ x - y + z = 2 \\ x - y - z = 0 \end{cases}$

43. $\begin{cases} x + y + 2z = 7 \\ x + 2y + z = 8 \\ 2x + y + z = 9 \end{cases}$ **44.** $\begin{cases} x + 2y + 2z = 10 \\ 2x + y + 2z = 9 \\ 2x + 2y + z = 1 \end{cases}$

45. $\begin{cases} 2x + y + z = 5 \\ x - 2y + 3z = 10 \\ x + y - 4z = -3 \end{cases}$ **46.** $\begin{cases} 3x + 2y - z = -8 \\ 2x - y + 7z = 10 \\ 2x + 2y - 3z = -10 \end{cases}$

47. $\begin{cases} 4x - 3y = 1 \\ 6x - 8z = 1 \\ 2y - 4z = 0 \end{cases}$ **48.** $\begin{cases} 4x + 3z = 4 \\ 2y - 6z = -1 \\ 8x + 4y + 3z = 9 \end{cases}$

49. $\begin{cases} 2x + 3y + 4z = 6 \\ 2x - 3y - 4z = -4 \\ 4x + 6y + 8z = 12 \end{cases}$

50. $\begin{cases} x - 3y + 4z - 2 = 0 \\ 2x + y + 2z - 3 = 0 \\ 4x - 5y + 10z - 7 = 0 \end{cases}$

51. $\begin{cases} 2x + y - z - 1 = 0 \\ x + 2y + 2z - 2 = 0 \\ 4x + 5y + 3z - 3 = 0 \end{cases}$

52. $\begin{cases} 2x - y + 4z + 2 = 0 \\ 5x + 8y + 7z = -8 \\ x + 3y + z + 3 = 0 \end{cases}$

53. $\begin{cases} x + y = 1 \\ \frac{1}{2}y + z = \frac{5}{2} \\ x - z = -3 \end{cases}$ **54.** $\begin{cases} \frac{1}{2}x + y + z + \frac{3}{2} = 0 \\ x + \frac{1}{2}y + z - \frac{1}{2} = 0 \\ x + y + \frac{1}{2}z + \frac{1}{2} = 0 \end{cases}$

APPLICATIONS   *Write a system of equations to solve each problem. Then use Cramer's rule to solve it.*

55. INVENTORIES The table below shows an end-of-the-year inventory report for a warehouse that supplies electronics stores. If the warehouse stocks two models of cordless telephones, one valued at $67 and the other at $100, how many of each model of phone did the warehouse have at the time of the inventory?

Item	Number	Merchandise value
Televisions	800	$1,005,450
Radios	200	$15,785
Cordless phones	360	$29,400

56. SIGNALING A system of sending signals uses two flags held in various positions to represent letters of the alphabet. The illustration shows how the letter U is signaled. Find x and y, if y is to be 30° more than x.

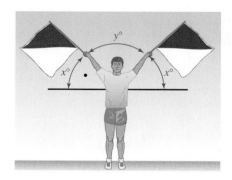

57. INVESTING A student wants to average a 6.6% return in the first year by investing $20,000 in the three stocks listed in the table. Because HiTech is a high-risk investment, he wants to invest three times as much in SaveTel and OilCo combined as he invests in HiTech. How much should he invest in each stock?

Stock	Rate of return
HiTech	10%
SaveTel	5%
OilCo	6%

58. INVESTING A woman wants to average a $7\frac{1}{3}\%$ return in the first year by investing $30,000 in three certificates of deposit. She wants to invest five times as much in the 8% CD as in the 6% CD. How much should she invest in each CD?

Type of CD	Rate of return
12-month	6%
24-month	7%
36-month	8%

Use a calculator with matrix capabilities to evaluate each determinant.

59. $\begin{vmatrix} 2 & -3 & 4 \\ -1 & 2 & 4 \\ 3 & -3 & 1 \end{vmatrix}$ **60.** $\begin{vmatrix} -3 & 2 & -5 \\ 3 & -2 & 6 \\ 1 & -3 & 4 \end{vmatrix}$

61. $\begin{vmatrix} 2 & 1 & -3 \\ -2 & 2 & 4 \\ 1 & -2 & 2 \end{vmatrix}$ **62.** $\begin{vmatrix} 4 & 2 & -3 \\ 2 & -5 & 6 \\ 2 & 5 & -2 \end{vmatrix}$

WRITING

63. Explain how to find the minor of an element of a determinant.

64. Explain how to find x when solving a system of three linear equations by Cramer's rule. Use the words *coefficients* and *constants* in your explanation.

REVIEW

65. Are the lines $y = 2x - 7$ and $x - 2y = 7$ perpendicular?

66. Are the lines $y = 2x - 7$ and $2x - y = 10$ parallel?

67. Are the equations $y = 2x - 7$ and $f(x) = 2x - 7$ the same?

68. How are the graphs of $f(x) = x^2$ and $g(x) = x^2 - 2$ related?

69. For the linear function $y = 2x - 7$, what variable is associated with the domain?

70. Is the graph of a circle the graph of a function?

71. The graph of a line passes through $(0, -3)$. Is this the x-intercept or the y-intercept of the line?

72. What is the name of the function $f(x) = |x|$?

73. For $y = 2x^2 + 6x + 1$, what is the independent variable and what is the dependent variable?

74. If $f(x) = x^3 - x$, find $f(-1)$.

Systems of Equations

In Chapter 3, we have solved problems involving two and three variables by writing and solving a **system of equations.**

Solutions of a System of Equations

A solution of a system of equations in two or three variables is an ordered pair or an ordered triple whose coordinates satisfy each equation of the system. In Exercises 1 and 2, decide whether the given ordered pair or ordered triple is a solution of the system.

1. $\begin{cases} 2x - y = 1 \\ 4x + 2y = 0 \end{cases} \left(\dfrac{1}{4}, -\dfrac{1}{2} \right)$

2. $\begin{cases} 2x - y + z = 9 \\ 3x + y - 4z = 8 \\ 2x - 7z = -1 \end{cases} (4, 0, 1)$

Methods of Solving Systems of Linear Equations

We have studied several methods for solving systems of two and three linear equations.

3. Solve $\begin{cases} 2x + 5y = 8 \\ y = 3x + 5 \end{cases}$ using the *graphing method.*

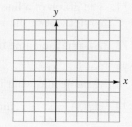

4. Solve $\begin{cases} 9x - 8y = 1 \\ 6x + 12y = 5 \end{cases}$ using the *addition method.*

5. Solve $\begin{cases} 4x - y - 10 = 0 \\ 3x + 5y = 19 \end{cases}$ using the *substitution method.*

6. Solve $\begin{cases} -x + 3y + 2z = 5 \\ 3x + 2y + z = -1 \\ 2x - y + 3z = 4 \end{cases}$ using the *addition method.*

7. Solve $\begin{cases} x - 6y = 3 \\ x + 3y = 21 \end{cases}$ using matrices.

8. Solve $\begin{cases} x + 2z = 7 \\ 2x - y + 3z = 9 \\ y - z = 1 \end{cases}$ using Cramer's rule.

Dependent Equations and Inconsistent Systems

If the equations in a system of two linear equations are dependent, the system has infinitely many solutions. An inconsistent system has no solutions.

9. Suppose you are solving a system of two equations by the addition method, and you obtain the following.

$$\begin{array}{r} 2x - 3y = 4 \\ -2x + 3y = -4 \\ \hline 0 = 0 \end{array}$$

What can you conclude?

10. Suppose you are solving a system of two equations by the substitution method, and you obtain

$$-2(x - 3) + 2x = 7$$
$$-2x + 6 + 2x = 7$$
$$6 = 7$$

What can you conclude?

ACCENT ON TEAMWORK

SECTION 3.1

LINE GRAPHS   Find five examples of line graphs where two lines intersect like that shown below. Your school library is a good resource. Ask to look through the library's collection of recent magazines and newspapers. For each example, explain the information given by the point of intersection of the graphs.

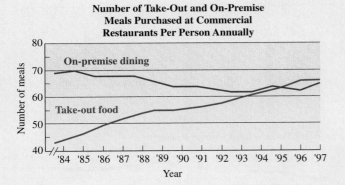

Number of Take-Out and On-Premise Meals Purchased at Commercial Restaurants Per Person Annually

SECTION 3.2

COMPARING SOLUTIONS   Example 8 of Section 3.2, involving water treatment, was solved using two variables. This problem can also be solved using one variable. Solve it using one variable and then comment on which method you like better. What are the advantages and the drawbacks of each method?

BREAK-POINT ANALYSIS   Suppose you are a financial analyst for the coathanger company mentioned in Example 10 of Section 3.2. It is your job to decide whether the company should purchase the new machine.

First, graph the equations

$$C = 1.5x + 400$$
$$C = 1.25x + 500$$

on the same coordinate system. Then write a report that could be given to company managers, explaining their options concerning the purchase of the new machine. Under what conditions should they keep the machine now in use? Under what conditions should they buy the new machine?

SECTION 3.3

GRAPHS OF SYSTEMS OF EQUATIONS   From your textbook, find and then graph examples of each of the following types of systems of two linear equations:

- Consistent system
- Inconsistent system
- Dependent equations

Then make cardboard models of each of the graphs of the systems of three linear equations illustrated in Figure 3-10 and Figure 3-11 of Section 3.3. Make a presentation to your class using the graphs and cardboard models as visual aids to help you explain whether each system has a solution, and if so, what form the solution takes.

SECTION 3.4

GAUSSIAN ELIMINATION   Use matrices and elementary row operations to show that the solution of this 4×4 system of linear equations is $(1, 1, 0, 1)$.

$$\begin{cases} a + b + c + d = 3 \\ a - b - c - d = -1 \\ a + b - c - d = 1 \\ a + b - c + d = 3 \end{cases}$$

SECTION 3.5

METHODS OF SOLUTION   Have each person in your group solve the system

$$\begin{cases} x - y = 4 \\ 2x + y = 5 \end{cases}$$

in a different way. The methods to use are graphing, addition, substitution, matrices, and Cramer's rule. Have each person briefly explain his or her method of solution. After everyone has presented a solution, discuss the advantages and drawbacks of each method. Can your group come to a consensus? Is there a favorite method?

SECTION 3.1	*Solving Systems by Graphing*

CONCEPTS

The graph of a linear equation is the graph of all points (x, y) on the rectangular coordinate system whose coordinates satisfy the equation.

To solve a system of two linear equations by the *graphing method*, find the coordinates of the point where the two graphs intersect.

REVIEW EXERCISES

See the illustration on the right.

1. Give three points that satisfy the equation $2x + y = 5$.

2. Give three points that satisfy the equation $x - y = 4$.

3. What is the solution of $\begin{cases} 2x + y = 5 \\ x - y = 4 \end{cases}$?

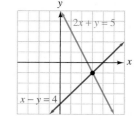

4. POLITICS Explain the importance of the points of intersection of the graphs shown below.

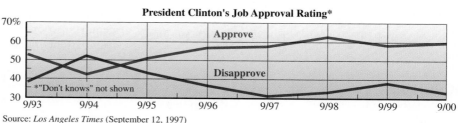

Source: *Los Angeles Times* (September 12, 1997)

A system of equations that has at least one solution, is called a *consistent system*. If the graphs are parallel lines, the system has no solution, and it is called an *inconsistent system*.

Equations with different graphs are called *independent equations*. If the graphs are the same line, the system has infinitely many solutions. The equations are called *dependent equations*.

Solve each system by the graphing method, if possible. If a system is inconsistent or if the equations are dependent, so indicate.

5. $\begin{cases} 2x + y = 11 \\ -x + 2y = 7 \end{cases}$

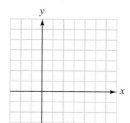

6. $\begin{cases} y = -\frac{3}{2}x \\ 2x - 3y + 13 = 0 \end{cases}$

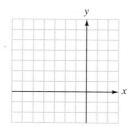

7. $\begin{cases} \frac{1}{2}x + \frac{1}{3}y = 2 \\ y = 6 - \frac{3}{2}x \end{cases}$

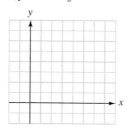

8. $\begin{cases} \frac{x}{3} - \frac{y}{2} = 1 \\ 6x - 9y = 3 \end{cases}$

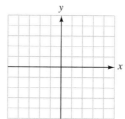

To solve a system by the *substitution method:*

1. Solve one equation for one of its variables.

2. Substitute the resulting expression for that variable into the other equation and solve that equation.

3. Find the value of the other variable by substituting the value of the variable found in step 2 into the equation from step 1.

To solve a system by the *addition method:*

1. Write both equations in general form: $Ax + By = C$.

2. Multiply the terms of one or both equations by constants so that the coefficients of one variable differ only in sign.

3. Add the equations from step 2 and solve the resulting equation.

4. Substitute the value obtained in step 3 into either original equation and solve for the remaining variable.

Solve each system using the substitution method, if possible. If a system is inconsistent or if the equations are dependent, so indicate.

9. $\begin{cases} x = y - 4 \\ 2x + 3y = 7 \end{cases}$

10. $\begin{cases} y = 2x + 5 \\ 3x - 5y = -4 \end{cases}$

11. $\begin{cases} 0.1x + 0.2y = 1.1 \\ 2x - y = 2 \end{cases}$

12. $\begin{cases} x = -2 - 3y \\ -2x - 6y = 4 \end{cases}$

Solve each system using the addition method, if possible.

13. $\begin{cases} x + y = -2 \\ 2x + 3y = -3 \end{cases}$

14. $\begin{cases} 2x - 3y = 5 \\ 2x - 3y = 8 \end{cases}$

15. $\begin{cases} x + \dfrac{1}{2}y = 7 \\ -2x = 3y - 6 \end{cases}$

16. $\begin{cases} y = \dfrac{x - 3}{2} \\ x = \dfrac{2y + 7}{2} \end{cases}$

17. To solve $\begin{cases} 5x - 2y = 19 \\ 3x + 4y = 1 \end{cases}$, which method, addition or substitution, would you use? Explain why.

18. Estimate the solution of the system

$$\begin{cases} y = -\tfrac{2}{3}x \\ 2x - 3y = -4 \end{cases}$$

from the graphs in the illustration. Then solve the system algebraically.

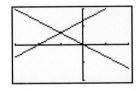

Use two equations to solve each problem.

19. MILEAGE MAPS The distance between Austin and Houston is 4 miles less than twice the distance between Austin and San Antonio. The round trip from Houston to Austin to San Antonio and back to Houston is 442 miles. Determine the mileages between Austin and Houston and between Austin and San Antonio.

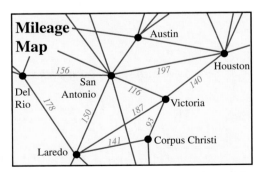

20. RIVERBOAT RIDES A Mississippi riverboat travels 30 miles downstream in 3 hours and then makes the return trip upstream in 5 hours. Find the speed of the riverboat in still water and the speed of the current.

21. BREAK POINTS A bottling company is considering purchasing a new piece of equipment for their production line. The machine they currently use has a setup cost of $250 and a cost of $0.04 per bottle. The new machine has a setup cost of $600 and a cost of $0.02 per bottle. Find the break point.

Systems of Three Equations

The solution of a system of three linear equations is an *ordered triple*.

To solve a system of linear equations with three variables:

1. Pick any two equations and eliminate a variable.

2. Pick a different pair of equations and eliminate the same variable.

3. Solve the resulting pair of equations.

4. Use substitution to find the value of the third variable.

22. Determine whether $(2, -1, 1)$ is a solution of the system $\begin{cases} x - y + z = 4 \\ x + 2y - z = -1. \\ x + y - 3z = -1 \end{cases}$

Solve each system, if possible.

23. $\begin{cases} 2x + y + z = -1 \\ 6x - 3y - 2z = 3 \\ 4x - y - z = 4 \end{cases}$

24. $\begin{cases} 2x + 3y + z = -5 \\ -x + 2y - z = -6 \\ 3x + y + 2z = 4 \end{cases}$

25. $\begin{cases} x + y - z = -3 \\ x + z = 2 \\ 2x - y + 2z = 3 \end{cases}$

26. $\begin{cases} 3x + 3y + 6z = -6 \\ -x - y - 2z = 2 \\ 2x + 2y + 4z = -4 \end{cases}$

27. A system of three linear equations in three variables is graphed on the right. Does the system have a solution? If so, how many solutions does it have?

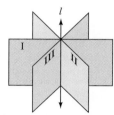

28. MIXING NUTS The owner of a produce store wanted to mix peanuts selling for $3 per pound, cashews selling for $9 per pound, and Brazil nuts selling for $9 per pound to get 50 pounds of a mixture that would sell for $6 per pound. She used 15 fewer pounds of cashews than peanuts. How many pounds of each did she use?

Solving Systems Using Matrices

A *matrix* is a rectangular array of numbers.

A system of linear equations can be represented by an *augmented matrix*.

Systems of linear equations can be solved using *Gaussian elimination* and *elementary row operations:*

1. Any two rows can be interchanged.

2. Any row can be multiplied by a nonzero constant.

3. Any row can be changed by adding a nonzero constant multiple of another row to it.

Represent each system of equations using an augmented matrix.

29. $\begin{cases} 5x + 4y = 3 \\ x - y = -3 \end{cases}$

30. $\begin{cases} x + 2y + 3z = 6 \\ x - 3y - z = 4 \\ 6x + y - 2z = -1 \end{cases}$

Solve each system using matrices, if possible.

31. $\begin{cases} x - y = 4 \\ 3x + 7y = -18 \end{cases}$

32. $\begin{cases} x + 2y - 3z = 5 \\ x + y + z = 0 \\ 3x + 4y + 2z = -1 \end{cases}$

33. $\begin{cases} 16x - 8y = 32 \\ -2x + y = -4 \end{cases}$

34. $\begin{cases} x + 2y + 2z = 2 \\ 4x + 5y + 3z = 3 \\ 2x + y - z = 1 \end{cases}$

35. INVESTING One year, a couple invested a total of $10,000 in two projects. The first investment, a mini-mall, made a 6% profit. The other investment, a skateboard park, made a 12% profit. If their investments made $960, how much was invested at each rate? To answer this question, write a system of two equations and solve it using matrices.

A *determinant of a square matrix* is a number.

To evaluate a 2 × 2 determinant:

$$\begin{vmatrix} a & b \\ c & d \end{vmatrix} = ad - bc$$

To evaluate a 3 × 3 determinant, we expand it by *minors* along any row or column using the *array of signs*.

Cramer's rule can be used to solve systems of linear equations.

Evaluate each determinant.

36. $\begin{vmatrix} 2 & 3 \\ -4 & 3 \end{vmatrix}$

37. $\begin{vmatrix} -3 & -4 \\ 5 & -6 \end{vmatrix}$

38. $\begin{vmatrix} -1 & 2 & -1 \\ 2 & -1 & 3 \\ 1 & -2 & 2 \end{vmatrix}$

39. $\begin{vmatrix} 3 & -2 & 2 \\ 1 & -2 & -2 \\ 2 & 1 & -1 \end{vmatrix}$

Use Cramer's rule to solve each system, if possible.

40. $\begin{cases} 3x + 4y = 10 \\ 2x - 3y = 1 \end{cases}$

41. $\begin{cases} -6x - 4y = -6 \\ 3x + 2y = 5 \end{cases}$

42. $\begin{cases} x + 2y + z = 0 \\ 2x + y + z = 3 \\ x + y + 2z = 5 \end{cases}$

43. $\begin{cases} 2x + 3y + z = 2 \\ x + 3y + 2z = 7 \\ x - y - z = -7 \end{cases}$

44. VETERINARY MEDICINE The daily requirements of a balanced diet for an animal are shown in the nutritional pyramid below. The number of grams per cup of nutrients in three food mixes are shown in the table. How many cups of each mix should be used to meet the daily requirements for protein, carbohydrates, and essential fatty acids in the animal's diet? To answer this problem, write a system of three equations and solve it using Cramer's rule.

	Grams per cup		
	Protein	Carbohydrates	Fatty acids
Mix A	5	2	1
Mix B	6	3	2
Mix C	8	3	1

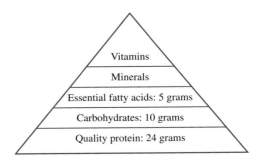

Vitamins

Minerals

Essential fatty acids: 5 grams

Carbohydrates: 10 grams

Quality protein: 24 grams

1. Solve $\begin{cases} 2x + y = 5 \\ y = 2x - 3 \end{cases}$ by graphing.

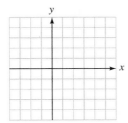

2. Use substitution to solve $\begin{cases} 2x - 4y = 14 \\ x + 2y = 7 \end{cases}$.

3. Use addition to solve $\begin{cases} 2x + 3y = -5 \\ 3x - 2y = 12 \end{cases}$.

4. Are the equations of the following system dependent or independent?

$$\begin{cases} 3(x + y) = x - 3 \\ -y = \dfrac{2x + 3}{3} \end{cases}$$

5. Is $(-1, -\frac{1}{2}, 5)$ a solution of $\begin{cases} x - 2y + z = 5 \\ 2x + 4y = -4 \\ -6y + 4z = 22 \end{cases}$?

6. Solve the following system using the addition method.

$$\begin{cases} x + y + z = 4 \\ x + y - z = 6 \\ 2x - 3y + z = -1 \end{cases}$$

Write a system of equations to solve each problem.

7. In the EXIT sign, find x and y, if y is 15 more than x.

8. ANTIFREEZE How much of a 40% antifreeze solution must a mechanic mix with an 80% antifreeze solution if 20 gallons of a 50% antifreeze solution are needed?

Use matrices to solve each system.

9. $\begin{cases} x + y = 4 \\ 2x - y = 2 \end{cases}$

10. $\begin{cases} x + y + 2z = -1 \\ x + 3y - 6z = 7 \\ 2x - y + 2z = 0 \end{cases}$

Evaluate each determinant.

11. $\begin{vmatrix} 2 & -3 \\ 4 & 5 \end{vmatrix}$

12. $\begin{vmatrix} 1 & 2 & 0 \\ 2 & 0 & 3 \\ 1 & -2 & 2 \end{vmatrix}$

Consider the system $\begin{cases} x - y = -6 \\ 3x + y = -6 \end{cases}$*, which is to be solved using Cramer's rule.*

13. When solving for x, what is the numerator determinant D_x? (**Don't evaluate it.**)

14. When solving for y, what is the denominator determinant D? (**Don't evaluate it.**)

15. Solve the system for x.

16. Solve the system for y.

17. Solve the following system for z only, using Cramer's rule.

$$\begin{cases} x + y + z = 4 \\ x + y - z = 6 \\ 2x - 3y + z = -1 \end{cases}$$

18. MOVIE TICKETS The receipts for one showing of a movie were $410 for an audience of 100 people. The ticket prices are given in the table. If twice as many children's tickets as general admission tickets were purchased, how many of each type of ticket were sold?

Ticket prices	
Children	$3.00
General Admission	$6.00
Seniors	$5.00

19. Which method, substitution or addition, would you use to solve the following system? Explain your reasoning.

$$\begin{cases} \dfrac{x}{2} - \dfrac{y}{4} = -4 \\ y = -2 - x \end{cases}$$

20. What does it mean to say that a system of two linear equations in two variables is an *inconsistent* system?

21. Suppose that two variables are used to solve an application problem. Why must two equations be written to solve the problem?

22. POPULATION PROJECTIONS If the population trends for the years 2005–2015 continue as projected, estimate the point of intersection of the graphs. Interpret your answer.

Children under age 18 and adults 65 and older as a percent of the U.S. population

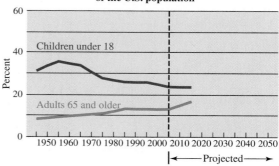

Source: U.S. Bureau of the Census

1. Complete the illustration by labeling the rational numbers, irrational numbers, integers, whole numbers, and natural numbers.

Real numbers

Natural numbers

2. FEDERAL BUDGET President Bush's proposed budget for the fiscal year 2005 was $2.4 trillion. The illustration shows how a typical dollar of the budget was to be spent. Determine the amount he proposed to spend on Social Security.

Social Security: 21¢ Defense: 18¢ Other entitlements: 16¢ Medicare: 12¢ Interest: 7¢ Medicaid: 7¢ Other spending: 19¢

Source: Budget of the United States Government FY 2005

Evaluate each expression for a = −3 and b = −5.

3. $-|b| - ab^2$

4. $\dfrac{14 + 2[2a - (b - a)]}{-b - 2}$

Simplify each expression.

5. $0.5x^2 - 6(2.1x^2 - x) + 6.7x$

6. $-(c + 2) - (2 - c)$

7. COMMUTING Use the following facts to determine a commuter's average speed when she drives to work.

- If she drives her car, it takes a quarter of an hour to get to work.
- If she rides the bus, it takes half an hour to get to work.
- When she drives, her average speed is 10 miles per hour faster than that of the bus.

8. DRIED FRUITS Dried apple slices cost $4.60 per pound, and dried banana chips sell for $3.40 per pound. How many pounds of each should be used to create a 10-pound mixture that sells for $4 per pound?

Solve each equation, if possible. If an equation is an identity, so indicate.

9. $\dfrac{3}{4}x + 1.5 = -19.5$

10. $7 - x - x - x = 8$

11. $\dfrac{x + 7}{3} = \dfrac{x - 2}{5} - \dfrac{x}{15} + \dfrac{7}{3}$

12. $3p - 6 = 4(p - 2) + 2 - p$

Solve each equation for the indicated variable.

13. $C = Ax + AB$ for B

14. $l = a + (n - 1)d$ for n

Graph each equation.

15. $3x = 4y - 11$

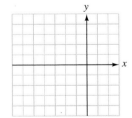

16. $y = -4$

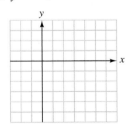

17. Write an equation of the line that passes through $(4, 5)$ and is parallel to the graph of $y = -3x$. Answer in slope–intercept form.

18. Find the slope of the line in the graph.

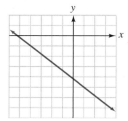

If $f(x) = -x^2 - \dfrac{x}{2}$, find each value.

19. $f(10)$ **20.** $f(-10)$

21. We can think of a function as a machine. Write a function that turns the given input in the illustration into the given output.

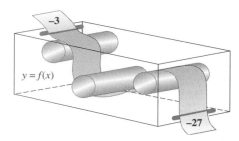

22. Determine whether the following graph is the graph of a function. Explain.

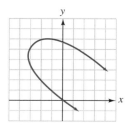

23. Does $x = y^2$ define a function?

24. COLLECTIBLES A collector buys the Hummel figurine shown in the illustration anticipating that it will be worth $650 in 20 years. Assuming straight-line appreciation, write an equation that gives the value v of the figurine x years after it is purchased.

Price: $300.00

25. Solve: $\begin{cases} y = \dfrac{-2x + 1}{3} \\ 3x - 2y = 8 \end{cases}$.

26. Solve: $\begin{cases} -x + 3y + 2z = 5 \\ 3x + 2y + z = -1 \\ 2x - y + 3z = 4 \end{cases}$.

Evaluate each determinant.

27. $\begin{vmatrix} 5 & -2 \\ -2 & 6 \end{vmatrix}$ **28.** $\begin{vmatrix} 2 & 1 & -3 \\ -2 & 2 & 4 \\ 1 & -2 & 2 \end{vmatrix}$

Inequalities

© C. McIntyre/Photolink/Getty Images

INTERMEDIATE
Algebra $f(x)$ Now™

Throughout the chapter, this icon introduces resources on the Intermediate AlgebraNow Web site, accessed through http://1pass.thomson.com, that will

Help you test your knowledge of the material with a pre-test and a post-test

Provide a personalized learning plan targeting areas you should study

There are many ways to measure distance. The above signpost in Maine measures the distances to these towns and lake areas. When building a staircase, carpenters use a tape measure to make sure the distances between the vertical posts are the same. Scientists use a beam of light to measure the distances from Earth to the planets. In mathematics, we use absolute value to measure distance. Recall that the absolute value of a real number is its distance from zero on the number line. In this chapter, we will define absolute value more formally and we will solve equations and inequalities that contain the absolute value of a variable expression.

To learn more about absolute value, visit *The Learning Equation* on the Internet at http://tle.brookscole.com. (The log-in instructions are in the Preface.) For Chapter 4, the online lesson is:

• *TLE* Lesson 6: Absolute Value Equations

Check Your Knowledge

1. To solve a linear inequality in one variable means to find all values of the variable that make the inequality _____.

2. ∞ is a symbol that represents positive _____.

3. The set of real numbers greater than -2 can be represented by the _____ $(-2, \infty)$.

4. If an edge is not included in the graph of an inequality, we draw it as a _____ line.

5. Is -3 a solution of $-5(x - 2) \le 4(3 - x)$?

6. Write interval notation for $\{x \mid -3 \le x < 2\}$, and then graph it.

Solve each inequality. Write the solution set in interval notation and then graph it.

7. $4x + 3 \le 7$

8. $3(4 - y) > 15$

Solve each compound inequality. Write the solution set in interval notation and then graph it.

9. $3x + 4 < 7$ and $-4x + 3 \le 7$

10. $7x + 2 \le -5$ or $2x - 3 \ge 5$

11. $0 < \dfrac{x + 3}{2} \le 1$

12. Find the value of each expression.
    a. $|3 - 4|$ b. $-|-17|$

Solve each equation.

13. $|x - 3| + 2 = 7$ 14. $|2x + 3| = |3x - 2|$

Solve each inequality. Write the solution set in interval notation (if possible) and then graph it.

15. $|2x - 7| < 3$

16. $|4 - x| > 2$

17. $|3 - 2x| - 3 \le 0$

18. $\left|\dfrac{x}{3} - 15\right| \le -5$

Graph each solution set.

19. $2x - y \le 4$ 20. $\begin{cases} x - y \le 3 \\ 2x + y > 2 \end{cases}$ 21. $\begin{cases} y \ge x \\ x - y > -2 \end{cases}$

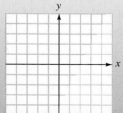

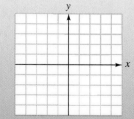

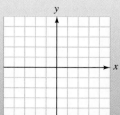

22. Stephanie has received exam scores of 75, 80, 85, and 90. What score must she get on the next exam in order to have an average between 70 and 80?

Study Skills Workshop

REWORKING NOTES TO USE AS STUDY AIDS

As mentioned in the first Study Skills Workshop, you should be taking notes during each of your class lectures. Your instructor will give you examples that he or she thinks are good illustrations of the basic concepts that you are learning. Often, when making up tests, instructors remember the examples that they presented and will include these as test questions. In class it is often difficult to organize the topics in a way that will be conducive to study so reworking your notes afterward is a good way to remind yourself of the material and to prepare for homework, quizzes, and tests. Reworking your notes should be done as soon as possible after class (no later than the next day is best), because the longer you wait to do this, the less chance you have of remembering clearly what was discussed.

How to Rework Notes. You can use a few different systems to rework notes as study aids, depending on personal preference and learning style. The main idea is break down ideas into three components:

1. the basic concept
2. an example that your instructor gave you that illustrates the concept
3. a breakdown of the example into steps, giving a justification or reason for each step

One way of doing this uses index cards stored in a small file box (or held together by a rubber band). If using this method, sort the index cards with tabs by basic concept. On the front of the card write an example that is not worked. On the back of the card show the worked example, citing the justification for each step. When you are studying for a test, look at the example and see whether you can work the problem without help of any kind. If you have trouble with this, you need only turn the card over to look the steps on the back. Continue to rework the problem until you can do it without looking at the solution.

Another way to do this on regular writing paper is to divide the paper into three columns, putting the basic concept in the leftmost column, the example in the middle column, and the worked problem in the last column. You can look over the work in the last column while you attempt to work the problem in the middle.

If you are an auditory learner (or if you are in a car for long stretches), record your voice on audiotape or CD by stating the original concept in your own words, then an example, and then a description of the steps taken in the example. Leave a pause between each phase of the problem and see if you can recite the steps before they are revealed on the tape.

ASSIGNMENT

1. Recall the results of your learning style assessment. Which method above do you think would work the best for your particular learning style and preference?
2. Rework your notes from your last class meeting. If you think of another method for doing this, feel free to experiment (for example, some people might think of creative ways of using their computer).
3. Poll your study group or 3 other classmates to see how they organized their reworked notes. Whose method did you like best? Why?

When working with unequal quantities, we use inequalities instead of equations to describe the situation mathematically.

4.1 Solving Linear Inequalities

- Inequalities   • Graphs, intervals, and set-builder notation
- Solving linear inequalities   • Problem solving

Traffic signs often appear in front of schools. From Figure 4-1, a motorist knows that

- A speed *greater than* 25 miles per hour breaks the law and could possibly result in a ticket for speeding.
- A speed *less than or equal to* 25 miles per hour is within the posted speed limit.

Statements such as these can be expressed using *inequality symbols.*

SPEED LIMIT **25**

FIGURE 4-1

Inequalities

Inequalities are statements indicating that two quantities are unequal. Inequalities contain one or more of the following symbols.

Inequality symbols		
$a \neq b$	means	"*a* is not equal to *b*."
$a < b$	means	"*a* is less than *b*."
$a > b$	means	"*a* is greater than *b*."
$a \leq b$	means	"*a* is less than or equal to *b*."
$a \geq b$	means	"*a* is greater than or equal to *b*."

By definition, $a < b$ means that "*a* is less than *b*," but it also means that $b > a$. Furthermore, if *a* is to the left of *b* on a number line, then $a < b$. If *a* is to the right of *b* on a number line, then $a > b$.

By definition, $a \leq b$ is true if *a* is less than *b* or if *a* is equal to *b*. For example, the inequality $-2 \leq 4$ is true, and so is $4 \leq 4$.

We can use a variable and inequality symbols to describe the warning that the traffic sign in Figure 4-1 gives to drivers. If *x* represents the motorist's speed in miles per hour, he or she is in danger of receiving a speeding ticket if $x > 25$, and he or she is observing the posted speed limit if $x \leq 25$.

Graphs, intervals, and set-builder notation

The graph of a set of real numbers that is a portion of a number line is called an **interval.** The graph shown in Figure 4-2 on the next page represents all real numbers that are greater than -5. This interval contains numbers that satisfy the inequality $x > -5$, such as -4.99, -3, -1.8, 0, $2\frac{3}{4}$, π, and 1,050. The left **parenthesis** at -5 indicates that -5 is not included in the interval.

We can also express this interval in **interval notation** as $(-5, \infty)$, where ∞ (read as **positive infinity**) indicates that the interval extends indefinitely to the right. The left parenthesis is used to show that the endpoint -5 is not included.

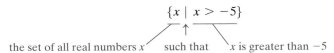

FIGURE 4-2

Set-builder notation is another way of describing the set of real numbers graphed in Figure 4-2. With this notation, the condition for membership in the set is specified using a variable. For example, the set of real numbers greater than -5 is written in set-builder notation as

$$\{x \mid x > -5\}$$

the set of all real numbers x such that x is greater than -5

Changing Values

THINK IT THROUGH

College students are more interested in making money than ever before and less interested in developing a personal take on life. The rising trend to be financially sound is attributed to several major changes, including a steady rise in students' desire to raise a family. Lindsey Bowers, *The Daily Cougar* (U. Houston), 2004

The graph below shows the results of the annual survey conducted by the Higher Education Research Institute. The study has been conducted for 38 years and is the longest-running survey of its kind. A total of 267,449 incoming freshmen at 413 colleges and universities were asked about their attitudes and priorities. For which years is the following true?

The percent citing that being well off financially is very important $>$ The percent citing that developing a meaningful philosophy of life is very important

The interval shown in Figure 4-3 on the next page is the graph of the real numbers less than or equal to 7. This interval contains numbers that satisfy the inequality $x \le 7$. The right **bracket** at 7 indicates that 7 is included in the interval. To express this interval in interval notation, we write $(-\infty, 7]$, where $-\infty$ (read as **negative infinity**) indicates that the interval extends indefinitely to the left. The bracket is used to show that 7 is included in the interval. To describe the interval using set-builder notation, we write $\{x \mid x \le 7\}$.

FIGURE 4-3

! COMMENT The symbol ∞ does not represent a number. It indicates that an interval extends indefinitely to the right. We always use a parenthesis after the symbol ∞. For similar reasons, a parenthesis is always used in front of the symbol $-\infty$ when writing interval notation.

INTERMEDIATE
Algebra $f(x)$ **Now™**

Self Check 1
Represent the set of negative real numbers using interval notation, with a graph, and using set-builder notation.

EXAMPLE 1   Represent the set of real numbers greater than or equal to 8 using interval notation, with a graph, and using set-builder notation.

Solution   All real numbers that are greater than or equal to 8 is the interval $[8, \infty)$. The graph is shown in Figure 4-4. Using set-builder notation, we write $\{x \mid x \ge 8\}$.

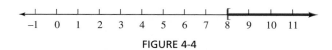

FIGURE 4-4

Answers $(-\infty, 0)$
$\{x \mid x < 0\}$

If an interval extends forever in one direction, as in the previous examples, it is called an **unbounded interval.** The following chart illustrates the types of unbounded intervals and shows how they are described using an inequality and a graph.

Unbounded intervals

The interval (a, ∞) includes all real numbers x such that $x > a$.

The interval $[a, \infty)$ includes all real numbers x such that $x \ge a$.

The interval $(-\infty, a)$ includes all real numbers x such that $x < a$.

The interval $(-\infty, a]$ includes all real numbers x such that $x \le a$.

The interval $(-\infty, \infty)$ includes all real numbers x. The graph of this interval is the entire number line.

When graphing intervals, an open circle can be used to show that a point is not included in a graph, and a solid circle can be used to show that a point *is* included. For example,

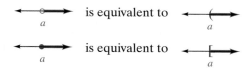

In this book, we will use parentheses and brackets when graphing intervals, because they are consistent with interval notation.

Solving linear inequalities

In this section, we will work with **linear inequalities** in one variable.

> ### Linear Inequalities
>
> A **linear inequality** in one variable (say, x) is any inequality that can be expressed in one of the following forms, where a, b, and c represent real numbers and $a \neq 0$.
>
> $$ax + b < c \quad ax + b \leq c \quad ax + b > c \quad \text{or} \quad ax + b \geq c$$

Some examples of linear inequalities are

$$3x < 0, \quad 3(2x - 9) < 9, \quad \text{and} \quad -12x - 18 \geq 16$$

To **solve a linear inequality** means to find all the values that, when substituted for the variable, make the inequality true. The set of all solutions of an inequality is called its **solution set.** Most of the inequalities we will solve have infinitely many solutions. We will use the following properties to solve inequalities.

> ### Addition and Subtraction Properties of Inequality
>
> Adding the same number to, or subtracting the same number from, both sides of an inequality does not change the solutions.
>
> For any real numbers a, b, and c,
>
> If $a < b$, then $a + c < b + c$.
>
> If $a < b$, then $a - c < b - c$.
>
> Similar statements can be made for the symbols $\leq$, $>$, or $\geq$.

As with equations, there are properties for multiplying and dividing both sides of an inequality by the same number. To develop what is called *the multiplication property of inequality,* consider the true statement $2 < 5$. If both sides are multiplied by a positive number, such as 3, another true inequality results.

$$2 < 5$$
$$3 \cdot 2 < 3 \cdot 5 \quad \text{Multiply both sides by 3.}$$
$$6 < 15 \quad \text{This is a true inequality.}$$

However, if we multiply both sides of $2 < 5$ by a negative number, such as -3, the direction of the inequality symbol is reversed to produce another true inequality.

$$2 < 5$$
$$-3 \cdot 2 > -3 \cdot 5 \quad \text{Multiply both sides by the negative number } -3 \text{ and reverse the direction of the inequality.}$$
$$-6 > -15 \quad \text{This is a true inequality.}$$

The inequality $-6 > -15$ is true because -6 is to the right of -15 on the number line.

Dividing both sides of an inequality by the same negative number also requires that the direction of the inequality symbol be reversed.

$$-4 < 6 \quad \text{This is a true inequality.}$$
$$\frac{-4}{-2} > \frac{6}{-2} \quad \text{Divide both sides by } -2 \text{ and change } < \text{ to } >.$$
$$2 > -3 \quad \text{This is a true inequality.}$$

These examples illustrate the multiplication and division properties of inequality.

> **Multiplication and Division Properties of Inequality**
>
> Multiplying or dividing both sides of an inequality by the same positive number does not change the solutions.
>
> For any real numbers a, b, and c, where c is positive,
>
> If $a < b$, then $ac < bc$.
>
> If $a < b$, then $\frac{a}{c} < \frac{b}{c}$.
>
> If we multiply or divide both sides of an inequality by a negative number, the direction of the inequality symbol must be reversed for the inequalities to have the same solutions.
>
> For any real numbers a, b, and c, where c is negative,
>
> If $a < b$, then $ac > bc$.
>
> If $a < b$, then $\frac{a}{c} > \frac{b}{c}$.
>
> Similar statements can be made for the symbols $\leq$, $>$, or $\geq$.

After applying one of the properties of inequality, the resulting inequality is equivalent to the original one. Like equivalent equations, **equivalent inequalities** have the same solution set.

INTERMEDIATE
Algebra $f(x)$ Now™

Self Check 2

Solve: $2(3x + 2) > -44$. Write the solution set in interval notation and then graph it.

EXAMPLE 2

Solve: $3(2x - 9) < 9$. Write the solution set in interval notation and then graph it.

Solution   To isolate x on the left-hand side of the inequality, we use the same strategy as we used to solve equations.

$$3(2x - 9) < 9$$
$$6x - 27 < 9 \qquad \text{Distribute the multiplication by 3.}$$
$$6x < 36 \qquad \text{To undo the subtraction of 27, add 27 to both sides.}$$
$$x < 6 \qquad \text{To undo the multiplication by 6, divide both sides by 6.}$$

The solution set is the interval $(-\infty, 6)$, whose graph is shown in Figure 4-5. We can also write the solution set using set-builder notation: $\{x \mid x < 6\}$.

The solution set contains infinitely many real numbers. We cannot check to see whether all of them satisfy the original inequality. As an informal check, we pick one number in the graph, such as 4, and see whether it satisfies the inequality.

Check: $3(2x - 9) < 9$ This is the original inequality.

$3[2(\mathbf{4}) - 9] \overset{?}{<} 9$ Substitute 4 for x. Read $\overset{?}{<}$ as "is possibly less than."

$3(8 - 9) \overset{?}{<} 9$

$3(-1) \overset{?}{<} 9$

$-3 < 9$ This is a true statement.

FIGURE 4-5

Since $-3 < 9$, 4 satisfies the inequality. The solution appears to be correct.

Answer $(-8, \infty)$

EXAMPLE 3 Solve: $-12x - 8 \leq 16$. Write the solution set in interval notation and then graph it.

Solution To solve this inequality, we need to isolate x.

$-12x - 8 \leq 16$

$-12x \leq 24$ To undo the subtraction of 8, add 8 to both sides.

$x \geq -2$ To undo the multiplication by -12, divide both sides by -12. Because we are dividing by a negative number, we reverse the $\leq$ symbol.

The solution set is $\{x \mid x \geq -2\}$ or the interval $[-2, \infty)$, whose graph is shown in Figure 4-6.

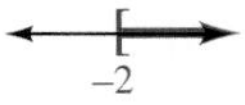

-2

FIGURE 4-6

Self Check 3
Solve: $-6x + 6 \leq 0$. Write the solution set in interval notation and then graph it.

Answer $[1, \infty)$

1

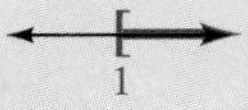

EXAMPLE 4 Solve: $\frac{2}{3}(x + 2) > \frac{4}{5}(x - 3)$. Write the solution set in interval notation and then graph it.

Solution It will be easier to solve the inequality if we clear it of fractions. We do that by multiplying both sides by the LCD of $\frac{2}{3}$ and $\frac{4}{5}$.

$\frac{2}{3}(x + 2) > \frac{4}{5}(x - 3)$

$\mathbf{15} \cdot \frac{2}{3}(x + 2) > \mathbf{15} \cdot \frac{4}{5}(x - 3)$ Multiply both sides by the LCD of $\frac{2}{3}$ and $\frac{4}{5}$, which is 15.

$10(x + 2) > 12(x - 3)$ Simplify: $15 \cdot \frac{2}{3} = 10$ and $15 \cdot \frac{4}{5} = 12$.

$10x + 20 > 12x - 36$ Distribute the multiplication by 10 and 12.

$-2x + 20 > -36$ To eliminate $12x$ on the right-hand side, subtract $12x$ from both sides.

$-2x > -56$ Subtract 20 from both sides.

$x < 28$ Divide both sides by -2 and reverse the $>$ symbol.

The solution set is the interval $(-\infty, 28)$, whose graph is shown in Figure 4-7.

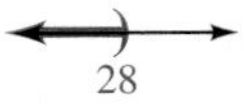

28

FIGURE 4-7

Self Check 4
Solve: $\frac{3}{2}(x + 2) < \frac{3}{5}(x - 3)$. Write the solution set in interval notation and then graph it.

Answer $\left(-\infty, -\frac{16}{3}\right)$

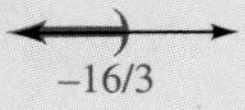

$-16/3$

! **COMMENT** When solving an inequality, the variable sometimes ends up on the right-hand side. For instance, suppose we solve an inequality and obtain $-3 < x$. This inequality can be expressed in the equivalent form $x > -3$, which most students find easier to graph and to express in interval notation.

CALCULATOR SNAPSHOT   Solving linear inequalities

There are several ways to solve linear inequalities graphically. For example, to solve $3(2x - 9) < 9$ we can subtract 9 from both sides and solve instead the equivalent inequality $3(2x - 9) - 9 < 0$. Using standard window settings of $[-10, 10]$ for x and $[-10, 10]$ for y, we graph $y = 3(2x - 9) - 9$ and then use TRACE. Moving the cursor closer and closer to the x-axis, as shown in Figure 4-8(a), we see that the graph is below the x-axis for x-values in the interval $(-\infty, 6)$. This interval is the solution, because in this interval, $3(2x - 9) - 9 < 0$.

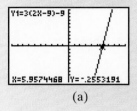
(a)

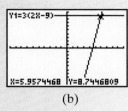
(b)

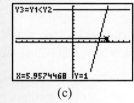
(c)

FIGURE 4-8

Another way to solve $3(2x - 9) < 9$ is to graph $y = 3(2x - 9)$ and $y = 9$. We can then trace to see that the graph of $y = 3(2x - 9)$ is below the graph of $y = 9$ for x-values in the interval $(-\infty, 6)$. See Figure 4-8(b). This interval is the solution, because in this interval, $3(2x - 9) < 9$.

A third approach is to enter and then graph

$Y_1 = 3(2x - 9)$

$Y_2 = 9$

$Y_3 = Y_1 < Y_2$ To do this, use the VARS key. Consult your owner's manual for the specific directions.

The graphs of $y = 3(2x - 9)$, $y = 9$, and a horizontal line 1 unit above the x-axis will be displayed, as shown in Figure 4-8(c). In the TRACE mode, we then move the cursor to the rightmost endpoint of the horizontal line to determine that the interval $(-\infty, 6)$ is the solution of $3(2x - 9) < 9$.

█ Problem solving

In previous chapters, we have used a five-step problem-solving strategy to solve problems. This process involved writing and then solving equations. We will now show how inequalities can be used to solve problems. To decide whether to use an equation or an inequality to solve a problem, you must look for key words and phrases. Here are some common statements that translate to inequalities.

The statement	Translates to
a does not exceed b.	$a \leq b$
a is at most b.	$a \leq b$
a is no more than b.	$a \leq b$

The statement	Translates to
a is at least b.	$a \geq b$
a is not less than b.	$a \geq b$
a will exceed b.	$a > b$

INTERMEDIATE
Algebra *f(x)* **Now**™

EXAMPLE 5   Translate the sentence to mathematical symbols: *The instructor said that the test would take no more than 50 minutes.*

Solution   Since the test will take no more than 50 minutes, it will take 50 minutes or less to complete. If we let t represent the time it takes to complete the test, then $t \leq 50$.

Self Check 5

Translate the sentence to mathematical symbols: *A PG-13 movie rating means that you must be at least 13 years old to see the movie.*

Answer $a \geq 13$

INTERMEDIATE
Algebra *f(x)* **Now**™

EXAMPLE 6   **Political contributions.**   Some volunteers are making long-distance telephone calls to solicit contributions for their candidate. The calls are billed at the rate of 25¢ for the first three minutes and 7¢ for each additional minute or part thereof. If the campaign chairperson has ordered that the cost of each call is not to exceed $1.00, for how many minutes can a volunteer talk to a prospective donor on the phone?

Analyze the problem

We are given the rate at which a call is billed. Since the cost of a call is not to exceed $1.00, the cost must be *less than or equal to* $1.00. This phrase indicates that we should write an inequality to find how long a volunteer can talk to a prospective donor.

Form an inequality

We will let x = the total number of minutes that a call can last. Then the cost of a call will be 25¢ for the first three minutes plus 7¢ times the number of additional minutes, where the number of *additional* minutes is $x - 3$ (the total number of minutes minus the first 3 minutes). With this information, we can form an inequality.

The cost of the first three minutes	plus	the cost of the additional minutes	is not to exceed	$1.00.
0.25	+	0.07(x − 3)	≤	1

Solve the inequality

To simplify the computations, we first clear the inequality of decimals.

$$0.25 + 0.07(x - 3) \leq 1$$

$$25 + 7(x - 3) \leq 100 \qquad \text{To eliminate the decimals, multiply both sides by 100.}$$

$$25 + 7x - 21 \leq 100 \qquad \text{Distribute the multiplication by 7.}$$

$$7x + 4 \leq 100 \qquad \text{Combine like terms.}$$

$$7x \leq 96 \qquad \text{Subtract 4 from both sides.}$$

$$x \leq 13.\overline{714285} \qquad \text{Divide both sides by 7.}$$

State the conclusion

Since the phone company doesn't bill for part of a minute, the longest time a call can last is 13 minutes. If a call lasts for $13.\overline{714285}$ minutes, it will be charged as a 14-minute call, and the cost will be $0.25 + $0.07(11) = $1.02.

Check the result

If the call lasts 13 minutes, the cost will be $0.25 + $0.07(10) = $0.95. This is less than $1.00. The result checks.

Section 4.1 STUDY SET

INTERMEDIATE
Algebra $f(x)$ Now™

VOCABULARY   *Fill in the blanks.*

1. $<$, $>$, $\leq$, and $\geq$ are _____ symbols.
2. $(-\infty, 5)$ is an example of an unbounded _____.
3. The _____ on the right of the interval notation $(-\infty, 5)$ indicates that 5 is not included in the interval.
4. To _____ an inequality means to find all values of the variable that make the inequality true.
5. $3x + 2 \geq 7$ is an example of a _____ inequality.
6. ∞ is a symbol representing positive _____.
7. The symbol for "_____" is $<$. The symbol for "_____" is $\geq$.
8. We read the _____ notation $\{x \mid x < 1\}$ as "the set of all real numbers x _____ _____ x is less than 1."

CONCEPTS

9. Describe each set of real numbers using interval notation and set-builder notation, and then graph it.
   a. All real numbers greater than 4
   b. All real numbers less than -4
   c. All real numbers less than or equal to 4

10. Match each interval with its graph.

   a. $(-\infty, -1]$
   b. $(-\infty, 1)$
   c. $[-1, \infty)$

   i.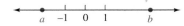
   ii.
   iii.

11. Classify each of the following as an equation, an expression, or an inequality.

   a. $-6 - 5x = 8$
   b. $5 - 2x$

   c. $7x - 5x > -4x$
   d. $-(7x - 9)$
   e. $\frac{x}{2} + 1 \leq 3(x + 7)$

12. In each case, determine what is wrong with the interval notation.
   a. $(\infty, -3)$
   b. $[-\infty, -3)$

13. In the illustration, which of the following are true?
   i. $b > 0$
   ii. $a - b < 0$
   iii. $ab > 0$

14. In the illustration above, which of the following are false?
   i. $b - a > 0$
   ii. $ab < 0$
   iii. $b - a < 0$

15. Perform each step listed below on the inequality $4 > -2$. **Do not reverse the inequality symbol.** Is the resulting statement true?
   a. Add 2 to both sides.
   b. Subtract 4 from both sides.
   c. Multiply both sides by 4.
   d. Divide both sides by -2.

16. Consider the linear inequality $3x + 6 \leq 6$. Determine whether each value is a solution of the inequality.
   a. 0   b. $\frac{2}{3}$
   c. -10 d. 1.5

17. What inequality symbol is suggested by each sentence?

   a. As many as 16 people were seriously injured.

   b. There are no fewer than 10 references to carpools in the speech.

   c. The car is at least 25 years old.

18. The solution set of a linear inequality in x is graphed in the illustration. For that inequality, determine whether a true or false statement results when

   a. -4 is substituted for x.

   b. -3 is substituted for x.

   c. 0 is substituted for x.

NOTATION *Complete each solution to solve the inequality.*

19. $-5x - 1 \geq -11$

$$-5x \geq \boxed{}$$

$$\frac{-5x}{-5} \; \boxed{} \; \frac{-10}{-5}$$

$$x \leq \boxed{}$$

Using interval notation, the result is $(\boxed{}, 2]$.
Using set-builder notation, the result is $\{x \mid \boxed{}\}$.

20. $3 - 6x < 17 + x$

$$3 - \boxed{} < 17$$

$$-7x < \boxed{}$$

$$\frac{-7x}{-7} \; \boxed{} \; \frac{14}{-7}$$

$$x > \boxed{}$$

Using interval notation, the result is $(-2, \infty)$.
Using set-builder notation, the result is $\{x \mid \boxed{}\}$.

21. Write an equivalent inequality with the variable on the left-hand side.

   a. $-10 > x$

   b. $\dfrac{7}{8} < x$

   c. $0 \leq x$

22. Translate to set-builder notation:

   the set of all real numbers x such that
   x is less than or equal to 40

PRACTICE *Solve each inequality. Write the solution set in interval notation and then graph it.*

23. $3x > -9$

24. $4x < -36$

25. $-30y \leq -600$

26. $-6y \geq -600$

27. $0.6x \geq 36$

28. $0.2x < 8$

29. $3 > -\dfrac{9}{10}x$

30. $-\dfrac{2}{5} < -\dfrac{4}{5}x$

31. $x + 4 < 5$

32. $x - 5 > 2$

33. $-5t + 3 \leq 5$

34. $-9t + 6 \geq 16$

35. $-3x - 1 \leq 5$

36. $-2y + 6 < 16$

37. $7 < \dfrac{5}{3}a - 3$

38. $5 > \dfrac{7}{2}a - 9$

39. $0.4x + 0.4 \leq 0.1x + 0.85$

40. $0.05 - 0.5x \leq -0.7 - 0.8x$

41. $3(z - 2) \leq 2(z + 7)$

42. $5(3 + z) > -3(z + 3)$

43. $-11(2 - b) < 4(2b + 2)$

44. $-9(h - 3) + 2h \leq 8(4 - h)$

45. $\dfrac{1}{2}y + 2 \geq \dfrac{1}{3}y - 4$

46. $\frac{1}{4}x - \frac{1}{3} \le x + 2$

47. $\frac{2}{3}x + \frac{3}{2}(x - 5) \le x$

48. $\frac{5}{9}(x + 3) - \frac{4}{3}(x - 3) \ge x - 1$

APPLICATIONS

49. REAL ESTATE Refer to the illustration below. For which regions of the country was the following inequality true in the year 2003?

Median sales price $<$ U.S. median price

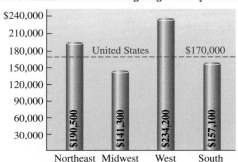

2003 Median Price of Existing Single-Family Homes

Source: National Association of Realtors

50. PUBLIC EDUCATION Refer to the illustration below. For which years is the following inequality true?

$$\text{Enrollment in grade 4} \ge \text{Enrollment in grade 1}$$

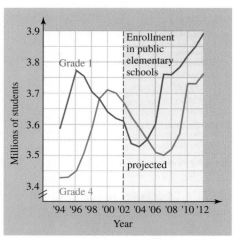

Source: National Center for Education Statistics

51. GEOMETRY The **triangle inequality** states an important relationship between the sides of any triangle:

The sum of the lengths of two sides of a triangle $>$ the length of the third side.

Use the triangle inequality to show that the dimensions of the shuffleboard court shown below must be mislabeled.

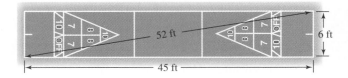

52. COMPUTER PROGRAMMING Flowcharts like the one below are used by programmers to show the step-by-step instructions of a computer program. For row 1 in the table, work through the steps of the flow chart using the values of a, b, and c, and determine what the computer printout would be. Now do the same for row 2, and then for row 3.

	a	b	c
Row 1	1	1	1
Row 2	9	-12	4
Row 3	11	-25	-24

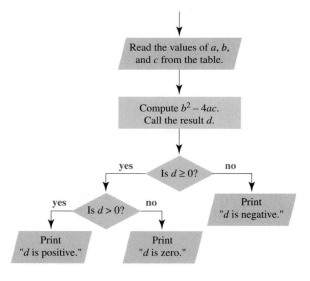

53. FUNDRAISING A school PTA wants to rent a dunking tank for its annual school fundraising carnival. The cost is $85.00 for the first three hours and then $19.50 for each additional hour or part thereof. How long can the tank be rented if up to $185 is budgeted for this expense?

54. INVESTMENTS If a woman has invested $10,000 at 8% annual interest, how much more must she invest at 9% so that her annual income will exceed $1,250?

55. BUYING COMPUTERS A student who can afford to spend up to $2,000 sees the ad shown below. If she decides to buy the computer, find the greatest number of CD-ROMs that she can also purchase. (Disregard sales tax.)

Big Sale!!!!

$1,695.95

All CD-ROMs
$19.95

56. GRADES A student has scores of 70, 77, and 85 on three government exams. What score does she need on a fourth exam to give her an average of 80 or better?

57. WORK SCHEDULES A student works two part-time jobs. He earns $7 an hour for working at the college library and $12 an hour for construction work. To save time for study, he limits his work to 20 hours a week. If he enjoys working at the library more, how many hours can he work at the library and still earn at least $175 a week?

58. SCHEDULING EQUIPMENT An excavating company charges $300 an hour for the use of a backhoe and $500 an hour for the use of a bulldozer. (Part of an hour counts as a full hour.) The company employs one operator for 40 hours per week to operate the machinery. If the company wants to bring in at least $18,500 each week from equipment rental, how many hours per week can it schedule the operator to use a backhoe?

59. MEDICAL PLANS A college provides its employees with a choice of the two medical plans, shown below. For what size hospital bills is Plan 2 better for the employee than Plan 1? (*Hint:* The cost to the employee includes both the deductible payment and the employee's coinsurance payment.)

Plan 1	Plan 2
Employee pays $100 Plan pays 70% of the rest	Employee pays $200 Plan pays 80% of the rest

60. MEDICAL PLANS To save costs, the college in Exercise 59 raised the employee deductible as shown below. For what size hospital bills is Plan 2 better for the employee than Plan 1? (*Hint:* The cost to the employee includes both the deductible payment and the employee's coinsurance payment.)

Plan 1	Plan 2
Employee pays $200 Plan pays 70% of the rest	Employee pays $400 Plan pays 80% of the rest

Use a graphing calculator to solve each inequality.

61. $2x + 3 < 5$

62. $3x - 2 > 4$

63. $5x + 2 \geq 4x - 2$

64. $3x - 4 \leq 2x + 4$

WRITING

65. The techniques for solving linear equations and linear inequalities are similar, yet different. Explain.

66. Explain how the symbol ∞ is used in this section. Is ∞ a real number?

67. Explain how to use the following graph to solve $2x + 1 < 3$.

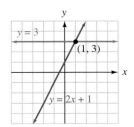

68. Explain each property of inequalities.

a. If $a < b$, and c is any real number, then $a + c < b + c$.

b. If $a < b$, and c is any negative real number, then $ac > bc$.

REVIEW *Use the graph to find f(−1), f(0), and f(2).*

69.

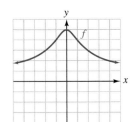

70.

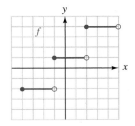

Complete each table.

71. $f(x) = x - x^3$

Input	Output
-2	
2	

72. $g(t) = \dfrac{t^2 - 1}{5}$

Input	Output
-6	
4	

4.2 Solving Compound Inequalities

- Solving compound inequalities containing the word *and*   • Double linear inequalities
- Compound inequalities containing the word *or*
- Solving compound inequalities containing the word *or*

The label on the tube of antibiotic ointment shown in Figure 4-9 advises the user about the temperature at which the medication should be stored. A careful reading of the statement reveals that the storage instruction consists of two parts:

> The storage temperature should be at least 59°F.

and

> The storage temperature should be at most 77°F.

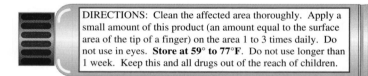

DIRECTIONS: Clean the affected area thoroughly. Apply a small amount of this product (an amount equal to the surface area of the tip of a finger) on the area 1 to 3 times daily. Do not use in eyes. **Store at 59° to 77°F.** Do not use longer than 1 week. Keep this and all drugs out of the reach of children.

FIGURE 4-9

When the words *and* or *or* are used to connect pairs of inequalities, we call these statements *compound inequalities*. In this section, we will discuss how to solve compound inequalities.

Solving compound inequalities containing the word *and*

When two inequalities are joined with the word *and*, we call the statement a **compound inequality.** Some examples are

$$x \geq -3 \quad \text{and} \quad x \leq 6$$

$$\frac{x}{2} + 1 > 0 \quad \text{and} \quad 2x - 3 < 5$$

$$x + 3 \leq 2x - 1 \quad \text{and} \quad 3x - 2 < 5x - 4$$

The solution set of these inequalities contains the numbers that make *both* of the inequalities true. For example, we can find the solution set of the compound inequality $x \geq -3$ and $x \leq 6$ by first graphing the solution sets of each inequality on the same number line and then looking for the numbers common to both graphs.

In Figure 4-10(a) on the next page, the graph of the solution set of $x \geq -3$ is shown in red, and the graph of the solution set of $x \leq 6$ is shown in blue. Figure 4-10(b) shows the graph of the solution of $x \geq -3$ and $x \leq 6$. The purple shaded interval in Figure 4-10(b) is where the red and blue graphs overlap in Figure 4-10(a). It represents the numbers common to the graphs of $x \geq -3$ and $x \leq 6$.

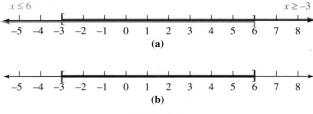

FIGURE 4-10

The solution set of $x \geq -3$ and $x \leq 6$ can be denoted by the **bounded interval** $[-3, 6]$, where the brackets indicate that the endpoints, -3 and 6, are included. It represents all real numbers between -3 and 6, including -3 and 6. Intervals such as this, which contain both endpoints, are called **closed intervals.**

When solving a compound inequality containing *and,* the solution set is the *intersection* of the solution sets of the two inequalities. The **intersection** of two sets is the set of elements that are common to both sets. We can denote the intersection of two sets using the symbol $\cap$, which is read as "intersection." For the compound inequality $x \geq -3$ and $x \leq 6$, we can write

$$[-3, \infty) \cap (-\infty, 6] = [-3, 6]$$

The solution set of the compound inequality $x \geq -3$ and $x \leq 6$ can be expressed in several ways:

1. *As a graph:*

$\begin{array}{c} \text{[———]} \\ \text{-3 \quad 6} \end{array}$

2. *In interval notation:* $[-3, 6]$

3. *In words:* all real numbers between -3 and 6, including -3 and 6

4. *Using set-builder notation:* $\{x \mid x \geq -3 \text{ and } x \leq 6\}$

EXAMPLE 1 Solve: $\dfrac{x}{2} + 1 > 0$ and $2x - 3 < 5$. Graph the solution set.

Solution We begin by solving each linear inequality separately.

$$\dfrac{x}{2} + 1 > 0 \qquad \text{and} \qquad 2x - 3 < 5$$

$$\dfrac{x}{2} > -1 \qquad\qquad\qquad 2x < 8$$

$$x > -2 \qquad\qquad\qquad\quad x < 4$$

Next, we graph the solutions of each inequality on the same number line and determine their intersection. See Figure 4-11.

FIGURE 4-11

The intersection of the graphs in Figure 4-11 is the set of all real numbers between -2 and 4. Using interval notation, the solution set is the interval $(-2, 4)$, whose graph

INTERMEDIATE
Algebra *f(x)* **Now**™
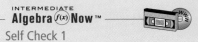

Self Check 1

Solve: $3x > -18$ and $\dfrac{x}{5} - 1 \leq 1$.

Graph the solution set.

is shown in purple in Figure 4-12. This bounded interval, which does not include either endpoint, is called an **open interval.**

Answer $(-6, 10]$

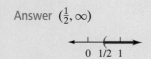

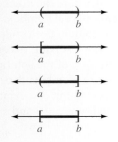

FIGURE 4-12

The solution of the compound inequality in the Self Check of Example 1 is the interval $(-6, 10]$. A bounded interval such as this, which includes only one endpoint, is called a **half-open interval.** The following chart shows the various types of bounded intervals, along with the inequalities and interval notation that describe them.

Intervals		
Open intervals	The interval (a, b) includes all real numbers x such that $a < x < b$.	
Half-open intervals	The interval $[a, b)$ includes all real numbers x such that $a \le x < b$.	
	The interval $(a, b]$ includes all real numbers x such that $a < x \le b$.	
Closed intervals	The interval $[a, b]$ includes all real numbers x such that $a \le x \le b$.	

Self Check 2
Solve: $2x + 3 < 4x + 2$ and $3x + 1 < 5x + 3$. Graph the solution set.

EXAMPLE 2   Solve: $x + 3 \le 2x - 1$ and $3x - 2 < 5x - 4$.

Solution   We solve each inequality separately.

$$x + 3 \le 2x - 1 \qquad \text{and} \qquad 3x - 2 < 5x - 4$$
$$4 \le x \qquad\qquad\qquad\qquad 2 < 2x$$
$$x \ge 4 \qquad\qquad\qquad\qquad 1 < x$$
$$\qquad\qquad\qquad\qquad\qquad\qquad x > 1$$

The graph of $x \ge 4$ is shown below in red and the graph of $x > 1$ is shown below in blue.

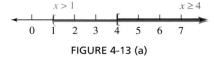

FIGURE 4-13 (a)

Only those x where $x \ge 4$ and $x > 1$ are in the solution set of the compound inequality. Since all numbers greater than or equal to 4 are also greater than 1, the solutions are the numbers x where $x \ge 4$. The solution set is the interval $[4, \infty)$, whose graph is shown below.

Answer $(\frac{1}{2}, \infty)$

FIGURE 4-13 (b)

Self Check 3
Solve: $2x - 3 < x - 2$ and $0 < x - 3.5$.

EXAMPLE 3   Solve: $x - 1 > -3$ and $2x < -8$.

Solution   We solve each inequality separately.

$$x - 1 > -3 \qquad \text{and} \qquad 2x < -8$$
$$x > -2 \qquad\qquad\qquad\qquad x < -4$$

We note that the graphs of the solution sets shown in Figure 4-14 do not intersect.

FIGURE 4-14

This means that there are no numbers that make both parts of the original compound inequality true. So the solution set is the empty set, which can be denoted $\varnothing$.

Answer **no solution**

◼ Double linear inequalities

Inequalities that contain two inequality symbols are called **double inequalities.** An example of a double inequality is

$-3 \le 2x + 5 < 7$ Read as "-3 is less than or equal to $2x + 5$ and $2x + 5$ is less than 7."

Any double linear inequality can be rewritten as a compound inequality containing the word *and.* In general, the following is true.

Double linear inequalities

The compound inequality $c < x < d$ is equivalent to $c < x$ and $x < d$.

EXAMPLE 4   Solve: $-3 \le 2x + 5 < 7$. Graph the solution set.

Solution   This double inequality $-3 \le 2x + 5 < 7$ means that

$-3 \le 2x + 5$ and $2x + 5 < 7$

We could solve each linear inequality separately, but we note that each solution would involve the same steps: subtracting 5 from both sides and dividing both sides by 2. We can solve the double inequality more efficiently by leaving it in its original form and applying these steps to each of its three parts to isolate x in the middle.

$$-3 \le 2x + 5 < 7$$

$$-3 - 5 \le 2x + 5 - 5 < 7 - 5 \qquad \text{To undo the addition of 5, subtract 5 from all three parts.}$$

$$-8 \le 2x < 2 \qquad \text{Perform the subtractions.}$$

$$\frac{-8}{2} \le \frac{2x}{2} < \frac{2}{2} \qquad \text{To undo the multiplication by 2, divide all three parts by 2.}$$

$$-4 \le x < 1 \qquad \text{Perform the divisions.}$$

The solution set is the half-open interval $[-4, 1)$, whose graph is shown in Figure 4-15.

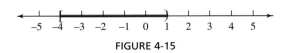

FIGURE 4-15

Self Check 4
Solve: $-5 \le 3x - 8 \le 7$.
Graph the solution set.

Answer $[1, 5]$

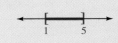

When multiplying or dividing all three parts of a double inequality by a negative number, don't forget to reverse the direction of *both* inequalities. As an example, we will solve $-15 < -5x \le 25$.

$$-15 < -5x \le 25$$

$$\frac{-15}{-5} > \frac{-5x}{-5} \ge \frac{25}{-5}$$ Divide all three parts by -5 to isolate x in the middle. Reverse both inequality signs.

$$3 > x \ge -5$$ Perform the divisions.

$$-5 \le x < 3$$ Write an equivalent compound inequality with the smaller number, -5, on the left.

▌ Compound inequalities containing the word *or*

A warning on the water temperature gauge of a commercial dishwasher, shown in Figure 4-16, cautions the operator to shut down the unit if

    The water temperature goes below 140°

or

    The water temperature goes above 160°

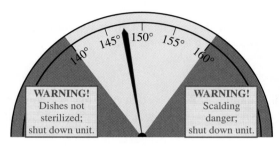

FIGURE 4-16

When two inequalities are joined with the word *or,* we also call the statement a compound inequality. Some examples are

$$x < 140 \quad \text{or} \quad x > 160$$
$$x \le -3 \quad \text{or} \quad x \ge 2$$
$$\frac{x}{3} > \frac{2}{3} \quad \text{or} \quad -(x - 2) > 3$$

▌ Solving compound inequalities containing the word *or*

The solution set of a compound inequality containing the word *or* contains the numbers that make *one or the other or both* inequalities true. For example, we can find the solution set of the compound inequality $x \le -3$ or $x \ge 2$ by putting the graphs of each inequality on the same number line.

In Figure 4-17(a), the graph of the solution set of $x \le -3$ is shown in red, and the graph of the solution set of $x \ge 2$ is shown in blue. Figure 4-17(b) shows the graph of the solution set of $x \le -3$ or $x \ge 2$. This graph is a combination of the graph of $x \le -3$ with the graph of $x \ge 2$.

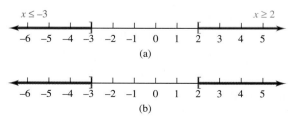

FIGURE 4-17

When solving a compound inequality containing *or*, the solution set is the *union* of the solution sets of the two inequalities. The **union** of two sets is the set of elements that are in either of the sets or both. We can denote the union of two sets using the symbol ∪, which is read as "union." For the compound inequality $x \leq -3$ or $x \geq 2$, we can write the solution set using interval notation:

$$(-\infty, -3] \cup [2, \infty)$$

We can express the solution set of the compound inequality $x \leq -3$ or $x \geq 2$ in several ways:

1. *As a graph:*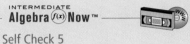
  $$\quad\quad\quad\quad\quad -3 \quad\quad 2$$

2. *In interval notation:* $(-\infty, -3] \cup [2, \infty)$

3. *In words:* all real numbers less than or equal to -3 or greater than or equal to 2

4. *In set-builder notation:* $\{x \mid x \leq -3 \text{ or } x \geq 2\}$

! COMMENT In the statement $x \leq -3$ or $x \geq 2$, it is incorrect to string the inequalities together as $2 \leq x \leq -3$, because that would imply that $2 \leq -3$, which is false.

EXAMPLE 5 Solve: $\dfrac{x}{3} > \dfrac{2}{3}$ or $-(x-2) > 3$. Graph the solution set.

Solution We solve each inequality separately.

$$\dfrac{x}{3} > \dfrac{2}{3} \quad\quad \text{or} \quad\quad -(x-2) > 3$$
$$x > 2 \quad\quad\quad\quad\quad\quad -x + 2 > 3$$
$$\quad\quad\quad\quad\quad\quad\quad\quad -x > 1$$
$$\quad\quad\quad\quad\quad\quad\quad\quad x < -1$$

The graph of the solution set is obtained by graphing the solution sets of each inequality on the same number line. See Figure 4-18.

FIGURE 4-18

The union of the two solution sets consists of all real numbers less than -1 or greater than 2. Using interval notation, the solution set is the interval $(-\infty, -1) \cup (2, \infty)$. Its graph appears in Figure 4-19.

FIGURE 4-19

EXAMPLE 6 Solve: $x + 3 \geq -3$ or $-x > 0$.

Solution We solve each inequality separately.

$$x + 3 \geq -3 \quad\quad \text{or} \quad\quad -x > 0$$
$$x \geq -6 \quad\quad\quad\quad\quad\quad x < 0$$

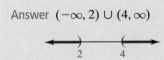

We graph the solution set of each inequality on the same number line in Figure 4-20.

FIGURE 4-20

The entire number line is shaded, which indicates that all real numbers satisfy the original compound inequality. Using interval notation, the solution set is denoted $(-\infty, \infty)$. Its graph is shown in Figure 4-21.

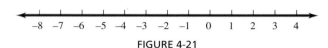

FIGURE 4-21

Answer $(-\infty, \infty)$

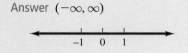

THINK IT THROUGH   Study Abroad

"American students are studying abroad in growing numbers despite economic and security concerns post-Sept 11." Open Doors Report 2003, Institute of International Education

Recent figures released by Institute of International Education show that the United Kingdom, Spain, and Italy are the top three destinations for American students studying abroad. For what year, or years, was

1. the number of American students studying in the United Kingdom greater than 30,000 *and* the number of American students studying in Spain greater than 13,000?

2. the number of American students studying in Spain greater than 15,000 *or* the number of American students studying in Italy less than 12,000?

3. the number of American students studying in Spain greater than the number studying in Italy *and* the number of American students studying in the United Kingdom at least twice the number studying in Italy?

American Students Studying in . . .

United Kingdom	Spain	Italy
2002: 31,143	2002: 17,176	2002: 17,169
2001: 30,289	2001: 16,016	2001: 16,127
2000: 29,289	2000: 13,974	2000: 12,930
1999: 27,720	1999: 12,292	1999: 11,281

Section 4.2 STUDY SET

INTERMEDIATE
Algebra $f(x)$ Now™

VOCABULARY   *Fill in the blanks.*

1. $x \geq 3$ and $x \leq 4$ is a _____ inequality.

2. $-6 \leq x + 1 < 1$ is a _____ linear inequality.

3. The bounded _____ (2, 8] includes all real numbers x such that $2 < x \leq 8$.

4. $x \leq 3$ or $x > 5$ is a compound _____.

CONCEPTS   *Fill in the blanks.*

5. The word *and* between two inequality statements requires that _____ of the inequalities must be true for the entire statement to be true.

6. The word *or* between two inequality statements requires that only _____ of the inequalities must be true for the entire statement to be true.

7. If the three parts of a double inequality are divided by a negative number, the direction of both inequality symbols must be _____.

8. The double inequality $-2 < 3x + 4 < 10$ can be written as $-2 < 3x + 4$ _____ $3x + 4 < 10$.

9. In each case, determine whether -3 is a solution of the compound inequality.

a. $\dfrac{x}{3} + 1 \geq 0$ and $2x - 3 < -10$

b. $2x \leq 0$ or $-3x < -5$

10. In each case, determine whether -3 is a solution of the double linear inequality.

a. $-1 < -3x + 4 < 12$

b. $-1 < -3x + 4 < 14$

11. Give the solution of each inequality in interval notation, if possible.

a. $x < -3$ and $x > 3$

b. $x < 3$ or $x > -3$

12. Give the solution of each inequality in interval notation, if possible.

a. $x < 0$ or $x \geq 0$

b. $x < 0$ and $x > 0$

13. Match each interval with its corresponding graph.

a. $[2, 3)$

i.
   2   3

b. $(2, 3)$

ii.
   2   3

c. $[2, 3]$

iii.
   2   3

14. Give the interval notation that describes each set. Then graph it.

a. The real numbers between -3 and 3

b. The real numbers less than -3 or greater than 3

c. The real numbers between -3 and 3, including 3

d. $\{x \mid x \geq -3 \text{ and } x \leq 3\}$

e. $\{x \mid x \leq -3 \text{ or } x \geq 3\}$

NOTATION

15. a. Graph: $(-\infty, 2) \cup [3, \infty)$.

b. Graph: $(-\infty, 3) \cap [-2, \infty)$.

c. Graph: $(0, 8) \cap (2, 10)$.

d. Graph: $(-5, 7] \cup [2, 9]$.

16. Classify each interval as open, half-open, or closed.

a. $(-2, 15]$

b. $[-2, 15]$

c. $(-2, 15)$

d. $[-2, 15)$

17. What is incorrect about the double inequality $3 < -3x + 4 < -3$?

18. What set is denoted by the interval notation $(-\infty, \infty)$? Graph it.

PRACTICE   *Solve each compound inequality. Write the solution set (if one exists) in interval notation and then graph it.*

19. $x > -2$ and $x \leq 5$

20. $x \leq -4$ and $x \geq -7$

21. $2.2x < -19.8$ and $-4x < 40$

22. $\frac{1}{2}x \le 2$ and $0.75x \ge -6$

23. $x + 3 < 3x - 1$ and $4x - 3 \le 3x$

24. $4x \ge -x + 5$ and $6 \ge 4x - 3$

25. $x + 2 < -\frac{1}{3}x$ and $-6x < 9x$

26. $5(x - 2) \ge 0$ and $-3x < 9$

27. $x - 1 \le 2(x + 2)$ and $x \le 2x - 5$

28. $5(x + 1) \le 4(x + 3)$ and $x + 12 < -3$

29. $4 \le x + 3 \le 7$

30. $-5.3 < x - 2.3 < -1.3$

31. $15 > 2x - 7 > 9$

32. $25 > 3x - 2 > 7$

33. $-2 < -b + 3 < 5$

34. $2 < -t - 2 < 9$

35. $-6 < -3(x - 4) \le 24$

36. $-4 \le -2(x + 8) < 8$

37. $2x + 1 \ge 5$ and $-3(x + 1) \ge -9$

38. $2(-2) \le 3x - 1$ and $3x - 1 \le -1 - 3$

39. $\frac{x}{0.7} + 5 > 4$ and $-4.8 \le \frac{3x}{-0.125}$

40. $-x < -2x$ and $3x > 2x$

41. $-4 > \frac{2}{3}x - 2 > -6$

42. $-6 \le \frac{1}{3}a + 1 < 0$

43. $0 \le \frac{4 - x}{3} \le 2$

44. $-2 \le \frac{5 - 3x}{2} \le 2$

45. $x \le -2$ or $x > 6$

46. $x \ge -1$ or $x \le -3$

47. $x - 3 < -4$ or $x - 2 > 0$

48. $4x < -12$ or $\frac{x}{2} > 4$

49. $3x + 2 < 8$ or $2x - 3 > 11$

50. $3x + 4 < -2$ or $3x + 4 > 10$

51. $x > 3$ or $x < 5$

52. $x < -15$ or $x > -100$

53. $-4(x + 2) \ge 12$ or $3x + 8 < 11$

54. $4.5x - 1 < -10$ or $6 - 2x \ge 12$

55. $4.5x - 2 > 2.5$ or $\frac{1}{2}x \le 1$

56. $0 < x$ or $3x - 5 > 4x - 7$

APPLICATIONS

57. BABY FURNITURE A company manufactures various sizes of playpens having perimeters between 128 and 192 inches, inclusive. (See the illustration on the next page.)

a. Complete the double inequality that mathematically describes the range of the perimeters of the playpens.

$$\boxed{} \le 4s \le \boxed{}$$

b. Solve the double inequality to find the range of the side lengths of the playpens.

58. TRUCKING The distance that a truck can travel in 8 hours, at a constant rate of r mph, is given by $8r$. A trucker wants to travel at least 350 miles, and company regulations don't allow him to exceed 450 miles in one 8-hour shift.

a. Complete the double inequality that describes the mileage range of the truck.

$$\boxed{} \le 8r \le \boxed{}$$

b. Solve the double inequality to find the range of the average rate (speed) of the truck for the 8-hour trip.

59. TREATING A FEVER Use the following flow chart to determine what action should be taken for a 13-month-old child who has had a 99.8° temperature for 3 days and is not suffering any other symptoms. T represents the child's temperature, A the child's age in months, and S the number of hours the child has experienced the symptoms.

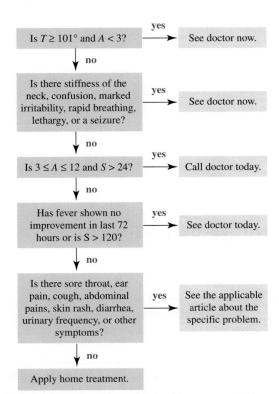

Based on information from *Take Care of Yourself* (Addison-Wesley, 1993)

60. THERMOSTATS The *Temp range* control on the thermostat shown below directs the heater to come on when the room temperature gets 5 degrees below the *Temp setting;* it directs the air conditioner to come on when the room temperature gets 5 degrees above the *Temp setting.* Use interval notation to describe

a. the temperature range for the room when neither the heater nor the air conditioner will be on.

b. the temperature range for the room when either the heater or the air conditioner will be on. (*Note:* The lowest temperature theoretically possible is $-460°$ F, called *absolute zero.*)

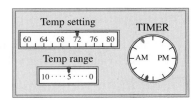

61. HEALTH CARE Refer to the illustration below. Let P represent the percent of children covered by private insurance, M the percent covered by Medicare/Medicaid, and N the percent not covered. For what years are the following true?

a. $P \ge 68$ and $M \ge 18$

b. $P \ge 68$ or $M \ge 18$

c. $P \ge 67$ and $N \le 12.5$

d. $P \ge 67$ and $N \le 12.5$

U.S. Health Care Coverage for Persons Under 18 Years of Age (in percent)

Private insurance   Medicaid

Not covered

	Private insurance	Medicaid	Not covered
1998	68.4	17.1	12.7
1999	68.8	18.1	11.9
2000	67.0	19.4	12.4
2001	66.7	21.2	11.0

Source: U.S. Department of Health and Human Services

62. POLLS For each response to the poll question shown in the illustration, the *margin of error* is +/− (read as "plus or minus") 3.2%. This means that for the statistical methods used to do the polling, the actual response could be as much as 3.2 points more or 3.2 points less than shown. Use interval notation to describe the possible interval (in percent) for each response.

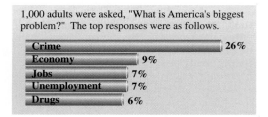

1,000 adults were asked, "What is America's biggest problem?" The top responses were as follows.

Crime	26%
Economy	9%
Jobs	7%
Unemployment	7%
Drugs	6%

WRITING

63. Explain how to find the union and how to find the intersection of $(-\infty, 5)$ and $(-2, \infty)$ graphically.

64. Explain why the double inequality

$$2 < x < 8$$

can be written in the form

$$2 < x \text{ and } x < 8$$

65. Each of the shaded regions in the **Venn diagram** in the illustration below represents a set. Describe the intersection of set A and set B.

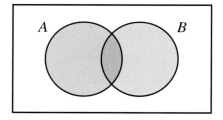

66. See Exercise 65. Describe the union of set A and set B.

REVIEW   *Refer to the illustration below, which shows the results of each of the games of the eventual champion, the University of Kentucky, in the 1998 NCAA Men's Basketball Tournament. Round to the nearest tenth when necessary.*

67. What are the mean, median, and mode of the set of Kentucky scores?

68. What are the mean and the median of the set of scores of Kentucky's opponents?

69. Find the margin of victory for Kentucky in each of its games. Then find the average (mean) margin of victory for Kentucky in the tournament.

70. What was the average (mean) combined score for Kentucky and its opponents in the tournament?

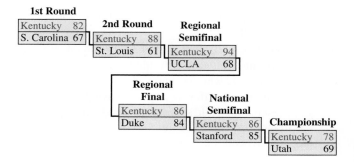

4.3 Solving Absolute Value Equations and Inequalities

- Absolute value • Equations of the form $|x| = k$ • Equations with two absolute values
- Inequalities of the form $|x| < k$ • Inequalities of the form $|x| > k$

Many quantities in mathematics, science, and engineering are expressed as positive numbers. To guarantee that a quantity is positive, we often use absolute value. In this section, we will work with equations and inequalities involving the absolute value of algebraic expressions. Using the definition of absolute value, we will develop procedures to solve absolute value equations and absolute value inequalities.

Absolute value

Recall that the absolute value of any real number is the distance between the number and zero on the number line. For example, the points shown in Figure 4-22 with coordinates of 4 and -4 both lie 4 units from 0. Thus, $|4| = |-4| = 4$.

The absolute value of a real number can be defined more formally as follows.

> **Absolute value**
>
> If $x \geq 0$, then $|x| = x$.
>
> If $x < 0$, then $|x| = -x$.

This definition gives a way to associate a nonnegative real number with any real number.

- If $x \geq 0$, then x (which is positive or 0) is its own absolute value.
- If $x < 0$, then $-x$ (which is positive) is the absolute value.

Either way, $|x|$ is positive or 0. That is, $|x| \geq 0$ for all real numbers x.

EXAMPLE 1   Find: **a.** $|9|$,   **b.** $|-5.68|$,   **c.** $|0|$,   and   **d.** $2|-8|$.

Solution

a. Since $9 \geq 0$, the number 9 is its own absolute value: $|9| = 9$.

b. Since $-5.68 < 0$, the opposite (negative) of -5.68 is the absolute value:

$$|-5.68| = -(-5.68) = 5.68$$

c. Since $0 \geq 0$, 0 is its own absolute value: $|0| = 0$.

d. $2|-8| = 2 \cdot |-8|$
$= 2 \cdot 8$
$= 16$

! COMMENT The placement of a $-$ sign in an expression containing an absolute value symbol is important. For example, $|-19| = 19$, but $-|19| = -19$.

Equations of the form $|x| = k$

The absolute value of a real number represents the distance on a number line from a point to the origin. To solve the **absolute value equation** $|x| = 5$, we must find the coordinates of all points on a number line that are exactly 5 units from zero. See Figure 4-23. The only two points that satisfy this condition have coordinates 5 and -5. That is, $x = 5$ or $x = -5$.

In general, the solution set of the absolute value equation $|x| = k$, where $k \geq 0$, includes the coordinates of the points on the number line that are k units from the origin. (See Figure 4-24.)

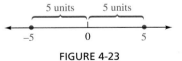

FIGURE 4-23

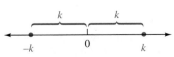

FIGURE 4-24

> **Absolute value equations**
>
> If $k \geq 0$, then
>
> $$|x| = k \qquad \text{is equivalent to} \qquad x = k \text{ or } x = -k$$

Self Check 2
Solve:

a. $|y| = 24$ **b.** $|x| = \dfrac{1}{2}$

c. $|a| = -1.1$

Answers **a.** 24, −24, **b.** $\dfrac{1}{2}, -\dfrac{1}{2}$,
c. no solution

EXAMPLE 2 Solve: **a.** $|x| = 8$, **b.** $|s| = 0.003$, and **c.** $|c| = -15$.

Solution

a. If $|x| = 8$, then $x = 8$ or $x = -8$.

b. If $|s| = 0.003$, then $s = 0.003$ or $s = -0.003$.

c. The absolute value of a number is either positive or zero but never negative. Therefore, there is no value for c for which $|c| = -15$. This equation has no solution.

The equation $|x - 3| = 7$ indicates that a point on a number line with a coordinate of $x - 3$ is 7 units from the origin. Thus, $x - 3$ can be either 7 or −7.

$$x - 3 = 7 \qquad \text{or} \qquad x - 3 = -7$$
$$x = 10 \qquad\qquad\quad x = -4$$

The solutions of the absolute value equation are 10 and −4. We can graph them on a number line, as shown in Figure 4-25. If either of these numbers is substituted for x in $|x - 3| = 7$, the equation is satisfied.

FIGURE 4-25

Check:

$$|x - 3| = 7 \qquad\qquad |x - 3| = 7$$
$$|10 - 3| \stackrel{?}{=} 7 \qquad\qquad |-4 - 3| \stackrel{?}{=} 7$$
$$|7| \stackrel{?}{=} 7 \qquad\qquad\quad |-7| \stackrel{?}{=} 7$$
$$7 = 7 \qquad\qquad\qquad\quad 7 = 7$$

Self Check 3
Solve:
a. $|2x - 3| = 7$

b. $\left|\dfrac{x}{4} - 1\right| = -3$

EXAMPLE 3 Solve: **a.** $|3x - 2| = 5$ and **b.** $|5 - x| = -10$.

Solution

a. We can solve $|3x - 2| = 5$ by writing and then solving an equivalent compound equation:

$$3x - 2 = 5 \qquad \text{or} \qquad 3x - 2 = -5$$

Now we solve each equation for x.

$$3x - 2 = 5 \qquad \text{or} \qquad 3x - 2 = -5$$
$$3x = 7 \qquad\qquad\qquad 3x = -3$$
$$x = \frac{7}{3} \qquad\qquad\qquad x = -1$$

Verify that both solutions, $\frac{7}{3}$ and −1, check.

Answers **a.** 5, −2,
b. no solution

b. For any real number x, $|5 - x|$, will be nonnegative. For this reason, $|5 - x| = -10$ has no solution.

When solving absolute value equations, we want the absolute value isolated on one side. To isolate an absolute value, we can use the equation-solving procedures studied earlier.

EXAMPLE 4　Solve: $\left|\dfrac{2}{3}x + 3\right| + 4 = 10$.

Self Check 4
Solve: $|0.4x - 2| - 0.6 = 0.4$.

Solution　We can isolate $\left|\frac{2}{3}x + 3\right|$ on the left-hand side of the equation by subtracting 4 from both sides.

$$\left|\frac{2}{3}x + 3\right| + 4 = 10$$

$$\left|\frac{2}{3}x + 3\right| = 6 \qquad \text{Subtract 4 from both sides.}$$

With the absolute value now isolated, we can solve $\left|\frac{2}{3}x + 3\right| = 6$ by writing and then solving an equivalent compound equation.

$$\frac{2}{3}x + 3 = 6 \qquad \text{or} \qquad \frac{2}{3}x + 3 = -6$$

Now we solve each equation for x.

$$\frac{2}{3}x + 3 = 6 \qquad \text{or} \qquad \frac{2}{3}x + 3 = -6$$

$$\frac{2}{3}x = 3 \qquad\qquad \frac{2}{3}x = -9$$

$$2x = 9 \qquad\qquad 2x = -27$$

$$x = \frac{9}{2} \qquad\qquad x = -\frac{27}{2}$$

Verify that both solutions, $\frac{9}{2}$ and $-\frac{27}{2}$, check.

Answer　7.5, 2.5

❗ COMMENT　Since the absolute value of a quantity cannot be negative, equations such as $\left|7x + \frac{1}{2}\right| = -4$ have no solution. Their solution sets are empty.

INTERMEDIATE
Algebra _f(x)_ Now™

EXAMPLE 5　Solve: $3\left|\dfrac{1}{2}x - 5\right| - 4 = -4$.

Self Check 5
Solve: $-5\left|\dfrac{2x}{3} + 4\right| + 1 = 1$.

Solution　We first isolate $\left|\frac{1}{2}x - 5\right|$ on the left-hand side.

$$3\left|\frac{1}{2}x - 5\right| - 4 = -4$$

$$3\left|\frac{1}{2}x - 5\right| = 0 \qquad \text{Add 4 to both sides.}$$

$$\left|\frac{1}{2}x - 5\right| = 0 \qquad \text{Divide both sides by 3.}$$

Since 0 is the only number whose absolute value is 0, the expression $\frac{1}{2}x - 5$ must be 0, and we have

$$\frac{1}{2}x - 5 = 0$$

$$\frac{1}{2}x = 5 \qquad \text{Add 5 to both sides.}$$

$$x = 10 \qquad \text{Multiply both sides by 2.}$$

Verify that 10 satisfies the original equation.

Answer　-6

Equations with two absolute values

The equation $|a| = |b|$ is true when $a = b$ or when $a = -b$. For example,

$$|3| = |3| \qquad \text{or} \qquad |3| = |-3|$$

These are the same number.   These numbers are opposites.

In general, the following statement is true.

Equations with two absolute values
If X and Y represent algebraic expressions, the equation $|X| = |Y|$ is equivalent to $X = Y \qquad \text{or} \qquad X = -Y$

INTERMEDIATE
Algebra $f(x)$ **Now**™

Self Check 6
Solve: $|2x - 3| = |4x + 9|$.

EXAMPLE 6   Solve: $|5x + 3| = |3x + 25|$.

Solution   This equation is true when $5x + 3 = 3x + 25$, or when $5x + 3 = -(3x + 25)$. We solve each equation for x.

$$
\begin{array}{rcl}
5x + 3 = 3x + 25 & \text{or} & 5x + 3 = -(3x + 25) \\
2x = 22 & & 5x + 3 = -3x - 25 \\
x = 11 & & 8x = -28 \\
& & x = -\dfrac{28}{8} \\
& & x = -\dfrac{7}{2}
\end{array}
$$

Answer $-1, -6$

Verify that both solutions, 11 and $-\frac{7}{2}$, check.

Inequalities of the form $|x| < k$

To solve the **absolute value inequality** $|x| < 5$, we must find the coordinates of all points on a number line that are less than 5 units from the origin. see Figure 4-26. Thus, x is between -5 and 5, and

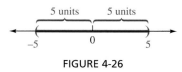

FIGURE 4-26

$$|x| < 5 \quad \text{is equivalent to} \quad -5 < x < 5$$

In general, the solution set of the absolute value inequality $|x| < k$ where $k > 0$ includes the coordinates of the points on the number line that are less than k units from the origin. See Figure 4-27.

FIGURE 4-27

Solving $|x| < k$ and $|x| \le k$			
$|x| < k$	is equivalent to	$-k < x < k$	where $k > 0$
$|x| \le k$	is equivalent to	$-k \le x \le k$	where $k \ge 0$

EXAMPLE 7   Solve $|2x - 3| < 9$ and graph the solution set.

Solution   To solve $|2x - 3| < 9$, we write and then solve an equivalent double inequality.

$|2x - 3| < 9$ is equivalent to $-9 < 2x - 3 < 9$

Now we solve for x.

$-9 < 2x - 3 < 9$
$-6 < 2x < 12$ Add 3 to all three parts.
$-3 < x < 6$ Divide all parts by 2.

Any number between -3 and 6 is in the solution set. This is the interval $(-3, 6)$; its graph is shown in Figure 4-28.

FIGURE 4-28

Answer $\left(-2, \frac{2}{3}\right)$

EXAMPLE 8   **Tolerances.**   When manufactured parts are inspected by a quality control engineer, they are classified as acceptable if each dimension falls within a given *tolerance range* of the dimensions listed on the blueprint. For the bracket shown in Figure 4-29, the distance between the two drilled holes is given as 2.900 inches. Because the tolerance is ± 0.015 inch, this distance can be as much as 0.015 inch longer or 0.015 inch shorter, and the part will be considered acceptable. The acceptable distance d between holes can be represented by the absolute value inequality $|d - 2.900| \leq 0.015$. Solve the inequality and explain the result.

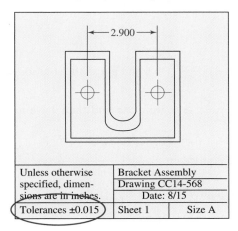

FIGURE 4-29

Solution   To solve the absolute value inequality, we write and then solve an equivalent double inequality.

$|d - 2.900| \leq 0.015$ is equivalent to $-0.015 \leq d - 2.900 \leq 0.015$

Now we solve for d.

$-0.015 \leq d - 2.900 \leq 0.015$
$2.885 \leq d \leq 2.915$ Add 2.900 to all three parts.

The solution set is the interval [2.885, 2.915]. This means that the distance between the two holes should be between 2.885 and 2.915 inches, inclusive. If the distance is less than 2.885 inches or more than 2.915 inches, the part should be rejected.

▪ Inequalities of the form $|x| > k$

To solve the absolute value inequality $|x| > 5$, we must find the coordinates of all points on a number line that are more than 5 units from the origin. See Figure 4-30.

FIGURE 4-30

Thus, $x < -5$ or $x > 5$.

In general, the solution set of $|x| > k$ includes the coordinates of the points on the number line that are more than k units from the origin. See Figure 4-31. Thus,

FIGURE 4-31

$$|x| > k \quad \text{is equivalent to} \quad x < -k \text{ or } x > k$$

The word *or* indicates an either/or situation. It is only necessary that x satisfy one of the two conditions to be in the solution set.

Solving $|x| > k$ and $|x| \geq k$

If $k \geq 0$, then

$$|x| > k \qquad \text{is equivalent to} \qquad x < -k \quad \text{or} \quad x > k$$
$$|x| \geq k \qquad \text{is equivalent to} \qquad x \leq -k \quad \text{or} \quad x \geq k$$

INTERMEDIATE
Algebra *f(x)* **Now**™

Self Check 9

Solve $\left| \dfrac{2 - x}{4} \right| \geq 1$ and graph the solution set.

EXAMPLE 9   Solve $\left| \dfrac{3 - x}{5} \right| \geq 6$ and graph the solution set.

Solution   To solve $\left| \dfrac{3 - x}{5} \right| \geq 6$, we write and then solve an equivalent compound inequality.

$$\left| \frac{3 - x}{5} \right| \geq 6 \quad \text{is equivalent to} \quad \frac{3 - x}{5} \leq -6 \quad \text{or} \quad \frac{3 - x}{5} \geq 6$$

Now we solve each inequality for x.

$\dfrac{3 - x}{5} \leq -6$	or	$\dfrac{3 - x}{5} \geq 6$	
$3 - x \leq -30$		$3 - x \geq 30$	Multiply both sides by 5.
$-x \leq -33$		$-x \geq 27$	Subtract 3 from both sides.
$x \geq 33$		$x \leq -27$	Divide both sides by -1 and reverse the direction of the inequality symbol.

The solution set is the interval $(-\infty, -27] \cup [33, \infty)$. Its graph appears in Figure 4-32.

Answer $(-\infty, -2] \cup [6, \infty)$

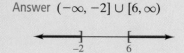

FIGURE 4-32

EXAMPLE 10   Solve $\left|\dfrac{2}{3}x - 2\right| - 3 > 6$ and graph the solution set.

Solution   We add 3 to both sides to isolate the absolute value on the left-hand side.

$$\left|\frac{2}{3}x - 2\right| - 3 > 6$$

$$\left|\frac{2}{3}x - 2\right| > 9 \quad \text{Add 3 to both sides to isolate the absolute value.}$$

We then proceed as follows:

$$\frac{2}{3}x - 2 < -9 \qquad \text{or} \qquad \frac{2}{3}x - 2 > 9$$

$$\frac{2}{3}x < -7 \qquad\qquad\qquad \frac{2}{3}x > 11 \quad \text{Add 2 to both sides.}$$

$$2x < -21 \qquad\qquad\qquad 2x > 33 \quad \text{Multiply both sides by 3.}$$

$$x < -\frac{21}{2} \qquad\qquad\qquad x > \frac{33}{2} \quad \text{Divide both sides by 2.}$$

The solution set is $\left(-\infty, -\frac{21}{2}\right) \cup \left(\frac{33}{2}, \infty\right)$. Its graph appears in Figure 4-33.

−21/2   33/2

FIGURE 4-33

INTERMEDIATE
Algebra *f(x)* **Now**™

Self Check 10

Solve $\left|\dfrac{3}{4}x + 2\right| - 1 > 3$ and graph the solution set.

Answer $(-\infty, -8) \cup \left(\dfrac{8}{3}, \infty\right)$

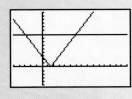

−8   8/3

The following summary shows how we can interpret absolute value in three ways. Assume $k > 0$.

Geometric description	Graphic description	Algebraic description
1. $\|x\| = k$ means that x is k units from 0 on the number line.		$\|x\| = k$ is equivalent to $x = k$ or $x = -k$.
2. $\|x\| < k$ means that x is less than k units from 0 on the number line.		$\|x\| < k$ is equivalent to $-k < x < k$.
3. $\|x\| > k$ means that x is more than k units from 0 on the number line.		$\|x\| > k$ is equivalent to $x > k$ or $x < -k$.

Solving absolute value inequalities

CALCULATOR SNAPSHOT

We can also solve absolute value inequalities using a graphing calculator. For example, to solve $|2x - 3| < 9$, we graph the equations $y = |2x - 3|$ and $y = 9$ on the same coordinate system. If we use settings of $[-5, 15]$ for x and $[-5, 15]$ for y, we get the graph shown in Figure 4-34.

The inequality $|2x - 3| < 9$ will be true for all x-coordinates of points that lie on the graph of $y = |2x - 3|$ and below the graph of $y = 9$. Using the TRACE or INTERSECT feature, we can see that these values of x are in the interval $(-3, 6)$.

FIGURE 4-34

Section 4.3 STUDY SET

VOCABULARY *Fill in the blanks.*

1. $|2x - 1| = 10$ is an absolute value _____.
2. $|2x - 1| > 10$ is an absolute value _____.
3. To _____ the absolute value in $|3 - x| - 4 = 5$, we add 4 to both sides.
4. $|x| = 2$ is _____ to $x = 2$ or $x = -2$.

CONCEPTS *Fill in the blanks.*

5. $|x| \geq$ ▢ for all real numbers x.
6. If $x < 0$, $|x| =$ ▢.
7. To solve $|x| > 5$, we must find the coordinates of all points on a number line that are _____ _____ 5 units from 0.
8. To solve $|x| < 5$, we must find the coordinates of all points on a number line that are _____ _____ 5 units from 0.
9. To solve $|x| = 5$, we must find the coordinates of all points on a number line that are ▢ units from 0.
10. The equation $|a| = |b|$ is true when $a =$ ▢ or when $a =$ ▢.

11. Determine whether -3 is a solution of the given equation or inequality.
    a. $|x - 1| = 4$ b. $|x - 1| > 4$
    c. $|x - 1| \leq 4$
    d. $|5 - x| = |x + 12|$

12. For each absolute value equation or inequality, write an equivalent compound equation or inequality.
    a. $|x| = 8$
    b. $|x| \geq 8$
    c. $|x| \leq 8$
    d. $|5x - 1| = |x + 3|$

NOTATION

13. Match each equation or inequality with its graph.
    a. $|x| = 1$ i.
    b. $|x| > 1$ ii.
    c. $|x| < 1$ iii.

14. Match each graph with its corresponding equation or inequality.
    a.   i. $|x| \geq 2$
    b.   ii. $|x| \leq 2$
    c.   iii. $|x| = 2$

Write each compound inequality as an inequality containing absolute value symbols.

15. $-4 < x < 4$
16. $x < -4$ or $x > 4$
17. $x + 3 < -6$ or $x + 3 > 6$
18. $-5 \leq x - 3 \leq 5$

PRACTICE *Find the value of each expression.*

19. $|8|$ 20. $|-18|$
21. $-|0.02|$ 22. $-|-3.14|$
23. $-\left|-\dfrac{31}{16}\right|$ 24. $-\left|\dfrac{25}{4}\right|$
25. $|\pi|$ 26. $\left|-\dfrac{\pi}{2}\right|$
27. $5|-5|$ 28. $9|-1|$
29. $-\dfrac{1}{2}|-4|$ 30. $-16\left|-\dfrac{1}{4}\right|$

Solve each equation, if possible.

31. $|x| = 23$ 32. $|x| = 90$
33. $|x - 3.1| = 6$ 34. $|x + 4.3| = 8.9$
35. $|3x + 2| = 16$ 36. $|5x - 3| = 22$
37. $\left|\dfrac{7}{2}x + 3\right| = -5$ 38. $\left|\dfrac{2x}{3} + 10\right| = 0$
39. $|3 - 4x| = 5$ 40. $|8 - 5x| = 18$
41. $2|3x + 24| = 0$ 42. $5|x - 21| = -8$
43. $\left|\dfrac{3x + 48}{3}\right| = 12$ 44. $\left|\dfrac{4x - 64}{4}\right| = 32$
45. $|x + 3| + 7 = 10$ 46. $|2 - x| + 3 = 5$
47. $8 = -1 + |0.3x - 3|$ 48. $-1 = 1 - |0.1x + 8|$

49. $|2x + 1| = |3x + 3|$ **50.** $|5x - 7| = |4x + 1|$

51. $|2 - x| = |3x + 2|$ **52.** $|4x + 3| = |9 - 2x|$

53. $\left|\dfrac{x}{2} + 2\right| = \left|\dfrac{x}{2} - 2\right|$ **54.** $|7x + 12| = |x - 6|$

55. $\left|x + \dfrac{1}{3}\right| = |x - 3|$ **56.** $\left|x - \dfrac{1}{4}\right| = |x + 4|$

Solve each inequality if possible. Write the solution set in interval notation and graph it.

57. $|x| < 4$

58. $|x| < 9$

59. $|x + 9| \le 12$

60. $|x - 8| \le 12$

61. $|3x - 2| < 10$

62. $|4 - 3x| \le 13$

63. $|3x + 2| \le -3$

64. $|5x - 12| < -5$

65. $|x| > 3$

66. $|x| > 7$

67. $|x - 12| > 24$

68. $|x + 5| \ge 7$

69. $|3x + 2| > 14$

70. $|2x - 5| > 25$

71. $|4x + 3| > -5$

72. $|7x + 2| > -8$

73. $|2 - 3x| \ge 8$

74. $|-1 - 2x| > 5$

75. $-|2x - 3| < -7$

76. $-|3x + 1| < -8$

77. $\left|\dfrac{x - 2}{3}\right| \le 4$

78. $\left|\dfrac{x - 2}{3}\right| > 4$

79. $|3x + 1| + 2 < 6$

80. $1 + \left|\dfrac{1}{7}x + 1\right| \le 1$

81. $\left|\dfrac{1}{3}x + 7\right| + 5 > 6$

82. $-2|3x - 4| < 16$

83. $|0.5x + 1| + 2 \le 0$

84. $15 \ge 7 - |1.4x + 9|$

APPLICATIONS

85. TEMPERATURE RANGES The temperatures on a summer day satisfied the inequality $|t - 78°| \le 8°$, where t is a temperature in degrees Fahrenheit. Solve this inequality and express the range of temperatures as a double inequality.

86. OPERATING TEMPERATURES A car CD player has an operating temperature of $|t - 40°| < 80°$, where t is a temperature in degrees Fahrenheit. Solve the inequality and express this range of temperatures as an interval.

87. AUTO MECHANICS On most cars, the bottoms of the front wheels are closer together than the tops, creating a *camber angle*. This lessens road shock to the steering system. The specifications for a certain car state that the camber angle c of its wheels should be $0.6° \pm 0.5°$.

   a. Express the range with an inequality containing absolute value symbols.

   b. Solve the inequality and express this range of camber angles as an interval.

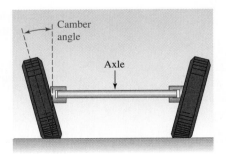

88. STEEL PRODUCTION A sheet of steel is to be 0.250 inch thick with a tolerance of 0.025 inch.

    a. Express this specification with an inequality containing absolute value symbols, using x to represent the thickness of a sheet of steel.

    b. Solve the inequality and express the range of thickness as an interval.

89. ERROR ANALYSIS In a lab, students measured the percent of copper p in a sample of copper sulfate. The students know that copper sulfate is actually 25.46% copper by mass. They are to compare their results to the actual value and find the amount of *experimental error*.

    a. Which measurements shown satisfy the absolute value inequality $|p - 25.46| \leq 1.00$?

    b. What can be said about the amount of error for each of the trials listed in part a?

> Lab 4   Section A
>
> Title:
> "Percent copper (Cu) in
> copper sulfate ($CuSO_4 \cdot 5H_2O$)"
>
> **Results**
>
> | | % Copper |
> |---|---|
> | Trial #1: | 22.91% |
> | Trial #2: | 26.45% |
> | Trial #3: | 26.49% |
> | Trial #4: | 24.76% |

90. ERROR ANALYSIS See Exercise 89.

    a. Which measurements satisfy the absolute value inequality $|p - 25.46| > 1.00$?

    b. What can be said about the amount of error for each of the trials listed in part a?

WRITING

91. Explain how to find the absolute value of a given number.

92. Explain why the equation $|x - 4| = -5$ has no solutions.

93. Explain the use of parentheses and brackets when graphing inequalities.

94. Explain the differences between the solution set of $|x| < 8$ and the solution set of $|x| > 8$.

95. Explain how to use the graph to solve $|x - 2| < 3$.

96. Explain how to use the graph to solve $|x - 2| \geq 3$.

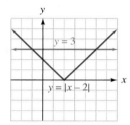

REVIEW

97. RAILROAD CROSSINGS The warning sign in the illustration is to be painted on the street in front of a railroad crossing. If y is 30° more than twice x, find x and y.

98. GEOMETRY Refer to the illustration. What is $2x + 2y$?

4.4   Linear Inequalities in Two Variables

- Graphing linear inequalities   • Horizontal and vertical boundary lines
- Problem solving

In the first three sections of this chapter, we have worked with linear inequalities in one variable. Some examples are

$$x \geq -7, \qquad 5 < \frac{7}{2}a - 9, \qquad \text{and} \qquad 5(3 + z) > -3(z + 3)$$

These inequalities have infinitely many solutions. When their solutions are graphed on a real number line, we obtain an interval.

In this section, we will discuss linear inequalities in two variables. Some examples are

$$y > 3x + 2, \qquad 2x - 3y \leq 6, \qquad \text{and} \qquad y < 2x$$

The solutions of these inequalities are ordered pairs. We can graph their solutions on a rectangular coordinate system.

■ Graphing linear inequalities

The **graph of a linear inequality** in x and y is the graph of all ordered pairs (x, y) that satisfy the inequality.

Linear inequalities

A **linear inequality** in x and y is any inequality that can be written in the form

$$Ax + By < C \quad \text{or} \quad Ax + By > C \quad \text{or} \quad Ax + By \leq C \quad \text{or} \quad Ax + By \geq C$$

where A, B, and C represent real numbers and A and B are not both 0.

To graph the linear inequality $y > 3x + 2$, we begin by graphing the linear equation $y = 3x + 2$. Its graph, shown in Figure 4-35(a), is a **boundary line** that separates the rectangular coordinate plane into two regions called **half-planes.** It is drawn with a dashed line to show that it is not part of the graph of $y > 3x + 2$.

To find which half-plane is the graph of $y > 3x + 2$, we can substitute the coordinates of any point in either half-plane. We will choose the origin as the **test point** because its coordinates, $(0, 0)$, make the computations easy. We substitute 0 for x and 0 for y into the inequality and simplify.

Check the test point (0, 0):

$$y > 3x + 2 \qquad \text{This is the original inequality.}$$

$$0 \overset{?}{>} 3(0) + 2 \qquad \text{Substitute 0 for } y \text{ and 0 for } x.$$

$$0 > 2 \qquad \text{This statement is false.}$$

Since the coordinates of the origin do not satisfy $y > 3x + 2$, the origin and all the other points on that side of the boundary are not part of the graph of the inequality. Thus, the half-plane on the other side of the dashed line is the graph. We shade that region, as shown in Figure 4-35(b).

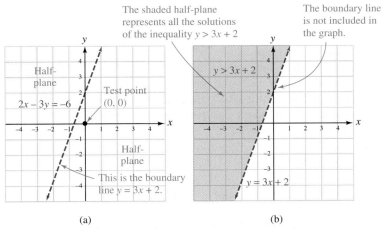

(a)   (b)

FIGURE 4-35

INTERMEDIATE
Algebra *f(x)* Now™

Self Check 1

Graph: $3x - 2y \geq 6$.

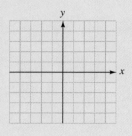

Answer

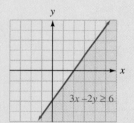

EXAMPLE 1   Graph: $2x - 3y \leq 6$.

Solution   This inequality is the combination of the inequality $2x - 3y < 6$ and the equation $2x - 3y = 6$.

We begin by graphing $2x - 3y = 6$ to find the boundary line that separates the two half-planes. We do so by noting that the line's *x*-intercept is $(3, 0)$ and its *y*-intercept is $(0, -2)$. This time, we draw the solid line shown in Figure 4-36(a), because equality is permitted by the symbol $\leq$. To decide which half-plane to shade, we check to see whether the coordinates of the origin satisfy the inequality.

Check the test point (0, 0):

$$2x - 3y \leq 6$$

$$2(0) - 3(0) \stackrel{?}{\leq} 6 \quad \text{Substitute 0 for } x \text{ and 0 for } y.$$

$$0 \leq 6 \quad \text{This statement is true.}$$

The coordinates of the origin satisfy the inequality. In fact, the coordinates of every point on the same side of the boundary line as the origin satisfy the inequality. We then shade that half-plane to complete the graph of $2x - 3y \leq 6$, shown in Figure 4-36(b).

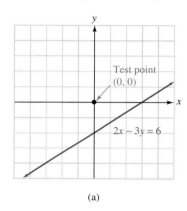

(a)

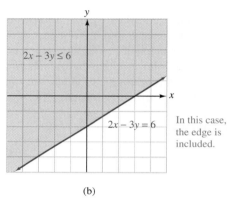

In this case, the edge is included.

(b)

FIGURE 4-36

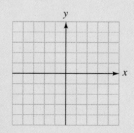

INTERMEDIATE
Algebra *f(x)* Now™

Self Check 2

Graph: $y > -x$.

EXAMPLE 2   Graph: $y < 2x$.

Solution   To graph $y = 2x$, we use the fact that the equation is in slope–intercept form and that $m = 2 = \frac{2}{1}$ and $b = 0$. Since the symbol $<$ does not include an equals symbol, the points on the graph of $y = 2x$ are not on the graph of $y < 2x$. We draw the boundary line as a dashed line to show this, as in Figure 4-37(a) on the next page.

To decide which half-plane is the graph of $y < 2x$, we check to see whether the coordinates of some fixed point satisfy the inequality. We cannot use the origin as a test point, because the boundary line passes through the origin. However, we can choose a different point — say, $(3, 1)$.

Check the test point (3, 1):

$$y < 2x$$

$$1 \stackrel{?}{<} 2(3) \quad \text{Substitute 1 for } y \text{ and 3 for } x.$$

$$1 < 6 \quad \text{This is a true statement.}$$

Since $1 < 6$ is a true inequality, the point $(3, 1)$ satisfies the inequality and is in the graph of $y < 2x$. We then shade the half-plane containing $(3, 1)$, as shown in Figure 4-37(b).

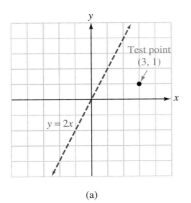

(a)

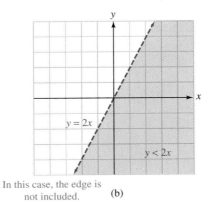

In this case, the edge is not included.　(b)

FIGURE 4-37

Answer

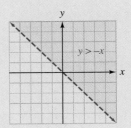

The following is a summary of the procedure for graphing linear inequalities.

Graphing linear inequalities in two variables

1. Graph the boundary line of the region. If the inequality allows the possibility of equality (the symbol is either $\leq$ or $\geq$), draw the boundary line as a solid line. If equality is not allowed ($<$ or $>$), draw the boundary line as a dashed line.

2. Pick a test point that is on one side of the boundary line. (Use the origin if possible.) Replace x and y in the original inequality with the coordinates of that point. If the inequality is satisfied, shade the side that contains that point. If the inequality is not satisfied, shade the other side.

▮ Horizontal and vertical boundary lines

Recall from Chapter 2 that the graph of the equation $x = a$ is a vertical line with x-intercept at $(a, 0)$, and the graph of the equation $y = b$ is a horizontal line with y-intercept at $(0, b)$.

EXAMPLE 3　Graph: $x \geq -1$.

Solution　The graph of the boundary $x = -1$ is a vertical line passing through $(-1, 0)$. We draw the boundary as a solid line to show that it is part of the solution. See Figure 4-38(a) on the next page.

In this case, we need not pick a test point. The inequality $x \geq -1$ is satisfied by points with an x-coordinate greater than or equal to -1. Points satisfying this condition lie to the right of the boundary. We shade that half-plane, as shown in Figure 4-38(b) on the next page, to complete the graph of $x \geq -1$.

Self Check 3
Graph: $y < 4$

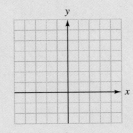

Answer

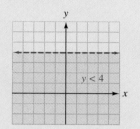

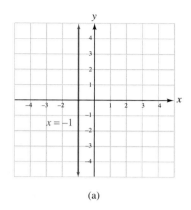

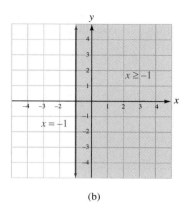

(a)   (b)

FIGURE 4-38

INTERMEDIATE
Algebra *f(x)* Now™

Problem solving

In the next example, we will solve a problem by writing a linear inequality in two variables to model a situation mathematically.

EXAMPLE 4   Social Security.   Retirees, ages 62–65, can earn as much as $11,640 and still receive their full Social Security benefits. If their annual earnings exceed $11,640, their benefits are reduced. A 64-year-old retired woman receiving Social Security works two part-time jobs: one at the library, paying $485 per week, and another at a pet store, paying $388 per week. Write an inequality representing the number of weeks the woman can work at each job during the year without losing any of her benefits.

Analyze the problem

We need to find the various combinations of weeks she can work at the library and at the pet store so that her annual income is less than or equal to $11,640.

Form an inequality

If we let x = the number of weeks she works at the library, she will earn $485x$ annually. If we let y = the number of weeks she works at the pet store, she will earn $388y$ annually. Combining the income from these jobs, the total is not to exceed $11,640.

The weekly rate on the library job		the weeks worked on the library job	plus	the weekly rate on the pet store job		the weeks worked on the pet store job	should not exceed	$11,640.
$485	·	x	+	$388	·	y	≤	$11,640

Solve the inequality

The graph of $485x + 388y \leq 11,640$ is shown in the figure on next page. Since she cannot work a negative number of weeks, the graph has no meaning when x or y is negative, so only the first quadrant is used. Any point in the shaded region indicates a way that she can schedule her work weeks and earn $11,640 or less annually. For example, if she works 8 weeks at the library and 16 weeks at the pet store, represented by the ordered pair (8, 16), she will earn

$$\$485(8) + \$388(16) = \$3,880 + \$6,208$$
$$= \$10,088$$

If she works 18 weeks at the library and 4 weeks at the pet store, represented by (18, 4), she will earn

$$\$485(18) + \$388(4) = \$8,730 + \$1,552$$
$$= \$10,282$$

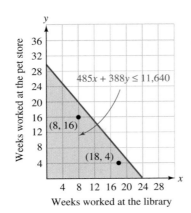

FIGURE 4-39

Graphing inequalities

Some graphing calculators (such as the TI-83 PLUS) have a graph style icon in the Y = editor. Some of the different graph styles are

\	line	A straight line or curved graph is shown.	\Y$_1$ =
◥	above	Shading covers the area above a graph.	◥Y$_1$ =
◣	below	Shading covers the area below a graph.	◣ Y$_1$ =

We can change the icon in front of Y$_1$ by placing the cursor on it and pressing the ENTER key.

To graph $2x - 3y \leq 6$ of Example 1, we first write it in an equivalent form, with y isolated on the left-hand side.

$$2x - 3y \leq 6$$

$$-3y \leq -2x + 6 \qquad \text{Subtract } 2x \text{ from both sides.}$$

$$y \geq \frac{2}{3}x - 2 \qquad \text{Divide both sides by } -3. \text{ Change the direction of the inequality symbol.}$$

We then change the graph style icon to above (◥), because the inequality $y \geq \frac{2}{3}x - 2$ contains a $\geq$ symbol. Using window settings of $[-10, 10]$ for x and $[-10, 10]$ for y, we enter the boundary equation $y = \frac{2}{3}x - 2$. See Figure 4-40(a). Finally, we press the GRAPH key to get Figure 4-40(b).

To graph $y < 2x$ from Example 2, we change the graph style icon to below (◣), because the inequality contains a $<$ symbol. Using window settings of $[-10, 10]$ for x and $[-10, 10]$ for y, we enter the boundary equation $y = 2x$ and press the GRAPH key to get Figure 4-40(c).

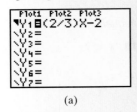

(a)

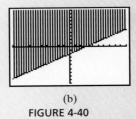

(b)

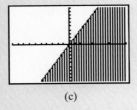

(c)

FIGURE 4-40

If your calculator does not have a graph style icon, you can graph linear inequalities with a SHADE feature. To do so, consult your owner's manual.

It is important to note that graphing calculators do not distinguish between solid and broken lines to show whether or not the edge of a region is included within the graph.

Section 4.4 STUDY SET

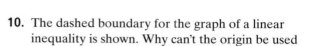

INTERMEDIATE
Algebra *f(x)* **Now**™

■ VOCABULARY   *Fill in the blanks.*

1. $4x - 2y \geq -8$ is an example of a _____ inequality in _____ variables.
2. Graphs of linear inequalities are _____.
3. The boundary line of a half-plane is called an _____.
4. The graph of a linear inequality in x and y contains the points (x, y) whose coordinates _____ the inequality.

■ CONCEPTS

5. Determine whether each ordered pair is a solution of $3x - 2y \geq 5$.

   a. $(3, 1)$ **b.** $(0, 3)$

   c. $(-1, -4)$ **d.** $\left(1, \dfrac{1}{2}\right)$

6. A linear inequality has been graphed below. Determine whether each point satisfies the inequality.

   a. $(-1, 4)$
   b. $(3, -2)$
   c. $(0, 0)$
   d. $(-3, -3)$

   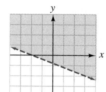

7. To graph the inequality $y > 3x - 1$, we begin by graphing the boundary line $y = 3x - 1$. What is the slope m of the line? What is its y-intercept?

8. To graph the inequality $2x + 3y \leq -6$ we begin by graphing the boundary line $2x + 3y = -6$. Find its x- and y-intercepts.

9. ZOOS To determine the allowable number of juvenile chimpanzees x and adult chimpanzees y that can live in an enclosure, a zookeeper refers to the following graph. Can 7 juvenile and 4 adult chimps be kept in the enclosure?

   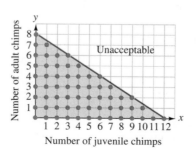

10. The dashed boundary for the graph of a linear inequality is shown. Why can't the origin be used as a test point to decide which side to shade?

    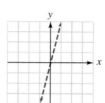

■ NOTATION

11. **a.** Solve the inequality in one variable and graph its solution set: $2x + 4 \geq 8$.

    b. Graph the inequality in two variables: $2x + 4y \geq 8$.

    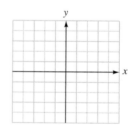

12. Decide whether the graph of each inequality includes the boundary line. In each case, would the boundary be a solid or a dashed line?

    a. $y < 3x - 1$

    b. $2x + 3y \geq -6$

    c. $y \leq -10$

    d. $x > 1$

■ PRACTICE   *Graph each inequality.*

13. $y > x + 1$ 14. $y < 2x - 1$

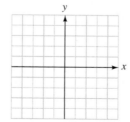   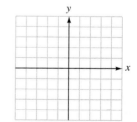

15. $y \geq x$

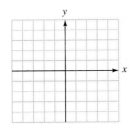

16. $y \leq 2x$

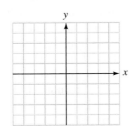

25. $3x + y > 2 + x$

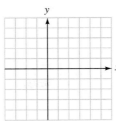

26. $3x - y > 6 + y$

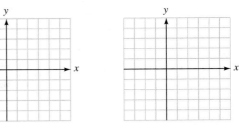

17. $2x + y \leq 6$

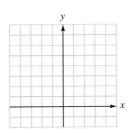

18. $x - 2y \geq 4$

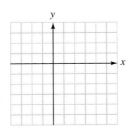

27. $2(x + 1) \geq 3\left(y - \dfrac{4}{3}\right)$

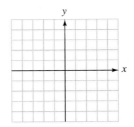

28. $3x + 1 < -2(y + 1)$

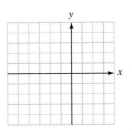

19. $3x \geq -y + 3$

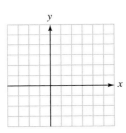

20. $2x \leq -3y - 12$

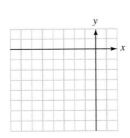

29. $\dfrac{x}{2} + \dfrac{y}{2} \leq 2$

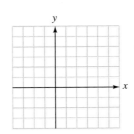

30. $\dfrac{x}{3} - \dfrac{y}{2} \geq 1$

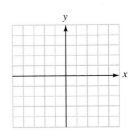

21. $y \geq 1 - \dfrac{3}{2}x$

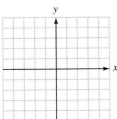

22. $y < \dfrac{x}{3} - 1$

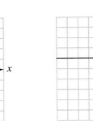

31. $x < 4$

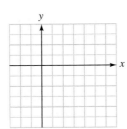

32. $y \geq -2$

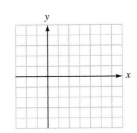

23. $2y < -x$

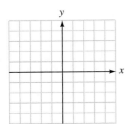

24. $3y > x$

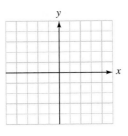

33. $y < 0$

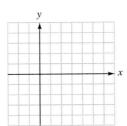

34. $x \geq 0$

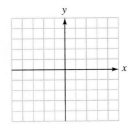

Find the equation of the boundary line and find the inequality whose graph is shown.

35.
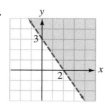

36.
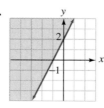

37.
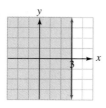

38.
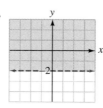

Use a graphing calculator to graph each inequality.

39. $y < 0.27x - 1$ **40.** $y > -3.5x + 2.7$

41. $y \geq -2.37x + 1.5$ **42.** $y < 3.37x - 1.7$

APPLICATIONS

43. GEOGRAPHY A region of the continental United States is shaded in the following map.

  a. What is the boundary that separates the shaded and unshaded regions?

  b. In words, describe the shaded area with respect to the boundary.

44. THE KOREAN WAR
After World War II, the 38th parallel of north latitude was established as the boundary between North Korea and South Korea. The Korean War began on June 25, 1950, when the North Korean army crossed this line and invaded South Korea. In the illustration, shade the region of the Korean Peninsula south of the 38th parallel.

Write a linear inequality that models the situation. Then graph each inequality for nonnegative values of x and y and give three ordered pairs that satisfy the inequality.

45. RESTAURANT SEATING As part of a remodeling project, a restaurant owner will install new booths that seat 4 persons, and new tables that seat 6 persons. The overall seating must conform to the sign shown below. Write an inequality that describes the possible combinations of booths (x) and tables (y) that the owner can install.

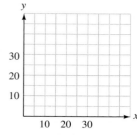

MAXIMUM OCCUPANCY
NOT TO EXCEED
120
By order of Clake County Fire Marshal

46. GARDENING During an Arbor Day sale, a garden store sold more than $2,000 worth of maple and pine trees. If a 6-foot maple costs $100 and a 5-foot pine costs $125, write an inequality that shows the possible ways that maple trees (x) and pine trees (y) were sold.

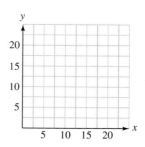

47. SPORTING GOODS
A sporting goods manufacturer allocates at least 1,200 units of time per day to make fishing rods and reels. If it takes 10 units of time to make a rod and 15 units of time to make a reel, write an inequality that describes the possible ways to schedule the time to make rods (x) and reels (y).

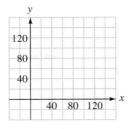

48. HOUSEKEEPING One housekeeper charges $12 per hour, and another charges $9 per hour. If Sarah can afford no more than $54 per month to clean her house, write an inequality that describes the possible number of hours that she can hire the first housekeeper (x) and the second housekeeper (y).

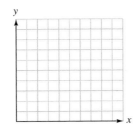

WRITING

49. Explain how to decide where to draw the boundary of the graph of a linear inequality, and whether to draw it as a solid or a dashed line.

50. Explain how to decide which side of the boundary of the graph of a linear inequality should be shaded.

REVIEW *Determine whether the ordered pair $(-4, 3)$ is a solution of the system of linear equations.*

51. $\begin{cases} 4x - y = -19 \\ 3x + 2y = -6 \end{cases}$

52. $\begin{cases} y = 2x + 11 \\ \frac{x}{2} + y = 0 \end{cases}$

Solve each system of equations.

53. $\begin{cases} x + y = 4 \\ x - y = 2 \end{cases}$

54. $\begin{cases} x - \frac{y}{2} = -2 \\ 0.01x + 0.02y = 0.03 \end{cases}$

4.5 Systems of Linear Inequalities

- Solving systems of linear inequalities • Compound inequalities • Problem solving

We have discussed how to solve systems of linear equations by the graphing method. For example, to solve

$$\begin{cases} y = -x + 1 \\ 2x - y = 2 \end{cases}$$

we graph both equations on the same set of coordinate axes and find the coordinates of the point of intersection of the straight lines.

In this section, we will discuss how to graphically **solve systems of linear inequalities,** such as

$$\begin{cases} y \leq -x + 1 \\ 2x - y > 2 \end{cases}$$

Solving systems of linear inequalities

When the solution of a linear inequality in x and y is graphed, the result is a half-plane. To solve a system of linear inequalities, we graph each of the inequalities on one set of coordinate axes and look for the intersection, or overlap, of the shaded half-planes.

Graph: $\begin{cases} x + y \geq 1 \\ 2x - y < 2 \end{cases}$.

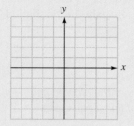

EXAMPLE 1   Graph the solution set of $\begin{cases} y \leq -x + 1 \\ 2x - y > 2 \end{cases}$.

Solution   To make it easy to see the intersection of the two half-planes, we graph one solution set in red and the other in blue.

To graph $y \leq -x + 1$, we graph the boundary $y = -x + 1$, as shown in Figure 4-41(a). Since the edge is to be included, we draw it as a solid line. To determine which half-plane to shade, we use the origin as a test point. Because the coordinates of the origin satisfy $y \leq -x + 1$, we shade (in red) the half-plane containing the origin.

In Figure 4-41(b), we superimpose the graph of $2x - y > 2$ on the graph of $y \leq -x + 1$ so that we can determine the points that the graphs have in common. To graph $2x - y > 2$, we graph the boundary $2x - y = 2$ as a dashed line. Since the test point $(0, 0)$ does not satisfy $2x - y > 2$, we then shade (in blue) the half-plane that does not contain $(0, 0)$.

The area that is shaded twice represents the solutions of the given system. Any point in the doubly shaded region in purple (including the purple portion of one of the boundaries) has coordinates that satisfy both inequalities.

$y = -x + 1$	$2x - y = 2$
We graph this line using the slope and y-intercept.	We graph this line using the intercept method.

$m = -1 = -\dfrac{1}{1}$

$b = 1$

y-intercept: $(0, 1)$

x	y	(x, y)
0	-2	$(0, -2)$
1	0	$(1, 0)$

Answer

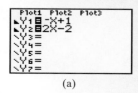

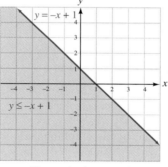

The graph of $y \leq -x + 1$
is shaded in red.

(a)

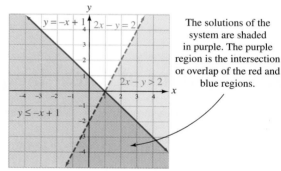

The solutions of the system are shaded in purple. The purple region is the intersection or overlap of the red and blue regions.

The graph of $2x - y > 2$ is shaded in blue.
It is drawn over the graph of $y \leq -x + 1$.

(b)

FIGURE 4-41

Solving systems of inequalities

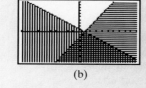

(a)

(b)

FIGURE 4-42

To solve the system of Example 1 with a graphing calculator, we can use window settings of $x = [-10, 10]$ and $y = [-10, 10]$. To graph $y \leq -x + 1$, we enter the boundary equation $y = -x + 1$ and change the graph style icon to below ($\blacktriangleright$). See Figure 4-42(a). To graph $2x - y > 2$, we first write it in equivalent form as $y < 2x - 2$. Then we enter the boundary equation $y = 2x - 2$ and change the graph style icon to below ($\blacktriangleright$). See Figure 4-42(a). Finally, we press the $\boxed{\text{GRAPH}}$ key to obtain Figure 4-42(b).

In general, to solve systems of linear inequalities, we will follow these steps.

> **Solving Systems of Linear Inequalities**
>
> 1. Graph each inequality on the same rectangular coordinate system.
> 2. Use shading to highlight the intersection of the graphs (the region where the graphs overlap). The points in this region are the solutions of the system.
> 3. As an informal check, pick a point from the region and verify that its coordinates satisfy each inequality of the original system.

EXAMPLE 2 Graph the solution set: $\begin{cases} x \geq 1 \\ y \geq x \\ 4x + 5y < 20 \end{cases}$.

Solution   We will find the graph of the solution set of the system in stages, using several graphs.

The graph of $x \geq 1$ includes the points that lie on the graph of $x = 1$ and to the right, as shown in red in Figure 4-43(a).

Figure 4-43(b) shows the graph of $x \geq 1$ and the graph of $y \geq x$. The graph of $y \geq x$, in blue, includes the points that lie on the graph of the boundary $y = x$ and above it.

INTERMEDIATE
Algebra $f(x)$ **Now** ™

Self Check 2

Graph: $\begin{cases} x \geq 0 \\ y \leq 0 \\ y \geq -2 \end{cases}$.

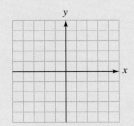

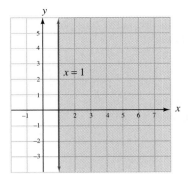

The graph of $x \geq 1$ is shaded in red.

(a)

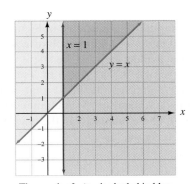

The graph of $y \geq x$ is shaded in blue.

(b)

The graph of $4x + 5y < 20$
is shaded in grey.

(c)

This is the graph of the
solution of the system.

(d)

FIGURE 4-43

Answer

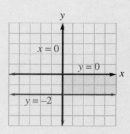

Figure 4-43(c) shows the graph of $x \geq 1$, $y \geq x$, and $4x + 5y < 20$. The graph of $4x + 5y < 20$ includes the points that lie below the graph of the boundary $4x + 5y = 20$.

The graph of the solution of the system includes the points that lie within the shaded triangle together with the points on the two sides of the triangle that are drawn with solid line segments, as shown in Figure 4-43(d).

Check: Pick a point in the shaded region, such as $(1.5, 2)$, and show that it satisfies each inequality of the system.

Compound inequalities

We have graphed the solution set of double linear inequalities, such as $2 < x \leq 5$, on a number line. These inequalities contained only one variable. In the next example, we will graph the solution set of $2 < x \leq 5$ in the context of two variables. In this case, we use the rectangular coordinate system.

Self Check 3

Graph: $-2 \leq y < 3$.

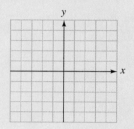

Answer

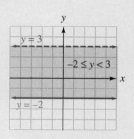

EXAMPLE 3   Graph $2 < x \leq 5$ on the rectangular coordinate plane.

Solution   The compound inequality $2 < x \leq 5$ is equivalent to the following system of two linear inequalities:

$$\begin{cases} 2 < x \\ x \leq 5 \end{cases}$$

The graph of $2 < x$, shown in Figure 4-44 in red, is the half-plane to the right of the vertical line $x = 2$. The graph of $x \leq 5$, shown in the figure in blue, includes the line $x = 5$ and the half-plane to its left. The graph of $2 < x \leq 5$ will contain all points in the plane that satisfy the inequalities $2 < x$ and $x \leq 5$ simultaneously (at the same time). These points are in the purple-shaded region of the figure.

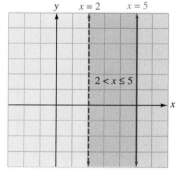

FIGURE 4-44

To graph a compound inequality containing the word *or* in the rectangular coordinate system, we sketch the *union* of the solution sets of the inequalities involved. For example, Figure 4-45 shows the graph of the compound inequality

$$x \leq -2 \text{ or } x > 3$$

in the rectangular coordinate system.

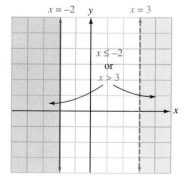

FIGURE 4-45

Problem solving

EXAMPLE 4 **Landscaping.** A homeowner has a budget of $300 to $600 for trees and bushes to landscape his yard. After some shopping, he finds that good trees cost $150 and mature bushes cost $75. What combinations of trees and bushes can he afford to buy?

Analyze the problem

We need to find the number of trees and the number of bushes that the homeowner can afford. This suggests we should use two variables. We know that he is willing to spend *at least* $300 and *at most* $600 for trees and bushes. These phrases suggest that we should write two inequalities that model the situation.

Form two inequalities

If x = the number of trees purchased, then $150x$ will be the cost of the trees. If y = the number of bushes purchased, then $75y$ will be the cost of the bushes. We know that the homeowner wants the sum of these costs to be from $300 to $600. We can then form the following system of linear inequalities.

The cost of a tree	·	the number of trees purchased	plus	the cost of a bush	·	the number of bushes purchased	should be at least	$300.
$150	·	x	+	$75	·	y	$\geq$	$300

The cost of a tree	·	the number of trees purchased	plus	the cost of a bush	·	the number of bushes purchased	should be at most	$600.
$150	·	x	+	$75	·	y	$\leq$	$600

Solve the system

We graph the system

$$\begin{cases} 150x + 75y \geq 300 \\ 150x + 75y \leq 600 \end{cases}$$

as in Figure 4-46. The coordinates of each point shown in the red-shaded graph give a possible combination of trees (x) and bushes (y) that can be purchased.

State the conclusion

The possible combinations of trees and bushes that can be purchased are given by

$(0, 4), (0, 5), (0, 6), (0, 7), (0, 8)$

$(1, 2), (1, 3), (1, 4), (1, 5), (1, 6)$ The ordered pair $(1, 6)$, for example, indicates that

$(2, 0), (2, 1), (2, 2), (2, 3), (2, 4)$ the homeowner can afford 1 tree and 6 bushes.

$(3, 0), (3, 1), (3, 2), (4, 0)$

Only these points can be used, because the homeowner cannot buy a portion of a tree or a bush.

Check the result

Check some of the ordered pairs to verify that they satisfy both inequalities.

Because the homeowner cannot buy a negative number of trees or bushes, we graph the system only for $x \geq 0$ and $y \geq 0$.

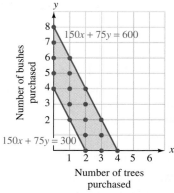

FIGURE 4-46

Section 4.5   STUDY SET

INTERMEDIATE
Algebra $f(x)$ Now™

VOCABULARY   *Fill in the blanks.*

1. $\begin{cases} x + y \leq 2 \\ x - 3y > 10 \end{cases}$ is a system of linear _____.

2. If an edge is included in the graph of an inequality, we draw it as a _____ line.

3. To solve a system of inequalities by graphing, we graph each inequality. The solution is the region where the graphs overlap or _____.

4. To determine which half-plane to shade when graphing a linear inequality, we see whether the coordinates of a test _____ satisfy the inequality.

CONCEPTS

5. Determine whether each ordered pair satisfies the system of linear inequalities $\begin{cases} x + y \leq 2 \\ x - 3y > 10 \end{cases}$.

   a. $(2, -3)$ **b.** $(12, -1)$
   c. $(0, -3)$ **d.** $(-0.5, -5)$

6. **a.** Determine whether $(-3, 10)$ satisfies the compound inequality $-5 < x \leq 8$ in the rectangular coordinate system.

   b. Determine whether $(-3, 3)$ satisfies the compound inequality $y \leq 0$ or $y > 4$ in the rectangular coordinate system.

7. In the illustration in the next column, the solution of one linear inequality is shaded in red, and the solution of a second linear inequality is shaded in blue. The intersection of these two regions is shaded in purple. Determine whether a true or false statement results if the coordinates of the given point are substituted into the given inequality.

   a. A, inequality 1   **b.** A, inequality 2
   c. B, inequality 1   **d.** B, inequality 2
   e. C, inequality 1   **f.** C, inequality 2

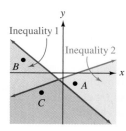

8. Match each equation, inequality, or system with the graph of its solution.

   a. $2x + y = 2$ **b.** $2x + y \geq 2$

   c. $\begin{cases} 2x + y = 2 \\ 2x - y = 2 \end{cases}$ **d.** $\begin{cases} 2x + y \geq 2 \\ 2x - y \leq 2 \end{cases}$

   i
   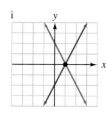

   ii
   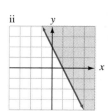

   iii
   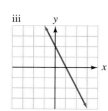

   iv
   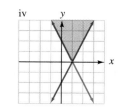

PRACTICE *Graph the solution set of each system.*

9. $\begin{cases} y < 3x + 2 \\ y < -2x + 3 \end{cases}$

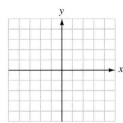

10. $\begin{cases} y \le x - 2 \\ y \ge 2x + 1 \end{cases}$

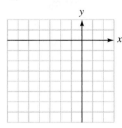

11. $\begin{cases} 3x + 2y > 6 \\ x + 3y \le 2 \end{cases}$

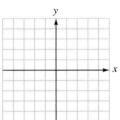

12. $\begin{cases} 3x + y \le 1 \\ -x + 2y \ge 6 \end{cases}$

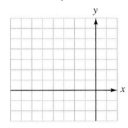

13. $\begin{cases} x + y < 2 \\ x + y \le 1 \end{cases}$

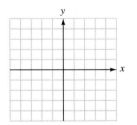

14. $\begin{cases} x + 2y < 3 \\ 2x + 4y < 8 \end{cases}$

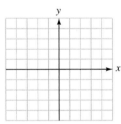

15. $\begin{cases} x > 0 \\ y > 0 \end{cases}$

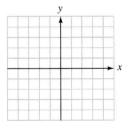

16. $\begin{cases} x \le 0 \\ y < 0 \end{cases}$

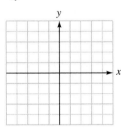

17. $\begin{cases} 2x + 3y \le 6 \\ 3x + y \le 1 \\ x \le 0 \end{cases}$

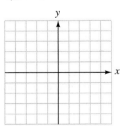

18. $\begin{cases} 2x + y \le 2 \\ y \ge x \\ x \ge 0 \end{cases}$

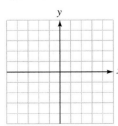

19. $\begin{cases} x - y < 4 \\ y \le 0 \\ x \ge 0 \end{cases}$

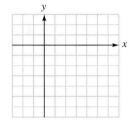

20. $\begin{cases} x \ge 0 \\ y \ge 0 \\ 9x + 3y \le 18 \\ 3x + 6y \le 18 \end{cases}$

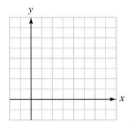

Graph each inequality.

21. $-2 \le x < 0$

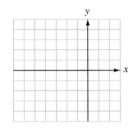

22. $-3 < y \le -1$

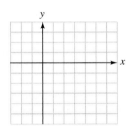

23. $y < -2 \text{ or } y > 3$

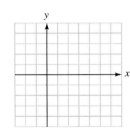

24. $-x \le 1 \text{ or } x \ge 2$

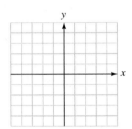

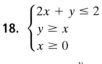 *Use a graphing calculator to solve each system.*

25. $\begin{cases} y < 3x + 2 \\ y < -2x + 3 \end{cases}$

26. $\begin{cases} y > -x + 2 \\ y < -x + 4 \end{cases}$

27. $\begin{cases} 2x + y \ge 6 \\ y \le 2(2x - 3) \end{cases}$

28. $\begin{cases} 3x + y < -2 \\ y > 3(1 - x) \end{cases}$

APPLICATIONS

29. FOOTBALL In 2003, the Green Bay Packers scored either a touchdown or a field goal 65.4% of the time when their offense was in the *red zone*. This was the best record in the NFL. If *x* represents the yard line the football is on, a team's red zone is an area on their opponent's half of the field that can be described by the system

$$\begin{cases} x > 0 \\ x \le 20 \end{cases}$$

Shade the red zone on the field shown below.

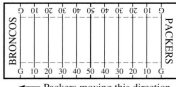

30. SHOT PUT In the shot put, the solid metal ball must land in a marked sector for it to be a fair throw. In the illustration, graph the system of inequalities that describes the region outside of the ring in which a shot must land.

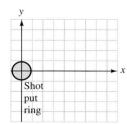

$$\begin{cases} y \le \dfrac{3}{8}x \\ y \ge -\dfrac{3}{8}x \\ x \ge 1 \end{cases}$$

31. NO-FLY ZONES After the Gulf War, U.S. and Allied forces enforced northern and southern no-fly zones over Iraq. Iraqi aircraft were prohibited from flying in this air space. If *y* represents the latitude parallel measurement, the no-fly zones can be described by

$$y \ge 36 \text{ or } y \le 33$$

On the map below, shade the regions of Iraq over which there was a no-fly zone.

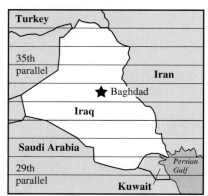

32. CARDIOVASCULAR FITNESS The following graph shows the range of pulse rates that persons ages 20–90 should maintain during aerobic exercise to get the most benefit from the training. The shaded region "Effective Training Heart Rate Zone" can be described by a system of linear inequalities. Determine what inequality symbol should be inserted in each blank.

$$\begin{cases} x \quad\rule{1em}{0.5pt}\quad 20 \\ x \quad\rule{1em}{0.5pt}\quad 90 \\ y \quad\rule{1em}{0.5pt}\quad -0.87x + 191 \\ y \quad\rule{1em}{0.5pt}\quad -0.72x + 158 \end{cases}$$

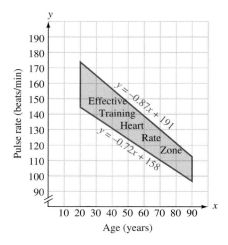

Graph each system and give two possible solutions.

33. COMPACT DISCS Melodic Music has compact discs on sale for either $10 or $15. If a customer wants to spend at least $30 but no more than $60 on CDs, use the illustration to graph a system of inequalities that will show the possible ways a customer can buy $10 CDs (*x*) and $15 CDs (*y*).

34. BOAT SALES Dry Boat Works wholesales aluminum boats for $800 and fiberglass boats for $600. Northland Marina wants to order at least $2,400 worth but no more than $4,800 worth of boats. Use the illustration to graph a system of inequalities that will show the

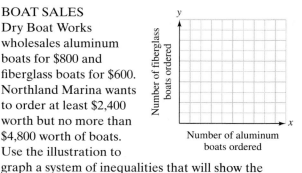

possible combination of aluminum boats (x) and fiberglass boats (y) that can be ordered.

35. FURNITURE SALES

A distributor wholesales desk chairs for $150 and side chairs for $100. Best Furniture wants to order no more than $900 worth of chairs, including more side chairs than desk chairs. Use the illustration to graph a system of inequalities that will show the possible combinations of desk chairs (x) and side chairs (y) that can be ordered.

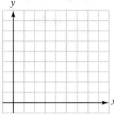

36. FURNACE EQUIPMENT

J. Bolden Heating Company wants to order no more than $2,000 worth of electronic air cleaners and humidifiers from a wholesaler that charges $500 for air cleaners and $200 for humidifiers. If Bolden wants more humidifiers than air cleaners, use the illustration to graph a system of inequalities that will show the possible combinations of air cleaners (x) and humidifiers (y) that can be ordered.

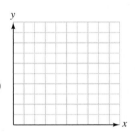

WRITING

37. Explain how to solve a system of two linear inequalities graphically.

38. Explain how a system of two linear inequalities might have no solution.

REVIEW *Use the given conditions to determine in which quadrant of a rectangular coordinate system each point (x, y) is located.*

39. $x > 0$ and $y < 0$

40. $x < 0$ and $y < 0$

41. $x < 0$ and $y > 0$

42. $x > 0$ and $y > 0$

KEY CONCEPT

Inequalities

Types of Inequalities

An **inequality** is a statement indicating that quantities are unequal. In Chapter 4, we have worked with several different types of inequalities and combinations of inequalities.

1. Classify each statement as one of the following: linear inequality in one variable, compound inequality, double linear inequality, absolute value inequality, linear inequality in two variables, system of linear inequalities.

a. $x - 3 < -4$ or $x - 2 > 0$

b. $\begin{cases} y < 3x + 2 \\ y < -2x + 3 \end{cases}$

c. $|x - 8| \leq 12$

d. $y < \dfrac{x}{3} - 1$

e. $\dfrac{1}{2}x + 2 \geq \dfrac{1}{3}x - 4$

f. $-6 < -3(x - 4) \leq 24$

g. $5(x - 2) \geq 0$ and $-3x < 9$

h. $|-1 - 2x| > 5$

i. $y > -x$

Solutions of Inequalities

A solution of a linear inequality in one variable is a value that, when substituted for the variable, makes the inequality true. A solution of a linear inequality in two variables (or a system of linear inequalities) is an ordered pair whose coordinates satisfy the inequality (or inequalities).

2. Determine whether -2 is a solution of the inequalities in one variable. Determine whether $(-1, 3)$ is a solution of the inequalities (or system of inequalities) in two variables.

a. $x - 3 < -4$ or $x - 2 > 0$

b. $\begin{cases} y < 3x + 2 \\ y < -2x + 3 \end{cases}$

c. $|x - 8| \leq 12$

d. $y < \dfrac{x}{3} - 1$

e. $\dfrac{1}{2}x + 2 \geq \dfrac{1}{3}x - 4$

f. $-6 < -3(x - 4) \leq 24$

g. $5(x - 2) \geq 0$ and $-3x < 9$

h. $|-1 - 2x| > 5$

i. $y > -x$

Graphs of Inequalities

To graph the solution set of an inequality in one variable, we use a number line. To graph the solution set of an inequality in two variables, we use a rectangular coordinate system.

3. Graph the solution set of the linear inequality in one variable: $2x + 1 > 4$.

4. Graph the solution set of the linear inequality in two variables: $2x + y \geq 4$.

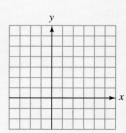

ACCENT ON TEAMWORK

SECTION 4.1

INEQUALITY STATEMENTS   The warning given on the street sign below can be described using an inequality. If we let w stand for the weight of a truck, then its weight must be such than $w \leq 3,000$ pounds if the truck is to use that street. Find six more real-life situations that can be described using an inequality.

TRIANGLE INEQUALITY   The triangle inequality is discussed in Exercise 51 of Study Set 4.1. To demonstrate it, cut two straws to lengths of 2 inches and 4 inches, to serve as two sides of a triangle. Cut a third straw to a length of 5 inches. Show that there is a triangle with sides of 2, 4, and 5 inches. Cut another straw to a length of 6 inches. Show that there is no triangle with sides of 2, 4, and 6 inches. How long must the third straw be for you to be able to form a triangle?

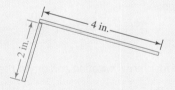

SECTION 4.2

POLLS   In Exercise 62 of Study Set 4.2, the margin of error for a poll of 1,000 adults was discussed. Find a published poll in a magazine or a newspaper in which the margin of error is given. Use interval notation to describe the possible intervals for each response in the poll. Taking the possible error into account, could the rankings of the responses possibly change?

ROOM TEMPERATURE   What temperature range do you like to keep inside your home? If t is the temperature, write a double linear inequality that describes the acceptable temperature range. Graph the solution set and express it as an interval. Then write a compound inequality containing the word *or* to describe the unacceptable temperature range inside your home. Graph the solution set and express it using interval notation.

VOCABULARY

a. How are the connecting words *and* and *or* used with inequalities? Give examples.

b. Explain the concepts of union and intersection of sets. Give some examples using a number line.

c. How are the symbols ∞ and $-\infty$ used in the context of a number line?

SECTION 4.3

TOLERANCES   Visit a machine shop or automobile repair shop and ask the employees to show you some examples of how they work with tolerances in their profession. Videotape their explanations and then play the video for your class. Show how the tolerances can be described using an absolute value inequality and interval notation.

SECTION 4.4

INEQUALITIES IN TWO VARIABLES

a. Can an inequality in two variables be an identity, one that is satisfied by all (x, y) pairs? Explain your answer.

b. Can an inequality in two variables have no solutions? Explain your answer.

SECTION 4.5

INTERSECTION   Sketch six examples of objects that intersect (overlap). Shade the region where they intersect. For instance, you could draw the intersection of two major streets in your city or the intersection of a chair and the floor.

CHAPTER REVIEW

INTERMEDIATE
Algebra $f(x)$ Now ™

| SECTION 4.1 | *Solving Linear Inequalities* |

CONCEPTS

To solve an inequality, apply the *properties of inequalities*.

If both sides of an inequality are multiplied (or divided) by a negative number, another inequality results, but with the opposite direction from the original inequality.

Set-builder notation is used to describe a set.

REVIEW EXERCISES

Solve each inequality. Give each solution set in interval notation and graph it.

1. $5(x - 2) \leq 5$

2. $0.3x - 0.4 \geq 1.2 - 0.1x$

3. $-16 < -\dfrac{4}{5}x$

4. $\dfrac{7}{4}(x + 3) < \dfrac{3}{8}(x - 3)$

5. Write interval notation for $\{x \mid x \geq 2\}$.

6. INVESTMENTS A woman has invested \$10,000 at 6% annual interest. How much more must she invest at 7% so that her annual income is at least \$2,000?

| SECTION 4.2 | *Solving Compound Inequalities* |

A solution of a compound inequality containing *and* makes both of the inequalities true.

The solution set of a compound inequality containing *and* is the *intersection* of the two solution sets.

Double linear inequalities:

$$c < x < d$$

is equivalent to

$$c < x \text{ and } x < d$$

A solution of a compound inequality containing the word *or* makes one, or the other, or both inequalities true.

The solution set of a compound inequality containing *or* is the *union* of the two solution sets.

Determine whether -4 is a solution of the compound inequality.

7. $x < 0$ and $x > -5$

8. $x + 3 < -3x - 1$ and $4x - 3 > 3x$

Solve each compound inequality. Give the result in interval notation and graph the solution set.

9. $-2x > 8$ and $x + 4 \geq -6$

10. $5(x + 2) \leq 4(x + 1)$ and $11 + x < 0$

Solve each compound inequality. Give the result in interval notation and graph the solution set.

11. $3 < 3x + 4 < 10$

12. $-2 \leq \dfrac{5 - x}{2} \leq 2$

Determine whether -4 is a solution of the compound inequality.

13. $x < 1.6$ or $x > -3.9$

14. $x + 1 < 2x - 1$ or $4x - 3 > 3x$

Solve each compound inequality. Give the result in interval notation and graph the solution set.

15. $x + 1 < -4$ or $x - 4 > 0$

16. $\dfrac{x}{2} + 3 > -2$ or $4 - x > 4$

17. **INTERIOR DECORATING** A manufacturer makes a line of rugs that are 4 feet wide and of varying lengths l (in feet). The floor area covered by the rugs ranges from 17 ft^2 to 25 ft^2. Write and then solve a double linear inequality to find the range of the lengths of the rugs.

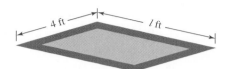

18. Match each word in Column I with *two* items in Column II.

    Column I

    a. or

    b. and

    Column II

    i. ∩

    ii. ∪

    iii. intersection

    iv. union

SECTION 4.3	*Solving Absolute Value Equations and Inequalities*

Definition of *absolute value*:

$$\begin{cases} \text{If } x \geq 0, |x| = x. \\ \text{If } x < 0, |x| = -x. \end{cases}$$

Evaluate each expression.

19. $|-7|$

20. $\left|\dfrac{5}{16}\right|$

21. $-|71.05|$

22. $-|-12|$

Absolute value equations:

If $k \geq 0$, $|x| = k$

is equivalent to

$x = k$ or $x = -k$

$|a| = |b|$ is equivalent to

$a = b$ or $a = -b$

Solve each absolute value equation.

23. $|4x| = 8$

24. $2|3x + 1| = 20$

25. $\left|\dfrac{3}{2}x - 4\right| - 10 = -1$

26. $\left|\dfrac{2 - x}{3}\right| = 4$

27. $|3x + 2| = |2x - 3|$

28. $\left|\dfrac{3 - 2x}{2}\right| = \left|\dfrac{3x - 2}{3}\right|$

Absolute value inequalities:

If $k > 0$, $|x| < k$

is equivalent to

$-k < x < k$

Solve each absolute value inequality. Give the solution in interval notation and graph it.

29. $|x| \leq 3$

30. $|2x + 7| < 3$

31. $|5 - 3x| \leq 14$

32. $\left|\dfrac{2}{3}x + 14\right| < 0$

Absolute value inequalities:

$|x| > k$

is equivalent to

$x < -k$ or $x > k$

33. $|x| > 1$

34. $\left|\dfrac{1 - 5x}{3}\right| \geq 7$

35. $|3x - 8| > 4$

36. $\left|\dfrac{3}{2}x - 14\right| \geq 0$

37. PRODUCE Before packing, freshly picked tomatoes are weighed on the scale shown. Tomatoes having a weight w (in ounces) that falls within the highlighted range are sold to grocery stores.

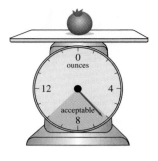

    a. Express this acceptable weight range using an absolute value inequality.

    b. Solve the inequality and express this range as an interval.

38. Explain the error.

Solve: $|x| + 3 = 9$.

$$x + 3 = 9 \quad \text{or} \quad x + 3 = -9$$
$$x = 6 \quad | \quad x = -12$$

SECTION 4.4	*Linear Inequalities in Two Variables*

To graph a linear inequality:

1. Graph the *boundary line*. Draw a solid line if the inequality contains $\leq$ or $\geq$ and a dashed line if it contains $<$ or $>$.

2. Pick a *test point* on one side of the boundary. Use the origin if possible. Replace x and y with the coordinate of that point. If the inequality is satisfied, shade the side that contains the point. If the inequality is not satisfied, shade the other side.

Graph each inequality.

39. $2x + 3y > 6$

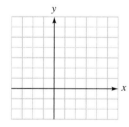

40. $y \leq 4 - x$

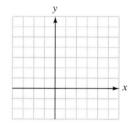

41. $y + 2 < \dfrac{x + 4}{2}$

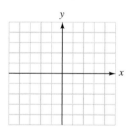

42. $x \geq -\dfrac{3}{2}$

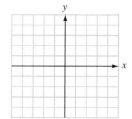

43. CONCERT TICKETS Tickets to a concert cost $6 for reserved seats and $4 for general admission. If receipts must be at least $10,200 to meet expenses, find an inequality that shows the possible ways that the box office can sell reserved seats (x) and general admission tickets (y). Then graph each inequality for nonnegative values of x and y and give three ordered pairs that satisfy the inequality.

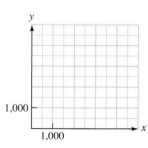

44. Find the equation of the boundary line. Then give the inequality whose graph is shown.

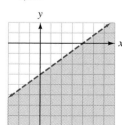

Systems of Linear Inequalities

To solve a *system of linear inequalities,* graph each of the inequalities on the same set of coordinate axes and look for the intersection of the shaded *half-planes.*

Graph the solution set of each system.

45. $\begin{cases} y \geq x + 1 \\ 3x + 2y < 6 \end{cases}$

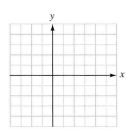

46. $\begin{cases} x - y < 3 \\ y \leq 0 \\ x \geq 0 \end{cases}$

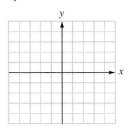

Compound inequalities can be graphed in the rectangular coordinate system.

Graph each compound inequality in the rectangular coordinate system.

47. $-2 < x < 4$

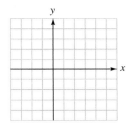

48. $y \leq -2$ or $y > 1$

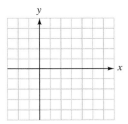

49. PETROLEUM EXPLORATION Organic matter converts to oil and gas within a specific range of temperature and depth called the *petroleum window.* The petroleum window in the illustration below can be described by a system of linear inequalities, where x is the temperature in °C of the soil at a depth of y meters. Determine which inequality symbol should be inserted in each blank.

$$\begin{cases} x \quad 35 \\ x \quad 130 \\ y \quad -56x + 280 \\ y \quad -18x + 90 \end{cases}$$

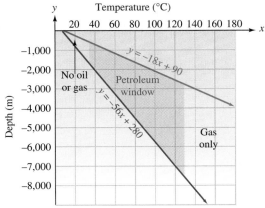

Based on data from *The Blue Planet* (Wiley, 1995)

311

50. In the illustration, the solution of one linear inequality is shaded in red, and the solution of a second is shaded in blue. The intersection of those regions is shaded in purple. Decide whether a true or false statement results if the coordinates of the given point are substituted into the given inequality.

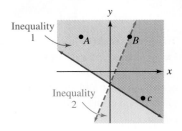

a. A, inequality 1   **b.** A, inequality 2

c. B, inequality 1   **d.** B, inequality 2

e. C, inequality 1   **f.** C, inequality 2

1. Determine whether the statement is true or false.

$$-5.67 \geq -5$$

2. Determine whether -2 is a solution of the inequality.

$$3(x - 2) \leq 2(x + 7)$$

3. Write interval notation for $\{x \mid x < -5\}$,

4. Suppose c and d represent real numbers. Translate the following statement to mathematical symbols:

c is at most d.

Graph the solution set of each inequality and give the solution in interval notation.

5. $7 < \dfrac{2}{3}t - 1$

6. $-2(2x + 3) \geq 14$

7. AVERAGING GRADES   Use the information from the gradebook to determine what score Karen Nelson-Sims needs on the fifth exam to keep her exam average (mean) above 80 in the class.

Sociology 101 8:00-10:00 pm MW	Exam 1	Exam 2	Exam 3	Exam 4	Exam 5
Nelson-Sims, Karen	70	79	85	88	

Solve each compound inequality. Give the result in interval notation and graph the solution set.

8. $3x \geq -2x + 5$ and $7 \geq 4x - 2$

9. $3x < -9$ or $-\dfrac{x}{4} < -2$

10. $-2 < \dfrac{x - 4}{3} < 4$

Evaluate each expression.

11. $|8|$

12. $-|-4.75|$

Solve each equation.

13. $|4 - 3x| = 19$

14. $|3x + 4| = |x + 12|$

Graph the solution set of each inequality and give the solution in interval notation.

15. $|x + 3| \leq 4$

16. $|2x - 4| > 22$

17. $|4 - 2x| + 1 > 3$

18. $2|2x - 4| \leq 4$

Graph each solution set in the rectangular coordinate system.

19. $3x + 2y \geq 6$

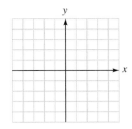

20. $y < x$

21. $\begin{cases} 2x - 3y \geq 6 \\ y \leq -x + 1 \end{cases}$

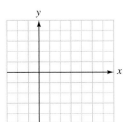

22. $-2 \leq y < 5$

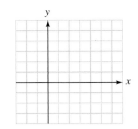

23. **ACCOUNTING** On average, it takes an accountant 1 hour to complete a simple tax return and 3 hours to complete a complicated return. If the accountant wants to work less than 9 hours per day, find an inequality that shows the number of possible ways that simple returns (x) and complicated returns (y) can be completed each day. Then graph the inequality and give three ordered pairs that satisfy it.

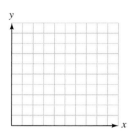

24. Two linear inequalities are graphed on the same coordinate axes as shown below. The solution set of the first inequality is shaded in red, and the solution set of the second in blue. The intersection of these regions is shaded in purple.

   a. Is $(3, -4)$ a solution of either inequality?

   b. Is $(3, -4)$ a solution of the system of two linear inequalities? Explain.

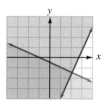

25. **INDOOR CLIMATE** The general zone of comfort acceptable to most people when working in an office can be described by a system of linear inequalities where x is the dry bulb temperature and y is the percent relative humidity, as shown in the illustration below. Determine what inequality symbol should be inserted in each blank.

$$\begin{cases} y \phantom{<} 60 \\ y \phantom{<} 27 \\ y \phantom{<} -11x + 852 \\ y \phantom{<} -5x + 445 \end{cases}$$

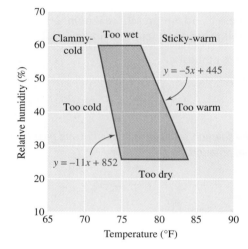

26. Explain why the inequality symbol must be reversed if both sides of $-2 < 5$ are multiplied by a negative number.

1. The following diagram shows the sets that compose the set of real numbers. Which of the indicated sets make up the *rational numbers* and the *irrational numbers?*

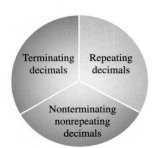

2. SCREWS The thread profile of a screw is determined by the distance between threads. This distance, indicated by the letter p, is known as the *pitch*. If $p = 0.125$, find each of the dimensions labeled in the illustration.

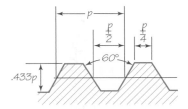

Evaluate each expression when $x = 2$ and $y = -4$.

3. $|x| - xy$

4. $\dfrac{x^2 - y^2}{3x + y}$

Simplify each expression.

5. $3p^2 - 6(5p^2 + p) + p^2$

6. $-(a + 2) - (a - b)$

7. PLASTIC WRAPS Estimate the number of *square feet* of plastic wrap on a roll if the dimensions printed on the box describe the roll as 205 feet long by $11\frac{3}{4}$ inches wide.

8. INVESTMENTS Find the amount of money that was invested at $8\frac{7}{8}\%$ if it earned \$1,775 in simple interest in one year.

Solve each equation, if possible.

9. $3x - 6 = 20$

10. $6(x - 1) = 2(x + 3)$

11. $\dfrac{5b}{2} - 10 = \dfrac{b}{3} + 3$

12. $2a - 5 = -2a + 4(a - 2) + 1$

Determine whether the lines represented by the equations are parallel or perpendicular.

13. $3x + 2y = 12, \ 2x - 3y = 5$

14. $3x = y + 4, \ y = 3(x - 4) - 1$

15. Write the equation of the line that passes through $(-2, 3)$ and is perpendicular to the graph of $3x + y = 8$. Answer in slope–intercept form.

16. Find the slope of the line that passes through $(0, -8)$ and $(-5, 0)$.

17. PRISONS The following graph shows the growth of the U.S. prison population from 1970 to 2000. Find the rate of change in the prison population from 1970 to 1975.

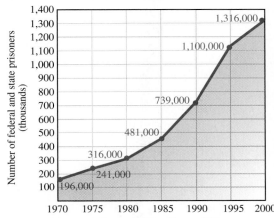

Source: *U.S. Statistical Abstract and Time Almanac 2004*

18. PRISONS Refer to the graph above. During what 5-year period was the rate of change in the U.S. prison population the greatest? Find the rate of change.

Let $f(x) = 3x^2 - x$ and find each value.

19. $f(2)$

20. $f(-2)$

21. Use graphing to solve: $\begin{cases} 2x + y = 5 \\ x - 2y = 0 \end{cases}$.

22. Use addition to solve: $\begin{cases} \dfrac{x}{10} + \dfrac{y}{5} = \dfrac{1}{2} \\ \dfrac{x}{2} - \dfrac{y}{5} = \dfrac{13}{10} \end{cases}$

23. Use substitution to solve: $\begin{cases} y = 4 - 3x \\ 2x - 3y = -1 \end{cases}$

24. Solve: $\begin{cases} x + y + z = 1 \\ 2x - y - z = -4 \\ x - 2y + z = 4 \end{cases}$

25. Use matrices to solve the system $\begin{cases} 4x - 3y = -1 \\ 3x + 4y = -7 \end{cases}$

26. Use Cramer's rule to solve the system
$\begin{cases} x - 2y - z = -2 \\ 3x + y - z = 6 \\ 2x - y + z = -1 \end{cases}$

27. U.S. WORKERS The following graph shows how the makeup of the U.S. workforce changed over the years 1900–2000. Estimate the coordinates of the points of intersection in the graph. Explain their significance.

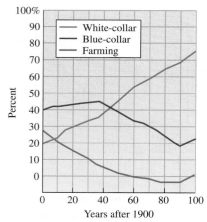

Source: *U.S. Statistical Abstract*

28. AGING The graph in the next column shows the effects of aging on cardiac output (the amount of blood that the heart can pump in one minute).

   a. Write the equation of the line.

   b. Use your answer to part a to determine the cardiac output at age 90.

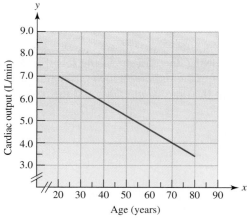

Based on data from *Cardiopulmonary Anatomy and Physiology, Essentials for Respiratory Care*, 2nd ed. (Delmar Publishers, 1994)

29. ENTREPRENEURSHIP A person invests $18,375 to set up a small business producing a piece of computer software that will sell for $29.95. If each piece can be produced for $5.45, how many pieces must be sold to break even?

30. CONCERT TICKETS Tickets for a concert cost $5, $3, and $2. Twice as many $5 tickets were sold as $2 tickets. The receipts for 750 tickets were $2,625. How many tickets were sold at each price?

Solve each equation.

31. $|4x - 3| = 9$

32. $|2x - 1| = |3x + 4|$

Solve each inequality, write the solution in interval notation, and graph the interval.

33. $-3(x - 4) \geq x - 32$

34. $-8 < -3x + 1 < 10$

35. $|3x - 2| \leq 4$

36. $|2x + 3| - 1 > 4$

Use graphing to solve each inequality or system of inequalities.

37. $2x - 3y \leq 12$

38. $\begin{cases} y < x + 2 \\ 3x + y \leq 6 \end{cases}$

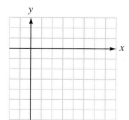

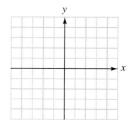

CHAPTER 5

Exponents, Polynomials, and Polynomial Functions

INTERMEDIATE
Algebra ƒ⁽ˣ⁾ Now™

Throughout the chapter, this icon introduces resources on the Intermediate AlgebraNow Web site, accessed through http://1pass .thomson.com, that will

- Help you test your knowledge of the material with a pre-test and a post-test

- Provide a personalized learning plan targeting areas you should study

© Getty Images/David Noton

TLE Over the past twenty years, the federal government, and many state and local governments, have increased their efforts to clean up hazardous waste sites that threaten public health and the environment. Environmental engineers are often hired to oversee these projects that deal with leaking underground storage tanks, chemical spills, asbestos, and lead paint. They use principles of biology, chemistry, and mathematics to develop plans to restore the sites to their original condition.

To learn more about the role of mathematics in environmental cleanup, visit *The Learning Equation* on the Internet at http://tle.brookscole.com. (The log-in instructions are in the Preface.) For Chapter 5, the online lessons are:

- *TLE* Lesson 7: The Greatest Common Factor and Factoring by Grouping
- *TLE* Lesson 8: Factoring Trinomials and the Difference of Squares

Check Your Knowledge

1. In the exponential expression x^n, x is called the _____, and n is called the _____.

2. 2.4×10^4 is written in _____ notation.

3. A _____ is the sum of one or more algebraic terms whose variables have whole-number exponents. No variable appears in a denominator.

4. Terms having the same variables with the same exponents are called _____ terms.

Simplify each expression. Write all answers without using negative exponents. Assume that no denominators are zero.

5. $(x^2)(x^5)(7x)(x^3)(x^2)$

6. $(-2\,x^3y^2)^3$

7. $(-3xyz^0)^{-2}$

8. $\left(\dfrac{x^3y^4}{5x^{-2}y^2}\right)^{-3}$

9. a. Write in scientific notation: 0.000093.
   b. Write in standard notation: 6.21×10^4.

10. The distance from the sun to the Earth is approximately 9.3×10^7 miles and the speed of light is approximately 1.86×10^5 miles per second. Use scientific notation to calculate the approximate time that the light from the sun takes to reach the Earth.

11. Consider the polynomial $-7x^5 + 5x^4 + 3x^3 + x^2 - 2$.
    a. What is the degree of the polynomial?
    b. What is the lead coefficient?

12. Consider the function $f(x) = x^3 - 4x$.
    a. Graph the function.
    b. Determine the solutions of the polynomial equation $x^3 - 4x = 0$ from the graph.

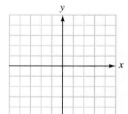

Perform the operations.

13. $(x^2 - 3x + 4) + (x^3 - 2x^2 + x + 4)$
14. $(y^2 - 4y + 7) - (2y^2 + 2y - 3)$

15. $(x^2yz^4)(-5x^{-3}y^{-2}z)$

16. $(2xy)(3x + 2y)$

17. $(y^4 + 7)(y^4 - 7)$

18. $(2x^2 - 3y)^2$

19. $(x + \tfrac{2}{3})^2$

20. $3z(x + y)(x - y)$

21. What is the greatest common factor (GCF) of $24a^2b^3c^3$, $6ab^3c^3$, and $40a^2b^2c^8$?

22. Write a polynomial that represents the area of a rectangular garden that is $(2x - 7)$ units long and $(x + 3)$ units wide.

Factor each expression, if possible.

23. $-16t^2 + 64t$
24. $x^2y - 3xy - 2x + 6$
25. $x^2z^2 - 64$

26. $3x^4 - 48$
27. $9x^2 + 25y^2$
28. $27x^3 + 64$
29. $5y^3 - 5$
30. $x^2 + 7xy + 6y^2$
31. $2x^2 + 7x + 3$
32. $y^2 + 2y + 1 - z^2$

Solve each equation.

33. $x^2 + x - 6 = 0$

34. $x^2 - 7x = 0$

35. $3x\left(x + \dfrac{4}{3}\right) = 4$

36. Solve: $b^2x^2 + a^2y^2 = a^2b^2$ for b^2

Study Skills Workshop

PREPARING FOR THE FIRST TEST

Preparation is essential for any test. When you properly review for a test, you establish the material firmly in your mind and increase your confidence. Be sure to allot the extra time necessary for test preparation on your study calendar and complete all homework that will be included on your test before studying. It is best to spread this study-time over a period of at least 4 days before the test to avoid feelings of frustration and anxiety, and to give yourself a chance to firmly embed the concepts in your mind.

Four Days Before the Test. Using your reworked notes, make a list of basic rules, formulas, and concepts that will be covered on your upcoming test on one $8\frac{1}{2} \times 11$ inch sheet of paper (or make an audio version if you are an auditory learner). When this is completed, go through the sections in your textbook that contain this same material to see whether anything was omitted; if so add it to your study sheet. Basic rules and concepts are usually found in colored boxes or in boldface type in your text. Carry this sheet or audiotape around with you all the time until you have taken your test. Then if you have some "down" time (waiting for a doctor's appointment, waiting during your kids' soccer practice, etc.), you will be able to use that time for a quick study review.

Three Days Before the Test. Go through the examples from your reworked notes. Try to work as many examples as possible without assistance from your notes or text. If you get stuck, look at the back of your index card or study sheet for help. Go over every problem until you can do all of them without assistance. Make a note of anything that you don't understand and see a tutor, a classmate, or your instructor during office hour for help.

Two Days Before the Test. Make a practice test for yourself with approximately the same number of questions that will appear on your real test. Pick problems for which a solution is available: examples in your text, odd problems from homework sections, or problems from chapter tests in your text are all good sources for practice test questions. The iLrn Web site also has online quizzes that you could use as practice tests. Once your practice test is complete, close your book, set a timer, and take the test under circumstances that closely resemble those in a real test-taking situation. When time is up, grade your test and make a list of problems that you got wrong.

One Day Before the Test. Correct all of the problems that you got wrong on your practice test and get help with any that you don't understand. Go over the examples from your reworked notes again. Get a good night's sleep on the night before your test.

Test Day. If possible, go over your study sheet before the test and review the material about which you feel confident. Don't try to learn new material right before the test, especially if you tend to suffer from test anxiety. Arrive at your test feeling confident and relaxed, knowing that you are well prepared.

ASSIGNMENT

1. One week before your test, check your study calendar to ensure that you have allotted extra time for test preparation (at least 1 to 2 hours for the 4 days before your test).
2. Four days before your test is scheduled, make up your study sheet or audiotape.
3. Two days before the test, make and take your practice test.
4. Make a list of problems that you missed on your practice test and get help with them if necessary.

Polynomials are algebraic expressions that can be used to model many real-world situations. They often contain terms in which the variables have exponents.

5.1 Exponents

- Exponents • Rules for exponents • Zero exponents • Negative exponents
- More rules for exponents

In Chapter 1, we evaluated exponential expressions having natural-number exponents. In this section, we will extend the definition of exponent to include negative-integer exponents, as in 3^{-2}, and zero exponents, as in 3^0. We will also use the definition of exponent to develop rules that can be used to simplify expressions.

Exponents

Exponents provide a way to write products of repeated factors in a concise form. For example,

$$y \cdot y = y^2 \qquad \text{Read } y^2 \text{ as "y to the second power" or "y squared."}$$
$$z \cdot z \cdot z = z^3 \qquad \text{Read } z^3 \text{ as "z to the third power" or "z cubed."}$$
$$x \cdot x \cdot x \cdot x = x^4 \qquad \text{Read } x^4 \text{ as "x to the fourth power."}$$

These examples illustrate the following definition.

> **Natural-number exponents**
> If n represents a natural number, then
> $$x^n = \overbrace{x \cdot x \cdot x \cdot \cdots \cdot x}^{n \text{ factors of } x}$$

The **exponential expression** x^n is called a **power of x,** and we read it as "x to the nth power." In this expression, x is called the **base,** and n is called the **exponent.**

$$\text{Base} \rightarrow x^n \leftarrow \text{Exponent}$$

A natural-number exponent tells how many times the base of an exponential expression is to be used as a factor in a product. When $n = 1$, the exponent is usually omitted. For example, $x^1 = x$.

INTERMEDIATE
Algebra $^{f(x)}$ **Now**™

Self Check 1
Identify the base and the exponent in each expression:
a. $(-kt)^4$ **b.** πr^2 **c.** $-h^8$

Answers **a.** $-kt, 4$, **b.** $r, 2$
c. $h, 8$

EXAMPLE 1   Identify the base and the exponent in each expression:
a. $(-a)^2$, **b.** $-a^2$, **c.** $5x^3$, and **d.** $(5x)^3$.

Solution
a. For $(-a)^2$, $-a$ is the base and the exponent is 2: $(-a)^2 = (-a)(-a)$.
b. For $-a^2$, the base is a and the exponent is 2: $-a^2 = -(a \cdot a)$.
c. For $5x^3$, x is the base and the exponent is 3: $5x^3 = 5 \cdot x \cdot x \cdot x$.
d. For $(5x)^3$, the base is $5x$ and the exponent is 3: $(5x)^3 = (5x)(5x)(5x)$.

Rules for exponents

Several rules for exponents come from the definition of exponent. The first rule, called the **product rule for exponents,** gives a way to find the result when multiplying exponential expressions that have the same base.

Since x^5 means that x is to be used as a factor five times, and since x^3 means that x is to be used as a factor three times, $x^5 \cdot x^3$ means that x will be used as a factor eight times.

$$x^5 x^3 = \overbrace{x \cdot x \cdot x \cdot x \cdot x}^{5 \text{ factors of } x} \cdot \overbrace{x \cdot x \cdot x}^{3 \text{ factors of } x} = \overbrace{x \cdot x \cdot x \cdot x \cdot x \cdot x \cdot x \cdot x}^{8 \text{ factors of } x} = x^8$$

In general,

$$x^m x^n = \overbrace{x \cdot x \cdot x \cdot \,\cdots\, \cdot x}^{m \text{ factors of } x} \cdot \overbrace{x \cdot x \cdot x \cdot \,\cdots\, \cdot x}^{n \text{ factors of } x} = \overbrace{x \cdot x \cdot x \cdot x \cdot \,\cdots\, \cdot x}^{m + n \text{ factors of } x} = x^{m+n}$$

Thus, *to multiply exponential expressions with the same base, keep the common base and add the exponents.*

> **The product rule for exponents**
>
> If m and n represent natural numbers, then
>
> $$x^m x^n = x^{m+n}$$

INTERMEDIATE
Algebra $f(x)$ **Now**™

EXAMPLE 2   Simplify each expression: **a.** $x^{11}x^5$,   **b.** $y^5 y^4 y$,   **c.** $a^2 b^3 a^3 b^2$, and **d.** $-8x^4(x^3)$.

Solution

a. $x^{11}x^5 = x^{11+5}$ Keep the common base x.
     $\quad\ \ = x^{16}$ Add the exponents.

b. $y^5 y^4 y = (y^5 y^4)y$
     $\qquad\ \ = y^9 y^1 \qquad y = y^1.$
     $\qquad\ \ = y^{10}$

c. $a^2 b^3 a^3 b^2 = a^2 a^3 b^3 b^2$
     $\qquad\quad\ = a^5 b^5$

d. $-8x^4(x^3) = -8(x^4 x^3)$
     $\qquad\qquad\ = -8x^7$

Self Check 2
Simplify each expression:
a. $2^3 2^5$
b. $k \cdot k^4$
c. $a^2 b^3 a^3 b^4$
d. $-8a^4(a^2 b)$

Answers **a.** $2^8 = 256$,   **b.** k^5, **c.** $a^5 b^7$,   **d.** $-8a^6 b$

! COMMENT   The product rule for exponents applies only to exponential expressions with the same base. The expression $x^5 y^3$, for example, cannot be simplified, because the bases of the exponential expressions are different.

To find another rule for exponents, we simplify $(x^4)^3$, which means x^4 cubed or $x^4 \cdot x^4 \cdot x^4$.

$$(x^4)^3 = \overbrace{x^4}^{x^4} \cdot \overbrace{x^4}^{x^4} \cdot \overbrace{x^4}^{x^4} = x \cdot x \cdot x \cdot x \cdot x \cdot x \cdot x \cdot x \cdot x \cdot x \cdot x \cdot x = x^{12}$$

In general, we have

$$(x^m)^n = \overbrace{x^m \cdot x^m \cdot x^m \cdot \,\cdots\, \cdot x^m}^{n \text{ factors of } x^m} = \overbrace{x \cdot x \cdot x \cdot x \cdot x \cdot \,\cdots\, \cdot x}^{mn \text{ factors of } x} = x^{mn}$$

Thus, *to raise an exponential expression to a power, keep the base and multiply the exponents.*

To find a third rule for exponents, we square $3x$ to get

$$(3x)^2 = (3x)(3x) = 3 \cdot 3 \cdot x \cdot x = 3^2x^2 = 9x^2$$

In general, we have

$$\overbrace{(xy)^n = (xy)(xy)(xy) \;\cdots\; (xy)}^{n \text{ factors of } xy} = \overbrace{xxx \;\cdots\; x}^{n \text{ factors of } x} \cdot \overbrace{yyy \;\cdots\; y}^{n \text{ factors of } y} = x^ny^n$$

To find a fourth rule for exponents, we cube $\dfrac{x}{3}$ to get

$$\left(\frac{x}{3}\right)^3 = \frac{x}{3} \cdot \frac{x}{3} \cdot \frac{x}{3} = \frac{x \cdot x \cdot x}{3 \cdot 3 \cdot 3} = \frac{x^3}{3^3} = \frac{x^3}{27}$$

In general, we have

$$\left(\frac{x}{y}\right)^n = \overbrace{\left(\frac{x}{y}\right)\left(\frac{x}{y}\right)\left(\frac{x}{y}\right) \;\cdots\; \left(\frac{x}{y}\right)}^{n \text{ factors of } \frac{x}{y}} \quad (y \neq 0)$$

$$= \frac{\overbrace{xxx \;\cdots\; x}^{n \text{ factors of } x}}{\underbrace{yyy \;\cdots\; y}_{n \text{ factors of } y}} \qquad \text{Multiply the numerators and multiply the denominators.}$$

$$= \frac{x^n}{y^n}$$

The previous results are called the **power rules for exponents.**

The power rules for exponents

If m and n represent natural numbers, then

$$(x^m)^n = x^{mn} \qquad (xy)^n = x^ny^n \qquad \left(\frac{x}{y}\right)^n = \frac{x^n}{y^n} \quad \text{where } y \neq 0$$

INTERMEDIATE
Algebra *f(x)* **Now**™

Self Check 3
Simplify each expression:
a. $(a^5)^8$

b. $(6^3)^5$

c. $(a^4a^3)^3$

d. $(-3a^3)^3(a^2)^3$

Answers **a.** a^{40}, **b.** 6^{15},
c. a^{21}, **d.** $-27a^{15}$

EXAMPLE 3 Simplify each expression: **a.** $(3^2)^3$, **b.** $(x^{11})^5$, **c.** $(x^2x^3)^6$, and
d. $(-2x^2)^4(x^3)^2$.

Solution
a. $(3^2)^3 = 3^{2 \cdot 3}$ Keep the base 3.   **b.** $(x^{11})^5 = x^{11 \cdot 5}$
                      Multiply the exponents. $= x^{55}$
$ = 3^6$
$ = 729$

c. $(x^2x^3)^6 = (x^5)^6$ **d.** $(-2x^2)^4(x^3)^2 = (-2)^4x^8x^6$
$ = x^{30}$ $ = 16x^{14}$

INTERMEDIATE
Algebra *f(x)* **Now**™

Self Check 4
Simplify each expression:
a. $(a^4b^5)^2$

EXAMPLE 4 Simplify each expression. Assume that no denominators are zero.
a. $(x^2y)^3$, **b.** $(x^3y^4)^4$, **c.** $\left(\dfrac{x}{y^2}\right)^4$, and **d.** $\left(\dfrac{6x^3}{y^4}\right)^2$.

Solution

a. $(x^2y)^3 = (x^2)^3y^3$ Raise each factor of the product

$\quad = x^6y^3$ x^2y to the 3rd power.

b. $(x^3y^4)^4 = (x^3)^4(y^4)^4$

$\quad\quad\quad\quad = x^{12}y^{16}$

c. $\left(\dfrac{x}{y^2}\right)^4 = \dfrac{x^4}{(y^2)^4}$ Raise the numerator and

$\quad = \dfrac{x^4}{y^8}$ denominator to the 4th power.

d. $\left(\dfrac{6x^3}{y^4}\right)^2 = \dfrac{6^2(x^3)^2}{(y^4)^2}$

$\quad\quad\quad\quad = \dfrac{36x^6}{y^8}$

b. $\left(\dfrac{-6a^5}{b^7}\right)^3$

Answers **a.** a^8b^{10},

b. $-\dfrac{216a^{15}}{b^{21}}$

Zero exponents

Since we want the rules for exponents to hold for exponents of 0, we have

$$x^0x^n = x^{0+n} = x^n = 1x^n$$

Because $x^0x^n = 1x^n$, it follows that $x^0 = 1$ where $x \neq 0$. In words, *a nonzero base raised to the 0 power is 1*.

Zero exponents

If $x \neq 0$, then $x^0 = 1$.

! COMMENT 0^0 is undefined.

Because of the previous definition, any nonzero base raised to the 0th power is 1. For example, if no variables are zero, then

$$3^0 = 1, \quad (-7)^0 = 1, \quad (3ax^3)^0 = 1, \quad \left(\tfrac{1}{2}x^5y^7z^9\right)^0 = 1$$

INTERMEDIATE
Algebra *f(x)* **Now**™

EXAMPLE 5   Simplify each expression: **a.** $(5x)^0$,   **b.** $5x^0$,   and   **c.** $-5x^0y$.

Self Check 5
Simplify each expression:
a. $2xy^0$

Solution

a. $(5x)^0 = 1$ The base is $5x$ and the exponent is 0.

b. $-(xy)^0$

b. $5x^0 = 5 \cdot x^0 = 5 \cdot 1 = 5$ The base is x and the exponent is 0.

c. $-5x^0y = -5 \cdot x^0 \cdot y = -5 \cdot 1 \cdot y = -5y$

Answers **a.** $2x$,   **b.** -1

Negative exponents

Since the rules for exponents are true for negative-integer exponents, we have

$$x^{-n}x^n = x^{-n+n} = x^0 = 1 \quad \text{where } x \neq 0$$

Because $x^{-n} \cdot x^n = 1$ and $\dfrac{1}{x^n} \cdot x^n = 1$, we define x^{-n} *to be the reciprocal of* x^n.

Negative exponents

If n represents an integer and $x \neq 0$,

$$x^{-n} = \dfrac{1}{x^n} \quad \text{and} \quad \dfrac{1}{x^{-n}} = x^n$$

Using this definition, we can write expressions containing negative exponents as expressions without negative exponents. For example,

$$3^{-2} = \frac{1}{3^2} = \frac{1}{9} \qquad 10^{-3} = \frac{1}{10^3} = \frac{1}{1,000} \qquad \frac{1}{4^{-4}} = 4^4 = 256$$

and if b, c, and x are not 0, we have

$$(2c)^{-3} = \frac{1}{(2c)^3} = \frac{1}{8c^3} \qquad 3x^{-1} = 3 \cdot \frac{1}{x^1} = \frac{3}{x} \qquad \frac{7}{b^{-2}} = 7 \cdot \frac{1}{b^{-2}} = 7b^2$$

INTERMEDIATE
Algebra $f(x)$ **Now**™

Self Check 6
Write each expression without negative exponents:
a. $-3.14t^{-7}$

b. $\dfrac{32j}{y^{-9}}$

Answers **a.** $-\dfrac{3.14}{t^7}$,   **b.** $32jy^9$

EXAMPLE 6   Write each expression without negative exponents: **a.** $-2m^{-8}$ and   **b.** $\dfrac{6a}{y^{-5}}$.

Solution

a. $-2m^{-8} = -2 \cdot \dfrac{1}{m^8} = -\dfrac{2}{m^8}$ The base is m. The original exponent is -8.

b. $\dfrac{6a}{y^{-5}} = 6a \cdot \dfrac{1}{y^{-5}} = 6a \cdot y^5 = 6ay^5$ Use the rule $\dfrac{1}{x^{-n}} = x^n$.

> **! COMMENT** By the definition of negative exponents, a base cannot be 0. Thus, an expression such as 0^{-5} is undefined.

INTERMEDIATE
Algebra $f(x)$ **Now**™

Self Check 7
Simplify each expression:
a. $a^{-7}a^3$

b. $(a^{-5})^{-3}$

Answers **a.** $\dfrac{1}{a^4}$,   **b.** a^{15}

EXAMPLE 7   Simplify each expression: **a.** $x^{-5}x^3$ and   **b.** $(x^{-3})^{-2}$.

Solution
a. $x^{-5}x^3 = x^{-5+3}$ Keep the common base x and add the exponents.

$\qquad\qquad = x^{-2}$

$\qquad\qquad = \dfrac{1}{x^2}$

b. $(x^{-3})^{-2} = x^{(-3)(-2)}$ Keep the base x and multiply the exponents.

$\qquad\qquad\quad = x^6$

▐ More rules for exponents

To develop a rule for dividing exponential expressions, we proceed as follows:

$$\frac{x^m}{x^n} = x^m \left(\frac{1}{x^n}\right) = x^m x^{-n} = x^{m+(-n)} = x^{m-n}$$

Thus, *to divide exponential expressions with the same nonzero base, keep the common base and subtract the exponent in the denominator from the exponent in the numerator.*

The quotient rule for exponents
If m and n represent integers and $x \neq 0$, then

$$\frac{x^m}{x^n} = x^{m-n}$$

EXAMPLE 8 Simplify each expression. Write each answer without using negative exponents. **a.** $\dfrac{a^5}{a^3}$ and **b.** $\dfrac{2x^{-5}}{x^{11}}$.

Solution

a. $\dfrac{a^5}{a^3} = a^{5-3}$ Keep the common base a.
      Subtract the exponents.

   $= a^2$

b. $\dfrac{2x^{-5}}{x^{11}} = 2x^{-5-11}$

   $= 2x^{-16}$

   $= \dfrac{2}{x^{16}}$

EXAMPLE 9 Simplify each expression. Write each answer without using negative exponents. **a.** $\dfrac{x^4 x^3}{x^{-5}}$, **b.** $\dfrac{(x^2)^3}{(x^3)^2}$, **c.** $\dfrac{x^2 y^3}{xy^4}$, and **d.** $\left(\dfrac{a^{-2} b^3}{a^5 b^4}\right)^3$.

Solution

a. $\dfrac{x^4 x^3}{x^{-5}} = \dfrac{x^7}{x^{-5}}$

   $= x^{7-(-5)}$

   $= x^{12}$

b. $\dfrac{(x^2)^3}{(x^3)^2} = \dfrac{x^6}{x^6}$

   $= x^{6-6}$

   $= x^0$

   $= 1$

c. $\dfrac{x^2 y^3}{xy^4} = x^{2-1} y^{3-4}$

   $= xy^{-1}$

   $= x \cdot \dfrac{1}{y}$

   $= \dfrac{x}{y}$

d. $\left(\dfrac{a^{-2} b^3}{a^5 b^4}\right)^3 = (a^{-2-5} b^{3-4})^3$

   $= (a^{-7} b^{-1})^3$

   $= \left(\dfrac{1}{a^7 b}\right)^3$

   $= \dfrac{1}{a^{21} b^3}$

   To illustrate another rule for exponents, we consider the following simplification of $\left(\frac{2}{3}\right)^{-4}$.

$$\left(\frac{2}{3}\right)^{-4} = \frac{1}{\left(\dfrac{2}{3}\right)^4} = \frac{1}{\dfrac{2^4}{3^4}} = 1 \div \frac{2^4}{3^4} = 1 \cdot \frac{3^4}{2^4} = \frac{3^4}{2^4} = \left(\frac{3}{2}\right)^4$$

The example illustrates that to raise a fraction to a negative power, we can invert the fraction and raise it to a positive power.

Fractions to negative powers

If n represents an integer and $x \neq 0$ and $y \neq 0$, then

$$\left(\frac{x}{y}\right)^{-n} = \left(\frac{y}{x}\right)^{n}$$

INTERMEDIATE
Algebra *f(x)* **Now**™

Self Check 10

Write $\left(\dfrac{3a^3b^2}{2aa^5b^{-2}} \right)^{-5}$ without using parentheses or negative exponents.

Answer $\dfrac{32a^{15}}{243b^{20}}$

EXAMPLE 10 Write each expression without using parentheses or negative exponents. **a.** $\left(\dfrac{2}{3} \right)^{-4}$, **b.** $\left(\dfrac{y^2}{x^3} \right)^{-3}$, **c.** $\left(\dfrac{2x^2}{3y^{-3}} \right)^{-4}$, and **d.** $\left(\dfrac{a^{-2}b^3}{a^2a^3b^4} \right)^{-3}$.

Solution

a. $\left(\dfrac{2}{3} \right)^{-4} = \left(\dfrac{3}{2} \right)^{4}$

$= \dfrac{3^4}{2^4}$

$= \dfrac{81}{16}$

b. $\left(\dfrac{y^2}{x^3} \right)^{-3} = \left(\dfrac{x^3}{y^2} \right)^{3}$

$= \dfrac{x^9}{y^6}$

c. $\left(\dfrac{2x^2}{3y^{-3}} \right)^{-4} = \left(\dfrac{3y^{-3}}{2x^2} \right)^{4}$

$= \dfrac{3^4 y^{-12}}{2^4 x^8}$

$= \dfrac{81}{16x^8} \cdot y^{-12}$

$= \dfrac{81}{16x^8} \cdot \dfrac{1}{y^{12}}$

$= \dfrac{81}{16x^8 y^{12}}$

d. $\left(\dfrac{a^{-2}b^3}{a^2a^3b^4} \right)^{-3} = \left(\dfrac{a^2a^3b^4}{a^{-2}b^3} \right)^{3}$

$= \left(\dfrac{a^5b^4}{a^{-2}b^3} \right)^{3}$

$= (a^{5-(-2)}b^{4-3})^3$

$= (a^7b)^3$

$= a^{21}b^3$

We summarize the rules for exponents as follows.

Rules for exponents

If there are no divisions by 0, then for all integers m and n,

$$x^m x^n = x^{m+n} \qquad (x^m)^n = x^{mn} \qquad (xy)^n = x^n y^n \qquad \left(\dfrac{x}{y} \right)^n = \dfrac{x^n}{y^n}$$

$$x^0 = 1 \quad (x \neq 0) \qquad x^{-n} = \dfrac{1}{x^n} \qquad \dfrac{x^m}{x^n} = x^{m-n} \qquad \left(\dfrac{x}{y} \right)^{-n} = \left(\dfrac{y}{x} \right)^n$$

Section 5.1   STUDY SET

INTERMEDIATE
Algebra *f(x)* **Now**™

VOCABULARY   *Fill in the blanks.*

1. x^n is read as "x to the nth _____."

2. In the exponential expression x^n, x is called the _____, and n is called the _____.

3. $3^4 \cdot 3^8$ is a _____ of exponential expressions with the same base and $\dfrac{x^4}{x^2}$ is a _____ of exponential expressions with the same base.

4. The exponential expression 3^{-2} has a _____ exponent.

CONCEPTS   *Complete the rules for exponents. Assume that $x \neq 0$ and $y \neq 0$.*

5. $x^m x^n = $

6. $(x^m)^n = $

7. $(xy)^n = $

8. $\left(\dfrac{x}{y} \right)^n = $

9. $x^0 = $

10. $x^{-n} = $

11. $\dfrac{x^m}{x^n} = $

12. $\left(\dfrac{x}{y} \right)^{-n} = $

13. An exponential expression with a negative exponent can be written as an equivalent expression with a positive exponent. Give an example.

14. Explain the difference between the two expressions.
  a. $2x$ and x^2
  b. $-2x$ and x^{-2}

15. a. To multiply exponential expressions with the same base, keep the common base and _____ the exponents.
  b. To divide exponential expressions with the same base, keep the common base and _____ the exponents.
  c. To raise an exponential expression to a power, keep the base and _____ the exponents.

16. a. To raise a product to a power, raise each _____ of the product to that power.
  b. To raise a quotient to a power, raise the _____ and the _____ to that power.
  c. Any nonzero base raised to the 0 power is ___.
  d. x^{-n} is the _____ of x^n.

NOTATION *Complete each simplification.*

17. $\dfrac{x^5 x^4}{x^{-2}} = \dfrac{x^\blacksquare}{x^{-2}}$

  $= x^{9-\blacksquare}$

  $= x^\blacksquare$

18. $\left(\dfrac{a^{-4}}{a^3}\right)^2 = (a^{-4-3})^2$

  $= (a^\blacksquare)^2$

  $= a^\blacksquare$

  $= \dfrac{1}{\blacksquare}$

PRACTICE *Identify the base and the exponent.*

19. 5^3
20. -7^2
21. $-x^5$
22. $(-t)^4$
23. $2b^6$
24. $(3xy)^5$
25. $\left(\dfrac{n}{4}\right)^3$
26. $(-pq)^2$

Simplify each expression. Assume that no denominators are zero. Write each answer without using negative exponents.

27. 3^2
28. 3^4
29. -3^2
30. -3^4
31. $(-3)^2$
32. $(-3)^3$
33. 5^{-2}
34. 5^{-4}
35. -5^{-2}
36. -5^{-4}
37. $(-5)^{-2}$
38. $(-5)^{-4}$
39. 8^0
40. -9^0
41. $(-8)^0$
42. $(-9)^0$
43. $(-2x)^5$
44. $(-3a)^3$
45. $x^2 x^3$
46. $y^3 y^4$
47. $x^2 x^3 x^5$
48. $y^3 y^7 y^2$
49. $k^0 k^7$
50. $x^8 x^{11}$
51. $2aba^3 b^4$
52. $2x^2 y^3 x^3 y^2$
53. $p^9 p p^0$
54. $z^7 z^0 z$
55. $(-x)^2 y^4 x^3$
56. $-x^2 y^7 y^3 x^{-2}$
57. $(b^{-8})^9$
58. $(z^{12})^2$
59. $(x^4)^7$
60. $(y^7)^5$
61. $(r^{-3}s)^3$
62. $(m^5 n^2)^{-3}$
63. $(a^2 a^3)^4$
64. $(bb^2 b^3)^4$
65. $(-d^2)^3 (d^{-3})^3$
66. $(c^3)^2 (c^4)^{-2}$
67. $(3x^3 y^4)^3$
68. $\left(\dfrac{1}{2} a^2 b^5\right)^4$
69. $\left(-\dfrac{1}{3} mn^2\right)^6$
70. $(-3p^2 q^3)^5$
71. $\left(\dfrac{a^3}{b^2}\right)^5$
72. $\left(\dfrac{a^2}{b^3}\right)^4$
73. $\left(\dfrac{a^{-3}}{b^{-2}}\right)^{-2}$
74. $\left(\dfrac{k^{-3}}{k^{-4}}\right)^{-1}$
75. $\dfrac{a^8}{a^3}$
76. $\dfrac{c^7}{c^2}$
77. $\dfrac{c^{12} c^5}{c^{10}}$
78. $\dfrac{a^{33}}{a^2 a^3}$
79. $\left(\dfrac{2}{3}\right)^{-2}$
80. $\left(\dfrac{4}{5}\right)^{-3}$
81. $\dfrac{1}{a^{-4}}$
82. $\dfrac{3}{b^{-5}}$
83. $\dfrac{(3x^2)^{-2}}{x^3 x^{-4} x^0}$
84. $\dfrac{y^{-3} y^{-4} y^0}{(2y^{-2})^3}$
85. $\left(\dfrac{4a^{-2} b}{3ab^{-3}}\right)^3$
86. $\left(\dfrac{2ab^{-3}}{3a^{-2} b^2}\right)^2$
87. $\left(\dfrac{3a^{-2} b^2}{17a^2 b^3}\right)^0$
88. $\dfrac{a^0 + b^0}{2(a+b)^0}$
89. $\left(\dfrac{-2a^4 b}{a^{-3} b^2}\right)^3$
90. $\left(\dfrac{-3x^4 y^2}{-9x^5 y^{-2}}\right)^2$

91. $\left(-\dfrac{2a^3b^2}{3a^{-3}b^2}\right)^{-3}$

92. $\left(\dfrac{3x^5y^2}{6x^5y^{-2}}\right)^{-4}$

93. $\left(\dfrac{-3pqr^{-4}}{2p^2q^{-3}r^2}\right)^{-2}$

94. $\left(\dfrac{4a^2b^3z^{-4}}{3a^{-2}b^{-7}z^3}\right)^{-3}$

95. $(2x^{-4}y^3)^3(3x^2y^{-2})^{-2}$

96. $(\frac{1}{2}a^2b^{-3})^{-2}(2ab^2)^2$

97. $\dfrac{(3x^2y^{-4})^{-2}}{(2x^3y^2)^{-3}}$

98. $\dfrac{(-2m^{-3}n^2)^2}{(3m^2n^3)^{-2}}$

Use a calculator to find each value.

99. 1.23^6

100. 0.0537^4

101. -6.25^3

102. $(-25.1)^5$

Use a calculator to verify that each statement is true by showing that the values on either side of the equation are equal.

103. $(3.68)^0 = 1$

104. $(2.1)^4(2.1)^3 = (2.1)^7$

105. $(7.2)^2(2.7)^2 = [(7.2)(2.7)]^2$

106. $\left(\dfrac{5.4}{2.7}\right)^{-4} = \left(\dfrac{2.7}{5.4}\right)^4$

107. $(3.2)^2(3.2)^{-2} = 1$

108. $(7.23)^{-3} = \dfrac{1}{(7.23)^3}$

APPLICATIONS

109. MICROSCOPES The illustration in the next column shows the relative sizes of some chemical and biological structures, expressed as fractions of a meter (m). Express each fraction shown as a power of 10, from the largest to the smallest.

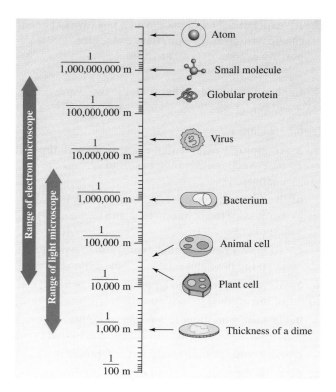

110. ASTRONOMY The distance d, in miles, of the nth planet from the sun is given by the formula

$$d = 9,275,200[3(2^{n-2}) + 4]$$

From the illustration below, determine n for Earth and Mars. Then find the distance of Earth and the distance of Mars from the sun.

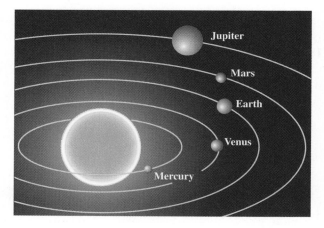

111. LICENSE PLATES The number of different license plates of the form three digits followed by three letters is $10 \cdot 10 \cdot 10 \cdot 26 \cdot 26 \cdot 26$. Write this expression using exponents. Then evaluate it.

112. PHYSICS Albert Einstein's work in relativity resulted in the observation that the total energy E of a body is equal to its total mass m times the square of the speed of light c. This relationship is given by the formula $E = mc^2$. Identify the base and exponent on the right-hand side of the equation.

113. GEOMETRY A cube is shown on the right.
   a. Find the area of its base.
   b. Find its volume.

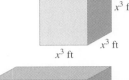

x^3 ft

x^3 ft

x^3 ft

114. GEOMETRY A rectangular solid is shown on the right.
   a. Find the area of its base.
   b. Find its volume.

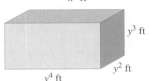

y^3 ft

y^2 ft

y^4 ft

WRITING

115. Explain how an exponential expression with a negative exponent can be expressed as an equivalent expression with a positive exponent. Give an example.

116. In the definition of x^{-n}, x cannot be 0. Why not?

117. Explain the error in the following solution.

$$-8ab^{-3} = \frac{a}{8b^3}$$

118. Is a positive number greater than 1 raised to a negative power greater than or less than 1? Explain your answer.

REVIEW *Solve each inequality. Give the result in interval notation and graph it.*

119. $a + 5 < 6$

120. $-9x + 5 \geq 15$

121. $6(t - 2) \leq 4(t + 7)$

122. $\frac{1}{4}p - \frac{1}{3} \leq p + 2$

5.2 Scientific Notation

- Writing numbers in scientific notation   • Converting from scientific notation
- Using scientific notation to simplify computations

Very large and small numbers occur in science and other disciplines. For example, the star nearest to Earth (excluding the sun) is Proxima Centauri, about 24,793,000,000,000 miles away, and the mass of a hydrogen atom is approximately 0.00000000000000000000000001673 gram.

Because these numbers contain many zeros, they are difficult to read and cumbersome to work with in computations. In this section, we will discuss a notation that enables us to express such numbers in a more manageable form.

Writing numbers in scientific notation

Scientific notation provides a compact way of writing large and small numbers.

> **Scientific notation**
>
> A positive number is written in **scientific notation** when it is written in the form $N \times 10^n$, where $1 \leq N < 10$ and n represents an integer.

Each of the following numbers is written in scientific notation.

3.67×10^6 2.24×10^{-4} 9.875×10^{22}

Every positive number written in scientific notation is the product of a decimal number between 1 (including 1) and 10 and an integer power of 10.

An integer exponent
↓

$$\boxed{}.\boxed{} \times 10^{\boxed{}}$$

↑
A decimal that is at least 1
but less than 10

Self Check 1
Write each italicized number in scientific notation.
a. In 1998, the country earning the most money from tourism was the United States: *$74,240,000,000.*

b. DNA molecules contain and transmit the information that allows cells to reproduce. They are only *0.000000002* meter wide.

Answers a. 7.424×10^{10}, **b.** 2×10^{-9}

EXAMPLE 1   Write each number in scientific notation: **a.** 24,793,000,000,000 and   **b.** 0.000000000000000000000001673.

Solution

a. The number 2.4793 is between 1 and 10. To get 24,793,000,000,000, the decimal point in 2.4793 must be moved 13 places to the *right*.

2.4,793,000,000,000.
⌣⌣⌣⌣⌣⌣⌣⌣⌣⌣⌣⌣⌣
13 places

We can move the decimal point 13 places to the right by multiplying 2.4793 by 10^{13}.

$$24{,}793{,}000{,}000{,}000 = 2.4793 \times 10^{13}$$

b. The number 1.673 is between 1 and 10. To get 0.000000000000000000000001673, the decimal point in 1.673 must be moved 24 places to the *left*.

0.000000000000000000000001.673
⌣⌣⌣⌣⌣⌣⌣⌣⌣⌣⌣⌣⌣⌣⌣⌣⌣⌣
24 places

We can move the decimal point 24 places to the left by multiplying 1.673 by 10^{-24}.

$$0.000000000000000000000001673 = 1.673 \times 10^{-24}$$

Numbers such as 47.2×10^3 and 0.063×10^{-2} appear to be written in scientific notation, because they are the product of a number and a power of 10. However, they are not. Their first factors (47.2 and 0.063) are not between 1 and 10.

Self Check 2
Write   **a.** 27.3×10^2 and **b.** 0.0025×10^{-3} in scientific notation.

Answers a. 2.73×10^3, **b.** 2.5×10^{-6}

EXAMPLE 2   Write **a.** 47.2×10^3 and   **b.** 0.063×10^{-2} in scientific notation.

Solution Since the first factors are not between 1 and 10, neither number is in scientific notation. However, we can change them to scientific notation as follows:

a. $47.2 \times 10^3 = (\mathbf{4.72 \times 10^1}) \times 10^3$ Write 47.2 in scientific notation.

$\qquad\qquad = 4.72 \times (10^1 \times 10^3)$ Group the powers of 10 together.

$\qquad\qquad = 4.72 \times 10^4$ Apply the product rule for exponents: $10^1 \times 10^3 = 10^{1+3} = 10^4$.

b. $0.063 \times 10^{-2} = (\mathbf{6.3 \times 10^{-2}}) \times 10^{-2}$ Write 0.063 in scientific notation.

$\qquad\qquad = 6.3 \times (10^{-2} \times 10^{-2})$

$\qquad\qquad = 6.3 \times 10^{-4}$

The American Educational System

Between 2001 and 2013, the number of high school graduates is projected to increase nationally by 11 percent. Increases are expected in each region of the country, especially the West.

National Center for Education Statistics, Projections to 2013

The figure on the right shows the three-level structure of the American educational system as illustrated in the *Digest for Education Statistics*, 2002. Write each italicized number in the following excerpt from the digest using scientific notation.

> For 2002, enrollment in U.S. elementary and secondary schools was estimated to be *53.6 million* and enrollment in U.S. postsecondary schools was estimated to be *15.6 million*. Total spending on all three levels of education was estimated to be *$745 billion*.

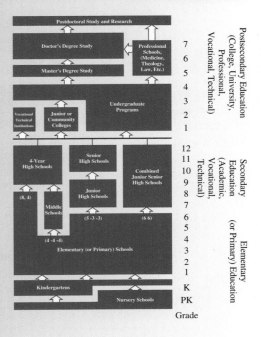

■ Converting from scientific notation

We can change a number written in scientific notation to **standard notation.** For example, to write 9.3×10^7 in standard notation, we multiply 9.3 by 10^7.

$9.3 \times 10^7 = 9.3 \times 10,000,000 = 93,000,000$ 10^7 is 1 followed by 7 zeros.

EXAMPLE 3   Change   **a.** 8.706×10^5 and   **b.** 1.1×10^{-3} to standard notation.

Solution

a. Since multiplication by 10^5 or 100,000 moves the decimal point 5 places to the right,

$8.706 \times 10^5 = 8\,7\,0\,6\,0\,0. = 870,600$

b. Since multiplication by 10^{-3} or 0.001 moves the decimal point 3 places to the left,

$1.1 \times 10^{-3} = 0.0\,0\,1\,1 = 0.0011$

INTERMEDIATE
Algebra *f(x)* Now™

Self Check 3
Change each number in scientific notation to standard notation.
a. Russia is the largest country in land area, with over 6.5×10^6 square miles.
b. The average distance between molecules of air in a room is 3.937×10^{-7} inch.

Answers **a.** 6,500,000,
b. 0.0000003937

Each of the following numbers is written in both scientific and standard notation. In each case, the exponent gives the number of places that the decimal point moves, and the sign of the exponent indicates the direction that it moves:

$5.32 \times 10^4 = 5\,3\,2\,0\,0.$
4 places to the right

$6.45 \times 10^7 = 6\,4\,5\,0\,0\,0\,0\,0.$
7 places to the right

$2.37 \times 10^{-4} = 0.\,0\,0\,0\,2\,3\,7$
4 places to the left

$9.234 \times 10^{-2} = 0.\,0\,9\,2\,3\,4$
2 places to the left

$4.89 \times 10^0 = 4.89$
No movement of the decimal point

Using scientific notation to simplify computations

Scientific notation is useful when multiplying and dividing very large or very small numbers.

Self Check 4

A light year is 5.88×10^{12} miles. What is the diameter of the Milky Way in miles?

EXAMPLE 4   Astronomy.

The galaxy in which we live is called the Milky Way. (See Figure 5-1.) This system of some 10^{11} stars, one of which is the sun, has a diameter of approximately 100,000 light years. (A light year is the distance light travels in a vacuum in one year: 9.46×10^{15} meters.) What is the diameter of the Milky Way in meters?

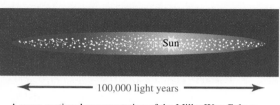

100,000 light years

A cross-sectional representation of the Milky Way Galaxy

FIGURE 5-1

Solution   To find the diameter of the Milky Way in meters, we will multiply the diameter of the Milky Way, expressed in light years, by the number of meters in a light year. To perform the calculation, we write 100,000 in scientific notation as 1.0×10^5.

$$1.0 \times 10^5 \cdot 9.46 \times 10^{15}$$

$$= (1.0 \cdot 9.46) \times (10^5 \cdot 10^{15})$$ Apply the commutative and associative properties of multiplication to group the first factors together and the powers of 10 together.

$$= 9.46 \times 10^{5+15}$$ Perform the multiplication: $1.0 \cdot 9.46 = 9.46$. For the powers of 10, keep the base and add the exponents.

$$= 9.46 \times 10^{20}$$ Perform the addition.

Answer 5.88×10^{17} mi

The Milky Way galaxy is about 9.46×10^{20} meters in diameter.

EXAMPLE 5   World oil reserves/production.

According to estimates in the *Oil and Gas Journal*, there were 1.03×10^{12} barrels of oil reserves in the ground at the start of 2003. At that time, world production was 2.89×10^{10} barrels per year. If annual production remains the same and if no new oil discoveries are made, when will the world's oil supply run out?

Solution   If we divide the estimated number of barrels of oil in reserve, 1.03×10^{12}, by the number of barrels produced each year, 2.89×10^{10}, we can find the number of years of oil supply left.

$$\frac{1.03 \times 10^{12}}{2.89 \times 10^{10}} = \frac{1.03}{2.89} \times \frac{10^{12}}{10^{10}}$$ Divide the first factors and the second factors in the numerator and denominator separately.

$$\approx 0.36 \times 10^{12-10}$$ Perform the division: $\frac{1.03}{2.89} \approx 0.36$. For the powers of 10, keep the base and subtract the exponents.

$$\approx 0.36 \times 10^2$$ Perform the subtraction.

$$\approx 36$$ Write 0.36×10^2 in standard notation.

According to industry estimates, as of 2003, there were 36 years of oil reserves left. Under these conditions, the world's oil supply will run out in the year 2039.

EXAMPLE 6   Use scientific notation to evaluate

$$\frac{(0.00000064)(24{,}000{,}000{,}000)}{(400{,}000{,}000)(0.0000000012)}$$

Solution   After writing each number in scientific notation, we can do the arithmetic on the numbers and the exponential expressions separately.

$$\frac{(0.00000064)(24{,}000{,}000{,}000)}{(400{,}000{,}000)(0.0000000012)} = \frac{(6.4 \times 10^{-7})(2.4 \times 10^{10})}{(4 \times 10^{8})(1.2 \times 10^{-9})}$$

$$= \frac{(6.4)(2.4)}{(4)(1.2)} \times \frac{10^{-7} 10^{10}}{10^{8} 10^{-9}}$$

$$= \frac{15.36}{4.8} \times 10^{-7+10-8-(-9)}$$

$$= 3.2 \times 10^{4}$$

The result is 3.2×10^{4}. In standard notation, this is 32,000.

INTERMEDIATE
Algebra $f(x)$ **Now**™

Self Check 6
Use scientific notation to evaluate

$$\frac{(320)(25{,}000)}{0.00004}$$

Answer
$2 \times 10^{11} = 200{,}000{,}000{,}000$

Using scientific notation

CALCULATOR SNAPSHOT

Scientific calculators and graphing calculators often give answers in scientific notation. For example, if we use a calculator to find 301.2^{8}, the display will read

> `6.77391496 ` `19`

On a scientific calculator

> `301.2^8`
> `        6.77391496E19`

On a graphing calculator

In either case, the answer is given in scientific notation and we interpret it as

$$6.77391496 \times 10^{19}$$

Numbers can also be entered into a calculator in scientific notation. For example, to enter 24,000,000,000 (which is 2.4×10^{10} in scientific notation), we enter these numbers and press these keys:

2.4 [EXP] 10   On most scientific calculators

2.4 [EE] 10   On a graphing calculator and on some scientific calculators

To use a scientific calculator to evaluate

$$\frac{(24{,}000{,}000{,}000)(0.00000006495)}{0.00000004824}$$

we must enter each number in scientific notation, because each number has too many digits to be entered directly. In scientific notation, the three numbers are

$$2.4 \times 10^{10} \qquad 6.495 \times 10^{-8} \qquad 4.824 \times 10^{-8}$$

Using a scientific calculator, we enter these numbers and press these keys:

2.4 [EXP] 10 [×] 6.495 [EXP] 8 [+/−] [÷] 4.824 [EXP] 8 [+/−] [=]

The display will read   `3.231343284 ` `10` . In standard notation, the answer is 32,313,432,840.

The keystrokes are similar on a graphing calculator.

Section 5.2 STUDY SET

VOCABULARY   *Fill in the blanks.*

1. 7.4×10^6 is written in _____ notation. 7,400,000 is written in _____ notation.

2. 10^{-3}, 10^0, 10^1, and 10^4 are _____ of 10.

CONCEPTS   *Fill in the blanks.*

3. A positive number is written in scientific notation when it is written in the form $N \times \boxed{}$, where $1 \leq N < 10$ and n is an _____.

4. Use > or <:
   The number $5.3 \times 10^2 \boxed{}$ the number 5.3×10^{-2}.

5. To change 6.31×10^{-4} to standard notation, we move the decimal point four places to the _____.

6. To change 9.7×10^3 to standard notation, we move the decimal point three places to the _____.

NOTATION

7. Explain why the number 60.22×10^{22} is not written in scientific notation.

8. Explain why the number 0.6022×10^{24} is not written in scientific notation.

PRACTICE   *Write each number in scientific notation.*

9. 3,900
10. 1,700
11. 0.0078
12. 0.068
13. 173,000,000,000,000
14. 89,800,000,000
15. 0.0000096
16. 0.000000046
17. 323×10^5
18. 689×10^9
19. $6,000 \times 10^{-7}$
20. 765×10^{-5}
21. 0.0527×10^5
22. 0.0298×10^3
23. 0.0317×10^{-2}
24. 0.0012×10^{-3}

Write each number in standard notation.

25. 2.7×10^2
26. 7.2×10^3
27. 3.23×10^{-3}
28. 6.48×10^{-2}
29. 7.96×10^5
30. 9.67×10^6
31. 3.7×10^{-4}
32. 4.12×10^{-5}
33. 5.23×10^0
34. 8.67×10^0
35. 23.65×10^6
36. 75.6×10^{-5}

Give all answers in scientific notation. **Use a calculator to check your results.**

37. $(7.9 \times 10^5)(2.3 \times 10^6)$
38. $(6.1 \times 10^8)(3.9 \times 10^5)$
39. $(9.1 \times 10^{-5})(5.5 \times 10^{12})$
40. $(8.4 \times 10^{-13})(4.8 \times 10^9)$
41. $\dfrac{4.2 \times 10^{-12}}{8.4 \times 10^{-5}}$
42. $\dfrac{1.21 \times 10^{-15}}{1.1 \times 10^2}$
43. $\dfrac{(3.9 \times 10^{-9})(9.5 \times 10^{-4})}{1.95 \times 10^{-2}}$
44. $\dfrac{(4.9 \times 10^{60})(2.7 \times 10^{30})}{6.3 \times 10^{40}}$

Write each numeral in scientific notation and perform the operations. Give all answers in scientific notation and in standard form. **Use a calculator to check your results.**

45. $(89,000,000,000)(4,500,000,000)$
46. $(0.000000061)(3,500,000,000)$
47. $\dfrac{0.00000129}{0.0003}$
48. $\dfrac{4,400,000,000,000}{0.0002}$
49. $\dfrac{(220,000)(0.000009)}{0.00033}$
50. $\dfrac{(640,000)(2,700,000)}{120,000}$
51. $\dfrac{(0.00024)(96,000,000)}{640,000,000}$
52. $\dfrac{(0.0000013)(0.00009)}{0.00039}$

APPLICATIONS

53. FIVE-CARD POKER The odds against being dealt the hand shown on the right are about 2.6×10^6 to 1. Express the odds using standard notation.

54. ENERGY See the illustration on the next page. Express each of the following using scientific notation. (1 quadrillion is 10^{15}.)

    a. 2001 U.S. energy consumption

    b. 2001 U.S. energy production

    c. The difference in 2001 consumption and production

2001 U.S. Energy Consumption and Production
(petroleum, natural gas, coal, hydroelectric, nuclear, geothermal, solar, wind)

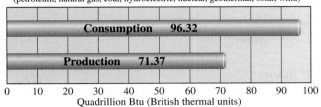

Consumption   96.32

Production   71.37

0   10   20   30   40   50   60   70   80   90   100
Quadrillion Btu (British thermal units)

Source: Energy Information Administration, United States Department of Energy

55. **THE YEAR 2000** Express in scientific notation each of the dollar amounts that appeared in the following excerpt from the *Federal Computer Week* Web page (February 16, 1998).

> President Clinton's fiscal 1999 budget proposal of $1.7 trillion includes expenditures of about $3.9 billion to ensure that federal computers can accept dates after Dec. 31, 1999. Clinton has proposed spending $275 million at the Defense Department and $312 million at the Treasury Department to fix the year 2000 problem.

56. **STAR TREK** In the science fiction series *Star Trek,* crew members talk of their spacecraft, the U.S.S. Enterprise, traveling at various warp speeds. To convert a warp speed, W, to an equivalent velocity in miles per second, v, we can use the equation

$$v = W^3 c$$

where c is the speed of light, 1.86×10^5 miles per second. Find the velocity of a spacecraft traveling at warp 2.

57. **ATOMS** A simple model of a helium atom is shown. If a proton has a mass of 1.7×10^{-24} grams, and if the mass of an electron is only about $\frac{1}{2,000}$ that of a proton, find the mass of an electron.

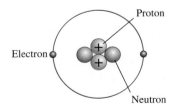

Proton

Electron

Neutron

58. **OCEANS** The mass of Earth's oceans is only about $\frac{1}{4,400}$ that of Earth. If the mass of Earth is 6.578×10^{21} tons, find the mass of the oceans.

59. **LIGHT YEARS** Light travels about 300,000,000 meters per second. A **light year** is the distance that light can travel in one year. Estimate the number of meters in one light year.

60. **AQUARIUMS** Express the volume of the fish tank shown below in scientific notation.

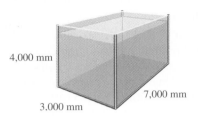

4,000 mm

7,000 mm

3,000 mm

61. **THE BIG DIPPER** One of the stars in the Big Dipper is named Merak. It is approximately 4.65×10^{14} miles from Earth.

   a. If light travels about 1.86×10^5 miles/sec, how many seconds does it take light emitted from Merak to reach Earth? (*Hint:* Use the formula $t = \frac{d}{r}$.)

   b. Convert your result from part a to years.

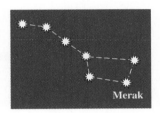

Merak

62. **BIOLOGY** A paramecium is a single-celled organism that propels itself with hair-like projections called *cilia.* Use the scale in the illustration below to estimate the length of the paramecium. Express the result in scientific and in standard notation.

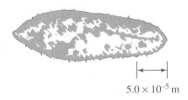

5.0×10^{-5} m

63. **COMET HALE-BOPP** On March 23, 1997, Comet Hale-Bopp made its closest approach to Earth, coming within 1.3 **astronomical units.** One astronomical unit (AU) is the distance from Earth to the sun — about 9.3×10^7 miles. Express this distance in miles, using scientific notation.

64. **DIAMONDS** The approximate number of atoms of carbon in a $\frac{1}{2}$-carat diamond can be found by computing

$$\frac{6.0 \times 10^{23}}{1.2 \times 10^2}$$

Express the number of carbon atoms in scientific and in standard notation.

65. ATOMS A hydrogen atom is so small that a single drop of water contains more than a million million billion hydrogen atoms. Express this number in scientific notation.

66. ASTRONOMY The American Physical Society recently honored first-year graduate student Gwen Bell for coming up with what it considers the most accurate estimate of the mass of the Milky Way. In pounds, her estimate is a 3 with 42 zeros after it. Express this number in scientific notation.

WRITING

67. Explain how to change a number from standard notation to scientific notation.

68. Explain how to change a number from scientific notation to standard notation.

69. Explain why 9.99×10^n represents a number less than 1 but greater than 0 if n is a negative integer.

70. Explain the advantages of writing very large and very small numbers in scientific notation.

REVIEW   *Solve each compound inequality. Give the result in interval notation and graph the solution set.*

71. $4x \geq -x + 5$ and $6 \geq 4x - 3$

72. $15 > 2x - 7 > 9$

73. $3x + 2 < 8$ or $2x - 3 > 11$

74. $-4(x + 2) \geq 12$ or $3x + 8 < 11$

5.3 Polynomials and Polynomial Functions

- Polynomials • Degree of a polynomial • Polynomial functions
- Evaluating polynomial functions • Graphing polynomial functions
- Combining like terms • Adding polynomials • Subtracting polynomials

In arithmetic, we add, subtract, multiply, divide, and find powers of numbers. In algebra, we perform these operations on algebraic expressions called *polynomials.*

Polynomials

A **term** is a number or a product of a number and a variable (or variables) raised to a power. Some examples are

$$17, \qquad 9x, \qquad \frac{15}{16}y^2, \qquad \text{and} \qquad -2.4x^4y^5$$

If a term contains only a number, such as 17, it is called a **constant term,** or simply a **constant.**

The **numerical coefficient,** or simply the **coefficient,** is the numerical factor of a term. For example, the coefficient of $9x$ is 9 and the coefficient of $-2.4x^4y^5$ is -2.4. The coefficient of a constant term is that constant.

> **Polynomials**
>
> A **polynomial** is an algebraic term or the sum of algebraic terms whose variables have whole-number exponents. No variable appears in a denominator.

The following expressions are polynomials in x:

$$-6x, \qquad 3x^2 + 2x, \qquad \frac{3}{2}x^5 - \frac{7}{3}x^4 - \frac{8}{3}x^3, \qquad \text{and} \qquad 19x^{20} + \sqrt{3}x^{14} + 4.5x^{11} - x^2$$

! COMMENT The following expressions are not polynomials:

$$\frac{2x}{x^2 + 1}, \qquad x^{1/2} - 8, \qquad \text{and} \qquad x^{-3} + 2x + 24$$

The first expression is a quotient and has a variable in the denominator. The last two have exponents that are not whole numbers.

If any terms of a polynomial contain more than one variable, we say that the polynomial is in more than one variable. Some examples are

$$3xy, \qquad 5x^2y^2 + 2xy - 3y, \qquad \text{and} \qquad u^2v^2w^2 + uv + 1$$

Polynomials can be classified according to the number of terms they have. A polynomial with one term is called a **monomial,** a polynomial with two terms is called a **binomial,** and a polynomial with three terms is called a **trinomial.**

Monomials	Binomials	Trinomials
$2x^3$	$2x + 5$	$2x^2 + 4x + 3$
a^2b	$-17x^4 - \dfrac{3}{5}x$	$3mn^3 - m^2n^3 + 7n$
$3x^3y^5z^2$	$32x^{13}y^5z^3 + 47x^3yz$	$-12x^5y^2 + 13x^4y^3 - 7x^3y^3$

■ Degree of a polynomial

Because the variable x occurs three times as a factor in the monomial $2x^3$, the monomial is called a *third-degree monomial* or a *monomial of degree 3.* The monomial $3x^3y^5z^2$ is called a *monomial of degree 10,* because the variables x, y, and z occur as factors a total of ten times $(3 + 5 + 2)$. These examples illustrate the following definition.

> **Degree of a monomial**
>
> The **degree of a monomial** in one variable is the exponent on the variable. The degree of a monomial in several variables is the sum of the exponents on those variables. If the monomial is a nonzero constant, its degree is 0. The constant 0 has no defined degree.

INTERMEDIATE
Algebra $f(x)$ **Now**™

EXAMPLE 1 Find the degree of **a.** $3x^4$, **b.** $-4x^2y^3$, and **c.** 3.

Solution
a. $3x^4$ is a monomial of degree 4, because the exponent on the variable is 4.
b. $-4x^2y^3$ is a monomial of degree 5, because the sum of the exponents on the variables is $2 + 3 = 5$.
c. 3 is a monomial of degree 0, because $3 = 3x^0$.

Self Check 1
Find the degree of
a. $-12a^2$
b. $8a^3b^2$
c. $\dfrac{1}{2}x^3y^2z^{12}$

Answers **a.** 2, **b.** 5, **c.** 17

We determine the degree of a polynomial by considering the degrees of each of its terms.

> **Degree of a polynomial**
>
> The **degree of a polynomial** is the same as the degree of the term in the polynomial with largest degree.

Self Check 2

Find the degree of
a. $x^2 - x + 1$

b. $x^2y^3 - 12x^7y^2 + 3x^9y^3 - 3$

Answers a. 2, b. 12

EXAMPLE 2   Find the degree of each polynomial: **a.** $3x^5 + 4x^2 + 7$, **b.** $7x^2y^8 - 3x^2y^2$, and **c.** $3x + 2y - xy$.

Solution

a. The terms of $3x^5 + 4x^2 + 7$ have degree 5, 2, and 0, respectively. This trinomial is of degree 5, because the largest degree of the three terms is 5.

b. $7x^2y^8 - 3x^2y^2$ is a binomial of degree 10.

c. $3x + 2y - xy$ is a trinomial of degree 2. (Recall that $xy = x^1y^1$.)

If the terms of a polynomial in one variable are written so that the exponents on the variable decrease as we move from left to right, we say that the terms are written with their exponents in *descending order*. If the terms are written so that the exponents on the variable increase as we move from left to right, we say that the terms are written with their exponents in *ascending order*.

$-5x^4 + 2x^3 + 7x^2 + 3x - 1$ This polynomial is written in descending powers of x.

$-1 + 3x + 7x^2 + 2x^3 - 5x^4$ The same polynomial is now written in ascending powers of x.

Polynomial functions

In Chapter 2, we saw that linear functions are defined by equations of the form $f(x) = mx + b$. Some examples of linear functions are

$$f(x) = 3x + 1 \qquad g(x) = -\frac{1}{2}x - 1 \qquad h(x) = 5x$$

In each case, the right-hand side of the equation is a polynomial. For this reason, linear functions are members of a larger class of functions known as **polynomial functions.**

Polynomial functions

A **polynomial function** is a function whose equation is defined by a polynomial in one variable.

Another example of a polynomial function is $f(x) = -x^2 + 6x - 8$. This is a second-degree polynomial function, called a **quadratic function.** Quadratic functions are of the form $f(x) = ax^2 + bx + c$, where $a \neq 0$.

An example of a third-degree polynomial function is $f(x) = x^3 - 3x^2 - 9x + 2$. Third-degree polynomial functions, also called **cubic functions,** are of the form $f(x) = ax^3 + bx^2 + cx + d$, where $a \neq 0$.

Evaluating polynomial functions

Polynomial functions can be used to model many real-life situations. If we are given a polynomial function model, we can learn more about the situation by evaluating the function at specific values. To *evaluate a polynomial function* at a specific value, we replace the variable in the defining equation with that value, called the **input.** Then we simplify the resulting expression to find the **output.**

INTERMEDIATE
Algebra *f(x)* Now™

EXAMPLE 3 **Rocketry.** If a toy rocket is shot straight up with an initial velocity of 128 feet per second, its height, in feet, t seconds after liftoff is given by the function

$$h(t) = -16t^2 + 128t$$

Find the height of the rocket at **a.** 0 second, **b.** 3 seconds, and **c.** 7.9 seconds.

Self Check 3
In Example 3, find the height of the rocket 4 seconds after it is shot upward.

Solution
a. To find the height of the rocket at 0 second, we substitute 0 for t and evaluate the right-hand side.

$$h(t) = -16t^2 + 128t \qquad \text{This is the given function.}$$
$$h(0) = -16(0)^2 + 128(0) \qquad \text{The input is 0.}$$
$$= 0 \qquad\qquad \text{The output is 0.}$$

At 0 second, the rocket's height is 0. It is on the ground waiting to be launched.

b. To find the height at 3 seconds, we substitute 3 for t and evaluate the right-hand side.

$$h(t) = -16t^2 + 128t \qquad \text{This is the given function.}$$
$$h(3) = -16(3)^2 + 128(3) \qquad \text{The input is 3.}$$
$$= -16(9) + 384$$
$$= -144 + 384$$
$$= 240 \qquad\qquad \text{The output is 240.}$$

At 3 seconds after liftoff, the height of the rocket is 240 feet.

c. To find the height at 7.9 seconds, we substitute 7.9 for t and evaluate the right-hand side.

$$h(t) = -16t^2 + 128t \qquad \text{This is the given function.}$$
$$h(7.9) = -16(7.9)^2 + 128(7.9) \qquad \text{The input is 7.9.}$$
$$= -16(62.41) + 1,011.2$$
$$= -998.56 + 1,011.2$$
$$= 12.64 \qquad\qquad \text{The output is 12.64.}$$

At 7.9 seconds, the height is 12.64 feet. The rocket has almost fallen back to Earth.

Answer 256 ft

INTERMEDIATE
Algebra *f(x)* Now™

EXAMPLE 4 **Packaging.** To make boxes, a manufacturer cuts equal-sized squares from each corner of a 10 in. × 12 in. piece of cardboard and then folds up the sides. (See Figure 5-2.) The polynomial function $f(x) = 4x^3 - 44x^2 + 120x$ gives the volume (in cubic inches) of the resulting box when a square with sides x inches long is cut from each corner. Find the volume of a box if 3-inch squares are cut out.

Self Check 4
In Example 4, find the volume of the resulting box if 2-inch squares are cut from each corner of the cardboard.

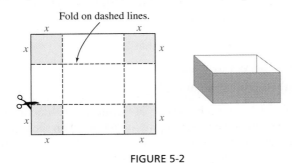

Fold on dashed lines.

FIGURE 5-2

Solution   To find the volume of the box, we evaluate the function for $x = 3$.

$$f(x) = 4x^3 - 44x^2 + 120x \qquad \text{This is the given function.}$$
$$f(3) = 4(3)^3 - 44(3)^2 + 120(3) \qquad \text{Substitute 3 for } x.$$
$$= 4(27) - 44(9) + 120(3) \qquad \text{Evaluate the right-hand side.}$$
$$= 108 - 396 + 360$$
$$= 72$$

Answer 96 in.³

If 3-inch squares are cut out, the box will have a volume of 72 in.³.

Graphing polynomial functions

We have previously graphed three basic polynomial functions. Figure 5-3 shows the graph of a linear function $f(x) = x$, the graph of the squaring function $f(x) = x^2$, and the graph of the cubing function $f(x) = x^3$. From the graphs in the figure, it is easy to determine the domain and range of each of these functions.

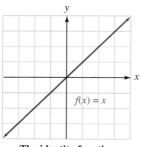

The identity function
The domain is $(-\infty, \infty)$.
The range is $(-\infty, \infty)$.

(a)

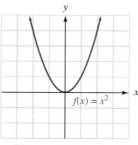

The squaring function
The domain is $(-\infty, \infty)$.
The range is $[0, \infty)$.

(b)

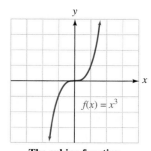

The cubing function
The domain is $(-\infty, \infty)$.
The range is $(-\infty, \infty)$.

(c)

FIGURE 5-3

When graphing a linear function, we need to plot only two points, because the graph is a straight line. The graphs of polynomial functions of degree greater than 1 are smooth, continuous curves. To graph them, we must plot more points.

In Example 3, we saw that the polynomial function $h(t) = -16t^2 + 128t$ gives the height of the rocket t seconds after it is shot upward. Since the height of the rocket depends on time, we say that the height is a function of time. To graph this function, we can make a table of values, plot the points, and join them with a smooth curve.

INTERMEDIATE
Algebra *f(x)* **Now™**

Self Check 5
Use the graph in Figure 5-4 to estimate the height of the rocket at 6.5 seconds.

EXAMPLE 5   Graph: $h(t) = -16t^2 + 128t$.

Solution   From Example 3, we have seen that

At 0 second, the rocket's height is 0: $h(0) = 0$.

At 3 seconds, the rocket's height is 240: $h(3) = 240$.

To graph the function, we select other values of t, evaluate the function at those values, and write the ordered pairs in the table shown in Figure 5-4 on the next page. Then we plot the pairs and join the resulting points to get the parabola shown in the figure. From the graph, we can see that 4 seconds into the flight, the rocket attains a maximum height of 256 feet.

b. $(3rt^2 + 4r^2t^2) - (8rt^2 - 4r^2t^2 + r^3t^2)$

$= 3rt^2 + 4r^2t^2 - 8rt^2 + 4r^2t^2 - r^3t^2$ Remove the parentheses. Change the sign of every term in the second polynomial.

$= 3rt^2 - 8rt^2 + 4r^2t^2 + 4r^2t^2 - r^3t^2$ Rearrange terms.

$= -5rt^2 + 8r^2t^2 - r^3t^2$ Combine like terms.

Answer $8a^2b^3 - 3a^2b^2$

To subtract polynomials in vertical form, we add the opposite (or the negative) of the polynomial that is being subtracted.

$$\begin{array}{r} 8x^3y + 2x^2y \\ - \underline{2x^3y - 3x^2y} \end{array} \Rightarrow \begin{array}{r} 8x^3y + 2x^2y \\ + \underline{-2x^3y + 3x^2y} \\ 6x^3y + 5x^2y \end{array}$$ This is the opposite of $2x^3y - 3x^2y$.

Section 5.3   STUDY SET

VOCABULARY   *Fill in the blanks.*

1. A _____ is the sum of one or more algebraic terms whose variables have whole-number exponents.

2. A _____ is a polynomial with one term.
   A _____ is a polynomial with two terms.
   A _____ is a polynomial with three terms.

3. The _____ of a monomial in one variable is the exponent on the variable.

4. A second-degree polynomial function is also called a _____ function.

5. A third-degree polynomial function is also called a _____ function.

6. The _____ of the term $-15x^2y^3$ is -15. The _____ of the term is 5.

7. Terms having the same variables with the same exponents are called _____ terms.

8. The _____ of $x^2 + x - 3$ is $-x^2 - x + 3$.

CONCEPTS   *Classify each polynomial as a monomial, a binomial, a trinomial, or none of these. Then determine its degree.*

9. $3x^2$

10. $2y^3 + 4y^2$

11. $3x^2y - 2x + 3y$

12. $a^2 + ab + b^2$

13. $x^2 - y^2$

14. $\dfrac{17}{2}x^3 + 3x^2 - x - 4$

15. 5

16. $8x^3y^5$

17. $9x^2y^4 - x - y^{10} + 1$

18. x^{17}

19. $4x^9 + 3x^2y^4$

20. -12

21. Write each polynomial with the exponents on x in descending order.

   a. $3x - 2x^4 + 7 - 5x^2$

   b. $a^2x - ax^3 + 7a^3x^5 - 5a^3x^2$

22. Write each polynomial with the exponents on y in ascending order.

   a. $4y^2 - 2y^5 + 7y - 5y^3$

   b. $x^3y^2 + x^2y^3 - 2x^3y + x^7y^6 - 3x^6$

Determine whether the terms are like or unlike terms. If they are like terms, combine them.

23. $3x, 7x$

24. $-8x, 3y$

25. $7x, 7y$

26. $3mn, 5mn$

27. $3r^2t^3, -8r^2t^3$

28. $9u^2v, 10u^2v$

29. $9x^2y^3, 3x^2y^2$

30. $27x^6y^4z, 8x^6y^4z^2$

31. Write a polynomial that represents the perimeter of the triangle.

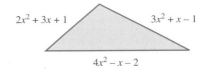

$2x^2 + 3x + 1$ $3x^2 + x - 1$

$4x^2 - x - 2$

32. Write a polynomial that represents the perimeter of the rectangle.

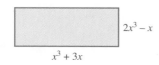

$2x^3 - x$

$x^3 + 3x$

NOTATION *Complete each solution.*

33. If $f(x) = 2x^2 + x + 2$, find $f(-1)$.

$$f(x) = 2x^2 + x + 2$$
$$f(-1) = 2()^2 + () + 2$$
$$= 2() + (-1) + 2$$
$$= $$

34. If $h(t) = -t^3 - t^2 + 2t + 1$, find $h(3)$.

$$h(t) = -t^3 - t^2 + 2t + 1$$
$$h(3) = -()^3 - ()^2 + 2(3) + 1$$
$$= - 9 + 6 + 1$$
$$= $$

PRACTICE *Complete each table, and graph each polynomial function.*

35. $f(x) = 2x^2 - 4x + 2$

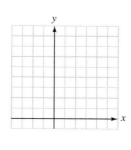

x	f(x)
−1	
0	
1	
2	
3	

36. $f(x) = -x^2 + 2x + 6$

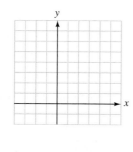

x	f(x)
−2	
−1	
0	
1	
2	
3	
4	

37. $f(x) = 2x^3 - 3x^2 - 11x + 6$

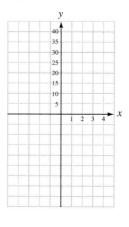

x	f(x)
−3	
−2	
−1	
0	
1	
2	
3	
4	

38. $f(x) = -x^3 - x^2 + 6x$

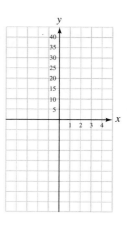

x	f(x)
−4	
−3	
−2	
−1	
0	
1	
2	
3	

Use a graphing calculator to graph each polynomial function. Use window settings of $[-4, 6]$ for x and $[-5, 5]$ for y.

39. $f(x) = 2.75x^2 - 4.7x + 1.5$

40. $f(x) = 0.37x^3 - 1.4x + 1.5$

Perform each operation.

41. $(3x^2 + 2x + 1) + (-2x^2 - 7x + 5)$

42. $(-2a^2 - 5a - 7) + (-3a^2 + 7a + 1)$

43. $(-a^2 + 2a + 3) - (4a^2 - 2a - 1)$

44. $(x^2 - 3x + 8) - (3x^2 + x + 3)$

45. $(2a^2 + 4a - 7) + (3a^2 - a - 2)$

46. $(6x^3 + 3x - 2) - (2x^3 + 3x^2 + 5)$

47. $(7y^3 + 4y^2 + y + 3) + (-8y^3 - y + 3)$

48. $(-8p^3 - 2p - 4) - (2p^3 + p^2 - p)$

49. $(3pq + p - q) + (-pq - p - q)$

50. $(-2mn + 2m - n) - (-2mn - 2m + n)$

51. $(-2x^2y^3 + 6xy + 5y^2) + (+4x^2y^3 + 7xy + 2y^2)$

52. $(3ax^3 - 2ax^2 + 3a^3) + (4ax^3 + 3ax^2 - 2a^3)$

53. $(3x^2 + 4x - 3) + (2x^2 - 3x - 1) + (x^2 + x + 7)$

54. $(-2x^2 + 6x + 5) - (-4x^2 - 7x + 2) -$
    $(4x^2 + 10x + 5)$

55. $\begin{array}{r} 3x^3 - 2x^2 + 4x - 3 \\ + -2x^3 + 3x^2 + 3x - 2 \\ \hline 5x^3 - 7x^2 + 7x - 12 \end{array}$

56. $\begin{array}{r} 7a^3 + 3a + 7 \\ + -2a^3 + 4a^2 - 13 \\ \hline 3a^3 - 3a^2 + 4a + 5 \end{array}$

57. $\begin{array}{r} 3x^2 - 4x + 17 \\ + -2x^2 + 4x + 5 \\ \hline \end{array}$

58. $\begin{array}{r} -2y^2 - 4y + 3 \\ - 3y^2 + 10y - 5 \\ \hline \end{array}$

59. $\begin{array}{r} -5y^3 + 4y^2 - 11y + 3 \\ - 2y^3 - 14y^2 + 17y - 32 \\ \hline \end{array}$

60. $\begin{array}{r} 17x^4 - 3x^2 - 65x - 12 \\ - 23x^4 + 14x^2 + 3x - 23 \\ \hline \end{array}$

61. $\begin{array}{r} 4x^2 + 0x \\ - 4x^2 - 2x \\ \hline \end{array}$

62. $\begin{array}{r} -2a^3 + 0a^2 \\ - 2a^3 + 3a^2 \\ \hline \end{array}$

63. Subtract $(-x^2 - 2x + 1)$ from the sum of
$(4x^2 + 5x - 3)$ and $(7x^2 + 2x - 10)$.

64. Subtract $(a^2b - 2ab + b)$ from the sum of
$(-6a^2b + 4ab - a)$ and $(8a^2b + 2ab - 10b)$.

APPLICATIONS

65. JUGGLING A juggler tosses one
ball straight upward while continuing
to juggle three others. The height
$f(t)$, in feet, of the ball is given
by the polynomial function
$f(t) = -16t^2 + 32t + 4$, where t is
the time in seconds since the ball was
thrown. Find the height of the ball
1 second after it is tossed upward.

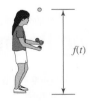

66. STOPPING DISTANCES The number of feet that
a car travels before stopping depends on the driver's
reaction time and the braking distance. (See the
illustration in the next column.) For one driver,

the stopping distance $d(v)$, in feet, is given by the
polynomial function $d(v) = 0.04v^2 + 0.9v$, where
v is the velocity of the car in mph. Find the stopping
distance at 60 mph.

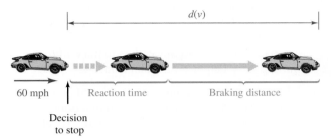

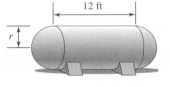

67. STORAGE TANKS
The volume $V(r)$ of the
gasoline storage tank,
in cubic feet, is given by
the polynomial function
$V(r) = 4.2r^3 + 37.7r^2$,
where r is the radius in feet of the cylindrical part of
the tank. What is the capacity of the tank shown on
the right if its radius is 4 feet?

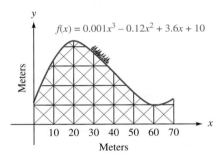

68. ROLLER COASTERS The polynomial function
$f(x) = 0.001x^3 - 0.12x^2 + 3.6x + 10$ models the path
of a portion of the track of a roller coaster, as shown
below. Find the height of the track for $x = 0, 20, 40$,
and 60.

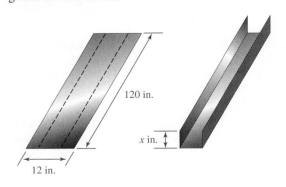

69. RAIN GUTTERS A rectangular sheet of metal will
be used to make a rain gutter by bending up its sides,
as shown below. If the ends are covered, the capacity
$f(x)$ of the gutter is a polynomial function of x:
$f(x) = -240x^2 + 1,440x$. Find the capacity of the
gutter if x is 3 inches.

70. CUSTOMER SERVICE A software company has found that on Mondays, the polynomial function $C(t) = -0.0625t^4 + t^3 - 6t^2 + 16t$ approximates the number of callers to its hotline at any one time. Here, t represents the time, in hours, since the hotline opened at 8:00 A.M. How many service technicians should be on duty on Mondays at noon if the company doesn't want any callers to the hotline waiting to be helped by a technician?

71. REAL ESTATE A computer analysis of two properties on the market generated functions to predict the value, in dollars, of each property after x years.

> Rental home: $R(x) = 1,100x + 125,000$
>
> Duplex: $D(x) = 1,400x + 150,000$

a. Find one polynomial function V that will give the combined value of the two properties after x years.

b. Use your answer to part a to find the combined value of the two properties after 20 years.

72. BUSINESS EXPENSES A company purchased two cars for its sales force to use. The following functions give the respective values of the vehicles after x years.

> Toyota Camry LE: $T(x) = -2,100x + 16,600$
>
> Ford Explorer Sport: $F(x) = -2,700x + 19,200$

a. Find one polynomial function V that will give the value of both cars after x years.

b. Use your answer to part a to find the combined value of the two cars after 3 years.

73. MATHEMATICS In calculus, an important polynomial function is

$$f(x) = 1 - \frac{x^2}{2} + \frac{x^4}{24} - \frac{x^6}{720}$$

Refer to the polynomial on the right-hand side of the equation.

a. How many terms does the polynomial have?

b. Are the terms written in descending or ascending powers of x?

c. What is the coefficient of the fourth term?

d. What is the degree of the third term?

e. What is the degree of the polynomial?

74. CALCULUS See Exercise 73. Another polynomial function used in advanced mathematics is

$$f(x) = 1 + x + \frac{x^2}{2} + \frac{x^3}{6} + \frac{x^4}{24}$$

a. Find $f(0)$.   **b.** Find $f(1)$.

WRITING

75. Explain why the terms x^2y and xy^2 are not like terms.

76. The family of polynomial functions contains linear, quadratic, and cubic functions. Explain.

77. Explain what is wrong in the following solution.

> Subtract $2x - 3$ from $3x + 4$.
>
> $(2x - 3) - (3x + 4) = 2x - 3 - 3x - 4$
> $= -x - 7$

78. Explain why $f(x) = \dfrac{1}{x + 1}$ is not a polynomial function.

79. Use the word *descending* in a sentence in which the context is not mathematical. Do the same for the word *ascending*.

80. Look up the meaning of the prefix *poly* in a dictionary. Why do you think the name *polynomial* was given to expressions such as $x^3 - x^2 + 2x + 15$?

REVIEW   *Solve each inequality. Write the solution set in interval notation.*

81. $|x| \leq 5$ **82.** $|x| > 7$

83. $|x - 4| < 5$ **84.** $|2x + 1| \geq 7$

5.4 Multiplying Polynomials

- Multiplying monomials • Multiplying a polynomial by a monomial
- Multiplying a polynomial by a polynomial • The FOIL method
- Multiplying three polynomials • Special products • Simplifying expressions
- Applications of multiplying polynomials

In this section, we will show how to multiply polynomials. These procedures involve the application of several algebraic concepts introduced in earlier chapters, such as the commutative and associative properties of multiplication, the rules for exponents, and the distributive property.

■ Multiplying monomials

We begin by considering the simplest case of polynomial multiplication, multiplying two monomials.

> **Multiplying monomials**
>
> To multiply two monomials, multiply the numerical factors (the coefficients) and then multiply the variable factors.

Self Check 1
Multiply:
a. $(-2a^3)(4a^2)$
b. $(-5b^3)(-3a)(a^2b)$

EXAMPLE 1 Perform each multiplication. **a.** $(3x^2)(6x^3)$, **b.** $(-8x)(2y)(xy)$, and **c.** $(2a^3b)(-7b^2c)(-12ac^4)$.

Solution
We can use the commutative and associative properties of multiplication to rearrange and regroup the factors.

a. $(3x^2)(6x^3) = 3 \cdot x^2 \cdot 6 \cdot x^3$
$\qquad\qquad = (3 \cdot 6)(x^2 \cdot x^3)$
$\qquad\qquad = 18x^5$ To simplify $x^2 \cdot x^3$, keep the base and add the exponents.

b. $(-8x)(2y)(xy) = -8 \cdot x \cdot 2 \cdot y \cdot x \cdot y$
$\qquad\qquad\quad = (-8 \cdot 2) \cdot x \cdot x \cdot y \cdot y$
$\qquad\qquad\quad = -16x^2y^2$

c. $(2a^3b)(-7b^2c)(-12ac^4) = 2 \cdot a^3 \cdot b \cdot (-7) \cdot b^2 \cdot c \cdot (-12) \cdot a \cdot c^4$
$\qquad\qquad\qquad\qquad\quad = 2(-7)(-12) \cdot a^3 \cdot a \cdot b \cdot b^2 \cdot c \cdot c^4$
$\qquad\qquad\qquad\qquad\quad = 168a^4b^3c^5$

Answers a. $-8a^5$, **b.** $15a^3b^4$

■ Multiplying a polynomial by a monomial

To multiply a polynomial by a monomial, we use the distributive property.

> **Multiplying polynomials by monomials**
>
> To multiply a monomial and a polynomial, multiply each term of the polynomial by the monomial.

Self Check 2
Multiply:
$-2a^2(a^2 - a + 3)$

EXAMPLE 2 Perform each multiplication. **a.** $3x^2(6xy + 3y^2)$, **b.** $5x^3y^2(xy^3 - 2x^2y)$, and **c.** $-2ab^2(3bz - 2az + 4z^3)$.

Solution
We can use the distributive property to remove parentheses.

a. $3x^2(6xy + 3y^2) = 3x^2 (6xy) + 3x^2 (3y^2)$ Distribute the multiplication by $3x^2$.
$\qquad\qquad\qquad\quad = 18x^3y + 9x^2y^2$ Perform the multiplications.

Since $18x^3y$ and $9x^2y^2$ are not like terms, we cannot add them.

b. $5x^3y^2(xy^3 - 2x^2y) = 5x^3y^2 (xy^3) - 5x^3y^2 (2x^2y)$ Distribute $5x^3y^2$.
$\qquad\qquad\qquad\qquad\quad = 5x^4y^5 - 10x^5y^3$

c. $-2ab^2(3bz - 2az + 4z^3)$

$= -2ab^2 \cdot 3bz - (-2ab^2) \cdot 2az + (-2ab^2) \cdot 4z^3$

$= -6ab^3z + 4a^2b^2z - 8ab^2z^3$

Answer $-2a^4 + 2a^3 - 6a^2$

■ Multiplying a polynomial by a polynomial

To multiply a polynomial by a polynomial, we use the distributive property repeatedly.

Self Check 3

Multiply:

$(2a + b)(3a - 2b)$

EXAMPLE 3   Perform each multiplication: **a.** $(3x + 2)(4x + 9)$ and

b. $(2a - b)(3a^2 - 4ab + b^2)$.

Solution

a. $(3x + 2)(4x + 9) = (3x + 2) \cdot 4x + (3x + 2) \cdot 9$ Distribute $3x + 2$.

$= 12x^2 + 8x + 27x + 18$ Distribute $4x$ and distribute 9.

$= 12x^2 + 35x + 18$ Combine like terms.

b. $(2a - b)(3a^2 - 4ab + b^2)$

$= (2a - b)3a^2 - (2a - b)4ab + (2a - b)b^2$ Distribute $2a - b$.

$= 6a^3 - 3a^2b - 8a^2b + 4ab^2 + 2ab^2 - b^3$ Distribute $3a^2$, $4ab$, and b^2.

$= 6a^3 - 11a^2b + 6ab^2 - b^3$ Combine like terms.

Answer $6a^2 - ab - 2b^2$

The results of Example 3 suggest the following rule.

> **Multiplying polynomials**
>
> To multiply two polynomials, multiply each term of one polynomial by each term of the other polynomial, and then combine like terms.

In the next example, we organize the work done in Example 3 vertically.

Self Check 4

Multiply:

$3x^2 + 2x - 5$

$\underline{ 2x + 1}$

EXAMPLE 4   Perform each multiplication vertically: **a.** $(3x + 2)(4x + 9)$

and   **b.** $(3a^2 - 4ab + b^2)(2a - b)$.

Solution

a.
$$
\begin{array}{r}
3x + 2 \\
\underline{4x + 9} \\
27x + 18 \\
\underline{12x^2 + 8x } \\
12x^2 + 35x + 18
\end{array}
$$
← This is the result of multiplying $3x + 2$ by 9.

← This is the result of multiplying $3x + 2$ by $4x$.

Combine like terms, column by column.

b.
$$
\begin{array}{r}
3a^2 - 4ab + b^2 \\
\underline{2a - b} \\
-3a^2b + 4ab^2 - b^3 \\
\underline{6a^3 - 8a^2b + 2ab^2 } \\
6a^3 - 11a^2b + 6ab^2 - b^3
\end{array}
$$
← Multiply $3a^2 - 4ab + b^2$ by $-b$.

← Multiply $3a^2 - 4ab + b^2$ by $2a$.

Combine like terms, column by column.

Answer $6x^3 + 7x^2 - 8x - 5$

INTERMEDIATE
Algebra $f(x)$ **Now**™

EXAMPLE 5   Find the product of $-2y^3 - 6y^2 + 12$ and $5y^2 - 10y - 10$.

Self Check 5
Find the product of $2a^2 + 6a - 12$
and $3a^2 + 9a - 9$.

Solution

To multiply these expressions, we must multiply each term of one polynomial by each term of the other polynomial.

$$(\mathbf{-2y^3 - 6y^2 + 12})(5y^2 - 10y - 10)$$

For lengthy multiplications like this, we can use vertical form. We begin by multiplying $-2y^3 - 6y^2 + 12$ by -10; then we multiply $-2y^3 - 6y^2 + 12$ by $-10y$; and finally we multiply $-2y^3 - 6y^2 + 12$ by $5y^2$. Then we combine like terms, column by column.

$$
\begin{array}{r}
-2y^3 - 6y^2 + 12 \\
5y^2 - 10y - 10 \\
\hline
20y^3 + 60y^2 - 120 \\
20y^4 + 60y^3 - 120y \\
-10y^5 - 30y^4 + 60y^2 \\
\hline
-10y^5 - 10y^4 + 80y^3 + 120y^2 - 120y - 120
\end{array}
$$

There is no y-term; leave a space.
There is no y^2-term; leave a space.
There is no y^3-term; leave a space.

Answer
$6a^4 + 36a^3 - 162a + 108$

■ The FOIL method

When we multiply two binomials, the distributive property requires that each term of one binomial be multiplied by each term of the other binomial. This fact can be emphasized by drawing arrows to show the indicated products. For example, to multiply $3x + 2$ and $x + 4$, we can write

First terms Last terms

$$
\begin{aligned}
(\mathbf{3x + 2})(\mathbf{x + 4}) &= \mathbf{3x(x) + 3x(4) + 2(x) + 2(4)} \\
&= 3x^2 + 12x + 2x + 8 \\
&= 3x^2 + 14x + 8
\end{aligned}
$$

Inner terms

Outer terms

Combine like terms:
$12x + 2x = 14x.$

We note that

- the product of the **F**irst terms is $3x \cdot x = 3x^2$,
- the product of the **O**uter terms is $3x \cdot 4 = 12x$,
- the product of the **I**nner terms is $2 \cdot x = 2x$, and
- the product of the **L**ast terms is $2 \cdot 4 = 8$.

The procedure is called the **FOIL** method of multiplying two binomials. Foil is an acronym for **F**irst terms, **O**uter terms, **I**nner terms, and **L**ast terms. Of course, the resulting terms of the product must be combined, if possible.

It is easy to multiply binomials by sight using the FOIL method. We find the product of the first terms, then find the products of the outer terms and the inner terms and add them (when possible), and then find the product of the last terms.

INTERMEDIATE
Algebra $f(x)$ **Now**™

EXAMPLE 6   Find each product: **a.** $(2x - 3)(3x + 2)$,   **b.** $(3x^2 + 1)(3x^2 + 4)$, and   **c.** $(4xy - 5)(2x^2 - 3y)$.

Self Check 6
Multiply:
a. $(3a + 4b)(2a - b)$
b. $(a^2b - 3)(a^2b - 1)$
c. $(6c^4 - d)(3c + d)$

Solution

a. $(2x - 3)(3x + 2) = 6x^2 - 5x - 6$

The product of the first terms is $2x \cdot 3x = 6x^2$. The middle term in the result comes from combining the outer and inner products of $4x$ and $-9x$:

$$4x + (-9x) = -5x$$

The product of the last terms is $-3 \cdot 2 = -6$.

b. $(3x^2 + 1)(3x^2 + 4) = 9x^4 + 15x^2 + 4$

The product of the first terms is $3x^2 \cdot 3x^2 = 9x^4$. The middle term in the result comes from combining the products $12x^2$ and $3x^2$:

$$12x^2 + 3x^2 = 15x^2$$

The product of the last terms is $1 \cdot 4 = 4$.

c. $(4xy - 5)(2x^2 - 3y) = 8x^3y - 12xy^2 - 10x^2 + 15y$

The product of the first terms is $4xy \cdot 2x^2 = 8x^3y$. The product of the outer terms is $4xy(-3y) = -12xy^2$. The product of the inner terms is $-5(2x^2) = -10x^2$. The product of the last terms is $-5(-3y) = 15y$. Since the terms of $8x^3y - 12xy^2 - 10x^2 + 15y$ are unlike, we cannot simplify this result.

Answers **a.** $6a^2 + 5ab - 4b^2$,
b. $a^4b^2 - 4a^2b + 3$,
c. $18c^5 + 6c^4d - 3cd - d^2$

▮ Multiplying three polynomials

When finding the product of three polynomials, we begin by multiplying *any* two of them, and then we multiply that result by the third polynomial.

Self Check 7
Multiply: $-2r(r - 2s)(5r - 4s)$.

EXAMPLE 7   Multiply: $3cd(c + 2d)(3c - d)$.

Solution   First, we find the product of the two binomials. Then we multiply that result by $3cd$.

$$3cd(c + 2d)(3c - d) = 3cd(3c^2 - cd + 6cd - 2d^2) \qquad \text{Use the FOIL method to find } (c + 2d)(3c - d).$$

$$= 3cd(3c^2 + 5cd - 2d^2) \qquad \text{Combine like terms: } -cd + 6cd = 5cd.$$

$$= 9c^3d + 15c^2d^2 - 6cd^3 \qquad \text{Distribute the multiplication by } 3cd.$$

Answer $-10r^3 + 28r^2s - 16rs^2$

▮ Special products

We often must find the square of a binomial. To do so, we can use the FOIL method. For example, to find $(x + y)^2$ and $(x - y)^2$, we proceed as follows.

$$(x + y)^2 = (x + y)(x + y) \qquad\qquad (x - y)^2 = (x - y)(x - y)$$

$$= x^2 + xy + xy + y^2 \qquad\qquad\quad = x^2 - xy - xy + y^2$$

$$= x^2 + 2xy + y^2 \qquad\qquad\qquad = x^2 - 2xy + y^2$$

In each case, we see that the square of the binomial is the square of its first term, twice the product of its two terms, and the square of its last term.

Figure 5-7 shows how $(x + y)^2$ can be found geometrically.

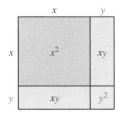

The area of the largest square is the product of its length and width: $(x + y)(x + y) = (x + y)^2$.

The area of the largest square is also the sum of its four parts: $x^2 + xy + xy + y^2 = x^2 + 2xy + y^2$.

Thus, $(x + y)^2 = x^2 + 2xy + y^2$.

FIGURE 5-7

Another common binomial product is the product of the sum and difference of the same two terms. An example of such a product is $(x + y)(x - y)$. To find this product, we use the FOIL method.

$$(x + y)(x - y) = x^2 - xy + xy + y^2$$
$$= x^2 - y^2 \qquad \text{Combine like terms: } -xy + xy = 0.$$

We see that the product of the sum and the difference of the same two terms is the square of the first term minus the square of the second term.

The square of a binomial and the product of the sum and the difference of the same two terms are called **special products.** Because special products occur so often, it is useful to learn their forms.

Special product formulas

$$(x + y)^2 = (x + y)(x + y) = x^2 + 2xy + y^2 \quad \text{The square of a sum}$$
$$(x - y)^2 = (x - y)(x - y) = x^2 - 2xy + y^2 \quad \text{The square of a difference}$$
$$(x + y)(x - y) = x^2 - y^2 \qquad\qquad\quad \text{The product of the sum and difference of two terms}$$

! COMMENT The squares $(x + y)^2$ and $(x - y)^2$ have trinomials for their products. Don't forget to write the middle terms in these products. Remember that

$$(x + y)^2 \neq x^2 + y^2 \qquad \text{and} \qquad (x - y)^2 \neq x^2 - y^2$$

Missing $2xy$ Missing $-2xy$ — Should be + symbol

Also remember that the product $(x + y)(x - y)$ is the binomial $x^2 - y^2$. And since $(x + y)(x - y) = (x - y)(x + y)$ by the commutative property of multiplication,

$$(x - y)(x + y) = x^2 - y^2$$

EXAMPLE 8 Multiply: **a.** $(5c + 3d)^2$, **b.** $\left(\dfrac{1}{2}a^4 - b^2\right)^2$, and

c. $(0.2m^3 + 2.5n)(0.2m^3 - 2.5n)$.

Solution
a. To find $(5c + 3d)^2$ using a special product formula, we begin by noting that the first term of the binomial is $5c$ and the last term is $3d$.

The square of the first term   Twice the product of the two terms   The square of the last term

$$(5c + 3d)^2 = (5c)^2 + 2(5c)(3d) + (3d)^2$$
$$= 25c^2 + 30cd + 9d^2$$

Self Check 8
Multiply:
a. $(8r + 2s)^2$

b. $\left(\dfrac{1}{3}a^3 - b^6\right)^2$

c. $(0.4x + 1.2y^4)(0.4x - 1.2y^4)$

b. To find $\left(\frac{1}{2}a^4 - b^2\right)^2$ using a special product formula, we begin by noting that the first term of the binomial is $\frac{1}{2}a^4$ and the last term is $-b^2$.

$$\left(\frac{1}{2}a^4 - b^2\right)^2 = \overset{\substack{\text{The square of}\\\text{the first term}\\\downarrow}}{\left(\frac{1}{2}a^4\right)^2} + \overset{\substack{\text{Twice the product}\\\text{of the two terms}\\\downarrow}}{2\left(\frac{1}{2}a^4\right)(-b^2)} + \overset{\substack{\text{The square of}\\\text{the last term}\\\downarrow}}{(-b^2)^2}$$

$$= \frac{1}{4}a^8 - a^4b^2 + b^4$$

c. $(0.2m^3 + 2.5n)(0.2m^3 - 2.5n)$ is the product of the sum and the difference of the same two terms: $0.2m^3$ and $2.5n$. Using a special product formula, we proceed as follows.

$$(0.2m^3 + 2.5n)(0.2m^3 - 2.5n) = \overset{\substack{\text{The square of}\\\text{the first term}\\\downarrow}}{(0.2m^3)^2} - \overset{\substack{\text{The square of}\\\text{the last term}\\\downarrow}}{(2.5n)^2}$$

$$= 0.04m^6 - 6.25n^2$$

Answers
a. $64r^2 + 32rs + 4s^2$,
b. $\dfrac{1}{9}a^6 - \dfrac{2}{3}a^3b^6 + b^{12}$,
c. $0.16x^2 - 1.44y^8$

◼ Simplifying expressions

The procedures discussed in this section are often useful when we simplify algebraic expressions that involve the multiplication of polynomials.

INTERMEDIATE
Algebra $f(x)$ **Now**™

Self Check 9
Simplify:
$(y - 7)(y + 7) - (4y + 3)^2$.

EXAMPLE 9   Simplify: $(5x - 4)^2 - (x - 7)(x + 1)$.

Solution   Before doing the subtraction, we use a special product formula to find $(5x - 4)^2$ and the FOIL method to find $(x - 7)(x + 1)$.

$(5x - 4)^2 - (x - 7)(x + 1)$

$= 25x^2 - 40x + 16 - (x^2 - 6x - 7)$ $(5x - 4)^2 = (5x)^2 + 2(5x)(-4) + (-4)^2$.

$= 25x^2 - 40x + 16 - x^2 + 6x + 7$ To subtract $(x^2 - 6x - 7)$, remove the parentheses and change the sign of each term within the parentheses.

Answer $-15y^2 - 24y - 58$

$= 24x^2 - 34x + 23$ Combine like terms.

◼ Applications of multiplying polynomials

Profit, revenue, and cost are terms used in the business world. The profit p earned on the sale of one or more items is given by the formula

Profit = revenue − cost

If a salesperson has 12 vacuum cleaners and sells them for \$225 each, the revenue will be $r = 12 \cdot \$225 = \$2,700$. This illustrates the following formula for finding the revenue r:

$$r = \boxed{\substack{\text{number of}\\\text{items sold } x}} \cdot \boxed{\substack{\text{selling price of}\\\text{each item } p}} = xp = px$$

EXAMPLE 10   Selling vacuum cleaners.   Over the years, a saleswoman has found that the number of vacuum cleaners she can sell depends on price. The lower the price, the more she can sell. She has determined that the number of vacuums x that she can sell at a price p is related by the equation $x = -\frac{2}{25}p + 28$.

a. Find a formula for the revenue r.

b. How much revenue will be taken in if the vacuums are priced at $250?

Solution

a. To find a formula for revenue, we substitute $-\frac{2}{25}p + 28$ for x in the formula $r = px$ and multiply.

$r = px$ 　　　　　　　　　This is the formula for revenue.

$r = p\left(-\dfrac{2}{25}p + 28\right)$ 　Substitute $-\frac{2}{25}p + 28$ for x.

$r = -\dfrac{2}{25}p^2 + 28p$ 　　Multiply the polynomials.

b. To find how much revenue will be taken in if the vacuums are priced at $250, we substitute 250 for p in the formula for revenue.

$r = -\dfrac{2}{25}p^2 + 28p$ 　　　　　This is the formula for revenue.

$r = -\dfrac{2}{25}(250)^2 + 28(250)$ 　Substitute 250 for p.

$= -5,000 + 7,000$

$= 2,000$

The revenue will be $2,000.

Section 5.4　STUDY SET

VOCABULARY   *Fill in the blanks.*

1. The expression $(x + 4)(x - 5)$ is the _____ of two binomials.

2. The expression $(x + 4)^2$ is the _____ of a binomial.

3. The polynomial $3x^2 - 3x + 2$ contains three _____.

4. To find $-2x(3x^2 - 2)$, we use the _____ property.

CONCEPTS   *Fill in the blanks.*

5. To multiply a monomial by a monomial, we multiply the numerical _____ and then multiply the variable factors.

6. To multiply a polynomial by a monomial, we multiply each _____ of the polynomial by the monomial.

7. To multiply a polynomial by a polynomial, we multiply each _____ of one polynomial by each term of the other polynomial.

8. FOIL is an acronym for _____ terms, _____ terms, _____ terms, and _____ terms.

9. $(x + y)^2 = (x + y)(x + y) =$

10. $(x - y)^2 = (x - y)(x - y) =$

11. $(x + y)(x - y) =$

12. a. The square of a binomial is the _____ of its first term, _____ the product of its two terms, and the _____ of its last term.

b. The product of the sum and difference of the same two terms is the _____ of the first term minus the _____ of the second term.

13. Write a polynomial that represents the area of the rectangle.

14. Write a polynomial that represents the area of the triangle.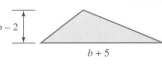

15. Write a polynomial that represents the area of the square.

$4a + 3$

16. Write a polynomial that represents the area of the rectangle.

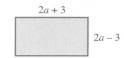

$2a + 3$

$2a - 3$

17. Consider $(2x + 4)(4x - 3)$. Give the

  a. First terms   **b.** Outer terms

  c. Inner terms   **d.** Last terms

18. Find

  a. $(4b - 1) + (2b - 1)$

  b. $(4b - 1)(2b - 1)$

  c. $(4b - 1) - (2b - 1)$

PRACTICE   *Find each product.*

19. $(2a^2)(-3ab)$ **20.** $(-3x^2y)(3xy)$

21. $(-3ab^2c)(5ac^2)$

22. $(-2m^2n)(-4mn^3)$

23. $(4a^2b)(-5a^3b^2)(6a^4)$

24. $(2x^2y^3)(4xy^5)(-5y^6)$

25. $(-5xx^2)(-3xy)^4$

26. $(-2a^2ab^2)^3(-3ab^2b^2)$

27. $3(x + 2)$ **28.** $-5(a + b)$

29. $3x(x^2 + 3x)$ **30.** $-2x(3x^2 - 2)$

31. $-2x(3x^2 - 3x + 2)$

32. $3a(4a^2 + 3a - 4)$

33. $7rst(r^2 + s^2 - t^2)$

34. $3x^2yz(x^2 - 2y + 3z^2)$

35. $4m^2n(-3mn)(m + n)$

36. $-3a^2b^3(2b)(3a + b)$

37. $(x + 2)(x + 3)$

38. $(y - 3)(y + 4)$

39. $(3t - 2)(2t + 3)$

40. $(p + 3)(3p - 4)$

41. $(3y - z)(2y - z)$

42. $(2m - n)(3m - n)$

43. $\left(\dfrac{1}{2}b + 8\right)(4b + 6)$

44. $\left(\dfrac{2}{3}x + 1\right)(15x - 9)$

45. $(0.4t - 3)(0.5t - 3)$

46. $(0.7d - 2)(0.1d + 3)$

47. $(b^3 - 1)(b + 1)$

48. $(c^3 + 1)(1 - c)$

49. $(3tu - 1)(-2tu + 3)$

50. $(-5st + 1)(10st - 7)$

51. $(9b^3 - c)(3b^2 - c)$

52. $(h^5 - k)(4h^3 - k)$

53. $(11m^2 + 3n^3)(5m + 2n^2)$

54. $(50m^4 - 3n^4)(2m + 2n^3)$

55. $6p^2(3p - 4)(p + 3)$

56. $4a(2a + 3)(3a - 2)$

57. $(3m - y)(4my)(2m - y)$

58. $(2h - z)(-3hz)(3h - z)$

59. $(x + 2)^2$ **60.** $(x - 3)^2$

61. $(a - 4)^2$ **62.** $(y + 5)^2$

63. $(2a + b)^2$

64. $(a - 2b)^2$

65. $(5r^2 + 6)^2$

66. $(6p^2 - 3)^2$

67. $(9ab - 4)^2$

68. $(2yz + 5)^2$

69. $\left(\dfrac{1}{4}b + 2\right)^2$

70. $\left(\dfrac{2}{3}y - 7\right)^2$

71. $(4k - 1.3)^2$

72. $(0.5k + 6)^2$

73. $(x + 2)(x - 2)$ **74.** $(z + 3)(z - 3)$

75. $(y^3 + 2)(y^3 - 2)$ **76.** $(y^4 + 3)(y^4 - 3)$

77. $(xy - 6)(xy + 6)$

78. $(a^4b - c)(a^4b + c)$

79. $\left(\dfrac{1}{2}x - 16\right)\left(\dfrac{1}{2}x + 16\right)$

80. $\left(\dfrac{3}{4}h^2 - \dfrac{2}{3}\right)\left(\dfrac{3}{4}h^2 + \dfrac{2}{3}\right)$

81. $(2.4 + y)(2.4 - y)$

82. $(3.5t + 4.1u)(3.5t - 4.1u)$

83. $(x - y)(x^2 + xy + y^2)$

84. $(x + y)(x^2 - xy + y^2)$

85. $(3y + 1)(2y^2 + 3y + 2)$

86. $(a + 2)(3a^2 + 4a - 2)$

87. $(2a - b)(4a^2 + 2ab + b^2)$

88. $(x - 3y)(x^2 + 3xy + 9y^2)$

89. $(a + b)(a - b)(a - 3b)$

90. $(x - y)(x + 2y)(x - 2y)$

91. $(a + b + c)(2a - b - 2c)$

92. $(x + 2y + 3z)^2$

93. $(r + s)^2(r - s)^2$

94. $r(r + s)(r - s)^2$

Simplify each expression.

95. $3x(2x + 4) - 3x^2$

96. $2y - 3y(y^2 + 4)$

97. $3pq - p(p - q)$

98. $-4rs(r - 2) + 4rs$

99. $(x + 3)(x - 3) + (2x - 1)(x + 2)$

100. $(2b + 3)(b - 1) - (b + 2)(3b - 1)$

101. $(3x - 4)^2 - (2x + 3)^2$

102. $(3y + 1)^2 + (2y - 4)^2$

Use a calculator to help find each product.

103. $(3.21x - 7.85)(2.87x + 4.59)$

104. $(7.44y + 56.7)(-2.1y - 67.3)$

105. $(-17.3y + 4.35)^2$

106. $(-0.31x + 29.3)(-0.31x - 29.3)$

APPLICATIONS

107. THE YELLOW PAGES Refer to the illustration below.

   a. Describe the area occupied by the ads for movers by using a product of two binomials.

   b. Describe the area occupied by the ad for Budget Moving Co. by using a product. Then perform the multiplication.

   c. Describe the area occupied by the ad for Snyder Movers by using a product. Then perform the multiplication.

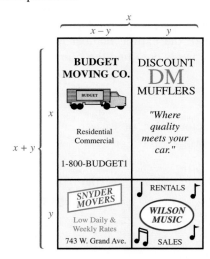

   d. Explain why your answer to part a is equal to the sum of your answers to parts b and c. What special product does this exercise illustrate?

108. HELICOPTER PADS To determine the amount of fluorescent paint needed to paint the circular ring on the landing pad design shown in the illustration, painters must find its area. The area of the ring is given by the expression $\pi(R + r)(R - r)$.

   a. Find the product $\pi(R + r)(R - r)$.

   b. If $R = 25$ feet and $r = 20$ feet, find the area to be painted. Round to the nearest tenth.

   c. If a quart of fluorescent paint covers 65 ft^2, how many quarts will be needed to paint the ring?

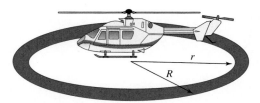

109. GIFT BOXES The corners of a 12-in.-by-12-in. piece of cardboard are creased, folded inward, and glued to make a gift box. Write a polynomial that gives the volume of the resulting box.

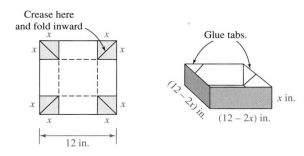

110. REVENUE A salesperson has found that the number x of televisions she can sell at a certain price p is related by the equation $x = -\frac{1}{5}p + 90$.

   a. Find the number of TVs she will sell if the price is $375.

   b. Write a formula for the revenue when x TVs are sold.

   c. Find the revenue generated by TV sales if they are priced at $400 each.

WRITING

111. Explain how to use the FOIL method.

112. Explain how you would multiply two trinomials.

113. On a test, when asked to find $(x - y)^2$, a student answered $x^2 - y^2$. What error did the student make?

114. Explain how the distributive property is used to find the following product: $2x^3(x^2 - 5x + 1)$.

REVIEW   *Graph each inequality or system of inequalities.*

115. $2x + y \leq 2$

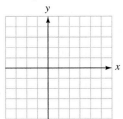

116. $x \geq 2$

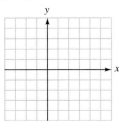

117. $\begin{cases} y - 2 < 3x \\ y + 2x < 3 \end{cases}$

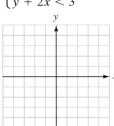

118. $\begin{cases} y < 0 \\ x < 0 \end{cases}$

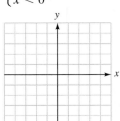

5.5 The Greatest Common Factor and Factoring by Grouping

- The prime-factored form of a natural number   • Factoring out the greatest common factor
- Factoring by grouping   • Formulas

We have discussed how to multiply polynomials. We will now reverse the operation of multiplication and show how to find the factors of a known product. The process of finding the individual factors of a known product is called **factoring.**

The prime-factored form of a natural number

If one number a divides a second number b, then a is called a **factor** of b. For example, because 3 divides 24, it is a factor of 24. Each number in the following list is a factor of 24, because each number divides 24.

1, 2, 3, 4, 6, 8, 12, and 24

To factor a natural number means to write it as a product of other natural numbers. If each factor is a prime number, the natural number is said to be written in **prime-factored form.** Example 1 shows how to find the prime-factored forms of 60, 84, and 180, respectively.

INTERMEDIATE
Algebra *f(x)* **Now™**

Self Check 1
Find the prime factorization of 120.

EXAMPLE 1   Find the prime factorization of each number: **a.** 60,   **b.** 84, and   **c.** 180.

Solution

a. $60 = 6 \cdot 10$
$= 2 \cdot 3 \cdot 2 \cdot 5$
$= 2^2 \cdot 3 \cdot 5$

b. $84 = 4 \cdot 21$
$= 2 \cdot 2 \cdot 3 \cdot 7$
$= 2^2 \cdot 3 \cdot 7$

c. $180 = 10 \cdot 18$
$= 2 \cdot 5 \cdot 3 \cdot 6$
$= 2 \cdot 5 \cdot 3 \cdot 3 \cdot 2$
$= 2^2 \cdot 3^2 \cdot 5$

Answer $2^3 \cdot 3 \cdot 5$

The largest natural number that divides 60, 84, and 180 is called the **greatest common factor (GCF)** of the numbers. Because 60, 84, and 180 all have two factors of 2 and one factor of 3, the GCF of these three numbers is $2^2 \cdot 3 = 12$. We note that

$$\frac{60}{12} = 5, \qquad \frac{84}{12} = 7, \qquad \text{and} \qquad \frac{180}{12} = 15$$

There is no natural number greater than 12 that divides 60, 84, and 180.

Algebraic monomials can also have greatest common factors.

EXAMPLE 2   Find the GCF of $6a^2b^3c$, $9a^3b^2c$, and $18a^4c^3$.

Solution   We begin by factoring each monomial.

$$6a^2b^3c = \mathbf{3 \cdot 2} \cdot \mathbf{a \cdot a} \cdot b \cdot b \cdot b \cdot \mathbf{c}$$
$$9a^3b^2c = \mathbf{3} \cdot 3 \cdot \mathbf{a \cdot a} \cdot a \cdot b \cdot b \cdot \mathbf{c}$$
$$18a^4c^3 = 2 \cdot \mathbf{3} \cdot 3 \cdot \mathbf{a \cdot a} \cdot a \cdot a \cdot \mathbf{c} \cdot c \cdot c$$

Since each monomial has one factor of 3, two factors of a, and one factor of c in common, the GCF is

$$3^1 \cdot a^2 \cdot c^1 = 3a^2c$$

INTERMEDIATE
Algebra $f(x)$ **Now** ™
Self Check 2
Find the GCF of $24x^2y^3$, $3x^3y$, and $18x^2y^2$.

Answer $3x^2y$

To find the GCF of several monomials, we follow these steps.

Steps for finding the GCF

1. Find the prime-factored form of each monomial.
2. Identify the prime factors that are common to each monomial.
3. Find the product of the factors found in Step 2, with each factor raised to the smallest power that occurs in any one monomial.

▮ Factoring out the greatest common factor

We have seen that the distributive property provides a method for multiplying a polynomial by a monomial. For example,

$$2x^3y^3(3x^2 - 4y^3) = \mathbf{2x^3y^3} \cdot 3x^2 - \mathbf{2x^3y^3} \cdot 4y^3$$
$$= 6x^5y^3 - 8x^3y^6$$

If the product of a multiplication is $6x^5y^3 - 8x^3y^6$, we can use the distributive property in reverse to find the individual factors.

$$6x^5y^3 - 8x^3y^6 = \mathbf{2x^3y^3} \cdot 3x^2 - \mathbf{2x^3y^3} \cdot 4y^3$$
$$= \mathbf{2x^3y^3}(3x^2 - 4y^3)$$

Since $2x^3y^3$ is the GCF of the terms of $6x^5y^3 - 8x^3y^6$, this process is called **factoring out the greatest common factor.**

EXAMPLE 3   Factor: $25a^3b + 15ab^3$.

Solution   We begin by factoring each monomial:

$$25a^3b = \mathbf{5} \cdot 5 \cdot \mathbf{a} \cdot a \cdot a \cdot \mathbf{b}$$
$$15ab^3 = \mathbf{5} \cdot 3 \cdot \mathbf{a} \cdot \mathbf{b} \cdot b \cdot b$$

INTERMEDIATE
Algebra $f(x)$ **Now** ™
Self Check 3
Factor: $9x^4y^2 - 12x^3y^3$.

Since each term has one factor of 5, one factor of a, and one factor of b in common, and there are no other common factors, $5ab$ is the GCF of the two terms. We can use the distributive property to factor it out.

$$25a^3b + 15ab^3 = \mathbf{5ab} \cdot 5a^2 + \mathbf{5ab} \cdot 3b^2$$
$$= \mathbf{5ab}(5a^2 + 3b^2)$$

Answer $3x^3y^2(3x - 4y)$

INTERMEDIATE
Algebra $f(x)$ **Now**™

Self Check 4
Factor: $2a^4b^2 + 6a^3b^2 - 4a^2b$.

EXAMPLE 4 Factor: $3xy^2z^3 + 6xyz^3 - 3xz^2$.

Solution We begin by factoring each monomial:

$$3xy^2z^3 = \mathbf{3} \cdot x \cdot y \cdot y \cdot z \cdot z \cdot z$$
$$6xyz^3 = \mathbf{3} \cdot 2 \cdot x \cdot y \cdot z \cdot z \cdot z$$
$$-3xz^2 = -\mathbf{3} \cdot x \cdot z \cdot z$$

Since each term has one factor of 3, one factor of x, and two factors of z in common, and because there are no other common factors, $3xz^2$ is the GCF of the three terms. We can use the distributive property to factor it out.

$$3xy^2z^3 + 6xyz^3 - 3xz^2 = \mathbf{3xz^2} \cdot y^2z + \mathbf{3xz^2} \cdot 2yz - \mathbf{3xz^2} \cdot 1$$
$$= \mathbf{3xz^2}(y^2z + 2yz - 1)$$

Answer $2a^2b(a^2b + 3ab - 2)$

! **COMMENT** In Example 4, when the $3xz^2$ is factored out from the third term, remember to write the -1.

A polynomial that cannot be factored is called a **prime polynomial** or an **irreducible polynomial.**

INTERMEDIATE
Algebra $f(x)$ **Now**™

Self Check 5
Factor: $6a^3 + 7b^2 + 5$.

EXAMPLE 5 Factor $3x^2 + 4y + 7$, if possible.

Solution We factor each monomial:

$$3x^2 = 3 \cdot x \cdot x \qquad 4y = 2 \cdot 2 \cdot y \qquad 7 = 7$$

Since there are no common factors other than 1, this polynomial cannot be factored. It is a prime polynomial.

Answer a prime polynomial

INTERMEDIATE
Algebra $f(x)$ **Now**™

Self Check 6
Factor out the opposite of the GCF from $-8a^2b^2 - 12ab^3$.

EXAMPLE 6 Factor the opposite of the GCF from $-6u^2v^3 + 8u^3v^2$.

Solution Because the GCF of the two terms is $2u^2v^2$, the opposite of the GCF is $-2u^2v^2$. To factor out $-2u^2v^2$, we proceed as follows:

$$-6u^2v^3 + 8u^3v^2 = -2u^2v^2 \cdot 3v + 2u^2v^2 \cdot 4u$$
$$= -2u^2v^2 \cdot 3v - (-2u^2v^2)4u$$
$$= -2u^2v^2(3v - 4u)$$

Answer $-4ab^2(2a + 3b)$

INTERMEDIATE
Algebra $f(x)$ **Now**™

EXAMPLE 7 Factor: $a(x - y + z) - b(x - y + z) + 3(x - y + z)$.

Solution We can factor out the GCF of the three terms, which is $(x - y + z)$.

$$a(x - y + z) - b(x - y + z) + 3(x - y + z)$$
$$= (x - y + z)a - (x - y + z)b + (x - y + z)3$$
$$= (x - y + z)(a - b + 3)$$

Self Check 7
Factor:
$x(a + b - c) - y(a + b - c)$.

Answer $(a + b - c)(x - y)$

▊ Factoring by grouping

Suppose that we wish to factor

$$ac + ad + bc + bd$$

Although there is no factor common to all four terms, there is a common factor of a in the first two terms and a common factor of b in the last two terms. We can factor out these common factors to get

$$ac + ad + bc + bd = a(c + d) + b(c + d)$$

We can now factor out the common factor of $c + d$ on the right-hand side:

$$ac + ad + bc + bd = (c + d)(a + b)$$

The grouping in this type of problem is not always unique. For example, if we write the expression $ac + ad + bc + bd$ in the form

$$ac + bc + ad + bd$$

and factor c from the first two terms and d from the last two terms, we obtain

$$ac + bc + ad + bd = c(a + b) + d(a + b)$$
$$= (a + b)(c + d) \qquad \text{This is equivalent to } (c + d)(a + b).$$

The method used in the previous examples is called **factoring by grouping.**

Factoring by Grouping

1. Group the terms of the polynomial so that each group has a common factor.

2. Factor out the common factor from each group.

3. Factor out the resulting common factor. If there is no common factor, regroup the terms of the polynomial and repeat steps 2 and 3.

INTERMEDIATE
Algebra $f(x)$ **Now**™

EXAMPLE 8 Factor: $3ax^2 + 3bx^2 + a + 5bx + 5ax + b$.

Solution Although there is no factor common to all six terms, $3x^2$ can be factored out of the first two terms, and $5x$ can be factored out of the fourth and fifth terms to get

$$3ax^2 + 3bx^2 + a + 5bx + 5ax + b = 3x^2(a + b) + a + 5x(b + a) + b$$

This result can be written in the form

$$3ax^2 + 3bx^2 + a + 5bx + 5ax + b = 3x^2(a + b) + 5x(a + b) + 1(a + b)$$

Since $a + b$ is common to all three terms, it can be factored out to get

$$3ax^2 + 3bx^2 + a + 5bx + 5ax + b = (a + b)(3x^2 + 5x + 1)$$

Self Check 8
Factor:
$2x^3 + 3x^2 + x + 2x^2y + 3xy + y$.

Answer $(x + y)(2x^2 + 3x + 1)$

A polynomial is **factored completely** when no factor can be factored further. To factor an expression completely, it is often necessary to factor more than once, as the following example illustrates.

Self Check 9
Factor:
$3a^3b + 3a^2b - 2a^2b^2 - 2ab^2$.

EXAMPLE 9   Factor: $3x^3y - 4x^2y^2 - 6x^2y + 8xy^2$.

Solution   We begin by factoring out the common factor xy.

$$3x^3y - 4x^2y^2 - 6x^2y + 8xy^2 = xy(3x^2 - 4xy - 6x + 8y)$$

We can now factor $3x^2 - 4xy - 6x + 8y$ by grouping:

$$3x^3y - 4x^2y^2 - 6x^2y + 8xy^2$$
$$= xy(3x^2 - 4xy - 6x + 8y)$$
$$= xy[x(3x - 4y) - 2(3x - 4y)] \quad \text{Factor } x \text{ from } 3x^2 - 4xy \text{ and } -2 \text{ from}$$
$$\qquad\qquad\qquad\qquad\qquad\qquad\qquad -6x + 8y.$$
$$= xy(3x - 4y)(x - 2) \qquad \text{Factor out } 3x - 4y.$$

Answer $ab(3a - 2b)(a + 1)$

Because no more factoring can be done, the factorization is complete.

> **! COMMENT** Whenever you factor an expression, always factor it completely. Each factor of a completely factored expression will be prime.

■ Formulas

Factoring is often required to solve a **literal equation** (an equation containing more than one variable) for one of its variables.

Self Check 10
Solve $A = p + prt$ for p.

EXAMPLE 10   Electronics.   The formula $r_1r_2 = rr_2 + rr_1$ is used in electronics to relate the combined resistance r of two resistors wired in parallel. The variable r_1 represents the resistance of the first resistor, and the variable r_2 represents the resistance of the second. Solve for r_2.

Solution   To isolate r_2 on one side of the equation, we get all terms involving r_2 on the left-hand side and all terms not involving r_2 on the right-hand side. We proceed as follows:

$$r_1r_2 = rr_2 + rr_1$$
$$r_1r_2 - rr_2 = rr_1 \qquad \text{Subtract } rr_2 \text{ from both sides.}$$
$$r_2(r_1 - r) = rr_1 \qquad \text{Factor out } r_2 \text{ on the left-hand side.}$$
$$r_2 = \frac{rr_1}{r_1 - r} \qquad \text{Divide both sides by } r_1 - r.$$

Answer $p = \dfrac{A}{1 + rt}$

Section 5.5 STUDY SET

■ **VOCABULARY**   *Fill in the blanks.*

1. When we write as $2x + 4$ as $2(x + 2)$, we say that we have _____ $2x + 4$.

2. When we write 100 as $2^2 \cdot 5^2$, we say that we have written 100 in _____ form.

3. Because 5 divides 20, we say that 5 is a ____ of 20.

4. The polynomial $2x^2y^3 - 4xy^2 + 6xy$ has three _____.

5. The abbreviation GCF stands for _____
_____ _____.

6. If a polynomial cannot be factored, it is called a
_____ polynomial or an irreducible polynomial.

CONCEPTS

7. The prime factorizations of three monomials are
shown here. Find their GCF.

$$2 \cdot 2 \cdot 3 \cdot x \cdot x \cdot y \cdot y \cdot y$$
$$2 \cdot 3 \cdot 3 \cdot x \cdot y \cdot y \cdot y \cdot y$$
$$2 \cdot 3 \cdot 3 \cdot 7 \cdot x \cdot x \cdot x \cdot y \cdot y$$

8. a. What property is illustrated below?
$$4a^2b(2ab^3 - 3a^2b^4) = 4a^2b \cdot 2ab^3 - 4a^2b \cdot 3a^2b^4$$

b. Explain how we use the distributive property in
reverse to factor $8a^3b^4 - 12a^4b^5$.

9. Explain why each factorization of $30t^2 - 20t^3$ is not
complete.

a. $5t^2(6 - 4t)$

b. $10t(3t - 2t^2)$

10. a. Factor $-5y^3 - 10y^2 + 15y$ by factoring out the
positive GCF.

b. Factor $-5y^3 - 10y^2 + 15y$ by factoring out the
opposite GCF.

NOTATION   *Complete each factorization.*

11. $3a - 12 = 3(a - \boxed{})$

12. $8z^3 + 4z^2 + 2z = 2z(4z^2 + 2z + \boxed{})$

13. $x^3 - x^2 + 2x - 2 = \boxed{}(x - 1) + \boxed{}(x - 1)$
$$= (x - 1)(x^2 + 2)$$

14. $-24a^3b^2 + 12ab^2 = -12ab^2(2a^2 \boxed{} 1)$

PRACTICE   *Find the prime-factored form of each number.*

15. 6   **16.** 10

17. 135   **18.** 98

19. 128   **20.** 357

21. 325   **22.** 288

Find the GCF of each set of monomials.

23. 36, 48   **24.** 45, 75

25. 42, 36, 98   **26.** 16, 40, 60

27. $4a^2b, 8a^3c$ **28.** $6x^3y^2z, 9xyz^2$

29. $18x^4y^3z^2, -12xy^2z^3$

30. $6x^2y^3, 24xy^3, 40x^2y^2z^3$

Factor each polynomial, if possible.

31. $2x + 8$ **32.** $3y - 9$

33. $2x^2 - 6x$ **34.** $3y^3 + 3y^2$

35. $5xy + 12ab^2$ **36.** $7x^2 + 14x$

37. $15x^2y - 10x^2y^2$

38. $11m^3n^2 - 12x^2y$ **39.** $14r^2s^3 + 15t^6$

40. $13ab^2c^3 - 26a^3b^2c$

41. $27z^3 + 12z^2 + 3z$

42. $25t^6 - 10t^3 + 5t^2$

43. $45x^{10}y^3 - 63x^7y^7 + 81x^{10}y^{10}$

44. $48u^6v^6 - 16u^4v^4 - 3u^6v^3$

Factor out the opposite of the greatest common factor.

45. $-3a - 6$ **46.** $-6b + 12$

47. $-3x^2 - x$ **48.** $-4a^3 + a^2$

49. $-6x^2 - 3xy$

50. $-15y^3 + 25y^2$

51. $-18a^2b - 12ab^2$

52. $-21t^5 + 28t^3$

53. $-63u^3v^6z^9 + 28u^2v^7z^2 - 21u^3v^3z^4$

54. $-56x^4y^3z^2 - 72x^3y^4z^5 + 80xy^2z^3$

Factor each expression.

55. $4(x + y) + t(x + y)$

56. $5(a - b) - t(a - b)$

57. $(a - b)r - (a - b)s$

58. $(x + y)u + (x + y)v$

59. $3(m + n + p) + x(m + n + p)$

60. $x(x - y - z) + y(x - y - z)$

61. $(u + v)^2 - (u + v)$

62. $a(x - y) - (x - y)^2$

63. $-a(x + y) + b(x + y)$

64. $-bx(a - b) - cx(a - b)$

65. $ax + bx + ay + by$

66. $ar - br + as - bs$

67. $x^2 + yx + 2x + 2y$

68. $2c + 2d - cd - d^2$

69. $3c - cd + 3d - c^2$

70. $x^2 + 4y - xy - 4x$

71. $a^2 - 4b + ab - 4a$

72. $7u + v^2 - 7v - uv$

73. $ax + bx - a - b$

74. $x^2y - ax - xy + a$

75. $x^2 + xy + xz + xy + y^2 + zy$

76. $ab - b^2 - bc + ac - bc - c^2$

77. $mpx + mqx + npx + nqx$

78. $abd - abe + acd - ace$

79. $x^2y + xy^2 + 2xyz + xy^2 + y^3 + 2y^2z$

80. $a^3 - 2a^2b + a^2c - a^2b + 2ab^2 - abc$

81. $2n^4p - 2n^2 - n^3p^2 + np + 2mn^3p - 2mn$

82. $a^2c^3 + ac^2 + a^3c^2 - 2a^2bc^2 - 2bc^2 + c^3$

Solve for the indicated variable.

83. $r_1r_2 = rr_2 + rr_1$ for r_1

84. $r_1r_2 = rr_2 + rr_1$ for r

85. $d_1d_2 = fd_2 + fd_1$ for f

86. $d_1d_2 = fd_2 + fd_1$ for d_1

87. $b^2x^2 + a^2y^2 = a^2b^2$ for a^2

88. $b^2x^2 + a^2y^2 = a^2b^2$ for b^2

89. $S(1 - r) = a - lr$ for r

90. $Sn = (n - 2)180°$ for n

APPLICATIONS

91. GEOMETRIC FORMULAS

   a. Write an expression that gives the area of the portion of the figure below that is shaded red.

   b. Do the same for the portion of the figure that is shaded blue.

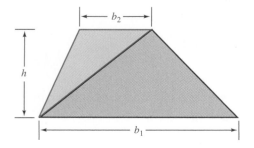

c. Add the results from parts a and b and then factor that expression. What important formula from geometry do you obtain?

92. PACKAGING The amount of cardboard needed to make the cereal box shown below can be found by computing the area A, which is given by the formula

$$A = 2wh + 4wl + 2lh$$

where w is the width, h the height, and l the length. Solve the equation for the width.

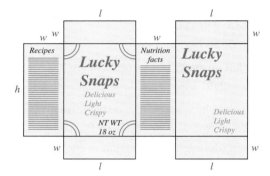

93. LANDSCAPING The combined area of the portions of the square lot that the sprinkler doesn't reach is given by $4r^2 - \pi r^2$, where r is the radius of the circular spray. Factor this expression.

94. CRAYONS The amount of colored wax used to make the crayon shown below can be found by computing its volume using the formula

$$V = \pi r^2 h_1 + \frac{1}{3}\pi r^2 h_2$$

Factor the expression on the right-hand side of this equation.

WRITING

95. One student commented, "Factoring undoes the distributive property." What do you think she meant?

96. Explain how to find the greatest common factor of two natural numbers.

97. Explain what is wrong with the following factorization.

$$5x^2 + x - 2 = 5(x^2 + x - 2)$$

98. What is a prime polynomial?

▮ REVIEW

99. What figure results when the function $f(x) = 3x + 1$ is graphed?

100. What figure results when the function $g(x) = x^2$ is graphed?

101. Solve the inequality $-x > 3$. Express the solution set using interval notation.

102. Evaluate: $2|-25 - (-6)(3)|$.

103. If two different lines are parallel, what can be said about their slopes?

104. Are $-3t^2$ and $12t^2$ like terms? If so, combine them.

5.6 The Difference of Two Squares; the Sum and Difference of Two Cubes

- Perfect squares • The difference of two squares • Perfect cubes
- The sum and difference of two cubes

In this section, we will discuss rules of factoring that apply to polynomials that can be written as the difference of two squares or as the sum or difference of two cubes. To use these methods, we must be able to recognize such polynomials. We begin with a discussion that will help you recognize polynomials with terms that are *perfect squares.*

▮ Perfect squares

To factor the difference of two squares, it is helpful to know the first 20 integers that are **perfect squares.**

1, 4, 9, 16, 25, 36, 49, 64, 81, 100, 121, 144, 169, 196, 225, 256, 289, 324, 361, 400

Expressions such as $x^6y^4z^2$ are also perfect squares, because they can be written as the square of another quantity:

$$x^6y^4z^2 = (x^3y^2z)^2$$

▮ The difference of two squares

In Section 5.4, we developed the special product formula

(1) $(x + y)(x - y) = x^2 - y^2$

The binomial $x^2 - y^2$ is called the **difference of two squares,** because x^2 represents the square of x, y^2 represents the square of y, and $x^2 - y^2$ represents the difference of these squares.

Equation 1 can be written in reverse order to give a formula for factoring the difference of two squares.

> **Factoring the difference of two squares**
>
> $x^2 - y^2 = (x + y)(x - y)$

If we think of the difference of two squares as the square of a **First** quantity minus the square of a **Last** quantity, we have the formula

$$F^2 - L^2 = (F + L)(F - L)$$

and we say: *To factor the square of a **First** quantity minus the square of a **Last** quantity, we multiply the **First** plus the **Last** by the **First** minus the **Last**.*

INTERMEDIATE
Algebra *(f(x))* **Now**™

Self Check 1
Factor: $81p^2 - 25$.

EXAMPLE 1   Factor: $49x^2 - 16$.

Solution   We begin by rewriting the binomial $49x^2 - 16$ as a difference of two squares: $(7x)^2 - (4)^2$. Then we use the formula for factoring the difference of two squares:

$$
\begin{array}{cccccc}
F^2 & - & L^2 & = & (F + L) & (F - L) \\
\downarrow & & \downarrow & & \downarrow\ \ \ \downarrow & \downarrow\ \ \ \downarrow \\
(7x)^2 & - & 4^2 & = & (7x + 4) & (7x - 4)
\end{array}
$$

We can verify this result using the FOIL method to perform the multiplication.

$$
\begin{aligned}
(7x + 4)(7x - 4) &= 49x^2 - 28x + 28x - 16 \\
&= 49x^2 - 16
\end{aligned}
$$

Answer $(9p + 5)(9p - 5)$

> **! COMMENT**   Many expressions that represent the sum of two squares, such as $(7x)^2 + 4^2$, cannot be factored in the real number system. The binomial $49x^2 + 16$ is a prime binomial.

INTERMEDIATE
Algebra *(f(x))* **Now**™

Self Check 2
Factor: $36r^4 - s^2$.

EXAMPLE 2   Factor: $64a^4 - 25b^2$.

Solution   We can write $64a^4 - 25b^2$ in the form $(8a^2)^2 - (5b)^2$ and use the formula for factoring the difference of two squares.

$$
\begin{array}{cccccc}
F^2 & - & L^2 & = & (F + L) & (F - L) \\
\downarrow & & \downarrow & & \downarrow\ \ \ \downarrow & \downarrow\ \ \ \downarrow \\
(8a^2)^2 & - & (5b)^2 & = & (8a^2 + 5b) & (8a^2 - 5b)
\end{array}
$$

Answer $(6r^2 + s)(6r^2 - s)$

Verify by multiplication.

Self Check 3
Factor: $a^4 - 81$.

EXAMPLE 3   Factor: $x^4 - 1$.

Solution   Because the binomial is the difference of the squares of x^2 and 1, it factors into the sum of x^2 and 1 and the difference of x^2 and 1.

$$
\begin{aligned}
x^4 - 1 &= (x^2)^2 - (1)^2 \\
&= (x^2 + 1)(x^2 - 1)
\end{aligned}
$$

The factor $x^2 + 1$ is the sum of two quantities and is prime. However, the factor $x^2 - 1$ is the difference of two squares and can be factored as $(x + 1)(x - 1)$. Thus,

$$
\begin{aligned}
x^4 - 1 &= (x^2 + 1)(x^2 - 1) \\
&= (x^2 + 1)(x + 1)(x - 1)
\end{aligned}
$$

Answer $(a^2 + 9)(a + 3)(a - 3)$

COMMENT When asked to factor a polynomial, we must be sure to factor it completely. After factoring a polynomial, always check to see whether any of the factors in the result can be factored further.

EXAMPLE 4 Factor: $(x + y)^4 - z^4$.

Solution This expression is the difference of two squares and can be factored:

$$(x + y)^4 - z^4 = [(x + y)^2]^2 - (z^2)^2$$
$$= [(x + y)^2 + z^2][(x + y)^2 - z^2]$$

The factor $(x + y)^2 + z^2$ is the sum of two squares and is prime. However, the factor $(x + y)^2 - z^2$ is the difference of two squares and can be factored as $(x + y + z)(x + y - z)$. Thus,

$$(x + y)^4 - z^4 = [(x + y)^2 + z^2][(x + y)^2 - z^2]$$
$$= [(x + y)^2 + z^2](x + y + z)(x + y - z)$$

Self Check 4
Factor: $(a - b)^4 - c^4$.

Answer $[(a - b)^2 + c^2](a - b + c)(a - b - c)$

When possible, we always factor out a common factor before factoring the difference of two squares. The factoring process is easier when all common factors are factored out first.

EXAMPLE 5 Factor: $2x^4y - 32y$.

Solution
$$2x^4y - 32y = 2y(x^4 - 16) \qquad \text{Factor out the GCF, which is } 2y.$$
$$= 2y(x^2 + 4)(x^2 - 4) \qquad \text{Factor } x^4 - 16.$$
$$= 2y(x^2 + 4)(x + 2)(x - 2) \qquad \text{Factor } x^2 - 4.$$

Self Check 5
Factor: $3a^4 - 3$.

Answer $3(a^2 + 1)(a + 1)(a - 1)$

EXAMPLE 6 Factor: $x^2 - y^2 + x - y$.

Solution If we group the first two terms and factor the difference of two squares, we have

$$x^2 - y^2 + x - y = (x + y)(x - y) + (x - y) \qquad \text{Factor } x^2 - y^2.$$
$$= (x - y)(x + y + 1) \qquad \text{Factor out } x - y.$$

Self Check 6
Factor: $a^2 - b^2 + a + b$.

Answer $(a + b)(a - b + 1)$

▉ Perfect cubes

The number 125 is called a perfect cube, because $5^3 = 125$. To factor the sum or difference of two cubes, it is helpful to know the first ten **perfect cubes:**

1, 8, 27, 64, 125, 216, 343, 512, 729, 1,000

Expressions such as $x^9y^6z^3$ are also perfect cubes, because they can be written as the cube of another quantity:

$$x^9y^6z^3 = (x^3y^2z)^3$$

■ The sum and difference of two cubes

To find formulas for factoring the sum or difference of two cubes, we use the following product formulas:

(2) $(x + y)(x^2 - xy + y^2) = x^3 + y^3$

(3) $(x - y)(x^2 + xy + y^2) = x^3 - y^3$

To verify Equation 2, we multiply $x^2 - xy + y^2$ by $x + y$.

$$(x + y)(x^2 - xy + y^2) = (x + y)x^2 - (x + y)xy + (x + y)y^2$$
$$= x \cdot x^2 + y \cdot x^2 - x \cdot xy - y \cdot xy + x \cdot y^2 + y \cdot y^2$$
$$= x^3 + x^2y - x^2y - xy^2 + xy^2 + y^3$$
$$= x^3 + y^3$$

Equation 3 can also be verified by multiplication.

   If we write Equations 2 and 3 in reverse order, we have the formulas for factoring the sum and difference of two cubes.

Factoring the sum and difference of two cubes

$$x^3 + y^3 = (x + y)(x^2 - xy + y^2)$$
$$x^3 - y^3 = (x - y)(x^2 + xy + y^2)$$

   If we think of the sum of two cubes as the sum of the cube of a **First** quantity plus the cube of a **Last** quantity, we have the formula

$$F^3 + L^3 = (F + L)(F^2 - FL + L^2)$$

*To factor the cube of a **First** quantity plus the cube of a **Last** quantity, we multiply the sum of the **First** and **Last** by*

- *the **First** squared*
- *minus the **First** times the **Last***
- *plus the **Last** squared.*

The formula for the difference of two cubes is

$$F^3 - L^3 = (F - L)(F^2 + FL + L^2)$$

*To factor the cube of a **First** quantity minus the cube of a **Last** quantity, we multiply the difference of the **First** and **Last** by*

- *the **First** squared*
- *plus the **First** times the **Last***
- *plus the **Last** squared.*

Self Check 7
Factor: $p^3 + 27$.

EXAMPLE 7   Factor: $a^3 + 8$.

Solution   Since $a^3 + 8$ can be written as $a^3 + 2^3$, we have the sum of two cubes, which factors as follows:

$$\mathbf{F^3 + L^3 = (F + L)(F^2 - FL + L^2)}$$
$$\downarrow \quad \downarrow \qquad \downarrow \quad \downarrow \ \downarrow \quad \downarrow\downarrow \quad \downarrow$$
$$a^3 + 2^3 = (a + 2)(a^2 - a\,2 + 2^2)$$
$$= (a + 2)(a^2 - 2a + 4)$$

Answer $(p + 3)(p^2 - 3p + 9)$

Thus, $a^3 + 8 = (a + 2)(a^2 - 2a + 4)$. Check by multiplication.

INTERMEDIATE
Algebra $f(x)$ Now™

EXAMPLE 8   Factor: $27a^3 - 64b^3$.

Self Check 8
Factor: $8p^3 - 125q^3$.

Solution   Since $27a^3 - 64b^3$ can be written as $(3a)^3 - (4b)^3$, we have the difference of two cubes, which factors as follows:

$$\mathbf{F^3} - \mathbf{L^3} = (\mathbf{F} - \mathbf{L})\ (\mathbf{F^2} + \mathbf{F}\ \mathbf{L} + \mathbf{L^2})$$
$$\downarrow \qquad \downarrow \qquad \downarrow \quad \downarrow \quad \downarrow \qquad \downarrow \quad \downarrow \qquad \downarrow$$
$$(3a)^3 - (4b)^3 = (3a - 4b)[(3a)^2 + (3a)(4b) + (4b)^2]$$
$$= (3a - 4b)(9a^2 + 12ab + 16b^2)$$

Thus, $27a^3 - 64b^3 = (3a - 4b)(9a^2 + 12ab + 16b^2)$. Check by multiplication.

Answer
$(2p - 5q)(4p^2 + 10pq + 25q^2)$

INTERMEDIATE
Algebra $f(x)$ Now™

EXAMPLE 9   Factor: $a^3 - (c + d)^3$.

Self Check 9
Factor: $(p + q)^3 - r^3$.

Solution

$$a^3 - (c + d)^3 = [a - (c + d)][a^2 + a(c + d) + (c + d)^2]$$

Now we simplify the expressions inside both sets of brackets.

$$a^3 - (c + d)^3 = (a - c - d)(a^2 + ac + ad + c^2 + 2cd + d^2)$$

Answer $(p + q - r)(p^2 + 2pq + q^2 + pr + qr + r^2)$

INTERMEDIATE
Algebra $f(x)$ Now™

EXAMPLE 10   Factor: $x^6 - 64$.

Self Check 10
Factor: $x^6 - 1$.

Solution   This expression is both the difference of two squares and the difference of two cubes. It is easier to factor it as the difference of two squares first. This expression factors into the product of a sum and a difference.

$$x^6 - 64 = (x^3)^2 - 8^2$$
$$= (x^3 + 8)(x^3 - 8)$$

Each of these factors further, however, for one is the sum of two cubes and the other is the difference of two cubes:

$$x^6 - 64 = (x + 2)(x^2 - 2x + 4)(x - 2)(x^2 + 2x + 4)$$

Answer $(x + 1)(x^2 - x + 1)(x - 1)(x^2 + x + 1)$

INTERMEDIATE
Algebra $f(x)$ Now™

EXAMPLE 11   Factor: $2a^5 + 128a^2$.

Self Check 11
Factor: $3x^5 + 24x^2$.

Solution   We first factor out the common monomial factor $2a^2$ to obtain

$$2a^5 + 128a^2 = 2a^2(a^3 + 64)$$

Then we factor $a^3 + 64$ as the sum of two cubes to obtain

$$2a^5 + 128a^2 = 2a^2(a + 4)(a^2 - 4a + 16)$$

Answer $3x^2(x + 2)(x^2 - 2x + 4)$

Section 5.6 STUDY SET

VOCABULARY *Fill in the blanks.*

1. When the polynomial $4x^2 - 25$ is rewritten as $(2x)^2 - (5)^2$, we see that it is the difference of two _____.

2. When the polynomial $8x^3 + 125$ is rewritten as $(2x)^3 + (5)^3$, we see that it is the sum of two _____.

CONCEPTS

3. Write the first ten integers that are perfect squares.

4. Write the first ten perfect cubes.

5. Use multiplication to verify that the sum of two squares $x^2 + 25$ does not factor as $(x + 5)(x + 5)$.

6. Use multiplication to verify that the difference of two squares $x^2 - 25$ factors as $(x + 5)(x - 5)$.

7. Explain why each factorization is not complete.
   a. $4g^2 - 16 = (2g + 4)(2g - 4)$

   b. $1 - t^8 = (1 + t^4)(1 - t^4)$

8. When asked to factor $81t^2 - 16$, one student answered $(9t - 4)(9t + 4)$, and another answered $(9t + 4)(9t - 4)$. Explain why both students are correct.

9. Factor each polynomial.
   a. $5p^2 + 20$

   b. $5p^2 - 20$

10. Factor each polynomial.
    a. $5p^3 + 20$

    b. $5p^3 + 40$

NOTATION *Complete each factorization.*

11. $p^3 + q^3 = (p + q)()$

12. $p^3 - q^3 = (p - q)()$

13. $p^2 - q^2 = (p + q)()$

14. $p^2q + pq^2 = (p + q)$

15. $36y^2 - 49m^2 = ()^2 - (7m)^2$
    $ = (6y 7m)(6y -)$

16. $h^3 - 27k^3 = (h)^3 - ()^3$
    $ = (h 3k)(h^2 + + 9k^2)$

PRACTICE *Factor each polynomial, if possible.*

17. $x^2 - 4$

18. $y^2 - 9$

19. $9y^2 - 64$

20. $16x^4 - 81y^2$

21. $x^2 + 25$

22. $144a^2 - b^4$

23. $625a^2 - 169b^4$

24. $4y^2 + 9z^4$

25. $81a^4 - 49b^2$

26. $64r^6 - 121s^2$

27. $36x^4y^2 - 49z^4$

28. $4a^2b^4c^6 - 9d^8$

29. $(x + y)^2 - z^2$

30. $a^2 - (b - c)^2$

31. $(a - b)^2 - c^2$

32. $(m + n)^2 - p^4$

33. $x^4 - y^4$

34. $16a^4 - 81b^4$

35. $256x^4y^4 - z^8$

36. $255a^4 - 16b^8c^{12}$

37. $2x^2 - 288$

38. $8x^2 - 72$

39. $2x^3 - 32x$

40. $3x^3 - 243x$

41. $5x^3 - 125x$

42. $6x^4 - 216x^2$

43. $r^2s^2t^2 - t^2x^4y^2$

44. $16a^4b^3c^4 - 64a^2bc^6$

45. $a^2 - b^2 + a + b$

46. $x^2 - y^2 - x - y$

47. $a^2 - b^2 + 2a - 2b$

48. $m^2 - n^2 + 3m + 3n$

49. $2x + y + 4x^2 - y^2$

50. $m - 2n + m^2 - 4n^2$

51. $r^3 + s^3$

52. $t^3 - v^3$

53. $x^3 - 8y^3$

54. $27a^3 + b^3$

55. $64a^3 - 125b^6$

56. $8x^6 + 125y^3$

57. $125x^3y^6 + 216z^9$

58. $1{,}000a^6 - 343b^3c^6$

59. $x^6 + y^6$

60. $x^9 + y^9$

61. $5x^3 + 625$

62. $2x^3 - 128$

63. $4x^5 - 256x^2$

64. $2x^6 + 54x^3$

65. $128u^2v^3 - 2t^3u^2$

66. $56rs^2t^3 + 7rs^2v^6$

67. $(a + b)x^3 + 27(a + b)$

68. $(c - d)r^3 - (c - d)s^3$

APPLICATIONS

69. CANDY To find the amount of chocolate used in the outer coating of the malted-milk ball shown below, we can find the volume V of the chocolate shell using the formula

$$V = \frac{4}{3}\pi r_1{}^3 - \frac{4}{3}\pi r_2{}^3$$

Factor the expression on the right-hand side of the formula.

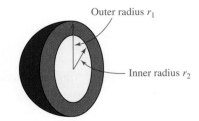

Outer radius r_1

Inner radius r_2

70. MOVIE STUNTS The function that gives the distance a stuntwoman is above the ground t seconds after she falls over the side of a 144-foot-tall building is

$$h(t) = 144 - 16t^2$$

Factor the right-hand side of the equation.

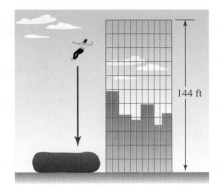

144 ft

WRITING

71. Describe the pattern used to factor the difference of two squares.

72. Describe the patterns used to factor the sum and the difference of two cubes.

REVIEW   *Graph the line with the given characteristics.*

73. Passing through $(-2, -1)$; slope $= -\dfrac{2}{3}$

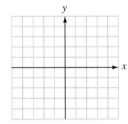

74. y-intercept $(0, -4)$; slope $= 3$

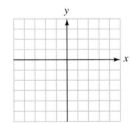

75. Horizontal; y-intercept $(0, -2)$

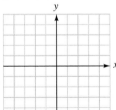

76. Parallel to the y-axis, passing through $(1, 4)$

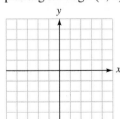

77. Write the equation of line l shown in the illustration below.

78. Write the equation of line r shown in the illustration to the right.

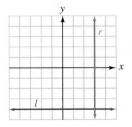

5.7 Factoring Trinomials

- Perfect square trinomials   • Factoring trinomials with lead coefficients of 1
- Factoring trinomials with lead coefficients other than 1   • Test for factorability
- Using substitution to factor trinomials   • Using grouping to factor trinomials

In this section, we will discuss several techniques for factoring trinomials. These techniques are based on the fact that the product of two binomials is often a trinomial. With that observation in mind, we begin the study of trinomial factoring by considering two special products.

Perfect square trinomials

Many trinomials can be factored by using the following special product formulas.

(1) $(x + y)(x + y) = x^2 + 2xy + y^2$

(2) $(x - y)(x - y) = x^2 - 2xy + y^2$

To factor $x^2 + 6x + 9$, we note that it can be written in the form $x^2 + 2(3)x + 3^2$. If $y = 3$, this form matches the right-hand side of Equation 1. Thus, $x^2 + 6x + 9$ factors as

$$x^2 + 6x + 9 = x^2 + 2(3)x + 3^2$$
$$= (x + 3)(x + 3)$$
$$= (x + 3)^2$$

Since $x^2 + 6x + 9$ is the square of $x + 3$, $x^2 + 6x + 9$, is called a **perfect square trinomial.** This result can be verified by multiplication:

$$(x + 3)(x + 3) = x^2 + 3x + 3x + 9$$
$$= x^2 + 6x + 9$$

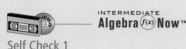
INTERMEDIATE
Algebra *f(x)* Now™

Self Check 1
Factor: $b^2 - 10b + 25$.

Answer $(b - 5)^2$

EXAMPLE 1   Factor: $x^2 - 4xz + 4z^2$.

Solution   To factor the perfect square trinomial $x^2 - 4xz + 4z^2$, we note that it can be written in the form $x^2 - 2x(2z) + (2z)^2$. If $y = 2z$, this form matches the right-hand side of Equation 2.

$$x^2 - 4xz + 4z^2 = x^2 - 2x(2z) + (2z)^2$$
$$= (x - 2z)(x - 2z)$$
$$= (x - 2z)^2$$

This result can be verified by multiplication.

We begin our discussion of *general trinomials* by considering trinomials with lead coefficients (the coefficient of the squared term) of 1.

Factoring trinomials with lead coefficients of 1

Since the product of two binomials is often a trinomial, we expect that many trinomials will factor as two binomials. For example, to factor $x^2 + 7x + 12$, we must find two binomials $x + a$ and $x + b$ such that

$$x^2 + 7x + 12 = (x + a)(x + b)$$

where $ab = 12$ and $ax + bx = 7x$.

To find the numbers a and b, we list the possible factorizations of 12 and find the one where the sum of the factors is 7.

This is the one to choose.
$$\downarrow$$

$$12(1) \quad 6(2) \quad 4(3) \quad -12(-1) \quad -6(-2) \quad -4(-3)$$

Thus, $a = 4$, $b = 3$, and

$$x^2 + 7x + 12 = (x + a)(x + b)$$

(3) $x^2 + 7x + 12 = (x + 4)(x + 3)$

This factorization can be verified by multiplying $x + 4$ and $x + 3$ and observing that the product is $x^2 + 7x + 12$.

Because of the commutative property of multiplication, the order of the factors in Equation 3 is not important.

To factor trinomials with lead coefficients of 1, we follow these steps.

Factoring trinomials with lead coefficients of 1

1. Write the trinomial in descending powers of one variable.
2. List the factorizations of the third term of the trinomial.
3. Pick the factorization where the sum of the factors is equal to the coefficient of the middle term.

INTERMEDIATE
Algebra $f(x)$ **Now**™
Self Check 2

EXAMPLE 2   Factor: $x^2 - 6x + 8$.

Factor: $a^2 - 7a + 12$.

Solution   Since the trinomial is written in descending powers of x, we can move to step 2 and list the possible factorizations of the third term of the trinomial, which is 8.

This is the one to choose.
$$\downarrow$$

$$8(1) \quad 4(2) \quad -8(-1) \quad -4(-2)$$

In the trinomial, the coefficient of the middle term is -6. The only factorization where the sum of the factors is -6 is $-4(-2)$. Thus, $a = -4$, $b = -2$, and

$$x^2 - 6x + 8 = (x + a)(x + b)$$
$$= (x - 4)(x - 2)$$

We can verify this result by multiplication:

$$(x - 4)(x - 2) = x^2 - 2x - 4x + 8 \quad \text{Use the FOIL method.}$$
$$= x^2 - 6x + 8$$

Answer $(a - 4)(a - 3)$

INTERMEDIATE
Algebra $f(x)$ **Now**™
Self Check 3

EXAMPLE 3   Factor: $-x + x^2 - 12$.

Factor: $-3a + a^2 - 10$.

Solution   We begin by writing the trinomial in descending powers of x:

$$-x + x^2 - 12 = x^2 - x - 12$$

The possible factorizations of the third term are

This is the one to choose.
$$\downarrow$$

$$12(-1) \quad 6(-2) \quad 4(-3) \quad 1(-12) \quad 2(-6) \quad 3(-4)$$

In the trinomial, the coefficient of the middle term is -1. The only factorization where the sum of the factors is -1 is $3(-4)$. Thus, $a = 3$, $b = -4$, and

$$-x + x^2 - 12 = (x + a)(x + b)$$
$$= (x + 3)(x - 4)$$

Answer $(a + 2)(a - 5)$

Self Check 4
Factor: $18a + 3ab - 3ab^2$.

EXAMPLE 4   Factor: $30x - 4xy - 2xy^2$.

Solution   We begin by writing the trinomial in descending powers of y:

$$30x - 4xy - 2xy^2 = -2xy^2 - 4xy + 30x$$

Each term in this trinomial has a common factor of $-2x$, which we will factor out.

$$30x - 4xy - 2xy^2 = -2x(y^2 + 2y - 15)$$

To factor $y^2 + 2y - 15$, we list the factors of -15 and find the pair whose sum is 2.

This is the one to choose.
↓

$$15(-1) \qquad 5(-3) \qquad 1(-15) \qquad 3(-5)$$

The only factorization where the sum of the factors is 2 (the coefficient of the middle term of $y^2 + 2y - 15$) is $5(-3)$. Thus, $a = 5$, $b = -3$, and

$$30x - 4xy - 2xy^2 = -2x(y^2 + 2y - 15)$$
$$= -2x(y + 5)(y - 3)$$

Answer $-3a(b + 2)(b - 3)$

! COMMENT In Example 4, be sure to include all factors in the final result. It is a common error to forget to write the $-2x$.

▮ Factoring trinomials with lead coefficients other than 1

There are more combinations of factors to consider when factoring trinomials with lead coefficients other than 1. To factor $5x^2 + 7x + 2$, for example, we must find two binomials of the form $ax + b$ and $cx + d$ such that

$$5x^2 + 7x + 2 = (ax + b)(cx + d)$$

Since the first term of the trinomial $5x^2 + 7x + 2$ is $5x^2$, the first terms of the binomial factors must be $5x$ and x.

$$5x^2$$
$$5x^2 + 7x + 2 = (5x + b)(x + d)$$

Since the product of the last terms must be 2, and the sum of the products of the outer and inner terms must be $7x$, we must find two numbers whose product is 2 that will give a middle term of $7x$.

$$2$$
$$5x^2 + 7x + 2 = (5x + b)(x + d)$$
$$O + I = 7x$$

Since $2(1)$ and $(-2)(-1)$ give a product of 2, there are four possible combinations to consider:

$(5x + 2)(x + 1)$ $(5x - 2)(x - 1)$

$(5x + 1)(x + 2)$ $(5x - 1)(x - 2)$

Of these possibilities, only the first one gives the correct middle term of $7x$. Thus,

(4) $5x^2 + 7x + 2 = (5x + 2)(x + 1)$

We can verify this result by multiplication:

$$(5x + 2)(x + 1) = 5x^2 + 5x + 2x + 2$$
$$= 5x^2 + 7x + 2$$

▮ Test for factorability

If a trinomial has the form $ax^2 + bx + c$, with integer coefficients and $a \neq 0$, we can test to see whether it is factorable.

- If the value of $b^2 - 4ac$ is a perfect square, the trinomial can be factored using only integers.
- If the value of $b^2 - 4ac$ is not a perfect square, the trinomial cannot be factored using only integers.

For example, $5x^2 + 7x + 2$ is a trinomial in the form $ax^2 + bx + c$ with

$a = 5,$ $b = 7,$ and $c = 2$

For this trinomial, the value of $b^2 - 4ac$ is

$$b^2 - 4ac = 7^2 - 4(5)(2)$$
$$= 49 - 40$$
$$= 9$$

Since 9 is a perfect square, the trinomial is factorable. Its factorization is shown in Equation 4.

> **Test for factorability**
>
> A trinomial of the form $ax^2 + bx + c$, with integer coefficients and $a \neq 0$, will factor into two binomials with integer coefficients if the value of $b^2 - 4ac$ is a perfect square. If $b^2 - 4ac = 0$, the factors will be the same.

INTERMEDIATE
Algebra $f(x)$ **Now**™

EXAMPLE 5   Factor: $3p^2 - 4p - 4$.

Solution   In the trinomial, $a = 3$, $b = -4$, and $c = -4$. To see whether it factors, we evaluate $b^2 - 4ac$.

$$b^2 - 4ac = (-4)^2 - 4(3)(-4)$$
$$= 16 + 48$$
$$= 64$$

Since 64 is a perfect square, the trinomial is factorable.

To factor the trinomial, we note that the first terms of the binomial factors must be $3p$ and p to give the first term of $3p^2$.

$$3p^2 - 4p - 4 = (3p + ?)(p + ?)$$

Self Check 5
Factor: $4q^2 - 9q - 9$.

The product of the last terms must be -4, and the sum of the products of the outer terms and the inner terms must be $-4p$.

$$3p^2 - 4p - 4 = (3p + ?)(p + ?)$$

$$O + I = -4p$$

Because $1(-4)$, $-1(4)$, and $-2(2)$ all give a product of -4, there are six possible combinations to consider:

$$(3p + 1)(p - 4) \qquad (3p - 4)(p + 1)$$
$$(3p - 1)(p + 4) \qquad (3p + 4)(p - 1)$$
$$(3p - 2)(p + 2) \qquad (3p + 2)(p - 2)$$

Of these possibilities, only the last gives the required middle term of $-4p$. Thus,

$$3p^2 - 4p - 4 = (3p + 2)(p - 2)$$

Answer $(4q + 3)(q - 3)$

Self Check 6
Factor $5a^2 - 8a + 2$, if possible.

Answer a prime polynomial

EXAMPLE 6   Factor $4t^2 - 3t - 5$, if possible.

Solution   In the trinomial, $a = 4$, $b = -3$, and $c = -5$. To see whether the trinomial is factorable, we evaluate $b^2 - 4ac$ by substituting the values of a, b, and c.

$$b^2 - 4ac = (-3)^2 - 4(4)(-5)$$
$$= 9 + 80$$
$$= 89$$

Since 89 is not a perfect square, the trinomial is not factorable using only integer coefficients.

It is not easy to give specific rules for factoring general trinomials. However, the following hints are helpful.

> **Factoring a general trinomial**
>
> 1. Write the trinomial in descending powers of one variable.
> 2. Factor out any greatest common factor (including -1, if that is necessary to make the coefficient of the first term positive).
> 3. Test the trinomial for factorability.
> 4. When the sign of the third term of the trinomial is $+$, the signs between the terms of each binomial factor are the same as the sign of the middle term of the trinomial.
>
>   When the sign of the third term of the trinomial is $-$, the signs between the terms of the binomials are opposite.
> 5. Try various combinations of the factors of the first terms and the last terms until you find the one that works.
> 6. Check the factorization by multiplication.

INTERMEDIATE
Algebra $f(x)$ **Now**™

EXAMPLE 7   Factor: $24y^2 + 10xy - 6x^2$.

Self Check 7
Factor: $-6x^2 - 15xy - 6y^2$.

Solution   We write the trinomial in descending powers of x and factor out -2.

$$24y^2 + 10xy - 6x^2 = -6x^2 + 10xy + 24y^2$$
$$= -2(3x^2 - 5xy - 12y^2)$$

In the trinomial $3x^2 - 5xy - 12y^2$, $a = 3$, $b = -5$, and $c = -12$.

$$b^2 - 4ac = (-5)^2 - 4(3)(-12)$$
$$= 25 + 144$$
$$= 169$$

Since 169 is a perfect square, the trinomial will factor.

The sign of the first term of $3x^2 - 5xy - 12y^2$ is positive and the sign of the third term is negative. Therefore, the signs between the binomial factors will be opposite. Because the first term is $3x^2$, the first terms of the binomial factors must be $3x$ and x.

$$3x^2$$

$$-2(3x^2 - 5xy - 12y^2) = -2(3x \qquad)(x \qquad)$$

The product of the last terms must be $-12y^2$, and the sum of the product of the outer terms and the product of the inner terms must be $-5xy$.

$$-12y^2$$

$$24y^2 + 10xy - 6x^2 = -2(3x \qquad ?y)(x \qquad ?y)$$

$$O + I = -5xy$$

Since $1(-12), 2(-6), 3(-4), 12(-1), 6(-2)$, and $4(-3)$ all give a product of -12, there are 12 possible combinations to consider.

$(3x + 1y)(x - 12y)$	$(3x - 12y)(x + 1y)$
$(3x + 2y)(x - 6y)$	$(3x - 6y)(x + 2y)$
$(3x + 3y)(x - 4y)$	$(3x - 4y)(x + 3y)$
$(3x + 12y)(x - 1y)$	$(3x - 1y)(x + 12y)$
$(3x + 6y)(x - 2y)$	$(3x - 2y)(x + 6y)$
This is the one to choose. → $(3x + 4y)(x - 3y)$	$(3x - 3y)(x + 4y)$

The combinations marked in color cannot work, because one of the binomial factors has a common factor. This implies that $3x^2 - 5xy - 12y^2$ would have a common factor, which it doesn't.

After mentally trying the remaining combinations, we find that only $(3x + 4y)(x - 3y)$ gives the proper middle term of $-5xy$.

$$24y^2 + 10xy - 6x^2 = -2(3x^2 - 5xy - 12y^2)$$
$$= -2(3x + 4y)(x - 3y)$$

Verify this result by multiplication.

Answer $-3(x + 2y)(2x + y)$

INTERMEDIATE
Algebra $f(x)$ **Now**™

EXAMPLE 8   Factor: $6y + 13x^2y + 6x^4y$.

Self Check 8
Factor: $4b + 11a^2b + 6a^4b$.

Solution   We write the trinomial in descending powers of x and factor out the common factor y to obtain

$$6y + 13x^2y + 6x^4y = 6x^4y + 13x^2y + 6y$$
$$= y(6x^4 + 13x^2 + 6)$$

A test for factorability will show that $6x^4 + 13x^2 + 6$ will factor.

Since the coefficients of the first and last terms of $6x^4 + 13x^2 + 6$ are positive, the signs between the terms in each binomial will be +.

Since the first term of the trinomial is $6x^4$, the first terms of the binomial factors must be either $2x^2$ and $3x^2$ or x^2 and $6x^2$.

Since the product of the last terms of the binomial factors must be 6, we must find two numbers whose product is 6 that will lead to a middle term of $7x^2$. After trying some combinations, we find the one that works.

$$6y + 13x^2y + 6x^4y = y(6x^4 + 13x^2 + 6)$$
$$= y(2x^2 + 3)(3x^2 + 2)$$

Answer $b(2a^2 + 1)(3a^2 + 4)$

Verify this result by multiplication.

Self Check 9
Factor: $a^2 + 4a + 4 - b^2$.

EXAMPLE 9   Factor: $x^2 + 6x + 9 - z^2$.

Solution   We group the first three terms together and factor the trinomial to get

$$x^2 + 6x + 9 - z^2 = (x + 3)(x + 3) - z^2$$
$$= (x + 3)^2 - z^2$$

We can now factor the difference of two squares to get

Answer $(a + 2 + b)(a + 2 - b)$

$$x^2 + 6x + 9 - z^2 = (x + 3 + z)(x + 3 - z)$$

Using substitution to factor trinomials

For more complicated expressions, a substitution sometimes helps to simplify the factoring process.

Self Check 10
Factor: $(a + b)^2 - 3(a + b) - 10$.

EXAMPLE 10   Factor: $(x + y)^2 + 7(x + y) + 12$.

Solution   We rewrite the trinomial $(x + y)^2 + 7(x + y) + 12$ as $z^2 + 7z + 12$, where $z = x + y$. The trinomial $z^2 + 7z + 12$ factors as $(z + 4)(z + 3)$.

To find the factorization of $(x + y)^2 + 7(x + y) + 12$, we substitute $x + y$ for z in the expression $(z + 4)(z + 3)$ to obtain

$$z^2 + 7z + 12 = (z + 4)(z + 3)$$

Answer $(a + b + 2)(a + b - 5)$

$(x + y)^2 + 7(x + y) + 12 = (x + y + 4)(x + y + 3)$ Replace z with $x + y$.

Using grouping to factor trinomials

The method of factoring by grouping can be used to help factor trinomials of the form $ax^2 + bx + c$. For example, to factor the trinomial $6x^2 + 7x - 3$, we proceed as follows:

1. First find the product ac: $6(-3) = -18$. This number is called the **key number.**

2. Find two factors of the key number -18 whose sum is $b = 7$:
$$9(-2) = -18 \quad \text{and} \quad 9 + (-2) = 7$$

3. Use the factors 9 and -2 as coefficients of two terms to be placed between $6x^2$ and -3:
$$6x^2 + 7x - 3 = 6x^2 + 9x - 2x - 3 \quad \text{Express } 7x \text{ as } 9x - 2x.$$

4. Factor by grouping:

$$6x^2 + 9x - 2x - 3 = 3x(2x + 3) - 1(2x + 3) \qquad \text{From } 6x^2 + 9x, \text{ factor out } 3x.$$
$$\text{From } -2x - 3, \text{ factor out } -1.$$
$$= (2x + 3)(3x - 1) \qquad \text{Factor out } 2x + 3.$$

We can verify this factorization by multiplication.

Factoring by grouping is especially useful when the lead coefficient, a, and the constant term, c, have many factors.

Factoring trinomials by grouping

To factor a trinomial by grouping:

1. Factor out any GCF (including -1 if that is necessary to make $a > 0$ in a trinomial of the form $ax^2 + bx + c$).

2. Identify a, b, and c, and find the key number ac.

3. Find two numbers whose product is the key number and whose sum is b.

4. Enter the two numbers as coefficients of x between the first and last terms and factor the polynomial by grouping.

The product of these numbers must be ac.

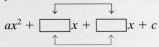

The sum of these numbers must be b.

5. Check by multiplying.

INTERMEDIATE
Algebra $f(x)$ **Now**™

EXAMPLE 11 Factor: $10x^2 + 17x - 6$.

Solution Since $a = 10$ and $c = -6$ in the trinomial, $ac = -60$. We now find two factors of -60 whose sum is 17. Two such factors are 20 and -3. We use these factors as coefficients of two terms to be placed between $10x^2$ and -6:

$$10x^2 + 17x - 6 = 10x^2 + 20x - 3x - 6 \qquad \text{Express } 17x \text{ as } 20x - 3x.$$

Finally, we factor by grouping.

$$= 10x(x + 2) - 3(x + 2) \qquad \text{From } 10x^2 + 20x, \text{ factor out } 10x.$$
$$\text{From } -3x - 6, \text{ factor out } -3.$$
$$= (x + 2)(10x - 3) \qquad \text{Factor out } x + 2.$$

Self Check 11
Factor: $15a^2 + 17a - 4$.

Answer $(3a + 4)(5a - 1)$

Section 5.7 STUDY SET

INTERMEDIATE
Algebra $f(x)$ **Now**™

VOCABULARY *Fill in the blanks.*

1. A polynomial with three terms, such as $3x^2 - 2x + 4$, is called a _____.

2. Since $y^2 + 2y + 1$ is the square of $y + 1$, we call $y^2 + 2y + 1$ a _____ square trinomial.

3. For $a^2 - a - 6$, the _____ coefficient (the coefficient of the a^2 term) is 1.

4. The trinomial $4a^2 - 5a - 6$ is written in _____ powers of a.

CONCEPTS

5. Consider $3x^2 - x + 16$. What is the sign of the
  a. First term?
  b. Middle term?
  c. Last term?

6. Explain what is meant when we say that the trinomial $h^2 - 12h + 27$ can be written as the product of two binomials.

7. If $b^2 - 4ac$ is a perfect square, the trinomial $ax^2 + bx + c$ can be factored using what type of coefficients?

8. Use the substitution $x = a + b$ to rewrite the trinomial $6(a + b)^2 - 17(a + b) - 3$.

NOTATION   *Find each product.*

9. $(x + y)(x + y) = x^2 + \boxed{}$

10. $(x - y)(x - y) = x^2 - \boxed{}$

11. $(x + y)(x - y) = \boxed{}$

12. $(a + b)(a + b) = \boxed{} + b^2$

13. The trinomial $4m^2 - 4m + 1$ is written in $ax^2 + bx + c$ form. Identify a, b, and c.

14. Consider the trinomial $15s^2 + 4s - 4$. Is $b^2 - 4ac$ a perfect square?

PRACTICE   *Complete each factorization.*

15. $x^2 + 5x + 6 = (x + 3)\boxed{}$

16. $x^2 - 6x + 8 = (x - 4)\boxed{}$

17. $x^2 + 2x - 15 = (x + 5)\boxed{}$

18. $x^2 - 3x - 18 = (x - 6)\boxed{}$

19. $2a^2 + 9a + 4 = \boxed{}(a + 4)$

20. $6p^2 - 5p - 4 = \boxed{}(2p + 1)$

Use a special product formula to factor each perfect square trinomial.

21. $x^2 + 2x + 1$ **22.** $y^2 - 2y + 1$

23. $a^2 - 18a + 81$ **24.** $b^2 + 12b + 36$

25. $4y^2 + 4y + 1$ **26.** $9x^2 + 6x + 1$

27. $9b^2 - 12b + 4$ **28.** $4a^2 - 12a + 9$

Test each trinomial for factorability and factor it, if possible.

29. $x^2 - 5x + 6$

30. $y^2 + 7y + 6$

31. $x^2 - 7x + 10$

32. $c^2 - 7c + 12$

33. $b^2 + 8b + 18$

34. $x^2 + 4x - 28$

35. $x^2 - x - 30$

36. $a^2 + 4a - 45$

37. $a^2 + 5a - 50$

38. $b^2 + 9b - 36$

39. $x^2 - 4xy - 21y^2$

40. $a^2 + 4ab - 5b^2$

Factor each trinomial. Factor out all common monomials first (including -1 if the lead coefficient is negative). If a trinomial is prime, so indicate.

41. $3x^2 + 12x - 63$

42. $2y^2 + 4y - 48$

43. $b^2x^2 - 12bx^2 + 35x^2$

44. $c^3x^2 + 11c^3x - 42c^3$

45. $-a^2 + 4a + 32$

46. $-x^2 - 2x + 15$

47. $-3x^2 + 15x - 18$

48. $-2y^2 - 16y + 40$

49. $-2p^2 - 2pq + 4q^2$

50. $-6m^2 + 3mn + 3n^2$

51. $6y^2 + 7y + 2$

52. $6x^2 - 11x + 3$

53. $8a^2 + 6a - 9$

54. $15b^2 + 4b - 4$

55. $6x^2 - 5xy - 4y^2$

56. $18y^2 - 3yz - 10z^2$

57. $5x^2 + 4x + 1$

58. $6z^2 + 17z + 12$

59. $8x^2 - 10x + 3$

60. $4a^2 + 20a + 3$

61. $a^2 - 3ab - 4b^2$

62. $b^2 + 2bc - 80c^2$

63. $3x^3 - 10x^2 + 3x$

64. $3t^3 - 3t^2 + t$

65. $-3a^2 + ab + 2b^2$

66. $-2x^2 + 3xy + 5y^2$

67. $5a^2 + 45b^2 - 30ab$

68. $-4x^2 - 9 + 12x$

69. $21x^4 - 10x^3 - 16x^2$

70. $16x^3 - 50x^2 + 36x$

71. $x^4 + 8x^2 + 15$

72. $x^4 + 11x^2 + 24$

73. $y^4 - 13y^2 + 30$

74. $y^4 - 13y^2 + 42$

75. $a^4 - 13a^2 + 36$

76. $b^4 - 17b^2 + 16$

Use a substitution to help factor each expression.

77. $(x + a)^2 + 2(x + a) + 1$

78. $(a + b)^2 - 2(a + b) + 1$

79. $3(a + b)^2 - 14(a + b) - 24$

80. $2(x - y)^2 + (x - y) - 10$

Factor each expression by using grouping.

81. $x^2 + 4x + 4 - y^2$

82. $x^2 - 6x + 9 - 4y^2$

83. $x^2 + 2x + 1 - 9z^2$

84. $x^2 + 10x + 25 - 16z^2$

85. $c^2 - 4a^2 + 4ab - b^2$

86. $4c^2 - a^2 - 6ab - 9b^2$

87. $2a^2 - 33a + 16$

88. $3b^2 + 2b - 21$

89. $2u^2 + 5u + 3$

90. $6y^2 + 5y - 6$

91. $20r^2 - 7rs - 6s^2$

92. $6s^2 + st - 12t^2$

APPLICATIONS

93. ICE The surface area of the cubical block of ice shown on the right is $6x^2 + 36x + 54$. Find the length of an edge of the block.

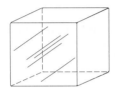

94. CHECKERS The area of the square checkerboard in the illustration is $25x^2 - 40x + 16$. Find the length of a side.

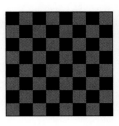

WRITING

95. Explain how you would factor -1 from a trinomial.

96. Explain how you would test the polynomial $ax^2 + bx + c$ for factorability.

REVIEW

97. If $f(x) = |2x - 1|$, find $f(-2)$.

98. If $g(x) = 2x^2 - 1$, find $g(-2)$.

99. Solve: $-3 = -\dfrac{9}{8}s$.

100. Solve: $2x + 3 = \dfrac{2}{3}x - 1$.

101. Simplify: $3p^2 - 6(5p^2 + p) + p^2$.

102. Solve: $\begin{cases} 2(2x + 3y) = 5 \\ 8x = 3(1 + 3y) \end{cases}$.

5.8 Summary of Factoring Techniques

• A general factoring strategy

Factoring some polynomials involves several steps in which two or more factoring techniques must be used. In this section, we will discuss a general factoring strategy — a step-by-step plan to follow when factoring any polynomial.

A general factoring strategy

In this section we will discuss ways to approach a randomly chosen factoring problem. For example, suppose we wish to factor the trinomial

$$x^2y^2z^3 + 7xy^2z^3 + 6y^2z^3$$

We begin by attempting to identify the problem type. The first possibility to look for is factoring out a common factor. Because the trinomial has a common factor y^2z^3, we factor it out:

$$x^2y^2z^3 + 7xy^2z^3 + 6y^2z^3 = y^2z^3(x^2 + 7x + 6)$$

We note that $x^2 + 7x + 6$ is a trinomial that can be factored as $(x + 6)(x + 1)$. Thus,

$$x^2y^2z^3 + 7xy^2z^3 + 6y^2z^3 = y^2z^3(x^2 + 7x + 6)$$
$$= y^2z^3(x + 6)(x + 1)$$

To identify the type of factoring problem, we follow these steps.

> **A general factoring strategy**
>
> **1.** Factor out all common factors.
> **2.** If an expression has two terms, check for the following problem types:
>   **a.** **The difference of two squares:** $x^2 - y^2 = (x + y)(x - y)$
>   **b.** **The sum of two cubes:** $x^3 + y^3 = (x + y)(x^2 - xy + y^2)$
>   **c.** **The difference of two cubes:** $x^3 - y^3 = (x - y)(x^2 + xy + y^2)$
> **3.** If an expression has three terms, attempt to factor it as a **trinomial.**
> **4.** If an expression has four or more terms, try factoring by **grouping.**
> **5.** Continue until each individual factor is prime.
> **6.** Check the results by multiplying.

INTERMEDIATE
Algebra $f(x)$ **Now**™

Self Check 1
Factor: $3p^4r^3 - 3q^4r^3$.

EXAMPLE 1   Factor: $48a^4c^3 - 3b^4c^3$.

Solution   We begin by factoring out the common factor $3c^3$:

$$48a^4c^3 - 3b^4c^3 = 3c^3(16a^4 - b^4)$$

Since the expression $16a^4 - b^4$ has two terms, we check to see whether it is the difference of two squares, which it is. As the difference of two squares, it factors as $(4a^2 + b^2)(4a^2 - b^2)$.

$$48a^4c^3 - 3b^4c^3 = 3c^3(\mathbf{16a^4 - b^4})$$
$$= 3c^3(\mathbf{4a^2 + b^2})(\mathbf{4a^2 - b^2})$$

The binomial $4a^2 + b^2$ is the sum of two squares and is prime. However, $4a^2 - b^2$ is the difference of two squares and factors as $(2a + b)(2a - b)$.

$$48a^4c^3 - 3b^4c^3 = 3c^3(16a^4 - b^4)$$
$$= 3c^3(4a^2 + b^2)(\mathbf{4a^2 - b^2})$$
$$= 3c^3(4a^2 + b^2)(\mathbf{2a + b})(\mathbf{2a - b})$$

Answer
$3r^3(p^2 + q^2)(p + q)(p - q)$

Since each of the individual factors is prime, the factorization is complete.

INTERMEDIATE
Algebra $f(x)$ **Now**™

Self Check 2
Factor:
$a^5p - a^3b^2p + a^2b^3p - b^5p$.

EXAMPLE 2   Factor: $x^5y + x^2y^4 - x^3y^3 - y^6$.

Solution   We begin by factoring out the common factor y:

$$x^5y + x^2y^4 - x^3y^3 - y^6 = y(x^5 + x^2y^3 - x^3y^2 - y^5)$$

Because the expression $x^5 + x^2y^3 - x^3y^2 - y^5$ has four terms, we try factoring by grouping to obtain

$$x^5y + x^2y^4 - x^3y^3 - y^6$$
$$= y(x^5 + x^2y^3 - x^3y^2 - y^5) \qquad \text{Factor out } y.$$
$$= y[x^2(x^3 + y^3) - y^2(x^3 + y^3)] \qquad \text{Factor by grouping.}$$
$$= y(x^3 + y^3)(x^2 - y^2) \qquad \text{Factor out } x^3 + y^3.$$

Finally, we factor $x^3 + y^3$ (the sum of two cubes) and $x^2 - y^2$ (the difference of two squares) to obtain

$$x^5y + x^2y^4 - x^3y^3 - y^6 = y(x + y)(x^2 - xy + y^2)(x + y)(x - y)$$

Because each of the individual factors is prime, the factorization is complete.

Answer $p(a + b)(a^2 - ab$
$+ \ b^2)(a + b)(a - b)$

EXAMPLE 3   Factor: $x^3 + 5x^2 + 6x + x^2y + 5xy + 6y$.

Solution   There are no common factors. Since there are more than three terms, we try factoring by grouping. We can factor x from the first three terms and y from the last three terms.

$$x^3 + 5x^2 + 6x + x^2y + 5xy + 6y$$
$$= x(x^2 + 5x + 6) + y(x^2 + 5x + 6)$$
$$= (x^2 + 5x + 6)(x + y) \qquad \text{Factor out } x^2 + 5x + 6.$$
$$= (x + 3)(x + 2)(x + y) \qquad \text{Factor } x^2 + 5x + 6.$$

INTERMEDIATE
Algebra $f(x)$ Now™
Self Check 3
Factor:
$a^3 - 5a^2 + 6a + a^2b - 5ab + 6b$.

Answer $(a - 2)(a - 3)(a + b)$

EXAMPLE 4   Factor: $x^4 + 2x^3 + x^2 + x + 1$.

Solution   There are no common factors. Since there are more than three terms, we try factoring by grouping. We can factor x^2 from the first three terms.

$$x^4 + 2x^3 + x^2 + x + 1 = x^2(x^2 + 2x + 1) + (x + 1)$$
$$= x^2(x + 1)(x + 1) + (x + 1) \qquad \text{Factor } x^2 + 2x + 1.$$
$$= (x + 1)[x^2(x + 1) + 1] \qquad \text{Factor out } x + 1.$$
$$= (x + 1)(x^3 + x^2 + 1)$$

INTERMEDIATE
Algebra $f(x)$ Now™
Self Check 4
Factor: $a^4 - a^3 - 2a^2 + a - 2$.

Answer $(a - 2)(a^3 + a^2 + 1)$

Section 5.8   STUDY SET

INTERMEDIATE
Algebra $f(x)$ Now™

VOCABULARY   *Fill in the blanks.*

1. The process of finding the individual factors of a known product is called _____.

2. $x^3 + y^3$ is called a sum of two _____.

3. $x^3 - y^3$ is called a difference of two _____.

4. $x^2 - y^2$ is called a _____ of two squares.

CONCEPTS   *Fill in the blanks.*

5. In any factoring problem, always factor out any _____ factors first.

6. When factoring, if an expression has two terms, check to see whether the problem type is the _____ of two squares, the sum of two _____, or the _____ of two cubes.

7. When factoring, if an expression has three terms, try to factor it as a _____.

8. When factoring, if an expression has four or more terms, try factoring it by _____.

9. Explain how to verify that $y^2z^3(x + 6)(x + 1)$ is the factored form of $x^2y^2z^3 + 7xy^2z^3 + 6y^2z^3$.

10. Why is the polynomial $x + 6$ classified as prime?

NOTATION   *Complete each factorization.*

11. $18a^3b + 3a^2b^2 - 6ab^3 = \quad (6a^2 + ab - 2b^2)$
$$= 3ab(3a + \quad)(\quad - b)$$

12. $2x^4 - 1,250 = 2()$

$ = 2()(x^2 - 25)$

$ = 2(x^2 + 25)(x + 5)()$

PRACTICE   *Factor each polynomial, if possible.*

13. $x^2 + 16 + 8x$

14. $20 + 11x - 3x^2$

15. $8x^3y^3 - 27$

16. $3x^2y + 6xy^2 - 12xy$

17. $xy - ty + xs - ts$

18. $bc + b + cd + d$

19. $25x^2 - 16y^2$

20. $27x^9 - y^3$

21. $12x^2 + 52x + 35$

22. $12x^2 + 14x - 6$

23. $6x^2 - 14x + 8$

24. $12x^2 - 12$

25. $4x^2y^2 + 4xy^2 + y^2$

26. $100z^2 - 81t^2$

27. $x^3 + (a^2y)^3$

28. $4x^2y^2z^2 - 26x^2y^2z^3$

29. $2x^3 - 54$

30. $4(xy)^3 + 256$

31. $ae + bf + af + be$

32. $a^2x^2 + b^2y^2 + b^2x^2 + a^2y^2$

33. $2(x + y)^2 + (x + y) - 3$

34. $(x - y)^3 + 125$

35. $625x^4 - 256y^4$

36. $2(a - b)^2 + 5(a - b) + 3$

37. $36x^4 - 36$

38. $6x^2 - 63 - 13x$

39. $a^4 - 13a^2 + 36$

40. $x^4 - 17x^2 + 16$

41. $x^2 + 6x + 9 - y^2$

42. $x^2 + 10x + 25 - y^8$

43. $4x^2 + 4x + 1 - 4y^2$

44. $9x^2 - 6x + 1 - 25y^2$

45. $x^2 - y^2 - 2y - 1$

46. $a^2 - b^2 + 4b - 4$

WRITING

47. What is your strategy for factoring a polynomial?

48. For the factorization below, explain why the polynomial is not factored completely.

$$48a^4c^3 - 3b^4c^3 = 3c^3(16a^4 - b^4)$$

REVIEW

49. Determine whether the graphs of $x + y = 2$ and $y = x + 5$ are parallel or perpendicular.

50. When expressed as a decimal, is $\frac{7}{8}$ a terminating or a repeating decimal?

51. Evaluate: $\begin{vmatrix} 1 & 15 \\ 15 & 0 \end{vmatrix}$.

52. If a triangle has exactly two sides with equal measures, what type of triangle is it?

5.9 Solving Equations by Factoring

• Solving quadratic equations   • Solving higher-degree polynomial equations   • Problem solving

Equations that involve *first-degree* polynomials, such as $3x + 6 = 0$, are called linear equations. Equations such as $3x^2 + 6x = 0$ that involve *second-degree* polynomials are called quadratic equations. The techniques that we have used to solve linear equations cannot be used to solve quadratic equations. However, we can solve many quadratic equations using factoring.

Solving quadratic equations

An equation such as $3x^2 + 4x - 7 = 0$ or $-5y^2 + 3y + 8 = 0$ is called a **quadratic** or **second-degree** equation.

> **Quadratic equations**
>
> A **quadratic equation** is any equation that can be written in the form
>
> $$ax^2 + bx + c = 0$$
>
> where a, b, and c represent real numbers and $a \neq 0$.

Many quadratic equations can be solved by factoring and then by using the **zero-factor property.**

> **Zero-factor property**
>
> If a and b represent real numbers, then
>
> If $ab = 0$, then $a = 0$ or $b = 0$.

The zero-factor property states that *if the product of two or more numbers is 0, then at least one of the numbers must be 0.*

To solve the quadratic equation $x^2 + 5x + 6 = 0$, we factor its left-hand side to obtain

$$(x + 3)(x + 2) = 0$$

Since the product of $x + 3$ and $x + 2$ is 0, at least one of the factors must be 0. Thus, we can set each factor equal to 0 and solve each resulting linear equation for x:

$$x + 3 = 0 \qquad \text{or} \qquad x + 2 = 0$$
$$x = -3 \qquad \qquad \qquad x = -2$$

To check these solutions, we substitute -3 and -2 for x in the equation and verify that each number satisfies the equation.

Check:

$$x^2 + 5x + 6 = 0 \qquad \text{or} \qquad x^2 + 5x + 6 = 0$$
$$(-3)^2 + 5(-3) + 6 \stackrel{?}{=} 0 \qquad \quad (-2)^2 + 5(-2) + 6 \stackrel{?}{=} 0$$
$$9 - 15 + 6 \stackrel{?}{=} 0 \qquad \qquad \quad 4 - 10 + 6 \stackrel{?}{=} 0$$
$$0 = 0 \qquad \qquad \qquad \qquad \quad 0 = 0$$

Both -3 and -2 are solutions, because both satisfy the equation.

INTERMEDIATE
Algebra $f(x)$ **Now** ™

EXAMPLE 1 Solve: $3x^2 + 6x = 0$.

Self Check 1
Solve: $4p^2 - 12p = 0$.

Solution To solve the equation, we factor the left-hand side, set each factor equal to 0, and solve each resulting equation for x.

$$3x^2 + 6x = 0$$
$$3x(x + 2) = 0 \qquad \text{Factor out the common factor of } 3x.$$
$$3x = 0 \quad \text{or} \quad x + 2 = 0 \qquad \text{By the zero-factor property, at least one of the factors must be equal to zero.}$$
$$x = 0 \qquad \qquad x = -2 \qquad \text{Solve each linear equation.}$$

Verify that both solutions, 0 and -2, check.

Answer $0, 3$

! COMMENT In Example 1, do not attempt to solve the equation by dividing both sides by $3x$, or you will lose the solution 0.

Self Check 2

Solve: $a^2 - 81 = 0$.

EXAMPLE 2 Solve: $x^2 - 16 = 0$.

Solution To solve the equation, we factor the difference of two squares on the left-hand side, set each factor equal to 0, and solve each resulting equation.

$$x^2 - 16 = 0$$
$$(x + 4)(x - 4) = 0$$
$$x + 4 = 0 \quad \text{or} \quad x - 4 = 0$$
$$x = -4 \quad \quad \quad x = 4$$

Answer 9, −9

Verify that both solutions, −4 and 4, check.

The following steps can be used to solve a quadratic equation by factoring.

> **Solving a quadratic equation by the factoring method**
> 1. Write the equation in $ax^2 + bx + c = 0$ form (called *quadratic* form).
> 2. Factor the polynomial.
> 3. Use the zero-factor property to set each factor equal to zero.
> 4. Solve each resulting equation.
> 5. Check the proposed solutions in the original equation.

Many equations that do not appear to be quadratic can be put into quadratic form and then solved by factoring.

Self Check 3

Solve: $x = \dfrac{6}{7}x^2 - \dfrac{3}{7}$.

EXAMPLE 3 Solve: $x = \dfrac{6}{5} - \dfrac{6}{5}x^2$.

Solution We must write the equation in quadratic form. To clear the equation of fractions, we multiply both sides by 5.

$$x = \frac{6}{5} - \frac{6}{5}x^2$$
$$5x = 6 - 6x^2 \quad \text{Multiply both sides by 5.}$$

To use factoring to solve this quadratic equation, one side of the equation must be 0. Since it is easier to factor a second-degree polynomial if the coefficient of the squared term is positive, we add $6x^2$ to both sides and subtract 6 from both sides to obtain

$$6x^2 + 5x - 6 = 0$$
$$(3x - 2)(2x + 3) = 0 \qquad \text{Factor the trinomial.}$$
$$3x - 2 = 0 \quad \text{or} \quad 2x + 3 = 0 \qquad \text{Set each factor equal to 0 and solve for } x.$$
$$3x = 2 \quad \quad \quad 2x = -3$$
$$x = \frac{2}{3} \quad \quad \quad x = -\frac{3}{2}$$

Answer $\dfrac{3}{2}, -\dfrac{1}{3}$

Verify that both solutions, $\dfrac{2}{3}$ and $-\dfrac{3}{2}$, check.

❗ COMMENT To solve a quadratic equation by factoring, be sure to set the quadratic polynomial equal to 0 before factoring and applying the zero-factor property. Do not make the following error:

$$6x^2 + 5x = 6$$

$$x(6x + 5) = 6$$

If the product of two numbers is 6, neither number need be 6. For example, $2 \cdot 3 = 6$.

$$x = 6 \qquad \text{or} \qquad 6x + 5 = 6$$

$$x = \frac{1}{6}$$

Neither solution checks.

Solving quadratic equations

To solve a quadratic equation such as $x^2 + 4x - 5 = 0$ with a graphing calculator, we can use standard window settings of $[-10, 10]$ for x and $[-10, 10]$ for y and graph the quadratic function $y = x^2 + 4x - 5$, as shown in Figure 5-8(a). We can then trace to find the x-coordinates of the x-intercepts of the parabola. See Figures 5-8(b) and 5-8(c). For better results, we can zoom in. Since these are the numbers x that make $y = 0$, they are the solutions of the equation.

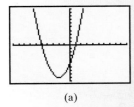 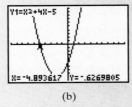 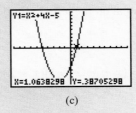

(a)   (b)   (c)

FIGURE 5-8

We can also find the x-intercepts of the graph of $y = x^2 + 4x - 5$ by using the ZERO feature found on most graphing calculators. Figures 5-9(a) and 5-9(b) show how this feature locates the x-intercept and displays its coordinates. (Consult your owner's manual for the specific instructions about how to use this feature.) From the displays, we can conclude that -5 and 1 are solutions of $x^2 + 4x - 5 = 0$. Verify this by checking them in the original equation.

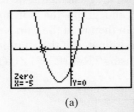 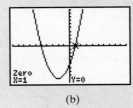

(a)   (b)

FIGURE 5-9

■ Solving higher-degree polynomial equations

We can solve many polynomial equations with degree greater than 2 by factoring and applying an extension of the zero-factor property.

EXAMPLE 4   Solve: $6x^3 - x^2 = 2x$.

Solution   First, we subtract $2x$ from both sides so that the right-hand side of the equation is 0.

$$6x^3 - x^2 - 2x = 0$$

INTERMEDIATE
Algebra *f(x)* **Now**™

Self Check 4
Solve: $5x^3 + 13x^2 = 6x$.

Then we factor x from the third-degree polynomial on the left-hand side and proceed as follows:

$$6x^3 - x^2 - 2x = 0$$

$$x(6x^2 - x - 2) = 0 \quad \text{Factor out } x.$$

$$x(3x - 2)(2x + 1) = 0 \quad \text{Factor } 6x^2 - x - 2.$$

$$x = 0 \quad \text{or} \quad 3x - 2 = 0 \quad \text{or} \quad 2x + 1 = 0 \quad \begin{array}{l}\text{Set each of the three}\\\text{factors equal to 0.}\end{array}$$

$$x = \frac{2}{3} \qquad\qquad x = -\frac{1}{2} \quad \text{Solve each equation.}$$

Answer $0, \dfrac{2}{5}, -3$

Verify that the three solutions, 0, $\dfrac{2}{3}$, and $-\dfrac{1}{2}$, check.

 INTERMEDIATE
Algebra $f(x)$ **Now**™

Self Check 5
Solve: $a^4 + 36 - 13a^2 = 0$.

EXAMPLE 5   Solve: $x^4 + 4 - 5x^2 = 0$.

Solution   First, we write the powers of x in descending order. Then we factor the trinomial on the left-hand side and proceed as follows:

$$x^4 - 5x^2 + 4 = 0$$

$$(x^2 - 1)(x^2 - 4) = 0$$

$$(x + 1)(x - 1)(x + 2)(x - 2) = 0 \quad \text{Factor } x^2 - 1 \text{ and } x^2 - 4.$$

$$x + 1 = 0 \quad \text{or} \quad x - 1 = 0 \quad \text{or} \quad x + 2 = 0 \quad \text{or} \quad x - 2 = 0$$

$$x = -1 \qquad x = 1 \qquad x = -2 \qquad x = 2$$

Answer $2, -2, 3, -3$

Verify that each solution checks.

CALCULATOR SNAPSHOT   Solving equations

To solve the equation $x^4 - 5x^2 + 4 = 0$ with a graphing calculator, we can use window settings of $[-6, 6]$ for x and $[-5, 10]$ for y and graph the polynomial function $y = x^4 - 5x^2 + 4$ as shown in Figure 5-10. We can then read the values of x that make $y = 0$. They are $x = -2, -1, 1$, and 2. If the x-coordinates of the x-intercepts were not obvious, we could approximate their values by using TRACE and ZOOM or by using the ZERO feature.

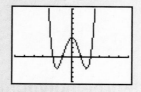

FIGURE 5-10

▇ Problem solving

EXAMPLE 6   Stained glass.   The triangular stained glass window shown in Figure 5-11 on the next page is to be installed in a chapel. The length of the base of the window is 3 times its height. The area of the window is 96 square feet. Find its base and height.

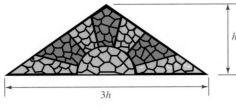

FIGURE 5-11

Analyze the problem

We are to find the length of the base and the height of the window. The formula that gives the area of a triangle is $A = \frac{1}{2}bh$, where b is the length of the base and h the height.

Form an equation

We can let h = the positive number that represents the height of the window. Then $3h$ = the length of the base. To form an equation in terms of h, we can substitute $3h$ for b and 96 for A in the formula for the area of a triangle.

$$A = \frac{1}{2}bh$$

$$96 = \frac{1}{2}(3h)h$$

Solve the equation

To solve this equation, we must write it in quadratic form.

$96 = \frac{1}{2}(3h)h$

$192 = 3h^2$ $(3h)h = 3h^2$. To clear the equation of the fraction, multiply both sides by 2.

$64 = h^2$ Divide both sides by 3.

$0 = h^2 - 64$ To obtain 0 on the left-hand side, subtract 64 from both sides.

$0 = (h + 8)(h - 8)$ Factor the difference of two squares.

$h + 8 = 0$ or $h - 8 = 0$

$h = -8$ | $h = 8$

State the conclusion

Since the height of a triangle cannot be negative, we must discard the negative solution. Thus, the height of the window is 8 feet, and the length of its base is 3(8), or 24 feet.

Check the result

The area of a triangle with a base of 24 feet and a height of 8 feet is 96 square feet:

$$A = \frac{1}{2}bh = \frac{1}{2}(24)(8) = 12(8) = 96$$

The result checks.

EXAMPLE 7 Ballistics. If the initial velocity of an object thrown straight up into the air is 176 feet per second, when will the object strike the ground?

Analyze the problem

The height, in feet, of an object thrown straight up into the air with an initial velocity of v feet per second is given by the formula

$$h = -16t^2 + vt$$

The height h is in feet, and t represents the number of seconds since the object was released. When the object hits the ground, its height will be 0.

Form an equation

In the formula, we set h equal to 0 and set v equal to 176.

$$h = -16t^2 + vt$$
$$0 = -16t^2 + 176t$$

Solve the equation

To solve this equation, we will use the factoring method.

$$0 = -16t^2 + 176t$$
$$0 = -16t(t - 11) \qquad \text{Factor out } -16t.$$
$$-16t = 0 \quad \text{or} \quad t - 11 = 0 \qquad \text{Set each factor equal to 0.}$$
$$t = 0 \qquad \qquad t = 11$$

State the conclusion

When t is 0, the object's height above the ground is 0 feet, because it has not been released. When t is 11, the height is again 0 feet, and the object has returned to the ground. The solution is 11 seconds.

Check the result

Verify that h is 0 when t is 11.

Section 5.9 STUDY SET

VOCABULARY *Fill in the blanks.*

1. A _____ equation is any equation that can be written in the form $ax^2 + bx + c = 0$ where $a \neq 0$.

2. To _____ an equation means to find all the values of the variable that make the equation true.

CONCEPTS

3. If the product of two numbers is 0, what must be true about at least one of the numbers?

4. Use a check to determine whether -5 and 4 are solutions of $a^2 - 9a + 20 = 0$.

5. Determine whether each equation is a quadratic equation.

a. $w^2 + 7w + 12 = 0$

b. $6t + 11 = 0$

c. $x(x + 3) = -2$

d. $k^3 - 4k^2 + k - 15 = 0$

6. What is wrong with the following work?

$$x(x + 2) = 8$$
$$x = 8 \quad \text{or} \quad x + 2 = 8$$
$$x = 8 \qquad \qquad x = 6$$

7. Use the graph below to solve the quadratic equation $x^2 - 2x - 3 = 0$.

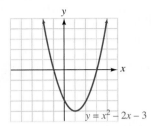

$y = x^2 - 2x - 3$

8. Use the graph below to solve the polynomial equation $x^3 - 4x^2 + 4x = 0$.

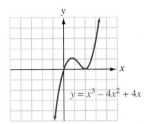

$y = x^3 - 4x^2 + 4x$

NOTATION *Complete each solution.*

9. Solve: $y^2 - 3y - 54 = 0$

$$(y - 9)() = 0$$

$$ = 0 \quad \text{or} \quad y + 6 = 0$$

$$y = 9 \quad \mid \quad y = $$

10. Solve: $x^2 - x = 12$.

$$x^2 - x - 12 = 0$$

$$(x - 4)() = 0$$

$$x - 4 = 0 \quad \text{or} \quad = 0$$

$$x = \quad \mid \quad x = $$

PRACTICE *Solve each equation.*

11. $4x^2 + 8x = 0$

12. $x^2 - 9 = 0$

13. $y^2 - 16 = 0$

14. $y^2 - 25 = 0$

15. $x^2 + x = 0$

16. $x^2 - 3x = 0$

17. $5y^2 - 25y = 0$

18. $y^2 - 36 = 0$

19. $z^2 + 8z + 15 = 0$

20. $w^2 + 7w + 12 = 0$

21. $x^2 + 6x + 8 = 0$

22. $x^2 + 9x + 20 = 0$

23. $3m^2 + 10m + 3 = 0$

24. $2r^2 + 5r + 3 = 0$

25. $2y^2 - 5y + 2 = 0$

26. $2x^2 - 3x + 1 = 0$

27. $2x^2 - x - 1 = 0$

28. $2x^2 - 3x - 5 = 0$

Write each equation in quadratic form and solve it by factoring.

29. $x(x - 6) + 9 = 0$

30. $x^2 + 8(x + 2) = 0$

31. $8a^2 = 3 - 10a$

32. $5z^2 = 6 - 13z$

33. $b(6b - 7) = 10$

34. $2y(4y + 3) = 9$

35. $\dfrac{3a^2}{2} = \dfrac{1}{2} - a$

36. $x^2 = \dfrac{1}{2}(x + 1)$

37. $x^2 + 1 = \dfrac{5}{2}x$

38. $\dfrac{3}{5}(x^2 - 4) = -\dfrac{9}{5}x$

39. $x\left(3x + \dfrac{22}{5}\right) = 1$

40. $x\left(\dfrac{x}{11} - \dfrac{1}{7}\right) = \dfrac{6}{77}$

Solve each equation.

41. $x^3 + x^2 = 0$

42. $2x^4 + 8x^3 = 0$

43. $y^3 - 49y = 0$

44. $2z^3 - 200z = 0$

45. $x^3 - 4x^2 - 21x = 0$

46. $x^3 + 8x^2 - 9x = 0$

47. $z^4 - 13z^2 + 36 = 0$

48. $y^4 - 10y^2 + 9 = 0$

49. $3a(a^2 + 5a) = -18a$

50. $7t^3 = 2t\left(t + \dfrac{5}{2}\right)$

51. $\dfrac{x^2(6x + 37)}{35} = x$

52. $x^2 = -\dfrac{4x^3(3x + 5)}{3}$

53. INTEGER PROBLEM The product of two consecutive even integers is 288. Find the integers. (*Hint:* Let x = the first even integer. Then represent the second even integer in terms of x.)

54. INTEGER PROBLEM The product of two consecutive odd integers is 143. Find the integers. (*Hint:* Let x = the first odd integer. Then represent the second odd integer in terms of x.)

Use a graphing calculator to find the solutions of each equation, if one exists. If an answer is not exact, give the answer to the nearest hundredth.

55. $2x^2 - 7x + 4 = 0$

56. $x^2 - 4x + 7 = 0$

57. $-3x^3 - 2x^2 + 5 = 0$

58. $-2x^3 - 3x - 5 = 0$

APPLICATIONS

59. COOKING The electric griddle shown below has a cooking surface of 160 square inches. Find the length and the width of the griddle.

60. STRUCTURAL ENGINEERING The formula for the area of a trapezoid is $A = \dfrac{h(B + b)}{2}$. The area of the trapezoidal truss in the illustration is 44 square feet. Find the height of the truss shown below if the shorter base is the same as the height.

shorter base: h ft

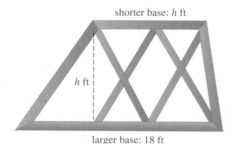

larger base: 18 ft

61. SWIMMING POOLS Building codes require that the rectangular swimming pool shown below be surrounded by a uniform-width walkway of at least 516 square feet. The length of the pool is 10 feet less than twice the width. How wide should the border be?

Surface area = 1,500 ft²

62. FINE ARTS An artist intends to paint a 60-square-foot mural on the large wall shown below. Find the dimensions of the mural if the artist leaves a border of uniform width around it.

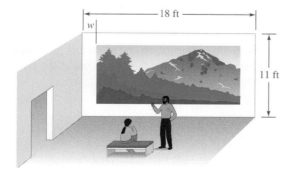

63. ARCHITECTURE The rectangular room shown below is twice as long as it is wide. It is divided into two rectangular parts by a partition, positioned as shown. If the larger part of the room contains 560 square feet, find the dimensions of the entire room.

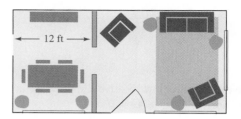

12 ft

64. WINTER RECREATION The length of the rectangular ice-skating rink shown below is 20 meters greater than twice its width. Find the width.

Area = 6,000 m² w m

65. BALLISTICS The muzzle velocity of a cannon is 480 feet per second. If a cannonball is fired vertically, at what times will it be at a height of 3,344 feet?

66. SLINGSHOTS A slingshot can provide an initial velocity of 128 feet per second. At what times will a stone, shot vertically upward, be 192 feet above the ground?

67. BUNGEE JUMPING The formula $h = -16t^2 + 212$ gives the distance a bungee jumper is from the ground for the free-fall portion of a jump, t seconds after leaping off a bridge as shown below. We can find the number of seconds it takes the jumper to reach the point in the fall where the 64-foot bungee cord starts to stretch by substituting 148 for h and solving for t. Find t.

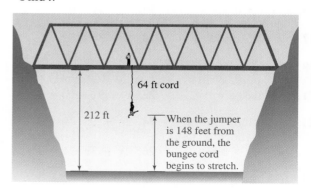

64 ft cord

212 ft

When the jumper is 148 feet from the ground, the bungee cord begins to stretch.

68. BASEBALL In 2001, pitcher Pedro Martinez of the Boston Red Sox threw a fastball that was clocked at 97 mph. This is a velocity of approximately 144 feet per second. If he could throw the baseball vertically into the air with this velocity, how long would it take for the ball to fall to the ground?

69. FORENSIC MEDICINE The kinetic energy E of a moving object is given by $E = \frac{1}{2}mv^2$, where m is the mass of the object (in kilograms) and v is the object's velocity (in meters per second). Kinetic energy is measured in joules. By measuring the damage done to a victim who has been struck by a 3-kilogram club, a police pathologist finds that the energy at impact was 54 joules. Find the velocity of the club at impact.

70. TRAFFIC ACCIDENTS Investigators at a traffic accident used the function $d(v) = 0.04v^2 + 0.8v$, where v is the velocity of the car (in mph) and $d(v)$ is the stopping distance of the car (in feet), to reconstruct the events leading up to a collision. From physical evidence, it was concluded that it took one car 32 feet to stop. At what velocity was the car traveling prior to the accident?

71. BREAK-EVEN POINT The cost for a guitar maker to hand-craft x guitars is given by the function $C(x) = \frac{1}{8}x^2 - x + 6$. The revenue taken in with the sale of x guitars is given by the function $R(x) = \frac{1}{4}x^2$. Find the number of guitars that must be sold so that the cost equals the revenue.

72. REVENUE Over the years, the manager of a crafts store has found that the number of scented candles x she can sell in a month depends on the price p according to the formula $x = 200 - 10p$. At what price should she sell the candles if she needs to bring in $750 in revenue a month from their sale? (*Hint:* Revenue = price · number sold = px.)

WRITING

73. Explain the zero-factor property.

74. In the work shown below, explain why the student has not solved for x.

$$\text{Solve: } x^2 + x - 6 = 0$$
$$x^2 + x = 6$$
$$\boxed{x = 6 - x^2}$$

75. Explain what is wrong with the following solution.

$$\text{Solve: } x^2 - x = 0$$
$$\frac{x^2}{x} - \frac{x}{x} = \frac{0}{x}$$
$$x - 1 = 0$$
$$\boxed{x = 1}$$

76. Explain what is wrong with the following solution.

$$\text{Solve: } \quad x^2 - x = 6$$
$$x(x - 1) = 6$$
$$\boxed{x = 6} \quad \text{or} \quad x - 1 = 6$$
$$\boxed{x = 7}$$

77. The graphs of the two polynomial functions, $f(x) = 2x^3 - 8x$ and $f(x) = 2x(x + 2)(x - 2)$, in the illustrations below appear to be the same. After examining the defining equations, explain why we know that they indeed are identical graphs.

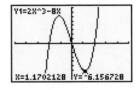 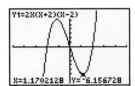

78. Explain why the x-coordinate of the x-intercept of the graph of $y = 8x^2 + 10x - 3$ (indicated as a *zero* in the illustration below) is a solution of $8x^2 + 10x - 3 = 0$.

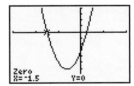

REVIEW

79. ALUMINUM FOIL Find the number of square feet of aluminum foil on a roll if it has dimensions of $8\frac{1}{3}$ yards × 12 inches.

80. HOCKEY A hockey puck is a vulcanized rubber disk 2.5 cm (1 in.) thick and 7.6 cm (3 in.) in diameter. Find the volume of a puck in cubic centimeters and cubic inches. Round to the nearest tenth.

Polynomials

A **polynomial** is an algebraic term or the sum of two or more algebraic terms whose variables have whole-number exponents. No variable appears in a denominator.

Operations with Polynomials

In arithmetic, we learned how to add, subtract, multiply, divide, and find powers of numbers. In algebra, we need to be able to perform these operations on polynomials.

Perform the operations.

1. $(-2x^2 - 5x - 7) + (-3x^2 + 7x + 1)$
2. $(6s^3 + 3s - 2) - (2s^3 + 3s^2 + 5)$
3. $(3m - 4)(m + 3)$
4. $3r^2st(r^2 - 2s + 3t^2)$
5. $(a - 2d)^2$
6. $(x - 3y)(x^2 + 3xy + 9y^2)$
7. $(3b + 1)(2b^2 + 3b + 2)$
8. $(2y + 3)(y - 1) - (y + 2)(3y - 1)$

Polynomial Functions

Polynomial functions can be used to model many real-world situations.

9. WINDOW WASHERS A man on a scaffold, washing the outside windows of a skyscraper, drops a squeegee. As it falls for t seconds, its distance in feet from the ground $d(t)$ after being dropped is given by the polynomial function $d(t) = -16t^2 + 576$. Find $d(6)$ and explain the result.

10. Write a polynomial function $V(x)$ that gives the volume of the ice chest shown below. Then find $V(3)$.

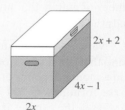

$2x + 2$

$4x - 1$

$2x$

Solving Equations by Factoring

Quadratic equations can be written in the form $ax^2 + bx + c = 0$. Many quadratic equations can be solved by factoring the polynomial $ax^2 + bx + c$ and using the zero-factor property. Some higher-degree equations can also be solved by using an extension of this procedure.

Solve each equation by factoring.

11. $x^2 - 81 = 0$
12. $5x^2 - 25x = 0$
13. $z^2 + 8z + 15 = 0$
14. $2r^2 + 5r + 3 = 0$
15. $\dfrac{3t^2}{2} + t = \dfrac{1}{2}$
16. $m^3 = 9m - 8m^2$

ACCENT ON TEAMWORK

SECTION 5.1

EXPONENTS   Have a student in your group write each of the eight rules for exponents on separate 3×5 cards. On another set of cards, write an explanation of each rule using words. On a third set of cards, write an example of the use of each rule for exponents. Shuffle the cards and work together to match the symbolic description, the word description, and the example for each of the eight rules for exponents.

SECTION 5.2

SCIENTIFIC NOTATION   Go to the library and find five examples of extremely large and five examples of extremely small numbers. Encyclopedias, government statistics books, and science books are good places to look. Write each number in scientific notation on a separate piece of paper. Include a brief explanation of what the number represents. Present the 10 examples in numerical order, beginning with the smallest number first.

SECTION 5.3

ADDING POLYNOMIALS   According to an old adage, "You can't add apples and oranges." Give some examples of how this concept applies when adding two polynomials.

SECTION 5.4

MULTIPLYING BINOMIALS   Use colored construction paper to make a model like that shown in the illustration. Use the model as part of a presentation to demonstrate why $(x + y)(x + y) = x^2 + 2xy + y^2$.

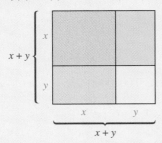

SECTION 5.5

SOLVING FORMULAS   Examine the student's work shown below. Write some comments to the student about what it means to *solve for the indicated variable*. Then write out the correct solution.

Solve for r_1: $r_1 r_2 = r r_2 + r r_1$.

$$\frac{r_1 r_2}{r_2} = \frac{r r_2 + r r_1}{r_2}$$

$$\boxed{r_1 = \frac{r r_2 + r r_1}{r_2}}$$

SECTION 5.6

FACTORING FORMULAS   Use multiplication to prove that in each statement given below, the expression on the left is not equal to the expression on the right.

$$x^2 - y^2 \neq (x - y)^2$$
$$x^3 - y^3 \neq (x - y)^3$$
$$x^3 + y^3 \neq (x + y)^3$$

For each statement, give the correct factoring formula.

SECTION 5.7

AUTHORING A TEXTBOOK   Assign each of the 11 examples in Section 5.7 to members of your group. Have them write a new but similar problem for each example and then write a solution, complete with an explanation and author notes, using the same format as in this book. They should also create an accompanying Self Check problem and include the answer. Compile all 11 examples into a booklet. Make copies of your booklet for the other members of the class.

SECTION 5.8

FACTORING FLOWCHART   Create a factoring flowchart that would lead a student through the correct steps to identify the type(s) of factoring necessary to factor any given polynomial. To get you started, the top of the flowchart should begin as shown below.

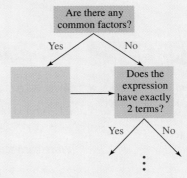

SECTION 5.9

QUADRATIC EQUATIONS   Suppose you know that the two solutions of a quadratic equation are 4 and -3. Consider the method used to solve quadratic equations studied in Section 5.9. Work backward to find the original equation.

CHAPTER REVIEW

SECTION 5.1	*Exponents*

CONCEPTS

If n represents a natural number,

$$x^n = \overbrace{x \cdot x \cdot x \cdots \cdot x}^{n \text{ factors of } x}$$

where x is the *base* and n the *exponent*.

Rules for exponents: If there are no divisions by 0, then for all integers m and n,

$$x^m x^n = x^{m+n} \qquad (x^m)^n = x^{mn}$$

$$(xy)^n = x^n y^n \qquad \left(\frac{x}{y}\right)^n = \frac{x^n}{y^n}$$

$$x^0 = 1 \qquad x^{-n} = \frac{1}{x^n}$$

$$\frac{x^m}{x^n} = x^{m-n}$$

$$\left(\frac{x}{y}\right)^{-n} = \left(\frac{y}{x}\right)^n$$

REVIEW EXERCISES

Evaluate each expression.

1. 3^6

2. -2^5

3. $(-4)^3$

4. 15^1

Simplify each expression and write all answers without negative exponents.

5. $x^4 \cdot x^2$

6. $a^3 b^5 a^2 b$

7. $(m^6)^3$

8. $(-t^2)^2 (t^3)^3$

9. $(3x^2 y^3)^2$

10. $\left(\dfrac{x^4}{b}\right)^4$

11. $-3x^0$

12. $(x^2)^{-5}$

13. -5^{-4}

14. $\dfrac{70}{x^{-4}}$

15. $(3x^{-3})^{-2}$

16. $2x^{-4} x^3$

17. $-\left(\dfrac{c^{-3}}{c^{-5}}\right)^5$

18. $\left(\dfrac{4}{5}\right)^{-2}$

19. $\dfrac{y^{-3}}{y^4 y}$

20. $\left(\dfrac{-2a^4 b}{a^{-3} b^2}\right)^{-3}$

SECTION 5.2	*Scientific Notation*

Scientific notation is a compact way of writing large and small numbers. Positive numbers are written in the form

$$N \times 10^n$$

where $1 \le N < 10$ and n is an integer.

Write each number in scientific notation.

21. 19,300,000,000

22. 0.00000002735

Write each number in standard notation.

23. 7.277×10^7

24. 8.3×10^{-9}

Write each number in scientific notation and perform the operations. Give answers in scientific notation.

25. THE SPEED OF LIGHT Light travels at about 300,000 kilometers per second. If the average distance from the sun to the planet Mars is approximately 228,000,000 kilometers, how long does it take light from the sun to reach Mars?

26. PROTONS If the mass of 1 proton is 0.00000000000000000000000167248 gram, find the mass of 1 million protons.

27. Evaluate: $\dfrac{(616{,}000{,}000)(0.000009)}{0.00066}$.

Polynomials and Polynomial Functions

A *polynomial* is an algebraic term or the sum of two or more algebraic terms whose variables have whole-number exponents. No variable appears in a denominator.

Determine whether each expression is a polynomial.

28. $\dfrac{2x^2}{x + 1}$

29. $-5x^3 + x^2 - 5x - 4$

30. $2.8y^{15} - y^{10} + y^8 - \dfrac{3}{2}y^6$

31. $x^{-3} + x^{-2} - x^{-1} - 1$

The *degree of a polynomial* is the *degree of the term* having the highest degree contained within the polynomial.

Classify each polynomial as a monomial, binomial, trinomial, or none of these. Then determine the degree of the polynomial.

32. $x^2 - 8$

33. $-15a^3b$

34. $x^4 + x^3 - x^2 + x - 4$

35. $9x^2y + 13x^3y^2 + 8x^4y^4$

To *evaluate a polynomial function*, we replace the variable in the defining equation with its value, called the *input*. Then we simplify to find the *output*.

36. SQUIRT GUNS The volume, in cubic inches, of the reservoir on top of the squirt gun shown on the right is given by the polynomial function $V(r) = 4.19r^3 + 25.12r^2$, where r is the radius in inches. Find $V(2)$ to the nearest cubic inch.

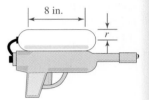

Graph each polynomial function.

37. $f(x) = x^2 - 2x$

38. $f(x) = x^3 - 3x^2 + 4$

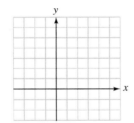

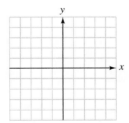

To add polynomials, remove parentheses and *combine like terms* (terms having the same variables with the same exponents).

Perform each operation.

39. $(3x^2 + 4x + 9) + (2x^2 - 2x + 7)$

40. $(2x^2y^3 - 5x^2y + 9y) + (x^2y^3 - 3x^2y - y)$

To subtract polynomials, add the first polynomial and the opposite (negative) of the second polynomial.

Write the opposite of each polynomial.

41. $x^2 - 3x$

42. $-2c^3d - 3c^2d + 1$

Subtract.

43. $(4x^3 + 4x^2 + 7) - (-2x^3 - x - 2)$

44. $\begin{array}{r} -10k^4 - 4k^3 + 5k^2 - k + 1 \\ - \underline{-16k^4 + 2k^3 - 4k^2 - k + 3} \end{array}$

Multiplying Polynomials

To *multiply monomials,* multiply their numerical factors and multiply their variable factors.

Find each product.

45. $(8a^2)\left(-\dfrac{1}{2}a\right)$

46. $(-3xy^2z)(-2xz^3)(xz)$

To *multiply a polynomial by a monomial,* multiply each term of the polynomial by the monomial.

Find each product.

47. $2xy^2(x^3y - 4xy^5)$

48. $-a^2b(-a^2 - 2ab + b^2)$

The *FOIL method* is used to multiply two binomials.

Find each product.

49. $(8x - 5)(2x + 3)$

50. $(3x^2 + 2)(2x - 4)$

51. $(5a - 6)^2$

52. $(0.7c^2 - d)(0.7c^2 + d)$

To *multiply polynomials,* multiply each term of one polynomial by each term of the other polynomial.

53. $\left(\dfrac{1}{3}a^3 - 1\right)^2$

54. $(ab + 1)(ab + 3)$

55. $(5x^2 - 4x)(3x^2 - 2x + 10)$

56. $(r + s)(r - s)(r - 3s)$

57. SHAVING A razor blade is made from a thin piece of platinum steel. Before its center is punched out, the blade has the shape shown. Write a polynomial that gives the area.

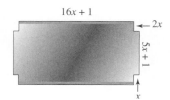

$16x + 1$

$2x$

$5x + 1$

x

The Greatest Common Factor and Factoring by Grouping

To *prime factor* a natural number means to write it as a product of prime numbers.

58. Find the prime factorization of 350.

The largest natural number that divides each number in a set of numbers is called their *greatest common factor (GCF).*

Find the GCF of each set of monomials.

59. $42, 36, 54$

60. $6x^2y^5, 15xy^3$

The process of finding the individual factors of a known product is called *factoring.*

Factor each polynomial, if possible.

61. $4x + 8$

62. $3x^3 - 6x^2 + 9x$

63. $5x^2y^3 - 11mn^2$

64. $7a^4b^2 + 49a^3b$

Always factor out *common factors* as the first step in a factoring problem. Use the distributive property to do this.

65. $5x^2(x + y) - 15x^3(x + y)$

66. $27x^3y^3z^3 + 81x^4y^5z^2 - 90x^2y^3z^7$

A polynomial that cannot be factored is a *prime polynomial.*

Factor out the opposite of the greatest common factor.

67. $-7b + 14$

68. $-49a^3b^2(a - b)^4 + 63a^2b^4(a - b)^3$

If an expression has four or more terms, try to factor the expression by *grouping*.

Factor each polynomial by grouping.

69. $xy + 2y + 4x + 8$

70. $r^2y - ar - ry + a$

71. Solve $m_1m_2 = mm_2 + mm_1$ for m_1.

SECTION 5.6
The Difference of Two Squares; the Sum and Difference of Two Cubes

Factoring the *difference of two squares*:

$$x^2 - y^2 = (x + y)(x - y)$$

Factor each expression, if possible.

72. $z^2 - 16$

73. $y^2 - 121$

74. $x^2y^4 - 64z^6$

75. $a^2b^2 + c^2$

76. $c^2 - (a + b)^2$

77. $3x^6 - 300x^2$

Factoring the *sum of two cubes*:

$$x^3 + y^3$$
$$= (x + y)(x^2 - xy + y^2)$$

Factoring the *difference of two cubes*:

$$x^3 - y^3$$
$$= (x - y)(x^2 + xy + y^2)$$

Factor each polynomial, if possible.

78. $t^3 + 64$

79. $2x^3y - 54yz^3$

SECTION 5.7
Factoring Trinomials

Perfect square trinomials are the squares of binomials:

$$x^2 + 2xy + y^2 = (x + y)^2$$
$$x^2 - 2xy + y^2 = (x - y)^2$$

Factor each trinomial, if possible.

80. $x^2 + 10x + 25$

81. $a^2 - 14a + 49$

Test for factorability: A trinomial of the form $ax^2 + bx + c$ will factor with integer coefficients if $b^2 - 4ac$ is a perfect square.

Factor each trinomial, if possible.

82. $y^2 + 21y + 20$

83. $z^2 - 11z + 30$

84. $-x^2 - 3x + 28$

85. $a^2 - 24b^2 - 5ab$

86. $4a^2 - 5a + 1$

87. $3b^2 + 2b + 1$

88. $y^3 + y^2 - 2y$

89. $15x^2 - 57xy - 12y^2$

To factor trinomials with a *lead coefficient of 1*, list the factorizations of the third term.

90. $2a^4 + 4a^3 - 6a^2$

91. $v^4 - 13v^2 + 42$

92. Use a substitution to factor $(s + t)^2 - 2(s + t) + 1$.

93. Use grouping to factor $k^2 + 2k + 1 - 9m^2$.

To factor trinomials with a *lead coefficient other than 1*, use the procedure for factoring a general trinomial.

14. Solve: $\begin{cases} 2(2x + 3y) = 5 \\ 8x = 3(1 + 3y) \end{cases}$.

15. Solve: $\begin{cases} 3x + 2y - z = -8 \\ 2x - y + 7z = 10 \\ 2x + 2y - 3z = -10 \end{cases}$.

16. Evaluate: $\begin{vmatrix} 2 & -3 & 4 \\ -1 & 2 & 4 \\ 3 & -3 & 1 \end{vmatrix}$.

Solve each inequality. Give the solution set in interval notation and then graph it.

17. $-9(t - 3) + 2t \leq 8(4 - t)$

18. $-6 \leq \dfrac{1}{3}h + 1 < 0$

19. $|m + 5| \geq 7$

20. $4.5x - 1 < -10$ or $6 - 2x \geq 12$

21. GROCERY SHOPPING Let x = the number of items in the grocery bag shown on the right. Which statement below mathematically describes the number of items this type of bag can hold?

i. $x > 10$ or $x < 15$ **ii.** $x < 10$ and $x > 15$

iii. $10 \leq x \leq 15$ **iv.** $x > 10$ and $x < 15$

22. Graph the solution set: $\begin{cases} x - y < 4 \\ y \leq 0 \\ x \geq 0 \end{cases}$.

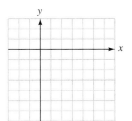

23. Simplify: $\left(\dfrac{-3a^4b^2}{-9a^5b^{-2}} \right)^{-2}$. Write the answer without using negative exponents.

24. Write 9.0895×10^{-8} in standard notation.

25. Find the product: $(2a - b)(4a^2 + 2ab + b^2)$.

26. Simplify: $(3k + 1)^2 + (2k - 4)^2$.

Factor each polynomial completely.

27. $x^2 + 4y - xy - 4x$

28. $6s^4 - 216s^2$

29. $8x^6 + 125y^3$

30. $-3a^2 + ab + 2b^2$

31. Solve: $x^2 = \dfrac{1}{2}(x + 1)$.

32. Solve $m_1m_2 = m_2g + m_1g$ for m_1.

CHAPTER 6

Rational Expressions, Equations, and Functions

INTERMEDIATE
Algebra *f(x)* Now™

Throughout the chapter, this icon introduces resources on the Intermediate AlgebraNow Web site, accessed through http://1pass .thomson.com, that will

- Help you test your knowledge of the material with a pre-test and a post-test

- Provide a personalized learning plan targeting areas you should study

© Bettmann /CORBIS

Air travel has come a long way since the Wright brothers made the world's first powered airplane flight on December 17, 1903. With Orville at the controls and Wilbur watching on the ground, the *Wright Flyer* traveled 120 feet in 12 seconds. That's an average rate of speed of $\frac{120 \text{ feet}}{12 \text{ seconds}}$, or 10 feet per second. Amazingly, today's passenger jets have cruising speeds of 500 mph, which is about 730 feet per second!

When calculating average rates of speed, we use the formula $r = \frac{d}{t}$, where r is the rate, d is the distance traveled, and t is the time traveled at that rate. This formula is an example of a *rational equation*. To learn more about rational equations, visit *The Learning Equation* on the Internet at http://tle.brookscole.com. (The log-in instructions are in the Preface.) For Chapter 6, the online lesson is:

- *TLE* Lesson 9: Solving Rational Equations

Check Your Knowledge

1. The _____ of a fraction can never be 0.

2. If two angles of one triangle have the same measure as two angles of a second triangle, the triangles are _____.

3. The equation $y = kx$ defines _____ variation, and $y = \dfrac{k}{x}$ defines _____ variation.

4. A proposed solution of an equation that does not satisfy the equation is called a(n) _____ solution.

Simplify each rational expression.

5. $\dfrac{6x^5y^4z^3}{3x^3y^4z^5}$

6. $\dfrac{a^2 - a}{1 - a}$

7. The time it takes to make a trip varies inversely with the average speed for the trip. If a trip takes 8 hours at an average speed of 50 miles per hour, how long will the trip take at 60 miles per hour?

8. Graph the rational function $f(x) = \dfrac{12}{x}$ for $x > 0$. Label the horizontal asymptote.

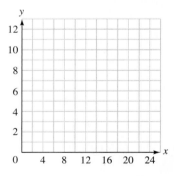

Perform the operations and simplify, if possible.

9. $\dfrac{x^2y^2z}{xy} \cdot \dfrac{x^3y^3}{z^3}$

10. $\dfrac{ax + ay + bx + by}{x^3 - 27} \div \dfrac{xc + xd + yc + yd}{x^2 + 3x + 9}$

11. $\dfrac{p^3 - q^3}{q^2 - p^2} \cdot \dfrac{q^2 + pq}{p^3 + p^2q + pq^2}$

12. $\dfrac{x}{x + y} + \dfrac{y}{x + y}$

13. $\dfrac{x^2 + x}{x - 2} + \dfrac{6}{2 - x}$

14. $\dfrac{x + 3}{2x^2 - 5x + 2} - \dfrac{3x - 1}{x^2 - x - 2}$

Simplify each complex fraction.

15. $\dfrac{x + \dfrac{1}{2x}}{x - \dfrac{1}{3x}}$

16. $\dfrac{\dfrac{h}{h^2 + 3h + 2}}{\dfrac{4}{h + 2} - \dfrac{4}{h + 1}}$

Solve each equation. If a solution is extraneous, so indicate.

17. $\dfrac{1}{3} + \dfrac{1}{x} = \dfrac{1}{2}$

18. $\dfrac{3 + 2a}{a^2 + 6 + 5a} + \dfrac{2 - 5a}{a^2 - 4} = \dfrac{2 - 3a}{a^2 - 6 + a}$

19. $\dfrac{3}{r} = \dfrac{12}{4r - r^2} - \dfrac{7}{r - 4}$

20. Solve $\dfrac{1}{a} + \dfrac{1}{b} = \dfrac{1}{c}$ for c.

21. John can type a 20-page term paper in 3 hours, but George takes 6 hours for the same job. How long will the job take if John and George work together?

22. Divide: $\dfrac{25x^3y^2 - 15x^2y^3 + 20xy}{10x^3y}$.

23. Long divide: $\dfrac{x^3 + 8}{x + 2}$.

24. Use synthetic division to divide $x^3 - 5x^2 + 6$ by $x - 2$.

Study Skills Workshop

HOW TO FORM A STUDY GROUP

In the Study Skills Workshop in Chapter 3, one assignment was to create an interest list about forming a study group. If you have not already done so, you may want to consider this idea again. Peer study groups allow you to ask questions in a nonthreatening environment and also to help fellow students by explaining a solution to a problem to them. Teaching someone else is one of the best ways to learn a topic. If you have already started a group, take some time to evaluate how well it's working and fine-tune the rules under which it operates.

Group Size. Ideally, a study group should be small. Three or four people per group is best, more than four can be too chaotic.

Time and Place. Establish a common meeting time once a week, usually two days before an assignment is due, for all group participants. This can be done by sharing your calendars and comparing study times. Find a meeting place that is convenient and practical for all members. There should be enough room to spread out and a place where you can talk (sometimes the conversation becomes quite animated!). Quiet sections in the library are *not* good meeting sites. Some schools have designated study areas that don't have to be kept quiet, like learning centers or math labs. Coffee shops or restaurants at off-peak times might be good places to meet, as are peoples' homes — a kitchen table can be an excellent place if you have your family's blessing. You may want to schedule an extra day to meet with your group during a test week.

How it Works. Don't expect to start and finish a complete homework assignment within your study group — at least attempt each problem before you meet with the group. During your private study time, make a list of the problems that you had trouble with, noting exactly where in the process you got stuck. Hopefully, someone in your group was able to do that problem correctly. If not, try verbalizing the problem with your group and brainstorm about possible ways to solve it. If there are still unanswered questions at the end of your meeting time, write these down. Plan to attend your instructor's office hour or see your group again the next day with this list.

As you become more familiar with your group, you may decide that it is working well without any formal rules: however, if you notice problems, it is okay to communicate your dissatisfaction to the group and ask whether a few ground-rules might make your group effective and organized.

▉ ASSIGNMENT

1. Are you currently involved with a study group?
   a. If so, evaluate how well it is working. Are there difficulties with the group that need to be addressed? Are there improvements that you would like to see made?
   b. If not, evaluate how well you are doing in the class at this point. If you're not sure, schedule an appointment with your instructor to find out. If you are not doing as well as expected, try forming a study group.
      i. Contact students either before or after your next class meeting to see whether you can find someone who can meet when you're free.
      ii. Find a meeting place and schedule a meeting with your group.
      iii. After the meeting, decide whether it helped. If it didn't, are there things that could be changed to make it work?

In this chapter, we extend the concept of fractions to include quotients of two polynomials, called rational expressions.

6.1 Rational Functions and Simplifying Rational Expressions

- Rational expressions   • Rational functions   • Graphing rational functions
- Finding the domain of a rational function   • Simplifying rational expressions
- Simplifying rational expressions by factoring out -1

We have seen that linear and polynomial functions can be used to model many real-world situations. In this section, we introduce another family of functions known as *rational functions*. Rational functions get their name from the fact that their defining equation contains a *ratio* (fraction) of two polynomials.

Rational expressions

Rational expressions are fractions that indicate the quotient (or ratio) of two polynomials, such as

$$\frac{-8y^3z^5}{6y^4z^3}, \qquad \frac{3x}{x-7}, \qquad \frac{5m+n}{8m+16}, \qquad \text{and} \qquad \frac{6a^2-13a+6}{3a^2+a-2}$$

! COMMENT Since division by 0 is undefined, the value of a polynomial in the denominator of a rational expression cannot be 0. For example, x cannot be 7 in the rational expression $\frac{3x}{x-7}$, because the value of the denominator would be 0. In the rational expression $\frac{5m+n}{8m+16}$, m cannot be -2, because the value of the denominator would be 0.

Rational functions

Rational expressions often define functions. For example, if the cost of subscribing to an online information network is $6 per month plus $1.50 per hour of access time, the average (mean) hourly cost of the service is the total monthly cost, divided by the number of hours of access time used that month:

$$\frac{C}{n} = \frac{1.50n+6}{n} \qquad \begin{array}{l}\text{C is the total monthly cost, and n is the number of} \\ \text{hours the service is used that month.}\end{array}$$

The right-hand side of this equation is a rational expression: the quotient of the binomial $1.50n + 6$ and the monomial n.

The rational function that gives the average hourly cost of using the online information network for n hours per month can be written as

$$f(n) = \frac{1.50n+6}{n}$$

We are assuming that at least one access call will be made each month, so the function is defined for $n > 0$.

Rational functions

A **rational function** is a function whose equation is defined by a rational expression in one variable, where the value of the polynomial in the denominator is never zero.

INTERMEDIATE
Algebra $f(x)$ Now™

EXAMPLE 1   Use the function $f(n) = \dfrac{1.50n + 6}{n}$ to find the average hourly cost when the network described earlier is used for   **a.** 1 hour   and   **b.** 9 hours.

Solution

a. To find the average hourly cost for 1 hour of access time, we find $f(1)$:

$$f(1) = \frac{1.50(1) + 6}{1} = 7.5 \quad \text{Input 1 for } n \text{ and simplify.}$$

The average hourly cost for 1 hour of access time is $7.50.

b. To find the average hourly cost for 9 hours of access time, we find $f(9)$:

$$f(9) = \frac{1.50(9) + 6}{9} = 2.166666666\ldots \quad \text{Input 9 for } n \text{ and simplify.}$$

The average hourly cost for 9 hours of access time is approximately $2.17.

Self Check 1
Find the average hourly cost when the network is used for
a. 3 hours
b. 100 hours

Answers **a.** $3.50,   **b.** $1.56

Graphing rational functions

To graph the rational function $f(n) = \frac{1.50n + 6}{n}$, we substitute values for n (the inputs) in the equation, compute the corresponding values of $f(n)$ (the outputs), and express the results as ordered pairs. From the evaluations in Example 1 and its Self Check, we know four such ordered pairs: $(1, 7.50)$, $(3, 3.50)$, $(9, 2.17)$, and $(100, 1.56)$. Those pairs and others are listed in the table in Figure 6-1. We then plot the points and draw a smooth curve through them to get the graph.

$f(n) = \dfrac{1.50n + 6}{n}$		
n	$f(n)$	$(n, f(n))$
1	7.50	(1, 7.50)
2	4.50	(2, 4.50)
3	3.50	(3, 3.50)
4	3.00	(4, 3.00)
5	2.70	(5, 2.70)
6	2.50	(6, 2.50)
7	2.36	(7, 2.36)
8	2.25	(8, 2.25)
9	2.17	(9, 2.17)
10	2.10	(10, 2.10)
100	1.56	(100, 1.56)

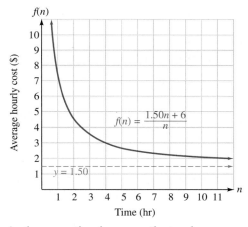

As the access time increases, the graph approaches the line $y = 1.50$, which indicates that the average hourly cost approaches $1.50 as the hours of use increase.

FIGURE 6-1

From the graph, we can see that the average hourly cost decreases as the number of hours of access time increases. Since the cost of each extra hour of access time is $1.50, the average hourly cost can approach $1.50 but never drop below it. Thus, the graph of the function approaches the line $y = 1.5$ as n increases. When a graph approaches a line, we call the line an **asymptote.** The line $y = 1.5$ is a **horizontal asymptote** of the graph.

As n gets smaller and approaches 0, the graph approaches the y-axis. The y-axis is a **vertical asymptote** of the graph.

Finding the domain of a rational function

Since division by 0 is undefined, any values that make the denominator 0 in a rational function must be excluded from the domain of the function.

Self Check 2
Find the domain of
$$f(x) = \frac{x^2 + 1}{x - 2}$$

EXAMPLE 2   Find the domain of $f(x) = \dfrac{3x + 2}{x^2 + x - 6}$.

Solution   From the set of real numbers, we must exclude any values of x that make the denominator 0. To find these values, we set $x^2 + x - 6$ equal to 0 and solve for x.

$$x^2 + x - 6 = 0$$

$(x + 3)(x - 2) = 0$ Factor the trinomial.

$x + 3 = 0$ or $x - 2 = 0$ Set each factor equal to 0.

$x = -3$ | $x = 2$ Solve each linear equation.

Thus, the domain of the function is the set of all real numbers except -3 and 2. In interval notation, the domain is $(-\infty, -3) \cup (-3, 2) \cup (2, \infty)$.

Answer $(-\infty, 2) \cup (2, \infty)$

CALCULATOR SNAPSHOT   Finding the domain and range of a rational function

We can find the domain and range of the function in Example 2 by looking at its graph. If we use window settings of $[-10, 10]$ for x and $[-10, 10]$ for y and graph the function

$$f(x) = \frac{3x + 2}{x^2 + x - 6}$$

we will obtain the graph in Figure 6-2(a) on the next page.

From the figure, we can see that

- As x approaches -3 from the left, the values of y decrease, and the graph approaches the vertical line $x = -3$.
- As x approaches -3 from the right, the values of y increase, and the graph approaches the vertical line $x = -3$.

From the figure, we can also see that

- As x approaches 2 from the left, the values of y decrease, and the graph approaches the vertical line $x = 2$.
- As x approaches 2 from the right, the values of y increase, and the graph approaches the vertical line $x = 2$.

The lines $x = -3$ and $x = 2$ are vertical asymptotes. Although the vertical lines in the graph appear to be the graphs of $x = -3$ and $x = 2$, they are not. Graphing calculators draw graphs by connecting dots whose x-coordinates are close together. Often when two such points straddle a vertical asymptote and their y-coordinates are far apart, the calculator draws a line between them anyway, producing what appears to be a vertical asymptote. If you set your calculator to dot mode instead of connected mode, the vertical lines will not appear.

From Figure 6-2(a), we can also see that

- As x increases to the right of 2, the values of y decrease and approach the line $y = 0$.
- As x decreases to the left of -3, the values of y increase and approach the line $y = 0$.

The line $y = 0$ (the x-axis) is a horizontal asymptote. Graphing calculators do not draw lines that appear to be horizontal asymptotes.

From the graph, we can see that every real number x, except -3 and 2, gives a value of y. This observation confirms that the domain of the function is $(-\infty, -3) \cup (-3, 2) \cup (2, \infty)$. We can also see that y can be any value. Thus, the range is $(-\infty, \infty)$.

To find the domain and range of the function $f(x) = \frac{2x + 1}{x - 1}$, we use a calculator to draw the graph shown in Figure 6-2(b). From this graph, we can see that the line $x = 1$ is a vertical asymptote and that the line $y = 2$ is a horizontal asymptote. Since x can be any real number except 1, the domain is the interval $(-\infty, 1) \cup (1, \infty)$. Since y can be any value except 2, the range is $(-\infty, 2) \cup (2, \infty)$.

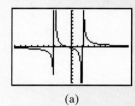　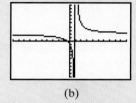

(a)　　　　　　　　　　　　　(b)

FIGURE 6-2

Learning and Remembering

THINK IT THROUGH

"Most students express a wish to be more efficient in their studies. Knowing how your brain takes in and processes information, and then working with this system, will greatly improve your efficiency."

From the *Study Skills Package,* University of Waterloo Counseling Services

The graph below is called the *curve of forgetting*. It shows how quickly a typical student forgets the new information presented in a one-hour lecture if he or she does not review the material later. Use the graph to estimate each of the following.

1. What percent of the information is retained by Day 2?

2. What percent of the information is forgotten by Day 7?

3. What percent of the information is retained by Day 30?

4. Do you think the curve of forgetting has an asymptote? Explain why or why not.

For ways to improve retention, visit the Web site
www.adm.**uwaterloo**.ca/infocs/study_skills/**curve**.html

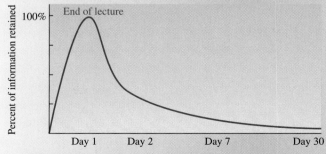

Source: The University of Waterloo, Canada, Counseling Service

Simplifying rational expressions

When working with rational expressions, we will use the familiar rules from arithmetic.

Properties of fractions

If a, b, c, d, and k represent real numbers, and if there are no divisions by 0, then

1. $\dfrac{a}{b} = \dfrac{c}{d}$ if and only if $ad = bc$ 2. $\dfrac{a}{1} = a$ and $\dfrac{a}{a} = 1$

3. $\dfrac{ak}{bk} = \dfrac{a}{b} \cdot \dfrac{k}{k} = \dfrac{a}{b}$ 4. $-\dfrac{a}{b} = \dfrac{-a}{b} = \dfrac{a}{-b}$

Property 3 of fractions is true because

$$\frac{ak}{bk} = \frac{a}{b} \cdot \frac{k}{k} = \frac{a}{b} \cdot 1 = \frac{a}{b} \qquad \text{Any number times 1 is the number.}$$

Property 3, which is known as the **fundamental property of fractions,** is used to simplify rational expressions. It enables us to divide out factors that are common to the numerator and the denominator of a fraction.

Simplifying a rational expression

To simplify a rational expression,

1. Factor the numerator and the denominator.
2. Divide out the common factors of the numerator and the denominator.

INTERMEDIATE
Algebra $f(x)$ **Now™**

Self Check 3

Simplify: $\dfrac{12a^4b^2}{3ab^4}$

Answer $\dfrac{4a^3}{b^2}$

EXAMPLE 3   Simplify: **a.** $\dfrac{10k}{25k^2}$ and **b.** $-\dfrac{8y^3z^5}{6y^4z^3}$.

Solution   To simplify these rational expressions, we factor each numerator and denominator and divide out all common factors.

a. $\dfrac{10k}{25k^2} = \dfrac{2 \cdot 5 \cdot k}{5 \cdot 5 \cdot k \cdot k}$

$= \dfrac{2 \cdot \overset{1}{\cancel{5}} \cdot \overset{1}{\cancel{k}}}{\underset{1}{\cancel{5}} \cdot 5 \cdot \underset{1}{\cancel{k}} \cdot k}$ Divide out the common factors of 5 and k. Show this using slashes and 1's.

$= \dfrac{2}{5k}$ Do the multiplications in the numerator and the denominator.

b. $-\dfrac{8y^3z^5}{6y^4z^3} = -\dfrac{2 \cdot 4 \cdot y \cdot y \cdot y \cdot z \cdot z \cdot z \cdot z \cdot z}{2 \cdot 3 \cdot y \cdot y \cdot y \cdot y \cdot z \cdot z \cdot z}$

$= -\dfrac{\overset{1}{\cancel{2}} \cdot 4 \cdot \overset{1}{\cancel{y}} \cdot \overset{1}{\cancel{y}} \cdot \overset{1}{\cancel{y}} \cdot \overset{1}{\cancel{z}} \cdot \overset{1}{\cancel{z}} \cdot \overset{1}{\cancel{z}} \cdot z \cdot z}{\underset{1}{\cancel{2}} \cdot 3 \cdot \underset{1}{\cancel{y}} \cdot \underset{1}{\cancel{y}} \cdot \underset{1}{\cancel{y}} \cdot y \cdot \underset{1}{\cancel{z}} \cdot \underset{1}{\cancel{z}} \cdot \underset{1}{\cancel{z}}}$

$= -\dfrac{4z^2}{3y}$

The fractions in Example 3 can also be simplified using the rules of exponents:

$$\frac{10k}{25k^2} = \frac{2 \cdot 5}{5 \cdot 5}k^{1-2}$$

$$= \frac{2}{5} \cdot k^{-1}$$

$$= \frac{2}{5} \cdot \frac{1}{k}$$

$$= \frac{2}{5k}$$

$$-\frac{8y^3z^5}{6y^4z^3} = -\frac{2 \cdot 4}{2 \cdot 3}y^{3-4}z^{5-3}$$

$$= -\frac{4}{3} \cdot y^{-1}z^2$$

$$= -\frac{4}{3} \cdot \frac{1}{y} \cdot \frac{z^2}{1}$$

$$= -\frac{4z^2}{3y}$$

INTERMEDIATE
Algebra $f(x)$ **Now**™

EXAMPLE 4   Simplify: $\dfrac{x^2 - 16}{x + 4}$.

Solution   We factor $x^2 - 16$ and use the fact that $\frac{x+4}{x+4} = 1$.

$$\frac{x^2 - 16}{x + 4} = \frac{(\overset{1}{\cancel{x + 4}})(x - 4)}{(\underset{1}{\cancel{x + 4}})} \qquad \text{Factor the difference of two squares.}$$

$$= \frac{x - 4}{1} \qquad \text{Divide out the common factor of } x + 4.$$

$$= x - 4$$

Self Check 4

Simplify: $\dfrac{x^2 - 9}{x - 3}$.

Answer $x + 3$

INTERMEDIATE
Algebra $f(x)$ **Now**™

EXAMPLE 5   Simplify: $\dfrac{6a^2 - 13a + 6}{3a^2 + a - 2}$.

Solution   We factor the trinomials in the numerator and the denominator and then divide out the common factor.

$$\frac{6a^2 - 13a + 6}{3a^2 + a - 2} = \frac{(\overset{1}{\cancel{3a - 2}})(2a - 3)}{(\underset{1}{\cancel{3a - 2}})(a + 1)}$$

$$= \frac{2a - 3}{a + 1}$$

Self Check 5

Simplify: $\dfrac{2b^2 + 7b - 15}{2b^2 + 13b + 15}$.

Answer $\dfrac{2b - 3}{2b + 3}$

We will encounter many fractions that are already in simplified form. For example, to attempt to simplify

$$\frac{x^2 + xa + 2x + 2a}{x^2 + x - 6}$$

we factor the numerator and denominator and divide out any common factors:

$$\frac{x^2 + xa + 2x + 2a}{x^2 + x - 6} = \frac{x(x + a) + 2(x + a)}{(x - 2)(x + 3)} = \frac{(x + a)(x + 2)}{(x - 2)(x + 3)}$$

Since there are no common factors in the numerator and denominator, the fraction is in *lowest terms*. It cannot be simplified.

! COMMENT Only *factors* that are common to the entire numerator and the entire denominator can be divided out. *Terms* common to both the numerator and denominator cannot be divided out. It is incorrect to divide out the common term of 3 in the following simplification, because it gives a wrong answer.

$$\frac{3 + 7}{3} = \frac{\overset{1}{\cancel{3}} + 7}{\underset{1}{\cancel{3}}} = \frac{1 + 7}{1} = 8 \qquad \text{The correct simplification is } \frac{3 + 7}{3} = \frac{10}{3}.$$

To be divided out, the 3 must be a factor of the entire numerator.

Similarly, it is not correct to divide out the y in the fraction $\frac{x^2 y + 6x}{y}$, because y is not a factor of the entire numerator.

■ Simplifying rational expressions by factoring out −1

To simplify $\frac{b - a}{a - b}$ where $a \neq b$, we factor −1 from the numerator and divide out any factors common to both the numerator and the denominator:

$$\frac{b - a}{a - b} = \frac{-a + b}{a - b} \qquad \text{Rewrite the numerator.}$$

$$= \frac{-1(a - b)}{(a - b)} \qquad \text{Factor out } -1 \text{ from each term in the numerator.}$$

$$= \frac{-1(a \overset{1}{\cancel{- b}})}{\underset{1}{a \cancel{- b}}} \qquad \text{Divide out the common factor.}$$

$$= \frac{-1}{1}$$

$$= -1$$

In general, we have the following principle.

> **Quotient of a quantity and its opposite**
>
> The quotient of any nonzero quantity and its negative (or opposite) is −1.

INTERMEDIATE
Algebra *f(x)* **Now**™

Self Check 6

Simplify: $\dfrac{2a^2 - 3ab - 9b^2}{3b^2 - ab}$.

EXAMPLE 6 Simplify: $\dfrac{3x^2 - 10xy - 8y^2}{4y^2 - xy}$.

Solution We factor the numerator and denominator. Because $x - 4y$ and $4y - x$ are negatives, their quotient is −1.

$$\frac{3x^2 - 10xy - 8y^2}{4y^2 - xy} = \frac{(3x + 2y)(x \overset{-1}{\cancel{- 4y}})}{y(\underset{1}{4y \cancel{- x}})}$$

$$= \frac{-(3x + 2y)}{y}$$

$$= \frac{-3x - 2y}{y}$$

Answer $-\dfrac{2a + 3b}{b}$ or $\dfrac{-2a - 3b}{b}$

Checking an algebraic simplification

After simplifying an expression, we can use a scientific calculator to check the answer. One way to check whether $\frac{x^2 - 16}{x + 4} = x - 4$ is correct in Example 4 is to evaluate $\frac{x^2 - 16}{x + 4}$ and $x - 4$ for a value of x (say, 3). The expressions should give identical results.

$$\boxed{(}\ \boxed{3}\ \boxed{x^2}\ \boxed{-}\ \boxed{16}\ \boxed{)}\ \boxed{\div}\ \boxed{(}\ \boxed{3}\ \boxed{+}\ \boxed{4}\ \boxed{)}\ \boxed{=}\qquad\boxed{-1\,}$$

Since $x - 4$ is -1 when $x = 3$, the results of the evaluations are the same. Evaluate the expressions for several other values of x. If the results differ for any given value, the original expression was not simplified correctly.

A graphing calculator can also be used to show that the simplification in Example 4 is correct. To do this, we enter the functions $f(x) = \frac{x^2 - 16}{x + 4}$ and $g(x) = x - 4$ as Y_1 and Y_2, respectively. See Figure 6-3(a). Then select the TABLE feature. Reading across the table, the values of Y_1 and Y_2 should be the same for each value of x as shown in Figure 6-3(b).

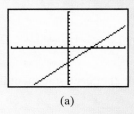 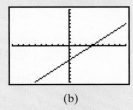

FIGURE 6-3

To use a third method, we can compare the graphs of $f(x) = \frac{x^2 - 16}{x + 4}$, shown in Figure 6-4(a), and $g(x) = x - 4$, shown in Figure 6-4(b). Except for the point where $x = -4$, the graphs are the same. The point where $x = -4$ is excluded from the graph of $f(x) = \frac{x^2 - 16}{x + 4}$, because -4 is not in the domain of f. However, graphing calculators do not show that this point is excluded. The point where $x = -4$ is included in the graph of $g(x) = x - 4$, because -4 is in the domain of g.

```
  (a)            (b)
```

FIGURE 6-4

Section 6.1 STUDY SET

INTERMEDIATE
Algebra *f(x)* Now™

VOCABULARY *Fill in the blanks.*

1. A fraction that is the quotient of two polynomials, such as $\frac{x^2 - x - 6}{x^3 - 8}$, is called a _____ expression.

2. In the rational expression $\frac{(x + 2)(3x - 1)}{(x + 2)(4x + 2)}$, $x + 2$ is a common _____ of the numerator and the denominator.

3. If a graph approaches a line, the line is called an
_____ .

4. The _____ of a fraction can
never be 0.

CONCEPTS

5. The graph of rational function f for $x > 0$ is shown
below. Find each of the following.

a. $f(1)$ **b.** $f(2)$

c. $f(4)$

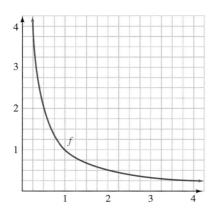

6. Show that $\dfrac{x - y}{y - x} = -1$ by first factoring out -1 from
each term in the numerator.

7. Simplify each expression.

a. $\dfrac{x + 8}{x + 8}$ **b.** $\dfrac{x + 8}{8 + x}$

c. $\dfrac{x - 8}{x - 8}$ **d.** $\dfrac{x - 8}{8 - x}$

8. Simplify each expression, if possible.

a. $\dfrac{x + 8}{x}$ **b.** $\dfrac{x + 8}{8}$

c. $\dfrac{a^3 + 8}{2}$ **d.** $\dfrac{x^2 + 5x + 6}{x^2 + x - 12}$

*Refer to the graphs below. Each graph shows the average cost
to manufacture a certain item for a given number of units
produced.*

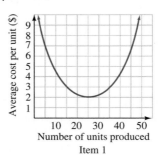

Item 1

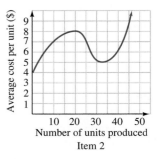

Item 2

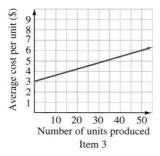

Item 3

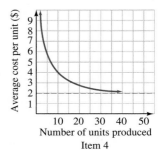

Item 4

9. MANUFACTURING For each graph, briefly
describe how the average cost per unit changes as
the number of units produced increases.

10. Which graph is best described as the graph of a

a. linear function?   **b.** quadratic function?

c. rational function?   **d.** polynomial function?

NOTATION

11. Determine whether each statement is true or false.

a. $-\dfrac{x - 4}{x + 4} = \dfrac{4 - x}{x + 4}$ **b.** $\dfrac{a - 3b}{2b - a} = \dfrac{3b - a}{a - 2b}$

12. Explain what the slashes and the 1's show.

$$\frac{t^2 - 4}{t^2 + 2t} = \frac{\overset{1}{(t + 2)}(t - 2)}{t\underset{1}{(t + 2)}} = \frac{t - 2}{t}$$

PRACTICE *The time t it takes to travel 600 miles is a
function of the average rate of speed, $t = \frac{600}{r}$. Find t for
each value of r.*

13. 30 mph   **14.** 40 mph

15. 50 mph   **16.** 60 mph

Complete the table for each rational function (round to the nearest hundredth when applicable). Then graph it. Each function is defined for $x > 0$. Identify the horizontal and vertical asymptotes.

17. $f(x) = \dfrac{6}{x}$

x	f(x)
1	
2	
4	
6	
8	
10	
12	

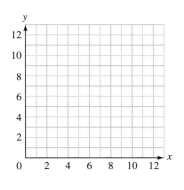

18. $f(x) = \dfrac{12}{x}$

x	f(x)
1	
4	
8	
12	
16	
20	
24	

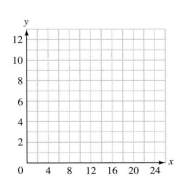

19. $f(x) = \dfrac{x + 2}{x}$

x	f(x)
1	
2	
4	
6	
8	
10	
12	

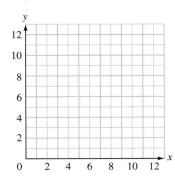

20. $f(x) = \dfrac{2x + 4}{x}$

x	f(x)
1	
4	
8	
12	
16	
20	
24	

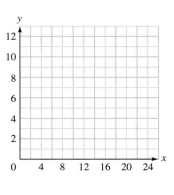

Find the domain of each rational function. Use interval notation.

21. $f(x) = \dfrac{2}{x}$

22. $f(x) = \dfrac{8}{x - 1}$

23. $f(x) = \dfrac{2}{x + 2}$

24. $f(x) = \dfrac{2}{x^2 - 2x}$

25. $f(x) = \dfrac{2}{x - x^2}$

26. $f(x) = \dfrac{2}{x^2 - 36}$

27. $f(x) = \dfrac{2}{x^2 - x - 56}$

28. $f(x) = \dfrac{2}{x^2 + 2x - 24}$

Simplify each rational expression when possible.

29. $\dfrac{12}{18}$

30. $\dfrac{25}{55}$

31. $-\dfrac{112}{36}$

32. $-\dfrac{49}{21}$

33. $\dfrac{12x^3}{3x}$

34. $-\dfrac{15a^2}{25a^3}$

35. $\dfrac{-24x^3y^4}{18x^4y^3}$

36. $\dfrac{15a^5b^4}{21b^3c^2}$

37. $-\dfrac{11x(x - y)}{22(x - y)}$

38. $\dfrac{x(x - 2)^2}{(x - 2)^3}$

39. $\dfrac{(a - b)(d - c)}{(c - d)(a - b)}$

40. $\dfrac{(p + q)(p - r)}{(r - p)(p + q)}$

41. $\dfrac{y + x}{x^2 - y^2}$

42. $\dfrac{x - y}{x^2 - y^2}$

68. $\dfrac{p^3 + p^2q - 2pq^2}{pq^2 + p^2q - 2p^3}$

69. $\dfrac{x^4 - y^4}{(x^2 + 2xy + y^2)(x^2 + y^2)}$

43. $\dfrac{5x - 10}{x^2 - 4x + 4}$

44. $\dfrac{y - xy}{xy - x}$

70. $\dfrac{(x^2 - 1)(x + 1)}{(x^2 - 2x + 1)^2}$

45. $\dfrac{12 - 3x^2}{x^2 - x - 2}$

46. $\dfrac{x^2 + 2x - 15}{25 - x^2}$

71. $\dfrac{6xy - 4x - 9y + 6}{6y^2 - 13y + 6}$

72. $\dfrac{x^2 + 2xy}{x + 2y + x^2 - 4y^2}$

47. $\dfrac{x^2 + y^2}{x + y}$

48. $\dfrac{3x + 6y}{2y + x}$

73. $\dfrac{(2x^2 + 3xy + y^2)(3a + b)}{(x + y)(2xy + 2bx + y^2 + by)}$

49. $\dfrac{x^3 + 8}{x^2 - 2x + 4}$

50. $\dfrac{x^2 + 3x + 9}{x^3 - 27}$

74. $\dfrac{(x - 1)(6ax + 9x + 4a + 6)}{(3x + 2)(2ax - 2a + 3x - 3)}$

75. $\dfrac{(x^2 + 2x + 1)(x^2 - 2x + 1)}{(x^2 - 1)^2}$

51. $\dfrac{x^2 + 2x + 1}{x^2 + 4x + 3}$

52. $\dfrac{6x^2 + x - 2}{8x^2 + 2x - 3}$

76. $\dfrac{2x^2 + 2x - 12}{x^3 + 3x^2 - 4x - 12}$

53. $\dfrac{sx + 4s - 3x - 12}{sx + 4s + 6x + 24}$

54. $\dfrac{ax + by + ay + bx}{a^2 - b^2}$

Use a graphing calculator to graph each function. From the graph, determine its domain and range.

55. $\dfrac{4x^2 + 24x + 32}{16x^2 + 8x - 48}$

56. $\dfrac{a^2 - 4}{a^3 - 8}$

77. $f(x) = \dfrac{x}{x - 2}$

78. $f(x) = \dfrac{x + 2}{x}$

57. $\dfrac{3x^2 - 3y^2}{x^2 + 2y + 2x + yx}$

58. $\dfrac{x^2 + x - 30}{x^2 - x - 20}$

79. $f(x) = \dfrac{x + 1}{x^2 - 4}$

80. $f(x) = \dfrac{x - 2}{x^2 - 3x - 4}$

59. $\dfrac{4x^2 + 8x + 3}{6 + x - 2x^2}$

60. $\dfrac{6x^2 + 13x + 6}{6 - 5x - 6x^2}$

APPLICATIONS

81. ENVIRONMENTAL CLEANUP Suppose the cost (in dollars) of removing $p\%$ of the pollution in a river is given by the rational function

$$f(p) = \dfrac{50{,}000p}{100 - p} \qquad (0 \le p < 100)$$

Find the cost of removing each percent of pollution.

a. 50%   **b.** 80%

61. $\dfrac{a^3 + 27}{4a^2 - 36}$

62. $\dfrac{a - b}{b^2 - a^2}$

63. $\dfrac{2x^2 - 3x - 9}{2x^2 + 3x - 9}$

64. $\dfrac{6x^2 - 7x - 5}{2x^2 + 5x + 2}$

82. DIRECTORY COSTS The average (mean) cost for a service club to publish a directory of its members is given by the rational function

$$f(x) = \dfrac{1.25x + 700}{x}$$

where x is the number of directories printed. Find the average cost per directory if

65. $\dfrac{(m + n)^3}{m^2 + 2mn + n^2}$

66. $\dfrac{x^3 - 27}{3x^2 - 8x - 3}$

a. 500 directories are printed.

b. 2,000 directories are printed.

67. $\dfrac{m^3 - mn^2}{mn^2 + m^2n - 2m^3}$

83. UTILITY COSTS An electric company charges $7.50 per month plus 9¢ for each kilowatt hour (kwh) of electricity used.

   a. Find a linear function that gives the total cost of n kwh of electricity.

   b. Find a rational function that gives the average cost per kwh when using n kwh.

   c. Find the average cost per kwh when 775 kwh are used.

84. SCHEDULING WORK CREWS The rational function

$$f(t) = \frac{t^2 + 2t}{2t + 2}$$

gives the number of days it would take two construction crews, working together, to frame a house that crew 1 (working alone) could complete in t days and crew 2 (working alone) could complete in $t + 2$ days.

   a. If crew 1 could frame a certain house in 15 days, how long would it take both crews working together?

   b. If crew 2 could frame a certain house in 20 days, how long would it take both crews working together?

85. FILLING A POOL The rational function

$$f(t) = \frac{t^2 + 3t}{2t + 3}$$

gives the number of hours it would take two pipes, working together, to fill a pool that the larger pipe (working alone) could fill in t hours and the smaller pipe (working alone) could fill in $t + 3$ hours.

   a. If the smaller pipe could fill a pool in 7 hours, how long would it take both pipes to fill the pool?

   b. If the larger pipe could fill a pool in 8 hours, how long would it take both pipes to fill the pool?

86. RETENTION STUDY After learning a list of words, two subjects were tested over a 28-day period to see what percent of the list they remembered. In both cases, their percent recall could be modeled by rational functions, as shown below.

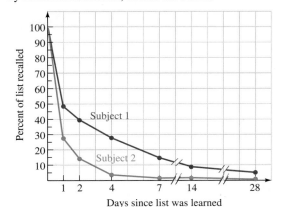

   a. Use the graphs to complete the table.

Days since learning	0	1	2	4	7	14	28
% recall — subject 1							
% recall — subject 2							

   b. After 28 days, which subject had the better recall?

WRITING

87. Explain why the domain of the rational function

$$f(x) = \frac{4x + 2}{x^2 - 25}$$

   does not include 5 or -5.

88. Explain why $\dfrac{x - 2}{2 - x} = -1$ and why $\dfrac{x + 2}{2 + x} = 1$.

REVIEW *Factor each expression.*

89. $3x^2 - 9x$

90. $-6t^2 + 5t + 6$

91. $27x^6 + 64y^3$

92. $x^2 + ax + 2x + 2a$

6.2 Multiplying and Dividing Rational Expressions

• Multiplying rational expressions • Finding powers of rational expressions
• Dividing rational expressions • Mixed operations

In this section, we begin with a review of the rules for multiplying and dividing arithmetic fractions — fractions whose numerators and denominators are integers. Then we use these rules, in combination with the simplification skills learned in Section 6.1, to find products and quotients of rational expressions.

▌ Multiplying rational expressions

In Section 6.1, we introduced four basic properties of fractions. We now discuss the rule for multiplying fractions.

Multiplying fractions

If a, b, c, and d represent real numbers and if no denominators are 0, then

$$\frac{a}{b} \cdot \frac{c}{d} = \frac{a \cdot c}{b \cdot d} = \frac{ac}{bd}$$

To multiply fractions, we multiply the numerators and multiply the denominators.

$$\frac{3}{5} \cdot \frac{2}{7} = \frac{3 \cdot 2}{5 \cdot 7} \qquad\qquad \frac{4}{7} \cdot \frac{5}{8} = \frac{4 \cdot 5}{7 \cdot 8}$$

$$= \frac{6}{35} \qquad\qquad\qquad = \frac{\overset{1}{2} \cdot \overset{1}{2} \cdot 5}{7 \cdot \underset{1}{2} \cdot \underset{1}{2} \cdot 2} \qquad \frac{2}{2} = 1.$$

$$\qquad\qquad\qquad\qquad\qquad = \frac{5}{14}$$

The same rule applies to rational expressions. If $t \neq 0$, then

$$\frac{x^2 y}{t} \cdot \frac{xy^3}{t^3} = \frac{x^2 y \cdot xy^3}{tt^3}$$

$$= \frac{x^2 x \cdot yy^3}{t^4}$$

$$= \frac{x^3 y^4}{t^4}$$

Self Check 1

Multiply: $\dfrac{a^2 + 6a + 9}{a} \cdot \dfrac{a^3}{a + 3}$.

Answer $a^2(a + 3)$

EXAMPLE 1   Find the product of $\dfrac{x^2 - 6x + 9}{x}$ and $\dfrac{x^2}{x - 3}$.

Solution   We multiply the numerators and multiply the denominators and then simplify the resulting fraction.

$$\frac{x^2 - 6x + 9}{x} \cdot \frac{x^2}{x - 3} = \frac{(x^2 - 6x + 9)x^2}{x(x - 3)} \qquad \text{Multiply the numerators and multiply the denominators.}$$

$$= \frac{(x - 3)(x - 3)xx}{x(x - 3)} \qquad \text{Factor in the numerator.}$$

$$= \frac{\overset{1}{(x - 3)}(x - 3)\overset{1}{x}x}{\underset{1}{x}\underset{1}{(x - 3)}} \qquad \text{Divide out common factors: } \frac{x - 3}{x - 3} = 1 \text{ and } \frac{x}{x} = 1.$$

$$= x(x - 3)$$

Checking an algebraic simplification

We can check the simplification in Example 1 by graphing the rational functions $f(x) = (\frac{x^2 - 6x + 9}{x})(\frac{x^2}{x - 3})$, shown in Figure 6-5(a), and $g(x) = x(x - 3)$, shown in Figure 6-5(b), and observing that the graphs are the same except that 0 and 3 are not included in the domain of the first function.

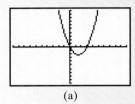
(a)

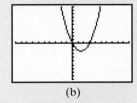
(b)

FIGURE 6-5

The split-screen G-T (graph, table) mode can also be used to check a simplification. If we enter $Y_3 = Y_1 - Y_2$, use the cursor to highlight the $=$ sign as shown in Figure 6-6(a), and then press $\boxed{\text{GRAPH}}$, we get the display shown in Figure 6-6(b). The zeros under the Y_3 column indicate that the value of $(\frac{x^2 - 6x + 9}{x})(\frac{x^2}{x - 3})$ and the value of $x(x - 3)$ are the same for different values of x. (The error message is given because when $x = 0$, $(\frac{x^2 - 6x + 9}{x})(\frac{x^2}{x - 3})$ is undefined.)

The graph of $Y_3 = Y_1 - Y_2$ is difficult to see because it lies on the x-axis. The graph indicates that for all x-values (except those that make the fractions undefined), $Y_3 = 0$. Or, more specifically, $(\frac{x^2 - 6x + 9}{x})(\frac{x^2}{x - 3}) = x(x - 3)$.

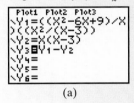
(a)

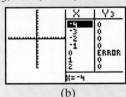
(b)

FIGURE 6-6

INTERMEDIATE
Algebra *f(x)* Now™

EXAMPLE 2 Multiply: $\dfrac{x^2 - x - 6}{x^2 - 4} \cdot \dfrac{x^2 + x - 6}{x^2 - 9}$.

Self Check 2

Solution

Multiply:
$$\dfrac{a^2 + a - 56}{a^2 - 49} \cdot \dfrac{a^2 - a - 56}{a^2 - 64}$$

$$\dfrac{x^2 - x - 6}{x^2 - 4} \cdot \dfrac{x^2 + x - 6}{x^2 - 9}$$

$$= \dfrac{(x^2 - x - 6)(x^2 + x - 6)}{(x^2 - 4)(x^2 - 9)} \qquad \text{Multiply the numerators and multiply the denominators.}$$

$$= \dfrac{(x - 3)(x + 2)(x + 3)(x - 2)}{(x + 2)(x - 2)(x + 3)(x - 3)} \qquad \text{Factor the polynomials.}$$

$$= \dfrac{\overset{1}{(\cancel{x - 3})}\overset{1}{(\cancel{x + 2})}\overset{1}{(\cancel{x + 3})}\overset{1}{(\cancel{x - 2})}}{\underset{1}{(\cancel{x + 2})}\underset{1}{(\cancel{x - 2})}\underset{1}{(\cancel{x + 3})}\underset{1}{(\cancel{x - 3})}} \qquad \text{Divide out common factors.}$$

$$= 1$$

Answer 1

⚠ COMMENT As in Example 2, when all factors divide out, the result is 1 and not 0.

Self Check 3
Multiply:

$$\frac{2a^2 + 5a - 12}{2a^2 + 11a + 12} \cdot \frac{2a^2 - 3a - 9}{2a^2 - a - 3}$$

EXAMPLE 3   Multiply: $\dfrac{6x^2 + 5x - 4}{2x^2 + 5x + 3} \cdot \dfrac{8x^2 + 6x - 9}{12x^2 + 7x - 12}$.

Solution

$$\frac{6x^2 + 5x - 4}{2x^2 + 5x + 3} \cdot \frac{8x^2 + 6x - 9}{12x^2 + 7x - 12}$$

$$= \frac{(6x^2 + 5x - 4)(8x^2 + 6x - 9)}{(2x^2 + 5x + 3)(12x^2 + 7x - 12)} \qquad \text{Multiply the numerators and multiply the denominators.}$$

$$= \frac{(3x + 4)(2x - 1)(4x - 3)(2x + 3)}{(2x + 3)(x + 1)(3x + 4)(4x - 3)} \qquad \text{Factor the polynomials.}$$

$$= \frac{\overset{1}{\cancel{(3x + 4)}}(2x - 1)\overset{1}{\cancel{(4x - 3)}}\overset{1}{\cancel{(2x + 3)}}}{\underset{1}{\cancel{(2x + 3)}}(x + 1)\underset{1}{\cancel{(3x + 4)}}\underset{1}{\cancel{(4x - 3)}}} \qquad \text{Divide out common factors.}$$

Answer $\dfrac{a - 3}{a + 1}$

$$= \frac{2x - 1}{x + 1}$$

Self Check 4
Multiply:

$$\frac{x^2 + 5x + 6}{(x^2 + 4x)(x + 2)}(x^3 + 4x^2)$$

EXAMPLE 4   Multiply: $(2x - x^2) \cdot \dfrac{x}{x^2 - 5x + 6}$.

Solution

$$(2x - x^2) \cdot \frac{x}{x^2 - 5x + 6}$$

$$= \frac{2x - x^2}{1} \cdot \frac{x}{x^2 - 5x + 6} \qquad \text{Write } 2x - x^2 \text{ as } \frac{2x - x^2}{1}.$$

$$= \frac{(2x - x^2)x}{1(x^2 - 5x + 6)} \qquad \text{Multiply the fractions.}$$

$$= \frac{x\overset{-1}{\cancel{(2 - x)}}x}{1\underset{1}{\cancel{(x - 2)}}(x - 3)} \qquad \begin{array}{l}\text{Factor out } x \text{ from } 2x - x^2 \text{ in the numerator and} \\ \text{factor the trinomial in the denominator. Recall that} \\ \text{the quotient of any nonzero quantity and its opposite} \\ \text{is } -1 \text{:} \frac{2 - x}{x - 2} = -1.\end{array}$$

$$= \frac{-x^2}{x - 3}$$

Since $\dfrac{-a}{b} = -\dfrac{a}{b}$, the $-$ sign can be written in front of the fraction. Thus, the final result can be written as

Answer $x(x + 3)$

$$-\frac{x^2}{x - 3}$$

In Examples 1–4, we would obtain the same answers if we had factored first and divided out the common factors before we multiplied.

Finding powers of rational expressions

EXAMPLE 5   Find $\left(\dfrac{x^2 + x - 1}{2x + 3}\right)^2$.

Solution   To square the rational expression, we write it as a factor twice and perform the multiplication.

$$\left(\frac{x^2 + x - 1}{2x + 3}\right)^2 = \left(\frac{x^2 + x - 1}{2x + 3}\right)\left(\frac{x^2 + x - 1}{2x + 3}\right)$$

$$= \frac{(x^2 + x - 1)(x^2 + x - 1)}{(2x + 3)(2x + 3)}$$

$$= \frac{x^4 + 2x^3 - x^2 - 2x + 1}{4x^2 + 12x + 9}$$

Self Check 5

Find $\left(\dfrac{x + 5}{x^2 - 6x}\right)^2$.

Answer $\dfrac{x^2 + 10x + 25}{x^4 - 12x^3 + 36x^2}$

Dividing rational expressions

Here is the rule for dividing fractions.

> **Dividing fractions**
>
> If a, b, c, and d represent real numbers and if no denominators are 0, then
>
> $$\frac{a}{b} \div \frac{c}{d} = \frac{a}{b} \cdot \frac{d}{c} = \frac{ad}{bc}$$

We can prove this rule as follows:

$$\frac{a}{b} \div \frac{c}{d} = \frac{\frac{a}{b}}{\frac{c}{d}} = \frac{\frac{a}{b}}{\frac{c}{d}} \cdot 1 = \frac{\frac{a}{b} \cdot \frac{d}{c}}{\frac{c}{d} \cdot \frac{d}{c}} = \frac{\frac{a}{b} \cdot \frac{d}{c}}{\frac{cd}{cd}} = \frac{\frac{a}{b} \cdot \frac{d}{c}}{1} = \frac{a}{b} \cdot \frac{d}{c} = \frac{ad}{bc}$$

Thus, *to divide two fractions, we can invert the divisor (the second fraction) and multiply.*

$$\frac{3}{5} \div \frac{2}{7} = \frac{3}{5} \cdot \frac{7}{2} \qquad\qquad \frac{4}{7} \div \frac{2}{21} = \frac{4}{7} \cdot \frac{21}{2}$$

$$= \frac{3 \cdot 7}{5 \cdot 2} \qquad\qquad\qquad = \frac{4 \cdot 21}{7 \cdot 2}$$

$$= \frac{21}{10} \qquad\qquad\qquad = \frac{\overset{1}{2} \cdot 2 \cdot 3 \cdot \overset{1}{7}}{\underset{1}{7} \cdot \underset{1}{2}}$$

$$\qquad\qquad\qquad\qquad = 6$$

We can state the rule for dividing two fractions in another way: *To divide two fractions, we multiply the first fraction and the reciprocal of the second.* The **reciprocal** of a fraction can be found by interchanging the numerator and denominator. For example, the reciprocal of $\frac{2}{7}$ is $\frac{7}{2}$. Similarly, the reciprocal of the rational expression $\frac{x^2 - 2x + 4}{2x^2 - 2}$ is $\frac{2x^2 - 2}{x^2 - 2x + 4}$.

The rule for division of fractions applies to rational expressions.

$$\frac{x^2}{y^3z^2} \div \frac{x^2}{yz^3} = \frac{x^2}{y^3z^2} \cdot \frac{yz^3}{x^2}$$ Invert the divisor and multiply (or multiply the first fraction by the reciprocal of the second).

$$= \frac{x^2yz^3}{x^2y^3z^2}$$ Multiply the numerators and the denominators.

$$= x^{2-2}y^{1-3}z^{3-2}$$ To divide exponential expressions with the same base, keep the base and subtract the exponents.

$$= x^0y^{-2}z^1$$ Simplify the exponents.

$$= 1 \cdot y^{-2} \cdot z$$ $x^0 = 1$.

$$= \frac{z}{y^2}$$ Write the result without the negative exponent.

Self Check 6

Divide: $\dfrac{x^3 - 8}{x - 1} \div \dfrac{x^2 + 2x + 4}{3x^2 - 3x}$.

EXAMPLE 6   Divide: $\dfrac{x^3 + 8}{x + 1} \div \dfrac{x^2 - 2x + 4}{2x^2 - 2}$.

Solution We invert the divisor, which is $\dfrac{x^2 - 2x + 4}{2x^2 - 2}$, and multiply.

$$\frac{x^3 + 8}{x + 1} \div \frac{x^2 - 2x + 4}{2x^2 - 2}$$

$$= \frac{x^3 + 8}{x + 1} \cdot \frac{2x^2 - 2}{x^2 - 2x + 4}$$

$$= \frac{(x^3 + 8)(2x^2 - 2)}{(x + 1)(x^2 - 2x + 4)}$$ Multiply the numerators and the denominators.

$$= \frac{(x + 2)(x^2 - 2x + 4)2(x + 1)(x - 1)}{(x + 1)(x^2 - 2x + 4)}$$ Factor $x^3 + 8$ and $2x^2 - 2$. The trinomial $x^2 - 2x + 4$ does not factor. Then divide out common factors.

Answer $3x(x - 2)$

$$= 2(x + 2)(x - 1)$$

Self Check 7

Divide: $\dfrac{m^4 - 9m^2}{a^2 - 3a} \div (m^2 + 3m)$.

EXAMPLE 7   Divide: $\dfrac{b^3 - 4b}{x - 1} \div (b - 2)$.

Solution

$$\frac{b^3 - 4b}{x - 1} \div (b - 2) = \frac{b^3 - 4b}{x - 1} \div \frac{b - 2}{1}$$ Write $b - 2$ as a fraction with a denominator of 1.

$$= \frac{b^3 - 4b}{x - 1} \cdot \frac{1}{b - 2}$$ Invert the divisor and multiply.

$$= \frac{b^3 - 4b}{(x - 1)(b - 2)}$$ Multiply the numerators and the denominators.

$$= \frac{b(b + 2)(b - 2)}{(x - 1)(b - 2)}$$ Factor $b^3 - 4b$ and then divide out common factors.

Answer $\dfrac{m(m - 3)}{a(a - 3)}$

$$= \frac{b(b + 2)}{x - 1}$$

Mixed operations

EXAMPLE 8 Simplify: $\dfrac{x^2 + 2x - 3}{6x^2 + 5x + 1} \div \dfrac{2x^2 - 2}{2x^2 - 5x - 3} \cdot \dfrac{6x^2 + 4x - 2}{x^2 - 2x - 3}$.

Solution The rules for the order of operations require multiplications and divisions to be done in order from left to right, so we begin by focusing on the division. We introduce grouping symbols to emphasize this. To divide the rational expressions within the parentheses, we invert $\dfrac{2x^2 - 2}{2x^2 - 5x - 3}$ and multiply.

$$\left(\dfrac{x^2 + 2x - 3}{6x^2 + 5x + 1} \div \dfrac{2x^2 - 2}{2x^2 - 5x - 3} \right) \dfrac{6x^2 + 4x - 2}{x^2 - 2x - 3}$$

$$= \left(\dfrac{x^2 + 2x - 3}{6x^2 + 5x + 1} \cdot \dfrac{2x^2 - 5x - 3}{2x^2 - 2} \right) \dfrac{6x^2 + 4x - 2}{x^2 - 2x - 3}$$

Next, we multiply the three fractions and simplify the result.

$$= \dfrac{(x^2 + 2x - 3)(2x^2 - 5x - 3)(6x^2 + 4x - 2)}{(6x^2 + 5x + 1)(2x^2 - 2)(x^2 - 2x - 3)}$$

$$= \dfrac{(x + 3)\overset{1}{\cancel{(x - 1)}}(2x + 1)\overset{1}{\cancel{(x - 3)}}\overset{1}{\cancel{2}}(3x - 1)\overset{1}{\cancel{(x + 1)}}}{(3x + 1)\cancel{(2x + 1)}\cancel{2}(x + 1)\underset{1}{\cancel{(x - 1)}}\underset{1}{\cancel{(x - 3)}}\underset{1}{\cancel{(x + 1)}}}$$

$$= \dfrac{(x + 3)(3x - 1)}{(3x + 1)(x + 1)}$$

Section 6.2 STUDY SET

VOCABULARY *Fill in the blanks.*

1. $\dfrac{a^2 - 9}{a^2 - 49} \cdot \dfrac{a - 7}{a + 3}$ is the product of two _____ expressions.

2. The _____ of $\dfrac{a + 3}{a + 7}$ is $\dfrac{a + 7}{a + 3}$.

3. In a fraction, the part above the fraction bar is called the _____. In a fraction, the part below the fraction bar is called the _____.

4. To simplify a rational expression, we divide out any _____ common to the numerator and denominator.

CONCEPTS *Fill in the blanks.*

5. To multiply two fractions, we _____ their numerators and multiply their denominators. In symbols, $\dfrac{a}{b} \cdot \dfrac{c}{d} = \boxed{}$.

6. To divide two fractions, we invert the divisor and _____. In symbols, $\dfrac{a}{b} \div \dfrac{c}{d} = \boxed{}$

7. The denominator of a fraction cannot be ▢.

8. $\dfrac{a + 1}{a + 1} = \boxed{}$, provided $a \neq -1$

NOTATION *Complete each solution.*

9. $\dfrac{x^2 + 3x}{5x - 25} \cdot \dfrac{x - 5}{x + 3} = \dfrac{(x^2 + 3x)(\boxed{})}{(\boxed{})(x + 3)}$

$$= \dfrac{\boxed{}(x - 5)}{\boxed{}(x + 3)}$$

$$= \dfrac{x}{5}$$

10. $\dfrac{x^2 - x - 6}{4x^2 + 16x} \div \dfrac{x - 3}{x + 4} = \dfrac{x^2 - x - 6}{4x^2 + 16x} \cdot \boxed{}$

$$= \dfrac{\boxed{}(x + 4)}{(4x^2 + 16x)\boxed{}}$$

$$= \dfrac{\boxed{}(x + 2)(x + 4)}{\boxed{}(x - 3)}$$

$$= \dfrac{x + 2}{4x}$$

11. A student checks her answers with those in the back of her textbook. Determine whether they are equivalent.

Student's answer	Book's answer	Equivalent?
$\dfrac{-x^{10}}{y^2}$	$-\dfrac{x^{10}}{y^2}$	
$\dfrac{x-3}{x+3}$	$\dfrac{3-x}{3+x}$	
$\dfrac{b+a}{(2-x)(d+c)}$	$\dfrac{a+b}{(x-2)(c+d)}$	

12. a. Write $5x^2 + 35x$ as a fraction.

   b. What is the reciprocal of $5x^2 + 35x$?

PRACTICE   *Perform the operations and simplify.*

13. $\dfrac{3}{4} \cdot \dfrac{5}{3}$

14. $-\dfrac{5}{6} \cdot \dfrac{3}{7}$

15. $-\dfrac{6}{11} \div \dfrac{36}{55}$

16. $\dfrac{17}{12} \div \dfrac{34}{3}$

17. $\dfrac{2x^2y^2}{cd} \cdot \dfrac{cd^2}{4c^2x}$

18. $\dfrac{b^2x}{6a^2y} \cdot \dfrac{3a^4b^4}{x^2y^3}$

19. $-\dfrac{x^3y^3}{y^2} \div \dfrac{y^3}{x^7}$

20. $(a^3)^2 b \div \dfrac{b}{(a^2)^3}$

21. $\dfrac{x^2 + 2x + 1}{x} \cdot \dfrac{x^2 - x}{x^2 - 1}$

22. $\dfrac{a+6}{16-a^2} \cdot \dfrac{3a-12}{3a+18}$

23. $\dfrac{2x^2 - x - 3}{x^2 - 1} \cdot \dfrac{x^2 + x - 2}{2x^2 + x - 6}$

24. $\dfrac{9x^2 + 3x - 20}{3x^2 - 7x + 4} \cdot \dfrac{3x^2 - 5x + 2}{9x^2 + 18x + 5}$

25. $\dfrac{x^2 - 16}{x^2 - 25} \div \dfrac{x+4}{x-5}$

26. $\dfrac{a^2 - 9}{a^2 - 49} \div \dfrac{a+3}{a+7}$

27. $-\dfrac{a^2 + 2a - 35}{12x} \div \dfrac{ax - 3x}{a^2 + 4a - 21}$

28. $\dfrac{x^2 - 4}{2b - bx} \div \dfrac{x^2 + 4x + 4}{2b + bx}$

29. $\dfrac{3t^2 - t - 2}{6t^2 - 5t - 6} \cdot \dfrac{4t^2 - 9}{2t^2 + 5t + 3}$

30. $\dfrac{2p^2 - 5p - 3}{p^2 - 9} \cdot \dfrac{2p^2 + 5p - 3}{2p^2 + 5p + 2}$

31. $\dfrac{3n^2 + 5n - 2}{12n^2 - 13n + 3} \div \dfrac{n^2 + 3n + 2}{4n^2 + 5n - 6}$

32. $\dfrac{8y^2 - 14y - 15}{6y^2 - 11y - 10} \div \dfrac{4y^2 - 9y - 9}{3y^2 - 7y - 6}$

33. $\dfrac{2x^2 + 5x + 3}{3x^2 - 5x + 2} \div \dfrac{2x^2 + x - 3}{-3x^2 + 5x - 2}$

34. $\dfrac{2p^2 - 5p - 3}{p^2 - 9} \div \dfrac{2p^2 + 5p + 2}{2p^2 + 5p - 3}$

35. $(x+1) \cdot \dfrac{1}{x^2 + 2x + 1}$ **36.** $-\dfrac{x-2}{x} \div (x^2 - 4)$

37. $(x^2 - x - 2) \cdot \dfrac{x^2 + 3x + 2}{x^2 - 4}$

38. $(2x^2 - 9x - 5) \cdot \dfrac{x}{2x^2 + x}$

39. $\dfrac{7b^5 - 7b^2}{5b^5 - 5b} \div \dfrac{14b^4 + 14b^3 + 14b^2}{20b^4 - 20b^2}$

40. $\dfrac{(s-t)^3}{s^3} \div \dfrac{(s-t)^4}{t^4}$

41. $\dfrac{x^3 + y^3}{x^3 - y^3} \div \dfrac{x^2 - xy + y^2}{x^2 + xy + y^2}$

42. $\dfrac{x^2 - 6x + 9}{4 - x^2} \div \dfrac{x^2 - 9}{x^2 - 8x + 12}$

43. $\dfrac{ax + ay + bx + by}{x^3 - 27} \div \dfrac{xc + xd + yc + yd}{x^2 + 3x + 9}$

44. $\dfrac{x^2 + 3x + yx + 3y}{x^2 - 9} \div \dfrac{x + 3}{x - 3}$

45. $\dfrac{x^2 - x - 6}{x^2 - 4} \cdot \dfrac{x^2 - x - 2}{9 - x^2}$

46. $\dfrac{p^3 - q^3}{q^2 - p^2} \cdot \dfrac{q^2 + pq}{p^3 + p^2q + pq^2}$

47. $(4x + 12) \cdot \dfrac{x^2}{2x - 6} \div \dfrac{2}{x - 3}$

48. $(4x^2 - 9) \div \dfrac{2x^2 + 5x + 3}{x + 2} \div (2x - 3)$

49. $(x^2 - x - 6) \div (x - 3) \div (x - 2)$

50. $(x^2 - x - 6) \div [(x - 3) \div (x - 2)]$

51. $\dfrac{2x^2 - 2x - 4}{x^2 + 2x - 8} \cdot \dfrac{3x^2 + 15x}{x + 1} \div \dfrac{4x^2 - 100}{x^2 - x - 20}$

52. $\dfrac{6a^2 - 7a - 3}{a^2 - 1} \div \dfrac{4a^2 - 12a + 9}{a^2 - 1} \cdot \dfrac{2a^2 - a - 3}{3a^2 - 2a - 1}$

53. $\dfrac{x^2 - x - 12}{x^2 + x - 2} \div \dfrac{x^2 - 6x + 8}{x^2 - 3x - 10} \cdot \dfrac{x^2 - 3x + 2}{x^2 - 2x - 15}$

54. $\dfrac{4a^2 - 10a + 6}{a^4 - 3a^3} \div \dfrac{3 - 2a}{2a^3} \cdot \dfrac{a - 3}{2a - 2}$

Find each power.

55. $\left(\dfrac{x - 3}{x^3 + 4}\right)^2$

56. $\left(\dfrac{2t^2 + t}{t - 1}\right)^2$

57. $\left(\dfrac{2m^2 - m - 3}{x^2 - 1}\right)^2$

58. $\left(\dfrac{-k - 3}{x^2 - x + 1}\right)^2$

APPLICATIONS

59. PHYSICS EXPERIMENTS The following table contains data from a physics experiment. k_1 and k_2 are constants. Complete the table.

Trial	Rate (m/sec)	Time (sec)	Distance (m)
1	$\dfrac{k_1^2 + 3k_1 + 2}{k_1 - 3}$	$\dfrac{k_1^2 - 3k_1}{k_1 + 1}$	
2	$\dfrac{k_2^2 + 6k_2 + 5}{k_2 + 1}$		$k_2^2 + 11k_2 + 30$

60. TRUNK CAPACITY The shape of the storage space in the trunk of the car shown is approximately a rectangular solid. Write a simplified rational expression that gives the number of cubic units of storage space in the trunk.

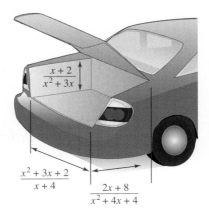

WRITING

61. Explain how to multiply two rational expressions.

62. Explain the error in the following work.

$$\dfrac{x^2 + x - 2}{x^2 - 4} \cdot \dfrac{x - 2}{x - 1} = \dfrac{(x + 2)(x - 1)}{(x + 2)(x - 2)} \cdot \dfrac{x - 2}{x - 1}$$

$$= \dfrac{\cancel{(x + 2)}\cancel{(x - 1)}\cancel{(x - 2)}}{\cancel{(x + 2)}\cancel{(x - 2)}\cancel{(x - 1)}}$$

$$= 0$$

REVIEW   *Perform each operation.*

63. $-2a^2(3a^3 - a^2)$

64. $(2t - 1)^2$

65. $(2g - n)(3g - n)$

66. $(2c - b)(4c^2 + 2cb + b^2)$

6.3 Adding and Subtracting Rational Expressions

- Adding and subtracting rational expressions with like denominators
- Adding and subtracting rational expressions with unlike denominators
- Finding the least common denominator

The methods used to add and subtract rational expressions are based on the rules for adding and subtracting arithmetic fractions. In this section, we will add and subtract rational expressions with *like* and *unlike* denominators.

Adding and subtracting rational expressions with like denominators

Fractions with like denominators are added and subtracted according to the following rules.

> **Adding and subtracting fractions**
>
> If a, b, and c represent real numbers and if there are no divisions by 0, then
>
> $$\frac{a}{b} + \frac{c}{b} = \frac{a+c}{b} \quad \text{and} \quad \frac{a}{b} - \frac{c}{b} = \frac{a-c}{b}$$

In words, *we add (or subtract) fractions with like denominators by adding (or subtracting) the numerators and keeping the common denominator.* Whenever possible, we should simplify the result. These two rules apply to addition and subtraction of rational expressions with like denominators.

INTERMEDIATE
Algebra $f(x)$ **Now**™

Self Check 1

Perform the operations:

a. $\dfrac{17}{22} + \dfrac{13}{22}$

b. $\dfrac{1}{6a} - \dfrac{7}{6a}$

c. $\dfrac{3a}{a-2} + \dfrac{2a}{a-2}$

Answers a. $\dfrac{15}{11}$,

b. $-\dfrac{1}{a}$, **c.** $\dfrac{5a}{a-2}$

EXAMPLE 1   Perform the operations:

a. $\dfrac{4}{3x} + \dfrac{7}{3x}$ and   **b.** $\dfrac{a^2}{a^2-1} - \dfrac{a}{a^2-1}$.

Solution

a. $\dfrac{4}{3x} + \dfrac{7}{3x} = \dfrac{4+7}{3x}$ The rational expressions have like denominators. Add the numerators and keep the common denominator.

$\qquad\qquad\quad = \dfrac{11}{3x}$ Perform the addition in the numerator.

b. $\dfrac{a^2}{a^2-1} - \dfrac{a}{a^2-1} = \dfrac{a^2-a}{a^2-1}$ The rational expressions have like denominators. Subtract the numerators and keep the common denominator.

We note that the polynomials factor in the numerator and the denominator of the result.

$$\frac{a^2}{a^2-1} - \frac{a}{a^2-1} = \frac{a\overset{1}{\cancel{(a-1)}}}{(a+1)\underset{1}{\cancel{(a-1)}}}$$ Simplify by dividing out the common factor of $(a-1)$.

$$= \frac{a}{a+1}$$

Checking algebra

We can check the subtraction in part b of Example 1 by graphing the rational functions $f(a) = \frac{a^2}{a^2 - 1} - \frac{a}{a^2 - 1}$, shown in Figure 6-7(a), and $g(a) = \frac{a}{a + 1}$, shown in Figure 6-7(b), and observing that the graphs are the same. Note that -1 and 1 are not in the domain of the first function and that -1 is not in the domain of the second function.

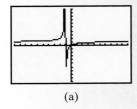
(a)

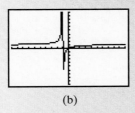
(b)

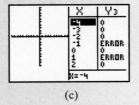
(c)

FIGURE 6-7

Figure 6-7(c) shows the display when the G-T mode is used to check the simplification. Here, $Y_3 = Y_1 - Y_2$, where $Y_1 = \frac{x^2}{x^2 - 1} - \frac{x}{x^2 - 1}$ and $Y_2 = \frac{x}{x + 1}$.

■ Adding and subtracting rational expressions with unlike denominators

To add or subtract fractions with unlike denominators, we change them into fractions with a common denominator. This is done using the **fundamental property of fractions.**

The fundamental property of fractions

If a, b, and k represent real numbers, and $b \neq 0$ and $k \neq 0$, then

$$\frac{a}{b} = \frac{a \cdot k}{b \cdot k}$$

In words, *multiplying the numerator and the denominator of a fraction by the same nonzero number does not change the value of the fraction.* This property is true because multiplying the numerator and denominator of a fraction by the same number is equivalent to multiplying it by 1. When a number is multiplied by 1, its value does not change. We use this property to add and subtract rational expressions with unlike denominators.

Recall that two polynomials are **opposites** if their terms are the same but are opposite in sign. When adding or subtracting two rational expressions with denominators that are opposites (negatives), we can multiply the numerator and the denominator of one of the expressions by -1 to get a common denominator.

EXAMPLE 2 Add: $\dfrac{x}{x - y} + \dfrac{y}{y - x}$.

Solution We note that the denominators, $x - y$ and $y - x$, are opposites. Either can serve as the LCD; we will choose $x - y$.

We must multiply the denominator of $\frac{y}{y - x}$ by -1 to obtain the LCD. By the fundamental property of fractions, we must also multiply the numerator by -1 to obtain an equivalent rational expression.

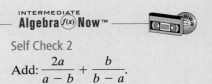

Self Check 2

Add: $\dfrac{2a}{a - b} + \dfrac{b}{b - a}$.

$$\frac{x}{x-y} + \frac{y}{y-x} = \frac{x}{x-y} + \frac{(-1)y}{(-1)(y-x)}$$

Multiply the numerator and the denominator of the second rational expression by -1.

$$= \frac{x}{x-y} + \frac{-y}{-y+x}$$

Perform the multiplication.

$$= \frac{x}{x-y} + \frac{-y}{x-y}$$

Rewrite the second denominator, $-y+x$, as $x-y$. The fractions now have a common denominator.

$$= \frac{x-y}{x-y}$$

Add the numerators and keep the common denominator.

$$= 1$$

Simplify.

Answer $\dfrac{2a-b}{a-b}$

When adding or subtracting two rational expressions with different denominators, we must often multiply one or both of them by an appropriate expression to get a common denominator. This process is called **building a fraction.**

Self Check 3

Subtract: $6 - \dfrac{5y}{6-y}$.

EXAMPLE 3   Subtract: $3 - \dfrac{7}{x-2}$.

Solution   If we write 3 as $\frac{3}{1}$ and then multiply its numerator and denominator by $x-2$, the fractions will have a common denominator of $x-2$.

$$3 - \frac{7}{x-2} = \frac{3}{1} - \frac{7}{x-2}$$

Write 3 as a fraction.

$$= \frac{3(x-2)}{1(x-2)} - \frac{7}{x-2}$$

Build $\frac{3}{1}$ to a fraction with denominator $x-2$ by multiplying its numerator and denominator by $x-2$.

$$= \frac{3x-6}{x-2} - \frac{7}{x-2}$$

Distribute the multiplication by 3.

$$= \frac{3x-6-7}{x-2}$$

Subtract the numerators and keep the common denominator.

$$= \frac{3x-13}{x-2}$$

Combine like terms in the numerator.

Answer $\dfrac{-11y+36}{6-y}$

Self Check 4

Add: $\dfrac{5}{a} + \dfrac{7}{b}$.

EXAMPLE 4   Add: $\dfrac{3}{x} + \dfrac{4}{y}$.

Solution   A common denominator for the fractions is xy. We multiply each numerator and denominator by the appropriate factor so that each denominator builds to xy.

$$\frac{3}{x} + \frac{4}{y} = \frac{3 \cdot y}{x \cdot y} + \frac{4 \cdot x}{y \cdot x}$$

$$= \frac{3y}{xy} + \frac{4x}{xy}$$

Multiply in the numerators and in the denominators.

$$= \frac{3y+4x}{xy}$$

Add the numerators and keep the common denominator.

Answer $\dfrac{5b+7a}{ab}$

EXAMPLE 5   Subtract: $\dfrac{4x}{x+2} - \dfrac{7x}{x-2}$.

Solution   By inspection, we see that a common denominator is $(x+2)(x-2)$. We multiply the numerator and denominator of each rational expression by the appropriate factor, so that each rational expression has a denominator of $(x+2)(x-2)$.

$$\frac{4x}{x+2} - \frac{7x}{x-2}$$

$$= \frac{4x(x-2)}{(x+2)(x-2)} - \frac{(x+2)7x}{(x+2)(x-2)}$$

$$= \frac{4x^2 - 8x}{(x+2)(x-2)} - \frac{7x^2 + 14x}{(x+2)(x-2)}$$

In each numerator, use the distributive property.

$$= \frac{4x^2 - 8x - (7x^2 + 14x)}{(x+2)(x-2)}$$

Subtract the numerators and keep the common denominator. Write $7x^2 + 14x$ within parentheses so that both of its terms will be subtracted from $4x^2 - 8x$.

$$= \frac{4x^2 - 8x - 7x^2 - 14x}{(x+2)(x-2)}$$

In the numerator, to subtract the polynomials, remove the parentheses and change the sign of every term in the second polynomial.

$$= \frac{-3x^2 - 22x}{(x+2)(x-2)}$$

Combine like terms.

If the common factor of $-x$ is factored out of the terms in the numerator, this result can be written in two other equivalent forms.

$$\frac{-3x^2 - 22x}{(x+2)(x-2)} = \frac{-x(3x+22)}{(x+2)(x-2)} = -\frac{x(3x+22)}{(x+2)(x-2)}$$

Finding the least common denominator

When adding or subtracting rational expressions with unlike denominators, it is easiest if we write the rational expressions in terms of the smallest common denominator possible, called the **least** (or lowest) **common denominator (LCD).** To find the least common denominator of several rational expressions, we follow these steps.

> **Finding the LCD**
> 1. Factor each denominator.
> 2. List the different factors of each denominator.
> 3. Write each factor found in Step 2 to the highest power that occurs in any one factorization.
> 4. The LCD is the product of the factors to the highest powers found in Step 3.

EXAMPLE 6   Find the LCD of   **a.** $\dfrac{5a}{24b}$ and $\dfrac{11a}{18b^2}$ and

b. $\dfrac{1}{x^2 - 12x + 36}$ and $\dfrac{3-x}{x^2 - 6x}$.

INTERMEDIATE
Algebra *f(x)* **Now**™
Self Check 6
Find the LCD of
a. $\dfrac{3y}{28z^3}$ and $\dfrac{5x}{21z}$

b. $\dfrac{a-1}{a^2-25}$ and $\dfrac{3-a^2}{a^2+7a+10}$

Solution

a. We write each denominator as the product of prime numbers and variables.

$$24b = 2 \cdot 2 \cdot 2 \cdot 3 \cdot b = 2^3 \cdot 3 \cdot b$$
$$18b^2 = 2 \cdot 3 \cdot 3 \cdot b \cdot b = 2 \cdot 3^2 \cdot b^2$$

To find the LCD, we form a product using each of these factors the greatest number of times it appears in any one factorization.

The greatest number of times the factor 2 appears is three times.
The greatest number of times the factor 3 appears is twice.
The greatest number of times the factor b appears is twice.

$$\text{LCD} = 2 \cdot 2 \cdot 2 \cdot 3 \cdot 3 \cdot b \cdot b$$
$$= 72b^2$$

b. We factor each denominator completely:

$$x^2 - 12x + 36 = (x-6)^2 \qquad x-6 \text{ occurs as a factor twice.}$$
$$x^2 - 6x = x(x-6)$$

To find the LCD, we form a product using the highest power of each of the factors:

$$\text{LCD} = x(x-6)^2$$

Answers **a.** $84z^3$,
b. $(a-5)(a+5)(a+2)$

To add or subtract rational expressions with unlike denominators, we follow these steps.

> **Adding or subtracting rational expressions with unlike denominators**
> 1. Find the LCD.
> 2. Express each rational expression with a denominator that is the LCD.
> 3. Add (or subtract) the resulting rational expressions.
> 4. Simplify the result if possible.

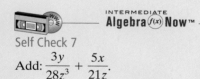
INTERMEDIATE
Algebra $f(x)$ **Now**™

Self Check 7

Add: $\dfrac{3y}{28z^3} + \dfrac{5x}{21z}$.

EXAMPLE 7   Add: $\dfrac{5a}{24b} + \dfrac{11a}{18b^2}$.

Solution   In Example 6, we saw that the LCD of these rational expressions is $72b^2$. We multiply each numerator and denominator by whatever it takes to build the denominator to $72b^2$.

$$\frac{5a}{24b} + \frac{11a}{18b^2} = \frac{5a \cdot 3b}{24b \cdot 3b} + \frac{11a \cdot 4}{18b^2 \cdot 4}$$

$$= \frac{15ab}{72b^2} + \frac{44a}{72b^2} \qquad \text{Peform the multiplications in the numerators and in the denominators.}$$

$$= \frac{15ab + 44a}{72b^2} \qquad \text{Add the numerators and keep the common denominator. The result does not simplify.}$$

Answer $\dfrac{9y + 20xz^2}{84z^3}$

INTERMEDIATE
Algebra $f(x)$ **Now**™

Self Check 8
Subtract:

$$\frac{a}{a^2-4a+4} - \frac{2}{a^2-4}$$

EXAMPLE 8   Subtract: $\dfrac{x}{x^2-2x+1} - \dfrac{4}{x^2-1}$.

Solution   We factor each denominator to find the LCD:

$$x^2 - 2x + 1 = (x-1)(x-1) = (x-1)^2$$
$$x^2 - 1 = (x+1)(x-1)$$

The LCD is $(x-1)^2(x+1)$ or $(x-1)(x-1)(x+1)$.

We now write each rational expression with its denominator in factored form. Then we multiply each numerator and denominator by the missing factor, so that each rational expression has a denominator of $(x - 1)(x - 1)(x + 1)$.

$$\frac{x}{x^2 - 2x + 1} - \frac{4}{x^2 - 1}$$

$$= \frac{x}{(x - 1)(x - 1)} - \frac{4}{(x + 1)(x - 1)}$$

$$= \frac{x(x + 1)}{(x - 1)(x - 1)(x + 1)} - \frac{4(x - 1)}{(x + 1)(x - 1)(x - 1)}$$

$$= \frac{x^2 + x}{(x - 1)(x - 1)(x + 1)} - \frac{4x - 4}{(x + 1)(x - 1)(x - 1)}$$

$$= \frac{(x^2 + x) - (4x - 4)}{(x - 1)(x - 1)(x + 1)} \qquad \text{Subtract the numerators and keep the common denominator.}$$

$$= \frac{x^2 + x - 4x + 4}{(x - 1)(x - 1)(x + 1)} \qquad \text{In the numerator, subtract the polynomials.}$$

$$= \frac{x^2 - 3x + 4}{(x - 1)^2(x + 1)} \qquad \text{Combine like terms. The result does not simplify.}$$

Answer $\dfrac{a^2 + 4}{(a - 2)^2(a + 2)}$

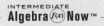

EXAMPLE 9   Combine: $\dfrac{2x}{x^2 - 4} - \dfrac{1}{x^2 - 3x + 2} + \dfrac{x + 1}{x^2 + x - 2}$.

Solution   We factor each denominator to find the LCD:

$$\left.\begin{array}{l} x^2 - 4 = (x - 2)(x + 2) \\ x^2 - 3x + 2 = (x - 2)(x - 1) \\ x^2 + x - 2 = (x - 1)(x + 2) \end{array}\right\} \text{ LCD} = (x - 2)(x + 2)(x - 1)$$

We then write each rational expression as an equivalent rational expression with the LCD as its denominator and do the subtraction and addition.

$$\frac{2x}{x^2 - 4} - \frac{1}{x^2 - 3x + 2} + \frac{x + 1}{x^2 + x - 2}$$

$$= \frac{2x}{(x - 2)(x + 2)} - \frac{1}{(x - 2)(x - 1)} + \frac{x + 1}{(x - 1)(x + 2)}$$

$$= \frac{2x(x - 1)}{(x - 2)(x + 2)(x - 1)} - \frac{1(x + 2)}{(x - 2)(x - 1)(x + 2)} + \frac{(x + 1)(x - 2)}{(x - 1)(x + 2)(x - 2)}$$

$$= \frac{2x(x - 1) - 1(x + 2) + (x + 1)(x - 2)}{(x + 2)(x - 2)(x - 1)}$$

$$= \frac{2x^2 - 2x - x - 2 + x^2 - x - 2}{(x + 2)(x - 2)(x - 1)}$$

$$= \frac{3x^2 - 4x - 4}{(x + 2)(x - 2)(x - 1)} \qquad \text{This result can be simplified.}$$

$$= \frac{(3x + 2)\overset{1}{\cancel{(x - 2)}}}{(x + 2)\underset{1}{\cancel{(x - 2)}}(x - 1)} \qquad \text{Factor the trinomial and divide out the common factor.}$$

$$= \frac{3x + 2}{(x + 2)(x - 1)}$$

Section 6.3 STUDY SET

INTERMEDIATE
Algebra $f(x)$ Now™

VOCABULARY *Fill in the blanks.*

1. The least common _____ of $\dfrac{3}{2a}$ and $\dfrac{7}{3a}$ is $6a$.

2. The rational expressions $\dfrac{x}{x-5}$ and $\dfrac{x+1}{x-5}$ have like _____. The rational expressions $\dfrac{2x}{x+5}$ and $\dfrac{5}{x-5}$ have _____ denominators.

3. The abbreviation for "least common denominator" is _____.

4. Two polynomials are _____ if their terms are the same but are opposite in sign.

CONCEPTS *Fill in the blanks.*

5. To subtract fractions with like denominators, we _____ the numerators and _____ the common denominator. In symbols,
$$\dfrac{a}{b} - \dfrac{c}{b} = \boxed{}$$

6. To add fractions with like denominators, we _____ the numerators and keep the _____ denominator. In symbols,
$$\dfrac{a}{b} + \dfrac{c}{b} = \boxed{}$$

7. Multiplying the numerator and the denominator of a fraction by the _____ nonzero number does not change the value of the fraction.

8. To find the LCD of several rational expressions, we _____ each denominator and use each factor to the _____ power that it appears in any one factorization.

9. Consider the two procedures.

i. $\dfrac{x^2 - 2x}{x^2 + 4x - 12} = \dfrac{x(\overset{1}{\cancel{x-2}})}{(x+6)(\underset{1}{\cancel{x-2}})} = \dfrac{x}{x+6}$

ii. $\dfrac{x}{x+6} = \dfrac{x(x-2)}{(x+6)(x-2)} = \dfrac{x^2-2x}{x^2+4x-12}$

a. In which of these procedures are we *building* a rational expression?

b. For what type of problem is this procedure often necessary?

c. What name is used to describe the other procedure?

10. Consider the following factorizations.
$$2 \cdot 3 \cdot 3 \cdot (x-2)$$
$$3(x-2)(x+1)$$

a. What is the greatest number of times the factor 3 appears in any one factorization?

b. What is the greatest number of times the factor $x-2$ appears in any one factorization?

NOTATION *Complete each solution.*

11. $\dfrac{6x-1}{3x-1} + \dfrac{3x-2}{3x-1} = \dfrac{6x-1+\boxed{}}{3x-1}$

$= \dfrac{9x-\boxed{}}{3x-1}$

$= \dfrac{3(\boxed{})}{3x-1}$

$= 3$

12. $\dfrac{8}{3v} - \dfrac{1}{4v^2} = \dfrac{8(\ \ )}{3v(4v)} - \dfrac{1(3)}{4v^2(\ \ )}$

$= \dfrac{\boxed{}}{12v^2} - \dfrac{3}{\boxed{}}$

$= \dfrac{32v-3}{12v^2}$

PRACTICE *Perform the operations and simplify the result when possible.*

13. $\dfrac{3}{4} + \dfrac{7}{4}$

14. $\dfrac{10}{33} - \dfrac{21}{33}$

15. $\dfrac{3}{4y} + \dfrac{8}{4y}$

16. $\dfrac{5}{3z^2} - \dfrac{6}{3z^2}$

17. $\dfrac{3x}{2x+2} + \dfrac{x+4}{2x+2}$

18. $\dfrac{4y}{y-4} - \dfrac{16}{y-4}$

19. $\dfrac{3x}{x-3} - \dfrac{9}{x-3}$

20. $\dfrac{9x}{x-y} - \dfrac{9y}{x-y}$

21. $\dfrac{5x}{x+1} + \dfrac{3}{x+1} - \dfrac{2x}{x+1}$

22. $\dfrac{4}{a+4} - \dfrac{2a}{a+4} + \dfrac{3a}{a+4}$

23. $\dfrac{3(x^2+x)}{x^2-5x+6} + \dfrac{-3(x^2-x)}{x^2-5x+6}$

24. $\dfrac{2x+4}{x^2+13x+12} - \dfrac{x+3}{x^2+13x+12}$

Find the LCD of the denominators listed.

25. $12x, 18x^2$

26. $15ab^2, 27a^2b$

27. $x^2 + 3x, x^2 - 9$

28. $3y^2 - 6y,\ 3y(y - 4)$

29. $x^3 + 27,\ x^2 + 6x + 9$

30. $x^3 - 8,\ x^2 - 4x + 4$

31. $2x^2 + 5x + 3,\ 4x^2 + 12x + 9,\ x^2 + 2x + 1$

32. $2x^2 + 5x + 3,\ 4x^2 + 12x + 9,\ 4x + 6$

Perform the operations and simplify the result when possible.

33. $\dfrac{1}{2} + \dfrac{1}{3}$

34. $\dfrac{8}{9} - \dfrac{5}{12}$

35. $\dfrac{3a}{2} - \dfrac{4b}{7}$

36. $\dfrac{a}{2} + \dfrac{2a}{5}$

37. $\dfrac{3}{4x} + \dfrac{2}{3x}$

38. $\dfrac{2}{5a} + \dfrac{3}{2b}$

39. $\dfrac{3a}{2b} - \dfrac{2b}{3a}$

40. $\dfrac{5m}{2n} - \dfrac{3n}{4m}$

41. $\dfrac{3}{ab^2} - \dfrac{5}{a^2 b}$

42. $\dfrac{1}{xy^3} - \dfrac{2}{x^2 y}$

43. $\dfrac{r}{4b^2} + \dfrac{s}{6b}$

44. $\dfrac{t}{12c^3} + \dfrac{t}{15c^2}$

45. $\dfrac{a + b}{3} + \dfrac{a - b}{7}$

46. $\dfrac{x - y}{2} + \dfrac{x + y}{3}$

47. $\dfrac{3}{x + 2} + \dfrac{5}{x - 4}$

48. $\dfrac{2}{a + 4} - \dfrac{6}{a + 3}$

49. $\dfrac{x + 2}{x + 5} - \dfrac{x - 3}{x + 7}$

50. $\dfrac{7}{x + 3} + \dfrac{4x}{x + 6}$

51. $4 + \dfrac{1}{x}$

52. $2 - \dfrac{1}{x + 1}$

53. $\dfrac{x + 8}{x - 3} - \dfrac{x - 14}{3 - x}$

54. $\dfrac{3 - x}{2 - x} + \dfrac{x - 1}{x - 2}$

55. $\dfrac{2a + 1}{3a - 2} - \dfrac{a - 4}{2 - 3a}$

56. $\dfrac{4}{x - 3} + \dfrac{5}{3 - x}$

57. $\dfrac{x}{x^2 + 5x + 6} + \dfrac{x}{x^2 - 4}$

58. $\dfrac{x}{3x^2 - 2x - 1} + \dfrac{4}{3x^2 + 10x + 3}$

59. $\dfrac{4}{x^2 - 2x - 3} - \dfrac{x}{3x^2 - 7x - 6}$

60. $\dfrac{2a}{a^2 - 2a - 8} + \dfrac{3}{a^2 - 5a + 4}$

61. $2x + 3 + \dfrac{1}{x + 1}$

62. $x + 1 + \dfrac{1}{x - 1}$

63. $1 + x - \dfrac{x}{x - 5}$

64. $2 - x + \dfrac{3}{x - 9}$

65. $\dfrac{8}{x^2 - 9} + \dfrac{2}{x - 3} - \dfrac{6}{x}$

66. $\dfrac{x}{x^2 - 4} - \dfrac{x}{x + 2} + \dfrac{2}{x}$

67. $\dfrac{3x}{2x - 1} + \dfrac{x + 1}{3x + 2} - \dfrac{2x}{6x^3 + x^2 - 2x}$

68. $\dfrac{x + 3}{2x^2 - 5x + 2} - \dfrac{3x - 1}{x^2 - x - 2}$

69. $\dfrac{3}{x + 1} - \dfrac{2}{x - 1} + \dfrac{x + 3}{x^2 - 1}$

70. $\dfrac{2}{x - 2} + \dfrac{3}{x + 2} - \dfrac{x - 1}{x^2 - 4}$

71. $\dfrac{a}{a - b} + \dfrac{b}{a + b} + \dfrac{a^2 + b^2}{b^2 - a^2}$

72. $\dfrac{1}{x + y} - \dfrac{1}{x - y} - \dfrac{2y}{y^2 - x^2}$

73. $\dfrac{7n^2}{m - n} + \dfrac{3m}{n - m} - \dfrac{3m^2 - n}{m^2 - 2mn + n^2}$

74. $\dfrac{3b}{2a - b} + \dfrac{2a - 1}{b - 2a} - \dfrac{3a^2 + b}{b^2 - 4ab + 4a^2}$

75. $\dfrac{m+1}{m^2+2m+1} + \dfrac{m-1}{m^2-2m+1} + \dfrac{2}{m^2-1}$

(*Hint:* Simplify first.)

76. $\dfrac{a+2}{a^2+3a+2} + \dfrac{a-1}{a^2-1} + \dfrac{3}{a+1}$

(*Hint:* Simplify first.)

APPLICATIONS

77. DRAFTING Among the tools used in drafting are the 45°–45°–90° and the 30°–60°–90° triangles shown. Find the perimeter of each triangle. Express each result as a single rational expression.

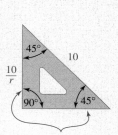

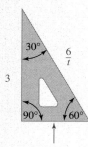

For a 45°-45°-90° triangle, these two sides are the same length.

For a 30°-60°-90° triangle, this side is half as long as the hypotenuse.

78. THE AMAZON The Amazon River flows in a generally eastern direction to the Atlantic Ocean. In Brazil, when the river is at low stage, the rate of flow is about 5 mph. Suppose that a river guide can canoe in still water at a rate of r mph.

  a. Complete the table to find rational expressions that represent the time it would take the guide to canoe 3 miles downriver and to canoe 3 miles upriver on the Amazon.

	Rate (mph)	Time (hr)	Distance (mi)
Downriver			3
Upriver			3

  b. Find the difference in the times for the trips upriver and downriver. Express the result as a single rational expression.

WRITING

79. Explain how to find the least common denominator of a set of fractions.

80. Explain how to add two rational expressions with unlike denominators.

81. Explain the error in the following work.

$$\frac{1}{x} \cdot \frac{3}{2} = \frac{1 \cdot 2}{x \cdot 2} \cdot \frac{3 \cdot x}{2 \cdot x}$$

$$= \frac{2}{2x} \cdot \frac{3x}{2x}$$

$$= \frac{6x}{2x}$$

82. Explain the error in the following work.

$$\frac{1}{x} - \frac{x+1}{x} = \frac{1-x+1}{x}$$

$$= \frac{2-x}{x}$$

REVIEW   *Solve each equation.*

83. $a(a-6) = -9$

84. $x^2 - \dfrac{1}{2}(x+1) = 0$

85. $y^3 + y^2 = 0$

86. $5x^2 = 6 - 13x$

6.4 Complex Fractions

- Simplifying complex fractions

A rational expression whose numerator and/or denominator contain rational expressions is called a **complex rational expression** or, more simply, a **complex fraction.** The expression above the main fraction bar of a complex fraction is the numerator, and the expression below the main fraction bar is the denominator. Two examples are:

$$\dfrac{\dfrac{3a}{b}}{\dfrac{6ac}{b^2}}$$ ← Numerator → $\dfrac{\dfrac{1}{x}+\dfrac{1}{y}}{\dfrac{1}{x}-\dfrac{1}{y}}$

$\leftarrow$ Main fraction bar $\rightarrow$

$\leftarrow$ Denominator $\rightarrow$

Simplifying complex fractions

To *simplify a complex fraction* means to express it as a single fraction. We can use two methods to simplify the complex fraction

$$\dfrac{\dfrac{3a}{b}}{\dfrac{6ac}{b^2}}$$

With the first method, we eliminate the fractions in the numerator and denominator by writing the complex fraction as a division and using the division rule for fractions:

$$\dfrac{\dfrac{3a}{b}}{\dfrac{6ac}{b^2}} = \dfrac{3a}{b} \div \dfrac{6ac}{b^2}$$ The main fraction bar of the complex fraction indicates division.

$$= \dfrac{3a}{b} \cdot \dfrac{b^2}{6ac}$$ Invert the divisor and multiply.

$$= \dfrac{b}{2c}$$ Multiply the fractions and simplify.

In the other method, we eliminate the fractions in the numerator and denominator by multiplying the fraction by 1, written in the form $\frac{b^2}{b^2}$. We use $\frac{b^2}{b^2}$ because b^2 is the LCD of $\frac{3a}{b}$ and $\frac{6ac}{b^2}$.

$$\dfrac{\dfrac{3a}{b}}{\dfrac{6ac}{b^2}} = \dfrac{b^2 \cdot \dfrac{3a}{b}}{b^2 \cdot \dfrac{6ac}{b^2}}$$

$$= \dfrac{\dfrac{3ab^2}{b}}{\dfrac{6acb^2}{b^2}}$$

$$= \dfrac{3ab}{6ac}$$ Simplify the fractions in the numerator and denominator of the complex fraction.

$$= \dfrac{b}{2c}$$ Divide out the common factor of $3a$.

With either method, the result is the same.

Methods for simplifying complex fractions

Method 1 Write the numerator and denominator of the complex fraction as single fractions. Then divide the fractions and simplify.

Method 2 Multiply the numerator and denominator of the complex fraction by the LCD of the fractions in its numerator and denominator. Then simplify the result, if possible.

! COMMENT Simplifying using division works well when a complex fraction is written, or can be easily written, as a quotient of two single rational expressions.

Simplifying using the LCD works well when the complex fraction has sums and/or differences in the numerator or denominator.

Self Check 1

Simplify: $\dfrac{\dfrac{3}{a} + 2}{2 + a}$.

EXAMPLE 1 Simplify: $\dfrac{\dfrac{2}{x} + 1}{3 + x}$.

Solution

Method 1: We add the fractions in the numerator and proceed as follows:

$$\frac{\dfrac{2}{x} + 1}{3 + x} = \frac{\dfrac{2}{x} + \dfrac{x}{x}}{\dfrac{3 + x}{1}} \qquad \text{Write 1 as } \frac{x}{x} \text{ and } 3 + x \text{ as } \frac{3 + x}{1}.$$

$$= \frac{\dfrac{2 + x}{x}}{\dfrac{3 + x}{1}} \qquad \text{Add } \frac{2}{x} \text{ and } \frac{x}{x} \text{ to get } \frac{2 + x}{x}.$$

$$= \frac{2 + x}{x} \div \frac{3 + x}{1} \qquad \text{The main fraction bar of the complex fraction indicates division.}$$

$$= \frac{2 + x}{x} \cdot \frac{1}{3 + x} \qquad \text{Multiply by the reciprocal of } \frac{x + 3}{1}.$$

$$= \frac{2 + x}{x^2 + 3x} \qquad \text{Multiply the numerators and multiply the denominators.}$$

Method 2: To eliminate the denominator of x, we multiply the numerator and the denominator of the complex fraction by x.

$$\frac{\dfrac{2}{x} + 1}{3 + x} = \frac{x\left(\dfrac{2}{x} + 1\right)}{x(3 + x)}$$

$$= \frac{2 + x}{x^2 + 3x} \qquad \begin{array}{l} \text{Use the distributive property to remove parentheses.} \\ \text{Write } 3x + x^2 \text{ as } x^2 + 3x. \end{array}$$

Answer $\dfrac{2a + 3}{a^2 + 2a}$

CALCULATOR SNAPSHOT   **Checking algebra**

An informal check of the simplification done in Example 1 can be performed using a scientific calculator. If

$$\frac{\dfrac{2}{x} + 1}{3 + x} = \frac{2 + x}{x^2 + 3x}$$

CALCULATOR SNAPSHOT   Checking algebra (continued)

then the expressions on each side will have identical values when evaluated for a given value of x (say, $x = 5$). To evaluate the expression on the left-hand side, we enter

$$\boxed{(}\; 2 \;\boxed{\div}\; 5 \;\boxed{+}\; 1 \;\boxed{)}\; \boxed{\div}\; \boxed{(}\; 3 \;\boxed{+}\; 5 \;\boxed{)}\; \boxed{=} \qquad \boxed{0.175}$$

To evaluate the expression on the right-hand side, we enter

$$\boxed{(}\; 2 \;\boxed{+}\; 5 \;\boxed{)}\; \boxed{\div}\; \boxed{(}\; 5 \;\boxed{x^2}\; \boxed{+}\; 3 \;\boxed{\times}\; 5 \;\boxed{)}\; \boxed{=} \qquad \boxed{0.175}$$

The results are the same, so it appears that the simplification is correct. We say "appears" because checking for only a single value of x is not definitive. The expressions should yield identical values when evaluated for any value of x for which the fractions are defined.

We can also check the simplification in Example 1 by graphing the functions

$$f(x) = \frac{\dfrac{2}{x} + 1}{3 + x} \qquad \text{shown in Figure 6-8(a),}$$

and

$$g(x) = \frac{2 + x}{x^2 + 3x} \qquad \text{shown in Figure 6-8(b),}$$

and observing that the graphs are the same. Each graph has window settings of $[-10, 10]$ for x and $[-10, 10]$ for y.

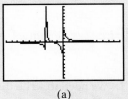

(a)   (b)

FIGURE 6-8

Self Check 2

Simplify: $\dfrac{\dfrac{1}{a} - \dfrac{1}{b}}{\dfrac{1}{a} + \dfrac{1}{b}}$.

EXAMPLE 2   Simplify: $\dfrac{\dfrac{1}{x} + \dfrac{1}{y}}{\dfrac{1}{x} - \dfrac{1}{y}}$.

Solution

Method 1: To write the numerator and denominator of the complex fraction as single fractions, we add the fractions in the numerator and subtract the fractions in the denominator.

$$\frac{\dfrac{1}{x} + \dfrac{1}{y}}{\dfrac{1}{x} - \dfrac{1}{y}} = \frac{\dfrac{1 \cdot y}{x \cdot y} + \dfrac{1 \cdot x}{y \cdot x}}{\dfrac{1 \cdot y}{x \cdot y} - \dfrac{1 \cdot x}{y \cdot x}}$$

Write the fractions in the numerator in terms of the LCD, xy. Do the same in the denominator.

$$= \frac{\dfrac{y + x}{xy}}{\dfrac{y - x}{xy}}$$

Add the fractions in the numerator, and subtract the fractions in the denominator.

$$= \frac{y + x}{xy} \div \frac{y - x}{xy}$$

Write the complex fraction as a division.

$$= \frac{y + x}{xy} \cdot \frac{xy}{y - x} \qquad \text{Invert the divisor and multiply.}$$

$$= \frac{(y + x)\overset{1}{\cancel{x}}\overset{1}{\cancel{y}}}{\underset{1}{\cancel{x}}\underset{1}{\cancel{y}}(y - x)} \qquad \begin{array}{l}\text{Multiply the fractions and divide out the common} \\ \text{factors.}\end{array}$$

$$= \frac{y + x}{y - x}$$

Method 2: We multiply the numerator and denominator by xy (the LCD of the fractions appearing in the complex fraction) and simplify.

$$\frac{\dfrac{1}{x} + \dfrac{1}{y}}{\dfrac{1}{x} - \dfrac{1}{y}} = \frac{xy\left(\dfrac{1}{x} + \dfrac{1}{y}\right)}{xy\left(\dfrac{1}{x} - \dfrac{1}{y}\right)} \qquad \text{The LCD of } \dfrac{1}{x} \text{ and } \dfrac{1}{y} \text{ is } xy.$$

$$= \frac{\dfrac{xy}{x} + \dfrac{xy}{y}}{\dfrac{xy}{x} - \dfrac{xy}{y}} \qquad \text{Use the distributive property to remove parentheses.}$$

Answer $\dfrac{b - a}{b + a}$
$$= \frac{y + x}{y - x} \qquad \text{Simplify each fraction.}$$

Self Check 3

Simplify: $\dfrac{a^{-2} + b^{-2}}{a^{-1} - b^{-1}}$.

EXAMPLE 3 Simplify: $\dfrac{x^{-1} + y^{-1}}{x^{-2} - y^{-2}}$.

Solution

Method 1: We need to write the numerator and denominator of the complex fraction as single fractions.

$$\frac{x^{-1} + y^{-1}}{x^{-2} - y^{-2}} = \frac{\dfrac{1}{x} + \dfrac{1}{y}}{\dfrac{1}{x^2} - \dfrac{1}{y^2}} \qquad \begin{array}{l}\text{Write the fraction without using negative} \\ \text{exponents.}\end{array}$$

$$= \frac{\dfrac{y}{xy} + \dfrac{x}{xy}}{\dfrac{y^2}{x^2 y^2} - \dfrac{x^2}{x^2 y^2}} \qquad \begin{array}{l}\text{Get a common denominator } (xy) \text{ in the} \\ \text{numerator and a common denominator} \\ (x^2 y^2) \text{ in the denominator.}\end{array}$$

$$= \frac{\dfrac{y + x}{xy}}{\dfrac{y^2 - x^2}{x^2 y^2}} \qquad \begin{array}{l}\text{Add the fractions in the numerator and} \\ \text{subtract the fractions in the denominator.}\end{array}$$

$$= \frac{y + x}{xy} \div \frac{y^2 - x^2}{x^2 y^2} \qquad \text{Write the fraction as a division.}$$

$$= \frac{y + x}{xy} \cdot \frac{xxyy}{(y - x)(y + x)} \qquad \text{Invert and multiply; factor } y^2 - x^2 \text{ and } x^2 y^2.$$

$$= \frac{\overset{1}{\cancel{(y + x)}}\underset{1}{\cancel{x}}x\underset{1}{\cancel{y}}y}{\underset{1}{\cancel{x}}\underset{1}{\cancel{y}}(y - x)\underset{1}{\cancel{(y + x)}}} \qquad \begin{array}{l}\text{Multiply the numerators and the} \\ \text{denominators. Divide out the common} \\ \text{factors.}\end{array}$$

$$= \frac{xy}{y - x}$$

Method 2: We multiply both numerator and denominator by x^2y^2, the LCD of the fractions, and proceed as follows:

$$\frac{x^{-1} + y^{-1}}{x^{-2} - y^{-2}} = \frac{\dfrac{1}{x} + \dfrac{1}{y}}{\dfrac{1}{x^2} - \dfrac{1}{y^2}}$$ Write the fraction without negative exponents.

$$= \frac{x^2y^2\left(\dfrac{1}{x} + \dfrac{1}{y}\right)}{x^2y^2\left(\dfrac{1}{x^2} - \dfrac{1}{y^2}\right)}$$ Multiply numerator and denominator by x^2y^2.

$$= \frac{xy^2 + yx^2}{y^2 - x^2}$$ Use the distributive property to remove parentheses.

$$= \frac{xy(y + x)}{(y + x)(y - x)}$$ Factor the numerator and denominator.

$$= \frac{xy}{y - x}$$ Divide out the common factor $y + x$.

Answer $\dfrac{b^2 + a^2}{ab(b - a)}$

❗ COMMENT $x^{-1} + y^{-1}$ means $\dfrac{1}{x} + \dfrac{1}{y}$, and $(x + y)^{-1}$ means $\dfrac{1}{x + y}$. Thus, $x^{-1} + y^{-1} \neq (x + y)^{-1}$.

EXAMPLE 4 Simplify: $\dfrac{\dfrac{1}{a^2 - 3a + 2}}{\dfrac{3}{a - 2} - \dfrac{2}{a - 1}}$.

Self Check 4

Simplify: $\dfrac{\dfrac{b}{b + 4} + \dfrac{3}{b + 3}}{\dfrac{b}{b^2 + 7b + 12}}$.

Solution We will use Method 2 to do the simplification. To determine the LCD for all the fractions appearing in the complex fraction, we must factor $a^2 - 3a + 2$.

$$\frac{\dfrac{1}{a^2 - 3a + 2}}{\dfrac{3}{a - 2} - \dfrac{2}{a - 1}} = \frac{\dfrac{1}{(a - 2)(a - 1)}}{\dfrac{3}{a - 2} - \dfrac{2}{a - 1}}$$

The LCD of the fractions in the numerator and denominator of the complex fraction is $(a - 2)(a - 1)$. We multiply the numerator and the denominator by the LCD.

$$= \frac{(a - 2)(a - 1)\left[\dfrac{1}{(a - 2)(a - 1)}\right]}{(a - 2)(a - 1)\left[\dfrac{3}{a - 2} - \dfrac{2}{a - 1}\right]}$$

$$= \frac{\dfrac{(a - 2)(a - 1)}{(a - 2)(a - 1)}}{\dfrac{3(a - 2)(a - 1)}{a - 2} - \dfrac{2(a - 2)(a - 1)}{a - 1}}$$ Perform the multiplication in the numerator. In the denominator, distribute the LCD.

$$= \frac{1}{3(a - 1) - 2(a - 2)}$$ Simplify each of the three rational expressions by dividing out the common factors.

$$= \frac{1}{3a - 3 - 2a + 4}$$ In the denominator, remove parentheses.

$$= \frac{1}{a + 1}$$ Combine like terms.

Answer $\dfrac{b^2 + 6b + 12}{b}$

If a fraction has a complex fraction in its numerator or denominator, it is often called a **continued fraction.**

Self Check 5

Simplify: $\dfrac{3 + \dfrac{2}{a}}{\dfrac{2a}{1 + \dfrac{1}{a}} + 3}$.

EXAMPLE 5   Simplify: $\dfrac{\dfrac{2x}{1 - \dfrac{1}{x}} + 3}{3 - \dfrac{2}{x}}$.

Solution   We begin by multiplying the numerator and denominator of the fraction

$$\frac{2x}{1 - \dfrac{1}{x}}$$

by x to eliminate the complex fraction in the numerator of the given fraction.

$$\frac{\dfrac{2x}{1 - \dfrac{1}{x}} + 3}{3 - \dfrac{2}{x}} = \frac{\dfrac{x(2x)}{x\left(1 - \dfrac{1}{x}\right)} + 3}{3 - \dfrac{2}{x}}$$

$$= \frac{\dfrac{2x^2}{x - 1} + 3}{3 - \dfrac{2}{x}}$$

We then multiply the numerator and denominator of the previous fraction by $x(x - 1)$, the LCD of $\frac{2x^2}{x - 1}$, 3, and $\frac{2}{x}$, and simplify.

$$\frac{\dfrac{2x}{1 - \dfrac{1}{x}} + 3}{3 - \dfrac{2}{x}} = \frac{x(x - 1)\left(\dfrac{2x^2}{x - 1} + 3\right)}{x(x - 1)\left(3 - \dfrac{2}{x}\right)}$$

$$= \frac{2x^3 + 3x(x - 1)}{3x(x - 1) - 2(x - 1)}$$

$$= \frac{2x^3 + 3x^2 - 3x}{3x^2 - 5x + 2}$$

Answer $\dfrac{(3a + 2)(a + 1)}{a(2a^2 + 3a + 3)}$

This result does not simplify.

Section 6.4   STUDY SET

▌ VOCABULARY   *Fill in the blanks.*

1. A _____ fraction is a fraction that has fractions (rational expressions) in its numerator and/or its denominator.

2. To _____ a complex fraction means to express it as a single fraction.

▌ CONCEPTS

3. The first step in simplifying a complex fraction using Method 2 is shown on the next page. With this method, the fractions in the numerator and denominator are to be eliminated by multiplying the complex fraction by 1. How is the 1 written in this case?

$$\frac{\dfrac{4}{t^2}}{\dfrac{3b}{t}} = \frac{t^2 \cdot \dfrac{4}{t^2}}{t^2 \cdot \dfrac{3b}{t}}$$

4. Determine the LCD of the rational expressions appearing in each complex fraction.

a. $\dfrac{1 + \dfrac{4}{c}}{\dfrac{2}{c} + c}$

b. $\dfrac{\dfrac{6}{m^2} + \dfrac{1}{2m}}{\dfrac{m^2 - 1}{4}}$

c. $\dfrac{\dfrac{p}{p + 2} + \dfrac{12}{p + 3}}{\dfrac{p - 1}{p^2 + 5p + 6}}$

d. $\dfrac{2 + \dfrac{3}{x + 1}}{\dfrac{1}{x} + x + x^2}$

NOTATION *Complete each solution to simplify the complex fraction.*

5. $\dfrac{\dfrac{5m^2}{6}}{\dfrac{25m}{3}} = \dfrac{5m^2}{6} \,\blacksquare\, \dfrac{25m}{3}$

$= \dfrac{5m^2}{6} \cdot \blacksquare$

$= \dfrac{5 \cdot \blacksquare \cdot m \cdot \blacksquare}{2 \cdot \blacksquare \cdot 5 \cdot 5 \cdot \blacksquare}$

$= \dfrac{m}{\blacksquare}$

6. $\dfrac{\dfrac{2}{a^2} - \dfrac{1}{b}}{\dfrac{2}{a} + \dfrac{1}{b^2}} = \dfrac{\blacksquare \left(\dfrac{2}{a^2} - \dfrac{1}{b} \right)}{\blacksquare \left(\dfrac{2}{a} + \dfrac{1}{b^2} \right)}$

$= \dfrac{\dfrac{2a^2b^2}{a^2} - \blacksquare}{\dfrac{2a^2b^2}{a} + \blacksquare}$

$= \dfrac{\blacksquare - a^2b}{2ab^2 + \blacksquare}$

7. The fraction $\dfrac{\dfrac{a}{b}}{\dfrac{c}{d}}$ is equivalent to $\dfrac{a}{b} \,\blacksquare\, \dfrac{c}{d}$.

8. What is the numerator and what is the denominator of the following complex fraction?

$$\frac{6 - k - \dfrac{5}{k}}{\dfrac{6}{k^2} + \dfrac{4}{k} - 4}$$

PRACTICE *Simplify each complex fraction.*

9. $\dfrac{\dfrac{1}{2}}{\dfrac{3}{4}}$

10. $-\dfrac{\dfrac{3}{4}}{\dfrac{1}{2}}$

11. $\dfrac{\dfrac{1}{2} - \dfrac{2}{3}}{\dfrac{2}{3} + \dfrac{1}{2}}$

12. $\dfrac{\dfrac{1}{4} - \dfrac{1}{5}}{\dfrac{1}{3}}$

13. $\dfrac{\dfrac{4x}{y}}{\dfrac{6xz}{y^2}}$

14. $\dfrac{\dfrac{5t^4}{9x}}{\dfrac{2t}{18x}}$

15. $\dfrac{\dfrac{5ab^2}{ab}}{25}$

16. $\dfrac{\dfrac{6a^2b}{4t}}{3a^2b^2}$

17. $\dfrac{\dfrac{x - y}{xy}}{\dfrac{y - x}{x}}$

18. $\dfrac{\dfrac{x^2 + 5x + 6}{3xy}}{\dfrac{x^2 - 9}{6xy}}$

19. $\dfrac{\dfrac{1}{x} - \dfrac{1}{y}}{xy}$

20. $\dfrac{xy}{\dfrac{1}{x} - \dfrac{1}{y}}$

21. $\dfrac{\dfrac{1}{a} + \dfrac{1}{b}}{\dfrac{1}{a}}$

22. $\dfrac{\dfrac{1}{b}}{\dfrac{1}{a} - \dfrac{1}{b}}$

23. $\dfrac{1 + \dfrac{x}{y}}{1 - \dfrac{x}{y}}$

24. $\dfrac{\dfrac{x}{y} + 1}{1 - \dfrac{x}{y}}$

25. $\dfrac{\dfrac{y}{x} - \dfrac{x}{y}}{\dfrac{1}{x} + \dfrac{1}{y}}$

26. $\dfrac{\dfrac{y}{x} - \dfrac{x}{y}}{\dfrac{1}{y} - \dfrac{1}{x}}$

27. $\dfrac{\dfrac{1}{a} - \dfrac{1}{b}}{\dfrac{a}{b} - \dfrac{b}{a}}$

28. $\dfrac{\dfrac{1}{a} + \dfrac{1}{b}}{\dfrac{a}{b} - \dfrac{b}{a}}$

29. $\dfrac{x + 1 - \dfrac{6}{x}}{\dfrac{1}{x}}$

30. $\dfrac{x - 1 - \dfrac{2}{x}}{\dfrac{x}{3}}$

31. $\dfrac{5xy}{1 + \dfrac{1}{xy}}$

32. $\dfrac{3a}{a + \dfrac{1}{a}}$

33. $\dfrac{a - 4 + \dfrac{1}{a}}{-\dfrac{1}{a} - a + 4}$

34. $\dfrac{a + 1 + \dfrac{1}{a^2}}{\dfrac{1}{a^2} + a - 1}$

53. $\dfrac{x - \dfrac{1}{1 - \dfrac{x}{2}}}{\dfrac{3}{x + \dfrac{2}{3}} - x}$

54. $\dfrac{3x - \dfrac{1}{3 - \dfrac{x}{2}}}{\dfrac{3}{\dfrac{x}{2} - 3} + x}$

35. $\dfrac{1 + \dfrac{6}{x} + \dfrac{8}{x^2}}{1 + \dfrac{1}{x} - \dfrac{12}{x^2}}$

36. $\dfrac{1 - x - \dfrac{2}{x}}{\dfrac{6}{x^2} + \dfrac{1}{x} - 1}$

APPLICATIONS

55. **ENGINEERING** The stiffness k of the shaft shown below is given by the formula

$$k = \dfrac{1}{\dfrac{1}{k_1} + \dfrac{1}{k_2}}$$

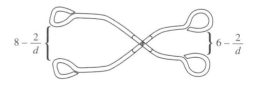

Section 1   Section 2

where k_1 and k_2 are the individual stiffnesses of each section. Simplify the complex fraction.

37. $\dfrac{\dfrac{1}{a + 1} + 1}{\dfrac{3}{a - 1} + 1}$

38. $\dfrac{2 + \dfrac{4}{y - 7}}{\dfrac{4}{y - 7}}$

39. $\dfrac{2 + \dfrac{3}{x + 1}}{\dfrac{1}{x} + x}$

40. $\dfrac{\dfrac{1}{x} - \dfrac{4}{x - 1}}{\dfrac{3}{x - 1} + \dfrac{2}{x}}$

56. **TRANSPORTATION** If a bus travels a distance d_1 at a speed s_1, and then travels a distance d_2 at a speed s_2, the average (mean) speed $\bar{s}$ is given by the formula

$$\bar{s} = \dfrac{d_1 + d_2}{\dfrac{d_1}{s_1} + \dfrac{d_2}{s_2}}$$

Simplify the complex fraction.

41. $\dfrac{y}{x^{-1} - y^{-1}}$

42. $\dfrac{x^{-1} + y^{-1}}{(x + y)^{-1}}$

43. $\dfrac{x - y^{-2}}{y - x^{-2}}$

44. $\dfrac{x^{-2} - y^{-2}}{x^{-1} - y^{-1}}$

57. **KITCHEN UTENSILS** What is the ratio of the width of the opening of the ice tongs shown below to the width of the opening of the handles? Express the result in simplest form.

45. $\dfrac{\dfrac{t}{x^2 - y^2}}{\dfrac{t}{x + y}}$

46. $\dfrac{\dfrac{7}{a - b}}{\dfrac{b}{a^3 - b^3}}$

47. $\dfrac{\dfrac{2}{x + 3} - \dfrac{1}{x - 3}}{\dfrac{3}{x^2 - 9}}$

48. $\dfrac{2 + \dfrac{1}{x^2 - 1}}{1 + \dfrac{1}{x - 1}}$

$8 - \dfrac{2}{d}$ $6 - \dfrac{2}{d}$

58. **DATA ANALYSIS** Use the data in the table to find the average measurement for the three-trial experiment. Express the answer as a rational expression.

49. $\dfrac{\dfrac{h}{h^2 + 3h + 2}}{\dfrac{4}{h + 2} - \dfrac{4}{h + 1}}$

50. $\dfrac{\dfrac{1}{r^2 + 4r + 4}}{\dfrac{r}{r + 2} + \dfrac{r}{r + 2}}$

	Trial 1	Trial 2	Trial 3
Measurement	$\dfrac{k}{3}$	$\dfrac{k}{5}$	$\dfrac{k}{6}$

51. $a + \dfrac{a}{1 + \dfrac{a}{a + 1}}$

52. $b + \dfrac{b}{1 - \dfrac{b + 1}{b}}$

59. What is a complex fraction?

60. Two methods can be used to simplify a complex fraction. Which method do you think is simpler? Why?

REVIEW *Solve each equation.*

61. $\dfrac{8(a - 5)}{3} = 2(a - 4)$

62. $\dfrac{3t^2}{5} + \dfrac{7t}{10} = \dfrac{3t + 6}{5}$

63. $a^4 - 13a^2 + 36 = 0$

64. $|2x - 1| = 9$

6.5 Equations Containing Rational Expressions

- Solving rational equations • Extraneous solutions
- Solving formulas for a specified variable • Problem solving

In this section, we will use the five-step problem-solving strategy to solve problems from disciplines such as business, photography, aviation, electronics, and publishing. We will encounter a new type of equation when we write mathematical models of each of these situations. The equations will contain rational expressions and are therefore called *rational equations.*

Solving rational equations

If an equation contains one or more rational expressions, it is called a **rational equation.** Some examples of rational equations are

$$\frac{3}{5} + \frac{7}{x + 2} = 2, \qquad \frac{x + 3}{x - 3} = \frac{2}{x^2 - 4}, \qquad \text{and} \qquad \frac{-x^2 + 10}{x^2 - 1} + \frac{3x}{x - 1} = \frac{2x}{x + 1}$$

To solve a rational equation, we can multiply both sides of the equation by the LCD of the rational expressions in the equation to clear it of fractions.

EXAMPLE 1   Solve: $\dfrac{3}{5} + \dfrac{7}{x + 2} = 2.$

Solution   We note that x cannot be -2, because this would give a 0 in the denominator of $\frac{7}{x + 2}$. If $x \neq -2$, we can multiply both sides of the equation by $5(x + 2)$, which is the LCD of $\frac{3}{5}$ and $\frac{7}{x + 2}$.

$$5(x + 2)\left(\frac{3}{5} + \frac{7}{x + 2}\right) = 5(x + 2)(2) \qquad \text{Multiply both sides by the LCD.}$$

$$5(x + 2)\left(\frac{3}{5}\right) + 5(x + 2)\left(\frac{7}{x + 2}\right) = 5(x + 2)2 \qquad \text{On the left-hand side, distribute } 5(x + 2).$$

$$\overset{1}{5}(x + 2)\left(\frac{3}{\underset{1}{5}}\right) + 5(\overset{1}{x + 2})\left(\frac{7}{\underset{1}{x + 2}}\right) = 5(x + 2)2 \qquad \text{On the left-hand side, divide out the common factors.}$$

$$3(x + 2) + 5(7) = 10(x + 2) \qquad \text{Simplify each side.}$$

The resulting equation does not contain any fractions. We now solve this *linear equation* for x.

INTERMEDIATE
Algebra $f(x)$ **Now™**

Self Check 1

Solve: $\dfrac{2}{5} + \dfrac{5}{x - 2} = \dfrac{29}{10}.$

$$3x + 6 + 35 = 10x + 20 \qquad \text{Use the distributive property and simplify.}$$
$$3x + 41 = 10x + 20 \qquad \text{Combine like terms.}$$
$$-7x = -21 \qquad\qquad\quad \text{Subtract } 10x \text{ and } 41 \text{ from both sides.}$$
$$x = 3 \qquad\qquad\qquad\ \text{Divide both sides by } -7.$$

The solution is 3. To check, we substitute 3 for x in the original equation and simplify:

Check: $\dfrac{3}{5} + \dfrac{7}{x + 2} = 2$

$$\dfrac{3}{5} + \dfrac{7}{3 + 2} \overset{?}{=} 2$$

$$\dfrac{3}{5} + \dfrac{7}{5} \overset{?}{=} 2$$

Answer 4

$$2 = 2$$

CALCULATOR SNAPSHOT   Solving rational equations graphically

To use a graphing calculator to approximate the solution of $\frac{3}{5} + \frac{7}{x+2} = 2$, we graph the functions $f(x) = \frac{3}{5} + \frac{7}{x+2}$ and $g(x) = 2$ using window settings of $[-10, 10]$ for x and $[-10, 10]$ for y. To find the point of intersection of the two graphs, we use the INTERSECT feature. In Figure 6-9, the display shows that the graphs intersect at the point $(3, 2)$. This implies that the solution of the rational equation is 3.

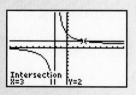

FIGURE 6-9

INTERMEDIATE
Algebra $f(x)$ **Now**™

EXAMPLE 2   Solve: $\dfrac{-x^2 + 10}{x^2 - 1} + \dfrac{3x}{x - 1} = \dfrac{2x}{x + 1}$.

Solution   We start by noting that x cannot be 1 or -1, because this would give a 0 in the denominator of a fraction. If $x \neq 1$ and $x \neq -1$, we can clear the equation of fractions by multiplying both sides by the LCD of the three rational expressions.

$$\dfrac{-x^2 + 10}{x^2 - 1} + \dfrac{3x}{x - 1} = \dfrac{2x}{x + 1}$$

$$\dfrac{-x^2 + 10}{(x + 1)(x - 1)} + \dfrac{3x}{x - 1} = \dfrac{2x}{x + 1} \qquad \begin{array}{l}\text{Factor the denominator } x^2 - 1. \text{ We determine}\\ \text{the LCD to be } (x + 1)(x - 1).\end{array}$$

Multiply both sides by the LCD. Distribute the multiplication by $(x + 1)(x - 1)$. Then simplify each rational expression by dividing out common factors.

$$(x + 1)(x - 1)\left(\dfrac{-x^2 + 10}{(x + 1)(x - 1)} + \dfrac{3x}{x - 1} \right) = (x + 1)(x - 1)\dfrac{2x}{x + 1}$$

Form an equation

We can let c = the speed of the current. Since the boat travels 12 mph and a current of c mph pushes the boat while it is going downstream, the speed of the boat going downstream is $(12 + c)$ mph. On the return trip, the current pushes against the boat, and its speed is $(12 - c)$ mph. Since time $= \frac{\text{distance}}{\text{rate}}$, the time required for the downstream leg of the trip is $\frac{9}{12 + c}$ hours, and the time required for the upstream leg of the trip is $\frac{9}{12 - c}$ hours. We can organize this information in the table shown in Figure 6-14.

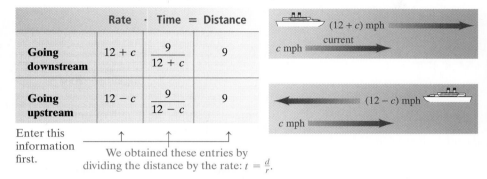

	Rate ·	Time =	Distance
Going downstream	$12 + c$	$\dfrac{9}{12 + c}$	9
Going upstream	$12 - c$	$\dfrac{9}{12 - c}$	9

Enter this information first.   We obtained these entries by dividing the distance by the rate: $t = \frac{d}{r}$.

FIGURE 6-14

Furthermore, we know that the total time required for the round trip is 1.6 or $\frac{8}{5}$ hours.

The time it takes to travel downstream	plus	the time it takes to travel upstream	is	the total time for the round trip.
$\dfrac{9}{12 + c}$	$+$	$\dfrac{9}{12 - c}$	$=$	$\dfrac{8}{5}$

Solve the equation

Multiply both sides of this equation by $5(12 + c)(12 - c)$ to clear it of fractions.

$$5(12 + c)(12 - c)\left(\frac{9}{12 + c} + \frac{9}{12 - c}\right) = 5(12 + c)(12 - c)\left(\frac{8}{5}\right)$$

$$\frac{5\overset{1}{\cancel{(12 + c)}}(12 - c)9}{\underset{1}{\cancel{12 + c}}} + \frac{5(12 + c)\overset{1}{\cancel{(12 - c)}}9}{\underset{1}{\cancel{12 - c}}} = \frac{\overset{1}{\cancel{5}}(12 + c)(12 - c)8}{\underset{1}{\cancel{5}}} \qquad \begin{array}{l}\text{Distribute, and} \\ \text{then divide out the} \\ \text{common factors.}\end{array}$$

$$45(12 - c) + 45(12 + c) = 8(12 + c)(12 - c) \qquad \text{Simplify.}$$

On the left-hand side, distribute. On the right-hand side, use the FOIL method.

$$540 - 45c + 540 + 45c = 8(144 - c^2)$$

Combine like terms and multiply. We see that the result is a quadratic equation.

$$1{,}080 = 1{,}152 - 8c^2$$

$$8c^2 - 72 = 0 \qquad \begin{array}{l}\text{Add } 8c^2 \text{ and} \\ \text{subtract } 1{,}152 \\ \text{from both sides.}\end{array}$$

$$c^2 - 9 = 0 \qquad \begin{array}{l}\text{Divide both sides} \\ \text{by 8.}\end{array}$$

$$(c + 3)(c - 3) = 0 \qquad \text{Factor } c^2 - 9.$$

$$c + 3 = 0 \quad \text{or} \quad c - 3 = 0 \qquad \begin{array}{l}\text{Set each factor} \\ \text{equal to 0.}\end{array}$$

$$c = -3 \qquad \qquad c = 3$$

State the conclusion

Since the current cannot be negative, the solution -3 must be discarded. The current in the Rock River is 3 mph.

Check the result

The downstream trip is at $12 + 3 = 15$ mph for $\frac{9}{12+3} = \frac{3}{5}$ hr. Thus, the distance traveled is $15 \cdot \frac{3}{5} = 9$ miles. The upstream trip is at $12 - 3 = 9$ mph for $\frac{9}{12-3} = 1$ hr. Thus, the distance traveled is $9 \cdot 1 = 9$ miles. Since both distances are 9 miles, the result checks.

Section 6.5   STUDY SET

VOCABULARY   *Fill in the blanks.*

1. An equation that contains rational expressions, such as $\frac{2}{x} + \frac{3}{2} = \frac{3}{4x}$, is called a _____ equation.

2. A proposed solution of an equation that does not satisfy the equation is called an _____ solution.

3. To _____ a rational equation of fractions, multiply both sides by the LCD of all rational expressions in the equation.

4. To _____ a rational equation means to find all the values of the variable that make the equation a true statement.

CONCEPTS

5. Is $x = 2$ a solution of the following equations?

 a. $\dfrac{x + 2}{x + 3} + \dfrac{1}{x^2 + 2x - 3} = 1$

 b. $\dfrac{x + 2}{x - 3} + \dfrac{1}{x^2 - 4} = 1$

6. To clear the equation

$$\frac{4}{10} + y = \frac{4y - 50}{5y - 25}$$

of fractions, by what should both sides be multiplied?

7. Complete the table.

	r	$\cdot$	t	$=$	d
Running	x				12
Bicycling	$x + 15$				12

8. The following table shows the length of time it takes each child in a family to wash their mother's minivan. Complete the table.

	Time to wash the van alone (min)	Amount of van washed in 1 minute
Glenn	25	
Brandon	30	
Kevin	x	

9. Consider the rational equation $\dfrac{x}{x - 3} = \dfrac{1}{x} + \dfrac{2}{x - 3}$.

 a. What values of x make a denominator 0?

 b. What values of x make a rational expression undefined?

 c. What numbers can't be solutions of the equation?

10. Perform each multiplication.

 a. $4x\left(\dfrac{3}{4x}\right)$ **b.** $(x + 6)(x - 2)\left(\dfrac{3}{x - 2}\right)$

NOTATION   *Complete each solution to solve the equation.*

11.
$$\frac{10}{3y} - \frac{7}{30} = \frac{9}{2y}$$

$$\boxed{}\left(\frac{10}{3y} - \frac{7}{30}\right) = 30y\left(\boxed{}\right)$$

$$30y\left(\boxed{}\right) - 30y\left(\frac{7}{30}\right) = 30y\left(\frac{9}{2y}\right)$$

$$100 - \boxed{} = 135$$

$$-7y = \boxed{}$$

$$y = -5$$

12.

$$\frac{2}{u-1} + \frac{1}{u} = \frac{1}{u^2 - u}$$

$$\boxed{}\left(\frac{2}{u-1} + \frac{1}{u}\right) = u(u-1)\left[\frac{1}{u(u-1)}\right]$$

$$u(u-1)\left(\frac{2}{u-1}\right) + u(u-1)\left(\boxed{}\right) = u(u-1)\left[\frac{1}{u(u-1)}\right]$$

$$2u + (\boxed{}) = \boxed{}$$

$$\boxed{} = 2$$

$$u = \boxed{}$$

▊ PRACTICE *Solve each equation. If a solution is extraneous, so indicate.*

13. $\dfrac{1}{4} + \dfrac{9}{x} = 1$

14. $\dfrac{1}{3} - \dfrac{10}{x} = -3$

15. $\dfrac{34}{x} - \dfrac{3}{2} = -\dfrac{13}{20}$

16. $\dfrac{1}{2} + \dfrac{7}{x} = 2 + \dfrac{1}{x}$

17. $\dfrac{3}{y} + \dfrac{7}{2y} = 13$

18. $\dfrac{2}{x} + \dfrac{1}{2} = \dfrac{7}{2x}$

19. $\dfrac{5x-1}{x-4} = \dfrac{6x-5}{x-4}$

20. $\dfrac{x-3}{x-1} - \dfrac{2x-4}{x-1} = 0$

21. $\dfrac{7}{5x} - \dfrac{1}{2} = \dfrac{5}{6x} + \dfrac{1}{3}$

22. $\dfrac{2}{x} + \dfrac{1}{2} = \dfrac{9}{4x} - \dfrac{1}{2x}$

23. $\dfrac{3-5y}{2+y} = \dfrac{3+5y}{2-y}$

24. $\dfrac{x}{x-2} = 1 + \dfrac{1}{x-3}$

25. $\dfrac{a+2}{a+1} - \dfrac{a-4}{a-3} = 0$

26. $\dfrac{z+2}{z+8} - \dfrac{z-3}{z-2} = 0$

27. $\dfrac{x+2}{x+3} - 1 = \dfrac{1}{3-2x-x^2}$

28. $\dfrac{x-3}{x-2} - \dfrac{1}{x} = \dfrac{x-3}{x}$

29. $\dfrac{x}{x+2} = 1 - \dfrac{3x+2}{x^2+4x+4}$

30. $\dfrac{3+2a}{a^2+6+5a} + \dfrac{2-5a}{a^2-4} = \dfrac{2-3a}{a^2-6+a}$

31. $\dfrac{2}{x-2} + \dfrac{1}{x+1} = \dfrac{1}{x^2-x-2}$

32. $\dfrac{5}{y-1} + \dfrac{3}{y-3} = \dfrac{8}{y-2}$

33. $\dfrac{a-1}{a+3} - \dfrac{1-2a}{3-a} = \dfrac{2-a}{a-3}$

34. $\dfrac{5}{2z^2+z-3} - \dfrac{2}{2z+3} = \dfrac{z+1}{z-1} - 1$

35. $\dfrac{5}{x+4} + \dfrac{1}{x+4} = x-1$

36. $\dfrac{2}{x-1} + \dfrac{x-2}{3} = \dfrac{4}{x-1}$

37. $\dfrac{3}{x+1} - \dfrac{x-2}{2} = \dfrac{x-2}{x+1}$

38. $\dfrac{x-4}{x-3} + \dfrac{x-2}{x-3} = x-3$

39. $\dfrac{2}{x-3} + \dfrac{3}{4} = \dfrac{17}{2x}$

40. $\dfrac{30}{y-2} + \dfrac{24}{y-5} = 13$

41. $\dfrac{x+4}{x+7} - \dfrac{x}{x+3} = \dfrac{3}{8}$

42. $\dfrac{5}{x+4} - \dfrac{1}{3} = \dfrac{x-1}{x}$

43. $\dfrac{2a}{a-3} + \dfrac{24}{a^2-9} = -\dfrac{4}{a+3}$

44. $\dfrac{3}{r} = \dfrac{12}{4r-r^2} - \dfrac{7}{r-4}$

Solve each formula for the indicated variable.

45. $I = \dfrac{E}{R_L + r}$ for r (from physics)

46. $P = \dfrac{R - C}{n}$ for C (from business)

47. $S = \dfrac{a - lr}{1 - r}$ for r (from mathematics)

48. $\mu_R = \dfrac{n_1(n_1 + n_2 + 1)}{2}$ for n_2 (from statistics)

49. $P = \dfrac{Q_1}{Q_2 - Q_1}$ for Q_1 (from refrigeration/heating)

50. $\dfrac{P_1 V_1}{T_1} = \dfrac{P_2 V_2}{T_2}$ for T_2 (from chemistry)

51. $\dfrac{1}{R} = \dfrac{1}{R_1} + \dfrac{1}{R_2} + \dfrac{1}{R_3}$ for R (from electronics)

52. $P + \dfrac{a}{V^2} = \dfrac{RT}{V - b}$ for b (from physics)

APPLICATIONS

53. PHOTOGRAPHY The illustration shows the relationship between distances when taking a photograph. The design of a camera lens uses the equation

$$\frac{1}{f} = \frac{1}{s_1} + \frac{1}{s_2}$$

which relates the focal length f of a lens to the image distance s_1 and the object distance s_2. Find the focal length of the lens in the illustration. (*Hint:* Convert feet to inches.)

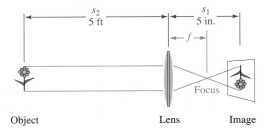

Object   Lens   Image

54. OPTICS The focal length f of a lens is given by the lensmaker's formula,

$$\frac{1}{f} = 0.6\left(\frac{1}{r_1} + \frac{1}{r_2}\right)$$

where f is the focal length of the lens and r_1 and r_2 are the radii of the two circular surfaces. Find the focal length of the lens in the illustration.

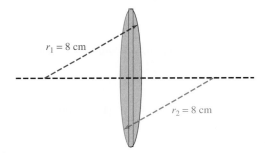

55. TAX ACCOUNTING As a piece of equipment gets older, its value usually lessens. One way to calculate *depreciation* is to use the formula

$$V = C - \left(\frac{C - S}{L}\right)N$$

where V denotes the value of the equipment at the end of year N, L is its useful lifetime (in years), C is its cost new, and S is its salvage value at the end of its useful life. Solve for L. Then determine what an accountant considered the useful lifetime of a forklift that cost $25,000 new, was worth $13,000 after 4 years, and has a salvage value of $1,000.

56. MECHANICAL ENGINEERING The equation

$$a = \frac{9.8m_2 - f}{m_2 + m_1}$$

models the system shown below where a is the acceleration of the suspended block, m_1 and m_2 are the masses of the blocks, and f is the friction force. Solve the equation for m_2.

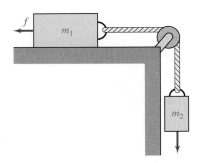

57. HOUSEPAINTING Two bids to paint a house are shown below.

  a. To get the job done quicker, the homeowner hired both the painters who submitted bids. How long will it take them to paint the house working together?

  b. What will the homeowner have to pay each painter?

Santos Painting
Residential
Bid:
  3 days
  @ $220 a day
Total: $660

Mays House Painting
Bid:
  $200 per day
  5 days work
Total: $1,000

58. ROOFING A HOUSE A homeowner estimates that it will take him 7 days to roof his house. A professional roofer estimates that he could roof the house in 4 days. How long will it take if the homeowner helps the roofer?

59. OYSTERS According to the *Guinness Book of World Records,* the record for opening oysters is 100 in 140 seconds by Mike Racz in Invercargill, New Zealand, on July 16, 1990. If it would take a novice $8\frac{1}{2}$ minutes to perform the same task, how long would it take them working together to open 100 oysters?

60. FARMING In 10 minutes, a conveyor belt can move 1,000 bushels of corn into the storage bin shown on the next page. A smaller belt can move 1,000 bushels to the storage bin in 14 minutes. If both belts are used, how long will it take to move 1,000 bushels to the storage bin?

61. FILLING A POND One pipe can fill a pond in 3 weeks, and a second pipe can fill it in 5 weeks. However, evaporation and seepage can empty the pond in 10 weeks. If both pipes are used, how long will it take to fill the pond?

62. HOUSECLEANING Sally can clean the house in 6 hours, and her father can clean the house in 4 hours. Sally's younger brother, Dennis, can completely mess up the house in 8 hours. If Sally and her father clean and Dennis plays, how long will it take to clean the house?

63. BOXING For his morning workout, a boxer bicycles for 8 miles and then jogs back to camp along the same route. If he bicycles 6 mph faster than he jogs, and the entire workout lasts 2 hours, how fast does he jog?

64. DELIVERIES A FedEx delivery van traveled from Rockford to Chicago in 3 hours less time than it took a second FedEx van to travel from Rockford to St. Louis. If the vans traveled at the same average speed, use the map below to help determine how long the first driver was on the road.

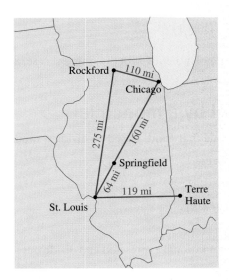

65. RATES OF SPEED Two trains made the same 315-mile run. Since one train traveled 10 mph faster than the other, it arrived 2 hours earlier. Find the speed of each train.

66. TRAIN TRAVEL A train traveled 120 miles from Freeport to Chicago and returned the same distance in a total time of 5 hours. If the train traveled 20 mph slower on the return trip, how fast did the train travel in each direction?

67. CROP DUSTING A helicopter spraying fertilizer over a field can fly 0.5 mile downwind in the same time as it can fly 0.4 mile upwind. Find the speed of the wind if the helicopter travels 45 mph in still air when dusting crops.

68. BOATING A man can drive a motorboat 45 miles down the Rock River in the same amount of time that he can drive 27 miles upstream. Find the speed of the current if the speed of the boat is 12 mph in still water.

69. UNIT COSTS One month, an appliance store manager bought several microwave ovens for a total of $1,800. The next month, because the unit cost of the same model of microwave increased by $25, she bought one oven fewer for the same total price. How many ovens did she buy the first month? (*Hint:* Write an expression for the unit cost of a microwave for the second month, then use the formula: Unit cost · number = total cost.)

70. VACATIONS Use the facts in the E-mail message below to determine how long the student had originally planned to stay in Europe. (*Hint:* Unit cost · number = total cost.)

E-Mail
Hi Mom and Dad,   After working so hard to earn $1,200 to take this trip to Europe, it was all worth it! I've been very frugal and been able to cut $20 from my daily expenses. Because of this, I'll be able to stay three extra days. Please pick me up on Friday instead. Love, Liz

WRITING

71. Why is it necessary to check the solutions of a rational equation?

72. Explain what it means to *clear* a rational equation of fractions.

73. Refer to the graph on the next page. Explain how to solve the rational equation graphically:

$$\frac{3x}{x-2} + \frac{1}{5} = 2$$

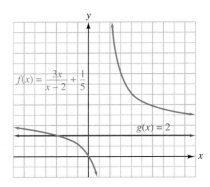

$$f(x) = \frac{3x}{x-2} + \frac{1}{5}$$

$$g(x) = 2$$

74. Would you use the same approach to answer the following problems? Explain why or why not.

Simplify: $\dfrac{x^2 - 10}{x^2 - 1} - \dfrac{3x}{x - 1} - \dfrac{2x}{x + 1}$

Solve: $\dfrac{x^2 - 10}{x^2 - 1} - \dfrac{3x}{x - 1} = -\dfrac{2x}{x + 1}$

REVIEW   *Write each italicized number in scientific notation.*

75. OIL The total cost of the Alaskan pipeline, running 800 miles from Prudhoe Bay to Valdez, was *$9,000,000,000*.

76. NATURAL GAS The TransCanada Pipeline transported a record *2,352,000,000,000* cubic feet of gas in 1995.

77. RADIOACTIVITY The least stable radioactive isotope is lithium 5, which decays in *0.0000000000000000000044* second.

78. BALANCES The finest balances in the world are made in Germany. They can weigh objects to an accuracy of *35×10^{-11}* ounce.

6.6 Dividing Polynomials

- Dividing a monomial by a monomial • Dividing a polynomial by a monomial
- Dividing a polynomial by a polynomial • Missing terms • Synthetic division
- The remainder theorem

We have discussed addition, subtraction, and multiplication of polynomials. We will now introduce the methods used to divide polynomials. This topic appears in a chapter about rational expressions because rational expressions indicate division of polynomials. For example,

$$\frac{x^2 - 3x + 7}{x + 1} = (x^2 - 3x + 7) \div (x + 1)$$

We begin the discussion of division of polynomials with the simplest case, a monomial divided by a monomial.

Dividing a monomial by a monomial

In Example 1, we review two methods that can be used to divide a monomial by a monomial.

INTERMEDIATE
Algebra *f(x)* **Now™**
Self Check 1
Simplify:
$\dfrac{6x^3 y^2}{8x^2 y^3}$

EXAMPLE 1   Simplify: $3a^2 b^3 \div (2a^3 b)$.

Solution
Method 1 We write $3a^2 b^3 \div (2a^3 b)$ as a rational expression and divide out all common factors:

$$\frac{3a^2 b^3}{2a^3 b} = \frac{3 \overset{1}{a} \overset{1}{a} b \overset{1}{b} b b}{2 \underset{1}{a} \underset{1}{a} a \underset{1}{b}}$$

$$= \frac{3b^2}{2a}$$

Method 2 We write $3a^2b^3 \div (2a^3b)$ as a rational expression and use the rules for exponents:

$$\frac{3a^2b^3}{2a^3b} = \frac{3}{2}a^{2-3}b^{3-1} \qquad \text{When dividing like bases, keep the base and subtract exponents.}$$

$$= \frac{3}{2}a^{-1}b^2 \qquad \text{Subtract.}$$

$$= \frac{3}{2}\left(\frac{1}{a}\right)\frac{b^2}{1} \qquad \text{We don't want negative exponents in the result.}$$

$$= \frac{3b^2}{2a} \qquad \text{Multiply the numerators. Multiply the denominators.}$$

Answer $\dfrac{3x}{4y}$

Dividing monomials

Method 1 Factor the numerator and denominator completely. Then divide out all common factors.

Method 2 Divide the coefficients. Then use the rules for exponents to simplify the divisions of variable factors with like bases.

◼ Dividing a polynomial by a monomial

In Section 6.3, we used the following rule to add and subtract fractions with like denominators.

Adding and subtracting fractions with like denominators

To add (or subtract) fractions with like denominators, add (or subtract) their numerators and keep the common denominator. In symbols, if a, b, c, and d represent real numbers, and d is not 0,

$$\frac{a}{d} + \frac{b}{d} = \frac{a+b}{d} \qquad\qquad \frac{a}{d} - \frac{b}{d} = \frac{a-b}{d}$$

We can use this rule in reverse to divide polynomials by monomials.

INTERMEDIATE
Algebra *f(x)* **Now**™

Self Check 2
Divide:
$$\frac{8a^3b^4 - 4a^4b^2 + a^2b^2}{4a^2b^2}$$

EXAMPLE 2 Divide: $4x^3y^2 + 3x^2y^5 - 12xy$ by $3x^2y^3$.

Solution We divide each term of the numerator by the denominator.

$$\frac{4x^3y^2 + 3x^2y^5 - 12xy}{3x^2y^3} = \frac{4x^3y^2}{3x^2y^3} + \frac{3x^2y^5}{3x^2y^3} - \frac{12xy}{3x^2y^3}$$

We then simplify each of the rational expressions on the right-hand side of the equals symbol to get

$$\frac{4x^3y^2 + 3x^2y^5 - 12xy}{3x^2y^3} = \frac{4x}{3y} + y^2 - \frac{4}{xy^2}$$

Answer $2ab^2 - a^2 + \dfrac{1}{4}$

> **Dividing a polynomial by a monomial**
> **1.** Write each term of the numerator over the denominator.
> **2.** Simplify each of the resulting rational expressions using the rules for simplifying a monomial divided by a monomial.

▮ Dividing a polynomial by a polynomial

There is an **algorithm** (a repeating series of steps) to use when the divisor is not a monomial. To use the division algorithm to divide $x^2 + 7x + 12$ (the **dividend**) by $x + 4$ (the **divisor**), we write the division in long division form and proceed as follows:

$$x + 4 \overline{)x^2 + 7x + 12} \qquad \overset{x}{}$$

How many times does x divide x^2?
$\dfrac{x^2}{x} = x$. Write x in the quotient.

$$\begin{array}{r} x \\ x + 4 \overline{)x^2 + 7x + 12} \\ \underline{x^2 + 4x} \\ 3x + 12 \end{array}$$

Multiply each term in the divisor by x to get $x^2 + 4x$, subtract $x^2 + 4x$ from $x^2 + 7x$, and bring down the 12.

$$\begin{array}{r} x + 3 \\ x + 4 \overline{)x^2 + 7x + 12} \\ \underline{x^2 + 4x} \\ 3x + 12 \end{array}$$

How many times does x divide $3x$?
$\dfrac{3x}{x} = 3$. Write 3 in the quotient.

$$\begin{array}{r} x + 3 \\ x + 4 \overline{)x^2 + 7x + 12} \\ \underline{x^2 + 4x} \\ 3x + 12 \\ \underline{3x + 12} \\ 0 \end{array}$$

Multiply each term in the divisor by 3 to get $3x + 12$, and subtract $3x + 12$ from $3x + 12$ to get 0.

The division process stops when the result of the subtraction is a constant or a polynomial with degree less than the degree of the divisor. Here, the quotient is $x + 3$ and the remainder is 0.

We can check the answer using the fact that for any division:

$$\text{divisor} \cdot \text{quotient} + \text{remainder} = \text{dividend}$$

Divisor · quotient + remainder = dividend

Check: $(x + 4)(x + 3) + \quad 0 \quad = x^2 + 7x + 12$ The quotient checks.

INTERMEDIATE
Algebra $f(x)$ Now™

EXAMPLE 3   Divide $\dfrac{2a^3 + 9a^2 + 5a - 6}{2a + 3}$ using long division.

Solution

$$\begin{array}{r} a^2 \\ 2a + 3 \overline{)2a^3 + 9a^2 + 5a - 6} \end{array}$$

How many times does $2a$ divide $2a^3$?
$\dfrac{2a^3}{2a} = a^2$. Write a^2 in the quotient.

$$\begin{array}{r} a^2 \\ 2a + 3 \overline{)2a^3 + 9a^2 + 5a - 6} \\ \underline{2a^3 + 3a^2} \\ 6a^2 + 5a \end{array}$$

Multiply each term in the divisor by a^2 to get $2a^3 + 3a^2$, subtract $2a^3 + 3a^2$ from $2a^3 + 9a^2$, and bring down the $5a$.

$$\begin{array}{r} a^2 + 3a \\ 2a + 3 \overline{)2a^3 + 9a^2 + 5a - 6} \\ \underline{2a^3 + 3a^2} \\ 6a^2 + 5a \end{array}$$

How many times does $2a$ divide $6a^2$?
$\dfrac{6a^2}{2a} = 3a$. Write $3a$ in the quotient.

$$
\begin{array}{r}
a^2 + 3a\phantom{{}-6} \\
2a + 3\overline{\smash{)}2a^3 + 9a^2 + 5a - 6} \\
\underline{2a^3 + 3a^2}\phantom{{}+ 5a - 6} \\
6a^2 + 5a\phantom{{}- 6} \\
\underline{6a^2 + 9a}\phantom{{}- 6} \\
-4a - 6
\end{array}
$$

Multiply each term in the divisor by $3a$ to get $6a^2 + 9a$, subtract $6a^2 + 9a$ from $6a^2 + 5a$, and bring down the -6.

$$
\begin{array}{r}
a^2 + 3a - 2 \\
2a + 3\overline{\smash{)}2a^3 + 9a^2 + 5a - 6} \\
\underline{2a^3 + 3a^2}\phantom{{}+ 5a - 6} \\
6a^2 + 5a\phantom{{}- 6} \\
\underline{6a^2 + 9a}\phantom{{}- 6} \\
-4a - 6
\end{array}
$$

How many times does $2a$ divide $-4a$? $\dfrac{-4a}{2a} = -2$. Write -2 in the quotient.

$$
\begin{array}{r}
a^2 + 3a - 2 \\
2a + 3\overline{\smash{)}2a^3 + 9a^2 + 5a - 6} \\
\underline{2a^3 + 3a^2}\phantom{{}+ 5a - 6} \\
6a^2 + 5a\phantom{{}- 6} \\
\underline{6a^2 + 9a}\phantom{{}- 6} \\
-4a - 6 \\
\underline{-4a - 6} \\
0
\end{array}
$$

Multiply each term in the divisor by -2 to get $-4a - 6$; subtract $-4a - 6$ from $-4a - 6$ to get 0.

Since the remainder is 0, the quotient is $a^2 + 3a - 2$. We can check the quotient by verifying that

$$
\underbrace{\text{Divisor} \cdot}_{} \quad \underbrace{\text{quotient}}_{} \quad + \text{ remainder} = \quad \underbrace{\text{dividend}}_{}
$$

$$
(2a + 3)(a^2 + 3a - 2) \quad + \quad 0 \quad = \quad 2a^3 + 9a^2 + 5a - 6
$$

INTERMEDIATE
Algebra $f(x)$**Now** ™

EXAMPLE 4 Divide: $\dfrac{3x^3 + 2x^2 - 3x + 8}{x - 2}$.

Solution

$$
\begin{array}{r}
3x^2 + 8x + 13 \\
x - 2\overline{\smash{)}3x^3 + 2x^2 - 3x + 8} \\
\underline{3x^3 - 6x^2}\phantom{{}- 3x + 8} \\
8x^2 - 3x\phantom{{}+ 8} \\
\underline{8x^2 - 16x}\phantom{{}+ 8} \\
13x + 8 \\
\underline{13x - 26} \\
34
\end{array}
$$

This division gives a quotient of $3x^2 + 8x + 13$ and a remainder of 34. It is common to form a fraction with the remainder as the numerator and the divisor as the denominator and to write the result as

$$
3x^2 + 8x + 13 + \dfrac{34}{x - 2}
$$

To check, we verify that

$$
(x - 2)(3x^2 + 8x + 13) + 34 = 3x^3 + 2x^2 - 3x + 8
$$

Self Check 4
Divide:
$$
\dfrac{2a^3 + 3a^2 - a + 2}{a - 3}
$$

Answer
$$
2a^2 + 9a + 26 + \dfrac{80}{a - 3}
$$

Self Check 5
Divide:
$$2 + 3a\overline{)-4a + 15a^2 + 18a^3 - 4}$$

EXAMPLE 5   Divide: $(-9x + 8x^3 + 10x^2 - 9) \div (3 + 2x)$.

Solution   The division algorithm works best when the polynomials in the dividend and the divisor are written in descending powers of x. We can use the commutative property of addition to rearrange the terms. Then the division is routine:

$$
\begin{array}{r}
4x^2 - x - 3 \\
2x + 3\overline{)8x^3 + 10x^2 - 9x - 9} \\
\underline{8x^3 + 12x^2} \\
-2x^2 - 9x \\
\underline{-2x^2 - 3x} \\
-6x - 9 \\
\underline{-6x - 9} \\
0
\end{array}
$$

Thus,

$$\frac{-9x + 8x^3 + 10x^2 - 9}{3 + 2x} = 4x^2 - x - 3$$

Answer $6a^2 + a - 2$

Missing terms

If a power of the variable is missing in the dividend, it is helpful to insert "placeholder" terms, because they aid in the subtraction step of the long division procedure.

Self Check 6
Divide $27a^3 - 1$ by $3a - 1$.

EXAMPLE 6   Divide $8x^3 + 1$ by $2x + 1$.

Solution   When we write the terms in the dividend in descending powers of x, we see that the terms involving x^2 and x are missing. We can introduce the terms $0x^2$ and $0x$ in the dividend or leave spaces for them. Then the division is routine.

$$
\begin{array}{r}
4x^2 - 2x + 1 \\
2x + 1\overline{)8x^3 + 0x^2 + 0x + 1} \\
\underline{8x^3 + 4x^2} \\
-4x^2 + 0x \\
\underline{-4x^2 - 2x} \\
2x + 1 \\
\underline{2x + 1} \\
0
\end{array}
$$

Thus,

$$\frac{8x^3 + 1}{2x + 1} = 4x^2 - 2x + 1$$

Answer $9a^2 + 3a + 1$

Self Check 7
Divide:
$$\frac{2a^2 + 3a^3 + a^4 - 7 + a}{a^2 - 2a + 1}$$

EXAMPLE 7   Divide $-17x^2 + 5x + x^4 + 2$ by $x^2 - 1 + 4x$.

Solution   We write the problem with the divisor and the dividend in descending powers of x. After introducing $0x^3$ for the missing term in the dividend, we proceed as follows:

$$
\begin{array}{r}
x^2 - 4x \\
x^2 + 4x - 1\overline{)x^4 + 0x^3 - 17x^2 + 5x + 2} \\
\underline{x^4 + 4x^3 - x^2} \\
-4x^3 - 16x^2 + 5x \\
\underline{-4x^3 - 16x^2 + 4x} \\
x + 2
\end{array}
$$

This division gives a quotient of $x^2 - 4x$ and a remainder of $x + 2$.

$$\frac{-17x^2 + 5x + x^4 + 2}{x^2 - 1 + 4x} = x^2 - 4x + \frac{x + 2}{x^2 + 4x - 1}$$

Write the denominator in descending powers of x.

Answer

$$a^2 + 5a + 11 + \frac{18a - 18}{a^2 - 2a + 1}$$

◼ Synthetic division

To divide a polynomial by a binomial of the form $x - r$, we can use a shortcut method called **synthetic division.** To see how synthetic division works, we consider the division of $4x^3 - 5x^2 - 11x + 20$ by $x - 2$.

$$
\begin{array}{r}
4x^2 + 3x - 5 \\
x - 2\,\overline{)\,4x^3 - 5x^2 - 11x + 20} \\
\underline{4x^3 - 8x^2} \\
3x^2 - 11x \\
\underline{3x^2 - 6x} \\
-5x + 20 \\
\underline{-5x + 10} \\
10 \quad \text{(remainder)}
\end{array}
$$

$$
\begin{array}{r}
4 \quad 3 \; -5 \\
1 - 2\,\overline{)\,4 - 5 - 11 \quad 20} \\
\underline{4 - 8} \\
3 - 11 \\
\underline{3 - 6} \\
-5 \quad 20 \\
\underline{-5 \quad 10} \\
10 \quad \text{(remainder)}
\end{array}
$$

On the left is the long division, and on the right is the same division with the variables and their exponents removed. The various powers of x can be remembered without actually writing them, because the exponents of the terms in the divisor, dividend, and quotient were written in descending order.

We can further shorten the version on the right. The numbers printed in color need not be written, because they are duplicates of the numbers above them. Thus, we can write the division in the following form:

$$
\begin{array}{r}
4 \quad 3 \; -5 \\
1 - 2\,\overline{)\,4 - 5 - 11 \quad 20} \\
\underline{-8} \\
3 \\
\underline{-6} \\
-5 \\
\underline{10} \\
10
\end{array}
$$

We can shorten the process further by compressing the work vertically and eliminating the 1 (the coefficient of x in the divisor):

$$
\begin{array}{r}
4 \quad\; 3 \; -5 \\
-2\,\overline{)\,4 \;\; -5 \;\; -11 \quad 20} \\
\underline{-8 \quad -6 \quad 10} \\
3 \quad -5 \quad 10
\end{array}
$$

If we write the 4 in the quotient on the left of the bottom line, the bottom line gives the coefficients of the quotient and the remainder. If we eliminate the top line, the division appears as follows:

$$
\begin{array}{r}
-2\,\rfloor\; 4 \;\; -5 \;\; -11 \quad 20 \\
\underline{-8 \;\; -6 \quad 10} \\
4 \quad\; 3 \;\; -5 \quad 10
\end{array}
$$

The bottom line was obtained by subtracting the middle line from the top line. If we replace the -2 in the divisor by 2, the division process will reverse the signs of every entry in the middle line, and then the bottom line can be obtained by addition. This gives the final form of the synthetic division.

$$
\begin{array}{r}
2\,\rfloor\; 4 \;\; -5 \;\; -11 \quad\; 20 \\
\underline{8 \quad\; 6 \quad -10} \\
4 \quad\; 3 \;\; -5 \quad\; 10
\end{array}
$$

These are the coefficients of the dividend.

These are the coefficients of the quotient and the remainder.

$$4x^2 + 3x - 5 + \frac{10}{x - 2}$$

Read the result from the bottom row.

Thus,

$$\frac{4x^3 - 5x^2 - 11x + 20}{x - 2} = 4x^2 + 3x - 5 + \frac{10}{x - 2}$$

INTERMEDIATE
Algebra $f(x)$ Now™

Self Check 8
Use synthetic division to divide
$2x^2 - 5x + 1$ by $x - 1$.

EXAMPLE 8   Use synthetic division to divide $6x^2 + 5x - 2$ by $x - 5$.

Solution   We write the coefficients of the dividend and the 5 in the divisor in the following form:

$$\underline{5|}\quad 6\quad 5\quad -2$$

Then we follow these steps:

$\underline{5|}\quad 6\quad\;\; 5\quad -2$ Begin by bringing down the 6.
$\qquad\;\;\downarrow$
$\qquad\;\; 6$

$\underline{5|}\quad 6\quad\;\; 5\quad -2$ Multiply 5 by 6 to get 30.
$\qquad\;\;\searrow\;\;\mathbf{30}$
$\qquad\;\; 6$

$\underline{5|}\quad 6\quad\;\; 5\quad -2$ Add 5 and 30 to get 35.
$\qquad\qquad 30$
$\qquad\;\; 6\quad \mathbf{35}$

$\underline{5|}\quad 6\quad\;\; 5\quad -2$ Multiply 35 by 5 to get 175.
$\qquad\qquad 30\quad \mathbf{175}$
$\qquad\;\; 6\quad 35$

$\underline{5|}\quad 6\quad\;\; 5\quad -2$ Add -2 and 175 to get 173.
$\qquad\qquad 30\quad 175$
$\qquad\;\; 6\quad 35\quad \mathbf{173}$

The numbers 6 and 35 represent the quotient $6x + 35$, and 173 is the remainder. Thus,

Answer $2x - 3 + \dfrac{-2}{x - 1}$

$$\frac{6x^2 + 5x - 2}{x - 5} = 6x + 35 + \frac{173}{x - 5}$$

! COMMENT Synthetic division is used only when dividing a polynomial by a binomial of the form $x - r$.

INTERMEDIATE
Algebra $f(x)$ Now™

Self Check 9
Use synthetic division to divide
$x^3 - 3x + 10$ by $x - 2$.

EXAMPLE 9   Use synthetic division to divide $5x^3 + x^2 - 3$ by $x - 2$.

Solution   We begin by writing

$\underline{2|}\quad 5\quad 1\quad \mathbf{0}\quad -3$ Write 0 for the coefficient of x, the missing term.

and complete the division as follows:

$\underline{2|}\quad 5\quad 1\quad 0\quad -3$ $\qquad$ $\underline{2|}\quad 5\quad 1\quad 0\quad -3$ $\qquad$ $\underline{2|}\quad 5\quad 1\quad 0\quad -3$
$\qquad\;\;\searrow\; 10$ $\qquad\qquad\qquad\;\; 10\quad 22$ $\qquad\qquad\qquad\; 10\quad 22\quad 44$
$\qquad\; 5\quad 11$ $\qquad\qquad\qquad 5\quad 11\quad 22$ $\qquad\qquad\quad 5\quad 11\quad 22\quad 41$

Multiply, then add. $\qquad$ Multiply, then add. $\qquad$ Multiply, then add.

Thus,

$$\frac{5x^3 + x^2 - 3}{x - 2} = 5x^2 + 11x + 22 + \frac{41}{x - 2}$$

Answer $x^2 + 2x + 1 + \dfrac{12}{x - 2}$

INTERMEDIATE
Algebra $f(x)$ **Now™**

EXAMPLE 10 Use synthetic division to divide $5x^2 + 6x^3 + 2 - 4x$ by $x + 2$.

Solution First, we write the dividend with the exponents in descending order.

$$6x^3 + 5x^2 - 4x + 2$$

Then we write the divisor in $x - r$ form: $x + 2 = x - (-2)$. Using synthetic division, we begin by writing

This represents
division by $x + 2$.

$$-2 \,\rvert\; \begin{array}{cccc} 6 & 5 & -4 & 2 \end{array}$$

and complete the division.

$$-2 \,\rvert\; \begin{array}{cccc} 6 & 5 & -4 & 2 \\ & -12 & 14 & -20 \\ \hline 6 & -7 & 10 & -18 \end{array}$$ The remainder is negative.

Thus,

$$\frac{5x^2 + 6x^3 + 2 - 4x}{x + 2} = 6x^2 - 7x + 10 + \frac{-18}{x + 2}$$

This result can also be written as $6x^2 - 7x + 10 - \dfrac{18}{x + 2}$.

Self Check 10
Use synthetic division to divide $6x^2 - 5 - 3x + x^3$ by $x + 3$.

Answer $x^2 + 3x - 12 + \dfrac{31}{x + 3}$

◼ The remainder theorem

In Example 8, when we divided $6x^2 + 5x - 2$ by $x - 5$, the remainder was 173. If we evaluate the polynomial function $P(x) = 6x^2 + 5x - 2$ for $x = 5$, we obtain an interesting result, 173.

$$\begin{aligned} P(x) &= 6x^2 + 5x - 2 \\ P(5) &= 6(5)^2 + 5(5) - 2 \\ &= 6(25) + 25 - 2 \\ &= 150 + 25 - 2 \\ &= 173 \end{aligned}$$

This is an illustration of the **remainder theorem.**

Remainder theorem

If a polynomial $P(x)$ is divided by $x - r$, the remainder is $P(r)$.

❗ COMMENT It is easier to find $P(r)$ by using synthetic division than by substituting r for x in $P(x)$. This is especially true if r is a decimal.

Self Check 11
Let $P(x) = x^3 + 3x^2 - x + 5$.
Find **a.** $P(-2)$ and
b. the remainder when $P(x)$ is
divided by $x + 2$.

EXAMPLE 11   Let $P(x) = 2x^3 - 3x^2 - 2x + 1$. Find **a.** $P(3)$ and **b.** the
remainder when $P(x)$ is divided by $x - 3$.

Solution

a. $P(3) = 2(3)^3 - 3(3)^2 - 2(3) + 1$ Substitute 3 for x.

$= 2(27) - 3(9) - 6 + 1$

$= 54 - 27 - 6 + 1$

$= \mathbf{22}$

b. We use synthetic division to find the remainder when $P(x) = 2x^3 - 3x^2 - 2x + 1$ is
divided by $x - 3$.

$$
\begin{array}{r|rrrr}
3 & 2 & -3 & -2 & 1 \\
  &   & 6 & 9 & 21 \\
\hline
  & 2 & 3 & 7 & \mathbf{22}
\end{array}
$$

The remainder is 22.

Answers a. 11,   **b.** 11

The results of parts a and b show that when $P(x)$ is divided by $x - 3$, the remainder
is $P(3)$.

Section 6.6   STUDY SET

INTERMEDIATE
Algebra $f^{(x)}$ **Now**™

VOCABULARY   *Fill in the blanks.*

1. In the division

$$
\begin{array}{r}
x + 3 \\
x - 4 \overline{)x^2 - x - 12}
\end{array}
$$

$x^2 - x - 12$ is the _____, $x - 4$ is the _____,
and $x + 3$ is the _____.

2. For the division shown below, the _____ is 1.

$$
\begin{array}{r}
x + 3 \\
x + 4 \overline{)x^2 + 7x + 13} \\
\underline{x^2 + 4x} \\
3x + 13 \\
\underline{3x + 12} \\
1
\end{array}
$$

3. The expression $5x^2 + 6$ is missing an x-term. We
can insert a _____ $0x$ term and write it as
$5x^2 + 0x + 6$.

4. In the polynomial $4x^4 + 2x^3 - x^2 + x + 7$, the powers
of x are written in _____ order.

CONCEPTS   *Fill in the blanks.*

5. $\dfrac{a + c}{b} = \dfrac{a}{} + \dfrac{c}{}$

6. Divisor · _____ + remainder = dividend

7. Suppose that after dividing $2x^3 + 5x^2 - 11x + 4$
by $2x - 1$, you obtain $x^2 + 3x - 4$. Show how
multiplication can be used to check the result.

8. Consider the first step of the division process for

$$
2x^2 - 1 \overline{)4x^4 + 0x^3 + 0x^2 + 0x - 1}
$$

How many times does $2x^2$ divide $4x^4$?

NOTATION   *Complete each solution.*

9.
$$
\begin{array}{r}
2x + 1 \\
x + 4 \overline{)2x^2 + 9x + 4} \\
\underline{ + 8x} \\
 + 4 \\
\underline{x + } \\
0
\end{array}
$$

10.
$$
\begin{array}{r}
2x - 1 \\
3x + 4 \overline{)6x^2 + 5x - 4} \\
\underline{6x^2 + } \\
 - 4 \\
\underline{-3x - } \\
0
\end{array}
$$

11. If a polynomial is divided by $3a - 2$ and the quotient
is $3a^2 + 5$ with a remainder of 6, how do we write the
result?

12. If a polynomial is divided by $3a - 2$ and the quotient is $3a^2 + 5$ with a remainder of -6, how do we write the result?

13. List three ways we can use symbols to write $x^2 - x - 12$ divided by $x - 4$.

14. Determine whether the statement below is true or false. Justify your answer.

$$2x^3 - 9 = 2x^3 + 0x^2 + 0x - 9$$

PRACTICE *Perform each division. Write answers without using negative exponents.*

15. $\dfrac{4x^2y^3}{8x^5y^2}$

16. $\dfrac{25x^4y^7}{5xy^9}$

17. $-\dfrac{33a^2b^2}{44a^4b^2}$

18. $\dfrac{-63a^4}{81a^6b^3}$

19. $\dfrac{4x + 6}{2}$

20. $\dfrac{11a^3 - 11a^2}{11}$

21. $\dfrac{4x^2 - x^3}{-6x}$

22. $\dfrac{5y^4 + 45y^3}{-15y^2}$

23. $\dfrac{12x^2y^3 + x^3y^2}{6xy}$

24. $\dfrac{54a^3y^2 - 18a^4y^3}{27a^2y^2}$

25. $\dfrac{24x^6y^7 - 12x^5y^{12} + 36xy}{48x^2y^3}$

26. $\dfrac{9x^4y^3 + 18x^2y - 27xy^4}{-9x^3y^3}$

27. $\dfrac{x^2 + 5x + 6}{x + 3}$

28. $\dfrac{x^2 - 5x + 6}{x - 3}$

29. $\dfrac{x^2 + 10x + 21}{x + 3}$

30. $\dfrac{x^2 + 10x + 21}{x + 7}$

31. $\dfrac{6x^2 - x - 12}{2x + 3}$

32. $\dfrac{6x^2 - x - 12}{2x - 3}$

33. $\dfrac{3x^3 - 2x^2 + x - 6}{x - 1}$

34. $\dfrac{4a^3 + a^2 - 3a + 7}{a + 1}$

35. $\dfrac{6x^3 + 11x^2 - x - 2}{3x - 2}$

36. $\dfrac{6x^3 + 11x^2 - 9x - 5}{2x + 3}$

37. $\dfrac{6x^3 - x^2 - 6x - 9}{2x - 3}$

38. $\dfrac{16x^3 + 16x^2 - 9x - 5}{4x + 5}$

39. $(2a + 1 + a^2) \div (a + 1)$

40. $(a - 15 + 6a^2) \div (2a - 3)$

41. $(6y - 4 + 10y^2) \div (5y - 2)$

42. $(-10x + x^2 + 16) \div (x - 2)$

43. $\dfrac{-18x + 12 + 6x^2}{x - 1}$

44. $\dfrac{27x + 23x^2 + 6x^3}{2x + 3}$

45. $\dfrac{13x + 16x^4 + 3x^2 + 3}{4x + 3}$

46. $\dfrac{3x^2 + 9x^3 + 4x + 4}{3x + 2}$

47. $a^3 + 1$ divided by $a - 1$

48. $27a^3 - 8$ divided by $3a - 2$

49. $\dfrac{15a^3 - 29a^2 + 16}{3a - 4}$

50. $\dfrac{4x^3 - 12x^2 + 17x - 12}{2x - 3}$

51. $y - 2 \overline{) -24y + 24 + 6y^2}$

52. $3 - a \overline{) 21a - a^2 - 54}$

53. $x^2 - 2 \overline{) x^6 - x^4 + 2x^2 - 8}$

54. $x^2 + 3 \overline{) x^6 + 2x^4 - 6x^2 - 9}$

55. $\dfrac{x^4 + 2x^3 + 4x^2 + 3x + 2}{x^2 + x + 2}$

56. $\dfrac{2x^4 + 3x^3 + 3x^2 - 5x - 3}{2x^2 - x - 1}$

57. $\dfrac{x^3 + 3x + 5x^2 + 6 + x^4}{x^2 + 3}$

58. $\dfrac{x^5 + 3x + 2}{x^3 + 1 + 2x}$

Use a calculator to help find each quotient.

59. $x - 2\overline{)9.8x^2 - 3.2x - 69.3}$

60. $2.5x - 3.7\overline{)-22.25x^2 - 38.9x - 16.65}$

Complete each synthetic division.

61. $2\rfloor$ 6   1 -23 2
        5   6
      6 13   3

62. $-3\rfloor$ 2 -4 -25 15
          -6 30
                      0

Use synthetic division to perform each division.

63. $(x^2 + x - 2) \div (x - 1)$

64. $(x^2 + x - 6) \div (x - 2)$

65. $(x^2 + 8 + 6x) \div (x + 4)$

66. $(x^2 - 15 - 2x) \div (x + 3)$

67. $(x^2 - 5x + 14) \div (x + 2)$

68. $(x^2 + 13x + 42) \div (x + 6)$

69. $(3x^3 - 10x^2 + 5x - 6) \div (x - 3)$

70. $(2x^3 - 9x^2 + 10x - 3) \div (x - 3)$

71. $(2x^3 - 5x - 6) \div (x - 2)$

72. $(4x^3 + 5x^2 - 1) \div (x + 2)$

73. $(5x^2 + 6x^3 + 4) \div (x + 1)$

74. $(4 - 3x^2 + x) \div (x - 4)$

Use a calculator and synthetic division to perform each division.

75. $(7.2x^2 - 2.1x + 0.5) \div (x - 0.2)$

76. $(8.1x^2 + 3.2x - 5.7) \div (x - 0.4)$

77. $(2.7x^2 + x - 5.2) \div (x + 1.7)$

78. $(1.3x^2 - 0.5x - 2.3) \div (x + 2.5)$

Let $P(x) = 2x^3 - 4x^2 + 2x - 1$. Evaluate $P(x)$ by substituting the given value of x into the polynomial and simplifying. Then evaluate the polynomial by using the remainder theorem and synthetic division.

79. $P(1)$ **80.** $P(2)$

81. $P(-2)$ **82.** $P(-1)$

Let $Q(x) = x^4 - 3x^3 + 2x^2 + x - 3$. Evaluate $Q(x)$ by substituting the given value of x into the polynomial and simplifying. Then evaluate the polynomial by using the remainder theorem and synthetic division.

83. $Q(-1)$ **84.** $Q(1)$

85. $Q(2)$ **86.** $Q(-2)$

Use the remainder theorem to find $P(r)$.

87. $P(x) = x^3 - 4x^2 + x - 2; r = 2$

88. $P(x) = x^3 - 3x^2 + x + 1; r = 1$

89. $P(x) = 2x^3 + x + 2; r = 3$

90. $P(x) = x^3 + x^2 + 1; r = -2$

91. $P(x) = x^4 - 2x^3 + x^2 - 3x + 2; r = -2$

92. $P(x) = x^5 + 3x^4 - x^2 + 1; r = -1$

93. $P(x) = 3x^5 + 1; r = -\frac{1}{2}$

94. $P(x) = 5x^7 - 7x^4 + x^2 + 1; r = 2$

APPLICATIONS

95. ADVERTISING Find the length of one of the longer sides of the billboard shown below if its area is given by $x^3 - 4x^2 + x + 6$.

96. MASONRY The steel trowel shown below is in the shape of an isosceles triangle. Find the height if the area is given by $6 + 18t + t^2 + 3t^3$.

97. WINTER TRAVEL Complete the table below, which lists the rate (mph), time traveled (hr), and distance traveled (mi) by an Alaskan trail guide.

	r	$\cdot$	t	$=$	d
Dog sled			$4x + 7$		$12x^2 + 13x - 14$
Snowshoes	$3x + 4$				$3x^2 + 19x + 20$

98. PRICING Complete the table below for two items sold at a produce store.

	Price per lb · Number of lb =		Value
Cashews	$x^2 + 2x + 4$		$x^4 + 4x^2 + 16$
Sunflower seeds		$x^2 + 6$	$x^4 - x^2 - 42$

WRITING

99. Explain how to divide a monomial by a monomial.

100. Explain how to check the result of a division problem if there is a nonzero remainder.

101. If you are given $P(x)$, explain how to use synthetic division to calculate $P(a)$.

102. Explain how synthetic division and long division of polynomials are related.

REVIEW   *Simplify each expression.*

103. $2(x^2 + 4x - 1) + 3(2x^2 - 2x + 2)$

104. $3(2a^2 - 3a + 2) - 4(2a^2 + 4a - 7)$

105. $-2(3y^3 - 2y + 7) - (y^2 + 2y - 4) + 4(y^3 + 2y - 1)$

106. $3(4y^3 + 3y - 2) + 2(3y^2 - y + 3) - 5(2y^3 - y^2 - 2)$

6.7 Proportion and Variation

- Ratios and rates • Proportions • Solving proportions • Similar triangles
- Direct variation • Inverse variation • Joint variation • Combined variation

In this section, we discuss five mathematical models that have a variety of applications. First, we show how a *ratio-proportion model* can be used to solve shopping problems and to determine the height of a tree given the length of its shadow. Then we introduce four types of *variation models,* each of which expresses a special relationship between two or more quantities. We use these models to solve problems involving travel, lighting, geometry, and highway construction.

Ratios and rates

The quotient of two numbers or two quantities with the same units is often referred to as a **ratio.** For example, $\frac{2}{3}$ can be read as "the ratio of 2 to 3." The notation $2 : 3$ (read as "2 is to 3") is another common way to denote a ratio. Some more examples of ratios are

$\dfrac{4x}{7y}$ The ratio of $4x$ to $7y$ and $\dfrac{x-2}{3x}$ The ratio of $x - 2$ to $3x$

When we compare two quantities having different units, we call the comparison a **rate,** and we can write it as a fraction. One example is an average rate of speed.

A distance traveled → $\dfrac{372 \text{ miles}}{6 \text{ hours}} = 62 \text{ mph}$ ← The average rate of speed
in a period of time →

Rates are often used to express **unit costs,** such as the cost per pound of ground beef.

The cost of a package of ground beef → $\dfrac{\$7.47}{5 \text{ lb}} \approx \$1.49 \text{ per lb}$ ← The cost per pound
The weight of the package →

Proportions

An equation indicating that two ratios or rates are equal is called a **proportion.** Two examples of proportions are

$\dfrac{1}{4} = \dfrac{2}{8}$ and $\dfrac{4}{7} = \dfrac{12}{21}$

In the proportion $\frac{a}{b} = \frac{c}{d}$, a and d are called the **extremes** of the proportion, and b and c are called the **means.**

To develop a fundamental property of proportions, we suppose that

$$\frac{a}{b} = \frac{c}{d}$$

is a proportion and multiply both sides by bd to obtain

$$bd\left(\frac{a}{b}\right) = bd\left(\frac{c}{d}\right)$$

$$\frac{\cancel{b}da}{\cancel{b}} = \frac{b\cancel{d}c}{\cancel{d}} \qquad \text{Divide out common factors.}$$

$$ad = bc$$

Since $ad = bc$, the product of the extremes equals the product of the means.

The same products ad and bc can be found by multiplying diagonally in the proportion $\frac{a}{b} = \frac{c}{d}$. We call ad and bc **cross products.**

$$\overset{ad}{\underset{}{}} \qquad \overset{bc}{}$$
$$\frac{a}{b} \diagup\!\!\!\!\diagdown \frac{c}{d}$$

> **The fundamental property of proportions**
>
> In a proportion, the product of the extremes is equal to the product of the means.
>
> If $\dfrac{a}{b} = \dfrac{c}{d}$, then $ad = bc$ and   if $ad = bc$, then $\dfrac{a}{b} = \dfrac{c}{d}$.

▮ Solving proportions

We can solve many problems by writing and then solving a proportion. To solve a proportion, we apply the fundamental property of proportions.

INTERMEDIATE
Algebra $f(x)$ Now™

Self Check 1
Solve for x:
$$\frac{x-1}{x} = \frac{x}{x+3}$$

EXAMPLE 1   Solve: $\dfrac{x+3}{x} = \dfrac{x}{x+6}$.

Solution

$$\frac{x+3}{x} = \frac{x}{x+6}$$

$$(x+3)(x+6) = x \cdot x \qquad \text{In a proportion, the product of the extremes equals the product of the means.}$$

$$x^2 + 9x + 18 = x^2 \qquad \text{Perform the multiplications.}$$

$$9x + 18 = 0 \qquad \text{Subtract } x^2 \text{ from both sides.}$$

$$x = -2 \qquad \text{Subtract 18 from both sides and then divide by 9.}$$

Thus, x is -2. To check, we substitute -2 for x in the proportion and simplify:

Check: $\dfrac{x+3}{x} = \dfrac{x}{x+6}$

$$\frac{-2+3}{-2} \overset{?}{=} \frac{-2}{-2+6}$$

$$\frac{1}{-2} \overset{?}{=} \frac{-2}{4}$$

$$-\frac{1}{2} = -\frac{1}{2}$$

Answer $\dfrac{3}{2}$

! COMMENT In Example 1, the expression $\frac{x+3}{x}$ is undefined if $x = 0$, because division by 0 would be indicated. Similarly, $\frac{x}{x+6}$ is undefined if $x = -6$. For these reasons, before beginning the solution process, we can rule out $x = 0$ and $x = -6$ as possible solutions of $\frac{x+3}{x} = \frac{x}{x+6}$.

EXAMPLE 2 Solve: $\frac{5a+2}{2a} = \frac{18}{a+4}$.

Solution Because $\frac{5a+2}{2a}$ is undefined if $a = 0$, we state the restriction that $a \neq 0$. Because $\frac{18}{a+4}$ is undefined if $a = -4$, we also note that $a \neq -4$.

$$\frac{5a+2}{2a} = \frac{18}{a+4}$$

$(5a+2)(a+4) = 2a(18)$ In a proportion, the product of the extremes equals the product of the means.

$5a^2 + 22a + 8 = 36a$ Multiply.

$5a^2 - 14a + 8 = 0$ Subtract 36*a* from both sides.

$(5a - 4)(a - 2) = 0$ Factor to solve the quadratic equation.

$5a - 4 = 0$ or $a - 2 = 0$ Set each factor equal to 0.

$5a = 4$ $a = 2$ Solve each linear equation.

$a = \frac{4}{5}$

Thus, $a = \frac{4}{5}$ or $a = 2$. **Check each solution.**

INTERMEDIATE
Algebra *f(x)* **Now**™

Self Check 2
Solve:
$\frac{3x+1}{12} = \frac{x}{x+2}$

Answer $\frac{2}{3}$, 1

EXAMPLE 3 **Gourmet cooking.** To make a dessert of Pears Hélène, a chef needs to purchase 14 pears. If they are on sale at 6 for $2.34, what will 14 cost?

Solution A proportion can be used to model this situation. First, we let c = the cost of 14 pears. The price per pear when purchasing 6 pears is $\frac{\$2.34}{6}$, and the price per pear when purchasing 14 pears is $\frac{\$c}{14}$. Since these ratios are equal, we have the following proportion: *$2.34 is to 6 pears as $c is to 14 pears.*

Cost of 6 pears → $\dfrac{2.34}{6} = \dfrac{c}{14}$ ← Cost of 14 pears
6 pears →   ← 14 pears

$14(2.34) = 6c$ In a proportion, the product of the extremes is equal to the product of the means.

$32.76 = 6c$ Multiply.

$\dfrac{32.76}{6} = c$ Divide both sides by 6.

$c = 5.46$ Simplify.

Fourteen pears will cost $5.46.

INTERMEDIATE
Algebra *f(x)* **Now**™

Self Check 3
A model railroad engine is 9 inches long. The scale is 87 feet to 1 foot. How long is a real engine? (*Hint:* Note the difference in units.)

Answer $65\frac{1}{4}$ ft

▌ Similar triangles

If two angles of one triangle have the same measure as two angles of a second triangle, the triangles will have the same shape. In this case, we call the triangles **similar triangles.** Here are some facts about similar triangles.

> **Similar triangles**
>
> If two triangles are similar, then
>
> 1. the three angles of the first triangle have the same measure, respectively, as the three angles of the second triangle.
> 2. the lengths of all corresponding sides are in proportion.

The triangles shown in Figure 6-15 are similar triangles.

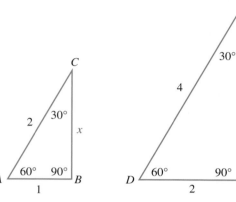

Corresponding angles have the same measure.

The corresponding sides are in proportion:

$$\frac{2}{4} = \frac{x}{2x}$$

$$\frac{x}{2x} = \frac{1}{2}$$

$$\frac{1}{2} = \frac{2}{4}$$

FIGURE 6-15

The properties of similar triangles often enable us to determine the lengths of the sides of triangles indirectly. For example, on a sunny day, we can find the height of a tree and stay safely on the ground.

INTERMEDIATE
Algebra *f(x)* Now™

EXAMPLE 4   Height of a tree.

A tree casts a shadow of 29 feet at the same time as a vertical yardstick casts a shadow of 2.5 feet. Find the height of the tree.

Solution   Refer to Figure 6-16, which shows the triangles determined by the tree and its shadow and the yardstick and its shadow. Because the triangles have the same shape, they are similar, and the measures of their corresponding sides are in proportion. If we let h represent the height of the tree, we can find h by setting up and solving the following proportion: h is to 3 as 29 is to 2.5.

$$\text{height of tree} \rightarrow \frac{h}{3} = \frac{29}{2.5} \leftarrow \text{length of tree's shadow}$$
$$\text{height of yardstick} \rightarrow \qquad\qquad \leftarrow \text{length of yardstick's shadow}$$

$2.5h = 3(29)$ In a proportion, the product of the extremes is equal to the product of the means.

$2.5h = 87$ Multiply.

$\quad h = 34.8$ Divide both sides by 2.5.

The tree is about 35 feet tall.

3 ft

2.5 ft

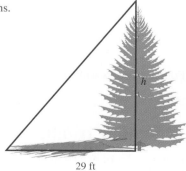

h

29 ft

FIGURE 6-16

Direct variation

To introduce direct variation, we consider the formula for the circumference of a circle

$$C = \pi D$$

where C is the circumference, D is the diameter, and $\pi \approx 3.14159$. If we double the diameter of a circle, we determine another circle with a larger circumference C_1 such that

$$C_1 = \pi(2D) = 2\pi D = 2C$$

Thus, doubling the diameter results in doubling the circumference. Likewise, if we triple the diameter, we will triple the circumference.

In this formula, we say that the variables C and D *vary directly,* or that they are *directly proportional.* This is because C is always found by multiplying D by a constant. In this example, the constant π is called the *constant of variation* or the *constant of proportionality.*

> **Direct variation**
>
> The words "y varies directly with x" or "y is directly proportional to x" mean that $y = kx$ for some nonzero constant k. The constant k is called the **constant of variation** or the **constant of proportionality.**

Since the formula for direct variation ($y = kx$) defines a linear function, its graph is always a line with a y-intercept at the origin. The graph of $y = kx$ appears in Figure 6-17 for three positive values of k.

One example of direct variation is Hooke's law from physics. Hooke's law states that the distance a spring will stretch varies directly with the force that is applied to it.

If d represents a distance and f represents a force, this verbal model of Hooke's law can be expressed mathematically as

$$d = kf \quad \text{This direct variation model can also be read as "}d\text{ is directly proportional to }f\text{."}$$

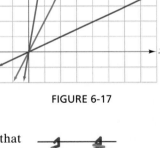

FIGURE 6-17

where k is the constant of variation. Suppose we know that a certain spring stretches 10 inches when a weight of 6 pounds is attached (see Figure 6-18). We can find k as follows:

$$d = kf$$
$$10 = k(6) \quad \text{Substitute 10 for } d \text{ and 6 for } f.$$
$$\frac{5}{3} = k$$

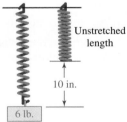

FIGURE 6-18

To find the force required to stretch the spring a distance of 35 inches, we can solve the equation $d = kf$ for f, with $d = 35$ and $k = \frac{5}{3}$.

$$d = kf$$
$$35 = \frac{5}{3}f \quad \text{Substitute 35 for } d \text{ and } \tfrac{5}{3} \text{ for } k.$$
$$105 = 5f \quad \text{Multiply both sides by 3.}$$
$$21 = f \quad \text{Divide both sides by 5.}$$

Thus, the force required to stretch the spring a distance of 35 inches is 21 pounds.

Self Check 5

When exchanging currencies, the number of British pounds received is directly proportional to the number of U.S. dollars to be exchanged. If $800 converts to 440 pounds, how many pounds will be received if $1,500 is exchanged?

EXAMPLE 5   Currency exchange.

The currency calculator shown on the right converts from U.S. dollars to Japanese yen. When exchanging these currencies, the number of yen received is directly proportional to the number of dollars to be exchanged. How many yen will an exchange of $1,200 bring?

Solution   The verbal model *the number of yen is directly proportional to the number of dollars* can be expressed by the equation

$$y = kd \qquad \text{This is a direct variation model.}$$

where y is the number of yen, k is the constant of variation, and d is the number of dollars. From the illustration, we see that an exchange of $500 brings 54,665 yen. To find k, we substitute 500 for d and 54,665 for y, and then we solve for k.

$$y = kd$$
$$54{,}665 = k(500)$$
$$109.33 = k \qquad \text{Divide both sides by 500.}$$

To find how many yen an exchange of $1,200 will bring, we substitute 109.33 for k and 1,200 for d in the direct variation model, and then we evaluate the right-hand side.

$$y = kd$$
$$y = 109.33\,(1{,}200)$$
$$y = 131{,}196$$

Answer 825 British pounds

An exchange of $1,200 will bring 131,196 yen.

Solving variation problems

To solve a variation problem:

1. Translate the verbal model into an equation.
2. Substitute the first set of values into the equation from step 1 to determine the value of k.
3. Substitute the value of k into the equation from step 1.
4. Substitute the remaining set of values into the equation from step 3 and solve for the unknown.

Inverse variation

In the formula $w = \frac{12}{l}$, w gets smaller as l gets larger, and w gets larger as l gets smaller. Since these variables vary in opposite directions in a predictable way, we say that the variables *vary inversely*, or that they are *inversely proportional*. The constant 12 is the constant of variation.

Inverse variation

The words "y varies inversely with x" or "y is inversely proportional to x" mean that $y = \frac{k}{x}$ for some nonzero constant k. The constant k is called the **constant of variation**.

The formula for inverse variation $\left(y = \frac{k}{x}\right)$ defines a rational function whose graph will have the x- and y-axes as asymptotes. The graph of $y = \frac{k}{x}$ appears in Figure 6-19 for three positive values of k.

In an elevator, the amount of floor space per person varies inversely with the number of people in the elevator. If f represents the amount of floor space per person and n the number of people in the elevator, the relationship between f and n can be expressed by the equation

$$f = \frac{k}{n}$$ This inverse variation model can also be read as "f is inversely proportional to n."

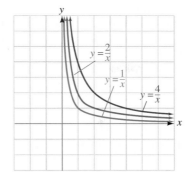

FIGURE 6-19

Figure 6-20 shows 6 people in an elevator; each has 8.25 square feet of floor space. To determine how much floor space each person would have if 15 people were in the elevator, we begin by determining k.

$$f = \frac{k}{n}$$

$$8.25 = \frac{k}{6}$$ Substitute 8.25 for f and 6 for n.

$$k = 49.5$$ Multiply both sides by 6 to solve for k.

FIGURE 6-20

To find the amount of floor space per person if 15 people are in the elevator, we proceed as follows:

$$f = \frac{k}{n}$$

$$f = \frac{49.5}{15}$$ Substitute 49.5 for k and 15 for n.

$$f = 3.3$$ Do the division.

If 15 people were in the elevator, each would have 3.3 square feet of floor space.

EXAMPLE 6   Photography.

The intensity I of light received from a light source varies inversely with the square of the distance from the light source. If a photographer, 16 feet away from his subject, has a light meter reading of 4 foot-candles of illuminance, what will the meter read if the photographer moves in for a close-up 4 feet away from the subject?

Solution   The words *intensity varies inversely with the square of the distance d* can be expressed by the equation

$$I = \frac{k}{d^2}$$ This inverse variation model can also be read as "I is inversely proportional to d^2."

To find k, we substitute 4 for I and 16 for d and solve for k.

$$I = \frac{k}{d^2}$$

$$4 = \frac{k}{16^2}$$

$$4 = \frac{k}{256}$$

$$1{,}024 = k$$

INTERMEDIATE
Algebra $f(x)$ Now™

Self Check 6
In Example 6, find the intensity when the photographer is 8 feet away from the subject.

To find the intensity when the photographer is 4 feet away from the subject, we substitute 4 for d and 1,024 for k and simplify.

$$I = \frac{k}{d^2}$$

$$I = \frac{1,024}{4^2}$$

$$= 64$$

Answer 16 foot-candles

The intensity at 4 feet is 64 foot-candles.

Joint variation

There are times when one variable varies with the product of several variables. For example, the area of a triangle varies directly with the product of its base and height:

$$A = \frac{1}{2}bh$$

Such variation is called *joint variation.*

Joint variation

If one variable varies directly with the product of two or more variables, the relationship is called **joint variation.** If y varies jointly with x and z, then $y = kxz$. The nonzero constant k is called the **constant of variation.**

EXAMPLE 7   The force of the wind on a billboard varies jointly with the area of the billboard and the square of the wind velocity. When the wind is blowing at 20 mph, the force on a billboard 30 feet wide and 18 feet high is 972 pounds. (Refer to Figure 6-21.) Find the force on a billboard having an area of 300 square feet caused by a 40-mph wind.

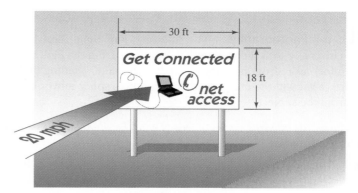

FIGURE 6-21

Solution   We will let f represent the force of the wind, A the area of the billboard, and v the velocity of the wind. The words *the force of the wind on a billboard varies jointly with the area of the billboard and the square of the wind velocity* mean that f varies directly as the product of A and v^2. Thus,

$$f = kAv^2 \qquad \text{The joint variation model can also be read as "} f \text{ is directly proportional to the product of } A \text{ and } v^2\text{."}$$

Since the billboard is 30 feet wide and 18 feet high, it has an area of $30 \cdot 18 = 540$ square feet. We can find k by substituting 972 for f, 540 for A, and 20 for v.

$$f = kAv^2$$
$$972 = k(540)(20)^2$$
$$972 = k(216,000) \qquad \text{Find the power: } (20)^2 = 400. \text{ Then perform the multiplication.}$$
$$0.0045 = k \qquad \text{Divide both sides by 216,000 to solve for } k.$$

To find the force exerted on a 300-square-foot billboard by a 40-mph wind, we use the formula $f = 0.0045Av^2$ and substitute 300 for A and 40 for v.

$$f = 0.0045Av^2$$
$$= 0.0045(300)(40)^2$$
$$= 2,160$$

The 40-mph wind exerts a force of 2,160 pounds on the billboard.

■ Combined variation

Many applied problems involve a combination of direct and inverse variation. Such variation is called **combined variation.**

INTERMEDIATE
Algebra $f(x)$ **Now**™

EXAMPLE 8   Highway construction.   The time it takes to build a highway varies directly with the length of the road, but inversely with the number of workers. If it takes 100 workers 4 weeks to build 2 miles of highway, how long will it take 80 workers to build 10 miles of highway?

Self Check 8
In Example 8, how long will it take 60 workers to build 6 miles of highway?

Solution   We can let t represent the time in weeks to build a highway, l represent the length of the highway in miles, and w represent the number of workers. The relationship between these variables can be expressed by the equation

$$t = \frac{kl}{w} \qquad \text{This is a combined variation model.}$$

We substitute 4 for t, 100 for w, and 2 for l to find k:

$$4 = \frac{k(2)}{100}$$
$$400 = 2k \qquad \text{Multiply both sides by 100.}$$
$$200 = k \qquad \text{Divide both sides by 2.}$$

We now substitute 80 for w, 10 for l, and 200 for k in the equation $t = \frac{kl}{w}$ and simplify:

$$t = \frac{kl}{w}$$
$$t = \frac{200(10)}{80}$$
$$= 25$$

It will take 25 weeks for 80 workers to build 10 miles of highway.

Answer 20 weeks

! COMMENT In variation problems, the constant of variation is usually positive, because most real-life applications involve only positive quantities. However, the definitions of direct, inverse, joint, and combined variation allow for a negative constant of variation.

Section 6.7 STUDY SET

VOCABULARY *Fill in the blanks.*

1. A _____ is the quotient of two numbers or two quantities with the same units.

2. An equation that states that two ratios are equal, such as $\frac{1}{2} = \frac{4}{8}$, is called a _____.

3. In a proportion, the product of the _____ is equal to the product of the _____.

4. If two angles of one triangle have the same measure as two angles of a second triangle, the triangles are _____.

5. The equation $y = kx$ defines _____ variation, and $y = \frac{k}{x}$ defines _____ variation.

6. The equation $y = kxz$ defines _____ variation, and $y = \frac{kx}{z}$ defines _____ variation.

7. _____ variation is represented by a rational function.

8. _____ variation is represented by a linear function.

CONCEPTS *Determine whether direct or inverse variation applies and sketch a reasonable graph for the situation.*

9.   **10.**

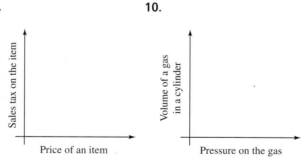

11.   **12.**

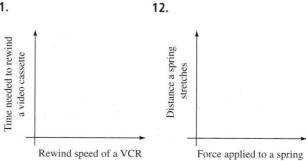

NOTATION *Complete each solution to solve the proportion.*

13. $\dfrac{-7}{6} = \dfrac{x + 3}{12}$

$\boxed{}(12) = \boxed{}(x + 3)$

$-84 = 6x + \boxed{}$

$\boxed{} = 6x$

$-17 = x$

14. $\dfrac{18}{2x + 1} = \dfrac{3}{14}$

$\boxed{}(14) = (\boxed{})3$

$252 = \boxed{} + 3$

$249 = \boxed{}$

$41.5 = x$

PRACTICE *Solve each proportion, if possible.*

15. $\dfrac{x}{5} = \dfrac{15}{25}$

16. $\dfrac{4}{y} = \dfrac{6}{27}$

17. $\dfrac{r - 2}{3} = \dfrac{r}{5}$

18. $\dfrac{x + 1}{x - 1} = \dfrac{6}{4}$

19. $\dfrac{5}{5z + 3} = \dfrac{2z}{2z^2 + 6}$

20. $\dfrac{9t + 6}{t} = \dfrac{7}{3}$

21. $\dfrac{2(y + 3)}{3} = \dfrac{4(y - 4)}{5}$

22. $\dfrac{b + 4}{5} = \dfrac{3(b - 2)}{3}$

23. $\dfrac{2}{3x} = \dfrac{6x}{36}$

24. $\dfrac{y}{4} = \dfrac{4}{y}$

25. $\dfrac{2}{c} = \dfrac{c - 3}{2}$

26. $\dfrac{2}{x + 6} = \dfrac{-2x}{5}$

27. $\dfrac{1}{x + 3} = \dfrac{-2x}{x + 5}$

28. $\dfrac{x - 1}{x + 1} = \dfrac{2}{3x}$

29. $\dfrac{9z + 6}{z(z + 3)} = \dfrac{7}{z + 3}$

30. $\dfrac{3}{n(n + 3)} = \dfrac{2}{(n + 1)(n + 3)}$

Express each verbal model in symbols.

31. A varies directly with the square of p.

32. z varies inversely with the cube of t.

33. v varies inversely with the square of r.

34. C varies jointly with x, y, and z.

35. P varies directly with the square of a and inversely with the cube of j.

36. M varies inversely with the cube of n and jointly with x and the square of z.

Express each variation model in words. In each equation, k is the constant of variation.

37. $L = kmn$

38. $P = \dfrac{km}{n}$

39. $R = \dfrac{kL}{d^2}$

40. $U = krs^2t$

■ **APPLICATIONS** *Set up a proportion and solve it.*

41. CAFFEINE Many convenience stores sell super-size 44-ounce soft drinks in refillable cups. For each of the products listed in the table, find the amount of caffeine contained in one of the large cups. Round to the nearest milligram.

Soft drink, 12 oz	Caffeine (mg)
Mountain Dew	55
Coca-Cola Classic	47
Pepsi	37

Based on data from *Los Angeles Times* (November 11, 1997) p. S4

42. TELEPHONES As of 2003, Iceland had 221 mobile cellular telephones per 250 inhabitants—the highest rate of any country in the world. If Iceland's population is about 280,000, how many mobile cellular telephones does the country have?

43. WALLPAPERING The instructions on the label of wallpaper adhesive read as shown. Estimate the amount of adhesive needed to paper 500 square feet of kitchen walls if a heavy wallpaper will be used.

> COVERAGE: One-half gallon will hang approximately 4 single rolls (140 sq ft), depending on the weight of the wall covering and the condition of the wall.

44. RECOMMENDED DOSAGES The recommended child's dose of the sedative hydroxine is 0.006 gram per kilogram of body mass. Find the dosage for a 30-kg child in milligrams.

45. ERGONOMICS The science of ergonomics coordinates the design of working conditions with the requirements of the worker. The illustration gives guidelines for the dimensions (in inches) of a computer workstation to be used by a person whose height is 69 inches. Find a set of workstation dimensions for a person 5 feet 11 inches tall. Round to the nearest tenth.

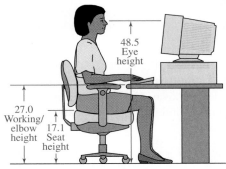

Based on information from the Anthropometric Survey of the U.S. Army personnel database

46. SHOPPING A recipe for guacamole dip calls for 5 avocados. If they are advertised at 3 for $1.98, what will 5 avocados cost?

47. DRAWING To make an enlargement of the sailboat, an artist drew a grid over the smaller picture and transferred the contents of each small box to its corresponding larger box on another sheet of paper. If the smaller picture is 3 in. × 5 in. and if the width of the enlargement is 7.5 in., what is the length of the enlargement?

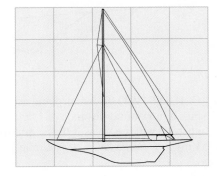

48. DRAFTING In a scale drawing, a 280-foot antenna tower is drawn $7\frac{1}{2}$ inches high. The building next to it is drawn $2\frac{1}{4}$ inches high. How tall is the actual building?

Use similar triangles to help solve each problem.

49. WASHINGTON, D.C. The Washington Monument casts a shadow of $166\frac{1}{2}$ feet at the same time as a 5-foot-tall tourist casts a shadow of $1\frac{1}{2}$ feet. Find the height of the monument.

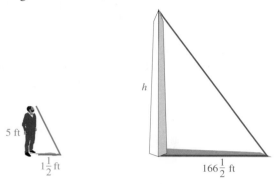

5 ft

h

$1\frac{1}{2}$ ft $166\frac{1}{2}$ ft

50. HEIGHT OF A FLAGPOLE A man places a mirror on the ground and sees the reflection of the top of a flagpole, as shown. The two triangles in the illustration are similar. Find the height h of the flagpole.

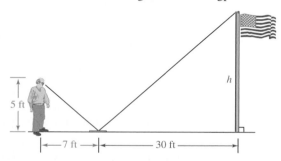

5 ft

h

← 7 ft → ← 30 ft →

51. WIDTH OF A RIVER Use the dimensions in the illustration below to find w, the width of the river. The two triangles in the illustration are similar.

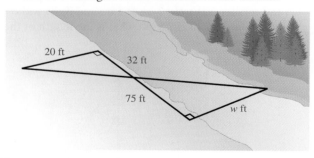

20 ft

32 ft

75 ft

w ft

52. FLIGHT PATHS An airplane ascends 150 feet as it flies a horizontal distance of 1,000 feet. How much altitude will it gain as it flies a horizontal distance of 1 mile? (*Hint:* 5,280 feet = 1 mile.)

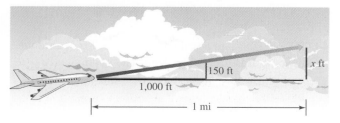

150 ft

x ft

1,000 ft

1 mi

53. SKI RUNS A ski course with $\frac{1}{2}$ mile of horizontal run falls 100 feet in every 300 feet of run. Find the height of the hill.

54. GRAPHIC ARTS The compass shown below is used to draw circles with different radii (plural for radius). For the setting shown, what radius will the resulting circle have?

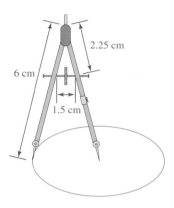

2.25 cm

6 cm

1.5 cm

Solve each problem by writing a variation model of the situation.

55. FREE FALL An object in free fall travels a distance s that is directly proportional to the square of the time t. If an object falls 1,024 feet in 8 seconds, how far will it fall in 10 seconds?

56. FINDING DISTANCES The distance that a car can go varies directly with the number of gallons of gasoline it consumes. If a car can go 288 miles on 12 gallons of gasoline, how far can it go on a full tank of 18 gallons?

57. FARMING The length of time that a given number of bushels of corn will last when feeding cattle varies inversely with the number of animals. If x bushels will feed 25 cows for 10 days, how long will the feed last for 10 cows?

58. ORGAN PIPES The frequency of vibration of air in an organ pipe is inversely proportional to the length of the pipe. If a pipe 2 feet long vibrates 256 times per second, how many times per second will a 6-foot pipe vibrate?

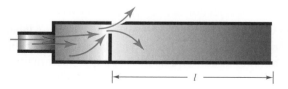

l

59. GAS PRESSURE Under constant temperature, the volume occupied by a gas varies inversely to the pressure applied. If the gas occupies a volume of 20 cubic inches under a pressure of 6 pounds per square inch, find the volume when the gas is subjected to a pressure of 10 pounds per square inch.

60. REAL ESTATE The table below shows the listing price for three homes in the same general locality. Write the variation model (direct or inverse) that describes the relationship between the listing price and the number of square feet of a house in this area.

Number of square feet	Listing price
1,720	$129,000
1,205	$90,375
1,080	$81,000

61. TRUCKING COSTS The costs incurred by a trucking company vary jointly with the number of trucks in service and the number of hours they are used. When 4 trucks are used for 6 hours each, the costs are $1,800. Find the costs of using 10 trucks, each for 12 hours.

62. OIL STORAGE The number of gallons of oil that can be stored in a cylindrical tank varies jointly with the height of the tank and the square of the radius of its base. The constant of proportionality is 23.5. Find the number of gallons that can be stored in the cylindrical tank below.

63. ELECTRONICS The voltage (in volts) measured across a resistor is directly proportional to the current (in amperes) flowing through the resistor. The constant of variation is the **resistance** (in ohms). If 6 volts is measured across a resistor carrying a current of 2 amperes, find the resistance.

64. ELECTRONICS The power (in watts) lost in a resistor (in the form of heat) varies directly with the square of the current (in amperes) passing through it. The constant of proportionality is the resistance (in ohms). What power is lost in a 5-ohm resistor carrying a 3-ampere current?

65. STRUCTURAL ENGINEERING The deflection of a beam is inversely proportional to its width and the cube of its depth. If the deflection of a 4-inch-by-4-inch beam is 1.1 inches, find the deflection of a 2-inch-by-8-inch beam positioned as shown in the next column.

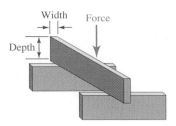

66. STRUCTURAL ENGINEERING Find the deflection of the beam in Exercise 65 when the beam is positioned as shown below.

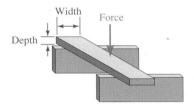

67. TENSION IN A STRING When playing with a Skip It toy, a child swings a weighted ball on the end of a string in a circular motion around one leg while jumping over the revolving string with the other leg. The tension T in the string is directly proportional to the square of the speed s of the ball and inversely proportional to the radius r of the circle. If the tension in the string is 6 pounds when the speed of the ball is 6 feet per second and the radius is 3 feet, find the tension when the speed is 8 feet per second and the radius is 2.5 feet.

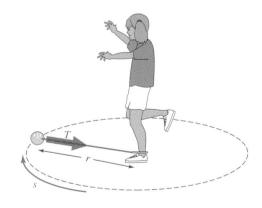

68. GAS PRESSURE The pressure of a certain amount of gas is directly proportional to the temperature (measured on the Kelvin scale) and inversely proportional to the volume. A sample of gas at a pressure of 1 atmosphere occupies a volume of 1 cubic meter at a temperature of 273 Kelvin. When heated, the gas expands to twice its volume, but the pressure remains constant. To what temperature is it heated?

WRITING

69. Distinguish between a *ratio* and a *proportion*.

70. From everyday life, give examples of two quantities that vary directly and two quantities that vary inversely.

REVIEW   *Simplify each expression.*

71. $(x^2x^3)^2$

72. $\left(\dfrac{a^3a^5}{a^{-2}}\right)^3$

73. $\dfrac{b^0 - 2b^0}{b^0}$

74. $\left(\dfrac{2r^{-2}r^{-3}}{4r^{-5}}\right)^{-3}$

Expressions and Equations

In this chapter, we have discussed procedures for working with **rational expressions** and procedures for solving **rational equations.**

■ Rational Expressions

The **fundamental property of fractions** is used when simplifying rational expressions and when multiplying and dividing rational expressions: *We can divide out factors that are common to the numerator and the denominator of a fraction.*

1. a. Simplify: $\dfrac{6x^2 + x - 2}{8x^2 + 2x - 3}$.

   b. What common factor was divided out?

2. a. Multiply: $\dfrac{3d^2 - d - 2}{6d^2 - 5d - 6} \cdot \dfrac{4d^2 - 9}{2d^2 + 5d + 3}$.

   b. What common factors were divided out?

The fundamental property of fractions also states that *multiplying the numerator and denominator of a fraction by the same nonzero number does not change the value of the fraction.* We use this concept to "build" fractions when adding or subtracting rational expressions with unlike denominators, and when simplifying complex fractions.

3. a. Add: $\dfrac{3}{x + 2} + \dfrac{5}{x - 4}$.

   b. By what did you multiply the first fraction to rewrite it in terms of the LCD? By what did you multiply the second fraction?

4. a. Use Method 2 to simplify $\dfrac{n - 1 - \dfrac{2}{n}}{\dfrac{n}{3}}$.

   b. By what did you multiply the numerator and denominator to simplify the complex fraction?

■ Rational Equations

The multiplication property of equality states that *if equal quantities are multiplied by the same nonzero number, the results will be equal quantities.* We use this property when solving rational equations. If we multiply both sides of the equation by the LCD of the rational expressions in the equation, we can clear it of fractions.

5. a. Solve: $\dfrac{t - 3}{t - 2} - \dfrac{t - 3}{t} = \dfrac{1}{t}$.

   b. By what did you multiply both sides to clear the equation of fractions?

6. a. Solve: $\dfrac{5}{2x^2 + x - 3} - \dfrac{x + 1}{x - 1} = \dfrac{2}{2x + 3} - 1$.

   b. By what did you multiply both sides to clear the equation of fractions?

7. a. Solve: $\dfrac{x + 1}{x + 2} = \dfrac{x - 3}{x - 4}$.

   b. By what did you multiply both sides to clear the equation of fractions?

8. a. Solve: $\dfrac{x^2}{a^2} - \dfrac{y^2}{b^2} = 1$ for a^2.

   b. By what did you multiply both sides to clear the equation of fractions?

ACCENT ON TEAMWORK

SECTION 6.1

VERTICAL ASYMPTOTES Consider the function $f(x) = \frac{1}{x}$. Use your calculator to complete the two tables of values.

x	3	2	1	0.5	0.25	0.1	0.01	0.001
$f(x)$								

x approaches 0 from the right

x	−3	−2	−1	−0.5	−0.25	−0.1	−0.01	−0.001
$f(x)$								

x approaches 0 from the left

Plot the points from each table on the same graph and draw a smooth curve through them. Explain what happens to the function values $f(x)$ as x approaches 0 from the right and approaches 0 from the left. Label the vertical asymptote and explain why the function is not defined for $x = 0$.

SECTION 6.2

MULTIPLYING RATIONAL EXPRESSIONS
Explain what is wrong with the student's work shown here. Then write a correct solution.

$$\frac{3}{4x} + \frac{2}{3x} = 12x\left(\frac{3}{4x} + \frac{2}{3x}\right)$$

$$= 12x \cdot \frac{3}{4x} + 12x \cdot \frac{2}{3x}$$

$$= 9 + 8$$

$$= 17$$

SECTION 6.3

ADDING RATIONAL EXPRESSIONS Add the fractions by expressing them in terms of a common denominator $24b^3$. (Note: This is not the LCD.)

$$\frac{r}{4b^2} + \frac{s}{6b}$$

An extra step had to be performed because the lowest common denominator was not used. What was it?

SECTION 6.4

COMPLEX FRACTIONS Each complex fraction in the list

$$1 + \frac{1}{2}, \quad 1 + \frac{1}{1 + \frac{1}{2}}, \quad 1 + \frac{1}{1 + \frac{1}{1 + \frac{1}{2}}}$$

$$1 + \frac{1}{1 + \frac{1}{1 + \frac{1}{1 + \frac{1}{2}}}}, \ldots$$

can be simplified by using the value of the expression preceding it. For example, to simplify the second expression in the list, replace $1 + \frac{1}{2}$ with $\frac{3}{2}$. Show that the expressions in the list simplify to the fractions $\frac{3}{2}, \frac{5}{3}, \frac{8}{5}, \frac{13}{8}, \frac{21}{13}, \frac{34}{21}, \ldots$. Do you see a pattern? Can you predict the next fraction?

SECTION 6.5

SHARED WORK Use a watch to time (in seconds) how long it takes to fill a pail using a garden hose. From a different faucet, with the water running at a different rate, determine how long it takes a second hose to fill the same pail. Use the concepts studied in Section 6.5 to determine how long it would take to fill the pail with both hoses. Then fill the pail using both hoses, keeping track of the time. How close is the actual time to the predicted time? What are some reasons for any difference between them? Make a video of this project and show it to your class.

SECTION 6.6

FACTORS Since 6 is a factor of 24, 6 divides 24 exactly with no remainder. Use this same reasoning to determine whether

a. $2x - 3$ is a factor of $10x^2 - x - 21$.
b. $x - 1$ is a factor of $x^5 - 1$.
c. $2x + y$ is a factor of $32x^5 + y^5$.

SECTION 6.7

COOKING Find a simple recipe for a treat that you can make for your classmates. Use a proportion to determine the amount of each ingredient needed to make enough of the recipe for the exact number of people in your class. Write the old recipe and the new recipe on separate pieces of poster board. Did the recipe serve the correct number of people? Tell the class how you made the calculations and what difficulties you encountered.

SECTION 6.1	*Rational Functions and Simplifying Rational Expressions*

CONCEPTS

Rational expressions are fractions that indicate the quotient of two polynomials.

A *rational function* is defined by a rational expression in one variable where the polynomial in the denominator is never zero.

To *simplify a rational expression:*

1. Factor the numerator and the denominator.

2. Divide out the common factors of the numerator and denominator by applying the *fundamental property of fractions:*

$$\frac{ak}{bk} = \frac{a}{b}$$

The quotient of any nonzero quantity and its opposite is -1.

REVIEW EXERCISES

1. Complete the table of values for the rational function $f(x) = \frac{4}{x}$ where $x > 0$. Then graph it. Label the horizontal asymptote.

x	f(x)
$\frac{1}{2}$	
1	
2	
3	
4	
5	
6	
7	
8	

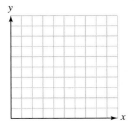

2. Use a graphing calculator to graph the rational function $f(x) = \frac{3x + 2}{x}$. From the graph, determine the equations of the horizontal and vertical asymptotes and the domain and range.

Simplify each rational expression.

3. $\dfrac{248x^2y}{576xy^2}$

4. $\dfrac{x^2 - 49}{x^2 + 14x + 49}$

5. $\dfrac{x^2 - 2x + 4}{2x^3 + 16}$

6. $\dfrac{x^2 + 6x + 36}{x^3 - 216}$

7. $\dfrac{ac - ad + bc - bd}{d^2 - c^2}$

8. $\dfrac{m^3 + m^2n - 2mn^2}{mn^2 + m^2n - 2m^3}$

9. $\dfrac{x - y}{y - x}$

10. $\dfrac{2m - 2n}{n - m}$

SECTION 6.2	*Multiplying and Dividing Rational Expressions*

To multiply two fractions, multiply the numerators and multiply the denominators:

$$\frac{a}{b} \cdot \frac{c}{d} = \frac{ac}{bd}$$

Perform the operations and simplify:

11. $\dfrac{x^2 + 4x + 4}{x^2 - x - 6} \cdot \dfrac{9 - x^2}{x^2 + 5x + 6}$

12. $\dfrac{2a^2 - 5a - 3}{a^2 - 9} \div \dfrac{2a^2 + 5a + 2}{2a^2 + 5a - 3}$

To divide two fractions, invert the divisor and multiply:

$$\frac{a}{b} \div \frac{c}{d} = \frac{a}{b} \cdot \frac{d}{c}$$

(Multiply the first fraction by the *reciprocal* of the second.)

13. $\left(\dfrac{h-2}{h^3+4}\right)^2$

14. $\dfrac{t^2-4}{t} \div (t+2)$

15. $\dfrac{x^2+3x+2}{x^2-x-6} \cdot \dfrac{3x^2-3x}{x^2-3x-4} \div \dfrac{x^2+3x+2}{x^2-2x-8}$

SECTION 6.3   *Adding and Subtracting Rational Expressions*

To add (or subtract) two rational expressions with *like* denominators, add (or subtract) the numerators and keep the common denominator.

Perform the operations and simplify.

16. $\dfrac{5y}{x-y} - \dfrac{3}{x-y}$

17. $\dfrac{3x-1}{x^2+2} + \dfrac{3(x-2)}{x^2+2}$

18. $\dfrac{4}{t-3} + \dfrac{6}{3-t}$

19. $\dfrac{p+3}{p^2+13p+12} - \dfrac{2p+4}{p^2+13p+12}$

To find the LCD of several rational expressions, factor each denominator and use each factor the greatest number of times that it appears in any one denominator. The product of these factors is the LCD.

The denominators of some rational expressions are given. Find the LCD.

20. $15a^2h, 20ah^3$

21. $ab^2 - ab, ab^2, b^2 - b$

22. $x^2 - 4x - 5, x^2 - 25$

23. $m^2 - 4m + 4, m^3 - 8$

To add or subtract rational expressions with *unlike* denominators, find the LCD and express each rational expression with a denominator that is the LCD. Add (or subtract) the resulting fractions and simplify the result if possible.

Perform the operations and simplify.

24. $9 - \dfrac{1}{a+1}$

25. $\dfrac{x}{4z^2} + \dfrac{y}{6z}$

26. $\dfrac{4x}{x-4} - \dfrac{3}{x+3}$

27. $\dfrac{4}{3xy-6y} - \dfrac{4}{10-5x}$

28. $\dfrac{-2(3+x)}{x^2+6x+9} + \dfrac{3(x+2)}{x^2-6x+9} - \dfrac{1}{x^2-9}$

SECTION 6.4   *Complex Fractions*

To *simplify a complex fraction:*

Method 1: Write the numerator and denominator as single fractions. Then divide the fractions and simplify.

Method 2: Multiply the numerator and denominator by the LCD of the fractions in the numerator and denominator. Then simplify the results.

Simplify each complex fraction.

29. $\dfrac{\dfrac{p^2-9}{6pt}}{\dfrac{p^2+5p+6}{3pt}}$

30. $\dfrac{\dfrac{1}{a}+\dfrac{2}{b}}{\dfrac{2}{a}-\dfrac{1}{b}}$

31. $\dfrac{1-\dfrac{1}{x}-\dfrac{2}{x^2}}{1+\dfrac{4}{x}+\dfrac{3}{x^2}}$

32. $\dfrac{(x-y)^{-2}}{x^{-2}-y^{-2}}$

33. $\dfrac{3x - \dfrac{1}{3-\dfrac{x}{2}}}{\dfrac{3}{x-6}+x}$

Equations Containing Rational Expressions

To solve a *rational equation*, multiply both sides by the LCD of the rational expressions in the equation to clear it of fractions.

Multiplying both sides of an equation by a quantity that contains a variable can lead to *extraneous* (false) solutions. All possible solutions of a rational equation must be checked.

Solve each equation. If a solution is extraneous, so indicate.

34. $\dfrac{4}{x} - \dfrac{1}{10} = \dfrac{7}{2x}$

35. $\dfrac{5}{7+t} - 1 = \dfrac{-4}{t+7}$

36. $\dfrac{q-3}{6} = \dfrac{10}{q+4}$

37. $\dfrac{2}{x+5} - \dfrac{1}{6} = \dfrac{1}{x+4}$

38. $\dfrac{2(x-5)}{x-2} = \dfrac{6x+12}{4-x^2}$

39. $\dfrac{x+3}{x-5} + \dfrac{6+2x^2}{x^2-7x+10} = \dfrac{3x}{x-2}$

Solve each formula for the indicated variable.

40. $\dfrac{x^2}{a^2} - \dfrac{y^2}{b^2} = 1$ for y^2

41. $H = \dfrac{2ab}{a+b}$ for b

42. ADVERTISING A small plane pulling a banner can fly at a rate of 75 mph in calm air. Flying down the coast, with a tailwind, the plane flew 40 miles in the same time that it took to fly 35 miles up the coast, into a headwind. Find the rate of the wind.

43. TRIP LENGTH Traffic reduced a driver's usual speed by 10 mph, which lengthened her 200-mile trip by 1 hour. Find the driver's usual speed.

44. DRAINING A TANK If one outlet pipe can drain a tank in 24 hours and another pipe can drain the tank in 36 hours, how long will it take for both pipes to drain the tank?

45. INSTALLING SIDING Two men have estimated that they can side a house in 8 days. If one of them, who could have sided the house alone in 14 days, gets sick, how long will it take the other man to side the house alone?

46. METALLURGY The stiffness of the flagpole is given by the formula

$$k = \dfrac{1}{\dfrac{1}{k_1} + \dfrac{1}{k_2}}$$

where k_1 and k_2 are the individual stiffnesses of each section. If the design specifications require that the stiffness k of the entire pole be 1,900,000 in. lb/rad, what must the stiffness of Section 1 be?

Section 1
Stiffness k_1

Section 2
Stiffness
$k_2 = 4{,}200{,}000$ in. lb/rad

SECTION 6.6   *Dividing Polynomials*

To divide two *monomials:* Factor the numerator and denominator. Then divide out common factors. Or divide the coefficients and apply the rules for exponents.

Perform each division. Write each answer without negative exponents.

47. $\dfrac{25h^4k^7}{5hk^9}$

48. $(-5x^6y^3) \div (10x^3y^6)$

To divide a *polynomial by a monomial,* write each term of the numerator over the denominator. Then simplify each resulting rational expression.

Find each quotient.

49. $\dfrac{36a + 32}{6}$

50. $\dfrac{30x^3y^2 - 15x^2y - 10xy^2}{-10xy}$

The *long division* algorithm can be used to divide a *polynomial by a polynomial.* Write the powers of the variable in descending order. If a power of a variable is missing, insert a placeholder term with a coefficient of 0.

Find each quotient using long division.

51. $b + 5\overline{)b^2 + 9b + 20}$

52. $\dfrac{-33v - 8v^2 + 3v^3 - 10}{1 + 3v}$

53. $x + 2\overline{)x^3 + 8}$

54. $(2x^3 + 7x^2 + 3 + 4x) \div (2x + 3)$

Synthetic division can be used to divide a polynomial by a binomial of the form $x - r$.

Use synthetic division to perform each division.

55. $x^3 - 13x - 12$ by $x - 4$

56 $x^2 + x^4 + 1$ by $x + 1$

The remainder theorem: If a polynomial $P(x)$ is divided by $x - r$, the remainder is $P(r)$.

57. Let $P(x) = 3x^2 - 2x + 3$. Use synthetic division and the remainder theorem to find $P(-1)$.

SECTION 6.7   *Proportion and Variation*

In a *proportion,* the product of the *extremes* is equal to the product of the *means.*

Solve each proportion.

58. $\dfrac{x + 1}{8} = \dfrac{4x - 2}{24}$

59. $\dfrac{1}{x + 6} = \dfrac{x + 10}{12}$

If two angles of one triangle have the same measure as two angles of a second triangle, the triangles are *similar.* The lengths of corresponding sides of similar triangles are proportional.

60. SIMILAR TRIANGLES Find the height of a tree if it casts a 44-foot shadow when a 4-foot shrub casts a $2\frac{1}{2}$-foot shadow.

Direct variation: $y = kx$, where k is the *constant of proportionality.*

61. PROPERTY TAXES The property tax in a certain county varies directly as assessed valuation. If a tax of \$1,575 is levied on a single-family home assessed at \$90,000, determine the property tax on an apartment complex assessed at \$312,000.

Inverse variation:

$$y = \frac{k}{x} \quad (k \text{ is a constant})$$

Joint variation: One variable varies with the product of several variables. For example, $y = kxz$ (k is a constant).

Combined variation: a combination of direct and inverse variation. For example,

$$y = \frac{kx}{z} \quad (k \text{ is a constant})$$

62. ELECTRICITY For a fixed voltage, the current in an electrical circuit varies inversely as the resistance in the circuit. If a certain circuit has a current of $2\frac{1}{2}$ amps when the resistance is 150 ohms, find the current in the circuit when the resistance is doubled.

63. HURRICANES The wind force on a vertical surface varies jointly as the area of the surface and the square of the wind's velocity. If a 10-mph wind exerts a force of 1.98 pounds on the sign shown, find the force on the sign if the wind is blowing at 80 mph.

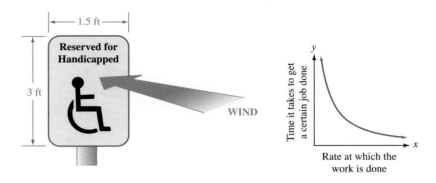

64. Does the graph above show direct or inverse variation?

65. Assume that x_1 varies directly with the third power of t and inversely with x_2. Find the constant of variation if $x_1 = 1.6$ when $t = 8$ and $x_2 = 64$.

Simplify each rational expression.

1. $-\dfrac{12x^2y^3z^2}{18x^3y^4z^2}$

2. $\dfrac{2x + 4}{x^2 - 4}$

3. $\dfrac{3y - 6z}{2z - y}$

4. $\dfrac{2x^2 + 7x + 3}{4x + 12}$

Perform the operations and simplify, if possible. Write all answers without negative exponents.

5. $\dfrac{x^2}{x^3z^2y^2} \cdot \dfrac{x^2z^4}{y^2z}$

6. $\dfrac{(x + 1)(x + 2)}{10} \cdot \dfrac{5}{x + 2}$

7. $\dfrac{u^2 + 5u + 6}{u^2 - 4} \cdot \dfrac{u^2 - 5u + 6}{u^2 - 9}$

8. $\dfrac{x^3 + y^3}{4} \div \dfrac{x^2 - xy + y^2}{2x + 2y}$

9. $\dfrac{xu + 2u + 3x + 6}{u^2 - 9} \cdot \dfrac{2u - 6}{x^2 + 3x + 2}$

10. $\dfrac{a^2 + 7a + 12}{a + 3} \div \dfrac{16 - a^2}{a - 4}$

11. $\dfrac{-3t + 4}{t^2 + t - 20} + \dfrac{6 + 5t}{t^2 + t - 20}$

12. $\dfrac{3w}{w - 5} + \dfrac{w + 10}{5 - w}$

13. $8b - 5 + \dfrac{5b + 4}{3b + 1}$

14. $\dfrac{x + 2}{x + 1} - \dfrac{x + 1}{x + 2}$

Simplify each complex fraction.

15. $\dfrac{\dfrac{2u^2w^3}{v^2}}{\dfrac{4uw^4}{uv}}$

16. $\dfrac{\dfrac{4}{3k} + \dfrac{k}{k + 1}}{\dfrac{k}{k + 1} - \dfrac{3}{k}}$

Solve each equation. If a solution is extraneous, so indicate.

17. $\dfrac{34}{x} + \dfrac{13}{20} = \dfrac{3}{2}$

18. $\dfrac{u - 2}{u - 3} + 3 = u + \dfrac{u - 4}{3 - u}$

19. $\dfrac{3}{x - 2} = \dfrac{x + 3}{2x}$

20. $\dfrac{4}{m^2 - 9} + \dfrac{5}{m^2 - m - 12} = \dfrac{7}{m^2 - 7m + 12}$

Solve each formula for the indicated variable.

21. $\dfrac{x^2}{a^2} + \dfrac{y^2}{b^2} = 1$ for a^2

22. $\dfrac{1}{r} = \dfrac{1}{r_1} + \dfrac{1}{r_2}$ for r_2

23. **ROOFING** One roofing crew can finish a 2,800-square-foot roof in 12 hours, and another crew can do the job in 10 hours. If they work together, can they finish before a predicted rain in 5 hours? If not, how long will they have to work in the rain?

24. **TOURING** A man bicycles 5 mph faster than he can walk. He bicycles 24 miles and then hikes back along the same route in 11 hours. How fast does he walk?

25. Divide: $\dfrac{18x^2y^3 - 12x^3y^2 + 9xy}{-3xy^4}$.

26. Divide: $(y^3 - 48) \div (y + 2)$.

27. Use synthetic division to divide $a^3 + 1$ by $a - 1$.

28. Use synthetic division to find the remainder when $4x^3 + 3x^2 + 2x - 1$ is divided by $x - 2$.

29. Explain why $x + 2$ can be divided out in the first expression and why it cannot be divided out in the second expression.

1st expression	*2nd expression*
$\dfrac{8(x + 2)(x - 3)}{x + 2}$	$\dfrac{8(x + 2) + (x - 3)}{x + 2}$

30. Explain what it means for two quantities to vary inversely. Give an example.

31. **ANNIVERSARY GIFTS** A florist sells a dozen long-stemmed red roses for $57.99. In honor of their 16th wedding anniversary, a man wants to buy 16 roses for his wife. What will the roses cost?

32. **SOUND** Sound intensity (loudness) varies inversely as the square of the distance from the source. If a rock band has a sound intensity of 100 decibels 30 feet away from the amplifier, find the sound intensity 60 feet away from the amplifier.

33. Graph the function $f(x) = \dfrac{2}{x}$ for $x > 0$. Label the horizontal asymptote.

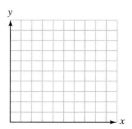

34. Draw a possible graph showing that the weekly salary of a person *varies directly* with the number of hours worked during the week. Label the axes.

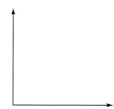

1. Solve: $-3 = -\dfrac{9}{8}t$.

2. Solve: $\dfrac{3x-4}{6} - \dfrac{x-2}{2} = \dfrac{-2x-3}{3}$.

3. AUTO SALES See the graph. An automobile dealership is going to order 80 new Ford Escorts. According to the survey, exactly how many green Escorts should be purchased to meet the expected customer demand?

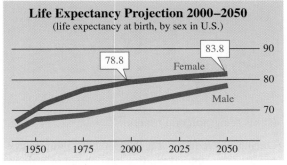

Vehicle Color Popularity Survey

Color	% of the market
Medium/dark green	17.5%
White	17.0%
Light brown	14.4%
Black	8.0%
Red	7.4%

Based on data from DuPont Automotive

4. LIFE EXPECTANCY Determine the predicted rate of change in the life expectancy of females during the years 2000–2050, as shown in the graph.

Life Expectancy Projection 2000–2050
(life expectancy at birth, by sex in U.S.)

Based on data from the Social Security Administration, Office of Chief Actuary

Write the equation of the line with the given properties. Express your answer in slope–intercept form.

5. Slope of -7, passing through $(7, 5)$

6. Passing through $(-4, 5)$ and $(2, -6)$

7. Draw the graph of the linear function $f(x) = \frac{2}{3}x - 2$. Then use interval notation to specify the domain and range.

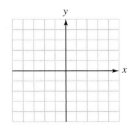

8. MECHANICAL ENGINEERING The tensions T_1 and T_2 (in pounds) in each of the ropes shown below can be found by solving the system

$$\begin{cases} 0.6T_1 - 0.8T_2 = 0 \\ 0.8T_1 + 0.6T_2 = 100 \end{cases}$$

Find T_1 and T_2.

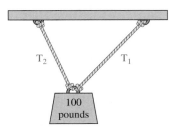

Solve each inequality and show the solution set as a graph on the number line.

9. $-2x \le -5$

10. $\left| \dfrac{3a}{5} - 2 \right| + 1 \ge \dfrac{6}{5}$

11. $5(x + 2) \le 4(x + 1)$ and $11 + x < 0$

12. $x + 1 < -4$ or $x - 4 > 0$

Simplify each expression.

13. $a^3b^2a^5b^2$

14. $\dfrac{a^3b^6}{a^7b^2}$

15. $\dfrac{1}{3^{-4}}$

16. $\left(\dfrac{2x^{-2}y^3}{x^2x^3y^4}\right)^{-3}$

Write each number in standard notation.

17. 4.25×10^4

18. 7.12×10^{-4}

19. Express as a formula: *y varies directly with the product of x and z, and inversely with r.*

20. Evaluate: $\begin{vmatrix} 3 & -2 \\ -2 & 4 \end{vmatrix}$.

21. Graph: $y < 4 - x$.

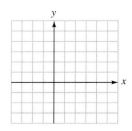

22. Graph: $f(x) = 2x^2 - 3$. Then use interval notation to specify the domain and range.

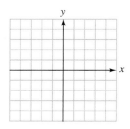

23. If $g(x) = -3x^3 + x - 4$, find $g(-2)$.

24. Find the degree of $3 + x^2y + 17x^3y^4$.

25. The graph of a line is shown on the graphing calculator screen below.

    a. Give the *x*- and *y*-intercepts of the line.

    b. What is the slope of the line?

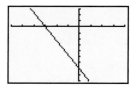

c. The line doesn't lie in one quadrant. Which quadrant is that?

d. What is the equation of the line?

e. Does the line pass through $(15, -42)$?

26. Express the dollar value of each type of U.S. coin and currency shown, using a power of 10. For example, one hundred dollars can be expressed as $\$10^2$.

Perform the operations and simplify, if possible.

27. $(x^3 + 3x^2 - 2x + 7) + (x^3 - 2x^2 + 2x + 5)$

28. $(-5x^2 + 3x + 4) - (-2x^2 + 3x + 7)$

29. $(3x + 4)(2x - 5)$

30. $(2x^3 - 1)^2$

The following graph shows the correction that must be made to a sundial reading to obtain accurate clock time. The difference is caused by Earth's orbit and tilted axis.

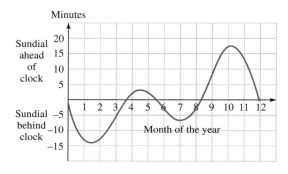

31. Is this the graph of a function?

32. During the year, what is the maximum number of minutes the sundial reading gets ahead of a clock?

33. During the year, what is the maximum number of minutes the sundial reading falls behind a clock?

34. How many times during a year is the sundial reading exactly the same as a clock?

Factor each expression.

35. $3r^2s^3 - 6rs^4$

36. $5(x - y) - a(x - y)$

37. $xu + yv + xv + yu$

38. $81x^4 - 16y^4$

39. $8x^3 - 27y^6$

40. $6x^2 + 5x - 6$

41. $9x^2 - 30x + 25$

42. $15x^2 - x - 6$

43. Solve: $6x^2 + 7 = -23x$.

44. Solve: $x^3 - 4x = 0$.

45. Solve $b^2x^2 + a^2y^2 = a^2b^2$ for b^2.

46. CAMPING The rectangular-shaped cooking surface of a small camping stove is 108 in.^{2}. If its length is 3 inches longer than its width, what are its dimensions?

Simplify each expression.

47. $\dfrac{2x^2y + xy - 6y}{3x^2y + 5xy - 2y}$

48. $\dfrac{p^3 - q^3}{q^2 - p^2} \cdot \dfrac{q^2 + pq}{p^3 + p^2q + pq^2}$

49. $\dfrac{2}{x + y} + \dfrac{3}{x - y} - \dfrac{x - 3y}{x^2 - y^2}$

50. $\dfrac{\dfrac{a}{b} + b}{a - \dfrac{b}{a}}$

51. Solve: $\dfrac{5x - 3}{x + 2} = \dfrac{5x + 3}{x - 2}$.

52. Solve: $\dfrac{3}{x - 2} + \dfrac{x^2}{(x + 3)(x - 2)} = \dfrac{x + 4}{x + 3}$.

Perform each division.

53. $(x^2 + 9x + 20) \div (x + 5)$

54. $(2x^2 + 4x - x^3 + 3) \div (x - 1)$

55. ECONOMICS The controversial Phillips curve depicts the tradeoff between unemployment and inflation as seen by one school of economists. If unemployment drops to very low levels, what does the theoretical model predict will happen to the inflation rate?

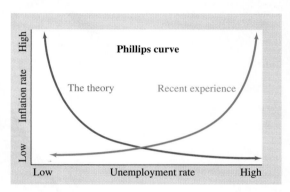

56. ECONOMICS See Exercise 55. The graph shows that economic factors have not followed the Phillips model in recent years. As unemployment dropped to very low levels, what happened to the inflation rate?

CHAPTER 7

Radical Expressions, Equations, and Functions

CORBIS

INTERMEDIATE Algebra _f(x)_ Now ™

Throughout the chapter, this icon introduces resources on the Inter-mediate AlgebraNow Web site, accessed through **http://1pass .thomson.com**, that will

- Help you test your knowledge of the material with a pre-test and a post-test

- Provide a personalized learning plan targeting areas you should study

When investigating an automobile accident, police and insurance professionals use clues from the scene to reconstruct the events that led to the collision. They estimate the speed of a vehicle prior to braking using a formula that involves the length of the skid marks and the condition of the road surface. This formula, $s = \sqrt{30Df}$, contains a square root. Square roots are more formally known as _radical expressions_.

To learn more about radical expressions and radical equations, visit _The Learning Equation_ on the Internet at http://tle.brookscole.com. (The log-in instructions are in the Preface.) For Chapter 7, the online lessons are:

- _TLE_ Lesson 10: Simplifying Radical Expressions
- _TLE_ Lesson 11: Radical Equations

Check Your Knowledge

1. The symbol $\sqrt{}$ is called a _____ symbol.

2. When n is an odd number, $\sqrt[n]{x}$ represents an _____ root. When n is an _____ number, $\sqrt[n]{x}$ represents an even root.

3. $\sqrt{x} + 1$ is the conjugate of _____ .

4. $32^{4/5}$ means the fourth _____ of the fifth _____ of 32.

5. In a right triangle, the side opposite the 90° angle is called the _____ .

6. Complete the table of values for $f(x) = -\sqrt{x - 1} + 1$. Then graph the function. Round to the nearest hundredth when necessary.

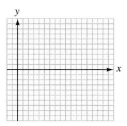

x	f(x)
1	
2	
3	
5	
10	
12	
17	

Simplify each expression, if possible. Assume that the variables represent positive numbers.

7. $\sqrt{8x^3}$

8. $\sqrt[3]{16y^4z^6}$

9. $\sqrt{\dfrac{9x^2}{16y^4}}$

10. $\sqrt[3]{-8y^6}$

11. $\sqrt[5]{-32}$

12. $\sqrt{-\dfrac{x^2}{9}}$

13. $\sqrt[3]{(x-1)^6}$

14. $\sqrt[3]{\dfrac{48x^4y^5}{81z^3}}$

15. $\sqrt{20x^5} + x\sqrt{80x^3}$

16. $\sqrt[3]{16} + \sqrt[3]{54} - \sqrt[3]{128}$

17. $(3\sqrt{5} - \sqrt{3})(\sqrt{5} + \sqrt{3})$

18. $(-27)^{1/3}$

19. $64^{-2/3}$

20. $\left(\dfrac{3x^{\frac{1}{2}}}{y^{\frac{1}{3}}}\right)^6$

Rationalize the denominator.

21. $\dfrac{x}{\sqrt{x} - 2}$

22. $\dfrac{\sqrt[3]{5}}{\sqrt[3]{9a}}$

Solve the equation.

23. $\sqrt{2x + 1} - 1 = -2\sqrt{x}$

24. $\sqrt[3]{y + 3} = \sqrt[3]{2y - 6}$

25. The hypotenuse of a right triangle is 13 meters and one leg is 12 meters. What is the length of the other leg?

26. Find the distance between the points $(5, 3)$ and $(12, 5)$. Write your answer using radical notation.

Study Skills Workshop

MIDTERM TUNE-UP

At midterm in your class, take a moment to evaluate your performance — there may still be time to make changes that will significantly impact your final grade.

What Is Your Overall Grade at Midterm? Many instructors give you information on how to calculate your grade based on test, homework, and other scores. Some instructors give formal progress reports to students when tests are returned. Meet with your instructor during office hours if you are not sure how to compute your grade and/or don't know how well you're doing. Based on your midterm grade, ask the following questions:

Should I Continue in the Class? If your midterm grade is a D or F, ask your instructor about your chances for passing the class. In many circumstances, it may be hard to raise your skills to a level that will allow you to be successful in your next math class. It may be best for you to drop the class at this point and spend extra time reviewing the material from your prerequisite class before re-enrolling the next term. (Remember that the most important indicator for success in a math class is the mastery of prerequisite skills!) You may be able to find help in reviewing your prerequisite skills in your college's tutorial center or with software such as that found on the iLrn Web site.

What Can I Do to Improve My Chances of Success? Especially if you find yourself at the C grade level at midterm, you want to do all that you can to improve your skills. Mastery of the skills that you are learning in this class is the most important indicator of success for your next math class! Ask yourself these questions:

1. Am I missing classes?
2. Am I adhering to my study calendar?
3. Am I taking good notes and reworking them to use as a study aid?
4. Am I using suggested techniques and spending the appropriate amount of time studying for tests?
5. Am I participating in a study group? If so, is it working well?
6. Am I seeking a tutor's or my instructor's help when I have unanswered questions?

Even if you have worked hard and are earning an A or B grade at midterm, you may still benefit by examining the above questions and fine-tuning your study routine. Strive to understand concepts as well as to perform routine tasks.

If you are doing well in the course so far, congratulate yourself and get ready to continue to work hard to maintain your grade. If your grade is not where you want it to be, you still have time to improve. Commit to improving your study habits, and schedule a meeting with your instructor to get ideas on how to raise your grade. Most important, don't give up — you have time to increase your success in the second half of the course.

ASSIGNMENT

1. What is your grade at midterm?
2. Respond to the six questions above. Where do you need to make adjustments?
3. Which of the above questions are you doing best with? Which one are you having the most trouble with?
4. What changes can you make that will enable you to do better?
5. What topics have you understood best so far in your class? Which topics are your weakest?

Radical expressions have the form $\sqrt[n]{a}$. In this chapter, we will see how they are used to model real-world situations.

7.1 Radical Expressions and Radical Functions

- Square roots   •   Square roots of expressions containing variables   •   The square root function
- Cube roots   •   The cube root function   •   *n*th roots

In this section, we will reverse the squaring process and learn how to find *square roots* of numbers. We will then generalize the concept of root and consider cube roots, fourth roots, fifth roots, and so on. We will also discuss a new family of functions, called *radical functions.*

■ Square roots

When solving problems, we must often find what number must be squared to obtain a second number *a*. If such a number can be found, it is called a **square root of *a*.** For example,

- 0 is a square root of 0, because $0^2 = 0$.
- 4 is a square root of 16, because $4^2 = 16$.
- -4 is a square root of 16, because $(-4)^2 = 16$.
- $7xy$ is a square root of $49x^2y^2$, because $(7xy)^2 = 49x^2y^2$.
- $-7xy$ is a square root of $49x^2y^2$, because $(-7xy)^2 = 49x^2y^2$.

The preceding examples illustrate the following definition.

Square root of *a*

The number *b* is a **square root of *a*** if $b^2 = a$.

All positive numbers have two real-number square roots, one that is positive and one that is negative.

EXAMPLE 1   Find the two square roots of 121.

Solution   The two square roots of 121 are 11 and -11, because

$$11^2 = 121 \quad \text{and} \quad (-11)^2 = 121$$

INTERMEDIATE
Algebra $f(x)$ Now™
Self Check 1
Find the square roots of 144.

Answer 12, -12

In the following definition, the symbol $\sqrt{}$ is called a **radical symbol,** and the number *x* within the radical symbol is called a **radicand.** An expression containing a radical is called a **radical expression.**

Principal square root

If $x > 0$, the **principal square root of *x*** is the positive square root of *x*, denoted as $\sqrt{x}$.

The principal square root of 0 is 0: $\sqrt{0} = 0$.

By definition, the principal square root of a positive number is always positive. Although 5 and -5 are both square roots of 25, only 5 is the principal square root. The radical expression $\sqrt{25}$ represents 5. The radical expression $-\sqrt{25}$ represents -5. When we write $\sqrt{25} = 5$ or $-\sqrt{25} = -5$, we say that we have *simplified the radical.*

INTERMEDIATE
Algebra *f(x)* **Now**™

Self Check 2
Simplify:
a. $-\sqrt{49}$

b. $\sqrt{\dfrac{25}{49}}$

Answers **a.** -7, **b.** $\dfrac{5}{7}$

EXAMPLE 2 Simplify each radical: **a.** $\sqrt{1}$, **b.** $\sqrt{81}$, **c.** $-\sqrt{81}$,

d. $-\sqrt{225}$, **e.** $\sqrt{\dfrac{1}{4}}$, **f.** $-\sqrt{\dfrac{16}{121}}$, **g.** $\sqrt{0.04}$, and **h.** $-\sqrt{0.0009}$.

Solution

a. $\sqrt{1} = 1$ **b.** $\sqrt{81} = 9$

c. $-\sqrt{81} = -9$ **d.** $-\sqrt{225} = -15$

e. $\sqrt{\dfrac{1}{4}} = \dfrac{1}{2}$ **f.** $-\sqrt{\dfrac{16}{121}} = -\dfrac{4}{11}$

g. $\sqrt{0.04} = 0.2$ **h.** $-\sqrt{0.0009} = -0.03$

Numbers such as 4, 9, 16, 49, and 1,600 are called **integer squares,** because each one is the square of an integer. The square root of every integer square is a rational number.

$$\sqrt{4} = 2, \qquad \sqrt{9} = 3, \qquad \sqrt{16} = 4, \qquad \sqrt{49} = 7, \qquad \sqrt{1,600} = 40$$

The square roots of many positive integers are not rational numbers. For example, $\sqrt{11}$ is an *irrational number.* To find an approximate value of $\sqrt{11}$, we enter 11 into a calculator and press the square root key $\boxed{\sqrt{}}$. (On some calculators, the $\boxed{\sqrt{}}$ key must be pressed first.)

$$\sqrt{11} \approx 3.31662479$$

! **COMMENT** Square roots of negative numbers are not real numbers. For example, $\sqrt{-9}$ is not a real number, because no real number squared equals -9. Square roots of negative numbers come from a set called the **imaginary numbers,** which we will discuss in the next chapter.

■ Square roots of expressions containing variables

If $x \neq 0$, the positive number x^2 has x and $-x$ for its two square roots. To denote the positive square root of $\sqrt{x^2}$, we must know whether x is positive or negative.

If x is positive, we can write

$$\sqrt{x^2} = x \qquad \sqrt{x^2} \text{ represents the positive square root of } x^2, \text{ which is } x.$$

If x is negative, then $-x > 0$, and we can write

$$\sqrt{x^2} = -x \qquad \sqrt{x^2} \text{ represents the positive square root of } x^2, \text{ which is } -x.$$

If we don't know whether x is positive or negative, we can use absolute value symbols to guarantee that $\sqrt{x^2}$ is positive.

Definition of $\sqrt{x^2}$

For any real number x,

$$\sqrt{x^2} = |x|$$

EXAMPLE 3 Simplify: **a.** $\sqrt{16x^2}$, **b.** $\sqrt{x^2 + 2x + 1}$, and **c.** $\sqrt{m^4}$.

Solution If x can be any real number, we have

a. $\sqrt{16x^2} = \sqrt{(4x)^2}$ Write the radicand $16x^2$ as $(4x)^2$.

$\quad\quad\quad\quad = |4x|$ Because $(|4x|)^2 = 16x^2$. Since x could be negative, absolute value symbols are needed.

$\quad\quad\quad\quad = 4|x|$ Since 4 is a positive constant in the product $4x$, we can write it outside the absolute value symbols.

b. $\sqrt{x^2 + 2x + 1}$

$\quad\quad = \sqrt{(x + 1)^2}$ Factor the radicand: $x^2 + 2x + 1 = (x + 1)^2$.

$\quad\quad = |x + 1|$ Since $x + 1$ can be negative (for example, when $x = -5$, $x + 1$ is -4), absolute value symbols are needed.

c. $\sqrt{m^4} = m^2$ Because $(m^2)^2 = m^4$. Since $m^2 \geq 0$, no absolute value symbols are needed.

Self Check 3
Simplify:
a. $\sqrt{25a^2}$
b. $\sqrt{16a^4}$

Answers **a.** $5|a|$, **b.** $4a^2$

If we are told that x represents a positive real number in parts a and b of Example 3, we do not need to use absolute value symbols to guarantee that the answers are positive.

$\sqrt{16x^2} = 4x$ If x is positive, $4x$ is positive.

$\sqrt{x^2 + 2x + 1} = x + 1$ If x is positive, $x + 1$ is positive.

▮ The square root function

Since there is one principal square root for every nonnegative real number x, the equation $f(x) = \sqrt{x}$ determines a function, called a **square root function.** Square root functions belong to a larger family of functions known as **radical functions.**

EXAMPLE 4 Graph $f(x) = \sqrt{x}$ and find its domain and range.

Solution To graph this square root function, we will evaluate it for several values of x. We begin with $x = 0$, since 0 is the smallest input for which $\sqrt{x}$ is defined.

$f(x) = \sqrt{x}$

$f(0) = \sqrt{0} = 0$

We enter the ordered pair $(0, 0)$ in the table of values in Figure 7-1 on the next page. Then we continue the evaluation process for $x = 1, 4, 9,$ and 16 and list the results in the table. After plotting all the ordered pairs, we draw a smooth curve through the points. This is the graph of function f (see Figure 7-1(a)). Since the equation defines a function, its graph passes the vertical line test.

We can use a graphing calculator with window settings of $[-1, 9]$ for x and $[-1, 9]$ for y to get the graph shown in Figure 7-1(b). From either graph, we can see that the domain and the range are the set of nonnegative real numbers. Expressed in interval notation, the domain is $[0, \infty)$, and the range is $[0, \infty)$.

Self Check 4
Graph: $f(x) = \sqrt{x} + 2$. Then give its domain and range and compare it to the graph of $f(x) = \sqrt{x}$.

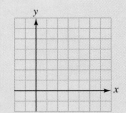

$f(x) = \sqrt{x}$		
x	$f(x)$	$(x, f(x))$
0	0	$(0, 0)$
1	1	$(1, 1)$
4	2	$(4, 2)$
9	3	$(9, 3)$
16	4	$(16, 4)$

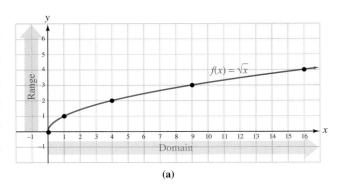

(a)

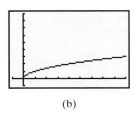

(b)

FIGURE 7-1

Answers

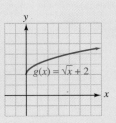

D: $[0, \infty)$, R: $[2, \infty)$; the graph is 2 units higher

The graphs of many radical functions are translations or reflections of the square root function, $f(x) = \sqrt{x}$. For example, if $k > 0$,

- The graph of $f(x) = \sqrt{x} + k$ is the graph of $f(x) = \sqrt{x}$ translated k units up.
- The graph of $f(x) = \sqrt{x} - k$ is the graph of $f(x) = \sqrt{x}$ translated k units down.
- The graph of $f(x) = \sqrt{x + k}$ is the graph of $f(x) = \sqrt{x}$ translated k units to the left.
- The graph of $f(x) = \sqrt{x - k}$ is the graph of $f(x) = \sqrt{x}$ translated k units to the right.
- The graph of $f(x) = -\sqrt{x}$ is the graph of $f(x) = \sqrt{x}$ reflected about the x-axis.

Self Check 5
Graph: $f(x) = \sqrt{x - 2} - 4$.
Then give the domain and the range.

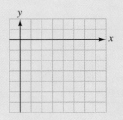

EXAMPLE 5   Graph $f(x) = -\sqrt{x + 4} - 2$ and find its domain and range.

Solution   This graph will be the reflection of $f(x) = \sqrt{x}$ about the x-axis, translated 4 units to the left and 2 units down. See Figure 7-2(a) on the next page. We can confirm this graph by using a graphing calculator with window settings of $[-5, 6]$ for x and $[-6, 2]$ for y to get the graph shown in Figure 7-2(b).

We can determine the domain of the function algebraically. Since the expression $\sqrt{x + 4}$ is not a real number when $x + 4$ is negative, we must require that

$$x + 4 \geq 0$$

Solving for x, we have

$$x \geq -4 \qquad \text{The } x\text{-inputs must be real numbers greater than or equal to } -4.$$

Thus, the domain of $f(x) = -\sqrt{x + 4} - 2$ is the interval $[-4, \infty)$.

From either graph, we can see that the domain is the interval $[-4, \infty)$ and that the range is the interval $(-\infty, -2]$.

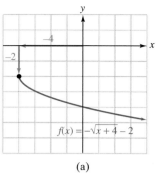

$f(x) = -\sqrt{x+4} - 2$

(a)

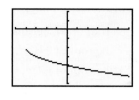

(b)

FIGURE 7-2

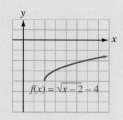

$f(x) = \sqrt{x-2} - 4$

D: $[2, \infty)$, R: $[-4, \infty)$

EXAMPLE 6 **Pendulums.** The **period of a pendulum** is the time required for the pendulum to swing back and forth to complete one cycle. The period (in seconds) is a function of the pendulum's length L (in feet) and is given by

$$f(L) = 2\pi\sqrt{\frac{L}{32}}$$

Find the period of the 5-foot-long pendulum of the clock shown in Figure 7-3.

Solution To determine the period, we substitute 5 for L.

$$f(L) = 2\pi\sqrt{\frac{L}{32}}$$

$$f(5) = 2\pi\sqrt{\frac{5}{32}}$$

$$\approx 2.483647066 \quad \text{Use a calculator to find an approximation.}$$

The period is approximately 2.5 seconds.

L

FIGURE 7-3

Self Check 6
To the nearest hundredth, find the period of a pendulum that is 3 feet long.

Answer 1.92 sec

Evaluating a square root function using its graph

CALCULATOR SNAPSHOT

To solve Example 6 with a graphing calculator with window settings of $[-2, 10]$ for x and $[-2, 10]$ for y, we graph the function $f(x) = 2\pi\sqrt{\frac{x}{32}}$, as in Figure 7-4(a). We then trace and move the cursor toward an x-value of 5 until we see the coordinates shown in Figure 7-4(b). The period is given by the y-value shown on the screen. By zooming in, we can get better results.

After entering $Y_1 = 2\pi\sqrt{\frac{x}{32}}$, we can also use the TABLE mode to evaluate the function. See Figure 7-4(c).

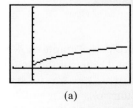

(a)

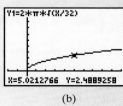

(b)

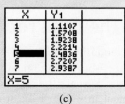

(c)

FIGURE 7-4

Cube roots

The **cube root of** x is any number whose cube is x. For example,

- 4 is a cube root of 64, because $4^3 = 64$.
- $3x^2y$ is a cube root of $27x^6y^3$, because $(3x^2y)^3 = 27x^6y^3$.
- $-2y$ is a cube root of $-8y^3$, because $(-2y)^3 = -8y^3$.

> **Cube roots**
>
> The **cube root of** x is denoted as $\sqrt[3]{x}$ and is defined by
>
> $$\sqrt[3]{x} = y \quad \text{if} \quad y^3 = x$$

We note that 64 has two real-number square roots, 8 and -8. However, 64 has only one real-number cube root 4, because 4 is the only real number whose cube is 64. Since every real number has exactly one real cube root, it is unnecessary to use absolute value symbols when simplifying cube roots.

> **Definition of** $\sqrt[3]{x^3}$
>
> For any real number x,
>
> $$\sqrt[3]{x^3} = x$$

INTERMEDIATE
Algebra *f(x)* **Now**™

Self Check 7
Simplify:
a. $\sqrt[3]{1{,}000}$

b. $\sqrt[3]{\dfrac{1}{27}}$

c. $\sqrt[3]{-125a^3}$

Answers **a.** 10, **b.** $\frac{1}{3}$, **c.** $-5a$

EXAMPLE 7   Simplify each expression: **a.** $\sqrt[3]{125}$, **b.** $\sqrt[3]{\dfrac{1}{8}}$, **c.** $\sqrt[3]{-27x^3}$, **d.** $\sqrt[3]{-\dfrac{8a^3}{27b^3}}$, and **e.** $\sqrt[3]{0.216x^3y^6}$.

Solution

a. $\sqrt[3]{125} = 5$ Because $5^3 = 5 \cdot 5 \cdot 5 = 125$.

b. $\sqrt[3]{\dfrac{1}{8}} = \dfrac{1}{2}$ Because $\left(\dfrac{1}{2}\right)^3 = \dfrac{1}{2} \cdot \dfrac{1}{2} \cdot \dfrac{1}{2} = \dfrac{1}{8}$.

c. $\sqrt[3]{-27x^3} = -3x$ Because $(-3x)^3 = (-3x)(-3x)(-3x) = -27x^3$.

d. $\sqrt[3]{-\dfrac{8a^3}{27b^3}} = -\dfrac{2a}{3b}$ Because $\left(-\dfrac{2a}{3b}\right)^3 = \left(-\dfrac{2a}{3b}\right)\left(-\dfrac{2a}{3b}\right)\left(-\dfrac{2a}{3b}\right) = -\dfrac{8a^3}{27b^3}$.

e. $\sqrt[3]{0.216x^3y^6} = 0.6xy^2$ Because $(0.6xy^2)^3 = (0.6xy^2)(0.6xy^2)(0.6xy^2) = 0.216x^3y^6$.

The cube root function

The equation $f(x) = \sqrt[3]{x}$ defines a **cube root function.** From the graph shown in Figure 7-5(a) on the next page, we can see that the domain and range of the function $f(x) = \sqrt[3]{x}$ are the set of real numbers. Note that the graph of $f(x) = \sqrt[3]{x}$ passes the vertical line test. Like square root functions, cube root functions are members of the family of radical functions. Figures 7-5(b) and 7-5(c) show several translations of the cube root function.

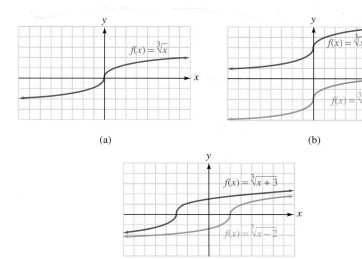

(a) (b)

(c)

FIGURE 7-5

*n*th roots

Just as there are square roots and cube roots, there are fourth roots, fifth roots, sixth roots, and so on.

When n is an odd natural number, the expression $\sqrt[n]{x}$ where $n > 1$ represents an **odd root.** Since every real number has just one real nth root when n is odd, we don't need to worry about absolute value symbols when finding odd roots. For example,

$$\sqrt[5]{243} = \sqrt[5]{3^5} = 3 \qquad \text{Because } 3^5 = 243.$$

$$\sqrt[7]{-128x^7} = \sqrt[7]{(-2x)^7} = -2x \quad \text{Because } (-2x)^7 = -128x^7.$$

When n is an even natural number, the expression $\sqrt[n]{x}$, where $n > 1$ and $x > 0$, represents an **even root.** In this case, there will be one positive and one negative real nth root. For example, the real sixth roots of 729 are 3 and -3, because $3^6 = 729$ and $(-3)^6 = 729$. When finding even roots, we can use absolute value symbols to guarantee that the nth root is positive.

$$\sqrt[4]{(-3)^4} = |-3| = 3 \qquad \begin{array}{l}\text{We could also simplify this as follows:}\\ \sqrt[4]{(-3)^4} = \sqrt[4]{81} = 3.\end{array}$$

$$\sqrt[6]{729x^6} = \sqrt[6]{(3x)^6} = |3x| = 3|x| \quad \begin{array}{l}\text{The absolute value symbols guarantee that the}\\ \text{sixth root is positive.}\end{array}$$

In general, we have the following rules.

Rules for $\sqrt[n]{x^n}$

If x represents a real number and $n > 1$, then

If n represents an odd natural number, $\sqrt[n]{x^n} = x$.

If n represents an even natural number, $\sqrt[n]{x^n} = |x|$.

In the radical expression $\sqrt[n]{x}$, n is called the **index** (or **order**) of the radical. When the index is 2, the radical is a square root, and we usually do not write the index.

$$\sqrt{x} = \sqrt[2]{x}$$

! COMMENT When n is even, where $n > 1$ and $x < 0$, $\sqrt[n]{x}$ is not a real number. For example, $\sqrt[4]{-81}$ is not a real number, because no real number raised to the fourth power is -81.

Self Check 8

Simplify:

a. $\sqrt[4]{\dfrac{1}{81}}$

b. $\sqrt[5]{10^5}$

c. $\sqrt[6]{-64}$

EXAMPLE 8 Simplify each radical, if possible: **a.** $\sqrt[4]{625}$, **b.** $\sqrt[4]{-1}$, **c.** $\sqrt[5]{-32}$, **d.** $\sqrt[6]{\dfrac{1}{64}}$, and **e.** $\sqrt[7]{10^7}$.

Solution

a. $\sqrt[4]{625} = 5$, because $5^4 = 625$ Read $\sqrt[4]{625}$ as "the fourth root of 625."

b. $\sqrt[4]{-1}$ is not a real number.   This is an even root of a negative number.

c. $\sqrt[5]{-32} = -2$, because $(-2)^5 = -32$ Read $\sqrt[5]{-32}$ as "the fifth root of -32."

d. $\sqrt[6]{\dfrac{1}{64}} = \dfrac{1}{2}$, because $\left(\dfrac{1}{2}\right)^6 = \dfrac{1}{64}$ Read $\sqrt[6]{\dfrac{1}{64}}$ as "the sixth root of $\dfrac{1}{64}$."

Answers a. $\frac{1}{3}$, **b.** 10, **c.** not a real number

e. $\sqrt[7]{10^7} = 10$, because $10^7 = 10^7$ Read $\sqrt[7]{10^7}$ as "the seventh root of 10^7."

CALCULATOR SNAPSHOT   Finding roots

The square root key $\boxed{\sqrt{}}$ on a scientific calculator can be used to evaluate square roots. To evaluate roots with an index greater than 2, we can use the root key $\boxed{\sqrt[x]{y}}$. For example, the function

$$r(V) = \sqrt[3]{\frac{3V}{4\pi}}$$

gives the radius of a sphere with volume V. To find the radius of the spherical propane tank shown in Figure 7-6, we substitute 113 for V to get

$$r(V) = \sqrt[3]{\frac{3V}{4\pi}}$$

$$r(113) = \sqrt[3]{\frac{3(113)}{4\pi}}$$

PROPANE
Capacity 113 ft³

FIGURE 7-6

To evaluate a root, we enter the radicand and press the root key $\boxed{\sqrt[x]{y}}$ followed by the index of the radical, which in this case is 3.

$$3 \boxed{\times} 113 \boxed{\div} \boxed{(} 4 \boxed{\times} \boxed{\pi} \boxed{)} \boxed{=} \boxed{\text{2nd}} \boxed{\sqrt[x]{y}} \boxed{3} \boxed{=}$$

$$\boxed{2.999139118}$$

To evaluate the cube root of $\frac{3(113)}{4\pi}$ using a graphing calculator, we enter these numbers and press these keys.

$$\boxed{\text{MATH}} \, 4 \boxed{(} \boxed{(} 3 \boxed{\times} 113 \boxed{)} \boxed{\div} \boxed{(} 4 \boxed{\times} \boxed{\text{2nd}} \boxed{\pi} \boxed{)} \boxed{)} \boxed{\text{ENTER}}$$

$$\boxed{\begin{array}{l} \sqrt[3]{}((3*113)/(4*\pi) \\ ) \\ \qquad\qquad 2.999139118 \end{array}}$$

The radius of the propane tank is about 3 feet.

EXAMPLE 9   Simplify each expression. Assume that x can be any real number.

a. $\sqrt[5]{x^5}$, **b.** $\sqrt[4]{16x^4}$, **c.** $\sqrt[6]{(x+4)^6}$, and **d.** $\sqrt[3]{(x+1)^3}$.

Self Check 9

Simplify:

a. $\sqrt[4]{16a^8}$

b. $\sqrt[5]{(a+5)^5}$

c. $\sqrt{(x^2+4x+4)^2}$

Solution

a. $\sqrt[5]{x^5} = x$ Since n is odd, absolute value symbols aren't needed.

b. $\sqrt[4]{16x^4} = |2x| = 2|x|$ Since n is even and x can be negative, absolute value symbols are needed to guarantee that the result is positive.

c. $\sqrt[6]{(x+4)^6} = |x+4|$ Absolute value symbols are needed to guarantee that the result is positive.

d. $\sqrt[3]{(x+1)^3} = x+1$ Since n is odd, absolute value symbols aren't needed.

Answers **a.** $2a^2$, **b.** $a+5$, **c.** $(x+2)^2$

If we are told that x represents a positive real number in parts b and c of Example 9, we do not need to use absolute value symbols to guarantee that the results are positive.

$$\sqrt[4]{16x^4} = 2x \qquad \text{If } x \text{ is positive, } 2x \text{ is positive.}$$
$$\sqrt[6]{(x+4)^6} = x+4 \qquad \text{If } x \text{ is positive, } x+4 \text{ is positive.}$$

We summarize the definitions concerning $\sqrt[n]{x}$ as follows.

Summary of the definitions of $\sqrt[n]{x}$

If n represents a natural number greater than 1 and x represents a real number, then

If $x > 0$, then $\sqrt[n]{x}$ is the positive number such that $\left(\sqrt[n]{x}\right)^n = x$.

If $x = 0$, then $\sqrt[n]{x} = 0$.

If $x < 0$ $\begin{cases} \text{and } n \text{ is odd, then } \sqrt[n]{x} \text{ is the real number such that } \left(\sqrt[n]{x}\right)^n = x. \\ \text{and } n \text{ is even, then } \sqrt[n]{x} \text{ is not a real number.} \end{cases}$

Section 7.1   STUDY SET

VOCABULARY   *Fill in the blanks.*

1. $5x^2$ is the _____ _____ of $25x^4$, because $(5x^2)^2 = 25x^4$.

2. $f(x) = \sqrt{x}$ and $g(t) = \sqrt[3]{t}$ are _____ functions.

3. The symbol $\sqrt{}$ is called a _____ symbol.

4. In the expression $\sqrt[3]{27x^6}$, 3 is the _____ and $27x^6$ is the _____.

5. When n is an odd number, $\sqrt[n]{x}$ represents an _____ root.

6. When n is an _____ number, $\sqrt[n]{x}$ represents an even root.

7. When we write $\sqrt{b^2 + 6b + 9} = |b+3|$, we say that we have _____ the radical.

8. 6 is the _____ _____ of 216 because $6^3 = 216$.

CONCEPTS   *Fill in the blanks.*

9. b is a square root of a if $b^2 =$ ▨ .

10. $\sqrt{0} =$ ▨ and $\sqrt[3]{0} =$ ▨ .

11. The number 25 has _____ square roots. The principal square root of 25 is the _____ square root of 25.

12. $\sqrt{-4}$ is not a real number, because no real number _____ equals -4.

13. $\sqrt[3]{x} = y$ if $y^3 =$ ▨ .

14. $\sqrt{x^2} =$ ▨ and $\sqrt[3]{x^3} =$ ▨ .

15. The graph of $f(x) = \sqrt{x} + 3$ is the graph of $f(x) = \sqrt{x}$ translated ▨ units _____.

16. The graph of $f(x) = \sqrt{x+5}$ is the graph of $f(x) = \sqrt{x}$ translated ▨ units to the _____.

Complete each table and graph the function.

17. $f(x) = -\sqrt{x}$

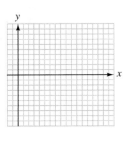

x	f(x)
0	
1	
4	
9	
16	

18. $f(x) = -\sqrt[3]{x}$

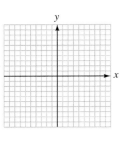

x	f(x)
-8	
-1	
0	
1	
8	

NOTATION   *Translate each sentence into mathematical symbols.*

19. The square root of x squared is the absolute value of x.

20. The cube root of x cubed is x.

21. f of x equals the square root of the quantity x minus five.

22. The fifth root of negative thirty-two is negative two.

PRACTICE   *Find each square root, if possible.*

23. $\sqrt{121}$ **24.** $\sqrt{144}$

25. $-\sqrt{64}$ **26.** $-\sqrt{1}$

27. $\sqrt{\dfrac{1}{9}}$ **28.** $-\sqrt{\dfrac{4}{25}}$

29. $\sqrt{0.25}$ **30.** $\sqrt{0.16}$

31. $\sqrt{-25}$ **32.** $-\sqrt{-49}$

33. $\sqrt{(-4)^2}$ **34.** $\sqrt{(-9)^2}$

Use a calculator to find each square root. Give the answer to four decimal places.

35. $\sqrt{12}$ **36.** $\sqrt{340}$

37. $\sqrt{679.25}$ **38.** $\sqrt{0.0063}$

Find each square root. Assume that all variables are unrestricted, and use absolute value symbols when necessary.

39. $\sqrt{4x^2}$ **40.** $\sqrt{16y^4}$

41. $\sqrt{(t+5)^2}$ **42.** $\sqrt{(a+6)^2}$

43. $\sqrt{(-5b)^2}$ **44.** $\sqrt{(-8c)^2}$

45. $\sqrt{a^2+6a+9}$ **46.** $\sqrt{x^2+10x+25}$

Find each value given that $f(x) = \sqrt{x-4}$ and $g(x) = \sqrt[3]{x-4}$.

47. $f(4)$ **48.** $f(8)$

49. $f(20)$ **50.** $f(29)$

51. $g(12)$ **52.** $g(-4)$

53. $g(-996)$ **54.** $g(1{,}004)$

 *Find each value given that $f(x) = \sqrt{x^2+4}$ and $g(x) = \sqrt[3]{x^2+1}$. Give answers to four decimal places.*

55. $f(4)$ **56.** $f(2.35)$

57. $g(6)$ **58.** $g(21.57)$

Graph each function and find its domain and range.

59. $f(x) = \sqrt{x+4}$ **60.** $f(x) = -\sqrt{x-1}$

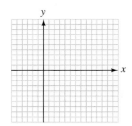   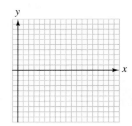

61. $f(x) = -\sqrt[3]{x} - 3$ **62.** $f(x) = \sqrt[3]{x} - 1$

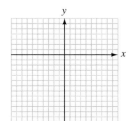   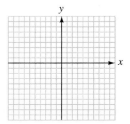

Simplify each cube root.

63. $\sqrt[3]{1}$ **64.** $\sqrt[3]{-8}$

65. $\sqrt[3]{-125}$ **66.** $\sqrt[3]{512}$

67. $\sqrt[3]{-\dfrac{8}{27}}$ **68.** $\sqrt[3]{\dfrac{125}{216}}$

69. $\sqrt[3]{0.064}$ **70.** $\sqrt[3]{0.001}$

71. $\sqrt[3]{8a^3}$ **72.** $\sqrt[3]{-27x^6}$

73. $\sqrt[3]{-1,000p^3q^3}$ **74.** $\sqrt[3]{343a^6b^3}$

75. $\sqrt[3]{-\dfrac{1}{8}m^6n^3}$ **76.** $\sqrt[3]{0.008z^9}$

77. $\sqrt[3]{-0.064s^9t^6}$ **78.** $\sqrt[3]{\dfrac{27}{1,000}a^6b^6}$

Simplify each radical, if possible. Assume that all variables represent positive real numbers.

79. $\sqrt[4]{81}$ **80.** $\sqrt[6]{64}$

81. $-\sqrt[5]{243}$ **82.** $-\sqrt[4]{625}$

83. $\sqrt[4]{-256}$ **84.** $\sqrt[6]{-729}$

85. $\sqrt[4]{\dfrac{16}{625}}$ **86.** $\sqrt[5]{-\dfrac{243}{32}}$

87. $-\sqrt[5]{-\dfrac{1}{32}}$ **88.** $-\sqrt[4]{\dfrac{81}{256}}$

89. $\sqrt[5]{32a^5}$ **90.** $\sqrt[5]{-32x^5}$

91. $\sqrt[4]{16a^4}$ **92.** $\sqrt[8]{x^{24}}$

93. $\sqrt[4]{k^{12}}$ **94.** $\sqrt[6]{64b^6}$

95. $\sqrt[4]{\dfrac{1}{16}m^4}$ **96.** $\sqrt[4]{\dfrac{1}{81}x^8}$

97. $\sqrt[25]{(x+2)^{25}}$ **98.** $\sqrt[44]{(x+4)^{44}}$

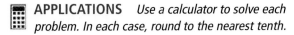 **APPLICATIONS** *Use a calculator to solve each problem. In each case, round to the nearest tenth.*

99. EMBROIDERY The radius r of a circle is given by the formula

$$r = \sqrt{\dfrac{A}{\pi}}$$

where A is its area. Find the diameter of the embroidery hoop shown on the right if there are 38.5 in.^{2} of stretched fabric on which to embroider.

100. BASEBALL The length of a diagonal of a square is given by the function $d(s) = \sqrt{2s^2}$, where s is the length of a side of the square. Find the distance from home plate to second base on a softball diamond and on a baseball diamond. The illustration in the next column gives the dimensions of each type of infield.

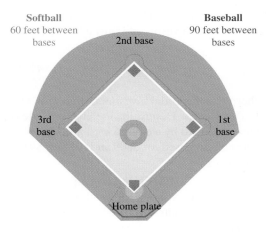

101. PULSE RATES The approximate pulse rate (in beats per minute) of an adult who is t inches tall is given by the function

$$p(t) = \dfrac{590}{\sqrt{t}}$$

The Guinness Book of World Records 1998 lists Ri Myong-hun of North Korea as the tallest living man, at 7 ft $8\frac{1}{2}$ in. Find his approximate pulse rate as predicted by the function.

102. THE GRAND CANYON The time t (in seconds) that it takes for an object to fall a distance of s feet is given by the formula

$$t = \dfrac{\sqrt{s}}{4}$$

In some places, the Grand Canyon is one mile (5,280 feet) deep. How long would it take a stone dropped over the edge of the canyon to hit bottom?

103. BIOLOGY Scientists will place five rats inside the controlled environment of a sealed hemisphere to study the rats' behavior. The function

$$d(V) = \sqrt[3]{12\left(\dfrac{V}{\pi}\right)}$$

gives the diameter of a hemisphere with volume V. Use the function to determine the diameter of the base of the hemisphere, if each rat requires 125 cubic feet of living space.

104. AQUARIUMS The function

$$s(g) = \sqrt[3]{\dfrac{g}{7.5}}$$

determines how long (in feet) an edge of a cube-shaped tank must be if it is to hold g gallons of water. What dimensions should a cube-shaped aquarium have if it is to hold 1,250 gallons of water?

105. COLLECTIBLES The *effective rate of interest r* earned by an investment is given by the formula

$$r = \sqrt[n]{\frac{A}{P}} - 1$$

where P is the initial investment that grows to value A after n years. Determine the effective rate of interest earned by a collector on a Lladró porcelain figurine purchased for \$800 and sold for \$950 five years later.

106. LAW ENFORCEMENT The graphs of the two radical functions shown in the illustration can be used to estimate the speed (in mph) of a car involved in an accident. Suppose a police accident report listed skid marks to be 220 feet long but failed to give the road conditions. Estimate the possible speeds the car was traveling prior to the brakes being applied.

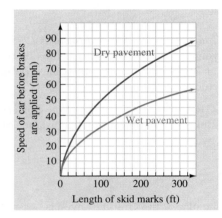

WRITING

107. Explain why 36 has two square roots, but $\sqrt{36}$ is just 6, not -6.

108. If x is any real number, then $\sqrt{x^2} = x$ is not correct. Explain.

109. Explain what is wrong with the graph below.

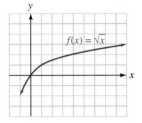

110. Explain how to estimate the domain and range of the radical function that is graphed on the right.

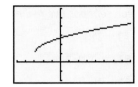

REVIEW   *Perform the operations.*

111. $\dfrac{x^2 - x - 6}{x^2 - 2x - 3} \cdot \dfrac{x^2 - 1}{x^2 + x - 2}$

112. $\dfrac{x^2 - 3x - 4}{x^2 - 5x + 6} \div \dfrac{x^2 - 2x - 3}{x^2 - x - 2}$

113. $\dfrac{3}{m + 1} + \dfrac{3m}{m - 1}$

114. $\dfrac{2x + 3}{3x - 1} - \dfrac{x - 4}{2x + 1}$

7.2 Simplifying and Combining Radical Expressions

- Properties of radicals • Simplifying radical expressions
- Adding and subtracting radical expressions

The square shown in Figure 7-7 has an area of 12 square centimeters. We can determine the length of one side of the square in the following way:

$$A = s^2$$

$$12 = s^2 \qquad \text{Substitute 12 for } A.$$

$$\sqrt{12} = \sqrt{s^2} \qquad \text{Take the positive square root of both sides.}$$

$$\sqrt{12} = s \qquad \text{Each side of the square is } \sqrt{12} \text{ centimeters long.}$$

Area = 12 cm²

FIGURE 7-7

The form in which we express the length of a side of the square depends upon the situation. If an approximation is acceptable, we can use a calculator to find that $\sqrt{12} \approx 3.464101615$, and then we can round to a specified degree of accuracy. For example, to the nearest hundredth, each side is 3.46 centimeters long.

If the situation calls for the exact length, we must use a radical expression. As you will see in this section, it is common practice to write a radical expression such as $\sqrt{12}$ in *simplified* form. To simplify radicals, we will use the multiplication and division properties of radicals.

Properties of radicals

We introduce the first of two properties of radicals with the following examples.

$$\sqrt{4 \cdot 25} = \sqrt{100} \qquad\qquad \sqrt{4}\sqrt{25} = 2 \cdot 5 \qquad \text{Read as "the square root of 4 times} \\ \text{the square root of 25."}$$

$$= 10 \qquad\qquad\qquad\qquad = 10$$

In each case, the answer is 10. Thus, $\sqrt{4 \cdot 25} = \sqrt{4}\sqrt{25}$. Likewise,

$$\sqrt[3]{8 \cdot 27} = \sqrt[3]{216} \qquad\qquad \sqrt[3]{8}\sqrt[3]{27} = 2 \cdot 3$$

$$= 6 \qquad\qquad\qquad\qquad = 6$$

In each case, the answer is 6. Thus, $\sqrt[3]{8 \cdot 27} = \sqrt[3]{8}\sqrt[3]{27}$. These results illustrate the **multiplication property of radicals**.

> **Multiplication property of radicals**
>
> If $\sqrt[n]{a}$ and $\sqrt[n]{b}$ represent real numbers, then
>
> $$\sqrt[n]{ab} = \sqrt[n]{a}\sqrt[n]{b}$$

As long as all radical expressions represent real numbers, *the nth root of the product of two numbers is equal to the product of their nth roots.*

COMMENT The multiplication property of radicals applies to the nth root of the product of two numbers. There is no such property for sums or differences. For example,

$$\sqrt{9 + 4} \neq \sqrt{9} + \sqrt{4} \qquad\qquad \sqrt{9 - 4} \neq \sqrt{9} - \sqrt{4}$$

$$\sqrt{13} \neq 3 + 2 \qquad\qquad\qquad \sqrt{5} \neq 3 - 2$$

$$\sqrt{13} \neq 5 \qquad\qquad\qquad\qquad \sqrt{5} \neq 1$$

Thus, $\sqrt{a + b} \neq \sqrt{a} + \sqrt{b}$ and $\sqrt{a - b} \neq \sqrt{a} - \sqrt{b}$.

To introduce the second property of radicals, we consider these examples.

$$\sqrt{\frac{100}{4}} = \sqrt{25} \qquad\qquad \frac{\sqrt{100}}{\sqrt{4}} = \frac{10}{2}$$

$$= 5 \qquad\qquad\qquad = 5$$

Since the answer is 5 in each case, $\sqrt{\dfrac{100}{4}} = \dfrac{\sqrt{100}}{\sqrt{4}}$. Likewise,

$$\sqrt[3]{\frac{64}{8}} = \sqrt[3]{8} \qquad\qquad \frac{\sqrt[3]{64}}{\sqrt[3]{8}} = \frac{4}{2}$$

$$= 2 \qquad\qquad\qquad = 2$$

Since the answer is 2 in each case, $\sqrt[3]{\dfrac{64}{8}} = \dfrac{\sqrt[3]{64}}{\sqrt[3]{8}}$. These results illustrate the **division property of radicals**.

Division property of radicals

If $\sqrt[n]{a}$ and $\sqrt[n]{b}$ represent real numbers, then

$$\sqrt[n]{\dfrac{a}{b}} = \dfrac{\sqrt[n]{a}}{\sqrt[n]{b}} \quad \text{where } b \neq 0$$

As long as all radical expressions represent real numbers, *the nth root of the quotient of two numbers is equal to the quotient of their nth roots.*

■ Simplifying radical expressions

A radical expression is said to be in simplified form when each of the following statements is true.

Simplified form of a radical expression

A radical expression is in simplest form when

1. No radicals appear in the denominator of a fraction.
2. The radicand contains no fractions or negative numbers.
3. Each factor in the radicand is to a power that is less than the index of the radical.

INTERMEDIATE
Algebra *f(x)* **Now**™ ──────

Self Check 1
Simplify:

a. $\sqrt{20}$

b. $\sqrt[3]{24}$

c. $\sqrt[5]{-128}$

EXAMPLE 1   Simplify: **a.** $\sqrt{12}$, **b.** $\sqrt{98}$, **c.** $\sqrt[3]{54}$, and **d.** $-\sqrt[4]{48}$.

Solution

a. Recall that numbers that are squares of integers, such as 1, 4, 9, 16, 25, and 36, are *perfect squares*. To simplify $\sqrt{12}$, we first factor 12 so that one factor is the largest perfect square that divides 12. Since 4 is the largest perfect square factor of 12, we write 12 as $4 \cdot 3$, use the multiplication property of radicals, and simplify.

$$\begin{aligned}
\sqrt{12} &= \sqrt{4 \cdot 3} &&\text{Write 12 in factored form as } 4 \cdot 3. \\
&= \sqrt{4}\sqrt{3} &&\text{By the multiplication property of radicals, the square root of a} \\
& &&\text{product is equal to the product of the square roots: } \sqrt{4 \cdot 3} = \sqrt{4}\sqrt{3}. \\
&= 2\sqrt{3} &&\text{Simplify: } \sqrt{4} = 2.
\end{aligned}$$

We can show that $2\sqrt{3}$ is a square root of 12 as follows:

$$\begin{aligned}
(2\sqrt{3})^2 &= (2)^2(\sqrt{3})^2 &&\text{Use a power rule of exponents: } (xy)^n = x^n y^n. \\
&= 4(3) &&\sqrt{3} \text{ is the number that, when squared, gives 3.} \\
&= 12
\end{aligned}$$

Thus, $\sqrt{12} = 2\sqrt{3}$.

b. The largest perfect square factor of 98 is 49. Thus,

$$\begin{aligned}
\sqrt{98} &= \sqrt{49 \cdot 2} &&\text{Write 98 in factored form: } 98 = 49 \cdot 2. \\
&= \sqrt{49}\sqrt{2} &&\text{Apply the multiplication property of radicals: } \sqrt{49 \cdot 2} = \sqrt{49}\sqrt{2}. \\
&= 7\sqrt{2} &&\text{Simplify: } \sqrt{49} = 7.
\end{aligned}$$

Check the result.

c. Numbers that are cubes of integers, such as 1, 8, 27, 64, 125, and 216, are called *perfect cubes.* Since the largest perfect cube factor of 54 is 27, we have

$$\sqrt[3]{54} = \sqrt[3]{27 \cdot 2}$$ Write 54 as $27 \cdot 2$.

$$= \sqrt[3]{27}\sqrt[3]{2}$$ By the multiplication property of radicals, the cube root of a product is equal to the product of the cube roots: $\sqrt[3]{27 \cdot 2} = \sqrt[3]{27}\sqrt[3]{2}$.

$$= 3\sqrt[3]{2}$$ Simplify: $\sqrt[3]{27} = 3$.

We can show that $3\sqrt[3]{2}$ is a cube root of 54 as follows:

$$(3\sqrt[3]{2})^3 = (3)^3(\sqrt[3]{2})^3 = 27(2) = 54$$

d. The largest perfect fourth-power factor of 48 is 16. Thus,

$$-\sqrt[4]{48} = -\sqrt[4]{16 \cdot 3}$$ Write 48 as $16 \cdot 3$.

$$= -\sqrt[4]{16}\sqrt[4]{3}$$ Apply the multiplication property of radicals.

$$= -2\sqrt[4]{3}$$ Simplify: $\sqrt[4]{16} = 2$.

Check the result.

Answers **a.** $2\sqrt{5}$, **b.** $2\sqrt[3]{3}$,
c. $-2\sqrt[5]{4}$

INTERMEDIATE
Algebra *f(x)* **Now**™

EXAMPLE 2   Simplify: **a.** $\sqrt{m^9}$, **b.** $\sqrt{128a^5}$, **c.** $\sqrt[3]{24x^7}$, and **d.** $\sqrt[5]{a^9b^5}$. Assume that all variables represent positive real numbers.

Self Check 2
Simplify:
a. $\sqrt{98b^3}$

b. $\sqrt[3]{54y^5}$

c. $\sqrt[4]{t^8u^{15}}$

Solution
a. The largest perfect square factor of m^9 is m^8.

$$\sqrt{m^9} = \sqrt{m^8 \cdot m}$$ Write m^9 in factored form as $m^8 \cdot m$.

$$= \sqrt{m^8}\sqrt{m}$$ Apply the multiplication property of radicals.

$$= m^4\sqrt{m}$$ Simplify: $\sqrt{m^8} = m^4$.

b. Since the largest perfect square factor of 128 is 64 and the largest perfect square factor of a^5 is a^4, the largest perfect square factor of $128a^5$ is $64a^4$. We write $128a^5$ as $64a^4 \cdot 2a$ and proceed as follows:

$$\sqrt{128a^5} = \sqrt{64a^4 \cdot 2a}$$

$$= \sqrt{64a^4}\sqrt{2a}$$ Use the multiplication property of radicals.

$$= 8a^2\sqrt{2a}$$ Simplify: $\sqrt{64a^4} = 8a^2$.

c. We write $24x^7$ as $8x^6 \cdot 3x$ and proceed as follows:

$$\sqrt[3]{24x^7} = \sqrt[3]{8x^6 \cdot 3x}$$ $8x^6$ is the largest perfect cube factor of $24x^7$.

$$= \sqrt[3]{8x^6}\sqrt[3]{3x}$$ Use the multiplication property of radicals.

$$= 2x^2\sqrt[3]{3x}$$ Simplify: $\sqrt[3]{8x^6} = 2x^2$.

d. The largest perfect fifth-power factor of a^9 is a^5, and b^5 is a fifth power.

$$\sqrt[5]{a^9b^5} = \sqrt[5]{a^5b^5 \cdot a^4}$$ a^5b^5 is the largest perfect fifth-power factor of a^9b^5.

$$= \sqrt[5]{a^5b^5}\sqrt[5]{a^4}$$ Use the multiplication property of radicals.

$$= ab\sqrt[5]{a^4}$$ Simplify: $\sqrt[5]{a^5b^5} = ab$.

Answers **a.** $7b\sqrt{2b}$,
b. $3y\sqrt[3]{2y^2}$, **c.** $t^2u^3\sqrt[4]{u^3}$

INTERMEDIATE
Algebra *f(x)* **Now**™

EXAMPLE 3   Simplify: **a.** $\sqrt{\dfrac{7}{64}}$, **b.** $\sqrt{\dfrac{15}{49x^2}}$, and **c.** $\sqrt[3]{\dfrac{10x^2}{27y^6}}$.

Assume that the variables represent positive real numbers.

Self Check 3

Simplify: **a.** $\sqrt{\dfrac{11}{36a^2}}$ where $a > 0$

b. $\sqrt[4]{\dfrac{a^3}{625y^{12}}}$

where $a > 0, y > 0$

Solution

a. We can write the square root of the quotient as the quotient of two square roots.

$$\sqrt{\dfrac{7}{64}} = \dfrac{\sqrt{7}}{\sqrt{64}} \qquad \text{Apply the division property of radicals.}$$

$$= \dfrac{\sqrt{7}}{8} \qquad \text{Simplify the denominator: } \sqrt{64} = 8.$$

b. $\sqrt{\dfrac{15}{49x^2}} = \dfrac{\sqrt{15}}{\sqrt{49x^2}} \qquad \text{Apply the division property of radicals.}$

$$= \dfrac{\sqrt{15}}{7x} \qquad \text{Simplify the denominator: } \sqrt{49x^2} = 7x.$$

c. We can write the cube root of the quotient as the quotient of two cube roots. Since $y \neq 0$, we have

$$\sqrt[3]{\dfrac{10x^2}{27y^6}} = \dfrac{\sqrt[3]{10x^2}}{\sqrt[3]{27y^6}} \qquad \text{Use the division property of radicals.}$$

$$= \dfrac{\sqrt[3]{10x^2}}{3y^2} \qquad \text{Simplify the denominator.}$$

Answers **a.** $\dfrac{\sqrt{11}}{6a}$, **b.** $\dfrac{\sqrt[4]{a^3}}{5y^3}$

Self Check 4
Simplify each expression
(assume that all variables
represent positive numbers):

a. $\dfrac{\sqrt{50ab^2}}{\sqrt{2a}}$

b. $\dfrac{\sqrt[3]{-2{,}000x^5v^3}}{\sqrt[3]{2x}}$

EXAMPLE 4 Simplify each expression. Assume that all variables represent positive numbers. **a.** $\dfrac{\sqrt{45xy^2}}{\sqrt{5x}}$ and **b.** $\dfrac{\sqrt[3]{-432x^5}}{\sqrt[3]{8x}}$.

Solution

a. We can write the quotient of the square roots as the square root of a quotient.

$$\dfrac{\sqrt{45xy^2}}{\sqrt{5x}} = \sqrt{\dfrac{45xy^2}{5x}} \qquad \text{Use the division property of radicals.}$$

$$= \sqrt{9y^2} \qquad \text{Simplify } \dfrac{45xy^2}{5x}.$$

$$= 3y \qquad \text{Simplify the radical.}$$

b. We can write the quotient of the cube roots as the cube root of a quotient.

$$\dfrac{\sqrt[3]{-432x^5}}{\sqrt[3]{8x}} = \sqrt[3]{\dfrac{-432x^5}{8x}} \qquad \text{Use the division property of radicals.}$$

$$= \sqrt[3]{-54x^4} \qquad \text{Simplify } \dfrac{-432x^5}{8x}.$$

$$= \sqrt[3]{-27x^3 \cdot 2x} \qquad -27x^3 \text{ is the largest perfect cube that divides } -54x^4.$$

$$= \sqrt[3]{-27x^3}\sqrt[3]{2x} \qquad \text{Use the multiplication property of radicals.}$$

$$= -3x\sqrt[3]{2x} \qquad \text{Simplify: } \sqrt[3]{-27x^3} = -3x.$$

Answers **a.** $5b$, **b.** $-10xv\sqrt[3]{x}$

■ Adding and subtracting radical expressions

Radical expressions with the same index and the same radicand are called **like** or **similar radicals.** For example, $3\sqrt{2}$ and $2\sqrt{2}$ are like radicals. However,

$3\sqrt{5}$ and $4\sqrt{2}$ are not like radicals, because the radicands are different.

$3\sqrt[3]{5}$ and $2\sqrt[3]{5}$ are not like radicals, because the indexes are different.

For a given expression containing two or more radical terms, we should attempt to combine like radicals, if possible. For example, to simplify the expression $3\sqrt{2} + 2\sqrt{2}$, we use the distributive property to factor out $\sqrt{2}$ and simplify.

$$3\sqrt{2} + 2\sqrt{2} = (3 + 2)\sqrt{2}$$
$$= 5\sqrt{2}$$

Radicals with the same index but different radicands can often be written as like radicals. For example, to simplify the expression $\sqrt{27} - \sqrt{12}$, we simplify both radicals first, and then we combine the like radicals.

$$\sqrt{27} - \sqrt{12} = \sqrt{9 \cdot 3} - \sqrt{4 \cdot 3} \qquad \text{Write 27 and 12 in factored form.}$$
$$= \sqrt{9}\sqrt{3} - \sqrt{4}\sqrt{3} \qquad \text{Use the multiplication property of radicals.}$$
$$= 3\sqrt{3} - 2\sqrt{3} \qquad \text{Simplify } \sqrt{9} \text{ and } \sqrt{4}.$$
$$= (3 - 2)\sqrt{3} \qquad \text{Factor out } \sqrt{3}.$$
$$= \sqrt{3} \qquad 1\sqrt{3} = \sqrt{3}.$$

As the previous examples suggest, we can use the following rule to add or subtract radicals.

Adding and subtracting radicals

To add or subtract radicals, simplify each radical expression and combine all like radicals. To add or subtract like radicals, combine the coefficients and keep the common radical.

INTERMEDIATE
Algebra *f(x)* **Now**™

EXAMPLE 5 Simplify: $2\sqrt{12} - 3\sqrt{48} + 3\sqrt{3}$.

Solution   We simplify $2\sqrt{12}$ and $3\sqrt{48}$ separately and then combine like radicals.

$$2\sqrt{12} - 3\sqrt{48} + 3\sqrt{3} = 2\sqrt{4 \cdot 3} - 3\sqrt{16 \cdot 3} + 3\sqrt{3}$$
$$= 2\sqrt{4}\sqrt{3} - 3\sqrt{16}\sqrt{3} + 3\sqrt{3}$$
$$= 2(2)\sqrt{3} - 3(4)\sqrt{3} + 3\sqrt{3}$$
$$= 4\sqrt{3} - 12\sqrt{3} + 3\sqrt{3} \qquad \text{All three expressions have the same index and radicand.}$$

$$= (4 - 12 + 3)\sqrt{3} \qquad \text{Combine the coefficients of these like radicals and keep } \sqrt{3}.$$

$$= -5\sqrt{3}$$

Self Check 5
Simplify:
$3\sqrt{75} - 2\sqrt{12} + 2\sqrt{48}$

Answer $19\sqrt{3}$

INTERMEDIATE
Algebra *f(x)* **Now**™
Self Check 6
Simplify:
$\sqrt[3]{24} - \sqrt[3]{16} + \sqrt[3]{54}$

EXAMPLE 6 Simplify: $\sqrt[3]{16} - \sqrt[3]{54} + \sqrt[3]{24}$.

Solution   We begin by simplifying each radical expression separately:

$$\sqrt[3]{16} - \sqrt[3]{54} + \sqrt[3]{24} = \sqrt[3]{8 \cdot 2} - \sqrt[3]{27 \cdot 2} + \sqrt[3]{8 \cdot 3}$$
$$= \sqrt[3]{8}\sqrt[3]{2} - \sqrt[3]{27}\sqrt[3]{2} + \sqrt[3]{8}\sqrt[3]{3}$$
$$= 2\sqrt[3]{2} - 3\sqrt[3]{2} + 2\sqrt[3]{3}$$

Now we combine the two radical expressions that have the same index and radicand.

$$\sqrt[3]{16} - \sqrt[3]{54} + \sqrt[3]{24} = -\sqrt[3]{2} + 2\sqrt[3]{3} \qquad 2\sqrt[3]{2} - 3\sqrt[3]{2} = -1\sqrt[3]{2} = -\sqrt[3]{2}.$$

Answer $2\sqrt[3]{3} + \sqrt[3]{2}$

! COMMENT Even though the radical expressions $-\sqrt[3]{2}$ and $2\sqrt[3]{3}$ in the last line of Example 6 have the same index, we cannot combine them, because their radicands are different. Neither can we combine radical expressions having the same radicand but a different index. For example, the expression $\sqrt[3]{2} + \sqrt[4]{2}$ cannot be simplified.

INTERMEDIATE
Algebra $f(x)$ Now™

Self Check 7
Simplify:

$$\sqrt{32x^3} + \sqrt{50x^3} - \sqrt{18x^3}$$

EXAMPLE 7   Simplify: $\sqrt[3]{16x^4} + \sqrt[3]{54x^4} - \sqrt[3]{-128x^4}$.

Solution   We simplify each radical expression separately, factor out $\sqrt[3]{2x}$, and simplify.

$$\sqrt[3]{16x^4} + \sqrt[3]{54x^4} - \sqrt[3]{-128x^4}$$
$$= \sqrt[3]{8x^3 \cdot 2x} + \sqrt[3]{27x^3 \cdot 2x} - \sqrt[3]{-64x^3 \cdot 2x}$$
$$= \sqrt[3]{8x^3}\sqrt[3]{2x} + \sqrt[3]{27x^3}\sqrt[3]{2x} - \sqrt[3]{-64x^3}\sqrt[3]{2x}$$
$$= 2x\sqrt[3]{2x} + 3x\sqrt[3]{2x} + 4x\sqrt[3]{2x} \quad \text{All three radicals have the same index and radicand.}$$
$$= (2x + 3x + 4x)\sqrt[3]{2x}$$
Answer $6x\sqrt{2x}$
$$= 9x\sqrt[3]{2x} \quad \text{Within the parentheses, combine like terms.}$$

Section 7.2   STUDY SET

INTERMEDIATE
Algebra $f(x)$ Now™

VOCABULARY   *Fill in the blanks.*

1. Radical expressions such as $\sqrt[3]{4}$ and $6\sqrt[3]{4}$ with the same index and the same radicand are called _____ radicals.

2. Numbers such as 1, 4, 9, 16, 25, and 36 are called perfect _____. Numbers such as 1, 8, 27, 64, and 125 are called perfect _____.

3. The largest perfect square _____ of 27 is 9.

4. "To _____ $\sqrt{24}$" means to write it as $2\sqrt{6}$.

CONCEPTS   *Fill in the blanks.*

5. $\sqrt[n]{ab} =$

In words, the nth root of the _____ of two numbers is equal to the product of their nth _____.

6. $\sqrt[n]{\dfrac{a}{b}} =$

In words, the nth root of the _____ of two numbers is equal to the quotient of their nth _____.

7. Consider the expressions $\sqrt{4 \cdot 5}$ and $\sqrt{4}\sqrt{5}$.

Which expression is

a. the square root of a product?

b. the product of square roots?

c. How are these two expressions related?

8. Consider the expressions

$$\dfrac{\sqrt[3]{a}}{\sqrt[3]{x^2}} \quad \text{and} \quad \sqrt[3]{\dfrac{a}{x^2}}$$

Which expression is

a. the cube root of a quotient?

b. the quotient of cube roots?

c. How are these two expressions related?

9. a. Write two radical expressions that have the same radicand but a different index. Can the expressions be added?

b. Write two radical expressions that have the same index but a different radicand. Can the expressions be added?

10. Explain the mistake in the student's solution.

$$\sqrt[3]{54} = \sqrt[3]{27 + 27}$$
$$= \sqrt[3]{27} + \sqrt[3]{27}$$
$$= 3 + 3$$
$$= 6$$

NOTATION *Complete each solution.*

11. $\sqrt[3]{32k^4} = \sqrt[3]{\boxed{} \cdot 4k}$

$$= \sqrt[3]{\boxed{}} \, \sqrt[3]{4k}$$

$$= 2k\sqrt[3]{\boxed{}}$$

12. $\dfrac{\sqrt{80s^2t^4}}{\sqrt{5s^2}} = \sqrt{\dfrac{80s^2t^4}{\boxed{}}}$

$$= \sqrt{\boxed{}}$$

$$= 4\boxed{}$$

PRACTICE *Use the multiplication or the division property of radicals to write each expression using one radical symbol. Then simplify. All variables represent positive numbers.*

13. $\sqrt{6}\sqrt{6}$ **14.** $\sqrt{11}\sqrt{11}$

15. $\sqrt{t}\sqrt{t}$ **16.** $-\sqrt{z}\sqrt{z}$

17. $\sqrt[3]{5x^2}\sqrt[3]{25x}$ **18.** $\sqrt[4]{25a}\sqrt[4]{25a^3}$

19. $\dfrac{\sqrt{500}}{\sqrt{5}}$ **20.** $\dfrac{\sqrt{128}}{\sqrt{2}}$

21. $\dfrac{\sqrt{98x^3}}{\sqrt{2x}}$ **22.** $\dfrac{\sqrt{75y^5}}{\sqrt{3y}}$

23. $\dfrac{\sqrt{180ab^4}}{\sqrt{5ab^2}}$ **24.** $\dfrac{\sqrt{112ab^3}}{\sqrt{7ab}}$

25. $\dfrac{\sqrt[3]{48}}{\sqrt[3]{6}}$ **26.** $\dfrac{\sqrt[3]{64}}{\sqrt[3]{8}}$

27. $\dfrac{\sqrt[3]{189a^4}}{\sqrt[3]{7a}}$ **28.** $\dfrac{\sqrt[3]{243x^7}}{\sqrt[3]{9x}}$

Simplify each radical expression. Assume that all variables represent postitive numbers.

29. $\sqrt{20}$ **30.** $\sqrt{8}$

31. $-\sqrt{200}$ **32.** $-\sqrt{250}$

33. $\sqrt[3]{80}$ **34.** $\sqrt[3]{270}$

35. $\sqrt[3]{-81}$ **36.** $\sqrt[3]{-72}$

37. $\sqrt[4]{32}$ **38.** $\sqrt[4]{48}$

39. $\sqrt[5]{96}$ **40.** $\sqrt[7]{256}$

41. $\sqrt{\dfrac{7}{9}}$ **42.** $\sqrt{\dfrac{3}{4}}$

43. $\sqrt[3]{\dfrac{7}{64}}$ **44.** $\sqrt[3]{\dfrac{4}{125}}$

45. $\sqrt[4]{\dfrac{3}{10,000}}$ **46.** $\sqrt[5]{\dfrac{4}{243}}$

47. $\sqrt[5]{\dfrac{3}{32}}$ **48.** $\sqrt[6]{\dfrac{5}{64}}$

49. $\sqrt{50x^2}$ **50.** $\sqrt{75a^2}$

51. $\sqrt{32b}$ **52.** $\sqrt{80c}$

53. $-\sqrt{112a^3}$ **54.** $\sqrt{147a^5}$

55. $\sqrt{175a^2b^3}$ **56.** $\sqrt{128a^3b^5}$

57. $-\sqrt{300xy}$ **58.** $\sqrt{200x^2y}$

59. $\sqrt[3]{-54x^6}$ **60.** $-\sqrt[3]{-81a^3}$

61. $\sqrt[3]{16x^{12}y^4}$ **62.** $\sqrt[3]{40a^3b^7}$

63. $\sqrt[4]{32x^{12}y^5}$ **64.** $\sqrt[5]{64x^{10}y^6}$

65. $\sqrt{\dfrac{z^2}{16x^2}}$ **66.** $\sqrt{\dfrac{b^4}{64a^8}}$

67. $\sqrt[4]{\dfrac{5x}{16z^4}}$ **68.** $\sqrt[3]{\dfrac{11a^2}{125b^6}}$

Simplify and combine like radicals. All variables represent positive numbers.

69. $4\sqrt{2x} + 6\sqrt{2x}$ **70.** $6\sqrt[3]{5y} + 3\sqrt[3]{5y}$

71. $8\sqrt[5]{7a^2} - 7\sqrt[5]{7a^2}$ **72.** $10\sqrt[6]{12xyz} - \sqrt[6]{12xyz}$

73. $\sqrt{2} - \sqrt{8}$ **74.** $\sqrt{20} - \sqrt{125}$

75. $\sqrt{98} - \sqrt{50}$ **76.** $\sqrt{72} - \sqrt{200}$

77. $3\sqrt{24} + \sqrt{54}$ **78.** $\sqrt{18} + 2\sqrt{50}$

79. $\sqrt[3]{24x} + \sqrt[3]{3x}$ **80.** $\sqrt[3]{16y} + \sqrt[3]{128y}$

81. $\sqrt[3]{32} - \sqrt[3]{108}$ **82.** $\sqrt[3]{80} - \sqrt[3]{10,000}$

83. $2\sqrt[3]{125} - 5\sqrt[3]{64}$ **84.** $3\sqrt[3]{27} + 12\sqrt[3]{216}$

85. $14\sqrt[4]{32} - 15\sqrt[4]{162}$ **86.** $23\sqrt[4]{768} + \sqrt[4]{48}$

87. $3\sqrt[4]{512} + 2\sqrt[4]{32}$ **88.** $4\sqrt[4]{243} - \sqrt[4]{48}$

89. $\sqrt{98} - \sqrt{50} - \sqrt{72}$

90. $\sqrt{20} + \sqrt{125} - \sqrt{80}$

91. $\sqrt{18t} + \sqrt{300t} - \sqrt{243t}$

92. $\sqrt{80m} - \sqrt{128m} + \sqrt{288m}$

93. $2\sqrt[3]{16} - \sqrt[3]{54} - 3\sqrt[3]{128}$

94. $\sqrt[4]{48} - \sqrt[4]{243} - \sqrt[4]{768}$

95. $\sqrt{25y^2z} - \sqrt{16y^2z}$ **96.** $\sqrt{25yz^2} + \sqrt{9yz^2}$

97. $\sqrt{36xy^2} + \sqrt{49xy^2}$ **98.** $3\sqrt{2x} - \sqrt{8x}$

99. $2\sqrt[3]{64a} + 2\sqrt[3]{8a}$ **100.** $3\sqrt[4]{x^4y} - 2\sqrt[4]{x^4y}$

101. $\sqrt{y^5} - \sqrt{9y^5} - \sqrt{25y^5}$

102. $\sqrt{8y^7} + \sqrt{32y^7} - \sqrt{2y^7}$

103. $\sqrt[5]{x^6y^2} + \sqrt[5]{32x^6y^2} + \sqrt[5]{x^6y^2}$

104. $\sqrt[3]{xy^4} + \sqrt[3]{8xy^4} - \sqrt[3]{27xy^4}$

■ **APPLICATIONS**   *First give the exact answer, expressed as a simplified radical expression. Then give an approximation, rounded to the nearest tenth.*

105. UMBRELLAS The surface area of a cone is given by the formula $S = \pi r \sqrt{r^2 + h^2}$, where r is the radius of the base and h is its height. Use this formula to find the number of square feet of waterproof cloth used to make the umbrella.

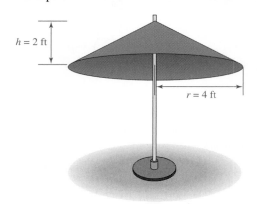

$h = 2$ ft

$r = 4$ ft

106. STRUCTURAL ENGINEERING Engineers have determined that two additional supports need to be added to strengthen a truss. Find the length L of each new support using the formula

$$L = \sqrt{\frac{b^2}{2} + \frac{c^2}{2} - \frac{a^2}{4}}$$

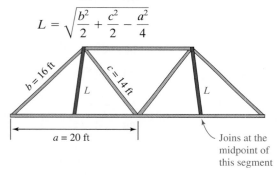

$b = 16$ ft $c = 14$ ft L L

$a = 20$ ft

Joins at the midpoint of this segment

107. BLOW DRYERS The current I (in amps), the power P (in watts), and the resistance R (in ohms) are related by the formula $I = \sqrt{\frac{P}{R}}$. What current is needed for a 1,200-watt hair dryer if the resistance is 16 ohms?

108. SATELLITES Engineers have determined that a spherical communications satellite needs to have a capacity of 565.2 cubic feet to house all of its operating systems. The volume V of a sphere is related to its radius r by the formula $r = \sqrt[3]{\frac{3V}{4\pi}}$. What radius must the satellite have to meet the engineer's specification? Use 3.14 for π.

109. DUCTWORK The pattern shown below is laid out on a sheet of galvanized tin. Then it is cut out with snips and bent along the dotted lines to make an air conditioning duct connection. Find the total length of the cut that must be made with the tin snips. (All measurements are in inches.)

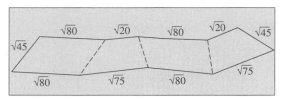

$\sqrt{80}$ $\sqrt{20}$ $\sqrt{80}$ $\sqrt{20}$ $\sqrt{45}$

$\sqrt{45}$

$\sqrt{80}$ $\sqrt{75}$ $\sqrt{80}$ $\sqrt{75}$

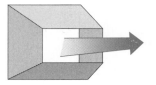

110. OUTDOOR COOKING The diameter of a circle is given by the function $d(A) = 2\sqrt{\frac{A}{\pi}}$, where A is the area of the circle. Find the difference between the diameters of the barbecue grills.

Cooking area
147π in.^{3}

Cooking area
48π in.^{3}

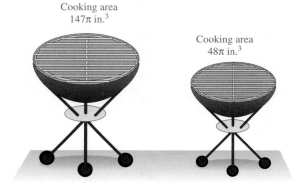

■ **WRITING**

111. Explain why $\sqrt[3]{9x^4}$ is not in simplified form.

112. How are the procedures used to simplify $3x + 4x$ and $3\sqrt{x} + 4\sqrt{x}$ similar?

113. Explain how the graphs of $Y_1 = 3\sqrt{24x} + \sqrt{54x}$ and $Y_2 = 9\sqrt{6x}$ can be used to verify the simplification $3\sqrt{24x} + \sqrt{54x} = 9\sqrt{6x}$. In each graph, settings of $[-5, 20]$ for x and $[-5, 100]$ for y were used.

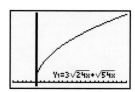
Y1=3√24x+√54x

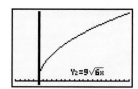
Y2=9√6x

114. Explain how to verify algebraically that $\sqrt{200x^3y^5} = 10xy^2\sqrt{2xy}$.

REVIEW *Perform each operation.*

115. $3x^2y^3(-5x^3y^{-4})$

116. $(2x^2 - 9x - 5) \cdot \dfrac{x}{2x^2 + x}$

117. $2p - 5\overline{)6p^2 - 7p - 25}$

118. $\dfrac{xy}{\dfrac{1}{x} - \dfrac{1}{y}}$

7.3 Multiplying and Dividing Radical Expressions

- Multiplying radical expressions   • Rationalizing denominators
- Rationalizing two-term denominators

In this section, we will discuss how to multiply and divide radical expressions. Solving these problems often requires the use of procedures and properties studied earlier, such as simplifying radical expressions, combining like radicals, the FOIL method, and the distributive property.

Multiplying radical expressions

Radical expressions with the same index can be multiplied. For example, to find the product of $\sqrt{5}$ and $\sqrt{10}$, we proceed as follows:

$$\sqrt{5}\sqrt{10} = \sqrt{5 \cdot 10} \quad \text{By the multiplication property of radicals: } \sqrt[n]{a}\sqrt[n]{b} = \sqrt[n]{ab}.$$
$$= \sqrt{50} \quad \text{Multiply the numbers under the radical. Note that } \sqrt{50} \text{ can be simplified.}$$
$$= \sqrt{25 \cdot 2} \quad \text{Begin the process of simplifying } \sqrt{50} \text{ by factoring 50.}$$
$$= 5\sqrt{2} \quad \sqrt{25 \cdot 2} = \sqrt{25}\sqrt{2} = 5\sqrt{2}.$$

EXAMPLE 1   Multiply and then simplify: **a.** $3\sqrt{6}\,(2\sqrt{3})$ and
b. $-2\sqrt[3]{7x} \cdot 6\sqrt[3]{49x^2}$.

Solution   We use the commutative and associative properties of multiplication to multiply the coefficients and the radicals separately. Then we simplify any radicals in the product, if possible.

a. $3\sqrt{6} \cdot 2\sqrt{3} = 3(2)\sqrt{6}\sqrt{3}$ Multiply the coefficients and multiply the radicals.
$$= 6\sqrt{18} \quad 3(2) = 6 \text{ and } \sqrt{6}\sqrt{3} = \sqrt{18}.$$
$$= 6\sqrt{9}\sqrt{2} \quad \text{Simplify: } \sqrt{18} = \sqrt{9 \cdot 2} = \sqrt{9}\sqrt{2}.$$
$$= 6(3)\sqrt{2} \quad \sqrt{9} = 3.$$
$$= 18\sqrt{2} \quad \text{Multiply.}$$

b. $-2\sqrt[3]{7x} \cdot 6\sqrt[3]{49x^2} = -2(6)\sqrt[3]{7x}\sqrt[3]{49x^2}$ Write the coefficients together and the radicals together.
$$= -12\sqrt[3]{7x \cdot 49x^2} \quad \text{Multiply the coefficients and multiply the radicals.}$$
$$= -12\sqrt[3]{7x \cdot 7^2x^2} \quad \text{Write 49 as } 7^2.$$
$$= -12\sqrt[3]{7^3x^3} \quad \text{Write } 7x \cdot 7^2x^2 \text{ as } 7^3x^3.$$
$$= -12(7x) \quad \text{Simplify: } \sqrt[3]{7^3x^3} = 7x.$$
$$= -84x \quad \text{Multiply.}$$

INTERMEDIATE
Algebra $f(x)$ **Now**™

Self Check 1
Multiply:
a. $-2\sqrt{14}\,(5\sqrt{2})$ and
b. $\sqrt[4]{4x^3} \cdot 9\sqrt[4]{8x^2}$.

Answers **a.** $-20\sqrt{7}$,   **b.** $18x\sqrt[4]{2x}$

Recall that to multiply a polynomial by a monomial, we use the distributive property. We use the same technique to multiply a radical expression that has two or more terms by a radical expression that has only one term.

INTERMEDIATE
Algebra $f(x)$ **Now**™

Self Check 2
Simplify:
$4\sqrt{2}(3\sqrt{5} - 2\sqrt{8})$

EXAMPLE 2   Simplify: $3\sqrt{3}(4\sqrt{8} - 5\sqrt{10})$.

Solution

$3\sqrt{3}(4\sqrt{8} - 5\sqrt{10})$
$= 3\sqrt{3}\cdot 4\sqrt{8} - 3\sqrt{3}\cdot 5\sqrt{10}$ Distribute the multiplication by $3\sqrt{3}$.
$= 12\sqrt{24} - 15\sqrt{30}$ Multiply the coefficients and multiply the radicals.
$= 12\sqrt{4}\sqrt{6} - 15\sqrt{30}$ Simplify: $\sqrt{24} = \sqrt{4\cdot 6} = \sqrt{4}\sqrt{6}$.
$= 12(2)\sqrt{6} - 15\sqrt{30}$ $\sqrt{4} = 2$.
$= 24\sqrt{6} - 15\sqrt{30}$

Answer $12\sqrt{10} - 32$

To multiply two binomials, we multiply each term of one binomial by each term of the other binomial and simplify. We multiply two radical expressions, each having two terms, in the same way.

INTERMEDIATE
Algebra $f(x)$ **Now**™

Self Check 3
Multiply:
$(\sqrt{5} + 2\sqrt{3})(\sqrt{5} - \sqrt{3})$

EXAMPLE 3   Multiply: $(\sqrt{7} + \sqrt{2})(\sqrt{7} - 3\sqrt{2})$.

Solution

$(\sqrt{7} + \sqrt{2})(\sqrt{7} - 3\sqrt{2})$
$= \sqrt{7}\sqrt{7} - 3\sqrt{7}\sqrt{2} + \sqrt{2}\sqrt{7} - 3\sqrt{2}\sqrt{2}$ Use the FOIL method.
$= 7 - 3\sqrt{14} + \sqrt{14} - 3(2)$ Perform each multiplication: $\sqrt{7}\sqrt{7} = \sqrt{49} = 7$ and $\sqrt{2}\sqrt{2} = \sqrt{4} = 2$.
$= 7 - 2\sqrt{14} - 6$ Combine like radicals: $-3\sqrt{14} + \sqrt{14} = -2\sqrt{14}$.

Answer $-1 + \sqrt{15}$

$= 1 - 2\sqrt{14}$ Combine like terms: $7 - 6 = 1$.

INTERMEDIATE
Algebra $f(x)$ **Now**™

Self Check 4
Multiply: $(\sqrt{x} + 1)(\sqrt{x} - 3)$.
Assume that $x > 0$.

EXAMPLE 4   Multiply: $(\sqrt{3x} - \sqrt{5})(\sqrt{2x} + \sqrt{10})$. Assume that $x > 0$.

Solution

$(\sqrt{3x} - \sqrt{5})(\sqrt{2x} + \sqrt{10})$
$= \sqrt{3x}\sqrt{2x} + \sqrt{3x}\sqrt{10} - \sqrt{5}\sqrt{2x} - \sqrt{5}\sqrt{10}$ Use the FOIL method.
$= \sqrt{6x^2} + \sqrt{30x} - \sqrt{10x} - \sqrt{50}$ Perform each multiplication.
$= \sqrt{6}\sqrt{x^2} + \sqrt{30x} - \sqrt{10x} - \sqrt{25}\sqrt{2}$ Simplify $\sqrt{6x^2}$ and $\sqrt{50}$.

Answer $x - 2\sqrt{x} - 3$

$= \sqrt{6}x + \sqrt{30x} - \sqrt{10x} - 5\sqrt{2}$

! COMMENT It is important to draw the radical sign carefully so that it completely covers the radicand, but no more than the radicand. To avoid confusion, we often write an expression such as $\sqrt{6}x$ in the form $x\sqrt{6}$.

Rationalizing denominators

In Section 7.2, we saw that a radical expression is in simplified form when each of the following statements is true.

1. No radicals appear in the denominator of a fraction.

2. The radicand contains no fractions or negative numbers.

3. Each factor in the radicand appears to a power that is less than the index of the radical.

We now consider radical expressions that do not satisfy requirement 1 and radical expressions that do not satisfy requirement 2 of this list. We will introduce an algebraic technique, called *rationalizing the denominator,* that is used to write such expressions in an equivalent simplified form.

To divide radical expressions, we **rationalize the denominator** of a fraction to replace the denominator with a rational number. For example, to divide $\sqrt{5}$ by $\sqrt{3}$, we write the division as the fraction

$$\frac{\sqrt{5}}{\sqrt{3}}$$ The denominator is the irrational number $\sqrt{3}$. This radical expression is not in simplified form, because a radical appears in the denominator.

To eliminate the radical in the denominator, we multiply the numerator and the denominator by a number that will give a perfect square *under the radical in the denominator.* Because $3 \cdot 3 = 9$ and 9 is a perfect square, $\sqrt{3}$ is such a number.

$$\frac{\sqrt{5}}{\sqrt{3}} = \frac{\sqrt{5} \cdot \sqrt{3}}{\sqrt{3} \cdot \sqrt{3}}$$ Multiply numerator and denominator of the fraction by $\sqrt{3}$.

$$= \frac{\sqrt{15}}{\sqrt{9}}$$ Multiply the radicals in the numerator and in the denominator.

$$= \frac{\sqrt{15}}{3}$$ Simplify: $\sqrt{9} = 3$. The denominator is now the rational number 3.

Thus, $\dfrac{\sqrt{5}}{\sqrt{3}} = \dfrac{\sqrt{15}}{3}$. Since there is no radical in the denominator and $\sqrt{15}$ cannot be simplified, the expression $\dfrac{\sqrt{15}}{3}$ is in simplest form, and the division is complete.

EXAMPLE 5   Rationalize each denominator: **a.** $\sqrt{\dfrac{20}{7}}$ and   **b.** $\dfrac{4}{\sqrt[3]{2}}$.

Solution

a. This radical expression is not in simplified form, because the radicand contains a fraction. We begin by writing the square root of the quotient as the quotient of two square roots:

$$\sqrt{\frac{20}{7}} = \frac{\sqrt{20}}{\sqrt{7}}$$ Apply the division property of radicals: $\sqrt[n]{\dfrac{a}{b}} = \dfrac{\sqrt[n]{a}}{\sqrt[n]{b}}$.

To rationalize the denominator, we proceed as follows:

$$\frac{\sqrt{20}}{\sqrt{7}} = \frac{\sqrt{20} \cdot \sqrt{7}}{\sqrt{7} \cdot \sqrt{7}}$$ Multiply the numerator and denominator by $\sqrt{7}$.

$$= \frac{\sqrt{140}}{\sqrt{49}}$$ Multiply the radicals.

$$= \frac{2\sqrt{35}}{7}$$ Simplify: $\sqrt{140} = \sqrt{4 \cdot 35} = \sqrt{4}\sqrt{35} = 2\sqrt{35}$ and $\sqrt{49} = 7$.

b. Here, we must rationalize a denominator that is a cube root. We multiply the numerator and the denominator by a number that will give a perfect cube under the radical sign. Since $2 \cdot 4 = 8$ is a perfect cube, $\sqrt[3]{4}$ is such a number.

INTERMEDIATE
Algebra *f(x)* **Now™**

Self Check 5
Rationalize each denominator:

a. $\sqrt{\dfrac{24}{5}}$ and   **b.** $\dfrac{5}{\sqrt[4]{3}}$

$$\frac{4}{\sqrt[3]{2}} = \frac{4 \cdot \sqrt[3]{4}}{\sqrt[3]{2} \cdot \sqrt[3]{4}} \qquad \text{Multiply numerator and denominator by } \sqrt[3]{4}.$$

$$= \frac{4\sqrt[3]{4}}{\sqrt[3]{8}} \qquad \text{Multiply the radicals in the denominator.}$$

$$= \frac{4\sqrt[3]{4}}{2} \qquad \text{Simplify: } \sqrt[3]{8} = 2.$$

$$= 2\sqrt[3]{4} \qquad \text{Simplify: } \frac{4\sqrt[3]{4}}{2} = \frac{\overset{1}{\cancel{2}} \cdot 2\sqrt[3]{4}}{\underset{1}{\cancel{2}}} = 2\sqrt[3]{4}.$$

Answers **a.** $\dfrac{2\sqrt{30}}{5}$, **b.** $\dfrac{5\sqrt[4]{27}}{3}$

EXAMPLE 6   Photography.   Many camera lenses (see Figure 7-8) have an adjustable opening called the **aperture,** which controls the amount of light passing through the lens. The **f-number** of a lens is its **focal length** divided by the diameter of its circular aperture:

FIGURE 7-8

$$f\text{-number} = \frac{f}{d} \qquad f \text{ is the focal length, and } d \text{ is the diameter of the aperture.}$$

A lens with a focal length of 12 centimeters and an aperture with a diameter of 6 centimeters has an *f*-number of $\frac{12}{6}$ and is an *f*/2 lens. If the area of the aperture is reduced to admit half as much light, the *f*-number of the lens will change. Find the new *f*-number to the nearest tenth.

Solution   We first find the area of the aperture when its diameter is 6 centimeters.

$$A = \pi r^2 \qquad \text{This is the formula for the area of a circle.}$$

$$A = \pi(3)^2 \qquad \text{Since a radius is half the diameter, substitute 3 for } r.$$

$$A = 9\pi$$

When the size of the aperture is reduced to admit half as much light, the area of the aperture will be $\frac{9\pi}{2}$ square centimeters. To find the diameter d of a circle with this area, we proceed as follows:

$$A = \pi r^2 \qquad \text{This is the formula for the area of a circle.}$$

$$\frac{9\pi}{2} = \pi\left(\frac{d}{2}\right)^2 \qquad \text{Substitute } \frac{9\pi}{2} \text{ for } A \text{ and } \frac{d}{2} \text{ for } r.$$

$$\frac{9\pi}{2} = \frac{\pi d^2}{4} \qquad \left(\frac{d}{2}\right)^2 = \frac{d^2}{4}.$$

$$18 = d^2 \qquad \text{Multiply both sides by 4 and divide both sides by } \pi.$$

$$d = 3\sqrt{2} \qquad \text{Take the positive square root of both sides. Then simplify:}$$
$$\sqrt{18} = \sqrt{9}\sqrt{2} = 3\sqrt{2}.$$

Since the focal length of the lens is still 12 centimeters and the diameter is now $3\sqrt{2}$ centimeters, the new *f*-number of the lens is

$$f\text{-number} = \frac{f}{d} = \frac{12}{3\sqrt{2}} \qquad \text{Substitute 12 for } f \text{ and } 3\sqrt{2} \text{ for } d.$$

$$= \frac{12\sqrt{2}}{3\sqrt{2}\sqrt{2}} \qquad \text{Rationalize the denominator.}$$

$$= \frac{12\sqrt{2}}{3(2)} \qquad \text{Simplify the denominator: } \sqrt{2}\sqrt{2} = 2.$$

$$= 2\sqrt{2} \qquad \text{Simplify: } \frac{12\sqrt{2}}{3(2)} = \frac{12\sqrt{2}}{6} = \frac{\overset{1}{\cancel{6}} \cdot 2\sqrt{2}}{\underset{1}{\cancel{6}}}.$$

$$\approx 2.828427125 \qquad \text{Use a calculator.}$$

The lens is now an $f/2.8$ lens.

EXAMPLE 7   Rationalize the denominator: $\dfrac{\sqrt{5xy^2}}{\sqrt{xy^3}}$. Assume that x and y are positive numbers.

Solution   In each case, we write the expression as a quotient of radicals and then simplify the radicand.

Method 1

$$\frac{\sqrt{5xy^2}}{\sqrt{xy^3}} = \sqrt{\frac{5xy^2}{xy^3}}$$

$$= \sqrt{\frac{5}{y}}$$

$$= \frac{\sqrt{5}}{\sqrt{y}}$$

$$= \frac{\sqrt{5}\sqrt{y}}{\sqrt{y}\sqrt{y}}$$

$$= \frac{\sqrt{5y}}{y}$$

Method 2

$$\frac{\sqrt{5xy^2}}{\sqrt{xy^3}} = \sqrt{\frac{5xy^2}{xy^3}}$$

$$= \sqrt{\frac{5}{y}}$$

$$= \sqrt{\frac{5 \cdot y}{y \cdot y}}$$

$$= \frac{\sqrt{5y}}{\sqrt{y^2}}$$

$$= \frac{\sqrt{5y}}{y}$$

EXAMPLE 8   Rationalize the denominator. Assume $q > 0$.

$$\sqrt{\frac{11}{20q^5}}$$

Solution   We write the expression as a quotient of two radicals. Then we simplify the radical in the denominator before rationalizing.

$$\sqrt{\frac{11}{20q^5}} = \frac{\sqrt{11}}{\sqrt{20q^5}} \qquad \text{The square root of a quotient is the quotient of the square roots.}$$

$$= \frac{\sqrt{11}}{\sqrt{4q^4 \cdot 5q}} \qquad \text{To simplify } \sqrt{20q^5}, \text{ write it as } \sqrt{4q^4 \cdot 5q}.$$

$$= \frac{\sqrt{11}}{2q^2\sqrt{5q}} \qquad \sqrt{4q^4 \cdot 5q} = \sqrt{4q^4}\sqrt{5q} = 2q^2\sqrt{5q}.$$

INTERMEDIATE
Algebra *f(x)* **Now** ™

Self Check 7
Rationalize the denominator:

$$\frac{\sqrt{4ab^3}}{\sqrt{2a^2b^2}}$$

Answer $\dfrac{\sqrt{2ab}}{a}$

Self Check 8
Rationalize the denominator:

$$\sqrt[3]{\frac{1}{16h^4}}$$

$$= \frac{\sqrt{11}\sqrt{5q}}{2q^2\sqrt{5q}\sqrt{5q}} \quad \text{Rationalize the denominator.}$$

$$= \frac{\sqrt{55q}}{2q^2(5q)} \quad \text{Multiply the radicals: } \sqrt{5q}\sqrt{5q} = 5q.$$

Answer $\dfrac{\sqrt[3]{4h^2}}{4h^2}$

$$= \frac{\sqrt{55q}}{10q^3} \quad \text{Multiply in the denominator.}$$

INTERMEDIATE
Algebra *f(x)* **Now**™
Self Check 9
Rationalize the denominator:

$$\frac{\sqrt{5}}{\sqrt{17b}}$$

EXAMPLE 9   Rationalize the denominator: $\dfrac{\sqrt[3]{5}}{\sqrt[3]{9m}}$.

Solution   We multiply the numerator and the denominator by $\sqrt[3]{3m^2}$, which will produce a perfect cube, $27m^3$, under the radical sign in the denominator.

$$\frac{\sqrt[3]{5}}{\sqrt[3]{9m}} = \frac{\sqrt[3]{5}\sqrt[3]{3m^2}}{\sqrt[3]{9m}\sqrt[3]{3m^2}} \quad \text{Multiply numerator and denominator by } \sqrt[3]{3m^2}.$$

$$= \frac{\sqrt[3]{15m^2}}{\sqrt[3]{27m^3}} \quad \text{Multiply the radicals.}$$

Answer $\dfrac{\sqrt{85b}}{17b}$

$$= \frac{\sqrt[3]{15m^2}}{3m} \quad \text{Simplify: } \sqrt[3]{27m^3} = 3m.$$

Rationalizing two-term denominators

So far, we have rationalized denominators that had only one term. We will now discuss a method to rationalize denominators that have two terms.

One-term denominators	**Two-term denominators**
$\dfrac{\sqrt{5}}{\sqrt{3}}, \quad \dfrac{11}{\sqrt{20q^5}}, \quad \dfrac{4}{\sqrt[3]{2}}$	$\dfrac{1}{\sqrt{2}+1}, \quad \dfrac{\sqrt{x}+\sqrt{2}}{\sqrt{x}-\sqrt{2}}$

To rationalize a denominator of $\dfrac{1}{\sqrt{2}+1}$, for example, we multiply the numerator and denominator by $\sqrt{2}-1$, because the product $(\sqrt{2}+1)(\sqrt{2}-1)$ contains no radicals.

$$(\sqrt{2}+1)(\sqrt{2}-1) = (\sqrt{2})^2 - (1)^2 \quad \text{Use a special product formula.}$$
$$= 2 - 1$$
$$= 1$$

Radical expressions that involve the sum and difference of the same two terms, such as $\sqrt{2}+1$ and $\sqrt{2}-1$, are called **conjugates.**

INTERMEDIATE
Algebra *f(x)* **Now**™
Self Check 10
Rationalize the denominator:

$$\frac{2}{\sqrt{3}+1}$$

EXAMPLE 10   Rationalize the denominator: $\dfrac{1}{\sqrt{2}+1}$.

Solution   We multiply the numerator and denominator of the fraction by $\sqrt{2}-1$, which is the conjugate of the denominator.

EXAMPLE 1   Solve: $\sqrt{x + 3} = 4$.

Solution   To eliminate the radical, we apply the power rule by squaring both sides of the equation and proceed as follows:

$$\sqrt{x + 3} = 4$$
$$(\sqrt{x + 3})^2 = (4)^2 \quad \text{Square both sides.}$$
$$x + 3 = 16$$
$$x = 13 \quad \text{Subtract 3 from both sides.}$$

We must check the proposed solution 13 to see whether it satisfies the original equation.

Check: $\sqrt{x + 3} = 4$
$$\sqrt{13 + 3} \overset{?}{=} 4 \quad \text{Substitute 13 for } x.$$
$$\sqrt{16} \overset{?}{=} 4$$
$$4 = 4$$

Since 13 satisfies the original equation, it is the solution.

To solve an equation containing radicals, we follow these steps.

Solving an equation containing radicals

1. Isolate one radical expression on one side of the equation.
2. Raise both sides of the equation to the power that is the same as the index of the radical.
3. Solve the resulting equation. If it still contains a radical, go back to step 1.
4. Check the results to eliminate extraneous solutions.

EXAMPLE 2   **Amusement park rides.**   The distance d in feet that an object will fall in t seconds is given by the formula

$$t = \sqrt{\frac{d}{16}}$$

If the designers of the amusement park attraction shown in Figure 7-9 want the riders to experience 3 seconds of vertical free fall, what length of vertical drop is needed?

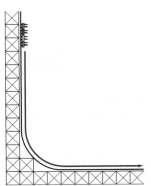

FIGURE 7-9

Solution   We substitute 3 for t in the formula and solve for d.

$$t = \sqrt{\dfrac{d}{16}}$$

$$3 = \sqrt{\dfrac{d}{16}} \qquad \text{Here the radical is isolated on the right-hand side.}$$

$$(3)^2 = \left(\sqrt{\dfrac{d}{16}}\right)^2 \qquad \text{Raise both sides to the second power.}$$

$$9 = \dfrac{d}{16} \qquad \text{Simplify.}$$

$$144 = d \qquad \text{Solve the resulting equation by multiplying both sides by 16.}$$

Answer 196 ft

The amount of vertical drop needs to be 144 feet.

Self Check 3
Solve: $\sqrt{4x + 1} + 1 = x$.

EXAMPLE 3   Solve: $\sqrt{3x + 1} + 1 = x$.

Solution   We first subtract 1 from both sides to isolate the radical. Then, to eliminate the radical, we square both sides of the equation and proceed as follows:

$$\sqrt{3x + 1} + 1 = x$$

$$\sqrt{3x + 1} = x - 1 \qquad \text{Subtract 1 from both sides.}$$

$$(\sqrt{3x + 1})^2 = (x - 1)^2 \qquad \text{Square both sides to eliminate the square root.}$$

$$3x + 1 = x^2 - 2x + 1 \qquad \begin{array}{l}\text{On the right-hand side, } (x - 1)^2 = (x - 1)(x - 1) = \\ x^2 - x - x + 1 = x^2 - 2x + 1.\end{array}$$

$$0 = x^2 - 5x \qquad \begin{array}{l}\text{Subtract } 3x \text{ and 1 from both sides. This is a} \\ \text{quadratic equation. Use factoring to solve it.}\end{array}$$

$$0 = x(x - 5) \qquad \text{Factor } x^2 - 5x.$$

$$x = 0 \quad \text{or} \quad x - 5 = 0 \qquad \text{Set each factor each to 0.}$$

$$x = 0 \quad \mid \quad x = 5$$

We must check each proposed solution to see whether it satisfies the original equation.

Check:
$$\sqrt{3x + 1} + 1 = x \qquad\qquad \sqrt{3x + 1} + 1 = x$$
$$\sqrt{3(0) + 1} + 1 \stackrel{?}{=} 0 \qquad\qquad \sqrt{3(5) + 1} + 1 \stackrel{?}{=} 5$$
$$\sqrt{1} + 1 \stackrel{?}{=} 0 \qquad\qquad \sqrt{16} + 1 \stackrel{?}{=} 5$$
$$2 \neq 0 \qquad\qquad\qquad 5 = 5$$

Answer 6, 0 is extraneous

Since 0 does not check, it must be discarded. The only solution of the original equation is 5.

CALCULATOR SNAPSHOT   **Solving equations containing radicals**

To find approximate solutions for $\sqrt{3x + 1} + 1 = x$ with a graphing calculator, we use window settings of $[-5, 10]$ for x and $[-2, 8]$ for y and graph the functions $f(x) = \sqrt{3x + 1} + 1$ and $g(x) = x$. We then use the INTERSECT feature to approximate the point of intersection of the graphs. See Figure 7-10. The intersection point of $(5, 5)$ implies that 5 is a solution of the radical equation. Check this result.

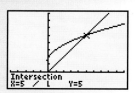

FIGURE 7-10

INTERMEDIATE
Algebra $f(x)$ **Now**™

EXAMPLE 4   Solve: $\sqrt[3]{x^3 + 7} = x + 1$.

Solution   To eliminate the radical, we cube both sides of the equation and proceed as follows:

$$\sqrt[3]{x^3 + 7} = x + 1$$
$$(\sqrt[3]{x^3 + 7})^3 = (x + 1)^3 \qquad \text{Cube both sides to eliminate the cube root.}$$
$$x^3 + 7 = x^3 + 3x^2 + 3x + 1 \quad (x + 1)^3 = (x + 1)(x + 1)(x + 1).$$
$$0 = 3x^2 + 3x - 6 \qquad \text{Subtract } x^3 \text{ and 7 from both sides.}$$
$$0 = x^2 + x - 2 \qquad \text{Divide both sides by 3. To solve this quadratic equation, use factoring.}$$
$$0 = (x + 2)(x - 1) \qquad \text{Factor the trinomial.}$$
$$x + 2 = 0 \qquad \text{or} \qquad x - 1 = 0$$
$$x = -2 \qquad \qquad x = 1$$

We check each proposed solution to see whether it satisfies the original equation.

Check:
$$\sqrt[3]{x^3 + 7} = x + 1 \qquad\qquad \sqrt[3]{x^3 + 7} = x + 1$$
$$\sqrt[3]{(-2)^3 + 7} \stackrel{?}{=} -2 + 1 \qquad\qquad \sqrt[3]{1^3 + 7} \stackrel{?}{=} 1 + 1$$
$$\sqrt[3]{-8 + 7} \stackrel{?}{=} -1 \qquad\qquad \sqrt[3]{1 + 7} \stackrel{?}{=} 2$$
$$\sqrt[3]{-1} \stackrel{?}{=} -1 \qquad\qquad \sqrt[3]{8} \stackrel{?}{=} 2$$
$$-1 = -1 \qquad\qquad 2 = 2$$

Both -2 and 1 are solutions of the original equation.

Self Check 4
Solve: $\sqrt[3]{x^3 + 8} = x + 2$.

Answer 0, -2

▮ Equations containing two radicals

EXAMPLE 5   Solve: $\sqrt{5x + 9} = 2\sqrt{3x + 4}$.

Solution   Each radical is isolated on one side of the equation, so we square both sides to eliminate them.

$$\sqrt{5x + 9} = 2\sqrt{3x + 4}$$
$$(\sqrt{5x + 9})^2 = (2\sqrt{3x + 4})^2 \qquad \text{Square both sides.}$$
$$5x + 9 = 4(3x + 4) \qquad \text{On the right-hand side:} \\ (2\sqrt{3x + 4})^2 = 2^2(\sqrt{3x + 4})^2 = 4(3x + 4).$$
$$5x + 9 = 12x + 16 \qquad \text{Distribute the multiplication by 4.}$$
$$-7 = 7x \qquad \text{Subtract } 5x \text{ and 16 from both sides.}$$
$$-1 = x \qquad \text{Divide both sides by 7.}$$

We check the proposed solution by substituting -1 for x in the original equation.

$$\sqrt{5x + 9} = 2\sqrt{3x + 4}$$
$$\sqrt{5(-1) + 9} \stackrel{?}{=} 2\sqrt{3(-1) + 4} \qquad \text{Substitute } -1 \text{ for } x.$$
$$\sqrt{4} \stackrel{?}{=} 2\sqrt{1}$$
$$2 = 2$$

The solution is -1.

INTERMEDIATE
Algebra $f(x)$ **Now**™
Self Check 5
Solve: $\sqrt{x - 4} = 2\sqrt{x - 16}$.

Answer 20

When more than one radical appears in an equation, it is often necessary to apply the power rule more than once.

Self Check 6

Solve: $\sqrt{a} + \sqrt{a+3} = 3$

EXAMPLE 6   Solve: $\sqrt{x} + \sqrt{x+2} = 2$.

Solution   To remove the radicals, we must square both sides of the equation. This is easier to do if one radical is on each side of the equation. So we subtract $\sqrt{x}$ from both sides to isolate $\sqrt{x+2}$ on the left-hand side of the equation.

$$\sqrt{x} + \sqrt{x+2} = 2$$

$$\sqrt{x+2} = 2 - \sqrt{x} \qquad \text{Subtract } \sqrt{x} \text{ from both sides.}$$

$$(\sqrt{x+2})^2 = (2 - \sqrt{x})^2 \qquad \text{Square both sides to eliminate the square root.}$$

$$x + 2 = 4 - 2\sqrt{x} - 2\sqrt{x} + x \qquad \begin{array}{l}(2 - \sqrt{x})^2 = (2 - \sqrt{x})(2 - \sqrt{x}) = \\ 4 - 2\sqrt{x} - 2\sqrt{x} + x.\end{array}$$

$$x + 2 = 4 - 4\sqrt{x} + x \qquad \begin{array}{l}\text{Combine like radicals:} \\ -2\sqrt{x} - 2\sqrt{x} = -4\sqrt{x}.\end{array}$$

$$2 = 4 - 4\sqrt{x} \qquad \text{Subtract } x \text{ from both sides.}$$

$$-2 = -4\sqrt{x} \qquad \text{Subtract 4 from both sides.}$$

$$\frac{1}{2} = \sqrt{x} \qquad \begin{array}{l}\text{Divide both sides by } -4 \text{ and} \\ \text{simplify.}\end{array}$$

$$\frac{1}{4} = x \qquad \text{Square both sides.}$$

Check: $\sqrt{x} + \sqrt{x+2} = 2$

$$\sqrt{\frac{1}{4}} + \sqrt{\frac{1}{4} + 2} \overset{?}{=} 2$$

$$\frac{1}{2} + \sqrt{\frac{9}{4}} \overset{?}{=} 2$$

$$\frac{1}{2} + \frac{3}{2} \overset{?}{=} 2$$

$$2 = 2$$

Answer 1

The solution is $\frac{1}{4}$.

CALCULATOR SNAPSHOT   **Solving equations containing radicals**

To find approximate solutions for $\sqrt{x} + \sqrt{x+2} = 5$ (an equation similar to that in Example 6) with a graphing calculator, we can use window settings of $[-2, 10]$ for x and $[-2, 8]$ for y and graph the functions $f(x) = \sqrt{x} + \sqrt{x+2}$ and $g(x) = 5$.

Figure 7-11 shows that the INTERSECT feature gives the approximate coordinates of the point of intersection of the two graphs as (5.29, 5). Therefore, an approximate solution of the radical equation is 5.29. Check its reasonableness.

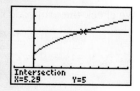

FIGURE 7-11

Solving formulas containing radicals

To *solve a formula for a variable* means to isolate that variable on one side of the equation, with all other quantities on the other side.

EXAMPLE 7 A piece of office equipment that is now worth V dollars originally cost C dollars 3 years ago. The rate r at which it has depreciated (lost value) is given by

$$r = 1 - \sqrt[3]{\frac{V}{C}}$$

Solve the formula for C.

Solution We begin by isolating the cube root on the right-hand side of the equation.

$$r = 1 - \sqrt[3]{\frac{V}{C}}$$

$$r - 1 = -\sqrt[3]{\frac{V}{C}} \qquad \text{Subtract 1 from both sides.}$$

$$(r - 1)^3 = \left[-\sqrt[3]{\frac{V}{C}}\right]^3 \qquad \text{Cube both sides.}$$

$$(r - 1)^3 = -\frac{V}{C} \qquad \text{Simplify the right-hand side.}$$

$$C(r - 1)^3 = -V \qquad \text{Multiply both sides by } C.$$

$$C = -\frac{V}{(r - 1)^3} \qquad \text{Divide both sides by } (r - 1)^3.$$

Self Check 7
A formula used in statistics to determine the necessary size of a sample to obtain the desired degree of accuracy is

$$e = z_0\sqrt{\frac{pq}{n}}$$

Solve the formula for n.

Answer $n = \dfrac{z_0^2 pq}{e^2}$

Section 7.4 STUDY SET

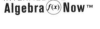

VOCABULARY *Fill in the blanks.*

1. Equations such as $\sqrt{x + 4} - 4 = 5$ and $\sqrt[3]{x + 1} = 12$ are called _____ equations.

2. When solving equations containing radicals, try to _____ one radical expression on one side of the equation.

3. Squaring both sides of an equation can introduce _____ solutions.

4. To _____ a proposed solution means to substitute it into the original equation and see whether a true statement results.

CONCEPTS

5. What is the first step in solving each equation?
   a. $\sqrt{x + 4} = 5$
   b. $\sqrt[3]{x + 4} = 2$

6. Fill in the blank to make a true statement.
   $\sqrt{x}$ represents the number that, when squared, gives ▨.

7. Simplify each expression.
   a. $(\sqrt{x})^2$ b. $(\sqrt{x - 5})^2$
   c. $(4\sqrt{2x})^2$ d. $(-\sqrt{x + 3})^2$

8. Simplify each expression.
   a. $(\sqrt[3]{x})^3$ b. $(\sqrt[4]{x})^4$
   c. $(-\sqrt[3]{2x})^3$ d. $(2\sqrt[3]{x} + 3)^3$

9. What is wrong with the work shown below?

$$\sqrt{x + 1} - 3 = 8$$
$$\sqrt{x + 1} = 11$$
$$(\sqrt{x + 1})^2 = 11$$
$$x + 1 = 11$$
$$x = 10$$

10. Solve $\sqrt{x-2}+2=4$ graphically, using the graphs below.

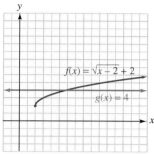

11. Use your own words to restate the power rule: If x, y, and n represent real numbers and $x=y$, then $x^n=y^n$.

12. The first step of a student's solution is shown below. What is a better way to begin the solution?

$$\text{Solve:}\quad \sqrt{x}+\sqrt{x+22}=12.$$
$$(\sqrt{x}+\sqrt{x+22})^2=12^2$$

NOTATION   *Complete each solution to solve the equation.*

13. $2\sqrt{x-2}=4$

$($ 　　　　$)^2=4^2$

　　$(x-2)=$ 　

$4x-$ 　$=16$

$4x=$ 　

$x=6$

14. $\sqrt{1-2x}=\sqrt{x+10}$

$($ 　　　　$)^2=(\sqrt{x+10})^2$

　　　　$=x+10$

　　　　$=9$

$x=-3$

PRACTICE   *Solve each equation and cross out all extraneous solutions.*

15. $\sqrt{5x-6}=2$ **16.** $\sqrt{7x-10}=12$

17. $\sqrt{6x+1}+2=7$ **18.** $\sqrt{6x+13}-2=5$

19. $2\sqrt{4x+1}=\sqrt{x+4}$

20. $\sqrt{3(x+4)}=\sqrt{5x-12}$

21. $\sqrt[3]{7n-1}=3$ **22.** $\sqrt[3]{12m+4}=4$

23. $\sqrt[4]{10p+1}=\sqrt[4]{11p-7}$

24. $\sqrt[4]{10y+6}=2\sqrt[4]{y}$

25. $x=\dfrac{\sqrt{12x-5}}{2}$ **26.** $x=\dfrac{\sqrt{16x-12}}{2}$

27. $\sqrt{x+2}-\sqrt{4-x}=0$

28. $\sqrt{6-x}-\sqrt{2x+3}=0$

29. $2\sqrt{x}=\sqrt{5x-16}$ **30.** $3\sqrt{x}=\sqrt{3x+54}$

31. $r-9=\sqrt{2r-3}$ **32.** $-s-3=2\sqrt{5-s}$

33. $\sqrt{-5x+24}=6-x$ **34.** $\sqrt{-x+2}=x-2$

35. $\sqrt{y+2}=4-y$ **36.** $\sqrt{22y+86}=y+9$

37. $\sqrt[3]{x^3-7}=x-1$ **38.** $\sqrt[3]{x^3+56}-2=x$

39. $\sqrt[4]{x^4+4x^2-4}=-x$ **40.** $\sqrt[4]{8x-8}+2=0$

41. $\sqrt[4]{12t+4}+2=0$ **42.** $u=\sqrt[4]{u^4-6u^2+24}$

43. $\sqrt{2y+1}=1-2\sqrt{y}$ **44.** $\sqrt{u}+3=\sqrt{u-3}$

45. $\sqrt{y+7}+3=\sqrt{y+4}$

46. $1+\sqrt{z}=\sqrt{z+3}$

47. $2+\sqrt{u}=\sqrt{2u+7}$

48. $5r+4=\sqrt{5r+20}+4r$

49. $\sqrt{6t+1}-3\sqrt{t}=-1$

50. $\sqrt{4s+1}-\sqrt{6s}=-1$

51. $\sqrt{2x+5}+\sqrt{x+2}=5$

52. $\sqrt{2x+5}+\sqrt{2x+1}+4=0$

53. $\sqrt{x-5}-\sqrt{x+3}=4$

54. $\sqrt{x+8}-\sqrt{x-4}=-2$

Solve each equation for the indicated variable.

55. $v=\sqrt{2gh}$ for h

56. $d=1.4\sqrt{h}$ for h

57. $T=2\pi\sqrt{\dfrac{l}{32}}$ for l

58. $d=\sqrt[3]{\dfrac{12V}{\pi}}$ for V

59. $r=\sqrt[3]{\dfrac{A}{P}}-1$ for A

60. $r=\sqrt[3]{\dfrac{A}{P}}-1$ for P

61. $L_A=L_B\sqrt{1-\dfrac{v^2}{c^2}}$ for v^2

62. $R_1=\sqrt{\dfrac{A}{\pi}-R_2^2}$ for A

APPLICATIONS

63. HIGHWAY DESIGNS A curved concrete road will accommodate traffic traveling s mph if the radius of the curve is r feet, according to the formula $s = 3\sqrt{r}$. If engineers expect 40-mph traffic, what radius should they specify? Give the result to the nearest foot.

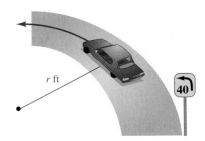

64. FORESTRY The higher a lookout tower is built, the farther an observer can see. That distance d (called the *horizon distance,* measured in miles) is related to the height h of the observer (measured in feet) by the formula $d = 1.4\sqrt{h}$. How tall must a lookout tower be to see the edge of the forest, 25 miles away? (Round to the nearest foot.)

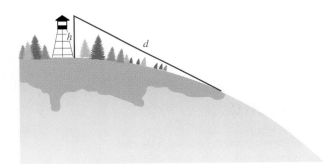

65. WIND POWER The power generated by a certain windmill is related to the velocity of the wind by the formula $v = \sqrt[3]{\dfrac{P}{0.02}}$ where P is the power (in watts) and v is the velocity of the wind (in mph). Find how much power the windmill is generating when the wind is 29 mph.

66. DIAMONDS The *effective rate of interest r* earned by an investment is given by the formula

$$r = \sqrt[n]{\dfrac{A}{P}} - 1$$

where P is the initial investment that grows to value A after n years. If a diamond buyer got \$4,000 for a 1.73-carat diamond that he had purchased 4 years earlier, and earned an annual rate of return of 6.5% on the investment, what did he originally pay for the diamond?

67. THEATER PRODUCTIONS The ropes, pulleys, and sandbags shown are part of a mechanical system used to raise and lower scenery for a stage play. For the scenery to be in the proper position, the following formula must apply:

$$w_2 = \sqrt{w_1^2 + w_3^2}$$

If $w_2 = 12.5$ lb and $w_3 = 7.5$ lb, find w_1.

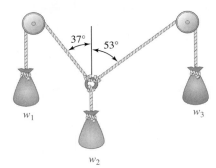

68. CARPENTRY During construction, carpenters often brace walls as shown, where the length of the brace is given by the formula

$$\ell = \sqrt{f^2 + h^2}$$

If a carpenter nails a 10-ft brace to the wall 6 feet above the floor, how far from the base of the wall should he nail the brace to the floor?

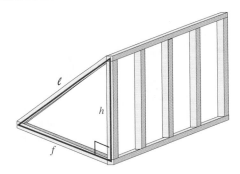

69. SUPPLY AND DEMAND The number of wrenches that will be produced at a given price can be predicted by the formula $s = \sqrt{5x}$, where s is the supply (in thousands) and x is the price (in dollars). The demand d for wrenches can be predicted by the formula $d = \sqrt{100 - 3x^2}$. Find the equilibrium price—that is, find the price at which supply will equal demand.

70. SUPPLY AND DEMAND The number of mirrors that will be produced at a given price can be predicted by the formula $s = \sqrt{23x}$, where s is the supply (in thousands) and x is the price (in dollars). The demand d for mirrors can be predicted by the formula $d = \sqrt{312 - 2x^2}$. Find the equilibrium price—that is, find the price at which supply will equal demand.

WRITING

71. If both sides of an equation are raised to the same power, the resulting equation might not be equivalent to the original equation. Explain.

72. Explain how the radical equation $\sqrt{2x - 1} = x$ can be solved graphically.

73. Explain how the table on the right can be used to solve $\sqrt{4x - 3} - 2 = \sqrt{2x - 5}$ if $Y_1 = \sqrt{4x - 3} - 2$ and $Y_2 = \sqrt{2x - 5}$.

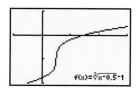

74. Explain how to use the graph shown on the right to approximate the solution of $\sqrt[3]{x - 0.5} = 1$.

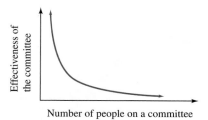

REVIEW

75. LIGHTING The intensity of the light reaching you from a light bulb varies inversely as the square of your distance from the bulb. If you are 5 feet away from a light bulb and the intensity is 40 foot-candles, what will the intensity be if you move 20 feet away from the bulb?

76. COMMITTEES What type of variation is shown below? As the number of people on this committee increased, what happened to its effectiveness?

77. TYPESETTING If 12-point type is 0.166044 inch tall, how tall is 30-point type?

78. GUITAR STRINGS The frequency of vibration of a string varies directly as the square root of the tension and inversely as the length of the string. Suppose a string 2.5 feet long, under a tension of 16 pounds, vibrates 25 times per second. Find k, the constant of variation.

7.5 Rational Exponents

- Rational exponents • Exponential expressions with variables in their bases
- Rational exponents with numerators other than 1 • Negative rational exponents
- Applying the rules for exponents • Simplifying radical expressions

We have worked with exponential expressions containing natural-number exponents, such as 5^3 and x^2. In Chapter 5, the definition of exponent was extended to include zero and negative integers, which gave meaning to expressions such as 8^{-3} and $(-9xy)^0$. In this section, we will again extend the definition of exponent — this time to include rational (fractional) exponents. We will see how expressions such as $9^{1/2}$, $\left(\frac{1}{16}\right)^{3/4}$, and $(-32x^5)^{-2/5}$ can be simplified by writing them in an equivalent radical form or by using the rules for exponents.

Rational exponents

We have seen that positive-integer exponents indicate the number of times that a base is to be used as a factor in a product. For example, x^5 means that x is to be used as a factor five times.

$$x^5 = \overbrace{x \cdot x \cdot x \cdot x \cdot x}^{5 \text{ factors of } x}$$

Furthermore, we recall the following rules for exponents.

Rules for exponents

If there are no divisions by 0, then for all integers m and n,

1. $x^m x^n = x^{m+n}$ **2.** $(x^m)^n = x^{mn}$ **3.** $(xy)^n = x^n y^n$ **4.** $\left(\dfrac{x}{y}\right)^n = \dfrac{x^n}{y^n}$

5. $x^0 = 1$ $(x \neq 0)$ **6.** $x^{-n} = \dfrac{1}{x^n}$ **7.** $\dfrac{x^m}{x^n} = x^{m-n}$ **8.** $\left(\dfrac{x}{y}\right)^{-n} = \left(\dfrac{y}{x}\right)^n$

It is possible to raise many bases to fractional powers. Since we want fractional exponents to obey the same rules as integer exponents, the square of $10^{1/2}$ must be 10, because

$$(10^{1/2})^2 = 10^{(1/2)2}$$ Keep the base and multiply the exponents.

$$= 10^1$$ $\frac{1}{2} \cdot 2 = 1$.

$$= 10$$ $10^1 = 10$

However, we have seen that

$$(\sqrt{10})^2 = 10$$

Since $(10^{1/2})^2$ and $(\sqrt{10})^2$ both equal 10, we define $10^{1/2}$ to be $\sqrt{10}$. Likewise, we define

$$10^{1/3} \text{ to be } \sqrt[3]{10} \quad \text{and} \quad 10^{1/4} \text{ to be } \sqrt[4]{10}$$

Rational exponents

If n represents a natural number greater than 1 and $\sqrt[n]{x}$ represents a real number, then

$$x^{1/n} = \sqrt[n]{x}$$

In words, *a rational exponent of $\frac{1}{n}$ indicates that the nth root of the base should be found.*

Using this definition, we can simplify the exponential expression $8^{1/3}$. The first step is to write it as an equivalent expression in radical form and proceed as follows:

$$8^{1/3} = \sqrt[3]{8}$$ The base of the exponential expression, 8, is the radicand. The denominator of the fractional exponent, 3, is the index of the radical.

$$= 2$$

EXAMPLE 1   Write each expression in radical form and simplify, if possible.

a. $9^{1/2}$,   **b.** $-\left(\dfrac{16}{9}\right)^{1/2}$,   **c.** $(-64)^{1/3}$,   **d.** $16^{1/4}$,   **e.** $\left(\dfrac{1}{32}\right)^{1/5}$,   **f.** $0^{1/8}$,   **g.** $y^{1/4}$,

and   **h.** $-(2x^2)^{1/5}$.

Solution

a. $9^{1/2} = \sqrt{9}$
$$= 3$$

b. $-\left(\dfrac{16}{9}\right)^{1/2} = -\sqrt{\dfrac{16}{9}}$
$$= -\dfrac{4}{3}$$

c. $(-64)^{1/3} = \sqrt[3]{-64}$
$$= -4$$

d. $16^{1/4} = \sqrt[4]{16}$
$$= 2$$

Self Check 1
Write each expression in radical form and simplify, if possible:

a. $16^{1/2}$

b. $\left(-\dfrac{27}{8}\right)^{1/3}$

c. $-(6x^3)^{1/4}$

e. $\left(\dfrac{1}{32}\right)^{1/5} = \sqrt[5]{\dfrac{1}{32}}$

$= \dfrac{1}{2}$

f. $0^{1/8} = \sqrt[8]{0}$

$= 0$

g. $y^{1/4} = \sqrt[4]{y}$

h. $-(2x^2)^{1/5} = -\sqrt[5]{2x^2}$

Answers **a.** 4, **b.** $-\dfrac{3}{2}$,

c. $-\sqrt[4]{6x^3}$

Self Check 2
Write the radical with a
fractional exponent: $\sqrt[6]{7ab}$.

Answer $(7ab)^{1/6}$

EXAMPLE 2   Write $\sqrt{5xyz}$ as an exponential expression with a rational exponent.

Solution   The radicand is $5xyz$, so the base of the exponential expression is $5xyz$. The index of the radical is an understood 2, so the denominator of the fractional exponent is 2.

$$\sqrt{5xyz} = (5xyz)^{1/2}$$

Rational exponents appear in formulas used in many disciplines, such as science and engineering.

EXAMPLE 3   **Satellites.**   See Figure 7-12. The formula

$$r = \left(\dfrac{GMP^2}{4\pi^2}\right)^{1/3}$$

gives the orbital radius (in meters) of a satellite circling Earth, where G and M are constants and P is the time in seconds for the satellite to make one complete revolution. Write the formula using a radical.

Solution   The fractional exponent $\frac{1}{3}$ has a denominator of 3, which indicates that we are to find the cube root of the base of the exponential expression. So we have

$$r = \sqrt[3]{\dfrac{GMP^2}{4\pi^2}}$$

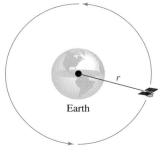

Earth

FIGURE 7-12

Exponential expressions with variables in their bases

As with radicals, when n represents an *odd natural number* in the expression $x^{1/n}$ where $n > 1$, there is exactly one real nth root, and we don't have to worry about absolute value symbols.

When n represents an *even natural number,* there are two nth roots. Since we want the expression $x^{1/n}$ to represent the positive nth root, we must often use absolute value symbols to guarantee that the simplified result is positive. Thus, if n is even,

$$(x^n)^{1/n} = |x|$$

When n is even and x is negative, the expression $x^{1/n}$ is not a real number.

EXAMPLE 4 Simplify each exponential expression. Assume that the variables can be any real number. **a.** $(-27x^3)^{1/3}$, **b.** $(256a^8)^{1/8}$, **c.** $[(y+4)^2]^{1/2}$, **d.** $(25b^4)^{1/2}$, and **e.** $(-256x^4)^{1/4}$.

Self Check 4
Simplify each expression:
a. $(625a^4)^{1/4}$
b. $(b^4)^{1/2}$

Solution

a. $(-27x^3)^{1/3} = -3x$ Because $(-3x)^3 = -27x^3$. Since n is odd, no absolute value symbols are needed.

b. $(256a^8)^{1/8} = 2|a|$ Because $(2|a|)^8 = 256a^8$. Since n is even and a can be any real number, $2a$ can be negative. Thus, absolute value symbols are needed.

c. $[(y+4)^2]^{1/2} = |y+4|$ Because $|y+4|^2 = (y+4)^2$. Since n is even and y can be any real number, $y + 4$ can be negative. Thus, absolute value symbols are needed.

d. $(25b^4)^{1/2} = 5b^2$ Because $(5b^2)^2 = 25b^4$. Since $b^2 \geq 0$, no absolute value symbols are needed.

e. $(-256x^4)^{1/4}$ is not a real number.   Because no real number raised to the 4th power is $-256x^4$.

Answers **a.** $5|a|$, **b.** b^2

If we are told that the variables represent positive real numbers in parts b and c of Example 4, the absolute value symbols in the answers are not needed.

$$(256a^8)^{1/8} = 2a \qquad \text{If } a \text{ represents a positive number, then } 2a \text{ is positive.}$$
$$[y+4)^2]^{1/2} = y + 4 \qquad \text{If } y \text{ represents a positive number, then } y + 4 \text{ is positive.}$$

We summarize the cases as follows.

Summary of the definitions of $x^{1/n}$

If n represents a natural number greater than 1 and x represents a real number,

If $x > 0$, then $x^{1/n}$ is the positive number such that $(x^{1/n})^n = x$.

If $x = 0$, then $x^{1/n} = 0$.

If $x < 0$ $\begin{cases} \text{and } n \text{ is odd, then } x^{1/n} \text{ is the real number such that } (x^{1/n})^n = x. \\ \text{and } n \text{ is even, then } x^{1/n} \text{ is not a real number.} \end{cases}$

▮ Rational exponents with numerators other than 1

We can extend the definition of $x^{1/n}$ to include fractional exponents with numerators other than 1. For example, since $8^{2/3}$ can be written as $(8^{1/3})^2$, we have

$$8^{2/3} = (8^{1/3})^2$$
$$= (\sqrt[3]{8})^2 \qquad \text{Write } 8^{1/3} \text{ in radical form.}$$
$$= 2^2 \qquad \text{Find the cube root first: } \sqrt[3]{8} = 2.$$
$$= 4 \qquad \text{Then find the power.}$$

Thus, we can simplify $8^{2/3}$ by finding the second power of the cube root of 8.

The numerator of the rational
exponent is the power.

$$8^{2/3} = (\sqrt[3]{8})^2$$ The base of the exponential expression is the radicand.

The denominator of the exponent
is the index of the radical.

We can also simplify $8^{2/3}$ by taking the cube root of 8 squared.

$$8^{2/3} = (8^2)^{1/3}$$

$\qquad = 64^{1/3}$ Find the power first: $8^2 = 64$.

$\qquad = \sqrt[3]{64}$ Write $64^{1/3}$ in radical form.

$\qquad = 4$ Now find the cube root.

In general, we have the following rule.

Changing from rational exponents to radicals

If m and n represent positive integers where $n \neq 1$ and $\sqrt[n]{x}$ represents a real number, then

$$x^{m/n} = \sqrt[n]{x^m} = \left(\sqrt[n]{x}\right)^m$$

Because of the previous definition, we can interpret $x^{m/n}$ in two ways:

1. $x^{m/n}$ means the nth root of the mth power of x.
2. $x^{m/n}$ means the mth power of the nth root of x.

Self Check 5
Simplify:
a. $16^{3/2}$

b. $(-27x^6)^{2/3}$

EXAMPLE 5   Simplify each expression in two ways: **a.** $9^{3/2}$,   **b.** $\left(\dfrac{1}{16}\right)^{3/4}$,   and   **c.** $(-8x^3)^{4/3}$.

Solution
a. $9^{3/2} = (\sqrt{9})^3$ or $9^{3/2} = \sqrt{9^3}$

$\qquad = 3^3 \qquad\qquad\qquad\quad = \sqrt{729}$

$\qquad = 27 \qquad\qquad\qquad\quad = 27$

b. $\left(\dfrac{1}{16}\right)^{3/4} = \left(\sqrt[4]{\dfrac{1}{16}}\right)^3$ or $\left(\dfrac{1}{16}\right)^{3/4} = \left[\left(\dfrac{1}{16}\right)^3\right]^{1/4}$

$\qquad\qquad = \left(\dfrac{1}{2}\right)^3 \qquad\qquad\qquad\qquad = \left(\dfrac{1}{4,096}\right)^{1/4}$

$\qquad\qquad = \dfrac{1}{8} \qquad\qquad\qquad\qquad\quad = \dfrac{1}{8}$

c. $(-8x^3)^{4/3} = (\sqrt[3]{-8x^3})^4$ or $(-8x^3)^{4/3} = \sqrt[3]{(-8x^3)^4}$

$\qquad\qquad = (-2x)^4 \qquad\qquad\qquad\qquad = \sqrt[3]{4,096x^{12}}$

$\qquad\qquad = 16x^4 \qquad\qquad\qquad\qquad\; = 16x^4$

Answers **a.** 64, **b.** $9x^4$

To avoid large numbers, it is usually better to find the root of the base first, as shown with the first solution of each part in Example 5.

Rational exponents

We can evaluate exponential expressions containing rational exponents using the exponential key $\boxed{y^x}$ or $\boxed{x^y}$ on a scientific calculator. For example, to evaluate $10^{2/3}$, we enter these numbers and press these keys:

$10 \boxed{y^x} \boxed{(} 2 \boxed{\div} 3 \boxed{)} \boxed{=}$ $\boxed{4.641588834}$

Note that parentheses were used when entering the power. Without them, the calculator would interpret the entry as $10^2 \div 3$.

To evaluate the exponential expression using a graphing calculator, we use the $\boxed{\wedge}$ key, which raises a base to a power. Again, we use parentheses when entering the power.

$10 \boxed{\wedge} \boxed{(} 2 \boxed{\div} 3 \boxed{)} \boxed{\text{ENTER}}$ $\boxed{\begin{array}{l} 10\wedge(2/3) \\ \qquad 4.641588834 \end{array}}$

To the nearest hundredth, $10^{2/3} \approx 4.64$.

Negative rational exponents

To be consistent with the definition of negative-integer exponents, we define $x^{-m/n}$ as follows.

Definition of $x^{-m/n}$

If m and n represent positive integers, $\frac{m}{n}$ is in simplified form, and $x^{1/n}$ represents a real number, then

$$x^{-m/n} = \frac{1}{x^{m/n}} \quad \text{and} \quad \frac{1}{x^{-m/n}} = x^{m/n} \quad \text{where } x \neq 0$$

EXAMPLE 6 Write each expression without using negative exponents and simplify, if possible: **a.** $64^{-1/2}$, **b.** $(-16)^{-3/4}$, **c.** $(-32x^5)^{-2/5}$, and **d.** $\dfrac{1}{16^{-3/2}}$.

a. $64^{-1/2} = \dfrac{1}{64^{1/2}}$

$= \dfrac{1}{\sqrt{64}}$

$= \dfrac{1}{8}$

b. $(-16)^{-3/4}$ is not a real number, because $(-16)^{1/4}$ is not a real number.

c. $(-32x^5)^{-2/5} = \dfrac{1}{(-32x^5)^{2/5}}$

$= \dfrac{1}{[(-32x^5)^{1/5}]^2}$

$= \dfrac{1}{(\sqrt[5]{-32x^5})^2}$

$= \dfrac{1}{(-2x)^2}$

$= \dfrac{1}{4x^2}$

d. $\dfrac{1}{16^{-3/2}} = 16^{3/2}$

$= (16^{1/2})^3$

$= (\sqrt{16})^3$

$= 4^3$

$= 64$

Self Check 6
Write without using negative exponents and simplify:
a. $16^{-1/4}$
b. $(-27a^3)^{-2/3}$

Answers **a.** $\dfrac{1}{2}$, **b.** $\dfrac{1}{9a^2}$

> **! COMMENT** By definition, 0^0 is undefined. A base of 0 raised to a negative power is also undefined. For example, 0^{-2} would equal $\frac{1}{0^2}$, which is undefined because we cannot divide by 0.

■ Applying the rules for exponents

We can use the rules for exponents to simplify many expressions with fractional exponents. If all variables represent positive numbers, no absolute value symbols are necessary.

INTERMEDIATE
Algebra $f(x)$ **Now**™

Self Check 7
Simplify:

a. $(x^{1/3}y^{3/2})^6$

b. $\dfrac{x^{5/3}x^{2/3}}{x^{1/3}}$

EXAMPLE 7 Assume that all variables represent positive numbers. Write all answers without using negative exponents. **a.** $5^{2/7}5^{3/7}$, **b.** $(5^{2/7})^3$, **c.** $(a^{2/3}b^{1/2})^6$, and **d.** $\dfrac{a^{8/3}a^{1/3}}{a^2}$.

Solution

a. $5^{2/7}5^{3/7} = 5^{2/7+3/7}$ Use the rule $x^m x^n = x^{m+n}$.

$\qquad\qquad = 5^{5/7}$ Add: $\frac{2}{7} + \frac{3}{7} = \frac{5}{7}$.

b. $(5^{2/7})^3 = 5^{(2/7)(3)}$ Use the rule $(x^m)^n = x^{mn}$.

$\qquad\quad = 5^{6/7}$ Multiply: $\frac{2}{7}(3) = \frac{6}{7}$.

c. $(a^{2/3}b^{1/2})^6 = (a^{2/3})^6(b^{1/2})^6$ Use the rule $(xy)^n = x^n y^n$.

$\qquad\qquad\quad = a^{12/3}b^{6/2}$ Use the rule $(x^m)^n = x^{mn}$ twice.

$\qquad\qquad\quad = a^4 b^3$ Simplify the exponents.

d. $\dfrac{a^{8/3}a^{1/3}}{a^2} = a^{8/3+1/3-2}$ Use the rules $x^m x^n = x^{m+n}$ and $\frac{x^m}{x^n} = x^{m-n}$.

$\qquad\qquad = a^{8/3+1/3-6/3}$ $2 = \frac{6}{3}$.

$\qquad\qquad = a^{3/3}$ $\frac{8}{3} + \frac{1}{3} - \frac{6}{3} = \frac{3}{3}$.

$\qquad\qquad = a$ $\frac{3}{3} = 1$.

Answers a. $x^2 y^9$, **b.** x^2

INTERMEDIATE
Algebra $f(x)$ **Now**™

EXAMPLE 8 Assume that all variables represent positive numbers, and perform the operations. Write all answers without negative exponents. **a.** $a^{4/5}(a^{1/5} + a^{3/5})$ and **b.** $x^{1/2}(x^{-1/2} + x^{1/2})$.

Solution

a. $a^{4/5}(a^{1/5} + a^{3/5}) = a^{4/5}a^{1/5} + a^{4/5}a^{3/5}$ Use the distributive property.

$\qquad\qquad\qquad\quad = a^{4/5+1/5} + a^{4/5+3/5}$ Use the rule $x^m x^n = x^{m+n}$.

$\qquad\qquad\qquad\quad = a^{5/5} + a^{7/5}$ Simplify the exponents.

$\qquad\qquad\qquad\quad = a + a^{7/5}$ We cannot add these terms because they are not like terms.

b. $x^{1/2}(x^{-1/2} + x^{1/2}) = x^{1/2}x^{-1/2} + x^{1/2}x^{1/2}$ Use the distributive property.

$\qquad\qquad\qquad\quad = x^{1/2+(-1/2)} + x^{1/2+1/2}$ Use the rule $x^m x^n = x^{m+n}$.

$\qquad\qquad\qquad\quad = x^0 + x^1$ Simplify each exponent.

$\qquad\qquad\qquad\quad = 1 + x$ $x^0 = 1$.

Simplifying radical expressions

We can simplify many radical expressions by using the following steps.

> **Using rational exponents to simplify radicals**
> 1. Change the radical expression into an exponential expression.
> 2. Simplify the rational exponents.
> 3. Change the exponential expression back into a radical.

INTERMEDIATE
Algebra *f(x)* **Now**™

EXAMPLE 9 Simplify: **a.** $\sqrt[4]{3^2}$, **b.** $\sqrt[8]{x^6}$, and **c.** $\sqrt[9]{27x^6y^3}$.

Self Check 9
Simplify:

a. $\sqrt[6]{3^3}$

b. $\sqrt[4]{64x^2y^2}$

Solution

a. $\sqrt[4]{3^2} = (3^2)^{1/4}$ Change the radical to an exponential expression.

      $= 3^{2/4}$ Use the rule $(x^m)^n = x^{mn}$.

      $= 3^{1/2}$ $\frac{2}{4} = \frac{1}{2}$.

      $= \sqrt{3}$ Change back to radical form.

b. $\sqrt[8]{x^6} = (x^6)^{1/8}$ Change the radical to an exponential expression.

      $= x^{6/8}$ Use the rule $(x^m)^n = x^{mn}$.

      $= x^{3/4}$ Simplify the fraction: $\frac{6}{8} = \frac{3}{4}$.

      $= (x^3)^{1/4}$ $\frac{3}{4} = 3(\frac{1}{4})$.

      $= \sqrt[4]{x^3}$ Change back to radical form.

c. $\sqrt[9]{27x^6y^3} = (3^3x^6y^3)^{1/9}$ Write 27 as 3^3 and change the radical to an exponential expression.

      $= 3^{3/9}x^{6/9}y^{3/9}$ Raise each factor to the $\frac{1}{9}$ power by multiplying the fractional exponents.

      $= 3^{1/3}x^{2/3}y^{1/3}$ Simplify each fractional exponent.

      $= (3x^2y)^{1/3}$ Use the rule $(xy)^n = x^ny^n$.

      $= \sqrt[3]{3x^2y}$ Change back to radical form.

Answers **a.** $\sqrt{3}$, **b.** $\sqrt{8xy}$

Section 7.5 STUDY SET

INTERMEDIATE
Algebra *f(x)* **Now**™

VOCABULARY *Fill in the blanks.*

1. The expressions $4^{1/2}$ and $(-8)^{-2/3}$ have _____ exponents.

2. In the exponential expression $27^{4/3}$, 27 is the _____, and 4/3 is the _____.

3. In the radical expression $\sqrt[3]{4{,}096x^{12}}$, 3 is the _____, and $4{,}096x^{12}$ is the _____.

4. $32^{4/5}$ means the fourth _____ of the fifth _____ of 32.

CONCEPTS

5. Complete the table by writing the given expression in the alternate form.

Radical form	Exponential form
$\sqrt[5]{25}$	
	$(-27)^{2/3}$
$(\sqrt[4]{16})^{-3}$	
	$81^{3/2}$
$-\sqrt{\frac{9}{64}}$	

6. Explain the two rules for rational exponents illustrated in the diagrams below.

   a. $(-32)^{1/5} = \sqrt[5]{-32}$

   b. $125^{4/3} = (\sqrt[3]{125})^4$

7. Graph each number on the number line.

$$\left\{ 8^{2/3}, (-125)^{1/3}, -16^{-1/4}, 4^{3/2}, -\left(\frac{9}{100}\right)^{-1/2} \right\}$$

-5 -4 -3 -2 -1 0 1 2 3 4 5 6 7 8

8. Evaluate $25^{3/2}$ in two ways. Which way is easier?

Complete each rule for exponents.

9. $x^m x^n = $

10. $(x^m)^n = $

11. $\dfrac{x^m}{x^n} = $

12. $x^{-n} = $

13. $x^{1/n} = $

14. $x^{m/n} = $ $= \sqrt[n]{x^m}$

Complete each table and graph the function.

15. $f(x) = x^{1/2}$

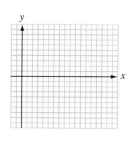

x	f(x)
0	
1	
4	
9	
16	

16. $f(x) = x^{1/3}$

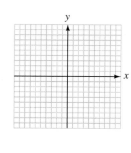

x	f(x)
-8	
-1	
0	
1	
8	

NOTATION *Complete each solution.*

17. $(100a^4)^{3/2} = ($ $)^3$

   $= ($ $)^3$

   $= 1{,}000a^6$

18. $(m^{1/3}n^{1/2})^6 = ($ $)^6(n^{1/2})^6$

   $= m$ $n^{6/2}$

   $= m^2 n^3$

PRACTICE *Write each expression in radical form.*

19. $x^{1/3}$

20. $b^{1/2}$

21. $(3x)^{1/4}$

22. $(4ab)^{1/6}$

23. $(17 x^3 y)^{1/4}$

24. $(34 a^2 b^2)^{1/5}$

25. $(x^2 + y^2)^{1/2}$

26. $(x^3 + y^3)^{1/3}$

Change each radical to an exponential expression.

27. $\sqrt{m}$

28. $\sqrt[3]{r}$

29. $\sqrt[4]{3a}$

30. $3\sqrt[5]{a}$

31. $\sqrt[6]{\dfrac{1}{7}abc}$

32. $\sqrt[7]{\dfrac{3}{8}p^2 q}$

33. $\sqrt[3]{a^2 - b^2}$

34. $\sqrt{x^2 + y^2}$

Evaluate each expression, if possible.

35. $4^{1/2}$

36. $25^{1/2}$

37. $125^{1/3}$

38. $8^{1/3}$

39. $16^{1/4}$

40. $625^{1/4}$

41. $32^{1/5}$

42. $0^{1/5}$

43. $\left(\dfrac{1}{16}\right)^{1/2}$

44. $\left(\dfrac{1}{4}\right)^{1/2}$

45. $-16^{1/4}$

46. $-125^{1/3}$

47. $(-64)^{1/2}$

48. $(-216)^{1/2}$

49. $(-27)^{1/3}$

50. $(-125)^{1/3}$

Simplify each expression, if possible. Assume that all variables are unrestricted and use absolute value symbols when necessary.

51. $(25y^2)^{1/2}$

52. $(-27x^3)^{1/3}$

53. $(16x^4)^{1/4}$

54. $(-16x^4)^{1/2}$

55. $(243x^5)^{1/5}$

56. $[(x + 1)^4]^{1/4}$

57. $(-64x^8)^{1/4}$

58. $[(x + 5)^3]^{1/3}$

Simplify each expression. Assume that all variables represent positive numbers.

59. $36^{3/2}$

60. $27^{2/3}$

61. $81^{3/4}$

62. $100^{3/2}$

63. $144^{3/2}$

64. $1{,}000^{2/3}$

65. $\left(\dfrac{1}{8}\right)^{2/3}$

66. $\left(\dfrac{4}{9}\right)^{3/2}$

67. $(25x^4)^{3/2}$

68. $(27a^3b^3)^{2/3}$

69. $\left(\dfrac{8x^3}{27}\right)^{2/3}$

70. $\left(\dfrac{27}{64y^6}\right)^{2/3}$

Write each expression without using negative exponents. Assume that all variables represent positive numbers.

71. $4^{-1/2}$

72. $8^{-1/3}$

73. $4^{-3/2}$

74. $25^{-5/2}$

75. $(16x^2)^{-3/2}$

76. $(81c^4)^{-3/2}$

77. $(-27y^3)^{-2/3}$

78. $(-8z^9)^{-2/3}$

79. $\left(\dfrac{27}{8}\right)^{-4/3}$

80. $\left(\dfrac{25}{49}\right)^{-3/2}$

81. $\left(-\dfrac{8x^3}{27}\right)^{-1/3}$

82. $\left(\dfrac{16}{81y^4}\right)^{-3/4}$

Use a calculator to evaluate each expression. Round to the nearest hundredth.

83. $\sqrt[3]{15}$

84. $\sqrt[4]{50.5}$

85. $\sqrt[5]{1.045}$

86. $\sqrt[7]{-1{,}000}$

Perform the operations. Write the answers without negative exponents. Assume that all variables represent positive numbers.

87. $5^{3/7}5^{2/7}$

88. $4^{2/5}4^{2/5}$

89. $(4^{1/5})^3$

90. $(3^{1/3})^5$

91. $\dfrac{9^{4/5}}{9^{3/5}}$

92. $\dfrac{7^{2/3}}{7^{1/2}}$

93. $6^{-2/3}6^{-4/3}$

94. $5^{1/3}5^{-5/3}$

95. $\dfrac{3^{4/3}3^{1/3}}{3^{2/3}}$

96. $\dfrac{2^{5/6}2^{1/3}}{2^{1/2}}$

97. $a^{2/3}a^{1/3}$

98. $b^{3/5}b^{1/5}$

99. $(a^{2/3})^{1/3}$

100. $(t^{4/5})^{10}$

101. $(a^{1/2}b^{1/3})^{3/2}$

102. $(mn^{-2/3})^{-3/5}$

103. $\dfrac{(4x^3y)^{1/2}}{(9xy)^{1/2}}$

104. $\dfrac{(27x^3y)^{1/3}}{(8xy^2)^{2/3}}$

105. $(27x^{-3})^{-1/3}$

106. $(16a^{-2})^{-1/2}$

107. $(2x^2y^{-1/4})^3(8y^{-2})^{2/3}$

108. $(27a^3b)^{-1/3}(9a^{-2}b^2)^{-1/2}$

Perform the multiplications. Assume that all variables are positive.

109. $y^{1/3}(y^{2/3} + y^{5/3})$

110. $y^{2/5}(y^{-2/5} + y^{3/5})$

111. $x^{3/5}(x^{7/5} - x^{2/5} + 1)$

112. $x^{4/3}(x^{2/3} + 3x^{5/3} - 4)$

Use rational exponents to simplify each radical. Assume that all variables represent positive numbers.

113. $\sqrt[6]{p^3}$

114. $\sqrt[8]{q^2}$

115. $\sqrt[4]{25b^2}$

116. $\sqrt[9]{-8x^6}$

APPLICATIONS

117. BALLISTIC PENDULUMS See the illustration below. The formula

$$v = \frac{m + M}{m}(2gh)^{1/2}$$

gives the velocity (in ft/sec) of a bullet with weight m fired into a block with weight M, that raises the height of the block h feet after the collision. The letter g represents the constant, 32. Find the velocity of the bullet to the nearest ft/sec.

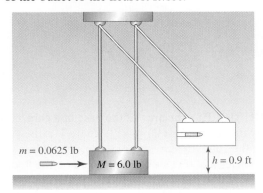

$m = 0.0625$ lb

$M = 6.0$ lb

$h = 0.9$ ft

118. GEOGRAPHY The formula

$$A = [s(s - a)(s - b)(s - c)]^{1/2}$$

gives the area of a triangle with sides of length a, b, and c, where s is one-half of the perimeter. Estimate the area of Virginia (to the nearest square mile) using the data given below.

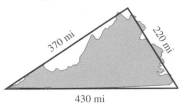

370 mi

220 mi

430 mi

119. RELATIVITY One of the concepts of relativity theory is that an object moving past an observer at a speed near the speed of light appears to have a larger mass because of its motion. If the mass of the object is m_0 when the object is at rest relative to the observer, its mass m will be given by the formula

$$m = m_0\left(1 - \frac{v^2}{c^2}\right)^{-1/2}$$

when it is moving with speed v (in miles per second) past the observer. The letter c is the speed of light, 186,000 mi/sec. If a proton with a rest mass of 1 unit is accelerated by a nuclear accelerator to a speed of 160,000 mi/sec, what mass will the technicians observe it to have? Round to the nearest hundredth.

120. LOGGING The width w and height h of the strongest rectangular beam that can be cut from a cylindrical log of radius a are given by

$$w = \frac{2a}{3}(3^{1/2}) \qquad h = a\left(\frac{8}{3}\right)^{1/2}$$

Find the width, height, and cross-sectional area of the strongest beam that can be cut from a log with *diameter* 4 feet. Round to the nearest hundredth.

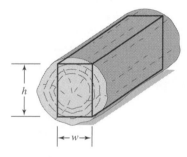

121. CUBICLES The area of the base of a cube is given by the function $A(V) = V^{2/3}$, where V is the volume of the cube. In a preschool room, 18 children's cubicles like that shown are placed on the floor around the room. Estimate how much floor space is lost to the cubicles. Give your answer in square inches and in square feet.

Mary S.

Storage capacity 4,096 in.^{3}

122. CARPENTRY The length L of the longest board that can be carried horizontally around the right-angle corner of two intersecting hallways is given by the formula

$$L = (a^{2/3} + b^{2/3})^{3/2}$$

where a and b represent the widths of the hallways. Find the longest shelf that a carpenter can carry around the corner if $a = 40$ in. and $b = 64$ in. Give your result in inches and in feet. In each case, round to the nearest tenth.

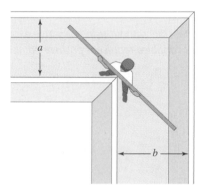

WRITING

123. What is a rational exponent? Give some examples.

124. Explain how the root key $\boxed{\sqrt[x]{y}}$ on a scientific calculator can be used in combination with other keys to evaluate the expression $16^{3/4}$.

REVIEW   *Solve each inequality. Write the solution set using interval notation.*

125. $5x - 4 < 11$

126. $-2(3t - 5) \geq 8$

127. $\frac{4}{5}(r - 3) > \frac{2}{3}(r + 2)$

128. $-4 < 2x - 4 \leq 8$

7.6 Geometric Applications of Radicals

- The Pythagorean theorem   •   45°–45°–90° triangles   •   30°–60°–90° triangles
- The distance formula

In this section, we will consider applications of square roots that occur in geometry. Then we will find the distance between two points on a rectangular coordinate system, using a formula that contains a square root. We begin by considering an important theorem (mathematical statement) about right triangles.

The Pythagorean theorem

If we know the lengths of two legs of a right triangle, we can find the length of the **hypotenuse** (the side opposite the 90° angle) by using the **Pythagorean theorem.**

> **Pythagorean theorem**
>
> If a and b represent the lengths of two legs of a right triangle and c represents the length of the hypotenuse, then
>
> $$a^2 + b^2 = c^2$$

In words, the Pythagorean theorem is expressed as follows:

> *In any right triangle, the square of the hypotenuse is equal to the sum of the squares of the two legs.*

Suppose the right triangle shown in Figure 7-13 has legs of length 3 and 4 units. To find the length of the hypotenuse, we use the Pythagorean theorem.

$$a^2 + b^2 = c^2$$
$$3^2 + 4^2 = c^2 \qquad \text{Substitute 3 for } a \text{ and 4 for } b.$$
$$9 + 16 = c^2$$
$$25 = c^2$$

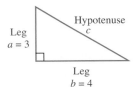

FIGURE 7-13

To solve for c, we need to "undo" the operation performed on it. Since c is squared, we take the positive square root of both sides of the equation.

$$\sqrt{25} = \sqrt{c^2} \qquad \text{Since } c \text{ represents the length of a side of a triangle, } c > 0.$$
$$5 = c \qquad \sqrt{c^2} = c, \text{ because } c \cdot c = c^2.$$

The length of the hypotenuse is 5 units.

EXAMPLE 1   Firefighting.   To fight a forest fire, the forestry department plans to clear a rectangular fire break around the fire, as shown in Figure 7-14 on the next page. Crews are equipped with mobile communications that have a 3,000-yard range. Can crews at points A and B remain in radio contact?

Self Check 1
In Example 1, can the crews communicate if $b = 1,500$ yards?

Solution   Points A, B, and C form a right triangle. To find the distance c from point A to point B, we can use the Pythagorean theorem, substituting 2,400 for a and 1,000 for b and solving for c.

$$a^2 + b^2 = c^2$$
$$2{,}400^2 + 1{,}000^2 = c^2$$
$$5{,}760{,}000 + 1{,}000{,}000 = c^2$$
$$6{,}760{,}000 = c^2$$
$$\sqrt{6{,}760{,}000} = \sqrt{c^2} \qquad \text{Take the positive square root of both sides.}$$
$$2{,}600 = c \qquad \text{Use a calculator to find the square root.}$$

The two crews are 2,600 yards apart. Because this distance is less than the range of the radios, they can communicate by radio.

Answer yes

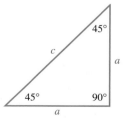

1,000 yd

c yd

C 2,400 yd B

FIGURE 7-14

THINK IT THROUGH   Pythagorean Triples

CORBIS

"Fraternity and sorority membership nationwide is declining, down about 30% in the last decade." Chronicle of Higher Education, 2002

The first college social fraternity, Phi Beta Kappa, was founded in 1776 on the campus of The College of William and Mary. However, secret societies have existed since ancient times, and from these roots the essence of today's fraternities and sororities have their foundation.

Pythagoras, the Greek mathematician of the 6th century B.C. was the leader of a secret fraternity/sorority called the Pythagoreans. They were a community of men and women that studied mathematics, and in particular, the "magic 3-4-5 triangle." This right triangle is special because the sum of the squares of the lengths of its legs is equal to the square of the length of its hypotenuse: $3^2 + 4^2 = 5^2$ or $9 + 16 = 25$. Today, we call a set of three natural numbers a, b, and c that satisfy $a^2 + b^2 = c^2$ a **Pythagorean triple.** Show that each list of numbers is a Pythagorean triple.

1. 5, 12, 13   **2.** 7, 24, 25   **3.** 8, 15, 17
4. 9, 40, 41   **5.** 28, 45, 53   **6.** 112, 15, 113

45°–45°–90° triangles

An **isosceles right triangle** is a right triangle with two legs of equal length. Isosceles right triangles have angle measures of 45°, 45°, and 90°. If we know the length of one leg of an isosceles right triangle, we can use the Pythagorean theorem to find the length of the hypotenuse. Since the triangle shown in Figure 7-15 is a right triangle, we have

45°

c

a

45°   90°

a

FIGURE 7-15

$$c^2 = a^2 + b^2$$
$$c^2 = a^2 + a^2 \qquad \text{Both legs are } a \text{ units long, so replace } b \text{ with } a.$$
$$c^2 = 2a^2 \qquad \text{Combine like terms.}$$
$$c = \sqrt{2a^2} \qquad \text{Take the positive square root of both sides.}$$
$$c = a\sqrt{2} \qquad \text{Simplify the radical: } \sqrt{2a^2} = \sqrt{2}\sqrt{a^2} = \sqrt{2}a = a\sqrt{2}.$$

Thus, *in an isosceles right triangle, the length of the hypotenuse is the length of one leg times $\sqrt{2}$.*

EXAMPLE 2 **45°–45°–90° triangles.** If one leg of the isosceles right triangle shown in Figure 7-15 is 10 feet long, find the length of the hypotenuse.

Solution Since the length of the hypotenuse is the length of a leg times $\sqrt{2}$, we have

$$c = 10\sqrt{2}$$

The length of the hypotenuse is $10\sqrt{2}$ units. To two decimal places, the length is 14.14 units.

If the length of the hypotenuse of an isosceles right triangle is known, we can use the Pythagorean theorem to find the length of each leg.

EXAMPLE 3 **Isosceles right triangle.** Find the exact length of each leg of the isosceles right triangle shown in Figure 7-16.

Solution We use the Pythagorean theorem.

$$c^2 = a^2 + b^2$$
$$\mathbf{25}^2 = a^2 + a^2 \qquad \text{Since both legs are } a \text{ units long,}$$

substitute a for b. The hypotenuse is 25 units long. Substitute 25 for c.

$$25^2 = 2a^2 \qquad \text{Combine like terms.}$$
$$\frac{625}{2} = a^2 \qquad \text{Square 25 and divide both sides by 2.}$$

$$\sqrt{\frac{625}{2}} = a$$

To solve for a, take the positive square root of both sides: $\sqrt{a^2} = a$.

$$\frac{\sqrt{625} \cdot \sqrt{2}}{\sqrt{2} \cdot \sqrt{2}} = a$$

Write $\sqrt{\frac{625}{2}}$ as $\frac{\sqrt{625}}{\sqrt{2}}$. Then rationalize the denominator.

$$\frac{25\sqrt{2}}{2} = a$$

In the numerator, simplify the radical: $\sqrt{625} = 25$. In the denominator, do the multiplication: $\sqrt{2} \cdot \sqrt{2} = 2$.

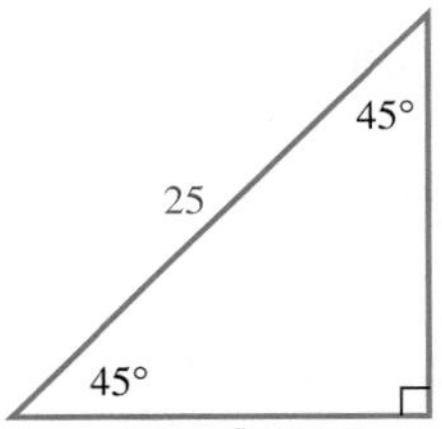

FIGURE 7-16

The exact length of each leg is $\dfrac{25\sqrt{2}}{2}$ units. To two decimal places, the length is 17.68 units.

30°–60°–90° triangles

From geometry, we know that an **equilateral triangle** is a triangle with three sides of equal length and three 60° angles. Each side of the equilateral triangle in Figure 7-17 is $2a$ units long. If an **altitude** is drawn to its base, the altitude bisects the base and divides the equilateral triangle into two 30°–60°–90° triangles. We can see that the shorter leg of each 30°–60°–90° triangle (the side *opposite* the 30° angle) is *a* units long. Thus,

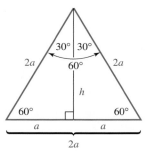

FIGURE 7-17

> *The length of the shorter leg of a 30°–60°–90° right triangle is half as long as the hypotenuse.*

We can discover another important relationship between the legs of a 30°–60°–90° triangle if we find the length of the altitude *h* in Figure 7-17. We begin by applying the Pythagorean theorem to one of the 30°–60°–90° triangles.

$$a^2 + b^2 = c^2$$
$$a^2 + h^2 = (2a)^2 \qquad \text{One leg is } h \text{ units long, so replace } b \text{ with } h. \text{ The hypotenuse is } 2a \text{ units long, so replace } c \text{ with } 2a.$$
$$a^2 + h^2 = 4a^2 \qquad (2a)^2 = (2a)(2a) = 4a^2.$$
$$h^2 = 3a^2 \qquad \text{Subtract } a^2 \text{ from both sides.}$$
$$h = \sqrt{3a^2} \qquad \text{Take the positive square root of both sides.}$$
$$h = a\sqrt{3} \qquad \text{Simplify the radical: } \sqrt{3a^2} = \sqrt{3}\sqrt{a^2} = a\sqrt{3}.$$

We see that the altitude (the longer side of the 30°–60°–90° triangle) is $\sqrt{3}$ times as long as the shorter leg. Thus,

> *The length of the longer leg of a 30°–60°–90° triangle is the length of the shorter leg times $\sqrt{3}$.*

INTERMEDIATE
Algebra $f(x)$ Now™

Self Check 4
Find the length of the hypotenuse and the longer leg of a 30°–60°–90° triangle if the shorter leg is 8 centimeters long.

EXAMPLE 4　30°–60°–90° triangles.　Find the length of the hypotenuse and the longer leg of the right triangle shown in Figure 7-18.

Solution　Since the shorter leg of a 30°–60°–90° triangle is half as long as the hypotenuse, the hypotenuse is 12 centimeters long.

Since the length of the longer leg is the length of the shorter leg times $\sqrt{3}$, the longer leg is $6\sqrt{3}$ (about 10.39) centimeters long.

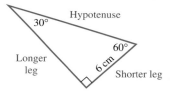

FIGURE 7-18

Answers　16 cm, $8\sqrt{3}$ cm

INTERMEDIATE
Algebra $f(x)$ Now™

EXAMPLE 5　**Stretching exercises.**　A doctor prescribed the back-strengthening exercise shown in Figure 7-19(a) for a patient. The patient was instructed to raise his leg to an angle of 60° and hold the position for 10 seconds. If the patient's leg is 36 inches long, how high off the floor will his foot be when his leg is held at the proper angle?

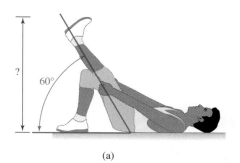

(a)

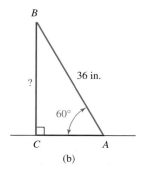

(b)

FIGURE 7-19

Solution In Figure 7-19(b), we see that a 30°–60°–90° triangle, which we will call triangle *ABC*, models the situation. Since the side opposite the 30° angle of a 30°–60°–90° triangle is half as long as the hypotenuse, side *AC* is 18 inches long.

Since the length of the side opposite the 60° angle is the length of the side opposite the 30° angle times $\sqrt{3}$, side *BC* is $18\sqrt{3}$, or about 31 inches long. So the patient's foot will be about 31 inches from the floor when his leg is in the proper stretching position.

The distance formula

With the *distance formula*, we can find the distance between any two points that are graphed on a rectangular coordinate system.

To find the distance d between points $P(x_1, y_1)$ and $Q(x_2, y_2)$ shown in Figure 7-20, we construct the right triangle *PRQ*. The distance between *P* and *R* is $|x_2 - x_1|$, and the distance between *R* and *Q* is $|y_2 - y_1|$. We apply the Pythagorean theorem to the right triangle *PRQ* to get

$$d^2 = |x_2 - x_1|^2 + |y_2 - y_1|^2$$
$$= (x_2 - x_1)^2 + (y_2 - y_1)^2 \qquad \text{Because } |x_2 - x_1|^2 = (x_2 - x_1)^2$$
$$\text{and } |y_2 - y_1|^2 = (y_2 - y_1)^2.$$

We take the positive square root of both sides:

(1) $\quad d = \sqrt{(x_2 - x_1)^2 + (y_2 - y_1)^2}$

Equation 1 is called the **distance formula.**

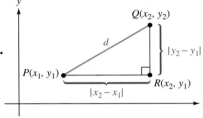

FIGURE 7-20

Distance formula

The distance between two points (x_1, y_1) and (x_2, y_2) is given by the formula

$$d = \sqrt{(x_2 - x_1)^2 + (y_2 - y_1)^2}$$

INTERMEDIATE
Algebra $f(x)$ **Now**™

EXAMPLE 6 Find the distance between points $(-2, 3)$ and $(4, -5)$.

Solution To find the distance, we can use the distance formula by substituting 4 for x_2, -2 for x_1, -5 for y_2, and 3 for y_1.

$$d = \sqrt{(x_2 - x_1)^2 + (y_2 - y_1)^2}$$
$$= \sqrt{[4 - (-2)]^2 + (-5 - 3)^2}$$
$$= \sqrt{(4 + 2)^2 + (-5 - 3)^2}$$
$$= \sqrt{6^2 + (-8)^2}$$
$$= \sqrt{36 + 64}$$
$$= \sqrt{100}$$
$$= 10$$

The distance between the points is 10 units.

Self Check 6
Find the distance between $(-2, -2)$ and $(3, 10)$.

Answer 13

EXAMPLE 7   Robotics.   Computerized robots are used to weld the parts of an automobile chassis together on an automated production line. To do this, an imaginary coordinate system is superimposed on the side of the vehicle, and the robot is programmed to move to specific positions to make each weld. See Figure 7-21, which is scaled in inches. If the welder unit moves from point to point at an average rate of speed of 48 in. per sec, how long will it take it to move from position 1 to position 2?

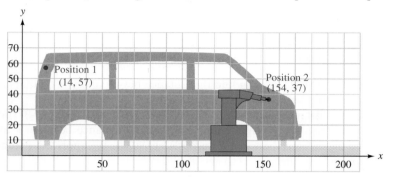

FIGURE 7-21

Solution   This is a uniform motion problem. We can use the formula $t = \frac{d}{r}$ to find the time it takes for the welder to move from position 1 coordinates $(14, 57)$ to position 2 with coordinates $(154, 37)$.

We can use the distance formula to find the distance d that the welder unit moves.

$$d = \sqrt{(x_2 - x_1)^2 + (y_2 - y_1)^2}$$

$$d = \sqrt{(154 - 14)^2 + (37 - 57)^2} \quad \text{Substitute 154 for } x_2, 14 \text{ for } x_1, 37 \text{ for } y_2, \text{ and } 57 \text{ for } y_1.$$

$$= \sqrt{140^2 + (-20)^2}$$

$$= \sqrt{20{,}000} \qquad\qquad 140^2 + (-20)^2 = 19{,}600 + 400 = 20{,}000.$$

$$= 100\sqrt{2} \qquad\qquad \text{Simplify: } \sqrt{20{,}000} = \sqrt{100 \cdot 100 \cdot 2} = \sqrt{100^2 \cdot 2} = 100\sqrt{2}.$$

The welder travels $100\sqrt{2}$ inches as it moves from position 1 to position 2. To find the time this will take, we divide the distance by the average rate of speed, 48 in. per sec.

$$t = \frac{d}{r}$$

$$t = \frac{100\sqrt{2}}{48} \qquad \text{Substitute } 100\sqrt{2} \text{ for } d \text{ and 48 for } r.$$

$$t \approx 2.9 \qquad\qquad \text{Use a calculator to find an approximation to the nearest tenth.}$$

It will take the welder about 2.9 seconds to travel from position 1 to position 2.

Section 7.6   STUDY SET

VOCABULARY   *Fill in the blanks.*

1. In a right triangle, the side opposite the 90° angle is called the _____.

2. An _____ right triangle is a right triangle with two legs of equal length.

3. The _____ theorem states that in any right triangle, the square of the hypotenuse is equal to the sum of the squares of the lengths of the two legs.

4. An _____ triangle has three sides of equal length and three 60° angles.

CONCEPTS *Fill in the blanks.*

5. If a and b are the lengths of two legs of a right triangle and c is the length of the hypotenuse, then ▓▓▓▓ .

6. In any right triangle, the square of the length of the hypotenuse is equal to the _____ of the squares of the lengths of the two _____ .

7. In an isosceles right triangle, the length of the hypotenuse is the length of one leg times ▓ .

8. The shorter leg of a $30°-60°-90°$ triangle is _____ as long as the hypotenuse.

9. The length of the longer leg of a $30°-60°-90°$ triangle is the length of the shorter leg times ▓ .

10. The formula to find the distance between two points (x_1, y_1) and (x_2, y_2) is $d =$ ▓▓▓▓ .

11. In a right triangle, the shorter leg is opposite the _____ angle, and the longer leg is opposite the _____ angle.

12. An isosceles triangle has _____ sides of equal length.

13. a. To solve the equation $c^2 = 20$, where c represents the length of the hypotenuse of a right triangle, how do we undo the operation performed on c?

b. What is the first step when solving the equation $25 + b^2 = 81$?

14. When the lengths of the sides of the triangle are substituted into the equation $a^2 + b^2 = c^2$, the result is a false statement. Explain why.

$$a^2 + b^2 = c^2$$
$$2^2 + 4^2 = 5^2$$
$$4 + 16 = 25$$
$$20 = 25$$

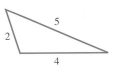

NOTATION *Complete each solution.*

15. Evaluate: $\sqrt{(-1-3)^2 + [2-(-4)]^2}$.

$$\sqrt{(-1-3)^2 + [2-(-4)]^2} = \sqrt{(-4)^2 + [\ \ ]^2}$$
$$= \sqrt{\ \ }$$
$$= \sqrt{\ \ } \cdot 13$$
$$= \ \ \sqrt{13}$$
$$\approx 7.21$$

16. Solve: $8^2 + 4^2 = c^2$.

$$\ \ + 16 = c^2$$
$$\ \ = c^2$$
$$\sqrt{\ \ } = \sqrt{c^2}$$
$$\sqrt{\ \ } \cdot 5 = c$$
$$\ \ \sqrt{5} = c$$
$$c \approx 8.94$$

PRACTICE *The lengths of two sides of the right triangle ABC are given. Find the length of the missing side.*

17. $a = 6$ ft and $b = 8$ ft

18. $a = 10$ cm and $c = 26$ cm

19. $b = 18$ m and $c = 82$ m

20. $a = 14$ in. and $c = 50$ in.

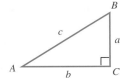

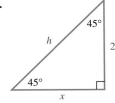 *Find the missing lengths in each triangle. Give the exact answer and then an approximation to two decimal places, when applicable.*

21.

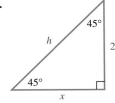

22.

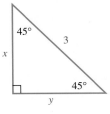

23.

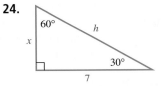

24.

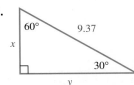 *Find the missing lengths in each triangle. Give the answer to two decimal places.*

25.

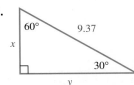

26.

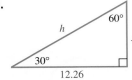

27.

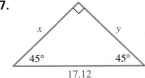

28.

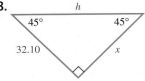

29. GEOMETRY Find the exact length of the diagonal (in blue) of one of the *faces* of the cube shown below.

30. GEOMETRY Find the exact length of the diagonal (in green) of the cube shown below.

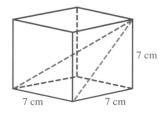

7 cm

7 cm　　7 cm

Find the distance between each pair of points.

31. $(0, 0)$, $(3, -4)$

32. $(0, 0)$, $(-12, 16)$

33. $(-2, -8)$, $(3, 4)$

34. $(-5, -2)$, $(7, 3)$

35. $(6, 8)$, $(12, 16)$

36. $(10, 4)$, $(2, -2)$

37. $(-3, 5)$, $(-5, -5)$

38. $(2, -3)$, $(4, -8)$

39. ISOSCELES TRIANGLES Use the distance formula to show that a triangle with vertices $(-2, 4)$, $(2, 8)$, and $(6, 4)$ is isosceles.

40. RIGHT TRIANGLES Use the distance formula and the Pythagorean theorem to show that a triangle with vertices $(2, 3)$, $(-3, 4)$, and $(1, -2)$ is a right triangle.

APPLICATIONS *Give the exact answer and round it to two decimal places.*

41. WASHINGTON, D.C. The square shows the 100-square-mile site selected by George Washington in 1790 to serve as a permanent capital for the United States. In 1847, the part of the district lying on the west bank of the Potomac was returned to Virginia. Find the coordinates of each corner of the original square that outlined the District of Columbia.

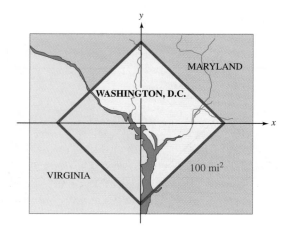

MARYLAND

WASHINGTON, D.C.

VIRGINIA

100 mi²

42. PAPER AIRPLANES The illustration below gives the directions for making a paper airplane from a square piece of paper with sides 8 inches long. Find the length l of the plane when it is completed.

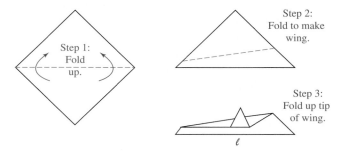

Step 1:
Fold up.

Step 2:
Fold to make wing.

Step 3:
Fold up tip of wing.

ℓ

43. HARDWARE The sides of the regular hexagonal nut shown below are 10 millimeters long. Find the height h of the nut.

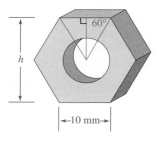

60°

h

←10 mm→

44. IRONING BOARDS Find the height h of the ironing board shown below.

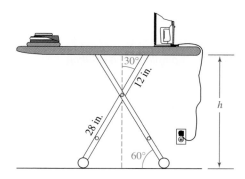

30°

12 in.

28 in.

60°

h

45. BASEBALL The baseball diamond shown on the next page is a square, 90 feet on a side. If the third baseman fields a ground ball 10 feet directly behind third base, how far must he throw the ball to throw a runner out at first base?

46. BASEBALL A shortstop fields a grounder at a point one-third of the way from second base to third base. (See the illustration on the next page.) How far will he have to throw the ball to make an out at first base?

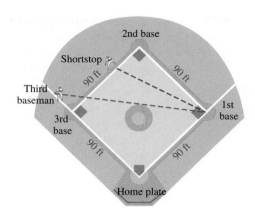

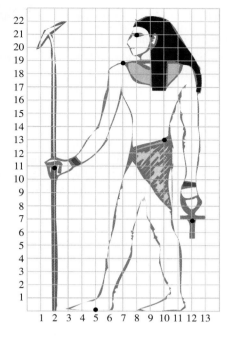

47. CLOTHESLINES A pair of damp jeans are hung on a clothesline to dry. They pull the center down 1 foot. By how much is the line stretched? Give the answer to the nearest hundredth.

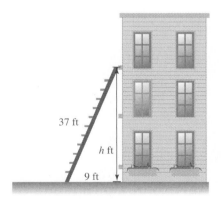

48. FIREFIGHTING The base of the 37-foot ladder shown below is 9 feet from the wall. Will the top reach a window ledge that is 35 feet above the ground? Explain how you arrived at your answer.

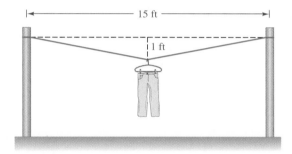

49. ART HISTORY A figure displaying some of the basic characteristics of Egyptian art is shown in the next column. Use the distance formula to find the following dimensions of the drawing. Round your answers to two decimal places.

   a. From the foot to the eye

   b. From the belt to the hand holding the staff

   c. From the shoulder to the symbol held in the hand

50. PACKAGING The diagonal *d* of a rectangular box with dimensions $a \times b \times c$ is given by

$$d = \sqrt{a^2 + b^2 + c^2}$$

Will the umbrella fit in the shipping carton below? Explain how you arrived at your answer.

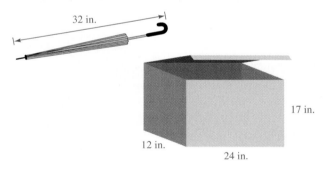

51. PACKAGING An archaeologist wants to ship a 34-inch femur bone. Will it fit in a 4-inch-tall box that has a 24-inch-square base? (See Exercise 50.) Explain how you arrived at your answer.

52. TELEPHONE SERVICE A telephone cable runs from *A* to *B* to *C* to *D*. How much cable is required to run from *A* to *D* directly?

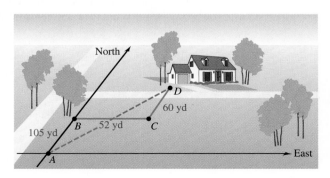

WRITING

53. State the Pythagorean theorem in words.

54. List the facts that you learned about special right triangles in this section.

REVIEW

55. DISCOUNT BUYING A repairman purchased some washing-machine motors for a total of $224. When the unit cost decreased by $4, he was able to buy one additional motor for the same total price. How many motors did he buy originally?

56. AVIATION An airplane can fly 650 miles with the wind in the same amount of time as it can fly 475 miles against the wind. If the wind speed is 40 mph, find the speed of the plane in still air.

57. Find the mean of 16, 6, 10, 4, 5, 13.

58. Find the median of 16, 6, 10, 4, 5.

Radicals

The expression $\sqrt[n]{a}$ is called a **radical expression.** In this chapter, we have discussed the properties and procedures used when simplifying radical expressions, solving radical equations, and writing radical expressions using rational exponents.

■ Expressions Containing Radicals

When working with expressions containing radicals, we must often apply the multiplication property and/or the division property of radicals to simplify the expression. Recall that

$$\sqrt[n]{ab} = \sqrt[n]{a}\sqrt[n]{b} \qquad\qquad \sqrt[n]{\frac{a}{b}} = \frac{\sqrt[n]{a}}{\sqrt[n]{b}} \quad \text{where } b \neq 0$$

Perform each operation and simplify the expression.

1. Simplify: $\sqrt[3]{-54h^6}$.

2. Add: $2\sqrt[3]{64e} + 3\sqrt[3]{8e}$.

3. Subtract: $\sqrt{72} - \sqrt{200}$.

4. Multiply: $-4\sqrt[3]{5r^2s}(5\sqrt[3]{2r})$.

5. Multiply: $(\sqrt{3s} - \sqrt{2t})(\sqrt{3s} + \sqrt{2t})$.

6. Multiply: $-\sqrt{3}(\sqrt{7} - \sqrt{5})$.

7. Find the power: $(3\sqrt{2n} - 2)^2$.

8. Rationalize the denominator: $\dfrac{\sqrt[3]{9j}}{\sqrt[3]{3jk}}$.

■ Equations Containing Radicals

When solving radical equations, our objective is to rid the equation of the radical. This is achieved by using the *power rule:*

If x, y, and n represent real numbers and $x = y$, then $x^n = y^n$.

If we raise both sides of an equation to the same power, the resulting equation might not be equivalent to the original equation. We must always check for extraneous solutions.

Solve each radical equation, if possible.

9. $\sqrt{1 - 2g} = \sqrt{g + 10}$

10. $4 - \sqrt[3]{4 + 12x} = 0$

11. $\sqrt{y + 2} - 4 = -y$

12. $\sqrt[4]{12t + 4} + 2 = 0$

■ Radicals and Rational Exponents

Radicals can be written using rational (fractional) exponents, and exponential expressions having fractional exponents can be written in radical form. To do this, we use two rules for exponents introduced in this chapter.

$$x^{1/n} = \sqrt[n]{x} \qquad\qquad x^{m/n} = \sqrt[n]{x^m} = \left(\sqrt[n]{x}\right)^m$$

13. Express using a rational exponent: $\sqrt[3]{3}$.

14. Express in radical form: $5a^{2/5}$.

ACCENT ON TEAMWORK

SECTION 7.1

A SPIRAL OF ROOTS   To do this project, you will need a piece of poster board, a protractor, a yardstick, and a pencil. Begin by drawing an isosceles right triangle near the right margin of the poster board. Label the length of each leg as 1 unit. (See the illustration.) Use the Pythagorean theorem to determine the length of the hypotenuse. Draw a second right triangle using the hypotenuse of the first triangle as one leg. Draw its second leg with a length of 1 unit. Find the length of the hypotenuse of triangle 2. Continue this process of creating right triangles, using the previous hypotenuse as one leg and drawing a new second leg of length 1 unit each time. Calculate the length of the resulting hypotenuse. What patterns, if any, do you see?

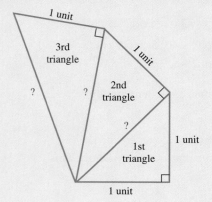

SECTION 7.2

COMMON ERRORS   In each addition or subtraction problem below, tell what mistake was made. Compare each problem to a similar one involving variables to help clarify your explanation. For example, compare Problem 1 to $2a + 3a$ to help explain the correct procedure that should be used to simplify the expression.

1. $2\sqrt{5x} + 3\sqrt{5x} = 5\sqrt{10x}$
2. $30 + 30\sqrt[4]{2} = 60\sqrt[4]{2}$
3. $7\sqrt[3]{y^2} - 5\sqrt[3]{y^2} = 2$
4. $6\sqrt{11ab} - 3\sqrt{5ab} = 3\sqrt{6ab}$

SECTION 7.3

MULTIPLYING RADICALS   In this chapter, when we were asked to find the product of two radical expressions, the radicals always had the same index. Brainstorm in your group to see if you can come up with a procedure that can be used to find

$$\sqrt[3]{3} \cdot \sqrt{3}$$

Keep in mind two things: The indices must be the same to apply the multiplication property of radicals, and radical expressions can be written using rational exponents.

SECTION 7.4

SOLVING RADICAL EQUATIONS   In this chapter, we solved equations containing two radicals. The radicals in those equations always had the same index. That is not the case for the following two equations:

$$\sqrt[3]{2x} = \sqrt{x} \qquad \sqrt[4]{x} = \sqrt{\frac{x}{4}}$$

Brainstorm in your group to see if you can come up with a procedure that can be used to solve these equations. What are their solutions?

SECTION 7.5

EXPONENTS   Use the $\boxed{y^x}$ key on your calculator to approximate each of the following exponential expressions. Then write them in order from least to greatest. Which of the exponents are not rational exponents? Could you have written the expressions in increasing order without having to approximate them? Explain.

1. $3^{0.999}$ **2.** $3^{3.9}$
3. $3^{1\frac{3}{4}}$ **4.** $3^{\sqrt{2}}$
5. $3^{1.7}$ **6.** 3^{π}
7. $3^{\frac{2}{3}}$ **8.** $3^{\frac{4}{3}}$
9. $3^{2\sqrt{3}}$ **10.** $3^{0.1}$

SECTION 7.6

GRAPHING IN THREE DIMENSIONS   A point is located in three-space by plotting ordered triples of numbers (x, y, z). The point $(3, 2, 4)$ is plotted below. In three-space, the distance formula is

$$d = \sqrt{(x_2 - x_1)^2 + (y_2 - y_1)^2 + (z_2 - z_1)^2}$$

Use this formula to find the distance between   **a.** $(3, 2, 1)$ and $(13, 12, 11)$ and   **b.** $(-1, -2, -4)$ and $(6, -2, 5)$.

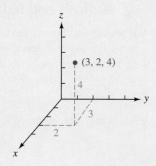

CHAPTER REVIEW

SECTION 7.1   *Radical Expressions and Radical Functions*

CONCEPTS

The number b is a *square root* of a if $b^2 = a$.

If $x > 0$, the *principal square root* of x is the positive square root of x, denoted $\sqrt{x}$. If x can be any real number, then $\sqrt{x^2} = |x|$.

The *cube root* of x is denoted as $\sqrt[3]{x}$ and is defined by

$$\sqrt[3]{x} = y \text{ if } y^3 = x$$

If n represents an even natural number,

$$\sqrt[n]{a^n} = |a|$$

If n represents an odd natural number,

$$\sqrt[n]{a^n} = a$$

If n represents a natural number greater than 1 and x represents a real number, then

- If $x > 0$, then $\sqrt[n]{x}$ is the positive number such that $(\sqrt[n]{x})^n = x$.
- If $x = 0$, then $\sqrt[n]{x} = 0$.
- If $x < 0$, and n is odd, $\sqrt[n]{x}$ is the real number such that $(\sqrt[n]{x})^n = x$.
- If $x < 0$, and n is even, $\sqrt[n]{x}$ is not a real number.

REVIEW EXERCISES

Simplify each radical expression, if possible. Assume that x can be any real number.

1. $\sqrt{49}$

2. $-\sqrt{121}$

3. $\sqrt{\dfrac{225}{49}}$

4. $\sqrt{-4}$

5. $\sqrt{0.01}$

6. $\sqrt{25x^2}$

7. $\sqrt{x^8}$

8. $\sqrt{x^2 + 4x + 4}$

Simplify each radical expression.

9. $\sqrt[3]{-27}$

10. $-\sqrt[3]{216}$

11. $\sqrt[3]{64a^6b^3}$

12. $\sqrt[3]{\dfrac{s^9}{125}}$

Simplify each radical expression, if possible. Assume that x and y can be any real number.

13. $\sqrt[4]{625}$

14. $\sqrt[5]{-32}$

15. $\sqrt[4]{256x^8y^4}$

16. $\sqrt{(-22y)^2}$

17. $-\sqrt[4]{\dfrac{1}{16}}$

18. $\sqrt[6]{-1}$

19. $\sqrt{0}$

20. $\sqrt[3]{0}$

21. GEOMETRY The side of a square with area A square feet is given by the function $s(A) = \sqrt{A}$. Find the *perimeter* of a square with an area of 144 square feet.

22. VOLUME OF A CUBE The total surface area of a cube is related to its volume V by the function $A(V) = 6\sqrt[3]{V^2}$. Find the surface area of a cube with a volume of 8 cm³.

Graph each radical function. Find the domain and range.

23. $f(x) = \sqrt{x + 2}$

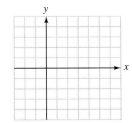

24. $f(x) = -\sqrt[3]{x} + 3$

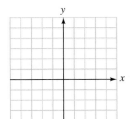

Simplifying and Combining Radical Expressions

A radical is in *simplest form* when:

1. No radicals appear in a denominator.

2. The radicand contains no fractions or negative numbers.

3. Each factor in the radicand appears to a power less than the index.

Properties of radicals:

1. Multiplication:

$$\sqrt[n]{ab} = \sqrt[n]{a}\sqrt[n]{b}$$

2. Division:

$$\sqrt[n]{\frac{a}{b}} = \frac{\sqrt[n]{a}}{\sqrt[n]{b}} \quad (b \neq 0)$$

Like radicals can be combined by addition and subtraction.

Radicals that are not similar can often be converted to radicals that are similar and then combined.

Simplify each expression. Assume that all variables represent positive real numbers.

25. $\sqrt{240}$

26. $\sqrt[3]{54}$

27. $\sqrt[4]{32}$

28. $-2\sqrt[5]{-96}$

29. $\sqrt{8x^5}$

30. $\sqrt[3]{r^{17}}$

31. $\sqrt[3]{16x^5y^4}$

32. $3\sqrt[3]{27j^7k}$

33. $\dfrac{\sqrt{32x^3}}{\sqrt{2x}}$

34. $\sqrt{\dfrac{17xy}{64a^4}}$

Simplify and combine like radicals. Assume that all variables represent positive real numbers.

35. $\sqrt{2} + 2\sqrt{2}$

36. $6\sqrt{20} - \sqrt{5}$

37. $2\sqrt[3]{3} - \sqrt[3]{24}$

38. $-\sqrt[4]{32} - 2\sqrt[4]{162}$

39. $2x\sqrt{8} + 2\sqrt{200x^2} + \sqrt{50x^2}$

40. $\sqrt[3]{54} - 3\sqrt[3]{16} + 4\sqrt[3]{128}$

41. SEWING A corner of fabric is folded over to form a collar and stitched down as shown on the right. From the dimensions given in the figure, determine the exact number of inches of stitching that must be made. Then give an approximation to one decimal place. (All measurements are in inches.)

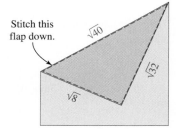

Stitch this flap down. $\sqrt{40}$ $\sqrt{32}$ $\sqrt{8}$

Multiplying and Dividing Radical Expressions

If two radicals have the same index, they can be multiplied:

$$\sqrt[n]{a}\sqrt[n]{b} = \sqrt[n]{ab}$$

If a radical appears in a denominator of a fraction, or if a radicand contains a fraction, we can write the radical in simplest form by *rationalizing the denominator.*

Simplify each expression. Assume that all variables represent positive real numbers.

42. $\sqrt{7}\sqrt{7}$

43. $(2\sqrt{5})(3\sqrt{2})$

44. $(-2\sqrt{8})^2$

45. $2\sqrt{6}\sqrt{216}$

46. $\sqrt{9x}\sqrt{x}$

47. $(\sqrt{33})^2$

48. $-\sqrt[3]{2x^2}\sqrt[3]{4x}$

49. $4\sqrt[3]{9}\sqrt[3]{9}$

50. $\sqrt{2}(\sqrt{8} - 3)$

51. $-\sqrt[4]{256x^5y^{11}}\sqrt[4]{625x^9y^3}$

52. $(\sqrt{3b} + \sqrt{3})^2$

53. $(2\sqrt{u} + 3)(3\sqrt{u} - 4)$

Rationalize each denominator.

54. $\dfrac{10}{\sqrt{3}}$

55. $\sqrt{\dfrac{3}{5}}$

56. $\dfrac{x}{\sqrt{xy}}$

57. $\dfrac{\sqrt[3]{uv}}{\sqrt[3]{u^5v^7}}$

58. $\dfrac{2}{\sqrt{2} - 1}$

59. $\dfrac{\sqrt{a} + 1}{\sqrt{a} - 1}$

To *rationalize a two-term denominator* of a fraction, multiply the numerator and the denominator by the conjugate of the binomial in the denominator.

Rationalize each numerator.

60. $\dfrac{3 - \sqrt{x}}{2}$

61. $\dfrac{\sqrt{a} - \sqrt{b}}{\sqrt{a}}$

62. VOLUME The formula relating the radius r of a sphere and its volume V is $r = \sqrt[3]{\dfrac{3V}{4\pi}}$. Write the radical in simplest form.

SECTION 7.4 *Radical Equations*

The power rule:

   If $x = y$, then $x^n = y^n$.

Solving equations containing radicals:

1. Isolate one radical expression on one side of the equation.

2. Raise both sides of the equation to the power that is the same as the index.

3. Solve the resulting equation. If it still contains a radical, go back to step 1.

4. Check the solutions to eliminate *extraneous* solutions.

Solve each equation. Write all solutions and cross out those that are extraneous.

63. $\sqrt{7x - 10} - 1 = 11$

64. $u = \sqrt{25u - 144}$

65. $2\sqrt{y - 3} = \sqrt{2y + 1}$

66. $\sqrt{z + 1} + \sqrt{z} = 2$

67. $\sqrt[3]{x^3 + 56} - 2 = x$

68. $\sqrt[4]{8x - 8} + 2 = 0$

69. Using the graphs in the illustration, estimate the solution of

$$\sqrt{2x - 3} = -2x + 5$$

Check the result.

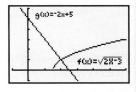

Solve each equation for the indicated variable.

70. $r = \sqrt{\dfrac{A}{P}} - 1$ for P

71. $h = \sqrt[3]{\dfrac{12I}{b}}$ for I

72. ELECTRONICS The current I (measured in amperes) and the power P (measured in watts) are related by the formula

$$I = \sqrt{\dfrac{P}{R}}$$

Find the resistance R in a circuit if the current used by an electrical appliance that is rated at 980 watts is 7.4 amperes.

SECTION 7.5 *Rational Exponents*

If n represents a natural number greater than 1 and $\sqrt[n]{x}$ represents a real number, then

   $x^{1/n} = \sqrt[n]{x}$

Write each expression in radical form.

73. $t^{1/2}$

74. $(5xy^3)^{1/4}$

Simplify each expression, if possible. Assume that all variables represent positive real numbers.

75. $25^{1/2}$

76. $-36^{1/2}$

77. $(-36)^{1/2}$

78. $1^{1/2}$

79. $\left(\dfrac{9}{x^2}\right)^{1/2}$

80. $\left(\dfrac{1}{27}\right)^{1/3}$

81. $(-8)^{1/3}$

82. $625^{1/4}$

83. $(27a^3b)^{1/3}$

84. $(81c^4d^4)^{1/4}$

If n represents a natural number greater than 1 and x represents a real number,

- If $x > 0$, then $x^{1/n}$ is the positive number such that $(x^{1/n})^n = x$.
- If $x = 0$, then $x^{1/n} = 0$.
- If $x < 0$, and n is odd, then $x^{1/n}$ is the real number such that $(x^{1/n})^n = x$.
- If $x < 0$ and n is even, then $x^{1/n}$ is not a real number.

If m and n are positive integers, $x > 0$, and $\frac{m}{n}$ is in simplest form,

$$x^{m/n} = \sqrt[n]{x^m} = \left(\sqrt[n]{x}\right)^m$$

$$x^{-m/n} = \frac{1}{x^{m/n}} \quad (x \neq 0)$$

$$\frac{1}{x^{-m/n}} = x^{m/n} \quad (x \neq 0)$$

The *rules for exponents* can be used to simplify expressions with fractional exponents.

Simplify each expression, if possible. Assume that all variables represent positive real numbers.

85. $9^{3/2}$

86. $8^{-2/3}$

87. $-49^{5/2}$

88. $\dfrac{1}{100^{-1/2}}$

89. $\left(\dfrac{4}{9}\right)^{-3/2}$

90. $\dfrac{1}{25^{5/2}}$

91. $(25x^2y^4)^{3/2}$

92. $(8u^6v^3)^{-2/3}$

Perform the operations. Write answers without negative exponents. Assume that all variables represent positive real numbers.

93. $5^{1/4}5^{1/2}$

94. $a^{3/7}a^{-2/7}$

95. $(k^{4/5})^{10}$

96. $\dfrac{(4g^3h)^{1/2}}{(9gh^{-1})^{1/2}}$

Perform the multiplications. Assume all variables represent positive real numbers.

97. $u^{1/2}(u^{1/2} - u^{-1/2})$

98. $v^{2/3}(v^{1/3} + v^{4/3})$

99. Simplify: $\sqrt[4]{\dfrac{a^2}{25b^2}}$. (The variables represent positive real numbers.)

100. Substitute the x- and y-coordinates of each point labeled in the graph into the equation

$$x^{2/3} + y^{2/3} = 32$$

Show that each one satisfies the equation.

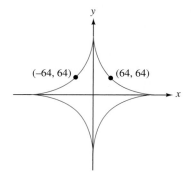

SECTION 7.6	*Geometric Applications of Radicals*

The Pythagorean theorem:

If a and b are the lengths of the *legs* of a right triangle and c is the length of the *hypotenuse,* then $a^2 + b^2 = c^2$.

101. CARPENTRY The gable end of the roof shown below is divided in half by a vertical brace 8 feet in height. Find the length of the roof line.

102. SAILING A technique called *tacking* allows a sailboat to make progress into the wind. A sailboat follows the course shown. Find d, the distance the boat advances into the wind after tacking.

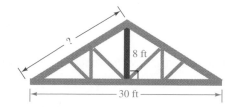

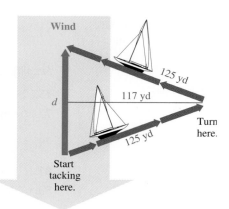

In an *isosceles right triangle,* the length of the hypotenuse is the length of one leg times $\sqrt{2}$.

The shorter leg of a $30°-60°-90°$ *triangle* (the side opposite the 30° angle) is half as long as the hypotenuse. The longer leg (the side opposite the 60° angle) is the length of the shorter leg times $\sqrt{3}$.

103. Find the length of the hypotenuse of an isosceles right triangle whose legs measure 7 meters.

104. The hypotenuse of a $30°-60°-90°$ triangle measures $12\sqrt{3}$ centimeters. Find the length of each leg.

Find x to two decimal places.

105.

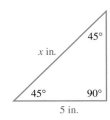

106.

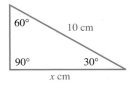

The distance formula:

$$d = \sqrt{(x_2 - x_1)^2 + (y_2 - y_1)^2}$$

Find the distance between each pair of points.

107. $(0, 0)$ and $(5, -12)$

108. $(-4, 6)$ and $(-2, 8)$

1. Complete the table for $f(x) = \sqrt{x-1}$. Then graph the function. Round to the nearest hundredth when necessary.

x	$f(x)$
1	
2	
3	
5	
10	
12	
17	

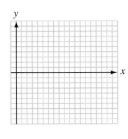

2. **SUBMARINES** The horizontal distance (measured in miles) that an observer can see is related to the height h (measured in feet) of the observer by the function $d(h) = 1.4\sqrt{h}$. If a submarine's periscope extends 4.7 feet above the surface of the ocean, how far is the horizon? Round to the nearest tenth.

Simplify each expression. Assume that the variables are unrestricted.

3. $\sqrt{x^2}$

4. $\sqrt{8x^2}$

5. $\sqrt[3]{54x^5}$

6. $\sqrt{18x^4y^8}$

Simplify each expression. Assume that all variables represent positive real numbers.

7. $\sqrt[3]{-64x^3y^6}$

8. $\sqrt{\dfrac{4a^2}{9}}$

9. $\sqrt[4]{-16}$

10. $\sqrt[5]{(t+8)^5}$

11. $\sqrt{48}$

12. $\sqrt{250x^3y^5}$

13. $\dfrac{\sqrt[3]{24x^{15}y^4}}{\sqrt[3]{y}}$

14. $\sqrt{\dfrac{3a^5}{48a^7}}$

Simplify and combine like radicals. Assume that all variables represent positive real numbers.

15. $\sqrt{12} - \sqrt{27}$

16. $2\sqrt[3]{40} - \sqrt[3]{5{,}000} + 4\sqrt[3]{625}$

17. $2\sqrt{48y^5} - 3y\sqrt{12y^3}$

18. $\sqrt[4]{768z^5} + z\sqrt[4]{48z}$

Perform each operation and simplify, if possible. All variables represent positive real numbers.

19. $-2\sqrt{xy}(3\sqrt{x} + \sqrt{xy^3})$

20. $(3\sqrt{2} + \sqrt{3})(2\sqrt{2} - 3\sqrt{3})$

21. $(\sqrt[3]{2a} + 9)^2$

Rationalize each denominator.

22. $\dfrac{1}{\sqrt{5}}$

23. $\dfrac{3t-1}{\sqrt{3t}-1}$

24. $\sqrt[3]{\dfrac{9}{4a^2}}$ ___

Solve and check each solution.

25. $2\sqrt{x} = \sqrt{x+1}$

26. $\sqrt[3]{6n+4} - 4 = 0$

27. $1 - \sqrt{u} = \sqrt{u - 3}$

28. Solve $r = \sqrt[3]{\dfrac{GMt^2}{4\pi^2}}$ for G.

Simplify each expression. Assume that all variables represent positive real numbers, and write answers without using negative exponents.

29. $16^{1/4}$

30. $27^{2/3}$

31. $36^{-3/2}$

32. $\left(-\dfrac{8}{27}\right)^{-2/3}$

33. $\dfrac{2^{5/3}2^{1/6}}{2^{1/2}}$

34. $\dfrac{(8x^3y)^{1/2}(8xy^5)^{1/2}}{(x^3y^6)^{1/3}}$

Find x to two decimal places.

35.

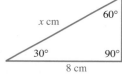

36.

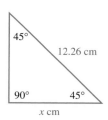

37. Find the distance between $(-2, 5)$ and $(22, 12)$.

38. PENDULUMS The time t, in seconds, it takes for a pendulum to swing back and forth to complete one period is given by the formula $t = 2\pi\sqrt{\dfrac{L}{32}}$, where L is the length of the pendulum in feet. Find the exact period of a 4-foot-long pendulum. Then approximate the period to the nearest tenth.

39. SHIPPING CRATE The diagonal brace on the shipping crate below is 53 inches. Find the height h of the crate.

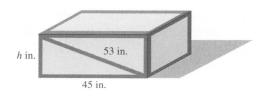

40. Explain why, without having to perform any algebraic steps, it is obvious that the equation $\sqrt{x - 8} = -10$ has no solutions.

1. a. What is a rational number?

   b. What is an irrational number?

   c. What is a real number?

2. Evaluate $\dfrac{|Ax_0 + By_0 + C|}{\sqrt{A^2 + B^2}}$
   for $A = 6$, $B = -8$, $C = -5$, and $x_0 = y_0 = -2$.

3. Solve $S = \dfrac{a - \ell r}{1 - r}$ for ℓ.

4. ELECTRONICS The illustration shows a closed circuit with two voltage sources and three resistors. The sum of the voltages in the loop must be 0. Find x.

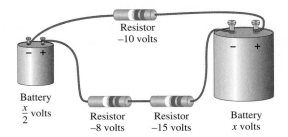

Resistor
−10 volts

Battery
$\dfrac{x}{2}$ volts

Resistor
−8 volts

Resistor
−15 volts

Battery
x volts

5. SALAD DRESSING A caterer is going to combine 10% vinegar–oil dressing with 18% vinegar–oil dressing to make 10 cups of a 15% vinegar–oil dressing. How many cups of each type of dressing will she need?

6. Refer to the illustration.

   a. What is the slope of the line?

   b. What is the y-intercept of the line?

   c. What is the equation of the line?

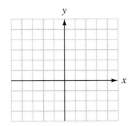

7. Refer to the illustration.

   a. What is the slope of line l?

   b. Write the equation of the line that passes through point A and is perpendicular to line l. Answer in slope–intercept form.

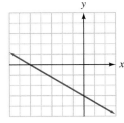

$A(4, -2)$

8. If $f(x) = 1.25 - x^5$, find $f(2)$.

9. Determine the domain and range of the function $f(x) = -|x| - 2$, which is graphed below. (The x- and y-axes are scaled in units of 1.)

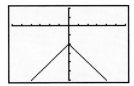

10. Graph: $g(x) = 1 + x^3$.

11. Solve $\begin{cases} x = \dfrac{3}{2}y + 5 \\ 2x - 3y = 8 \end{cases}$, if possible.

12. BUDGETS The family budget guidelines below were given on the Web site www.thefamily.com. It is recommended that housing, utilities, and transportation costs should be 60% of the budget; and food, clothing, and savings should be 25% of the budget. Find x, y, and z. (*Hint:* The sum of the percents from all categories in a circle graph is what number?)

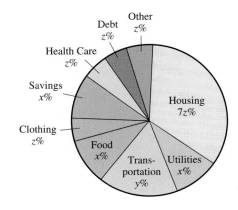

Other
$z\%$

Debt
$z\%$

Health Care
$z\%$

Savings
$x\%$

Clothing
$z\%$

Food
$x\%$

Trans-
portation
$y\%$

Utilities
$x\%$

Housing
$7z\%$

Solve each inequality or compound inequality and graph the solution set. Then express the solution set using interval notation.

13. $5(x + 1) \leq 4(x + 3)$ and $x + 12 < -3$

14. $|-1 - 2x| > 5$

15. Graph: $\begin{cases} x + 2y < 3 \\ 2x + 4y < 8 \end{cases}$.

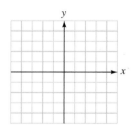

16. Simplify: $\left(\dfrac{4a^{-2}b}{3ab^{-3}} \right)^3$.

17. Find $(6.1 \times 10^8)(3.9 \times 10^5)$. Give the answer in scientific notation.

18. Subtract:
$(-2x^2y^3 + 6xy + 5y^2) - (-4x^2y^3 - 7xy + 2y^2)$.

19. Multiply: $(3y + 1)(2y^2 + 3y + 2)$.

20. Simplify: $(x + 3)(x - 3) + (2x - 1)(x + 2)$.

Factor each polynomial completely.

21. $3c - cd + 3d - c^2$

22. $x^3 - 8y^3$

23. $(a + b)^2 - 2(a + b) + 1$

24. $x^4 - 17x^2 + 16$

Solve each equation and check the result.

25. $2z^3 - 200z = 0$

26. $3m^2 + 10m = -3$

27. Simplify: $\dfrac{3x^2 - 10xy - 8y^2}{4y^2 - xy}$.

28. For what values of x is $\dfrac{2}{x^2 - x - 56}$ undefined?

Perform the operations.

29. $(2x^2 - 9x - 5) \cdot \dfrac{x}{2x^2 + x}$

30. $\dfrac{2x}{x^2 - 4} - \dfrac{1}{x^2 - 3x + 2} + \dfrac{x + 1}{x^2 + x - 2}$

31. SHARED WORK One pipe fills a tank in 4 hours, and another fills it in 6 hours. How long will it take to fill the tank using both pipes?

32. Explain the difference between direct variation and inverse variation. Assume a positive constant of variation.

Simplify each expression.

33. $\left(-\dfrac{8x^3}{27} \right)^{-1/3}$

34. $\sqrt{200x^4y^3z}$

35. $\sqrt[3]{16} + \sqrt[3]{128}$

36. $(\sqrt{5z} + \sqrt{3})(\sqrt{5z} + \sqrt{3})$

37. Solve: $\sqrt{-5x + 24} = 6 - x$.

38. Rationalize the denominator: $\dfrac{\sqrt{3}}{\sqrt{50}}$.

CHAPTER 8

Quadratic Equations, Functions, and Inequalities

© Bob Krist /CORBIS

INTERMEDIATE
Algebra $f(x)$ Now™

Throughout the chapter, this icon introduces resources on the Intermediate AlgebraNow Web site, accessed through **http://1pass .thomson.com**, that will

- Help you test your knowledge of the material with a pre-test and a post-test

- Provide a personalized learning plan targeting areas you should study

TLE   In a watercolor class, students learn that light, shadow, color, and perspective are fundamental components of an attractive painting. They also learn that the appropriate matting and frame can enhance their work. In this section, we will use mathematics to determine the dimensions of a uniform matting that is to have the same area as the picture it frames. To do this, we will write a quadratic equation and then solve it by *completing the square.* The technique of completing the square can be used to derive *the quadratic formula.* This formula is a valuable algebraic tool for solving any quadratic equation.

To learn more about quadratic equations, visit *The Learning Equation* on the Internet at http://tle.brookscole.com. (The log-in instructions are in the Preface.) For Chapter 8, the online lesson is:

- *TLE* Lesson 12: The Quadratic Formula

Check Your Knowledge

1. An equation of the form $ax^2 + bx + c = 0$ where $a \neq 0$ is called a _____ equation.

2. The lowest point on a parabola that opens upward is called the _____.

3. The _____ part of $3 + 4i$ is 3. The _____ part is 4.

4. The complex conjugate of $4 - 5i$ is _____.

5. For the quadratic equation $ax^2 + bx + c = 0$, the discriminant is _____.

Solve each equation by factoring or using the square root property.

6. $-16t^2 + 32t + 48 = 0$ 7. $5x^2 - 49 = 0$

8. Determine what number must be added to $y^2 - 10y$ to make a perfect square trinomial.

9. Solve $x^2 + 6x + 4 = 0$ by completing the square.

10. Solve $2x^2 = 4x + 1$ using the quadratic formula.

11. A rectangular window is 4 feet taller than it is wide and its area is 20 ft^2. Find the window's dimensions to the nearest tenth of a foot.

12. Simplify: $\sqrt{-75}$.

13. Simplify: i^{77}.

Perform the operations. Write all answers in the form $a + bi$.

14. $(2 - 3i) + (3 + 2i)$ 15. $(3 - 4i) - (2 + 2i)$

16. $3i(4 - 3i)$ 17. $(2 - 5i)(2 + 5i)$

18. $\dfrac{1}{i^9}$ 19. $\dfrac{2}{1 - i}$

20. Use the discriminant to determine whether the solutions of $3x^2 - 5x + 4 = 0$ are real or nonreal.

21. Solve: $(x - 2)^2 = -4$.

22. Solve: $x - 5\sqrt{x} + 6 = 0$.

Determine the vertex and the axis of symmetry of the graph of the function. Then graph it.

23. $f(x) = x^2 - x - 6$

24. $f(x) = -(x + 2)^2 + 2$

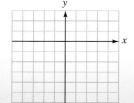
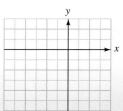

Solve the inequality and graph the solution set.

25. $\dfrac{x + 1}{x - 2} > 0$ 26. $x^2 - 3x + 2 > 0$

Study Skills Workshop
CAREER EXPLORATION

Your college major can have a direct influence on the math course(s) that you need to take after Intermediate Algebra. If you have plans to major in math, engineering, or the sciences, you will choose a sequence of courses leading to calculus and beyond; if you have a major that is more liberal studies oriented, you will most probably choose a different course or sequence of courses than the math/science majors. Ultimately, your choice of career will affect the math sequence that follows this course. Before the end of this term, it would be wise to have at least a general idea of which direction you would like to go.

How Do You Decide? If you aren't sure of your ultimate career goal, you may want to seek the advice of a counselor, pay a visit to your college's career center, or search the Internet for tools that will help you discover things about yourself and the careers, that might suit your interests. Even if you think you know what career you are aiming for, it's still a good idea to search a little. Personality tests tell you things about your basic makeup but aren't necessarily job or career-oriented; one such test is the Myers-Briggs test which may be available through your college counseling office. Once you know your basic personality type from this test, you may use books such as *Do What You Are*[1] to explore various career choices.

It's probably not a good idea to put all of your hopes on one type of test, though, and there are various free online tests that you can take to help determine your personality type. For example, see: http://www.personalitytype.com/, http://www.keirsey.com/, and http://www.9types.com/. There are also career or vocational tests that you can take online: http://www.review.com/career/careerquizhome.cfm?menuID=0&careers=6, and http://career.missouri.edu/ are examples. If you would rather read than use a computer, a classic book on making career choices is *What Color is Your Parachute? A Practical Manual for Job-Hunters and Career-Changers*.[2] There is also a Web site designed to be used with the book at http://www.jobhuntersbible.com/.

Once You've Decided. After you have a general idea of the career you would like to pursue, talk to a college counselor to find out what type of education is necessary to achieve your goal. Your counselor can tell you which classes you need to take at your particular school and may help you choose a college to which you can transfer, if that is necessary.

ASSIGNMENT

1. Do you have a general career goal in mind? If so, what is it?
2. Take at least two personality tests and two career tests. List the tests that you took and where you found them. Did taking the tests help you in making any decisions about a career choice?
3. Make an appointment and visit a counselor to discuss which classes you should take during your next term and beyond. If you are still unclear about which direction you should take, ask your counselor for a good all-purpose plan. Make a list of classes that your counselor suggests and keep a copy for yourself.

---

[1] Paul D. Tieger and Barbara Barron-Tieger, Little, Brown and Company.
[2] by Dick Bolles, Ten Speed Press Publishers.

We have previously solved quadratic equations by factoring. In this chapter, we will discuss other methods for solving quadratic equations, and we will consider the graphs of quadratic functions.

8.1 Completing the Square

- A review of solving quadratic equations by factoring • The square root property
- Completing the square • Solving equations by completing the square • Problem solving

We have seen that equations involving first-degree polynomials, such as $12x - 4 = 0$, are called *linear equations*. We have also seen that equations involving second-degree polynomials, such as $12x^2 - 4x = 0$, are called *quadratic equations*. In Chapter 5, we learned how to solve quadratic equations by factoring. However, as we shall see, the factoring method has its limitations. In this section, we will introduce a more general method that enables us to solve any quadratic equation.

A review of solving quadratic equations by factoring

A *quadratic equation* is an equation of the form $ax^2 + bx + c = 0$, where a, b, and c are real numbers and $a \neq 0$. We have discussed how to solve quadratic equations by factoring. For example, to solve $6x^2 - 7x - 3 = 0$, we proceed as follows:

$$6x^2 - 7x - 3 = 0$$
$$(2x - 3)(3x + 1) = 0 \qquad \text{Factor.}$$
$$2x - 3 = 0 \quad \text{or} \quad 3x + 1 = 0 \qquad \text{Set each factor equal to 0.}$$
$$x = \frac{3}{2} \qquad\qquad x = -\frac{1}{3} \qquad \text{Solve each linear equation.}$$

Many expressions do not factor as easily as $6x^2 - 7x - 3$. For example, it would be difficult to solve $2x^2 + 4x + 1 = 0$ by factoring, because $2x^2 + 4x + 1$ cannot be factored by using only integers. With this in mind, we will now develop another method of solving quadratic equations. It is based on the *square root property*.

The square root property

To develop general methods for solving all quadratic equations, we first consider the equation $x^2 = c$. If $c \geq 0$, we can find the real solutions of $x^2 = c$ as follows:

$$x^2 = c$$
$$x^2 - c = 0 \qquad \text{Subtract } c \text{ from both sides.}$$
$$x^2 - (\sqrt{c})^2 = 0 \qquad \text{Replace } c \text{ with } (\sqrt{c})^2, \text{ since } c = (\sqrt{c})^2.$$
$$(x + \sqrt{c})(x - \sqrt{c}) = 0 \qquad \text{Factor the difference of two squares.}$$
$$x + \sqrt{c} = 0 \quad \text{or} \quad x - \sqrt{c} = 0 \qquad \text{Set each factor equal to 0.}$$
$$x = -\sqrt{c} \qquad\qquad x = \sqrt{c} \qquad \text{Solve each linear equation.}$$

The two solutions of $x^2 = c$ are $x = \sqrt{c}$ and $x = -\sqrt{c}$.

> **Square root property**
>
> For any nonnegative real number c, if $x^2 = c$, then
>
> $$x = \sqrt{c} \qquad \text{or} \qquad x = -\sqrt{c}$$

EXAMPLE 1   Solve: $x^2 - 12 = 0$.

Solution   We can write the equation as $x^2 = 12$ and use the square root property to solve it.

$$x^2 - 12 = 0$$
$$x^2 = 12 \qquad \text{Add 12 to both sides.}$$
$$x = \sqrt{12} \quad \text{or} \quad x = -\sqrt{12} \qquad \text{Use the square root property.}$$
$$x = 2\sqrt{3} \quad | \quad x = -2\sqrt{3} \qquad \text{Simplify: } \sqrt{12} = \sqrt{4}\sqrt{3} = 2\sqrt{3}.$$

Verify that $2\sqrt{3}$ and $-2\sqrt{3}$ satisfy the original equation.

We can use **double-sign notation** $\pm$ to write the solutions in more compact form as $\pm 2\sqrt{3}$. Read $\pm$ as "positive or negative."

Self Check 1
Solve: $x^2 - 18 = 0$.

Answer $\pm 3\sqrt{2}$

EXAMPLE 2   **Phonograph records.**   Before compact disc (CD) technology, one way of recording music was by engraving grooves on thin vinyl discs called records. (See Figure 8-1.) The vinyl discs used for long-playing records had a surface area of about 111 square inches per side and were played at $33\frac{1}{3}$ revolutions per minute on a turntable. What is the radius of a long-playing record?

CD

FIGURE 8-1

Solution   The relationship between the area of a circle and its radius is given by the formula $A = \pi r^2$. We can find the radius of a record by substituting 111 for A and solving for r.

$$A = \pi r^2 \qquad \text{This is the formula for the area of a circle.}$$
$$111 = \pi r^2 \qquad \text{Substitute 111 for } A.$$
$$\frac{111}{\pi} = r^2 \qquad \text{To undo the multiplication by } \pi, \text{ divide both sides by } \pi.$$
$$r = \sqrt{\frac{111}{\pi}} \quad \text{or} \quad r = -\sqrt{\frac{111}{\pi}} \qquad \text{Use the square root property. Since the radius of the record cannot be negative, discard the second solution.}$$

The radius of a record is $\sqrt{\frac{111}{\pi}}$ inches — to the nearest tenth, 5.9 inches.

INTERMEDIATE
Algebra $f(x)$ Now™

Self Check 3
Solve: $(x + 2)^2 = 9$.

EXAMPLE 3 Solve: $(x - 3)^2 = 16$.

Solution

$$(x - 3)^2 = 16$$

$x - 3 = \sqrt{16}$ or	$x - 3 = -\sqrt{16}$	Use the square root property.
$x - 3 = 4$	$x - 3 = -4$	Simplify: $\sqrt{16} = 4$.
$x = 3 + 4$	$x = 3 - 4$	Add 3 to both sides.
$x = 7$	$x = -1$	Simplify.

Answer 1, −5

Verify that 7 and −1 satisfy the original equation.

Completing the square

All quadratic equations can be solved by **completing the square.** This method involves the special products

$$x^2 + 2ax + a^2 = (x + a)^2 \qquad \text{and} \qquad x^2 - 2ax + a^2 = (x - a)^2$$

The trinomials $x^2 + \mathbf{2a}x + a^2$ and $x^2 - \mathbf{2a}x + a^2$ are both perfect square trinomials, because both factor as the square of a binomial. In each case, the coefficient of the first term is 1, and if we take one-half of the coefficient of x in the middle term and square it, we obtain the third term.

$$\left[\frac{1}{2}(2a)\right]^2 = a^2 \qquad \left[\frac{1}{2}(-2a)\right]^2 = (-a)^2 = a^2$$

INTERMEDIATE
Algebra $f(x)$ Now™

Self Check 4
Add a number to $a^2 - 5a$ to make it a perfect trinomial square.

EXAMPLE 4 Add a number to make each binomial a perfect square trinomial: **a.** $x^2 + 10x$, **b.** $x^2 - 6x$, and **c.** $x^2 - 11x$.

Solution
a. To make $x^2 + 10x$ a perfect square trinomial, we find one-half of 10, square it, and add that result to $x^2 + 10x$.

$$x^2 + \mathbf{10x} + \left[\frac{1}{2}\mathbf{(10)}\right]^2 = x^2 + 10x + (5)^2 \quad \text{Simplify: } \frac{1}{2}(10) = 5.$$

$$= x^2 + 10x + 25 \quad \text{Note that } x^2 + 10x + 25 = (x + 5)^2.$$

b. To make $x^2 - 6x$ a perfect square trinomial, we find one-half of −6, square it, and add that result to $x^2 - 6x$.

$$x^2 - \mathbf{6x} + \left[\frac{1}{2}\mathbf{(-6)}\right]^2 = x^2 - 6x + (-3)^2 \quad \text{Simplify: } \frac{1}{2}(-6) = -3.$$

$$= x^2 - 6x + 9 \quad \text{Note that } x^2 - 6x + 9 = (x - 3)^2.$$

c. To make $x^2 - 11x$ a perfect square trinomial, we find one-half of −11, square it, and add that result to $x^2 - 11x$.

$$x^2 - \mathbf{11x} + \left[\frac{1}{2}\mathbf{(-11)}\right]^2$$

$$= x^2 - 11x + \left(-\frac{11}{2}\right)^2 \quad \text{Simplify: } \frac{1}{2}(-11) = -\frac{11}{2}.$$

Answer $a^2 - 5a + \dfrac{25}{4}$

$$= x^2 - 11x + \frac{121}{4} \quad \text{Note that } x^2 - 11x + \frac{121}{4} = \left(x - \frac{11}{2}\right)^2.$$

Solving equations by completing the square

We will now discuss how the method of completing the square and the square root property can be used to solve *any* quadratic equation.

Solving by completing the square

To solve a quadratic equation of the form $ax^2 + bx + c = 0$ by completing the square, we use the following steps.

1. Make sure that the coefficient of x^2 is 1. If it is not, make it 1 by dividing both sides of the equation by the coefficient of x^2.

2. If necessary, add or subtract a number from both sides of the equation so that all of the variable terms are on one side of the equation and constants on the other.

3. Complete the square:

   a. Find one-half of the coefficient of x and square it.

   b. Add that square to both sides of the equation.

4. Factor the trinomial square on one side of the equation. Combine like terms on the other side.

5. Solve the resulting equation using the square root property.

6. Check the answers in the original equation.

INTERMEDIATE
Algebra $f(x)$ **Now**™

EXAMPLE 5 Use completing the square to solve $x^2 + 8x + 7 = 0$.

Solution

Step 1: In this example, the coefficient of x^2 is understood to be 1.

Step 2: We subtract 7 from both sides of the equation so that only terms with variables are on the left-hand side.

$$x^2 + 8x + 7 = 0$$
$$x^2 + 8x = -7$$

Step 3: The coefficient of x is 8, one-half of 8 is 4, and $4^2 = 16$. To complete the square, we add 16 to both sides.

$$x^2 + 8x + \mathbf{16} = \mathbf{16} - 7$$

(1) $x^2 + 8x + 16 = 9$ Simplify: $16 - 7 = 9$.

Step 4: Since the left-hand side of Equation 1 is a perfect square trinomial, we can factor it to get $(x + 4)^2$.

$$x^2 + 8x + 16 = 9$$

(2) $(x + 4)^2 = 9$

Step 5: We then solve Equation 2 by using the square root property.

$$x + 4 = \pm\sqrt{9}$$

$$x + 4 = 3 \qquad \text{or} \qquad x + 4 = -3$$
$$x = -1 \qquad | \qquad x = -7$$

Step 6: Verify that -1 and -7 satisfy the original equation.

INTERMEDIATE
Algebra $f(x)$ **Now™**

EXAMPLE 6   Solve: $6x^2 + 5x - 6 = 0$.

Solution

Step 1: To make the coefficient of x^2 equal to 1, we divide both sides of the equation by 6.

$$6x^2 + 5x - 6 = 0$$

$$\frac{6x^2}{6} + \frac{5}{6}x - \frac{6}{6} = \frac{0}{6} \quad \text{Divide both sides by 6.}$$

$$x^2 + \frac{5}{6}x - 1 = 0 \quad \text{Simplify.}$$

Step 2: We add 1 to both sides so that only terms with variables are on the left-hand side of the equation.

$$x^2 + \frac{5}{6}x = 1$$

Step 3: The coefficient of x is $\frac{5}{6}$, one-half of $\frac{5}{6}$ is $\frac{5}{12}$, and $\left(\frac{5}{12}\right)^2 = \frac{25}{144}$. To complete the square, we add $\frac{25}{144}$ to both sides.

$$x^2 + \frac{5}{6}x + \frac{25}{144} = 1 + \frac{25}{144}$$

(3) $$x^2 + \frac{5}{6}x + \frac{25}{144} = \frac{169}{144} \quad \text{Simplify: } 1 + \frac{25}{144} = \frac{144}{144} + \frac{25}{144} = \frac{169}{144}.$$

Step 4: Since the left-hand side of Equation 3 is a perfect square trinomial, we can factor it to get $\left(x + \frac{5}{12}\right)^2$.

(4) $$\left(x + \frac{5}{12}\right)^2 = \frac{169}{144}$$

Step 5: We can solve Equation 4 using the square root property.

$$x + \frac{5}{12} = \pm\sqrt{\frac{169}{144}}$$

$$x + \frac{5}{12} = \frac{13}{12} \qquad \text{or} \qquad x + \frac{5}{12} = -\frac{13}{12} \qquad \text{Simplify: } \sqrt{\frac{169}{144}} = \frac{13}{12}.$$

$$x = -\frac{5}{12} + \frac{13}{12} \qquad\qquad x = -\frac{5}{12} - \frac{13}{12} \qquad \text{Subtract } \frac{5}{12} \text{ from both sides.}$$

$$x = \frac{8}{12} \qquad\qquad\qquad x = -\frac{18}{12} \qquad \text{Simplify.}$$

$$x = \frac{2}{3} \qquad\qquad\qquad x = -\frac{3}{2} \qquad \text{Simplify each fraction.}$$

Step 6: Verify that $\frac{2}{3}$ and $-\frac{3}{2}$ satisfy the original equation.

INTERMEDIATE
— **Algebra** $f(x)$ **Now™**

Self Check 7
Solve: $3x^2 + 6x + 1 = 0$.

EXAMPLE 7   Solve: $2x^2 + 4x + 1 = 0$.

Solution

$$2x^2 + 4x + 1 = 0$$

$$x^2 + 2x + \frac{1}{2} = 0 \qquad \text{Divide both sides by 2 to make the coefficient of } x^2 \text{ equal to 1.}$$

$$x^2 + 2x = -\frac{1}{2}$$ Subtract $\frac{1}{2}$ from both sides.

$$x^2 + 2x + \mathbf{1} = \mathbf{1} - \frac{1}{2}$$ Square one-half of the coefficient of x and add it to both sides.

$$(x + 1)^2 = \frac{1}{2}$$ Factor and combine like terms.

$$x + 1 = \pm\sqrt{\frac{1}{2}}$$ Apply the square root property.

To write $\sqrt{\frac{1}{2}}$ in simplified radical form, we write it as a quotient of square roots and then rationalize the denominator.

$$x + 1 = \frac{\sqrt{2}}{2}$$ $\Big|$ $$x + 1 = -\frac{\sqrt{2}}{2}$$ $\sqrt{\frac{1}{2}} = \frac{\sqrt{1}}{\sqrt{2}} = \frac{1 \cdot \sqrt{2}}{\sqrt{2}\sqrt{2}} = \frac{\sqrt{2}}{2}.$

$$x = -1 + \frac{\sqrt{2}}{2}$$ $\Big|$ $$x = -1 - \frac{\sqrt{2}}{2}$$ Subtract 1 from both sides.

We can express each solution in an alternate form if we write -1 as a fraction with a denominator of 2.

$$x = -\frac{2}{2} + \frac{\sqrt{2}}{2}$$ $\Big|$ $$x = -\frac{2}{2} - \frac{\sqrt{2}}{2}$$ Write -1 as $-\frac{2}{2}$.

$$x = \frac{-2 + \sqrt{2}}{2}$$ $\Big|$ $$x = \frac{-2 - \sqrt{2}}{2}$$ Add (subtract) the numerators and keep the common denominator of 2.

The exact solutions are $\dfrac{-2 + \sqrt{2}}{2}$ and $\dfrac{-2 - \sqrt{2}}{2}$, or more concisely, $x = \dfrac{-2 \pm \sqrt{2}}{2}$.

(Read $\pm$ as "plus or minus.") We can use a calculator to approximate them. To the nearest hundredth, they are -0.29 and -1.71.

Answer $\dfrac{-3 \pm \sqrt{6}}{3}$

! COMMENT Recall that to simplify a fraction, we divide out common *factors* of the numerator and denominator. In Example 7, since -2 is a *term* of the numerator of $\frac{-2 \pm \sqrt{2}}{2}$, no further simplification of this expression can be made.

Checking solutions of quadratic equations

CALCULATOR SNAPSHOT

We can use a graphing calculator to check the solutions of the quadratic equation $2x^2 + 4x + 1 = 0$ found in Example 7. After entering $Y_1 = 2x^2 + 4x + 1$, we call up the home screen by pressing $\boxed{\text{2nd}}$ QUIT. Then we press the $\boxed{\text{VARS}}$ key, arrow $\boxed{\blacktriangleright}$ to Y-VARS, and enter 1 and enter 1 again to get the display shown in Figure 8-2(a). We evaluate $2x^2 + 4x + 1$ for $x = \frac{-2 + \sqrt{2}}{2}$ by inputting the solution using function notation, as shown in Figure 8-2(b). When $\boxed{\text{ENTER}}$ is pressed, the result of 0 is confirmation that $x = \frac{-2 + \sqrt{2}}{2}$ is a solution of the equation.

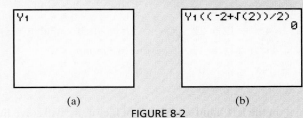

(a)   (b)

FIGURE 8-2

◼ Problem solving

INTERMEDIATE
Algebra *(x)* Now™

EXAMPLE 8   Graduation announcements.   In creating the announcement shown in Figure 8-3, the graphic artist wants to follow two design criteria:

- A border of uniform width should surround the text.
- Equal areas should be devoted to the text and to the border.

To meet these requirements, how wide should the border be?

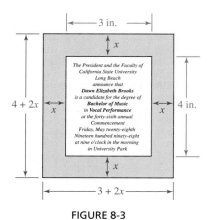

The President and the Faculty of
California State University
Long Beach
announce that
Dawn Elizabeth Brooks
is a candidate for the degree of
Bachelor of Music
in Vocal Performance
at the forty-sixth annual
Commencement
Friday, May twenty-eighth
Nineteen hundred ninety-eight
at nine o'clock in the morning
in University Park

FIGURE 8-3

Analyze the problem

The text occupies $4 \cdot 3 = 12$ in.^{2} of space. The border must also have an area of 12 in.^{2}.

Form an equation

If we let x = the width of the border, the length of the announcement is $(4 + 2x)$ inches and the width is $(3 + 2x)$ inches. We can now form the equation.

The area of the announcement	minus	the area of the text	equals	the area of the border.
$(4 + 2x)(3 + 2x)$	−	12	=	12

Solve the equation

$$(4 + 2x)(3 + 2x) - 12 = 12$$

$12 + 8x + 6x + 4x^2 - 12 = 12$ On the left-hand side, multiply the binomials.

$\qquad\qquad 4x^2 + 14x = 12$ Combine like terms.

$\qquad\quad 4x^2 + 14x - 12 = 0$ Subtract 12 from both sides.

$\qquad\qquad 2x^2 + 7x - 6 = 0$ Divide both sides by 2.

Since the trinomial on the left-hand side does not factor, we will solve the equation by completing the square.

$$x^2 + \frac{7}{2}x - 3 = 0$$

Divide both sides by 2 so that the coefficient of x^2 is 1.

$$x^2 + \frac{7}{2}x = 3$$

Add 3 to both sides.

$$x^2 + \frac{7}{2}x + \frac{49}{16} = 3 + \frac{49}{16}$$

One-half of $\frac{7}{2}$ is $\frac{7}{4}$. Square $\frac{7}{4}$, which is $\frac{49}{16}$, and add it to both sides.

$$\left(x + \frac{7}{4}\right)^2 = \frac{97}{16}$$

On the left-hand side, factor the trinomial. On the right-hand side, $3 = \frac{3 \cdot 16}{1 \cdot 16} = \frac{48}{16}$ and $\frac{48}{16} + \frac{49}{16} = \frac{97}{16}$.

$$x + \frac{7}{4} = \pm\frac{\sqrt{97}}{4}$$

Apply the square root property. On the right-hand side, $\sqrt{\frac{97}{16}} = \frac{\sqrt{97}}{\sqrt{16}} = \frac{\sqrt{97}}{4}$.

$$x = -\frac{7}{4} + \frac{\sqrt{97}}{4} \quad \text{or} \quad x = -\frac{7}{4} - \frac{\sqrt{97}}{4}$$

Subtract $\frac{7}{4}$ from both sides.

$$x = \frac{-7 + \sqrt{97}}{4} \qquad x = \frac{-7 - \sqrt{97}}{4}$$

Write each expression as a single fraction.

State the conclusion

The width of the border should be $\dfrac{-7 + \sqrt{97}}{4} \approx 0.71$ inch. (We discard the solution $\dfrac{-7 - \sqrt{97}}{4}$, since it is negative.)

Check the result

If the border is 0.71 inch wide, the announcement has an area of about $5.42 \cdot 4.42 \approx 23.96$ in.^{2}. If we subtract the area of the text from the area of the announcement, we get $23.96 - 12 = 11.96$ in.^{2}. This represents the area of the border, which was to be 12 in.^{2}. The answer seems reasonable.

Section 8.1 STUDY SET

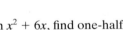

INTERMEDIATE
Algebra $f(x)$ Now™

VOCABULARY *Fill in the blanks.*

1. An equation of the form $ax^2 + bx + c = 0$ where $a \neq 0$ is called a _____ equation.

2. In the expression $2 \pm \sqrt{3}$, the symbol $\pm$ is read as "_____ ____ _____."

3. $x^2 + 6x + 9$ is called a _____ square trinomial because it factors as $(x + 3)^2$.

4. The _____ of x^2 in $x^2 - 12x + 36 = 0$ is 1, and the _____ term is 36.

CONCEPTS *Fill in the blanks.*

5. The solutions of $x^2 = c$, where $c > 0$, are $x = \boxed{}$ and $x = \boxed{}$.

6. To complete the square on x in $x^2 + 6x$, find one-half of $\boxed{}$, square it to get $\boxed{}$, and add $\boxed{}$ to get $\boxed{}$.

7. Check to see whether $3\sqrt{2}$ is a solution of $x^2 - 18 = 0$.

8. Check to see whether $-2 + \sqrt{2}$ is a solution of $x^2 + 4x + 2 = 0$.

9. Find one-half of the coefficient of x and then square it. What is the result?

    a. $x^2 + 12x$

    b. $x^2 - 5x$

    c. $x^2 - \dfrac{x}{2}$

10. Add a number to make each binomial a perfect square trinomial. Then factor the result.

  a. $x^2 + 8x$

  b. $x^2 - 8x$

  c. $x^2 - x$

11. What is the first step in solving the equation $x^2 + 12x = 35$

  a. by the factoring method?

  b. by completing the square?

12. Solve the equation: $x^2 = 16$.

  a. by the factoring method.

  b. by the square root method.

13. Write the expression on the right-hand side of the equation in simplified form by applying the quotient rule for radicals and then rationalizing the denominator.

$$x - 1 = \pm\sqrt{\frac{5}{2}}$$

14. Write the expression on the right-hand side of the equation as a single fraction by first expressing 1 as a fraction with a denominator of 2.

$$x = 1 \pm \frac{\sqrt{10}}{2}$$

15. Explain the error in the work below.

$$\frac{4 \pm \sqrt{3}}{8} = \frac{\overset{1}{\cancel{4}} \pm \sqrt{3}}{\underset{1}{\cancel{4} \cdot 2}}$$
$$= \frac{1 \pm \sqrt{3}}{2}$$

16. Explain the error in the work below.

$$\frac{1 \pm \sqrt{5}}{5} = \frac{1 \pm \sqrt{\cancel{5}}^{\,1}}{\underset{1}{\cancel{5}}}$$
$$= \frac{1 \pm 1}{1}$$

NOTATION

17. Explain why the following is true:

$$3 \pm \frac{\sqrt{10}}{5} \neq \frac{3 \pm \sqrt{10}}{5}$$

18. Isolate x on the left-hand side of the equation by subtracting 3 from both sides of the equation.

$$x + 3 = \pm\sqrt{5}$$

19. In solving a quadratic equation, a student obtains $x = \pm 2\sqrt{5}$.

  a. How many solutions are represented by this notation? List them.

  b. Approximate the solutions to the nearest hundredth.

20. In solving a quadratic equation, a student obtains

$$x = \frac{-5 \pm \sqrt{7}}{3}.$$

  a. How many solutions are represented by this notation? List them

  b. Approximate the solutions to the nearest hundredth.

PRACTICE   *Use factoring to solve each equation.*

21. $6x^2 + 12x = 0$ **22.** $5x^2 + 11x = 0$

23. $2y^2 - 50 = 0$ **24.** $4y^2 - 64 = 0$

25. $r^2 + 6r + 8 = 0$ **26.** $x^2 + 9x + 20 = 0$

27. $2z^2 = -2 + 5z$ **28.** $3x^2 = 8 - 10x$

Use the square root property to solve each equation.

29. $x^2 = 36$ **30.** $x^2 = 144$

31. $z^2 = 5$ **32.** $u^2 = 24$

33. $3x^2 - 16 = 0$ **34.** $5x^2 - 49 = 0$

35. $(x + 1)^2 = 1$ **36.** $(x - 1)^2 = 4$

37. $(s - 7)^2 - 9 = 0$ **38.** $(t + 4)^2 = 16$

39. $(x + 5)^2 - 3 = 0$ **40.** $(x + 3)^2 - 7 = 0$

Use the square root property to solve for the indicated variable. Assume that all variables represent positive numbers. Express all radicals in simplified form.

41. $2d^2 = 3h$ for d **42.** $2x^2 = d^2$ for d

43. $E = mc^2$ for c **44.** $A = \pi r^2$ for r

Use completing the square to solve each equation.

45. $x^2 + 2x - 8 = 0$ **46.** $x^2 + 6x + 5 = 0$

47. $k^2 - 8k + 12 = 0$ **48.** $p^2 - 4p + 3 = 0$

49. $g^2 + 5g - 6 = 0$ **50.** $s^2 + 5s - 14 = 0$

51. $x^2 - 3x - 4 = 0$ **52.** $x^2 - 7x + 12 = 0$

53. $2x^2 - x - 1 = 0$ **54.** $2x^2 - 5x + 2 = 0$

55. $6x^2 + x - 2 = 0$ **56.** $9 - 6r = 8r^2$

57. $x^2 + 8x + 6 = 0$ **58.** $x^2 + 6x + 4 = 0$

59. $x^2 - 2x - 17 = 0$ **60.** $x^2 + 10x - 7 = 0$

61. $b^2 - 3b = 5$ **62.** $a^2 - a = 3$

63. $3x^2 - 6x = 1$ **64.** $2x^2 - 6x = -3$

65. $4x^2 - 4x = 7$ **66.** $2x^2 - 8x = -5$

67. $2x^2 + 5x - 2 = 0$ **68.** $4x^2 - 4x - 1 = 0$

69. $\dfrac{7x + 1}{5} = -x^2$ **70.** $\dfrac{3x^2}{8} = \dfrac{1}{8} - x$

APPLICATIONS

71. FLAGS In 1912, an executive order by President Taft fixed the overall width and length of the U.S. flag in the ratio 1 to 1.9. If 100 square feet of cloth are to be used to make a U.S. flag, estimate its dimensions to the nearest $\frac{1}{4}$ foot.

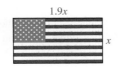

72. MOVIE STUNTS According to the *Guinness Book of World Records, 1998,* stuntman Dan Koko fell a distance of 312 feet into an airbag after jumping from the Vegas World Hotel and Casino. The distance d in feet traveled by a free-falling object in t seconds is given by the formula $d = 16t^2$. To the nearest tenth of a second, how long did the stuntman's free fall last?

73. ACCIDENTS The height h (in feet) of an object that is dropped from a height of s feet is given by the formula $h = s - 16t^2$, where t is the time the object has been falling. A 5-foot-tall woman on a sidewalk looks directly overhead and sees a window washer drop a bottle from 4 stories up. How long does she have to get out of the way? Round to the nearest tenth. (A story is 12 feet.)

74. GEOGRAPHY The surface area S of a sphere is given by the formula $S = 4\pi r^2$, where r is the radius of the sphere. An almanac lists the surface area of Earth as 196,938,800 square miles. Assuming Earth to be spherical, what is its radius to the nearest mile?

75. AUTOMOBILE ENGINES As the piston shown in the illustration below moves upward, it pushes a cylinder of a gasoline/air mixture that is ignited by the spark plug. The formula that gives the volume of a cylinder is $V = \pi r^2 h$, where r is the radius and h the height. Find the radius of the piston (to the nearest hundredth of an inch) if it displaces 47.75 cubic inches of gasoline/air mixture as it moves from its lowest to its highest point.

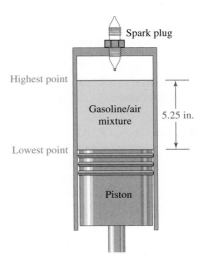

76. INVESTMENTS If P dollars are deposited in an account that pays an annual rate of interest r, then in n years, the amount of money A in the account is given by the formula $A = P(1 + r)^n$. A savings account was opened on January 3, 1999, with a deposit of $10,000 and closed on January 2, 2001, with an ending balance of $11,772.25. Find r, the rate of interest.

77. PICTURE FRAMING The matting around the picture below has a uniform width. How wide is the matting if its area equals the area of the picture? Round to the nearest hundredth of an inch.

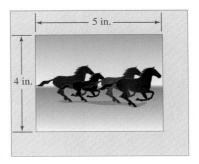

78. SWIMMING POOLS See the advertisement on the next page. How wide will the free concrete decking be if a uniform width is constructed around the perimeter of the pool? Round to the nearest hundredth of a yard. (*Hint:* Note the difference in units.)

SAHARA POOL & SPA
SUMMER SPECIAL

This 18 ft x 30 ft pool: only $18,500

Buy now and receive 28 square yards of concrete decking FREE!

79. DIMENSIONS OF A RECTANGLE A rectangle is 4 feet longer than it is wide, and its area is 20 square feet. Find its dimensions to the nearest tenth of a foot.

80. DIMENSIONS OF A TRIANGLE The height of a triangle is 4 meters longer than twice its base.

Find the base and height if the area of the triangle is 10 square meters. Round to the nearest hundredth of a meter.

■ WRITING

81. Explain how to complete the square.

82. Explain why a cannot be 0 in the quadratic equation $ax^2 + bx + c = 0$.

■ REVIEW *Simplify each expression. All variables represent positive real numbers.*

83. $\sqrt[3]{40a^3b^6}$

84. $\sqrt[3]{-27x^6}$

85. $\sqrt[8]{x^{24}}$

86. $\sqrt[4]{\dfrac{16}{625}}$

87. $\sqrt{175a^2b^3}$

88. $\sqrt{\dfrac{z^2}{16x^2}}$

8.2 The Quadratic Formula

- The quadratic formula
- Solving quadratic equations using the quadratic formula
- Solving an equivalent equation
- Problem solving

We can solve any quadratic equation by completing the square, but the work is often tedious. In this section, we will develop a formula, called the *quadratic formula,* that lets us solve quadratic equations with much less effort.

■ The quadratic formula

To develop a formula we can use to solve quadratic equations, we solve the general quadratic equation $ax^2 + bx + c = 0$ for x, where $a > 0$.

$$ax^2 + bx + c = 0$$

$$\frac{ax^2}{a} + \frac{bx}{a} + \frac{c}{a} = \frac{0}{a}$$
Since $a > 0$, we can divide both sides by a.

$$x^2 + \frac{bx}{a} = -\frac{c}{a}$$
$\frac{0}{a} = 0$; subtract $\frac{c}{a}$ from both sides.

$$x^2 + \frac{b}{a}x + \left(\frac{b}{2a}\right)^2 = \left(\frac{b}{2a}\right)^2 - \frac{c}{a}$$
Complete the square on x. One-half of $\frac{b}{a}$ is $\frac{b}{2a}$. Add $\left(\frac{b}{2a}\right)^2$ to both sides.

$$x^2 + \frac{b}{a}x + \frac{b^2}{4a^2} = \frac{b^2}{4a^2} - \frac{4ac}{4aa}$$
Find both powers. Get a common denominator of $4a^2$ on the right-hand side.

(1)
$$\left(x + \frac{b}{2a}\right)^2 = \frac{b^2 - 4ac}{4a^2}$$
Factor the left-hand side and add the fractions on the right-hand side.

We can solve Equation 1 using the square root property.

$$x + \frac{b}{2a} = \sqrt{\frac{b^2 - 4ac}{4a^2}} \qquad \text{or} \qquad x + \frac{b}{2a} = -\sqrt{\frac{b^2 - 4ac}{4a^2}}$$

$$x + \frac{b}{2a} = \frac{\sqrt{b^2 - 4ac}}{\sqrt{4a^2}} \qquad\qquad x + \frac{b}{2a} = -\frac{\sqrt{b^2 - 4ac}}{\sqrt{4a^2}}$$

$$x + \frac{b}{2a} = \frac{\sqrt{b^2 - 4ac}}{2a} \qquad\qquad x + \frac{b}{2a} = -\frac{\sqrt{b^2 - 4ac}}{2a}$$

$$x = -\frac{b}{2a} + \frac{\sqrt{b^2 - 4ac}}{2a} \qquad\qquad x = -\frac{b}{2a} - \frac{\sqrt{b^2 - 4ac}}{2a}$$

$$x = \frac{-b + \sqrt{b^2 - 4ac}}{2a} \qquad\qquad x = \frac{-b - \sqrt{b^2 - 4ac}}{2a}$$

If $a < 0$, it can be shown that the same two solutions are obtained. The result is called the **quadratic formula.**

The quadratic formula

The solutions of $ax^2 + bx + c = 0$ where $a \neq 0$ are given by the formula

$$x = \frac{-b \pm \sqrt{b^2 - 4ac}}{2a} \qquad \text{Read the symbol } \pm \text{ as "plus or minus."}$$

! COMMENT Be sure to draw the fraction bar under both parts of the numerator, and be sure to draw the radical symbol only over $b^2 - 4ac$. Do not write the quadratic formula as

$$x = -b \pm \frac{\sqrt{b^2 - 4ac}}{2a} \qquad \text{or as} \qquad x = -b \pm \sqrt{\frac{b^2 - 4ac}{2a}}$$

■ Solving quadratic equations using the quadratic formula

INTERMEDIATE
Algebra *f(x)* Now™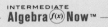

EXAMPLE 1   Solve: $6x^2 - x - 5 = 0$.

Solution   This equation is written in $ax^2 + bx + c = 0$ form, where $a = 6$, $b = -1$, and $c = -5$.

$$x = \frac{-b \pm \sqrt{b^2 - 4ac}}{2a}$$

$$= \frac{-(-1) \pm \sqrt{(-1)^2 - 4(6)(-5)}}{2(6)} \qquad \text{Substitute 6 for } a, -1 \text{ for } b, \text{ and } -5 \text{ for } c.$$

$$= \frac{1 \pm \sqrt{1 + 120}}{12} \qquad \text{Simplify: } -(-1) = 1. \text{ Do the operations within the radical.}$$

$$= \frac{1 \pm \sqrt{121}}{12} \qquad \text{Simplify within the radical.}$$

$$= \frac{1 \pm 11}{12} \qquad \text{Simplify: } \sqrt{121} = 11.$$

This result represents two solutions. To find the first solution, we add in the numerator. To find the second, we subtract in the numerator.

Self Check 1

Solve: $3x^2 - 5x - 2 = 0$.

$$x = \frac{1 + 11}{12} \quad \text{or} \quad x = \frac{1 - 11}{12}$$

$$x = \frac{12}{12} \qquad\qquad x = \frac{-10}{12}$$

$$x = 1 \qquad\qquad x = -\frac{5}{6}$$

Answer $2, -\dfrac{1}{3}$

Verify that 1 and $-\frac{5}{6}$ satisfy the original equation.

! COMMENT You may have noticed in Example 1 that $6x^2 - x - 5$ factors as $(6x + 5)(x - 1)$. Therefore, we could have solved $6x^2 - x - 5 = 0$ using the factoring method.

When using the quadratic formula, we should write the equation in $ax^2 + bx + c = 0$ form (called **quadratic form**) so that a, b, and c can be determined.

Self Check 2
Solve: $3x^2 = 2x + 3$.
Approximate the solutions to two decimal places.

EXAMPLE 2   Solve: $2x^2 = -4x - 1$.

Solution   We begin by writing the equation in quadratic form:

$$2x^2 + 4x + 1 = 0 \quad \text{Add } 4x \text{ and 1 to both sides.}$$

In this equation, $a = 2$, $b = 4$, and $c = 1$.

$$x = \frac{-b \pm \sqrt{b^2 - 4ac}}{2a}$$

$$= \frac{-4 \pm \sqrt{4^2 - 4(2)(1)}}{2(2)} \qquad \text{Substitute 2 for } a, \text{ 4 for } b, \text{ and 1 for } c.$$

$$= \frac{-4 \pm \sqrt{16 - 8}}{4} \qquad \text{Simplify within the radical.}$$

$$= \frac{-4 \pm \sqrt{8}}{4}$$

$$x = \frac{-4 \pm 2\sqrt{2}}{4} \qquad \text{Simplify: } \sqrt{8} = \sqrt{4 \cdot 2} = 2\sqrt{2}.$$

We can write the solutions in simpler form by factoring out 2 from the terms in the numerator and then dividing out the common factor.

$$x = \frac{-4 \pm 2\sqrt{2}}{4} = \frac{2(-2 \pm \sqrt{2})}{4} = \frac{\overset{1}{2}(-2 \pm \sqrt{2})}{\underset{1}{2} \cdot 2} = \frac{-2 \pm \sqrt{2}}{2}$$

Answer $\dfrac{1 \pm \sqrt{10}}{3}$; -0.72, 1.39

The solutions are $\dfrac{-2 + \sqrt{2}}{2}$ and $\dfrac{-2 - \sqrt{2}}{2}$. We can approximate the solutions using a calculator. To two decimal places, they are -0.29 and -1.71.

■ Solving an equivalent equation

When solving a quadratic equation by the quadratic formula, we can often simplify the computations by solving an equivalent equation.

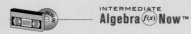
INTERMEDIATE
Algebra $f(x)$ Now™

EXAMPLE 3   For each equation, write an equivalent equation so that the quadratic formula computations will be easier:
a. $-2x^2 + 4x - 1 = 0$,   **b.** $x^2 + \frac{4}{5}x - \frac{1}{3} = 0$,   and   **c.** $20x^2 - 60x - 40 = 0$.

Solution

a. It is often easier to solve a quadratic equation using the quadratic formula if a is positive. If we multiply (or divide) both sides of $-2x^2 + 4x - 1 = 0$ by -1, we obtain an equivalent equation with $a > 0$.

$$-2x^2 + 4x - 1 = 0 \qquad \text{Here, } a = -2.$$

$$(-1)(-2x^2 + 4x - 1) = (-1)(0)$$

$$2x^2 - 4x + 1 = 0 \qquad \text{Now } a = 2.$$

b. For $x^2 + \frac{4}{5}x - \frac{1}{3} = 0$, two coefficients are fractions: $b = \frac{4}{5}$ and $c = -\frac{1}{3}$. We can multiply both sides of the equation by their least common denominator, 15, to obtain an equivalent equation having coefficients that are integers.

$$x^2 + \frac{4}{5}x - \frac{1}{3} = 0 \qquad \text{Here, } a = 1, b = \tfrac{4}{5}, \text{ and } c = -\tfrac{1}{3}.$$

$$15\left(x^2 + \frac{4}{5}x - \frac{1}{3} \right) = 15(0)$$

$$15x^2 + 12x - 5 = 0 \qquad \text{Now } a = 15, b = 12, \text{ and } c = -5.$$

c. For $20x^2 - 60x - 40 = 0$, the coefficients 20, -60, and -40 have a common factor of 20. If we divide both sides of the equation by their GCF, we obtain an equivalent equation having smaller coefficients.

$$20x^2 - 60x - 40 = 0 \qquad \text{Here, } a = 20, b = -60, \text{ and } c = -40.$$

$$\frac{20x^2}{20} - \frac{60x}{20} - \frac{40}{20} = \frac{0}{20}$$

$$x^2 - 3x - 2 = 0 \qquad \text{Now } a = 1, b = -3, \text{ and } c = -2.$$

Self Check 3
For each equation, write an equivalent equation so that the quadratic formula computations will be simpler:
a. $-6x^2 + 7x - 9 = 0$
b. $\frac{1}{3}x^2 - \frac{2}{3}x - \frac{5}{6} = 0$
c. $44x^2 + 66x - 55 = 0$

Answers **a.** $6x^2 - 7x + 9 = 0$,
b. $2x^2 - 4x - 5 = 0$,
c. $4x^2 + 6x - 5 = 0$

Problem solving

EXAMPLE 4 **Taking shortcuts.** Instead of using the existing hallways, students are wearing a path through a planted quad area to walk 195 feet directly from the classrooms to the cafeteria. See Figure 8-4. If the length of the hallway from the office to the cafeteria is 105 feet longer than the hallway from the office to the classrooms, how much walking are the students saving by taking the shortcut?

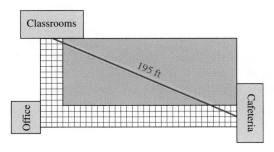

FIGURE 8-4

Analyze the problem

The two hallways and the shortcut form a right triangle with a hypotenuse 195 feet long. We will use the Pythagorean theorem to solve this problem.

Form an equation

If we let x = the length (in feet) of the hallway from the classrooms to the office, then the length of the hallway from the office to the cafeteria is $(x + 105)$ feet. Substituting these lengths into the Pythagorean theorem, we have

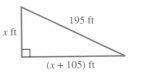

x ft   195 ft   (x + 105) ft

$$a^2 + b^2 = c^2 \qquad \text{This is the Pythagorean theorem.}$$

$$x^2 + (x + 105)^2 = 195^2 \qquad \text{Substitute } x \text{ for } a, (x + 105) \text{ for } b, \text{ and } 195 \text{ for } c.$$

$$x^2 + x^2 + 105x + 105x + 11{,}025 = 38{,}025 \qquad \text{Find } (x + 105)^2.$$

$$2x^2 + 210x + 11{,}025 = 38{,}025 \qquad \text{Combine like terms.}$$

$$2x^2 + 210x - 27{,}000 = 0 \qquad \text{Subtract 38,025 from both sides.}$$

$$x^2 + 105x - 13{,}500 = 0 \qquad \text{Divide both sides by 2.}$$

Solve the equation

To solve $x^2 + 105x - 13{,}500 = 0$, we will use the quadratic formula with $a = 1$, $b = 105$, and $c = -13{,}500$.

$$x = \frac{-b \pm \sqrt{b^2 - 4ac}}{2a}$$

$$x = \frac{-105 \pm \sqrt{(105)^2 - 4(1)(-13{,}500)}}{2(1)}$$

$$x = \frac{-105 \pm \sqrt{65{,}025}}{2} \qquad \text{Simplify: } (105)^2 - 4(1)(-13{,}500) = 11{,}025 + 54{,}000 = 65{,}025.$$

$$x = \frac{-105 \pm 255}{2} \qquad \text{Use a calculator: } \sqrt{65{,}025} = 255.$$

$$x = \frac{150}{2} \quad \text{or} \quad x = \frac{-360}{2}$$

$$x = 75 \qquad\qquad x = -180 \qquad \text{Since the length of the hallway can't be negative, discard the solution } x = -180.$$

State the conclusion

The length of the hallway from the classrooms to the office is 75 feet. The length of the hallway from the office to the cafeteria is $75 + 105 = 180$ feet. Instead of using the hallways, a distance of $75 + 180 = 255$ feet, the students are taking the 195-foot short-cut to the cafeteria, a savings of $(255 - 195)$, or 60 feet.

Check the result

The length of the 180-foot hallway is 105 feet longer than the length of the 75-foot hallway. The sum of the squares of the lengths of the hallways is $75^2 + 180^2 = 38{,}025$. This equals the square of the length of the 195-foot shortcut. The answer checks.

EXAMPLE 5   Mass transit.   A bus company has 4,000 passengers daily, each currently paying a 75¢ fare. For each 15¢ fare increase, the company estimates that it will lose 50 passengers. If the company needs to bring in $6,570 per day to stay in business, what fare must be charged to produce this amount of revenue?

Analyze the problem

To understand how a fare increase affects the number of passengers, let's consider what happens if there are two fare increases. We organize the data in a table. The fares are expressed in terms of dollars.

Number of increases	New fare	Number of passengers
One $0.15 increase	$0.75 + $0.15(1) = $0.90	4,000 − 50(1) = 3,950
Two $0.15 increases	$0.75 + $0.15(2) = $1.05	4,000 − 50(2) = 3,900

In general, the new fare will be the old fare ($0.75) plus the number of fare increases times $0.15. The number of passengers who will pay the new fare is 4,000 minus 50 times the number of $0.15 fare increases.

Form an equation

If we let $x =$ the number of $0.15 fare increases necessary to bring in $6,570 daily, then $(0.75 + 0.15x)$ is the fare that must be charged. The number of passengers who will pay this fare is $(4,000 - 50x)$. We can now form the equation.

The bus fare	times	the number of passengers who will pay that fare	equals	$6,570.
$(0.75 + 0.15x)$	·	$(4,000 - 50x)$	=	6,570

Solve the equation

$$(0.75 + 0.15x)(4,000 - 50x) = 6,570$$

$3,000 - 37.5x + 600x - 7.5x^2 = 6,570$ Multiply the binomials.

$-7.5x^2 + 562.5x + 3,000 = 6,570$ Combine like terms: $-37.5x + 600x = 562.5x$.

$-7.5x^2 + 562.5x - 3,570 = 0$ Subtract 6,570 from both sides.

$7.5x^2 - 562.5x + 3,570 = 0$ Multiply both sides by -1 so that a, 7.5, is positive.

To solve this equation, we will use the quadratic formula.

$$x = \frac{-b \pm \sqrt{b^2 - 4ac}}{2a}$$

$$x = \frac{-(-562.5) \pm \sqrt{(-562.5)^2 - 4(7.5)(3,570)}}{2(7.5)}$$ Substitute 7.5 for a, -562.5 for b, and 3,570 for c.

$$x = \frac{562.5 \pm \sqrt{209,306.25}}{15}$$ Simplify: $(-562.5)^2 - 4(7.5)(3,570) = 316,406.25 - 107,100 = 209,306.25$.

$$x = \frac{562.5 \pm 457.5}{15}$$ Use a calculator: $\sqrt{209,306.25} = 457.5$.

$$x = \frac{1,020}{15} \quad \text{or} \quad x = \frac{105}{15}$$

$$x = 68 \qquad\qquad x = 7$$

State the conclusion

If there are 7 fifteen-cent increases in the fare, the new fare will be $0.75 + $0.15(7) = $1.80. If there are 68 fifteen-cent increases in the fare, the new fare will be $0.75 + $0.15(68) = $10.95. Although this fare would bring in the necessary revenue, a $10.95 bus fare is unreasonable, so we discard it.

Check the result

A fare of $1.80 will be paid by $[4{,}000 - 50(7)] = 3{,}650$ bus riders. The amount of revenue brought in would be $\$1.80(3{,}650) = \$6{,}570$. The answer checks.

EXAMPLE 6 Lawyers.

The number of lawyers N in the United States each year from 1980 to 2002 is approximated by $N = 222x^2 + 17{,}630x + 571{,}178$, where $x = 0$ corresponds to the year 1980, $x = 1$ corresponds to 1981, $x = 2$ corresponds to 1982, and so on. (Thus, $0 \le x \le 22$.) In what year does this model indicate that the United States had one million lawyers?

Solution   We will substitute 1,000,000 for N in the equation. Then we can solve for x, which will give the number of years after 1980 that the United States had approximately 1,000,000 lawyers.

$$N = 222x^2 + 17{,}630x + 571{,}178$$

$$\mathbf{1{,}000{,}000} = 222x^2 + 17{,}630x + 571{,}178 \qquad \text{Replace } N \text{ with 1,000,000.}$$

$$0 = 222x^2 + 17{,}630x - 428{,}822 \qquad \begin{array}{l}\text{Subtract 1,000,000 from both sides so that}\\\text{the equation is in quadratic form.}\end{array}$$

We can simplify the computations by dividing both sides of the equation by 2, which is the greatest common factor of 222, 17,630, and 428,822.

$$111x^2 + 8{,}815x - 214{,}411 = 0 \qquad \text{Divide both sides by 2.}$$

We solve this equation using the quadratic formula.

$$x = \frac{-b \pm \sqrt{b^2 - 4ac}}{2a}$$

$$x = \frac{-8{,}815 \pm \sqrt{(8{,}815)^2 - 4(111)(-214{,}411)}}{2(111)} \qquad \begin{array}{l}\text{Substitute 111 for } a, \text{ 8,815 for } b,\\\text{and } -214{,}411 \text{ for } c.\end{array}$$

$$x = \frac{-8{,}815 \pm \sqrt{172{,}902{,}709}}{222} \qquad \begin{array}{l}\text{Evaluate the expression within}\\\text{the radical.}\end{array}$$

$$x \approx \frac{4{,}334}{222} \quad \text{or} \quad x \approx \frac{-21{,}964}{222} \qquad \text{Use a calculator.}$$

$$x \approx 19.5 \qquad\qquad x \approx -98.9 \qquad \begin{array}{l}\text{Since the model is defined only}\\\text{for } 0 \le x \le 22, \text{ we discard the}\\\text{second solution.}\end{array}$$

In 19.5 years after 1980, or midway through 1999, the United States had approximately 1,000,000 lawyers.

Section 8.2 STUDY SET

■ VOCABULARY   *Fill in the blanks.*

1. An equation of the form $ax^2 + bx + c = 0$ where $a \ne 0$ is a _____ equation.

2. $x = \frac{-b \pm \sqrt{b^2 - 4ac}}{2a}$ is called the _____ formula.

■ CONCEPTS

3. Write each equation in quadratic form.

   a. $x^2 + 2x = -5$.   **b.** $3x^2 = -2x + 1$.

4. For each quadratic equation, find a, b, and c.

   a. $x^2 + 5x + 6 = 0$ b. $8x^2 - x = 10$

5. Determine whether each statement is true or false.

   a. Any quadratic equation can be solved by using the quadratic formula.

   b. Any quadratic equation can be solved by completing the square.

6. What is wrong with the beginning of the solution shown below?

   Solve: $x^2 - 3x = 2$.

   $a = 1$ $b = -3$ $c = 2$

7. What form would the quadratic formula take if, in developing it, we began with the general quadratic equation expressed as $rc^2 + sc + t = 0$ where $r \neq 0$?

8. A student used the quadratic formula to solve a quadratic equation and obtained $x = \dfrac{-2 \pm \sqrt{3}}{2}$.

   a. How many solutions does the equation have? What are they exactly?

   b. Graph the solutions on a number line.

9. Simplify each of the following solutions.

   a. $x = \dfrac{3 \pm 6\sqrt{2}}{3}$

   b. $x = \dfrac{-12 \pm 4\sqrt{7}}{8}$

10. For each of the following, write an equivalent equation so that the quadratic formula computations will be easier to perform.

    a. $-5x^2 + 9x - 2 = 0$

    b. $\dfrac{1}{8}x^2 + \dfrac{1}{2}x - \dfrac{3}{4} = 0$

    c. $45x^2 + 30x - 15 = 0$

NOTATION

11. On a quiz, students were asked to write the quadratic formula. What is wrong with each answer.

    a. $x = -b \pm \dfrac{\sqrt{b^2 - 4ac}}{2a}$ b. $x = \dfrac{-b\sqrt{b^2 - 4ac}}{2a}$

12. In reading $\dfrac{-b \pm \sqrt{b^2 - 4ac}}{2a}$, we say, "The _____ of b, plus or _____ the square _____ of b _____ minus ▇ times a times c, all _____ $2a$."

PRACTICE *Use the quadratic formula to solve each equation.*

13. $x^2 + 3x + 2 = 0$

14. $x^2 - 3x + 2 = 0$

15. $x^2 + 12x = -36$

16. $y^2 - 18y = -81$

17. $2x^2 + 5x - 3 = 0$

18. $6x^2 - x - 1 = 0$

19. $5x^2 + 5x + 1 = 0$

20. $4w^2 + 6w + 1 = 0$

21. $8u = -4u^2 - 3$

22. $4t + 3 = 4t^2$

23. $-16y^2 - 8y + 3 = 0$

24. $-16x^2 - 16x - 3 = 0$

25. $x^2 - \dfrac{14}{15}x = \dfrac{8}{15}$

26. $x^2 = -\dfrac{5}{4}x + \dfrac{3}{2}$

27. $\dfrac{x^2}{2} + \dfrac{5}{2}x = -1$

28. $\dfrac{x^2}{8} - \dfrac{x}{4} = \dfrac{1}{2}$

29. $2x^2 - 1 = 3x$

30. $-9x = 2 - 3x^2$

31. $-x^2 + 10x = 18$

32. $-3x = \dfrac{x^2}{2} + 2$

33. $x^2 - 6x = 391$

34. $-x^2 + 27x = -280$

35. $x^2 - \dfrac{5}{3} = -\dfrac{11}{6}x$

36. $x^2 - \dfrac{1}{2} = \dfrac{2}{3}x$

37. $50x^2 + 30x - 10 = 0$

38. $120b^2 + 120b - 40 = 0$

39. $900x^2 - 8,100x = 1,800$

40. $-14x^2 + 21x = -49$

41. $-0.6x^2 - 0.03 = -0.4x$

42. $2x^2 + 0.1x = 0.04$

Use the quadratic formula and a calculator to solve each equation. Round answers to the nearest hundredth.

43. $x^2 + 8x + 5 = 0$

44. $2x^2 - x - 9 = 0$

45. $3x^2 - 2x - 2 = 0$

46. $81x^2 + 12x - 80 = 0$

47. $0.7x^2 - 3.5x - 25 = 0$

48. $-4.5x^2 + 0.2x + 3.75 = 0$

APPLICATIONS

49. THEATER SCREENS The largest permanent movie screen is in the Panasonic Imax theater at Darling Harbor, Sydney, Australia. The rectangular screen has an area of 11,349 square feet. Find the dimensions of the screen if it is 20 feet longer than it is wide.

50. ROCK CONCERTS During a 1997 tour, the rock group U2 used the world's largest LED (light emitting diode) electronic screen as part of the stage backdrop. The rectangular screen, with an area of 9,520 square feet, had a length that was 2 feet more than three times its width. Find the dimensions of the LED screen.

51. CENTRAL PARK Central Park is one of New York's best-known landmarks. Rectangular in shape, its length is 5 times its width. When measured in miles, its perimeter numerically exceeds its area by 4.75. Find the dimensions of Central Park if we know that its width is less than 1 mile.

52. HISTORY One of the most important cities of the ancient world was Babylon. Greek historians wrote that the city was square-shaped. Measured in miles, its area numerically exceeded its perimeter by about 124. Find its dimensions. (Round to the nearest tenth.)

53. BADMINTON The person who wrote the instructions for setting up the badminton net shown below forgot to give the specific dimensions for securing the pole. How long is the support string?

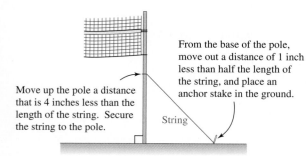

Move up the pole a distance that is 4 inches less than the length of the string. Secure the string to the pole.

From the base of the pole, move out a distance of 1 inch less than half the length of the string, and place an anchor stake in the ground.

String

54. RIGHT TRIANGLES The hypotenuse of a right triangle is 2.5 units long. The longer leg is 1.7 units longer than the shorter leg. Find the lengths of the sides of the triangle.

55. SCHOOL DANCES Tickets to a school dance cost $4, and the projected attendance is 300 persons. It is further projected that for every 10¢ increase in ticket price, the average attendance will decrease by 5. At what ticket price will the receipts from the dance be $1,248?

56. TICKET SALES A carnival at a county fair normally sells three thousand 25¢ ride tickets on a Saturday. For each 5¢ increase in price, management estimates that 80 fewer tickets will be sold. What increase in ticket price will produce $994 of revenue on Saturday?

57. MAGAZINE SALES The *Gazette's* profit is $20 per year for each of its 3,000 subscribers. Management estimates that the profit per subscriber will increase by 1¢ for each additional subscriber over the current 3,000. How many subscribers will bring a total profit of $120,000?

58. POLYGONS The five-sided polygon called a *pentagon,* shown in the illustration, has 5 diagonals. The number of diagonals d of a polygon of n sides is given by the formula

$$d = \frac{n(n-3)}{2}$$

Find the number of sides of a polygon if it has 275 diagonals.

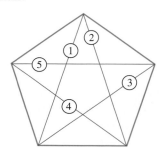

59. INVESTMENT RATES A woman invests $1,000 in a mutual fund for which interest is compounded annually at a rate r. After one year, she deposits an additional $2,000. After two years, the balance in the account is

$$\$1,000(1 + r)^2 + \$2,000(1 + r)$$

If this amount is $3,368.10, find r.

60. METAL FABRICATION A box with no top is to be made by cutting a 2-inch square from each corner of the square sheet of metal shown on the next page. After bending up the sides, the volume of the box is to be 220 cubic inches. How large should the piece of metal be? Round to the nearest hundredth.

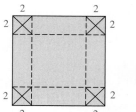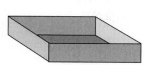

61. RETIREMENT The labor force participation rate P (in percent) for men ages 55–64 from 1970 to 2000 is approximated by the quadratic equation $P = 0.03x^2 - 1.37x + 82.51$, where $x = 0$ corresponds to the year 1970, $x = 1$ corresponds to 1971, $x = 2$ corresponds to 1972, and so on. (Thus, $0 \le x \le 30$.) When does the model indicate that 75% of the men ages 55–64 were part of the workforce?

62. SPACE PROGRAM The yearly budget B (in billions of dollars) for the National Aeronautics and Space Administration (NASA) is approximated by the equation $B = 0.0596x^2 - 0.3811x + 14.2709$, where x is the number of years since 1995 and $0 \le x \le 9$. In what year does the model indicate that NASA's budget was about $15 billion?

WRITING

63. Explain why the quadratic formula, in most cases, is less tedious to use in solving a quadratic equation than is the method of completing the square.

64. On an exam, a student was asked to solve the equation $-4w^2 - 6w - 1 = 0$. Her first step was to multiply both sides of the equation by -1. She then used the quadratic formula to solve $4w^2 + 6w + 1 = 0$ instead. Is this a valid approach? Explain.

REVIEW *Change each radical to an exponential expression.*

65. $\sqrt{n}$

66. $\sqrt[7]{\dfrac{3}{8}r^2s}$

67. $\sqrt[4]{3b}$

68. $3\sqrt[3]{c^2 - d^2}$

Write each expression in radical form.

69. $t^{1/3}$

70. $\left(\dfrac{3}{4}m^2n^2\right)^{1/5}$

71. $(3t)^{1/4}$

72. $(c^2 + d^2)^{1/2}$

8.3 Quadratic Functions and Their Graphs

- Quadratic functions • Graphing $f(x) = ax^2$ • Graphing $f(x) = ax^2 + k$
- Graphing $f(x) = a(x - h)^2$ • Graphing $f(x) = a(x - h)^2 + k$
- Graphing $f(x) = ax^2 + bx + c$ • Determining minimum and maximum values
- Solving quadratic equations graphically

The graph in Figure 8-5 on the next page shows a trampolinist's distance from the ground (in relation to time) as she bounds into the air and then falls back down to the trampoline.

From the graph, we can see that the trampolinist is 14 feet above the ground 0.5 second after bounding upward and that her height above the ground after 1.75 seconds is 9 feet.

The parabola shown in Figure 8-5 is the graph of the *quadratic function* $s(t) = -16t^2 + 32t + 2$. In this section, we will discuss two forms in which quadratic functions are written and how to graph them.

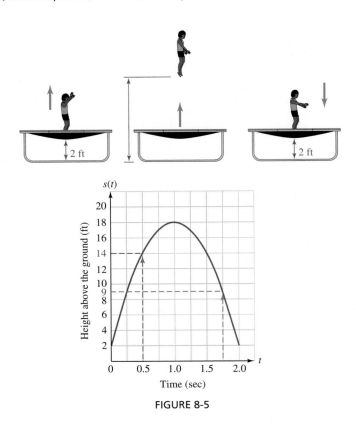

FIGURE 8-5

Quadratic functions

Quadratic functions

A quadratic function is a second-degree polynomial function of the form

$$f(x) = ax^2 + bx + c$$

where a, b, and c represent real numbers and $a \neq 0$.

INTERMEDIATE
Algebra $f(x)$ Now™

Self Check 1
Find the distance the trampolinist is from the ground after being in the air for $1\frac{1}{2}$ seconds.

EXAMPLE 1   The quadratic function $s(t) = -16t^2 + 32t + 2$ gives the distance (in feet) that the trampolinist shown in Figure 8-5 is from the ground, t seconds after bounding upward. How far is she from the ground after being in the air for $\frac{3}{4}$ second?

Solution   To find her distance from the ground, we find the value of the function for $t = \frac{3}{4} = 0.75$.

$$s(t) = -16t^2 + 32t + 2$$
$$s(0.75) = -16(0.75)^2 + 32(0.75) + 2 \quad \text{Replace } t \text{ with } 0.75.$$
$$= -9 + 24 + 2$$
$$= 17$$

Answer 14 feet

The trampolinist is 17 feet off the ground $\frac{3}{4}$ second after bounding upward.

We have seen that important information can be gained from the graph of a quadratic function. We now begin a discussion of how to graph such a function by considering the simplest case, quadratic functions of the form $f(x) = ax^2$.

Graphing $f(x) = ax^2$

EXAMPLE 2 Graph: **a.** $f(x) = x^2$, **b.** $g(x) = 3x^2$, and **c.** $h(x) = \dfrac{1}{3}x^2$.

Solution We can make a table of values for each function, plot each point, and join them with a smooth curve, as in Figure 8-6. We note that the graph of $h(x) = \frac{1}{3}x^2$ is wider than the graph of $f(x) = x^2$, and that the graph of $g(x) = 3x^2$ is narrower than the graph of $f(x) = x^2$. In the function $f(x) = ax^2$, the smaller the value of $|a|$, the wider the graph.

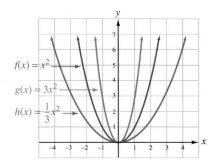

FIGURE 8-6

$f(x) = x^2$		
x	$f(x)$	$(x, f(x))$
-2	4	$(-2, 4)$
-1	1	$(-1, 1)$
0	0	$(0, 0)$
1	1	$(1, 1)$
2	4	$(2, 4)$

$g(x) = 3x^2$		
x	$g(x)$	$(x, g(x))$
-2	12	$(-2, 12)$
-1	3	$(-1, 3)$
0	0	$(0, 0)$
1	3	$(1, 3)$
2	12	$(2, 12)$

$h(x) = \frac{1}{3}x^2$		
x	$h(x)$	$(x, h(x))$
-2	$\frac{4}{3}$	$(-2, \frac{4}{3})$
-1	$\frac{1}{3}$	$(-1, \frac{1}{3})$
0	0	$(0, 0)$
1	$\frac{1}{3}$	$(1, \frac{1}{3})$
2	$\frac{4}{3}$	$(2, \frac{4}{3})$

EXAMPLE 3 Graph: $f(x) = -3x^2$.

Solution We make a table of values for the function, plot each point, and join them with a smooth curve, as in Figure 8-7. We see that the parabola opens downward and has the same shape as the graph of $g(x) = 3x^2$ in Example 2.

$f(x) = -3x^2$		
x	$f(x)$	$(x, f(x))$
-2	-12	$(-2, -12)$
-1	-3	$(-1, -3)$
0	0	$(0, 0)$
1	-3	$(1, -3)$
2	-12	$(2, -12)$

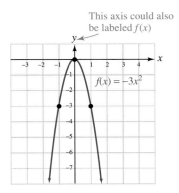

This axis could also be labeled $f(x)$

$f(x) = -3x^2$

FIGURE 8-7

Self Check 3

Graph: $f(x) = -\dfrac{1}{3}x^2$.

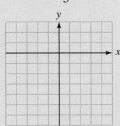

Answer

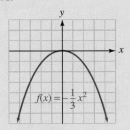

$f(x) = -\dfrac{1}{3}x^2$

The graphs of quadratic functions of the form $f(x) = ax^2$ are **parabolas.** The lowest point of a parabola that opens upward, or the highest point of a parabola that opens downward, is called the **vertex** of the parabola. For example, the vertex of the parabola shown in Figure 8-7 is the point $(0, 0)$. The vertical line, called an **axis of symmetry,** that passes through the vertex divides the parabola into two congruent halves. If we fold the paper along the axis of symmetry, the two sides of the parabola will match. The axis of symmetry of the parabola shown in Figure 8-7 is the line $x = 0$ (the y-axis).

The results of Examples 2 and 3 confirm the following facts.

> **The graph of $f(x) = ax^2$**
>
> The graph of $f(x) = ax^2$ is a parabola opening upward when $a > 0$ and downward when $a < 0$, with vertex at the point $(0, 0)$ and axis of symmetry the line $x = 0$.

■ Graphing $f(x) = ax^2 + k$

EXAMPLE 4   Graph: **a.** $f(x) = 2x^2$,   **b.** $g(x) = 2x^2 + 3$,   and **c.** $h(x) = 2x^2 - 3$.

Solution   We make a table of values for each function, plot each point, and join them with a smooth curve, as in Figure 8-8. We note that the graph of $g(x) = 2x^2 + 3$ is identical to the graph of $f(x) = 2x^2$, except that it has been translated 3 units upward. The graph of $h(x) = 2x^2 - 3$ is identical to the graph of $f(x) = 2x^2$, except that it has been translated 3 units downward. In each case, the axis of symmetry is the line $x = 0$.

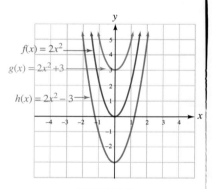

FIGURE 8-8

$f(x) = 2x^2$		
x	$f(x)$	$(x, f(x))$
-2	8	$(-2, 8)$
-1	2	$(-1, 2)$
0	0	$(0, 0)$
1	2	$(1, 2)$
2	8	$(2, 8)$

$g(x) = 2x^2 + 3$		
x	$g(x)$	$(x, g(x))$
-2	11	$(-2, 11)$
-1	5	$(-1, 5)$
0	3	$(0, 3)$
1	5	$(1, 5)$
2	11	$(2, 11)$

$h(x) = 2x^2 - 3$		
x	$h(x)$	$(x, h(x))$
-2	5	$(-2, 5)$
-1	-1	$(-1, -1)$
0	-3	$(0, -3)$
1	-1	$(1, -1)$
2	5	$(2, 5)$

The results of Example 4 confirm the following facts.

> **The graph of $f(x) = ax^2 + k$**
>
> The graph of $f(x) = ax^2 + k$ is a parabola having the same shape as $f(x) = ax^2$ but translated upward k units if k is positive and downward $|k|$ units if k is negative. The vertex is at the point $(0, k)$, and the axis of symmetry is the line $x = 0$.

Graphing $f(x) = a(x - h)^2$

EXAMPLE 5 Graph: **a.** $f(x) = 2x^2$, **b.** $g(x) = 2(x - 3)^2$, and
c. $h(x) = 2(x + 3)^2$.

Solution We make a table of values for each function, plot each point, and join them with a smooth curve, as in Figure 8-9. We note that the graph of $g(x) = 2(x - 3)^2$ is identical to the graph of $f(x) = 2x^2$, except that it has been translated 3 units to the right. The graph of $h(x) = 2(x + 3)^2$ is identical to the graph of $f(x) = 2x^2$, except that it has been translated 3 units to the left.

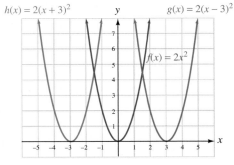

FIGURE 8-9

$f(x) = 2x^2$		
x	$f(x)$	$(x, f(x))$
-2	8	$(-2, 8)$
-1	2	$(-1, 2)$
0	0	$(0, 0)$
1	2	$(1, 2)$
2	8	$(2, 8)$

$g(x) = 2(x - 3)^2$		
x	$g(x)$	$(x, g(x))$
1	8	$(1, 8)$
2	2	$(2, 2)$
3	0	$(3, 0)$
4	2	$(4, 2)$
5	8	$(5, 8)$

$h(x) = 2(x + 3)^2$		
x	$h(x)$	$(x, h(x))$
-5	8	$(-5, 8)$
-4	2	$(-4, 2)$
-3	0	$(-3, 0)$
-2	2	$(-2, 2)$
-1	8	$(-1, 8)$

The results of Example 5 confirm the following facts.

The graph of $f(x) = a(x - h)^2$

The graph of $f(x) = a(x - h)^2$ is a parabola having the same shape as $f(x) = ax^2$ but translated h units to the right if h is positive and $|h|$ units to the left if h is negative. The vertex is at the point $(h, 0)$, and the axis of symmetry is the line $x = h$.

Graphing $f(x) = a(x - h)^2 + k$

EXAMPLE 6 Graph: $f(x) = 2(x - 3)^2 - 4$.

Solution The graph of $f(x) = 2(x - 3)^2 - 4$ is identical to the graph of $g(x) = 2(x - 3)^2$, except that it has been translated 4 units downward. The graph of $g(x) = 2(x - 3)^2$ is identical to the graph of $h(x) = 2x^2$, except that it has been translated 3 units to the right. Thus, to graph $f(x) = 2(x - 3)^2 - 4$, we can graph $h(x) = 2x^2$ and shift it 3 units to the right and then 4 units downward, as shown in Figure 8-10 on the next page.

Self Check 6
Graph: $f(x) = 2(x - 1)^2 - 2$. Label the vertex and draw the axis of symmetry.

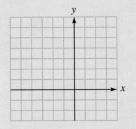

The vertex of the graph is the point $(3, -4)$, and the axis of symmetry is the line $x = 3$.

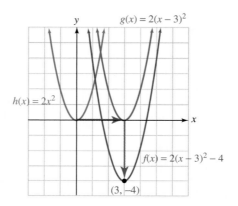

FIGURE 8-10

Answer

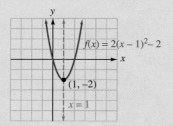

The results of Example 6 confirm the following facts.

Graphing a quadratic function in standard form

The graph of the quadratic function

$$f(x) = a(x - h)^2 + k \quad \text{where } a \neq 0$$

is a parabola with vertex at (h, k). The axis of symmetry is the line $x = h$. The parabola opens upward when $a > 0$ and downward when $a < 0$.

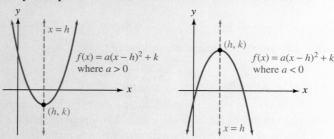

INTERMEDIATE
Algebra $f(x)$ **Now**™

Self Check 7

Answer parts a, b, and c of Example 7 for the graph of $f(x) = 6(x - 5)^2 + 1$.

EXAMPLE 7 Consider the graph of $f(x) = -3(x + 1)^2 - 4$.

a. Does the graph open upward or downward?

b. What are the coordinates of the vertex?

c. What is the axis of symmetry?

Solution Rewriting the given function in $f(x) = a(x - h)^2 + k$ form, we have

Standard form requires a minus sign here. Standard form requires a plus sign here.

$$f(x) = \underset{a}{-3} \, [x - \underset{h}{(-1)}]^2 + \underset{k}{(-4)}$$

a. Since $a = -3 < 0$, the parabola opens downward.

b. The vertex is $(h, k) = (-1, -4)$.

c. The axis of symmetry is the line $x = h$. In this case, $x = -1$.

Answer **a.** upward,
b. $(5, 1)$,   **c.** $x = 5$

Graphing $f(x) = ax^2 + bx + c$

To graph functions of the form $f(x) = ax^2 + bx + c$, we complete the square to write
the function in the form $f(x) = a(x - h)^2 + k$.

EXAMPLE 8   Graph: $f(x) = 2x^2 - 4x - 1$.

Self Check 8

Graph: $f(x) = 3x^2 - 12x + 8$.

Solution

Step 1: We complete the square on x to write the given function in the form
$f(x) = a(x - h)^2 + k$.

$$f(x) = 2x^2 - 4x - 1$$
$$f(x) = 2(x^2 - 2x) - 1 \quad \text{Factor 2 from } 2x^2 - 4x.$$

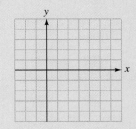

Now we complete the square on x by adding 1 within the parentheses. Since this adds 2
to the right-hand side, we also subtract 2 from the right-hand side.

By the distributive property,
when 1 is added to the
expression within the
parentheses, $2 \cdot 1 = 2$ is
added to the right-hand side.

Subtract 2 to
counteract the
addition of 2.

$$f(x) = 2(x^2 - 2x + 1) - 1 - 2 \quad \begin{array}{l} \text{To complete the square within the parentheses,} \\ \text{one-half of } -2 = -1 \text{ and } (-1)^2 = 1. \end{array}$$

(1) $f(x) = 2(x - 1)^2 - 3 \quad\quad\quad \text{Factor } x^2 - 2x + 1 \text{ and combine like terms.}$

Step 2: From Equation 1, we can see that $h = 1$ and $k = -3$, so the vertex will be at
the point $(1, -3)$, and the axis of symmetry is $x = 1$. We plot the vertex and axis of
symmetry on the coordinate system in Figure 8-11 shown below.

Step 3: Finally, we construct a table of values to determine several points on the
parabola. Since the x-coordinate of the vertex is 1, we choose values for x close to 1
and on the same side of the axis of symmetry. After plotting $(2, -1)$ and $(3, 5)$, we use
symmetry to locate two other points on the parabola: $(-1, 5)$ and $(0, -1)$. Then we
draw the graph, as in Figure 8-11.

$f(x) = 2x^2 - 4x - 1$		
x	**$f(x)$**	**$(x, f(x))$**
2	-1	$(2, -1)$
3	5	$(3, 5)$

↑
The x-coordinate of the vertex is 1.
Choose values for x close to 1 and on the
same side of the axis of symmetry.

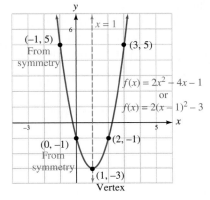

FIGURE 8-11

Answer

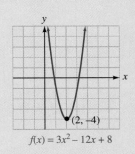

$f(x) = 3x^2 - 12x + 8$

We can derive a formula for the vertex of the graph of $f(x) = ax^2 + bx + c$ by completing the square in the same manner as we did in Example 8. After using similar steps, the result is

$$f(x) = a\left[x - \left(-\frac{b}{2a}\right)\right]^2 + \frac{4ac - b^2}{4a}$$

$$\uparrow \qquad\qquad \uparrow$$
$$h \qquad\qquad k$$

The x-coordinate of the vertex is $-\dfrac{b}{2a}$. The y-coordinate of the vertex is $\dfrac{4ac - b^2}{4a}$. However, we can also find the y-coordinate of the vertex by substituting the x-coordinate, $-\dfrac{b}{2a}$, for x in the quadratic function.

Formula for the vertex of a parabola

The vertex of the graph of the quadratic function $f(x) = ax^2 + bx + c$ is

$$\left(-\frac{b}{2a}, f\left(-\frac{b}{2a}\right)\right)$$

and the axis of symmetry of the parabola is the line $x = -\frac{b}{2a}$.

In Example 8, for the function $f(x) = 2x^2 - 4x - 1$, $a = 2$ and $b = -4$. To find the vertex of its graph, we compute

$$-\frac{b}{2a} = -\frac{-4}{2(2)} \qquad\qquad f\left(-\frac{b}{2a}\right) = f(1)$$

$$= -\frac{-4}{4} \qquad\qquad\qquad = 2(1)^2 - 4(1) - 1$$

$$= 1 \qquad\qquad\qquad\qquad = -3$$

The vertex is the point $(1, -3)$. This agrees with the result we obtained in Example 8 by completing the square.

CALCULATOR SNAPSHOT

Finding the vertex

To graph the function $f(x) = 2x^2 + 6x - 3$ and find the coordinates of the vertex and the axis of symmetry of the parabola, we can use a graphing calculator with window settings of $[-10, 10]$ for x and $[-10, 10]$ for y. If we enter the function, we will obtain the graph shown in Figure 8-12(a).

We then trace to move the cursor to the lowest point on the graph, as shown in Figure 8-12(b). By zooming in, we can see that the vertex is the point $(-1.5, -7.5)$, or $\left(-\frac{3}{2}, -\frac{15}{2}\right)$, and that the line $x = -\frac{3}{2}$ is the axis of symmetry.

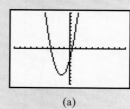
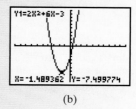

(a)   (b)

FIGURE 8-12

Some calculators have an fmin or fmax feature that can also be used to find the vertex. Consult your owner's manual for details.

Determining minimum and maximum values

It is often useful to know the smallest or largest possible value a quantity can assume. For example, companies try to minimize their costs and maximize their profits. If the quantity is expressed by a quadratic function, the vertex of the graph of the function gives its minimum or maximum value.

EXAMPLE 9 **Minimizing costs.** A glassworks that makes lead crystal vases has daily production costs given by the function $C(x) = 0.2x^2 - 10x + 650$, where x is the number of vases made each day. How many vases should be produced to minimize the per-day costs? What will the costs be?

Solution The graph of $C(x) = 0.2x^2 - 10x + 650$ is a parabola opening upward. The vertex is the lowest point on the graph. To find the vertex, we compute

$$-\frac{b}{2a} = -\frac{-10}{2(0.2)} \quad \begin{matrix} b = -10 \text{ and} \\ a = 0.2. \end{matrix} \qquad f\left(-\frac{b}{2a}\right) = f(25)$$

$$= -\frac{-10}{0.4} \qquad\qquad\qquad\qquad = 0.2(25)^2 - 10(25) + 650$$

$$= 25 \qquad\qquad\qquad\qquad\qquad = 525$$

The vertex is the point (25, 525), and it indicates that the costs are a minimum of \$525 when 25 vases are made daily.

To solve this problem with a graphing calculator with window settings of [0, 50] for x and [0, 1,000] for y, we graph the function $C(x) = 0.2x^2 - 10x + 650$. By using TRACE and ZOOM, we can locate the vertex of the graph. See Figure 8-13. The coordinates of the vertex indicate that the minimum cost is \$525 when the number of vases produced is 25.

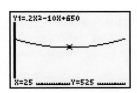

FIGURE 8-13

EXAMPLE 10 **Maximizing area.** A kennel owner wants to build the rectangular pen shown in Figure 8-14(a) on the next page to house his dog. If he uses one side of his barn, find the maximum area that he can enclose with 80 feet of fencing.

Solution We can let the width of the area be represented by w. Then the length is represented by $80 - 2w$. The function that gives the area the pen encloses is

$$A(w) = (80 - 2w)w \qquad \text{To find the area of a rectangle, multiply its length and width.}$$

We can find the maximum value of A by determining the vertex of the graph of the function. This time, we will find the vertex by completing the square.

$$A(w) = 80w - 2w^2 \qquad\qquad\qquad \text{Distribute the multiplication by } w.$$

$$= -2(w^2 - 40w) \qquad\qquad\quad \text{Factor out } -2.$$

$$= -2(w^2 - 40w + 400) + 800 \qquad \text{Complete the square on } w. \text{ Within the parentheses, add } \left(\frac{-40}{2}\right)^2 = 400. \text{ Since this subtracts 800 from the right-hand side, add 800 to the right-hand side.}$$

$$= -2(w - 20)^2 + 800 \qquad\qquad \text{Factor } w^2 - 40w + 400.$$

Thus, the coordinates of the vertex of the graph of the quadratic function are (20, 800), and the maximum area is 800 square feet.

To solve this problem with a graphing calculator with window settings of [0, 50] for x and [0, 1,000] for y, we graph the function $A(w) = -2w^2 + 80w$ to get the graph in Figure 8-14(b). By using TRACE and ZOOM, we can determine the vertex of the graph, which shows that the maximum area is 800 square feet when the width is 20 feet.

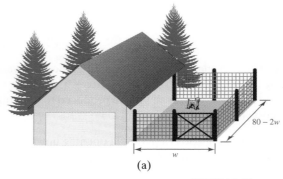
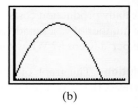

(a)   (b)

FIGURE 8-14

THINK IT THROUGH   Automobile Accidents

"Motor vehicle accidents can have many causes, but they can usually be divided into negligence, intentional misconduct, or product liability."
Allen L. Rothenberg, InjuryLawyer.com

The graph below shows the results of a study that explored the relationship between the age of drivers and the number of traffic accidents in which they were involved. For example, we see that drivers 16–20 years old were involved in 28 reported accidents for every one million miles driven. On the graph, sketch a parabola that could be used to model the data and locate its vertex. What information about age of the driver and traffic accidents do the coordinates of the vertex give?

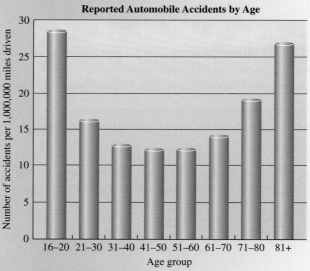

Source: Quality Planning Corporation

Solving quadratic equations graphically

We can solve quadratic equations graphically. For example, the solutions of $x^2 - 6x + 8 = 0$ are the numbers x that will make y equal to 0 in the quadratic function $f(x) = x^2 - 6x + 8$. To find these numbers, we carefully inspect the graph of $f(x) = x^2 - 6x + 8$ and locate the points on the graph with a y-coordinate of 0. In Figure 8-15, these points are $(2, 0)$ and $(4, 0)$, the x-intercepts of the graph. We can conclude that the x-coordinates of the x-intercepts, $x = 2$ and $x = 4$, are the solutions of $x^2 - 6x + 8 = 0$.

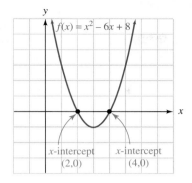

FIGURE 8-15

Solving quadratic equations graphically

CALCULATOR SNAPSHOT

We can use a graphing calculator to find approximate solutions of quadratic equations. For example, the solutions of $0.7x^2 + 2x - 3.5 = 0$ are the numbers x that will make $y = 0$ in the quadratic function $f(x) = 0.7x^2 + 2x - 3.5$. To approximate these numbers, we graph the quadratic function and read the x-intercepts from the graph using the ZERO feature. In Figure 8-16, we see that the x-coordinate of the left-most x-intercept of the graph is given as -4.082025. This means that an approximate solution of the equation is -4.08. To find the positive x-intercept, we use similar steps.

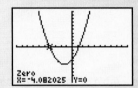

FIGURE 8-16

When solving quadratic equations graphically, we must consider three possibilities. If the graph of the associated quadratic function has two x-intercepts, the quadratic equation has two real-number solutions. Figure 8-17(a) shows an example of this. If the graph has one x-intercept, as shown in Figure 8-17(b), the equation has one real-number solution. Finally, if the graph does not have an x-intercept, as shown in Figure 8-17(c), the equation does not have any real-number solutions.

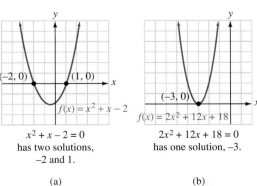

$$x^2 + x - 2 = 0$$
has two solutions,
-2 and 1.

(a)

$$2x^2 + 12x + 18 = 0$$
has one solution, -3.

(b)

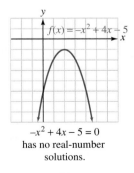

$$-x^2 + 4x - 5 = 0$$
has no real-number
solutions.

(c)

FIGURE 8-17

Section 8.3   STUDY SET

INTERMEDIATE
Algebra $f(x)$ Now™

VOCABULARY   *Refer to the illustration and fill in the blanks.*

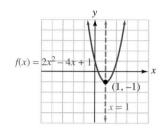

$f(x) = 2x^2 - 4x + 1$

$(1, -1)$

$x = 1$

1. The function graphed in the illustration is called a _____ function.

2. The graph is called a _____.

3. The lowest point on the graph, $(1, -1)$, is called the _____ of the parabola.

4. The vertical line $x = 1$ divides the parabola into two halves. This line is called the _____ ____ _____.

CONCEPTS

5. Refer to the graph below.

   a. What do we call the curve shown there?

   b. What are the x-intercepts of the graph?

   c. What is the y-intercept of the graph?

   d. What is the vertex?

   e. What is the axis of symmetry?

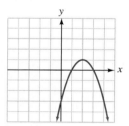

6. The vertex of a parabola is at $(1, -3)$, its y-intercept is $(0, -2)$, and it passes through the point $(3, 1)$, as shown below. Draw the axis of symmetry and use it to help determine two other points on the parabola.

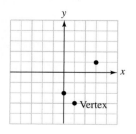

Vertex

Make a table of values for each function and graph them on the same coordinate system.

7. $f(x) = x^2,\ g(x) = 2x^2,\ h(x) = \dfrac{1}{2}x^2$

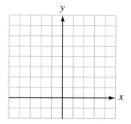

8. $f(x) = -x^2,\ g(x) = -\dfrac{1}{4}x^2,\ h(x) = -4x^2$

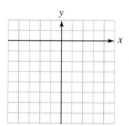

Make a table of values to graph function f. Then use a translation to graph the other two functions on the same coordinate system.

9. $f(x) = 4x^2,\ g(x) = 4x^2 + 3,\ h(x) = 4x^2 - 2$

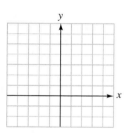

10. $f(x) = 3x^2,\ g(x) = 3(x + 2)^2,\ h(x) = 3(x - 3)^2$

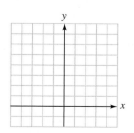

Make a table of values to graph function f. Then use a series of translations to graph function g on the same coordinate system.

11. $f(x) = -3x^2$, $g(x) = -3(x - 2)^2 - 1$

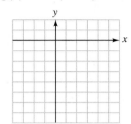

12. $f(x) = -\dfrac{1}{2}x^2$, $g(x) = -\dfrac{1}{2}(x + 1)^2 + 2$

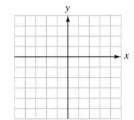

13. Use the graph of $f(x) = \frac{x^2}{10} - \frac{x}{5} - \frac{3}{2}$, shown below, to estimate the solutions of the quadratic equation $\frac{x^2}{10} - \frac{x}{5} - \frac{3}{2} = 0$.

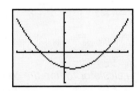

14. Three quadratic equations are to be solved graphically. The graphs of their associated quadratic functions are shown below. Determine which graph indicates that the equation has

  a. two real solutions.

  b. one real solution.

  c. no real solutions.

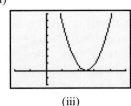
        (i)   (ii)

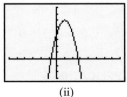
        (iii)

NOTATION

15. The function $f(x) = 2(x + 1)^2 + 6$ is written in the form $f(x) = a(x - h)^2 + k$. Is $h = -1$ or is $h = 1$? Explain.

16. The vertex of a quadratic function $f(x) = ax^2 + bx + c$ is given by the formula $\left(-\frac{b}{2a}, f\left(-\frac{b}{2a}\right)\right)$. Explain what is meant by the notation $f\left(-\frac{b}{2a}\right)$.

PRACTICE
Find the coordinates of the vertex and the axis of symmetry of the graph of each function. If necessary, complete the square on x to write the equation in the form $f(x) = a(x - h)^2 + k$. Do not graph the equation, but determine whether the graph will open upward or downward.

17. $f(x) = (x - 1)^2 + 2$

18. $f(x) = 2(x - 2)^2 - 1$

19. $f(x) = 2(x + 3)^2 - 4$

20. $f(x) = -3(x + 1)^2 + 3$

21. $f(x) = 2x^2 - 4x$

22. $f(x) = 3x^2 - 3$

23. $f(x) = -4x^2 + 16x + 5$

24. $f(x) = 5x^2 + 20x + 25$

25. $f(x) = 3x^2 + 4x + 2$

26. $f(x) = -6x^2 + 5x - 7$

First determine the vertex and the axis of symmetry of the graph of the function. Then plot several points and complete the graph. (See Example 8.)

27. $f(x) = (x - 3)^2 + 2$ **28.** $f(x) = (x + 1)^2 - 2$

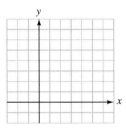   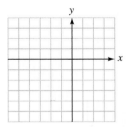

29. $f(x) = -(x - 2)^2$ **30.** $f(x) = -(x + 2)^2$

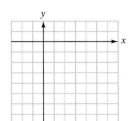   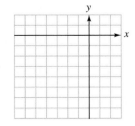

31. $f(x) = -2(x + 3)^2 + 4$ **32.** $f(x) = 2(x - 2)^2 - 4$

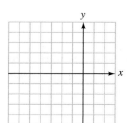   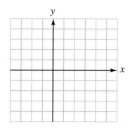

41. $f(x) = -x^2 + 2x + 3$ **42.** $f(x) = -x^2 + x + 2$

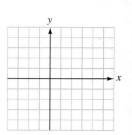   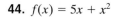

43. $f(x) = -3x^2 + 2x$ **44.** $f(x) = 5x + x^2$

33. $f(x) = -(x + 4)^2 - 1$ **34.** $f(x) = -(x - 3)^2 + 1$

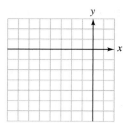   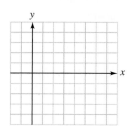

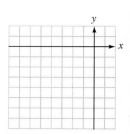   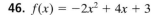

45. $f(x) = -12x^2 - 6x + 6$ **46.** $f(x) = -2x^2 + 4x + 3$

35. $f(x) = x^2 + x - 6$ **36.** $f(x) = x^2 - x - 6$

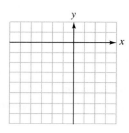   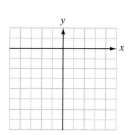

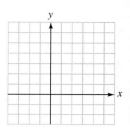   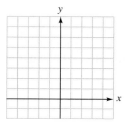

Use a graphing calculator to find the coordinates of the vertex of the graph of each quadratic function. Round to the nearest hundredth.

47. $f(x) = 2x^2 - x + 1$ **48.** $f(x) = x^2 + 5x - 6$

49. $f(x) = 7 + x - x^2$ **50.** $f(x) = 2x^2 - 3x + 2$

37. $f(x) = 3x^2 - 12x + 10$ **38.** $f(x) = 4x^2 + 16x + 9$

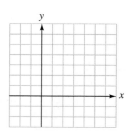   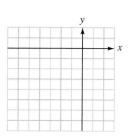

Use a graphing calculator to solve each equation. If an answer is not exact, round to the nearest hundredth.

51. $x^2 + x - 6 = 0$ **52.** $2x^2 - 5x - 3 = 0$
53. $0.5x^2 - 0.7x - 3 = 0$ **54.** $2x^2 - 0.5x - 2 = 0$

APPLICATIONS

55. FIREWORKS A fireworks shell is shot straight up with an initial velocity of 120 feet per second. Its height s after t seconds is given by the equation $s = 120t - 16t^2$. If the shell is designed to explode when it reaches its maximum height, how long after being fired, and at what height, will the fireworks appear in the sky?

39. $f(x) = 4x^2 + 24x + 37$ **40.** $f(x) = 6x^2 + 18x + 16.5$

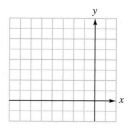   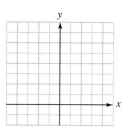

56. BALLISTICS From the top of the building, a ball is thrown straight up with an initial velocity of 32 feet per second. The equation

$$s = -16t^2 + 32t + 48$$

gives the height s of the ball t seconds after it is thrown. Find the maximum height reached by the ball and the time it takes for the ball to hit the ground.

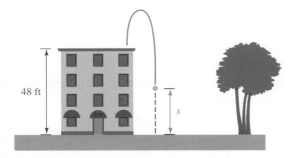

57. FENCING A FIELD A farmer wants to fence in three sides of a rectangular field shown below with 1,000 feet of fencing. The other side of the rectangle will be a river. If the enclosed area is to be maximum, find the dimensions of the field.

58. POLICE INVESTIGATIONS A police officer seals off the scene of a car collision using a roll of yellow police tape that is 300 feet long, as shown below. What dimensions should be used to seal off the maximum rectangular area around the collision? What is the maximum area?

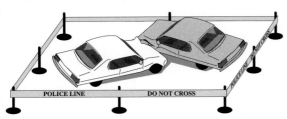

59. OPERATING COSTS The cost C in dollars of operating a certain concrete-cutting machine is related to the number of minutes n the machine is run by the function

$$C(n) = 2.2n^2 - 66n + 655$$

For what number of minutes is the cost of running the machine a minimum? What is the minimum cost?

60. WATER USAGE The height (in feet) of the water level in a reservoir over a 1-year period is modeled by the function

$$H(t) = 3.3(t - 9)^2 + 14$$

where $t = 1$ represents January, $t = 2$ represents February, and so on. How low did the water level get that year, and when did it reach the low mark?

61. U.S. ARMY The function

$$N(x) = -0.0534x^2 + 0.337x + 0.969$$

gives the number of active-duty military personnel in the United States Army (in millions) for the years 1965–1972, where $x = 0$ corresponds to 1965, $x = 1$ corresponds to 1966, and so on. For this period, when was the army's personnel strength level at its highest, and what was it? Historically, can you explain why?

62. SCHOOL ENROLLMENTS The total annual enrollment (in millions) in U.S. elementary and secondary schools for the years 1975–1996 is given by the model

$$E(x) = 0.058x^2 - 1.162x + 50.604$$

where $x = 0$ corresponds to 1975, $x = 1$ corresponds to 1976, and so on.

a. For this period, when was enrollment the lowest? What was it?

b. Use the model to complete the bar graph below.

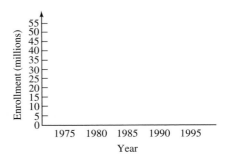

63. MAXIMIZING REVENUE The revenue R received for selling x stereos is given by the formula

$$R = -\frac{x^2}{5} + 80x - 1,000$$

How many stereos must be sold to obtain the maximum revenue? Find the maximum revenue.

64. MAXIMIZING REVENUE When priced at $30 each, a toy has annual sales of 4,000 units. The manufacturer estimates that each $1 increase in cost will decrease sales by 100 units. Find the unit price that will maximize total revenue. (*Hint*: Total revenue = price · the number of units sold.)

WRITING

65. Use the example of a stream of water from a drinking fountain to explain the concepts of the vertex and the axis of symmetry of a parabola.

66. What are some quantities that are good to maximize? What are some quantities that are good to minimize?

67. A table of values for $f(x) = 2x^2 - 4x + 3$ is shown on the right. Explain why it appears that the vertex of the graph of f is the point $(1, 1)$.

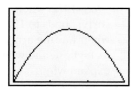

X	Y1
.25	2.125
.5	1.5
.75	1.125
1	1
1.25	1.125
1.5	1.5
1.75	2.125

X=.25

68. The illustration on the right shows the graph of the quadratic function $f(x) = -4x^2 + 12x$ with domain $[0, 3]$. Explain how the value of $f(x)$ changes as the value of x increases from 0 to 3.

69. A mirror is held against the y-axis of the graph of a quadratic function. What fact about parabolas does this illustrate?

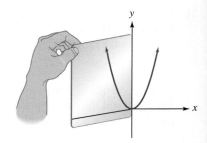

70. Give a definition of the vertex of a parabola that opens upward.

REVIEW *Perform the operations. All variables represent positive numbers.*

71. $\sqrt{8a}\,\sqrt{2a^3b}$

72. $(\sqrt{23})^2$

73. $\dfrac{\sqrt{3}}{\sqrt{50}}$

74. $\dfrac{3}{\sqrt[3]{9}}$

75. $3(\sqrt{5b} - \sqrt{3})^2$

76. $-2\sqrt{5b}(4\sqrt{2b} - 3\sqrt{3})$

8.4 Complex Numbers

- Imaginary numbers • Simplifying imaginary numbers • Complex numbers
- Arithmetic of complex numbers • Complex conjugates
- Rationalizing the denominator • Powers of i

We have seen that negative numbers do not have real-number square roots. For years, people believed that numbers like

$$\sqrt{-1}, \qquad \sqrt{-3}, \qquad \sqrt{-4}, \qquad \text{and} \qquad \sqrt{-9}$$

were nonsense. In the 17th century, René Descartes (1596–1650) called them *imaginary numbers*. Today, imaginary numbers have many uses, such as describing the behavior of alternating current in electronics.

In this section, we will introduce imaginary numbers and discuss a broader set of numbers called *complex numbers*.

■ Imaginary numbers

So far, all of our work with quadratic equations has involved only real numbers. However, the solutions of many quadratic equations are not real numbers.

INTERMEDIATE
Algebra *(f(x))* **Now**™
Self Check 1
Solve: $a^2 + 3a + 5 = 0$.

EXAMPLE 1   Solve: $x^2 + x + 1 = 0$.

Solution   Because $x^2 + x + 1$ is prime and cannot be factored, we will use the quadratic formula, with $a = 1$, $b = 1$, and $c = 1$, to solve it.

$$x = \frac{-b \pm \sqrt{b^2 - 4ac}}{2a}$$

$$= \frac{-1 \pm \sqrt{1^2 - 4(1)(1)}}{2(1)}$$ Substitute 1 for a, 1 for b, and 1 for c.

$$= \frac{-1 \pm \sqrt{1 - 4}}{2}$$ Simplify within the radical.

$$= \frac{-1 \pm \sqrt{-3}}{2}$$

$$x = \frac{-1 + \sqrt{-3}}{2} \quad \text{or} \quad x = \frac{-1 - \sqrt{-3}}{2}$$

Each solution contains the number $\sqrt{-3}$. Since no real number squared is -3, $\sqrt{-3}$ is not a real number, and these two solutions are not real numbers.

To solve equations such as $x^2 + x + 1 = 0$, we must define the square root of a negative number.

Answer $\dfrac{-3 \pm \sqrt{-11}}{2}$

Square roots of negative numbers are called **imaginary numbers.** The imaginary number $\sqrt{-1}$ is often denoted by the letter i:

$$i = \sqrt{-1}$$

Because i represents the square root of -1, it follows that

$$i^2 = -1$$

◼ Simplifying imaginary numbers

We can use extensions of the multiplication and division properties of radicals discussed in Chapter 6 to simplify imaginary numbers. For example,

$$\sqrt{-25} = \sqrt{-1 \cdot 25} = \sqrt{-1} \cdot \sqrt{25} = i \cdot 5 = 5i$$

$$\sqrt{\frac{-100}{49}} = \sqrt{-1 \cdot \frac{100}{49}} = \sqrt{-1} \cdot \frac{\sqrt{100}}{\sqrt{49}} = i \cdot \frac{10}{7} = \frac{10}{7}i$$

$$\sqrt{-3} = \sqrt{-1 \cdot 3} = \sqrt{-1} \cdot \sqrt{3} = i \cdot \sqrt{3} = i\sqrt{3}$$

These examples illustrate a rule for simplifying square roots of negative numbers.

Square root of a negative number

For any positive real number b,

$$\sqrt{-b} = i\sqrt{b}$$

To justify this rule, we use the fact that $\sqrt{-1} = i$:

$$\sqrt{-b} = \sqrt{(-1)b}$$
$$= \sqrt{-1}\sqrt{b}$$
$$= i\sqrt{b}$$

INTERMEDIATE
Algebra *f(x)* Now™

EXAMPLE 2 Write each expression in terms of i: **a.** $\sqrt{-81}$, **b.** $\sqrt{-5}$, **c.** $-\sqrt{-8}$, and **d.** $3\sqrt{-\dfrac{98}{16}}$.

Self Check 2
Write each expression in terms of i:

a. $-\sqrt{-4}$

b. $\sqrt{-11}$

c. $2\sqrt{-28}$

d. $\sqrt{-\dfrac{27}{100}}$

Answers **a.** $-2i$,
b. $i\sqrt{11}$ or $\sqrt{11}i$,

c. $4i\sqrt{7}$ or $4\sqrt{7}i$, **d.** $\dfrac{3\sqrt{3}}{10}i$

Solution

a. $\sqrt{-81} = \sqrt{-1 \cdot 81} = \sqrt{-1} \cdot \sqrt{81} = i \cdot 9 = 9i$

b. $\sqrt{-5} = \sqrt{-1 \cdot 5} = \sqrt{-1} \cdot \sqrt{5} = i \cdot \sqrt{5}$ or $\sqrt{5}i$

! COMMENT Since is easy to confuse $\sqrt{5}i$ with $\sqrt{5i}$, we write i first so that it is clear that the i is not under the radical. However, both $i\sqrt{5}$ and $\sqrt{5}i$ are correct.

c. $-\sqrt{-8} = -\sqrt{-1 \cdot 8} = -\sqrt{-1} \cdot \sqrt{8} = -i \cdot \sqrt{8}$

Now we simplify $\sqrt{8}$:

$$-i \cdot \sqrt{8} = -i \cdot 2\sqrt{2} = -2i\sqrt{2} \text{ or } -2\sqrt{2}i$$

d. $3\sqrt{-\dfrac{98}{16}} = 3\sqrt{-1 \cdot \dfrac{98}{16}} = \dfrac{3\sqrt{-1}\sqrt{49}\sqrt{2}}{\sqrt{16}} = \dfrac{21\sqrt{2}}{4}i$

INTERMEDIATE
Algebra $f(x)$ **Now**™

Self Check 3
Solve: $x^2 + 75 = 0$.

EXAMPLE 3 Solve: $x^2 + 21 = 0$.

Solution We can write the equation as $x^2 = -21$ and use the square root property to solve it.

$$x^2 + 21 = 0$$
$$x^2 = -21 \qquad \text{Subtract 21 from both sides.}$$
$$x = \pm\sqrt{-21} \qquad \text{Apply the square root property. Note the square root of a negative number.}$$

This equation has imaginary solutions, which we can express in terms of i.

$$x = \pm\sqrt{-1 \cdot 21}$$
$$x = \pm i\sqrt{21}$$

Answer $\pm 5i\sqrt{3}$

Verify that $i\sqrt{21}$ and $-i\sqrt{21}$ satisfy the original equation.

■ Complex numbers

The imaginary numbers are a subset of a set of numbers called the **complex numbers.**

> **Complex numbers**
>
> A **complex number** is any number that can be written in the form $a + bi$, where a and b represent real numbers and $i = \sqrt{-1}$.
>
> In the complex number $a + bi$, a is called the **real part,** and b is called the **imaginary part.**

Some examples of complex numbers are

$3 + 4i$ $\qquad a = 3$ and $b = 4$.

$-6 - 5i$ $\qquad$ Think of this as $-6 + (-5)i$, where $a = -6$ and $b = -5$. It is common to use $a - bi$ form as a substitute for $a + (-b)i$.

$\dfrac{1}{2} + 0i$ $\qquad a = \dfrac{1}{2}$ and $b = 0$.

$0 + i\sqrt{2}$ $\qquad a = 0$ and $b = \sqrt{2}$.

If $b = 0$, the complex number $a + bi$ is a real number. If $b \neq 0$ and $a = 0$, the complex number $0 + bi$ (or just bi) is an imaginary number. The relationship between the real numbers, the imaginary numbers, and the complex numbers is shown in Figure 8-18.

COMPLEX NUMBERS

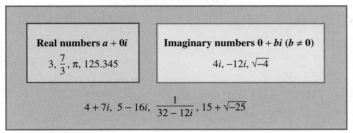

Real numbers $a + 0i$	Imaginary numbers $0 + bi$ $(b \neq 0)$
$3, \dfrac{7}{3}, \pi, 125.345$	$4i, -12i, \sqrt{-4}$

$$4 + 7i,\ 5 - 16i,\ \dfrac{1}{32 - 12i},\ 15 + \sqrt{-25}$$

FIGURE 8-18

INTERMEDIATE
Algebra $f(x)$ Now™

EXAMPLE 4 Solve: $4t^2 - 6t + 3 = 0$, and write the solutions in $a + bi$ form.

Solution We use the quadratic formula to solve the equation.

$$t = \frac{-b \pm \sqrt{b^2 - 4ac}}{2a}$$

$$= \frac{-(-6) \pm \sqrt{(-6)^2 - 4(4)(3)}}{2(4)} \qquad \text{Substitute 4 for } a, -6 \text{ for } b, \text{ and 3 for } c.$$

$$= \frac{6 \pm \sqrt{36 - 48}}{8} \qquad \text{Simplify within the radical.}$$

$$= \frac{6 \pm \sqrt{-12}}{8}$$

$$= \frac{6 \pm 2i\sqrt{3}}{8} \qquad \sqrt{-12} = i\sqrt{12} = i\sqrt{4}\sqrt{3} = 2i\sqrt{3}.$$

$$= \frac{3 \pm i\sqrt{3}}{4} \qquad \frac{6 \pm 2i\sqrt{3}}{8} = \frac{\overset{1}{2}(3 \pm i\sqrt{3})}{\underset{1}{2} \cdot 4} = \frac{3 \pm i\sqrt{3}}{4}.$$

Writing each solution as a complex number in the form $a + bi$, they are

$$\frac{3}{4} + \frac{\sqrt{3}}{4}i \qquad \text{and} \qquad \frac{3}{4} - \frac{\sqrt{3}}{4}i$$

Self Check 4
Solve: $a^2 + 2a + 3 = 0$.

Answer $-1 \pm i\sqrt{2}$

Arithmetic of complex numbers

We now consider the rules used to add, subtract, multiply, and divide complex numbers.

Addition and subtraction of complex numbers

To add (or subtract) two complex numbers, add (or subtract) their real parts and add (or subtract) their imaginary parts.

$$(a + bi) + (c + di) = (a + c) + (b + d)i$$
$$(a + bi) - (c + di) = (a - c) + (b - d)i$$

Self Check 5
Perform the operations:

a. $(3 - 5i) + (-2 + 7i)$

b. $(3 - \sqrt{-25}) - (-2 + \sqrt{-49})$

EXAMPLE 5 Perform the operations: **a.** $(8 + 4i) + (12 + 8i)$, **b.** $(7 - \sqrt{-16}) + (9 + \sqrt{-4})$, and **c.** $(-6 + i) - (3 + 2i)$.

a. $(8 + 4i) + (12 + 8i) = (8 + 12) + (4 + 8)i$ Add the real parts. Add the imaginary parts.

$$= 20 + 12i$$

b. $(7 - \sqrt{-16}) + (9 + \sqrt{-4})$

$$= (7 - 4i) + (9 + 2i)$$ Write $\sqrt{-16}$ and $\sqrt{-4}$ in terms of i.

$$= (7 + 9) + (-4 + 2)i$$ Add the real parts. Add the imaginary parts.

$$= 16 - 2i$$ Write $16 + (-2i)$ in the form $16 - 2i$.

c. $(-6 + i) - (3 + 2i) = (-6 - 3) + (1 - 2)i$ Subtract the real parts. Subtract the imaginary parts.

$$= -9 - i$$

Answers **a.** $1 + 2i$, **b.** $5 - 12i$

Imaginary numbers are not real numbers; some properties of real numbers do not apply to imaginary numbers. For example, *we cannot use the product rule for radicals to find* $\sqrt{-2}\sqrt{-20}$.

To multiply two imaginary numbers, they should first be expressed in terms of i.

$$\sqrt{-2}\sqrt{-20} = (i\sqrt{2})(2i\sqrt{5})$$ Simplify: $\sqrt{-20} = i\sqrt{20} = 2i\sqrt{5}$.

$$= 2i^2\sqrt{10}$$ $i \cdot 2i = 2i^2$.

$$= -2\sqrt{10}$$ Simplify: $i^2 = -1$.

! COMMENT If a and b are both negative, then $\sqrt{ab} \neq \sqrt{a}\sqrt{b}$. For example, if $a = -16$ and $b = -4$,

$$\sqrt{(-16)(-4)} = \sqrt{64} = 8 \quad \text{but} \quad \sqrt{-16}\sqrt{-4} = (4i)(2i) = 8i^2 = 8(-1) = -8$$

To multiply a complex number by a real number, we use the distributive property to remove parentheses and then simplify. For example,

$$6(2 + 9i) = 6(2) + 6(9i)$$ Use the distributive property.

$$= 12 + 54i$$ Simplify.

To multiply a complex number by an imaginary number, we use the distributive property to remove parentheses and then simplify. For example,

$$-5i(4 - 8i) = -5i(4) - (-5i)8i$$ Use the distributive property.

$$= -20i + 40i^2$$ Simplify.

$$= -40 - 20i$$ Since $i^2 = -1$, $40i^2 = 40(-1) = -40$.

To multiply two complex numbers, we use the FOIL method to develop the following definition.

Multiplying complex numbers

Complex numbers are multiplied as if they were binomals, with $i^2 = -1$:

$$(a + bi)(c + di) = ac + adi + bci + bdi^2$$

$$= (ac - bd) + (ad + bc)i$$

INTERMEDIATE
Algebra *f(x)* **Now**™

EXAMPLE 6   Multiply the complex numbers: **a.** $(2 + 3i)(3 - 2i)$ and
b. $(-4 + 2i)(2 + i)$.

Self Check 6
Multiply:
$(-2 + 3i)(3 - 2i)$

a. $(2 + 3i)(3 - 2i) = 6 - 4i + 9i - 6i^2$ Use the FOIL method.

$\qquad\qquad\qquad\quad = 6 + 5i - (-6)$ Combine the imaginary terms: $-4i + 9i = 5i$.
$\qquad\qquad\qquad\qquad\qquad\qquad\quad$ Simplify: $i^2 = -1$, so $6i^2 = -6$.

$\qquad\qquad\qquad\quad = 6 + 5i + 6$

$\qquad\qquad\qquad\quad = 12 + 5i$ Combine like terms.

b. $(-4 + 2i)(2 + i) = -8 - 4i + 4i + 2i^2$ Use the FOIL method.

$\qquad\qquad\qquad\quad = -8 + 0i - 2$ $-4i + 4i = 0i$. Since $i^2 = -1$, $2i^2 = -2$.

$\qquad\qquad\qquad\quad = -10 + 0i$

Answer $0 + 13i$

▍Complex conjugates

Complex conjugates
The complex numbers $a + bi$ and $a - bi$ are called **complex conjugates.**

For example,

$3 + 4i$ and $3 - 4i$ are complex conjugates.

$5 - 7i$ and $5 + 7i$ are complex conjugates.

$8 + 17i$ and $8 - 17i$ are complex conjugates.

INTERMEDIATE
Algebra *f(x)* **Now**™

EXAMPLE 7 Find the product of $3 + i$ and its complex conjugate.

Self Check 7
Multiply:
$(2 + 3i)(2 - 3i)$

Solution   The complex conjugate of $3 + i$ is $3 - i$. We can find the product as follows:

$(3 + i)(3 - i) = 9 - 3i + 3i - i^2$ Use the FOIL method.

$\qquad\qquad\quad = 9 - i^2$ Combine like terms.

$\qquad\qquad\quad = 9 - (-1)$ Simplify: $i^2 = -1$.

$\qquad\qquad\quad = 10$

Answer 13

The product of the complex number $a + bi$ and its complex conjugate $a - bi$ is
the real number $a^2 + b^2$, as the following work shows:

$(a + bi)(a - bi) = a^2 - abi + abi - b^2i^2$ Use the FOIL method.

$\qquad\qquad\qquad = a^2 - b^2(-1)$ Combine like terms. $i^2 = -1$.

$\qquad\qquad\qquad = a^2 + b^2$

▍Rationalizing the denominator

To divide complex numbers, we often have to rationalize a denominator.

Self Check 8
Divide 1 by $5 - i$.

EXAMPLE 8   Divide and write the result in $a + bi$ form:

$$\frac{1}{3 + i}$$

Solution   The denominator of the fraction $\frac{1}{3+i}$ is not a rational number. We can rationalize the denominator by multiplying both the numerator and the denominator of the fraction by the complex conjugate of the denominator, which is $3 - i$.

$$\frac{1}{3 + i} = \frac{1}{3 + i} \cdot \frac{3 - i}{3 - i}$$ Multiply by 1: $\frac{3 - i}{3 - i} = 1$.

$$= \frac{3 - i}{9 - 3i + 3i - i^2}$$ Multiply the numerators and multiply the denominators.

$$= \frac{3 - i}{9 - (-1)}$$ Simplify: $i^2 = -1$.

$$= \frac{3 - i}{10}$$ Simplify in the denominator.

Answer $\dfrac{5}{26} + \dfrac{1}{26}i$

$$= \frac{3}{10} - \frac{1}{10}i$$ Write the complex number in $a + bi$ form.

Self Check 9
Divide:
$$\frac{5 + 4i}{3 + 2i}$$

EXAMPLE 9   Divide and write in $a + bi$ form: $\dfrac{3 - i}{2 + i}$.

Solution   We multiply both the numerator and the denominator of the fraction by the complex conjugate of the denominator.

$$\frac{3 - i}{2 + i} = \frac{3 - i}{2 + i} \cdot \frac{2 - i}{2 - i}$$ Multiply by 1: $\frac{2 - i}{2 - i} = 1$.

$$= \frac{6 - 3i - 2i + i^2}{4 - 2i + 2i - i^2}$$ Multiply the numerators and multiply the denominators.

$$= \frac{5 - 5i}{4 - (-1)}$$ $i^2 = -1$.

$$= \frac{5(1 - i)}{5}$$ Factor out 5 in the numerator and divide out the common factor of 5.

Answer $\dfrac{23}{13} + \dfrac{2}{13}i$

$$= 1 - i$$ Simplify.

Self Check 10
Write
$$\frac{3 + \sqrt{-9}}{4 + \sqrt{-16}}$$
in $a + bi$ form.

EXAMPLE 10   Write $\dfrac{4 + \sqrt{-16}}{2 + \sqrt{-4}}$ in $a + bi$ form.

Solution

$$\frac{4 + \sqrt{-16}}{2 + \sqrt{-4}} = \frac{4 + 4i}{2 + 2i}$$ Write the numerator and denominator in $a + bi$ form.

$$= \frac{2(\overset{1}{\cancel{2 + 2i}})}{\underset{1}{\cancel{2 + 2i}}}$$ Factor out 2 in the numerator and divide out the common factor of $2 + 2i$.

Answer $\dfrac{3}{4} + 0i$

$$= 2 + 0i$$

EXAMPLE 11 Find the quotient: $\dfrac{7}{2i}$. Express the result in $a + bi$ form.

Solution The denominator can be expressed as $0 + 2i$. Its conjugate is $0 - 2i$, or just $-2i$.

$$\frac{7}{2i} = \frac{7}{2i} \cdot \frac{-2i}{-2i} \qquad \text{Multiply by 1: } \frac{-2i}{-2i} = 1.$$

$$= \frac{-14i}{-4i^2}$$

$$= \frac{-14i}{4} \qquad i^2 = -1.$$

$$= \frac{-7i}{2} \qquad \text{Simplify the fraction.}$$

$$= 0 - \frac{7}{2}i \qquad \text{Write in } a + bi \text{ form.}$$

INTERMEDIATE
Algebra *f(x)* **Now**™
Self Check 11
Divide:
$\dfrac{5}{-i}$

Answer $0 + 5i$

Powers of *i*

The powers of i produce an interesting pattern:

$$i = \sqrt{-1} = i \qquad\qquad i^5 = i^4 i = 1i = i$$
$$i^2 = (\sqrt{-1})^2 = -1 \qquad i^6 = i^4 i^2 = 1(-1) = -1$$
$$i^3 = i^2 i = -1i = -i \qquad i^7 = i^4 i^3 = 1(-i) = -i$$
$$i^4 = i^2 i^2 = (-1)(-1) = 1 \qquad i^8 = i^4 i^4 = (1)(1) = 1$$

As we find larger powers of i, the pattern of results continues: $i, -1, -i, 1, \ldots$.

Larger powers of i can be simplified by using the fact that $i^4 = 1$. For example, to simplify i^{29}, we note that 29 divided by 4 gives a quotient of 7 and a remainder of 1. Thus, $29 = 4 \cdot 7 + 1$ and

$$i^{29} = i^{4 \cdot 7 + 1} \qquad 4 \cdot 7 = 28.$$

$$= (i^4)^7 \cdot i^1$$

$$= 1^7 \cdot i \qquad i^4 = 1.$$

$$= i \qquad 1 \cdot i = i.$$

The result of this example illustrates the following fact.

Powers of *i*

If n represents a natural number that has a remainder of r when divided by 4, then

$$i^n = i^r$$

EXAMPLE 12 Simplify: i^{55}.

Solution We divide 55 by 4 and get a remainder of 3. Therefore,

$$i^{55} = i^3 = -i$$

INTERMEDIATE
Algebra *f(x)* **Now**™
Self Check 12
Simplify: i^{62}.

Answer -1

Section 8.4 STUDY SET

VOCABULARY *Fill in the blanks.*

1. $\sqrt{-1}$, $\sqrt{-3}$, and $\sqrt{-4}$ are examples of _____ numbers.

2. $3 + 5i$, $2 - 7i$, and $5 - \frac{1}{2}i$ are examples of _____ numbers.

3. The _____ part of $5 + 7i$ is 5. The _____ part is 7.

4. $6 + 3i$ and $6 - 3i$ are called complex _____.

5. _____ is a number whose square is -1.

6. i^{25} is called a _____ of i.

CONCEPTS *Fill in the blanks.*

7. $i = $ ▨

8. $i^2 = $ ▨

9. $i^3 = $ ▨

10. $i^4 = $ ▨

11. We multiply two complex numbers by using the _____ method.

12. To divide two complex numbers, we _____ the denominator.

13. Use a check to see whether $-5i\sqrt{2}$ is a solution of $x^2 + 50 = 0$.

14. Give the complex conjugate of each number.

  a. $2 - 3i$ **b.** 2 **c.** $-3i$

15. Complete the illustration below to show the relationship between the real numbers, the imaginary numbers, the complex numbers, the rational numbers, and the irrational numbers.

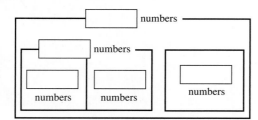

16. Determine whether each statement is true or false.

  a. Every complex number is a real number.

  b. Every real number is a complex number.

  c. Imaginary numbers can always be denoted using i.

  d. The square root of a negative number is an imaginary number.

17. Solve each equation.

  a. $x^2 - 1 = 0$ **b.** $x^2 + 1 = 0$

18. Is $\sqrt[3]{-64}$ an imaginary number? Explain.

NOTATION *Complete each solution.*

19. $(3 + 2i)(3 - i) = ▨ - 3i + ▨ - 2i^2$

                 $= 9 + 3i + ▨$

                 $= 11 + 3i$

20. $\dfrac{3}{2 - i} = \dfrac{3}{2 - i} \cdot \dfrac{▨}{}$

        $= \dfrac{6 + ▨}{4 + ▨ - 2i - i^2}$

        $= \dfrac{6 + 3i}{▨}$

        $= \dfrac{6}{5} + \dfrac{3}{5}i$

21. Determine whether each statement is true or false.

  a. $\sqrt{6}i = i\sqrt{6}$ **b.** $\sqrt{8}i = \sqrt{8i}$

  c. $\sqrt{-25} = -\sqrt{25}$ **d.** $-i = i$

  e. $-i\sqrt{18} = -3i\sqrt{2}$ **f.** $\dfrac{2 \pm 4i}{8} = \dfrac{1 \pm 4i}{4}$

22. Determine whether each statement is true or false.

  a. $\sqrt{-3}\sqrt{-2} = \sqrt{6}$

  b. $\sqrt{-36} + \sqrt{-25} = \sqrt{-61}$

PRACTICE *Express each number in terms of i.*

23. $\sqrt{-9}$ **24.** $\sqrt{-4}$

25. $\sqrt{-7}$ **26.** $\sqrt{-11}$

27. $\sqrt{-24}$ **28.** $\sqrt{-28}$

29. $-\sqrt{-24}$ **30.** $-\sqrt{-72}$

31. $5\sqrt{-81}$ **32.** $6\sqrt{-49}$

33. $\sqrt{-\dfrac{25}{9}}$ **34.** $-\sqrt{-\dfrac{121}{144}}$

Simplify each expression.

35. $\sqrt{-1}\sqrt{-36}$ **36.** $\sqrt{-9}\sqrt{-100}$

37. $\sqrt{-2}\sqrt{-6}$ **38.** $\sqrt{-3}\sqrt{-6}$

39. $\dfrac{\sqrt{-25}}{\sqrt{-64}}$ **40.** $\dfrac{\sqrt{-4}}{\sqrt{-1}}$

41. $-\dfrac{\sqrt{-400}}{\sqrt{-1}}$ **42.** $-\dfrac{\sqrt{-225}}{\sqrt{-16}}$

Solve each equation. Write all solutions in bi or a + bi form.

43. $x^2 + 9 = 0$ **44.** $x^2 + 100 = 0$

45. $3x^2 = -16$ **46.** $2x^2 = -25$

47. $x^2 + 2x + 2 = 0$ **48.** $x^2 - 2x + 6 = 0$

49. $2x^2 + x + 1 = 0$ **50.** $3x^2 + 2x + 1 = 0$

51. $3x^2 - 4x = -2$ **52.** $2x^2 + 3x = -3$

53. $3x^2 - 2x = -3$ **54.** $5x^2 = 2x - 1$

Perform the operations. Write all answers in a + bi form.

55. $(3 + 4i) + (5 - 6i)$ **56.** $(5 + 3i) - (6 - 9i)$

57. $(7 - 3i) - (4 + 2i)$ **58.** $(8 + 3i) + (-7 - 2i)$

59. $(6 - i) + (9 + 3i)$ **60.** $(5 - 4i) - (3 + 2i)$

61. $(8 + \sqrt{-25}) + (7 + \sqrt{-4})$

62. $(-7 + \sqrt{-81}) - (-2 - \sqrt{-64})$

63. $3(2 - i)$ **64.** $-4(3 + 4i)$

65. $-5i(5 - 5i)$ **66.** $2i(7 + 2i)$

67. $(2 + i)(3 - i)$ **68.** $(4 - i)(2 + i)$

69. $(3 - 2i)(2 + 3i)$ **70.** $(3 - i)(2 + 3i)$

71. $(4 + i)(3 - i)$ **72.** $(1 - 5i)(1 - 4i)$

73. $(2 - \sqrt{-16})(3 + \sqrt{-4})$

74. $(3 - \sqrt{-4})(4 - \sqrt{-9})$

75. $(2 + \sqrt{2}i)(3 - \sqrt{2}i)$ **76.** $(5 + \sqrt{3}i)(2 - \sqrt{3}i)$

77. $(2 + i)^2$ **78.** $(3 - 2i)^2$

Write each expression in a + bi form.

79. $\dfrac{1}{i}$ **80.** $\dfrac{1}{i^3}$

81. $\dfrac{4}{5i^3}$ **82.** $\dfrac{3}{2i}$

83. $\dfrac{3i}{8\sqrt{-9}}$ **84.** $\dfrac{5i^3}{2\sqrt{-4}}$

85. $\dfrac{-3}{5i^5}$ **86.** $\dfrac{-4}{6i^7}$

87. $\dfrac{5}{2 - i}$ **88.** $\dfrac{26}{3 - 2i}$

89. $\dfrac{3}{5 + i}$ **90.** $\dfrac{-4}{7 - 2i}$

91. $\dfrac{-12}{7 - \sqrt{-1}}$ **92.** $\dfrac{4}{3 + \sqrt{-1}}$

93. $\dfrac{5i}{6 + 2i}$ **94.** $\dfrac{-4i}{2 - 6i}$

95. $\dfrac{-2i}{3 + 2i}$ **96.** $\dfrac{3i}{6 - i}$

97. $\dfrac{3 - 2i}{3 + 2i}$ **98.** $\dfrac{2 + 3i}{2 - 3i}$

99. $\dfrac{3 + 2i}{3 + i}$ **100.** $\dfrac{2 - 5i}{2 + 5i}$

101. $\dfrac{\sqrt{5} - \sqrt{3}i}{\sqrt{5} + \sqrt{3}i}$ **102.** $\dfrac{\sqrt{3} + \sqrt{2}i}{\sqrt{3} - \sqrt{2}i}$

Simplify each expression.

103. i^{21} **104.** i^{19}

105. i^{27} **106.** i^{22}

107. i^{100} **108.** i^{42}

109. i^{97} **110.** i^{200}

▮ APPLICATIONS

111. FRACTAL GEOMETRY Complex numbers are fundamental in the creation of the intricate geometric shape shown below, called a **fractal.** Fractal geometry is a rapidly expanding field with applications in science, medicine, and computer graphics. The process of creating this image is based on the following sequence of steps, which begins by picking any complex number, which we will call z.

1. Square z, and then add that result to z.

2. Square the result from step 1, and then add it to z.

3. Square the result from step 2, and then add it to z.

If we begin with the complex number i, what is the result after performing steps 1, 2, and 3?

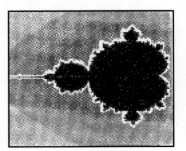

112. ELECTRONICS The impedance Z in an AC (alternating current) circuit is a measure of how much the circuit impedes (hinders) the flow of current through it. The impedance is related to the voltage V and the current I by the formula

$$V = IZ$$

If a circuit has a current of $(0.5 + 2.0i)$ amps and an impedance of $(0.4 - 3.0i)$ ohms, find the voltage.

WRITING

113. What is an imaginary number?

114. In Example 1 of this section, what unusual situation illustrated the need to define the square root of a negative number?

REVIEW

115. What are the lengths of the two legs of a $30°-60°-90°$ triangle if the hypotenuse is 30 units long?

116. What are the lengths of the other two sides of a $45°-45°-90°$ triangle if one leg is 30 units long?

117. WIND SPEEDS A plane that can fly 200 mph in still air makes a 330-mile flight with a tailwind and returns, flying into the same wind. Find the speed of the wind if the total flying time is $3\frac{1}{3}$ hours.

118. FINDING RATES A student drove a distance of 135 miles at an average speed of 50 mph. How much faster would she have to drive on the return trip to save 30 minutes of driving time?

8.5 The Discriminant and Equations That Can Be Written in Quadratic Form

- The discriminant • Equations that can be written in quadratic form • Problem solving

In this section, we will discuss how to predict what type of solutions a quadratic equation will have without actually solving the equation. We will then solve some special equations that can be written in quadratic form. Finally, we will use the equation-solving methods of this chapter to solve a shared-work problem.

The discriminant

We can predict what type of solutions a particular quadratic equation will have without solving it. To see how, we suppose that the coefficients a, b, and c in the equation $ax^2 + bx + c = 0$ represent real numbers and $a \neq 0$. Then the solutions of the equation are given by the quadratic formula

$$x = \frac{-b \pm \sqrt{b^2 - 4ac}}{2a}$$

If $b^2 - 4ac \geq 0$, the solutions are real numbers. If $b^2 - 4ac < 0$, the solutions are nonreal complex numbers. Thus, the value of $b^2 - 4ac$, called the **discriminant,** determines the type of solutions for a particular quadratic equation.

The discriminant

If a, b, and c represent real numbers and

if $b^2 - 4ac$ is ...	the solutions are ...
positive,	two different real numbers.
0,	two real numbers that are equal.
negative,	two different nonreal complex numbers that are complex conjugates.

If a, b, and c represent rational numbers and

if $b^2 - 4ac$ is ...	the solutions are ...
a perfect square,	two different rational numbers.
positive and not a perfect square,	two different irrational numbers.

EXAMPLE 1 Determine the type of solutions for each equation:
a. $x^2 + x + 1 = 0$ and **b.** $3x^2 + 5x + 2 = 0$.

Solution
a. We calculate the discriminant for $x^2 + x + 1 = 0$:

$$b^2 - 4ac = 1^2 - 4(1)(1) \quad a = 1, b = 1, \text{ and } c = 1.$$
$$= -3 \quad \text{The result is a negative number.}$$

Since $b^2 - 4ac < 0$, the solutions of $x^2 + x + 1 = 0$ are two nonreal complex numbers that are complex conjugates.

b. For $3x^2 + 5x + 2 = 0$,

$$b^2 - 4ac = 5^2 - 4(3)(2) \quad a = 3, b = 5, \text{ and } c = 2.$$
$$= 25 - 24$$
$$= 1 \quad \text{The result is a positive number.}$$

Since $b^2 - 4ac > 0$ and $b^2 - 4ac$ is a perfect square, the two solutions of $3x^2 + 5x + 2 = 0$ are rational and unequal.

INTERMEDIATE
Algebra $f(x)$ **Now**™
Self Check 1
Determine the type of
solutions for
a. $x^2 + x - 1 = 0$
b. $4x^2 - 10x + 25 = 0$

Answers **a.** real numbers that are
irrational and unequal,
b. nonreal numbers that are
complex conjugates

Equations that can be written in quadratic form

Many nonquadratic equations can be written in quadratic form ($ax^2 + bx + c = 0$) and solved using the techniques discussed in previous sections. For example, a careful inspection of the equation $x^4 - 5x^2 + 4 = 0$ leads to the following observations:

In the lead term, the exponent on the variable is the square of the exponent on the variable in the middle term. $x^4 - 5x^2 + 4 = 0$

The last term is a constant.

Equations having the above characteristics are said to be *quadratic in form*. One way to solve them is to make a substitution. We will let $y = x^2$.

$$x^4 - 5x^2 + 4 = 0 \quad \text{This is the given equation.}$$
$$(x^2)^2 - 5(x^2) + 4 = 0 \quad \text{Write } x^4 \text{ as } (x^2)^2.$$
$$y^2 - 5y + 4 = 0 \quad \text{Replace each } x^2 \text{ with } y.$$

We can solve this quadratic equation by factoring.

$$(y - 4)(y - 1) = 0 \qquad \text{Factor } y^2 - 5y + 4.$$

$$y - 4 = 0 \quad \text{or} \quad y - 1 = 0 \quad \text{Set each factor equal to 0.}$$

$$y = 4 \qquad \qquad y = 1$$

These are not the solutions for x. To find x, we undo the earlier substitutions by replacing each y with x^2 and we solve for x.

$$x^2 = 4 \quad \text{or} \quad x^2 = 1$$

$$x = \pm\sqrt{4} \qquad x = \pm\sqrt{1} \quad \text{Use the square root property.}$$

$$x = \pm 2 \qquad \quad x = \pm 1$$

This equation has four solutions: 1, −1, 2, and −2. Verify that each one satisfies the original equation.

Self Check 2

Solve: $x^4 - 3x^2 - 4 = 0$.

EXAMPLE 2   Solve: $x^4 - 8x^2 - 9 = 0$.

Solution   This equation can be written in quadratic form because the exponent on the variable of the lead term is the square of the exponent of the middle term. We can write the equation in the form

$$(x^2)^2 - 8(x^2) - 9 = 0$$

which can be solved using substitution by letting $y = x^2$.

$$(x^2)^2 - 8(x^2) - 9 = 0$$

$$y^2 - 8y - 9 = 0 \qquad \text{Substitute } y \text{ for } x^2.$$

$$(y - 9)(y + 1) = 0 \qquad \text{Factor } y^2 - 8y - 9.$$

$$y - 9 = 0 \quad \text{or} \quad y + 1 = 0 \quad \text{Set each factor equal to 0.}$$

$$y = 9 \qquad \qquad y = -1 \quad \text{Solve each equation.}$$

Since $y = x^2$, we have

$$x^2 = 9 \quad \text{or} \quad x^2 = -1$$

$$x = \pm 3 \qquad x = \pm i \qquad \text{Use the square root property.}$$

Answer 2, −2, i, −i

The equation has four solutions: −3, 3, i, and −i. Check each one.

Self Check 3

Solve: $x + x^{1/2} - 6 = 0$.

EXAMPLE 3   Solve: $x - 7\sqrt{x} + 12 = 0$.

Solution   This equation can be written in quadratic form, because the power of the variable of the lead term is the square of the variable factor of the middle term: $x = (\sqrt{x})^2$. If we let $y = \sqrt{x}$, then $y^2 = x$. With this substitution, the equation

$$x - 7\sqrt{x} + 12 = 0$$

becomes a quadratic equation that can be solved by factoring.

$$y^2 - 7y + 12 = 0 \qquad \text{Substitute } y^2 \text{ for } x \text{ and } y \text{ for } \sqrt{x}.$$

$$(y - 3)(y - 4) = 0 \qquad \text{Factor } y^2 - 7y + 12 = 0.$$

$$y - 3 = 0 \quad \text{or} \quad y - 4 = 0 \quad \text{Set each factor equal to 0.}$$

$$y = 3 \qquad \qquad y = 4$$

Replace each y with $\sqrt{x}$ and solve the radical equations by squaring both sides.

$$\sqrt{x} = 3 \quad \text{or} \quad \sqrt{x} = 4$$

$$x = 9 \qquad \quad x = 16 \quad \text{Square both sides.}$$

Answer 4

Verify that 9 and 16 satisfy the original equation.

EXAMPLE 4 **Solving equations that are quadratic in form.** Solve: $2m^{2/3} - 2 = 3m^{1/3}$.

INTERMEDIATE
Algebra $f(x)$ Now™
Self Check 4
Solve: $a^{2/3} = -3a^{1/3} + 10$.

Solution After writing the equation in descending powers of m, we see that

$$2m^{2/3} - 3m^{1/3} - 2 = 0$$

can be written in quadratic form, because $m^{2/3} = (m^{1/3})^2$. We will use the substitution $y = m^{1/3}$.

$$2m^{2/3} - 3m^{1/3} - 2 = 0$$
$$2(m^{1/3})^2 - 3m^{1/3} - 2 = 0$$
$$2y^2 - 3y - 2 = 0 \qquad \text{Replace } m^{1/3} \text{ with } y.$$
$$(2y + 1)(y - 2) = 0 \qquad \text{Factor } 2y^2 - 3y - 2 = 0.$$
$$2y + 1 = 0 \quad \text{or} \quad y - 2 = 0 \quad \text{Set each factor equal to 0.}$$
$$y = -\frac{1}{2} \qquad\qquad y = 2$$

Replace each y with $m^{1/3}$ and solve for m.

$$m^{1/3} = -\frac{1}{2} \qquad \text{or} \qquad m^{1/3} = 2$$
$$(m^{1/3})^3 = \left(-\frac{1}{2}\right)^3 \qquad (m^{1/3})^3 = (2)^3 \quad \begin{array}{l}\text{Recall that } m^{1/3} = \sqrt[3]{m}. \text{ To solve} \\ \text{for } m, \text{ cube both sides.}\end{array}$$
$$m = -\frac{1}{8} \qquad\qquad m = 8$$

Verify that $-\frac{1}{8}$ and 8 satisfy the original equation.

Answer $-125, 8$

INTERMEDIATE
Algebra $f(x)$ Now™

EXAMPLE 5 Solve: $(4t + 2)^2 - 30(4t + 2) + 224 = 0$.

Self Check 5
Solve: $(n + 3)^2 - 6(n + 3) = -8$.

Solution If we make the substitution $y = 4t + 2$, the given equation becomes

$$y^2 - 30y + 224 = 0$$

which can be solved by using the quadratic formula.

$$y = \frac{-b \pm \sqrt{b^2 - 4ac}}{2a}$$

$$= \frac{-(-30) \pm \sqrt{(-30)^2 - 4(1)(224)}}{2(1)} \qquad \text{Substitute 1 for } a, -30 \text{ for } b, \text{ and 224 for } c.$$

$$= \frac{30 \pm \sqrt{900 - 896}}{2} \qquad\qquad \text{Simplify within the radical.}$$

$$= \frac{30 \pm 2}{2} \qquad\qquad\qquad \sqrt{900 - 896} = \sqrt{4} = 2.$$

$$y = 16 \quad \text{or} \quad y = 14$$

To find t, we replace y with $4t + 2$ and solve for t.

$$4t + 2 = 16 \quad \text{or} \quad 4t + 2 = 14$$
$$4t = 14 \qquad\qquad 4t = 12$$
$$t = 3.5 \qquad\qquad t = 3$$

Verify that 3.5 and 3 satisfy the original equation.

Answer $-1, 1$

Problem solving

EXAMPLE 6 Household appliances.

A water temperature control on a washing machine is shown in Figure 8-19. When the warm setting is selected, both the hot and cold water inlets open to fill the tub in 2 minutes 15 seconds. When the cold temperature setting is chosen, the cold water inlet fills the tub 45 seconds faster than when the hot setting is used. How long does it take to fill the washing machine with hot water?

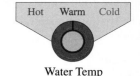

Water Temp

FIGURE 8-19

Analyze the problem

The key to solving this problem is to determine how much of the tub is filled by each water temperature setting in 1 second. On the warm setting, when the hot and cold inlets are working together, the tub is filled in 2 minutes 15 seconds, or 135 seconds. So in 1 second, they fill $\frac{1}{135}$ of the tub.

Form an equation

Let x = the number of seconds it takes to fill the tub when the hot temperature is chosen. In 1 second, the hot water inlet fills $\frac{1}{x}$ of the tub. Since the cold water inlet fills the tub in 45 seconds less time, the washing machine can be filled with cold water in $(x - 45)$ seconds. In 1 second, $\frac{1}{x - 45}$ of the tub is filled by the cold water inlet. We can now form an equation.

What the hot water inlet pipe can do in 1 second	plus	what the cold water inlet pipe can do in 1 second	equals	what they can do together in 1 second.
$\dfrac{1}{x}$	$+$	$\dfrac{1}{x - 45}$	$=$	$\dfrac{1}{135}$

Solve the equation

$$\frac{1}{x} + \frac{1}{x - 45} = \frac{1}{135}$$

$$135x(x - 45)\left(\frac{1}{x} + \frac{1}{x - 45}\right) = 135x(x - 45)\left(\frac{1}{135}\right)$$

Multiply both sides by $135x(x - 45)$ to clear the equation of fractions.

$$135(x - 45) + 135x = x(x - 45)$$

Simplify.

$$135x - 6{,}075 + 135x = x^2 - 45x$$

Distribute the multiplication by 135 and by x.

$$270x - 6{,}075 = x^2 - 45x$$

Combine like terms.

$$0 = x^2 - 315x + 6{,}075$$

Subtract $270x$ from both sides. Add 6,075 to both sides.

To solve this equation, we can use the quadratic formula, with $a = 1$, $b = -315$, and $c = 6{,}075$.

$$x = \frac{-b \pm \sqrt{b^2 - 4ac}}{2a}$$

$$= \frac{-(-315) \pm \sqrt{(-315)^2 - 4(1)(6{,}075)}}{2(1)}$$ Substitute 1 for a, -315 for b, and 6,075 for c.

$$= \frac{315 \pm \sqrt{99{,}225 - 24{,}300}}{2}$$ Simplify within the radical.

$$= \frac{315 \pm \sqrt{74{,}925}}{2}$$

$$x \approx \frac{589}{2} \quad \text{or} \quad x \approx \frac{41}{2}$$

$$x \approx 294 \quad \bigg| \quad x \approx 21$$

State the conclusion

We disregard the solution of 21 seconds, because this would imply that the cold water inlet fills the tub in a negative number of seconds ($21 - 45 = -24$). Therefore, the hot water inlet fills the washing machine tub in about 294 seconds, which is 4 minutes 54 seconds.

Check the result

Use estimation to check the result.

Section 8.5   STUDY SET

VOCABULARY   *Fill in the blanks.*

1. For the quadratic equation $ax^2 + bx + c = 0$, the discriminant is _____ .

2. When an equation is written in the form $ax^2 + bx + c = 0$, we say that it is written in _____ form.

CONCEPTS   *Consider the quadratic equation $ax^2 + bx + c = 0$, where a, b, and c represent rational numbers, and fill in the blanks to make the statements true.*

3. If $b^2 - 4ac < 0$, the solutions of the equation are nonreal complex _____ .

4. If $b^2 - 4ac = $ ▮, the solutions of the equation are equal real numbers.

5. If $b^2 - 4ac$ is a perfect square, the solutions are _____ numbers and _____ .

6. If $b^2 - 4ac$ is positive and not a perfect square, the solutions are _____ numbers and _____ .

7. Consider $x^4 - 3x^2 + 2 = 0$.

  a. What is the relationship between the powers of x in the first two terms on the left-hand side?

  b. Is this equation quadratic in form?

8. Consider $x^{2/3} + 4x^{1/3} - 5 = 0$.

  a. What is the relationship between the powers of x in the first two terms on the left-hand side?

  b. Is this equation quadratic in form?

NOTATION   *Complete each solution.*

9. To find the type of solutions for the equation $x^2 + 5x + 6 = 0$, we compute the discriminant.

$$b^2 - 4ac = \text{▮}^2 - 4(1)(\text{ })$$

$$= 25 - \text{▮}$$

$$= 1$$

Since a, b, and c are rational numbers and the value of the discriminant is a perfect square, the solutions are _____ numbers and unequal.

10. Change $\dfrac{3}{4} + x = \dfrac{3x - 50}{4(x - 6)}$ to quadratic form.

$$\text{▮}\left(\frac{3}{4} + x\right) = \text{▮} \,\frac{3x - 50}{4(x - 6)}$$

$$3(x - 6) + 4x(\text{▮}) = 3x - 50$$

$$3x - \text{▮} + 4x^2 - \text{▮} = 3x - 50$$

$$4x^2 - 24x + \text{▮} = 0$$

$$\text{▮} - 6x + 8 = 0$$

PRACTICE *Use the discriminant to determine what type of solutions exist for each quadratic equation.* **Do not solve the equation.**

11. $4x^2 - 4x + 1 = 0$ **12.** $6x^2 - 5x - 6 = 0$

13. $5x^2 + x + 2 = 0$ **14.** $3x^2 + 10x - 2 = 0$

15. $2x^2 = 4x - 1$ **16.** $9x^2 = 12x - 4$

17. $x(2x - 3) = 20$ **18.** $x(x - 3) = -10$

19. Use the discriminant to determine whether the solutions of $1{,}492x^2 + 1{,}776x - 2{,}000 = 0$ are real numbers.

20. Use the discriminant to determine whether the solutions of $1{,}776x^2 - 1{,}492x + 2{,}000 = 0$ are real numbers.

Solve each equation.

21. $x^4 - 17x^2 + 16 = 0$ **22.** $x^4 - 10x^2 + 9 = 0$

23. $x^4 = 6x^2 - 5$ **24.** $2x^4 + 24 = 26x^2$

25. $t^4 + 3t^2 = 28$ **26.** $3h^4 + h^2 - 2 = 0$

27. $2x + \sqrt{x} - 3 = 0$ **28.** $2x - \sqrt{x} - 1 = 0$

29. $3x + 5\sqrt{x} + 2 = 0$ **30.** $3x - 4\sqrt{x} + 1 = 0$

31. $x - 6\sqrt{x} = -8$ **32.** $x - 5x^{1/2} + 4 = 0$

33. $x^{2/3} + 5x^{1/3} + 6 = 0$ **34.** $x^{2/3} - 7x^{1/3} + 12 = 0$

35. $a^{2/3} - 2a^{1/3} - 3 = 0$ **36.** $r^{2/3} + 4r^{1/3} - 5 = 0$

37. $2(2x + 1)^2 - 7(2x + 1) + 6 = 0$

38. $3(2 - x)^2 + 10(2 - x) - 8 = 0$

39. $(c + 1)^2 - 4(c + 1) - 8 = 0$

40. $(k - 7)^2 + 6(k - 7) + 10 = 0$

41. $x + 5 + \dfrac{4}{x} = 0$ **42.** $x - 4 + \dfrac{3}{x} = 0$

43. $\dfrac{1}{x + 2} + \dfrac{24}{x + 3} = 13$ **44.** $\dfrac{3}{x} + \dfrac{4}{x + 1} = 2$

45. $\dfrac{2}{x - 1} + \dfrac{1}{x + 1} = 3$ **46.** $\dfrac{3}{x - 2} - \dfrac{1}{x + 2} = 5$

47. $x^{-4} - 2x^{-2} + 1 = 0$ **48.** $4x^{-4} + 1 = 5x^{-2}$

49. $x + \dfrac{2}{x - 2} = 0$ **50.** $x + \dfrac{x + 5}{x - 3} = 0$

APPLICATIONS

51. FLOWER ARRANGEMENTS A florist needs to determine the height h of the flowers shown. The radius r, the width w, and the height h of the circular-shaped arrangement are related by the formula

$$r = \frac{4h^2 + w^2}{8h}$$

If w is to be 34 inches and r is to be 18 inches, find h to the nearest tenth of an inch.

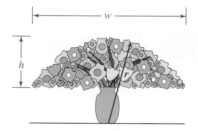

52. ARCHITECTURE A **golden rectangle** is said to be one of the most visually appealing of all geometric forms. The Parthenon, built by the Greeks in the 5th century B.C. and shown in the illustration below, fits into a golden rectangle once its ruined triangular pediment is drawn in.

   In a golden rectangle, the length l and width w must satisfy the equation

$$\frac{l}{w} = \frac{w}{l - w}$$

If a rectangular billboard is to have a width of 20 feet, what should its length be so that it is a golden rectangle? Round to the nearest tenth.

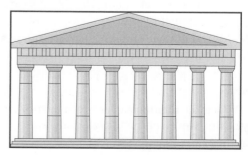

53. SNOWMOBILING A woman drives her snowmobile 150 miles at a rate of r mph. She could have gone the same distance in 2 hours less time if she had increased her speed by 20 mph. Find r.

54. BICYCLING Tina bicycles 160 miles at the rate of r mph. The same trip would have taken 2 hours longer if she had decreased her speed by 4 mph. Find r.

55. CROWD CONTROL After a sold-out performance at a county fair, security guards have found that the grandstand area can be emptied in 6 minutes if both the east and west exits are opened. If just the east exit is used, it takes 4 minutes longer to clear the grandstand than it does if just the west exit is opened. How long does it take to clear the grandstand if everyone must file through the east exit?

56. PAPER ROUTES When a father, in a car, and his son, on a bicycle, work together to distribute the morning edition, it takes them 35 minutes to complete a paper route. Working alone, it takes the son 25 minutes longer than the father. To the nearest minute, how long does it take the son to cover the paper route on his bicycle?

WRITING

57. Describe how to predict what type of solutions the equation $3x^2 - 4x + 5 = 0$ will have.

58. Explain how the method of substitution is used in this section to solve equations.

REVIEW *Solve each equation.*

59. $\dfrac{1}{4} + \dfrac{1}{t} = \dfrac{1}{2t}$

60. $\dfrac{p-3}{3p} + \dfrac{1}{2p} = \dfrac{1}{4}$

61. Find the slope of the line passing through $(-2, -4)$ and $(3, 5)$.

62. Write the equation of the line passing through $(-2, -4)$ and $(3, 5)$ in general form.

8.6 Quadratic and Other Nonlinear Inequalities

- Solving quadratic inequalities • Solving rational inequalities
- Graphs of nonlinear inequalities in two variables

If $a \neq 0$, inequalities of the form $ax^2 + bx + c < 0$ and $ax^2 + bx + c > 0$ are called *quadratic inequalities in one variable*. We will begin this section by showing how to solve them by making a sign chart. We will then show how to solve other nonlinear inequalities using the same technique. To conclude, we will show how to find graphical solutions of nonlinear inequalities in two variables.

Solving quadratic inequalities

To solve the inequality $x^2 + x - 6 < 0$, we must find the values of x that make the inequality true. This can be done using a number line. We begin by factoring the trinomial to obtain

$(x + 3)(x - 2) < 0$

Since the product of $x + 3$ and $x - 2$ must be less than 0, the values of $x + 3$ and $x - 2$ must be opposite in sign. To find the intervals where this is true, we keep track of their signs by constructing the chart in Figure 8-20. The chart shows that

- $x - 2$ is 0 when $x = 2$, is positive when $x > 2$ (indicated with $+$ signs in the figure), and is negative when $x < 2$ (indicated with $-$ signs in the figure).
- $x + 3$ is 0 when $x = -3$, is positive when $x > -3$, and is negative when $x < -3$.

The only place where the values of the binomials are opposite in sign is in the interval $(-3, 2)$. Therefore, the solution set of the inequality can be denoted as

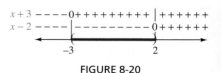

$$-3 < x < 2$$

The graph of the solution set is shown on the number line in Figure 8-20.

FIGURE 8-20

Self Check 1
Solve $x^2 + 2x - 15 > 0$ and
graph the solution set.

EXAMPLE 1   Solve: $x^2 + 2x - 3 \geq 0$.

Solution   We factor the trinomial to get $(x - 1)(x + 3)$ and construct a sign chart, as in Figure 8-21.

- $x - 1$ is 0 when $x = 1$, is positive when $x > 1$, and is negative when $x < 1$.
- $x + 3$ is 0 when $x = -3$, is positive when $x > -3$, and is negative when $x < -3$.

The product of $x - 1$ and $x + 3$ will be greater than 0 when the signs of the binomial factors are the same. This occurs in the intervals $(-\infty, -3)$ and $(1, \infty)$. The numbers -3 and 1 are also included, because they make the product equal to 0. Thus, the solution set is

$$(-\infty, -3] \cup [1, \infty) \qquad \text{or} \qquad x \leq -3 \text{ or } x \geq 1$$

The graph of the solution set is shown on the number line in Figure 8-21.

Answer $(-\infty, -5) \cup (3, \infty)$

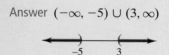

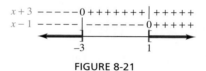

FIGURE 8-21

■ Solving rational inequalities

Making a sign chart is useful for solving many inequalities that are neither linear nor quadratic. In the next three examples, we will use a sign chart to solve *rational inequalities*.

INTERMEDIATE
Algebra $f(x)$ **Now**™

Self Check 2

Solve: $\dfrac{3}{x} > 5$.

EXAMPLE 2   Solve: $\dfrac{1}{x} < 6$.

Solution   We subtract 6 from both sides to make the right-hand side equal to 0. We then find a common denominator and subtract:

$$\frac{1}{x} < 6$$

$$\frac{1}{x} - 6 < 0 \qquad \text{Subtract 6 from both sides.}$$

$$\frac{1}{x} - \frac{6x}{x} < 0 \qquad \text{Get a common denominator.}$$

$$\frac{1 - 6x}{x} < 0 \qquad \text{Subtract the numerators and keep the common denominator.}$$

We now make a sign chart, as shown in Figure 8-22.

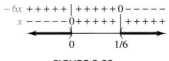

- The denominator x is 0 when $x = 0$, is positive when $x > 0$, and is negative when $x < 0$.

FIGURE 8-22

- The numerator $1 - 6x$ is 0 when $x = \frac{1}{6}$, is positive when $x < \frac{1}{6}$, and is negative when $x > \frac{1}{6}$.

The fraction $\frac{1 - 6x}{x}$ will be less than 0 when the numerator and denominator are opposite in sign. This occurs in the interval

$$(-\infty, 0) \cup \left(\frac{1}{6}, \infty\right) \qquad \text{or} \qquad x < 0 \text{ or } x > \frac{1}{6}$$

The graph of this interval is shown in Figure 8-22.

Answer $\left(0, \dfrac{3}{5}\right)$

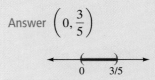

!**COMMENT** Since we don't know whether x is positive, 0, or negative, multiplying both sides of the inequality $\frac{1}{x} < 6$ by x is a three-case situation:

- If $x > 0$, then $1 < 6x$.
- If $x = 0$, then $\frac{1}{x}$ is undefined.
- If $x < 0$, then $1 > 6x$.

If we multiply both sides by x and solve the linear inequality $1 < 6x$, we are considering only one case and will get only part of the answer.

INTERMEDIATE
Algebra *f(x)* **Now**™

EXAMPLE 3 Solve: $\dfrac{x^2 - 3x + 2}{x - 3} \geq 0$.

Self Check 3
Solve
$$\frac{x + 2}{x^2 - 2x - 3} > 0$$
and graph the solution set.

Solution We write the fraction with the numerator in factored form.

$$\frac{(x - 2)(x - 1)}{x - 3} \geq 0$$

To keep track of the signs of $x - 2$, $x - 1$, and $x - 3$, we construct the sign chart shown in Figure 8-23. The fraction will be positive in the intervals where all factors are positive, or where exactly two factors are negative. The numbers 1 and 2 are included, because they make the numerator (and thus the fraction) equal to 0. The number 3 is not included, because it gives a 0 in the denominator.

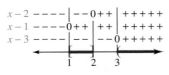

FIGURE 8-23

The solution is the interval $[1, 2] \cup (3, \infty)$. The graph appears in Figure 8-23.

Answer $(-2, -1) \cup (3, \infty)$

INTERMEDIATE
Algebra *f(x)* **Now**™

EXAMPLE 4 Solve: $\dfrac{3}{x - 1} < \dfrac{2}{x}$.

Solution We subtract $\frac{2}{x}$ from both sides to get 0 on the right-hand side and proceed as follows:

Self Check 4
Solve

$$\frac{2}{x + 1} > \frac{1}{x}$$

and graph the solution set.

$$\frac{3}{x - 1} < \frac{2}{x}$$

$$\frac{3}{x - 1} - \frac{2}{x} < 0 \qquad \text{Subtract } \frac{2}{x} \text{ from both sides.}$$

$$\frac{3x}{(x - 1)x} - \frac{2(x - 1)}{x(x - 1)} < 0 \qquad \text{Get a common denominator.}$$

$$\frac{3x - 2x + 2}{x(x - 1)} < 0 \qquad \text{Keep the denominator and subtract the numerators.}$$

$$\frac{x + 2}{x(x - 1)} < 0 \qquad \text{Combine like terms.}$$

We can keep track of the signs of $x + 2$, x, and $x - 1$ with the sign chart shown in Figure 8-24. The fraction will be negative in the intervals with either one or three negative factors. The numbers 0 and 1 are not included, because they give a 0 in the denominator, and the number -2 is not included, because it does not satisfy the inequality.

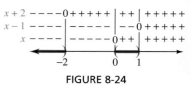

FIGURE 8-24

The solution is the interval $(-\infty, -2) \cup (0, 1)$, as shown in Figure 8-24.

Answer $(-1, 0) \cup (1, \infty)$

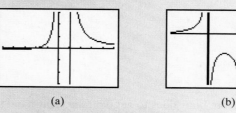

Solving inequalities graphically

To approximate the solutions of $x^2 + 2x - 3 \geq 0$ (Example 1) by graphing, we can use the standard window settings of $[-10, 10]$ for x and $[-10, 10]$ for y and graph the quadratic function $y = x^2 + 2x - 3$, as in Figure 8-25. The solution of the inequality will be those numbers x for which the graph of $y = x^2 + 2x - 3$ lies above or on the x-axis. We can trace to find that this interval is $(-\infty, -3] \cup [1, \infty)$.

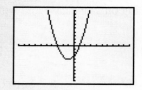

FIGURE 8-25

To approximate the solutions of $\frac{3}{x - 1} < \frac{2}{x}$ (Example 4), we first write the inequality in the form

$$\frac{x + 2}{x(x - 1)} < 0$$

We use window settings of $[-5, 5]$ for x and $[-3, 3]$ for y and graph the function $y = \frac{x + 2}{x(x - 1)}$, as in Figure 8-26(a). The solution of the inequality will be those numbers x for which the graph lies below the x-axis.

We can trace to see that the graph is below the x-axis when x is less than -2. Since we cannot see the graph in the interval $0 < x < 1$, we redraw the graph using window settings of $[-1, 2]$ for x and $[-25, 10]$ for y. See Figure 8-26(b).

We can now see that the graph is below the x-axis in the interval $(0, 1)$. Thus, the solution of the inequality is the union of two intervals: $(-\infty, -2) \cup (0, 1)$.

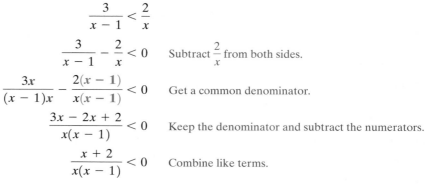

(a)   (b)

FIGURE 8-26

Graphs of nonlinear inequalities in two variables

We now consider the graphs of nonlinear inequalities in two variables.

INTERMEDIATE
Algebra $f(x)$ **Now**™

EXAMPLE 5   Graph: $y < -x^2 + 4$.

Solution   The graph of $y = -x^2 + 4$ is the parabolic boundary separating the region representing $y < -x^2 + 4$ and the region representing $y > -x^2 + 4$.

We graph the quadratic function $y = -x^2 + 4$ as a dashed parabola, because equality is not permitted. Since the coordinates of the origin satisfy the inequality $y < -x^2 + 4$, the test point $(0, 0)$ is in the graph. The complete graph is shown in Figure 8-27.

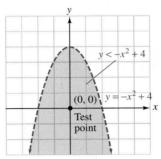

FIGURE 8-27

Self Check 5
Graph: $y \geq -x^2 + 4$.

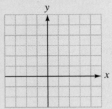

Answer

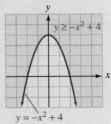

INTERMEDIATE
Algebra $f(x)$ **Now**™

EXAMPLE 6   Graph: $x \leq |y|$.

Solution   We first graph $x = |y|$ as in Figure 8-28(a). We use a solid line, because equality is permitted. Because the origin is on the graph, we cannot use the origin as a test point. However, any other point, such as $(1, 0)$, will do. We substitute 1 for x and 0 for y into the inequality to get

$x \leq |y|$

$1 \leq |0|$

$1 \leq 0$

Since $1 \leq 0$ is a false statement, the test point $(1, 0)$ does not satisfy the inequality and is not part of the graph. Thus, the graph of $x \leq |y|$ is to the left of the boundary.

The complete graph and the test point are shown in Figure 8-28(b).

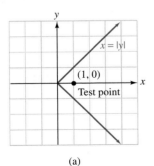

(a)   (b)

FIGURE 8-28

Self Check 6
Graph: $x \geq -|y|$.

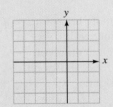

Answer

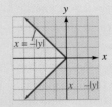

Section 8.6 STUDY SET

VOCABULARY   *Fill in the blanks.*

1. Any inequality of the form $ax^2 + bx + c > 0$ where $a \neq 0$ is called a _____ inequality.

2. The inequality $y < x^2 - 2x + 3$ is a nonlinear inequality in _____ variables.

3. The _____ (3, 5) represents the real numbers between 3 and 5.

4. To decide which side of the boundary to shade when solving inequalities in two variables, we pick a _____ point.

CONCEPTS   *Fill in the blanks.*

5. When $x > 3$, the binomial $x - 3$ is _____ than zero.

6. When $x < 3$, the binomial $x - 3$ is _____ than zero.

7. If $x = 0$, the fraction $\frac{1}{x}$ is _____.

8. To keep track of the signs of factors in a product or quotient, we can use a _____ chart.

9. Estimate the solution of the inequality $x^2 - x - 6 > 0$ using the graph of $y = x^2 - x - 6$, shown on the right.

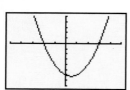

10. Estimate the solution of the inequality $\frac{x-3}{x} \leq 0$ using the graph of $y = \frac{x-3}{x}$, shown on the right.

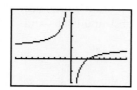

NOTATION   *Consider the inequality*
$$(x + 2)(x - 3) > 0$$

11. Since the product of $x + 2$ and $x - 3$ is positive, the values of $x + 2$ and $x - 3$ are both positive or both negative. Fill in each blank with a number.

  a. $x + 2 = 0$ when $x = $ ▢

  b. $x + 2 > 0$ when $x > $ ▢

  c. $x + 2 < 0$ when $x < $ ▢

  d. $x - 3 = 0$ when $x = $ ▢

  e. $x - 3 > 0$ when $x > $ ▢

  f. $x - 3 < 0$ when $x < $ ▢

12. Use the information in Exercise 11 to make a sign chart of the data and graph the solution set.

13. Match each inequality with the corresponding interval notation.

  a. $-4 \leq x < 5$ **i.** $(-\infty, \infty)$

  b. $x \leq -4$ or $x > 5$ **ii.** $[-4, \infty)$

  c. $x \geq -4$ **iii.** $[-4, 5)$

  d. $x < 5$ or $x \geq -4$ **iv.** $(-\infty, -4] \cup (5, \infty)$

14. What is the meaning of the symbol $\cup$?

PRACTICE   *Solve each inequality. Give each result in interval notation and graph the solution set.*

15. $x^2 - 5x + 4 < 0$

16. $x^2 - 3x - 4 > 0$

17. $x^2 - 8x + 15 > 0$

18. $x^2 + 2x - 8 < 0$

19. $x^2 + x - 12 \leq 0$

20. $x^2 - 8x \leq -15$

21. $x^2 + 8x < -16$

22. $x^2 + 6x \geq -9$

23. $x^2 \geq 9$

24. $x^2 \geq 16$

25. $2x^2 - 50 < 0$

26. $3x^2 - 243 < 0$

27. $\frac{1}{x} < 2$

28. $\frac{1}{x} > 3$

29. $-\dfrac{5}{x} < 3$

30. $\dfrac{4}{x} \geq 8$

31. $\dfrac{x^2 - x - 12}{x - 1} < 0$

32. $\dfrac{x^2 + x - 6}{x - 4} \geq 0$

33. $\dfrac{6x^2 - 5x + 1}{2x + 1} > 0$

34. $\dfrac{6x^2 + 11x + 3}{3x - 1} < 0$

35. $\dfrac{3}{x - 2} < \dfrac{4}{x}$

36. $\dfrac{-6}{x + 1} \geq \dfrac{1}{x}$

37. $\dfrac{7}{x - 3} \geq \dfrac{2}{x + 4}$

38. $\dfrac{-5}{x - 4} < \dfrac{3}{x + 1}$

39. $\dfrac{x}{x + 4} \leq \dfrac{1}{x + 1}$

40. $\dfrac{x}{x + 9} \geq \dfrac{1}{x + 1}$

41. $(x + 2)^2 > 0$

42. $(x - 3)^2 < 0$

 Use a graphing calculator to solve each inequality. Give the answer in interval notation.

43. $x^2 - 2x - 3 < 0$ **44.** $x^2 + x - 6 > 0$

45. $\dfrac{x + 3}{x - 2} > 0$ **46.** $\dfrac{3}{x} < 2$

Graph each inequality.

47. $y < x^2 + 1$ **48.** $y > x^2 - 3$

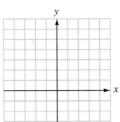 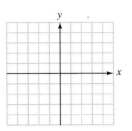

49. $y \leq x^2 + 5x + 6$ **50.** $y \geq x^2 + 5x + 4$

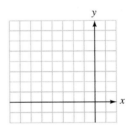 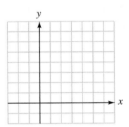

51. $y < |x + 4|$ **52.** $y \geq |x - 3|$

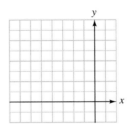 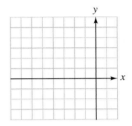

53. $y \leq -|x| + 2$ **54.** $y > |x| - 2$

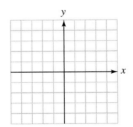 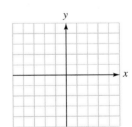

APPLICATIONS

55. SUSPENSION BRIDGES If an x-axis is superimposed over the roadway of the Golden Gate Bridge, with the origin at the center of the bridge as shown on the next page, the length L in feet of a vertical support cable can be approximated by the formula

$$L = \dfrac{1}{9,000}x^2 + 5$$

For the Golden Gate Bridge, $-2{,}100 < x < 2{,}100$. For what intervals along the x-axis are the vertical cables more than 95 feet long?

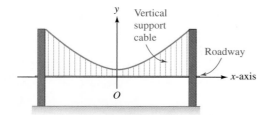

56. MALLS The number of people n in a mall is modeled by the formula

$$n = -100x^2 + 1{,}200x$$

where x is the number of hours since the mall opened. If the mall opened at 9 A.M., when were there 2,000 or more people in it?

WRITING

57. Explain why $(x - 4)(x + 5)$ will be positive only when the signs of $x - 4$ and $x + 5$ are the same.

58. Explain how to find the graph of $y \geq x^2$.

59. The graph of $f(x) = x^2 - 3x + 4$ is shown. Explain why the quadratic inequality $x^2 - 3x + 4 < 0$ has no solution.

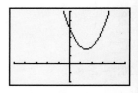

60. What three important facts about the expression $x - 1$ are indicated below?

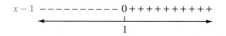

REVIEW   *Translate each statement into an equation.*

61. x varies directly with y.

62. y varies inversely with t.

63. t varies jointly with x and y.

64. d varies directly with t and inversely with u^2.

Find the slope of the graph of each linear function.

65. $f(x) = 3x - 4$

66. $f(x) = -x$

Solving Quadratic Equations

We have discussed five methods for solving **quadratic equations.** Let's review each of them and list an advantage and a drawback of each method.

▮ Factoring

- It can be very quick and simple if the factoring pattern is evident.
- Much of the time, $ax^2 + bx + c$ cannot be factored or is not easily factored.

Solve each equation by factoring.

1. $4k^2 + 8k = 0$ **2.** $z^2 + 8z + 15 = 0$ **3.** $2r^2 + 5r = -3$

▮ The square root method

- If the equation can be written in the form $x^2 = a$ or $(x + d)^2 = a$, where a is a constant, the square root method is a fast method, requiring few computations.
- Most quadratic equations that we must solve are not written in either of these forms.

Solve each equation by the square root method.

4. $u^2 = 24$ **5.** $(s - 7)^2 - 9 = 0$ **6.** $3x^2 - 16 = 0$

▮ Completing the square

- It can be used to solve any quadratic equation.
- Most often, it involves more steps than the other methods.

Solve each equation by completing the square.

7. $x^2 + 10x - 7 = 0$ **8.** $4x^2 - 4x - 1 = 0$ **9.** $x^2 + 2x + 2 = 0$

▮ The quadratic formula

- It simply involves an evaluation of the expression $\dfrac{-b \pm \sqrt{b^2 - 4ac}}{2a}$ for values of a, b, and c.

- If applicable, the factoring method and the square root method are usually faster than the formula method.

Solve each equation by using the quadratic formula.

10. $2x^2 - 1 = 3x$ **11.** $x^2 - 6x - 391 = 0$ **12.** $3x^2 + 2x + 1 = 0$

▮ The graphing method

- We can solve the equation using a graphing calculator. It doesn't require any computations.
- It usually gives only approximations of the solutions.

13. Use the graph of $y = x^2 + x - 2$ to solve $x^2 + x - 2 = 0$.

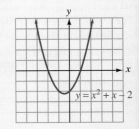

ACCENT ON TEAMWORK

SECTION 8.1

PICTURE FRAMES Each person in your group is to make an 8 in. $\times$ 10 in. collage that illustrates an important aspect of his or her life. Photographs, magazine pictures, drawings, and the like can be used to highlight family, hobbies, jobs, pets, etc. Assign each person the task of making a matting of uniform width to frame his or her collage. Some can make a matting whose area equals that of the picture. Others can make a matting that is half the area or double the area. When completed, let each person briefly explain his or her collage to the other group members. See Example 8, Section 8.1, for some hints on how to do the mathematics.

SECTION 8.2

TEAM RELAYS Have your group make a presentation to the class showing the derivation of the quadratic formula. (See page 582 of the text.) Begin with the general quadratic equation, $ax^2 + bx + c = 0$, and have each member of your group perform and explain several steps. Make a baton from a paper towel roll, like that shown below. As one student finishes with his or her segment of the derivation, the baton is passed to the next student to pick up where he or she left off.

QUADRATIC FORMULA RELAY

SECTION 8.3

DETERMINING THE NUMBER OF SOLUTIONS GRAPHICALLY Experiment with various combinations of a's, b's, and c's to write quadratic functions of the form $f(x) = ax^2 + bx + c$ that, when graphed using a calculator, intersect the x-axis **a.** two times, **b.** one time, and **c.** no times. Then give the associated quadratic equation in each case and tell how many real-number solutions the equation has.

SECTION 8.4

COMPLEX NUMBERS Write a report about complex numbers that answers these questions: When were they first used and by whom? Why were they "invented"? What are some of their applications?

A book about the history of mathematics will be helpful. The school library or perhaps one of your mathematics instructors may have one you can borrow. Some colleges offer a course about complex numbers called Complex Variables. Try to find a copy of the textbook. Germany once issued a postage stamp honoring Carl Gauss and featuring complex numbers. Try to find a picture of it. Physics instructors or electronics instructors are two other possible resources for material.

SECTION 8.5

SOLUTIONS OF A QUADRATIC EQUATION Consider the following property about the solutions of a quadratic equation: If

$$r_1 + r_2 = -\frac{b}{a} \quad \text{and} \quad r_1 r_2 = \frac{c}{a}$$

then r_1 and r_2 are the solutions of a quadratic equation $ax^2 + bx + c = 0$, with $a \neq 0$.

Use this property to show that

a. $\frac{3}{2}$ and $-\frac{1}{3}$ are solutions of $6x^2 - 7x - 3 = 0$.

b. $i\sqrt{51}$ and $-i\sqrt{51}$ are solutions of $x^2 + 51 = 0$.

c. $1 + 3\sqrt{2}$ and $1 - 3\sqrt{2}$ are solutions of $x^2 - 2x - 17 = 0$.

SECTION 8.6

QUADRATIC INEQUALITIES The fraction shown below has two factors in the numerator and two factors in the denominator.

$$\frac{(x - 1)(x + 4)}{(x + 2)(x + 1)}$$

a. With the four factors in mind, under what conditions will the fraction be positive?

b. With the four factors in mind, under what conditions will the fraction be negative?

c. Solve: $\dfrac{(x - 1)(x + 4)}{(x + 2)(x + 1)} > 0$.

CHAPTER REVIEW

SECTION 8.1 — *Completing the Square*

CONCEPTS

The square root property:
If $c > 0$, the equation $x^2 = c$ has two real solutions:

$$x = \sqrt{c} \qquad x = -\sqrt{c}$$

To *complete the square:*

1. Make sure the coefficient of x^2 is 1.

2. Make sure the constant term is on the right-hand side of the equation.

3. Add the square of one-half of the coefficient of x to both sides.

4. Factor the trinomial.

5. Use the square root property.

REVIEW EXERCISES

Solve each equation by factoring or using the square root property.

1. $x^2 + 9x + 20 = 0$

2. $6x^2 + 17x + 5 = 0$

3. $x^2 = 28$

4. $(t + 2)^2 = 36$

5. $5a^2 + 11a = 0$

6. $5x^2 - 49 = 0$

7. What number must be added to $x^2 - x$ to make a perfect square trinomial?

Solve each equation by completing the square.

8. $x^2 + 6x + 8 = 0$

9. $2x^2 - 6x + 3 = 0$

10. HAPPY NEW YEAR As part of a New Year's Eve celebration, a huge ball is to be dropped from the top of a 605-foot-tall building at the proper moment so that it strikes the ground at exactly 12:00 midnight. The distance d in feet traveled by a free-falling object in t seconds is given by the formula $d = 16t^2$. To the nearest second, when should the ball be dropped from the building?

SECTION 8.2 — *The Quadratic Formula*

The quadratic formula:
The solutions of
$ax^2 + bx + c = 0$ where
$a \neq 0$ are given by

$$x = \frac{-b \pm \sqrt{b^2 - 4ac}}{2a}$$

Solve each equation using the quadratic formula.

11. $-x^2 + 10x - 18 = 0$

12. $x^2 - 10x = 0$

13. $2x^2 + 13x = 7$

14. $26y - 3y^2 = 2$

15. SPORTS POSTERS The design specifications for a poster of tennis star Lindsay Davenport shown below call for a 615-square-inch photograph to be surrounded by a blue border. The borders on the sides of the poster are to be half as wide as those at the top and bottom. Find the width of each border.

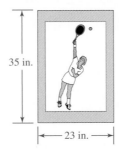

35 in.

23 in.

16. ACROBATS To begin his routine, an acrobat is catapulted upward as shown above. His distance d (in feet) from the arena floor during this maneuver is given by the formula $d = -16t^2 + 40t + 5$, where t is the time in seconds since being launched. If the trapeze bar is 25 feet in the air, at what two times will he be able to grab it? Round to the nearest tenth.

A *quadratic function* is a second-degree polynomial function of the form $f(x) = ax^2 + bx + c$.

The graph of $f(x) = ax^2$ is a *parabola* opening upward when $a > 0$ and downward when $a < 0$, with *vertex* at the point $(0, 0)$ and *axis of symmetry* the line $x = 0$.

Each of the following functions has a graph that is the same shape as $f(x) = ax^2$ but involves a vertical or horizontal translation.

1. $f(x) = ax^2 + k$: translated upward if $k > 0$, downward if $k < 0$.

2. $f(x) = a(x - h)^2$: translated right if $h > 0$, and left if $h < 0$.

If $a \neq 0$, the graph of $f(x) = a(x - h)^2 + k$ is a parabola with vertex at (h, k). It opens upward when $a > 0$ and downward when $a < 0$.

The vertex of the graph of $f(x) = ax^2 + bx + c$ is

$$\left(-\frac{b}{2a}, f\left(-\frac{b}{2a}\right)\right)$$

and the axis of symmetry is the line $x = -\dfrac{b}{2a}$

The vertex of the graph of a quadratic function gives the *minimum* or *maximum* value of the function.

17. AEROSPACE INDUSTRY The annual sales of the Boeing Company in billions of dollars for the years 1993–1998 can be modeled by the quadratic function

$$S(x) = 2.4375x^2 - 8.225x + 40.7$$

where x is the number of years since 1993. What were the annual sales for 1997?

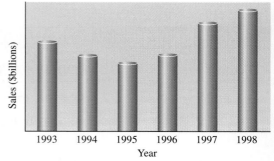

Source: The Boeing Company Annual Report, 1999 and 2002

Make a table of values to graph function f. Then use a series of translations to graph function g on the same coordinate system.

18. $f(x) = 2x^2$
$g(x) = 2x^2 - 3$

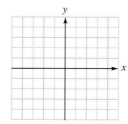

19. $f(x) = -4x^2$
$g(x) = -4(x - 2)^2 + 1$

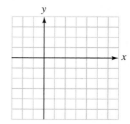

First determine the vertex and the axis of symmetry of the graph of each function. Then plot several points and complete the graph.

20. $f(x) = -\left(x + \dfrac{3}{2}\right)^2 + \dfrac{5}{2}$

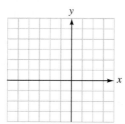

21. $f(x) = 5x^2 + 10x - 1$

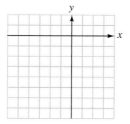

22. FARMING The number of farms in the United States for the years 1870–1970 is approximated by

$$N(x) = -1{,}526x^2 + 155{,}652x + 2{,}500{,}200$$

where $x = 0$ represents 1870, $x = 1$ represents 1871, and so on. For this period, when was the number of U.S. farms a maximum? How many farms were there?

23. Estimate the solutions of $-3x^2 - 5x + 2 = 0$ from the graph of $f(x) = -3x^2 - 5x + 2$, shown on the right.

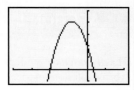

Complex Numbers

Square roots of negative numbers are called *imaginary numbers*.

$$i = \sqrt{-1} \quad and \quad i^2 = -1$$

For any positive real number b,

$$\sqrt{-b} = i\sqrt{b}$$

A *complex number* is any number that can be written in the form $a + bi$, where a and b are real numbers and $i^2 = -1$.

Adding complex numbers:
$(a + bi) + (c + di) =$
$\qquad (a + c) + (b + d)i$

Subtracting complex numbers:
$(a + bi) - (c + di) =$
$\qquad (a - c) + (b - d)i$

Multiplying complex numbers:
$(a + bi)(c + di) =$
$\qquad (ac - bd) + (ad + bc)i$

To divide complex numbers, we often have to *rationalize* a denominator by multiplying the numerator and the denominator by the *complex conjugate* of the denominator.

The *powers of i* rotate through a cycle of four numbers: $i = i$, $i^2 = -1$, $i^3 = -i$, $i^4 = 1$

Simplify each expression.

24. $\sqrt{-4}$

25. $\sqrt{-7}$

26. $-3\sqrt{-20}$

27. $\sqrt{-\dfrac{36}{49}}$

28. $\sqrt{-3}\,\sqrt{-3}$

29. $\dfrac{\sqrt{-64}}{\sqrt{-9}}$

Solve each equation.

30. $a^2 = -25$

31. $x^2 - 2x + 13 = 0$

Peform the operations and give all answers in $a + bi$ form.

32. $(5 + 4i) + (7 - 12i)$

33. $(-6 - 40i) - (-8 + 28i)$

34. $(-8 + \sqrt{-8}) + (6 - \sqrt{-32})$

35. $2i(64 + 9i)$

36. $(2 - 7i)(-3 + 4i)$

37. $(5 - \sqrt{-27})(-6 + \sqrt{-12})$

Write each expression in $a + bi$ form.

38. $\dfrac{6}{2 + i}$

39. $\dfrac{4 + i}{4 - i}$

40. $\dfrac{\sqrt{3} + \sqrt{-4}}{\sqrt{3} - \sqrt{-4}}$

41. $\dfrac{-2}{5i^3}$

Simplify each power of i.

42. i^{65}

43. i^{48}

SECTION 8.5

The Discriminant and Equations That Can Be Written in Quadratic Form

The *discriminant* predicts the type of solutions of $ax^2 + bx + c = 0$:

1. If $b^2 - 4ac > 0$, the solutions are unequal real numbers.

2. If $b^2 - 4ac = 0$, the solutions are equal real numbers.

3. If $b^2 - 4ac < 0$, the solutions are complex conjugates.

Use the discriminant to determine what types of solutions exist for each equation.

44. $3x^2 + 4x - 3 = 0$

45. $4x^2 - 5x + 7 = 0$

46. $9x^2 - 12x + 4 = 0$

Many equations that are not quadratic can be written in quadratic form.

Solve each equation.

47. $x - 13\sqrt{x} + 12 = 0$

48. $a^{2/3} + a^{1/3} - 6 = 0$

49. $6x^4 - 19x^2 + 3 = 0$

50. $\dfrac{6}{x + 2} + \dfrac{6}{x + 1} = 5$

51. $(x - 3)^2 - 8(x - 3) + 7 = 0$

52. WEEKLY CHORES Working together, two sisters can do the yard work at their house in 45 minutes. When the older girl does it all herself, she can complete the job in 20 minutes less time than it takes the younger girl working alone. How long does it take the older girl to do the yard work?

SECTION 8.6

Quadratic and Other Nonlinear Inequalities

To graph a *quadratic inequality in one variable*, get 0 on the right-hand side. Then factor the polynomial on the left-hand side. Use a *sign chart* to determine the solution set.

Solve each inequality. Give each result in interval notation and graph the solution set.

53. $x^2 + 2x - 35 > 0$

54. $x^2 - 81 \leq 0$

To solve rational inequalities, get 0 on the right-hand side and a single fraction on the left-hand side. Factor the numerator and denominator. Then use a sign chart to determine the solution set.

55. $\dfrac{3}{x} \leq 5$

56. $\dfrac{2x^2 - x - 28}{x - 1} > 0$

Use a graphing calculator to solve each inequality. Compare the results with those in Exercises 53 and 56.

57. $x^2 + 2x - 35 > 0$

58. $\dfrac{2x^2 - x - 28}{x - 1} > 0$

To graph a *nonlinear inequality in two variables,* first graph the boundary. Then use a test point to determine which half-plane to shade.

Graph each inequality.

59. $y < \dfrac{1}{2}x^2 - 1$

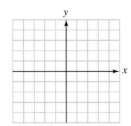

60. $y \geq -|x|$

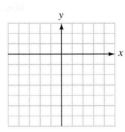

Solve each equation by factoring.

1. $3x^2 + 18x = 0$

2. $x(6x + 19) = -15$

3. Determine what number must be added to $x^2 + 24x$ to make it a perfect square.

4. Solve $x^2 - x - 1 = 0$ by completing the square.

5. Solve the equation $2x^2 - 8x = -5$ using the quadratic formula.

6. TABLECLOTHS In 1990, Sportex of Highland, Illinois, made what was at the time the world's longest tablecloth. Find the dimensions of the rectangular tablecloth if it covered an area of 6,759 square feet and its length was 8 feet more than 332 times its width.

7. Simplify: $\sqrt{-48}$.

8. Simplify: i^{54}.

Perform the operations. Give all answers in $a + bi$ form.

9. $(2 + 4i) + (-3 + 7i)$

10. $(3 - \sqrt{-9}) - (-1 + \sqrt{-16})$

11. $2i(3 - 4i)$

12. $(3 + 2i)(-4 - i)$

13. $\dfrac{1}{i^3}$

14. $\dfrac{2 + i}{3 - i}$

15. Use the discriminant to determine whether the solutions of $3x^2 + 5x + 17 = 0$ are real or nonreal.

Solve each equation.

16. $x^2 = -12$.

17. $13 = 4t - t^2$.

18. $2y - 3\sqrt{y} + 1 = 0$.

19. $x^4 - x^2 - 12 = 0$.

20. $4\left(\dfrac{x + 2}{3x}\right)^2 - 4\left(\dfrac{x + 2}{3x}\right) - 3 = 0$.

Determine the vertex and the axis of symmetry of the graph of the function. Then graph it.

21. $f(x) = 2x^2 + x - 1$

22. $f(x) = -3(x - 1)^2 - 2$

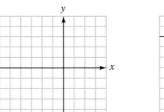

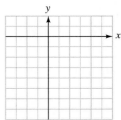

Solve the inequality and graph the solution set.

23. $x^2 - 2x - 8 > 0$

24. $\dfrac{x - 2}{x + 3} \le 0$

25. Graph: $y \le -x^2 + 3$.

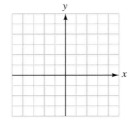

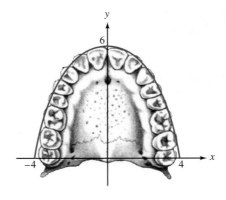

26. DRAWING An artist uses four equal-sized right triangles to block out a perspective drawing of an old hotel. For each triangle, one leg is 14 inches longer than the other, and the hypotenuse is 26 inches. On the centerline of the drawing, what is the length of the segment extending from the ground to the top of the building?

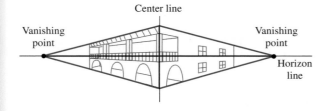

27. ANTHROPOLOGY Anthropologists refer to the shape of the human jaw as a *parabolic dental arcade*. Which function is the best mathematical model of the parabola shown in the illustration in the next column?

i. $f(x) = -\dfrac{3}{8}(x - 4)^2 + 6$

ii. $f(x) = -\dfrac{3}{8}(x - 6)^2 + 4$

iii. $f(x) = -\dfrac{3}{8}x^2 + 6$

iv. $f(x) = \dfrac{3}{8}x^2 + 6$

28. COOKING Working together, a chef and his assistant can make a pastry dessert in 25 minutes. When the chef makes it himself, it takes him 8 minutes less time than it takes his assistant working alone. How long does it take the chef to make the dessert?

29. DISTRESS SIGNALS A flare is fired directly upward into the air from a boat that is experiencing engine problems. The height of the flare (in feet) above the water, t seconds after being fired, is given by the formula $h = -16t^2 + 112t + 15$. If the flare is designed to explode when it reaches its highest point, at what height will this occur?

30. What is an imaginary number? Give some examples.

31. The graph of a quadratic function of the form $f(x) = ax^2 + bx + c$ is shown below. Estimate the solutions of the corresponding quadratic equation $ax^2 + bx + c = 0$.

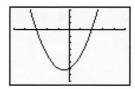

32. See Problem 31. Estimate the solution of the quadratic inequality $ax^2 + bx + c \le 0$. Write the solution set in interval notation.

Write an equation of the line with the given properties. Express the result in slope–intercept form.

1. $m = 3$, passing through $(-2, -4)$

2. Parallel to the graph of $2x + 3y = 6$ and passing through $(0, -2)$

3. AIRPORT TRAFFIC From the graph shown below, determine the projected average rate of change in the number of takeoffs and landings at Los Angeles International Airport for the years 2000–2015.

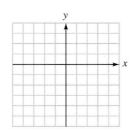

4. Solve by graphing:
$$\begin{cases} y = -\dfrac{5}{2}x + \dfrac{1}{2} \\ 2x - \dfrac{3}{2}y = 5 \end{cases}$$

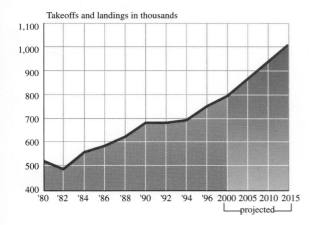

5. Solve using Cramer's rule:
$$\begin{cases} x - y + z = 4 \\ x + 2y - z = -1 \\ x + y - 3z = -2 \end{cases}$$

6. Graph the solution set of the system
$$\begin{cases} 3x + 2y > 6 \\ x + 3y \le 2 \end{cases}$$

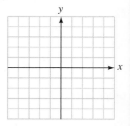

7. Solve the inequality $-9x + 6 > 16$. Give the result in interval notation and graph the solution set.

8. Solve: $|2x - 5| \ge 25$. Give the result in interval notation and graph the solution set.

Find the domain and range of each function.

9. $f(x) = 2x^2 - 3$

10. $y = -|x - 4|$

Perform each operation.

11. $(2a^2 + 4a - 7) - 2(3a^2 - 4a)$

12. $8(3x + 2)(2x - 3)$

Factor each expression.

13. $x^4 - 16y^4$

14. $15x^2 - 2x - 8$

15. $x^2z + 4yz - xyz - 4xz$

16. $8x^6 + 125y^3$

Solve each equation.

17. $x^2 - 5x - 6 = 0$

18. $6a^3 - 2a = a^2$

19. $\dfrac{x - 4}{x - 3} + \dfrac{x - 2}{x - 3} = x - 3$

20. Solve for b: $P + \dfrac{a}{V^2} = \dfrac{RT}{V - b}$.

Simplify each expression.

21. $\dfrac{x^3 + y^3}{x^3 - y^3} \div \dfrac{x^2 - xy + y^2}{x^2 + xy + y^2}$

22. $\dfrac{1}{x + y} - \dfrac{1}{x - y} - \dfrac{2y}{y^2 - x^2}$

Graph each function and determine its domain and range.

23. $f(x) = x^3 + x^2 - 6x$

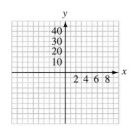

24. $f(x) = \dfrac{4}{x}$ for $x > 0$

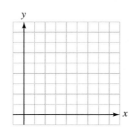

25. LIGHT As light energy radiates away from its source, its intensity varies inversely as the square of the distance from the source. The illustration below shows that the light energy passing through an area 1 foot from the source spreads out over 4 units of area 2 feet from the source. That energy is therefore less intense 2 feet from the source than it was 1 foot from the source. Over how many units of area will the light energy spread out 3 feet from the source?

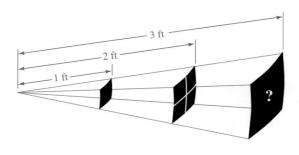

26. Graph the function $f(x) = \sqrt{x - 2}$ and give its domain and range.

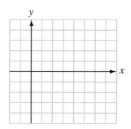

Simplify each expression. All variables represent positive real numbers.

27. $\sqrt[3]{-27x^3}$

28. $\sqrt{48t^3}$

29. $64^{-2/3}$

30. $\dfrac{x^{5/3} x^{1/2}}{x^{3/4}}$

Simplify each expression.

31. $-3\sqrt[4]{32} - 2\sqrt[4]{162} + 5\sqrt[4]{48}$

32. $3\sqrt{2}(2\sqrt{3} - 4\sqrt{12})$

33. $\dfrac{\sqrt{x} + 2}{\sqrt{x} - 1}$

34. $\dfrac{5}{\sqrt[3]{x}}$

Solve each equation.

35. $5\sqrt{x + 2} = x + 8$

36. $\sqrt{x} + \sqrt{x + 2} = 2$

37. Find the length of the hypotenuse of the right triangle shown below.

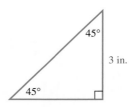

38. Find the length of the hypotenuse of the right triangle shown below.

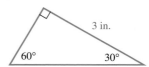

39. Find the distance between $(-2, 6)$ and $(4, 14)$.

40. What number must be added to $x^2 + 6x$ to make a perfect square trinomial?

41. Use the method of completing the square to solve the equation $2x^2 + x - 3 = 0$.

42. Use the quadratic formula to solve the equation $3x^2 + 4x - 1 = 0$.

43. Graph $f(x) = \frac{1}{2}x^2 - x + 1$ and find the coordinates of its vertex.

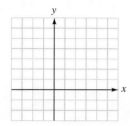

44. Graph $f(x) = -x^2 - 4x$ and find the coordinates of its vertex.

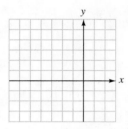

Perform the operations. Give all answers in the form $a + bi$.

45. $(3 + 5i) + (4 - 3i)$

46. $\dfrac{5}{3 - i}$

47. Simplify: $\sqrt{-64}$.

48. Solve: $a - 7a^{1/2} + 12 = 0$.

49. Solve: $2x^2 + 25 = 0$.

50. Simplify: i^{42}.

51. The graph of $f(x) = 16x^2 + 24x + 9$ is shown below. Estimate the solution(s) of $16x^2 + 24x + 9 = 0$.

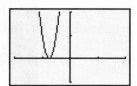

52. Use the graph above to determine the solution, if any, of $16x^2 + 24x + 9 < 0$.

Exponential and Logarithmic Functions

INTERMEDIATE
Algebra *f(x)* **Now**™

Throughout the chapter, this icon introduces resources on the Inter-mediate AlgebraNow Web site, accessed through http://1pass .thomson.com, that will

• Help you test your knowledge of the material with a pre-test and a post-test

• Provide a personalized learning plan targeting areas you should study

Financial planners advise us that we should begin saving for our retirement at a young age. However, it's difficult to make saving a budgeting priority with today's high cost of living. Fortunately, we don't have to choose between paying our current financial obligations and saving for retirement. Thanks to the power of compounding, a modest amount of money invested wisely can turn into a large sum over time. That is because compounding calculates interest on the principal and on the prior period's interest.

To learn more about compound interest, visit *The Learning Equation* on the Internet at http://tle.brookscole.com. (The log-in instructions are in the Preface.) For Chapter 9, the online lessons are:

• *TLE* Lesson 13: Exponential Functions
• *TLE* Lesson 14: Properties of Logarithms

Check Your Knowledge

1. If any horizontal line that intersects the graph of a function does so more than once, the function is not _____.

2. The graphs of a function and its inverse are symmetric about the line ____.

3. If $b > 0$ and $b \neq 1$, the function $f(x) = b^x$ is a(n) _____ function.

4. The functions $f(x) = \log_b(x)$ and $f(x) = b^x$ are _____ functions.

5. Base-e logarithms are often called _____ logarithms.

In Problems 6–9, $f(x) = x^2$ and $g(x) = 2x - 1$. Find each function or value.

6. $f \cdot g$

7. $f - g$

8. $g(f(x))$

9. $f \circ g\,(0)$

Find the inverse of each function.

10. $f(x) = 2x + 3$

11. $f(x) = x^2 \quad (x \geq 0)$

12. Graph: $f(x) = 3^{-x}$.

13. Graph: $f(x) = e^x$.

14. Graph: $f(x) = \log x$.

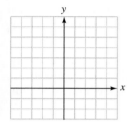
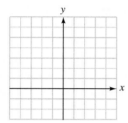
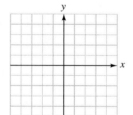

15. If $6000 is deposited in an account paying 5% interest compounded quarterly, what amount will be in the account after 6 years? Round your answer to the nearest cent.

16. The number of grams A of a certain radioactive substance present after t years is given by the formula $A = 80e^{-0.05t}$. Find the amount present (in grams) after 10 years. Round your answer to two decimal places.

17. An account contains $2,000 and has been earning 8% interest, compounded continuously. How long will it take the $2,000 dollars in the account to double? Round your answer to the nearest tenth of a year.

Find x.

18. $\log_4 x = 3$

19. $\log_x 4 = -2$

20. $\log_3 27 = x$

21. $\ln x = 0$

22. Write the statement $\log_5 125 = 3$ in exponential form.

23. Write the statement $e^0 = 1$ in logarithmic form.

24. Write the expression $\log 2x^2y^3$ in terms of the logarithms of 2, x, and y.

25. Write the expression $2 \ln x + 4 \ln y - 3 \ln z$ as a logarithm of a single quantity.

26. Use the change-of-base formula to find $\log_3 13$ to four decimal places.

Solve each equation. Round to four decimal places when necessary.

27. $2^x = 5$

28. $5^{2x-1} = 125$

29. $\log x = \log(2x - 1)$

30. $\log x - \log 3 = \log 6$

Study Skills Workshop

FINDING MATH IN THE WORLD AROUND YOU

This chapter in your text contains many problems that have applications that directly affect your life, so project-based assignments are particularly relevant here. However, some students, even those who have completed research papers in an English class, may feel a little out of their element when doing a research project for a math class. The good news is that researching a project in math follows the same basic steps as writing a paper for any other class. The main idea is that the paper you hand in is your product, but there is a process that you must go through to get your final results.

Process

- *Know your topic*   Clearly state your idea and set the bounds you will explore in your project (you might adjust the bounds after doing some research). Decide whether you wish to write a report (a restatement of information that you found from a variety of sources) or an analysis of a topic. An analysis-based research paper is one in which you use others' ideas to make an argument of your own.
- *Do research*   A wealth of material is available to explore in both libraries and on the Internet. In fact, once you start looking, you may find lots more information than you can use to write your project!
- *Write your project*   This step requires organization and analysis, and may make use of math skills on your part. Examination will help you decide which topics you found in your research belong within the bounds of your project.

Tools for Research

- Explore how to research and write a paper. A good resource for getting started can be found at http://www.ipl.org/div/aplus/.
- Visit your college library or your library's Web site to see what subject matter resources are available there.
- Use Internet resources. Some good ones are "The Electric Library," (http://www.highbeam.com/library/index.asp?) and "Internet Mathematics Library" (http://mathforum.org/library/). Use an Internet search engine to look up specific math topics. To determine whether what you find is a good resource, use "How to Critically Analyze Information Sources" at http://www.library.cornell.edu/olinuris/ref/research/skill26.htm.

Most important, when doing any research project, give yourself ample time to be able to think about the information that you have found and how you want to use it. Depending on the size of the project, give yourself at least a few days after you have found information so that you can analyze and incorporate it into your work.

ASSIGNMENT

1. Choose a topic that you learned this semester that especially interests you. Your project could be a report, an analysis, or a relevant application.
2. Use at least three sources to gather information on your project.
3. Write the details of your project in a paper (3–5 pages).

In this chapter, we discuss the concept of function in more depth. We also introduce exponential and logarithmic functions, which have applications in many areas.

9.1 Algebra and Composition of Functions

- Algebra of functions   • Composition of functions   • The identity function
- Writing composite functions

Just as it is possible to perform operations on real numbers, it is possible to perform operations on functions. We will begin by showing how to add, subtract, multiply, and divide functions. Then we will consider another method of combining functions, called *composition of functions*.

◼ Algebra of functions

We now consider how functions can be added, subtracted, multiplied, and divided.

Operations on functions

If the domains and ranges of functions f and g are subsets of the real numbers, then

The **sum** of f and g, denoted as $f + g$, is defined by

$$(f + g)(x) = f(x) + g(x)$$

The **difference** of f and g, denoted as $f - g$, is defined by

$$(f - g)(x) = f(x) - g(x)$$

The **product** of f and g, denoted as $f \cdot g$, is defined by

$$(f \cdot g)(x) = f(x)g(x)$$

The **quotient** of f and g, denoted as f/g, is defined by

$$(f/g)(x) = \frac{f(x)}{g(x)} \quad \text{where } g(x) \neq 0$$

The domain of each of these functions is the set of real numbers x that are in the domain of both f and g. In the case of the quotient, there is the further restriction that $g(x) \neq 0$.

EXAMPLE 1   Let $f(x) = 2x^2 + 1$ and $g(x) = 5x - 3$. Find each function and its domain: **a.** $(f + g)(x)$ and   **b.** $(f - g)(x)$.

Solution

a. $(f + g)(x) = f(x) + g(x)$

$\qquad\qquad = (2x^2 + 1) + (5x - 3)$

$\qquad\qquad = 2x^2 + 5x - 2$ 　　　　　Combine like terms.

The domain of $f + g$ is the set of real numbers that are in the domain of both f and g. Since the domain of both f and g is the interval $(-\infty, \infty)$, the domain of $f + g$ is also the interval $(-\infty, \infty)$.

INTERMEDIATE
Algebra $f(x)$ Now™

Self Check 1

Let $f(x) = 3x - 2$ and $g(x) = 2x^2 + 3x$. Find:

a. $(f + g)(x)$

b. $(f - g)(x)$

b. $(f - g)(x) = f(x) - g(x)$

$$= (2x^2 + 1) - (5x - 3)$$

$$= 2x^2 + 1 - 5x + 3$$

$$= 2x^2 - 5x + 4 \qquad \text{Combine like terms.}$$

Since the domain of both f and g is $(-\infty, \infty)$, the domain of $f - g$ is also the interval $(-\infty, \infty)$.

Answers **a.** $2x^2 + 6x - 2$,
b. $-2x^2 - 2$

INTERMEDIATE
Algebra *f(x)* Now™

Self Check 2
Let $f(x) = 2x^2 - 3$ and
$g(x) = x^2 + 1$. Find:

a. $(f \cdot g)(x)$

b. $(f/g)(x)$

EXAMPLE 2   Let $f(x) = 2x^2 + 1$ and $g(x) = 5x - 3$. Find each function and its domain: **a.** $(f \cdot g)(x)$ and   **b.** $(f/g)(x)$.

Solution

a. $(f \cdot g)(x) = f(x)g(x)$

$$= (2x^2 + 1)(5x - 3)$$

$$= 10x^3 - 6x^2 + 5x - 3 \qquad \text{Multiply the binomials.}$$

The domain of $f \cdot g$ is the set of real numbers that are in the domain of both f and g. Since the domain of both f and g is the interval $(-\infty, \infty)$, the domain of $f \cdot g$ is also the interval $(-\infty, \infty)$.

b. $(f/g)(x) = \dfrac{f(x)}{g(x)}$

$$= \dfrac{2x^2 + 1}{5x - 3}$$

Answers **a.** $2x^4 - x^2 - 3$,

b. $\dfrac{2x^2 - 3}{x^2 + 1}$

Since the denominator of the fraction cannot be 0, $x \neq \frac{3}{5}$. Thus, the domain of f/g is the interval $\left(-\infty, \frac{3}{5}\right) \cup \left(\frac{3}{5}, \infty\right)$.

THINK IT THROUGH   Black Colleges and Universities

"Historically Black Colleges and Universities (HBCUs) provide an academic atmosphere which recognizes, responds, and appreciates the diverse background and qualifications of thousands of students that annually enroll at these institutions." Wilvena T. McDowell, Assistant Direction of Admissions, North Carolina A&T State University

In the illustration, the graph of function f gives the number of men and the graph of function g gives the number of women enrolled in historically Black colleges and universities. The variable x represents the number of years since 1980. Sketch the graph of the function $f + g$ on the same coordinate system and explain its significance.

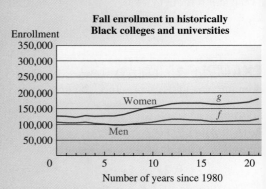

Source: National Center for Educational Statistics, 2004

Composition of functions

We have seen that a function can be represented by a machine: We input a number from the domain, and a number from the range comes out. For example, if we put the number 2 into the machine shown in Figure 9-1(a), the number $f(2) = 8$ comes out. In general, if we put x into the machine shown in Figure 9-1(b), the value $f(x)$ comes out.

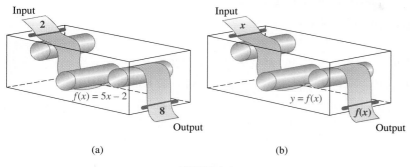

FIGURE 9-1

Often one quantity is a function of a second quantity that depends, in turn, on a third quantity. For example, the cost of a car trip is a function of the gasoline consumed. The amount of gasoline consumed, in turn, is a function of the number of miles driven. Such chains of dependence can be analyzed mathematically as **compositions of functions.**

Suppose that $y = f(x)$ and $y = g(x)$ define two functions. Any number x in the domain of g will produce the corresponding value $g(x)$ in the range of g. If $g(x)$ is in the domain of function f, then $g(x)$ can be substituted into f, and a corresponding value $f(g(x))$ will be determined. This two-step process defines a new function, called a **composite function,** denoted by $f \circ g$. (This is read as "f composed with g.")

The function machines shown in Figure 9-2 illustrate the composition $f \circ g$. When we put a number x into the function g, $g(x)$ comes out. The value $g(x)$ goes into function f, which transforms $g(x)$ into $f(g(x))$. (This is read as "f of g of x.") If the function machines for g and f were connected to make a single machine, that machine would be named $f \circ g$.

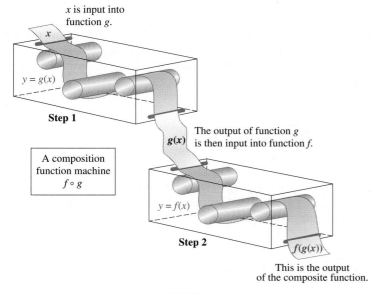

FIGURE 9-2

To be in the domain of the composite function $f \circ g$, a number x has to be in the domain of g. Also, the output of g must be in the domain of f. Thus, the domain of $f \circ g$ consists of those numbers x that are in the domain of g, and for which $g(x)$ is in the domain of f.

Composite functions

The **composite function $f \circ g$** is defined by

$$(f \circ g)(x) = f(g(x))$$

For example, if $f(x) = 4x$ and $g(x) = 3x + 2$, then

$$
\begin{aligned}
(f \circ g)(x) &= f(g(x)) & (g \circ f)(x) &= g(f(x)) \\
&= f(3x + 2) & &= g(4x) \\
&= 4(3x + 2) & &= 3(4x) + 2 \\
&= 12x + 8 & &= 12x + 2
\end{aligned}
$$

Different results

!COMMENT Note that in the previous example, $(f \circ g)(x) \neq (g \circ f)(x)$. This shows that the composition of functions is not commutative. Also note that $(f \circ g)(x)$ is not equal to $(f \cdot g)(x)$: $(f \circ g)(x) = 12x + 8$, but $(f \cdot g)(x) = 12x^2 + 8x$.

INTERMEDIATE
Algebra $f(x)$ Now™

Self Check 3
Let $f(x) = 3x - 1$ and
$g(x) = x + 5$. Find:

a. $(f \circ g)(4)$

b. $(f \circ g)(x)$

c. $(g \circ f)(-2)$

EXAMPLE 3 Let $f(x) = 2x + 1$ and $g(x) = x - 4$. Find **a.** $(f \circ g)(9)$, **b.** $(f \circ g)(x)$, and **c.** $(g \circ f)(-2)$.

Solution

a. $(f \circ g)(9)$ means $f(g(9))$. In Figure 9-3(a), function g receives the number 9, subtracts 4, and releases the number $g(9) = 5$. Then 5 goes into the f function, which doubles 5 and adds 1. The final result, 11, is the output of the composite function $f \circ g$:

Read as "f of g of 9."
↓
$$(f \circ g)(9) = f(g(9)) = f(5) = 2(5) + 1 = 11$$

Thus, $(f \circ g)(9) = 11$.

b. $(f \circ g)(x)$ means $f(g(x))$. In Figure 9-3(a), function g receives the number x, subtracts 4, and releases the number $x - 4$. Then $x - 4$ goes into the f function, which doubles $x - 4$ and adds 1. The final result, $2x - 7$, is the output of the composite function $f \circ g$.

Read as "f of g of x."
↓
$$(f \circ g)(x) = f(g(x)) = f(x - 4) = 2(x - 4) + 1 = 2x - 7$$

Thus, $(f \circ g)(x) = 2x - 7$.

c. $(g \circ f)(-2)$ means $g(f(-2))$. In Figure 9-3(b), function f receives the number -2, doubles it and adds 1, and releases -3 into the g function. Function g subtracts 4 from -3 and outputs a final result of -7. Thus,

Read as "g of f of -2."
↓
$$(g \circ f)(-2) = g(f(-2)) = g(-3) = -3 - 4 = -7$$

Answers **a.** 26, **b.** $3x + 14$, **c.** -2

Thus, $(g \circ f)(-2) = -7$.

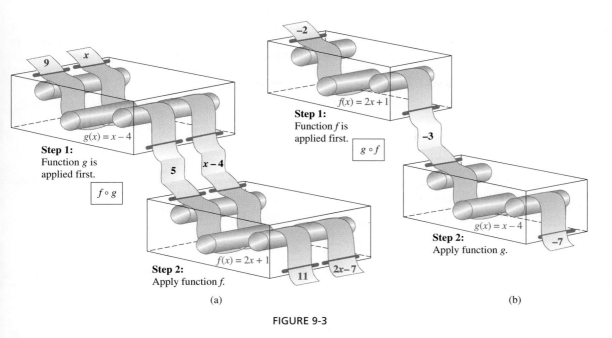

FIGURE 9-3

The identity function

The **identity function** is defined by the equation $I(x) = x$. Under this function, the value that is assigned to any real number x is x itself. For example $I(2) = 2$, $I(-3) = -3$, and $I(7.5) = 7.5$. If f is any function, the composition of f with the identity function is just the function f:

$$(f \circ I)(x) = (I \circ f)(x) = f(x)$$

EXAMPLE 4   Let f be any function and let I be the identity function, $I(x) = x$. Show that   **a.** $(f \circ I)(x) = f(x)$ and   **b.** $(I \circ f)(x) = f(x)$.

Solution
a. $(f \circ I)(x)$ means $f(I(x))$. Because $I(x) = (x)$, we have

$$(f \circ I)(x) = f(I(x)) = f(x)$$

b. $(I \circ f)(x)$ means $I(f(x))$. Because I passes any number through unchanged, we have $I(f(x)) = f(x)$ and

$$(I \circ f)(x) = I(f(x)) = f(x)$$

Writing composite functions

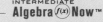

EXAMPLE 5   Biological research.   A laboratory specimen is stored in a refrigeration unit at a temperature of 15° Fahrenheit. Biologists remove the specimen and warm it at the rate of 3° F per hour. Express the sample's Celsius temperature as a function of the time t since it was removed from refrigeration.

Solution   The temperature of the specimen is 15°F when the time $t = 0$. Because it warms at a rate of 3°F per hour, its initial temperature of 15° increases by $3t$°F in t hours. The Fahrenheit temperature of the specimen is given by the function

$$F(t) = 3t + 15$$

The Celsius temperature C is a function of this Fahrenheit temperature F, given by the function

$$C(F) = \frac{5}{9}(F - 32)$$

To express the specimen's Celsius temperature as a function of *time*, we find the composite function

$$
\begin{aligned}
(C \circ F)(t) &= C(F(t)) \\
&= C(3t + 15) &&\text{Substitute } 3t + 15 \text{ for } F(t). \\
&= \frac{5}{9}[(3t + 15) - 32] &&\text{Substitute } 3t + 15 \text{ for } F \text{ in } \frac{5}{9}(F - 32). \\
&= \frac{5}{9}(3t - 17) &&\text{Simplify.}
\end{aligned}
$$

The composite function, $C(t) = \frac{5}{9}(3t - 17)$, finds the temperature of the specimen in degrees Celsius t hours after it is removed from refrigeration.

Section 9.1   STUDY SET

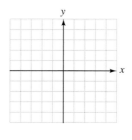

VOCABULARY   *Fill in the blanks.*

1. The _____ of f and g, denoted as $f + g$, is defined by $(f + g)(x) = $ _____ .

2. The _____ of f and g, denoted as $f - g$, is defined by $(f - g)(x) = $ _____ .

3. The _____ of f and g, denoted as $f \cdot g$, is defined by $(f \cdot g)(x) = $ _____ .

4. The _____ of f and g, denoted as f/g, is defined by $(f/g)(x) = $ _____ .

5. In Exercises 1–3, the _____ of each function is the set of real numbers x that are in the domain of both f and g.

6. The _____ function $f \circ g$ is defined by $(f \circ g)(x) = f(g(x))$.

7. Under the _____ function, the value that is assigned to any real number x is x itself.

8. When reading the notation $f(g(x))$, we say "f _____ g _____ x."

CONCEPTS

9. Fill in the blanks.
   a. $(f \circ g)(x) = f($ _____ $)$
   b. To find $f(g(x))$, we first find _____ and then substitute that value for x in $f(x)$.

10. a. If $f(x) = 3x + 1$ and $g(x) = 1 - 2x$, find $f(g(3))$ and $g(f(3))$.
   b. Is the composition of functions commutative? Explain.

11. Complete the table of values for the identity function, $I(x) = x$. Then graph it.

x	$I(x)$
-3	
-2	
-1	
0	
1	
2	
3	

12. Fill in the blanks in the drawing of the function machines below that show how to compute $g(f(-2))$.

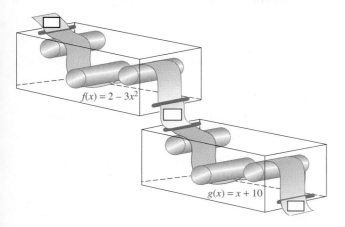

$f(x) = 2 - 3x^2$

$g(x) = x + 10$

■ **NOTATION**   *Complete each solution.*

13. Let $f(x) = 3x - 1$ and $g(x) = 2x + 3$. Find $f \cdot g$.

$$(f \cdot g)(x) = f(x) \cdot g(x)$$
$$= \boxed{}\,(2x + 3)$$
$$= 6x^2 + \boxed{} - \boxed{} - 3$$
$$= 6x^2 + 7x - 3$$

14. Let $f(x) = 3x - 1$ and $g(x) = 2x + 3$. Find $f \circ g$.

$$(f \circ g)(x) = f(g(x))$$
$$= f(\boxed{})$$
$$= 3(\boxed{}) - 1$$
$$= \boxed{} + \boxed{} - 1$$
$$= 6x + 8$$

■ **PRACTICE**   *Let $f(x) = 3x$ and $g(x) = 4x$. Find each function and its domain.*

15. $f + g$ **16.** $f - g$

17. $f \cdot g$ **18.** f/g

19. $g - f$ **20.** $g + f$

21. g/f **22.** $g \cdot f$

Let $f(x) = 2x + 1$ and $g(x) = x - 3$. Find each function and its domain.

23. $f + g$ **24.** $f - g$

25. $f \cdot g$ **26.** f/g

27. $g - f$ **28.** $g + f$

29. g/f

30. $g \cdot f$

Let $f(x) = 3x - 2$ and $g(x) = 2x^2 + 1$. Find each function and its domain.

31. $f - g$ **32.** $f + g$

33. f/g **34.** $f \cdot g$

Let $f(x) = x^2 - 1$ and $g(x) = x^2 - 4$. Find each function and its domain.

35. $f - g$ **36.** $f + g$

37. g/f **38.** $g \cdot f$

Let $f(x) = 2x + 1$ and $g(x) = x^2 - 1$. Find each value.

39. $(f \circ g)(2)$ **40.** $(g \circ f)(2)$

41. $(g \circ f)(-3)$ **42.** $(f \circ g)(-3)$

43. $(f \circ g)(0)$ **44.** $(g \circ f)(0)$

45. $(f \circ g)\left(\dfrac{1}{2}\right)$ **46.** $(g \circ f)\left(\dfrac{1}{3}\right)$

47. $(f \circ g)(x)$ **48.** $(g \circ f)(x)$

49. $(g \circ f)(2x)$ **50.** $(f \circ g)(2x)$

Let $f(x) = 3x - 2$ and $g(x) = x^2 + x$. Find each value.

51. $(f \circ g)(4)$ **52.** $(g \circ f)(4)$

53. $(g \circ f)(-3)$ **54.** $(f \circ g)(-3)$

55. $(g \circ f)(0)$ **56.** $(f \circ g)(0)$

57. $(g \circ f)(x)$ **58.** $(f \circ g)(x)$

59. If $f(x) = x + 1$ and $g(x) = 2x - 5$, show that $(f \circ g)(x) \neq (g \circ f)(x)$.

60. If $f(x) = x^2 + 1$ and $g(x) = 3x^2 - 2$, show that $(f \circ g)(x) \neq (g \circ f)(x)$.

■ **APPLICATIONS**

61. METALLURGY A molten alloy must be cooled slowly to control crystallization. When removed from the furnace, its temperature is $2{,}700°$ F, and it will be cooled at $200°$ per hour. Express the Celsius temperature as a function of the number of hours t since cooling began.

62. WEATHER FORECASTING A high-pressure area promises increasingly warmer weather for the next 48 hours. The temperature is now $34°$ Celsius and is expected to rise $1°$ every 6 hours. Express the Fahrenheit temperature as a function of the number of hours from now. (*Hint:* $F = \frac{9}{5}C + 32$.)

63. MILEAGE COSTS

    a. Use the graphs below to determine the cost of the gasoline consumed if a family drove 500 miles on a summer vacation.

    b. Write a composition function that expresses the cost of the gasoline consumed on the vacation as a function of the miles driven.

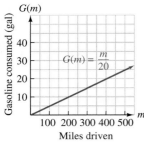
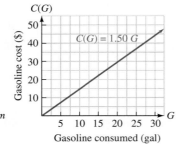

64. COSTUMES The tables on the back of the pattern package shown below can be used to determine the number of yards of material needed to make a rabbit costume for a child.

    a. How many yards of material are needed if the child's chest measures 29 inches?

    b. In this exercise, one quantity is a function of a second quantity that depends, in turn, on a third quantity. Explain this dependence.

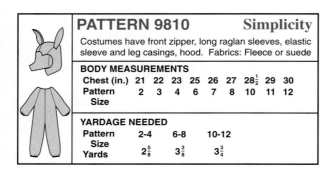

PATTERN 9810   **Simplicity**

Costumes have front zipper, long raglan sleeves, elastic sleeve and leg casings, hood. Fabrics: Fleece or suede

BODY MEASUREMENTS

Chest (in.)	21	22	23	25	26	27	28½	29	30
Pattern Size	2	3	4	6	7	8	10	11	12

YARDAGE NEEDED

Pattern Size	2-4	6-8	10-12
Yards	$2\frac{5}{8}$	$3\frac{3}{8}$	$3\frac{3}{4}$

WRITING

65. Problem 63 illustrates a chain of dependence between the cost of the gasoline, the gasoline consumed, and the miles driven. Describe another chain of dependence that could be represented by a composition function.

66. Explain how to add, subtract, multiply, and divide two functions.

67. Write out in words how to say each of the following:

$$(f \circ g)(2) \qquad g(f(-8))$$

68. If $Y_1 = f(x)$ and $Y_2 = g(x)$, explain how to use the tables shown below to find $g(f(2))$.

X	Y₁
-2	-2
0	2
2	6
4	10
6	14
8	18
10	22

X=-2

X	Y₂
-2	-7
0	-6
2	-5
4	-4
6	-3
8	-2
10	-1

X=-2

REVIEW   *Simplify each expression.*

69. $\dfrac{3x^2 + x - 14}{4 - x^2}$

70. $\dfrac{2x^3 + 14x^2}{3 + 2x - x^2} \cdot \dfrac{x^2 - 3x}{x}$

71. $\dfrac{x^2 - 2x - 8}{3x^2 - x - 12} \div \dfrac{3x^2 + 5x - 2}{3x - 1}$

72. $\dfrac{x - 1}{1 + \dfrac{x}{x - 2}}$

9.2 Inverses of Functions

   • One-to-one functions   • The horizontal line test   • Finding inverses of functions

The function defined by $C(F) = \frac{5}{9}(F - 32)$ can be used to convert degrees Fahrenheit to degrees Celsius. If we input a Fahrenheit reading, the output is a Celsius reading. For example, if we substitute 41° for F, we will obtain a Celsius reading of 5°:

$$C(F) = \frac{5}{9}(F - 32)$$

$$C(41) = \frac{5}{9}(41 - 32) \quad \text{Substitute 41 for } F.$$

$$= \frac{5}{9}(9)$$

$$= 5$$

If we want to find a Fahrenheit reading from a Celsius reading, we need a function into which we can input a Celsius reading and have a Fahrenheit reading come out. Such a function is $F(C) = \frac{9}{5}C + 32$, which takes the Celsius reading of 5° and turns it back into a Fahrenheit reading of 41°.

$$F(C) = \frac{9}{5}C + 32$$

$$F(5) = \frac{9}{5}(5) + 32 \quad \text{Substitute 5 for } C.$$

$$= 41$$

These two functions do opposite things. The first turns 41° F into 5° C, and the second turns 5° C back into 41° F. For this reason, we say that the functions are *inverses* of each other. In this section, we will show how to find inverses of one-to-one functions.

One-to-one functions

We have seen that a function assigns to each input exactly one output. For some functions, different inputs are assigned different outputs, as shown in Figure 9-4(a). For other functions, different inputs are assigned the *same* output, as in Figure 9-4(b). When each output corresponds to exactly one input, as in Figure 9-4(a), we say the function is *one-to-one*.

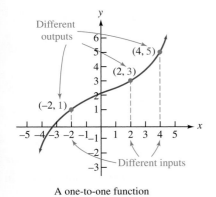

A one-to-one function
(a)

Not a one-to-one function
(b)

FIGURE 9-4

One-to-one functions

A function is called **one-to-one** if each input value of x in the domain determines a different output value of y in the range.

Self Check 1

Determine whether $f(x) = 2x + 3$ is one-to-one. Explain why or why not.

Answer yes, because different input values determine different output values.

EXAMPLE 1   Determine whether the functions   **a.** $f(x) = x^2$ and **b.** $f(x) = x^3$ are one-to-one.

Solution

a. The function $f(x) = x^2$ is not one-to-one, because different input values x can determine the same output value y. For example, inputs of 3 and -3 produce the same output value of 9.

$$f(3) = 3^2 = 9   \text{and}   f(-3) = (-3)^2 = 9$$

b. The function $f(x) = x^3$ is one-to-one, because different input values x determine different output values y. This is because different numbers have different cubes.

■ The horizontal line test

The **horizontal line test** can be used to decide whether the graph of a function represents a one-to-one function.

> **The horizontal line test**
>
> If every horizontal line that intersects the graph of a function does so only once, the function is one-to-one. Otherwise, the function is not one-to-one.

In Figure 9-5(a), since the graph of the function (shown in red) passes the horizontal line test, the function is one-to-one. In Figure 9-5(b), the graph of the function does not pass the horizontal line test. Therefore, the function is not one-to-one.

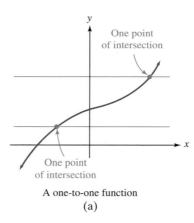

One point of intersection

One point of intersection

A one-to-one function

(a)

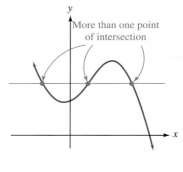

More than one point of intersection

Not a one-to-one function

(b)

FIGURE 9-5

Self Check 2

Determine whether the following graphs represent one-to-one functions.

a.

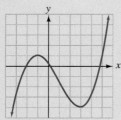

EXAMPLE 2   Use the horizontal line test to decide whether the graphs in Figure 9-6 on the next page represent one-to-one functions.

Solution

a. Refer to the graph in Figure 9.6(a). Because at least one horizontal line intersects the graph twice, the graph does not represent a one-to-one function.

b. Refer to Figure 9.6(b). Because every horizontal line that can be drawn intersects the graph exactly once, the graph represents a one-to-one function.

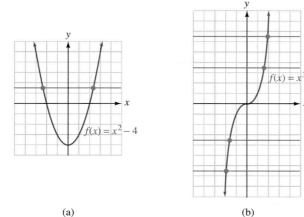

b.

(a)   (b)

FIGURE 9-6

Answers **a.** no,   **b.** yes

Finding inverses of functions

If f is the function determined by the ordered pairs in the table shown in Figure 9-7(a), it turns the number 1 into 10, 2 into 20, and 3 into 30. Since the inverse of f must turn 10 back into 1, 20 back into 2, and 30 back into 3, it consists of the ordered pairs shown in Figure 9-7(b).

FUNCTION f		
x	y	(x, y)
1	10	(1, 10)
2	20	(2, 20)
3	30	(3, 30)

INVERSE OF f		
x	y	(x, y)
10	1	(10, 1)
20	2	(20, 2)
30	3	(30, 3)

The x- and y-coordinates
are interchanged.

(a)   (b)

FIGURE 9-7

We note that the domain of f and the range of its inverse is $\{1, 2, 3\}$. The range of f and the domain of its inverse is $\{10, 20, 30\}$.

This example illustrates that to form the inverse of a function f, we simply interchange the coordinates of each ordered pair that determines f. When the inverse of a function f is also a function, we call it f **inverse** and denote it with the symbol f^{-1}.

! COMMENT The symbol $f^{-1}(x)$ is read as "the inverse of $f(x)$" or just "f inverse." The -1 in the notation $f^{-1}(x)$ is not an exponent. Remember that $f^{-1}(x) \neq \frac{1}{f(x)}$.

Finding the inverse of a one-to-one function

If a function is one-to-one, we find its inverse as follows:

1. If the function is written using function notation, replace $f(x)$ with y.

2. Interchange the variables x and y.

3. Solve the resulting equation for y.

4. We can then substitute $f^{-1}(x)$ for y to denote the inverse function.

INTERMEDIATE
Algebra *f(x)* Now™

Self Check 3

If $f(x) = -5x - 3$, find the inverse of f and tell whether it is a function.

EXAMPLE 3   If $f(x) = 4x + 2$, find the inverse of f and tell whether it is a function.

Solution   We proceed as follows:

$$f(x) = 4x + 2$$

$$y = 4x + 2 \quad \text{Step 1: Replace } f(x) \text{ with } y.$$

$$x = 4y + 2 \quad \text{Step 2: Interchange the variables } x \text{ and } y.$$

To decide whether the inverse $x = 4y + 2$ is a function, we solve for y (step 3).

$$x = 4y + 2$$

$$x - 2 = 4y \qquad \text{Subtract 2 from both sides.}$$

(1) $\qquad y = \dfrac{x - 2}{4} \qquad$ Divide both sides by 4 and write y on the left-hand side.

Because each input x that is substituted into Equation 1 gives one output y, the inverse of f is a function, so we can express it in the form

Answers $f^{-1}(x) = \dfrac{-x - 3}{5}$,

yes

$$f^{-1}(x) = \dfrac{x - 2}{4} \qquad \text{Step 4: Substitute } f^{-1}(x) \text{ for } y.$$

To emphasize an important relationship between a function and its inverse, we substitute some number x, such as $x = 3$, into the function $f(x) = 4x + 2$ of Example 3. The corresponding value of y produced is

$$f(3) = 4(3) + 2 = 14$$

If we substitute 14 into the inverse function, f^{-1}, the corresponding value of y that is produced is

$$f^{-1}(14) = \dfrac{14 - 2}{4} = 3$$

Thus, the function f turns 3 into 14, and the inverse function f^{-1} turns 14 back into 3. In general, the composition of a function and its inverse is the identity function.

To prove that $f(x) = 4x + 2$ and $f^{-1}(x) = \frac{x-2}{4}$ are inverse functions, we must show that their composition (in both directions) is the identity function:

$$(f \circ f^{-1})(x) = f(f^{-1}(x)) \qquad\qquad (f^{-1} \circ f)(x) = f^{-1}(f(x))$$

$$= f\left(\dfrac{x - 2}{4}\right) \qquad\qquad\qquad = f^{-1}(4x + 2)$$

$$= 4\left(\dfrac{x - 2}{4}\right) + 2 \qquad\qquad = \dfrac{4x + 2 - 2}{4}$$

$$= x - 2 + 2 \qquad\qquad\qquad = \dfrac{4x}{4}$$

$$= x \qquad\qquad\qquad\qquad\qquad = x$$

Thus, $(f \circ f^{-1})(x) = (f^{-1} \circ f)(x) = x$, which is the identity function $I(x)$.

Self Check 4

Find the inverse of the function defined by $f(x) = \frac{2}{3}x - 2$. Graph the function and its inverse on one coordinate system. Also graph $y = x$.

EXAMPLE 4   Find the inverse function of $f(x) = -\frac{3}{2}x + 3$. Graph the function and its inverse on one coordinate system.

Solution   To find the inverse function of $f(x) = -\frac{3}{2}x + 3$, we begin by replacing $f(x)$ with y.

$$y = -\dfrac{3}{2}x + 3$$

Then we interchange the variables x and y and solve the equation for y.

$$x = -\frac{3}{2}y + 3$$

$$x - 3 = -\frac{3}{2}y \qquad \text{Subtract 3 from both sides.}$$

$$2x - 6 = -3y \qquad \text{Multiply both sides by 2.}$$

$$-\frac{2}{3}x + 2 = y \qquad \text{Divide both sides by } -3.$$

$$-\frac{2}{3}x + 2 = f^{-1}(x) \qquad \text{Substitute } f^{-1}(x) \text{ for } y.$$

The inverse of $f(x) = -\frac{3}{2}x + 3$ is $f^{-1}(x) = -\frac{2}{3}x + 2$. The graphs of the function, its inverse, and the line $y = x$ appear in Figure 9-8.

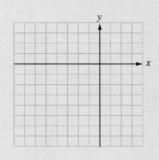

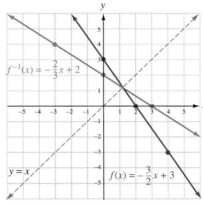

FIGURE 9-8

Answer $f^{-1}(x) = \frac{3}{2}x + 3$

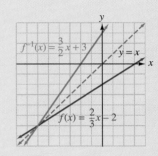

In Figure 9-8, we see that the graphs of $f(x) = -\frac{3}{2}x + 3$ and $f^{-1}(x) = -\frac{2}{3}x + 2$ are symmetric about the line $y = x$. That is, the graphs are mirror images (reflections) of each other with reference to the line $y = x$. This is always the case with a function and its inverse, because when the coordinates (a, b) satisfy an equation, the coordinates (b, a) will satisfy its inverse.

Graphing the inverse of a function   CALCULATOR SNAPSHOT

We can use a graphing calculator for an informal check of the result found in Example 4. First, we enter $f(x) = -\frac{3}{2}x + 3$ as $Y_1 = -\frac{3}{2}x + 3$. Then we enter what we believe to be the inverse function, as $Y_2 = -\frac{2}{3}x + 2$, as well as the equation $y = x$ as $Y_3 = x$. See Figure 9-9(a). Before graphing, we adjust the display so that the graphing grid will be composed of squares. (The standard window display is rectangular, having a grid with a unit width different from the unit height.) Check your owner's manual for directions on how to program the calculator to graph in the square mode.

In Figure 9-9(b), it appears that the two graphs are symmetric about the line $y = x$. Although it is not definitive, this visual check does help to validate that $f^{-1}(x) = -\frac{2}{3}x + 2$ is the inverse of the function $f(x) = -\frac{3}{2}x + 3$.

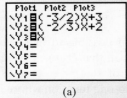

(a)   (b)

FIGURE 9-9

In each example so far, the inverse of a function has been another function. This is not always true, as the following example will show.

Self Check 5
Find the inverse of the function determined by $f(x) = 4x^2$.

EXAMPLE 5   Find the inverse of the function determined by $f(x) = x^2$.

Solution

$$y = x^2 \qquad \text{Replace } f(x) \text{ with } y.$$
$$x = y^2 \qquad \text{Interchange } x \text{ and } y.$$
$$y = \pm\sqrt{x} \qquad \text{Use the square root property and write } y \text{ on the left-hand side.}$$

When the inverse $y = \pm\sqrt{x}$ is graphed, as in Figure 9-10, we see that the graph (in blue) does not pass the vertical line test. Thus, it is not a function.

The graph of $f(x) = x^2$ is also shown in the figure. Note that it is not one-to-one. The graphs of $f(x) = x^2$ and $y = \pm\sqrt{x}$ are symmetric about the line $y = x$.

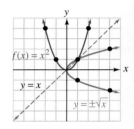

FIGURE 9-10

Answer $y = \pm\dfrac{\sqrt{x}}{2}$

One-to-one functions and inverses

If a function is one-to-one, its inverse is a function.

Self Check 6
Find the inverse of the $f(x) = x^5$.

EXAMPLE 6   Find the inverse of $f(x) = x^3$.

Solution   Since $f(x) = x^3$ is a one-to-one function (see Example 2(b)), its inverse is a function. To find the inverse, we proceed as follows:

$$y = x^3 \qquad \text{Replace } f(x) \text{ with } y.$$
$$x = y^3 \qquad \text{Interchange the variables } x \text{ and } y.$$
$$\sqrt[3]{x} = y \qquad \text{Take the cube root of both sides.}$$

We note that to each number x there corresponds one real cube root. Thus, $y = \sqrt[3]{x}$ represents a function. Using $f^{-1}(x)$ notation, the inverse of $f(x) = x^3$ is

$$f^{-1}(x) = \sqrt[3]{x}$$

Answer $f^{-1}(x) = \sqrt[5]{x}$

If a function is not one-to-one, it is often possible to make it one-to-one by restricting its domain. For example, we have seen that $f(x) = x^2$ is not a one-to-one function. However, if we restrict the domain of $f(x) = x^2$ to $x \geq 0$ (only 0 and positive input values for x), the result is a one-to-one function whose inverse is also a function.

EXAMPLE 7 Find the inverse of the function defined by $f(x) = x^2$ with $x \geq 0$. Graph the function and its inverse on one set of coordinate axes.

Solution The inverse of the function $f(x) = x^2$ with $x \geq 0$ is

$x = y^2$ with $y \geq 0$ Interchange the variables x and y.

Then we solve for y.

$y = \pm\sqrt{x}$ with $y \geq 0$

Since $y \geq 0$, we can discard the possibility that $y = -\sqrt{x}$. Thus, the inverse of $f(x) = x^2$ with $x \geq 0$ is $f^{-1}(x) = \sqrt{x}$.

The graphs of the function and its inverse appear in Figure 9-11. The line $y = x$ is included so that we can see that the graphs are symmetric about the line $y = x$.

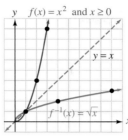

FIGURE 9-11

Section 9.2 STUDY SET

VOCABULARY *Fill in the blanks.*

1. A function is called _____ if each input determines a different output.

2. The _____ line test can be used to decide whether the graph of a function represents a one-to-one function.

3. The functions f and f^{-1} are _____.

4. An input value is an element of a function's _____. An output value is an element of a function's _____.

CONCEPTS *Fill in the blanks.*

5. If every horizontal line that intersects the graph of a function does so only _____, the function is one-to-one.

6. If any horizontal line that intersects the graph of a function does so more than once, the function is not _____.

7. If a function turns an input of 2 into an output of 5, the inverse function will turn an input of 5 into an output of ____.

8. The graphs of a function and its inverse are symmetrical about the line $y = $ ▮.

9. $(f \circ f^{-1})(x) = $ ▮.

10. To find the inverse of the function $f(x) = 2x - 3$, we begin by replacing $f(x)$ with ▮, and then we _____ x and y.

11. Use the table of values of the one-to-one function f to complete a table of values for f^{-1}.

x	f(x)		x	f⁻¹(x)
−6	−3		−3	
−4	−2		−2	
0	0		0	
2	1		1	
8	4		4	

12. How can we tell that function f is not one-to-one from the table of values?

x	f(x)
−2	4
−1	1
0	0
2	4
3	9

13. Is the inverse of a function always a function?

14. Name four points that the line $y = x$ passes through.

15. If f is a one-to-one function, and if $f(2) = 6$, then what is $f^{-1}(6)$?

16. If the point $(2, -4)$ is on the graph of the one-to-one function f, then what point is on the graph of f^{-1}?

17. Two functions are graphed on the calculator display along with the line $y = x$. Explain why we know that the functions are not inverses of each other.

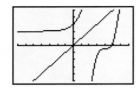

18. A table of values for a function f is shown in illustration (a). A table of values for f^{-1} is shown in illustration (b). Explain how to use the tables to find $f^{-1}(f(4))$ and $f(f^{-1}(2))$.

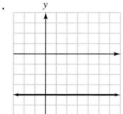

      (a)   (b)

▓ **NOTATION** *Complete each solution.*

19. Find the inverse of $f(x) = 2x - 3$.

$$y = \boxed{} - 3 \qquad \text{Replace } f(x) \text{ with } y.$$
$$x = \boxed{} - 3 \qquad \text{Interchange the variables } x \text{ and } y.$$
$$x + \boxed{} = 2y \qquad \text{Add 3 to both sides.}$$
$$\frac{x + 3}{2} = \boxed{} \qquad \text{Divide both sides by 2.}$$

The inverse of $f(x) = 2x - 3$ is $\boxed{} = \dfrac{x + 3}{2}$

20. Find the inverse of $f(x) = \sqrt[3]{x} + 2$.

$$\boxed{} = \sqrt[3]{x} + 2 \qquad \text{Replace } f(x) \text{ with } y.$$
$$x = \sqrt[3]{\boxed{}} + 2 \qquad \text{Interchange the variables } x \text{ and } y.$$
$$x - \boxed{} = \sqrt[3]{y} \qquad \text{Subtract 2 from both sides.}$$
$$(x - 2)^3 = \boxed{} \qquad \text{Cube both sides.}$$

The inverse of $f(x) = \sqrt[3]{x} + 2$ is $\boxed{} = (x - 2)^3$.

21. The symbol $f^{-1}(x)$ is read as "_____ f" or "f _____."

22. Explain the difference in the meaning of the -1 in the notation $f^{-1}(x)$ as compared to x^{-1}.

▓ **PRACTICE** *Determine whether each function is one-to-one.*

23. $f(x) = 2x$ **24.** $f(x) = |x|$

25. $f(x) = x^4$ **26.** $f(x) = x^3 + 1$

Each graph represents a function. Use the horizontal line test to decide whether each function is one-to-one.

27.
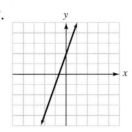

28.
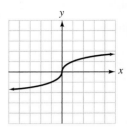

29.
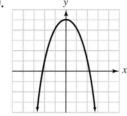

30.
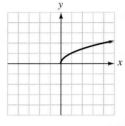

31.
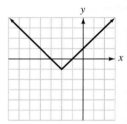

32.

Determine whether the set of ordered pairs (x, y) is a function. Then find the inverse of the set of ordered pairs (x, y) and determine whether the inverse is a function.

33. $\{(3, 2), (2, 1), (1, 0)\}$

34. $\{(4, 1), (5, 1), (6, 1), (7, 1)\}$

35. $\{(1, 1), (2, 1), (3, 1), (4, 1)\}$

36. $\{(1, 2), (2, 3), (1, 3), (1, 5)\}$

Find the inverse of the function and express it using $f^{-1}(x)$ notation.

37. $f(x) = \dfrac{4}{5}x - 4$ **38.** $f(x) = 5x - 1$

39. $f(x) = \dfrac{x}{5} + \dfrac{4}{5}$ **40.** $f(x) = \dfrac{x}{3} - \dfrac{1}{3}$

41. $f(x) = \dfrac{x - 4}{5}$ **42.** $f(x) = \dfrac{2x + 6}{3}$

Find the inverse of each function and express it in terms of x and y. Then graph the function and its inverse on one coordinate system. Show the line of symmetry on the graph.

43. $f(x) = 4x + 3$

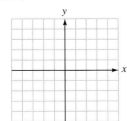

44. $f(x) = \frac{x}{3} + \frac{1}{3}$

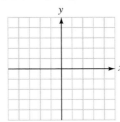

45. $f(x) = -\frac{2}{3}x + 3$

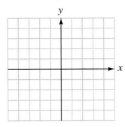

46. $f(x) = -\frac{1}{3}x + \frac{4}{3}$

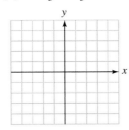

47. $f(x) = x^3$

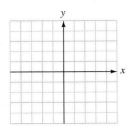

48. $f(x) = x^3 + 1$

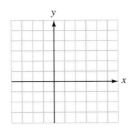

49. $f(x) = x^2 - 1 \ (x \geq 0)$

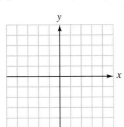

50. $f(x) = x^2 + 1 \ (x \geq 0)$

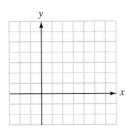

Find the inverse of the function and determine whether it is a function. If it is a function, express it using $f^{-1}(x)$ notation.

51. $f(x) = x^2 + 4$

52. $f(x) = x^2 + 5$

53. $f(x) = x^3$

54. $xy = 4$

55. $f(x) = |x|$

56. $f(x) = \sqrt[3]{x}$

57. $f(x) = 2x^3 - 3$

58. $f(x) = \frac{3}{x^3} - 1$

APPLICATIONS

59. INTERPERSONAL RELATIONSHIPS Feelings of anxiety in a relationship can increase or decrease, depending on what is going on in the relationship. The following graph shows how a person's anxiety might vary as a relationship develops over time.

   a. Is this the graph of a function? Is its inverse a function?

   b. Does each anxiety level correspond to exactly one point in time? Use the dashed lined labeled *Maximum threshold* to explain.

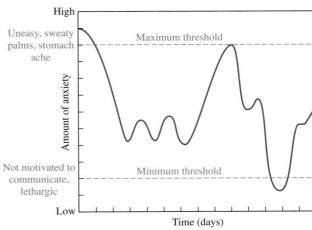

Source: Gudykunst, *Building Bridges: Interpersonal Skills for a Changing World* (Houghton Mifflin, 1994)

60. LIGHTING LEVELS The ability of the eye to see detail increases as the level of illumination increases. This relationship can be modeled by a function E, whose graph is shown on the right.

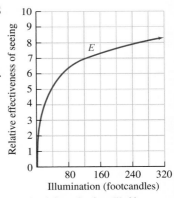

Based on information from *World Book Encyclopedia*

   a. From the graph, determine $E(240)$.

   b. Is function E one-to-one? Does E have an inverse?

   c. If the effectiveness of seeing in an office is 7, what is the illumination in the office? How can this question be asked using inverse function notation?

WRITING

61. What does it mean when we say that one graph is symmetric to the other about the line $y = x$?

62. Explain the purpose of the horizontal line test.

63. In the illustration below, a function f and its inverse f^{-1} have been graphed on the same coordinate system. Explain what concept can be demonstrated by folding the graph paper on the dotted line.

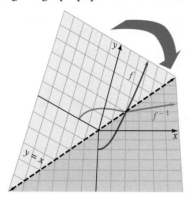

64. Write in words how to read the notation
$$f^{-1}(x) = \frac{1}{2}x - 3.$$

65. $3 - \sqrt{-64}$

66. $(2 - 3i) + (4 + 5i)$

67. $(3 + 4i)(2 - 3i)$

68. $\dfrac{6 + 7i}{3 - 4i}$

69. $(6 - 8i)^2$

70. i^{100}

9.3 Exponential Functions

- Irrational exponents • Exponential functions • Graphing exponential functions
- Vertical and horizontal translations • Compound interest • Exponential functions as models

The graph in Figure 9-12 shows the balance in a bank account in which $10,000 was invested in 1998 at 9%, compounded monthly. The graph shows that in the year 2008, the value of the account will be approximately $25,000, and in the year 2028, the value will be approximately $147,000. The curve shown in Figure 9-12 is the graph of a function called an *exponential function*.

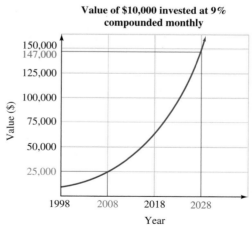

Value of $10,000 invested at 9% compounded monthly

FIGURE 9-12

Exponential functions are also suitable for modeling many other situations, such as population growth, the spread of an epidemic, the temperature of a heated object as it cools, and radioactive decay. Before we can discuss exponential functions in more detail, we must define irrational exponents.

Irrational exponents

We have discussed expressions of the form b^x, where x is a rational number.

$8^{1/2}$ means "the square root of 8."

$5^{1/3}$ means "the cube root of 5."

$3^{-2/5} = \dfrac{1}{3^{2/5}}$ means "the reciprocal of the fifth root of 3^2."

To give meaning to b^x when x is an irrational number, we consider the expression $5^{\sqrt{2}}$, where $\sqrt{2}$ is the irrational number $1.414213562\ldots$

Each number in the following list is defined, because each exponent is a rational number.

$5^{1.4}, \quad 5^{1.41}, \quad 5^{1.414}, \quad 5^{1.4142}, \quad 5^{1.41421}, \quad \ldots$

Since the exponents are getting closer to $\sqrt{2}$, the numbers in this list are successively better approximations of $5^{\sqrt{2}}$. We can use a calculator to obtain a very good approximation.

Evaluating exponential expressions

CALCULATOR SNAPSHOT

To find the value of $5^{\sqrt{2}}$ with a scientific calculator, we enter these numbers and press these keys:

$5\;\boxed{y^x}\;2\;\boxed{\sqrt{}}\;\boxed{=}$
$$\boxed{9.738517742}$$

With a graphing calculator, we enter these numbers and press these keys:

$5\;\boxed{\wedge}\;\boxed{\sqrt{}}\;2\;\boxed{)}\;\boxed{\text{ENTER}}$
$$\boxed{\begin{array}{l} 5\wedge\sqrt{}(2) \\ \quad\quad 9.738517742 \end{array}}$$

In general, if $b > 0$ and x represents a real number, then b^x represents a positive number. It can be shown that all of the familiar rules of exponents are also true for irrational exponents.

EXAMPLE 1   Use the rules of exponents to simplify   **a.** $(5^{\sqrt{2}})^{\sqrt{2}}$ and **b.** $b^{\sqrt{3}} \cdot b^{\sqrt{12}}$.

Solution
a. $(5^{\sqrt{2}})^{\sqrt{2}} = 5^{\sqrt{2}\sqrt{2}}$ Keep the base and multiply the exponents.

$\qquad\qquad\;\; = 5^2$ $\sqrt{2}\sqrt{2} = \sqrt{4} = 2.$

$\qquad\qquad\;\; = 25$

b. $b^{\sqrt{3}} \cdot b^{\sqrt{12}} = b^{\sqrt{3}+\sqrt{12}}$ Keep the base and add the exponents.

$\qquad\qquad\;\; = b^{\sqrt{3}+2\sqrt{3}}$ Simplify: $\sqrt{12} = \sqrt{4}\sqrt{3} = 2\sqrt{3}.$

$\qquad\qquad\;\; = b^{3\sqrt{3}}$ Combine like radicals: $\sqrt{3} + 2\sqrt{3} = 3\sqrt{3}.$

INTERMEDIATE
Algebra $f(x)$ Now™

Self Check 1
Simplify:
a. $(3^{\sqrt{2}})^{\sqrt{8}}$

b. $b^{\sqrt{2}} \cdot b^{\sqrt{18}}$

Answers **a.** 81, **b.** $b^{4\sqrt{2}}$

Exponential functions

If $b > 0$ and $b \neq 1$, the function $f(x) = b^x$ is an **exponential function.** Since x can be any real number, its domain is the set of real numbers. This is the interval $(-\infty, \infty)$.

Because b is positive, the value of $f(x)$ is positive, and the range is the set of positive numbers. This is the interval $(0, \infty)$.

Since $b \neq 1$, an exponential function cannot be the constant function $f(x) = 1^x$, in which $f(x) = 1$ for every real number x.

Exponential functions

An **exponential function with base b** is defined by the equation

$$f(x) = b^x \qquad \text{or} \qquad y = b^x \quad \text{where } b > 0, b \neq 1, \text{ and } x \text{ is a real number}$$

The **domain of any exponential function** is the interval $(-\infty, \infty)$. The **range** is the interval $(0, \infty)$.

▮ Graphing exponential functions

Since the domain and range of $f(x) = b^x$ are sets of real numbers, we can graph exponential functions on a rectangular coordinate system.

Self Check 2

Graph: $f(x) = 4^x$.

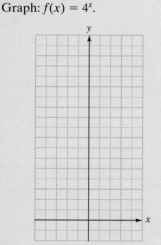

Answer

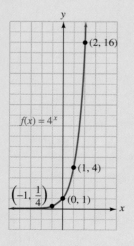

EXAMPLE 2   Graph: $f(x) = 2^x$.

Solution   To graph $f(x) = 2^x$, we make a table of values for the function, plot the points, and join them with a smooth curve, as shown in Figure 9-13. For example, if $x = -1$, we have

$$f(x) = 2^x$$
$$f(-1) = 2^{-1} \quad \text{Substitute } -1 \text{ for } x.$$
$$= \frac{1}{2}$$

The point $\left(-1, \frac{1}{2}\right)$ is on the graph of $f(x) = 2^x$.

$f(x) = 2^x$		
x	$f(x)$	$(x, f(x))$
-1	$\frac{1}{2}$	$\left(-1, \frac{1}{2}\right)$
0	1	$(0, 1)$
1	2	$(1, 2)$
2	4	$(2, 4)$
3	8	$(3, 8)$
4	16	$(4, 16)$

The graph steadily approaches the x-axis, but never touches or crosses it.

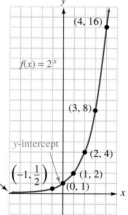

FIGURE 9-13

From the graph, we can verify that the domain of $f(x) = 2^x$ is the interval $(-\infty, \infty)$ and the range is the interval $(0, \infty)$. Since the graph passes the horizontal line test, the function is one-to-one.

Note that as x decreases, the values of $f(x)$ decrease and approach 0. Thus, the x-axis is a horizontal asymptote of the graph. The graph does not have an x-intercept, the y-intercept is $(0, 1)$, and the graph passes through the point $(1, 2)$.

EXAMPLE 3   Graph: $f(x) = \left(\dfrac{1}{2}\right)^x$.

Solution   We make a table of values for the function. For example, if $x = -4$, we have

$$f(x) = \left(\frac{1}{2}\right)^x$$

$$f(-4) = \left(\frac{1}{2}\right)^{-4} \qquad \text{Substitute } -4 \text{ for } x.$$

$$= \left(\frac{2}{1}\right)^4$$

$$= 16$$

The point $(-4, 16)$ is on the graph of $f(x) = \left(\frac{1}{2}\right)^x$. The graph of $f(x) = \left(\frac{1}{2}\right)^x$ appears in Figure 9-14.

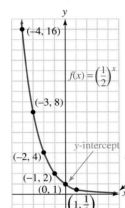

$f(x) = \left(\dfrac{1}{2}\right)^x$		
x	$f(x)$	$(x, f(x))$
-4	16	$(-4, 16)$
-3	8	$(-3, 8)$
-2	4	$(-2, 4)$
-1	2	$(-1, 2)$
0	1	$(0, 1)$
1	$\frac{1}{2}$	$(1, \frac{1}{2})$

FIGURE 9-14

From the graph, we can verify that the domain of $f(x) = \left(\frac{1}{2}\right)^x$ is the interval $(-\infty, \infty)$ and the range is the interval $(0, \infty)$. Since the graph passes the horizontal line test, the function is one-to-one.

Note that as x increases, the values of $f(x)$ decrease and approach 0. Thus, the x-axis is a horizontal asymptote of the graph. The graph does not have an x-intercept, the y-intercept is $(0, 1)$, and the graph passes through the point $\left(1, \frac{1}{2}\right)$.

Examples 2 and 3 illustrate the following properties of exponential functions.

Properties of exponential functions

The domain of the exponential function $f(x) = b^x$ is the interval $(-\infty, \infty)$.

The range is the interval $(0, \infty)$.

The graph has a y-intercept of $(0, 1)$.

The x-axis (the line $y = 0$) is an asymptote of the graph.

The graph of $f(x) = b^x$ passes through the point $(1, b)$.

In Example 2 (where $b = 2$), the values of y increase as the values of x increase. Since the graph rises as we move to the right, we call the function an *increasing function*. When $b > 1$, the larger the value of b, the steeper the curve.

INTERMEDIATE
Algebra *f(x)* **Now**™

Self Check 3

Graph: $g(x) = \left(\dfrac{1}{3}\right)^x$.

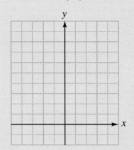

Answer

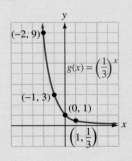

In Example 3 (where $b = \frac{1}{2}$), the values of y decrease as the values of x increase. Since the graph drops as we move to the right, we call the function a *decreasing function*. When $0 < b < 1$, the smaller the value of b, the steeper the curve.

In general, the following is true.

Increasing and decreasing functions

If $b > 1$, then $f(x) = b^x$ is an **increasing function.**

If $0 < b < 1$, then $f(x) = b^x$ is a **decreasing function.**

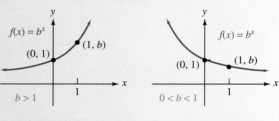

Increasing function   Decreasing function

An exponential function with base b is either increasing (for $b > 1$) or decreasing ($0 < b < 1$). Since different real numbers x determine different values of b^x, exponential functions are one-to-one.

Graphing exponential functions

To use a graphing calculator to graph $f(x) = \left(\frac{2}{3}\right)^x$ and $g(x) = \left(\frac{3}{2}\right)^x$, we enter the right-hand sides of the equations after the symbols $Y_1 =$ and $Y_2 =$. The screen will show the following equations.

$$Y_1 = (2/3)\,{}^\wedge X$$
$$Y_2 = (3/2)\,{}^\wedge X$$

If we use window settings of $[-10, 10]$ for x and $[-2, 10]$ for y and press the $\boxed{\text{GRAPH}}$ key, we will obtain the graph shown in Figure 9-15.

We note that the graph of $f(x) = \left(\frac{2}{3}\right)^x$ passes through the points $(0, 1)$ and $\left(1, \frac{2}{3}\right)$. Since $\frac{2}{3} < 1$, the function is decreasing.

The graph of $g(x) = \left(\frac{3}{2}\right)^x$ passes through the points $(0, 1)$ and $\left(1, \frac{3}{2}\right)$. Since $\frac{3}{2} > 1$, the function is increasing.

Since both graphs pass the horizontal line test, each function is one-to-one.

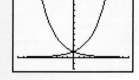

FIGURE 9-15

■ Vertical and horizontal translations

INTERMEDIATE
Algebra $f(x)$ **Now™**

Self Check 4
Graph $f(x) = 4^x$ and $g(x) = 4^x - 3$, and describe the translation.

EXAMPLE 4   On one set of axes, graph $f(x) = 2^x$ and $g(x) = 2^x + 3$, and describe the translation.

Solution   The graph of $g(x) = 2^x + 3$ is identical to the graph of $f(x) = 2^x$, except that it is translated 3 units upward. (See Figure 9-16.) We note that the asymptote of the graph of $g(x) = 2^x + 3$ is the line $y = 3$.

$f(x) = 2^x$		
x	$f(x)$	$(x, f(x))$
-4	$\frac{1}{16}$	$(-4, \frac{1}{16})$
0	1	$(0, 1)$
2	4	$(2, 4)$

$g(x) = 2^x + 3$		
x	$g(x)$	$(x, g(x))$
-4	$3\frac{1}{16}$	$(-4, 3\frac{1}{16})$
0	4	$(0, 4)$
2	7	$(2, 7)$

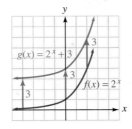

FIGURE 9-16

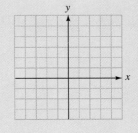

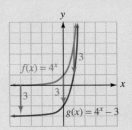

Answer The graph of $f(x) = 4^x$ is translated 3 units downward.

INTERMEDIATE
Algebra _f(x)_ **Now**™

EXAMPLE 5   On one set of axes, graph $f(x) = 2^x$ and $g(x) = 2^{x+3}$, and describe the translation.

Solution   The graph of $g(x) = 2^{x+3}$ is identical to the graph of $f(x) = 2^x$, except that it is translated 3 units to the left. (See Figure 9-17.)

$f(x) = 2^x$		
x	$f(x)$	$(x, f(x))$
-1	$\frac{1}{2}$	$(-1, \frac{1}{2})$
0	1	$(0, 1)$
1	2	$(1, 2)$

$g(x) = 2^{x+3}$		
x	$g(x)$	$(x, g(x))$
-1	4	$(-1, 4)$
0	8	$(0, 8)$
1	16	$(1, 16)$

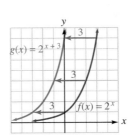

FIGURE 9-17

Self Check 5
On one set of axes, graph $f(x) = 4^x$ and $g(x) = 4^{x-3}$, and describe the translation.

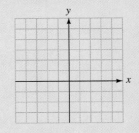

Answer The graph of $f(x) = 4^x$ is translated 3 units to the right.

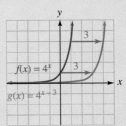

The graphs of $f(x) = kb^x$ and $f(x) = b^{kx}$ are vertical and horizontal stretchings of the graph of $f(x) = b^x$. To graph these functions, we can plot several points and join them with a smooth curve, or we can use a graphing calculator.

Graphing exponential functions

To use a graphing calculator to graph the exponential function $f(x) = 3(2^{x/3})$, we enter the right-hand side of the equation after the symbol $Y_1 =$. The display will show the equation

$$Y_1 = 3(2 \wedge (X/3))$$

If we use window settings of $[-10, 10]$ for x and $[-2, 18]$ for y and press the graph key, we will obtain the graph shown in Figure 9-18.

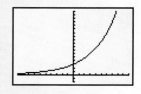

FIGURE 9-18

Compound interest

If we deposit $\$P$ in an account paying an annual interest rate r, we can find the amount A in the account at the end of t years by using the formula $A = P + Prt$, or $A = P(1 + rt)$.

Suppose that we deposit $\$500$ in such an account that pays interest every six months. Then $P = 500$, and after six months ($\frac{1}{2}$ year), the amount in the account will be

$$A = 500(1 + rt)$$
$$= 500\left(1 + r \cdot \frac{1}{2}\right) \quad \text{Substitute } \tfrac{1}{2} \text{ for } t.$$
$$= 500\left(1 + \frac{r}{2}\right)$$

The account will begin the second six-month period with a value of $\$500(1 + \frac{r}{2})$. After the second six-month period, the amount in the account will be

$$A = P(1 + rt)$$
$$A = \left[500\left(1 + \frac{r}{2}\right)\right]\left(1 + r \cdot \frac{1}{2}\right) \quad \text{Substitute } 500\left(1 + \tfrac{r}{2}\right) \text{ for } P \text{ and } \tfrac{1}{2} \text{ for } t.$$
$$= 500\left(1 + \frac{r}{2}\right)\left(1 + \frac{r}{2}\right)$$
$$= 500\left(1 + \frac{r}{2}\right)^2$$

At the end of a third six-month period, the amount in the account will be

$$A = 500\left(1 + \frac{r}{2}\right)^3$$

In this discussion, the earned interest is deposited back in the account and also earns interest. When this is the case, we say that the account is earning **compound interest.** This example illustrates the following formula for compound interest.

> **Formula for compound interest**
>
> If $\$P$ is deposited in an account and interest is paid k times a year at an annual rate r, the amount A in the account after t years is given by
>
> $$A = P\left(1 + \frac{r}{k}\right)^{kt}$$

EXAMPLE 6 Saving for college.

To save for college, parents invest $12,000 for their newborn child in a mutual fund that should average a 10% annual return. If the quarterly dividends are reinvested, how much will be available in 18 years?

Solution   We substitute 12,000 for P, 0.10 for r, and 18 for t in the formula for compound interest and find A. Since interest is paid quarterly, $k = 4$.

$$A = P\left(1 + \frac{r}{k}\right)^{kt}$$

$$A = 12{,}000\left(1 + \frac{0.10}{4}\right)^{4(18)} \qquad \text{Express } r = 10\% \text{ as a decimal.}$$

$$= 12{,}000(1 + 0.025)^{72}$$

$$= 12{,}000(1.025)^{72}$$

$$= 71{,}006.74 \qquad \text{Use a scientific calculator and press these keys:}$$
$$\qquad\qquad\qquad\qquad 1.025 \;\boxed{y^x}\; 72 \;\boxed{=}\; \boxed{\times}\; 12{,}000 \;\boxed{=}\; .$$

In 18 years, the account will be worth $71,006.74.

Self Check 6
In Example 6, how much would be available if the parents invested $20,000?

Answer $118,344.56

Solving investment problems

CALCULATOR SNAPSHOT

Suppose $1 is deposited in an account earning 6% annual interest, compounded monthly. To use a graphing calculator to estimate how much will be in the account in 100 years, we can substitute 1 for P, 0.06 for r, and 12 for k in the formula

$$A = P\left(1 + \frac{r}{k}\right)^{kt}$$

$$A = 1\left(1 + \frac{0.06}{12}\right)^{12t}$$

and simplify to get

$$A = (1.005)^{12t}$$

We now graph $A = (1.005)^{12t}$ using window settings of $[0, 120]$ for t and $[0, 400]$ for A to obtain the graph shown in Figure 9-19. We can then trace and zoom to estimate that $1 grows to be approximately $397 in 100 years. From the graph, we can see that the money grows slowly in the early years and rapidly in the later years.

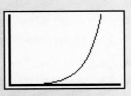

FIGURE 9-19

   In business applications, the initial amount of money deposited is often called the **present value** (PV). The amount to which the money will grow is called the **future value** (FV). The interest rate used for each compounding period is the **periodic interest rate** (i), and the number of times interest is compounded is the number of **compounding periods** (n). Using these definitions, we have an alternate formula for compound interest.

Formula for compound interest

$$FV = PV(1 + i)^n$$

This alternate formula appears on business calculators. To use this formula to solve Example 6, we proceed as follows:

$$FV = PV(1 + i)^n$$

$$FV = \mathbf{12{,}000}(1 + \mathbf{0.025})^{72} \qquad i = \frac{0.10}{4} = 0.025 \text{ and } n = 4(18) = 72.$$

$$= 71{,}006.74 \qquad \text{Use a calculator to evaluate the expression.}$$

Exponential functions as models

EXAMPLE 7 **Cellular phones.** For the years 1990–1997, the U.S. cellular telephone industry experienced exponential growth. (See Figure 9-20.) The exponential function $S(n) = 5.74(1.39)^n$ approximates the number of cellular telephone subscribers in millions, where n is the number of years since 1990 and $0 \le n \le 7$.

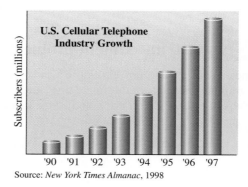

Source: *New York Times Almanac*, 1998

FIGURE 9-20

a. How many subscribers were there in 1990?

b. How many subscribers were there in 1997?

Solution

a. The year 1990 is 0 years after 1990. To find the number of subscribers in 1990, we substitute 0 for n in the function and find $S(0)$.

$$S(n) = 5.74(1.39)^n$$

$$S(0) = 5.74(1.39)^0$$

$$= 5.74 \cdot 1 \qquad (1.39)^0 = 1.$$

$$= 5.74$$

In 1990, there were approximately 5.74 million cellular telephone subscribers.

b. The year 1997 is 7 years after 1990. We need to find $S(7)$.

$$S(n) = 5.74(1.39)^n$$

$$S(7) = 5.74(1.39)^7 \qquad \text{Substitute 7 for } n.$$

$$\approx 57.54604675 \qquad \text{Use a calculator to find an approximation.}$$

In 1997, there were approximately 57.54 million cellular telephone subscribers.

Section 9.3 STUDY SET

VOCABULARY *Refer to the graph of f(x) = 3ˣ shown below.*

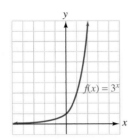

$f(x) = 3^x$

1. What type of function is $f(x) = 3^x$?

2. What is the domain of the function?

3. What is the range of the function?

4. What is the y-intercept of the graph?

5. What is the x-intercept of the graph?

6. What is an asymptote of the graph?

7. Is f an increasing or a decreasing function?

8. The graph passes through the point $(1, y)$. What is y?

CONCEPTS

9. Graph the functions $f(x) = x^2$ and $g(x) = 2^x$ on the same set of coordinate axes.

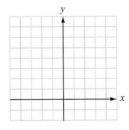

10. Graph $y = x^{\frac{1}{2}}$ and $y = \left(\frac{1}{2}\right)^x$ on the same set of coordinate axes.

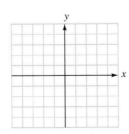

11. What are the two formulas that are used to determine the amount of money in a savings account that is earning compound interest?

12. Explain the order in which the expression $20{,}000(1.036)^{72}$ should be evaluated.

13. The illustration shows the graph of $f(x) = 2^x$, as well as two vertical translations of that graph. Using the notation $g(x)$ for one translation and $h(x)$ for the other, write the defining equation for each function.

14. The illustration shows the graph of $f(x) = 2^x$, as well as a horizontal translation of that graph. Using the notation $g(x)$ for the translation, write its defining equation.

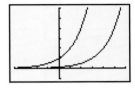

NOTATION

15. In $A = P\left(1 + \frac{r}{k}\right)^{kt}$, what is the base, and what is the exponent?

16. For an exponential function of the form $f(x) = b^x$, what are the restrictions on b?

PRACTICE *Find each value to four decimal places.*

17. $2^{\sqrt{2}}$ **18.** $7^{\sqrt{2}}$

19. $5^{\sqrt{5}}$ **20.** $6^{\sqrt{3}}$

Simplify each expression.

21. $\left(2^{\sqrt{3}}\right)^{\sqrt{3}}$ **22.** $3^{\sqrt{2}}3^{\sqrt{18}}$

23. $7^{\sqrt{3}}7^{\sqrt{12}}$ **24.** $\left(3^{\sqrt{5}}\right)^{\sqrt{5}}$

Graph each exponential function.

25. $f(x) = 5^x$ **26.** $f(x) = 6^x$

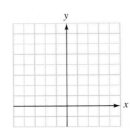
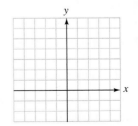

27. $y = \left(\frac{1}{4}\right)^x$

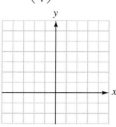

28. $y = \left(\frac{1}{5}\right)^x$

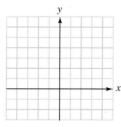

29. $f(x) = 3^x - 2$

30. $f(x) = 2^x + 1$

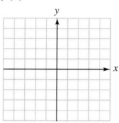

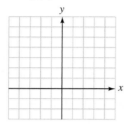

31. $f(x) = 3^{x-1}$

32. $f(x) = 2^{x+1}$

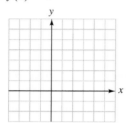

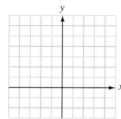

Use a graphing calculator to graph each function. Determine whether the function is an increasing or a decreasing function.

33. $f(x) = \frac{1}{2}(3^{x/2})$

34. $f(x) = -3(2^{x/3})$

35. $f(x) = 2(3^{-x/2})$

36. $f(x) = -\frac{1}{4}(2^{-x/2})$

APPLICATIONS   *Assume that there are no deposits or withdrawals.*

37. COMPOUND INTEREST An initial deposit of $10,000 earns 8% interest, compounded quarterly. How much will be in the account after 10 years?

38. COMPOUND INTEREST An initial deposit of $10,000 earns 8% interest, compounded monthly. How much will be in the account after 10 years?

39. COMPARING INTEREST RATES How much more interest could $1,000 earn in 5 years, compounded quarterly, if the annual interest rate were $5\frac{1}{2}$% instead of 5%?

40. COMPARING SAVINGS PLANS Which institution in the illustration provides the better investment?

Fidelity Savings & Loan

Earn 5.25%
compounded monthly

Union Trust

Money Market Account
paying 5.35%
compounded annually

41. COMPOUND INTEREST If $1 had been invested on July 4, 1776, at 5% interest, compounded annually, what would it be worth on July 4, 2076?

42. FREQUENCY OF COMPOUNDING $10,000 is invested in each of two accounts, both paying 6% annual interest. In the first account, interest compounds quarterly, and in the second account, interest compounds daily. Find the difference between the accounts after 20 years.

43. WORLD POPULATION See the illustration below.

a. Estimate when the world's population reached $\frac{1}{2}$ billion and when it reached 1 billion.

b. Estimate the world's population in the year 2000.

c. What type of function does it appear could be used to model the population growth?

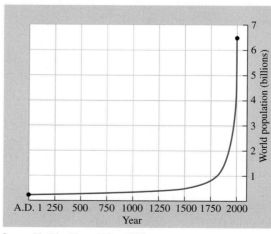

Source: *The Blue Planet* (Wiley, 1995)

44. THE STOCK MARKET The Dow Jones Industrial Average is a measure of how well the stock market is doing. Plot the following Dow milestones as ordered pairs of the form (year, average) on the following graph. What type of function does it appear could be used to model the growth of the stock market?

Year	Average	Year	Average
Jan. 1906	100	Nov. 1995	5,000
Mar. 1956	500	Oct. 1996	6,000
Nov. 1972	1,000	Feb. 1997	7,000
Jan. 1987	2,000	July 1997	8,000
Apr. 1991	3,000	Apr. 1998	9,000
Feb. 1995	4,000	Mar. 1999	10,000
		May 1999	11,000

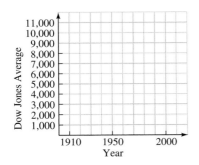

45. VALUE OF A CAR The following graph shows how the value of the average car depreciates as a percent of its original value over a 10-year period. It also shows the yearly maintenance costs as a percent of the car's value.

a. When is the car worth half of its purchase price?

b. When is the car worth a quarter of its purchase price?

c. When do the average yearly maintenance costs surpass the value of the car?

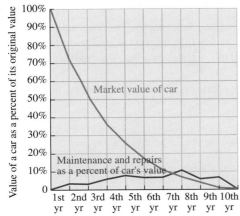

Source: U.S. Department of Transportation

46. DIVING *Bottom time* is the time a scuba diver spends descending plus the actual time spent at a certain depth. Complete the graph of the bottom time limits given in the table.

Bottom time limits			
Depth (ft)	Bottom time (min)	Depth (ft)	Bottom time (min)
30	no limit	80	40
35	310	90	30
40	200	100	25
50	100	110	20
60	60	120	15
70	50	130	10

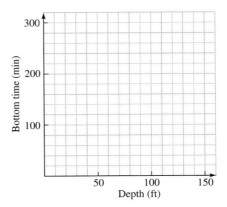

47. BACTERIAL CULTURE A colony of 6 million bacteria is growing in a culture medium as shown below. The population P after t hours is given by the formula $P = (6 \times 10^6)(2.3)^t$. Find the population after 4 hours.

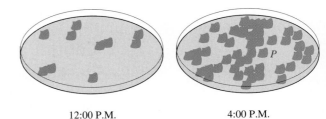

12:00 P.M.　　　　　　　　4:00 P.M.

48. RADIOACTIVE DECAY A radioactive material decays according to the formula $A = A_0\left(\frac{2}{3}\right)^t$, where A_0 is the initial amount present and t is measured in years. Find the amount present in 5 years.

49. DISCHARGING A BATTERY The charge remaining in a battery decreases as the battery discharges. The charge C (in coulombs) after t days is given by the formula $C = (3 \times 10^{-4})(0.7)^t$. Find the charge after 5 days.

50. POPULATION GROWTH The population of North Rivers is decreasing exponentially according to the formula $P = 3,745(0.93)^t$, where t is measured in years from the present date. Find the population in 6 years, 9 months.

51. SALVAGE VALUE A small business purchased a computer for $4,700. It is expected that its value each year will be 75% of its value in the preceding year. If the business disposes of the computer after 5 years, find its salvage value (the value after 5 years).

52. LOUISIANA PURCHASE In 1803, the United States negotiated the Louisiana Purchase with France. The country doubled its territory by adding 827,000 square miles of land for $15 million. If the land appreciated at the rate of 6% each year, what would one square mile of land be worth in 2005?

WRITING

53. If world population is increasing exponentially, why is there cause for concern?

54. How do the graphs of $f(x) = 3^x$ and $g(x) = \left(\frac{1}{3}\right)^x$ differ? How are they similar?

REVIEW   *Refer to the illustration below, in which lines r and s are parallel.*

55. Find x.

56. Find the measure of $\angle 1$.

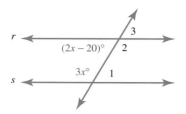

57. Find the measure of $\angle 2$.

58. Find the measure of $\angle 3$.

9.4 Base-*e* Exponential Functions

- Continuous compound interest   • The natural exponential function
- Graphing the natural exponential function   • Vertical and horizontal translations
- Malthusian population growth   • Base-*e* exponential function models

An exponential function that appears in applications is the base-*e* exponential function. In this section, we will show how to evaluate e, graph the base-*e* exponential function, and discuss one of its applications in analyzing population growth.

Continuous compound interest

If a bank pays interest twice a year, we say that interest is compounded semiannually. If it pays interest four times a year, we say that interest is compounded quarterly. If it pays interest continuously (infinitely many times in a year), we say that interest is compounded continuously.

   To develop the formula for continuous compound interest, we start with the formula

$$A = P\left(1 + \frac{r}{k}\right)^{kt}$$ The formula for compound interest: r is the annual rate and k is the number of times per year interest is paid.

and let $rn = k$. Since r and k are positive numbers, so is n.

$$A = P\left(1 + \frac{r}{rn}\right)^{rnt}$$

We can then simplify the fraction $\dfrac{r}{rn}$ and use the commutative property of multiplication to change the order of the factors in the exponent.

$$A = P\left(1 + \frac{1}{n}\right)^{nrt}$$

Finally, we can use a property of exponents to write the formula as

(1) $A = P\left[\left(1 + \dfrac{1}{n}\right)^{n}\right]^{rt}$ Use the property $a^{mn} = (a^m)^n$.

To find the value of $\left(1 + \frac{1}{n}\right)^{n}$, we evaluate it for several values of n, as shown in Table 9-1.

n	$\left(1 + \frac{1}{n}\right)^{n}$
1	2
2	2.25
4	2.44140625...
12	2.61303529...
365	2.71456748...
1,000	2.71692393...
100,000	2.71826830...
1,000,000	2.71828137...

TABLE 9-1

The results illustrate that as n gets larger, the value of $\left(1 + \frac{1}{n}\right)^{n}$ approaches the number 2.71828.... This number is called e, which has the following value.

$e = \mathbf{2.718281828459...}$

In continuous compound interest, k (the number of compoundings) is infinitely large. Since $k, r,$ and n are all positive, r is a fixed rate, and $k = rn$, as k gets very large (approaches infinity), so does n. Therefore, we can replace $\left(1 + \frac{1}{n}\right)^{n}$ in Equation 1 with e to get

$$A = Pe^{rt}$$

Formula for exponential growth

If a quantity P increases or decreases at an annual rate r, compounded continuously, the amount A after t years is given by

$$A = Pe^{rt}$$

If time is measured in years, then r is called the **annual growth rate.** If r is negative, the growth represents a decrease.

■ The natural exponential function

The number e is often used as the base in applications of exponential functions.

> **The natural exponential function**
>
> The function defined by $f(x) = e^x$ or $y = e^x$ is the **natural exponential function** (or the **base-*e* exponential function**), where $e = 2.71828. \ldots$

The $\boxed{e^x}$ key on a calculator is used to enter the natural exponential function.

CALCULATOR SNAPSHOT   The natural exponential function key

To compute the amount to which $12,000 will grow if invested for 18 years at 10% annual interest, compounded continuously, we substitute 12,000 for P, 0.10 for r, and 18 for t in the formula for exponential growth:

$$A = Pe^{rt}$$
$$A = 12{,}000e^{0.10(18)}$$
$$= 12{,}000e^{1.8}$$

To evaluate this expression, we enter these numbers and press these keys on a scientific calculator:

1.8 $\boxed{e^x}$ $\boxed{\times}$ 12000 $\boxed{=}$ $\qquad$ $\boxed{72595.76957}$

Using a graphing calculator, we enter these numbers and press these keys:

12000 $\boxed{\times}$ $\boxed{2nd}$ $\boxed{e^x}$ 1.8 $\boxed{)}$ $\boxed{ENTER}$ $\qquad$ $\boxed{\begin{array}{l} 12000*e\char94(1.8) \\ \qquad\qquad 72595.76957 \end{array}}$

After 18 years, the account will contain $72,595.77. This is $1,589.03 more than the result in Example 6 in the previous section, where interest was compounded quarterly.

INTERMEDIATE
Algebra $f(x)$ **Now™**

Self Check 1
In Example 1, find the balance in 60 years.

EXAMPLE 1   If $25,000 accumulates interest at an annual rate of 8%, compounded continuously, find the balance in the account in 50 years.

Solution

$$A = Pe^{rt} \qquad \text{This is the exponential growth formula.}$$
$$A = 25{,}000e^{0.08(50)} \qquad 8\% = 0.08.$$
$$= 25{,}000e^4$$
$$\approx 1{,}364{,}953.75 \qquad \text{Use a calculator.}$$

Answer $3,037,760.44

In 50 years, the balance will be $1,364,953.75 — more than a million dollars.

▇ Graphing the natural exponential function

To graph $f(x) = e^x$, we plot several points and join them with a smooth curve, as shown in Figure 9-21 on the next page. By looking at the graph, we can verify that it is an increasing function. The domain is the interval $(-\infty, \infty)$, and the range is the interval $(0, \infty)$. Note that as x decreases, the values of $f(x)$ decrease and approach 0. Thus, the x-axis is a horizontal asymptote of the graph.

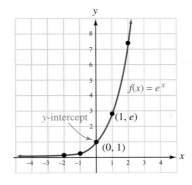

$f(x) = e^x$		
x	$f(x)$	$(x, f(x))$
-2	0.1	$(-2, 0.1)$
-1	0.4	$(-1, 0.4)$
0	1	$(0, 1)$
1	2.7	$(1, 2.7)$
2	7.4	$(2, 7.4)$

↑
The outputs can be found using the e^x key on a calculator. They are rounded to the nearest tenth to make point plotting easier.

FIGURE 9-21

■ Vertical and horizontal translations

We can illustrate the effects of vertical and horizontal translations of the natural exponential function by using a graphing calculator.

Translations of the natural exponential function	CALCULATOR SNAPSHOT

Figure 9-22(a) shows the calculator graphs of $f(x) = e^x$, $g(x) = e^x + 5$, and $h(x) = e^x - 3$. To graph these functions with window settings of $[-3, 6]$ for x and $[-5, 15]$ for y, we enter the right-hand sides of the equations after the symbols $Y_1 =$, $Y_2 =$, and $Y_3 =$. The display will show

$$Y_1 = e^{\wedge}(X)$$
$$Y_2 = e^{\wedge}(X) + 5$$
$$Y_3 = e^{\wedge}(X) - 3$$

After graphing these functions, we can see that the graph of $g(x) = e^x + 5$ is 5 units above the graph of $f(x) = e^x$, and that the graph of $h(x) = e^x - 3$ is 3 units below the graph of $f(x) = e^x$.

Figure 9-22(b) shows the calculator graphs of $f(x) = e^x$, $g(x) = e^{x+5}$, and $h(x) = e^{x-3}$. To graph these functions with window settings of $[-7, 10]$ for x and $[-5, 15]$ for y, we enter the right-hand sides of the equations after the symbols $Y_1 =$, $Y_2 =$, and $Y_3 =$. The display will show

$$Y_1 = e^{\wedge}(X)$$
$$Y_2 = e^{\wedge}(X + 5)$$
$$Y_3 = e^{\wedge}(X - 3)$$

After graphing these functions, we can see that the graph of $g(x) = e^{x+5}$ is 5 units to the left of the graph of $f(x) = e^x$, and that the graph of $h(x) = e^{x-3}$ is 3 units to the right of the graph of $f(x) = e^x$.

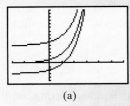
(a)

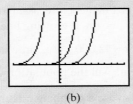
(b)

FIGURE 9-22

Graphing exponential functions

Figure 9-23 shows the calculator graph of $f(x) = 3e^{-x/2}$. To graph this function with window settings of $[-7, 10]$ for x and $[-5, 15]$ for y, we enter the right-hand side of the equation after the symbol $Y_1 =$. The display will show the equation

$$Y_1 = 3(e^\wedge(-X/2))$$

Explain why the graph has a y-intercept of $(0, 3)$.

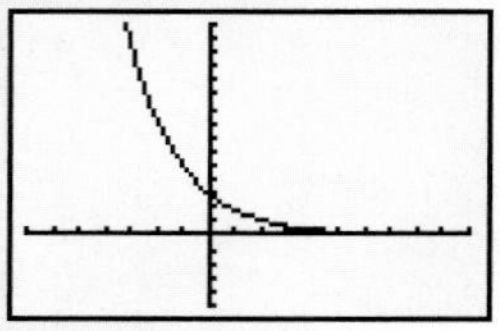

FIGURE 9-23

■ Malthusian population growth

An equation based on the natural exponential function provides a model for **population growth.** In the **Malthusian model for population growth,** the future population of a colony is related to the present population by the formula $A = Pe^{rt}$.

INTERMEDIATE
Algebra $f(x)$ **Now**™

Self Check 2
In Example 2, find the population in 50 years.

EXAMPLE 2   **City planning.**   The population of a city is currently 15,000, but changing economic conditions are causing the population to decrease 3% each year. If this trend continues, find the population in 30 years.

Solution   Since the population is decreasing 3% each year, the annual growth rate is -3%, or -0.03. We can substitute -0.03 for r, 30 for t, and 15,000 for P in the formula for exponential growth and find A.

$$A = Pe^{rt}$$
$$A = 15{,}000e^{-0.03(30)}$$
$$= 15{,}000e^{-0.9}$$
$$\approx 6.099$$

Answer 3,347

In 30 years, the expected population will be 6,099.

The English economist Thomas Robert Malthus (1766–1834) pioneered in population study. He believed that poverty and starvation were unavoidable, because the human population tends to grow exponentially, but the food supply tends to grow linearly.

Self Check 3
In Example 3, suppose that the population grows at a 3% rate. Use a graphing calculator to determine for how many years the food supply will be adequate.

EXAMPLE 3   **The Malthusian model.**   Suppose that a country with a population of 1,000 people is growing exponentially according to the formula

$$P = 1{,}000e^{0.02t}$$

where t is in years. Furthermore, assume that the food supply F, measured in adequate food per day per person, is growing linearly according to the formula

$$F = 30.625t + 2{,}000 \quad (t \text{ is time in years})$$

In how many years will the population outstrip the food supply?

Solution   We can use a graphing calculator with window settings of $[0, 100]$ for x and $[0, 10{,}000]$ for y. After graphing the functions, we obtain Figure 9-24(a) on the next page. If we trace, as in Figure 9-24(b), we can find the point where the two graphs intersect. From the graph, we can see that the food supply will be adequate for about 71 years. At that time, the population of approximately 4,200 people will begin to have problems.

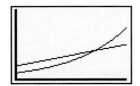

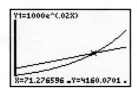

FIGURE 9-24

Answer about 38 years

■ Base-*e* exponential function models

EXAMPLE 4   Baking.   A mother takes a cake out of the oven and sets it on a rack to cool. The function $T(t) = 68 + 220e^{-0.18t}$ gives the cake's temperature in degrees Fahrenheit after it has cooled for t minutes. If her children will be home from school in 20 minutes, will the cake have cooled enough for the children to eat it?

Solution   When the children arrive home, the cake will have cooled for 20 minutes. To find the temperature of the cake at that time, we need to find $T(20)$.

$$T(t) = 68 + 220e^{-0.18t}$$
$$T(20) = 68 + 220e^{-0.18(20)} \quad \text{Substitute 20 for } t.$$
$$= 68 + 220e^{-3.6}$$
$$\approx 74.0 \qquad\qquad \text{Use a calculator.}$$

When the children return home, the temperature of the cake will be about 74°, and it can be eaten.

Section 9.4 STUDY SET

■ **VOCABULARY**   *Refer to the graph of f(x) = e^x in the illustration below.*

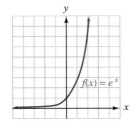

$f(x) = e^x$

1. What is the name of the function $f(x) = e^x$?

2. What is the domain of the function?

3. What is the range of the function?

4. What is the y-intercept of the graph?

5. What is the x-intercept of the graph?

6. What is an asymptote of the graph?

7. Is f an increasing or a decreasing function?

8. The graph passes through the point $(1, y)$. What is y?

■ **CONCEPTS**   *Fill in the blanks.*

9. In _____ compound interest, the number of compoundings is infinitely large.

10. The formula for continuous compound interest is $A =$ [ ].

11. To two decimal places, the value of e is [ ].

12. If n gets larger and larger, the value of $\left(1 + \frac{1}{n}\right)^n$ approaches the value of [ ].

13. Graph each irrational number on the number line. $\left\{\pi, e, \sqrt{2}, \frac{\sqrt{3}}{2}\right\}$

14. Complete the table of values. Round to the nearest hundredth.

x	-2	-1	0	1	2
e^x					

15. POPULATION OF THE UNITED STATES
Complete the graph of the U.S. census population figures shown in the table (in millions). What type of function does it appear could be used to model the population?

Year	Population	Year	Population
1790	3.9	1900	76.0
1800	5.3	1910	92.2
1810	7.2	1920	106.0
1820	9.6	1930	123.2
1830	12.9	1940	132.1
1840	17.0	1950	151.3
1850	23.1	1960	179.3
1860	31.4	1970	203.3
1870	38.5	1980	226.5
1880	50.1	1990	248.7
1890	62.9	2000	281.4

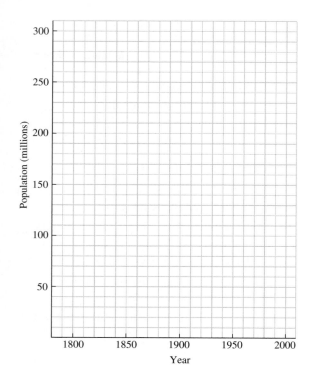

16. What is the Malthusian population growth formula?

17. The function $f(x) = e^x$ is graphed, and the TRACE feature is used. What is the y-coordinate of the point on the graph having an x-coordinate of 1? What is the name given this number?

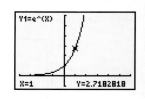

18. The calculator display shows a table of values for $f(x) = e^x$. As x decreases, what happens to the values of $f(x)$ listed in the Y_1 column? Will the value of $f(x)$ ever be 0 or negative?

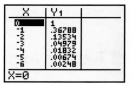

NOTATION   *Evaluate A in the formula $A = Pe^{rt}$ for the following values of r and t.*

19. $P = 1,000$, $r = 0.09$ and $t = 10$

$$A = \boxed{}\, e^{(0.09)()}$$
$$= 1,000e^{\boxed{}}$$
$$\approx \boxed{}\,(2.459603111)$$
$$\approx 2,459.603111$$

20. $P = 1,000$, $r = 0.12$ and $t = 50$

$$A = 1,000e^{(\boxed{})(50)}$$
$$= \boxed{}\, e^6$$
$$\approx 1,000(\boxed{})$$
$$\approx 403,428.7935$$

PRACTICE   *Graph each function. Check your work with a graphing calculator. Compare each graph to the graph of $f(x) = e^x$.*

21. $f(x) = e^x + 1$ **22.** $f(x) = e^x - 2$

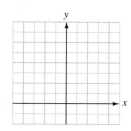
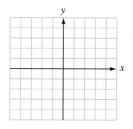

23. $y = e^{x+3}$

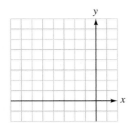

24. $y = e^{x-5}$

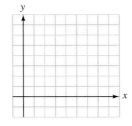

25. $f(x) = -e^x$

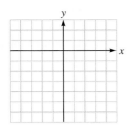

26. $f(x) = -e^x + 1$

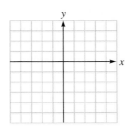

27. $f(x) = 2e^x$

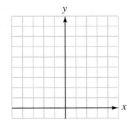

28. $f(x) = \dfrac{1}{2}e^x$

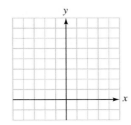

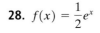 **APPLICATIONS** *In Exercises 29–34, assume that there are no deposits or withdrawals.*

29. CONTINUOUS COMPOUND INTEREST
An initial investment of $5,000 earns 8.2% interest, compounded continuously. What will the investment be worth in 12 years?

30. CONTINUOUS COMPOUND INTEREST
An initial investment of $2,000 earns 8% interest, compounded continuously. What will the investment be worth in 15 years?

31. COMPARING COMPOUNDING METHODS
An initial deposit of $5,000 grows at an annual rate of 8.5% for 5 years. Compare the final balances resulting from annual compounding and continuous compounding.

32. COMPARING COMPOUNDING METHODS
An initial deposit of $30,000 grows at an annual rate of 8% for 20 years. Compare the final balances resulting from annual compounding and continuous compounding.

33. DETERMINING THE INITIAL DEPOSIT
An account now contains $11,180 and has been accumulating interest at 7% annual interest, compounded continuously, for 7 years. Find the initial deposit.

34. DETERMINING THE PREVIOUS BALANCE
An account now contains $3,610 and has been accumulating interest at 8% annual interest, compounded continuously. How much was in the account 4 years ago?

35. WORLD POPULATION GROWTH The population of the Earth is approximately 6.1 billion people and is growing at an annual rate of 1.4%. Assuming a Malthusian growth model, find the world population in 30 years.

36. HIGHS AND LOWS Somalia, in eastern Africa, has one of the greatest population growth rates in the world. Bulgaria, in southeastern Europe, has one of the smallest. Assuming a Malthusian growth model, complete the table.

Country	Population 2003	Annual growth rate	Estimated population 2015
Somalia	8,025,190	3.43%	
Bulgaria	7,537,929	−1.09%	

Source: nationmaster.com

37. POPULATION GROWTH The growth of a population is modeled by

$$P = 173e^{0.03t}$$

How large will the population be when $t = 20$?

38. POPULATION DECLINE The decline of a population is modeled by

$$P = (1.2 \times 10^6)e^{-0.008t}$$

How large will the population be when $t = 30$?

39. OCEANOGRAPHY The width w (in millimeters) of successive growth spirals of the sea shell *Catapulus voluto,* shown in the illustration below, is given by the exponential function

$$w = 1.54e^{0.503n}$$

where n is the spiral number. Find the width, to the nearest tenth of a millimeter, of the sixth spiral.

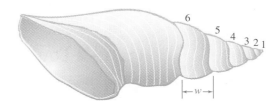

40. EPIDEMICS The spread of hoof and mouth disease through a herd of cattle can be modeled by the formula

$$P = P_0 e^{0.27t} \qquad (t \text{ is in days})$$

If a rancher does not act quickly to treat two cases, how many cattle will have the disease in 12 days?

41. HALF-LIFE OF A DRUG The quantity of a prescription drug in the bloodstream of a patient t hours after it is administered can be modeled by an exponential function. (See the graph in the illustration.) Use the graph to determine the time it takes to eliminate half of the initial dose from the body.

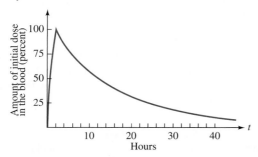

42. MEDICINE The concentration x of a certain prescription drug in an organ after t minutes is given by

$$x = 0.08(1 - e^{-0.1t})$$

Find the concentration of the drug at 30 minutes.

43. SKYDIVING Before the parachute opens, a skydiver's velocity v in meters per second is given by

$$v = 50(1 - e^{-0.2t})$$

Find the velocity after 20 seconds of free fall.

44. FREE FALL After t seconds a certain falling object has a velocity v given by

$$v = 50(1 - e^{-0.3t})$$

Which is falling faster after 2 seconds — the object or the skydiver in Exercise 43?

Use a graphing calculator to solve each problem.

45. THE MALTHUSIAN MODEL In Example 3, suppose that better farming methods changed the formula for food growth to $y = 31x + 2{,}000$. How long would the food supply be adequate?

46. THE MALTHUSIAN MODEL In Example 3, suppose that a birth-control program changed the formula for population growth to $P = 1{,}000e^{0.01t}$. How long would the food supply be adequate?

WRITING

47. Explain why the graph of $y = e^x - 5$ is five units below the graph of $y = e^x$.

48. A feature article in a newspaper stated that the sport of snowboarding was growing *exponentially*. Explain what the author of the article meant by that.

REVIEW   *Simplify each expression. Assume that all variables represent positive numbers.*

49. $\sqrt{240x^5}$

50. $\sqrt[3]{-125x^5y^4}$

51. $4\sqrt{48y^3} - 3y\sqrt{12y}$

52. $\sqrt[4]{48z^5} + \sqrt[4]{768z^5}$

9.5 Logarithmic Functions

- Logarithms • Graphs of logarithmic functions • Vertical and horizontal translations
- Base-10 logarithms • Applications of logarithms

A function that is closely related to the exponential function is the *logarithmic function*. It can be used to solve many application problems from fields such as electronics, seismology (the study of earthquakes), business, and population growth.

Logarithms

Because an exponential function defined by $f(x) = b^x$ (or $y = b^x$) is one-to-one, it has an inverse function that is defined by the equation $x = b^y$. To express this inverse function in the form $y = f^{-1}(x)$, we must solve the equation $x = b^y$ for y. For this, we need the following definition.

Logarithmic functions

If $b > 0$ and $b \neq 1$, the **logarithmic function with base b** is defined by

$$y = \log_b x \quad \text{if and only if} \quad x = b^y$$

Read $\log_b x$ as "the logarithm base b of x" or "log base b of x."

The **domain of the logarithmic function** is the interval $(0, \infty)$. The **range** is the interval $(-\infty, \infty)$. We can also write a logarithmic function in the form $f(x) = \log_b x$.

Since the function $y = \log_b x$ is the inverse of the one-to-one exponential function $y = b^x$, the logarithmic function is also one-to-one.

! COMMENT Because the domain of the logarithmic function is the set of positive numbers, it is impossible to find the logarithm of 0 or the logarithm of a negative number.

The previous definition guarantees that any pair (x, y) that satisfies the equation $y = \log_b x$ also satisfies the exponential equation $x = b^y$. Because of this relationship, a statement written in logarithmic form can be written in an equivalent exponential form.

Logarithmic form		**Exponential form**	
$\log_7 1 = 0$	because	$1 = 7^0$	Read as "log base 7 of 1 is 0."
$\log_5 25 = 2$	because	$25 = 5^2$	
$\log_5 \dfrac{1}{25} = -2$	because	$\dfrac{1}{25} = 5^{-2}$	
$\log_{16} 4 = \dfrac{1}{2}$	because	$4 = 16^{1/2}$	
$\log_2 \dfrac{1}{8} = -3$	because	$\dfrac{1}{8} = 2^{-3}$	
$\log_b x = y$	because	$x = b^y$	

In each of these examples, the logarithm of a number is an exponent. In fact,

$\log_b x$ is the exponent to which b is raised to get x.

Translating this statement into symbols, we have

$$b^{\log_b x} = x$$

INTERMEDIATE
Algebra $f(x)$ Now™

EXAMPLE 1 Find y in each equation: **a.** $\log_6 1 = y$, **b.** $\log_3 27 = y$, and **c.** $\log_5 \dfrac{1}{5} = y$.

Solution
a. We can change the equation $\log_6 1 = y$ into the equivalent exponential equation $1 = 6^y$. Since $1 = 6^0$, it follows that $y = 0$. Thus,

$$\log_6 1 = 0$$

Self Check 1
Find y in each equation:

a. $\log_3 9 = y$

b. $\log_2 16 = y$

c. $\log_5 \dfrac{1}{125} = y$

b. $\log_3 27 = y$ is equivalent to $27 = 3^y$. Since $27 = 3^3$, it follows that $3^y = 3^3$, and $y = 3$. Thus,

$$\log_3 27 = 3$$

c. $\log_5 \frac{1}{5} = y$ is equivalent to $\frac{1}{5} = 5^y$. Since $\frac{1}{5} = 5^{-1}$, it follows that $5^y = 5^{-1}$, and $y = -1$. Thus,

$$\log_5 \frac{1}{5} = -1$$

Answers **a.** 2, **b.** 4, **c.** −3

INTERMEDIATE
Algebra ⨍(x) Now™

Self Check 2
Find x in each equation:

a. $\log_2 32 = x$

b. $\log_x 8 = 3$

c. $\log_5 x = 2$

Answers **a.** 5, **b.** 2, **c.** 25

EXAMPLE 2 Find the value of x in each equation: **a.** $\log_3 81 = x$, **b.** $\log_x 125 = 3$, and **c.** $\log_4 x = 3$.

Solution

a. $\log_3 81 = x$ is equivalent to $3^x = 81$. Because $3^4 = 81$, it follows that $3^x = 3^4$. Thus, $x = 4$.

b. $\log_x 125 = 3$ is equivalent to $x^3 = 125$. Because $5^3 = 125$, it follows that $x^3 = 5^3$. Thus, $x = 5$.

c. $\log_4 x = 3$ is equivalent to $4^3 = x$. Because $4^3 = 64$, it follows that $x = 64$.

INTERMEDIATE
Algebra ⨍(x) Now™

Self Check 3
Find x in each equation:

a. $\log_{1/4} x = 3$

b. $\log_{1/4} x = -2$

Answers **a.** $\dfrac{1}{64}$, **b.** 16

EXAMPLE 3 Find the value of x in each equation: **a.** $\log_{1/3} x = 2$, **b.** $\log_{1/3} x = -2$, and **c.** $\log_{1/3} \dfrac{1}{27} = x$.

Solution

a. $\log_{1/3} x = 2$ is equivalent to $\left(\frac{1}{3}\right)^2 = x$. Thus, $x = \frac{1}{9}$.

b. $\log_{1/3} x = -2$ is equivalent to $\left(\frac{1}{3}\right)^{-2} = x$. Thus,

$$x = \left(\frac{1}{3}\right)^{-2} = 3^2 = 9$$

c. $\log_{1/3} \frac{1}{27} = x$ is equivalent to $\left(\frac{1}{3}\right)^x = \frac{1}{27}$. Because $\left(\frac{1}{3}\right)^3 = \frac{1}{27}$, it follows that $x = 3$.

▌ Graphs of logarithmic functions

To graph the logarithmic function $y = \log_2 x$ (or $f(x) = \log_2 x$), we calculate and plot several points with coordinates (x, y) that satisfy the equivalent equation $x = 2^y$. After joining these points with a smooth curve, we have the graph shown in Figure 9-25(a) on the next page.

To graph $y = \log_{1/2} x$ (or $g(x) = \log_{1/2} x$), we calculate and plot several points with coordinates (x, y) that satisfy the equation $x = \left(\frac{1}{2}\right)^y$. After joining these points with a smooth curve, we have the graph shown in Figure 9-25(b).

$y = \log_2 x$		
x	y	(x, y)
$\frac{1}{4}$	-2	$(\frac{1}{4}, -2)$
$\frac{1}{2}$	-1	$(\frac{1}{2}, -1)$
1	0	$(1, 0)$
2	1	$(2, 1)$
4	2	$(4, 2)$
8	3	$(8, 3)$

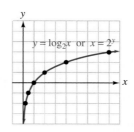

↑
Choose values for x that
are integer powers of 2.

(a)

$y = \log_{1/2} x$		
x	y	(x, y)
$\frac{1}{4}$	2	$(\frac{1}{4}, 2)$
$\frac{1}{2}$	1	$(\frac{1}{2}, 1)$
1	0	$(1, 0)$
2	-1	$(2, -1)$
4	-2	$(4, -2)$
8	-3	$(8, -3)$

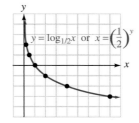

↑
Choose values for x that
are integer powers of $\frac{1}{2}$.

(b)

FIGURE 9-25

The graphs of all logarithmic functions are similar to those in Figure 9-25. If $b > 1$, the logarithmic function is increasing, as in Figure 9-26(a). If $0 < b < 1$, the logarithmic function is decreasing, as in Figure 9-26(b).

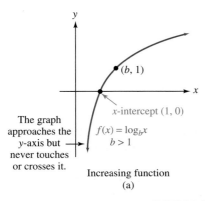

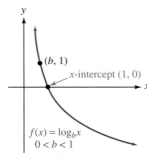

FIGURE 9-26

> **Properties of logarithmic functions**
>
> The graph of $y = \log_b x$ (or $f(x) = \log_b x$) has the following properties.
>
> **1.** It passes through the point $(1, 0)$.
> **2.** It passes through the point $(b, 1)$.
> **3.** The y-axis (the line $x = 0$) is an asymptote.
> **4.** The domain is $(0, \infty)$ and the range is $(-\infty, \infty)$.

The exponential and logarithmic functions are inverses of each other, so their graphs have symmetry about the line $y = x$. The graphs of $f(x) = \log_b x$ and $g(x) = b^x$ are shown in Figure 9-27(a) when $b > 1$ and in Figure 9-27(b) when $0 < b < 1$.

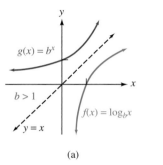
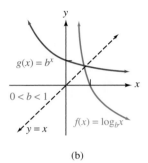

(a)   (b)

FIGURE 9-27

■ Vertical and horizontal translations

The graphs of many functions involving logarithms are translations of the basic logarithmic graphs.

Self Check 4
Graph $g(x) = \log_3 x - 2$ and describe the translation.

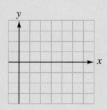

Answer The graph of $y = \log_3 x$ is translated 2 units downward.

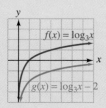

EXAMPLE 4   Graph the function $g(x) = 3 + \log_2 x$ and describe the translation.

Solution   The graph of $g(x) = 3 + \log_2 x$ is identical to the graph of $f(x) = \log_2 x$, except that it is translated 3 units upward. (See Figure 9-28.)

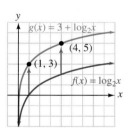

FIGURE 9-28

EXAMPLE 5   Graph $y = \log_{1/2}(x - 1)$ and describe the translation.

Solution   The graph of $y = \log_{1/2}(x - 1)$ is identical to the graph of $y = \log_{1/2} x$, except that it is translated 1 unit to the right. (See Figure 9-29.) We note that the asymptote of the graph of $y = \log_{1/2}(x - 1)$ is the line $x = 1$.

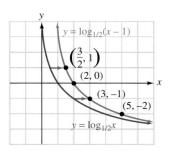

FIGURE 9-29

Self Check 5
Graph $y = \log_{1/3}(x + 2)$ and describe the translation.

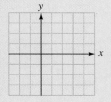

Answer The graph of $y = \log_{1/3} x$ is translated 2 units to the left.

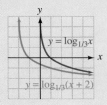

Graphing logarithmic functions

CALCULATOR SNAPSHOT

Graphing calculators can draw graphs of logarithmic functions. To use a calculator to graph the logarithmic function $f(x) = -2 + \log_{10}\left(\frac{1}{2}x\right)$, we enter the right-hand side of the equation after the symbol $Y_1 =$. The display will show the equation

$$Y_1 = -2 + \log(X/2)$$

If we use window settings of $[-1, 5]$ for x and $[-4, 1]$ for y and press the $\boxed{\text{GRAPH}}$ key, we will obtain the graph shown in Figure 9-30.

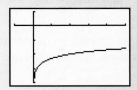

FIGURE 9-30

■ Base-10 logarithms

For computational purposes and in many applications, we will use base-10 logarithms (also called **common logarithms**). When the base b is not indicated in the notation $\log x$, we assume that $b = 10$:

log x means $\log_{10} x$

Because base-10 logarithms appear so often, it is a good idea to become familiar with the following base-10 logarithms:

Logarithmic form		**Exponential form**	
$\log \dfrac{1}{100} = -2$	because	$10^{-2} = \dfrac{1}{100}$	Read as "log of $\dfrac{1}{100}$ is -2."
$\log \dfrac{1}{10} = -1$	because	$10^{-1} = \dfrac{1}{10}$	
$\log 1 = 0$	because	$10^{0} = 1$	
$\log 10 = 1$	because	$10^{1} = 10$	
$\log 100 = 2$	because	$10^{2} = 100$	
$\log 1{,}000 = 3$	because	$10^{3} = 1{,}000$	

In general, we have

$$\log 10^{x} = x$$

CALCULATOR SNAPSHOT   **Finding logarithms**

Before calculators, extensive tables provided logarithms of numbers. Today, logarithms are easy to find with a calculator. For example, to find log 2.34 with a scientific calculator, we enter these numbers and press these keys:

2.34 $\boxed{\text{LOG}}$ $\boxed{\text{.369215857}}$

To use a graphing calculator, we enter these numbers and press these keys:

$\boxed{\text{LOG}}$ 2.34 $\boxed{\text{)}}$ $\boxed{\text{ENTER}}$ $\boxed{\begin{array}{l}\text{log(2.34)}\\ \qquad\text{.3692158574}\end{array}}$

To four decimal places, log 2.34 = 0.3692.

INTERMEDIATE
Algebra $f(x)$ **Now**™

Self Check 6
Solve: $\log x = 1.87737$. Round to four decimal places.

EXAMPLE 6   Find x in the equation $\log x = 0.3568$. Round to four decimal places.

Solution   The equation $\log x = 0.3568$ is equivalent to $10^{0.3568} = x$. To find x with a scientific calculator, we enter these numbers and press these keys:

10 $\boxed{y^x}$ .3568 $\boxed{=}$

The display will read $\boxed{\text{2.274049951}}$. To four decimal places,

$$x = 2.2740$$

Answer 75.3998

If your calculator has a $\boxed{10^x}$ key, enter .3568 and press it to get the same result.

Applications of logarithms

Common logarithms are used in electrical engineering to express the voltage gain (or loss) of an electronic device such as an amplifier. The unit of gain (or loss), called the **decibel,** is defined by a logarithmic relation.

Decibel voltage gain

If E_O represents the output voltage of a device and E_I the input voltage, the decibel voltage gain of the device (db gain) is given by

$$\text{db gain} = 20 \log \frac{E_O}{E_I}$$

INTERMEDIATE
Algebra $f(x)$ Now™

EXAMPLE 7   Finding db gain.   If the input to an amplifier is 0.5 volt and the output is 40 volts, find the decibel voltage gain of the amplifier.

Solution   We can find the decibel voltage gain by substituting 0.5 for E_I and 40 for E_O into the formula for db gain:

$$\text{db gain} = 20 \log \frac{E_O}{E_I}$$

$$\text{db gain} = 20 \log \frac{40}{0.5}$$

$$= 20 \log 80 \qquad \text{Perform the division.}$$

$$\approx 38 \qquad \text{Use a calculator: } 20 \log 80 \text{ means } 20 \cdot \log 80.$$

The amplifier provides a 38-decibel voltage gain.

In seismology, common logarithms are used to measure the intensity of earthquakes on the **Richter scale.** The intensity of an earthquake is given by the following logarithmic function.

Richter scale

If R represents the intensity of an earthquake, A the amplitude (measured in micrometers), and P the period (the time of one oscillation of Earth's surface measured in seconds), then

$$R = \log \frac{A}{P}$$

INTERMEDIATE
Algebra $f(x)$ Now™

EXAMPLE 8   Earthquakes.   Find the measure on the Richter scale of an earthquake with an amplitude of 5,000 micrometers (0.5 centimeter) and a period of 0.1 second.

Solution   We substitute 5,000 for A and 0.1 for P in the Richter scale formula and simplify:

$$R = \log \frac{A}{P}$$

$$R = \log \frac{5,000}{0.1}$$

$$= \log 50,000 \qquad \text{Perform the division.}$$

$$\approx 4.698970004 \qquad \text{Use a calculator.}$$

The earthquake measures about 4.7 on the Richter scale.

Section 9.5 STUDY SET

INTERMEDIATE
Algebra $f(x)$ Now™

VOCABULARY   *Refer to the graph of $f(x) = \log_4 x$.*

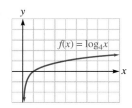

1. What type of function is $f(x) = \log_4 x$?

2. What is the domain of the function?

3. What is the range of the function?

4. What is the y-intercept of the graph?

5. What is the x-intercept of the graph?

6. What is an asymptote of the graph?

7. Is f an increasing or a decreasing function?

8. The graph passes through the point $(4, y)$. What is y?

CONCEPTS   *Fill in the blanks.*

9. The equation $y = \log_b x$ is equivalent to the exponential equation $x = \boxed{}$.

10. $\log_b x$ is the _____ to which b is raised to get x.

11. The functions $f(x) = \log_b x$ and $f(x) = b^x$ are _____ functions.

12. The inverse of an exponential function is called a _____ function

Complete each table of values. If an evaluation is not possible, write "none."

13. $y = \log x$

x	y
$\frac{1}{100}$	
$\frac{1}{10}$	
1	
10	
100	

14. $f(x) = \log_5 x$

x	f(x)
$\frac{1}{25}$	
$\frac{1}{5}$	
1	
5	
25	

15. $f(x) = \log_6 x$

x	f(x)
-6	
0	
$\frac{1}{216}$	
$\sqrt{6}$	
6^8	

16. $f(x) = \log_8 x$

x	f(x)
-8	
0	
$\frac{1}{8}$	
$\sqrt{8}$	
64	

17. Use a calculator to complete the table of values for $y = \log x$. Round to the nearest hundredth.

x	y
0.5	
1	
2	
3	
4	
5	
6	
7	
8	
9	
10	

18. Graph $y = \log x$ on the illustration below. (See Exercise 17.) Note that the units on the x- and y-axes are different.

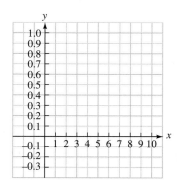

19. A table of values for $f(x) = \log x$ is shown below. As x decreases and gets close to 0, what happens to the values of $f(x)$?

X	Y₁
1	0
.9	-.0458
.8	-.0969
.7	-.1549
.6	-.2218
.5	-.301
.4	-.3979

X=1

20. For each function, determine $f^{-1}(x)$.

a. $f(x) = 10^x$ **b.** $f(x) = 3^x$

c. $f(x) = \log x$ **d.** $f(x) = \log_2 x$

NOTATION   *Fill in the blanks.*

21. The notation $\log x$ means $\log$ ___ x.

22. $\log_{10} 10^x =$ ___.

PRACTICE   *Write the equation in exponential form.*

23. $\log_3 81 = 4$ **24.** $\log_7 7 = 1$

25. $\log_{12} 12 = 1$ **26.** $\log_6 36 = 2$

27. $\log_4 \dfrac{1}{64} = -3$ **28.** $\log_6 \dfrac{1}{36} = -2$

29. $\log 0.001 = -3$ **30.** $\log_3 243 = 5$

Write each equation in logarithmic form.

31. $8^2 = 64$ **32.** $10^3 = 1,000$

33. $4^{-2} = \dfrac{1}{16}$ **34.** $3^{-4} = \dfrac{1}{81}$

35. $\left(\dfrac{1}{2}\right)^{-5} = 32$ **36.** $\left(\dfrac{1}{3}\right)^{-3} = 27$

37. $x^y = z$ **38.** $m^n = p$

Find each value of x.

39. $\log_2 8 = x$ **40.** $\log_3 9 = x$

41. $\log_4 64 = x$ **42.** $\log_6 216 = x$

43. $\log_{1/2} \dfrac{1}{8} = x$ **44.** $\log_{1/3} \dfrac{1}{81} = x$

45. $\log_9 3 = x$ **46.** $\log_{125} 5 = x$

47. $\log_8 x = 2$ **48.** $\log_7 x = 0$

49. $\log_{25} x = \dfrac{1}{2}$ **50.** $\log_4 x = \dfrac{1}{2}$

51. $\log_5 x = -2$ **52.** $\log_3 x = -4$

53. $\log_{36} x = -\dfrac{1}{2}$ **54.** $\log_{27} x = -\dfrac{1}{3}$

55. $\log_{100} \dfrac{1}{1,000} = x$ **56.** $\log_{5/2} \dfrac{4}{25} = x$

57. $\log_{27} 9 = x$ **58.** $\log_{12} x = 0$

59. $\log_x 5^3 = 3$ **60.** $\log_x 5 = 1$

61. $\log_x \dfrac{9}{4} = 2$ **62.** $\log_x \dfrac{\sqrt{3}}{3} = \dfrac{1}{2}$

63. $\log_x \dfrac{1}{64} = -3$ **64.** $\log_x \dfrac{1}{100} = -2$

65. $\log_8 x = 0$ **66.** $\log_4 8 = x$

Use a calculator to find each value. Give answers to four decimal places.

67. $\log 3.25$ **68.** $\log 0.57$

69. $\log 0.00467$ **70.** $\log 375.876$

Use a calculator to find each value of x. Give answers to two decimal places.

71. $\log x = 1.4023$ **72.** $\log x = 0.926$

73. $\log x = -1.71$ **74.** $\log x = -0.5$

Graph each function. Determine whether it is an increasing or a decreasing function.

75. $f(x) = \log_3 x$ **76.** $f(x) = \log_{1/3} x$

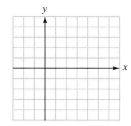
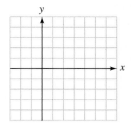

77. $y = \log_{1/2} x$ **78.** $y = \log_4 x$

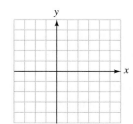
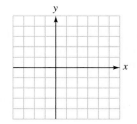

Graph each function.

79. $f(x) = 3 + \log_3 x$ **80.** $f(x) = \log_{1/3} x - 1$

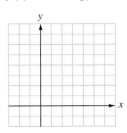   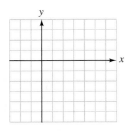

81. $y = \log_{1/2} (x - 2)$ **82.** $y = \log_4 (x + 2)$

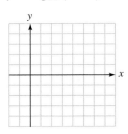   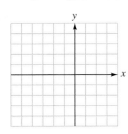

Graph each pair of inverse functions on a single coordinate system. Draw the axis of symmetry.

83. $f(x) = 6^x$ **84.** $f(x) = 3^x$

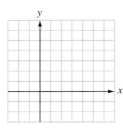   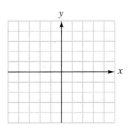

85. $f(x) = 5^x$ **86.** $f(x) = 8^x$

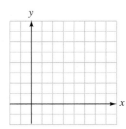   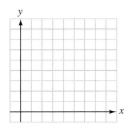

■ **APPLICATIONS**

87. FINDING INPUT VOLTAGE Find the db gain of an amplifier if the input voltage is 0.71 volts when the output voltage is 20 volts.

88. FINDING OUTPUT VOLTAGE Find the db gain of an amplifier if the output voltage is 2.8 volts when the input voltage is 0.05 volt.

89. db GAIN OF AN AMPLIFIER Find the db gain of the amplifier shown below.

90. GAIN OF AN AMPLIFIER An amplifier produces an output of 80 volts when driven by an input of 0.12 volts. Find the amplifier's db gain.

91. THE RICHTER SCALE An earthquake has amplitude of 5,000 micrometers and a period of 0.2 second. Find its measure on the Richter scale.

92. EARTHQUAKES Find the period of an earthquake with amplitude of 80,000 micrometers that measures 6 on the Richter scale.

93. EARTHQUAKES An earthquake with a period of $\frac{1}{4}$ second measures 4 on the Richter scale. Find its amplitude.

94. EARTHQUAKES In 1985, Mexico City experienced an earthquake of magnitude 8.1 on the Richter scale. In 1989, the San Francisco Bay area was rocked by an earthquake measuring 7.1. By what factor must the amplitude of an earthquake change to increase its severity by 1 point on the Richter scale? (Assume that the period remains constant.)

95. DEPRECIATION In business, equipment is often depreciated using the double declining-balance method. In this method, a piece of equipment with a life expectancy of N years, costing $\$C$, will depreciate to a value of $\$V$ in n years, where n is given by the formula

$$n = \frac{\log V - \log C}{\log \left(1 - \frac{2}{N}\right)}$$

A computer that cost $\$37,000$ has a life expectancy of 5 years. If it has depreciated to a value of $\$8,000$, how old is it?

96. DEPRECIATION A typewriter worth $\$470$ when new had a life expectancy of 12 years. If it is now worth $\$189$, how old is it? (See Exercise 95.)

97. TIME FOR MONEY TO GROW If $\$P$ is invested at the end of each year in an annuity earning annual interest at a rate r, the amount in the account will be $\$A$ after n years, where

$$n = \frac{\log \left[\dfrac{Ar}{P} + 1\right]}{\log (1 + r)}$$

If $\$1,000$ is invested each year in an annuity earning 12% annual interest, how long will it take for the account to be worth $\$20,000$?

98. TIME FOR MONEY TO GROW If $5,000 is invested each year in an annuity earning 8% annual interest, how long will it take for the account to be worth $50,000? (See Exercise 97.)

REVIEW *Solve each equation.*

101. $\sqrt[3]{6x + 4} = 4$

102. $\sqrt{3x - 4} = \sqrt{-7x + 2}$

103. $\sqrt{a + 1} - 1 = 3a$

104. $3 - \sqrt{t - 3} = \sqrt{t}$

WRITING

99. Explain the mathematical relationship between $y = \log x$ and $y = 10^x$.

100. Explain why it is impossible to find the logarithm of a negative number.

9.6 Base-*e* Logarithms

- Base-*e* logarithms
- Graphing the natural logarithmic function
- An application of base-*e* logarithms

A special logarithmic function is the base-*e* logarithmic function. In this section, we will show how to evaluate base-*e* logarithms, graph the base-*e* logarithmic function, and solve some problems that involve the base-*e* logarithmic function.

Base-*e* logarithms

We have seen the importance of the number *e* in mathematical models of events in nature. Base-*e* logarithms are just as important. They are called **natural logarithms** or **Napierian logarithms** after John Napier (1550–1617). They are usually written as ln *x* rather than $\log_e x$:

> **ln *x***   **means** $\log_e x$ Read ln *x* letter-by-letter as "ℓ-n of *x*."

Like all logarithmic functions, the domain of $f(x) = \ln x$ is the interval $(0, \infty)$, and the range is the interval $(-\infty, \infty)$.

To find the base-*e* logarithms of numbers, we can use a calculator.

Finding base-*e* (natural) logarithms   CALCULATOR SNAPSHOT

To use a scientific calculator to find the value of ln 2.34, we enter these numbers and press these keys:

2.34 $\boxed{\text{LN}}$ $\boxed{.850150929}$

To use a graphing calculator, we enter these numbers and press these keys:

$\boxed{\text{LN}}$ 2.34 $\boxed{)}$ $\boxed{\text{ENTER}}$ $\boxed{\begin{array}{l}\ln(2.34)\\ \quad\quad .8501509294\end{array}}$

To four decimal places, ln 2.34 = 0.8502.

INTERMEDIATE
Algebra $f(x)$ Now™

Self Check 1
Find each value to four decimal places:

a. $\ln \pi$

b. $\ln 0$

EXAMPLE 1 Use a calculator to find each value: **a.** $\ln 17.32$ and **b.** $\ln (-0.05)$.

Solution
a. Enter these numbers and press these keys:

Scientific calculator	**Graphing calculator**
17.32 $\boxed{\text{LN}}$	$\boxed{\text{LN}}$ 17.32 $\boxed{)}$ $\boxed{\text{ENTER}}$

Either way, the result is 2.851861903.

b. Enter these numbers and press these keys:

Scientific calculator	**Graphing calculator**
0.05 $\boxed{+/-}$ $\boxed{\text{LN}}$	$\boxed{\text{LN}}$ $\boxed{(-)}$ 0.05 $\boxed{)}$ $\boxed{\text{ENTER}}$

Either way, we obtain an error, because we cannot take the logarithm of a negative number.

Answers **a.** 1.1447, **b.** no value

INTERMEDIATE
Algebra $f(x)$ Now™

Self Check 2
Solve:

a. $\ln x = 1.9344$

b. $-3 = \ln x$

Give each result to four decimal places.

EXAMPLE 2 Solve each equation: **a.** $\ln x = 1.335$ and **b.** $\ln x = -5.5$. Give each result to four decimal places.

Solution
a. Since the base of the natural logarithmic function is e, the equation $\ln x = 1.335$ is equivalent to $e^{1.335} = x$. To use a scientific calculator to find x, press these keys:

1.335 $\boxed{e^x}$

The display will read 3.799995946. To four decimal places,

$x = 3.8000$

b. The equation $\ln x = -5.5$ is equivalent to $e^{-5.5} = x$. To use a scientific calculator to find x, press these keys:

5.5 $\boxed{+/-}$ $\boxed{e^x}$

The display will read 0.004086771. To four decimal places,

$x = 0.0041$

Answers **a.** 6.9199, **b.** 0.0498

Graphing the natural logarithmic function

The equation $y = \ln x$ is equivalent to the equation $x = e^y$. To get the graph of $\ln x$, we can plot points that satisfy the equation $x = e^y$ and join them with a smooth curve, as shown in Figure 9-31(a). Figure 9-31(b) shows the calculator graph of $y = \ln x$.

$y = \ln x$		
x	y	(x, y)
$\frac{1}{e}$	-1	$\left(\frac{1}{e}, -1\right)$
1	0	$(1, 0)$
e	1	$(e, 1)$
e^2	2	$(e^2, 2)$

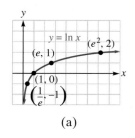
(a)

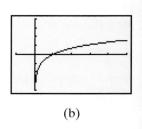
(b)

FIGURE 9-31

The exponential function and the natural logarithmic function are inverse functions. Figure 9-32 shows that their graphs are symmetric to the line $y = x$.

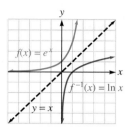

FIGURE 9-32

Graphing base-*e* logarithmic functions

CALCULATOR SNAPSHOT

Many graphs of logarithmic functions involve translations of the graph of $f(x) = \ln x$. For example, Figure 9-33 shows calculator graphs of the functions $f(x) = \ln x$, $g(x) = \ln x + 2$, and $h(x) = \ln x - 3$.

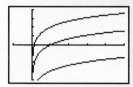

The graph of $g(x) = \ln x + 2$ is 2 units above the graph of $f(x) = \ln x$.

The graph of $h(x) = \ln x - 3$ is 3 units below the graph of $f(x) = \ln x$.

FIGURE 9-33

Figure 9-34 shows the calculator graph of the functions $f(x) = \ln x$, $g(x) = \ln (x - 2)$, and $h(x) = \ln (x + 3)$.

The graph of $h(x) = \ln (x + 3)$ is 3 units to the left of the graph of $f(x) = \ln x$.

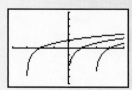

The graph of $g(x) = \ln (x - 2)$ is 2 units to the right of the graph of $f(x) = \ln x$.

FIGURE 9-34

An application of base-*e* logarithms

Base-*e* logarithms have many applications. If a population grows exponentially at a certain annual rate, the time required for the population to double is called the **doubling time.** It is given by the following formula.

Formula for doubling time

If r represents the annual rate, compounded continuously, and t represents time required for a population to double, then

$$t = \frac{\ln 2}{r}$$

INTERMEDIATE
Algebra $f(x)$ Now™

Self Check 3
See Example 3. If the population's annual growth rate could be reduced to 1.5% per year, what would be the doubling time?

EXAMPLE 3 **Doubling time.** The population of Earth is growing at the approximate rate of 2% per year. If this rate continues, how long will it take for the population to double?

Solution Because the population is growing at the rate of 2% per year, we substitute 0.02 for r in the formula for doubling time and simplify.

$$t = \frac{\ln 2}{r}$$

$$t = \frac{\ln 2}{0.02}$$

$$\approx 34.65735903 \quad \text{Use a calculator. Find ln 2 first. Then divide the result by 0.02.}$$

Answer about 46 years

The population will double in about 35 years.

INTERMEDIATE
Algebra $f(x)$ Now™

Self Check 4
In Example 4, how long will it take to double at 9%, compounded continuously?

EXAMPLE 4 **Doubling time.** How long will it take $1,000 to double at an annual rate of 8%, compounded continuously?

Solution We substitute 0.08 for r and simplify:

$$t = \frac{\ln 2}{r}$$

$$t = \frac{\ln 2}{0.08}$$

$$\approx 8.664339757 \quad \text{Use a calculator. Find ln 2 first. Then divide the result by 0.08.}$$

Answer about 7.7 years

It will take about $8\frac{2}{3}$ years for the money to double.

Section 9.6 STUDY SET

INTERMEDIATE
Algebra $f(x)$ Now™

VOCABULARY   *Fill in the blanks.*

1. Base-e logarithms are often called _____ logarithms.

2. $f(x) = \ln x$ and $f(x) = e^x$ are _____ functions.

CONCEPTS

3. Use a calculator to complete the table of values for $f(x) = \ln x$. Round to the nearest hundredth.

x	0.5	1	2	3	4	5	6	7	8	9	10
$f(x)$											

4. Graph $f(x) = \ln x$ on the illustration. (See Exercise 3.) Note that the units on the x- and y-axes are different.

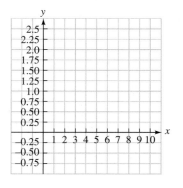

Fill in the blanks.

5. The graph of $f(x) = \ln x$ has the _____ as an asymptote.

6. The domain of the function $f(x) = \ln x$ is the interval ____.

7. The range of the function $f(x) = \ln x$ is the interval _____.

8. The graph of $f(x) = \ln x$ passes through the point (__, 0).

9. The statement $y = \ln x$ is equivalent to the exponential statement $e^y = $ __.

10. The logarithm of a negative number is _____.

11. A table of values for $f(x) = \ln x$ is shown below. Explain why ERROR appears in the Y_1 column for the first three entries.

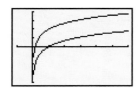

12. The illustration shows the graph of $f(x) = \ln x$, as well as a vertical translation of that graph. Using the notation $g(x)$ for the translation, write the defining equation for the function.

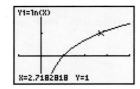

13. In the illustration, $f(x) = \ln x$ was graphed, and the TRACE feature was used. What is the x-coordinate of the point on the graph having a y-coordinate of 1? What is the name given this number?

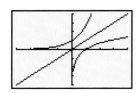

14. The graphs of $f(x) = \ln x$, $f(x) = e^x$, and $y = x$ are shown. What phrase is used to describe the relationship between the graphs?

NOTATION   *Fill in the blanks.*

15. $\ln 2$ means $\log_{} 2$.

16. $\log 2$ means $\log_{} 2$.

17. If a population grows exponentially at a rate r, the time it will take the population to double is given by the formula $t = $ __.

18. To evaluate a base-10 logarithm with a calculator, use the _____ key. To evaluate a base-e logarithm, use the _____ key.

PRACTICE   *Use a calculator to find each value, if possible. Express all answers to four decimal places.*

19. $\ln 35.15$ 20. $\ln 0.675$

21. $\ln 0.00465$ 22. $\ln 378.96$

23. $\ln 1.72$ 24. $\ln 2.7$

25. $\ln (-0.1)$ 26. $\ln (-10)$

Use a calculator to find x. Express all answers to four decimal places.

27. $\ln x = 1.4023$ 28. $\ln x = 2.6490$

29. $\ln x = 4.24$ 30. $\ln x = 0.926$

31. $\ln x = -3.71$ 32. $\ln x = -0.28$

33. $1.001 = \ln x$ 34. $\ln x = -0.001$

Use a graphing calculator to graph each function.

35. $y = \ln \left(\dfrac{1}{2}x \right)$ 36. $y = \ln x^2$

37. $f(x) = \ln (-x)$ 38. $f(x) = \ln (3x)$

APPLICATIONS

39. POPULATION GROWTH How long will it take the population of River City to double?

40. DOUBLING MONEY How long will it take $1,000 to double if it is invested at an annual rate of 5% compounded continuously?

41. POPULATION GROWTH A population growing continuously at an annual rate r will triple in a time t given by the formula

$$t = \frac{\ln 3}{r}$$

How long will it take the population of a town to triple if it is growing at the rate of 12% per year?

42. TRIPLING MONEY Find the length of time for $25,000 to triple when it is invested at 6% annual interest, compounded continuously. See Exercise 41.

43. FORENSIC MEDICINE To estimate the number of hours t that a murder victim had been dead, a coroner used the formula

$$t = \frac{1}{0.25} \ln \frac{98.6 - T_s}{82 - T_s}$$

where T_s is the temperature of the surroundings where the body was found. If the crime took place in an apartment where the thermostat was set at 70°F, approximately how long ago did the murder occur?

44. MAKING JELL-O® After the contents of a package of JELL-O® are combined with boiling water, the mixture is placed in a refrigerator whose temperature remains a constant 42°F. Estimate the number of hours t that it will take for the JELL-O® to cool to 50°F using the formula

$$t = -\frac{1}{0.9} \ln \frac{50 - T_r}{200 - T_r}$$

where T_r is the temperature of the refrigerator.

WRITING

45. Explain the difference between the functions $f(x) = \log x$ and $f(x) = \ln x$.

46. How are the functions $f(x) = \ln x$ and $f(x) = e^x$ related?

REVIEW   *Write an equation of the required line.*

47. Parallel to $y = 5x - 8$ and passing through the origin

48. Having a slope of 7 and a y-intercept of 3

49. Passing through the point $(3, 2)$ and perpendicular to the line $y = \frac{2}{3}x - 12$

50. Parallel to the line $3x + 2y = 9$ and passing through the point $(-3, 5)$

51. Vertical line through the point $(2, 3)$

52. Horizontal line through the point $(2, 3)$

9.7 Properties of Logarithms

- Properties of logarithms   • The change-of-base formula   • An application from chemistry

In this section, we will discuss eight properties of logarithms and use them to simplify logarithmic expressions. We will then show how to change a logarithm from one base to another. We conclude the section by solving some problems from the field of chemistry.

Properties of logarithms

Since logarithms are exponents, the properties of exponents have counterparts in the theory of logarithms. We begin with four basic properties.

> **Properties of logarithms**
>
> If b represents a positive number and $b \neq 1$, then
>
> **1.** $\log_b 1 = 0$ **2.** $\log_b b = 1$
> **3.** $\log_b b^x = x$ **4.** $b^{\log_b x} = x$ where $x > 0$

Properties 1 through 4 follow directly from the definition of a logarithm.

1. $\log_b 1 = 0$, because $b^0 = 1$.

2. $\log_b b = 1$, because $b^1 = b$.

3. $\log_b b^x = x$, because $b^x = b^x$

4. $b^{\log_b x} = x$, because $\log_b x$ is the exponent to which b is raised to get x.

Properties 3 and 4 also indicate that the composition of the exponential and logarithmic functions (in both directions) is the identity function. This is expected, because the exponential and logarithmic functions are inverse functions.

EXAMPLE 1   Simplify each expression: **a.** $\log_5 1$,   **b.** $\log_3 3$,   **c.** $\ln e^3$,   and   **d.** $9^{\log_9 7}$.

Solution

a. By property 1, $\log_5 1 = 0$, because $5^0 = 1$.

b. By property 2, $\log_3 3 = 1$, because $3^1 = 3$.

c. By property 3, $\ln e^3 = 3$, because $e^3 = e^3$.

d. By property 4, $9^{\log_9 7} = 7$, because $\log_9 7$ is the power to which 9 is raised to get 7.

INTERMEDIATE
Algebra $f(x)$ **Now**™

Self Check 1

Simplify   **a.** $\log_4 1$,   **b.** $\log_4 4$, **c.** $\log_2 2^4$,   and   **d.** $5^{\log_5 2}$.

Answers **a.** 0,   **b.** 1,   **c.** 4, **d.** 2

The next two properties state that

The logarithm of a product is the sum of the logarithms.

The logarithm of a quotient is the difference of the logarithms.

Properties of logarithms

If M, N, and b represent positive numbers and $b \neq 1$, then

5. $\log_b MN = \log_b M + \log_b N$ **6.** $\log_b \dfrac{M}{N} = \log_b M - \log_b N$

Proof   To prove property 5, we let $x = \log_b M$ and $y = \log_b N$. We use the definition of logarithm to write each equation in exponential form.

$$M = b^x \qquad \text{and} \qquad N = b^y$$

Then $MN = b^x b^y$, and a property of exponents gives

$$MN = b^{x+y} \qquad \text{Keep the base and add the exponents: } b^x b^y = b^{x+y}.$$

We write this exponential equation in logarithmic form as

$$\log_b MN = x + y$$

Substituting the values of x and y completes the proof.

$$\log_b MN = \log_b M + \log_b N \qquad\qquad \square$$

The proof of property 6 is similar.

In the following examples, we will use these properties to write a logarithm of a product as the sum of logarithms and a logarithm of a quotient as the difference of logarithms. Assume that all variables are positive and that $b \neq 1$.

INTERMEDIATE
Algebra *f(x)* **Now**™

Self Check 2
Rewrite:
a. $\log_3 (4 \cdot 3)$
b. $\log (1,000y)$

Answers **a.** $\log_3 4 + 1$,
b. $3 + \log y$

EXAMPLE 2   Use the product rule for logarithms to rewrite each of the following: **a.** $\log_2 (2 \cdot 7)$ and   **b.** $\log (100x)$.

Solution
a. $\log_2 (2 \cdot 7) = \log_2 2 + \log_2 7$ The log of a product is the sum of the logs.
$\qquad\qquad\quad = 1 + \log_2 7$ $\log_2 2 = 1$.

b. $\log (100x) = \log 100 + \log x$ The log of a product is the sum of the logs.
$\qquad\qquad\ = 2 + \log x$ $\log 100 = 2$.

INTERMEDIATE
Algebra *f(x)* **Now**™

Self Check 3
Rewrite:
a. $\log_6 \dfrac{6}{5}$ **b.** $\ln \dfrac{y}{100}$

Answers **a.** $1 - \log_6 5$,
b. $\ln y - \ln 100$

EXAMPLE 3   Use the quotient rule for logarithms to rewrite each of the following: **a.** $\ln \dfrac{10}{7}$ and   **b.** $\log_4 \dfrac{x}{64}$.

Solution
a. $\ln \dfrac{10}{7} = \ln 10 - \ln 7$ The log of a quotient is the difference of the logs.

b. $\log_4 \dfrac{x}{64} = \log_4 x - \log_4 64$ The log of a quotient is the difference of the logs.
$\qquad\qquad\ = \log_4 x - 3$ $\log_4 64 = 3$.

INTERMEDIATE
Algebra *f(x)* **Now**™

Self Check 4

Rewrite $\log_b \dfrac{x}{yz}$.

Answer $\log_b x - \log_b y - \log_b z$

EXAMPLE 4   Use logarithm properties to rewrite each expression: **a.** $\log_b xyz$ and   **b.** $\log \dfrac{xy}{z}$.

Solution
a. We observe that $\log_b xyz$ is the logarithm of a product.

$\log_b xyz = \log_b (xy)z$ Group the first two factors together.
$\qquad\quad = \log_b (xy) + \log_b z$ The log of a product is the sum of the logs.
$\qquad\quad = \log_b x + \log_b y + \log_b z$ The log of a product is the sum of the logs.

b. In the expression $\log \dfrac{xy}{z}$, we have the logarithm of a quotient.

$\log \dfrac{xy}{z} = \log (xy) - \log z$ The log of a quotient is the difference of the logs.
$\qquad\quad = (\log x + \log y) - \log z$ The log of a product is the sum of the logs.
$\qquad\quad = \log x + \log y - \log z$ Remove parentheses.

❗ COMMENT By property 5 of logarithms, the logarithm of a *product* is equal to the *sum* of the logarithms. The logarithm of a sum or a difference usually does not simplify. In general,

$$\log_b (M + N) \neq \log_b M + \log_b N \qquad \text{and} \qquad \log_b (M - N) \neq \log_b M - \log_b N$$

By property 6, the logarithm of a *quotient* is equal to the *difference* of the logarithms. The logarithm of a quotient is not the quotient of the logarithms:

$$\log_b \dfrac{M}{N} \neq \dfrac{\log_b M}{\log_b N}$$

Verifying properties of logarithms

We can use a calculator to illustrate property 5 of logarithms by showing that

$$\ln[(3.7)(15.9)] = \ln 3.7 + \ln 15.9$$

We calculate the left- and right-hand sides of the equation separately and compare the results. To use a scientific calculator to find $\ln[(3.7)(15.9)]$, we enter these numbers and press these keys:

3.7 $\boxed{\times}$ 15.9 $\boxed{=}$ $\boxed{\text{LN}}$ $\boxed{\text{4.074651929}}$

To find $\ln 3.7 + \ln 15.9$, we enter these numbers and press these keys:

3.7 $\boxed{\text{LN}}$ $\boxed{+}$ 15.9 $\boxed{\text{LN}}$ $\boxed{=}$ $\boxed{\text{4.074651929}}$

Since the left- and right-hand sides are equal, the equation $\ln[(3.7)(15.9)] = \ln 3.7 + \ln 15.9$ is true.

Two more properties of logarithms state that

The logarithm of a power is the power times the logarithm.

If the logarithms of two numbers are equal, the numbers are equal.

Properties of logarithms

If M, p, and b represent positive numbers and $b \neq 1$, then

7. $\log_b M^p = p \log_b M$ **8.** If $\log_b x = \log_b y$, then $x = y$.

Proof   To prove property 7, we let $x = \log_b M$, write the expression in exponential form, and raise both sides to the pth power:

$$M = b^x$$
$$(M)^p = (b^x)^p \quad \text{Raise both sides to the } p\text{th power.}$$
$$M^p = b^{px} \quad \text{Keep the base and multiply the exponents.}$$

Using the definition of logarithms gives

$$\log_b M^p = px$$

Substituting the value for x completes the proof.

$$\log_b M^p = p \log_b M$$

Property 8 follows from the fact that the logarithmic function is a one-to-one function. Property 8 will be used in the next section, when we solve logarithmic equations.

EXAMPLE 5   Use the power rule for logarithms to rewrite each of the following:
a. $\log_5 6^2$ and   **b.** $\log \sqrt{10}$.

Solution
a. $\log_5 6^2 = 2 \log_5 6$ The log of a power is the power times the log.

b. $\log \sqrt{10} = \log(10)^{1/2}$ Write $\sqrt{10}$ using a fractional exponent: $\sqrt{10} = (10)^{1/2}$.

$$= \frac{1}{2} \log 10 \qquad \text{The log of a power is the power times the log.}$$

$$= \frac{1}{2} \qquad\qquad \text{Simplify: } \log 10 = 1.$$

INTERMEDIATE
Algebra $f(x)$ Now™

Self Check 5
Rewrite:
a. $\ln x^4$,   **b.** $\log_2 \sqrt[3]{3}$

Answers **a.** $4 \ln x$,

b. $\frac{1}{3}\log_2 3$

Self Check 6

Rewrite $\log \sqrt[4]{\dfrac{x^3 y}{z}}$.

EXAMPLE 6 Use logarithm properties to rewrite each expression:
a. $\log_b (x^2 y^3 z)$ and **b.** $\ln \dfrac{y^3 \sqrt{x}}{z}$.

Solution
a. We begin by recognizing that $\log_b (x^2 y^3 z)$ is the logarithm of a product.

$$\log_b (x^2 y^3 z) = \log_b x^2 + \log_b y^3 + \log_b z \qquad \text{The log of a product is the sum of the logs.}$$

$$= 2 \log_b x + 3 \log_b y + \log_b z \qquad \text{The log of a power is the power times the log.}$$

b. The expression $\ln \dfrac{y^3 \sqrt{x}}{z}$ is the logarithm of a quotient.

$$\ln \frac{y^3 \sqrt{x}}{z} = \ln (y^3 \sqrt{x}) - \ln z \qquad \text{The log of a quotient is the difference of the logs.}$$

$$= \ln y^3 + \ln \sqrt{x} - \ln z \qquad \text{The log of a product is the sum of the logs.}$$

$$= \ln y^3 + \ln x^{1/2} - \ln z \qquad \text{Write } \sqrt{x} \text{ as } x^{1/2}.$$

$$= 3 \ln y + \frac{1}{2} \ln x - \ln z \qquad \text{The log of a power is the power times the log.}$$

Answer

$\dfrac{1}{4}(3 \log x + \log y - \log z)$

We can use the properties of logarithms to combine several logarithms into one logarithm.

Self Check 7

Write the expression as one logarithm:

$2 \log_a x + \frac{1}{2} \log_a y - 2 \log_a (x - y)$

EXAMPLE 7 Write each of the given expressions as one logarithm:
a. $3 \log_a x + \frac{1}{2} \log_a y$ and **b.** $\frac{1}{2} \log_b (x - 2) - \log_b y + 3 \log_b z$.

Solution
a. We begin by applying the power rule to each term of the expression.

$$3 \log_a x + \frac{1}{2} \log_a y$$

$$= \log_a x^3 + \log_a y^{1/2} \qquad \text{A power times a log is the log of the power.}$$

$$= \log_a (x^3 y^{1/2}) \qquad \text{The sum of two logs is the log of the product.}$$

b. The first and third terms of this expression can be rewritten using the power rule of logarithms.

$$\frac{1}{2} \log_b (x - 2) - \log_b y + 3 \log_b z$$

$$= \log_b (x - 2)^{1/2} - \log_b y + \log_b z^3 \qquad \text{A power times a log is the log of the power.}$$

$$= \log_b \frac{(x - 2)^{1/2}}{y} + \log_b z^3 \qquad \text{The difference of two logs is the log of the quotient.}$$

$$= \log_b \frac{z^3 \sqrt{x - 2}}{y} \qquad \text{The sum of two logs is the log of the product. Write } (x - 2)^{1/2} \text{ as } \sqrt{x - 2}.$$

Answer $\log_a \dfrac{x^2 \sqrt{y}}{(x - y)^2}$

We summarize the properties of logarithms as follows.

Properties of logarithms

If b, M, and N represent positive numbers and $b \neq 1$, then

1. $\log_b 1 = 0$ **2.** $\log_b b = 1$

3. $\log_b b^x = x$ **4.** $b^{\log_b x} = x$

5. $\log_b MN = \log_b M + \log_b N$ **6.** $\log_b \dfrac{M}{N} = \log_b M - \log_b N$

7. $\log_b M^p = p \log_b M$ **8.** If $\log_b x = \log_b y$, then $x = y$.

EXAMPLE 8 Given that $\log 2 \approx 0.3010$ and $\log 3 \approx 0.4771$, find approximations for **a.** $\log 6$ and **b.** $\log 18$.

Solution

a. $\log 6 = \log (2 \cdot 3)$ Write 6 using the factors 2 and 3.

$\qquad = \log 2 + \log 3$ The log of a product is the sum of the logs.

$\qquad \approx 0.3010 + 0.4771$ Substitute the value of each logarithm.

$\qquad \approx 0.7781$

b. $\log 18 = \log (2 \cdot 3^2)$ Write 18 using the factors 2 and 3.

$\qquad = \log 2 + \log 3^2$ The log of a product is the sum of the logs.

$\qquad = \log 2 + 2 \log 3$ The log of a power is the power times the log.

$\qquad \approx 0.3010 + 2(0.4771)$ Substitute the value of each logarithm.

$\qquad \approx 1.2552$

INTERMEDIATE
Algebra $f(x)$ **Now**™

Self Check 8
Find:
a. $\log 1.5$
b. $\log 0.75$

Answers **a.** 0.1761,
b. -0.1249

◼ The change-of-base formula

Most calculators can find common logarithms and natural logarithms. If we need to find a logarithm with some other base, we use a conversion formula.

  If we know the base-a logarithm of a number, we can find its logarithm to some other base b by using a formula called the **change-of-base formula.**

Change-of-base formula

If a, b, and x represent positive numbers, $a \neq 1$, and $b \neq 1$, then

$$\log_b x = \frac{\log_a x}{\log_a b}$$

Proof To prove this formula, we begin with the equation $\log_b x = y$.

(1) $\qquad y = \log_b x$

$\qquad\qquad x = b^y$ Change the equation from logarithmic to exponential form.

$\qquad \log_a x = \log_a b^y$ Take the base-a logarithm of both sides.

$\qquad \log_a x = y \log_a b$ The log of a power is the power times the log.

$\qquad\qquad y = \dfrac{\log_a x}{\log_a b}$ Divide both sides by $\log_a b$.

$\qquad \log_b x = \dfrac{\log_a x}{\log_a b}$ Refer to Equation 1 and substitute $\log_b x$ for y. $\square$

If we know logarithms to base a (for example, $a = 10$), we can find the logarithm of x to a new base b. We simply divide the base-a logarithm of x by the base-a logarithm of b.

! COMMENT $\dfrac{\log_a x}{\log_a b}$ means that one logarithm is to be divided by the other. They are not to be subtracted.

INTERMEDIATE
Algebra $f(x)$ **Now**™

Self Check 9
Find $\log_5 3$ to four decimal places.

EXAMPLE 9 Find $\log_3 5$.

Solution We can use base-10 logarithms to find a base-3 logarithm. To do this, we substitute 3 for b, 10 for a, and 5 for x in the change-of-base formula:

$$\log_b x = \frac{\log_a x}{\log_a b}$$

$$\log_3 5 = \frac{\log_{10} 5}{\log_{10} 3} \qquad b = 3, x = 5, \text{ and } a = 10.$$

$$\approx 1.464973521 \qquad \text{Use a calculator.}$$

To four decimal places, $\log_3 5 = 1.4650$.

We can also use the natural logarithm function (base e) in the change-of-base formula to find a base-3 logarithm.

$$\log_b x = \frac{\log_a x}{\log_a b}$$

$$\log_3 5 = \frac{\ln 5}{\ln 3} \qquad \begin{array}{l} b = 3, x = 5, \text{ and } a = e. \\ \log_e 5 = \ln 5 \text{ and } \log_e 3 = \ln 3. \end{array}$$

$$\approx 1.464973521 \qquad \text{Use a calculator.}$$

Answer 0.6826

We obtain the same result.

▌ An application from chemistry

In chemistry, common logarithms are used to express the acidity of solutions. The more acidic a solution, the greater the concentration of hydrogen ions. This concentration is indicated indirectly by the *pH scale,* or *hydrogen ion index.* The pH of a solution is defined as follows.

pH of a solution

If $[H^+]$ represents the hydrogen ion concentration in gram-ions per liter, then

$$pH = -\log [H^+]$$

EXAMPLE 10 pH meters.
One of the most accurate ways to measure pH is with a probe and meter. What reading should the meter give for pure water if water has a hydrogen ion concentration [H⁺] of approximately 10^{-7} gram-ions per liter?

Solution Since pure water has approximately 10^{-7} gram-ions per liter, its pH is

$$pH = -\log [\mathbf{H^+}]$$
$$pH = -\log \mathbf{10^{-7}}$$
$$= -(-7) \log 10 \qquad \text{The log of a power is the power times the log.}$$
$$= -(-7) \cdot 1 \qquad \text{Simplify: } \log 10 = 1.$$
$$= 7$$

The meter should give a reading of 7.

EXAMPLE 11 Hydrogen ion concentration.
Find the hydrogen ion concentration of seawater if its pH is 8.5.

Solution To find its hydrogen ion concentration, we solve the following equation for [H⁺].

$$\mathbf{pH} = -\log [\mathbf{H^+}]$$
$$\mathbf{8.5} = -\log [\mathbf{H^+}]$$
$$-8.5 = \log [\mathbf{H^+}] \qquad \text{Multiply both sides by } -1.$$
$$[\mathbf{H^+}] = 10^{-8.5} \qquad \text{Change the equation to exponential form.}$$

We can use a calculator to find that

$$[\mathbf{H^+}] \approx 3.2 \times 10^{-9} \text{ gram-ions per liter}$$

Section 9.7 STUDY SET

VOCABULARY Fill in the blanks.

1. The expression $\log_3 (4x)$ is the logarithm of a _____.

2. The expression $\log_2 \frac{5}{x}$ is the logarithm of a _____.

3. The expression $\log 4^x$ is the logarithm of a _____.

4. In the expression $\log_5 4$, the number 5 is the _____ of the logarithm.

CONCEPTS Fill in the blanks.

5. $\log_b 1 =$ ▢

6. $\log_b b =$ ▢

7. $\log_b MN = \log_b$ ▢ $+ \log_b$ ▢

8. $b^{\log_b x} =$ ▢

9. If $\log_b x = \log_b y$, then ▢ $=$ ▢.

10. $\log_b \dfrac{M}{N} = \log_b M$ ▢ $\log_b N$

11. $\log_b x^p = p \cdot \log_b$ ▢

12. $\log_b b^x =$ ▢

13. $\log_b (A + B)$ ▢ $\log_b A + \log_b B$

14. $\log_b A + \log_b B$ ▢ $\log_b AB$

15. $\log_b x = \dfrac{\log_a x}{▢}$

16. $pH =$ ▢

17. Three logarithmic expressions have been evaluated, and the results are shown on the calculator display on the right. Show that each result is correct by writing the equivalent base-10 exponential statement.

```
log(1)
            0
log(10)
            1
log(10²)
            2
```

18. Three logarithmic expressions have been evaluated, and the results are shown on the calculator display on the right. Show that each result is correct by writing the equivalent base-e exponential statement. (The notation $\ln(e\wedge(2))$ means $\ln e^2$.)

```
ln(e^(2))
            2
ln(e^(3))
            3
ln(e^(4))
            4
```

Evaluate each expression.

19. $\log_4 1$

20. $\log_4 4$

21. $\log_4 4^7$

22. $\ln e^8$

23. $5^{\log_5 10}$

24. $8^{\log_8 10}$

25. $\log_5 5^2$

26. $\log_4 4^2$

27. $\ln e$

28. $\log_7 1$

29. $\log_3 3^7$

30. $5^{\log_5 8}$

■ **NOTATION**   *Complete each solution.*

31. $\log_b rst = \log_b (\quad)t$

$\qquad = \log_b (rs) + \log_b \underline{\quad}$

$\qquad = \log_b \underline{\quad} + \log_b \underline{\quad} + \log_b t$

32. $\log \dfrac{r}{st} = \log r - \log (\underline{\quad})$

$\qquad = \log r - (\log \underline{\quad} + \log t)$

$\qquad = \log r - \log s \underline{\quad} \log t$

▦ **PRACTICE**   *Use a calculator to verify each equation.*

33. $\log [(2.5)(3.7)] = \log 2.5 + \log 3.7$

34. $\log 45.37 = \dfrac{\ln 45.37}{\ln 10}$

35. $\ln (2.25)^4 = 4 \ln 2.25$

36. $\ln \dfrac{11.3}{6.1} = \ln 11.3 - \ln 6.1$

37. $\log \sqrt{24.3} = \dfrac{1}{2} \log 24.3$

38. $\ln 8.75 = \dfrac{\log 8.75}{\log e}$

Use the properties of logarithms to rewrite each expression. Assume that x, y, and z are positive numbers.

39. $\log_2 (4 \cdot 5)$

40. $\log_3 (27 \cdot 5)$

41. $\log_6 \dfrac{x}{36}$

42. $\log_8 \dfrac{y}{8}$

43. $\ln y^4$

44. $\ln z^9$

45. $\log \sqrt{5}$

46. $\log \sqrt[3]{7}$

Assume that x, y, z, and b are positive numbers and $b \neq 1$. Use the properties of logarithms to write each expression in terms of the logarithms of x, y, and z.

47. $\log xyz$

48. $\log 4xz$

49. $\log_2 \dfrac{2x}{y}$

50. $\log_3 \dfrac{x}{yz}$

51. $\log x^3 y^2$

52. $\log xy^2 z^3$

53. $\log_b (xy)^{1/2}$

54. $\log_b x^3 y^{1/2}$

55. $\log_a \dfrac{\sqrt[3]{x}}{\sqrt[4]{yz}}$

56. $\log_b \sqrt[4]{\dfrac{x^3 y^2}{z^4}}$

57. $\ln x\sqrt{z}$

58. $\ln \sqrt{xy}$

Assume that x, y, z, and b are positive numbers and $b \neq 1$. Use the properties of logarithms to write each expression as the logarithm of a single quantity.

59. $\log_2 (x + 1) - \log_2 x$

60. $\log_3 x + \log_3 (x + 2) - \log_3 8$

61. $2 \log x + \dfrac{1}{2} \log y$

62. $-2 \log x - 3 \log y + \log z$

63. $-3 \log_b x - 2 \log_b y + \dfrac{1}{2} \log_b z$

64. $3 \log_b (x + 1) - 2 \log_b (x + 2) + \log_b x$

65. $\ln \left(\dfrac{x}{z} + x \right) - \ln \left(\dfrac{y}{z} + y \right)$

66. $\ln (xy + y^2) - \ln (xz + yz) + \ln z$

Determine whether the given statement is true. If a statement is false, explain why.

67. $\log xy = (\log x)(\log y)$

68. $\log ab = \log a + 1$

69. $\log_b (A - B) = \dfrac{\log_b A}{\log_b B}$

70. $\dfrac{\log_b A}{\log_b B} = \log_b A - \log_b B$

71. $\log_b \dfrac{A}{B} = \log_b A - \log_b B$

72. $\log_b 2 = \log_2 b$

Assume that $\log_b 4 = 0.6021$, $\log_b 7 = 0.8451$, and $\log_b 9 = 0.9542$. Use these values and the properties of logarithms to find each value.

73. $\log_b 28$

74. $\log_b \dfrac{7}{4}$

75. $\log_b \dfrac{4}{63}$

76. $\log_b 36$

77. $\log_b \dfrac{63}{4}$

78. $\log_b 2.25$

79. $\log_b 64$

80. $\log_b 49$

Use the change-of-base formula to find each logarithm to four decimal places.

81. $\log_3 7$

82. $\log_7 3$

83. $\log_{1/3} 3$

84. $\log_{1/2} 6$

85. $\log_3 8$

86. $\log_5 10$

87. $\log_{\sqrt{2}} \sqrt{5}$

88. $\log_\pi e$

APPLICATIONS

89. pH OF A SOLUTION Find the pH of a solution with a hydrogen ion concentration of 1.7×10^{-5} gram-ions per liter.

90. HYDROGEN ION CONCENTRATION Find the hydrogen ion concentration of a saturated solution of calcium hydroxide whose pH is 13.2.

91. AQUARIUMS To test for safe pH levels in a freshwater aquarium, a test strip is compared with the scale shown. Find the corresponding range in the hydrogen ion concentration.

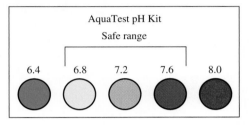

AquaTest pH Kit

Safe range

6.4 6.8 7.2 7.6 8.0

92. pH OF SOUR PICKLES The hydrogen ion concentration of sour pickles is 6.31×10^{-4}. Find the pH.

WRITING

93. Explain the difference between a logarithm of a product and the product of logarithms.

94. How can the $\boxed{\text{LOG}}$ key on a calculator be used to find $\log_2 7$?

REVIEW *Consider the line that passes through $P(-2, 3)$ and $Q(4, -4)$.*

95. Find the slope of line PQ.

96. Find the distance PQ.

97. Find the midpoint of segment PQ.

98. Write the equation of line PQ.

9.8 Exponential and Logarithmic Equations

- Solving exponential equations • Solving logarithmic equations
- Radioactive decay • Population growth

An **exponential equation** is an equation that contains a variable in one of its exponents. Some examples of exponential equations are

$$3^x = 5, \qquad 6^{x-3} = 2^x, \qquad \text{and} \qquad 2^{x^2+2x} = \dfrac{1}{2}$$

A **logarithmic equation** is an equation with a logarithmic expression that contains a variable. Some examples of logarithmic equations are

$$\log 5x = 1, \qquad \log(3x + 2) - \log(2x - 3) = 0, \qquad \text{and} \qquad \frac{\log_2(5x - 6)}{\log_2 x} = 2$$

In this section, we will learn how to solve many of these equations.

■ Solving exponential equations

If both sides of an exponential equation can be expressed as a power of the same base, we can use the following property to solve the equation.

> **Exponent property of equality**
>
> If two exponential expressions with the same base are equal, their exponents are equal.
>   For any real number b, where $b \neq -1, 0,$ or 1.
>
> $b^x = b^y$ is equivalent to $x = y$

Self Check 1

Solve: $3^{x^2 - 2x} = \dfrac{1}{3}$.

EXAMPLE 1   Solve: $2^{x^2 + 2x} = \dfrac{1}{2}$.

Solution   Since $\frac{1}{2} = 2^{-1}$, we can write the equation in the form

$2^{x^2 + 2x} = 2^{-1}$ Each side of the equation can be written as an exponential expression with base 2.

Since equal quantities with equal bases have equal exponents, we have

$$
\begin{aligned}
x^2 + 2x &= -1 & &\text{Equate the exponents.} \\
x^2 + 2x + 1 &= 0 & &\text{Add 1 to both sides.} \\
(x + 1)(x + 1) &= 0 & &\text{Factor the trinomial.} \\
x + 1 = 0 \qquad \text{or} \qquad x + 1 &= 0 & &\text{Set each factor equal to 0.} \\
x = -1 \qquad\qquad\quad x &= -1
\end{aligned}
$$

Answer 1, 1

Verify that -1 satisfies the original equation.

CALCULATOR SNAPSHOT   **Solving exponential equations graphically**

To use a graphing calculator to approximate the solutions of $2^{x^2 + 2x} = \frac{1}{2}$ (see Example 1), we can subtract $\frac{1}{2}$ from both sides of the equation to get

$$2^{x^2 + 2x} - \frac{1}{2} = 0 \quad \text{and graph the corresponding function} \quad y = 2^{x^2 + 2x} - \frac{1}{2}$$

If we use window settings of $[-4, 4]$ for x and $[-1, 2]$ for y, we obtain the graph shown in Figure 9-35(a) on the next page.

The solutions of $2^{x^2 + 2x} - \frac{1}{2} = 0$ are the x-coordinates of the x-intercepts of the graph of $y = 2^{x^2 + 2x} - \frac{1}{2}$. Using the ZERO feature, we see that the graph has only one x-intercept, $(-1, 0)$. Therefore, -1 is the only solution of $2^{x^2 + 2x} - \frac{1}{2} = 0$.

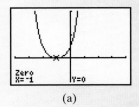
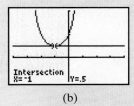

(a)   (b)

FIGURE 9-35

We can also solve $2^{x^2+2x} = \frac{1}{2}$ using the INTERSECT feature found on most graphing calculators. After graphing $Y_1 = 2^{x^2+2x}$ and $Y_2 = \frac{1}{2}$, we select INTERSECT, which approximates the coordinates of the point of intersection of the two graphs. From the display shown in Figure 9-35(b), we can conclude that the solution is -1. Verify this by checking.

EXAMPLE 2   Solve: $3^x = 5$.

Self Check 2

Solve: $5^x = 4$. Give the answer to four decimal places.

Solution   Since the logarithms of equal numbers are equal, we can take the common logarithm of each side of the equation. The power rule of logarithms then provides a way of moving the variable x from its position as an exponent to a position as a coefficient.

$$3^x = 5$$

$$\log 3^x = \log 5 \qquad \text{Take the common logarithm of each side.}$$

$$x \log 3 = \log 5 \qquad \text{The log of a power is the power times the log: } \log 3^x = x \log 3.$$

(1) $$x = \frac{\log 5}{\log 3} \qquad \text{Divide both sides by } \log 3.$$

$$x \approx 1.464973521 \qquad \text{Use a calculator.}$$

The exact solution is $\dfrac{\log 5}{\log 3}$. To four decimal places, the solution is 1.4650.

We can also take the natural logarithm of each side of the equation to solve for x.

$$3^x = 5$$

$$\ln 3^x = \ln 5 \qquad \text{Take the natural logarithm of each side.}$$

$$x \ln 3 = \ln 5 \qquad \text{Use the power rule of logarithms: } \ln 3^x = x \ln 3.$$

$$x = \frac{\ln 5}{\ln 3} \qquad \text{Divide both sides by } \ln 3.$$

$$x \approx 1.464973521 \qquad \text{Use a calculator.}$$

To check the solution, we substitute 1.4650 for x in 3^x and see if the result is close to 5. We can compute $3^{1.4650}$ by entering $3 \boxed{y^x} 1.4650 \boxed{=}$ on a scientific calculator. The result of 5.000145454 verifies that $x \approx 1.4650$ is an approximate solution of $3^x = 5$.

Answer   0.8614

❗ COMMENT   A careless reading of Equation 1 in the Example 2 solution leads to a common error. The right-hand side of Equation 1 calls for a division, not a subtraction.

$$\frac{\log 5}{\log 3} \qquad \text{means} \qquad (\log 5) \div (\log 3)$$

It is the expression $\log \frac{5}{3}$ that means $\log 5 - \log 3$.

INTERMEDIATE
Algebra *f(x)* Now™

Self Check 3
Solve: $5^{x-2} = 3^x$.

EXAMPLE 3   Solve: $6^{x-3} = 2^x$.

Solution

$$6^{x-3} = 2^x$$

$$\log 6^{x-3} = \log 2^x \qquad \text{Take the common logarithm of each side.}$$

$$(x-3)\log 6 = x \log 2 \qquad \text{The log of a power is the power times the log.}$$

$$x \log 6 - 3 \log 6 = x \log 2 \qquad \text{Use the distributive property.}$$

$$x \log 6 - x \log 2 = 3 \log 6 \qquad \text{On both sides, add 3 log 6 and subtract } x \log 2.$$

$$x (\log 6 - \log 2) = 3 \log 6 \qquad \text{Factor out } x \text{ on the left-hand side.}$$

$$x = \frac{3 \log 6}{\log 6 - \log 2} \qquad \text{Divide both sides by } \log 6 - \log 2.$$

$$x \approx 4.892789261 \qquad \text{Use a calculator.}$$

To four decimal places, the solution is 4.8928.

Answer
$$\frac{2 \log 5}{\log 5 - \log 3} \approx 6.3013$$

INTERMEDIATE
Algebra *f(x)* Now™

Self Check 4
Solve: $e^{2.1t} = 35$.

EXAMPLE 4   Solve: $e^{0.9t} = 8$.

Solution   The exponential expression on the left-hand side has base e. In such cases, the computations are somewhat simpler if we take the natural logarithm of each side.

$$e^{0.9t} = 8$$

$$\ln e^{0.9t} = \ln 8 \qquad \text{Take the natural logarithm of each side.}$$

$$0.9t \ln e = \ln 8 \qquad \text{Use the power rule of logarithms: } \ln e^{0.9t} = 0.9t \ln e.$$

$$0.9t \cdot 1 = \ln 8 \qquad \text{Simplify: } \ln e = 1.$$

$$0.9t = \ln 8$$

$$t = \frac{\ln 8}{0.9} \qquad \text{Divide both sides by 0.9.}$$

$$t \approx 2.310490602 \qquad \text{Use a calculator.}$$

Answer $\dfrac{\ln 35}{2.1} \approx 1.6930$

To four decimal places, the solution is 2.3105.

◼ Solving logarithmic equations

We can solve many logarithmic equations using properties of logarithms.

INTERMEDIATE
Algebra *f(x)* Now™

Self Check 5
Solve: $\log_2 (x - 3) = -1$

EXAMPLE 5   Solve: $\log 5x = 3$.

Solution   Recall that $\log 5x = \log_{10} 5x$. We can change the equation $\log 5x = 3$ into the equivalent base-10 exponential equation $10^3 = 5x$ and solve for x.

$$\log 5x = 3$$

$$10^3 = 5x$$

$$1,000 = 5x \qquad \text{Simplify: } 10^3 = 1,000.$$

$$200 = x \qquad \text{Divide both sides by 5.}$$

Answer $\dfrac{7}{2}$

The solution is 200. Check the result.

INTERMEDIATE
Algebra (f(x)) **Now**™

EXAMPLE 6   Solve: $\log (3x + 2) - \log (2x - 3) = 0$.

Self Check 6
Solve:
$\log(5x + 2) - \log(7x - 2) = 0$

Solution   We isolate each logarithmic expression on one side of the equation.

$$\log (3x + 2) - \log (2x - 3) = 0$$
$$\mathbf{log}\,(3x + 2) = \mathbf{log}\,(2x - 3) \quad \text{Add } \log (2x - 3) \text{ to both sides.}$$

In the previous section, we saw that if the logarithms of two numbers are equal, the numbers are equal. So we have

$$(3x + 2) = (2x - 3) \quad \text{If } \log r = \log s, \text{ then } r = s.$$
$$x = -5 \quad \text{Subtract } 2x \text{ and } 2 \text{ from both sides.}$$

Check: $\log (3x + 2) - \log (2x - 3) = 0$
$$\log [3(-5) + 2] - \log [2(-5) - 3] \overset{?}{=} 0$$
$$\log (-13) - \log (-13) \overset{?}{=} 0$$

Since the logarithm of a negative number does not exist, the proposed solution -5 must be discarded. This equation has no solutions.

Answer 2

INTERMEDIATE
Algebra (f(x)) **Now**™

EXAMPLE 7   Solve: $\log x + \log (x - 3) = 1$.

Self Check 7
Solve: $\log x + \log (x + 3) = 1$.

Solution

$$\log x + \log (x - 3) = 1$$

$\log x\,(x - 3) = 1$	Use the product rule of logarithms.
$\log_{10} x\,(x - 3) = 1$	The base is 10.
$x(x - 3) = \mathbf{10^1}$	Use the definition of logarithms to change the equation to exponential form: $\log x(x - 3) = \log_{10} x(x - 3)$.
$x^2 - 3x - 10 = 0$	Distribute the multiplication by x and subtract 10 from both sides.
$(x + 2)(x - 5) = 0$	Factor the trinomial.
$x + 2 = 0 \quad \text{or} \quad x - 5 = 0$	Set each factor equal to 0.
$x = -2 \qquad\quad x = 5$	

Check: The number -2 is not a solution, because it does not satisfy the equation (a negative number does not have a logarithm). We will check the remaining number, 5.

$$\log x + \log (x - 3) = 1$$
$$\log 5 + \log (5 - 3) \overset{?}{=} 1 \quad \text{Substitute 5 for } x.$$
$$\log 5 + \log 2 \overset{?}{=} 1$$
$$\log 10 \overset{?}{=} 1 \quad \text{Use the product rule of logarithms:}$$
$$\qquad\qquad\qquad \log 5 + \log 2 = \log (5 \cdot 2) = \log 10.$$
$$1 = 1 \quad \text{Simplify: } \log 10 = 1.$$

Since 5 satisfies the equation, it is the solution.

Answer 2

! COMMENT Examples 6 and 7 illustrate that we must check the solutions of a logarithmic equation.

CALCULATOR SNAPSHOT   **Solving logarithmic equations graphically**

To use a graphing calculator to approximate the solutions of the logarithmic equation $\log x + \log (x - 3) = 1$ (see Example 7), we can subtract 1 from both sides of the equation to get

$$\log x + \log (x - 3) - 1 = 0$$

and graph the corresponding function

$$y = \log x + \log (x - 3) - 1$$

If we use window settings of $[0, 20]$ for x and $[-2, 2]$ for y, we obtain the graph shown in Figure 9-36(a). Since the solution of the equation is the x-coordinate of the x-intercept, we can find the solution using the ZERO feature. The solution is $x = 5$.

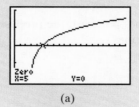 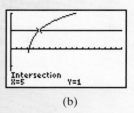

(a)   (b)

FIGURE 9-36

We can also solve $\log x + \log (x - 3) = 1$ using the INTERSECT feature. After graphing $Y_1 = \log x + \log (x - 3)$ and $Y_2 = 1$, we select INTERSECT, which approximates the coordinates of the point of intersection of the two graphs. From the display shown in Figure 9-36(b), we can conclude that the solution is $x = 5$. Verify this by checking.

INTERMEDIATE
Algebra *f(x)* **Now**™
Self Check 8

Solve:
$$\frac{\log_3 (5x + 6)}{\log_3 x} = 2$$

EXAMPLE 8   Solve: $\dfrac{\log_2 (5x - 6)}{\log_2 x} = 2$.

Solution   We can multiply both sides of the equation by $\log_2 x$ to get

$$\log_2 (5x - 6) = 2 \log_2 x$$

and apply the power rule of logarithms to get

$$\log_2 (5x - 6) = \log_2 x^2$$

By property 8 of logarithms, $5x - 6 = x^2$, because they have equal logarithms. Thus,

$$5x - 6 = x^2$$
$$0 = x^2 - 5x + 6$$
$$0 = (x - 3)(x - 2)$$

$$x - 3 = 0 \quad \text{or} \quad x - 2 = 0$$
$$x = 3 \qquad\qquad x = 2$$

Answer 6

Verify that both 2 and 3 satisfy the equation.

Radioactive decay

Experiments have determined the time it takes for half of a sample of a given radioactive material to decompose. This time is a constant, called the material's **half-life.**

When living organisms die, the oxygen/carbon dioxide cycle common to all living things ceases, and carbon-14, a radioactive isotope with a half-life of 5,700 years, is no longer absorbed. By measuring the amount of carbon-14 present in an ancient object, archaeologists can estimate the object's age by using the radioactive decay formula.

> **Radioactive decay formula**
>
> If A is the amount of radioactive material present at time t, A_0 was the amount present at $t = 0$, and h is the material's half-life, then
>
> $$A = A_0 2^{-t/h}$$

EXAMPLE 9   Carbon-14 dating.   How old is a wooden statue that retains only one-third of its original carbon-14 content?

Solution   To find the time t when $A = \frac{1}{3}A_0$, we substitute $\frac{A_0}{3}$ for A and 5,700 for h in the radioactive decay formula and solve for t:

$$A = A_0 2^{-t/h}$$

$$\frac{A_0}{3} = A_0 2^{-t/5,700} \qquad \text{The half-life of carbon-14 is 5,700 years.}$$

$$1 = 3(2^{-t/5,700}) \qquad \text{Divide both sides by } A_0 \text{ and multiply both sides by 3.}$$

$$\log 1 = \log 3(2^{-t/5,700}) \qquad \text{Take the common logarithm of each side.}$$

$$0 = \log 3 + \log 2^{-t/5,700} \qquad \log 1 = 0, \text{ and use the product rule of logarithms.}$$

$$-\log 3 = -\frac{t}{5,700}\log 2 \qquad \text{Subtract log 3 from both sides and use the power rule of logarithms.}$$

$$5,700\left(\frac{\log 3}{\log 2}\right) = t \qquad \text{Multiply both sides by } -\frac{5,700}{\log 2}.$$

$$t \approx 9,034.286254 \qquad \text{Use a calculator.}$$

The statue is approximately 9,000 years old.

Self Check 9
In Example 9, how old is a statue that retains 25% of its original carbon-14 content?

Answer about 11,400 years

Population growth

Recall that when there is sufficient food and space, populations of living organisms tend to increase exponentially according to the Malthusian growth model.

> **Malthusian growth model**
>
> If P is the population at some time t, P_0 is the initial population at $t = 0$, and k depends on the rate of growth, then
>
> $$P = P_0 e^{kt}$$

Self Check 10
In Example 10, how long will it take the population to reach 20,000?

EXAMPLE 10 Population growth.   The bacteria in a laboratory culture increased from an initial population of 500 to 1,500 in 3 hours. How long will it take for the population to reach 10,000?

Solution   We substitute 500 for P_0, 1,500 for P, and 3 for t and simplify to find k:

$$P = P_0 e^{kt}$$

$$1{,}500 = 500(e^{k3})$$ Substitute 1,500 for P, 500 for P_0, and 3 for t.

$$3 = e^{3k}$$ Divide both sides by 500.

$$3k = \ln 3$$ Change the equation from exponential to logarithmic form.

$$k = \frac{\ln 3}{3}$$ Divide both sides by 3.

To find when the population will reach 10,000, we substitute 10,000 for P, 500 for P_0, and $\frac{\ln 3}{3}$ for k in the equation $P = P_0 e^{kt}$ and solve for t:

$$P = P_0 e^{kt}$$

$$10{,}000 = 500 e^{[(\ln 3)/3]t}$$

$$20 = e^{[(\ln 3)/3]t}$$ Divide both sides by 500.

$$\left(\frac{\ln 3}{3}\right) t = \ln 20$$ Change the equation to logarithmic form.

$$t = \frac{3 \ln 20}{\ln 3}$$ Multiply both sides by $\frac{3}{\ln 3}$.

$$\approx 8.180499084$$ Use a calculator.

Answer about 10 hours

The culture will reach 10,000 bacteria in about 8 hours.

INTERMEDIATE
Algebra $f(x)$ Now™

EXAMPLE 11 Generation time.   If a medium is inoculated with a bacterial culture that contains 1,000 cells per milliliter, how many generations will pass by the time the culture has grown to a population of 1 million cells per milliliter?

Solution   During bacterial reproduction, the time required for a population to double is called the *generation time*. If b bacteria are introduced into a medium, then after the generation time of the organism has elapsed, there are $2b$ cells. After another generation, there are $2(2b)$ or $4b$ cells, and so on. After n generations, the number of cells present will be

(1) $B = b \cdot 2^n$

To find the number of generations that have passed while the population grows from b bacteria to B bacteria, we solve Equation 1 for n. To do this, we can take the common logarithm or the natural logarithm of each side.

$$\ln B = \ln (b \cdot 2^n)$$ Take the natural logarithm of each side.

$$\ln B = \ln b + n \ln 2$$ Apply the product and power rules of logarithms.

$$\ln B - \ln b = n \ln 2$$ Subtract $\ln b$ from both sides.

$$n = \frac{1}{\ln 2}(\ln B - \ln b)$$ Multiply both sides by $\frac{1}{\ln 2}$.

(2) $$n = \frac{1}{\ln 2}\left(\ln \frac{B}{b}\right)$$ Use the quotient rule of logarithms.

Equation 2 is a formula that gives the number of generations that will pass as the population grows from b bacteria to B bacteria.

To find the number of generations that have passed while a population of 1,000 cells per milliliter has grown to a population of 1 million cells per milliliter, we substitute 1,000 for b and 1,000,000 for B in Equation 2 and solve for n.

$$n = \frac{1}{\ln 2} \ln \frac{1,000,000}{1,000}$$

$$= \frac{1}{\ln 2} \ln 1,000 \qquad \text{Perform the division.}$$

$$n \approx 9.965784285 \qquad \text{Use a calculator.}$$

Approximately 10 generations will have passed.

Section 9.8   STUDY SET

VOCABULARY   *Fill in the blanks.*

1. An equation with a variable in its exponent, such as $3^{2x} = 8$, is called a(n) _____ equation.

2. An equation with a logarithmic expression that contains a variable, such as $\log_5 (2x - 3) = \log_5 (x + 4)$, is a(n) _____ equation.

CONCEPTS   *Fill in the blanks.*

3. The formula for radioactive decay is $A =$ ▮.

4. The formula for population growth is $P =$ ▮.

5. Use the graphs in the illustration to estimate the solution of $2^x = 3^{-x+3}$.

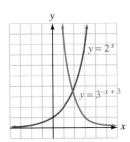

6. Use the graphs in the illustration to estimate the solution of $3 \log (x - 1) = 2 \log x$.

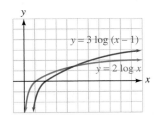

7. Fill in the blanks to make the statements true. To solve $5^x = 21$, we can take the _____ of each side of the equation to get

$$\log 5^x = \log 21$$

The power rule for logarithms then provides a way of moving the variable x from its position as an _____ to a position as a coefficient.

8. Use a calculator to determine whether $x \approx 2.5646$ is a solution of $2^{2x+1} = 70$.

9. Find $\dfrac{\log 8}{\log 5}$. Round to four decimal places.

10. Find $\dfrac{2 \ln 12}{\ln 9}$. Round to four decimal places.

11. Simplify: $\ln e$.

12. Does $\dfrac{\log 7}{\log 3} = \log 7 - \log 3$?

13. Write the corresponding base-10 exponential equation for $\log (x + 1) = 2$.

14. Write the corresponding base-e exponential equation for $\ln (x + 1) = 2$.

15. Solve each equation. Round to the nearest hundredth.

    a. $x^2 = 12$ **b.** $2^x = 12$

16. Solve each equation.

    a. $\log (x - 1) = 3$ **b.** $\log (x - 1) = \log 3$

17. Perform a check to see whether -4 is a solution of $\log_5 (x + 3) = \frac{1}{5}$.

18. Perform a check to see whether -2 is a solution of $5^{2x+3} = \frac{1}{5}$.

NOTATION *Complete each solution to solve the equation.*

19. $2^x = 7$

        $\boxed{}\,2^x = \log 7$

        $x\,\boxed{} = \log 7$

        $x = \dfrac{\log 7}{\boxed{}}$

20. $\log_2 (2x - 3) = \log_2 (x + 4)$

      $\boxed{} = x + 4$

      $x = \boxed{}$

PRACTICE *Solve each exponential equation. Give answers to four decimal places when necessary.*

21. $4^x = 5$

22. $7^x = 12$

23. $13^{x-1} = 2$

24. $5^{x+1} = 3$

25. $2^{x+1} = 3^x$

26. $5^{x-3} = 3^{2x}$

27. $2^x = 3^x$

28. $3^{2x} = 4^x$

29. $7^{x^2} = 10$

30. $8^{x^2} = 11$

31. $8^{x^2} = 9^x$

32. $5^{x^2} = 2^{5x}$

33. $e^{3x} = 9$

34. $e^{4x} = 60$

35. $e^{-0.2t} = 14.2$

36. $e^{0.3t} = 9.1$

37. $2^{x-2} = 64$

38. $3^{-3x+1} = 243$

39. $5^{4x} = \dfrac{1}{125}$

40. $8^{-x+1} = \dfrac{1}{64}$

41. $2^{x^2-2x} = 8$

42. $3^{x^2-3x} = 81$

43. $3^{x^2+4x} = \dfrac{1}{81}$

44. $7^{x^2+3x} = \dfrac{1}{49}$

Use a graphing calculator to solve each equation. Give all answers to the nearest tenth.

45. $2^{x+1} = 7$

46. $3^{x-1} = 2^x$

47. $4(2^{x^2}) = 8^{3x}$

48. $3^x - 10 = 3^{-x}$

Solve each logarithmic equation.

49. $\log (x + 2) = 4$

50. $\log 5x = 4$

51. $\log (7 - x) = 2$

52. $\log (2 - x) = 3$

53. $\ln x = 1$

54. $\ln x = 5$

55. $\ln (x + 1) = 3$

56. $\ln 2x = 5$

57. $\log 2x = \log 4$

58. $\log 3x = \log 9$

59. $\ln (3x + 1) = \ln (x + 7)$

60. $\ln (x^2 + 4x) = \ln (x^2 + 16)$

61. $\log (3 - 2x) - \log (x + 24) = 0$

62. $\log (3x + 5) - \log (2x + 6) = 0$

63. $\log \dfrac{4x + 1}{2x + 9} = 0$

64. $\log \dfrac{2 - 5x}{2(x + 8)} = 0$

65. $\log x^2 = 2$

66. $\log x^3 = 3$

67. $\log x + \log (x - 48) = 2$

68. $\log x + \log (x + 9) = 1$

69. $\log x + \log (x - 15) = 2$

70. $\log x + \log (x + 21) = 2$

71. $\log (x + 90) = 3 - \log x$

72. $\log (x - 90) = 3 - \log x$

73. $\log (x - 6) - \log (x - 2) = \log \dfrac{5}{x}$

74. $\log (3 - 2x) - \log (x + 9) = 0$

75. $\dfrac{\log (3x - 4)}{\log x} = 2$

76. $\dfrac{\log (8x - 7)}{\log x} = 2$

77. $\dfrac{\log (5x + 6)}{2} = \log x$

78. $\dfrac{1}{2} \log (4x + 5) = \log x$

79. $\log_3 x = \log_3 \left(\dfrac{1}{x}\right) + 4$

80. $\log_5 (7 + x) + \log_5 (8 - x) - \log_5 2 = 2$

81. $2 \log_2 x = 3 + \log_2 (x - 2)$

82. $2 \log_3 x - \log_3 (x - 4) = 2 + \log_3 2$

83. $\log (7y + 1) = 2 \log (y + 3) - \log 2$

84. $2 \log (y + 2) = \log (y + 2) - \log 12$

Use a graphing calculator to solve each equation. If an answer is not exact, round to the nearest tenth.

85. $\log x + \log (x - 15) = 2$

86. $\log x + \log (x + 3) = 1$

87. $\ln (2x + 5) - \ln 3 = \ln (x - 1)$

88. $2 \log (x^2 + 4x) = 1$

APPLICATIONS

89. TRITIUM DECAY The half-life of tritium is 12.4 years. How long will it take for 25% of a sample of tritium to decompose?

90. RADIOACTIVE DECAY In two years, 20% of a radioactive element decays. Find its half-life.

91. THORIUM DECAY An isotope of thorium, ^{227}Th, has a half-life of 18.4 days. How long will it take for 80% of the sample to decompose?

92. LEAD DECAY An isotope of lead, ^{201}Pb, has a half-life of 8.4 hours. How many hours ago was there 30% more of the substance?

93. CARBON-14 DATING A bone fragment analyzed by archaeologists contains 60% of the carbon-14 that it is assumed to have had initially. How old is it?

94. CARBON-14 DATING Only 10% of the carbon-14 in a small wooden bowl remains. How old is the bowl?

95. COMPOUND INTEREST If $500 is deposited in an account paying 8.5% annual interest, compounded semiannually, how long will it take for the account to increase to $800?

96. CONTINUOUS COMPOUND INTEREST In Exercise 95, how long will it take if the interest is compounded continuously?

97. COMPOUND INTEREST If $1,300 is deposited in a savings account paying 9% interest, compounded quarterly, how long will it take the account to increase to $2,100?

98. COMPOUND INTEREST A sum of $5,000 deposited in an account grows to $7,000 in 5 years. Assuming annual compounding, what interest rate is being paid?

99. RULE OF SEVENTY A rule of thumb for finding how long it takes an investment earning continuously compounded interest to double is called the **rule of seventy.** To apply the rule, divide 70 by the interest rate written as a percent. At 5%, doubling requires $\frac{70}{5} = 14$ years to double an investment. At 7%, it takes $\frac{70}{7} = 10$ years. Explain why this formula works.

100. BACTERIAL GROWTH A bacterial culture grows according to the formula

$$P = P_0 a^r$$

If it takes 5 days for the culture to triple in size, how long will it take to double in size?

101. RODENT CONTROL The rodent population in a city is currently estimated at 30,000. If it is expected to double every 5 years, when will the population reach 1 million?

102. POPULATION GROWTH The population of a city is expected to triple every 15 years. When can the city planners expect the present population of 140 persons to double?

103. BACTERIAL CULTURES A bacterial culture doubles in size every 24 hours. By how much will it have increased in 36 hours?

104. OCEANOGRAPHY The intensity I of a light a distance x meters beneath the surface of a lake decreases exponentially. From the illustration below, find the depth at which the intensity will be 20%.

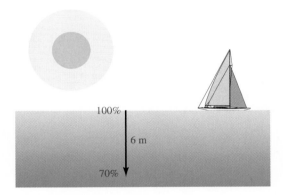

105. MEDICINE If a medium is inoculated with a bacterial culture containing 500 cells per milliliter, how many generations will have passed by the time the culture contains 5×10^6 cells per milliliter?

106. MEDICINE If a medium is inoculated with a bacterial culture containing 800 cells per milliliter, how many generations will have passed by the time the culture contains 6×10^7 cells per milliliter?

107. NEWTON'S LAW OF COOLING Water initially at 100°C is left to cool in a room at temperature 60°C. After 3 minutes, the water temperature is 90°. The water temperature T is a function of time t given by

$$T = 60 + 40e^{kt}$$

Find k.

108. NEWTON'S LAW OF COOLING Refer to Exercise 107 and find the time for the water temperature to reach 70°C.

WRITING

109. Explain how to solve $2^{x+1} = 31$.
110. Explain how to solve $2^{x+1} = 32$.

REVIEW   *Solve each equation.*

111. $5x^2 - 25x = 0$ **112.** $4y^2 - 25 = 0$

113. $3p^2 + 10p = 8$ **114.** $4t^2 + 1 = -6t$

115. Find the length of leg AC in the triangle.

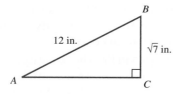

116. MEDICATIONS The amount of medicine a patient should take is often proportional to his or her weight. If a patient weighing 83 kilograms needs 150 milligrams of medicine, how much will be needed by a person weighing 99.6 kilograms?

KEY CONCEPT

Inverse Functions

One-to-One Functions

A function is *one-to-one* if each input value x in the domain determines a different output value y in the range. In Exercises 1–3, determine whether the function is one-to-one.

1. $f(x) = x^2$

2. $f(x) = |x|$

3.

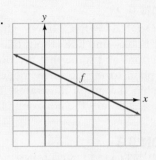

The Inverse of a Function

If a function is one-to-one, its inverse is a function. To find the inverse of a function, interchange x and y and solve for y.

4. Find $f^{-1}(x)$ if $f(x) = -2x - 1$.

5. Given the table of values for a one-to-one function f, complete the table of values for f^{-1}.

x	-2	1	3
$f(x)$	4	-2	-6

x	4	-2	-6
$f^{-1}(x)$			

Exponential and Logarithmic Functions

The exponential functions $f(x) = b^x$ and $g(x) = \log_b x$, where $b > 0$ and $b \neq 1$, are inverse functions.

6. Write the exponential statement $10^3 = 1,000$ in logarithmic form.

7. Write the logarithmic statement $\log_2 \frac{1}{8} = -3$ in exponential form.

8. If $\log_4 x = \frac{1}{2}$, what is x?

9. If $\log_b \frac{9}{4} = 2$, what is b?

The Natural Exponential and Natural Logarithmic Functions

A special exponential function that is used in many real-life applications involving growth and decay is the base-e exponential function, $f(x) = e^x$. Its inverse is the natural logarithm function $g(x) = \ln x$.

10. What is an approximate value of e?

11. What is the base of the logarithmic function $f(x) = \ln x$?

12. Use a calculator to find x: $\ln x = -0.28$.

13. Graph $f(x) = e^x$ and $g(x) = \ln x$ on the coordinate system. Label the axis of symmetry.

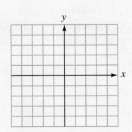

ACCENT ON TEAMWORK

SECTION 9.1

COMPOSITION OF FUNCTIONS Consider the functions $f(x) = x^2$, $g(x) = 2x + 1$, and $h(x) = 1 - x$. Determine whether the composition of functions f, g, and h is associative. That is, is the following true?

$$[f \circ (g \circ h)](x) \overset{?}{=} [(f \circ g) \circ h](x)$$

SECTION 9.2

ONE-TO-ONE FUNCTIONS In newspapers, magazines, or books, find line graphs that are graphs of one-to-one functions and some that are not one-to-one. In each case, explain to the other members of your group what relationship the graph illustrates. Then tell whether the graph passes or fails the horizontal line test.

SECTION 9.3

EXPONENTIAL FUNCTIONS On a piece of poster board, draw a rectangular coordinate system made up of a grid of 1-inch squares. Then graph each of the following exponential functions. How are the graphs similar? How are they different? For positive values of x, as the base increases, what happens to the steepness of the graph?

$$f(x) = 2^x \qquad f(x) = 3^x \qquad f(x) = 4^x$$
$$f(x) = 5^x \qquad f(x) = 6^x \qquad f(x) = 7^x$$

SECTION 9.4

GROWTH Make two copies of each of the graph shapes shown in the illustration below.

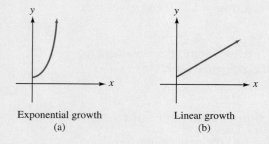

Exponential growth
(a)

Linear growth
(b)

Think of two examples of quantities that, in general, have grown exponentially over the years. Label the horizontal and vertical axes of the graphs with the proper titles. (You do not have to scale the axes in terms of units.) Do the same for two quantities that, in general, have grown

linearly over the years. Share your observations with the other members of your group. See if they agree with your models.

SECTION 9.5

LOGARITHMIC FUNCTIONS On a piece of poster board, draw a rectangular coordinate system made up of a grid of 1-inch squares. Then graph each of the following logarithmic functions. How are the graphs similar? How are they different? For positive values of x, as the base increases, what happens to the steepness of the graph?

$$f(x) = \log_2 x \qquad f(x) = \log_3 x \qquad f(x) = \log_4 x$$
$$f(x) = \log_5 x \qquad f(x) = \log_6 x \qquad f(x) = \log_7 x$$

SECTION 9.6

THE NUMBER e The value of e can be calculated to any degree of accuracy by adding the terms of the following pattern:

$$1, \; 1, \; \frac{1}{2}, \; \frac{1}{2 \cdot 3}, \; \frac{1}{2 \cdot 3 \cdot 4}, \; \frac{1}{2 \cdot 3 \cdot 4 \cdot 5}, \; \cdots$$

The more terms that are added, the closer the sum will be to e. Write each of the first six terms in simplified form and then add them. To how many decimal places is the sum accurate?

SECTION 9.7

The properties of logarithms discussed in Section 9.7 hold for logarithms of any base. Use these properties to simplify each of the following natural logarithmic expressions.

a. $\ln x^y + \ln x^z$ **b.** $\ln x^y - \ln x^z$

c. $\ln e^{10}$ **d.** $\ln \dfrac{x}{y} + \ln \dfrac{y}{x}$

SECTION 9.8

To solve exponential equations involving e, it is convenient to take the natural logarithm (instead of the common logarithm) of both sides of the equation. Use this technique to solve each of the following equations. Give all answers to four decimal places.

a. $e^x = 10$ **b.** $e^{2x} = 15$

c. $e^{x+1} = 50.5$ **d.** $3e^{5x-2} = 3$

SECTION 9.1 — *Algebra and Composition of Functions*

CONCEPTS

Functions can be added, subtracted, multiplied, and divided:

$(f + g)(x) = f(x) + g(x)$

$(f - g)(x) = f(x) - g(x)$

$(f \cdot g)(x) = f(x)g(x)$

$(f/g)(x) = \dfrac{f(x)}{g(x)} \quad (g(x) \neq 0)$

Composition of functions:

$(f \circ g)(x) = f(g(x))$

REVIEW EXERCISES

Let $f(x) = 2x$ and $g(x) = x + 1$. Find each function or value.

1. $f + g$　　　　　　　　　**2.** $f - g$

3. $f \cdot g$　　　　　　　　　**4.** f/g

5. $(f \circ g)(1)$　　　　　　　**6.** $g(f(1))$

7. $(f \circ g)(x)$　　　　　　　**8.** $(g \circ f)(x)$

SECTION 9.2 — *Inverses of Functions*

A function is *one-to-one* if each input value x in the domain determines a different output value y in the range.

Determine whether the function is one-to-one.

9. $f(x) = x^2 + 3$　　　　　　**10.** $f(x) = x + 3$

Horizontal line test: If every horizontal line that intersects the graph of a function does so only once, the function is one-to-one.

Use the horizontal line test to decide whether the function is one-to-one.

11.

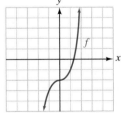

12.

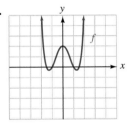

The graph of a function and its inverse are symmetric about the line $y = x$.

13. Given the graph of function f shown, graph f^{-1} on the same coordinate axes. Label the axis of symmetry.

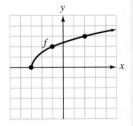

To *find the inverse of a function,* replace $f(x)$ with y, interchange the variables x and y, and solve for y.

Find the inverse of each function.

14. $f(x) = 6x - 3$　　　　　　**15.** $f(x) = 4x + 5$

16. $f(x) = x^3$　　　　　　　**17.** $f(x) = 2x^2 - 1 \quad (x \geq 0)$

Exponential Functions

An *exponential function* with base b is defined by the equation

$$f(x) = b^x \quad \text{or} \quad y = b^x$$

where $b > 0, b \neq 1$

If $b > 1$, then $f(x) = b^x$ is an *increasing function.*

If $0 < b < 1$, then $f(x) = b^x$ is a *decreasing function.*

Use properties of exponents to simplify each expression.

18. $5^{\sqrt{2}} \cdot 5^{\sqrt{2}}$

19. $(2^{\sqrt{5}})^{\sqrt{2}}$

Graph each function. Then give the domain and the range.

20. $y = 3^x$

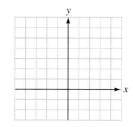

21. $f(x) = \left(\dfrac{1}{3}\right)^x$

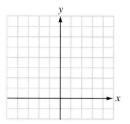

Graph each function by using a translation.

22. $h(x) = \left(\dfrac{1}{2}\right)^x - 2$

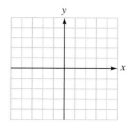

23. $g(x) = \left(\dfrac{1}{2}\right)^{x+2}$

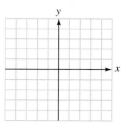

Exponential functions are suitable models for describing many situations involving the *growth* or *decay* of a quantity.

24. COAL PRODUCTION The table gives the number of tons of coal produced in the United States for the years 1800–1920. Graph the data on the illustration on the next page. What type of function does it appear could be used to model coal production over this period?

Year	Tons	Year	Tons
1800	108,000	1870	40,429,000
1810	178,000	1880	79,407,000
1820	881,000	1890	157,771,000
1830	1,334,000	1900	269,684,000
1840	2,474,000	1910	501,596,000
1850	8,356,000	1920	658,265,000
1860	20,041,000		

Based on information from *World Book Encyclopedia.*

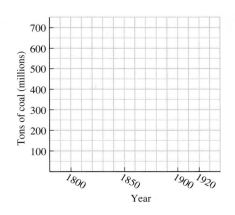

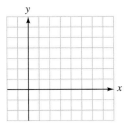

Compound interest: If $\$P$ is the deposit, and interest is paid k times a year at an annual rate r, the amount A in the account after t years is given by

$$A = P\left(1 + \frac{r}{k}\right)^{kt}$$

25. COMPOUND INTEREST How much will $\$10,500$ become if it earns 9% annual interest, compounded quarterly, for 60 years?

26. DEPRECIATION The value (in dollars) of a certain model car is given by the function $V(t) = 12,000\left(10^{-0.155t}\right)$, where t is the number of years from the present. What will the value of the car be in 5 years?

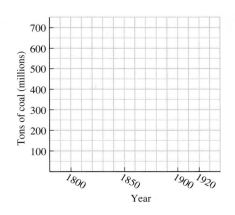

| SECTION 9.4 | *Base-e Exponential Functions* |

The function defined by $f(x) = e^x$ is the *natural exponential function* where $e = 2.718281828459\ldots.$

If a quantity increases or decreases at an annual rate r, *compounded continuously*, then the amount A after t years is given by

$$A = Pe^{rt}$$

27. MORTGAGE RATES The average annual interest rate on a 30-year fixed-rate home mortgage for the years 1980–1996 can be approximated by the function $r(t) = 13.9e^{-0.035t}$, where t is the number of years since 1980. To the nearest hundredth of a percent, what does this model predict was the 30-year fixed rate in 1995?

28. INTEREST COMPOUNDED CONTINUOUSLY If $\$10,500$ accumulates interest at an annual rate of 9%, compounded continuously, how much will be in the account in 60 years?

Graph each function and give the domain and the range.

29. $f(x) = e^x + 1$

30. $y = e^{x-3}$

Malthusian population growth is modeled by the formula

$$P = P_0 e^{kt}$$

31. POPULATION GROWTH As of July 2003, the population of the United States was estimated to be 290,340,000, with an annual growth rate of 0.92%. If the growth rate remains the same, how large will the population be in 50 years?

32. MEDICAL TESTS A radioactive dye is injected into a patient as part of a test to detect heart disease. The amount of dye remaining in his bloodstream t hours after the injection is given by the function $f(t) = 10e^{-0.27t}$. How can you tell from the function that the amount of dye in the bloodstream is decreasing?

If $b > 0$ and $b \neq 1$, then the *logarithmic function with base b* is defined by

$$y = \log_b x \qquad \text{or}$$
$$f(x) = \log_b x$$

$y = \log_b x$ means $x = b^y$.

$\log_b x$ is the exponent to which b is raised to get x.

$$b^{\log_b x} = x$$

33. Give the domain and range of the logarithmic function $f(x) = \log x$.

34. Write the statement $\log_4 64 = 3$ in exponential form.

35. Write the statement $7^{-1} = \frac{1}{7}$ in logarithmic form.

Find each value, if possible.

36. $\log_3 9$

37. $\log_9 \dfrac{1}{81}$

38. $\log_8 1$

39. $\log_5 (-25)$

40. $\log_6 \sqrt{6}$

41. $\log 1{,}000$

Find the value of x in each equation.

42. $\log_2 x = 5$

43. $\log_3 x = -4$

44. $\log_x 16 = 2$

45. $\log_x \dfrac{1}{100} = -2$

46. $\log_9 3 = x$

47. $\log_{27} x = \dfrac{2}{3}$

Use a calculator to find the value of x to four decimal places.

48. $\log 4.51 = x$

49. $\log x = 1.43$

If $b > 1$, then $f(x) = \log_b x$ is an *increasing function*. If $0 < b < 1$, then $f(x) = \log_b x$ is a *decreasing function*.

The exponential function $f(x) = b^x$ and the logarithmic function $f(x) = \log_b x$ are inverses of each other.

Graph each pair of equations on one set of coordinate axes.

50. $y = 4^x$ and $y = \log_4 x$

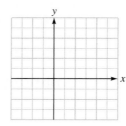

51. $f(x) = \left(\dfrac{1}{3}\right)^x$ and $g(x) = \log_{1/3} x$

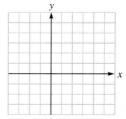

Graph each function.

52. $f(x) = \log (x - 2)$

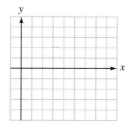

53. $f(x) = 3 + \log x$

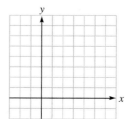

Decibel voltage gain:

$$\text{db gain} = 20 \log \dfrac{E_O}{E_I}$$

54. ELECTRICAL ENGINEERING An amplifier has an output of 18 volts when the input is 0.04 volt. Find the db gain.

The Richter scale:

$$R = \log \dfrac{A}{P}$$

55. EARTHQUAKES An earthquake had a period of 0.3 second and an amplitude of 7,500 micrometers. Find its measure on the Richter scale.

Base-e Logarithms

Natural logarithms:

$\ln x$ means $\log_e x$

Use a calculator to find each value to four decimal places.

56. $\ln 452$

57. $\ln 0.85$

Use a calculator to find the value of x to four decimal places.

58. $\ln x = 2.336$

59. $\ln x = -8.8$

60. What function is the inverse of $f(x) = \ln x$?

Graph each function.

61. $f(x) = 1 + \ln x$

62. $y = \ln (x + 1)$

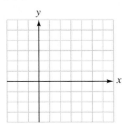

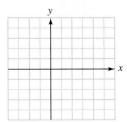

Population doubling time:

$t = \dfrac{\ln 2}{r}$

63. POPULATION GROWTH How long will it take the population of Mexico to double if the growth rate is currently 1.43% per year?

64. BOTANY The height (in inches) of a certain plant is approximated by the function $H(a) = 13 + 20.03 \ln a$, where a is its age in years. How tall will it be when it is 19 years old?

Properties of Logarithms

Properties of logarithms: If b is a positive number and $b \neq 1$,

1. $\log_b 1 = 0$ **2.** $\log_b b = 1$

3. $\log_b b^x = x$ **4.** $b^{\log_b x} = x$

5. $\log_b MN = \log_b M + \log_b N$

6. $\log_b \dfrac{M}{N} = \log_b M - \log_b N$

7. $\log_b M^p = p \log_b M$

8. If $\log_b x = \log_b y$, then $x = y$.

Simplify each expression.

65. $\log_7 1$

66. $\log_7 7$

67. $\log_7 7^3$

68. $7^{\log_7 4}$

69. $\ln e^4$

70. $\ln e$

Use the properties of logarithms to rewrite each expression.

71. $\log_3 27x$

72. $\log \dfrac{100}{x}$

73. $\log_5 \sqrt{27}$

74. $\log 10ab$

Write each expression in terms of the logarithms of x, y, and z.

75. $\log_b \dfrac{x^2 y^3}{z}$

76. $\ln \sqrt{\dfrac{x}{yz^2}}$

Write each expression as the logarithm of one quantity.

77. $3 \log_2 x - 5 \log_2 y + 7 \log_2 z$

78. $-3 \log_b y - 7 \log_b z + \dfrac{1}{2} \log_b (x + 2)$

Change-of-base formula:

$$\log_b x = \frac{\log_a x}{\log_a b}$$

Assume that $\log_b 5 = 1.1609$ and $\log_b 8 = 1.50000$. Use these values and the properties of logarithms to find each value to four decimal places.

79. $\log_b 40$ **80.** $\log_b 64$

81. Find $\log_5 17$ to four decimal places.

pH scale:

$$pH = -\log [H^+]$$

82. pH OF GRAPEFRUIT The pH of grapefruit juice is about 3.1 Find its hydrogen ion concentration.

SECTION 9.8   *Exponential and Logarithmic Equations*

An *exponential equation* is an equation that contains a variable in one of its exponents.

Solve each equation for x. Give answers to four decimal places when necessary.

83. $3^x = 7$ **84.** $5^{x+2} = 625$

85. $2^x = 3^{x-1}$ **86.** $2^{x^2+4x} = \dfrac{1}{8}$

87. $e^x = 7$ **88.** $e^{-0.4t} = 25$

A *logarithmic equation* is an equation with a logarithmic expression that contains a variable.

Solve each equation for x.

89. $\log (x - 4) = 2$ **90.** $\ln (2x - 3) = \ln 15$

91. $\log x + \log (29 - x) = 2$ **92.** $\log_2 x + \log_2 (x - 2) = 3$

93. $\dfrac{\log (7x - 12)}{\log x} = 2$ **94.** $\log_2 (x + 2) + \log_2 (x - 1) = 2$

95. $\log x + \log (x - 5) = \log 6$ **96.** $\log 3 - \log (x - 1) = -1$

Carbon dating:

$$A = A_0 2^{-t/h}$$

97. CARBON-14 DATING A wooden statue found in Egypt has a carbon-14 content that is two-thirds of that found in living wood. If the half-life of carbon-14 is 5,700 years, how old is the statue?

98. ANTS The number of ants in a colony is estimated to be 800. If the ant population is expected to triple every 14 days, how long will it take for the population to reach one million?

99. The approximate coordinates of the points of intersection of the graphs of $f(x) = \log x$ and $g(x) = 1 - \log (7 - x)$ are shown in parts a and b of the illustration below. Estimate the solutions of the logarithmic equation $\log x = 1 - \log (7 - x)$. Then check your answer.

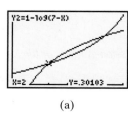   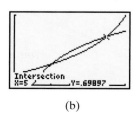

(a)   (b)

Let $f(x) = 4x$ and $g(x) = x - 1$. Find each function or value.

1. $g + f$

2. $g \cdot f$

3. Find $(g \circ f)(1)$.

4. Find $f(g(x))$.

Find the inverse of each function.

5. $f(x) = -\dfrac{3}{2}x + 6$

6. $f(x) = 3x^2 + 4 \quad (x \geq 0)$

Graph each function.

7. $f(x) = 2^x + 1$

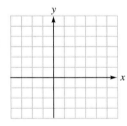

8. $y = 3^{-x}$

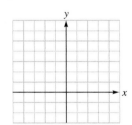

9. RADIOACTIVE DECAY A radioactive material decays according to the formula $A = A_0(2)^{-t}$. How much of a 3-gram sample will be left in 6 years?

10. COMPOUND INTEREST An initial deposit of $1,000 earns 6% interest, compounded twice a year. How much will be in the account in one year?

11. Graph: $f(x) = e^x$.

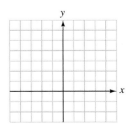

12. CONTINUOUS COMPOUNDING An account contains $2,000 and has been earning 8% interest, compounded continuously. How much will be in the account in 10 years?

Find x.

13. $\log_4 16 = x$

14. $\log_x 81 = 4$

15. $\log_3 x = -3$

16. $\ln x = 1$

17. Write the statement $\log_6 \frac{1}{36} = -2$ in exponential form:

18. Give the domain and range of the function $f(x) = \log x$.

Graph each function.

19. $f(x) = -\log_3 x$

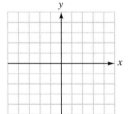

20. $f(x) = \ln x$

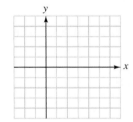

21. Write the expression $\log a^2bc^3$ in terms of the logarithms of a, b, and c.

22. Write the expression $\frac{1}{2} \ln (a + 2) + \ln b - 3 \ln c$ as a logarithm of a single quantity.

23. Use the change-of-base formula to find $\log_7 3$ to four decimal places.

24. What function is the inverse of $y = 10^x$?

25. pH Find the pH of a solution with a hydrogen ion concentration of 3.7×10^{-7}.
(*Hint:* pH $= -\log [\mathrm{H}^+]$.)

26. ELECTRONICS Find the db gain of an amplifier when $E_O = 60$ volts and $E_I = 0.3$ volt.
Hint: db gain $= 20 \log \left(\dfrac{E_O}{E_I}\right)$.

Solve each equation. Round to four decimal places when necessary.

27. $5^x = 3$

28. $3^{x-1} = 27$

29. $\ln(5x + 2) = \ln(2x + 5)$

30. $\log x + \log(x - 9) = 1$

31. The illustration shows the graphs of $y = \frac{1}{2}\ln(x - 1)$ and $y = \ln 2$ and the approximate coordinates of their point of intersection. Estimate the solution of the logarithmic equation $\frac{1}{2}\ln(x - 1) = \ln 2$.

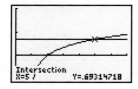

Intersection
X=5 Y=.69314718

32. Show a check of your answer to Problem 31.

33. Give an example of a situation studied in this chapter that is modeled by a function with a graph that has the shape shown. Label the axes. You do not have to scale the axes.

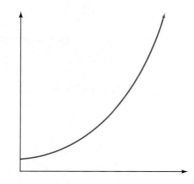

34. Consider the graph shown below.

   a. Is it the graph of a function?

   b. Is its inverse a function?

   c. What is $f^{-1}(260)$? What information about temperature and tire tread does it give?

Relationship between car speed and tire temperature

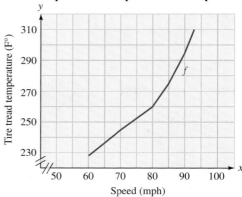

35. POPULATION GROWTH As of July 2003, the population of India was estimated to be 1,050,000,000, with an annual growth rate of 1.47%. If the growth rate remains the same, how large will the population be in 30 years?

36. INSECTS The number of insects attracted to a bright light is currently 5. If the number is expected to quadruple every 6 minutes, how long will it take for the number to reach 500?

CHAPTERS 1–9 CUMULATIVE REVIEW EXERCISES

1. Write the formula associated with each concept.

  a. Perimeter of a rectangle

  b. Area of a circle

  c. Area of a triangle

  d. Volume of a cube

  e. Simple interest

  f. Distance (uniform motion)

  g. Midpoint of a line segment

  h. Slope–intercept form of the equation of a line

  i. Point–slope form of the equation of a line

  j. Slope of a line

  k. Pythagorean theorem

  l. Distance between two points

  m. Direct variation

  n. Inverse variation

  o. Quadratic formula

  p. Exponential growth

  q. Change-of-base formula for logarithms

2. Fill in the blanks to complete the rules for exponents.

  a. $x^1 =$

  b. $x^m x^n =$

  c. $(x^m)^n =$

  d. $(xy)^n =$

  e. $x^0 =$

  f. $\left(\dfrac{x}{y}\right)^n =$

  g. $\dfrac{x^m}{x^n} =$

  h. $x^{-n} =$

  i. $\dfrac{1}{x^{-n}} =$

  j. $\left(\dfrac{x}{y}\right)^{-n} =$

  k. $x^{1/n} =$

  l. $x^{m/n} =$

3. Complete each factorization formula.

  a. $x^2 - y^2 =$

  b. $x^3 - y^3 =$

  c. $x^3 + y^3 =$

4. Complete each special product formula.

  a. $(x + y)^2 =$

  b. $(x - y)^2 =$

  c. $(x + y)(x - y) =$

5. Complete each property of radicals.

  a. $\sqrt[n]{ab} =$

  b. $\sqrt[n]{\dfrac{a}{b}} =$ $(b \neq 0)$

6. Approximate each irrational number to the nearest hundredth.

  a. π **b.** $\sqrt{2}$ **c.** e

7. To graph $y \geq x$, we first graph the boundary line $y = x$ as in the illustration. Explain how we determine which side of the boundary to shade.

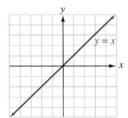

8. Complete each property of logarithms.

  a. $\log_b 1 =$

  b. $\log_b b =$

  c. $\log_b b^x =$

  d. $b^{\log_b x} =$

  e. $\log_b MN =$

  f. $\log_b \dfrac{M}{N} =$

  g. $\log_b M^p =$

9. Fill in the blanks to complete the fundamental property of fractions

  If a, b, and k represent real numbers, and $b \neq 0$ and $k \neq 0$, then

$$\frac{a}{b} = \frac{a \cdot }{b \cdot }$$

10. The following partial solutions show three important applications of the fundamental property of fractions. In each case, explain why it was used.

a. $\dfrac{5a}{24b} + \dfrac{11a}{18b^2} = \dfrac{5a \cdot 3b}{24b \cdot 3b} + \dfrac{11a \cdot 4}{18b^2 \cdot 4}$

b. $\dfrac{\sqrt{70}}{\sqrt{3}} = \dfrac{\sqrt{70} \cdot \sqrt{3}}{\sqrt{3} \cdot \sqrt{3}}$

c. $\dfrac{\dfrac{1}{x} + \dfrac{1}{y}}{\dfrac{1}{x} - \dfrac{1}{y}} = \dfrac{xy\left(\dfrac{1}{x} + \dfrac{1}{y}\right)}{xy\left(\dfrac{1}{x} - \dfrac{1}{y}\right)}$

11. The following partial solution shows an important application of the fundamental property of fractions. Explain why it was used and what the slashes and 1's mean.

$$\dfrac{6a^2 - 13a + 6}{3a^2 + a - 2} = \dfrac{\overset{1}{\cancel{(3a - 2)}}(2a - 3)}{\underset{1}{\cancel{(3a - 2)}}(a + 1)}$$

12. Fill in the blanks to complete the fundamental property of proportions.

If $\dfrac{a}{b} = \dfrac{c}{d}$, then $ad = \boxed{}$

13. What is a quadratic equation?

14. Consider the equation $(x + 1)(x - 7) = 0$. How is the zero-factor property used to solve this equation?

15. What is the first step used to solve the following equation?

$$\dfrac{x - 3}{x - 2} - \dfrac{1}{x} = \dfrac{x - 3}{x}$$

16. Which method would be the most efficient to solve the following system?

$$\begin{cases} 3x - 2y = -10 \\ 6x + 5y = 25 \end{cases}$$

Fill in the blanks to complete each definition.

17. For any real number x,

$$\begin{cases} \text{if } x \geq 0, \text{ then } |x| = \boxed{} \\ \text{if } x < 0, \text{ then } |x| = \boxed{} \end{cases}$$

18. The number b is a square root of a if $\boxed{} = a$.

19. If x can be any real number, then $\sqrt{x^2} = \boxed{}$.

20. If x is any real number, then $\sqrt[3]{x^3} = \boxed{}$.

21. If a and b are real numbers, $a - b = a \boxed{} (-b)$.

22. $i = \boxed{}$

Fill in the blanks

23. A function is a correspondence between a set of input values x (called the _____) and a set of output values y (called the _____), in which exactly _____ y-value in the range is assigned to each number x in the domain.

24. a. The solution set of a compound inequality containing the word *and* consists of all the numbers that make _____ inequalities true.

   b. The solution set of a compound inequality containing the word *or* consists of all the numbers that make _____ or the other or _____ inequalities true.

Conic Sections;
More Graphing

Getty Images /Image Bank

INTERMEDIATE
Algebra $f(x)$ **Now**™
Throughout the chapter, this icon
introduces resources on the Intermediate AlgebraNow Web site,
accessed through **http://1pass**
.thomson.com, that will

• Help you test your knowledge of
the material with a pre-test and
a post-test

• Provide a personalized learning
plan targeting areas you should
study

TLE The solar system consists of the sun, the nine planets, more than
130 satellites of the planets, and a large number of comets and asteroids.
The orbits of the planets are ellipses, though all except Mercury and Pluto are very nearly
circular. The orbit of a comet can be an ellipse or a hyperbola, depending on the total
energy of the comet. Circles, ellipses, parabolas, and hyperbolas belong to a family of
curves called *conic sections*. They are so named because they can be formed by the
intersection of a plane and a pair of right-circular cones.

To learn more about conic sections, visit *The Learning Equation* on the Internet at
http://tle.brookscole.com. (The log-in instructions are in the Preface.) For Chapter 10, the
online lesson is:

• *TLE* Lesson 15: Introduction to Conic Sections

Check Your Knowledge

1. A _____ is the set of all points in a plane that are a fixed distance from a point.

2. An _____ is the set of all points in a plane for which the sum of the distances from two fixed points is a constant.

3. A _____ is the set of all points in a plane for which the difference of the distances from two fixed points is a constant.

4. The curves formed by the intersection of a plane with an infinite right circular cone are called _____ _____.

5. Find the center and radius of the circle $(x + 2)^2 + (y + 1)^2 = 9$.

6. Find the center and radius of the circle $x^2 + y^2 - 6x + 2y = 6$.

Graph each equation.

7. $x^2 + y = 2$

8. $(x - 3)^2 + (y - 2)^2 = 4$

9. $x^2 - y^2 = 9$

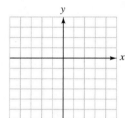 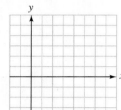 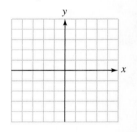

10. $\dfrac{(x - 1)^2}{4} + \dfrac{y^2}{9} = 1$

11. Write $4x^2 + y^2 + 16x - 2y = -13$ in standard form and graph the equation.

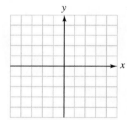 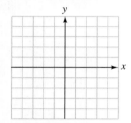

12. Find the vertex and the axis of symmetry of the graph of $x = -3y^2 - 12y - 13$.

13. Solve the system $\begin{cases} x^2 + y^2 = 9 \\ x - y = 3 \end{cases}$ by graphing.

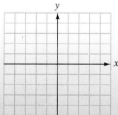

14. Solve the system $\begin{cases} x^2 + y^2 = 25 \\ 2x^2 - 3y^2 = 5 \end{cases}$.

15. The product of two numbers is 24 and the sum is 10. Use two variables to find the two numbers.

16. The area of a rectangle is 48 ft^2 and the perimeter is 28 ft. Use two variables to find the dimensions of the rectangle.

Study Skills Workshop
MASTERING THE FINAL EXAM

It can be a little overwhelming to think of studying for a comprehensive math final. Often a great deal of material has been covered in class and there usually are many topics that are somewhat hard to recall without review. The good thing about a comprehensive final, though, is that it gives you an opportunity to see the whole scope of the class, from beginning to end. It also provides an excellent start to your next math class. That is why it is so important to be organized and thorough in your final exam preparation. If you allow enough time and go about your studying in an organized, systematic way, you will feel more confident in your abilities as you approach the final. Be sure to set up a new calendar for finals week and incorporate study time for your finals for all of your classes. Allow yourself at least a week to prepare for your math final, even if you must start studying before classes actually end.

Review All Tests. The best way to start studying is to review all of your old tests. Look through each test carefully and make sure that you can do the problems that you did correctly on your tests as well as fix any errors that you made. Make a list of the concepts that your instructor tested and write down appropriate rules or formulas that would be good to memorize for your exam. Make another list of test questions that you couldn't answer correctly, even after your tried to find the correct solution.

Make a New Study Sheet (or Tape) for Your Final. Review your old study sheets (or tapes) for each test and incorporate them into one sheet, being sure to include any important rules or formulas that you found when you reviewed your old tests. Look at this list every day until the final, updating it as necessary.

Prepare a Practice Final. Like the practice tests that you prepared throughout the term, prepare a practice final exam at least three days before your final. Take this practice final without using books or notes and time yourself. When you are finished, make a note of problems that you did incorrectly.

Meet with Your Study Group. After taking your practice final, meet with your study group to see whether anyone can help you resolve unanswered questions. You may want to exchange practice finals with others in your group so you can try more problems. Make a note of any questions that you still don't know how to do.

Meet with a Tutor or Your Instructor. If you still need help, visit your tutor or instructor a day or two before the final. Continue to work problems, trying not to look in your text or notes unless absolutely necessary.

On the Day of the Exam. Review problems that you feel confident about, but don't try to learn new material right before your exam. Know the starting time of your final, because often the time is different from your class meeting time. Bring any materials that you will need, such as a calculator (check the battery!) or Scantron form. Relax as much as possible and feel confident that you have prepared well!

ASSIGNMENT

1. Review your old tests and prepare a study sheet (or audiotape).
2. Prepare and take a practice final at least three days before your exam.
3. Meet with your study group at least two days before the final to review.
4. The evening before your final, gather the materials you will need and make a note of what you may need to purchase the next day. Get a good night's sleep!
5. On the day of the final, arrive a little early so you can relax before the test begins.

In this chapter, we will discuss graphs that do not represent functions. Since these curves are cross sections of cones, they are called conic sections.

10.1 The Circle and the Parabola

● Conic sections ● The circle ● Problem solving ● The parabola ● Problem solving

We have previously graphed first-degree equations in two variables such as $y = 3x + 8$ and $4x - 3y = 12$. Their graphs are lines. In this section, we will graph second-degree equations in two variables such as $x^2 + y^2 = 25$ and $x = -3y^2 - 12y - 13$. The graphs of these equations are *conic sections.*

■ Conic sections

The curves formed by the intersection of a plane with an infinite right-circular cone are called **conic sections.** Those curves have four basic shapes, called **circles, parabolas, ellipses,** and **hyperbolas,** as shown below.

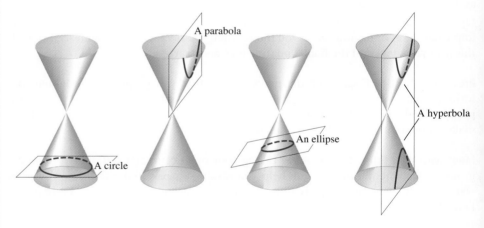

FIGURE 10-1

Conic sections have many applications. For example, everyone knows the importance of circular wheels and gears, pizza cutters, and Ferris wheels.

Parabolas can be rotated to generate dish-shaped surfaces called **paraboloids.** Any light or sound placed at the **focus** of a paraboloid is reflected outward in parallel paths, as shown in Figure 10-2(a) on the next page. This property makes parabolic surfaces ideal for flashlight and headlight reflectors. It also makes parabolic surfaces good antennas, because signals captured by such antennas are concentrated at a single point. Parabolic mirrors are capable of concentrating the rays of the sun at a single point and thereby generating tremendous heat. This property is used in the design of certain solar furnaces.

Any object thrown upward and outward travels in a parabolic path. An example of this is a stream of water flowing from a drinking fountain, as shown in Figure 10-2(b). In architecture, many arches are parabolic in shape, because this gives strength. Cables that support suspension bridges hang in the form of a parabola. (See Figure 10-2(c).)

Parabolas

Radar dish

Stream of water

Support cables

FIGURE 10-2

Ellipses have optical and acoustical properties that are useful in architecture and engineering. For example, many arches are portions of an ellipse, because the shape is pleasing to the eye. (See Figure 10-3(a).) The planets and many comets have elliptical orbits, as shown in Figure 10-3(b). Gears are often cut into elliptical shapes to provide nonuniform motion. (See Figure 10-3(c).)

Ellipses

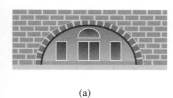

(a)

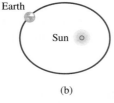

Earth

Sun

(b)

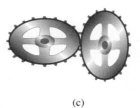

(c)

FIGURE 10-3

Hyperbolas serve as the basis of a navigational system known as LORAN (LOng RAnge Navigation). (See Figure 10-4.) They are also used to find the source of a distress signal, provide the basis for the design of hypoid gears, and describe the orbits of some comets.

Hyperbolas

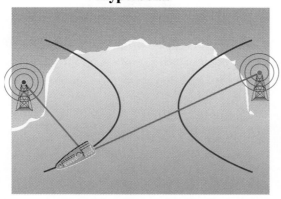

FIGURE 10-4

The circle

The most common conic section is the circle.

The circle

A **circle** is the set of all points in a plane that are a fixed distance from a point called its **center.** The fixed distance is called the **radius** of the circle.

To develop the equation of a circle, we will write the equation of a circle with a radius of r and with a center at some point $C(h, k)$, as in Figure 10-5. This task is equivalent to finding all points $P(x, y)$ such that the length of line segment CP is r. We can use the distance formula to find r.

$$r = \sqrt{(x - h)^2 + (y - k)^2}$$

We then square both sides to obtain

(1) $\quad r^2 = (x - h)^2 + (y - k)^2$

Equation 1 is called the **standard form of the equation of a circle** with a radius of r and center at the point with coordinates (h, k).

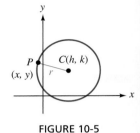

FIGURE 10-5

> **Standard equation of a circle with center at (h, k)**
>
> Any equation that can be written in the form
>
> $$(x - h)^2 + (y - k)^2 = r^2$$
>
> has a graph that is a circle with radius r and center at point (h, k).

If $r^2 = 0$, the graph reduces to a point called a **point circle.** If $r^2 < 0$, a circle does not exist. If both coordinates of the center are 0, the center of the circle is the origin.

> **Standard equation of a circle with center at (0, 0)**
>
> Any equation that can be written in the form
>
> $$x^2 + y^2 = r^2$$
>
> has a graph that is a circle with radius r and center at the origin.

INTERMEDIATE
Algebra $f(x)$ **Now**™

Self Check 1
Graph: $x^2 + y^2 = 4$.

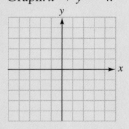

Answer

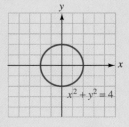

EXAMPLE 1　Graph: $x^2 + y^2 = 25$.

Solution　Because this equation can be written in the form $x^2 + y^2 = r^2$, its graph is a circle with center at the origin. Since $r^2 = 25 = 5^2$, the circle has a radius of 5. The graph appears in Figure 10-6.

$x^2 + y^2 = 25$		
x	y	(x, y)
-5	0	$(-5, 0)$
-4	± 3	$(-4, \pm 3)$
-3	± 4	$(-3, \pm 4)$
0	± 5	$(0, \pm 5)$
3	± 4	$(3, \pm 4)$
4	± 3	$(4, \pm 3)$
5	0	$(5, 0)$

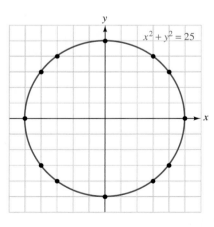

FIGURE 10-6

INTERMEDIATE
Algebra *f(x)* **Now**™

EXAMPLE 2   Find the equation of the circle with radius 7 and center at (3, 2).

Self Check 2
Find the equation of the circle
with radius 6 and center at (2, 3).

Solution   We substitute 7 for r, 3 for h, and 2 for k in standard form and simplify.

$$(x - h)^2 + (y - k)^2 = r^2$$
$$(x - 3)^2 + (y - 2)^2 = 7^2$$
$$x^2 - 6x + 9 + y^2 - 4y + 4 = 49$$
$$x^2 + y^2 - 6x - 4y + 9 + 4 - 49 = 0 \qquad \text{Subtract 49 from both sides.}$$
$$x^2 + y^2 - 6x - 4y - 36 = 0 \qquad \text{Combine like terms.}$$

The equation is $x^2 + y^2 - 6x - 4y - 36 = 0$.

Answer
$x^2 + y^2 - 4x - 6y - 23 = 0$

Ultimate Frisbee   THINK IT THROUGH

"Ultimate Frisbee combines speed, grace and powerful hurling with a grueling pace." *THE WALL STREET JOURNAL*

College students have tossed the plastic discs called Frisbees on campuses since the WHAM-O toy company first introduced them in 1958. Recently, the popularity of the Frisbee has been growing dramatically with a game called Ultimate Frisbee. Ultimate, as the players call it, is a high-endurance sport with few basic rules that combines the nonstop movement of soccer, the defensive strategies of basketball, and the passing of football.

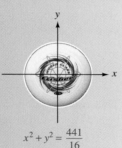

   The illustration shows a drawing of an official ultimate Frisbee centered on a rectangular coordinate system. Determine the diameter of the disc from the given equation. The units are inches.

$$x^2 + y^2 = \frac{441}{16}$$

INTERMEDIATE
Algebra *f(x)* **Now**™

EXAMPLE 3   Graph: $x^2 + y^2 - 4x + 2y = 11$.

Self Check 3
Write
$x^2 + y^2 + 2x - 4y - 11 = 0$ in
standard form and graph it.

Solution   To identify the curve, we complete the square on x and y and write the equation in standard form.

$$x^2 + y^2 - 4x + 2y = 11$$
$$x^2 - 4x + y^2 + 2y = 11$$

To complete the square on x and y, add 4 and 1 to both sides.

$$x^2 - 4x + 4 + y^2 + 2y + 1 = 11 + 4 + 1$$
$$(x - 2)^2 + (y + 1)^2 = 16 \qquad \text{Factor } x^2 - 4x + 4 \text{ and } y^2 + 2y + 1.$$
$$(x - 2)^2 + [y - (-1)]^2 = 4^2$$
$$\uparrow \qquad\qquad \uparrow \qquad \uparrow$$
$$(x - h)^2 \;+\; (y - k)^2 \;=\; r^2 \qquad h = 2, k = -1, \text{ and } r^2 = 4^2. \text{ Since the}$$
$$\text{radius of a circle must be positive. } r = 4.$$

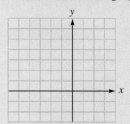

The center of the circle is the ordered pair $(h, k) = (2, -1)$ and the radius is 4. If we plot the center and draw a circle with a radius of 4 units, we will obtain the circle shown in Figure 10-7 on the next page.

Answer $(x + 1)^2 + (y - 2)^2 = 16$

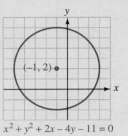

$x^2 + y^2 + 2x - 4y - 11 = 0$

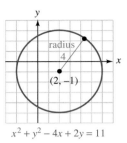

$x^2 + y^2 - 4x + 2y = 11$

FIGURE 10-7

Graphing circles

Since the graphs of circles fail the vertical line test, their equations do not represent functions. It is somewhat more difficult to use a graphing calculator to graph equations that are not functions. For example, to graph the circle described by $(x - 1)^2 + (y - 2)^2 = 4$, we must split the equation into two functions and graph each one separately. We begin by solving the equation for y.

$$(x - 1)^2 + (y - 2)^2 = 4$$

$$(y - 2)^2 = 4 - (x - 1)^2 \qquad \text{Subtract } (x - 1)^2 \text{ from both sides.}$$

$$y - 2 = \pm\sqrt{4 - (x - 1)^2} \qquad \text{Take the square root of both sides.}$$

$$y = 2 \pm \sqrt{4 - (x - 1)^2} \qquad \text{Add 2 to both sides.}$$

This equation defines two functions. If we use window settings of $[-3, 5]$ for x and $[-3, 5]$ for y and graph the functions

$$y = 2 + \sqrt{4 - (x - 1)^2} \qquad \text{and} \qquad y = 2 - \sqrt{4 - (x - 1)^2}$$

we get the distorted circle shown in Figure 10-8(a). To get a better circle, we can use the graphing calculator's squaring feature, which gives an equal unit distance on both the x- and y-axes. Using this feature, we get the circle shown in Figure 10-8(b). Sometimes the two arcs will not join because of approximations made by the calculator at each endpoint.

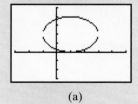

(a)

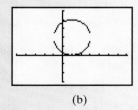

(b)

FIGURE 10-8

▍Problem solving

EXAMPLE 4 Radio translators. The broadcast area of a television station is bounded by the circle $x^2 + y^2 = 3{,}600$, where x and y are measured in miles. A translator station picks up the signal and retransmits it from the center of a circular area bounded by $(x + 30)^2 + (y - 40)^2 = 1,600$. Find the location of the translator and the greatest distance from the main transmitter that the signal can be received.

Solution The coverage of the television station is bounded by $x^2 + y^2 = 60^2$, a circle centered at the origin with a radius of 60 miles, as shown in yellow in Figure 10-9. Because the translator is at the center of the circle $(x + 30)^2 + (y - 40)^2 = 1,600$, it is located at $(-30, 40)$, a point 30 miles west and 40 miles north of the television station. The radius of the translator's coverage is $\sqrt{1,600}$, or 40 miles.

As shown in Figure 10-9, the greatest distance of reception is the sum of d, the distance from the translator to the television station, and 40 miles, the radius of the translator's coverage.

To find d, we use the distance formula to find the distance between $(x_1, y_1) = (-30, 40)$ and the origin, $(x_2, y_2) = (0, 0)$.

$$d = \sqrt{(x_1 - x_2)^2 + (y_1 - y_2)^2}$$
$$= \sqrt{(-30 - 0)^2 + (40 - 0)^2}$$
$$= \sqrt{(-30)^2 + 40^2}$$
$$= \sqrt{900 + 1,600}$$
$$= \sqrt{2,500}$$
$$= 50$$

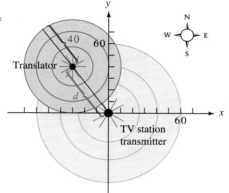

FIGURE 10-9

The translator is located 50 miles from the television station, and it broadcasts the signal an additional 40 miles. The greatest reception distance is 50 + 40, or 90 miles.

The parabola

We have seen that equations of the form $y = a(x - h)^2 + k$ where $a \neq 0$ represent parabolas with vertex at point (h, k). They open upward when $a > 0$ and downward when $a < 0$.

Equations of the form $x = a(y - k)^2 + h$ also represent parabolas with vertex at point (h, k). However, they open to the right when $a > 0$ and to the left when $a < 0$. Parabolas that open to the right or left do not represent functions, because their graphs fail the vertical line test.

Several types of parabolas are summarized in the following chart.

Equations of parabolas

Parabola opening	Vertex at origin	Vertex at (h, k)
If $a > 0$:		
Up	$y = ax^2$	$y = a(x - h)^2 + k$
Right	$x = ay^2$	$x = a(y - k)^2 + h$
If $a < 0$:		
Down	$y = ax^2$	$y = a(x - h)^2 + k$
Left	$x = ay^2$	$x = a(y - k)^2 + h$

Self Check 5

Graph: $x = \dfrac{1}{2}(y - 1)^2 - 2$.

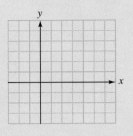

EXAMPLE 5   Graph: **a.** $x = \dfrac{1}{2}y^2$ and   **b.** $x = -2(y - 2)^2 + 3$.

Solution

a. We make a table of ordered pairs that satisfy the equation, plot each pair, and draw the parabola, as in Figure 10-10(a). Because the equation is of the form $x = ay^2$ with $a = \frac{1}{2} > 0$, the parabola opens to the right and has its vertex at the origin.

b. We make a table of ordered pairs that satisfy the equation, plot each pair, and draw the parabola, as in Figure 10-10(b). Because the equation is of the form $x = a(y - k)^2 + h$ with $a = -2$, the parabola opens to the left and has its vertex at the point with coordinates $(3, 2)$.

$x = \frac{1}{2}y^2$		
x	y	(x, y)
0	0	$(0, 0)$
2	2	$(2, 2)$
2	-2	$(2, -2)$
8	4	$(8, 4)$
8	-4	$(8, -4)$

$x = -2(y - 2)^2 + 3$		
x	y	(x, y)
-5	0	$(-5, 0)$
1	1	$(1, 1)$
3	2	$(3, 2)$
1	3	$(1, 3)$
-5	4	$(-5, 4)$

Answer

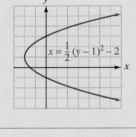

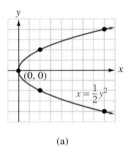

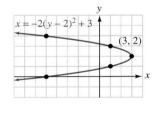

(a)   (b)

FIGURE 10-10

Self Check 6

Graph: $y = \dfrac{1}{2}x^2 - x - 1$.

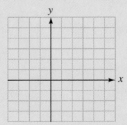

EXAMPLE 6   Graph: $y = -2x^2 + 12x - 15$.

Solution   Because the equation is not in standard form, the coordinates of the vertex are not obvious. To write the equation in standard form, we complete the square on x.

$$y = -2x^2 + 12x - 15$$
$$y = -2(x^2 - 6x) - 15 \qquad \text{Factor out } -2 \text{ from } -2x^2 + 12x.$$

This step adds $-2 \cdot 9$ or -18 to this side.   Add 18 to counteract the addition of -18.

$$y = -2(x^2 - 6x + 9) - 15 + 18 \qquad \text{Subtract and add 18; } -2(9) = -18.$$
$$y = -2(x - 3)^2 + 3$$

Since the equation is written in the form $y = a(x - h)^2 + k$ with $a = -2$, we can see that the parabola opens downward and has its vertex at $(3, 3)$. The graph of the function is shown in Figure 10-11 on the next page.

$$y = -2x^2 + 12x - 15$$

x	y	(x, y)
1	-5	(1, -5)
2	1	(2, 1)
3	3	(3, 3)
4	1	(4, 1)
5	-5	(5, -5)

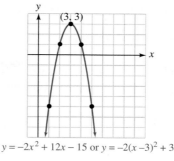

$$y = -2x^2 + 12x - 15 \text{ or } y = -2(x - 3)^2 + 3$$

FIGURE 10-11

Answer

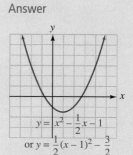

$$y = \left| x^2 - \frac{1}{2}x - 1 \right|$$

$$\text{or } y = \frac{1}{2}(x - 1)^2 - \frac{3}{2}$$

■ Problem solving

INTERMEDIATE
Algebra $f(x)$ **Now** ™

EXAMPLE 7   Gateway Arch.   The shape of the Gateway Arch in St. Louis is approximately a parabola, as shown in Figure 10-12(a). How high is the arch 100 feet from its foundation?

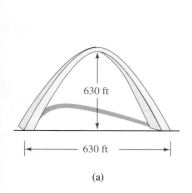

630 ft

630 ft

(a)

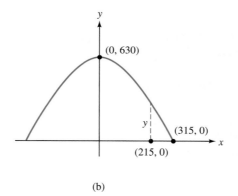

(0, 630)

(315, 0)

(215, 0)

(b)

FIGURE 10-12

Solution   We place the parabola on a coordinate system as in Figure 10-12(b), with ground level on the x-axis and the vertex of the parabola at the point $(h, k) = (0, 630)$. The equation of this downward-opening parabola has the following form, where $a < 0$.

$y = a(x - h)^2 + k$

$y = a(x - 0)^2 + 630$ Substitute 0 for h and 630 for k.

$y = ax^2 + 630$ Simplify.

Because the Gateway Arch is 630 feet wide at its base, the parabola passes through the point $\left(\frac{630}{2}, 0\right)$, or $(315, 0)$. To find a in the equation of the parabola, we proceed as follows:

$y = ax^2 + 630$

$0 = a(315)^2 + 630$ Substitute 315 for x and 0 for y.

$\dfrac{-630}{315^2} = a$ Subtract 630 from both sides and divide both sides by 315^2.

$-\dfrac{2}{315} = a$ Simplify the fraction:

$$-\frac{630}{315^2} = -\frac{2 \cdot \overset{1}{\cancel{315}}}{\underset{1}{\cancel{315}} \cdot 315} = -\frac{2}{315}.$$

The equation of the parabola that approximates the shape of the Gateway Arch is

$$y = -\frac{2}{315}x^2 + 630$$

To find the height of the arch at a point 100 feet from its foundation, we substitute $315 - 100$, or 215, for x in the equation of the parabola and solve for y.

$$y = -\frac{2}{315}x^2 + 630$$

$$y = -\frac{2}{315}(\mathbf{215})^2 + 630$$

$$\approx 336.5079365 \qquad \text{Use a calculator.}$$

At a point 100 feet from the foundation, the height of the arch is about 337 feet.

Section 10.1 STUDY SET

VOCABULARY *Fill in the blanks.*

1. The curves formed by the intersection of a plane with an infinite right-circular cone are called _____ _____.

2. A _____ is the set of all points in a plane that are a fixed distance from a point.

3. The fixed distance in Exercise 2 is called the _____ of the circle, and the point is called its _____.

4. The line that divides a parabola into two identical halves is called the axis of _____.

CONCEPTS *Fill in the blanks.*

5. The graph of $x = -5y^2$ is a parabola that has its _____ at the origin and opens to the _____.

6. The graph of $y = 6x^2$ is a _____ that has its vertex at the _____ and opens _____.

7. The graph of $x = 7(y - 2)^2 + 3$ is a _____ that has its vertex at _____ and opens to the _____.

8. The graph of $x = -4(y - 1)^2 - 3$ is a _____ that has its vertex at _____ and opens to the _____.

9. The standard equation of a circle with center at (h, k) is _____.

10. The standard equation of a circle with center at $(0, 0)$ is _____.

11. The standard equation of a parabola that opens upward and has its vertex at the origin is $y =$ _____, where $a >$ ___.

12. The standard equation of a parabola that opens to the left and has its vertex at (h, k) is $x =$ _____, where $a <$ ___.

NOTATION *Complete each solution.*

13. Find an equation of a circle with radius 6 and center at $(2, 3)$.

$$(x - h)^2 + (y - k)^2 = r^2$$
$$(x - \;\;)^2 + (y - 3)^2 = \;\;^2$$
$$x^2 - \;\;x + 4 + y^2 - 6y + 9 = \;\;$$
$$x^2 + y^2 - 4x - 6y - 23 = 0$$

14. Write an equation of a parabola with vertex at $(2, 3)$, opening to the left, and with $a = -2$.

$$x = a(y - k)^2 + h$$
$$x = \;\;(y - \;\;)^2 + 2$$
$$x = -2(y^2 - \;\;y + 9) + 2$$
$$x = -2y^2 + 12y - 16$$

PRACTICE *Graph each equation.*

15. $x^2 + y^2 = 9$ 16. $x^2 + y^2 = 16$

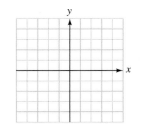   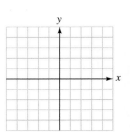

17. $(x - 2)^2 + y^2 = 9$

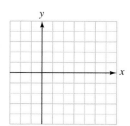

18. $x^2 + (y - 3)^2 = 4$

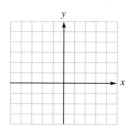

19. $(x - 2)^2 + (y - 4)^2 = 4$

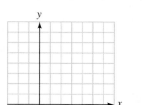

20. $(x - 3)^2 + (y - 2)^2 = 4$

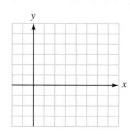

21. $(x + 3)^2 + (y - 1)^2 = 16$

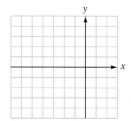

22. $(x - 1)^2 + (y + 4)^2 = 9$

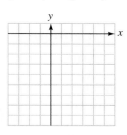

23. $x^2 + (y + 3)^2 = 1$

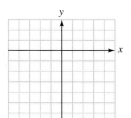

24. $(x + 4)^2 + y^2 = 1$

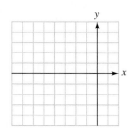

Use a graphing calculator to graph each equation.

25. $3x^2 + 3y^2 = 16$

26. $2x^2 + 2y^2 = 9$

27. $(x + 1)^2 + y^2 = 16$

28. $x^2 + (y - 2)^2 = 4$

Write the equation of the circle with the following properties.

29. Center at the origin; radius 1

30. Center at the origin; radius 4

31. Center at $(6, 8)$; radius 5

32. Center at $(5, 3)$; radius 2

33. Center at $(-2, 6)$; radius 12

34. Center at $(5, -4)$; radius 6

35. Center at the origin; diameter $4\sqrt{2}$

36. Center at the origin; diameter $8\sqrt{3}$

Graph each circle and give the coordinates of the center.

37. $x^2 + y^2 + 2x - 8 = 0$

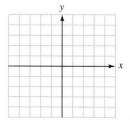

38. $x^2 + y^2 - 4y = 12$

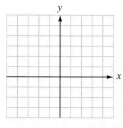

39. $x^2 + y^2 - 2x + 4y = -1$

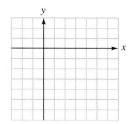

40. $x^2 + y^2 + 4x + 2y = 4$

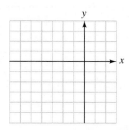

41. $x^2 + y^2 + 6x - 4y = -12$

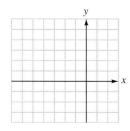

42. $x^2 + y^2 + 8x + 2y = -13$

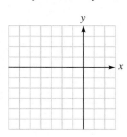

43. $4x^2 + 4y^2 + 4y = 15$ (*Hint:* Divide both sides by 4.)

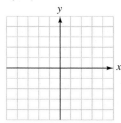

44. $9x^2 + 9y^2 - 12y = 5$ (*Hint:* Divide both sides by 9.)

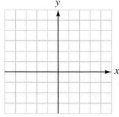

Find the vertex of each parabola and graph it.

45. $x = y^2$

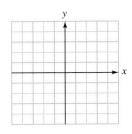

46. $x = -y^2 + 1$

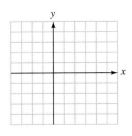

47. $x = -\dfrac{1}{4}y^2$

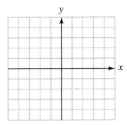

48. $x = 4y^2$

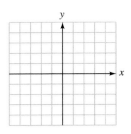

49. $y = x^2 + 4x + 5$

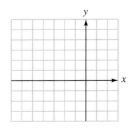

50. $y = -x^2 - 2x + 3$

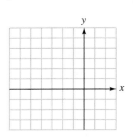

51. $y = -x^2 - x + 1$

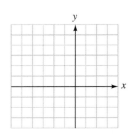

52. $x = \dfrac{1}{2}y^2 + 2y$

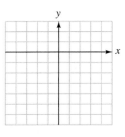

53. $y^2 + 4x - 6y = -1$

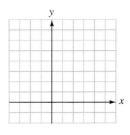

54. $x^2 - 2y - 2x = -7$

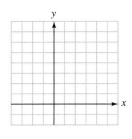

55. $y = 2(x - 1)^2 + 3$

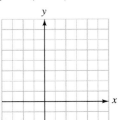

56. $y = -2(x + 1)^2 + 2$

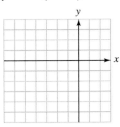

 *Use a graphing calculator to graph each equation.*

57. $x = 2y^2$

58. $x = y^2 - 4$

59. $x^2 - 2x + y = 6$

60. $x = -2(y - 1)^2 + 2$

APPLICATIONS

61. MESHING GEARS For design purposes, the large gear in the illustration is the circle $x^2 + y^2 = 16$. The smaller

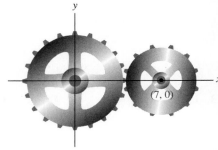

gear is a circle centered at $(7, 0)$ and tangent to the larger circle. Find the equation of the smaller gear.

62. WIDTH OF A WALKWAY The walkway below is bounded by the two circles $x^2 + y^2 = 2,500$ and $(x - 10)^2 + y^2 = 900$, measured in feet. Find the largest and the smallest width of the walkway.

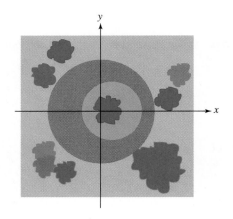

63. BROADCAST RANGES Radio stations applying for licensing may not use the same frequency if their broadcast areas overlap. One station's coverage is bounded by $x^2 + y^2 - 8x - 20y + 16 = 0$, and another's by $x^2 + y^2 + 2x + 4y - 11 = 0$. May they be licensed for the same frequency?

64. HIGHWAY DESIGN Engineers want to join two sections of highway with a curve that is one-quarter of a circle. The equation of the circle is $x^2 + y^2 - 16x - 20y + 155 = 0$, where distances are measured in kilometers. Find the locations (relative to the center of town) of the intersections of the highway with State and with Main.

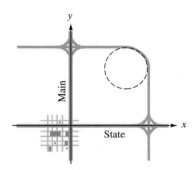

65. PROJECTILES The cannonball in the illustration below follows the parabolic trajectory $y = 30x - x^2$. How far short of the castle does it land?

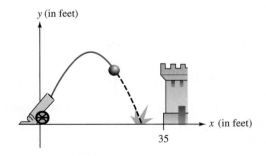

66. FLIGHT OF A PROJECTILE In Exercise 65, how high does the cannonball get?

67. ORBIT OF A COMET If the orbit of the comet shown is approximated by the equation $2y^2 - 9x = 18$, how far is it from the sun at the vertex of the orbit? Distances are measured in astronomical units (AU).

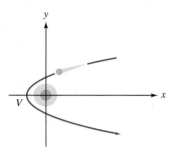

68. SATELLITE ANTENNA The cross section of the satellite antenna shown below is a parabola given by the equation $y = \frac{1}{16}x^2$, with distances measured in feet. If the dish is 8 feet wide, how deep is it?

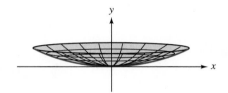

▮ **WRITING**

69. Explain how to decide from its equation whether the graph of a parabola opens up, down, right, or left.

70. From the equation of a circle, explain how to determine the radius and the coordinates of the center.

71. On the day of an election, the following warning was posted in front of a school. Explain what it means.

> *No electioneering within a 1,000-foot radius of this polling place.*

72. What is meant by the *turning radius* of a truck?

▮ **REVIEW** *Solve each equation.*

73. $|3x - 4| = 11$

74. $\left|\dfrac{4 - 3x}{5}\right| = 12$

75. $|3x + 4| = |5x - 2|$

76. $|6 - 4x| = |x + 2|$

10.2 The Ellipse

• The ellipse • Constructing an ellipse • Graphing ellipses • Problem solving

A third conic section is an oval-shaped curve called an *ellipse*. Ellipses can be nearly round and look almost like a circle, or they can be long and narrow. In this section, we will learn how to construct ellipses and how to graph equations that represent ellipses.

The ellipse

The ellipse

An **ellipse** is the set of all points P in a plane for which the sum of the distances from two fixed points is a constant. See Figure 10-13, in which $d_1 + d_2$ is a constant.

Each of the two points is called a **focus.** Midway between the **foci** (plural of "focus") is the **center** of the ellipse.

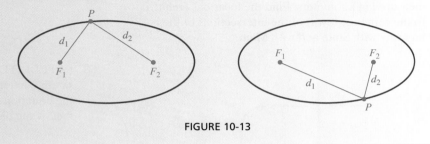

FIGURE 10-13

Constructing an ellipse

We can construct an ellipse by placing two thumbtacks fairly close together, as in Figure 10-14. We then tie each end of a piece of string to a thumbtack, catch the loop with the point of a pencil, and (keeping the string taut) draw the ellipse.

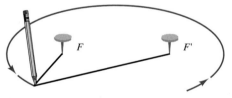

FIGURE 10-14

Using this method, we can construct an ellipse of any specific size. For example, to construct an ellipse that is 10 inches wide and 6 inches high, we must find the length of string to use and the distance between the thumbtacks.

To do this, we will let a represent the distance between the center and vertex V, as shown in Figure 10-15(a). We will also let c represent the distance between the center of the ellipse and either focus. When the pencil is at vertex V, the length of the string is $c + a + (a - c)$, or just $2a$. Because $2a$ is the 10-inch width of the ellipse, the string needs to be 10 inches long. The distance $2a$ is constant for any point on the ellipse, including point B shown in Figure 10-15(b).

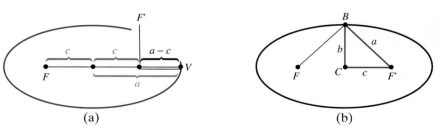

(a)   (b)

FIGURE 10-15

From right triangle CBF' (read F' as "F prime") and the Pythagorean theorem, we can find c as follows:

$$a^2 = b^2 + c^2 \quad \text{or} \quad c = \sqrt{a^2 - b^2}$$

Since distance b is one-half of the height of the ellipse, $b = 3$. Since $2a = 10$, $a = 5$. We can now substitute $a = 5$ and $b = 3$ into the formula to find c:

$$
\begin{aligned}
c &= \sqrt{5^2 - 3^2} \\
&= \sqrt{25 - 9} \\
&= \sqrt{16} \\
&= 4
\end{aligned}
$$

Since c is 4, the distance between the thumbtacks must be 8 inches. We can construct the ellipse by tying a 10-inch string to thumbtacks that are 8 inches apart.

■ Graphing ellipses

To graph ellipses, we can make a table of ordered pairs that satisfy the equation, plot them, and join the points with a smooth curve.

EXAMPLE 1   Graph: $\dfrac{x^2}{36} + \dfrac{y^2}{9} = 1$.

Solution   First we note that the equation can be written in the form

$$\frac{x^2}{6^2} + \frac{y^2}{3^2} = 1 \quad \text{Write 36 as } 6^2 \text{ and 9 as } 3^2.$$

After making a table of ordered pairs that satisfy the equation, plotting each of them, and joining the points with a curve, we obtain the ellipse shown in Figure 10-16.

     We note that the center of the ellipse is the origin, the ellipse intersects the x-axis at points $(6, 0)$ and $(-6, 0)$, and the ellipse intersects the y-axis at points $(0, 3)$ and $(0, -3)$.

$\dfrac{x^2}{36} + \dfrac{y^2}{9} = 1$		
x	y	(x, y)
-6	0	$(-6, 0)$
-4	± 2.2	$(-4, \pm 2.2)$
-2	± 2.8	$(-2, \pm 2.8)$
0	± 3	$(0, \pm 3)$
2	± 2.8	$(2, \pm 2.8)$
4	± 2.2	$(4, \pm 2.2)$
6	0	$(6, 0)$

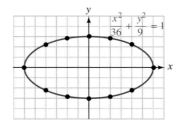

FIGURE 10-16

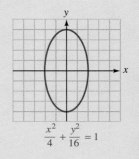

Example 1 illustrates that the graph of

$$\frac{x^2}{a^2} + \frac{y^2}{b^2} = 1$$

is an ellipse centered at the origin. To find the x-intercepts of the graph, we can let $y = 0$ and solve for x.

$$\frac{x^2}{a^2} + \frac{0^2}{b^2} = 1 \qquad \text{Substitute 0 for } y.$$

$$\frac{x^2}{a^2} + 0 = 1 \qquad \frac{0^2}{b^2} = 0.$$

$$x^2 = a^2 \qquad \text{Simplify and multiply both sides by } a^2.$$

$$x = a \qquad \text{or} \qquad x = -a$$

The x-intercepts are $(a, 0)$ and $(-a, 0)$.

To find the y-intercepts of the graph, we can let $x = 0$ and solve for y.

$$\frac{0^2}{a^2} + \frac{y^2}{b^2} = 1$$

$$0 + \frac{y^2}{b^2} = 1$$

$$y^2 = b^2$$

$$y = b \qquad \text{or} \qquad y = -b$$

The y-intercepts are $(0, b)$ and $(0, -b)$.

In general, we have the following results.

Equations of an ellipse centered at the origin

The equation of an ellipse centered at the origin with x-intercepts at $V_1(a, 0)$ and $V_2(-a, 0)$ and with y-intercepts at $(0, b)$ and $(0, -b)$ is

$$\frac{x^2}{a^2} + \frac{y^2}{b^2} = 1 \quad \text{where } a > b > 0 \qquad \text{See Figure 10-17(a).}$$

The equation of an ellipse centered at the origin with y-intercepts at $V_1(0, a)$ and $V_2(0, -a)$ and with x-intercepts at $(b, 0)$ and $(-b, 0)$ is

$$\frac{x^2}{b^2} + \frac{y^2}{a^2} = 1 \quad \text{where } a > b > 0 \qquad \text{See Figure 10-17(b).}$$

In Figure 10-17, the points V_1 and V_2 are the **vertices** of the ellipse, the midpoint of $\overline{V_1V_2}$ is the **center** of the ellipse, and the distance between the center of the ellipse and either vertex is a. The segment $\overline{V_1V_2}$ is called the **major axis,** and the segment joining either $(0, b)$ and $(0, -b)$ or $(b, 0)$ and $(-b, 0)$ is called the **minor axis.**

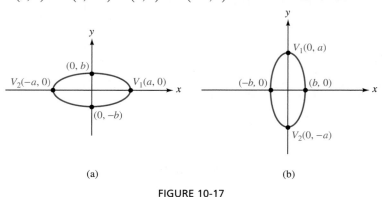

(a)   (b)

FIGURE 10-17

The equations for ellipses centered at (h, k) are as follows.

Equations of an ellipse centered at (h, k)

The equation of an ellipse centered at (h, k), with major axis parallel to the x-axis, is

(1) $\dfrac{(x - h)^2}{a^2} + \dfrac{(y - k)^2}{b^2} = 1$ where $a > b > 0$

The equation of an ellipse centered at (h, k), with major axis parallel to the y-axis, is

(2) $\dfrac{(x - h)^2}{b^2} + \dfrac{(y - k)^2}{a^2} = 1$ where $a > b > 0$

EXAMPLE 2   Graph: $\dfrac{(x - 2)^2}{16} + \dfrac{(y + 3)^2}{25} = 1$.

Solution
We write the equation in the form

$$\dfrac{(x - 2)^2}{4^2} + \dfrac{[y - (-3)]^2}{5^2} = 1$$

This is the equation of an ellipse centered at $(h, k) = (2, -3)$, with major axis parallel to the y-axis and with $b = 4$ and $a = 5$. We first plot the center, as shown in Figure 10-18. Since a is the distance from the center to the vertex, we can locate the vertices by counting 5 units above and 5 units below the center. The vertices are at points $(2, 2)$ and $(2, -8)$.

Since $b = 4$, we can locate two more points on the ellipse by counting 4 units to the left and 4 units to the right of the center. The points $(-2, -3)$ and $(6, -3)$ are also on the graph.

Using these four points as guides, we can draw the ellipse.

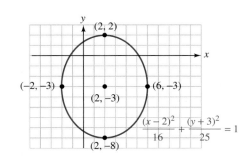

$\dfrac{(x-2)^2}{16} + \dfrac{(y+3)^2}{25} = 1$		
x	**y**	**(x, y)**
2	2	(2, 2)
2	−8	(2, −8)
6	−3	(6, −3)
−2	−3	(−2, −3)

FIGURE 10-18

INTERMEDIATE
Algebra $f(x)$ Now™

Self Check 2
Graph:
$$\dfrac{(x - 1)^2}{9} + \dfrac{(y + 2)^2}{16} = 1$$

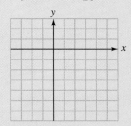

Answer

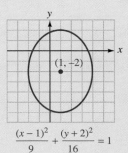

$$\dfrac{(x - 1)^2}{9} + \dfrac{(y + 2)^2}{16} = 1$$

CALCULATOR SNAPSHOT   Graphing ellipses

To use a graphing calculator to graph

$$\frac{(x + 2)^2}{4} + \frac{(y - 1)^2}{25} = 1$$

we first clear the equation of fractions by multiplying both sides by 100 and solving for y.

$$25(x + 2)^2 + 4(y - 1)^2 = 100$$ Multiply both sides by 100.

$$4(y - 1)^2 = 100 - 25(x + 2)^2$$ Subtract $25(x + 2)^2$ from both sides

$$(y - 1)^2 = \frac{100 - 25(x + 2)^2}{4}$$ Divide both sides by 4.

$$y - 1 = \pm\frac{\sqrt{100 - 25(x + 2)^2}}{2}$$ Take the square root of both sides.

$$y = 1 \pm \frac{\sqrt{100 - 25(x + 2)^2}}{2}$$ Add 1 to both sides.

If we use window settings $[-6, 6]$ for x and $[-6, 6]$ for y and graph the functions

$$y = 1 + \frac{\sqrt{100 - 25(x + 2)^2}}{2} \quad \text{and} \quad y = 1 - \frac{\sqrt{100 - 25(x + 2)^2}}{2}$$

we will obtain the ellipse shown in Figure 10-19.

As we saw with circles, the two portions of the ellipse do not quite connect. This is because the graphs are nearly vertical there.

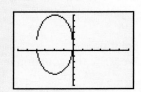

FIGURE 10-19

We can complete the square to write the equations of many ellipses in standard form.

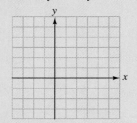

INTERMEDIATE
Algebra *f(x)* Now™

Self Check 3
Graph:
$4x^2 - 8x + 9y^2 - 36y = -4$

EXAMPLE 3   Write $4x^2 + 9y^2 - 16x - 18y = 11$ in standard form to show that the equation represents an ellipse.

Solution   We write the equation in standard form by completing the square on x and y:

$$4x^2 + 9y^2 - 16x - 18y = 11$$

$$4x^2 - 16x + 9y^2 - 18y = 11$$ Use the commutative property to rearrange terms.

$$4(x^2 - 4x) + 9(y^2 - 2y) = 11$$ Factor 4 from $4x^2 - 16x$ and factor 9 from $9y^2 - 18y$ to get coefficients of 1 for the squared terms.

Now we complete the square twice to make $x^2 - 4x$ and $y^2 - 2y$ perfect square trinomials.

$$4(x^2 - 4x + 4) + 9(y^2 - 2y + 1) = 11 + 16 + 9$$ Since 16 and 9 are added to the left-hand side, add 16 and 9 to the right-hand side.

$$4(x - 2)^2 + 9(y - 1)^2 = 36$$ Factor $x^2 - 4x + 4$ and $y^2 - 2y + 1$.

$$\frac{(x - 2)^2}{9} + \frac{(y - 1)^2}{4} = 1$$ Divide both sides by 36 and simplify.

Since this equation matches Equation 1 on page 751 it represents an ellipse. Its graph is shown in Figure 10-20.

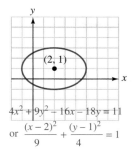

$4x^2 + 9y^2 - 16x - 18y = 11$

or $\dfrac{(x-2)^2}{9} + \dfrac{(y-1)^2}{4} = 1$

FIGURE 10-20

Answer

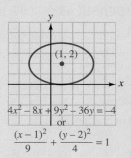

$4x^2 - 8x + 9y^2 - 36y = -4$

or

$\dfrac{(x-1)^2}{9} + \dfrac{(y-2)^2}{4} = 1$

◼ Problem solving

EXAMPLE 4 Landscape design.   A landscape architect is designing an elliptical pool that will fit in the center of a 20-by-30-foot rectangular garden, leaving at least 5 feet of space on all sides. Find the equation of the ellipse.

Solution   We place the rectangular garden in a coordinate system, as in Figure 10-21. To maintain 5 feet of clearance at the ends of the ellipse, the vertices must be the points $V_1(10, 0)$ and $V_2(-10, 0)$. Similarly, the y-intercepts are the points $(0, 5)$ and $(0, -5)$.

The equation of the ellipse has the form

$$\frac{x^2}{a^2} + \frac{y^2}{b^2} = 1$$

with $a = 10$ and $b = 5$. Thus, the equation of the boundary of the pool is

$$\frac{x^2}{100} + \frac{y^2}{25} = 1$$

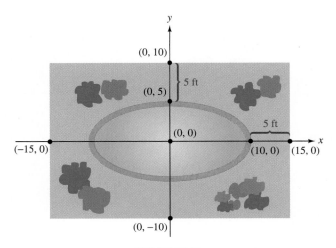

FIGURE 10-21

Section 10.2   STUDY SET

INTERMEDIATE
Algebra *f(x)* **Now**™

VOCABULARY   *Fill in the blanks.*

1. An _____ is the set of all points in a plane for which the _____ of the distances from two fixed points is a constant.

2. The fixed points in Exercise 1 are the _____ of the ellipse.

3. The midpoint of the line segment joining the foci of an ellipse is called the _____ of the ellipse.

4. The segment joining the vertices of an ellipse is called the _____ axis.

CONCEPTS   *Fill in the blanks.*

5. The elliptical graph of $\dfrac{x^2}{a^2} + \dfrac{y^2}{b^2} = 1$ has x-intercepts of ▮▮▮▮ .

6. The elliptical graph of $\dfrac{x^2}{a^2} + \dfrac{y^2}{b^2} = 1$ has y-intercepts of ▮▮▮▮ .

7. The center of the ellipse with an equation of
$$\frac{x^2}{a^2} + \frac{y^2}{b^2} = 1$$
is the point ▮▮▮▮ .

8. The center of the ellipse with an equation of
$$\frac{(x - h)^2}{a^2} + \frac{(y - k)^2}{b^2} = 1$$
is the point ▮▮▮▮ .

NOTATION   *Complete each solution.*

9. Write $16x^2 + 9y^2 - 144 = 0$ in standard form.
$$16x^2 + 9y^2 - 144 = 0$$
$$16x^2 + 9y^2 = \boxed{}$$
$$\frac{x^2}{9} + \frac{y^2}{\boxed{}} = 1$$

10. Write $9x^2 + 4y^2 - 18x + 8y = 23$ in standard form.
$$9x^2 + 4y^2 - 18x + 8y = 23$$
$$9x^2 - 18x + \boxed{}\, y^2 + 8y = 23$$
$$9(x^2 - \boxed{}\, x) + 4(y^2 + \boxed{}\, y) = 23$$
$$9(x^2 - 2x + \boxed{}\,) + 4(y^2 + 2y + 1) = 23 + 9 + 4$$
$$9(x - 1)^2 + 4(y + 1)^2 = \boxed{}$$
$$\frac{(x - 1)^2}{4} + \frac{(y + 1)^2}{9} = 1$$

PRACTICE   *Graph each equation.*

11. $\dfrac{x^2}{4} + \dfrac{y^2}{9} = 1$

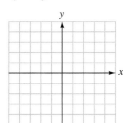

12. $\dfrac{x^2}{1} + \dfrac{y^2}{9} = 1$

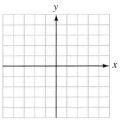

13. $\dfrac{x^2}{9} + \dfrac{y^2}{1} = 1$

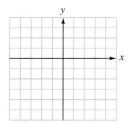

14. $\dfrac{x^2}{9} + \dfrac{y^2}{25} = 1$

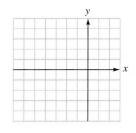

15. $16x^2 + 4y^2 = 64$
(*Hint:* Divide both sides by 64.)

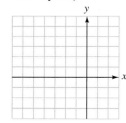

16. $4x^2 + 9y^2 = 36$
(*Hint:* Divide both sides by 36.)

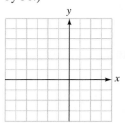

17. $\dfrac{(x - 2)^2}{9} + \dfrac{(y - 1)^2}{4} = 1$

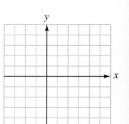

18. $\dfrac{(x - 1)^2}{9} + \dfrac{(y - 3)^2}{4} = 1$

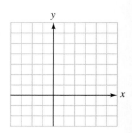

19. $\dfrac{(x+1)^2}{4} + \dfrac{(y+2)^2}{1} = 1$

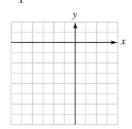

20. $\dfrac{(x+1)^2}{9} + \dfrac{y^2}{25} = 1$

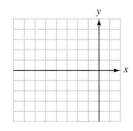

Use a graphing calculator to graph each equation.

21. $\dfrac{x^2}{9} + \dfrac{y^2}{4} = 1$ **22.** $x^2 + 16y^2 = 16$

23. $\dfrac{x^2}{4} + \dfrac{(y-1)^2}{4} = 1$ **24.** $\dfrac{(x+1)^2}{9} + \dfrac{(y-2)^2}{4} = 1$

Write each equation in standard form and graph it.

25. $x^2 + 4y^2 - 4x + 8y + 4 = 0$

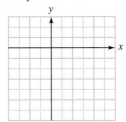

26. $x^2 + 4y^2 - 2x - 16y = -13$

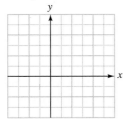

27. $9x^2 + 4y^2 - 18x + 16y = 11$

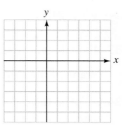

28. $16x^2 + 25y^2 - 160x - 200y + 400 = 0$

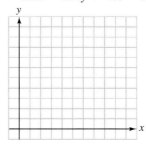

APPLICATIONS

29. DESIGNING AN UNDERPASS The arch of the underpass is a part of an ellipse. Find the equation of the arch.

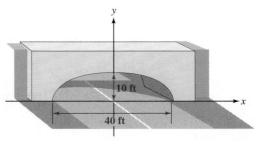

30. ARCHES Find the height of the elliptical arch in Exercise 29 at a point 10 feet from the center of the roadway.

31. AREA OF AN ELLIPSE The area A of the ellipse described by $\dfrac{x^2}{a^2} + \dfrac{y^2}{b^2} = 1$ is given by $A = \pi ab$. Find the area of the ellipse $\dfrac{x^2}{16} + \dfrac{y^2}{9} = 1$.

32. AREA OF A TRACK The elliptical track below is bounded by the ellipses $4x^2 + 9y^2 = 576$ and $9x^2 + 25y^2 = 900$. Find the area of the track. (See Exercise 31.)

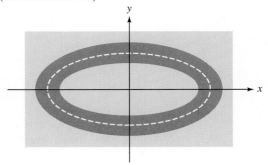

WRITING

33. Explain how to find the x- and y-intercepts of the graph of the ellipse $\dfrac{x^2}{a^2} + \dfrac{y^2}{b^2} = 1$.

34. Explain the relationship between the center, focus, and vertex of an ellipse.

REVIEW   *Find each product.*

35. $3x^{-2}y^2(4x^2 + 3y^{-2})$

36. $(2a^{-2} - b^{-2})(2a^{-2} + b^{-2})$

Write each expression as a complex fraction without using negative exponents. Then simplify.

37. $\dfrac{x^{-2} + y^{-2}}{x^{-2} - y^{-2}}$

38. $\dfrac{2x^{-3} - 2y^{-3}}{4x^{-3} + 4y^{-3}}$

10.3   The Hyperbola

• The hyperbola   • Problem solving

The final conic section, called a *hyperbola,* is a curve with two branches. In this section, we will learn how to graph equations that represent hyperbolas.

The hyperbola

> ### The hyperbola
>
> A **hyperbola** is the set of all points P in a plane for which the difference of the distances of each point from two fixed points is a constant. See Figure 10-22, where $d_1 - d_2$ is a constant.
>
>   Each of the two points is called a **focus.** Midway between the foci (plural of "focus") is the **center** of the hyperbola.
>
>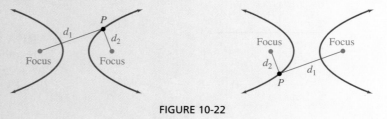
>
> FIGURE 10-22

    The graph of the equation

$$\frac{x^2}{25} - \frac{y^2}{9} = 1$$

is a hyperbola. To graph the equation, we make a table of ordered pairs that satisfy the equation, plot each pair, and join the points with a smooth curve, as in Figure 10-23 on the next page.

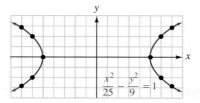

$\dfrac{x^2}{25} - \dfrac{y^2}{9} = 1$		
x	y	(x, y)
-7	± 2.9	$(-7, \pm 2.9)$
-6	± 2.0	$(-6, \pm 2.0)$
-5	0	$(-5, 0)$
5	0	$(5, 0)$
6	± 2.0	$(6, \pm 2.0)$
7	± 2.9	$(7, \pm 2.9)$

FIGURE 10-23

This graph is centered at the origin and intersects the x-axis at $(\sqrt{25}, 0)$ and $(-\sqrt{25}, 0)$. After simplifying, these points are $(5, 0)$ and $(-5, 0)$. We also note that the graph does not intersect the y-axis.

It is possible to draw a hyperbola without plotting points. For example, if we want to graph the hyperbola with an equation of

$$\frac{x^2}{a^2} - \frac{y^2}{b^2} = 1$$

we first look at the x- and y-intercepts. To find the x-intercepts, we let $y = 0$ and solve for x:

$$\frac{x^2}{a^2} - \frac{0^2}{b^2} = 1$$
$$x^2 = a^2$$
$$x = \pm a$$

The hyperbola crosses the x-axis at the points $V_1(a, 0)$ and $V_2(-a, 0)$, called the **vertices** of the hyperbola. See Figure 10-24.

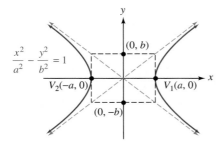

FIGURE 10-24

To attempt to find the y-intercepts, we let $x = 0$ and solve for y:

$$\frac{0^2}{a^2} - \frac{y^2}{b^2} = 1$$
$$y^2 = -b^2$$
$$y = \pm\sqrt{-b^2}$$

Since b^2 is always positive, $\sqrt{-b^2}$ is an imaginary number. This means that the hyperbola does not cross the y-axis.

If we construct a rectangle, called the **fundamental** or **central rectangle,** whose sides pass horizontally through $\pm b$ on the y-axis and vertically through $\pm a$ on the x-axis, the extended diagonals of the rectangle will be asymptotes of the hyperbola.

Equation of a hyperbola centered at the origin

Any equation that can be written in the form

$$\frac{x^2}{a^2} - \frac{y^2}{b^2} = 1$$

has a graph that is a hyperbola centered at the origin, as in Figure 10-25. The x-intercepts are the vertices $V_1(a, 0)$ and $V_2(-a, 0)$. There are no y-intercepts.

The asymptotes of the hyperbola are the extended diagonals of the rectangle in the figure.

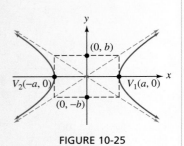

FIGURE 10-25

The branches of the hyperbola in previous discussions open to the left and to the right. It is possible for hyperbolas to have different orientations with respect to the x- and y-axes. For example, the branches of a hyperbola can open upward and downward. In that case, the following equation applies.

Equation of a hyperbola centered at the origin

Any equation that can be written in the form

$$\frac{y^2}{a^2} - \frac{x^2}{b^2} = 1$$

has a graph that is a hyperbola centered at the origin, as in Figure 10-26. The y-intercepts are the vertices $V_1(0, a)$ and $V_2(0, -a)$. There are no x-intercepts.

The asymptotes of the hyperbola are the extended diagonals of the rectangle shown in the figure.

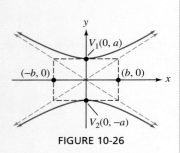

FIGURE 10-26

Self Check 1

Graph: $16y^2 - x^2 = 16$.

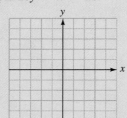

EXAMPLE 1   Graph: $9y^2 - 4x^2 = 36$.

Solution   To write the equation in standard form, we divide both sides by 36 to obtain

$$\frac{9y^2}{36} - \frac{4x^2}{36} = \frac{36}{36}$$

$$\frac{y^2}{4} - \frac{x^2}{9} = 1 \quad \text{Simplify each fraction.}$$

We then find the y-intercepts by letting $x = 0$ and solving for y:

$$\frac{y^2}{4} - \frac{0^2}{9} = 1$$

$$y^2 = 4$$

Thus, $y = \pm 2$, and the vertices of the hyperbola are $V_1(0, 2)$ and $V_2(0, -2)$. (See Figure 10-27 on the next page.)

Since $\pm\sqrt{9} = \pm 3$, we use the points $(3, 0)$ and $(-3, 0)$ on the x-axis to help draw the fundamental rectangle. We then draw its extended diagonals and sketch the hyperbola.

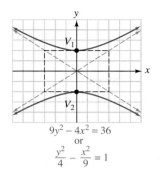

$$9y^2 - 4x^2 = 36$$
or
$$\frac{y^2}{4} - \frac{x^2}{9} = 1$$

FIGURE 10-27

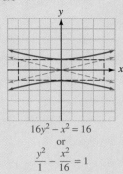
If a hyperbola is centered at a point with coordinates (h, k), the following equations apply.

Equations of hyperbolas centered at (h, k)

Any equation that can be written in the form

$$\frac{(x - h)^2}{a^2} - \frac{(y - k)^2}{b^2} = 1$$

is a hyperbola that has its center at (h, k) and opens left and right.
    Any equation of the form

$$\frac{(y - k)^2}{a^2} - \frac{(x - h)^2}{b^2} = 1$$

is a hyperbola that has its center at (h, k) and opens up and down.

EXAMPLE 2   Graph: $\dfrac{(x - 3)^2}{16} - \dfrac{(y + 1)^2}{4} = 1$.

Solution   We write the equation in the form

$$\frac{(x - 3)^2}{4^2} - \frac{[y - (-1)]^2}{2^2} = 1$$

to see that its graph will be a hyperbola centered at $(h, k) = (3, -1)$. Its vertices are located at $a = 4$ units to the right and left of the center, at $(7, -1)$ and $(-1, -1)$. Since $b = 2$, we can count 2 units above and below the center to locate points $(3, 1)$ and $(3, -3)$. With these points, we can draw the fundamental rectangle along with its extended diagonals. We can then sketch the hyperbola, as shown in Figure 10–28.

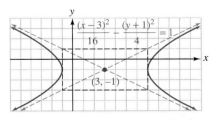

FIGURE 10-28

INTERMEDIATE
Algebra $f(x)$ Now™

Self Check 2

Graph: $\dfrac{(x + 2)^2}{9} - \dfrac{(y - 1)^2}{4} = 1$

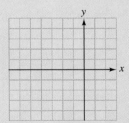

Answer

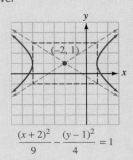

$$\frac{(x + 2)^2}{9} - \frac{(y - 1)^2}{4} = 1$$

INTERMEDIATE
Algebra *f(x)* **Now**™

Self Check 3
Graph:
$$x^2 - 4y^2 + 2x - 8y = 7$$

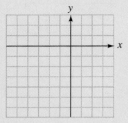

Answer

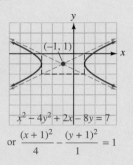

$$x^2 - 4y^2 + 2x - 8y = 7$$
or $\dfrac{(x+1)^2}{4} - \dfrac{(y+1)^2}{1} = 1$

EXAMPLE 3   Write the equation $x^2 - y^2 - 2x + 4y = 12$ in standard form to show that the equation represents a hyperbola. Then graph it.

Solution   We proceed as follows.

$$x^2 - y^2 - 2x + 4y = 12$$
$$x^2 - 2x - y^2 + 4y = 12 \quad \text{Use the commutative property to rearrange terms.}$$
$$x^2 - 2x - (y^2 - 4y) = 12 \quad \text{Factor } -1 \text{ from } -y^2 + 4y.$$

We then complete the square on x and y to make $x^2 - 2x$ and $y^2 - 4y$ perfect square trinomials.

$$x^2 - 2x + \mathbf{1} - (y^2 - 4y + 4) = 12 + \mathbf{1} - 4$$

We then factor $x^2 - 2x + 1$ and $y^2 - 4y + 4$ to get

$$(x-1)^2 - (y-2)^2 = 9$$
$$\frac{(x-1)^2}{9} - \frac{(y-2)^2}{9} = 1 \quad \text{Divide both sides by 9.}$$

This is the equation of a hyperbola with center at $(1, 2)$. Its graph is shown in Figure 10-29.

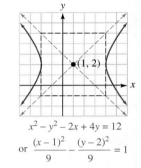

$$x^2 - y^2 - 2x + 4y = 12$$
or $\dfrac{(x-1)^2}{9} - \dfrac{(y-2)^2}{9} = 1$

FIGURE 10-29

There is a special type of hyperbola (also centered at the origin) that does not intersect either the x- or the y-axis. These hyperbolas have equations of the form $xy = k$, where $k \neq 0$.

Self Check 4
Graph: $xy = 6$

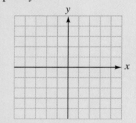

Answer

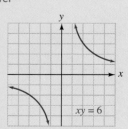

$xy = 6$

EXAMPLE 4   Graph: $xy = -8$.

Solution   We make a table of ordered pairs, plot each pair, and join the points with a smooth curve to obtain the hyperbola in Figure 10-30.

\multicolumn{3}{c}{$xy = -8$}
x	y	(x, y)
1	-8	$(1, -8)$
2	-4	$(2, -4)$
4	-2	$(4, -2)$
8	-1	$(8, -1)$
-1	8	$(-1, 8)$
-2	4	$(-2, 4)$
-4	2	$(-4, 2)$
-8	1	$(-8, 1)$

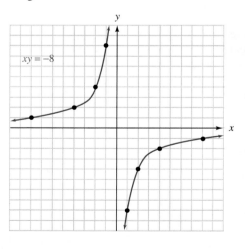

$xy = -8$

FIGURE 10-30

The result in Example 4 illustrates the following general equation.

Equations of hyperbolas of the form $xy = k$

Any equation of the form $xy = k$, where $k \neq 0$, has a graph that is a **hyperbola,** which does not intersect either the x- or the y-axis

---

Problem solving

EXAMPLE 5   Atomic structure.   In an experiment that led to the discovery of the atomic structure of matter, Lord Rutherford (1871–1937) shot high-energy alpha particles toward a thin sheet of gold. Many of them were reflected, and Rutherford showed the existence of the nucleus of a gold atom. The alpha particle in Figure 10–31 is repelled by the nucleus at the origin; it travels along the hyperbolic path given by $4x^2 - y^2 = 16$. How close does the particle come to the nucleus?

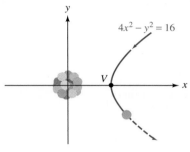

$4x^2 - y^2 = 16$

FIGURE 10-31

Solution   To find how far the particle is from the nucleus at the origin, we must find the coordinates of the vertex V. To do so, we write the equation of the particle's path in standard form:

$$4x^2 - y^2 = 16$$

$$\frac{4x^2}{16} - \frac{y^2}{16} = \frac{16}{16} \qquad \text{Divide both sides by 16.}$$

$$\frac{x^2}{4} - \frac{y^2}{16} = 1 \qquad \text{Simplify.}$$

$$\frac{x^2}{2^2} - \frac{y^2}{4^2} = 1 \qquad \text{Write 4 as } 2^2 \text{ and 16 as } 4^2.$$

This equation is in the form

$$\frac{x^2}{a^2} - \frac{y^2}{b^2} = 1$$

with $a = 2$. Thus, the vertex of the path is $(2, 0)$. The particle gets as close as 2 units from the nucleus.

Section 10.3   STUDY SET

VOCABULARY   *Fill in the blanks.*

1. A _____ is the set of all points in a plane for which the _____ of the distances from two fixed points is a constant.

2. The fixed points in Exercise 1 are the _____ of the hyperbola.

3. The midpoint of the line segment joining the foci of a hyperbola is called the _____ of the hyperbola.

4. The rectangle whose sides pass horizontally through $\pm b$ on the y-axis and vertically through $\pm a$ on the x-axis is called the _____ rectangle.

CONCEPTS *Fill in the blanks.*

5. The graph of

$$\frac{x^2}{a^2} - \frac{y^2}{b^2} = 1$$

is a hyperbola with x-intercepts of ▭.

6. The hyperbolic graph of

$$\frac{x^2}{a^2} - \frac{y^2}{b^2} = 1$$

has _____ y-intercepts.

7. The center of the hyperbola with an equation of

$$\frac{x^2}{a^2} - \frac{y^2}{b^2} = 1$$

is the point ▭.

8. The center of the hyperbola with an equation of

$$\frac{(x-h)^2}{a^2} - \frac{(y-k)^2}{b^2} = 1$$

is the point ▭.

NOTATION *Complete each solution.*

9. Write the equation $4x^2 - 9y^2 - 36 = 0$ in standard form.

$$4x^2 - 9y^2 - 36 = 0$$
$$4x^2 - 9y^2 = \;▭$$
$$\frac{4x^2}{▭} - \frac{9y^2}{▭} = 1$$
$$\frac{x^2}{9} - \frac{y^2}{4} = 1$$

10. Write the equation $x^2 - 4y^2 + 2x - 63 = 0$ in standard form.

$$x^2 - 4y^2 + 2x - 63 = 0$$
$$x^2 + 2x - 4y^2 = \;▭$$
$$x^2 + 2x + ▭ - 4y^2 = 63 + ▭$$
$$(\;▭\;)^2 - 4y^2 = 64$$
$$\frac{(x+1)^2}{64} - \frac{y^2}{16} = 1$$

PRACTICE *Graph each hyperbola.*

11. $\dfrac{x^2}{9} - \dfrac{y^2}{4} = 1$

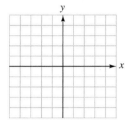

12. $\dfrac{x^2}{4} - \dfrac{y^2}{4} = 1$

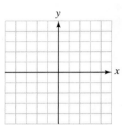

13. $\dfrac{y^2}{4} - \dfrac{x^2}{9} = 1$

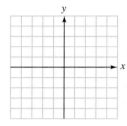

14. $\dfrac{y^2}{4} - \dfrac{x^2}{64} = 1$

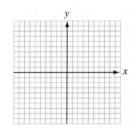

15. $25x^2 - y^2 = 25$

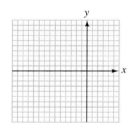

16. $9x^2 - 4y^2 = 36$

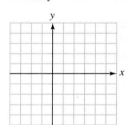

17. $\dfrac{(x-2)^2}{9} - \dfrac{y^2}{16} = 1$

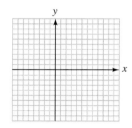

18. $\dfrac{(x+2)^2}{16} - \dfrac{(y-3)^2}{25} = 1$

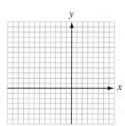

19. $4(x+3)^2 - (y-1)^2 = 4$

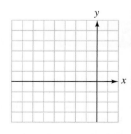

20. $(x + 5)^2 - 16y^2 = 16$

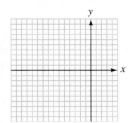

28. $x^2 - 9y^2 - 4x - 54y = 86$

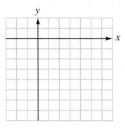

21. $xy = 8$

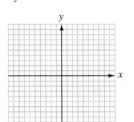

22. $xy = -10$

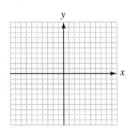

29. $4y^2 - x^2 + 8y + 4x = 4$

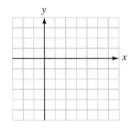

Use a graphing calculator to graph each equation.

23. $\dfrac{x^2}{9} - \dfrac{y^2}{4} = 1$

24. $y^2 - 16x^2 = 16$

30. $y^2 - 4x^2 - 4y - 8x = 4$

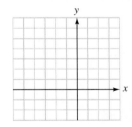

25. $\dfrac{x^2}{4} - \dfrac{(y - 1)^2}{9} = 1$

26. $\dfrac{(y + 1)^2}{9} - \dfrac{(x - 2)^2}{4} = 1$

APPLICATIONS

31. ALPHA PARTICLES The particle in the illustration approaches the nucleus at the origin along the path $\dfrac{y^2}{9} - \dfrac{x^2}{81} = 1$ in the coordinate system shown. How close does the particle come to the nucleus?

Write each equation in standard form and graph it.

27. $4x^2 - y^2 + 8x - 4y = 4$

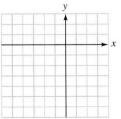

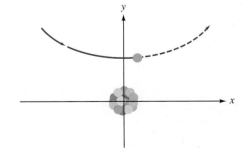

32. LORAN By determining the difference of the distances between the ship in the illustration below and two radio transmitters, the LORAN system places the ship on the hyperbola $x^2 - 4y^2 = 576$ in the coordinate system shown. If the ship is also 5 miles out to sea, find its coordinates.

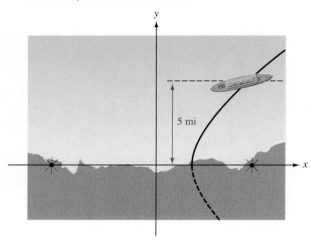

33. SONIC BOOM The position of the sonic boom caused by the faster-than-sound aircraft is the hyperbola $y^2 - x^2 = 25$ in the coordinate system shown. How wide is the hyperbola 5 units from its vertex?

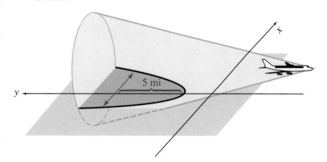

34. FLUIDS See the illustration below. Two glass plates in contact at the left, and separated by about 5 millimeters on the right, are dipped in beet juice, which rises by capillary action to form a hyperbola. The hyperbola is modeled by an equation of the form $xy = k$. If the curve passes through the point $(12, 2)$. what is k?

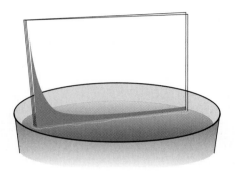

■ **WRITING**

35. What is a hyperbola?

36. Explain why the graph of the hyperbola

$$\frac{x^2}{a^2} - \frac{y^2}{b^2} = 1$$

has no y-intercept.

■ **REVIEW**   *Factor each expression.*

37. $-6x^4 + 9x^3 - 6x^2$

38. $4a^2 - b^2$

39. $15a^2 - 4ab - 4b^2$

40. $8p^3 - 27q^3$

10.4 Solving Simultaneous Second-Degree Equations

• Solving systems by graphing • Solving systems by substitution • Solving systems by addition

We have discussed how to solve systems of linear equations. To do this, we used the graphing, substitution, and addition methods. In this section, we will apply these methods to solve systems where at least one of the equations is nonlinear.

Solving systems by graphing

We now discuss ways to solve systems of two equations in two variables where at least one of the equations is of second degree.

EXAMPLE 1 Solve $\begin{cases} x^2 + y^2 = 25 \\ 2x + y = 10 \end{cases}$ by graphing.

Solution The graph of $x^2 + y^2 = 25$ is a circle with center at the origin and radius of 5. The graph of $2x + y = 10$ is a line. Depending on whether the line is a **secant** (intersecting the circle at two points) or a **tangent** (intersecting the circle at one point) or does not intersect the circle at all, there are two, one, or no solutions to the system, respectively.

After graphing the circle and the line, it appears that the points of intersection are $(5, 0)$ and $(3, 4)$. To verify that they are solutions of the system, we need to check each one.

Check: **For (5, 0)**   **For (3, 4)**

$2x + y = 10$	$x^2 + y^2 = 25$	$2x + y = 10$	$x^2 + y^2 = 25$
$2(5) + 0 \overset{?}{=} 10$	$5^2 + 0^2 \overset{?}{=} 25$	$2(3) + 4 \overset{?}{=} 10$	$3^2 + 4^2 \overset{?}{=} 25$
$10 = 10$	$25 = 25$	$10 = 10$	$25 = 25$

The ordered pair $(5, 0)$ satisfies both equations of the system, and so does $(3, 4)$. Thus, there are two solutions, $(5, 0)$ and $(3, 4)$, and the solution set is $\{(5, 0), (3, 4)\}$.

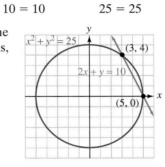

FIGURE 10-32

INTERMEDIATE
Algebra $f(x)$ **Now**™

Self Check 1

Solve $\begin{cases} x^2 + y^2 = 13 \\ y = -\frac{1}{5}x + \frac{13}{5} \end{cases}$ by graphing.

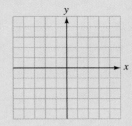

Answer $(3, 2), (-2, 3)$

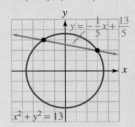

Solving systems of equations   **CALCULATOR SNAPSHOT**

To solve Example 1 with a graphing calculator, we graph the circle and the line on one set of coordinate axes. (See Figure 10-33(a).) We then trace to find the coordinates of the intersection points of the graphs. (See Figure 10-33(b) and Figure 10-33(c).)

We can zoom for better results.

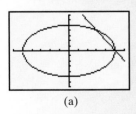

(a)

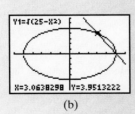

(b)

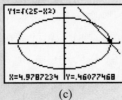

(c)

FIGURE 10-33

■ Solving systems by substitution

Algebraic methods can also be used to solve systems of equations.

EXAMPLE 2 Solve $\begin{cases} x^2 + y^2 = 25 \\ 2x + y = 10 \end{cases}$.

Solution This system is the one we solved in Example 1. We can also solve this type of system by substitution. Solving the linear equation for y gives

INTERMEDIATE
Algebra $f(x)$ **Now**™

Self Check 2

Solve $\begin{cases} x^2 + y^2 = 13 \\ y = -\frac{1}{5}x + \frac{13}{5} \end{cases}$ by substitution.

$$2x + y = 10 \qquad \text{This is the second equation of the given system.}$$
$$\textbf{(1)} \qquad y = -2x + 10$$

We can substitute $-2x + 10$ for y in the second-degree equation and solve the resulting quadratic equation for x:

$$x^2 + y^2 = 25$$
$$x^2 + (\mathbf{-2x + 10})^2 = 25$$
$$x^2 + 4x^2 - 40x + 100 = 25 \qquad \begin{array}{l}(-2x + 10)(-2x + 10) = \\ 4x^2 - 40x + 100.\end{array}$$
$$5x^2 - 40x + 75 = 0 \qquad \begin{array}{l}\text{Combine like terms and subtract 25} \\ \text{from both sides.}\end{array}$$
$$x^2 - 8x + 15 = 0 \qquad \text{Divide both sides by 5.}$$
$$(x - 5)(x - 3) = 0 \qquad \text{Factor } x^2 - 8x + 15.$$
$$x - 5 = 0 \quad \text{or} \quad x - 3 = 0 \qquad \text{Set each factor equal to 0.}$$
$$x = 5 \quad | \quad \qquad x = 3$$

If we substitute 5 for x in Equation 1, we get $y = 0$. If we substitute 3 for x in Equation 1, we get $y = 4$. The two solutions are

$$\begin{cases} x = 5 \\ y = 0 \end{cases} \quad \text{or} \quad \begin{cases} x = 3 \\ y = 4 \end{cases}$$

Answer $(3, 2), (-2, 3)$

which can be written as $(5, 0)$ and $(3, 4)$.

Self Check 3

Solve $\begin{cases} x^2 + y^2 = 20 \\ y = x^2 \end{cases}$

by substitution.

EXAMPLE 3   Solve $\begin{cases} 4x^2 + 9y^2 = 5 \\ y = x^2 \end{cases}$.

Solution   We can solve this system by substitution.

$$4x^2 + 9y^2 = 5$$
$$4y + 9y^2 = 5 \qquad \text{Substitute } y \text{ for } x^2.$$
$$9y^2 + 4y - 5 = 0 \qquad \text{Subtract 5 from both sides.}$$
$$(9y - 5)(y + 1) = 0 \qquad \text{Factor } 9y^2 + 4y - 5.$$
$$9y - 5 = 0 \quad \text{or} \quad y + 1 = 0 \qquad \text{Set each factor equal to 0.}$$
$$y = \frac{5}{9} \quad \Big| \quad \qquad y = -1$$

Since $y = x^2$, the values of x are found by solving the equations

$$x^2 = \frac{5}{9} \quad \text{and} \quad x^2 = -1$$

Because $x^2 = -1$ has no real solutions, this possibility is discarded. The solutions of $x^2 = \frac{5}{9}$ are

$$x = \frac{\sqrt{5}}{3} \quad \text{or} \quad x = -\frac{\sqrt{5}}{3}$$

The solutions of the system are

$$\left(\frac{\sqrt{5}}{3}, \frac{5}{9} \right) \quad \text{and} \quad \left(-\frac{\sqrt{5}}{3}, \frac{5}{9} \right)$$

Answer $(2, 4), (-2, 4)$

Solving systems by addition

INTERMEDIATE
Algebra $f(x)$ Now™

EXAMPLE 4 Solve $\begin{cases} 3x^2 + 2y^2 = 36 \\ 4x^2 - y^2 = 4 \end{cases}$.

Solution Since both equations are in the form $ax^2 + by^2 = c$, we can solve the system by addition. We copy the first equation and multiply the second equation by 2 to obtain the equivalent system

$$\begin{cases} 3x^2 + 2y^2 = 36 \\ 8x^2 - 2y^2 = 8 \end{cases}$$

We add the equations to eliminate the y^2-terms and solve the resulting equation for x:

$$11x^2 = 44$$
$$x^2 = 4$$
$$x = 2 \quad \text{or} \quad x = -2$$

To find y, we substitute 2 for x and then -2 for x in the first equation:

For $x = 2$	**For $x = -2$**
$3x^2 + 2y^2 = 36$	$3x^2 + 2y^2 = 36$
$3(2)^2 + 2y^2 = 36$	$3(-2)^2 + 2y^2 = 36$
$12 + 2y^2 = 36$	$12 + 2y^2 = 36$
$2y^2 = 24$	$2y^2 = 24$
$y^2 = 12$	$y^2 = 12$
$y = \sqrt{12} \quad \text{or} \quad y = -\sqrt{12}$	$y = \sqrt{12} \quad \text{or} \quad y = -\sqrt{12}$
$y = 2\sqrt{3} \quad \mid \quad y = -2\sqrt{3}$	$y = 2\sqrt{3} \quad \mid \quad y = -2\sqrt{3}$

The four solutions of this system are: $(2, 2\sqrt{3})$, $(2, -2\sqrt{3})$, $(-2, 2\sqrt{3})$, and $(-2, -2\sqrt{3})$.

Self Check 4

Solve: $\begin{cases} x^2 + 4y^2 = 16 \\ x^2 - y^2 = 1 \end{cases}$.

Answer $(2, \sqrt{3})$, $(2, -\sqrt{3})$, $(-2, \sqrt{3})$, $(-2, -\sqrt{3})$

Section 10.4 STUDY SET

INTERMEDIATE
Algebra $f(x)$ Now™

VOCABULARY *Fill in the blanks.*

1. We can solve systems of equations by _____ addition, or _____.

2. A _____ is a line that intersects a circle at two points.

3. A _____ is a line that intersects a circle at one point.

4. _____ equations have graphs that are not straight lines.

CONCEPTS *Fill in the blanks.*

5. At most, a line can intersect an ellipse at _____ points. At most, an ellipse can intersect a parabola at _____ points.

6. At most, an ellipse can intersect a circle at _____ points. At most, a hyperbola can intersect a circle at _____ points.

7. Check to determine whether $(1, -1)$ is a solution of the system $\begin{cases} 2x + y - 1 = 0 \\ x^2 + y^2 = 3 \end{cases}$.

8. What should be used as the substitution equation to solve the system $\begin{cases} 2x^2 - y^2 = 6 \\ x^2 - y = 3 \end{cases}$?

NOTATION *Complete each solution.*

9. Solve: $\begin{cases} x^2 + y^2 = 5 \\ y = 2x \end{cases}$.

$$x^2 + y^2 = 5$$
$$x^2 + (\quad)^2 = 5$$
$$x^2 + 4x^2 = \boxed{}$$
$$\boxed{} x^2 = 5$$
$$x^2 = \boxed{}$$
$$x = 1 \quad \text{or} \quad x = -1 \qquad \text{(continued)}$$

If $x = 1$, then
$$y = 2(\ \ ) = 2$$
If $x = -1$, then
$$y = 2(\ \ ) = -2$$
The solutions are $(1, \ \ )$ and $(-1, \ \ )$.

10. Solve: $\begin{cases} x^2 + y^2 = 13 \\ x^2 - y^2 = 5 \end{cases}$.

$$2x^2 = \ \ \quad\quad \text{Add the equations.}$$
$$x^2 = \ \ $$
$$x = \ \ \quad \text{or} \quad x = \ \ $$
$$2y^2 = \ \ \quad\quad \text{Subtract the}$$
$$y^2 = \ \ \quad\quad\quad \text{equations.}$$
$$y = \ \ \quad \text{or} \quad y = \ \ $$

The solutions are

$$(3, \ \ ), \quad (3, \ \ ), \quad (-3, \ \ ), \quad (-3, \ \ )$$

PRACTICE *Solve each system of equations by graphing.*

11. $\begin{cases} 8x^2 + 32y^2 = 256 \\ x = 2y \end{cases}$

12. $\begin{cases} x^2 + y^2 = 2 \\ x + y = 2 \end{cases}$

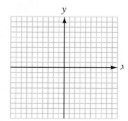

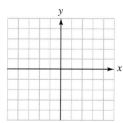

13. $\begin{cases} x^2 + y^2 = 10 \\ y = 3x^2 \end{cases}$

14. $\begin{cases} x^2 + y^2 = 5 \\ x + y = 3 \end{cases}$

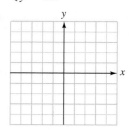

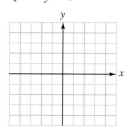

15. $\begin{cases} x^2 + y^2 = 25 \\ 12x^2 + 64y^2 = 768 \end{cases}$

16. $\begin{cases} x^2 + y^2 = 13 \\ y = x^2 - 1 \end{cases}$

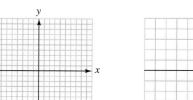

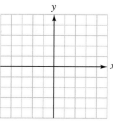

17. $\begin{cases} x^2 - 13 = -y^2 \\ y = \frac{2}{3}x \end{cases}$

18. $\begin{cases} x^2 + y^2 = 20 \\ y = x^2 \end{cases}$

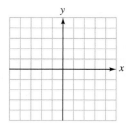

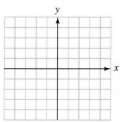

Use a graphing calculator to solve each system.

19. $\begin{cases} x^2 - 6x - y = -5 \\ x^2 - 6x + y = -5 \end{cases}$

20. $\begin{cases} x^2 - y^2 = -5 \\ 3x^2 + 2y^2 = 30 \end{cases}$

Solve each system of equations algebraically for real values of x and y.

21. $\begin{cases} 25x^2 + 9y^2 = 225 \\ 5x + 3y = 15 \end{cases}$

22. $\begin{cases} x^2 + y^2 = 20 \\ y = x^2 \end{cases}$

23. $\begin{cases} x^2 + y^2 = 2 \\ x + y = 2 \end{cases}$

24. $\begin{cases} x^2 + y^2 = 36 \\ 49x^2 + 36y^2 = 1{,}764 \end{cases}$

25. $\begin{cases} x^2 + y^2 = 5 \\ x + y = 3 \end{cases}$

26. $\begin{cases} x^2 - x - y = 2 \\ 4x - 3y = 0 \end{cases}$

27. $\begin{cases} x^2 + y^2 = 13 \\ y = x^2 - 1 \end{cases}$

28. $\begin{cases} x^2 + y^2 = 25 \\ 2x^2 - 3y^2 = 5 \end{cases}$

29. $\begin{cases} x^2 + y^2 = 30 \\ y = x^2 \end{cases}$

30. $\begin{cases} 9x^2 - 7y^2 = 81 \\ x^2 + y^2 = 9 \end{cases}$

31. $\begin{cases} x^2 + y^2 = 13 \\ x^2 - y^2 = 5 \end{cases}$

32. $\begin{cases} 2x^2 + y^2 = 6 \\ x^2 - y^2 = 3 \end{cases}$

33. $\begin{cases} x^2 + y^2 = 20 \\ x^2 - y^2 = -12 \end{cases}$

34. $\begin{cases} xy = -\dfrac{9}{2} \\ 3x + 2y = 6 \end{cases}$

35. $\begin{cases} y^2 = 40 - x^2 \\ y = x^2 - 10 \end{cases}$

36. $\begin{cases} x^2 - 6x - y = -5 \\ x^2 - 6x + y = -5 \end{cases}$

37. $\begin{cases} y = x^2 - 4 \\ x^2 - y^2 = -16 \end{cases}$ **38.** $\begin{cases} 6x^2 + 8y^2 = 182 \\ 8x^2 - 3y^2 = 24 \end{cases}$

39. $\begin{cases} x^2 - y^2 = -5 \\ 3x^2 + 2y^2 = 30 \end{cases}$ **40.** $\begin{cases} \frac{1}{x} + \frac{1}{y} = 5 \\ \frac{1}{x} - \frac{1}{y} = -3 \end{cases}$

41. $\begin{cases} \frac{1}{x} + \frac{2}{y} = 1 \\ \frac{2}{x} - \frac{1}{y} = \frac{1}{3} \end{cases}$ **42.** $\begin{cases} \frac{1}{x} + \frac{3}{y} = 4 \\ \frac{2}{x} - \frac{1}{y} = 7 \end{cases}$

43. $\begin{cases} 3y^2 = xy \\ 2x^2 + xy - 84 = 0 \end{cases}$ **44.** $\begin{cases} x^2 + y^2 = 10 \\ 2x^2 - 3y^2 = 5 \end{cases}$

45. $\begin{cases} xy = \dfrac{1}{6} \\ y + x = 5xy \end{cases}$ **46.** $\begin{cases} xy = \dfrac{1}{12} \\ y + x = 7xy \end{cases}$

47. INTEGER PROBLEM The product of two integers is 32, and their sum is 12. Find the integers.

48. NUMBER PROBLEM The sum of the squares of two numbers is 221, and the sum of the numbers is 9. Find the numbers.

49. GEOMETRY The area of a rectangle is 63 square centimeters, and its perimeter is 32 centimeters. Find the dimensions of the rectangle.

50. INVESTING Grant receives $225 annual income from one investment. Jeff invested $500 more than Grant, but at an annual rate of 1% less. Jeff's annual income is $240. What is the amount and rate of Grant's investment?

51. INVESTING Carol receives $67.50 annual income from one investment. John invested $150 more than Carol at an annual rate of $1\frac{1}{2}$% more. John's annual income is $94.50. What is the amount and rate of Carol's investment? (*Hint:* There are two answers.)

52. ARTILLERY The shell fired from the base of the hill in the illustration follows the parabolic path $y = -\frac{1}{6}x^2 + 2x$, with distances measured in miles. The hill has a slope of $\frac{1}{3}$. How far from the gun is the point of impact? (*Hint:* Find the coordinates of the point and then the distance.)

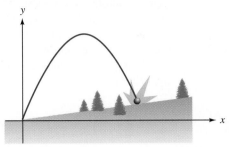

53. DRIVING RATES Jim drove 306 miles. Jim's brother made the same trip at a speed 17 mph slower than Jim did and required an extra $1\frac{1}{2}$ hours. What was Jim's rate and time?

54. FENCING PASTURES A rectangular pasture is to be fenced in along a riverbank, as shown. If 260 feet of fencing is to enclose an area of 8,000 square feet, find the dimensions of the pasture.

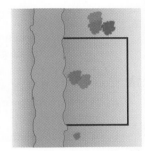

55. Describe the benefits of the graphical method for solving a system of equations.

56. Describe the drawbacks of the graphical method for solving a system of equations.

57. $\sqrt{200x^2} - 3\sqrt{98x^2}$

58. $a\sqrt{112a} - 5\sqrt{175a^3}$

59. $\dfrac{3t\sqrt{2t} - 2\sqrt{2t^3}}{\sqrt{18t} - \sqrt{2t}}$

60. $\sqrt[3]{\dfrac{x}{4}} + \sqrt[3]{\dfrac{x}{32}} - \sqrt[3]{\dfrac{x}{500}}$

KEY CONCEPT

Conic Sections

In this chapter, we have studied four **conic sections:** the circle, the parabola, the ellipse, and the hyperbola. They are called conic sections because they are formed by the intersection of a plane and a right-circular cone. These curves are graphs of second-degree equations in two variables.

Classify the graph of each equation as a circle, a parabola, an ellipse, or a hyperbola.

1. $\dfrac{(x-2)^2}{9} + \dfrac{(y-1)^2}{4} = 1$

2. $(x-3)^2 + (y-2)^2 = 4$

3. $y = 4x^2 - 2x + 3$

4. $\dfrac{x^2}{4} - \dfrac{(y-1)^2}{9} = 1$

5. $4x^2 - y^2 + 8x - 4y = 4$

6. $\dfrac{x^2}{4} + \dfrac{y^2}{9} = 1$

7. $9x^2 + 4y^2 - 18x + 16y = 11$

8. $x^2 - 2y - 2x = -7$

9. $x^2 + y^2 + 8x + 2y = -13$

Equations of Conic Sections

When the equation of a conic section is written in standard form, important features of its graph are apparent.

10. Consider $(x+1)^2 + (y-2)^2 = 16$.

  a. What are the coordinates of the center of the circle?

  b. What is the radius of the circle?

11. Consider $x = \frac{1}{2}(y-1)^2 - 2$.

  a. What are the coordinates of the vertex of the parabola?

  b. In which direction does the parabola open?

12. Consider $\dfrac{x^2}{4} + \dfrac{y^2}{16} = 1$.

  a. What are the coordinates of the center of the ellipse?

  b. To which axis is the major axis of the ellipse parallel?

  c. How long is the major axis?

  d. How long is the minor axis?

13. Consider $\dfrac{(x+2)^2}{9} - \dfrac{(y-1)^2}{4} = 1$.

  a. What are the coordinates of the center of the hyperbola?

  b. In which direction do the branches of the hyperbola open?

  c. What are the dimensions of the fundamental rectangle?

Graphing Conic Sections

Use the results from Problems 10–13 to graph each conic section.

14. $(x+1)^2 + (y-2)^2 = 16$

15. $x = \frac{1}{2}(y-1)^2 - 2$

16. $\dfrac{x^2}{4} + \dfrac{y^2}{16} = 1$

17. $\dfrac{(x+2)^2}{9} - \dfrac{(y-1)^2}{4} = 1$

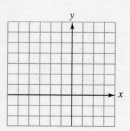
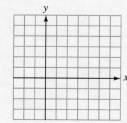
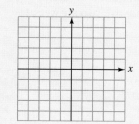
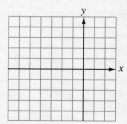

ACCENT ON TEAMWORK

SECTION 10.1

CONSTRUCTING PARABOLAS To construct a parabola, secure one end of a piece of string that is as long as a T-square to a large piece of paper. Attach the other end of the string to the upper end of the T-square. Then, hold the string taut against the T-square with a pencil and slide the T-square along the edge of the table. As the T-square moves, the pencil will trace a parabola.

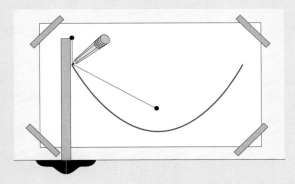

SECTION 10.2

CONSTRUCTING ELLIPSES We can construct an ellipse with two thumbtacks, a piece of string, and a pencil. Place the two thumbtacks fairly close together, as shown. Catch the loop of the string with the point of a pencil and (keeping the string taut) draw the ellipse. Experiment by moving one of the thumbtacks farther away and then closer to the other thumbtack. How does the shape of the ellipse change?

For each point on an ellipse, the sum of the distances of the point from two given points is a constant. With this method of construction, what are the two given points? What is the constant?

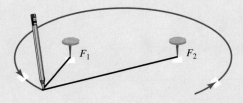

SECTION 10.3

CONIC SECTION MODELS Mold some clay into the shape of a right-circular cone, as shown. With both hands, pull a thin wire through the clay to slice it in such a way that a circular shape results. Experiment with ways to slice the clay to obtain an elliptical shape, a parabolic shape, and one branch of a hyperbolic shape.

SECTION 10.4

SIMULTANEOUS SYSTEMS Sketch the graphs of a system of second-degree equations that meet the following conditions, if possible.

The graph of one equation is a circle, the graph of the other equation is a parabola, and the system has

a. no solution.   b. one solution.
c. two solutions.   d. three solutions.
e. four solutions.   f. five solutions.

SIMULTANEOUS SYSTEMS Sketch the graphs of a system of second-degree equations that meet the following conditions, if possible.

The graph of one equation is an ellipse, the graph of the other equation is a hyperbola, and the system has

a. no solution.   b. one solution.
c. two solutions.   d. three solutions.
e. four solutions.   f. five solutions.

SIMULTANEOUS SYSTEMS Sketch the graphs of a system of second-degree equations that meet the following conditions, if possible.

The graph of one equation is a parabola, the graph of the other equation is a hyperbola, and the system has

a. no solution.   b. one solution.
c. two solutions.   d. three solutions.
e. four solutions.   f. five solutions.

CHAPTER REVIEW

INTERMEDIATE
Algebra $f(x)$ **Now**™

SECTION 10.1   *The Circle and the Parabola*

CONCEPTS

Equations of a circle:

$(x - h)^2 + (y - k)^2 = r^2$
center (h, k), radius r

$x^2 + y^2 = r^2$
center $(0, 0)$, radius r

REVIEW EXERCISES

Graph each equation.

1. $(x - 1)^2 + (y + 2)^2 = 9$

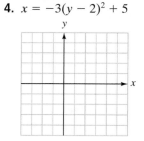

2. $x^2 + y^2 = 16$

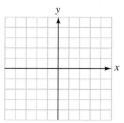

3. Write the equation in standard form and graph it.

$x^2 + y^2 + 4x - 2y = 4$

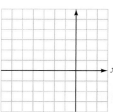

Equations of parabolas:

Parabola opening	Vertex at origin
If $a > 0$	
Up	$y = ax^2$
Right	$x = ay^2$
If $a < 0$	
Down	$y = ax^2$
Left	$x = ay^2$

Parabola opening	Vertex at (h, k)
If $a > 0$	
Up	$y = a(x - h)^2 + k$
Right	$x = a(y - k)^2 + h$
If $a < 0$	
Down	$y = a(x - h)^2 + k$
Left	$x = a(y - k)^2 + h$

Graph each equation.

4. $x = -3(y - 2)^2 + 5$

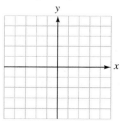

5. $x = 2(y + 1)^2 - 2$

6. LONG JUMP   The equation describing the flight path of the long jumper is $y = -\frac{5}{121}(x - 11)^2 + 5$. Show that she will land at a point 22 feet away from the take-off board.

The Ellipse

Equations of an ellipse:

Center at $(0, 0)$

$$\frac{x^2}{a^2} + \frac{y^2}{b^2} = 1 \quad (a > b > 0)$$

$$\frac{x^2}{b^2} + \frac{y^2}{a^2} = 1 \quad (a > b > 0)$$

Center at (h, k)

$$\frac{(x - h)^2}{a^2} + \frac{(y - k)^2}{b^2} = 1$$

$$\frac{(x - h)^2}{b^2} + \frac{(y - k)^2}{a^2} = 1$$

Graph each ellipse.

7. $9x^2 + 16y^2 = 144$

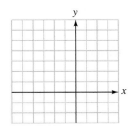

8. $\dfrac{(x - 2)^2}{4} + \dfrac{(y - 1)^2}{9} = 1$

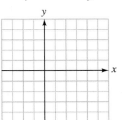

9. Write the equation in standard form and graph it.

$$4x^2 + 9y^2 + 8x - 18y = 23$$

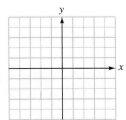

10. SALAMI When a delicatessen slices a cylindrical salami at an angle, the results are elliptical pieces that are larger than circular pieces. Write the equation of the shape of the slice of salami shown in the illustration if it was centered at the origin of a coordinate system.

6 cm
10 cm

The Hyperbola

Equations of a hyperbola:

Center at $(0, 0)$

$$\frac{x^2}{a^2} - \frac{y^2}{b^2} = 1$$

$$\frac{y^2}{b^2} - \frac{x^2}{a^2} = 1$$

Center at (h, k)

$$\frac{(x - h)^2}{a^2} - \frac{(y - k)^2}{b^2} = 1$$

$$\frac{(y - k)^2}{a^2} - \frac{(x - h)^2}{b^2} = 1$$

Graph each hyperbola.

11. $9x^2 - y^2 = -9$

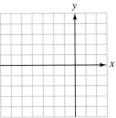

12. $xy = 9$

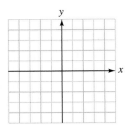

13. Write the equation $4x^2 - 2y^2 + 8x - 8y = 8$ in standard form and tell whether its graph will be an ellipse or a hyperbola.

14. Write the equation in standard form and graph it.

$$9x^2 - 4y^2 - 18x - 8y = 31$$

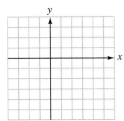

15. ELECTROSTATIC REPULSION
Two similarly charged particles are shot together for an almost head-on collision, as in the illustration. They repel each other and travel the two branches of the hyperbola given by $x^2 - 4y^2 = 4$ on the given coordinate system. How close do they get?

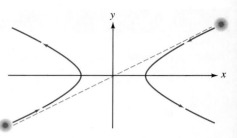

16. Determine whether the equation, when graphed, will be a circle, parabola, ellipse, or hyperbola.

a. $\dfrac{(x-4)^2}{16} + \dfrac{y^2}{49} = 1$

b. $x^2 + 6x - y^2 + 2y - 16 = 0$

c. $x = -4y^2 - y + 1$

d. $x^2 + 2x + y^2 - 4y = 40$

SECTION 10.4	*Solving Simultaneous Second-Degree Equations*

17. The graphs of $y^2 - x^2 = 9$ and $x^2 + y^2 = 9$ are shown. Estimate the solutions of the system

$$\begin{cases} y^2 - x^2 = 9 \\ x^2 + y^2 = 9 \end{cases}$$

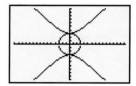

Solve each system.

18. $\begin{cases} 3x^2 + y^2 = 52 \\ x^2 - y^2 = 12 \end{cases}$

19. $\begin{cases} \dfrac{x^2}{16} + \dfrac{y^2}{12} = 1 \\ x^2 - \dfrac{y^2}{3} = 1 \end{cases}$

20. $\begin{cases} y = -x^2 + 2 \\ x^2 - y - 2 = 0 \end{cases}$

21. $\begin{cases} x^2 + 2y^2 = 12 \\ 2x - y = 2 \end{cases}$

1. Find the center and the radius of the circle $(x - 2)^2 + (y + 3)^2 = 4$.

6. $\dfrac{(x - 2)^2}{9} - y^2 = 1$

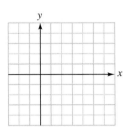

2. Find the center and the radius of the circle $x^2 + y^2 + 4x - 6y = 3$.

Graph each equation.

3. $(x + 1)^2 + (y - 2)^2 = 9$

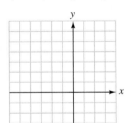

7. Graph $xy = -4$.

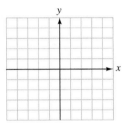

Write each equation in standard form and graph the equation.

8. $x^2 + 4y^2 + 6x + 16y = -9$

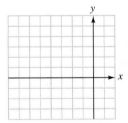

4. $x = (y - 2)^2 - 1$

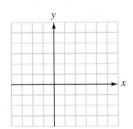

9. $4y^2 - 9x^2 - 8y = 32$

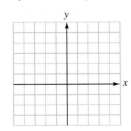

5. $9x^2 + 4y^2 = 36$

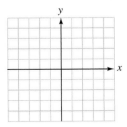

10. Solve the system graphically.

$$\begin{cases} x^2 + y^2 = 25 \\ y - x = 1 \end{cases}$$

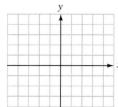

Solve each system.

11. $\begin{cases} 2x - y = -2 \\ x^2 + y^2 = 16 + 4y \end{cases}$

12. $\begin{cases} x^2 + y^2 = 25 \\ 4x^2 - 9y = 0 \end{cases}$

13. $\begin{cases} 5x^2 - y^2 - 3 = 0 \\ x^2 + 2y^2 = 5 \end{cases}$

14. TV HISTORY   In the early days of television, stations broadcast a black-and-white test pattern like that shown below during the early morning hours. Use the given coordinate system to write an equation of the large, bold circle in the center of the pattern.

15. SOUND   The equation $x = -\frac{1}{10}y^2$ defines a cross-section view of a parabolic dish. Construct a table of solutions for the equation, plot the ordered pairs, and connect the points with a smooth curve. Then plot the point $(-2.5, 0)$, which locates the microphone that picks up reflected sound waves. Draw a line parallel

to the axis of symmetry coming into the dish and striking the dish at $(-0.9, 3)$ and reflecting into the microphone.

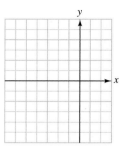

16. Write the equation in standard form of the ellipse graphed below.

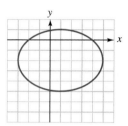

17. What is the equation in standard form of the hyperbola graphed below?

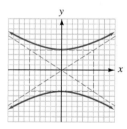

18. Determine whether the equation, when graphed, will be a circle, a parabola, an ellipse, or a hyperbola.

  a. $25x^2 + 100y^2 = 400$

  b. $x^2 - y^2 = 1$

  c. $x^2 + 8x + y^2 - 16y - 1 = 0$

  d. $x = 8y^2 - 9y + 4$

Consider the set $\{-\frac{4}{3}, \pi, 5.6, \sqrt{2}, 0, -23, e, 7i\}$. List the elements in the set that are

1. whole numbers

2. rational numbers

3. irrational numbers

4. real numbers

5. **FINANCIAL PLANNING** Ana has some money to invest. Her financial planner tells her that if she can come up with $3,000 more, she will qualify for an 11% annual interest rate. Otherwise, she will have to invest the money at 7.5% annual interest. The financial planner urges her to invest the larger amount, because the 11% investment would yield twice as much annual income as the 7.5% investment. How much does she originally have on hand to invest?

6. **BOATING** Use the graph below to determine the average rate of change in the sound level of the engine of a boat in relation to rpm of the engine.

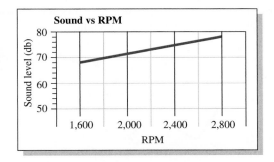

Determine whether the graphs of the equations are parallel or perpendicular.

7. $3x - 4y = 12$, $y = \frac{3}{4}x - 5$

8. $y = 3x + 4$, $x = -3y + 4$

Write an equation of the line with the given properties in slope–intercept form.

9. $m = -2$, passing through $(0, 5)$

10. Passing through $(8, -5)$ and $(-5, 4)$

11. Use substitution to solve $\begin{cases} 3x + y = 4 \\ 2x - 3y = -1 \end{cases}$.

12. Use addition to solve $\begin{cases} x + 2y = -2 \\ 2x - y = 6 \end{cases}$.

13. Solve using Cramer's rule: $\begin{cases} 4x - 3y = -1 \\ 3x + 4y = -7 \end{cases}$.

14. Solve: $\begin{cases} b + 2c = 7 - a \\ a + c = 8 - 2b \\ 2a + b + c = 9 \end{cases}$.

15. The graphs of $y = 4(x - 5) - x - 2$ and $y = -(2x + 6) - 1$ are shown below. Use the information in the display to solve $4(x - 5) - x - 2 = -(2x + 6) - 1$ graphically.

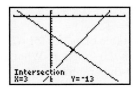

16. Evaluate: $\begin{vmatrix} 3 & -2 \\ 1 & -1 \end{vmatrix}$.

17. **MARTIAL ARTS** Find the measure of each angle of the triangle shown.

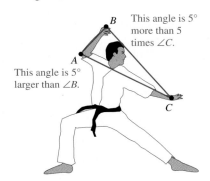

This angle is 5° more than 5 times $\angle C$.

This angle is 5° larger than $\angle B$.

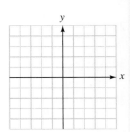

18. Solve: $\begin{cases} 3x - 2y \le 6 \\ y < -x + 2 \end{cases}$.

Give the solution set in interval notation and then graph it.

19. Solve: $|5 - 3x| \le 14$.

20. Solve: $4.5x - 1 < -10$ or $6 - 2x \ge 12$.

Perform the operations.

21. $(4x - 3y)(3x + y)$

22. $(-2x^2y^3 + 6xy + 5y^2) - (-4x^2y^3 - 7xy + 2y^2)$

23. $(a - 2b)^2$

24. $(a + 2)(3a^2 + 4a - 2)$

Factor the expression completely.

25. $3x^3y - 4x^2y^2 - 6x^2y + 8xy^2$

26. $256x^4y^4 - z^8$

27. Solve for λ: $\dfrac{A\lambda}{2} + 1 = 2d + 3\lambda$.

28. Complete the table of values for $f(x) = -x^3 - x^2 + 6x$ and then graph the function. What are the x- and y-intercepts of the graph?

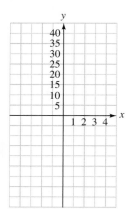

x	$f(x)$
-4	
-3	
-2	
-1	
0	
1	
2	
3	

Simplify.

29. $\left(\dfrac{4a^{-2}b}{3ab^{-3}}\right)^3$

30. $\dfrac{6x^2 + 13x + 6}{6 - 5x - 6x^2}$

31. $\dfrac{p^3 - q^3}{q^2 - p^2} \cdot \dfrac{q^2 + pq}{p^3 + p^2q + pq^2}$

32. $\dfrac{2}{a - 2} + \dfrac{3}{a + 2} - \dfrac{a - 1}{a^2 - 4}$

33. Solve: $\dfrac{x - 4}{x - 3} + \dfrac{x - 2}{x - 3} = x - 3$.

34. Solve $\dfrac{1}{R} = \dfrac{1}{R_1} + \dfrac{1}{R_2} + \dfrac{1}{R_3}$ for R.

35. TIRE WEAR See the illustration below.

　　a. What type of function does it appear would model the relationship between the inflation of a tire and the percent of service it gives?

　　b. At what percent(s) of inflation will a tire offer only 90% of its possible service?

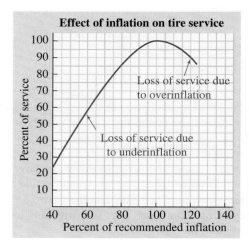

Effect of inflation on tire service

Loss of service due to overinflation

Loss of service due to underinflation

Percent of service / Percent of recommended inflation

36. CHANGING DIAPERS The illustration shows how to put a diaper on a baby. If the diaper is a square with sides 16 inches long, what is the largest waist size that this diaper can wrap around, assuming an overlap of 1 inch to pin the diaper?

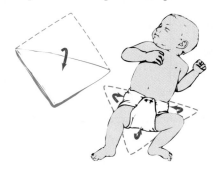

37. Use the long division method to find

$$(2x^2 + 4x - x^3 + 3) \div (x - 1)$$

38. The graph of $f(x) = \sqrt{2x + 5} + \sqrt{x + 2} - 5$ is shown. Use the information in the display to determine the solution of $\sqrt{2x + 5} = -\sqrt{x + 2} + 5$.

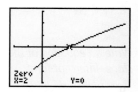

Zero
X=2　　　　Y=0

Simplify each expression.

39. $\sqrt{98} + \sqrt{8} - \sqrt{32}$

40. $12\sqrt[3]{648x^4} + 3\sqrt[3]{81x^4}$

41. Evaluate: $\left(\dfrac{25}{49}\right)^{-3/2}$.

42. Rationalize the denominator: $\dfrac{3t-1}{\sqrt{3t}+1}$.

Simplify: Write the result in the form a + bi.

43. $(-7+\sqrt{-81})-(-2-\sqrt{-64})$

44. $\dfrac{2-5i}{2+5i}$

Solve each equation.

45. $\sqrt{3a+1}=a-1$

46. $\sqrt{x+3}-\sqrt{3}=\sqrt{x}$

47. $6a^2+5a-6=0$

48. $4w^2+6w+1=0$

49. $2(2x+1)^2-7(2x+1)+6=0$

50. $3x^2-4x=-2$

51. If $f(x)=x^2-2$ and $g(x)=2x+1$, find $(f\circ g)(x)$.

52. Find the inverse function of $f(x)=2x^3-1$.

53. Graph: $f(x)=\left(\dfrac{1}{2}\right)^x$.

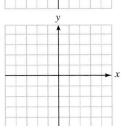

54. Graph $y=e^x$ and its inverse on the same coordinate system. Label the axis of symmetry.

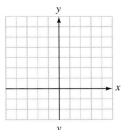

55. Write $y=\log_2 x$ as an exponential equation.

56. Apply properties of logarithms to simplify $\log_6 \dfrac{x}{36}$.

Find x.

57. $\log_x 25=2$

58. $\log_5 125=x$

59. $\log_3 x=-3$

60. $\ln e=x$

61. Find the inverse of $f(x)=\log_2 x$.

62. If $\log_{10} 10^x=y$, then y equals what quantity?

Let $\log 7=0.8451$ *and* $\log 14=1.1461$. *Evaluate each expression without using a calculator or tables.*

63. $\log 98$

64. $\log 2$

Solve each equation. Round to four decimal places when necessary.

65. $2^{x+2}=3^x$

66. $\log x+\log(x+9)=1$

67. $5^{4x}=\dfrac{1}{125}$

68. $\log_3 x=\log_3\left(\dfrac{1}{x}\right)+4$

69. BOAT DEPRECIATION How much will a $9,000 boat be worth after 9 years if it depreciates 12% per year?

70. Find $\log_6 8$.

71. Write the equation of the circle that has its center at $(1, 3)$ and passes through $(-2, -1)$.

72. Write the equation in standard form and then graph it.
$$x^2-9y^2+2x+36y=44$$

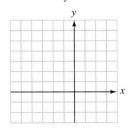

73. Graph: $\dfrac{(x-1)^2}{9}+\dfrac{(y-3)^2}{4}=1$.

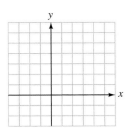

74. Graph: $y^2+4x-6y=-1$.

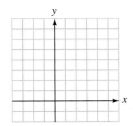

Fractions and Decimals

- Fractions • Simplifying fractions • Multiplying and dividing fractions
- Adding and subtracting fractions • Mixed numbers • Decimals
- Writing fractions as decimals

It is often said, "A building is only as strong as its foundation." The same is true of studying mathematics. A thorough understanding of arithmetic is essential to your success in algebra.

In arithmetic, we learned how to add, subtract, multiply, and divide with **whole numbers:** 0, 1, 2, 3, 4, 5, 6, 7, ... (The dots indicate that the numbers continue indefinitely in the same pattern.) Assuming that you have mastered those skills, we will now strengthen your mathematical foundation by reviewing the rules for arithmetic operations with fractions and decimals.

Fractions

In the **fractions**

$$\frac{1}{2}, \quad \frac{3}{5}, \quad \frac{2}{17}, \quad \text{and} \quad \frac{37}{7}$$

the number above the bar is called the **numerator,** and the number below it is called the **denominator.**

Fractions can be used to indicate parts of a whole. In Figure I-1(a), a rectangle has been divided into 3 equal parts, with 1 of the parts shaded. The fraction $\frac{1}{3}$ indicates how much of the figure is shaded. In Figure I-1(b), $\frac{4}{5}$ of the circle is shaded. In either example, the denominator of the fraction shows the total number of equal parts into which the whole is divided, and the numerator shows the number of these equal parts that are shaded.

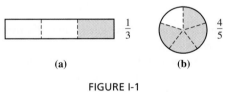

FIGURE I-1

Fractions with a numerator that is less than the denominator, such as $\frac{1}{3}$ and $\frac{4}{5}$, are called **proper fractions.** A proper fraction is less than 1. Fractions with a numerator that is equal to or greater than the denominator, such as $\frac{47}{47}$ and $\frac{5}{4}$, are called **improper fractions.** An improper fraction is greater than or equal to 1.

Fractions can also be used to indicate division. For example, the fraction $\frac{8}{2}$ indicates that 8 is to be divided by 2:

$$\frac{8}{2} = 8 \div 2 = 4$$

We note that $\frac{8}{2} = 4$, because $4 \cdot 2 = 8$, and that $\frac{0}{7} = 0$, because $0 \cdot 7 = 0$. However, the fraction $\frac{6}{0}$ is undefined, because no number multiplied by 0 gives 6. The fraction $\frac{0}{0}$ is indeterminate, because every number multiplied by 0 gives 0.

! **COMMENT** Remember that the denominator of a fraction cannot be zero.

Simplifying fractions

A fraction is in **lowest terms** when no natural number greater than 1 will divide both the numerator and the denominator exactly (with no remainder). The fraction $\frac{6}{11}$ is in lowest terms, because only 1 divides both 6 and 11 exactly. The fraction $\frac{6}{8}$ is not in lowest terms, because 2 divides both 6 and 8 exactly.

We can **simplify** (or **reduce**) a fraction that is not in lowest terms by dividing both its numerator and denominator by the same number. For example, to simplify the fraction $\frac{6}{8}$, we divide both numerator and denominator by 2:

$$\frac{6}{8} = \frac{6 \div 2}{8 \div 2} = \frac{3}{4}$$

From Figure I-2, we see that $\frac{6}{8}$ and $\frac{3}{4}$ represent identical shaded amounts of the rectangle. Two fractions that represent the same number are called **equivalent fractions.** Equivalent fractions may look different, but they have the same value.

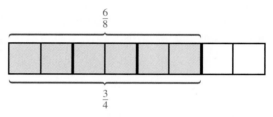

FIGURE I-2

When simplifying a fraction, it is often helpful to write the numerator and denominator in factored form. To **factor** a number means to express it as the product of two or more numbers. For example, 8 can be written as

$$1 \cdot 8, \quad 4 \cdot 2, \quad \text{and} \quad 2 \cdot 2 \cdot 2$$

In each case, we say that we have **factored** 8. The numbers 1, 2, 4, and 8 are called the **factors** of 8.

Some numbers have only two factors. For example, the only factors of 7 are 1 and 7, and the only factors of 23 are 1 and 23. We call such numbers **prime numbers.**

Prime numbers

A **prime number** is a whole number greater than 1 that has only 1 and itself as factors. The prime numbers are 2, 3, 5, 7, 11, 13, 17, 19, 23, 29, 31, ...

When a number is written as the product of prime numbers, we say that it is written in **prime-factored** form.

Self Check 1
Write 189 in prime-factored form.

EXAMPLE 1   Write 210 in prime-factored form.

Solution   We can write 210 as the product of 21 and 10 and proceed as follows:

$$210 = 21 \cdot 10$$
$$210 = 3 \cdot 7 \cdot 2 \cdot 5 \qquad \text{Factor 21 as } 3 \cdot 7 \text{ and factor 10 as } 2 \cdot 5.$$

The *factor tree* at the right shows an alternate way to prime factor 210.

Since 210 is now written as the product of prime numbers, its prime-factored form is $3 \cdot 7 \cdot 2 \cdot 5$.

Answer $7 \cdot 3 \cdot 3 \cdot 3$

To simplify a fraction, we factor its numerator and denominator and then divide out all common factors that appear in both the numerator and denominator. To simplify the fractions $\frac{6}{8}$ and $\frac{189}{210}$, for example, we proceed as follows:

$$\frac{6}{8} = \frac{3 \cdot 2}{4 \cdot 2} = \frac{3 \cdot \overset{1}{\cancel{2}}}{4 \cdot \underset{1}{\cancel{2}}} = \frac{3}{4} \quad \text{and} \quad \frac{189}{210} = \frac{\overset{1}{\cancel{7}} \cdot \overset{1}{\cancel{3}} \cdot 3 \cdot 3}{\underset{1}{\cancel{3}} \cdot \underset{1}{\cancel{7}} \cdot 2 \cdot 5} = \frac{9}{10}$$

Slashes and small 1's are used to show that common factors of the numerator and denominator have been divided out.

❗ COMMENT When working with fractions, you should always give your answers in lowest terms. Remember that a fraction is in lowest terms only when its numerator and denominator have no common factors other than 1.

EXAMPLE 2 Simplify each fraction, if possible: **a.** $\dfrac{6}{30}$ and **b.** $\dfrac{33}{40}$.

Solution

a. To simplify $\frac{6}{30}$, we factor the numerator and denominator and divide out the common factor of 6.

$$\frac{6}{30} = \frac{6 \cdot 1}{6 \cdot 5} = \frac{\overset{1}{\cancel{6}} \cdot 1}{\underset{1}{\cancel{6}} \cdot 5} = \frac{1}{5} \qquad \text{Perform the multiplications: } 1 \cdot 1 = 1 \text{ and } 1 \cdot 5 = 5.$$

b. To try to simplify $\frac{33}{40}$, we write the numerator and denominator in prime-factored form.

$$\frac{33}{40} = \frac{3 \cdot 11}{2 \cdot 2 \cdot 2 \cdot 5}$$

Since the numerator and denominator have no common factors, $\frac{33}{40}$ is in lowest terms.

Self Check 2
Simplify each fraction, if possible.

a. $\dfrac{24}{56}$

b. $\dfrac{16}{125}$

Answers **a.** $\dfrac{3}{7}$,

b. in lowest terms

The previous examples illustrate an important application of the **fundamental property of fractions.**

> **The fundamental property of fractions**
>
> Multiplying or dividing the numerator and the denominator of a fraction by the same nonzero number does not change the value of the fraction. In symbols, if a, b, and c represent numbers (and b and c are not zero),
>
> $$\frac{a}{b} = \frac{a \cdot c}{b \cdot c} \qquad \text{and} \qquad \frac{a}{b} = \frac{a \div c}{b \div c}$$

▮ Multiplying and dividing fractions

> ### Multiplying fractions
>
> To multiply two fractions, we multiply their numerators and multiply their denominators. In symbols, if a, b, c, and d represent numbers,
>
> $$\frac{a}{b} \cdot \frac{c}{d} = \frac{a \cdot c}{b \cdot d} \quad \text{where } b \neq 0 \text{ and } d \neq 0$$

❗ COMMENT After multiplying fractions, we should simplify the result, if possible.

Self Check 3
Find

a. $\dfrac{5}{9} \cdot \dfrac{2}{3}$

b. $\dfrac{6}{25}\left(\dfrac{5}{6}\right)$

EXAMPLE 3 Find **a.** $\dfrac{7}{8} \cdot \dfrac{3}{5}$ and **b.** $\dfrac{5}{8}\left(\dfrac{4}{5}\right)$.

Solution

a. $\dfrac{7}{8} \cdot \dfrac{3}{5} = \dfrac{7 \cdot 3}{8 \cdot 5}$ Multiply the numerators and multiply the denominators.

$\qquad\qquad = \dfrac{21}{40}$ Perform the multiplications: $7 \cdot 3 = 21$ and $8 \cdot 5 = 40$.

This result cannot be simplified.

b. $\dfrac{5}{8}\left(\dfrac{4}{5}\right) = \dfrac{5 \cdot 4}{8 \cdot 5}$ Multiply the numerators and multiply the denominators.

Here, we can simplify the result by dividing out common factors. The numerator and denominator have a common factor of 5. Furthermore, if we factor 8 as $4 \cdot 2$, there will also be a common factor of 4.

$\dfrac{5}{8}\left(\dfrac{4}{5}\right) = \dfrac{5 \cdot 4}{4 \cdot 2 \cdot 5}$ Factor: $8 = 4 \cdot 2$.

$\qquad\qquad = \dfrac{\overset{1}{\cancel{5}} \cdot \overset{1}{\cancel{4}}}{\underset{1}{\cancel{4}} \cdot 2 \cdot \underset{1}{\cancel{5}}}$ Divide out the common factors of 4 and 5.

$\qquad\qquad = \dfrac{1}{2}$ Perform the multiplications: $1 \cdot 1 = 1$ and $1 \cdot 2 \cdot 1 = 2$.

Answers a. $\dfrac{10}{27}$, **b.** $\dfrac{1}{5}$

One number is called the **reciprocal** of another if their product is 1. For example, $\frac{3}{5}$ is the reciprocal of $\frac{5}{3}$, because

$$\frac{3}{5} \cdot \frac{5}{3} = \frac{15}{15} = 1$$

> ### Dividing fractions
>
> To divide two fractions, we multiply the first fraction by the reciprocal of the second fraction. In symbols, if a, b, c, and d represent numbers,
>
> $$\frac{a}{b} \div \frac{c}{d} = \frac{a}{b} \cdot \frac{d}{c} \quad \text{where } b \neq 0, c \neq 0, \text{ and } d \neq 0$$

EXAMPLE 4   Find $\dfrac{1}{3} \div \dfrac{4}{5}$.

Solution

$$\dfrac{1}{3} \div \dfrac{4}{5} = \dfrac{1}{3} \cdot \dfrac{5}{4} \qquad \text{Multiply } \dfrac{1}{3} \text{ by the reciprocal of } \dfrac{4}{5}, \text{ which is } \dfrac{5}{4}.$$

$$= \dfrac{1 \cdot 5}{3 \cdot 4} \qquad \text{Multiply the numerators and multiply the denominators.}$$

$$= \dfrac{5}{12} \qquad \text{Multiply in the numerator and denominator.}$$

Self Check 4

Find $\dfrac{2}{3} \div \dfrac{7}{8}$.

Answer $\dfrac{16}{21}$

■ Adding and subtracting fractions

Adding and subtracting fractions with like denominators

To add (or subtract) two fractions with the same denominator, we add (or subtract) the numerators and keep the common denominator. In symbols, if a, b, and d represent numbers,

$$\dfrac{a}{d} + \dfrac{b}{d} = \dfrac{a+b}{d} \quad \text{and} \quad \dfrac{a}{d} - \dfrac{b}{d} = \dfrac{a-b}{d} \quad \text{provided that } d \neq 0$$

For example,

$$\dfrac{3}{7} + \dfrac{2}{7} = \dfrac{3+2}{7} \quad \text{and} \quad \dfrac{7}{9} - \dfrac{2}{9} = \dfrac{7-2}{9}$$

$$= \dfrac{5}{7} \qquad\qquad\qquad\quad = \dfrac{5}{9}$$

! COMMENT Remember that only *factors* that are common to the entire numerator and the entire denominator can be divided out. To simplify $\dfrac{5+8}{5}$, it would be incorrect to divide out the 5. Doing so would give an incorrect answer of 9.

Correct   **Incorrect**

$$\dfrac{5+8}{5} = \dfrac{13}{5} \qquad\qquad \dfrac{5+8}{5} = \dfrac{\overset{1}{\cancel{5}}+8}{\underset{1}{\cancel{5}}} = \dfrac{1+8}{1} = 9$$

To add fractions with unlike denominators, we rewrite the fractions so that they have a common denominator. The smallest common denominator, called the **least** or **lowest common denominator,** is usually the easiest common denominator to use.

Least common denominator

The **least common denominator (LCD)** for a set of fractions is the smallest number each denominator will divide exactly.

In the problem $\dfrac{3}{5} + \dfrac{1}{3}$, the denominators of the fractions are 5 and 3. The numbers 5 and 3 divide many numbers exactly (30, 45, and 60, to name a few), but the smallest number that 5 and 3 divide exactly is 15. Thus, 15 is the LCD for $\dfrac{3}{5}$ and $\dfrac{1}{3}$.

To find $\dfrac{3}{5} + \dfrac{1}{3}$, we express each fraction as an equivalent fraction with a denominator of 15. This process, known as **expressing a fraction in higher terms,** is an application of the fundamental property of fractions. To **build** each fraction so that it has a denominator of 15, we multiply its numerator and denominator by the same number.

$$\frac{3}{5} + \frac{1}{3} = \frac{3 \cdot 3}{5 \cdot 3} + \frac{1 \cdot 5}{3 \cdot 5}$$

↑ We need to multiply this denominator by 3 to obtain 15. We must also multiply the numerator by 3.

↑ We need to multiply this denominator by 5 to obtain 15. We must also multiply the numerator by 5.

$$= \frac{9}{15} + \frac{5}{15}$$ Perform the multiplications in the numerators and denominators. Note that the denominators are now the same.

$$= \frac{9 + 5}{15}$$ Add the numerators and write the sum over the common denominator 15.

$$= \frac{14}{15}$$ Perform the addition: $9 + 5 = 14$.

Self Check 5

Find $\dfrac{11}{48} - \dfrac{7}{40}$.

EXAMPLE 5 Find $\dfrac{3}{10} - \dfrac{5}{28}$.

Solution To find the LCD, we find the prime factorization of both denominators and use each prime factor the *greatest* number of times it appears in any one factorization:

$$\left. \begin{array}{l} 10 = 2 \cdot 5 \\ 28 = 2 \cdot 2 \cdot 7 \end{array} \right\} \quad \text{LCD} = 2 \cdot 2 \cdot 5 \cdot 7 = 140$$

2 appears twice in the factorization of 28.
5 appears once in the factorization of 10.
7 appears once in the factorization of 28.

Since 140 is the smallest number that 10 and 28 divide exactly, we write both fractions as fractions with the LCD of 140.

$$\frac{3}{10} - \frac{5}{28} = \frac{3 \cdot 14}{10 \cdot 14} - \frac{5 \cdot 5}{28 \cdot 5}$$ We must multiply 10 by 14 to obtain 140. We must multiply 28 by 5 to obtain 140.

$$= \frac{42}{140} - \frac{25}{140}$$ Perform the four multiplications.

$$= \frac{42 - 25}{140}$$ Subtract the numerators and keep the common denominator.

$$= \frac{17}{140}$$ Perform the subtraction.

Answer $\dfrac{13}{240}$

Since 17 is a prime number, it has no common factor with 140. Thus, $\frac{17}{140}$ is in lowest terms and cannot be simplified.

■ Mixed numbers

A **mixed number** is the sum of a whole number and a proper fraction. For example, the mixed number $3\frac{3}{4}$ means $3 + \frac{3}{4}$.

Self Check 6

Find $2\dfrac{3}{4} \cdot 6$.

EXAMPLE 6 Find $3\dfrac{3}{4} \div 2$.

Solution To multiply or divide with mixed numbers, we first change the mixed numbers to improper fractions.

$$3\frac{3}{4} = \frac{3(4) + 3}{4} = \frac{15}{4}$$ To write a mixed number as an improper fraction, multiply the whole-number part by the denominator of the fraction and add the result to the numerator. Then write that sum over the denominator.

Now we replace $3\frac{3}{4}$ with $\frac{15}{4}$ and divide.

$$3\frac{3}{4} \div 2 = \frac{15}{4} \div \frac{2}{1}$$ Write $3\frac{3}{4}$ as $\frac{15}{4}$ and write 2 as $\frac{2}{1}$.

$$= \frac{15}{4} \cdot \frac{1}{2}$$ Multiply by the reciprocal of $\frac{2}{1}$, which is $\frac{1}{2}$.

$$= \frac{15}{8}$$ Multiply the numerators and multiply the denominators.

Answer $\dfrac{33}{2}$

! COMMENT In algebra, you will see that it is usually preferable to work with an improper fraction rather than the equivalent mixed number.

EXAMPLE 7   Find $2\frac{1}{4} + 1\frac{1}{3}$. Write the answer as a mixed number.

Solution   We change the mixed numbers to improper fractions and proceed as follows.

$$2\frac{1}{4} + 1\frac{1}{3} = \frac{9}{4} + \frac{4}{3}$$ Write $2\frac{1}{4}$ as $\frac{9}{4}$ and $1\frac{1}{3}$ as $\frac{4}{3}$.

$$= \frac{9 \cdot 3}{4 \cdot 3} + \frac{4 \cdot 4}{3 \cdot 4}$$ The LCD is 12.

$$= \frac{27}{12} + \frac{16}{12}$$ Perform the four multiplications.

$$= \frac{43}{12}$$ Perform the addition: $27 + 16 = 43$.

Finally, we change $\frac{43}{12}$ to a mixed number.

$$\begin{array}{r} 3 \\ 12\overline{)43} \\ -36 \\ \hline 7 \end{array}$$ To write an improper fraction as a mixed number, divide the numerator by the denominator to obtain the whole-number part. The remainder over the divisor is the fractional part.

$$\frac{43}{12} = 3\frac{7}{12}$$

Self Check 7
Find $4\frac{1}{6} + 1\frac{1}{5}$. Write the answer as a mixed number.

Answer $5\frac{11}{30}$

▪ Decimals

In the **decimal numeration system,** columns to the left of the decimal point have a value greater than or equal to 1. Columns to the right of the decimal point have a value less than 1. (See Figure I-3.)

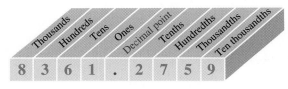

FIGURE I-3

We can use a vertical column format to add or subtract decimals. For example,

$$
\begin{array}{r}
25.568 \\
+\ 2.74 \\
\hline
28.308
\end{array}
\qquad
\begin{array}{r}
25.568 \\
-\ 2.74 \\
\hline
22.828
\end{array}
$$

First, line up the decimal points. Then add (or subtract) the numbers as if they were whole numbers. Finally, write the decimal point in the result directly below the decimal points in the problem.

To multiply decimals, begin by multiplying the numbers as if they were whole numbers. Then we place the decimal point in the result so that the number of decimal places in the answer is equal to the sum of the decimal places in the factors.

$$
\begin{array}{r}
3.453 \quad \leftarrow \text{Here there are three decimal places.}\\
\times\ \ 9.25 \quad \leftarrow \text{Here there are two decimal places.}\\
\hline
17265 \\
6906\ \ \\
31077\ \ \ \ \\
\hline
31.94025 \quad \leftarrow \text{The product has } 3 + 2 = 5 \text{ decimal places.}
\end{array}
$$

To multiply a decimal by 10, 100, 1,000, and so on (such numbers are called **powers of 10**), we move the decimal point the same number of places to the right as the number of zeros in the power of 10.

One zero in 10

$$8.675 \cdot 10 = 86.75$$

Move the decimal point
1 place to the right.

Two zeros in 100

$$8.675 \cdot 100 = 867.5$$

Move the decimal point
2 places to the right.

To divide decimals, we move the decimal point in the divisor (the number that we are dividing by) to the right so that it becomes a whole number. We then move the decimal point in the dividend (the number being divided) the same number of places to the right.

$$1.23\overline{)30.258}$$

Move the decimal point in both the divisor and the dividend two places to the right.

We align the decimal point in the quotient with the repositioned decimal point in the dividend and use long division.

$$
\begin{array}{r}
24.6 \\
123\overline{)3025.8} \\
246 \\
\hline
565 \\
492 \\
\hline
73\ 8 \\
73\ 8 \\
\hline
0
\end{array}
$$

■ Writing fractions as decimals

To write a fraction as a decimal, we divide the numerator by the denominator. For example, to write $\frac{1}{4}$ and $\frac{5}{22}$ as decimals, we proceed as follows:

$$
\begin{array}{r}
0.25 \\
4\overline{)1.00} \\
8 \\
\hline
20 \\
20 \\
\hline
0
\end{array}
$$

$$
\begin{array}{r}
0.22727\ldots \\
22\overline{)5.00000} \\
4\ 4 \\
\hline
60 \\
44 \\
\hline
160 \\
154 \\
\hline
60 \\
44 \\
\hline
160
\end{array}
$$

The decimal 0.25 is called a **terminating decimal** because it terminates, or ends. The decimal 0.227272727 . . . is called a **repeating decimal** because it repeats the block of digits 27 indefinitely. We can use an **overbar** to write repeating decimals in a more concise form: $0.227272727 \ldots = 0.2\overline{27}$.

! COMMENT When using an overbar to designate a repeating decimal, place the bar over the smallest repeating block of digits. For example, write 0.2272727 . . . as $0.2\overline{27}$ not $0.22\overline{727}$.

Terminating decimals

$\frac{1}{2} = 0.5$

$\frac{5}{8} = 0.625$

$\frac{3}{4} = 0.75$

Repeating decimals

$\frac{1}{6} = 0.16666 \ldots$ or $0.1\overline{6}$

$\frac{1}{3} = 0.3333 \ldots$ or $0.\overline{3}$

$\frac{2}{3} = 0.6666 \ldots$ or $0.\overline{6}$

Appendix I STUDY SET

VOCABULARY *Fill in the blanks.*

1. The _____ of the fraction $\frac{3}{4}$ is 3, and the _____ is 4.

2. When we express $\frac{6}{8}$ as $\frac{3}{4}$, we say that we have _____ $\frac{6}{8}$ to lowest terms. A fraction is in _____ terms when no whole number greater than 1 will divide its numerator and denominator exactly.

3. Two fractions that represent the same number, such as $\frac{1}{2}$ and $\frac{2}{4}$, are called _____ fractions.

4. Numbers that have only 1 and themselves as factors, such as 23, 37, and 41, are called _____ numbers.

5. When we write 60 as $20 \cdot 3$, we say that we have _____ 60. When we write 60 as $5 \cdot 3 \cdot 2 \cdot 2$, we say that we have written 60 in _____ form.

6. The number $\frac{2}{3}$ is the _____ of the number $\frac{3}{2}$, because their product is 1.

7. The number 0.75 is called a _____ decimal, because it terminates, or ends. The number 0.111 . . . is called a _____ decimal.

8. The _____ common denominator for a set of fractions is the smallest number each denominator will divide exactly.

CONCEPTS

9. What equivalent fractions are shown below?

10. The prime factorization of a number is $2 \cdot 2 \cdot 3 \cdot 5$. What is the number?

11. To express $\frac{3}{8}$ as an equivalent fraction with a denominator of 40, by what number must we multiply the numerator and denominator?

12. **a.** Give three numbers that 4 and 6 divide exactly.

    b. What is the smallest number that 4 and 6 divide exactly?

13. The prime factorizations of 24 and 36 are

    $24 = 2 \cdot 2 \cdot 2 \cdot 3$

    $36 = 2 \cdot 2 \cdot 3 \cdot 3$

    a. What is the greatest number of times 2 appears in any one factorization?

    b. What is the greatest number of times 3 appears in any one factorization?

14. **a.** Write $2\frac{15}{16}$ as an improper fraction.

    b. Write $\frac{49}{12}$ as a mixed number.

NOTATION

15. Consider

    $$\frac{70}{175} = \frac{\overset{1}{\cancel{7}} \cdot \overset{1}{\cancel{5}} \cdot 2}{\cancel{7} \cdot \cancel{5} \cdot 5}$$
    $$\phantom{\frac{70}{175} = \frac{}{}}{\scriptstyle 1 \quad 1}$$

    What do the slashes and the 1's show?

16. Consider 2,345.6789.

    a. What digit is in the thousands column?

    b. What digit is in the thousandths column?

▮ PRACTICE   *List the factors of each number.*

17. 20   **18.** 50
19. 28   **20.** 36

Give the prime factorization of each number.

21. 75   **22.** 20
23. 28   **24.** 54
25. 117   **26.** 147
27. 220   **28.** 270

Build up each fraction or whole number to an equivalent fraction having the indicated denominator.

29. $\frac{1}{3}$, denominator 9   **30.** $\frac{3}{8}$, denominator 24

31. $\frac{4}{9}$, denominator 54   **32.** $\frac{9}{16}$, denominator 64

33. 7, denominator 5   **34.** 12, denominator 3

Write each fraction in lowest terms. If the fraction is already in lowest terms, so indicate.

35. $\frac{6}{12}$ **36.** $\frac{3}{9}$

37. $\frac{24}{18}$ **38.** $\frac{35}{14}$

39. $\frac{15}{20}$ **40.** $\frac{22}{77}$

41. $\frac{72}{64}$ **42.** $\frac{26}{21}$

43. $\frac{36}{225}$ **44.** $\frac{175}{490}$

Perform each multiplication. Simplify the result when possible.

45. $\frac{1}{2} \cdot \frac{3}{5}$ **46.** $\frac{3}{4} \cdot \frac{5}{7}$

47. $\frac{4}{3}\left(\frac{6}{5}\right)$ **48.** $\frac{7}{8}\left(\frac{6}{15}\right)$

49. $\frac{5}{12} \cdot \frac{18}{5}$ **50.** $\frac{5}{4} \cdot \frac{12}{10}$

51. $\frac{10}{21} \cdot 14$ **52.** $\frac{5}{24} \cdot 16$

53. $7\frac{1}{2} \cdot 1\frac{2}{5}$ **54.** $3\frac{1}{4}\left(1\frac{1}{5}\right)$

Perform each division. Simplify the result when possible.

55. $\frac{3}{5} \div \frac{2}{3}$ **56.** $\frac{4}{5} \div \frac{3}{7}$

57. $\frac{3}{4} \div \frac{6}{5}$ **58.** $\frac{3}{8} \div \frac{15}{28}$

59. $\frac{21}{35} \div \frac{3}{14}$ **60.** $\frac{23}{25} \div \frac{46}{5}$

61. $6 \div \frac{3}{14}$ **62.** $23 \div \frac{46}{5}$

63. $3\frac{1}{3} \div 1\frac{5}{6}$ **64.** $2\frac{1}{2} \div 1\frac{1}{2}$

Perform each addition or subtraction. Simplify the result when possible.

65. $\frac{3}{5} + \frac{3}{5}$ **66.** $\frac{4}{13} - \frac{3}{13}$

67. $\frac{1}{6} + \frac{1}{24}$ **68.** $\frac{17}{25} - \frac{2}{5}$

69. $\frac{3}{5} + \frac{2}{3}$ **70.** $\frac{4}{3} + \frac{7}{2}$

71. $\frac{9}{4} - \frac{5}{6}$ **72.** $\frac{2}{15} + \frac{7}{9}$

73. $\frac{7}{10} - \frac{1}{14}$ **74.** $\frac{7}{25} + \frac{3}{10}$

75. $\frac{5}{14} - \frac{4}{21}$ **76.** $\frac{2}{33} + \frac{3}{22}$

77. $3 - \frac{3}{4}$ **78.** $\frac{17}{3} + 4$

79. $3\frac{3}{4} - 2\frac{1}{2}$ **80.** $15\frac{5}{6} + 11\frac{5}{8}$

81. $8\frac{2}{9} - 7\frac{2}{3}$ **82.** $3\frac{4}{5} - 3\frac{1}{10}$

Perform each operation.

83. $23.45 + 135.2$ **84.** $345.213 - 27.35$
85. $67.235 - 22.45$ **86.** $12.17 + 3.457$
87. $3.4 \cdot 13.2$ **88.** $4.21 \cdot 2.73$
89. $0.23\overline{)1.0465}$ **90.** $4.7\overline{)10.857}$
91. $2.9517(1,000)$ **92.** $100(333.614)$
93. $100 \cdot 0.05$ **94.** $1,000 \cdot 0.085$

Write each fraction as a decimal. If the result is a repeating decimal, use an overbar.

95. $\frac{5}{8}$ **96.** $\frac{3}{32}$

97. $\frac{1}{30}$ **98.** $\frac{7}{9}$

99. $\dfrac{21}{50}$ **100.** $\dfrac{2}{125}$

101. $\dfrac{5}{11}$ **102.** $\dfrac{1}{60}$

APPLICATIONS

103. BOTANY To assess the effects of smog on tree development, botanists cut down a pine tree and measured the width of the growth rings for the last 2 years.

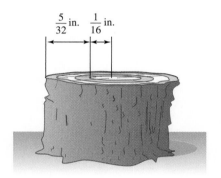

$\dfrac{5}{32}$ in. $\dfrac{1}{16}$ in.

   a. What was the growth over this 2-year period (in inches)?
   b. What is the difference in the widths of the rings?

104. HARDWARE Refer to the illustration below. How long is the threaded part of the bolt?

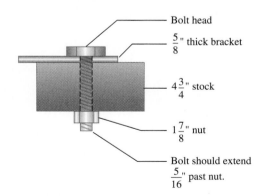

- Bolt head
- $\dfrac{5}{8}$" thick bracket
- $4\dfrac{3}{4}$" stock
- $1\dfrac{7}{8}$" nut
- Bolt should extend $\dfrac{5}{16}$" past nut.

105. FRAMES How much molding is needed to produce the square picture frame?

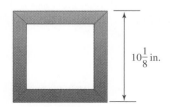

$10\dfrac{1}{8}$ in.

106. PRODUCTION PLANNING The materials used to make a pillow are shown. Examine the inventory list to decide how many pillows can be manufactured in one production run with the materials in stock.

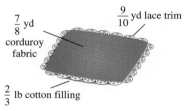

$\dfrac{7}{8}$ yd corduroy fabric

$\dfrac{9}{10}$ yd lace trim

$\dfrac{2}{3}$ lb cotton filling

Factory Inventory List

Materials	Amount in stock
Lace trim	135 yd
Corduroy fabric	154 yd
Cotton filling	98 lb

107. VEHICLE SPECIFICATIONS Certain dimensions of a compact car are shown. What is the wheelbase of the car?

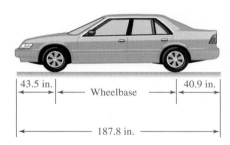

43.5 in.   Wheelbase   40.9 in.

187.8 in.

108. RETROFITTING FREEWAYS The illustration shows the width of the three columns of an existing freeway overpass. A computer analysis indicates that each column needs to be increased in width by a factor of 1.4 to ensure stability during an earthquake. According to the analysis, how wide should each of the columns be?

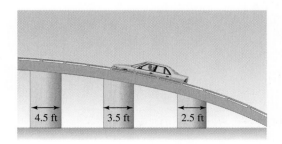

4.5 ft   3.5 ft   2.5 ft

▌ WRITING

109. What does it mean to say that an employee is *terminated*?

110. What are equivalent fractions?

111. How does the set of factors of a number differ from the prime factorization of the number? Give an example.

112. Explain the error in the following work.

$$\frac{4}{3} + \frac{3}{2} = \frac{4}{\overset{}{\underset{1}{3}}} + \frac{\overset{1}{3}}{5}$$

$$= \frac{4}{1} + \frac{1}{5}$$

$$= 4 + \frac{1}{5}$$

$$= 4\frac{1}{5}$$

Cumulative Review of Graphing

Graph each equation.

1. $y = 2x + 1$

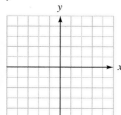

2. $-4y + 5x = -15$

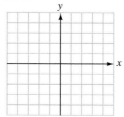

3. $x = 5$

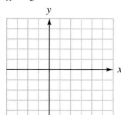

4. $y = -3$

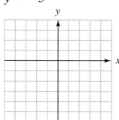

5. Solve the system using the graphing method.
$\begin{cases} 2x + y = 1 \\ x - 2y = -7 \end{cases}$

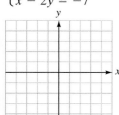

6. Graph $y > \dfrac{x}{3}$.

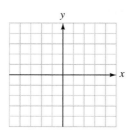

Graph the solution set.

7. $\begin{cases} 3x + 2y > 6 \\ x + 3y \le 2 \end{cases}$

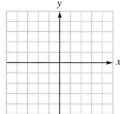

8. $y < -2$ or $y > 3$

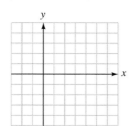

Graph each function.

9. $f(x) = x$

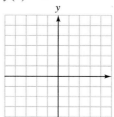

10. $f(x) = |x|$

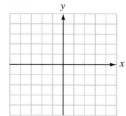

11. $f(x) = x^2$

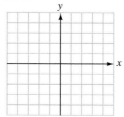

12. $f(x) = x^3$

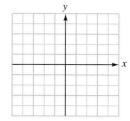

13. $f(x) = x^3 + 4x^2 + 4x$

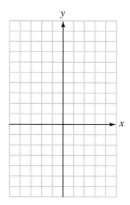

14. $f(x) = \dfrac{12}{x}$ for $x > 0$

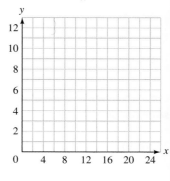

15. $f(x) = \sqrt{x}$

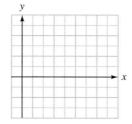

16. $f(x) = \sqrt[3]{x}$

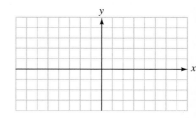

17. $f(x) = 2(x - 2)^2 - 4$

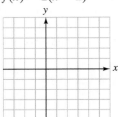

18. $f(x) = -12x^2 - 6x + 6$

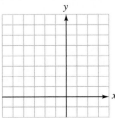

Graph each conic section.

27. $x^2 + y^2 = 16$

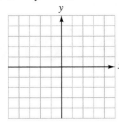

28. $x^2 + y^2 + 8x + 2y = -13$

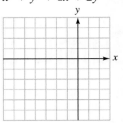

19. $f(x) = 5^x$

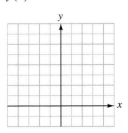

20. $y = \left(\dfrac{1}{4}\right)^x$

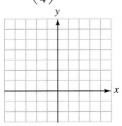

29. $x = -y^2 + 1$

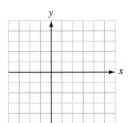

30. $\dfrac{x^2}{4} + \dfrac{y^2}{9} = 1$

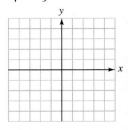

21. $f(x) = e^x$

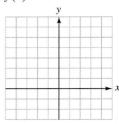

22. $f(x) = \log_3 x$

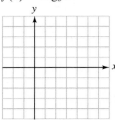

31. $\dfrac{(x + 2)^2}{16} - \dfrac{(y - 3)^2}{25} = 1$

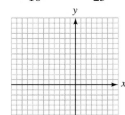

32. $xy = -10$

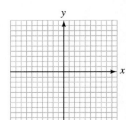

23. $f(x) = \log x$

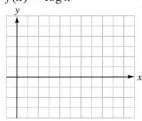

24. $f(x) = \ln x$

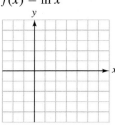

For each function, first sketch the associated function. Then draw the graph using a translation or a reflection.

33. $f(x) = |x + 2| - 3$

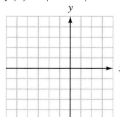

34. $f(x) = -x^3$

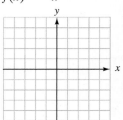

Graph the given function, its inverse, and the line $y = x$ on the same coordinate system.

25. $f(x) = \dfrac{1}{3}x + \dfrac{1}{3}$

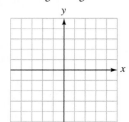

26. $f(x) = x^2 \quad (x \geq 0)$

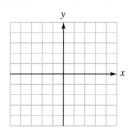

Answers to Selected Exercises

Chapter 1 Check Your Knowledge (page 2)

1. equation **2.** perimeter, area, volume **3.** real
4. like (or similar) **5.** $12 - x$
6. a.

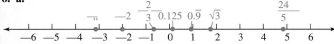

b. -2 **c.** $-2, -\frac{2}{3}, 0.125, 0.\overline{9}, \frac{24}{5}$ **d.** $-\pi, \sqrt{3}$

7. -2 **8.** -1 **9.** -125 **10.** $-\frac{1}{3}$ **11.** $-\frac{11}{40}$

12. 62 **13.** $\frac{8}{7}$ **14.** -3 **15. a.** $C = \frac{5}{9}(F - 32)$ **b.** -40.

16. a. distributive property **b.** associative property of
addition **17.** $-39x - \frac{y}{6}$ **18.** 3 **19.** 2 **20. a.** 11 **b.** 12
c. 12 **21.** 12 pounds $12, 24 pounds $6 **22.** 35 mph, 50 mph

Study Set Section 1.1 (page 8)

1. equation **3.** expressions **5.** formula **7.** multiplication
9. expression **11.** equation **13.** equation **15.** expression
17. a. a line graph **b.** 1 hour; 2 inches **c.** 7 in.; 0 in.
19. $c = 13u + 24$ **21.** $w = \frac{c}{75}$ **23.** $A = t + 15$ **25.** $c = 12b$
27. $h = t - 10$ **29. a.** D **b.** 2, 4 **c.** multiplication, addition
31. 2, 6, 15 **33.** 22.44, 21.43, 0 **35.** 37 in. **37.** 8 yd
39. a. $C = 10h + 20$ **b.** 30, 40, 50, 60, 100

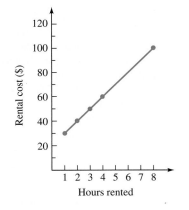

41. 150, 135, 90, 45, 30

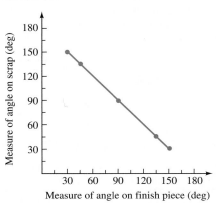

Study Set Section 1.2 (page 19)

1. rational **3.** absolute value **5.** Irrational **7.** composite
9. 1, 2, 9 **11.** $-3, 0, 1, 2, 9$ **13.** $\sqrt{3}, \pi$ **15.** 2 **17.** 2
19. 9 **21.** nonrepeating, irrational **23.** repeating, rational
25. $7 = \frac{7}{1}, -7\frac{3}{5} = \frac{-38}{5}, 0.007 = \frac{7}{1,000}, 700.1 = \frac{7,001}{10}$ **27.** 3.5
or -3.5 **29.**

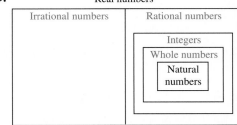

31. is less than **33.** braces **35.** 0.875, terminating
37. $-0.7\overline{3}$, repeating
39.

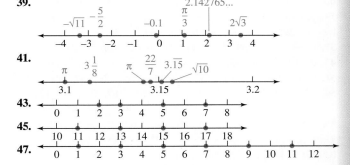

41.

43.

45.

47.

49. < **51.** > **53.** < **55.** > **57.** $12 < 19$ **59.** $-5 \geq -6$
61. 20 **63.** -6 **65.** 5.9 **67.** $\frac{5}{4}$ **69.** $3\frac{2}{25} = 3.0800, \frac{77}{50} = 1.5400,$
$\frac{15}{16} = 0.9375, 2\frac{5}{8} = 2.6250, \frac{\pi}{4} = 0.7854, \sqrt{8} = 2.8284$
71. 113.10 ft **77.** expression **79.** 2.2, 8.5, 29.1

Think It Through (page 27)

2.86

Study Set   Section 1.3 (page 32)

1. sum, difference **3.** evaluate **5.** squared, cubed **7.** base,
exponent **9.** opposite **11. a.** add first to get 18 or multiply
first to get 12 **b.** 12; multiplication is to be done before addition
13. a. area **b.** volume **c.** perimeter **15.** 8, 160
17. radical symbol
19. a. -6 **b.** 6 **21.** -8
23. -4.3 **25.** -7 **27.** 0
29. -12 **31.** -1.5 **33.** 60
35. -2 **37.** $\frac{1}{6}$ **39.** $\frac{11}{10}$
41. $-\frac{6}{7}$ **43.** $\frac{24}{25}$
45. 144 **47.** -25 **49.** 32
51. 1.69 **53.** 8 **55.** $-\frac{3}{4}$
57. -17 **59.** 64 **61.** 13
63. -2 **65.** -8 **67.** 10,000
69. -39 **71.** -32 **73.** 5
75. $-\frac{1}{2}$ **77.** 2 **79.** 91,985
81. -24 **83.** -2 **85.** 61 **87.** 1 **89.** 10 **91.** 11.3 cm^2
93. 94.35 cm^3 **95.** 775.73 m^3 **97.** 25 ft^2 **99.** 1st term: area
of bottom flap; 2nd term: area of left and right flaps; 3rd term:
area of top flap; 4th term: area of face; 42.5 in.^{2} **101.** \$(967)
103. yes **105.** 187 in.^{2} **107.** 9 units

Volume of cube (in.^{3}) — y-axis labeled 10 to 70; Length of edge (in.) — x-axis labeled 1 to 4

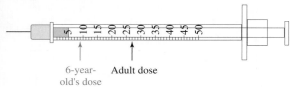

6-year-old's dose   Adult dose

113. -7 and 3 **115.** $\{\ldots, -2, -1, 0, 1, 2, \ldots\}$ **117.** true

Study Set   Section 1.4 (page 44)

1. like **3.** term **5.** simplify **7.** undefined
9. a. $(x + y) + z = x + (y + z)$ **b.** $xy = yx$ **c.** $r(s + t) = rs + rt$
11. a. 5 **b.** -5 **13. a.** $\frac{16}{15}$ **b.** $-\frac{1}{20}$ **c.** 2 **d.** $\frac{1}{x}$ **15. a.** $2x$
b. $-3a$ **c.** $-9t$ **d.** $22x$ **17. a.** $2x^2, -x, 6$ **b.** 2, $-1, 6$
19. yes, $8x$ **21.** no **23.** yes, $-2x^2$ **25.** no **27.** multiplication
by -1 **29.** $7 + 3$ **31.** $3 \cdot 2 + 3d$ **33.** c **35.** 1
37. $(8 + 7) + a$ **39.** $2(x + y)$ **41.** assoc. prop. of add.
43. distrib. prop. **45.** $-4t + 12$ **47.** $-t + 3$ **49.** $2s - 6$
51. $0.7s + 1.4$ **53.** $4x - 5y + 1$ **55.** $72m + 8n$ **57.** $24r + 18s$
59. $72m$ **61.** $-45q$ **63.** $30bp$ **65.** $80ry$ **67.** $18x$ **69.** $13x^2$
71. 0 **73.** 0 **75.** $6x$ **77.** 0 **79.** $3.1h$ **81.** $\frac{9}{10}ab$ **83.** $\frac{14}{15}t$
85. $-4y + 36$ **87.** $7z - 20$ **89.** $14c + 62$ **91.** $-2x^2 + 15x$
93. $-2a + b - 2$ **95.** $-6p + 17$ **97. a.** $20(x + 6)$ m^2
b. $(20x + 120)$ m^2 **c.** $20(x + 6) = 20x + 120$; distrib. prop.
99. $(9x - 54)$ in. **105.** 0 **107.** $-\frac{7}{8}$ **109.** 1,000 **111.** -3

Study Set   Section 1.5 (page 56)

1. equation **3.** satisfies **5.** contradiction **7.** c, c
9. a. $2y + 2$ **b.** 3 **c.** 18 **11.** $\mathbb{R}$ (all real numbers).
13. It clears the equation of fractions. **15.** $-2x$, 14, 14, 34,
$-2, -2$ **17. a.** -1 **b.** $\frac{2}{3}$ **19.** yes **21.** no **23.** 28 **25.** -20
27. $\frac{1}{2}$ **29.** $-\frac{6}{5}$ **31.** 0 **33.** 0 **35.** 2 **37.** -9 **39.** -166
41. 2.52 **43.** $\frac{2}{3}$ **45.** -8 **47.** 1.395 **49.** 3 **51.** 13 **53.** -11
55. -8 **57.** -2 **59.** 24 **61.** 0 **63.** 6 **65.** $\frac{21}{19}$ **67.** 3 **69.** 24
71. all real numbers, identity **73.** no solution, contradiction
75. all real numbers, identity
77. $B = \dfrac{3V}{h}$ **79.** $t = \dfrac{I}{Pr}$ **81.** $w = \dfrac{P - 2l}{2}$
83. $B = \dfrac{2A}{h} - b$ or $B = \dfrac{2A - bh}{h}$ **85.** $x = \dfrac{y - b}{m}$
87. $v_{\text{o}} = 2\overline{v} - v$ **89.** $l = \dfrac{a - S + Sr}{r}$ **91.** $l = \dfrac{2S - na}{n}$ or
$l = \dfrac{2S}{n} - a$ **93.** $C = \dfrac{5}{9}(F - 32)$; 432, -179; 58, -89; 17, -66
95. $d = \dfrac{360A}{\pi(r_1^2 - r_2^2)}$; 140, 160 **97.** $n = \dfrac{PV}{R(T + 273)}$; 0.008, 0.090
99. $n = \dfrac{C - 6.50}{0.07}$; 621, 1,000, about 1,692.9 kwh
101. $h = \dfrac{A - 2\pi r^2}{2\pi r}$ **107.** $t - 4$ **109.** $-4b + 32$ **111.** $2.9b$
113. t

Study Set   Section 1.6 (page 67)

1. acute **3.** complementary **5.** right **7.** angles **9.** $d + 15$,
$2d - 10, 2d + 20, \frac{d}{2} - 10, 2d$ **11. a.** $\frac{2}{3}x$ **b.** $2x$ **c.** $x + 2x + \frac{2}{3}x$
13. $26.5 - x$ **15.** Cheerios: \$663.5 million; Frosted Flakes:
\$339.5 million **17.** 20 **19.** 305 mi **21.** 300 shares of BB,
200 shares of SS **23.** 35 scientific, 50 graphing **25.** 7 ft, 15 ft
27. $30°, 150°$ **29.** $10°$ **31.** $50°$ **33.** $10°$ **35.** $60°$ **37.** ii
39. 156 ft by 312 ft **41.** 10 ft **43.** 6 in. **47.** repeating
49. $\{\ldots, -4, -3, -2, -1, 0, 1, 2, 3, 4, \ldots\}$ **51.** 0

Think It Through (page 74)

Medical assistant, about 59%

Study Set   Section 1.7 (page 81)

1. principal **3.** median **5.** amount, base **7. a.** 272,616
b. 272,616, what, 2,172,923 **9. a.** \$25,000 **b.** \$(30,000 − x)
11. 74.70, 2.79p, 2.59$(p + 30)$ **13.** $x = 0.05 \cdot 10.56$
15. $32.5 = 0.74x$ **17.** 100, 100, 9x, 40,000, 16,000 **19.** $I = Prt$
21. $v = pn$ **23.** about 415 quadrillion Btu **25.** 20% **27.** \$50
29. 9% **31.** -7.9%, 4.3% **33.** city: mean 37, median 32,
mode 32; hwy: mean 43.3, median 41, mode 41 **35.** 84
37. \$2,000 at 8%, \$10,000 at 9% **39.** \$45,000 **41.** \$100,000
43. $\frac{1}{4}$ hr = 15 min **45.** $\frac{2}{3}$ hr **47.** 3:30 P.M. **49.** $1\frac{1}{2}$ hr
51. 20 lb of \$1.90 candy; 10 lb of \$2.20 candy **53.** 4,000 ft^3 of
the premium mix, 2,000 ft^3 of sawdust **55.** 10 oz
57. 2 gal **63.** 0 **65.** 8

Key Concept (page 87)

1. given: 48 states, 4 more lie east of the Miss. River than west;
find: how many states lie west **3.** Let x = the length of the
shortest piece. **5. a.** subtraction **b.** multiplication
c. addition **d.** division **7.** 0.15(15), 0.50x, 0.40$(15 + x)$

Chapter Review (page 89)

1. $C = 2t + 15$ **2.** $l = \frac{25}{w}$ **3.** $P = u - 3$ **4.** 180, 195, 210, 225, 240 **5. a.**

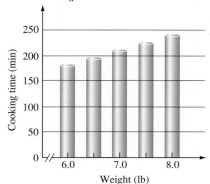

Weight (lb)

b.

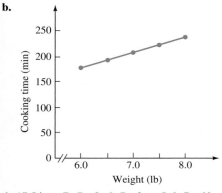

Weight (lb)

6. 17.5 in. **7.** 7 **8.** 0, 7 **9.** $-5, 0, 7$ **10.** $-5, 0, 2.4, 7, -\frac{2}{3},$ $-3.\overline{6}, \frac{15}{4}$ **11.** $-\sqrt{3}, \pi, 0.13242368\ldots$ **12.** all **13.** $-5, -\sqrt{3},$ $-\frac{2}{3}, -3.\overline{6}$ **14.** $2.4, 7, \pi, \frac{15}{4}, 0.13242368\ldots$ **15.** 7 **16.** none

17. 0 **18.** $-5, 7$ **19.** $>$ **20.** $<$ **21.** false **22.** true

23.

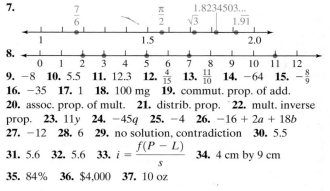

24.

25. 18 **26.** -6.26 **27.** -7 **28.** 10.1 **29.** $-\frac{3}{4}$ **30.** 2
31. 12.6 **32.** $-\frac{1}{32}$ **33.** 0.2 **34.** $-\frac{3}{56}$ **35.** -33 **36.** -5.7
37. -120 **38.** 1 **39.** -243 **40.** $\frac{4}{9}$ **41.** 0.027 **42.** -25
43. 2 **44.** -10 **45.** $\frac{3}{5}$ **46.** 0.8 **47.** 44 **48.** 1 **49.** -12
50. 58 **51.** 8 **52.** 3 **53.** 3,000 **54.** -64 **55.** 56 **56.** $-\frac{1}{2}$
57. 100 in.^{2} **58.** 251.3 in.^{3} **59.** $3x + 21$ **60.** $5t$
61. 0 **62.** $27 + (1 + 99)$ **63.** 1 **64.** m **65.** 1 **66.** 0
67. $-3(5 \cdot 2)$ **68.** $(z + t) \cdot t$ **69.** 1 **70.** 0 **71.** -25
72. undefined **73.** $8x + 48$ **74.** $-6x + 12$ **75.** $4 - 3y$
76. $3.6x - 2.4y$ **77.** $20c^2 - 10c + 10$ **78.** $2t + 6$ **79.** $48k$
80. $70xy$ **81.** $-189p$ **82.** $45a + 7$ **83.** 0 **84.** $3m - 48$
85. x **86.** $-24.54l$ **87.** $6t^3 + 4t^2$ **88.** $11h + 77$ **89.** yes
90. no **91.** $\{-225\}$ **92.** $\{7.9\}$ **93.** $\{0.014\}$ **94.** $\{-4\}$
95. $-\frac{12}{5}$ **96.** -9 **97.** 8 **98.** $\frac{11}{7}$ **99.** 8 **100.** 12 **101.** 0.06
102. -8 **103.** 0 **104.** 3 **105.** no solution, contradiction
106. all real numbers, identity **107.** $h = \dfrac{V}{\pi r^2}$

108. $g = \dfrac{m - Y}{2}$ **109.** $x = \dfrac{T}{ab} - y$ or $x = \dfrac{T - aby}{ab}$

110. $r^3 = \dfrac{3V}{4\pi}$ **111.** O'Hare: 66.5 million;

Atlanta: 76.5 million **112.** $245 - 5c$ **113.** \$5,000; \$1,000x
114. 42 ft, 45 ft, 48 ft, 51 ft **115.** 50°, 130° **116.** \$18,000 at
10%, \$7,000 at 9% **117. a.** 11.2% **b.** 3.8%
118. 504, 505, 505 **119.** 2 min **120.** 10 gal **121.** $3.95(x + 3)$

Chapter 1 Test (page 95)

1. $s = T + 10$ **2.** $A = \frac{1}{2}bh$ **3.** $-2, 0, 5$ **4.** $-2, 0, -3\frac{3}{4}, 9.2, \frac{14}{5}, 5$
5. $\pi, -\sqrt{7}$ **6.** all
7.

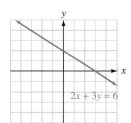

8.

9. -8 **10.** 5.5 **11.** 12.3 **12.** $\frac{4}{15}$ **13.** $\frac{11}{10}$ **14.** -64 **15.** $-\frac{8}{9}$
16. -35 **17.** 1 **18.** 100 mg **19.** commut. prop. of add.
20. assoc. prop. of mult. **21.** distrib. prop. **22.** mult. inverse
prop. **23.** $11y$ **24.** $-45q$ **25.** -4 **26.** $-16 + 2a + 18b$
27. -12 **28.** 6 **29.** no solution, contradiction **30.** 5.5
31. 5.6 **32.** 5.6 **33.** $i = \dfrac{f(P - L)}{s}$ **34.** 4 cm by 9 cm

35. 84% **36.** \$4,000 **37.** 10 oz

Chapter 2 Check Your Knowledge (page 98)

1. x, y **2.** satisfy **3.** linear **4.** perpendicular, parallel
5. range, domain **6. a.** $-4, 2$ **b.** $(0, -8)$ **c.** -9

7. $\left(-\dfrac{3}{2}, -\dfrac{5}{2}\right)$

8. $(3, 0), (0, 2)$

9. undefined **10.** -1 **11.** 2 **12.** -2
13. $y = -2x + 11$ **14.** $x + y = 3$ **15.** $y = -x$
16. no **17.** D: all real numbers except 0,
R: all real numbers except 0 **18. a.** 1 **b.** 2
19. a. not a function **b.** not a function **c.** a function
20.

Study Set Section 2.1 (page 108)

1. ordered **3.** origin **5.** rectangular **7.** midpoint
9. origin, right, down **11.** II **13.** $-3, -1, 1, 3$

15. one **17.** t **19.** x sub 1 **21–28.**
29. $(2, 4)$ **31.** $(-2.5, -1.5)$
33. $(3, 0)$ **35.** $(0, 0)$
37. a. on the surface **b.** diving
c. 1,000 ft **d.** 500 ft
39. a. 1993 **b.** Imports exceeded
production by about 3.5 million
barrels per day. **41.** $(3, 4)$
43. $(9, 12)$ **45.** $(\frac{7}{2}, 6)$ **47.** $(\frac{1}{2}, -2)$
49. $(-4, 0)$ **51.** $(4, 1)$ **53.** $(-20, -3)$
55. Jonesville $(5, B)$, Easley $(1, B)$,
Hodges $(2, E)$, Union $(6, C)$ **57. a.** $(2, -1)$ **b.** no
c. yes **59. a.** 6 **b.** 7 strokes **c.** the 16th hole **d.** the 18th
hole **61. a.** \$2 **b.** \$4 **c.** \$7 **d.** \$9 **63.** tip of tail, front of
engine, tip of wing **67.** 20 **69.** $\frac{1}{2}$ **71.** 0.7

Study Set   Section 2.2   (page 120)

1. satisfy **3.** x-intercept, y-intercept **5.** vertical **7. a.** yes
b. no **9.** x-intercept: $(-6, 0)$; y-intercept: $(0, 3)$
11. a. $(-3, 0)$; $(0, 4)$ **b.** false **13.** $y = -4x - 1$ **15.** $-3, -2$
17. the y-axis **19.** $5, 4, 2$ **21.** $0, -1, -2$

23.

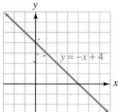

25.

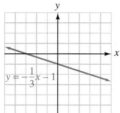

27.

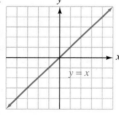

29.

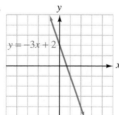

31.

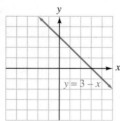

33.

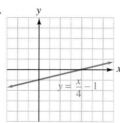

35.

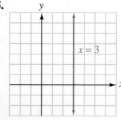

37.

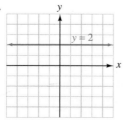

39.

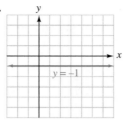

41.

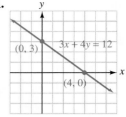

43.

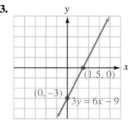

45.

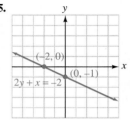

47.

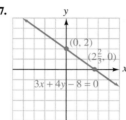

49.

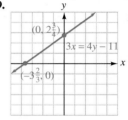

51. 1.22 **53.** 4.67 **55.** \$60
57. a. In 1990,
there were
65.5 million
swimmers
b. about 58.2
million

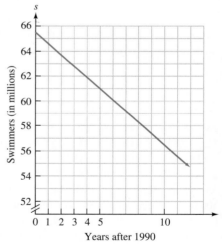

59. \$162,500 **61.** 12.5 yrs **63. a.** $c = 10t + 2$
b. 12, 22, 32, 42

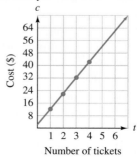

c. \$62 **67.** 11, 13, 17, 19, 23, 29 **69.** III **71.** 80s **73.** $3x + 8$

Think It Through   (page 130)

1960–1970; an increase of about 50 community colleges per yr

Study Set   Section 2.3   (page 133)

1. slope   **3.** change   **5.** reciprocals   **7. a.** l_3; 0   **b.** l_2; undefined   **c.** l_1; 2   **d.** l_4; −3   **9. a.** an increase of 73 million units/yr   **b.** a decrease of 35 million units/yr

11. a. $-\frac{4}{3}$ **b.** $-\frac{2}{3}$ **13.** $m = \frac{y_2 - y_1}{x_2 - x_1}$ **15. a.** 6   **b.** 8   **c.** $\frac{3}{4}$

17. $-\frac{8}{3}$ **19.** $\frac{7}{8}$ **21.** 3   **23.** −1   **25.** $-\frac{1}{3}$ **27.** 0   **29.** undefined

31. −1   **33.** $-\frac{3}{2}$ **35.** $\frac{3}{4}$ **37.** $\frac{1}{2}$ **39.** 0   **41.** perpendicular

43. neither   **45.** neither   **47.** parallel   **49.** perpendicular

51. neither   **53.** $\frac{3}{140}, \frac{1}{15}, \frac{1}{20}$; part 2   **55.** $\frac{1}{10}; \frac{1}{4}$ **57.** $\frac{1}{25}$; 4%

59. brace: $\frac{1}{2}$; support 1: −2; support 2: −1; yes, to support 1

65. 40 lb licorice; 20 lb gumdrops   **67.** 4 hr

Study Set   Section 2.4   (page 145)

1. $y - y_1 = m(x - x_1)$ **3.** perpendicular   **5.** no

7. $m = \frac{2}{3}$; $y + 3 = \frac{2}{3}(x + 2)$ **9.** $m = -\frac{2}{3}$; (0, 1)   **11.** yes

13. a. (0, 0)   **b.** none   **15.** No; the slopes are not negative reciprocals. Their product is not −1: 1(−0.9) = −0.9.

17. $\frac{1}{3}x, 2, 2, 1, \frac{1}{3}, -1$ **19.** $y = 5x + 7$ **21.** $y = -3x + 6$

23. $y = x$ **25.** $y = \frac{7}{3}x - 3$ **27.** $y = \frac{2}{3}x + \frac{11}{3}$

29. $y = 3x + 17$ **31.** $y = -7x + 54$ **33.** $y = -4$

35. $y = -\frac{1}{2}x + 11$ **37.** $\frac{3}{2}, (0, -4)$ **39.** $-\frac{1}{3}, (0, -\frac{5}{6})$

41. $1, (0, -1)$ **43.** $\frac{2}{3}, (0, 2)$

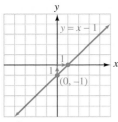

45. $-\frac{3}{4}, (0, -2)$

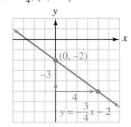

47. parallel   **49.** perpendicular

51. neither   **53.** perpendicular

55. $y = 4x$ **57.** $y = 4x - 3$

59. $y = \frac{4}{5}x - \frac{26}{5}$ **61.** $y = -\frac{1}{4}x$

63. $y = -\frac{1}{4}x + \frac{11}{2}$

65. $y = -\frac{5}{4}x + 3$

67. $y = -\frac{950}{3}x + 1,750$

69. $y = 1,811,250x + 36,225,000$

71. a. $B = \frac{1}{100}p - 195$ **b.** 905

73. a. $E = -\frac{1}{2}t + \frac{21}{2}$ **b.** The number of errors is reduced by 1 for every 2 trials.   **c.** On the 21st trial, the rat should make no errors.   **75. a.** $y = 1.35x - 31.25$

b. When the actual temperature is 0°F, the wind-chill temperature is about −31°F.   **81.** $29,100   **83.** 0

Think It Through   (page 154)

$28,000, $61,250

Study Set   Section 2.5   (page 158)

1. function   **3.** range   **5.** x **7.** y **9.** $f(-1)$

11. a.

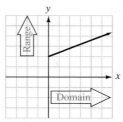

b. D: all real numbers greater than or equal to 0, R: all real numbers greater than or equal to 2   **13.** −5, 25   **15.** of

17. yes   **19.** yes   **21.** yes

23. no   **25.** 9, −3   **27.** 3, −5

29. 22, 2   **31.** 3, 11   **33.** 4, 9

35. 7, 26   **37.** 9, 16   **39.** 6, 15

41. 4, 4   **43.** 2, 2   **45.** $\frac{1}{5}$, 1

47. $-2, \frac{2}{5}$ **49.** 3.7, 1.1, 3.4   **51.** $-\frac{27}{64}, \frac{1}{216}, \frac{125}{8}$

53. $g(w) = 2w, g(w + 1) = 2w + 2$ **55.** $g(w) = 3w - 5,$ $g(w + 1) = 3w - 2$ **57.** D: the set of all real numbers except 6

59. D: the set of all real numbers   **61.** D: the set of all real numbers   **63.** D: the set of all real numbers   **65.** D: {−2, 4, 6}, R: {3, 5, 7}   **67.** D: the set of all real numbers except 4, R: the set of all real numbers except 0   **69.** not a function

71. a function   **73.** a function   **75.** a function

77.

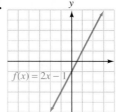

D: the set of all real numbers, R: the set of all real numbers

79.

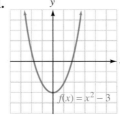

D: the set of all real numbers, R: the set of all real numbers

81. no   **83.** yes   **85.** between 20°C and 25°C   **87. a.** $I(b) =$ $1.75b - 50$ **b.** $142.50   **89. a.** (200, 25), (200, 90), (200, 105)   **b.** It doesn't pass the vertical line test.   **91. a.** 3,400; the tax on an income of $25,000 is $3,400   **b.** $T(a) = 3,910 + 0.25(a - 28,400)$ **93. a.** 624 ft   **b.** 0; the rocket strikes the ground 16 seconds after being shot   **97.** $\frac{-15}{4}$ **99.** $\frac{1}{3}$

Study Set   Section 2.6   (page 170)

1. squaring   **3.** absolute value   **5.** vertical   **7.** reflection   **9.** 4, left   **11.** 5, up   **13. a.** 2   **b.** 0   **15.** −6   **17.** −3, 1, 2   **19. a.** 3   **b.** 0   **c.** 1.5

21.

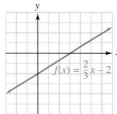

D: the set of real numbers, R: the set of all real numbers greater than or equal to −3

23.

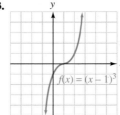

D: the set of real numbers, R: the set of real numbers

25.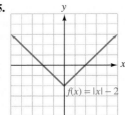

D: the set of real numbers,
R: the set of all real numbers
greater than or equal to −2

27.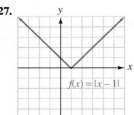

D: the set of real numbers,
R: the set of real numbers greater
than or equal to 0

29.   **31.**

33.   **35.**

37. 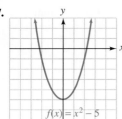   **39.**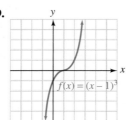

41. 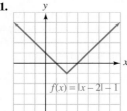   **43.**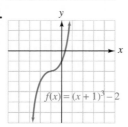

45. 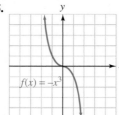   **47.**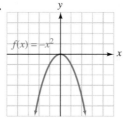

49. 4   **51.** −3   **53.** $f(x) = |x|$ **55.** a parabola

61. $W = T - ma$ **63.** $g = \dfrac{2(s - vt)}{t^2}$ **65.** \$5.4 million

Key Concept   (page 174)

1. a. correspondence, input, range, one, domain
b. dependent, independent   **3.** −17   **5.** 60 ft
7. $f(x) = 2x + 3, 3$ **9.** 0; the projectile will strike the ground
4 seconds after being shot into the air

Chapter Review   (page 176)

1–5.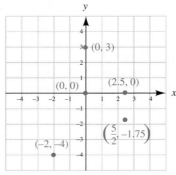

6. −5, −5; 0, 0; −1, −1; 4, 4; 3, 3; 2, 2; 3, 3
7. 1 ft below its normal level
8. decreased by 3 ft
9. from day 3 to the beginning of day 4
10. \$10 increments
11. \$800   **12.** (2, 2)
13. (2, −6)

14. 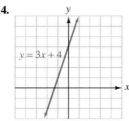   **15.**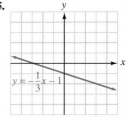

16. 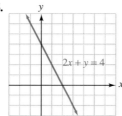   **17.**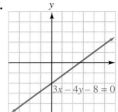

18. 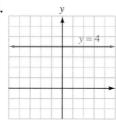   **19.**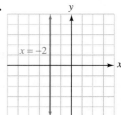

20. 9, 0, −9   **21.** −4, $-\frac{5}{2}$, −1   **22.** slope of $l_1 = \frac{4}{5}$;
slope of $l_2 = -\frac{8}{5}$ **23.** 1.37% per year   **24.** 1   **25.** $-\frac{14}{9}$
26. 0   **27.** undefined   **28.** $\frac{2}{3}$ **29.** −2   **30.** undefined
31. 0   **32.** perpendicular   **33.** parallel   **34.** $y = 3x + 29$
35. $y = -\frac{13}{8}x + \frac{3}{4}$ **36.** $3x - 2y = 1$ **37.** $2x + 3y = -21$
38. $y = -\frac{3}{4}x - 3; m = -\frac{3}{4}, (0, -3)$

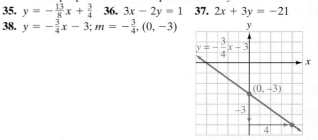

39. $y = -1,720x + 8,700$
40. yes   **41.** yes   **42.** no   **43.** no   **44.** −7   **45.** 18   **46.** 8

47. $3t + 2$ **48.** D: the set of real numbers, R: the set of real numbers   **49.** D: the set of real numbers, R: the set of all real numbers greater than or equal to 1   **50.** D: the set of all real numbers except 2, R: the set of all real numbers except 0   **51.** D: the set of real numbers, R: the set of nonpositive real numbers   **52.** function   **53.** not a function   **54.** $C(x) = 105x + 175$ **55.** \$805   **56.** yes   **57.** no

58. 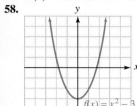   **59.**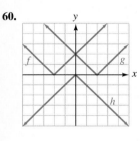

60. 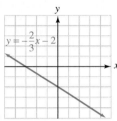   **61. a.** -4 **b.** 3   **62.** 4   **63.** -1 **64.** 2

Chapter 2 Test   *(page 181)*

1. 240 ft   **2.** 1 sec and 7 sec   **3.** about 260 ft   **4.** 8 sec
5. $\left(\frac{5}{2}, \frac{3}{2}\right)$ **6.**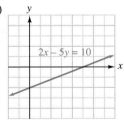

7. $(5, 0), (0, -2)$

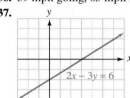

8.   **9.** 3   **10.** -1.5 degree/hr
11. $\frac{1}{2}$ **12.** $\frac{2}{3}$ **13.** undefined
14. 0   **15.** $y = \frac{2}{3}x - \frac{23}{3}$
16. $8x - y = -22$
17. $m = -\frac{1}{3}, \left(0, -\frac{3}{2}\right)$
18. neither   **19.** $y = \frac{3}{2}x$
20. a. $v = -600x + 4,000$
   b. $(0, 4,000)$; it gives the value
of the copier when new: \$4,000   **21.** no   **22.** yes
23. D: the set of real numbers, R: the set of nonnegative real numbers   **24.** D: the set of real numbers, R: the set of real numbers   **25.** 10   **26.** -1 **27.** 3   **28.** $r^2 - 2r - 1$
29. function   **30.** not a function

31. 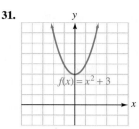   **32.**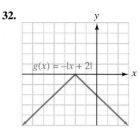

33. Find the x-coordinate of the x-intercept of the graph of $y = 3(x - 2) - 2(-2 + x)$; 2
34. Find the x-coordinate of the point of intersection of the graph of $y = 3(x - 2) - 2(-2 + x)$ and $y = 1$; 3

Cumulative Review Exercises   *(page 183)*

1. 1, 2, 6, 7   **2.** 0, 1, 2, 6, 7   **3.** $-2, 0, 1, 2, \frac{13}{12}, 6, 7$ **4.** $\sqrt{5}, \pi$
5. -2 **6.** $-2, 0, 1, 2, \frac{13}{12}, 6, 7, \sqrt{5}, \pi$ **7.** 2, 7   **8.** 6
9. $-2, 0, 2, 6$ **10.** 1, 7   **11.** -2 **12.** -2 **13.** 22   **14.** -2
15. $\frac{24}{25}$ **16.** 2   **17.** 4   **18.** -5 **19.** assoc. prop. of add.
20. distrib. prop.   **21.** commut. prop. of add.   **22.** assoc. prop. of mult.   **23.** $-5y$ **24.** $-28st$ **25.** 0   **26.** $z - 4$
27. 8   **28.** -27 **29.** -1 **30.** 6   **31.** $\frac{8}{3}$ **32.** 24
33. $a = \frac{2S}{n} - l$ or $a = \frac{2S - ln}{n}$ **34.** $h = \frac{2A}{b_1 + b_2}$ **35.** \$14,000
36. 39 mph going, 65 mph returning
37. 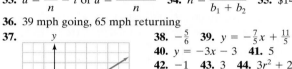   **38.** $-\frac{5}{6}$ **39.** $y = -\frac{7}{5}x + \frac{11}{5}$
40. $y = -3x - 3$ **41.** 5
42. -1 **43.** 3   **44.** $3r^2 + 2$

45. a function; D: the set of real numbers, R: the set of all real numbers less than or equal to 1

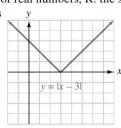

46. a function; D: the set of real numbers, R: the set of nonnegative real numbers

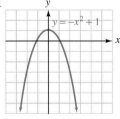

47. The points $(2, 20), (3, 40)$, and $(6, 60)$ do not lie on a straight line.   **48. a.** Find the x-coordinate of the x-intercept of the graph of $y = -7 - 5(x - 1) + 4x$; -2 **b.** Find the x-coordinate of the point of intersection of the graph of $y = -7 - 5(x - 1) + 4x$ and $y = -4$; 2

Chapter 3   Check Your Knowledge   (page 186)

1. consistent, inconsistent   **2.** independent, dependent
3. intersection   **4.** matrix   **5.** Cramer's   **6.** no
7. $(1, -2)$

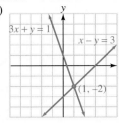

8. $(4, 1)$ **9.** $(-1, 3)$
10. consistent, dependent
11. yes   **12.** $(3, -2, 1)$
13. $4,000 at 4%, $11,000
at 6%   **14.** 12 gal 5%,
8 gal 10%   **15.** $(2, -3)$
16. $(-2, 3, -1)$ **17.** 7
18. 0   **19.** $(-1, 2)$
20. -1

Think It Through   (page 191)

$(1981, 50)$; in 1981, 50% of the bachelor's degrees that were
awarded went to men and 50% went to women.

Study Set   Section 3.1   (page 193)

1. system   **3.** inconsistent   **5.** dependent   **7. a.** true
b. false   **c.** true   **d.** true   **9. a.** $-4, (-4, 0), 2, (0, 2); 3, (2, 3)$
b. $(-4, 0); (0, 2)$ **11. a.** $\begin{cases} x + y = 5 \\ x - y = -1 \end{cases}$ (answers may vary)

b. $\begin{cases} x + y = 5 \\ 2x + 2y = 10 \end{cases}$ (answers may vary)   **c.** $\begin{cases} x + y = 5 \\ x + y = 4 \end{cases}$
(answers may vary)   **13.** brace   **15.** yes   **17.** no

19.
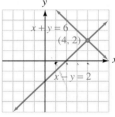

21.
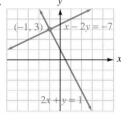

23.
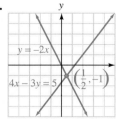

25.
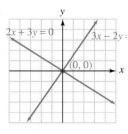

27.
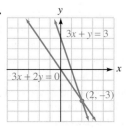

29.
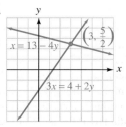

31.
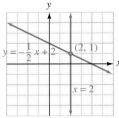

33.
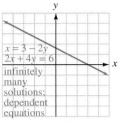

35.
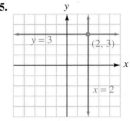

37.
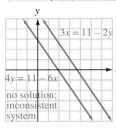

39.
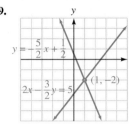

41.
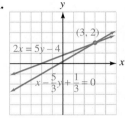

43.
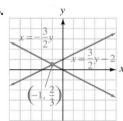

45. $(-0.37, -2.69)$
47. $(-7.64, 7.04)$ **49.** Gallup,
Grants, Albuquerque,
Tucumcari; Las Vegas, Santa Fe,
Albuquerque, Socorro, Las
Cruces; Albuquerque
51. $(2,000, 50)$

53. a.
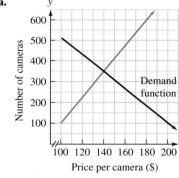

b. $140   **c.** Supply
increases, and
demand decreases.
55. a. yes
b. $(3.75, -0.5)$
c. no   **59.** -3
61. 0   **63.** D: the
set of real numbers,
R: the set of all real
numbers greater
than or equal to -2
65. 40.5 cm²

Study Set   Section 3.2   (page 207)

1. general   **3.** eliminated   **5. a.** 3; -4 (answers may vary)
b. 2, -3 (answers may vary)   **7.** addition method
9. a. ii   **b.** iii   **c.** i   **11.** $(2, 2)$ **13.** $(5, 3)$ **15.** $(-2, 4)$
17. no solution, inconsistent system   **19.** $(5, 2)$ **21.** $(-4, -2)$
23. $(1, 2)$ **25.** $\left(\frac{1}{2}, \frac{2}{3}\right)$ **27.** $\left(5, \frac{3}{2}\right)$ **29.** $\left(-2, \frac{3}{2}\right)$ **31.** infinitely
many solutions, dependent equations   **33.** no solution,
inconsistent system   **35.** $(4, 8)$ **37.** $(20, -12)$ **39.** $\left(\frac{2}{3}, \frac{3}{2}\right)$
41. $\left(\frac{1}{2}, -3\right)$ **43.** $(2, -3)$ **45.** $(9, -1)$ **47.** $(2, 3)$ **49.** $\left(-\frac{1}{3}, 1\right)$
51. $475, $800   **53.** dogs: 60 million; cats: 75 million

55. 16 m by 20 m   **57.** 75°, 25°   **59.** $3,000 at 10%, $5,000 at 12%   **61.** 45 mi   **63.** 85 racing bikes, 120 mountain bikes
65. 4,031   **67.** $4,666\frac{2}{3}$ books   **69.** 103   **71.** 148 g of the 0.2%, 37 g of the 0.7%   **73. a.** 590 units per month   **b.** 620 units per month   **c.** A (smaller loss)   **81.** $-\frac{5}{2}$ **83.** $-\frac{8}{5}$ **85.** $\frac{4}{3}$

Think It Through   (page 215)

$53,100; $52,500; $50,000

Study Set   Section 3.3   (page 219)

1. system   **3.** three   **5.** dependent   **7. a.** no solution
b. no solution   **9.** $x + 2y - 3z = -6$ **11.** yes   **13.** $(1, 1, 2)$
15. $(0, 2, 2)$ **17.** $(3, 2, 1)$ **19.** no solution, inconsistent system
21. $(60, 30, 90)$ **23.** $(2, 4, 8)$ **25.** infinitely many solutions, dependent equations   **27.** $(2, 6, 9)$ **29.** 30 expensive,
50 middle-priced, 100 inexpensive   **31.** 2, 3, 1   **33.** 3 poles,
2 bears, 4 deer   **35.** 78%, 21%, 1%   **37. a.** infinitely many
solutions, all lying on the line running down the binding
b. 3 parallel planes (shelves); no solution   **c.** each pair of
planes (cards) intersect; no solution   **d.** 3 planes (faces of die)
intersect at a corner; 1 solution   **39.** $y = \frac{1}{2}x^2 - 2x - 1$
41. $x^2 + y^2 - 2x - 2y - 2 = 0$ **43.** $A = 40°, B = 60°, C = 80°$
45. 12, 15, 21
49. 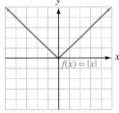   **51.**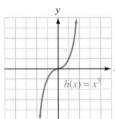

Study Set   Section 3.4   (page 229)

1. matrix   **3.** rows, columns   **5.** augmented   **7. a.** 2×3
b. 3×4 **9.** $\begin{cases} x - y = -10 \\ y = 6 \end{cases}$; $(-4, 6)$ **11.** It has no solution.
The system is inconsistent.

13. a. multiply row 1 by $\frac{1}{3}$; $\begin{bmatrix} 1 & 2 & -3 & | & 0 \\ 1 & 5 & -2 & | & 1 \\ -2 & 2 & -2 & | & 5 \end{bmatrix}$

b. to row 2, add -1 times row 1; $\begin{bmatrix} 1 & 2 & -3 & | & 0 \\ 0 & 3 & 1 & | & 1 \\ -2 & 2 & -2 & | & 5 \end{bmatrix}$

15. $-1, 1, -5, 2, y, 4$ **17.** $(1, 1)$ **19.** $(2, -3)$ **21.** $(-1, -1)$
23. $(0, -3)$ **25.** $(1, 2, 3)$ **27.** $(4, 5, 4)$ **29.** $(2, 1, 0)$
31. $(-1, -1, 2)$ **33.** no solution, inconsistent system
35. infinitely many solutions, dependent equations   **37.** $(0, 1, 3)$
39. no solution, inconsistent system   **41.** $(-4, 8, 5)$
43. infinitely many solutions, dependent equations   **45.** 22°, 68°
47. 40°, 65°, 75°   **49.** 76°, 104°   **51.** founder's circle: 100;
box seats: 300; promenade: 400   **55.** $m = \dfrac{y_2 - y_1}{x_2 - x_1}$ where $x_2 \neq x_1$
57. $y - y_1 = m(x - x_1)$

Study Set   Section 3.5   (page 239)

1. determinant   **3.** minor   **5.** rows, columns   **7.** dependent,
inconsistent   **9.** $ad - bc$ **11.** $\begin{vmatrix} 3 & 4 \\ 2 & -3 \end{vmatrix}$ **13.** $(\frac{7}{11}, -\frac{5}{11})$
17. 8   **19.** -2 **21.** 200   **23.** 6   **25.** 1   **27.** 26   **29.** 0
31. -79 **33.** $(4, 2)$ **35.** $(-\frac{1}{2}, \frac{1}{3})$ **37.** no solution,
inconsistent system   **39.** $(2, -1)$ **41.** $(1, 1, 2)$ **43.** $(3, 2, 1)$
45. $(3, -2, 1)$ **47.** $(-\frac{1}{2}, -1, -\frac{1}{2})$ **49.** infinitely many solutions,
dependent equations   **51.** no solution, inconsistent system
53. $(-2, 3, 1)$ **55.** 200 of the $67 phones, 160 of the $100 phones
57. $5,000 in HiTech, $8,000 in SaveTel, $7,000 in OilCo
59. -23 **61.** 26   **65.** no   **67.** yes   **69.** x **71.** y-intercept
73. $x; y$

Key Concept   (page 242)

1. yes   **3.** 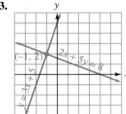   **5.** $(3, 2)$ **7.** $(15, 2)$
9. The equations of the
system are dependent.
There are infinitely many
solutions.

Chapter Review   (page 244)

1. $(1, 3), (2, 1), (4, -3)$ (answers may vary)   **2.** $(0, -4), (2, -2),$
$(4, 0)$ (answers may vary)   **3.** $(3, -1)$ **4.** President Clinton's
job approval and disapproval ratings were the same:
approximately 47% in 5/94 and approximately 48% in 5/95.

5. 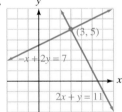   **6.**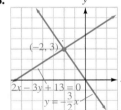

7. 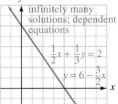   **8.**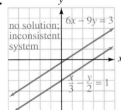

9. $(-1, 3)$ **10.** $(-3, -1)$ **11.** $(3, 4)$ **12.** infinitely many
solutions, dependent equations   **13.** $(-3, 1)$ **14.** no solution,
inconsistent system   **15.** $(9, -4)$ **16.** $(4, \frac{1}{2})$ **17.** Using the
addition method, the computations are easier.   **18.** $(-1, 0.7)$
(answers may vary); $(-1, \frac{2}{3})$ **19.** A-H: 162 mi, A-SA: 83 mi
20. 8 mph, 2 mph   **21.** 17,500 bottles   **22.** no   **23.** $(\frac{1}{2}, 4, -6)$
24. no solution, inconsistent system   **25.** $(-1, 1, 3)$
26. infinitely many solutions, dependent equations

27. yes; infinitely many solutions   **28.** 25 lb peanuts,
10 lb cashews, 15 lb Brazil nuts   **29.** $\begin{bmatrix} 5 & 4 & 3 \\ 1 & -1 & -3 \end{bmatrix}$

30. $\begin{bmatrix} 1 & 2 & 3 & 6 \\ 1 & -3 & -1 & 4 \\ 6 & 1 & -2 & -1 \end{bmatrix}$ **31.** $(1, -3)$ **32.** $(5, -3, -2)$

33. infinitely many solutions, dependent equations
34. no solution, inconsistent system   **35.** \$4,000 at 6%,
\$6,000 at 12%   **36.** 18   **37.** 38   **38.** -3 **39.** 28   **40.** $(2, 1)$
41. no solution, inconsistent system   **42.** $(1, -2, 3)$
43. $(-3, 2, 2)$ **44.** 2 cups mix A, 1 cup mix B, 1 cup mix C

Chapter 3 Test   (page 249)

1.
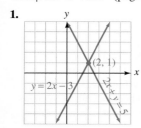

2. $(7, 0)$ **3.** $(2, -3)$
4. dependent   **5.** no
6. $(3, 2, -1)$ **7.** 55, 70
8. 15 gal 40%, 5 gal 80%
9. $(2, 2)$ **10.** $(1, 0, -1)$
11. 22   **12.** 4
13. $\begin{vmatrix} -6 & -1 \\ -6 & 1 \end{vmatrix}$ **14.** $\begin{vmatrix} 1 & -1 \\ 3 & 1 \end{vmatrix}$
15. -3 **16.** 3   **17.** -1

18. C: 60, GA: 30, S: 10   **20.** The system has no solution.
22. $(2035, 25)$

Cumulative Review Exercises   (page 251)

1.
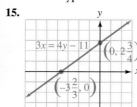

2. 504,000,000,000   **3.** 70   **4.** 2   **5.** $-12.1x^2 + 12.7x$
6. -4 **7.** 20 mph   **8.** 5 lb apple slices, 5 lb banana chips
9. -28 **10.** $-\frac{1}{3}$ **11.** -2 **12.** all real numbers, identity
13. $B = \dfrac{C - Ax}{A}$ **14.** $n = \dfrac{l - a + d}{d}$ or $n = \dfrac{l - a}{d} + 1$

15.
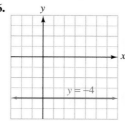
16.

17. $y = -3x + 17$ **18.** $-\frac{4}{5}$ **19.** -105 **20.** -95
21. $f(x) = x^3$ (answers may vary)   **22.** no   **23.** no
24. $v = 17.5x + 300$ **25.** $(2, -1)$ **26.** $(-1, 0, 2)$ **27.** 26   **28.** 26

Chapter 4   Check Your Knowledge   (page 254)

1. true   **2.** infinity   **3.** interval   **4.** dashed   **5.** no
6. $[-3, 2)$

7. $(-\infty, 1]$
8. $(-\infty, -1)$
9. $[-1, 1)$
10. $(-\infty, 1] \cup [4, \infty)$
11. $(-3, -1]$
12. a. 1   **b.** -17 **13.** $-2, 8$ **14.** $-\frac{1}{5}, 5$
15. $(2, 5)$
16. $(-\infty, 2) \cup (6, \infty)$
17. $[0, 3]$ **18.** no solution

19.
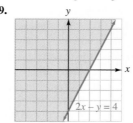
20.
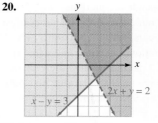

21.
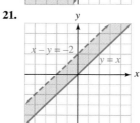
22. between 20 and 70

Think It Through   (page 257)

From 1979 to 2003

Study Set   Section 4.1   (page 264)

1. inequality   **3.** parenthesis   **5.** linear   **7.** is less than;
is greater than or equal to
9. a. $(4, \infty)$; $\{x \mid x > 4\}$
b. $(-\infty, -4)$; $\{x \mid x < -4\}$
c. $(-\infty, 4]$; $\{x \mid x \le 4\}$
11. a. equation   **b.** expression   **c.** inequality   **d.** expression
e. inequality   **13.** i and ii   **15. a.** yes   **b.** yes   **c.** yes   **d.** no
17. a. the number of seriously injured ≤ 16 **b.** the number of
references to carpools ≥ 10 **c.** The car is at least 25 years old.
19. $-10, \le, 2, -\infty, x \le 2$ **21. a.** $x < -10$ **b.** $x > \frac{7}{8}$ **c.** $x \ge 0$
23. $(-3, \infty)$
25. $[20, \infty)$ **27.** $[60, \infty)$
29. $\left(-\frac{10}{3}, \infty\right)$ **31.** $(-\infty, 1)$
33. $\left[-\frac{2}{5}, \infty\right)$ **35.** $[-2, \infty)$

37. $(6, \infty)$

39. $(-\infty, 1.5]$

41. $(-\infty, 20]$

43. $(-\infty, 10)$

45. $[-36, \infty)$

47. $(-\infty, \frac{45}{7}]$

49. Midwest, South   **51.** $6 + 45 \not> 52$ **53.** 8 hr   **55.** 15
57. 13 hr   **59.** anything over \$900   **61.** $x < 1$ **63.** $x \geq -4$
69. 4, 5, 3   **71.** 6, −6

Think It Through   (page 274)

1. 2001, 2002   **2.** 1999, 2001, 2002   **3.** 1999, 2000

Study Set   Section 4.2   (page 274)

1. compound   **3.** interval   **5.** both   **7.** reversed   **9. a.** no
b. yes   **11. a.** no solution   **b.** $(-\infty, \infty)$ **13. a.** ii   **b.** iii
c. i   **15. a.**   **b.**

c.   **d.**

17. $3 \not< -3$ **19.** $(-2, 5]$

21. $(-10, -9)$

23. $(2, 3]$

25. no solution   **27.** $[5, \infty)$

29. $[1, 4]$ **31.** $(8, 11)$

33. $(-2, 5)$

35. $[-4, 6)$

37. $[2, 2]$ **39.** $(-0.7, 0.2]$

41. $(-6, -3)$

43. $[-2, 4]$

45. $(-\infty, -2] \cup (6, \infty)$

47. $(-\infty, -1) \cup (2, \infty)$

49. $(-\infty, 2) \cup (7, \infty)$

51. $(-\infty, \infty)$ **53.** $(-\infty, 1)$

55. $(-\infty, \infty)$

57. a. 128, 192   **b.** $32 \leq s \leq 48$ **59.** See doctor today.
61. a. 1999   **b.** 1998, 1999, 2000, 2001   **c.** 1999, 2000
d. 1998, 1999, 2000, 2001   **67.** 85.7, 86, 86   **69.** 13.3 pts/game

Study Set   Section 4.3   (page 286)

1. equation   **3.** isolate   **5.** 0   **7.** more than   **9.** 5   **11. a.** yes
b. no   **c.** yes   **d.** no   **13. a.** ii   **b.** iii   **c.** i   **15.** $|x| < 4$
17. $|x + 3| > 6$ **19.** 8   **21.** −0.02   **23.** $-\frac{31}{16}$ **25.** π
27. 25   **29.** −2   **31.** 23, −23   **33.** 9.1, −2.9   **35.** $\frac{14}{3}$, −6
37. no solution   **39.** 2, $-\frac{1}{2}$ **41.** −8   **43.** −4, −28   **45.** 0, −6
47. 40, −20   **49.** −2, $-\frac{4}{5}$ **51.** 0, −2   **53.** 0   **55.** $\frac{4}{3}$

57. $(-4, 4)$

59. $[-21, 3]$

61. $(-\frac{8}{3}, 4)$

63. no solution   **65.** $(-\infty, -3) \cup (3, \infty)$

67. $(-\infty, -12) \cup (36, \infty)$

69. $(-\infty, -\frac{16}{3}) \cup (4, \infty)$

71. $(-\infty, \infty)$

73. $(-\infty, -2] \cup [\frac{10}{3}, \infty)$

75. $(-\infty, -2) \cup (5, \infty)$

77. $[-10, 14]$

79. $(-\frac{5}{3}, 1)$

81. $(-\infty, -24) \cup (-18, \infty)$

83. no solution   **85.** $70° \leq t \leq 86°$ **87. a.** $|c - 0.6°| \leq 0.5°$
b. $[0.1°, 1.1°]$ **89. a.** 26.45%, 24.76%   **b.** It is less than or
equal to 1%.   **97.** 50°, 130°

Study Set   Section 4.4   (page 294)

1. linear, two   **3.** edge   **5. a.** yes   **b.** no   **c.** yes   **d.** no
7. $m = 3$; $(0, -1)$ **9.** no   **11. a.** $x \geq 2$

b.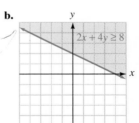

13.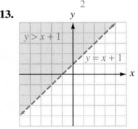

15.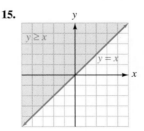

17.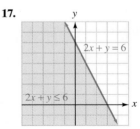

19.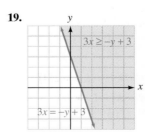

21.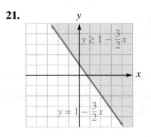

23.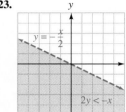

25.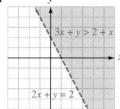

9.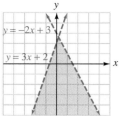

11.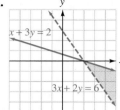

27.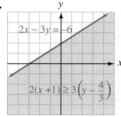

29.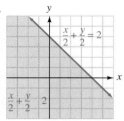

13.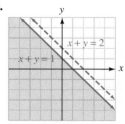

15.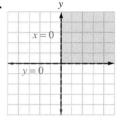

31.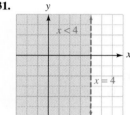

33.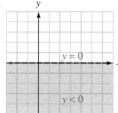

17.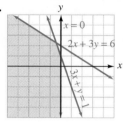

19.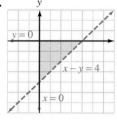

35. $3x + 2y > 6$ **37.** $x \le 3$

39.   **41.**

43. a. the Mississippi River   **b.** the area of the U.S. west of the Mississippi River

21.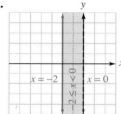

23.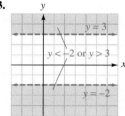

45. $(5, 15)$, $(15, 10)$, $(20, 5)$

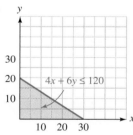

25.   **27.**

29.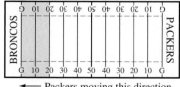

← Packers moving this direction

47. $(40, 80)$, $(80, 80)$, $(120, 40)$

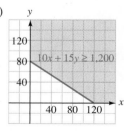

51. yes   **53.** $(3, 1)$

Study Set   Section 4.5   *(page 302)*

1. inequalities   **3.** intersect   **5. a.** yes   **b.** no   **c.** no   **d.** yes
7. a. false   **b.** true   **c.** true   **d.** false   **e.** true   **f.** true

31.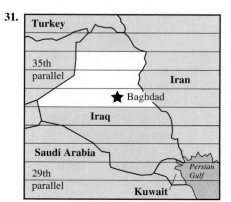

33. 1 $10 CD and 2 $15 CDs, 4 $10 CDs and 1 $15 CD

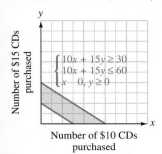

35. 2 desk chairs and 4 side chairs, 1 desk chair and 5 side chairs.

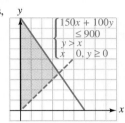

39. IV   **41.** II

Key Concept   (page 306)

1. a. compound inequality   **b.** system of linear inequalities
c. absolute value inequality   **d.** linear inequality in two
variables   **e.** linear inequality in one variable   **f.** double linear
inequality   **g.** compound inequality   **h.** absolute value
inequality   **i.** linear inequality in two variables   **3.** ⟵———(——⟶
                                                                3/2

Chapter Review   (page 308)

1. $(-\infty, 3]$ ⟵———]——⟶
                         3

2. $[4, \infty)$ ⟵———[——⟶
                       4

3. $(-\infty, 20)$ ⟵———)——⟶
                          20

4. $(-\infty, -\frac{51}{11})$ ⟵———)——⟶
                                    −51/11

5. $[2, \infty)$ **6.** $20,000 or more   **7.** yes   **8.** no

9. $[-10, -4)$ ⟵——[———)——⟶
                     −10   −4

10. $(-\infty, 11)$ ⟵——)——⟶
                           −11

11. $\left(-\frac{1}{3}, 2\right)$ ⟵——(———)——⟶
                                        −1/3   2

12. $[1, 9]$ ⟵——[———]——⟶   **13.** yes   **14.** no
                    1   9

15. $(-\infty, -5) \cup (4, \infty)$ ⟵—)——|——(—⟶
                                          −5 0 4

16. $(-\infty, \infty)$ ⟵——|——⟶
                               0

17. $17 \le 4l \le 25$, 4.25 ft $\le l \le$ 6.25 ft   **18. a.** ii, iv   **b.** i, iii
19. 7   **20.** $\frac{5}{16}$ **21.** −71.05   **22.** −12   **23.** 2, −2   **24.** 3, $-\frac{11}{3}$
25. $\frac{26}{3}, -\frac{10}{3}$ **26.** 14, −10   **27.** $\frac{1}{5}, -5$ **28.** $\frac{13}{12}$

29. $[-3, 3]$ ⟵——[———]——⟶
                    −3   3

30. $(-5, -2)$ ⟵——(———)——⟶
                     −5   −2

31. $[-3, \frac{19}{3}]$ ⟵——[———]——⟶   **32.** no solution
                          −3   19/3

33. $(-\infty, -1) \cup (1, \infty)$ ⟵—)——|——(—⟶
                                         −1 0 1

34. $(-\infty, -4] \cup [\frac{22}{5}, \infty)$ ⟵——]———[—⟶
                                                 −4   22/5

35. $(-\infty, \frac{4}{3}) \cup (4, \infty)$ ⟵——)———(—⟶
                                               4/3   4

36. $(-\infty, \infty)$ ⟵——|——⟶
                               0

37. a. $|w - 8| \le 2$ **b.** $[6, 10]$ **38.** We must isolate $|x|$ first.

39.
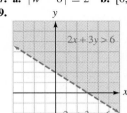

40.
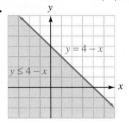

41.
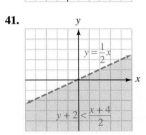

42.
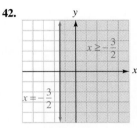

43. (1,800, 0), (1,000, 1,500), (2,000, 2,000) (answers may vary)

44. $3x - 4y > 12$

45.
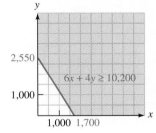

46.
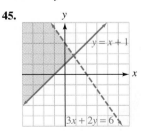

47.
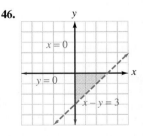

48.
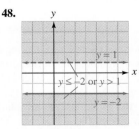
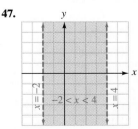

49. $\ge, \le, \ge, \le$ **50. a.** true   **b.** false   **c.** true   **d.** false
e. true   **f.** true

Chapter 4 Test (page 313)

1. false   **2.** yes   **3.** $(-\infty, -5)$ **4.** $c \le d$
5. $(12, \infty)$ 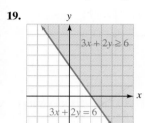   **6.** $(-\infty, -5]$

7. more than 78   **8.** $[1, \frac{9}{4}]$

9. $(-\infty, -3) \cup (8, \infty)$

10. $(-2, 16)$

11. 8   **12.** -4.75 **13.** $-5, \frac{23}{3}$ **14.** 4, -4
15. $[-7, 1]$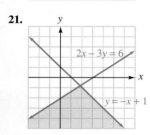

16. $(-\infty, -9) \cup (13, \infty)$

17. $(-\infty, 1) \cup (3, \infty)$

18. $[1, 3]$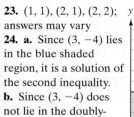

19. 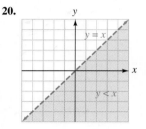   **20.**

21. 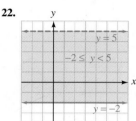   **22.**

23. $(1, 1), (2, 1), (2, 2)$; answers may vary
24. a. Since $(3, -4)$ lies in the blue shaded region, it is a solution of the second inequality.
b. Since $(3, -4)$ does not lie in the doubly-shaded purple region, it is not a solution of the system of inequalities.   **25.** $\le, \ge, \ge, \le$

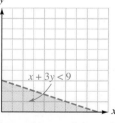

Cumulative Review Exercises (page 315)

1. rational numbers: terminating and repeating decimals; irrational numbers: nonterminating, nonrepeating decimals
2. 0.125, 0.0625, 0.03125, 0.054125   **3.** 10   **4.** -6
5. $-26p^2 - 6p$ **6.** $-2a + b - 2$ **7.** 201 ft^2 **8.** $20,000
9. $\frac{26}{3}$ **10.** 3   **11.** 6   **12.** no solution   **13.** perpendicular
14. parallel   **15.** $y = \frac{1}{3}x + \frac{11}{3}$ **16.** $-\frac{8}{5}$ **17.** 9,000 prisoners/yr
18. 1990–1995, 72,200 prisoners/yr   **19.** 10   **20.** 14
21. $(2, 1)$ **22.** $(3, 1)$ **23.** $(1, 1)$ **24.** $(-1, -1, 3)$

25. $(-1, -1)$ **26.** $(1, 2, -1)$ **27.** $(7, 23)$, in 1907 the percent of U.S. workers in white-collar and farming jobs was the same (23%); $(45, 42)$, in 1945 the percent of U.S. workers in white-collar and blue-collar jobs was the same (42%)
28. a. $y = -0.06x + 8.2$ **b.** 2.8 L/min
29. 750   **30.** 250 $5 tickets, 375 $3 tickets, 125 $2 tickets
31. 3, $-\frac{3}{2}$ **32.** $-5, -\frac{3}{5}$
33. $(-\infty, 11]$ **34.** $(-3, 3)$

35. $[-\frac{2}{3}, 2]$

36. $(-\infty, -4) \cup (1, \infty)$

37.   **38.**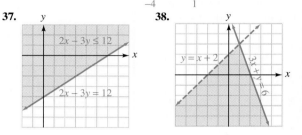

Chapter 5 Check Your Knowledge (page 318)

1. base, exponent   **2.** scientific   **3.** polynomial   **4.** like
5. $7x^{13}$ **6.** $-8x^9y^6$ **7.** $\frac{1}{9x^2y^2}$ **8.** $\frac{125}{x^{15}y^6}$
9. a. 9.3×10^{-5} **b.** 62,100   **10.** 500 seconds, 8 min, 20 secs
11. a. 5   **b.** -7 **12. a.**   **b.** $-2, 0, 2$

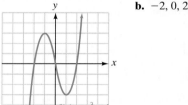

13. $x^3 - x^2 - 2x + 8$ **14.** $-y^2 - 6y + 10$ **15.** $\frac{-5z^5}{xy}$
16. $6x^2y + 4xy^2$ **17.** $y^2 - 49$ **18.** $4x^4 - 12x^2y + 9y^2$
19. $x^2 + \frac{4}{3}x + \frac{4}{9}$ **20.** $3x^2z - 3y^2z$ **21.** $2ab^2c^3$
22. $2x^2 - x - 21$ square units   **23.** $-16t(t - 4)$
24. $(x - 3)(xy - 2)$ **25.** $(xz + 8)(xz - 8)$
26. $3(x^2 + 4)(x + 2)(x - 2)$ **27.** prime
28. $(3x + 4)(9x^2 - 12x + 16)$ **29.** $5(y - 1)(y^2 + y + 1)$
30. $(x + 6y)(x + y)$ **31.** $(2x + 1)(x + 3)$
32. $(y + 1 + z)(y + 1 - z)$ **33.** $-3, 2$
34. 0, 7   **35.** $-2, \frac{2}{3}$ **36.** $b^2 = \frac{a^2y^2}{a^2 - x^2}$

Study Set   Section 5.1 (page 326)

1. power   **3.** product, quotient   **5.** x^{m+n} **7.** x^ny^n **9.** 1
11. x^{m-n} **13.** $4^{-2} = \frac{1}{4^2}$ (answers may vary)   **15. a.** add
b. subtract   **c.** multiply   **17.** 9, (-2), 11   **19.** base is 5, exponent is 3   **21.** base is x, exponent is 5   **23.** base is b, exponent is 6   **25.** base is $\frac{n}{4}$, exponent is 3   **27.** 9
29. -9 **31.** 9   **33.** $\frac{1}{25}$ **35.** $-\frac{1}{25}$ **37.** $\frac{1}{25}$ **39.** 1
41. 1   **43.** $-32x^5$ **45.** x^5

47. x^{10} **49.** k^7 **51.** $2a^4b^5$ **53.** p^{10} **55.** x^5y^4
57. $\dfrac{1}{b^{72}}$ **59.** x^{28} **61.** $\dfrac{s^3}{r^9}$ **63.** a^{20} **65.** $-\dfrac{1}{d^3}$ **67.** $27x^9y^{12}$
69. $\dfrac{1}{729}m^6n^{12}$ **71.** $\dfrac{a^{15}}{b^{10}}$ **73.** $\dfrac{a^6}{b^4}$ **75.** a^5 **77.** c^7 **79.** $\dfrac{9}{4}$
81. a^4 **83.** $\dfrac{1}{9x^3}$ **85.** $\dfrac{64b^{12}}{27a^9}$ **87.** 1 **89.** $-\dfrac{8a^{21}}{b^3}$
91. $-\dfrac{27}{8a^{18}}$ **93.** $\dfrac{4p^2r^{12}}{9q^8}$ **95.** $\dfrac{8y^{13}}{9x^{16}}$ **97.** $\dfrac{8x^5y^{14}}{9}$
99. 3.462825992 **101.** -244.140625 **109.** $10^{-2}, 10^{-3}, 10^{-4},$
$10^{-5}, 10^{-6}, 10^{-7}, 10^{-8}, 10^{-9}$ **111.** $10^3 \cdot 26^3$; 17,576,000
113. a. $x^6 \text{ ft}^2$ **b.** $x^9 \text{ ft}^3$
119. $(-\infty, 1)$ 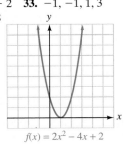 **121.** $(-\infty, 20]$

Think It Through   (page 331)

$5.36 \times 10^7, 1.56 \times 10^7, 7.45 \times 10^{11}$

Study Set   Section 5.2   (page 334)

1. scientific, standard **3.** 10^n, integer **5.** left **7.** 60.22 is not between 1 and 10. **9.** 3.9×10^3 **11.** 7.8×10^{-3}
13. 1.73×10^{14} **15.** 9.6×10^{-6} **17.** 3.23×10^7 **19.** 6.0×10^{-4}
21. 5.27×10^3 **23.** 3.17×10^{-4} **25.** 270 **27.** 0.00323
29. 796,000 **31.** 0.00037 **33.** 5.23 **35.** 23,650,000
37. 1.817×10^{12} **39.** 5.005×10^8 **41.** 5×10^{-8}
43. 1.9×10^{-10} **45.** 4.005×10^{20}; 400,500,000,000,000,000,000
47. 4.3×10^{-3}; 0.0043 **49.** 6×10^3; 6,000
51. 3.6×10^{-5}; 0.000036 **53.** 2,600,000 to 1 **55.** $\$1.7 \times 10^{12}$,
$\$3.9 \times 10^9, \$2.75 \times 10^8, \$3.12 \times 10^8$ **57.** 8.5×10^{-28} g
59. about 9.5×10^{15} m **61. a.** 2.5×10^9 sec = 2,500,000,000 sec
b. about 79 years **63.** 1.209×10^8 mi
65. 1.0×10^{21} **71.** $[1, \frac{9}{4}]$

73. $(-\infty, 2) \cup (7, \infty)$

Think It Through   (page 342)

1. about 21.9 yr and 25.2 yr; about 3.2 yr **2.** about 1959

Study Set   Section 5.3   (page 345)

1. polynomial **3.** degree **5.** cubic **7.** like **9.** monomial; 2
11. trinomial; 3 **13.** binomial; 2 **15.** monomial; 0 **17.** none of these; 10 **19.** binomial; 9 **21. a.** $-2x^4 - 5x^2 + 3x + 7$
b. $7a^3x^5 - ax^3 - 5a^3x^2 + a^2x$ **23.** like terms, $10x$ **25.** unlike terms **27.** like terms, $-5r^2t^3$ **29.** unlike terms
31. $9x^2 + 3x - 2$ **33.** $-1, -1, 1, 3$
35. 8, 2, 0, 2, 8

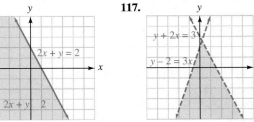
$f(x) = 2x^2 - 4x + 2$

37. $-42, 0, 12, 6, -6, -12, 0, 42$

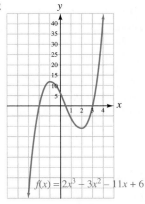
$f(x) = 2x^3 - 3x^2 - 11x + 6$

39.
41. $x^2 - 5x + 6$ **43.** $-5a^2 + 4a + 4$
45. $5a^2 + 3a - 9$ **47.** $-y^3 + 4y^2 + 6$
49. $2pq - 2q$ **51.** $2x^2y^3 + 13xy + 3y^2$
53. $4x^2 - 11$ **55.** $6x^3 - 6x^2 + 14x - 17$
57. $x^2 - 8x + 22$ **59.** $-3y^3 + 18y^2 - 28y + 35$ **61.** $2x$
63. $12x^2 + 9x - 14$ **65.** 20 ft **67.** 872 ft^3 **69.** 2,160 in.^{3}
71. a. $V(x) = 2,500x + 275,000$ **b.** \$325,000 **73. a.** 4
b. ascending **c.** $-\dfrac{1}{720}$ **d.** fourth **e.** sixth **81.** $[-5, 5]$
83. $(-1, 9)$

Study Set   Section 5.4   (page 355)

1. product **3.** terms **5.** factors **7.** term **9.** $x^2 + 2xy + y^2$
11. $x^2 - y^2$ **13.** $x^2 + 2x - 8$ **15.** $16a^2 + 24a + 9$
17. a. $2x, 4x$ **b.** $2x, -3$ **c.** $4, 4x$ **d.** $4, -3$ **19.** $-6a^3b$
21. $-15a^2b^2c^3$ **23.** $-120a^9b^3$ **25.** $-405x^7y^4$ **27.** $3x + 6$
29. $3x^3 + 9x^2$ **31.** $-6x^3 + 6x^2 - 4x$ **33.** $7r^3st + 7rs^3t - 7rst^3$
35. $-12m^4n^2 - 12m^3n^3$ **37.** $x^2 + 5x + 6$ **39.** $6t^2 + 5t - 6$
41. $6y^2 - 5yz + z^2$ **43.** $2b^2 + 35b + 48$ **45.** $0.2t^2 - 2.7t + 9$
47. $b^4 + b^3 - b - 1$ **49.** $-6t^2u^2 + 11tu - 3$
51. $27b^5 - 9b^3c - 3b^2c + c^2$ **53.** $55m^3 + 22m^2n^2 + 15mn^3 + 6n^5$
55. $18p^4 + 30p^3 - 72p^2$ **57.** $24m^3y - 20m^2y^2 + 4my^3$
59. $x^2 + 4x + 4$ **61.** $a^2 - 8a + 16$ **63.** $4a^2 + 4ab + b^2$
65. $25r^4 + 60r^2 + 36$ **67.** $81a^2b^2 - 72ab + 16$
69. $\frac{1}{16}b^2 + b + 4$ **71.** $16k^2 - 10.4k + 1.69$ **73.** $x^2 - 4$
75. $y^6 - 4$ **77.** $x^2y^2 - 36$ **79.** $\frac{1}{4}x^2 - 256$ **81.** $5.76 - y^2$
83. $x^3 - y^3$ **85.** $6y^3 + 11y^2 + 9y + 2$ **87.** $8a^3 - b^3$
89. $a^3 - 3a^2b - ab^2 + 3b^3$ **91.** $2a^2 + ab - b^2 - 3bc - 2c^2$
93. $r^4 - 2r^2s^2 + s^4$ **95.** $3x^2 + 12x$ **97.** $-p^2 + 4pq$
99. $3x^2 + 3x - 11$ **101.** $5x^2 - 36x + 7$
103. $9.2127x^2 - 7.7956x - 36.0315$
105. $299.29y^2 - 150.51y + 18.9225$ **107. a.** $(x + y)(x - y)$
b. $x(x - y); x^2 - xy$ **c.** $y(x - y); xy - y^2$
d. They represent the same area. $(x + y)(x - y) = x^2 - y^2$
109. $x(12 - 2x)(12 - 2x)$ in.^{3} = $(144x - 48x^2 + 4x^3)$ in.^{3}
115.
117.

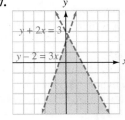

Study Set   Section 5.5   (page 362)

1. factored   **3.** factor   **5.** greatest common factor   **7.** $6xy^2$
9. a. The terms inside the parentheses have a common factor 2.
b. The terms inside the parentheses have a common factor t.
11. 4   **13.** $x^2, 2$ **15.** $2 \cdot 3$ **17.** $3^3 \cdot 5$ **19.** 2^7 **21.** $5^2 \cdot 13$
23. 12   **25.** 2   **27.** $4a^2$ **29.** $6xy^2z^2$ **31.** $2(x+4)$
33. $2x(x-3)$ **35.** prime   **37.** $5x^2y(3-2y)$ **39.** prime
41. $3z(9z^2+4z+1)$ **43.** $9x^7y^3(5x^3-7y^4+9x^3y^7)$
45. $-3(a+2)$ **47.** $-x(3x+1)$ **49.** $-3x(2x+y)$
51. $-6ab(3a+2b)$ **53.** $-7u^2v^3z^2(9uv^3z^7-4v^4+3uz^2)$
55. $(x+y)(4+t)$ **57.** $(a-b)(r-s)$
59. $(m+n+p)(3+x)$ **61.** $(u+v)(u+v-1)$
63. $-(x+y)(a-b)$ **65.** $(x+y)(a+b)$ **67.** $(x+2)(x+y)$
69. $(3-c)(c+d)$ **71.** $(a+b)(a-4)$ **73.** $(a+b)(x-1)$
75. $(x+y)(x+y+z)$ **77.** $x(m+n)(p+q)$
79. $y(x+y)(x+y+2z)$ **81.** $n(2n-p+2m)(n^2p-1)$
83. $r_1 = \dfrac{rr_2}{r_2 - r}$ **85.** $f = \dfrac{d_1 d_2}{d_2 + d_1}$ **87.** $a^2 = \dfrac{b^2 x^2}{b^2 - y^2}$
89. $r = \dfrac{S-a}{S-l}$ **91. a.** $\frac{1}{2}b_1h$ **b.** $\frac{1}{2}b_2h$ **c.** $\frac{1}{2}h(b_1+b_2)$;
the formula for the area of a trapezoid   **93.** $r^2(4-\pi)$
99. a line   **101.** $(-\infty, -3)$ **103.** They are the same.

Study Set   Section 5.6   (page 370)

1. squares   **3.** 1, 4, 9, 16, 25, 36, 49, 64, 81, 100
5. $(x+5)(x+5) = x^2 + 10x + 25$ **7. a.** A common factor of 2
can be factored out of each binomial.   **b.** $(1-t^4)$ factors as the
difference of two squares.   **9. a.** $5(p^2+4)$ **b.** $5(p+2)(p-2)$
11. $p^2 - pq + q^2$ **13.** $p-q$ **15.** $6y, +, 7m$
17. $(x+2)(x-2)$ **19.** $(3y+8)(3y-8)$ **21.** prime
23. $(25a+13b^2)(25a-13b^2)$ **25.** $(9a^2+7b)(9a^2-7b)$
27. $(6x^2y+7z^2)(6x^2y-7z^2)$ **29.** $(x+y+z)(x+y-z)$
31. $(a-b+c)(a-b-c)$ **33.** $(x^2+y^2)(x+y)(x-y)$
35. $(16x^2y^2+z^4)(4xy+z^2)(4xy-z^2)$ **37.** $2(x+12)(x-12)$
39. $2x(x+4)(x-4)$ **41.** $5x(x+5)(x-5)$
43. $t^2(rs+x^2y)(rs-x^2y)$ **45.** $(a+b)(a-b+1)$
47. $(a-b)(a+b+2)$ **49.** $(2x+y)(1+2x-y)$
51. $(r+s)(r^2-rs+s^2)$ **53.** $(x-2y)(x^2+2xy+4y^2)$
55. $(4a-5b^2)(16a^2+20ab^2+25b^4)$
57. $(5xy^2+6z^3)(25x^2y^4-30xy^2z^3+36z^6)$
59. $(x^2+y^2)(x^4-x^2y^2+y^4)$ **61.** $5(x+5)(x^2-5x+25)$
63. $4x^2(x-4)(x^2+4x+16)$ **65.** $2u^2(4v-t)(16v^2+4tv+t^2)$
67. $(a+b)(x+3)(x^2-3x+9)$ **69.** $\frac{4}{3}\pi(r_1-r_2)(r_1^2+r_1r_2+r_2^2)$
73.

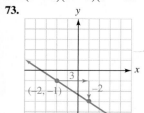

75.

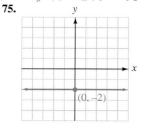

77. $y = -4$

Study Set   Section 5.7   (page 379)

1. trinomial   **3.** lead   **5. a.** positive   **b.** negative   **c.** positive
7. integers   **9.** $2xy + y^2$ **11.** $x^2 - y^2$ **13.** 4, −4, 1   **15.** $(x+2)$
17. $(x-3)$ **19.** $(2a+1)$ **21.** $(x+1)^2$ **23.** $(a-9)^2$
25. $(2y+1)^2$ **27.** $(3b-2)^2$ **29.** $(x-3)(x-2)$

31. $(x-2)(x-5)$ **33.** prime   **35.** $(x+5)(x-6)$
37. $(a+10)(a-5)$ **39.** $(x+3y)(x-7y)$ **41.** $3(x+7)(x-3)$
43. $x^2(b-7)(b-5)$ **45.** $-(a-8)(a+4)$
47. $-3(x-3)(x-2)$ **49.** $-2(p+2q)(p-q)$
51. $(3y+2)(2y+1)$ **53.** $(4a-3)(2a+3)$
55. $(3x-4y)(2x+y)$ **57.** prime   **59.** $(4x-3)(2x-1)$
61. $(a+b)(a-4b)$ **63.** $x(3x-1)(x-3)$
65. $-(3a+2b)(a-b)$ **67.** $5(a-3b)^2$ **69.** $x^2(7x-8)(3x+2)$
71. $(x^2+5)(x^2+3)$ **73.** $(y^2-10)(y^2-3)$
75. $(a+3)(a-3)(a+2)(a-2)$ **77.** $(x+a+1)^2$
79. $(3a+3b+4)(a+b-6)$ **81.** $(x+2+y)(x+2-y)$
83. $(x+1+3z)(x+1-3z)$ **85.** $(c+2a-b)(c-2a+b)$
87. $(a-16)(2a-1)$ **89.** $(2u+3)(u+1)$
91. $(5r+2s)(4r-3s)$ **93.** $x+3$ **97.** 5   **99.** $\frac{8}{3}$
101. $-26p^2 - 6p$

Study Set   Section 5.8   (page 383)

1. factoring   **3.** cubes   **5.** common   **7.** trinomial   **9.** Multiply
the factors of $y^2z^3(x+6)(x+1)$ to see if the product is $x^2y^2z^3 + 7xy^2z^3 + 6y^2z^3$.   **11.** $3ab, 2b, 2a$ **13.** $(x+4)^2$
15. $(2xy-3)(4x^2y+6xy+9)$ **17.** $(x-t)(y+s)$
19. $(5x+4y)(5x-4y)$ **21.** $(6x+5)(2x+7)$
23. $2(3x-4)(x-1)$ **25.** $y^2(2x+1)^2$
27. $(x+a^2y)(x^2-a^2xy+a^4y^2)$ **29.** $2(x-3)(x^2+3x+9)$
31. $(a+b)(f+e)$ **33.** $(2x+2y+3)(x+y-1)$
35. $(25x^2+16y^2)(5x+4y)(5x-4y)$
37. $36(x^2+1)(x+1)(x-1)$ **39.** $(a+3)(a-3)(a+2)(a-2)$
41. $(x+3+y)(x+3-y)$ **43.** $(2x+1+2y)(2x+1-2y)$
45. $(x+y+1)(x-y-1)$ **49.** perpendicular   **51.** -225

Study Set   Section 5.9   (page 390)

1. quadratic   **3.** At least one is 0.   **5. a.** yes   **b.** no   **c.** yes
d. no   **7.** 3, −1   **9.** $y+6, y-9, -6$ **11.** 0, −2   **13.** 4, −4
15. 0, −1   **17.** 0, 5   **19.** −3, −5   **21.** −2, −4   **23.** $-\frac{1}{3}, -3$
25. $\frac{1}{2}, 2$ **27.** $1, -\frac{1}{2}$ **29.** 3, 3   **31.** $\frac{1}{4}, -\frac{3}{2}$ **33.** $2, -\frac{5}{6}$ **35.** $\frac{1}{3}, -1$
37. $2, \frac{1}{2}$ **39.** $\frac{1}{5}, -\frac{5}{3}$ **41.** 0, 0, −1   **43.** 0, 7, −7   **45.** 0, 7, −3
47. 3, −3, 2, −2   **49.** 0, −2, −3   **51.** $0, \frac{5}{6}, -7$ **53.** 16, 18 or
−18, −16   **55.** 2.78, 0.72   **57.** 1   **59.** 10 in., 16 in.   **61.** 3 ft
63. 20 ft by 40 ft   **65.** 11 sec and 19 sec   **67.** 2 sec   **69.** 6 m/sec
71. 4   **79.** 25 ft^2

Key Concept   (page 394)

1. $-5x^2 + 2x - 6$ **3.** $3m^2 + 5m - 12$ **5.** $a^2 - 4ad + 4d^2$
7. $6b^3 + 11b^2 + 9b + 2$ **9.** 0; after falling for 6 seconds,
the squeegee will strike the ground   **11.** 9, −9   **13.** −3, −5
15. $\frac{1}{3}, -1$

Chapter Review   (page 396)

1. 729   **2.** −32   **3.** −64   **4.** 15   **5.** x^6 **6.** a^5b^6
7. m^{18} **8.** t^{13} **9.** $9x^4y^6$ **10.** $\dfrac{x^{16}}{b^4}$ **11.** −3   **12.** $\dfrac{1}{x^{10}}$ **13.** $-\dfrac{1}{625}$
14. $70x^4$ **15.** $\dfrac{x^6}{9}$ **16.** $\dfrac{2}{x}$ **17.** $-c^{10}$ **18.** $\dfrac{25}{16}$ **19.** $\dfrac{1}{y^8}$
20. $-\dfrac{b^3}{8a^{21}}$ **21.** 1.93×10^{10} **22.** 2.735×10^{-8} **23.** 72,770,000
24. 0.0000000083   **25.** 7.6×10^2 sec   **26.** 1.67248×10^{-18} g
27. 8.4×10^6 **28.** no   **29.** yes   **30.** yes   **31.** no
32. binomial, 2   **33.** monomial, 4   **34.** none of these, 4

35. trinomial, 8   **36.** 134 in.³   **37.**

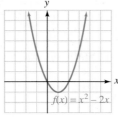

$f(x) = x^2 - 2x$

38.

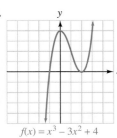

$f(x) = x^3 - 3x^2 + 4$

39. $5x^2 + 2x + 16$
40. $3x^2y^3 - 8x^2y + 8y$
41. $-x^2 + 3x$
42. $2c^3d + 3c^2d - 1$
43. $6x^3 + 4x^2 + x + 9$
44. $6k^4 - 6k^3 + 9k^2 - 2$
45. $-4a^3$ **46.** $6x^3y^2z^5$
47. $2x^4y^3 - 8x^2y^7$
48. $a^4b + 2a^3b^2 - a^2b^3$
49. $16x^2 + 14x - 15$
50. $6x^3 - 12x^2 + 4x - 8$ **51.** $25a^2 - 60a + 36$
52. $0.49c^4 - d^2$ **53.** $\frac{1}{9}a^6 - \frac{2}{3}a^3 + 1$ **54.** $a^2b^2 + 4ab + 3$
55. $15x^4 - 22x^3 + 58x^2 - 40x$ **56.** $r^3 - 3r^2s - rs^2 + 3s^3$
57. $154x^2 + 27x + 1$ **58.** $2 \cdot 5^2 \cdot 7$ **59.** 6 **60.** $3xy^3$
61. $4(x + 2)$ **62.** $3x(x^2 - 2x + 3)$ **63.** prime **64.** $7a^3b(ab + 7)$
65. $5x^2(x + y)(1 - 3x)$ **66.** $9x^2y^3z^2(3xz + 9x^2y^2 - 10z^5)$
67. $-7(b - 2)$ **68.** $-7a^2b^2(a - b)^3(7a^2 - 7ab - 9b^2)$
69. $(x + 2)(y + 4)$ **70.** $(ry - a)(r - 1)$ **71.** $m_1 = \dfrac{mm_2}{m_2 - m}$
72. $(z + 4)(z - 4)$ **73.** $(y + 11)(y - 11)$
74. $(xy^2 + 8z^3)(xy^2 - 8z^3)$ **75.** prime **76.** $(c + a + b)(c - a - b)$
77. $3x^2(x^2 + 10)(x^2 - 10)$ **78.** $(t + 4)(t^2 - 4t + 16)$
79. $2y(x - 3z)(x^2 + 3xz + 9z^2)$ **80.** $(x + 5)^2$ **81.** $(a - 7)^2$
82. $(y + 20)(y + 1)$ **83.** $(z - 5)(z - 6)$ **84.** $-(x + 7)(x - 4)$
85. $(a - 8b)(a + 3b)$ **86.** $(4a - 1)(a - 1)$ **87.** prime
88. $y(y + 2)(y - 1)$ **89.** $3(5x + y)(x - 4y)$
90. $2a^2(a + 3)(a - 1)$ **91.** $(v^2 - 7)(v^2 - 6)$ **92.** $(s + t - 1)^2$
93. $(k + 1 + 3m)(k + 1 - 3m)$ **94.** $x(x - 1)(x + 6)$
95. $3y(x + 3)(x - 7)$ **96.** $(z - 2)(z + x + 2)$
97. $(x + 1 + p)(x + 1 - p)$ **98.** $(x + 2 + 2p^2)(x + 2 - 2p^2)$
99. $(y + 2)(y + 1 + x)$ **100.** $4(ab + 4)(a^2b^2 - 4ab + 16)$
101. $36(z^2 + 1)(z + 1)(z - 1)$ **102.** $\frac{\pi}{2}h(r_1 + r_2)(r_1 - r_2)$
103. $0, \frac{3}{4}$ **104.** $6, -6$ **105.** $\frac{1}{2}, -\frac{5}{6}$ **106.** $\frac{2}{7}, 5$ **107.** $0, -\frac{2}{3}, \frac{4}{5}$
108. $-\frac{2}{3}, 7, 0$ **109.** 17 m by 20 m **110.** $-\frac{1}{2}, 1$

Chapter 5 Test   (page 401)

1. x^9 **2.** $-8x^6y^9$ **3.** $\dfrac{1}{m^5}$ **4.** $\dfrac{m^4}{9n^{10}}$ **5.** 4.706×10^{12}
6. 0.000245 **7.** 1.45×10^{19} **8.** 1.116×10^7 mi/min
9. 5 **10.** 13 **11.** 110 ft **12.** $3x^2 + 4x - 4$
13.

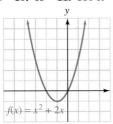

$f(x) = x^2 + 2x$

14. a.

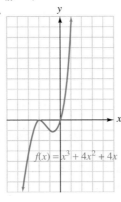

$f(x) = x^3 + 4x^2 + 4x$

14. b. $-2, 0$ **15.** $5y^2 + y - 1$ **16.** $-4uv^2 + 2uv - 14v$
17. $-6x^4y^3z^4$ **18.** $-15a^3b^4 + 10a^3b^5$ **19.** $z^6 - 0.16$
20. $\frac{9}{16}s^2t^2 + 12st + 64$ **21.** $16t^4 - 72t^2y + 81y^2$ **22.** $2s^3 - 2st^2$
23. $3x^5(5 + 2x^2)$ **24.** $3abc(4a^2b - abc + 2c^2)$
25. $(u - v)(r + s)$ **26.** $(a - y)(x + y)$ **27.** $(ax + 7)(ax - 7)$
28. $4(y^2 + 4)(y + 2)(y - 2)$ **29.** prime
30. $(b + 5)(b^2 - 5b + 25)$ **31.** $3(u - 2)(u^2 + 2u + 4)$
32. $(b + c - 6)(b + c + 1)$ **33.** $(3b + 2c)(2b - c)$
34. $(x + 3 + y)(x + 3 - y)$ **35.** $\frac{3}{4}, -\frac{3}{2}$ **36.** $0, -4$
37. $1, 0, -9$ **38.** $v = \dfrac{v_1 v_3}{v_3 + v_1}$ **39.** 11 ft by 22 ft

Cumulative Review Exercises   (page 403)

1. a. true **b.** false **c.** false **d.** true **2.** -27 **3.** 1
4. $C = \frac{5}{9}(F - 32)$ **5.** 1,687.22 cm³ **6.** 12 oz **7.** $y = 8x + 21$
8. $y = -\frac{10}{13}x + \frac{2}{13}$ **9.** -1.2% per year **10.** -3
11. $f(x) = 0.025x + 95$ **12.** D: the set of real numbers,
R: the set of all real numbers greater than or equal to 2

13.

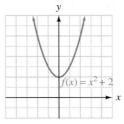

$f(x) = x^2 + 2$

$y = -\frac{5}{2}x + \frac{1}{2}$
$2x - \frac{3}{2}y = 5$
$(1, -2)$

14. $(\frac{3}{4}, \frac{1}{3})$ **15.** $(-2, 0, 2)$ **16.** -23 **17.** $(-\infty, 5]$

18. $[-21, -3)$

19. $(-\infty, -12] \cup [2, \infty)$

20. $(-\infty, -2)$ **21.** iii

22.

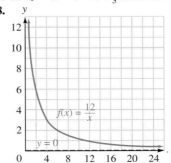

$y = 0$
$x - y = 4$
$x = 0$

23. $\dfrac{9a^2}{b^8}$ **24.** 0.000000090895
25. $8a^3 - b^3$
26. $13k^2 - 10k + 17$
27. $(x - y)(x - 4)$
28. $6s^2(s + 6)(s - 6)$
29. $(2x^2 + 5y)(4x^4 - 10x^2y + 25y^2)$
30. $-(3a + 2b)(a - b)$
31. $1, -\frac{1}{2}$ **32.** $m_1 = \dfrac{m_2 g}{m_2 - g}$

Chapter 6   Check Your Knowledge   (page 406)

1. denominator **2.** similar **3.** direct, inverse **4.** extraneous
5. $2x^2/z^2$ **6.** $-a$ **7.** $6\frac{2}{3}$ hours or 6 hours 40 minutes
8.

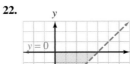

$f(x) = \dfrac{12}{x}$
$y = 0$

9. x^4y^4/z^2
10. $\dfrac{a + b}{(x - 3)(c + d)}$
11. $-\dfrac{q}{p}$
12. 1   **13.** $x + 3$
14. $-\dfrac{5x + 1}{(2x - 1)(x + 1)}$
15. $\dfrac{6x^2 + 3}{6x^2 - 2}$
16. $-\dfrac{h}{4}$

17. 6 **18.** $-\dfrac{2}{5}$ **19.** no solution; 0 is extraneous

20. $c = ab/(a + b)$ **21.** 2 hours **22.** $\dfrac{5y}{2} - \dfrac{3y^2}{2x} + \dfrac{2}{x^2}$

23. $x^2 - 2x + 4$ **24.** $x^2 - 3x - 6 - \dfrac{6}{x - 2}$

Think It Through (page 411)

1. about 40% **2.** about 85% **3.** about 5%

Study Set   Section 6.1 (page 415)

1. rational **3.** asymptote **5. a.** 1 **b.** 0.5 **c.** 0.25 **7. a.** 1
b. 1 **c.** 1 **d.** −1 **9.** 1: decreases then steadily increases;
2: increases, decreases, then steadily increases; 3: steadily
increases; 4: steadily decreases, approaching a cost of $2.00 per
unit **11. a.** true **b.** true **13.** 20 hr **15.** 12 hr **17.** 6, 3, 1.5,
1, 0.75, 0.6, 0.5

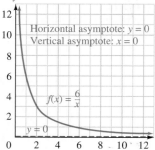

19. 3, 2, 1.5, 1.33, 1.25, 1.2, 1.17

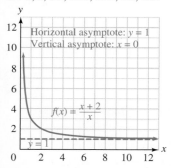

21. $(-\infty, 0) \cup (0, \infty)$ **23.** $(-\infty, -2) \cup (-2, \infty)$
25. $(-\infty, 0) \cup (0, 1) \cup (1, \infty)$
27. $(-\infty, -7) \cup (-7, 8) \cup (8, \infty)$

29. $\dfrac{2}{3}$ **31.** $-\dfrac{28}{9}$ **33.** $4x^2$ **35.** $-\dfrac{4y}{3x}$ **37.** $-\dfrac{x}{2}$ **39.** −1

41. $\dfrac{1}{x - y}$ **43.** $\dfrac{5}{x - 2}$ **45.** $\dfrac{-3(x + 2)}{x + 1}$ **47.** in lowest terms

49. $x + 2$ **51.** $\dfrac{x + 1}{x + 3}$ **53.** $\dfrac{s - 3}{s + 6}$ **55.** $\dfrac{x + 4}{2(2x - 3)}$

57. $\dfrac{3(x - y)}{x + 2}$ **59.** $\dfrac{2x + 1}{2 - x}$ **61.** $\dfrac{a^2 - 3a + 9}{4(a - 3)}$

63. in lowest terms **65.** $m + n$ **67.** $-\dfrac{m + n}{2m + n}$ or $\dfrac{-m - n}{2m + n}$

69. $\dfrac{x - y}{x + y}$ **71.** $\dfrac{2x - 3}{2y - 3}$ **73.** $\dfrac{3a + b}{y + b}$ **75.** 1

77. D: $(-\infty, 2) \cup (2, \infty)$; R: $(-\infty, 1) \cup (1, \infty)$
79. D: $(-\infty, -2) \cup (-2, 2) \cup (2, \infty)$; R: $(-\infty, \infty)$
81. a. $50,000 **b.** $200,000 **83. a.** $c(n) = 0.09n + 7.50$
b. $c(n) = \dfrac{0.09n + 7.50}{n}$ **c.** about 10¢ **85. a.** about 2.5 hr
b. about 4.6 hr **89.** $3x(x - 3)$
91. $(3x^2 + 4y)(9x^4 - 12x^2y + 16y^2)$

Study Set   Section 6.2 (page 425)

1. rational **3.** numerator, denominator **5.** multiply, $\frac{ac}{bd}$ **7.** 0
9. $x - 5, 5x - 25, x(x + 3), 5(x - 5)$ **11.** yes, no, yes

13. $\dfrac{5}{4}$ **15.** $-\dfrac{5}{6}$ **17.** $\dfrac{xy^2d}{2c^2}$ **19.** $-\dfrac{x^{10}}{y^2}$ **21.** $x + 1$ **23.** 1

25. $\dfrac{x - 4}{x + 5}$ **27.** $-\dfrac{(a + 7)^2(a - 5)}{12x^2}$ **29.** $\dfrac{t - 1}{t + 1}$ **31.** $\dfrac{n + 2}{n + 1}$

33. $-\dfrac{x + 1}{x - 1}$ **35.** $\dfrac{1}{x + 1}$ **37.** $(x + 1)^2$ **39.** $\dfrac{2b(b - 1)}{b^2 + 1}$

41. $\dfrac{x + y}{x - y}$ **43.** $\dfrac{a + b}{(x - 3)(c + d)}$ **45.** $-\dfrac{x + 1}{x + 3}$ or $\dfrac{-x - 1}{x + 3}$

47. $x^2(x + 3)$ **49.** $\dfrac{x + 2}{x - 2}$ **51.** $\dfrac{3x}{2}$ **53.** 1

55. $\dfrac{x^2 - 6x + 9}{x^6 + 8x^3 + 16}$ **57.** $\dfrac{4m^4 - 4m^3 - 11m^2 + 6m + 9}{x^4 - 2x^2 + 1}$

59. $k_1(k_1 + 2), k_2 + 6$ **63.** $-6a^5 + 2a^4$ **65.** $6g^2 - 5gn + n^2$

Study Set   Section 6.3 (page 434)

1. denominator **3.** LCD **5.** subtract, keep, $\dfrac{a - c}{b}$ **7.** same
9. a. ii **b.** adding or subtracting rational expressions
c. simplifying a rational expression **11.** $3x - 2, 3, 3x - 1$

13. $\dfrac{5}{2}$ **15.** $\dfrac{11}{4y}$ **17.** 2 **19.** 3 **21.** 3 **23.** $\dfrac{6x}{(x - 3)(x - 2)}$

25. $36x^2$ **27.** $x(x + 3)(x - 3)$ **29.** $(x + 3)^2(x^2 - 3x + 9)$

31. $(2x + 3)^2(x + 1)^2$ **33.** $\dfrac{5}{6}$ **35.** $\dfrac{21a - 8b}{14}$ **37.** $\dfrac{17}{12x}$

39. $\dfrac{9a^2 - 4b^2}{6ab}$ **41.** $\dfrac{3a - 5b}{a^2b^2}$ **43.** $\dfrac{3r + 2bs}{12b^2}$ **45.** $\dfrac{2(5a + 2b)}{21}$

47. $\dfrac{2(4x - 1)}{(x + 2)(x - 4)}$ **49.** $\dfrac{7x + 29}{(x + 5)(x + 7)}$ **51.** $\dfrac{4x + 1}{x}$

53. 2 **55.** $\dfrac{3(a - 1)}{3a - 2}$ **57.** $\dfrac{x(2x + 1)}{(x + 3)(x + 2)(x - 2)}$

59. $\dfrac{-x^2 + 11x + 8}{(3x + 2)(x + 1)(x - 3)}$ **61.** $\dfrac{2x^2 + 5x + 4}{x + 1}$

63. $\dfrac{x^2 - 5x - 5}{x - 5}$ **65.** $\dfrac{-2(2x^2 - 7x - 27)}{x(x + 3)(x - 3)}$

67. $\dfrac{11x^2 + 7x - 3}{(2x - 1)(3x + 2)}$ **69.** $\dfrac{2}{x + 1}$ **71.** $\dfrac{2b}{a + b}$

73. $\dfrac{7mn^2 - 7n^3 - 6m^2 + 3mn + n}{(m - n)^2}$ **75.** $\dfrac{2}{m - 1}$

77. $\dfrac{10r + 20}{r}; \dfrac{3t + 9}{t}$ **83.** 3, 3 **85.** 0, 0, −1

Study Set   Section 6.4 (page 442)

1. complex **3.** $\dfrac{t^2}{t^2}$ **5.** $\div, \dfrac{3}{25m}, m, 3, 3, m, 10$ **7.** ÷ **9.** $\dfrac{2}{3}$

11. $-\dfrac{1}{7}$ **13.** $\dfrac{2y}{3z}$ **15.** $125b$ **17.** $-\dfrac{1}{y}$ **19.** $\dfrac{y - x}{x^2y^2}$

21. $\dfrac{b + a}{b}$ **23.** $\dfrac{y + x}{y - x}$ **25.** $y - x$ **27.** $-\dfrac{1}{a + b}$

29. $x^2 + x - 6$ **31.** $\dfrac{5x^2y^2}{xy + 1}$ **33.** −1 **35.** $\dfrac{x + 2}{x - 3}$ **37.** $\dfrac{a - 1}{a + 1}$

39. $\dfrac{2x^2 + 5x}{x^3 + x^2 + x + 1}$ **41.** $\dfrac{xy^2}{y - x}$ **43.** $\dfrac{x^2(xy^2 - 1)}{y^2(x^2y - 1)}$ **45.** $\dfrac{1}{x - y}$

47. $\dfrac{x - 9}{3}$ **49.** $-\dfrac{h}{4}$ **51.** $\dfrac{3a^2 + 2a}{2a + 1}$

53. $\dfrac{(-x^2 + 2x - 2)(3x + 2)}{(2 - x)(-3x^2 - 2x + 9)}$ **55.** $\dfrac{k_1k_2}{k_2 + k_1}$ **57.** $\dfrac{4d - 1}{3d - 1}$

61. 8 **63.** 2, −2, 3, −3

Study Set Section 6.5 (page 454)

1. rational **3.** clear **5. a.** yes **b.** no

7. $\frac{12}{x}, \frac{12}{x+15}$ **9. a.** 3, 0 **b.** 3, 0 **c.** 3, 0 **11.** $30y, \frac{9}{2y}, \frac{10}{3y}, 7y, 35$

13. 12 **15.** 40 **17.** $\frac{1}{2}$ **19.** no solution; 4 is extraneous **21.** $\frac{17}{25}$

23. 0 **25.** 1 **27.** 2 **29.** 2 **31.** $\frac{1}{3}$ **33.** 0 **35.** 2, -5

37. $-4, 3$ **39.** $6, \frac{17}{3}$ **41.** $1, -11$ **43.** $-2; -3$ is extraneous

45. $r = \dfrac{E - IR_L}{I}$ **47.** $r = \dfrac{S - a}{S - l}$ **49.** $Q_1 = \dfrac{PQ_2}{1 + P}$

51. $R = \dfrac{R_1 R_2 R_3}{R_2 R_3 + R_1 R_3 + R_1 R_2}$ **53.** $4\frac{8}{13}$ in. **55.** $L = \frac{SN - CN}{V - C}$,

8 years **57. a.** $1\frac{7}{8}$ days **b.** Santos: \$412.50, Mays: \$375

59. about 110 sec **61.** $2\frac{4}{13}$ weeks **63.** 6 mph **65.** 35 mph and

45 mph **67.** 5 mph **69.** 9 **75.** 9.0×10^9 **77.** 4.4×10^{-22}

Study Set Section 6.6 (page 466)

1. dividend, divisor, quotient **3.** placeholder **5.** b, b

7. $(2x - 1)(x^2 + 3x - 4) = 2x^3 + 5x^2 - 11x + 4$

9. $2x^2, x, 4, 8x, -3x, -4$ **11.** $3a^2 + 5 + \dfrac{6}{3a - 2}$

13. $\dfrac{x^2 - x - 12}{x - 4}, x - 4\overline{)x^2 - x - 12}$,

$(x^2 - x - 12) \div (x - 4)$ **15.** $\dfrac{y}{2x^3}$ **17.** $-\dfrac{3}{4a^2}$

19. $2x + 3$ **21.** $-\dfrac{2x}{3} + \dfrac{x^2}{6}$ **23.** $2xy^2 + \dfrac{x^2 y}{6}$

25. $\dfrac{x^4 y^4}{2} - \dfrac{x^3 y^9}{4} + \dfrac{3}{4xy^2}$ **27.** $x + 2$ **29.** $x + 7$

31. $3x - 5 + \dfrac{3}{2x + 3}$ **33.** $3x^2 + x + 2 - \dfrac{4}{x - 1}$

35. $2x^2 + 5x + 3 + \dfrac{4}{3x - 2}$ **37.** $3x^2 + 4x + 3$ **39.** $a + 1$

41. $2y + 2$ **43.** $6x - 12$ **45.** $4x^3 - 3x^2 + 3x + 1$

47. $a^2 + a + 1 + \dfrac{2}{a - 1}$ **49.** $5a^2 - 3a - 4$ **51.** $6y - 12$

53. $x^4 + x^2 + 4$ **55.** $x^2 + x + 1$ **57.** $x^2 + x + 2$

59. $9.8x + 16.4 - \dfrac{36.5}{x - 2}$ **61.** $12, 26, 8$ **63.** $x + 2$

65. $x + 2$ **67.** $x - 7 + \dfrac{28}{x + 2}$ **69.** $3x^2 - x + 2$

71. $2x^2 + 4x + 3$ **73.** $6x^2 - x + 1 + \dfrac{3}{x + 1}$

75. $7.2x - 0.66 + \dfrac{0.368}{x - 0.2}$ **77.** $2.7x - 3.59 + \dfrac{0.903}{x + 1.7}$ **79.** -1

81. -37 **83.** 2 **85.** -1 **87.** -8 **89.** 59 **91.** 44 **93.** $\frac{29}{32}$

95. $x^2 - 5x + 6$ **97.** $3x - 2, x + 5$ **103.** $8x^2 + 2x + 4$

105. $-2y^3 - y^2 + 10y - 14$

Study Set Section 6.7 (page 478)

1. ratio **3.** extremes, means **5.** direct, inverse **7.** inverse

9.

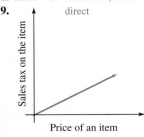

direct

Sales tax on the item

Price of an item

11.

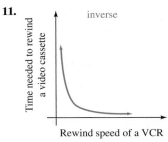

inverse

Time needed to rewind a video cassette

Rewind speed of a VCR

13. $-7, 6, 18, -102$ **15.** 3

17. 5 **19.** 5 **21.** 39

23. $2, -2$ **25.** $4, -1$

27. $-\frac{5}{2}, -1$ **29.** no solution

31. $A = kp^2$

33. $v = \dfrac{k}{r^2}$ **35.** $P = \dfrac{ka^2}{j^3}$

37. L varies jointly with m and n. **39.** R varies directly with L and inversely with d^2. **41.** 202, 172, 136 **43.** about 2 gal **45.** eye: 49.9 in.; seat: 17.6 in.; elbow: 27.8 in. **47.** 12.5 in. **49.** 555 ft

51. $46\frac{7}{8}$ ft **53.** 880 ft **55.** 1,600 ft **57.** 25 days **59.** 12 in.^{3}

61. \$9,000 **63.** 3 ohms **65.** 0.275 in. **67.** 12.8 lb **71.** x^{10}

73. -1

Key Concept (page 483)

1. a. $\dfrac{3x + 2}{4x + 3}$ **b.** $2x - 1$ **3. a.** $\dfrac{2(4x - 1)}{(x + 2)(x - 4)}$

b. $\dfrac{x - 4}{x - 4}, \dfrac{x + 2}{x + 2}$ **5. a.** 4 **b.** $t(t - 2)$

7. a. 1 **b.** $(x + 2)(x - 4)$

Chapter Review (page 485)

1. $8, 4, 2, 1.33, 1, 0.8, 0.67, 0.57, \frac{1}{2}$

2. $y = 3, x = 0$; D: $(-\infty, 0) \cup (0, \infty)$, R: $(-\infty, 3) \cup (3, \infty)$

3. $\dfrac{31x}{72y}$ **4.** $\dfrac{x - 7}{x + 7}$ **5.** $\dfrac{1}{2(x + 2)}$

6. $\dfrac{1}{x - 6}$ **7.** $\dfrac{-a - b}{c + d}$

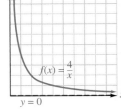

$f(x) = \dfrac{4}{x}$

$y = 0$

8. $\dfrac{-m - 2n}{n + 2m}$ **9.** -1 **10.** -2

11. -1 **12.** $\dfrac{2a - 1}{a + 2}$ **13.** $\dfrac{h^2 - 4h + 4}{h^6 + 8h^3 + 16}$ **14.** $\dfrac{t - 2}{t}$

15. $\dfrac{3x(x - 1)}{(x - 3)(x + 1)}$ **16.** $\dfrac{5y - 3}{x - y}$ **17.** $\dfrac{6x - 7}{x^2 + 2}$

18. $-\dfrac{2}{t - 3}$ **19.** $-\dfrac{1}{p + 12}$ **20.** $60a^2 h^3$ **21.** $ab^2(b - 1)$

22. $(x - 5)(x + 5)(x + 1)$ **23.** $(m^2 + 2m + 4)(m - 2)^2$

24. $\dfrac{9a + 8}{a + 1}$ **25.** $\dfrac{3x + 2yz}{12z^2}$ **26.** $\dfrac{4x^2 + 9x + 12}{(x - 4)(x + 3)}$

27. $\dfrac{4(3y + 5)}{15y(x - 2)}$ **28.** $\dfrac{x^2 + 26x + 3}{(x + 3)(x - 3)^2}$ **29.** $\dfrac{p - 3}{2(p + 2)}$

30. $\dfrac{b + 2a}{2b - a}$ **31.** $\dfrac{x - 2}{x + 3}$ **32.** $\dfrac{x^2 y^2}{(x - y)^2 (y^2 - x^2)}$

33. $\dfrac{3x^2 - 18x + 2}{x^2 - 6x + 3}$ **34.** 5 **35.** 2 **36.** $8, -9$ **37.** $-1, -2$

38. no solution; 2 and -2 are extraneous **39.** 0

40. $y^2 = \dfrac{x^2 b^2 - a^2 b^2}{a^2}$ **41.** $b = \dfrac{Ha}{2a - H}$ **42.** 5 mph

43. 50 mph **44.** $14\frac{2}{5}$ hr **45.** $18\frac{2}{3}$ days

46. about 3,500,000 in. lb/rad **47.** $\dfrac{5h^3}{k^2}$ **48.** $-\dfrac{x^3}{2y^3}$

49. $6a + \dfrac{16}{3}$ **50.** $-3x^2 y + \dfrac{3x}{2} + y$ **51.** $b + 4$

52. $v^2 - 3v - 10$ **53.** $x^2 - 2x + 4$ **54.** $x^2 + 2x - 1 + \dfrac{6}{2x + 3}$

55. $x^2 + 4x + 3$ **56.** $x^3 - x^2 + 2x - 2 + \dfrac{3}{x+1}$ **57.** 8

58. 5   **59.** $-4, -12$ **60.** 70.4 ft   **61.** \$5,460   **62.** 1.25 amps

63. 126.72 lb   **64.** inverse variation   **65.** 0.2

Chapter 6 Test   (page 491)

1. $-\dfrac{2}{3xy}$ **2.** $\dfrac{2}{x-2}$ **3.** -3 **4.** $\dfrac{2x+1}{4}$ **5.** $\dfrac{xz}{y^4}$

6. $\dfrac{x+1}{2}$ **7.** 1   **8.** $\dfrac{(x+y)^2}{2}$ **9.** $\dfrac{2}{x+1}$ **10.** -1 **11.** $\dfrac{2}{t-4}$

12. 2   **13.** $\dfrac{24b^2 - 2b - 1}{3b + 1}$ **14.** $\dfrac{2x+3}{(x+1)(x+2)}$ **15.** $\dfrac{u^2}{2vw}$

16. $\dfrac{3k^2 + 4k + 4}{3k^2 - 9k - 9}$ **17.** 40   **18.** 5; 3 is extraneous.

19. $6, -1$ **20.** 26   **21.** $a^2 = \dfrac{x^2b^2}{b^2 - y^2}$ **22.** $r_2 = \dfrac{rr_1}{r_1 - r}$

23. no, $\frac{5}{11}$ of an hour   **24.** 3 mph   **25.** $-\dfrac{6x}{y} + \dfrac{4x^2}{y^2} - \dfrac{3}{y^3}$

26. $y^2 - 2y + 4 - \dfrac{56}{y+2}$ **27.** $a^2 + a + 1 + \dfrac{2}{a-1}$ **28.** 47

31. \$77.32   **32.** 25 decibels

33.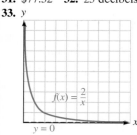

34.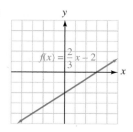

Cumulative Review Exercises   (page 493)

1. $\frac{8}{3}$ **2.** -2 **3.** 14   **4.** Life expectancy will increase 0.1 year each year during this period.   **5.** $y = -7x + 54$

6. $y = -\frac{11}{6}x - \frac{7}{3}$

7. D: $(-\infty, \infty)$; R: $(-\infty, \infty)$

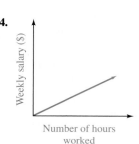

8. $T_1 = 80, T_2 = 60$ **9.**

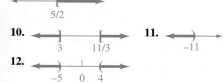

10.   **11.**

12.

13. a^8b^4 **14.** $\dfrac{b^4}{a^4}$ **15.** 81   **16.** $\dfrac{x^{21}y^3}{8}$ **17.** 42,500

18. 0.000712   **19.** $y = \dfrac{kxz}{r}$ **20.** 8

21.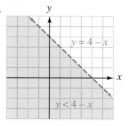

22. D: $(-\infty, \infty)$; R: $[-3, \infty)$

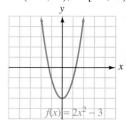

23. 18   **24.** 7   **25. a.** $(-3, 0), (0, -7)$ **b.** $-\frac{7}{3}$ **c.** I

d. $y = -\frac{7}{3}x - 7$ **e.** yes   **26.** penny: $\$10^{-2}$, dime: $\$10^{-1}$, one dollar bill: $\$10^0$, one hundred thousand dollar bill: $\$10^5$

27. $2x^3 + x^2 + 12$ **28.** $-3x^2 - 3$ **29.** $6x^2 - 7x - 20$

30. $4x^6 - 4x^3 + 1$ **31.** yes   **32.** about 17   **33.** about 14

34. 4   **35.** $3rs^3(r - 2s)$ **36.** $(x - y)(5 - a)$

37. $(x + y)(u + v)$ **38.** $(9x^2 + 4y^2)(3x + 2y)(3x - 2y)$

39. $(2x - 3y^2)(4x^2 + 6xy^2 + 9y^4)$ **40.** $(2x + 3)(3x - 2)$

41. $(3x - 5)^2$ **42.** $(5x + 3)(3x - 2)$ **43.** $-\dfrac{1}{3}, -\dfrac{7}{2}$

44. $0, 2, -2$ **45.** $b^2 = \dfrac{a^2y^2}{a^2 - x^2}$ **46.** 9 in. by 12 in.

47. $\dfrac{2x-3}{3x-1}$ **48.** $-\dfrac{q}{p}$ **49.** $\dfrac{4}{x-y}$ **50.** $\dfrac{a^2 + ab^2}{a^2b - b^2}$ **51.** 0

52. -17 **53.** $x + 4$ **54.** $-x^2 + x + 5 + \dfrac{8}{x-1}$

55. It will rise sharply.   **56.** It has dropped.

Chapter 7   Check Your Knowledge   (page 497)

1. radical   **2.** odd, even   **3.** $\sqrt{x} - 1$ **4.** power, root

5. hypotenuse   **6.** $1, 0, -0.41, -1, -2,$

$-2.32, -3$

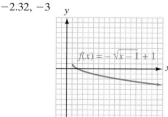

7. $2x\sqrt{2x}$

8. $2yz^2\sqrt{2y}$

9. $\dfrac{3x}{4y^2}$ **10.** $-2y^2$

11. -2

12. not real

13. $(x-1)^2$ or $x^2 - 2x + 1$

14. $\dfrac{2xy\sqrt[3]{2xy^2}}{3z}$

15. $6x^2\sqrt{5x}$ **16.** $\sqrt[3]{2}$ **17.** $12 + 2\sqrt{15}$

18. -3 **19.** $\dfrac{1}{16}$ **20.** $\dfrac{729x^3}{y^2}$ **21.** $\dfrac{2x + x\sqrt{x}}{x - 4}$ **22.** $\dfrac{\sqrt[3]{15a^2}}{3a}$

23. $0; 4$ **24.** 9   **25.** 5 meters   **26.** $\sqrt{53}$

Study Set   Section 7.1   (page 507)

1. square root   **3.** radical   **5.** odd   **7.** simplified   **9.** a

11. two, positive   **13.** x **15.** 3, up

17. $0, -1, -2, -3, -4$

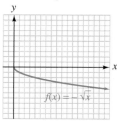

19. $\sqrt{x^2} = |x|$ **21.** $f(x) = \sqrt{x - 5}$ **23.** 11   **25.** -8

27. $\frac{1}{3}$ **29.** 0.5   **31.** not real   **33.** 4   **35.** 3.4641   **37.** 26.0624

39. $2|x|$ **41.** $|t + 5|$ **43.** $5|b|$ **45.** $|a + 3|$ **47.** 0   **49.** 4
51. 2   **53.** -10 **55.** 4.4721   **57.** 3.3322
59. D: $[-4, \infty)$, R: $[0, \infty)$

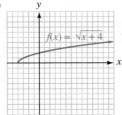

61. D: $(-\infty, \infty)$, R: $(-\infty, \infty)$

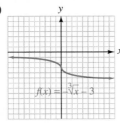

63. 1   **65.** -5 **67.** $-\frac{2}{3}$ **69.** 0.4   **71.** $2a$ **73.** $-10pq$
75. $-\frac{1}{2}m^2n$ **77.** $-0.4s^3t^2$ **79.** 3   **81.** -3 **83.** not real
85. $\frac{2}{5}$ **87.** $\frac{1}{2}$ **89.** $2a$ **91.** $2a$ **93.** k^3 **95.** $\frac{1}{2}m$ **97.** $x + 2$
99. 7.0 in.   **101.** about 61.3 beats/min   **103.** 13.4 ft   **105.** 3.5%
111. 1   **113.** $\dfrac{3(m^2 + 2m - 1)}{(m + 1)(m - 1)}$

Study Set   Section 7.2   (page 516)

1. like   **3.** factor   **5.** $\sqrt[n]{a}\sqrt[n]{b}$, product, roots   **7. a.** $\sqrt{4 \cdot 5}$
b. $\sqrt{4}\sqrt{5}$ **c.** $\sqrt{4 \cdot 5} = \sqrt{4}\sqrt{5}$ **9. a.** $\sqrt{5}$, $\sqrt[3]{5}$ (answers
may vary); no   **b.** $\sqrt{5}$, $\sqrt{6}$ (answers may vary); no
11. $8k^3$, $8k^3$, $4k$ **13.** 6   **15.** t **17.** $5x$ **19.** 10   **21.** $7x$ **23.** $6b$
25. 2   **27.** $3a$ **29.** $2\sqrt{5}$ **31.** $-10\sqrt{2}$ **33.** $2\sqrt[3]{10}$ **35.** $-3\sqrt[3]{3}$
37. $2\sqrt[4]{2}$ **39.** $2\sqrt[5]{3}$ **41.** $\dfrac{\sqrt{7}}{3}$ **43.** $\dfrac{\sqrt[3]{7}}{4}$ **45.** $\dfrac{\sqrt[4]{3}}{10}$ **47.** $\dfrac{\sqrt[5]{3}}{2}$
49. $5x\sqrt{2}$ **51.** $4\sqrt{2b}$ **53.** $-4a\sqrt{7a}$ **55.** $5ab\sqrt{7b}$
57. $-10\sqrt{3xy}$ **59.** $-3x^2\sqrt[3]{2}$ **61.** $2x^4y\sqrt[3]{2y}$ **63.** $2x^3y\sqrt[4]{2y}$
65. $\dfrac{z}{4x}$ **67.** $\dfrac{\sqrt[4]{5x}}{2z}$ **69.** $10\sqrt{2x}$ **71.** $\sqrt[5]{7a^2}$ **73.** $-\sqrt{2}$
75. $2\sqrt{2}$ **77.** $9\sqrt{6}$ **79.** $3\sqrt[3]{3x}$ **81.** $-\sqrt[3]{4}$ **83.** -10
85. $-17\sqrt[4]{2}$ **87.** $16\sqrt[4]{2}$ **89.** $-4\sqrt{2}$ **91.** $3\sqrt{2t} + \sqrt{3t}$
93. $-11\sqrt[3]{2}$ **95.** $y\sqrt{z}$ **97.** $13y\sqrt{x}$ **99.** $12\sqrt[3]{a}$ **101.** $-7y^2\sqrt{y}$
103. $4x\sqrt[5]{xy^2}$ **105.** $8\pi\sqrt{5}$ ft²; 56.2 ft²   **107.** $5\sqrt{3}$ amps;
8.7 amps   **109.** $(26\sqrt{5} + 10\sqrt{3})$ in.; 75.5 in.   **115.** $-\dfrac{15x^5}{y}$

117. $3p + 4 - \dfrac{5}{2p - 5}$

Study Set   Section 7.3   (page 525)

1. multiply   **3.** irrational   **5.** perfect   **7. a.** $6\sqrt{6}$ **b.** 48
c. can't be simplified   **d.** $-6\sqrt{6}$ **9.** When the numerator
and the denominator of a fraction are multiplied by the same
nonzero number, the value of the fraction is not changed.
11. A radical appears in the denominator.   **13.** $\sqrt{6}$, 48, 16, 4
15. 11   **17.** 7   **19.** 4   **21.** $5\sqrt{2}$ **23.** $6\sqrt{2}$ **25.** 5   **27.** 18
29. 8   **31.** 18   **33.** $2\sqrt[3]{3}$ **35.** ab^2 **37.** $5a\sqrt{b}$ **39.** $-20r\sqrt[3]{10s}$
41. $x^2(x + 3)$ **43.** $12\sqrt{5} - 15$ **45.** $24\sqrt{3} + 6\sqrt{14}$
47. $-8x\sqrt{10} + 6\sqrt{15x}$ **49.** $-1 - 2\sqrt{2}$ **51.** $5z + 2\sqrt{15z} + 3$

53. $3x - 2y$ **55.** $6a + 5\sqrt{3ab} - 3b$ **57.** $18r - 12\sqrt{2r} + 4$
59. $-6x - 12\sqrt{x} - 6$ **61.** $\dfrac{\sqrt{7}}{7}$ **63.** $\dfrac{\sqrt{30}}{5}$ **65.** $\dfrac{\sqrt{10}}{4}$
67. 2   **69.** $\dfrac{\sqrt[3]{4}}{2}$ **71.** $\sqrt[3]{3}$ **73.** $\dfrac{\sqrt[3]{6}}{3}$ **75.** $\dfrac{2\sqrt{2xy}}{xy}$ **77.** $\dfrac{\sqrt{5y}}{y}$
79. $\dfrac{\sqrt[3]{2ab^2}}{b}$ **81.** $\dfrac{\sqrt[4]{4}}{2}$ **83.** $\dfrac{\sqrt[5]{2}}{2}$ **85.** $\sqrt{2} + 1$ **87.** $\dfrac{3\sqrt{2} - \sqrt{10}}{4}$
89. $2 + \sqrt{3}$ **91.** $\dfrac{9 - 2\sqrt{14}}{5}$ **93.** $\dfrac{2(\sqrt{x} - 1)}{x - 1}$ **95.** $\sqrt{2z} + 1$
97. $\dfrac{x - 2\sqrt{xy} + y}{x - y}$ **99.** $\dfrac{\sqrt{2\pi}}{2\pi\sigma}$ **101.** $\dfrac{\sqrt{2}}{2}$ **103.** $f/4$ **107.** 1
109. $\frac{1}{3}$

Study Set   Section 7.4   (page 533)

1. radical   **3.** extraneous   **5. a.** Square both sides.   **b.** Cube
both sides.   **7. a.** x **b.** $x - 5$ **c.** $32x$ **d.** $x + 3$ **9.** Only one
side of the equation was squared.   **11.** If we raise two equal
quantities to the same power, the results are equal quantities.
13. $2\sqrt{x - 2}$, 4, 16, 8, 24   **15.** 2   **17.** 4   **19.** 0   **21.** 4   **23.** 8
25. $\frac{5}{2}$, $\frac{1}{2}$ **27.** 1   **29.** 16   **31.** 14, $\cancel{6}$ **33.** 4, 3   **35.** 2, $\cancel{7}$
37. 2, -1 **39.** -1, $\cancel{1}$ **41.** $\cancel{1}$, no solution   **43.** 0, $\cancel{4}$ **45.** $\cancel{-3}$,
no solution   **47.** 1, 9   **49.** 4, $\cancel{0}$ **51.** 2, $\cancel{142}$
53. $\cancel{6}$, no solution   **55.** $h = \dfrac{v^2}{2g}$ **57.** $l = \dfrac{8T^2}{\pi^2}$
59. $A = P(r + 1)^3$ **61.** $v^2 = c^2\left(1 - \dfrac{L_A^2}{L_B^2}\right)$ **63.** 178 ft
65. about 488 watts   **67.** 10 lb   **69.** \$5   **75.** 2.5 foot-candles
77. 0.41511 in.

Study Set   Section 7.5   (page 543)

1. rational (or fractional)   **3.** index, radicand
5. $25^{1/5}$, $(\sqrt[3]{-27})^2$, $16^{-3/4}$, $(\sqrt{81})^3$, $-\left(\frac{9}{64}\right)^{1/2}$
7.

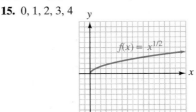

9. x^{m+n} **11.** x^{m-n} **13.** $\sqrt[n]{x}$
15. 0, 1, 2, 3, 4   **17.** $\sqrt{100a^4}$, $10a^2$
19. $\sqrt[3]{x}$ **21.** $\sqrt[4]{3x}$
23. $\sqrt[4]{17x^3y}$
25. $\sqrt{x^2 + y^2}$
27. $m^{1/2}$ **29.** $(3a)^{1/4}$
31. $\left(\frac{1}{7}abc\right)^{1/6}$
33. $(a^2 - b^2)^{1/3}$
35. 2   **37.** 5   **39.** 2
41. 2   **43.** $\frac{1}{4}$ **45.** -2 **47.** not real   **49.** -3 **51.** $5|y|$
53. $2|x|$ **55.** $3x$ **57.** not real   **59.** 216   **61.** 27   **63.** 1,728
65. $\dfrac{1}{4}$ **67.** $125x^6$ **69.** $\dfrac{4x^2}{9}$ **71.** $\dfrac{1}{2}$ **73.** $\dfrac{1}{8}$ **75.** $\dfrac{1}{64x^3}$ **77.** $\dfrac{1}{9y^2}$
79. $\dfrac{16}{81}$ **81.** $-\dfrac{3}{2x}$ **83.** 2.47   **85.** 1.01   **87.** $5^{5/7}$ **89.** $4^{3/5}$
91. $9^{1/5}$ **93.** $\dfrac{1}{36}$ **95.** 3   **97.** a **99.** $a^{2/9}$ **101.** $a^{3/4}b^{1/2}$
103. $\dfrac{2x}{3}$ **105.** $\dfrac{1}{3}x$ **107.** $\dfrac{32x^6}{y^{25/12}}$ **109.** $y + y^2$ **111.** $x^2 - x + x^{3/5}$
113. $\sqrt{p}$ **115.** $\sqrt{5b}$ **117.** 736 ft/sec   **119.** 1.96 units
121. 4,608 in.², 32 ft²   **125.** $(-\infty, 3)$ **127.** $(28, \infty)$

Study Set   Section 7.6   (page 552)

1. hypotenuse   **3.** Pythagorean   **5.** $a^2 + b^2 = c^2$ **7.** $\sqrt{2}$
9. $\sqrt{3}$ **11.** $30°, 60°$ **13. a.** Take the positive square root of
both sides.   **b.** Subtract 25 from both sides.   **15.** 6, 52, 4, 2
17. 10 ft   **19.** 80 m   **21.** $h = 2\sqrt{2} \approx 2.83, x = 2$
23. $x = 5\sqrt{3} \approx 8.66, h = 10$ **25.** $x = 4.69, y = 8.11$
27. $x = 12.11, y = 12.11$ **29.** $7\sqrt{2}$ cm   **31.** 5   **33.** 13   **35.** 10
37. $2\sqrt{26}$ **41.** $(5\sqrt{2}, 0), (0, 5\sqrt{2}), (-5\sqrt{2}, 0), (0, -5\sqrt{2})$;
(7.07, 0), (0, 7.07), (−7.07, 0) (0, −7.07)   **43.** $10\sqrt{3}$ mm,
17.32 mm   **45.** $10\sqrt{181}$ ft, 134.54 ft   **47.** about 0.13 ft
49. a. 21.21 units   **b.** 8.25 units   **c.** 13.00 units   **51.** yes
55. 7   **57.** 9

Key Concept   (page 557)

1. $-3h^2\sqrt[3]{2}$ **3.** $-4\sqrt{2}$ **5.** $3s - 2t$ **7.** $18n - 12\sqrt{2n} + 4$
9. -3 **11.** 2; 7 is extraneous   **13.** $3^{1/3}$

Chapter Review   (page 559)

1. 7   **2.** -11 **3.** $\frac{15}{7}$ **4.** not real   **5.** 0.1   **6.** $5|x|$ **7.** x^4
8. $|x + 2|$ **9.** -3 **10.** -6 **11.** $4a^2b$ **12.** $\frac{s^3}{5}$ **13.** 5
14. -2 **15.** $4x^2|y|$ **16.** $22|y|$ **17.** $-\frac{1}{2}$ **18.** not real
19. 0   **20.** 0   **21.** 48 ft   **22.** 24 cm^2
23. D: $[-2, \infty)$, R: $[0, \infty)$

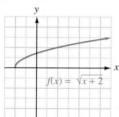

$f(x) = \sqrt{x + 2}$

24. D: $(-\infty, \infty)$, R: $(-\infty, \infty)$

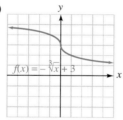

$f(x) = -\sqrt[3]{x} + 3$

25. $4\sqrt{15}$ **26.** $3\sqrt[3]{2}$ **27.** $2\sqrt[4]{2}$ **28.** $4\sqrt[5]{3}$ **29.** $2x^2\sqrt{2x}$
30. $r^5\sqrt[3]{r^2}$ **31.** $2xy\sqrt[3]{2x^2y}$ **32.** $9j^2\sqrt[3]{jk}$ **33.** $4x$ **34.** $\dfrac{\sqrt{17xy}}{8a^2}$
35. $3\sqrt{2}$ **36.** $11\sqrt{5}$ **37.** 0   **38.** $-8\sqrt[4]{2}$ **39.** $29x\sqrt{2}$
40. $13\sqrt[3]{2}$ **41.** $(6\sqrt{2} + 2\sqrt{10})$ in., 14.8 in.   **42.** 7   **43.** $6\sqrt{10}$
44. 32   **45.** 72   **46.** $3x$ **47.** 33   **48.** $-2x$ **49.** $12\sqrt[3]{3}$
50. $4 - 3\sqrt{2}$ **51.** $-20x^3y^3\sqrt{xy}$ **52.** $3b + 6\sqrt{b} + 3$
53. $6u - 12 + \sqrt{u}$ **54.** $\dfrac{10\sqrt{3}}{3}$ **55.** $\dfrac{\sqrt{15}}{5}$ **56.** $\dfrac{\sqrt{xy}}{y}$
57. $\dfrac{\sqrt[3]{u^2}}{u^2v^2}$ **58.** $2(\sqrt{2} + 1)$ **59.** $\dfrac{a + 2\sqrt{a} + 1}{a - 1}$ **60.** $\dfrac{9 - x}{2(3 + \sqrt{x})}$
61. $\dfrac{a - b}{a + \sqrt{ab}}$ **62.** $r = \dfrac{\sqrt[3]{6\pi^2V}}{2\pi}$ **63.** 22   **64.** 16, 9
65. $\frac{13}{2}$ **66.** $\frac{9}{16}$ **67.** 2, -4 **68.** 3, no solutions   **69.** 2
70. $P = \dfrac{A}{(r + 1)^2}$ **71.** $I = \dfrac{h^3b}{12}$ **72.** about 17.9 ohms   **73.** $\sqrt{t}$

74. $\sqrt[4]{5xy^3}$ **75.** 5   **76.** -6 **77.** not real   **78.** 1   **79.** $\frac{3}{x}$
80. $\frac{1}{3}$ **81.** -2 **82.** 5   **83.** $3ab^{1/3}$ **84.** $3cd$ **85.** 27   **86.** $\frac{1}{4}$
87. $-16,807$ **88.** 10   **89.** $\dfrac{27}{8}$ **90.** $\dfrac{1}{3,125}$ **91.** $125x^3y^6$
92. $\dfrac{1}{4u^4v^2}$ **93.** $5^{3/4}$ **94.** $a^{1/7}$ **95.** k^8 **96.** $\dfrac{2gh}{3}$
97. $u - 1$ **98.** $v + v^2$ **99.** $\sqrt{\dfrac{a}{5b}} = \dfrac{\sqrt{5ab}}{5b}$
100. Two true statements result: $16 = 16$ **101.** 17 ft
102. 88 yd   **103.** $7\sqrt{2}$ m   **104.** $6\sqrt{3}$ cm, 18 cm   **105.** 7.07 in.
106. 8.66 cm   **107.** 13   **108.** $2\sqrt{2}$

Chapter 7 Test   (page 565)

1. 0, 1, 1.41, 2, 3, 3.32, 4 y **2.** 3.0 mi

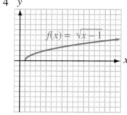

$f(x) = \sqrt{x - 1}$

3. $|x|$ **4.** $2|x|\sqrt{2}$ **5.** $3x\sqrt[3]{2x^2}$ **6.** $3x^2y^4\sqrt{2}$ **7.** $-4xy^2$
8. $\frac{2}{3}a$ **9.** not real   **10.** $t + 8$ **11.** $4\sqrt{3}$ **12.** $5xy^2\sqrt{10xy}$
13. $2x^5y\sqrt[3]{3}$ **14.** $\frac{1}{4a}$ **15.** $-\sqrt{3}$ **16.** $14\sqrt[3]{5}$ **17.** $2y^2\sqrt{3y}$
18. $6z\sqrt[4]{3z}$ **19.** $-6x\sqrt{y} - 2xy^2$ **20.** $3 - 7\sqrt{6}$
21. $\sqrt[3]{4a^2} + 18\sqrt[3]{2a} + 81$ **22.** $\dfrac{\sqrt{5}}{5}$ **23.** $\sqrt{3t} + 1$
24. $\dfrac{\sqrt[3]{18a}}{2a}$ **25.** $\frac{1}{3}$ **26.** 10   **27.** no solution; 4 is
extraneous   **28.** $G = \dfrac{4\pi^2r^3}{Mt^2}$ **29.** 2   **30.** 9   **31.** $\frac{1}{216}$
32. $\frac{9}{4}$ **33.** $2^{4/3}$ **34.** $8xy$ **35.** 9.24 cm   **36.** 8.67 cm
37. 25   **38.** $\dfrac{\pi\sqrt{2}}{2}$ sec, 2.2 sec   **39.** 28 in.

Cumulative Review Exercises   (page 567)

1. a. A rational number is any number that can be written
as a fraction with an integer numerator and a nonzero integer
denominator.   **b.** An irrational number is a nonterminating,
nonrepeating decimal.   **c.** A real number is any number that
is either a rational number or an irrational number.   **2.** $\frac{1}{10}$
3. $\ell = \dfrac{a - S + Sr}{r}$ **4.** 22   **5.** 10%: $3\frac{3}{4}$ cups, 18%: $6\frac{1}{4}$ cups
6. a. $-\frac{3}{5}$ **b.** $(0, -3)$ **c.** $y = -\frac{3}{5}x - 3$ **7. a.** 3
b. $y = -\frac{1}{3}x - \frac{2}{3}$ **8.** -30.75 **9.** D: $(-\infty, \infty)$; R: $(-\infty, -2]$
10. y **11.** no solution   **12.** 10, 15, 5

$g(x) = 1 + x^3$

13. $(-\infty, -15)$ (arrow) -15
14. $(-\infty, -3) \cup (2, \infty)$ (arrows) -3 2

15.

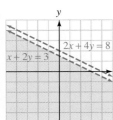

16. $\dfrac{64b^{12}}{27a^9}$ **17.** 2.379×10^{14}

18. $2x^2y^3 + 13xy + 3y^2$

19. $6y^3 + 11y^2 + 9y + 2$

20. $3x^2 + 3x - 11$

21. $(3 - c)(c + d)$

22. $(x - 2y)(x^2 + 2xy + 4y^2)$

23. $(a + b - 1)^2$

24. $(x + 1)(x - 1)(x + 4)(x - 4)$

25. $0, 10, -10$ **26.** $-\frac{1}{3}, -3$ **27.** $\dfrac{-3x - 2y}{y}$ **28.** $-7, 8$

29. $x - 5$ **30.** $\dfrac{3x + 2}{(x + 2)(x - 1)}$ **31.** $2\frac{2}{5}$ hr

32. Direct variation: As one quantity increases, the other increases, in a predictable way. Inverse variation: As one quantity increases, the other decreases, in a predictable way.

33. $-\dfrac{3}{2x}$ **34.** $10x^2y\sqrt{2yz}$ **35.** $6\sqrt[3]{2}$ **36.** $5z + 2\sqrt{15z} + 3$

37. $4, 3$ **38.** $\dfrac{\sqrt{6}}{10}$

Chapter 8   Check Your Knowledge   (page 570)

1. quadratic   **2.** vertex   **3.** real, imaginary   **4.** $4 + 5i$

5. $b^2 - 4ac$ **6.** $-1, 3$ **7.** $\pm\dfrac{7\sqrt{5}}{5}$ **8.** 25 **9.** $-3 \pm \sqrt{5}$

10. $\dfrac{2 \pm \sqrt{6}}{2}$ **11.** 6.9 ft high by 2.9 ft wide   **12.** $5i\sqrt{3}$ **13.** i

14. $5 - i$ **15.** $1 - 6i$ **16.** $9 + 12i$ **17.** $29 + 0i$ **18.** $0 - i$

19. $1 + i$ **20.** nonreal   **21.** $2 \pm 2i$ **22.** $4, 9$ **23.** vertex $(\frac{1}{2}, -\frac{25}{4})$ axis of symmetry $x = \frac{1}{2}$

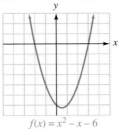

$f(x) = x^2 - x - 6$

24. vertex $(-2, 2)$ axis of symmetry $x = -2$

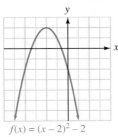

$f(x) = (x - 2)^2 - 2$

25. $(-\infty, -1) \cup (2, \infty)$

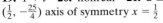

26. $(-\infty, 1) \cup (2, \infty)$

Study Set   Section 8.1   (page 579)

1. quadratic   **3.** perfect   **5.** $\sqrt{c}, -\sqrt{c}$ **7.** It is a solution   **9. a.** 36 **b.** $\frac{25}{4}$ **c.** $\frac{1}{16}$ **11. a.** Subtract 35 from both sides.   **b.** Add 36 to both sides.   **13.** $\pm\dfrac{\sqrt{10}}{2}$

15. 4 is not a factor of the numerator. It cannot be divided out.

17. $\dfrac{3 \pm \sqrt{10}}{5} = \dfrac{3}{5} \pm \dfrac{\sqrt{10}}{5}$ **19. a.** $2; 2\sqrt{5}, -2\sqrt{5}$

b. ± 4.47 **21.** $0, -2$ **23.** $5, -5$ **25.** $-2, -4$ **27.** $2, \frac{1}{2}$

29. ± 6 **31.** $\pm\sqrt{5}$ **33.** $\pm\dfrac{4\sqrt{3}}{3}$ **35.** $0, -2$ **37.** $4, 10$

39. $-5 \pm \sqrt{3}$ **41.** $d = \dfrac{\sqrt{6h}}{2}$ **43.** $c = \dfrac{\sqrt{Em}}{m}$ **45.** $2, -4$

47. $2, 6$ **49.** $1, -6$ **51.** $-1, 4$ **53.** $1, -\dfrac{1}{2}$ **55.** $\dfrac{1}{2}, -\dfrac{2}{3}$

57. $-4 \pm \sqrt{10}$ **59.** $1 \pm 3\sqrt{2}$ **61.** $\dfrac{3 \pm \sqrt{29}}{2}$

63. $\dfrac{3 \pm 2\sqrt{3}}{3}$ **65.** $\dfrac{1 \pm 2\sqrt{2}}{2}$ **67.** $\dfrac{-5 \pm \sqrt{41}}{4}$

69. $\dfrac{-7 \pm \sqrt{29}}{10}$ **71.** width: $7\frac{1}{4}$ ft; length: $13\frac{3}{4}$ ft   **73.** 1.6 sec

75. 1.70 in.   **77.** 0.92 in.   **79.** 2.9 ft, 6.9 ft   **83.** $2ab^2\sqrt[3]{5}$

85. x^3 **87.** $5ab\sqrt{7b}$

Study Set   Section 8.2   (page 588)

1. quadratic   **3. a.** $x^2 + 2x + 5 = 0$ **b.** $3x^2 + 2x - 1 = 0$

5. a. true   **b.** true   **7.** $c = \dfrac{-s \pm \sqrt{s^2 - 4rt}}{2r}$

9. a. $x = 1 \pm 2\sqrt{2}$ **b.** $x = \dfrac{-3 \pm \sqrt{7}}{2}$ **11. a.** The fraction bar wasn't drawn under both parts of the numerator.

b. A $\pm$ symbol wasn't written between b and the radical.

13. $-1, -2$ **15.** $-6, -6$ **17.** $\dfrac{1}{2}, -3$ **19.** $\dfrac{-5 \pm \sqrt{5}}{10}$

21. $-\dfrac{3}{2}, -\dfrac{1}{2}$ **23.** $\dfrac{1}{4}, -\dfrac{3}{4}$ **25.** $\dfrac{4}{3}, -\dfrac{2}{5}$ **27.** $\dfrac{-5 \pm \sqrt{17}}{2}$

29. $\dfrac{3 \pm \sqrt{17}}{4}$ **31.** $5 \pm \sqrt{7}$ **33.** $23, -17$ **35.** $\dfrac{2}{3}, -\dfrac{5}{2}$

37. $\dfrac{-3 \pm \sqrt{29}}{10}$ **39.** $\dfrac{9 \pm \sqrt{89}}{2}$ **41.** $\dfrac{10 \pm \sqrt{55}}{30}$

43. $-0.68, -7.32$ **45.** $1.22, -0.55$ **47.** $8.98, -3.98$

49. 97 ft by 117 ft   **51.** 0.5 mi by 2.5 mi   **53.** 34 in.

55. \$4.80 or \$5.20   **57.** 4,000   **59.** 9%   **61.** early 1976

65. $n^{1/2}$ **67.** $(3b)^{1/4}$ **69.** $\sqrt[3]{t}$ **71.** $\sqrt[4]{3t}$

Think It Through   (page 600)

We estimate that 50-yr-old drivers are involved in the least number of accidents, 12, per million miles driven.

Study Set   Section 8.3   (page 602)

1. quadratic   **3.** vertex   **5. a.** a parabola   **b.** $(1, 0), (3, 0)$

c. $(0, -3)$ **d.** $(2, 1)$ **e.** $x = 2$

7.

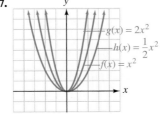

$g(x) = 2x^2$

$h(x) = \dfrac{1}{2}x^2$

$f(x) = x^2$

9.

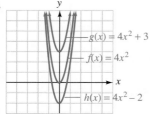

$g(x) = 4x^2 + 3$
$f(x) = 4x^2$
$h(x) = 4x^2 - 2$

11.

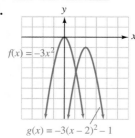

$f(x) = -3x^2$
$g(x) = -3(x - 2)^2 - 1$

13. $-3, 5$　**15.** $h = -1; f(x) = 2[x - (-1)]^2 + 6$
17. $(1, 2)$, $x = 1$, upward　**19.** $(-3, -4)$, $x = -3$, upward
21. $(1, -2)$, $x = 1$, upward　**23.** $(2, 21)$, $x = 2$, downward
25. $\left(-\frac{2}{3}, \frac{2}{3}\right)$, $x = -\frac{2}{3}$, upward

27. axis: $x = 3$

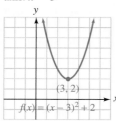

$f(x) = (x - 3)^2 + 2$
$(3, 2)$

29. axis: $x = 2$

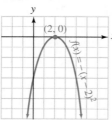

$(2, 0)$
$f(x) = -(x - 2)^2$

31. axis: $x = -3$

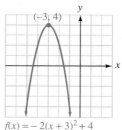

$(-3, 4)$
$f(x) = -2(x + 3)^2 + 4$

33. axis: $x = -4$

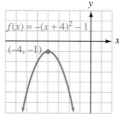

$f(x) = -(x + 4)^2 - 1$
$(-4, -1)$

35. axis: $x = -\frac{1}{2}$

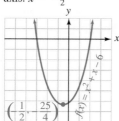

$\left(\frac{1}{2}, -\frac{25}{4}\right)$
$f(x) = x^2 + x - 6$

37. axis: $x = 2$

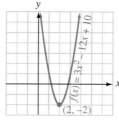

$f(x) = 3x^2 - 12x + 10$
$(2, -2)$

39. axis: $x = -3$

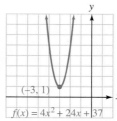

$(-3, 1)$
$f(x) = 4x^2 + 24x + 37$

41. axis: $x = 1$

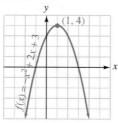

$(1, 4)$
$f(x) = -x^2 + 2x + 3$

43. axis: $x = \frac{1}{3}$

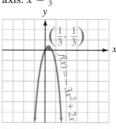

$\left(\frac{1}{3}, \frac{1}{3}\right)$
$f(x) = -3x^2 + 2x$

45. axis: $x = -\frac{1}{4}$

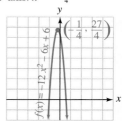

$\left(-\frac{1}{4}, \frac{27}{4}\right)$
$f(x) = -12x^2 - 6x + 6$

47. $(0.25, 0.88)$　**49.** $(0.50, 7.25)$　**51.** $2, -3$　**53.** $-1.85, 3.25$
55. 3.75 sec, 225 ft　**57.** 250 ft by 500 ft　**59.** 15 min, \$160
61. 1968, 1.5 million; the U.S. involvement in the war in
Vietnam was at its peak　**63.** 200, \$7,000　**71.** $4a^2\sqrt{b}$
73. $\dfrac{\sqrt{6}}{10}$　**75.** $15b - 6\sqrt{15b} + 9$

Study Set　Section 8.4　(page 614)

1. imaginary　**3.** real, imaginary　**5.** i or $\sqrt{-1}$　**7.** $\sqrt{-1}$
9. $-i$　**11.** FOIL　**13.** It is a solution.
15.

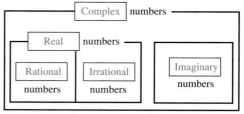

17. a. ± 1　**b.** $\pm i$　**19.** $9, 6i, 2$　**21. a.** true　**b.** false
c. false　**d.** false　**e.** true　**f.** false　**23.** $3i$　**25.** $i\sqrt{7}$ or $\sqrt{7}i$
27. $2i\sqrt{6}$ or $2\sqrt{6}i$　**29.** $-2i\sqrt{6}$ or $-2\sqrt{6}i$　**31.** $45i$
33. $\frac{5}{3}i$　**35.** -6　**37.** $-2\sqrt{3}$　**39.** $\frac{5}{8}$　**41.** -20　**43.** $\pm 3i$
45. $\pm\dfrac{4\sqrt{3}}{3}i$　**47.** $-1 \pm i$　**49.** $-\dfrac{1}{4} \pm \dfrac{\sqrt{7}}{4}i$　**51.** $\dfrac{2}{3} \pm \dfrac{\sqrt{2}}{3}i$
53. $\dfrac{1}{3} \pm \dfrac{2\sqrt{2}}{3}i$　**55.** $8 - 2i$　**57.** $3 - 5i$　**59.** $15 + 2i$
61. $15 + 7i$　**63.** $6 - 3i$　**65.** $-25 - 25i$
67. $7 + i$　**69.** $12 + 5i$　**71.** $13 - i$　**73.** $14 - 8i$
75. $8 + i\sqrt{2}$ or $8 + \sqrt{2}i$　**77.** $3 + 4i$　**79.** $0 - i$　**81.** $0 + \dfrac{4}{5}i$
83. $\dfrac{1}{8} + 0i$　**85.** $0 + \dfrac{3}{5}i$　**87.** $2 + i$　**89.** $\dfrac{15}{26} - \dfrac{3}{26}i$
91. $-\dfrac{42}{25} - \dfrac{6}{25}i$　**93.** $\dfrac{1}{4} + \dfrac{3}{4}i$　**95.** $-\dfrac{4}{13} - \dfrac{6}{13}i$　**97.** $\dfrac{5}{13} - \dfrac{12}{13}i$
99. $\dfrac{11}{10} + \dfrac{3}{10}i$　**101.** $\dfrac{1}{4} - \dfrac{\sqrt{15}}{4}i$　**103.** i　**105.** $-i$　**107.** 1
109. i　**111.** $-1 + i$　**115.** 15 units, $15\sqrt{3}$ units　**117.** 20 mph

Study Set　Section 8.5　(page 621)

1. $b^2 - 4ac$　**3.** conjugates　**5.** rational, unequal
7. a. $x^4 = (x^2)^2$　**b.** yes　**9.** 5, 6, 24　**11.** rational, equal
13. complex conjugates　**15.** irrational, unequal　**17.** rational,
unequal　**19.** yes　**21.** $1, -1, 4, -4$　**23.** $1, -1, \sqrt{5}, -\sqrt{5}$
25. $2, -2, i\sqrt{7}, -i\sqrt{7}$　**27.** 1　**29.** no solution　**31.** 16, 4
33. $-8, -27$　**35.** $-1, 27$　**37.** $\dfrac{1}{4}, \dfrac{1}{2}$　**39.** $1 - 2\sqrt{3}, 1 + 2\sqrt{3}$
41. $-1, -4$　**43.** $-1, -\dfrac{27}{13}$　**45.** $\dfrac{3 + \sqrt{57}}{6}, \dfrac{3 - \sqrt{57}}{6}$
47. $1, 1, -1, -1$　**49.** $1 \pm i$　**51.** 12.1 in.　**53.** 30 mph
55. 14.3 min　**59.** -2　**61.** $\dfrac{9}{5}$

Study Set   Section 8.6   (page 628)

1. quadratic   **3.** interval   **5.** greater   **7.** undefined
9. $(-\infty, -2) \cup (3, \infty)$ **11. a.** -2 **b.** -2 **c.** -2 **d.** 3
e. 3   **f.** 3   **13. a.** iii   **b.** iv   **c.** ii   **d.** i
15. $(1, 4)$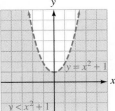

17. $(-\infty, 3) \cup (5, \infty)$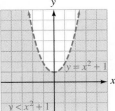

19. $[-4, 3]$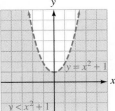

21. no solutions   **23.** $(-\infty, -3] \cup [3, \infty)$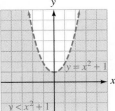

25. $(-5, 5)$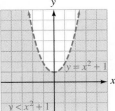

27. $(-\infty, 0) \cup (\frac{1}{2}, \infty)$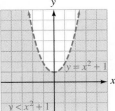

29. $(-\infty, -\frac{5}{3}) \cup (0, \infty)$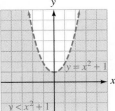

31. $(-\infty, -3) \cup (1, 4)$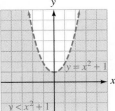

33. $(-\frac{1}{2}, \frac{1}{3}) \cup (\frac{1}{2}, \infty)$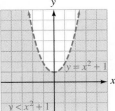

35. $(0, 2) \cup (8, \infty)$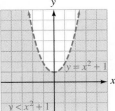

37. $[-\frac{34}{5}, -4) \cup (3, \infty)$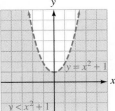

39. $(-4, -2] \cup (-1, 2]$

41. $(-\infty, -2) \cup (-2, \infty)$

43. $(-1, 3)$ **45.** $(-\infty, -3) \cup (2, \infty)$

47.   **49.**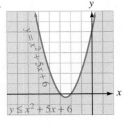

51.   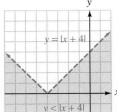   **53.**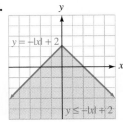

55. $(-2,100, -900) \cup (900, 2,100)$ **61.** $x = ky$
63. $t = kxy$ **65.** 3

Key Concept   (page 631)

1. $0, -2$ **3.** $-\frac{3}{2}, -1$ **5.** $4, 10$ **7.** $-5 \pm 4\sqrt{2}$ **9.** $-1 \pm i$
11. $23, -17$ **13.** $1, -2$

Chapter Review   (page 633)

1. $-5, -4$ **2.** $-\frac{1}{3}, -\frac{5}{2}$ **3.** $\pm 2\sqrt{7}$ **4.** $4, -8$ **5.** $0, -\frac{11}{5}$
6. $\pm\frac{7\sqrt{5}}{5}$ **7.** $\frac{1}{4}$ **8.** $-4, -2$ **9.** $\frac{3 \pm \sqrt{3}}{2}$
10. 6 seconds before midnight   **11.** $5 \pm \sqrt{7}$ **12.** $0, 10$
13. $\frac{1}{2}, -7$ **14.** $\frac{13 \pm \sqrt{163}}{3}$ **15.** sides: 1.25 in. wide;
top/bottom: 2.5 in. wide   **16.** 0.7 sec, 1.8 sec   **17.** \$46.8 billion
18. 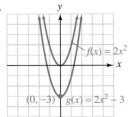   **19.**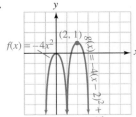

20. 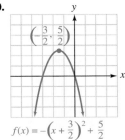   **21.**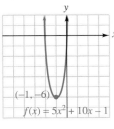

axis: $x = -\frac{3}{2}$
22. $1921, 6,469,326$ **23.** $-2, \frac{1}{3}$ **24.** $2i$ **25.** $i\sqrt{7}$
26. $-6i\sqrt{5}$ **27.** $\frac{6}{7}i$ **28.** -3 **29.** $\frac{8}{3}$ **30.** $\pm 5i$ **31.** $1 \pm 2i\sqrt{3}$
32. $12 - 8i$ **33.** $2 - 68i$ **34.** $-2 - 2i\sqrt{2}$ **35.** $-18 + 128i$
36. $22 + 29i$ **37.** $-12 + 28i\sqrt{3}$ **38.** $\frac{12}{5} - \frac{6}{5}i$ **39.** $\frac{15}{17} + \frac{8}{17}i$
40. $-\frac{1}{7} + \frac{4\sqrt{3}}{7}i$ **41.** $0 - \frac{2}{5}i$ **42.** i **43.** 1   **44.** irrational,
unequal   **45.** complex conjugates   **46.** equal rational numbers
47. $1, 144$ **48.** $8, -27$ **49.** $\sqrt{3}, -\sqrt{3}, \frac{\sqrt{6}}{6}, -\frac{\sqrt{6}}{6}$
50. $1, -\frac{8}{5}$ **51.** $10, 4$ **52.** about 81 min
53. $(-\infty, -7) \cup (5, \infty)$

54. $[-9, 9]$,

55. $(-\infty, 0) \cup [3/5, \infty)$

56. $(-7/2, 1) \cup (4, \infty)$

57.   **58.**

59. 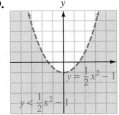   **60.**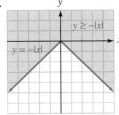

Chapter 8 Test (page 639)

1. $0, -6$ **2.** $-\frac{3}{2}, -\frac{5}{3}$ **3.** 144 **4.** $\dfrac{1 \pm \sqrt{5}}{2}$ **5.** $\dfrac{4 \pm \sqrt{6}}{2}$

6. 4.5 ft by 1,502 ft **7.** $4i\sqrt{3}$ **8.** -1 **9.** $-1 + 11i$

10. $4 - 7i$ **11.** $8 + 6i$ **12.** $-10 - 11i$ **13.** $0 + i$ **14.** $\frac{1}{2} + \frac{1}{2}i$

15. nonreal **16.** $\pm 2i\sqrt{3}$ **17.** $2 \pm 3i$ **18.** $1, \frac{1}{4}$

19. $2, -2, i\sqrt{3}, -i\sqrt{3}$ **20.** $-\frac{4}{5}, \frac{4}{7}$

21.
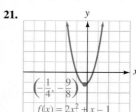

22.
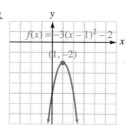

23. $(-\infty, -2) \cup (4, \infty)$
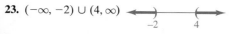

24. $(-3, 2]$

25.
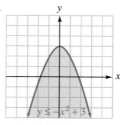

26. 20 in. **27.** iii **28.** about 46 min **29.** 211 ft **31.** $-3, 2$
32. $[-3, 2]$

Cumulative Review Exercises (page 641)

1. $y = 3x + 2$ **2.** $y = -\frac{2}{3}x - 2$ **3.** an increase of about 13,333 a year **4.**
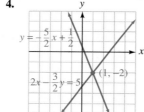

5. $(2, -1, 1)$ **6.**
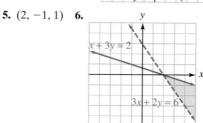

7. $\left(-\infty, -\frac{10}{9}\right)$
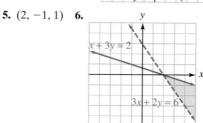

8. $(-\infty, -10] \cup [15, \infty)$
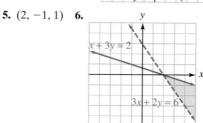

9. $D = (-\infty, \infty)$, $R = [-3, \infty)$ **10.** $D = (-\infty, \infty)$, $R = (-\infty, 0]$
11. $-4a^2 + 12a - 7$ **12.** $48x^2 - 40x - 48$
13. $(x^2 + 4y^2)(x + 2y)(x - 2y)$ **14.** $(3x + 2)(5x - 4)$
15. $z(x - y)(x - 4)$ **16.** $(2x^2 + 5y)(4x^4 - 10x^2y + 25y^2)$

17. $6, -1$ **18.** $0, \frac{2}{3}, -\frac{1}{2}$ **19.** 5; 3 is extraneous
20. $b = \dfrac{-RTV^2 + aV + PV^3}{PV^2 + a}$ **21.** $\dfrac{x + y}{x - y}$ **22.** 0

23. D: $(-\infty, \infty)$, R: $(-\infty, \infty)$
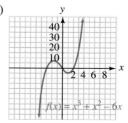

24. D: $(0, \infty)$, R: $(0, \infty)$
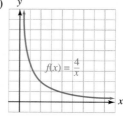

25. 9 **26.** D: $[2, \infty)$, R: $[0, \infty)$
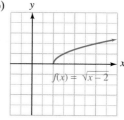

27. $-3x$ **28.** $4t\sqrt{3t}$ **29.** $\frac{1}{16}$ **30.** $x^{17/12}$
31. $-12\sqrt[4]{2} + 10\sqrt[4]{3}$ **32.** $-18\sqrt{6}$ **33.** $\dfrac{x + 3\sqrt{x} + 2}{x - 1}$

34. $\dfrac{5\sqrt[3]{x^2}}{x}$ **35.** $2, 7$ **36.** $\frac{1}{4}$ **37.** $3\sqrt{2}$ in. **38.** $2\sqrt{3}$ in.

39. 10 **40.** 9 **41.** $1, -\frac{3}{2}$ **42.** $\dfrac{-2 \pm \sqrt{7}}{3}$

43.
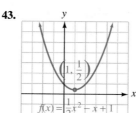

44.
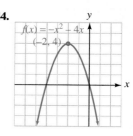

45. $7 + 2i$ **46.** $\frac{3}{2} + \frac{1}{2}i$ **47.** $8i$ **48.** $9, 16$ **49.** $\pm \dfrac{5\sqrt{2}}{2}i$
50. -1 **51.** $-\frac{3}{4}$ **52.** no solution

Chapter 9   Check Your Knowledge (page 645)

1. one-to-one **2.** $y = x$ **3.** exponential **4.** inverse
5. natural **6.** $(f \circ g)(x) = 2x^3 - x^2$
7. $(f - g)(x) = x^2 - 2x + 1$ **8.** $g(f(x)) = 2x^2 - 1$
9. 1 **10.** $f^{-1}(x) = \dfrac{x - 3}{2}$ **11.** $f^{-1}(x) = \sqrt{x}$

12.
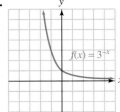

13.
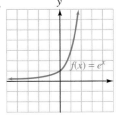

14.
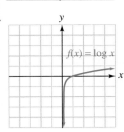

15. $8,084.11   **16.** 48.52 grams
17. 8.7 yr   **18.** 64   **19.** $\frac{1}{2}$
20. 3   **21.** 1   **22.** $5^3 = 125$
23. ln 1 = 0
24. log 2 + 2 log x + 3 log y
25. ln$\frac{x^2 y^4}{z^3}$ **26.** 2.3347
27. 2.3219   **28.** 2   **29.** 1
30. 18

21. the inverse of, inverse   **23.** yes   **25.** no

27.
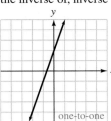

29.
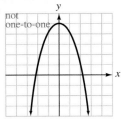

31.
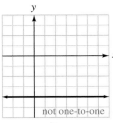

33. yes, {(2, 3), (1, 2), (0, 1)}, yes

35. yes, {(1, 1), (1, 2), (1, 3), (1, 4)}, no   **37.** $f^{-1}(x) = \frac{5}{4}x + 5$
39. $f^{-1}(x) = 5x - 4$ **41.** $f^{-1}(x) = 5x + 4$

43.
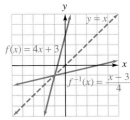

45.
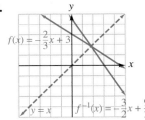

47.
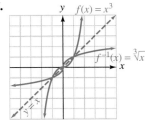

49.
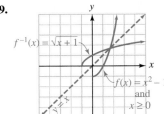

51. $y = \pm\sqrt{x - 4}$, no

Think It Through   (page 648)

The graph of $f + g$ gives the total number of students enrolled in historically Black colleges and universities.

Study Set   Section 9.1   (page 652)

1. sum, $f(x) + g(x)$ **3.** product, $f(x)g(x)$ **5.** domain
7. identity   **9. a.** $g(x)$ **b.** $g(x)$
11. −3, −2, −1, 0, 1, 2, 3

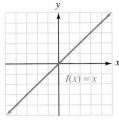

13. $(3x - 1), 9x, 2x$ **15.** $7x, (-\infty, \infty)$ **17.** $12x^2, (-\infty, \infty)$
19. $x, (-\infty, \infty)$ **21.** $\frac{4}{3}, (-\infty, 0) \cup (0, \infty)$ **23.** $3x - 2, (-\infty, \infty)$
25. $2x^2 - 5x - 3, (-\infty, \infty)$ **27.** $-x - 4, (-\infty, \infty)$
29. $\frac{x - 3}{2x + 1}, \left(-\infty, -\frac{1}{2}\right) \cup \left(-\frac{1}{2}, \infty\right)$ **31.** $-2x^2 + 3x - 3,$
$(-\infty, \infty)$ **33.** $\frac{3x - 2}{2x^2 + 1}, (-\infty, \infty)$ **35.** $3, (-\infty, \infty)$
37. $\frac{x^2 - 4}{x^2 - 1}, (-\infty, -1) \cup (-1, 1) \cup (1, \infty)$ **39.** 7   **41.** 24
43. −1   **45.** $-\frac{1}{2}$ **47.** $2x^2 - 1$ **49.** $16x^2 + 8x$ **51.** 58
53. 110   **55.** 2   **57.** $9x^2 - 9x + 2$ **61.** $C(t) = \frac{5}{9}(2,668 - 200t)$
63. a. about $37.50   **b.** $C(m) = \frac{1.50m}{20}$ **69.** $-\frac{3x + 7}{x + 2}$
71. $\frac{x - 4}{3x^2 - x - 12}$

Study Set   Section 9.2   (page 661)

1. one-to-one   **3.** inverses   **5.** once   **7.** 2   **9.** x
11. −6, −4, 0, 2, 8   **13.** no   **15.** 2   **17.** The graphs are not symmetric about the line $y = x$.   **19.** $2x, 2y, 3, y, f^{-1}(x)$

53. $f^{-1}(x) = \sqrt[3]{x}$, yes   **55.** $x = |y|$, no   **57.** $f^{-1}(x) = \sqrt[3]{\frac{x+3}{2}}$, yes   **59. a.** yes, no   **b.** No. Twice during this period, the person's anxiety level was at the maximum threshold value.
65. $3 - 8i$ **67.** $18 - i$ **69.** $-28 - 96i$

Study Set   Section 9.3   (page 673)

1. exponential   **3.** $(0, \infty)$ **5.** none   **7.** increasing
9.

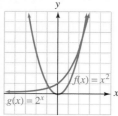

11. $A = P\left(1 + \frac{r}{k}\right)^{kt}, FV = PV(1 + i)^n$ **13.** $g(x) = 2^x + 3$;
$h(x) = 2^x - 2$ **15.** $\left(1 + \frac{r}{k}\right), kt$ **17.** 2.6651   **19.** 36.5548   **21.** 8
23. $7^{3\sqrt{3}}$ **25.**

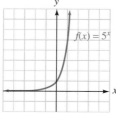

27.

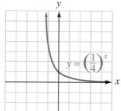

29.

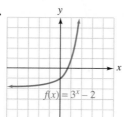

31.

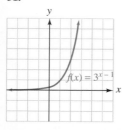

33. increasing

35. decreasing   **37.** \$22,080.40   **39.** \$32.03
41. \$2,273,996.13

43. a. about 1500, about 1825   **b.** 6.5 billion   **c.** exponential
45. a. at the end of the 2nd year   **b.** at the end of the 4th year   **c.** during the 7th year   **47.** 1.679046×10^8
49. 5.0421×10^{-5} coulombs   **51.** \$1,115.33   **55.** 40
57. 120°

Study Set   Section 9.4   (page 681)

1. the natural exponential function   **3.** $(0, \infty)$ **5.** none
7. increasing   **9.** continuous   **11.** 2.72

13.

15. an exponential function   **17.** $2.7182818\ldots$; e
19. 1,000, 10, 0.9, 1,000

21.

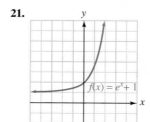

23.

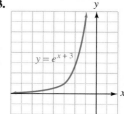

25.

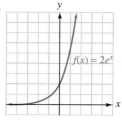

27.

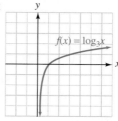

29. \$13,375.68   **31.** \$7,518.28 from annual compounding, \$7,647.95 from continuous compounding   **33.** \$6,849.16
35. about 9.3 billion   **37.** 315   **39.** 31.5 mm   **41.** 12 hr
43. 49 mps   **45.** about 72 yr   **49.** $4x^2\sqrt{15x}$ **51.** $10y\sqrt{3y}$

Study Set   Section 9.5   (page 692)

1. logarithmic   **3.** $(-\infty, \infty)$ **5.** $(1, 0)$ **7.** increasing
9. b^y **11.** inverse   **13.** $-2, -1, 0, 1, 2$ **15.** none, none,
$-3, \frac{1}{2}, 8$ **17.** $-0.30, 0, 0.30, 0.48, 0.60, 0.70, 0.78, 0.85, 0.90, 0.95, 1$
19. They decrease.   **21.** 10   **23.** $3^4 = 81$ **25.** $12^1 = 12$
27. $4^{-3} = \frac{1}{64}$ **29.** $10^{-3} = 0.001$ **31.** $\log_8 64 = 2$
33. $\log_4 \frac{1}{16} = -2$ **35.** $\log_{1/2} 32 = -5$ **37.** $\log_x z = y$ **39.** 3
41. 3   **43.** 3   **45.** $\frac{1}{2}$ **47.** 64   **49.** 5   **51.** $\frac{1}{25}$ **53.** $\frac{1}{6}$ **55.** $-\frac{3}{2}$
57. $\frac{2}{3}$ **59.** 5   **61.** $\frac{3}{2}$ **63.** 4   **65.** 1   **67.** 0.5119   **69.** -2.3307
71. 25.25   **73.** 0.02

75. increasing

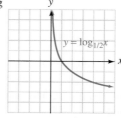

77. decreasing

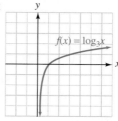

79.

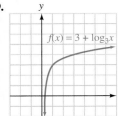

81.

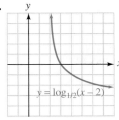

83.

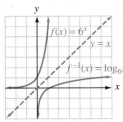

85.

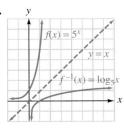

87. 29 db   **89.** 49.5 db   **91.** 4.4   **93.** 2,500 micrometers
95. 3 yr old   **97.** 10.8 yr   **101.** 10   **103.** $0; -\frac{5}{9}$ is extraneous

Study Set   Section 9.6   (page 698)

1. natural   **3.** $-0.69, 0, 0.69, 1.10, 1.39, 1.61, 1.79, 1.95, 2.08,$
$2.20, 2.30$ **5.** y-axis   **7.** $(-\infty, \infty)$ **9.** x
11. The logarithm of a negative number or zero is not defined.
13. $2.7182818\ldots, e$ **15.** e **17.** $\frac{\ln 2}{r}$ **19.** 3.5596
21. -5.3709 **23.** 0.5423   **25.** undefined   **27.** 4.0645
29. 69.4079   **31.** 0.0245   **33.** 2.7210   **35.**

37.

39. 5.8 yr   **41.** 9.2 yr   **43.** about 3.5 hr

47. $y = 5x$ **49.** $y = -\frac{3}{2}x + \frac{13}{2}$ **51.** $x = 2$

Study Set   Section 9.7   (page 707)

1. product   **3.** power   **5.** 0   **7.** M, N **9.** x, y **11.** x **13.** $\neq$
15. $\log_a b$ **17.** $10^0 = 1, 10^1 = 10, 10^2 = 10^2$ **19.** 0   **21.** 7
23. 10   **25.** 2   **27.** 1   **29.** 7   **31.** rs, t, r, s **39.** $2 + \log_2 5$
41. $\log_6 x - 2$ **43.** $4 \ln y$ **45.** $\frac{1}{2} \log 5$ **47.** $\log x + \log y + \log z$
49. $1 + \log_2 x - \log_2 y$ **51.** $3 \log x + 2 \log y$
53. $\frac{1}{2}(\log_b x + \log_b y)$ **55.** $\frac{1}{3}\log_a x - \frac{1}{4}\log_a y - \frac{1}{4}\log_a z$
57. $\ln x + \frac{1}{2}\ln z$ **59.** $\log_2 \frac{x+1}{x}$ **61.** $\log x^2 y^{1/2}$ **63.** $\log_b \frac{z^{1/2}}{x^3 y^2}$
65. $\ln \frac{\frac{x}{z}+x}{\frac{y}{z}+y} = \ln \frac{x}{y}$ **67.** false   **69.** false   **71.** true
73. 1.4472   **75.** -1.1972 **77.** 1.1972   **79.** 1.8063
81. 1.7712   **83.** -1.0000 **85.** 1.8928   **87.** 2.3219
89. 4.8   **91.** from 2.5×10^{-8} to 1.6×10^{-7} **95.** $-\frac{7}{6}$
97. $(1, -\frac{1}{2})$

Study Set   Section 9.8   (page 717)

1. exponential   **3.** $A_0 2^{-t/h}$ **5.** about 1.8   **7.** logarithm,
exponent   **9.** 1.2920   **11.** 1   **13.** $10^2 = x + 1$ **15. a.** ± 3.46
b. 3.58   **17.** not a solution   **19.** $\log, \log 2, \log 2$ **21.** 1.1610

23. 1.2702   **25.** 1.7095   **27.** 0   **29.** ± 1.0878 **31.** 0, 1.0566
33. 0.7324   **35.** -13.2662 **37.** 8   **39.** $-\frac{3}{4}$ **41.** 3, -1
43. $-2, -2$ **45.** 1.8   **47.** 8.8, 0.2   **49.** 9,998   **51.** -93
53. $e \approx 2.7183$ **55.** 19.0855   **57.** 2   **59.** 3   **61.** -7 **63.** 4
65. 10, -10 **67.** 50   **69.** 20   **71.** 10   **73.** 10   **75.** no solution
77. 6   **79.** 9   **81.** 4   **83.** 1, 7   **85.** 20   **87.** 8   **89.** 5.1 yr
91. 42.7 days   **93.** about 4,200 yr   **95.** 5.6 yr   **97.** 5.4 yr
99. because $\ln 2 \approx 0.7$ **101.** 25.3 yr   **103.** 2.828 times larger
105. 13.3   **107.** $\frac{1}{3}\ln 0.75$ **111.** 0, 5   **113.** $\frac{2}{3}, -4$
115. $\sqrt{137}$ in.

Key Concept   (page 721)

1. no   **3.** yes   **5.** $-2, 1, 3$ **7.** $2^{-3} = \frac{1}{8}$ **9.** $\frac{3}{2}$
11. e **13.**

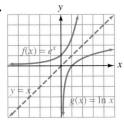

Chapter Review   (page 723)

1. $(f + g)(x) = 3x + 1$ **2.** $(f - g)(x) = x - 1$
3. $(f \cdot g)(x) = 2x^2 + 2x$ **4.** $(f/g)(x) = \frac{2x}{x+1}$ **5.** 4   **6.** 3
7. $(f \circ g)(x) = 2(x + 1)$ **8.** $(g \circ f)(x) = 2x + 1$ **9.** no
10. yes   **11.** yes   **12.** no   **13.**

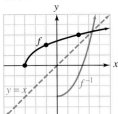

14. $f^{-1}(x) = \dfrac{x+3}{6}$ **15.** $f^{-1}(x) = \dfrac{x-5}{4}$ **16.** $f^{-1}(x) = \sqrt[3]{x}$

17. $f^{-1}(x) = \sqrt{\dfrac{x+1}{2}}$ **18.** $5^{2\sqrt{2}}$ **19.** $2^{\sqrt{10}}$

20. D: $(-\infty, \infty)$; R: $(0, \infty)$

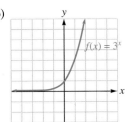

21. D: $(-\infty, \infty)$; R: $(0, \infty)$

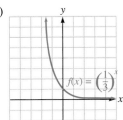

22.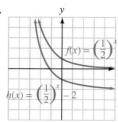

23.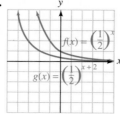

51.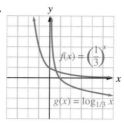

52.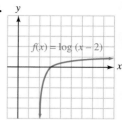

24. an exponential function

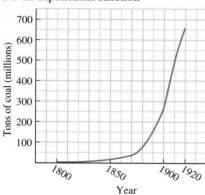

54. about 53   **55.** about 4.4
56. 6.1137   **57.** −0.1625
58. 10.3398   **59.** 0.0002
60. $f(x) = e^x$

25. $2,189,703.45
26. about $2,015   **27.** 8.22%
28. $2,324,767.37

29. D: $(-\infty, \infty)$, R: $(1, \infty)$

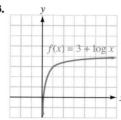

53.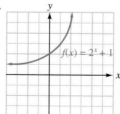

30. D: $(-\infty, \infty)$, R: $(0, \infty)$

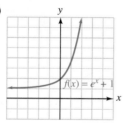

61.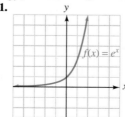

62.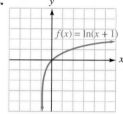

63. about $48\frac{1}{2}$ yr   **64.** about 72 in. (6 ft.)   **65.** 0   **66.** 1   **67.** 3
68. 4   **69.** 4   **70.** 1   **71.** $3 + \log_3 x$ **72.** $2 - \log x$
73. $\frac{1}{3} \log_5 27$ **74.** $1 + \log a + \log b$
75. $2 \log_b x + 3 \log_b y - \log_b z$ **76.** $\frac{1}{2}(\ln x - \ln y - 2 \ln z)$
77. $\log_2 \dfrac{x^3 z^7}{y^5}$ **78.** $\log_b \dfrac{\sqrt{x + 2}}{y^3 z^7}$ **79.** 2.6609   **80.** 3.0000
81. 1.7604   **82.** about 7.9×10^{-4} gram-ions/liter   **83.** 1.7712
84. 2   **85.** 2.7095   **86.** −3, −1   **87.** 1.9459   **88.** −8.0472
89. 104   **90.** 9   **91.** 25, 4   **92.** 4   **93.** 4, 3   **94.** 2   **95.** 6
96. 31   **97.** about 3,300 yr   **98.** about 91 days   **99.** 2, 5

Chapter 9 Test   (page 729)

1. $(g + f)(x) = 5x - 1$ **2.** $(g \cdot f)(x) = 4x^2 - 4x$ **3.** 3
4. $4(x - 1)$ **5.** $f^{-1}(x) = \dfrac{12 - 2x}{3}$ **6.** $f^{-1}(x) = \sqrt{\dfrac{x - 4}{3}}$

7.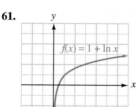

8.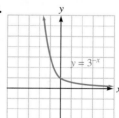

31. about 459,920,041   **32.** The exponent on the base e is
negative.   **33.** D: $(0, \infty)$, R: $(-\infty, \infty)$ **34.** $4^3 = 64$
35. $\log_7 \frac{1}{7} = -1$ **36.** 2   **37.** −2   **38.** 0   **39.** not possible
40. $\frac{1}{2}$ **41.** 3   **42.** 32   **43.** $\frac{1}{81}$ **44.** 4   **45.** 10   **46.** $\frac{1}{2}$ **47.** 9
48. 0.6542   **49.** 26.9153

50.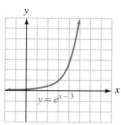

9. $\frac{3}{64}$ g = 0.046875 g   **10.** $1,060.90
11.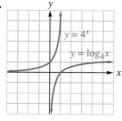

12. $4,451.08   **13.** 2   **14.** 3
15. $\frac{1}{27}$ **16.** e **17.** $6^{-2} = \frac{1}{36}$
18. D: $(0, \infty)$, R: $(-\infty, \infty)$

19. 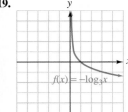   **20.**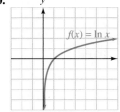

21. $2 \log a + \log b + 3 \log c$ **22.** $\ln \dfrac{b\sqrt{a + 2}}{c^3}$ **23.** 0.5646

24. $y = \log x$ **25.** 6.4   **26.** 46   **27.** 0.6826   **28.** 4   **29.** 1
30. 10   **31.** 5   **32.** $\frac{1}{2}\ln(5 - 1) = \frac{1}{2}\ln 4 = 0.69314718\ldots,$
which is ln 2.   **34. a.** yes   **b.** yes   **c.** 80; when the temperature
of the tire tread is 260°, the vehicle is traveling 80 mph
35. about 1,631,973,737   **36.** about 20 min

Cumulative Review Exercises   (page 731)

1. a. $P = 2l + 2w$ **b.** $A = \pi r^2$ **c.** $A = \frac{1}{2}bh$ **d.** $V = s^3$
e. $I = Prt$ **f.** $d = rt$ **g.** $\left(\dfrac{x_1 + x_2}{2}, \dfrac{y_1 + y_2}{2}\right)$ **h.** $y = mx + b$
i. $y - y_1 = m(x - x_1)$ **j.** $m = \dfrac{y_2 - y_1}{x_2 - x_1}$ **k.** $a^2 + b^2 = c^2$
l. $d = \sqrt{(x_2 - x_1)^2 + (y_2 - y_1)^2}$ **m.** $y = kx$ **n.** $y = \dfrac{k}{x}$
o. $x = \dfrac{-b \pm \sqrt{b^2 - 4ac}}{2a}$ **p.** $A = Pe^{rt}$ **q.** $\log_b x = \dfrac{\log_a x}{\log_a b}$
2. a. x **b.** x^{m+n} **c.** x^{mn} **d.** $x^n y^n$ **e.** 1   **f.** $\dfrac{x^n}{y^n}$ **g.** x^{m-n}
h. $\dfrac{1}{x^n}$ **i.** x^n **j.** $\left(\dfrac{y}{x}\right)^n$ **k.** $\sqrt[n]{x}$ **l.** $(\sqrt[n]{x})^m$
3. a. $(x + y)(x - y)$ **b.** $(x - y)(x^2 + xy + y^2)$
c. $(x + y)(x^2 - xy + y^2)$ **4. a.** $x^2 + 2xy + y^2$ **b.** $x^2 - 2xy + y^2$
c. $x^2 - y^2$ **5. a.** $\sqrt[n]{a}\sqrt[n]{b}$ **b.** $\dfrac{\sqrt[n]{a}}{\sqrt[n]{b}}$ **6. a.** 3.14   **b.** 1.41
c. 2.72   **7.** Pick a test point on one side of the boundary line. In
$y \geq x$, replace x and y with the coordinates of that point. If the
inequality is satisfied, shade the side that contains that point.
If the inequality is not satisfied, shade the other side.   **8. a.** 0
b. 1   **c.** x **d.** x **e.** $\log_b M + \log_b N$ **f.** $\log_b M - \log_b N$
g. $p \log_b M$ **9.** k, k **10. a.** to build up the fractions so they
have the same denominator   **b.** to rationalize the denominator
c. to simplify a complex fraction   **11.** It was used to simplify a
rational expression. The slashes and 1's show a common factor
of $3a - 2$ being divided out.   **12.** bc **13.** an equation of the
form $ax^2 + bx + c = 0$, where $a \neq 0$ **14.** $x + 1 = 0$ or $x - 7 = 0$
15. Multiply both sides of the equation by the LCD, which is
$x(x - 2)$.   **16.** the addition method   **17.** $x, -x$ **18.** b^2
19. $|x|$ **20.** x **21.** $+$ **22.** $\sqrt{-1}$ **23.** domain, range, one
24. a. both   **b.** one, both

Chapter 10   Check Your Knowledge   (page 734)

1. circle   **2.** ellipse   **3.** hyperbola   **4.** conic sections
5. $(-2, -1), 3$ **6.** $(3, -1), 4$ **7.**
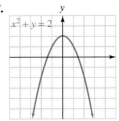

8. 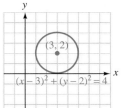   **9.**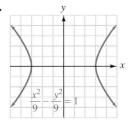

10. 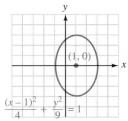   **11.**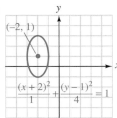

12. $(-1, -2), y = -2$
13. $(0, -3)$ and $(3, 0)$

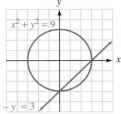

14. $(4, 3), (-4, 3), (4, -3), (-4, -3)$ **15.** 4, 6   **16.** 6 ft by 8 ft

Think It Through   (page 739)

$\dfrac{21}{2}$ in. = 10.5 in.

Study Set   Section 10.1   (page 744)

1. conic sections   **3.** radius, center   **5.** vertex, left
7. parabola, $(3, 2)$, right
9. $(x - h)^2 + (y - k)^2 = r^2$ **11.** $ax^2, 0$ **13.** 2, 6, 4, 36
15. 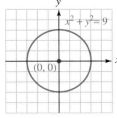   **17.**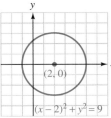

19. 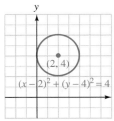   **21.**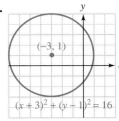

23. 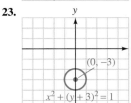   **25.**

27.

29. $x^2 + y^2 = 1$

59.

61. $(x - 7)^2 + y^2 = 9$ **63.** no
65. 5 ft   **67.** 2 AU   **73.** $5, -\frac{7}{3}$
75. $3, -\frac{1}{4}$

31. $(x - 6)^2 + (y - 8)^2 = 25$ **33.** $(x + 2)^2 + (y - 6)^2 = 144$
35. $x^2 + y^2 = 8$ **37.** $(x + 1)^2 + y^2 = 9$

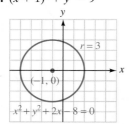

39. $(x - 1)^2 + (y + 2)^2 = 4$ **41.** $(x + 3)^2 + (y - 2)^2 = 1$

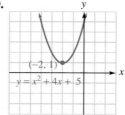

43.
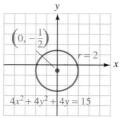

45.
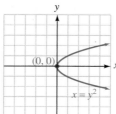

47.
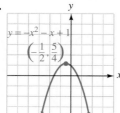

49.
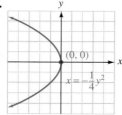

Study Set   Section 10.2   (page 754)

1. ellipse, sum   **3.** center   **5.** $(\pm a, 0)$ **7.** $(0, 0)$ **9.** 144, 16

11. 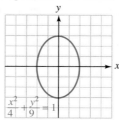   **13.**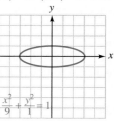

15. 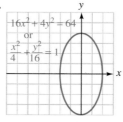   **17.**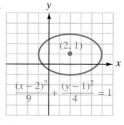

19. 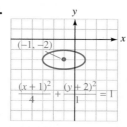   **21.**

23.   **25.**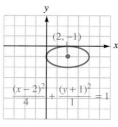

27.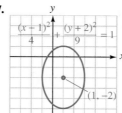

29. $y = \frac{1}{2}\sqrt{400 - x^2}$

31. 12π sq. units

35. $12y^2 + \dfrac{9}{x^2}$

37. $\dfrac{y^2 + x^2}{y^2 - x^2}$

51.
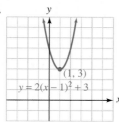

53.
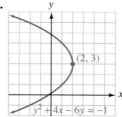

55.

57.
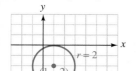

Study Set   Section 10.3   (page 761)

1. hyperbola, difference   **3.** center   **5.** $(\pm a, 0)$ **7.** $(0, 0)$
9. 36, 36, 36

11.

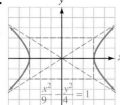

13.

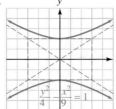

15.

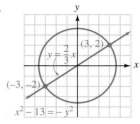

17.

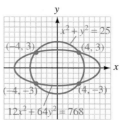

15.

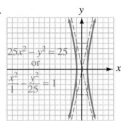

17.

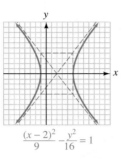

$\dfrac{(x-2)^2}{9} - \dfrac{y^2}{16} = 1$

19.

21. $(3, 0), (0, 5)$ **23.** $(1, 1)$
25. $(1, 2), (2, 1)$ **27.** $(-2, 3), (2, 3)$
29. $(\sqrt{5}, 5), (-\sqrt{5}, 5)$ **31.** $(3, 2),$
$(3, -2), (-3, 2), (-3, -2)$
33. $(2, 4), (2, -4), (-2, 4), (-2, -4)$ **35.** $(-\sqrt{15}, 5),$
$(\sqrt{15}, 5), (-2, -6), (2, -6)$ **37.** $(0, -4), (-3, 5), (3, 5)$
39. $(-2, 3), (2, 3), (-2, -3), (2, -3)$ **41.** $(3, 3)$
43. $(6, 2), (-6, -2), (\sqrt{42}, 0), (-\sqrt{42}, 0)$ **45.** $(\frac{1}{2}, \frac{1}{3}), (\frac{1}{3}, \frac{1}{2})$
47. $4, 8$ **49.** 7 cm by 9 cm   **51.** either \$750 at 9% or
\$900 at 7.5%   **53.** 68 mph, 4.5 hr   **57.** $-11x\sqrt{2}$ **59.** $\dfrac{t}{2}$

19.

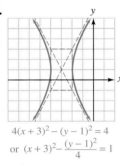

21.

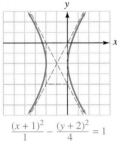

$4(x + 3)^2 - (y - 1)^2 = 4$
or $(x + 3)^2 - \dfrac{(y-1)^2}{4} = 1$

Key Concept   (page 770)

1. ellipse   **3.** parabola   **5.** hyperbola   **7.** ellipse
9. circle   **11. a.** $(-2, 1)$ **b.** right   **13. a.** $(-2, 1)$ **b.** left and
right   **c.** 6 units horizontally, 4 units vertically

23.

25.

15.

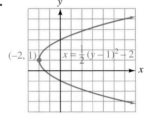

17.

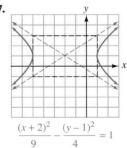

$x = \frac{1}{2}(y - 1)^2 - 2$

$\dfrac{(x+2)^2}{9} - \dfrac{(y-1)^2}{4} = 1$

27.

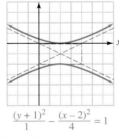

29.

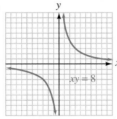

$\dfrac{(x+1)^2}{1} - \dfrac{(y+2)^2}{4} = 1$

$\dfrac{(y+1)^2}{1} - \dfrac{(x-2)^2}{4} = 1$

31. 3 units   **33.** $10\sqrt{3}$ units   **37.** $-3x^2(2x^2 - 3x + 2)$
39. $(5a + 2b)(3a - 2b)$

Chapter Review   (page 772)

1.

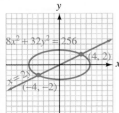

$(x - 1)^2 + (y + 2)^2 = 9$

2.

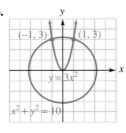

$x^2 + y^2 = 16$

Study Set   Section 10.4   (page 767)

1. graphing, substitution   **3.** tangent   **5.** two, four   **7.** no
9. $2x, 5, 5, 1, 1, -1, 2, -2$

11.

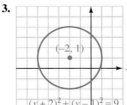

$8x^2 + 32y^2 = 256$

$x = 2y$

13.

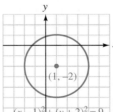

$y = 3x^2$

$x^2 + y^2 = 10$

3.

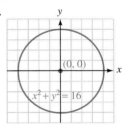

$(x + 2)^2 + (y - 1)^2 = 9$

4.

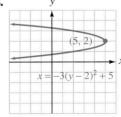

$x = -3(y - 2)^2 + 5$

5.
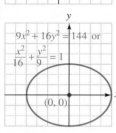
$x = 2(y+1)^2 - 2$
$(-2, -1)$

6. when $x = 22$, $y = 0$:
$-\frac{5}{121}(22 - 11)^2 + 5 = 0$

7.
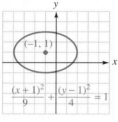
$9x^2 + 16y^2 = 144$ or
$\frac{x^2}{16} + \frac{y^2}{9} = 1$
$(0, 0)$

8.
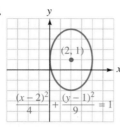
$(2, 1)$
$\frac{(x-2)^2}{4} + \frac{(y-1)^2}{9} = 1$

9.
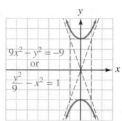
$(-1, 1)$
$\frac{(x+1)^2}{9} + \frac{(y-1)^2}{4} = 1$

10. $\dfrac{x^2}{25} + \dfrac{y^2}{9} = 1$

11.
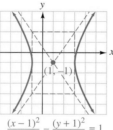
$9x^2 - y^2 = -9$
or
$\frac{y^2}{9} - x^2 = 1$

12.
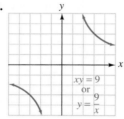
$xy = 9$
or
$y = \frac{9}{x}$

13. hyperbola

14.
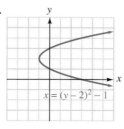
$(1, -1)$
$\frac{(x-1)^2}{4} - \frac{(y+1)^2}{9} = 1$

15. 4 units

16. a. ellipse
b. hyperbola
c. parabola
d. circle

17. $(0, 3)$, $(0, -3)$

18. $(4, 2)$, $(4, -2)$,
$(-4, 2)$, $(-4, -2)$

19. $(2, 3)$, $(2, -3)$,
$(-2, 3)$, $(-2, -3)$

20. $(\sqrt{2}, 0)$, $(-\sqrt{2}, 0)$ **21.** $(2, 2)$, $\left(-\dfrac{2}{9}, -\dfrac{22}{9}\right)$

Chapter 10 Test   (page 775)

1. $(2, -3)$, 2   **2.** $(-2, 3)$, 4

3.
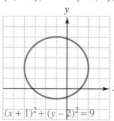
$(x+1)^2 + (y-2)^2 = 9$

4.
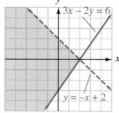
$x = (y-2)^2 - 1$

5.
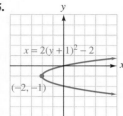
$9x^2 + 4y^2 = 36$
or
$\frac{x^2}{4} + \frac{y^2}{9} = 1$

6.
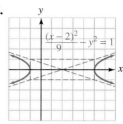
$\frac{(x-2)^2}{9} - y^2 = 1$

7.
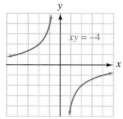
$xy = -4$

8.
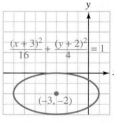
$\frac{(x+3)^2}{16} + \frac{(y+2)^2}{4} = 1$
$(-3, -2)$

9.
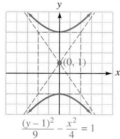
$(0, 1)$
$\frac{(y-1)^2}{9} - \frac{x^2}{4} = 1$

10.
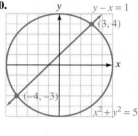
$y - x = 1$
$(3, 4)$
$(-4, -3)$
$x^2 + y^2 = 5$

11. $(2, 6)$, $(-2, -2)$ **12.** $(3, 4)$, $(-3, 4)$
13. $(1, \sqrt{2})$, $(1, -\sqrt{2})$, $(-1, \sqrt{2})$, $(-1, -\sqrt{2})$
14. $(x - 4)^2 + (y - 3)^2 = 9$

15.
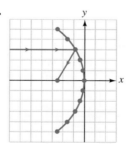

16. $\dfrac{(x-1)^2}{16} + \dfrac{(y+2)^2}{9} = 1$

17. $\dfrac{x^2}{16} - \dfrac{y^2}{36} = 1$

18. a. ellipse   **b.** hyperbola
c. circle   **d.** parabola

Cumulative Review Exercises   (page 777)

1. 0   **2.** $-\frac{4}{3}$, 5.6, 0, -23 **3.** π, $\sqrt{2}$, e **4.** $-\frac{4}{3}$, π, 5.6, $\sqrt{2}$, 0, -23, e **5.** \$8,250   **6.** $\frac{1}{120}$ db/rpm   **7.** parallel
8. perpendicular   **9.** $y = -2x + 5$ **10.** $y = -\frac{9}{13}x + \frac{7}{13}$
11. $(1, 1)$ **12.** $(2, -2)$ **13.** $(-1, -1)$ **14.** $(3, 2, 1)$
15. 3   **16.** -1 **17.** $85°$, $80°$, $15°$

18.
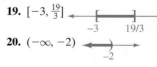
$3x - 2y = 6$
$y = -x + 2$

19. $[-3, \frac{19}{3}]$

20. $(-\infty, -2)$

21. $12x^2 - 5xy - 3y^2$
22. $2x^2y^3 + 13xy + 3y^2$
23. $a^2 - 4ab + 4b^2$
24. $3a^3 + 10a^2 + 6a - 4$
25. $xy(3x - 4y)(x - 2)$
26. $(16x^2y^2 + z^4)(4xy + z^2)(4xy - z^2)$ **27.** $\lambda = \dfrac{4d - 2}{A - 6}$

28. 24, 0, −8, −6, 0, 4, 0, −18; (−3, 0), (0, 0), (2, 0); (0, 0)

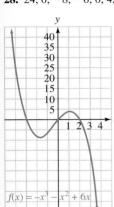

$f(x) = -x^3 - x^2 + 6x$

29. $\dfrac{64b^{12}}{27a^9}$ **30.** $-\dfrac{3x + 2}{3x - 2}$

31. $-\dfrac{q}{p}$ **32.** $\dfrac{4a - 1}{(a + 2)(a - 2)}$

33. 5; 3 is extraneous

34. $R = \dfrac{R_1 R_2 R_3}{R_2 R_3 + R_1 R_3 + R_1 R_2}$

35. a. a quadratic function
b. at about 85% and 120% of the suggested inflation
36. about $21\frac{1}{2}$ in.
37. $-x^2 + x + 5 + \dfrac{8}{x - 1}$
38. 2   **39.** $5\sqrt{2}$ **40.** $81x\sqrt[3]{3x}$
41. $\frac{343}{125}$ **42.** $\sqrt{3t} - 1$

43. $-5 + 17i$ **44.** $-\frac{21}{29} - \frac{20}{29}i$ **45.** 5, 0 is extraneous

46. 0   **47.** $\frac{2}{3}, -\frac{3}{2}$ **48.** $\dfrac{-3 \pm \sqrt{5}}{4}$ **49.** $\frac{1}{4}, \frac{1}{2}$

50. $\dfrac{2 \pm i\sqrt{2}}{3}$ **51.** $4x^2 + 4x - 1$ **52.** $f^{-1}(x) = \sqrt[3]{\dfrac{x + 1}{2}}$

53.

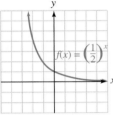

$f(x) = \left(\frac{1}{2}\right)^x$

54.

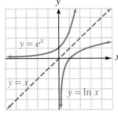

$y = e^x$
$y = x$
$y = \ln x$

55. $2^y = x$ **56.** $(\log_6 x) - 2$ **57.** 5   **58.** 3   **59.** $\frac{1}{27}$ **60.** 1
61. $f^{-1}(x) = 2^x$ **62.** x **63.** 1.9912   **64.** 0.301   **65.** 3.4190
66. 1, −10 is extraneous   **67.** $-\frac{3}{4}$ **68.** 9   **69.** $2,848.31
70. 1.16056   **71.** $x^2 + y^2 - 2x - 6y - 15 = 0$

72.

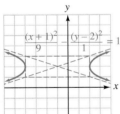

$\dfrac{(x + 1)^2}{9} - \dfrac{(y - 2)^2}{1} = 1$

73.

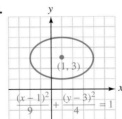

$(1, 3)$
$\dfrac{(x - 1)^2}{9} + \dfrac{(y - 3)^2}{4} = 1$

74.

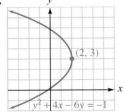

$(2, 3)$
$y^2 + 4x - 6y = -1$

Study Set   Appendix I   (page A-9)

1. numerator, denominator   **3.** equivalent   **5.** factored, prime-factored   **7.** terminating, repeating   **9.** $\frac{4}{12} = \frac{1}{3}$ **11.** 5

13. a. 3 times   **b.** 2 times   **15.** common factors of 7 and 5 in the numerator and denominator have been divided out.   **17.** 1, 2, 4, 5, 10, 20   **19.** 1, 2, 4, 7, 14, 28   **21.** $5 \cdot 5 \cdot 3$ **23.** $7 \cdot 2 \cdot 2$
25. $13 \cdot 3 \cdot 3$ **27.** $11 \cdot 5 \cdot 2 \cdot 2$ **29.** $\frac{3}{9}$ **31.** $\frac{24}{54}$ **33.** $\frac{35}{5}$ **35.** $\frac{1}{2}$
37. $\frac{4}{3}$ **39.** $\frac{3}{4}$ **41.** $\frac{9}{8}$ **43.** $\frac{4}{25}$ **45.** $\frac{3}{10}$ **47.** $\frac{8}{5}$ **49.** $\frac{3}{2}$ **51.** $\frac{20}{3}$
53. $10\frac{1}{2}$ **55.** $\frac{9}{10}$ **57.** $\frac{5}{8}$ **59.** $\frac{14}{5}$ **61.** 28   **63.** $1\frac{9}{11}$ **65.** $\frac{6}{5}$
67. $\frac{5}{24}$ **69.** $\frac{19}{15}$ **71.** $\frac{17}{12}$ **73.** $\frac{22}{35}$ **75.** $\frac{1}{6}$ **77.** $\frac{9}{4}$ **79.** $1\frac{1}{4}$ **81.** $\frac{5}{9}$
83. 158.65   **85.** 44.785   **87.** 44.88   **89.** 4.55   **91.** 2,951.7
93. 5   **95.** 0.625   **97.** $0.0\overline{3}$ **99.** 0.42   **101.** $0.\overline{45}$
103. a. $\frac{7}{32}$ in.   **b.** $\frac{3}{32}$ in.   **105.** $40\frac{1}{2}$ in.   **107.** 103.4 in.

Cumulative Review of Graphing   (page A-13)

1.

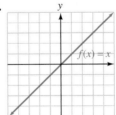

$y = 2x + 1$

3.

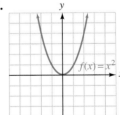

$x = 5$

5.

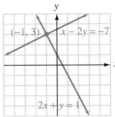

$(-1, 3)$
$x - 2y = -7$
$2x + y = 1$

7.

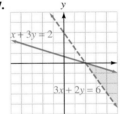

$x + 3y = 2$
$3x + 2y = 6$

9.

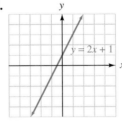

$f(x) = x$

11.

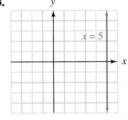

$f(x) = x^2$

13.

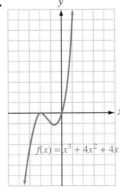

$f(x) = x^3 + 4x^2 + 4x$

15.

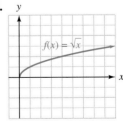

$f(x) = \sqrt{x}$

17.

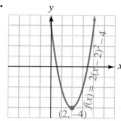

19.

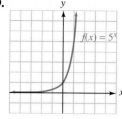

29.

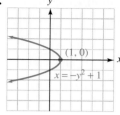

31.

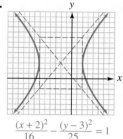

$$\frac{(x+2)^2}{16} - \frac{(y-3)^2}{25} = 1$$

21.

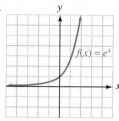

23.

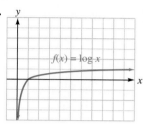

33.

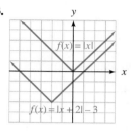

25.

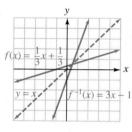

27.

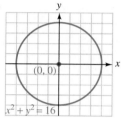

INDEX